Lecture Notes in Networks a

Volume 308

The series "Lecture Notes in Networks and Systems" publishes the latest developments in Networks and Systems—quickly, informally and with high quality. Original research reported in proceedings and post-proceedings represents the core of LNNS.

Volumes published in LNNS embrace all aspects and subfields of, as well as new challenges in, Networks and Systems.

The series contains proceedings and edited volumes in systems and networks, spanning the areas of Cyber-Physical Systems, Autonomous Systems, Sensor Networks, Control Systems, Energy Systems, Automotive Systems, Biological Systems, Vehicular Networking and Connected Vehicles, Aerospace Systems, Automation, Manufacturing, Smart Grids, Nonlinear Systems, Power Systems, Robotics, Social Systems, Economic Systems and other. Of particular value to both the contributors and the readership are the short publication timeframe and the world-wide distribution and exposure which enable both a wide and rapid dissemination of research output.

The series covers the theory, applications, and perspectives on the state of the art and future developments relevant to systems and networks, decision making, control, complex processes and related areas, as embedded in the fields of interdisciplinary and applied sciences, engineering, computer science, physics, economics, social, and life sciences, as well as the paradigms and methodologies behind them.

Indexed by SCOPUS, INSPEC, WTI Frankfurt eG, zbMATH, SCImago.

All books published in the series are submitted for consideration in Web of Science.

More information about this series at http://www.springer.com/series/15179

Cengiz Kahraman · Selcuk Cebi ·
Sezi Cevik Onar · Basar Oztaysi ·
A. Cagri Tolga · Irem Ucal Sari
Editors

Intelligent and Fuzzy Techniques for Emerging Conditions and Digital Transformation

Proceedings of the INFUS 2021 Conference, held August 24–26, 2021. Volume 2

Editors
Cengiz Kahraman
Department of Industrial Engineering
Istanbul Technical University
Istanbul, Turkey

Selcuk Cebi
Industrial Engineering Department
Yildiz Technical University
Istanbul, Turkey

Sezi Cevik Onar
Department of Industrial Engineering
Istanbul Technical University
Istanbul, Turkey

Basar Oztaysi
Department of Industrial Engineering
Istanbul Technical University
Istanbul, Turkey

A. Cagri Tolga
Industrial Engineering Department
Galatasaray University
Istanbul, Turkey

Irem Ucal Sari
Department of Industrial Engineering
Istanbul Technical University
Istanbul, Turkey

ISSN 2367-3370 ISSN 2367-3389 (electronic)
Lecture Notes in Networks and Systems
ISBN 978-3-030-85576-5 ISBN 978-3-030-85577-2 (eBook)
https://doi.org/10.1007/978-3-030-85577-2

This Springer imprint is published by the registered company Springer Nature Switzerland AG
The registered company address is: Gewerbestrasse 11, 6330 Cham, Switzerland

Preface

INFUS is an acronym for intelligent and fuzzy systems. It is a well-established international research forum to advance the foundations and applications of intelligent and fuzzy systems, computational intelligence, and soft computing for applied research in general and for complex engineering and decision support systems. The principal mission of INFUS is to construct a bridge between fuzzy and intelligence systems and real complex systems via joint research between universities and international research institutions, encouraging interdisciplinary research and bringing multidiscipline researchers together.

INFUS 2021 has been organized in August 24–26, 2021 in Istanbul, Turkey. The number of participants after the review process is about 200. The proceedings of INFUS 2019 and 2020 conferences have been published by Springer publishing house with an excellent quality before the conference began. INFUS 2021 proceedings is again published by Springer under "Advances in Intelligent Systems and Computing Series" and will be indexed in Scopus. Its title is Intelligent and Fuzzy Techniques: Emerging Conditions and Digital Transformation. Moreover, special issues of indexed journals will be devoted to a strictly refereed selection of extended papers presented at INFUS 2021. Journal of Intelligent and Fuzzy Systems (SCI-E) and Journal of Multi-Valued Logic and Soft Computing (SCI-E) are the two main journals that we publish their special issues for the papers presented at INFUS conferences. In addition to these journals, there are several journals that we are given their special issues having other indices.

Emerging conditions such as pandemic, wars, natural disasters, and various high technologies force us for significant changes in our business and social life. Pandemic caused all of us to live under quarantine for a certain period of time and serious restrictions in our business and social life. We clearly saw how important digital technologies are and how great the need for them is during this period. Digital transformation is the adoption of digital technologies to transform services or businesses, through replacing non-digital or manual processes with digital processes or replacing older digital technology with newer digital technologies. This may enable—in addition to efficiency via automation—new types of innovation and creativity, rather than simply enhancing and supporting traditional methods. INFUS

2021 focuses on revealing the reflection of digital transformation in our business and social life under emerging conditions through intelligent and fuzzy systems.

Our invited speakers this year are Prof. Krassimir Atanassov, Prof. Vicenc Torra, Prof. Janusz Kacprzyk, Prof. Ahmet Fahri Özok, and Prof. Ajith Abraham. It is an honor to include their speeches in our conference program. We appreciate their voluntary contributions to INFUS 2021, and we hope to see them at INFUS conferences for many years.

In the beginning of the planning process, we had planned to organize INFUS 2021 in Izmir as we did for INFUS 2020. Unfortunately, coronavirus pandemic prevented it as all of you know. We hope to organize an interactive conference in 2022 with your participation in Izmir. Our social program of INFUS 2021 in Izmir will be exactly realized at INFUS 2022, if everything goes on its way. We thank all of you very much since you did not give up your participation to INFUS 2021. We appreciate your sincerity and fidelity.

This year, the number of submitted papers became 355. After the review process, about 43% of these papers have been rejected. More than 50% of the accepted papers are from other countries outside Turkey. The distribution of the admitted papers with respect to their source countries is as follows from the most to the least: Turkey, Russia, China, Iran, Poland, India, Azerbaijan, Bulgaria, Morocco, Spain, Algeria, Serbia, Ukraine, Pakistan, Canada, Japan, South Korea, UK, Indonesia, USA, Vietnam, Finland, Romania, France, Uzbekistan, Italy, and Austria. We again thank all the representatives of their countries for selecting INFUS 2021 as an international scientific arena.

We also thank the anonymous reviewers for their hard works in selecting high-quality papers of INFUS 2021. Each of the organizing committee members provided invaluable contributions to INFUS 2021. INFUS conferences would be impossible without their efforts. We hope meeting you all next year in Turkey at an interactive, face-to-face conference.

Cengiz Kahraman
A. Cagri Tolga
Selcuk Cebi
Basar Oztaysi
Sezi Cevik Onar
Irem Ucal Sari

Organization

Program Chair

Cengiz Kahraman	Istanbul Technical University, Turkey

Program Committee

A. Cagri Tolga	Galatasaray University, Turkey
Irem Ucal Sari	Istanbul Technical University, Turkey
Basar Oztaysi	Istanbul Technical University, Turkey
Selcuk Cebi	Yildiz Technical University, Turkey
Sezi Cevik Onar	Istanbul Technical University, Turkey

Contents

Intuitionistic Fuzzy Sets

Temporal-Level Operators Over Intuitionistic Fuzzy Sets

Krassimir T. Atanassov[1,2(✉)]

[1] Department of Bioinformatics and Mathematical Modelling IBPhBME, Bulgarian Academy of Sciences, Acad. G. Bonchev Str. Bl. 105, 1113 Sofia, Bulgaria
krat@bas.bg

[2] Intelligent Systems Laboratory Prof. Asen Zlatarov University, 8010 Bourgas, Bulgaria

Abstract. The Intuitionistic Fuzzy Set (IFS) was introduced in 1983 as one of the first extensions of Zadeh's fuzzy set. In the following years, on the one hand, it was extended to intuitionistic L-fuzzy set (1984), interval valued IFS (1989), IFS of second type (1989) (although some authors have incorrectly called it Pythagorean fuzzy set) and more generally of n-th type, Temporal IFS (TIFS) and others. On the other hand, different relations, operations and operators have been introduced over IFSs. The operators over IFS are of modal, topological, level and other types.

In the present paper, as a continuation and fusion of the ideas of TIFS and of level operations over IFSs, temporal-level operators are introduced and some of their basic properties are studied.

Keywords: Intuitionistic fuzzy operator · Intuitionistic fuzzy set · Temporal intuitionistic fuzzy operator · Temporal intuitionistic fuzzy set

AMS Classification: 03E72

1 Introduction

In 1983, the Intuitionistic Fuzzy Sets (IFSs, see [4–6]) were introduced as extensions of the Zadeh's fuzzy sets (see[11]). Different operations, relations and operators were defined over IFSs. A part of them have analogues in ordinary fuzzy sets theory, but the majority of them do not have archetype in the standard case.

In a series of papers, 190 intuitionistic fuzzy implications were introduced (see [9]). In [1–3], to each one of these implications three (in some cases - different) intuitionistic fuzzy unions and intersections were juxtaposed.

In [5,6], the two level operators $P_{\alpha,\beta}$ and $Q_{\alpha,\beta}$ are defined for each IFS

$$A\{\langle x, \mu_A(x), \nu_A(x)\rangle | x \in E\},$$

C. Kahraman et al. (Eds.): INFUS 2021, LNNS 308, pp. 3–11, 2022.
https://doi.org/10.1007/978-3-030-85577-2_1

where $\mu_A, \nu_A : E \to [0,1]$ and $\mu_A(x) + \nu_A(x) \leq 1$ for each $x \in E$, by

$$P_{\alpha,\beta}(A) = \{\langle x, \max(\alpha, \mu_A(x)), \min(\beta, \nu_A(x))\rangle | x \in E\},$$
$$Q_{\alpha,\beta}(A) = \{\langle x, \min(\alpha, \mu_A(x)), \max(\beta, \nu_A(x))\rangle | x \in E\},$$

for $\alpha, \beta \in [0,1]$ and $\alpha + \beta \leq 1$, where $\alpha, \beta \in [0,1]$ and $\alpha + \beta \leq 1$, and their basic properties are studied.

Let us define for each $x \in E$:

$$\pi_A(x) = 1 - \mu_A(x) - \nu_A(x).$$

In the present paper, we will extend the idea for both level operators, changing their parameters $\alpha, \beta \in [0,1]$ with two functions $\alpha, \beta : T \to [0,1]$, where T is a fixed time-scale and for each $t \in T$:

$$\alpha(t) + \beta(t) \leq 1. \tag{1}$$

The basic properties of the new operators will be studied and some open problems will be formulated.

2 Preliminaries

Following [6], for every two IFSs A and B, we will define the following relations and operations (everywhere below "iff" means "if and only if"):

$$A \subseteq B \text{ iff } (\forall x \in E)(\mu_A(x) \leq \mu_B(x) \ \&\ \nu_A(x) \geq \nu_B(x));$$
$$A \supseteq B \text{ iff } B \subseteq A;$$
$$A = B \text{ iff } (\forall x \in E)(\mu_A(x) = \mu_B(x) \ \&\ \nu_A(x) = \nu_B(x));$$
$$A \cap B = \{\langle x, \min(\mu_A(x), \mu_B(x)), \max(\nu_A(x), \nu_B(x))\rangle | x \in E\};$$
$$A \cup B = \{\langle x, \max(\mu_A(x), \mu_B(x)), \min(\nu_A(x), \nu_B(x))\rangle | x \in E\};$$
$$A + B = \{\langle x, \mu_A(x) + \mu_B(x) - \mu_A(x)\mu_B(x), \nu_A(x)\nu_B(x)\rangle \mid x \in E\};$$
$$A.B = \{\langle x, \mu_A(x)\mu_B(x), \nu_A(x) + \nu_B(x) - \nu_A(x)\nu_B(x)\rangle \mid x \in E\};$$
$$A@B = \{\langle x, \tfrac{\mu_A(x)+\mu_B(x)}{2}, \tfrac{\nu_A(x)+\nu_B(x)}{2}\rangle | x \in E\}.$$

We must mention that operation @ plays simultaneously the role of both operations "union" and "intersection".

Following [6], we introduce 8 topological-type operators. The first two of them are analogous to the topological operators of closure and interior (see, e.g. [8,10]), while the rest ones are extensions of the first two.

For every IFS A,

$$\mathcal{C}(A) = \{\langle x, K, L\rangle | x \in E\},$$
$$\mathcal{I}(A) = \{\langle x, k, l\rangle | x \in E\},$$
$$\mathcal{C}_\mu(A) = \{\langle x, K, \min(1 - K, \nu_A(x))\rangle | x \in E\},$$
$$\mathcal{C}_\nu(A) = \{\langle x, \mu_A(x), L\rangle | x \in E\},$$
$$\mathcal{I}_\mu(A) = \{\langle x, k, \nu_A(x)\rangle | x \in E\},$$
$$\mathcal{I}_\nu(A) = \{\langle x, \min(1 - l, \mu_A(x)), l\rangle | x \in E\},$$
$$\mathcal{C}^*_\mu(A) = \{\langle x, \min(K, 1 - \nu_A(x)), \min(1 - K, \nu_A(x))\rangle | x \in E\},$$
$$\mathcal{I}^*_\nu(A) = \{\langle x, \min(1 - l, \mu_A(x)), \min(l, 1 - \mu_A(x))\rangle | x \in E\},$$

where

$$K = \sup_{y \in E} \mu_A(y),$$
$$L = \inf_{y \in E} \nu_A(y),$$
$$k = \inf_{y \in E} \mu_A(y),$$
$$l = \sup_{y \in E} \nu_A(y).$$

For every IFS A, following [4,6] we define 9 modal-type operators:

$$\Box A = \{\langle x, \mu_A(x), 1 - \mu_A(x)\rangle | x \in E\},$$
$$\Diamond A = \{\langle x, 1 - \nu_A(x), \nu_A(x)\rangle | x \in E\},$$
$$D_\alpha(A) = \{\langle x, \mu_A(x) + \alpha.\pi_A(x), \nu_A(x) + (1 - \alpha).\pi_A(x)\rangle | x \in E\},$$
$$F_{\alpha,\beta}(A) = \{\langle x, \mu_A(x) + \alpha.\pi_A(x), \nu_A(x) + \beta.\pi_A(x)\rangle | x \in E\},$$
$$G_{\alpha,\beta}(A) = \{\langle x, \alpha.\mu_A(x), \beta.\nu_A(x)\rangle | x \in E\},$$
$$H_{\alpha,\beta}(A) = \{\langle x, \alpha.\mu_A(x), \nu_A(x) + \beta.\pi_A(x)\rangle | x \in E\},$$
$$H^*_{\alpha,\beta}(A) = \{\langle x, \alpha.\mu_A(x), \nu_A(x) + \beta.(1 - \alpha.\mu_A(x) - \nu_A(x))\rangle | x \in E\},$$
$$J_{\alpha,\beta}(A) = \{\langle x, \mu_A(x) + \alpha.\pi_A(x), \beta.\nu_A(x)\rangle | x \in E\},$$
$$J^*_{\alpha,\beta}(A) = \{\langle x, \mu_A(x) + \alpha.(1 - \mu_A(x) - \beta.\nu_A(x)), \beta.\nu_A(x)\rangle | x \in E\}.$$

The first two operators correspond to the standard modal operators in the modal logic (see, e.g. [7]), while the rest operators are extensions or modifications of the first two ones.

The second type of modal operators, defined in IFSs theory are the following (see [6]):

$$\boxplus A = \{\langle x, \frac{\mu_A(x)}{2}, \frac{\nu_A(x) + 1}{2}\rangle | x \in E\},$$
$$\boxtimes A = \{\langle x, \frac{\mu_A(x) + 1}{2}, \frac{\nu_A(x)}{2}\rangle | x \in E\},$$
$$\boxplus_\alpha A = \{\langle x, \alpha.\mu_A(x), \alpha.\nu_A(x) + 1 - \alpha\rangle | x \in E\},$$
$$\boxtimes_\alpha A = \{\langle x, \alpha.\mu_A(x) + 1 - \alpha, \alpha.\nu_A(x)\rangle | x \in E\},$$

where $\alpha \in [0, 1]$,

$$\boxplus_{\alpha,\beta} A = \{\langle x, \alpha.\mu_A(x), \alpha.\nu_A(x) + \beta\rangle | x \in E\},$$
$$\boxtimes_{\alpha,\beta} A = \{\langle x, \alpha.\mu_A(x) + \beta, \alpha.\nu_A(x)\rangle | x \in E\},$$

where $\alpha, \beta, \alpha + \beta \in [0, 1]$,

$$\boxplus_{\alpha,\beta,\gamma} A = \{\langle x, \alpha.\mu_A(x), \beta.\nu_A(x) + \gamma\rangle | x \in E\},$$
$$\boxtimes_{\alpha,\beta,\gamma} A = \{\langle x, \alpha.\mu_A(x) + \gamma, \beta.\nu_A(x)\rangle | x \in E\},$$

where $\alpha, \beta, \gamma \in [0, 1]$ and $\max(\alpha, \beta) + \gamma \leq 1$.

$$\boxdot_{\alpha,\beta,\gamma,\delta} A = \{\langle x, \alpha.\mu_A(x) + \gamma, \beta.\nu_A(x) + \delta\rangle | x \in E\},$$

where $\alpha, \beta, \gamma, \delta \in [0, 1]$ and $\max(\alpha, \beta) + \gamma + \delta \leq 1$,

$$\boxcircle_{\alpha,\beta,\gamma,\delta,\varepsilon,\zeta} A = \{\langle x, \alpha.\mu_A(x) - \varepsilon.\nu_A(x) + \gamma,$$
$$\beta.\nu_A(x) - \zeta.\mu_A(x) + \delta\rangle | x \in E\},$$

where $\alpha, \beta, \gamma, \delta, \varepsilon, \zeta \in [0, 1]$ and

$$\max(\alpha - \zeta, \beta - \varepsilon) + \gamma + \delta \leq 1,$$
$$\min(\alpha - \zeta, \beta - \varepsilon) + \gamma + \delta \geq 0.$$

It is seen easy that operator $\boxcircle_{\alpha,\beta,\gamma,\delta,\varepsilon,\zeta}$ includes as a partial case operator $\boxdot_{\alpha,\beta,\gamma,\delta}$, that includes as a partial case operators $\boxplus_{\alpha,\beta,\gamma}$ and $\boxtimes_{\alpha,\beta,\gamma}$; they include as a partial case operators $\boxplus_{\alpha,\beta}$ and $\boxtimes_{\alpha,\beta}$, respectively; the rest ones include as partial case operators $\boxplus_{\alpha}$ and $\boxtimes_{\alpha}$, respectively, and finally, they include as partial case operators $\boxplus$ and $\boxtimes$, respectively.

Finally, following [6], we define the concept of the Temporal IFS (TIFS) by

$$A(T)\{\langle x, \mu_A(x, t), \nu_A(x, t)\rangle | x \in E \;\&\; t \in T\},$$

where $\mu_A, \nu_A : E \times T \to [0, 1]$, $\mu_A(x, t) + \nu_A(x, t) \leq 1$ for each $x \in E, t \in T$, and T is the above mentioned time-scale.

3 Temporal-Level Operators over Standard Intuitionistic Fuzzy Sets

Here, for a first time, we combine the ideas for the level operators $P_{\alpha,\beta}, Q_{\alpha,\beta}$ with the one for TIFS and in a result we obtain the two Temporal-level Operators (TLOs) $P^*_{\alpha,\beta}$ and $Q^*_{\alpha,\beta}$.

Let A be a fixed (standard) IFS. Then the new operators are defined by:

$$P^*_{\alpha,\beta}(A,T) = \{\langle x, \max(\alpha(t), \mu_A(x)), \min(\beta(t), \nu_A(x))\rangle | x \in E, t \in T\},$$
$$Q^*_{\alpha,\beta}(A,T) = \{\langle x, \min(\alpha(t), \mu_A(x)), \max(\beta(t), \nu_A(x))\rangle | x \in E, t \in T\}.$$

We must check that definitions of the TLOs are correct. Really, by the definitions of functions $\alpha, \beta, \mu_A, \nu_A$ it follows that for every $x \in E, t \in T$:

$$0 \le \max(\alpha(t), \mu_A(x)) \le 1,$$
$$0 \le \min(\beta(t), \nu_A(x)) \le 1.$$

Let

$$X \equiv \max(\alpha(t), \mu_A(x)) + \min(\beta(t), \nu_A(x)).$$

If $\alpha(t) \ge \mu_A(x)$, then

$$X = \alpha(t) + \min(\beta(t), \nu_A(x)) \le \alpha(t) + \beta(t) \le 1.$$

If $\alpha(t) < \mu_A(x)$, then

$$X = \mu_A(x) + \min(\beta(t), \nu_A(x)) \le \mu_A(x) + \nu_A(x) \le 1.$$

Therefore, the definition of operator $P^*_{\alpha,\beta}$ is correct. By the same manner we check that the definition of operator $Q^*_{\alpha,\beta}$ is correct, too.

The geometrical interpretation of the result of applying of operators $P^*_{\alpha,\beta}$ and $Q^*_{\alpha,\beta}$ over element $x \in E$ at time-moment $t \in T$ (we denote it by $P^*_{\alpha(t),\beta(t)}(A,T)(x)$ and $Q^*_{\alpha(t),\beta(t)}(A,T)(x)$) is shown on Figs. 1 and 2, respectively.

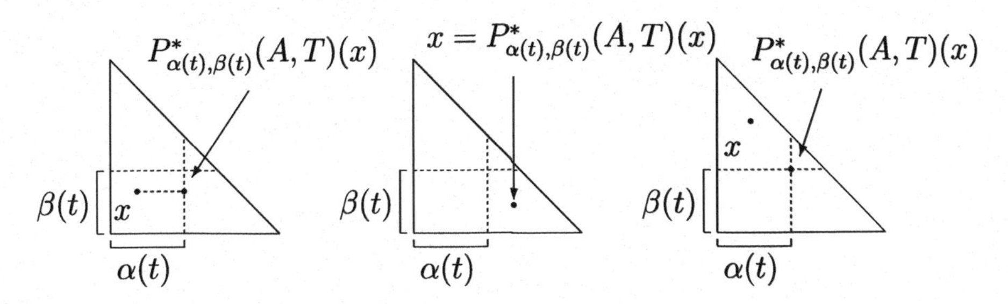

Fig. 1. Geometric interpretation of operator P^* (in three scenarios depending on the location of element x).

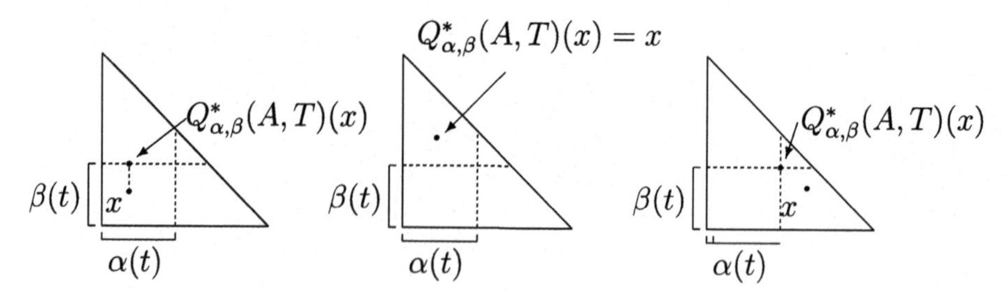

Fig. 2. Geometric interpretation of operator Q^* (in three scenarios depending on the location of element x).

The new operators satisfy the following properties.

Theorem 1. *For every IFS A, for every time-scale T and for every $\alpha, \beta, \gamma, \delta : T \to [0,1]$, such that for each $t \in T$: $\alpha(t) + \beta(t) \leq 1, \gamma(t) + \delta(t) \leq 1$:*

(a) $\neg P^*_{\alpha,\beta}(\neg A, T) = Q^*_{\beta,\alpha}(A, T)$;
(b) $\neg Q^*_{\alpha,\beta}(\neg A, T) = P^*_{\beta,\alpha}(A, T)$;
(c) $P^*_{\alpha,\beta}(P^*_{\gamma,\delta}(A, T), T) = P^*_{\max(\alpha,\gamma),\min(\beta,\delta)}(A, T)$;
(d) $P^*_{\alpha,\beta}(Q^*_{\gamma,\delta}(A, T), T) = Q^*_{\max(\alpha,\gamma),\min(\beta,\delta)}(P^*_{\alpha,\beta}(A, T), T)$;
(e) $Q^*_{\alpha,\beta}(P^*_{\gamma,\delta}(A, T), T) = P^*_{\min(\alpha,\gamma),\max(\beta,\delta)}(Q^*_{\alpha,\beta}(A, T), T)$;
(f) $Q^*_{\alpha,\beta}(Q^*_{\gamma,\delta}(A, T), T) = Q^*_{\min(\alpha,\gamma),\max(\beta,\delta)}(A, T)$.

Proof: Let A be a fixed IFS and T - a fixed time-scale. Then for (a) and (d) we obtain, respectively the following equalities:

$$\begin{aligned}
\neg P^*_{\alpha,\beta}(\neg A, T) &= \neg P^*_{\alpha,\beta}(A\{\langle x, \nu_A(x), \mu_A(x)\rangle | x \in E\}, T) \\
&= \neg\{\langle x, \max(\alpha(t), \nu_A(x)), \min(\beta(t), \mu_A(x))\rangle | x \in E, t \in T\} \\
&= \{\langle x, \min(\beta(t), \mu_A(x)), \max(\alpha(t), \nu_A(x))\rangle | x \in E, t \in T\}
\end{aligned}$$

$$Q^*_{\beta,\alpha}(A, T);$$

$$P^*_{\alpha,\beta}(Q^*_{\gamma,\delta}(A, T), T)$$

$$\begin{aligned}
&= P^*_{\alpha,\beta}(\{\langle x, \min(\gamma(t), \mu_A(x)), \max(\delta(t), \nu_A(x))\rangle | x \in E, t \in T\}, T) \\
&= \{\langle x, \max(\alpha(t), \min(\gamma(t), \mu_A(x)), \min(\beta(t), \max(\delta(t), \nu_A(x))\rangle | x \in E, t \in T\} \\
&= \{\langle x, \min(\max(\alpha(t), \gamma(t)), \max(\alpha(t), \mu_A(x))), \max(\min(\beta(t), \delta(t)), \max(\beta(t), \nu_A(x)))\rangle \\
&\qquad | x \in E, t \in T\} \\
&= Q^*_{\max(\alpha,\gamma),\min(\beta,\delta)}(\{\langle x, \max(\alpha(t), \mu_A(x)), \max(\beta(t), \nu_A(x))\rangle | x \in E, t \in T\}, T) \\
&= Q^*_{\max(\alpha,\gamma),\min(\beta,\delta)}(P^*_{\alpha,\beta}(A, T), T).
\end{aligned}$$

The rest assertions, as well as the following ones, are proved by the same manner.

Theorem 2. *For every two IFSs A and B, for each time-scale T and for every two functions $\alpha, \beta : T \to [0,1]$ satisfying (1) the following equalities hold:*

(a) $P^*_{\alpha,\beta}(A \cap B, T) = P^*_{\alpha,\beta}(A, T) \cap P^*_{\alpha,\beta}(B, T)$,
(b) $P^*_{\alpha,\beta}(A \cup B, T) = P^*_{\alpha,\beta}(A, T) \cup P^*_{\alpha,\beta}(B, T)$,
(c) $Q^*_{\alpha,\beta}(A \cap B, T) = Q^*_{\alpha,\beta}(A, T) \cap Q^*_{\alpha,\beta}(B, T)$,
(d) $Q^*_{\alpha,\beta}(A \cup B, T) = Q^*_{\alpha,\beta}(A, T) \cup Q^*_{\alpha,\beta}(B, T)$.

It can be checked that there is not any relation similar to (a) - (d) from Theorem 2, when the opertion is @.

Theorem 3. *For each IFS A, for each time-scale T and for every two functions $\alpha, \beta : T \to [0, 1]$ satisfying (1) the following equalities hold:*

(a) $\mathcal{C}(P^*_{\alpha,\beta}(A, T)) = P^*_{\alpha,\beta}(\mathcal{C}(A), T)$,
(b) $\mathcal{I}(P^*_{\alpha,\beta}(A, T)) = P^*_{\alpha,\beta}(\mathcal{I}(A), T)$,
(s) $\mathcal{C}(Q^*_{\alpha,\beta}(A, T)) = Q^*_{\alpha,\beta}(\mathcal{C}(A), T)$,
(d) $\mathcal{I}(Q^*_{\alpha,\beta}(A, T)) = Q^*_{\alpha,\beta}(\mathcal{I}(A), T)$.

An **Open Question** is: which relations will be valid, if operators $\mathcal{C}$ and $\mathcal{I}$ in (a) - (d) of Theorem 3 are changed with the rest topological operators described in Sect. 2.

Theorem 4. *For each IFS A, for each time-scale T and for every two functions $\alpha, \beta : T \to [0, 1]$ satisfying (1) the following equalities hold:*

(a) $\Box P^*_{\alpha,\beta}(A, T) \subseteq P^*_{\alpha,\beta}(\Box A, T)$,
(b) $\Diamond P^*_{\alpha,\beta}(A, T) \supseteq P^*_{\alpha,\beta}(\Diamond A, T)$,
(s) $\Box Q^*_{\alpha,\beta}(A, T) \supseteq Q^*_{\alpha,\beta}(\Box A, T)$,
(d) $\Diamond Q^*_{\alpha,\beta}(A, T) \subseteq Q^*_{\alpha,\beta}(\Diamond A, T)$.

An **Open Question** is: which relations will be valid, if operators $\Box$ and $\Diamond$ in (a) - (d) of Theorem 4 are changed with the rest modal operators from first type, described in Sect. 2.

Theorem 5. *For each IFS A, for each time-scale T, for every two functions $\alpha, \beta : T \to [0, 1]$ satisfying (1) and for every $a, b, c, d \in [0, 1]$ for which $\max(a, b) + c + d \leq 1$, the following equalities hold:*

(a) $\boxdot_{a,b,c,d} P^*_{\alpha,\beta}(A, T) = P^*_{a\alpha+c, b\beta+d}(\boxdot_{a,b,c,d} A)$,
(b) $\boxdot_{a,b,c,d} Q^*_{\alpha,\beta}(A, T) = Q^*_{a\alpha+c, b\beta+d}(\boxdot_{a,b,c,d} A)$.

Therefore, equalities (a) - (b) from Theorem 5 will be valid for the first 8 modal operators from the second type. An **Open Question** is: which relations will be valid, if operator $\boxdot_{a,b,c,d}$ in (a) - (b) of Theorem 5 is changed with the operator $\boxed{\circ}_{a,b,c,d,e,f}$, described in Sect. 2.

4 Temporal-Level Operators over Temporal Intuitionistic Fuzzy Sets

First, following [6], we can introduce two new operators defined over a TIFS $A(T)$ that will be simultaneously topological and temporal:

$$\mathcal{C}^*(A(T)) = \{\langle x, \sup_{t\in T} \mu_{A(T)}(x,t), \inf_{t\in T} \nu_{A(T)}(x,t)\rangle | x \in E\},$$
$$\mathcal{I}^*(A(T)) = \{\langle x, \inf_{t\in T} \mu_{A(T)}(x,t), \sup_{t\in T} \nu_{A(T)}(x,t)\rangle | x \in E\}.$$

For them, Theorem 3 obtains the form

Theorem 6. *For each IFS A, for each time-scale T and for every two functions* $\alpha, \beta : T \to [0,1]$ *satisfying (1) the following equalities hold:*

(a) $\mathcal{C}^*(P^*_{\alpha,\beta}(A,T)) = P^*_{\alpha,\beta}(\mathcal{C}^*(A),T)$,
(b) $\mathcal{I}^*(P^*_{\alpha,\beta}(A,T)) = P^*_{\alpha,\beta}(\mathcal{I}^*(A),T)$,
(s) $\mathcal{C}^*(Q^*_{\alpha,\beta}(A,T)) = Q^*_{\alpha,\beta}(\mathcal{C}^*(A),T)$,
(d) $\mathcal{I}^*(Q^*_{\alpha,\beta}(A,T)) = Q^*_{\alpha,\beta}(\mathcal{I}^*(A),T)$.

Now, if A is a fixed TIFS over universe E and time-scale T, then the new temporal-level operators (defined over the same time-scale) are defined by:

$$P^*_{\alpha,\beta}(A(T),T) = \{\langle x, \max(\alpha(t), \mu_A(x)), \min(\beta(t), \nu_A(x))\rangle | x \in E, t \in T\},$$
$$Q^*_{\alpha,\beta}(A(T),T) = \{\langle x, \min(\alpha(t), \mu_A(x)), \max(\beta(t), \nu_A(x))\rangle | x \in E, t \in T\}.$$

By the same manner as above, we can formulate and prove Theorem 1, 2, 4 and 5 for the case, when the TIFS A and the two temporal-level operators are defined over one time-scale T.

5 Conclusion: Next Extensions

In near future, we will study the situation when the TIFS A is defined over universe E and time-scale T, while the two temporal-level operators described here, are defined over another time-scale U.

On the other hand, it is clear that the new operators are extensions of the older P- and Q-operators, because when both functions α and β of the new functions give as a result only two constant, we obtain the older operators. Now, we will discuss a next step of extension, having in mind that in some sense P- and Q-operators are associated with operations "union" and "intersection", respectively. Therefore, we can change these operators with one O-operator, so that for both functions α and β:

$$O^{\cup}_{\alpha,\beta} = P^*_{\alpha,\beta},$$
$$O^{\cap}_{\alpha,\beta} = Q^*_{\alpha,\beta}$$

and for them the above formulated assertions, obviously, will be valid. Now, let $*$ and $\circ$ be two operations so that the first one is an analogous of operations "union" and the second one - of operation "intersection". Now, the O-operator will have the forms $O^*_{\alpha,\beta}$ and $O^\circ_{\alpha,\beta}$.

Acknowledgement. This research was funded by Bulgarian National Science Fund, grant number KP-06-N22/1/2018 "Theoretical research and applications of InterCriteria Analysis".

References

1. Angelova, N., Stoenchev, M.: Intuitionistic fuzzy conjunctions and disjunctions from first type. Ann. "Informatics" Sec., Union of Scientists in Bulgaria **8**, 1–17 (2015/2016)
2. Angelova, N., Stoenchev, M., Todorov, V.: Intuitionistic fuzzy conjunctions and disjunctions from second type. Issues Intuitionistic Fuzzy Sets Generalized Nets **13**, 143–170 (2017)
3. Angelova, N., Stoenchev, M.: Intuitionistic fuzzy conjunctions and disjunctions from third type. Notes Intuitionistic Fuzzy Sets **23**(5), 29–41 (2017)
4. Atanassov, K. Intuitionistic fuzzy sets, In: VII ITKR's Session, Sofia, June 1983 (Deposed in Central Sci. - Techn. Library of Bulg. Acad. of Sci., 1697/84) (in Bulg.). Reprinted: Int. J. Bioautomation, Vol. 20, S1–S6 (2016). (in English) (1983)
5. Atanassov, K.: Intuitionistic Fuzzy Sets. Springer, Heidelberg (1999). https://doi.org/10.1007/978-3-7908-1870-3
6. Atanassov, K.: On Intuitionistic Fuzzy Sets Theory. Springer, Berlin (2012). https://doi.org/10.1007/978-3-642-29127-2
7. Feys, R.: Modal Logics. Gauthier, Paris (1965)
8. Kuratowski, K.: Topology, vol. 1. Acad. Press, New York (1966)
9. Vassilev, P., Atanassov, K.: Extensions and Modifications of Intuitionistic Fuzzy Sets. "Prof. Marin Drinov" Academic Publishing House, Sofia (2019)
10. Yosida, K.: Functional Analysis. Springer, Berlin (1965)
11. Zadeh, L.: Fuzzy sets. Inf. Control **8**, 338–353 (1965)

Investigation of Cotton Production Companies Transition to Industry 4.0 Using the Intuitionistic Fuzzy TOPSIS

Merve Çil[1,2(✉)] and Babak Daneshvar Rouyendegh[2]

[1] Department of Industrial Engineering, Ankara Yıldırım Beyazıt University (AYBU), 06010 Ankara, Turkey

[2] Department of Mechanical and Industrial Engineering, Sultan Qaboos University, Muscut, Oman

Abstract. Manufacturing companies are looking for solutions for their current problems due to the latest developments in technology, changed expectations of consumers, decreased efficiency in production, changed conditions, and demands to increase product quality. Companies are aware that they cannot solve the problems of their current technologies. For this reason, companies have turned to industry 4.0. In this study, a survey consisting of 9 main criteria was conducted for experts working in the cotton manufacturing sector in Turkey. After testing the reliability of the survey data, they were evaluated using the intuitionistic fuzzy TOPSIS method. As a result, the readiness of cotton manufacturing firms in Turkey for the transition to the industry 4.0 model was examined, as well as the difficulties and deficiencies that firms will face during the transition to the model.

Keywords: Industry 4.0 · Intuitionistic fuzzy TOPSIS · Agriculture 4.0 · Production · Manufacturing

1 Introduction

Cotton, which is cultivated in 69 countries today, has not lost its relevance and importance in the world textile industry despite the increase in the production of different raw materials. 60% of woven raw material used in the textile industry still derives from cotton [1]. By-products obtained from cotton fiber and cottonseed are highly valued in the market. Cotton fiber is preferred due to many properties such as strength, conductivity, and hygiene. Vegetable oil, the raw material of the cellulose industry, and animal pulp are obtained from cotton seeds [2].

Turkey produces one-fourth of the cotton in the cotton processing plant established many years between 1860–1960. Cotton ginning factories opened after the establishment of simple weaving facilities in the 1860s; In the 1950s, large weaving factories took their place in the sector. Facilities with different capacities made significant contributions to the economy of the region and the cotton weaving industry [3].

One of the most important problems of cotton production in Turkey is the high cost of production. Most of the cotton we use in the industry is imported. The cost of cotton

C. Kahraman et al. (Eds.): INFUS 2021, LNNS 308, pp. 12–19, 2022.
https://doi.org/10.1007/978-3-030-85577-2_2

processing increases due to many reasons. To make improvements in costs, modern production technologies and tools should be used, necessary training should be given to producers and awareness should be created. At the same time, technical, economic, and administrative measures should be taken to forestall productivity and quality losses by preventing mishaps that may occur during pre-harvest and post-harvest transportation, storage, and ginning operations [4].

Traditional agriculture negatively affects the environment due to the methods it applies. The harmful effects of outdated methods used in cotton farming rise day by day. These methods cause soil congestion, soil degradation, and erosion. Also, due to improper spraying, groundwater mixes and pollution occurs. Soil operations without using advanced and developing state-of-the-art technology increase the emission of greenhouse gases to the atmosphere, causing global warming [5].

Turkey's industry is located at a point between the 2nd and 3rd Industrial Revolution. To switch to Industry 4.0, it is necessary to establish a technological infrastructure and to train well-equipped experts. To make the final determination of the case to capture Turkey's 4th industrial revolution and then must create a plan [6]. If manufacturing companies switch to industry 4.0, they will reach a high technological production level.

This paper is organized as follows. Section 2, survey studies in the manufacturing sector, intuitionistic fuzzy TOPSIS method and industry 4.0 are presented. Section 3, intuitionistic fuzzy TOPSIS is formulated. Section 4, the case study is briefly explained. The result and evaluation of this paper are explained in Sect. 5. Finally, a conclusion is made in Sect. 6.

2 Literature Review

Today, with the influence of information and communication technologies, both of which are in a state of rapid "change and development"; consumers can get any product they want and they can purchase it anywhere; in other words, competition between businesses now goes beyond the limits. Whereas rapid growth and diversification of consumer demands as well as consumers' tendency to perceive many products as similar to each other, on the other hand, provoke industries to develop their abilities and techniques further to gain ground among others [7].

2.1 Survey Studies in the Manufacturing Sector

In this study, a questionnaire consisting of a series of questions was used. The questionnaire can reach a larger audience easier than other data collection techniques. It can also achieve its purpose much faster and at less cost [8]. Many surveys have been conducted for improvement and development purposes in the manufacturing sector. The questionnaire has many advantages over other data collection methods. Surveys are the easiest to obtain data, reliable and lowest cost methods. Since the subjects will accept the questionnaire solution, it is more useful than other methods in obtaining information [9].

2.2 Intuitionistic Fuzzy TOPSIS Method

The traditional TOPSIS method is based on sorting alternative evaluations by calculating Euclidean (Euclidean) distances to ideal positive and negative solutions. Many researchers in the literature have expanded the traditional TOPSIS method with system theories and numbers other than crisp numbers and proposed different approaches using different distance measurement methods. In the TOPSIS method, the approach in which analyses are performed while using intuitionistic fuzzy numbers (IFN) is called the intuitionistic fuzzy TOPSIS (IF-TOPSIS) method.

2.3 Industry 4.0

The concept of industry 4.0 defined as "the integration of complex physical machinery and devices with networked sensors and software, used to predict, control and plan for better business and societal outcomes", or "a new level of value chain organization and management across the lifecycle of products" or "a collective term for technologies and concepts of value chain organization" [10].

3 Methodology

[11] defined the degree of belonging of the element "s" to the set of "B" as $\boldsymbol{\mu_B(s)}$, the degree of not being as $\mathbf{v_B(s)}$ and the index of hesitation as $\boldsymbol{\pi_B(s)}$ in the Intuitionistic Fuzzy Set. In the Intuitionistic fuzzy set theory, the sum of the degree of belonging and non-belonging is less than 1.

3.1 Intuitionistic Fuzzy TOPSIS (IF-TOPSIS)

The IF-TOPSIS method proposed by [12] based on the above definitions consists of the following steps. Linguistic terms are expressed in intuitively fuzzy numbers to define the weight of DMs. $D_j = [\mu_j, v_j, \pi_j]$ is the intuitionistic fuzzy number for jth DM ranking. The weight of decision-maker can be defined by the following formula:

$$\lambda_j = \frac{\left[\mu_j + \pi_j\left(\frac{\mu_j}{(\mu_j + v_j)}\right)\right]}{\sum_{j=1}^{p}\left[\mu_j + \pi_j\left(\frac{\mu_j}{(\mu_j + v_j)}\right)\right]} \tag{1}$$

$$\lambda_j \in [0, 1],\ j = 1, 2 \ldots, p \text{ and } \sum_{j=1}^{p} \lambda_j = 1 \tag{2}$$

The importance of the criteria is at different levels for each DM. $w_n^{\,j} = (\mu_n^{\,j}, v_n^{\,j}, \pi_n^{\,j})$, denote the intuitionistic fuzzy number about the nth criterion of jth DM. The weights of the criteria are calculated using the intuitionistic fuzzy weighted averaging (IFWA).

$$\begin{aligned} w_n &= IFWAr_\lambda\left(w_n^{(1)}, w_n^{(2)}, \ldots, w_n^{(p)}\right) = \lambda_1 w_n^{(1)} \oplus \lambda_2 w_n^{(2)} \oplus, \ldots, \oplus \lambda_p w_n^{(p)} \\ &= \left[1 - \prod_{j=1}^{p}\left(1 - \mu_n^{(j)}\right)^{\lambda_j}, \prod_{j=1}^{p}\left(v_n^{(j)}\right)^{\lambda_j}, \prod_{j=1}^{p}\left(1 - \mu_n^{(j)}\right)^{\lambda_j} - \prod_{j=1}^{p}\left(1 - v_n^{(j)}\right)^{\lambda_j}\right] \end{aligned} \tag{3}$$

Each view of the decision maker group must be combined into a single view to create the aggregated intuitionistic fuzzy decision matrix (AIFDM) model.

$m = (1, 2..., t)$, $n = (1, 2..., d)$, $R^{(j)} = (r_{mn}{}^{(j)})_{t*d}$ be the IFDM of each decision maker.

$\lambda = \{\lambda_1, \lambda_2, \lambda_3, ..., \lambda_p\}$ is set of weight of the decision maker

$$R = (r_{mm})_{t'*d'} \tag{4}$$

Later the weights of the criteria and the combined decision matrix was created, the aggregated weighted intuitionistic fuzzy decision matrix (AWIFDM) was obtained. While creating the AWIFDM equation below, the article of [12] was utilized.

$$\begin{aligned} r_{mn} &= IFWAr_{\lambda}\left(r_{mn}^{(1)}, r_{mn}^{(2)}, \ldots, r_{mn}^{(p)}\right) = \lambda_1 r_{mn}^{(1)} \oplus \lambda_2 r_{mn}^{(2)} \oplus, \ldots, \oplus \lambda_p r_{mn}^{(p)} \\ &= \left[1 - \prod_{j=1}^{p}\left(1 - \mu_{mn}^{(j)}\right)^{\lambda_j}, \prod_{j=1}^{p}\left(\nu_{mn}^{(j)}\right)^{\lambda_j}, \prod_{j=1}^{p}\left(1 - \mu_{mn}^{(j)}\right)^{\lambda_j} - \prod_{j=1}^{p}\left(1 - \nu_{mn}^{(j)}\right)^{\lambda_j}\right] \end{aligned} \tag{5}$$

$$R \oplus W = \left(\mu'_{mn}, \nu'_{mn}\right) = \{B, \mu_{mn} * \mu_n, \nu_{mn} + \nu_n - \nu_{mn} * \nu_n\} \tag{6}$$

$$\pi'_{mn} = 1 - \mu_{mn} * \mu_n - \nu_{mn} - \nu_n + \nu_{mn} * \nu_n \tag{7}$$

Let N_1 be benefit criterion and N_2 be cost criterion. A^* represents the intuitive fuzzy positive ideal solution while A-intuitive fuzzy negative represents the ideal solution.

$$A^* = \left(r_1'^*, r_2'^*, \ldots, r_d'^*\right) \quad r_n'^* = \left(\mu_n'^*, \nu_n'^*, \pi_n'^*\right), \; n = 1, 2, \ldots, d \tag{8}$$

$$A^- = \left(r_1'^-, r_2'^-, \ldots, r_d'^-\right) \quad r_n'^- = \left(\mu_n'^-, \nu_n'^-, \pi_n'^-\right) n = 1, 2, \ldots, d \tag{9}$$

$$\mu_n'^* = \left\{\left(\max m \left\{\mu'_{mn}\right\} n \in N_1\right), \left(\min m \left\{\mu'_{mn}\right\} n \in N_2\right)\right\}, \tag{10}$$

$$\nu_n'^* = \left\{\left(\max m \left\{\nu'_{mn}\right\} n \in N_1\right), \left(\min m \left\{\nu'_{mn}\right\} n \in N_2\right)\right\} \tag{11}$$

$$\pi_n'^- = \left\{\begin{array}{l} \left(1 - \max m\left\{\mu'_{mn}\right\} - \min m\left\{\nu'_{mn}\right\} n \in N_1\right), \\ \left(1 - \min m\left\{\mu'_{mn}\right\} - \max m\left\{\nu'_{mn}\right\} n \in N_2\right) \end{array}\right\}, \tag{12}$$

$$\mu_n'^- = \left\{\left(\max m \left\{\mu'_{mn}\right\} n \in N_1\right), \left(\min m \left\{\mu'_{mn}\right\} n \in N_2\right)\right\}, \tag{13}$$

$$\nu_n'^- = \left\{\left(\max m \left\{\nu'_{mn}\right\} n \in N_1\right), \left(\min m \left\{\nu'_{mn}\right\} n \in N_2\right)\right\}, \tag{14}$$

$$\pi_n'^- = \left\{\begin{array}{l} \left(1 - \max m \left\{\mu'_{mn}\right\} - \min m \left\{\nu'_{mn}\right\} n \in N_1\right), \\ \left(1 - \min m \left\{\mu'_{mn}\right\} - \max m \left\{\nu'_{mn}\right\} n \in N_2\right) \end{array}\right\}, \tag{15}$$

In this study, using the Euclidean distance measure, the separation measures of each alternative S_m^* and S_m^- are calculated in Eq. (16) and (17).

$$S_m^* = \sqrt{\frac{1}{2d}\sum_{n=1}^{d}\left[\left(\mu'_{mn} - \mu_n'^-\right)^2 + \left(\nu'_{mn} - \nu_n'^-\right)^2 + \left(\pi'_{mn} - \pi_n'^-\right)^2\right]} \tag{16}$$

$$S_m^- = \sqrt{\frac{1}{2d}\sum_{n=1}^{d}\left[\left(\mu'_{mn} - \mu'^{-}_{n}\right)^2 + \left(v'_{mn} - v'^{-}_{n}\right)^2 + \left(\pi'_{mn} - \pi'^{-}_{n}\right)^2\right]} \tag{17}$$

The relative closeness coefficient of an alternative A_m is calculated as in Eq. (18).

$$CC_m^* = \frac{S_m^-}{S_m^* + S_m^-} \text{ where } 0 \geq C_m^* \geq 1 \tag{18}$$

Later the relative closeness coefficient of each alternative is defined alternatives are ranked according to descending order of CC_y^*.

4 Case Study

The criteria were established as 9 main topics, using the literature to examine the compliance of the manufacturing companies in Adana with industry 4.0 (Schumacher, Erol and Sihn 2016). Under 9 criteria, 3 companies were evaluated; 77 questions were solved by 4 decision makers. Due to confidentiality, the firms were named X, U, and V. The intuitionistic fuzzy TOPSIS method was used to obtain more accurate results when evaluating questions by decision- makers. 3 companies that were involved in this study will contribute to determining the place in the industry 4.0 scenario.

Strategy (C_{01}); The strategy criterion allows the company to find different and innovative solutions when determining the company's roadmap and moving to implementation. Leadership (C_{02}); The leadership criterion shows that not only can one person catch up with rapidly developing technology with challenges, but also succeed with a management group consisting of many people. Customers (C_{03}); The customer criterion shows that it will meet the needs and expectations of consumers when it switches to a customer -oriented system using developing technology. Products (C_{04}); The product criteria show that the focus is on customer requests and that all processes from the production stage to the recycling of the product must be integrated. Operations (C_{05}); The operation criterion shows that with improvements made in the production process, efficiency will be increased in time and cost. Culture (C_{06}); The culture criterion shows that when firms have an organizational culture, mission and strategy are also more successful. People (C_{07}); The people criterion shows the importance of investing in employees. Governance (C_{08}); The governance criterion shows the importance of planning, the firm's predisposition to technology, coordination, and strategies. Technology (C_{09}); Technology criterion shows that with communication between machines and the use of the latest technology in all company departments, technology affects quality and efficiency simultaneously. The intuitionistic fuzzy TOPSIS method used for evaluation covers the following steps:

Step 1: Decision-makers' degree of importance are considered as linguistic variables. Then, the linguistic terms were transformed into intuitionistic fuzzy numbers and importance levels of DMs were determined (Table 1).

Table 1. Decision makers' weight

	DM-1	DM-2	DM-3	DM-4
Weights	0.28	0.24	0.23	0.25

Step 2: linguistic variables are given in Table 2.

Table 2. Linguistic terms for rating the importance of criteria and the decision makers

Linguistic terms	Intuitionistic fuzzy numbers		
Extremely important (EI)	0.85	0.15	0
Important (I)	0.65	0.25	0
Medium (M)	0.5	0.45	0
Unimportant (U)	0.3	0.7	0
Extremely unimportant (EU)	0.1	0.9	0

Table 3. Criteria evaluations of decision makers

	C_{01}	C_{02}	C_{03}	C_{04}	C_{05}	C_{06}	C_{07}	C_{08}	C_{09}
DM-1	EI	I	I	EI	EI	M	EU	EI	EI
DM-2	I	I	EI	I	I	M	U	I	I
DM-3	EI	EI	M	I	EI	U	EU	EI	I
DM-4	I	M	EI	EI	EI	U	M	I	EI

Step 3: Linguistic terms are defined so that four decision-makers can evaluate alternatives according to each criterion. Using IFWA, were combined to form a combined decision matrix. **Step 4:** The weighted decision matrix was obtained by multiplying the decision matrix values with the weight vector. **Step 5:** Keeping in mind the criteria, the positive and negative ideal solutions obtained as a result of the calculations were obtained. **Step 6:** It was calculated using the formulas in Eq. (16) and (17) to determine positive and negative separation measurements. **Step 7:** The relative closeness coefficient of the alternatives was calculated is given in Fig. 1 (Table 3).

Step 8: The order was organized from large to small. The company with the highest relative closeness coefficient means that it is the most suitable for industry 4.0, and the company with the lowest relative closeness coefficient implies that it is the one that is not the most suitable for industry 4.0 among the alternatives.

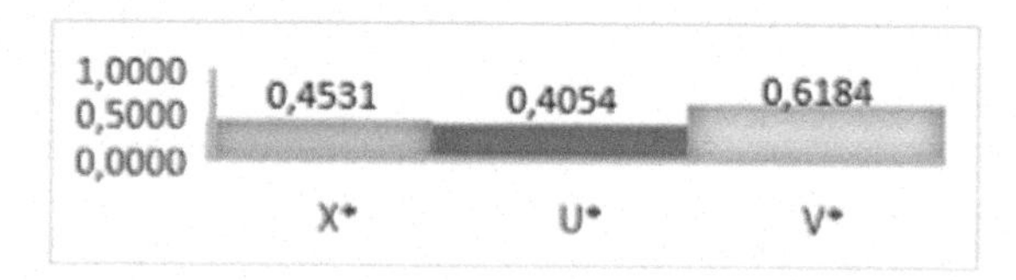

Fig. 1. The relative closeness coefficient of each alternative

5 Result and Evaluation

Based on the information obtained from 4 decision-makers, it was determined whether the companies comply with the industry 4.0 criteria or not. Strategy, leadership, customer, product, operation, culture, people, management, and technology criteria were taken into consideration to determine the compliance with industry 4.0. Company V ranked first by taking the highest relative closeness coefficient value. The feature that distinguishes V firm from other companies is its higher potential to comply with the operation, management, and technology criteria compared to others.

6 Conclusion

For the transition to industry 4.0, which requires a high budget, companies must use the latest technology in all departments. Companies do not want to take risks due to the high budget, so they continue their production in the same old order. Moreover, companies are afraid to switch to industry 4.0 because they have the fear of failing and coming to the brink of bankruptcy. Companies that switch to industry 4.0 show that a radical change at once is not a healthy step. It will be beneficial to switch to industry 4.0 gradually and in a way that is most suitable for the company, in a planned, scheduled, and time-consuming manner to eliminate risks and deficiencies. Every single company finds it difficult to give up its production and management habits. Including employees and customers, at first, it takes some time to get used to the new order but as time goes on, success in quality and efficiency satisfies everyone.

Industry 4.0 has many advantages. Some of these can be listed as being environmentally friendly systems, increasing quality, providing efficiency, reducing costs as well as waste, and saving energy. In addition to its advantages, industry 4.0 also has several disadvantages. Since industry 4.0 will mostly use automatic systems, the need for manpower will decrease. Employees will not want the transition to industry 4.0 as the effect of manpower on production will decrease and the effect of machines, on the contrary, will increase over time. Small businesses will begin to lose their effectiveness in the market, so many of them will have to withdraw from the sector. Some situations initially appear to be a disadvantage but later turn into advantages. Many professions will disappear with industry 4.0 whereas different professions will begin to emerge. Large companies will enter into a competitive environment and, therefore, will have to constantly improve themselves, which will be a major factor in increasing quality and productivity.

This study will guide companies that want to switch to Industry 4.0. Companies will gain awareness thanks to these and similar studies.

References

1. Çukobirlik Homepage (2013). http://www.cukobirlik.com.tr/?tekd=777&ikid=1&syf=*PAMUK*. Accessed 02 Sep 2020
2. TMMOB Homepage. http://www.taris.com.tr/pamukweb/t_pamuk_hak.asp. Accessed 08 Sep 2020
3. Özüdoğru, A.: Adana'da Dokuma Sanayi Yapılarının Endüstri Mirası Kapsamında İncelenmesi, vol. 29, no. 14, pp. 235–246 (2010)
4. Evcim, Ü.: Türkiye Pamuk Sorunları ve Çözüm Önerileri (2019). http://www.upk.org.tr/User_Files/editor/file/TRPamuk_25112019.pdf. Accessed 09 Sep 2020
5. Topdemir, T., Coşkun, M.B.: Menemen Koşullarında Pamuk Yetiştiriciliğinde Uygulanan Farklı Toprak İşleme Yöntemlerinin Enerji Verimliliği ve Kullanım Etkinliğinin Belirlenmesi. Adnan Menderes Üniversitesi Ziraat Fakültesi Derg **16**(1), 7–12 (2019)
6. Çetin, H.: Endüstri 4.0 Süreci ve Türkiye'nin Durumu. https://www.kirmizilar.com/tr/index.php/guncel-yazilar3/4826-endustri-4-0-sureci-ve-turkiye-nin-durumu. Accessed 09 Feb 2020
7. Taşkın, Ç., Emel, G.: İşletme Lojistiği. Alfa Aktüel (2010)
8. Büyüköztürk, Ş.: Anket Geliştirme. Türk Eğitim Bilim. Derg. **3**(2), 133–151 (2005)
9. Karaca, A., Hüsrev Turnagöl, H., Üniversitesi, H., Bilimleri, S., Yüksekokulu, T.: Reliability and validity of three different questionnaires in employees. Ankara (2007)
10. Mrugalska, B., Wyrwicka, M.K.: Towards lean production in Industry 4.0. Procedia Eng. **182**, 466–473 (2017)
11. Atanassov, K.T.: Intuitionistic fuzzy sets. Int. J. Bioautomation **20**, S1–S6 (2016)
12. Daneshvar Rouyendegh, B., Yildizbasi, A., Arikan, Ü.Z.B.: Using intuitionistic fuzzy TOPSIS in site selection of wind power plants in Turkey (2018)

A Novel Ranking Method of the Generalized Intuitionistic Fuzzy Numbers Based on Possibility Measures

Totan Garai(✉)

Department of Mathematics, Syamsundar College, Syamsundar, Purba Bardhaman 713424, West Bengal, India

Abstract. In the real line, generalized intuitionistic fuzzy numbers (GIFNs) are a special type of fuzzy sets (FSs). In this paper, we have developed a novel raking technique of GIFNs. Additionally we have defined possibility mean and standard deviation of GIFNs. Then formulates the magnitude of membership and non-membership function of GIFNs.

Keywords: Possibility mean · Possibility variance and standard deviation · Generalized intuitionistic fuzzy number

1 Introduction

In 1965 by Zadeh [1] proposed the FS theory. FS theory has been well exhibited and applied in many real science problem. FS information carried by membership function, for this purpose some difficulties aeries in some real problem. Thereafter, Atanassov [2] introduced the intuitionistic fuzzy set (IFS) by adding an additional non-membership function. This may fleet more enormous and lenient intimation as compared with the FS. Intuitionistic fuzzy numbers (IFNs) are a generalization of of fuzzy numbers [9]. Recently researchers work on three kinds IFNs which are triangular IFN (TrIFN) [5,6], trapezoidal IFN (TIFN) [4,7] and inter-valued trapezoidal IFN [3,8].

Now a days possibility mean and variance are the important mathematical tool of uncertain theory. First time the notations of lower and upper possibilistic mean was proposed by Carlsson and Fuller [11]. The weighted possibilistic mean of interval-valued fuzzy number introduced by Fuller and Majlender [10]. Garai et al. [13] proposed the possibility mean, variance and covariance of TIFNs. Recently, a correlation coefficient of generalized intuitionistic fuzzy sets was considered by Park et al. [12].

In decision making problem fuzzy set theory has a exceptional role. Jain [14] was introduced the idea of ranking fuzzy numbers (FNs). Based on the centroidal and degree of fragmentation, Zeng and Cao [15] formulated a new ranking method. The index technique of ranking fuzzy numbers was presented by Zeng and Cao [16]. Recently, Qiupeng and Zuxing [17] was formulated a novel ranking technique of FNs based on possibility theory.

C. Kahraman et al. (Eds.): INFUS 2021, LNNS 308, pp. 20–27, 2022.
https://doi.org/10.1007/978-3-030-85577-2_3

In spite of the above narrated improvements, following mergers can also be formed in the possibility mean and standard deviation of generalized intuitionistic numbers.

- Possibility mean and standard deviation of generalized intuitionistic numbers.
- The magnitude of generalized intuitionistic numbers for membership and non-membership function.

The rest of the paper is organized as follows: In Sect. 2, we present some basic knowledge of generalized intuitionistic numbers, In Sect. 3, we formulate the magnitude of generalized intuitionistic numbers. Finally, the conclusion and scope of future work plan consider in Sect. 4.

2 Basic Preliminaries

Definition 1. For any two real number ($\mathbb{R}$) $w_{\tilde{c}} \in [0,1]$ and $u_{\tilde{c}} \in [0,1]$, generalized intuitionistic fuzzy number (GIFN) number $\tilde{c}^I$ is a special kind FS on $\mathbb{R}$ where $0 \leq w_{\tilde{c}} + u_{\tilde{c}} \leq 1$. The membership function of GIFN is $\mu_{\tilde{c}^I} : \mathbb{R} \longrightarrow [0, w_{\tilde{c}}]$ and non membership function is $\nu_{\tilde{c}^I} : \mathbb{R} \longrightarrow [u_{\tilde{c}}, 1]$. Which satisfies the followings:

(i) For there exit at least two $x_1, x_2 \in \mathbb{R}$ satisfies that condition $\mu_{\tilde{c}^I}(x_1) = w_{\tilde{c}}$ and $\nu_{\tilde{c}^I}(x_2) = u_{\tilde{c}}$.
(ii) $\mu_{\tilde{c}^I}$ is quasi concave function on $\mathbb{R}$, and $\mu_{\tilde{c}^I}$ upper semi continuous function on $\mathbb{R}$.
(iii) $\nu_{\tilde{c}^I}$ is quasi convex function on $\mathbb{R}$, and $\nu_{\tilde{c}^I}$ lower semi continuous function on $\mathbb{R}$.
(iv) The support of $\tilde{c}^I$ ($i.e.$ $\tilde{c}^I_{<0,1>} = \{\mu_{\tilde{c}^I}(x) \geq 0, \nu_{\tilde{c}^I}(x) \leq 1; \forall x \in \mathbb{R}\}$) is bonded.

From the above definition of the GIFN, we can defined membership and non-membership function of GIFN $\tilde{c}^I = \{(\underline{c_1}, c_{1l}, c_{1r}, \overline{c_1}); w_{\tilde{c}}, (\underline{c_2}, c_{2l}, c_{2r}, \overline{c_2}); u_{\tilde{c}}\}$ as follows

$$\mu_{\tilde{c}^I}(x) = \begin{cases} 0 & \text{if } x < \underline{c_1} \\ g_{\mu l}(x) & \text{if } \underline{c_1} \leq x < a_{1l} \\ w_{\tilde{c}} & \text{if } c_{1l} \leq x \leq c_{1r} \\ g_{\mu r}(x) & \text{if } c_{1r} < x \leq \overline{c_1} \\ 0 & \text{if } x > \overline{c_1} \end{cases}$$

and

$$\nu_{\tilde{c}^I}(x) = \begin{cases} 1 & \text{if } x < \underline{c_2} \\ g_{\nu l}(x) & \text{if } \underline{c_2} \leq x < c_{2l} \\ u_{\tilde{c}} & \text{if } c_{2l} \leq x \leq c_{2r} \\ g_{\nu r}(x) & \text{if } c_{2r} < x \leq \overline{c_2} \\ 1 & \text{if } x > \overline{c_2} \end{cases}$$

respectively. Where the functions $g_{\mu l} : [\underline{c_1}, c_{1l}] \rightarrow [0, w_{\tilde{c}}]$ and $g_{\nu r} : [c_{2r}, \overline{c_2}] \rightarrow [u_{\tilde{c}}, 1]$ are continuous, non decreasing and satisfy the conditions $g_{\mu l}(\underline{c_1}) = 0$, $g_{\mu l}(c_{1l}) = w_{\tilde{c}}$, $g_{\nu r}(c_{2r}) = u_{\tilde{c}}$ and $g_{\nu r}(\overline{c_2}) = 1$; the functions $g_{\mu r} : [c_{1r}, \overline{c_1}] \rightarrow$

$[0, w_{\tilde{c}}]$ and $g_{\nu l} : [\underline{c_2}, c_{2l}] \to [u_{\tilde{c}}, 1]$ are continuous, non increasing and satisfy the conditions $g_{\mu r}(c_{1r}) = w_{\tilde{c}}, g_{\mu r}(\overline{c_1}) = 0$, $g_{\nu l}(\underline{c_2}) = 1$ and $g_{\nu l}(c_{2l}) = u_{\tilde{c}}$. $\overline{c_1}$ and $\underline{c_1}$ are called upper and lower limits and $[c_{1l}, c_{1r}]$ be called the mean interval of the GIFN $\tilde{c}^I$ for the membership function respectively.

Definition 2. Let $\tilde{c}^I$ be an GIFN [7] in $\mathbb{R}$. Membership function and non-membership function of $\tilde{c}^I$ defined as:

$$\mu_{\tilde{c}^I}(x) = \begin{cases} \dfrac{x - c_1}{c_2 - c_1} w_{\tilde{c}}, & \text{if } c_1 \leq x < c_2 \\ w_{\tilde{c}}, & \text{if } c_2 \leq x \leq c_3 \\ \dfrac{c_4 - x}{c_4 - c_3} w_{\tilde{c}}, & \text{if } c_3 < x \leq c_4 \\ 0 & \text{if } x < c_1 \text{ or } x > c_4 \end{cases}$$

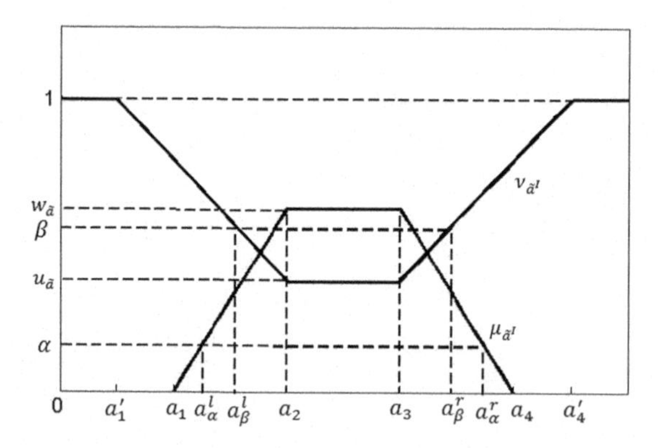

Fig. 1. α-cut and β-cut set of GTIFNs($\tilde{a}^I = (a_1, a_2, a_3, a_4; w_{\tilde{a}})(\acute{a}_1, a_2, a_3, \acute{a}_4; u_{\tilde{a}})$)

and

$$\nu_{\tilde{c}^I}(x) = \begin{cases} \dfrac{(c_2 - x) + (x - \acute{c}_1)u_{\tilde{c}}}{c_2 - \acute{c}_1}, & \text{if } \acute{c}_1 \leq x < c_2 \\ u_{\tilde{c}}, & \text{if } c_2 \leq x \leq c_3 \\ \dfrac{(x - c_3) + (\acute{c}_4 - x)u_{\tilde{c}}}{\acute{c}_4 - c_3}, & \text{if } c_3 < x \leq \acute{c}_4 \\ 0 & \text{if } x < \acute{c}_1 \text{ or } x > \acute{c}_4 \end{cases}$$

respectively, GIFN pictured in Fig. 1. $w_{\tilde{c}}$ represent the maximum degree of membership function and $u_{\tilde{c}}$ minimum degree of non-membership function. $w_{\tilde{c}}$ and $u_{\tilde{c}}$ satisfying the conditions: $0 \leq w_{\tilde{c}} \leq 1$, $0 \leq u_{\tilde{c}} \leq 1$ and $0 \leq w_{\tilde{c}} + u_{\tilde{c}} \leq 1$. Then, the IFN $\tilde{a}^I$ is called the generalized trapezoidal intuitionistic fuzzy number (GTIFN), denoted by $\tilde{c}^I = (c_1, c_2, c_3, c_4; w_{\tilde{c}})(\acute{c}_1, c_2, c_3, \acute{c}_4; u_{\tilde{c}})$. When $c_2 = c_3$, a GTIFN reduce to generalized triangular intuitionistic fuzzy number (GTrIFN), denoted by $\tilde{c}^I = (c_1, c_2, c_3; w_{\tilde{c}})(\acute{c}_1, c_2, \acute{c}_3; u_{\tilde{c}c})$.

Definition 3. If $\acute{c_1}, c_1 \geq 0$, and one of six [13] values $\acute{c_1}, c_1, c_2, c_3, c_4$ and $\acute{c_4}$ is not equal to zero, then the GTIFN $\tilde{c}^I = (c_1, c_2, c_3, c_4; w_{\tilde{c}})(\acute{c_1}, c_2, c_3, \acute{c_4}; u_{\tilde{c}})$ is called a positive GTIFN and its noted by $\tilde{c}^I > 0$.

Definition 4. Let $\tilde{c}^I = (c_1, c_2, c_3, c_4; w_{\tilde{c}})(\acute{c_1}, c_2, c_3, \acute{c_4}; u_{\tilde{c}})$ and $\tilde{e}^I = (e_1, e_2, e_3, e_4; w_{\tilde{e}})$ $(\acute{e_1}, e_2, e_3, \acute{e_4}; u_{\tilde{e}})$ be two GTIFNs and $k \geq 0$. Then the arithmetic operations of two [13] GTIFNs defined as:

(*i*) $\tilde{c}^I + \tilde{e}^I = (c_1 + e_1, c_2 + e_2, c_3 + e_3, c_4 + e_4; w_{\tilde{c}} \vee w_{\tilde{e}})(\acute{c_1} + \acute{e_1}, c_2 + e_2, c_3 + e_3, \acute{c_4} + \acute{e_4}; u_{\tilde{c}} \wedge e_{\tilde{b}})$, where the symbols '$\wedge$=min' and '$\vee$=max'.
(*ii*) $k\tilde{c}^I = (kc_1, kc_2, kc_3, kc_4; w_{\tilde{c}})(k\acute{c_1}, kc_2, kc_3, k\acute{c_4}; u_{\tilde{c}})$

Definition 5. Let $w_{\tilde{c}}$ and $u_{\tilde{c}}$ be two weights of $\tilde{c}^I$, then the (α, β)-cut set, α-cut set and β-cut set [13] are defined as: $\tilde{c}^I_{\alpha,\beta} = \{x : \mu_{\tilde{c}^I}(x) \geq \alpha, \nu_{\tilde{c}^I}(x) \leq \beta\}$, $\tilde{c}^I_\alpha = \{x : \mu_{\tilde{c}^I}(x) \geq \alpha\}$ and $\tilde{a}^I_\beta = \{x : \nu_{\tilde{c}^I}(x) \leq \beta\}$, respectively, where $0 \leq \alpha + \beta \leq 1$, $0 \leq \alpha \leq w_{\tilde{c}}$ and $u_{\tilde{c}} \leq \beta \leq 1$.

Definition 6. Let $\tilde{c}^I = (c_1, c_2, c_3, c_4; w_{\tilde{c}})(\acute{c_1}, c_2, c_3, \acute{c_4}; u_{\tilde{c}})$ be a GTIFN. By the Definition 6, the α, β-cuts set of GTIFN $\tilde{c}^I$ calculated as

$$\tilde{c}^I_\alpha = [c^l_\alpha, c^r_\alpha] = \left[c_1 + \frac{\alpha(c_2 - c_1)}{w_{\tilde{c}}}, c_4 - \frac{\alpha(c_4 - c_3)}{w_{\tilde{c}}}\right] \tag{1}$$

and

$$\tilde{c}^I_\beta = [c^l_\beta, c^I_\beta] = \left[\acute{c_1} + \frac{(1-\beta)(c_2 - \acute{c_1})}{1 - u_{\tilde{c}}}, \acute{c_4} - \frac{(1-\beta)(\acute{c_4} - c_3)}{1 - u_{\tilde{c}}}\right] \tag{2}$$

Definition 7. Let $\tilde{c}^I_\alpha = [c^l_\alpha, c^r_\alpha]$ be the α-cut and $\tilde{c}^I_\beta = [c^l_\beta, c^r_\beta]$ be the β-cut set of a GIFN $\tilde{c}^I$ with $0 \leq \alpha \leq w_{\tilde{c}}$ and $u_{\tilde{c}} \leq \beta \leq 1$. The possibility mean for membership function of GIFN $\tilde{c}^I$ defined as

$$M_\mu(\tilde{c}^I) = \frac{\int_0^{w_{\tilde{c}}} \alpha \frac{c^l_\alpha + c^r_\alpha}{2} d\alpha}{\int_0^{w_{\tilde{c}}} \alpha d\alpha} = \frac{1}{w^2_{\tilde{c}}} \int_0^{w_{\tilde{c}}} \alpha[c^l_\alpha + c^r_\alpha] d\alpha \tag{3}$$

and the possibility mean of non-membership function of GIFN $\tilde{c}^I$ defined as

$$M_\nu(\tilde{c}^I) = \frac{\int_{u_{\tilde{c}}}^1 \beta \frac{c^l_\beta + c^r_\beta}{2} d\beta}{\int_{u_{\tilde{c}}}^1 \beta d\beta} = \frac{1}{u^2_{\tilde{c}}} \int_{u_{\tilde{c}}}^1 \beta[c^l_\beta + c^r_\beta] d\beta \tag{4}$$

Definition 8. Let $\tilde{c}^I_\alpha = [c^l_\alpha, c^r_\alpha]$ be the α-cut and $\tilde{c}^I_\beta = [c^l_\beta, c^r_\beta]$ be the β-cut set of a GIFN $\tilde{c}^I$ with $0 \leq \alpha \leq w_{\tilde{c}}$ and $u_{\tilde{c}} \leq \beta \leq 1$. The possibility variance for membership function of GIFN $\tilde{c}^I$ defined as

$$V_\mu(\tilde{c}^I) = \frac{1}{2} \int_0^{w_{\tilde{c}}} \alpha(c^r_\alpha - c^l_\alpha)^2 d\alpha \tag{5}$$

and the possibility variance of non-membership function of GIFN $\tilde{a}^I$ defined as

$$V_\nu(\tilde{c}^I) = \frac{1}{2}\int_{u_{\tilde{c}}}^{1} \beta(c_\beta^r - c_\beta^l)^2 d\beta \tag{6}$$

Definition 9. Let $\tilde{c}_\alpha^I = [c_\alpha^l, c_\alpha^r]$ be the α-cut and $\tilde{c}_\beta^I = [c_\beta^l, c_\beta^r]$ be the β-cut set of a GIFN $\tilde{c}^I$ with $0 \le \alpha \le w_{\tilde{c}}$ and $u_{\tilde{c}} \le \beta \le 1$. The possibility standard deviation of membership function of GIFN $\tilde{c}^I$ defined as

$$\sigma_\mu(\tilde{c}^I) = \sqrt{V_\mu(\tilde{c}^I)} = \sqrt{\frac{1}{2}\int_0^{w_{\tilde{c}}} \alpha(c_\alpha^r - c_\alpha^l)^2 d\alpha} \tag{7}$$

and the possibility standard deviation of non-membership function of GIFN $\tilde{c}^I$ defined as

$$\sigma_\nu(\tilde{c}^I) = \sqrt{V_\nu(\tilde{c}^I)} = \sqrt{\frac{1}{2}\int_{u_{\tilde{c}}}^{1} \beta(c_\beta^r - c_\beta^l)^2 d\beta} \tag{8}$$

3 A Novel Approach Ranking of Generalized Intuitionistic Fuzzy Numbers

Let $M_\mu(\tilde{c}^I)$ and $M_\nu(\tilde{c}^I)$ are the average values for membership and non-membership functions of GIFN $\tilde{c}^I$. $\sigma_\mu(\tilde{c}^I)$ is the degree deviation from membership $M_\mu(\tilde{c}^I)$ and $\sigma_\nu(\tilde{c}^I)$ is the degree deviation from non-membership $M_\nu(\tilde{c}^I)$. Then, for GIFN $\tilde{c}^I$, we have define the magnitude of membership and non-membership functions of $\tilde{c}^I$ as

$$Mag_\mu(\tilde{c}^I) = M_\mu(\tilde{c}^I) + \sigma_\mu(\tilde{c}^I) \tag{9}$$

and

$$Mag_\nu(\tilde{c}^I) = M_\nu(\tilde{c}^I) + \sigma_\nu(\tilde{c}^I) \tag{10}$$

The magnitudes of a GIFN $\tilde{a}^I$ defined in equation (9) & (10) using the possibility mean, variance and standard deviation. The numerical value of $Mag_\mu(\tilde{c}^I)$ and $Mag_\nu(\tilde{c}^I)$ are used to rank the GIFNs.

The GIFN will be greater as one of $Mag_\mu(.)$ and $Mag_\mu(.)$ is getter. Therefore, for any two GIFNs $\tilde{c}^I$ and $\tilde{e}^I$, we define the ranking of $\tilde{c}^I$ and $\tilde{e}^I$ by the $Mag(.)$ value as follows:

(i) If $Mag_\mu(\tilde{c}^I) > Mag_\mu(\tilde{e}^I)$, then $\tilde{c}^I$ is bigger than $\tilde{e}^I$ denoted by $\tilde{c}^I \succ \tilde{e}^I$.
(ii) If $Mag_\mu(\tilde{c}^I) < Mag_\mu(\tilde{e}^I)$, then $\tilde{c}^I$ is smaller than $\tilde{e}^I$ denoted by $\tilde{c}^I \prec \tilde{e}^I$.
(iii) If $Mag_\mu(\tilde{c}^I) = Mag_\mu(\tilde{e}^I)$, then
  - (a) If $Mag_\nu(\tilde{c}^I) < Mag_\nu(\tilde{e}^I)$ then $\tilde{c}^I \prec \tilde{e}^I$;
  - (b) If $Mag_\nu(\tilde{c}^I) = Mag_\nu(\tilde{e}^I)$ then $\tilde{c}^I$ and $\tilde{e}^I$ represent the same information, denoted by $\tilde{c}^I \approx \tilde{e}^I$;
  - (c) If $Mag_\nu(\tilde{c}^I) > Mag_\nu(\tilde{e}^I)$ then $\tilde{c}^I \succ \tilde{e}^I$;

Then we formulate the order $\geq$ and $\leq$ as $\tilde{c}^I \geq \tilde{e}^I$ if and only if $\tilde{c}^I \succ \tilde{e}^I$ or $\tilde{c}^I \sim \tilde{e}^I$, $\tilde{c}^I \leq \tilde{e}^I$ if only if $\tilde{c}^I \prec \tilde{e}^I$ or $\tilde{c}^I \sim \tilde{e}^I$. Thus we define the ranking of $\tilde{c}^I$ and $\tilde{e}^I$ by the $Mag(.)$ as follows.

(i) If $Mag_\mu(\tilde{c}^I) \geq Mag_\mu(\tilde{e}^I)$ or $Mag_\nu(\tilde{c}^I) \geq Mag_\nu(\tilde{e}^I)$ then $\tilde{c}^I \geq \tilde{e}^I$.
(ii) If $Mag_\mu(\tilde{c}^I) \leq Mag_\mu(\tilde{e}^I)$ or $Mag_\nu(\tilde{c}^I) \leq Mag_\nu(\tilde{e}^I)$ then $\tilde{c}^I \leq \tilde{e}^I$.

Theorem 1. *For a GTIFNs* $\tilde{c}^I = (c_1, c_2, c_3, c_4; w_{\tilde{c}})(\acute{c}_1, c_2, c_3, \acute{c}_4; u_{\tilde{c}})$, $\tilde{e}^I = (e_1, e_2, e_3, e_4; w_{\tilde{e}})(\acute{e}_1, e_2, e_3, \acute{e}_4; u_{\tilde{e}}) \in F(\mathbb{R})$. $\tilde{c}^I \leq \tilde{e}^I$ *if*

$$\begin{aligned} &\frac{c_1 + 2c_2 + 2c_3 + c_4}{6} + \sqrt{\frac{1}{4}(c_4 - c_1)^2 + \frac{1}{8}(c_4 - c_3 + c_2 - c_1)^2 - \frac{1}{3}(c_4 - c_1)(c_4 - c_3 + c_2 - c_1)w_{\tilde{c}}} \\ \leq &\frac{e_1 + 2e_2 + 2e_3 + e_4}{6} + \sqrt{\frac{1}{4}(e_4 - c_1)^2 + \frac{1}{8}(e_4 - e_3 + e_2 - e_1)^2 - \frac{1}{3}(e_4 - e_1)(e_4 - e_3 + e_2 - e_1)w_{\tilde{e}}} \end{aligned} \quad (11)$$

Or,

$$\begin{aligned} &\frac{\acute{c}_1 + 2c_2 + 2c_3 + \acute{c}_4}{6} + \sqrt{\frac{1}{4}(\acute{c}_4 - \acute{c}_1)^2 + \frac{1}{8}(\acute{c}_4 - c_3 + c_2 - \acute{a}_1)^2 - \frac{1}{3}(\acute{c}_4 - \acute{c}_1)(\acute{c}_4 - c_3 + c_2 - \acute{c}_1)u_{\tilde{c}}} \\ \leq &\frac{\acute{e}_1 + 2e_2 + 2e_3 + \acute{e}_4}{6} + \sqrt{\frac{1}{4}(\acute{e}_4 - \acute{e}_1)^2 + \frac{1}{8}(\acute{e}_4 - e_3 + e_2 - \acute{e}_1)^2 - \frac{1}{3}(\acute{e}_4 - \acute{e}_1)(\acute{e}_4 - e_3 + e_2 - \acute{e}_1)u_{\tilde{e}}} \end{aligned} \quad (12)$$

Proof. Let $\tilde{c}^I = (c_1, c_2, c_3, c_4; w_{\tilde{c}})(\acute{c}_1, c_2, c_3, \acute{c}_4; u_{\tilde{c}})$, $\tilde{e}^I = (e_1, e_2, e_3, e_4; w_{\tilde{e}})(\acute{e}_1, e_2, e_3, \acute{e}_4; u_{\tilde{e}}) \in F(\mathbb{R})$. $\tilde{c}^I \leq \tilde{e}^I$ be two generalized trapezoidal intuitionistic fuzzy numbers . Then the possibility mean and variance of $\tilde{c}^I \& \tilde{e}^I$ calculate as follows

$$\begin{aligned} M(\tilde{c}^I) &= \frac{1}{w_{\tilde{c}}^2} \int_0^{w_{\tilde{c}}} \alpha \left(c_1 + \frac{\alpha(c_2 - c_1)}{w_{\tilde{c}}} + c_4 - \frac{\alpha(c_4 - c_3)}{w_{\tilde{c}}} \right) d\alpha \\ &= \frac{c_1 + 2c_2 + 2c_3 + c_4}{6} \end{aligned} \quad (13)$$

and

$$\begin{aligned} M(\tilde{e}^I) &= \frac{1}{w_{\tilde{e}}^2} \int_0^{w_{\tilde{e}}} \alpha \left(e_1 + \frac{\alpha(e_2 - e_1)}{w_{\tilde{e}}} + e_4 - \frac{\alpha(e_4 - e_3)}{w_{\tilde{e}}} \right) d\alpha \\ &= \frac{e_1 + 2e_2 + 2e_3 + e_4}{6} \end{aligned} \quad (14)$$

$$\begin{aligned} V_\mu(\tilde{c}^I) &= \frac{1}{2} \int_0^{w_{\tilde{c}}} \alpha (c_\alpha^r - c_\alpha^l)^2 d\alpha \\ &= \frac{1}{2} \int_0^{w_{\tilde{c}}} \alpha \left(c_4 - \frac{\alpha(c_4 - c_3)}{w_{\tilde{c}}} - c_1 - \frac{\alpha(c_2 - c_1)}{w_{\tilde{c}}} \right)^2 d\alpha \\ &= \frac{1}{4}(c_4 - c_1)^2 w_{\tilde{c}}^2 + \frac{1}{8}(c_4 - c_3 + c_2 - c_1)^2 w_{\tilde{c}}^2 - \frac{1}{3}(c_4 - c_1)(c_4 - c_3 + c_2 - c_1) w_{\tilde{c}}^2 \end{aligned} \quad (15)$$

and

$$\begin{aligned} V_\mu(\tilde{e}^I) &= \frac{1}{2} \int_0^{w_{\tilde{e}}} \alpha (e_\alpha^r - e_\alpha^l)^2 d\alpha \\ &= \frac{1}{2} \int_0^{w_{\tilde{e}}} \alpha \left(e_4 - \frac{\alpha(e_4 - e_3)}{w_{\tilde{e}}} - e_1 - \frac{\alpha(e_2 - e_1)}{w_{\tilde{e}}} \right)^2 d\alpha \\ &= \frac{1}{4}(e_4 - e_1)^2 w_{\tilde{e}}^2 + \frac{1}{8}(e_4 - e_3 + e_2 - e_1)^2 w_{\tilde{e}}^2 - \frac{1}{3}(e_4 - e_1)(e_4 - e_3 + e_2 - e_1) w_{\tilde{e}}^2 \end{aligned} \quad (16)$$

The possibilistic standard deviation of $\tilde{c}^I, \tilde{e}^I$ are

$$\sigma_\mu(\tilde{c}^I) = \sqrt{V_\mu(\tilde{c}^I)}$$
$$= \sqrt{\frac{1}{4}(c_4 - c_1)^2 + \frac{1}{8}(c_4 - c_3 + c_2 - c_1)^2 - \frac{1}{3}(c_4 - c_1)(c_4 - c_3 + c_2 - c_1)w_{\tilde{c}}} \quad (17)$$

and

$$\sigma_\mu(\tilde{e}^I) = \sqrt{V_\mu(\tilde{e}^I)}$$
$$= \sqrt{\frac{1}{4}(e_4 - e_1)^2 + \frac{1}{8}(e_4 - e_3 + e_2 - e_1)^2 - \frac{1}{3}(e_4 - e_1)(e_4 - e_3 + e_2 - e_1)w_{\tilde{e}}} \quad (18)$$

Then

$$Mag_\mu(\tilde{c}^I) = M_\mu(\tilde{c}^I) + \sigma_\mu(\tilde{c}^I) = \frac{c_1 + 2c_2 + 2c_3 + c_4}{6}$$
$$+ \sqrt{\frac{1}{4}(c_4 - c_1)^2 + \frac{1}{8}(c_4 - c_3 + c_2 - c_1)^2 - \frac{1}{3}(c_4 - c_1)(c_4 - c_3 + c_2 - c_1)w_{\tilde{c}}} \quad (19)$$

and

$$Mag_\mu(\tilde{e}^I) = M_\mu(\tilde{e}^I) + \sigma_\mu(\tilde{e}^I) = \frac{e_1 + 2e_2 + 2e_3 + e_4}{6}$$
$$+ \sqrt{\frac{1}{4}(e_4 - e_1)^2 + \frac{1}{8}(e_4 - e_3 + e_2 - e_1)^2 - \frac{1}{3}(e_4 - e_1)(e_4 - e_3 + e_2 - e_1)w_{\tilde{e}}} \quad (20)$$

So, $\tilde{c}^I \geq \tilde{e}^I$, if $Mag_\mu(\tilde{c}^I) \geq Mag_\mu(\tilde{e}^I)$.
Similarly, we can also prove that $\tilde{c}^I \geq \tilde{e}^I$, if $Mag_\nu(\tilde{c}^I) \geq Mag_\nu(\tilde{e}^I)$.
The proof is complete.

4 Conclusion

Possibility measures is the useful and compatible tools for ranking an GIFNs. The concepts possibility mean, standard deviation and magnitude of GIFNs are proposed here. Then we have introduced a novel ranking method of GIFNs considering the magnitude of decision maker, which is based on possibility mean and standard deviation. However, this decision making method can be applied in different ways like as: MADM, MGDM and so on.

References

1. Zadeh, A.L.: Fuzzy sets. Inf. Control **8**, 338–356 (1965)
2. Atanassov, K.T.: Intuitionistic fuzzy sets. Fuzzy Sets Syst. **20**, 87–96 (1986)

3. Dubois, D., Prade, H.: Fuzzy Sets and Systems: Theory and Applications. Academic press, New York (1980)
4. Li, F.D.: A note on "using intuitionistic fuzzy sets for fault-tree analysis on printed circuit board assembly." Microelectron. Reliab. **48**, 17–41 (2008)
5. Wan, P.S.: Multi-attribute decision making method based on possibility variance coefficient of triangular intuitionistic fuzzy numbers. Int. J. Uncertain. Fuzziness Knowl. Based Syst. **21**, 223–243 (2013)
6. Wan, P.S., Dong, Y.J.: Possibility method for triangular intuitionistic fuzzy multi-attribute group decision making with incomplete weight information. Int. J. Comput. Intell. Syst. **7**, 65–79 (2015)
7. Wang, Q.J., Zhang, Z.: Aggregation operators on intuitionistic trapezoidal fuzzy number and its application to multi-criteria decision making problems. J. Syst. Eng. Electron. **20**, 321–326 (2009)
8. Wu, J., Liu, J.Y.: An approach for multiple attribute group decision making problems with interval-valued intuitionistic trapezoidal fuzzy numbers. Comput. Indust. Engi. **66**, 311–324 (2013)
9. Xu, Z.S., Yager, R.R.: Some geometric aggregation operators based on intuitionistic fuzzy sets. Int. J. General Syst. **35**, 417–433 (2006)
10. Fuller, R., Majlender, P.: On weighted possibilistic mean and variance of fuzzy numbers. Fuzzy sets Syst. **136**, 363–374 (2003)
11. Carlsson, C., Fuller, R.: On possibilistic mean value and variance of fuzzy numbers. Fuzzy sets. Syst. **122**, 315–326 (2001)
12. Park, J.H., Park, Y., Lim, K.M.: Correlation coefficient of generalized intuitionistic fuzzy sets by statistical method. Honam Math. J. **28**, 317–326 (2006)
13. Garai, T., Chakraborty, D., Roy, T.K.: Possibility mean, variance and covariance of generalized intuitionistic fuzzy numbers and its application to multi-item inventory model with inventory level dependent demand. J. Intell. Fuzzy Syst. **35**, 1021–1036 (2018)
14. Jain, R.: A procedure for multi-aspect decision making using fuzzy sets. Int. J. Syst. Sci. **8**, 1–7 (1978)
15. Zeng, F.H., Cao, J.: Ranking method of fuzzy numbers based on center of fuzzy numbers. Fuzy Sets Syst. **22**, 142–147 (2011)
16. Zeng, F.H., Cao, J.: Exponential method of ranking fuzzy numbers. Statist. Decis. **341**, 134–160 (2011)
17. Qiupeng, Gu., Zuxinng, X.: A new approach for ranking fuzzy numbers based on possibility theory. J. Comput. Appl. Math. **309**, 674–682 (2017)

Prioritization of Factors Affecting the Digitalization of Quality Management Using Interval-Valued Intuitionistic Fuzzy Best -Worst Method

Nurşah Alkan(✉) and Cengiz Kahraman

Department of Industrial Engineering, Istanbul Technical University, Besiktas, 34367 Istanbul, Turkey
nalkan@itu.edu.tr

Abstract. With the adoption of new generation technologies, developments in quality management, which is an integral part of Industry 4.0, are increasing rapidly. The digitalization process in quality management provides significant advantages for enterprises with both an increase in performance and a reduction in costs. Digital quality management, which includes technologies that change the ways enterprises develop, manage and protect their quality standards, integrates new technologies into traditional systems and thus manages the quality by providing continuous improvement and improving overall business performance. For organizations to start and continue digitalization in their quality management, they should take into account many factors and be able to manage them correctly. For this aim, the factors that will ensure digitalization in the quality management of organizations should be evaluated and prioritized at the same time. In this study, the factors that provide digitalization in quality management are prioritized by using the best-worst method (BWM), which is one of the multi-criteria decision-making (MCDM) methods. After the criteria are first determined by a comprehensive literature review and expert opinions, the best-worst method based on interval-valued intuitionistic fuzzy sets is used to prioritize the related factors enabling the digitization in quality management to better address the uncertainty in decision makers' opinions.

Keywords: Digitalization · Quality management · Industry 4.0 · MCDM · IVIF-BWM

1 Introduction

Quality management ensures to consistent the tools used to meet the present and future needs and expectations of the products and services offered to the customer by controlling the different activities and tasks within the organization and aims to improve all processes, materials and services, suppliers, and all employees by continuously developing [1]. Traditional quality management practices face various challenges as customer needs are constantly changing and maintaining a high level of quality is difficult [2]. As

C. Kahraman et al. (Eds.): INFUS 2021, LNNS 308, pp. 28–39, 2022.
https://doi.org/10.1007/978-3-030-85577-2_4

new technologies emerge to overcome the difficulties in quality management, quality is increasingly important for organizations. The use by technology providers of Industry 4.0 to show that quality should be an organization-wide strategy has led to it being called Quality 4.0 or digital quality management [3]. The implementation of Industry 4.0 technologies to quality management has been important roles such as real-time process monitoring and data collection and analytic supported predictive maintenance in the factory of the future [4]. Technology providers of Industry 4.0 align and harmonize data between key processes such as design control, document control, non-compliance, complaint, change, and audit management throughout a product's lifecycle [4]. Therefore, Quality 4.0 or digital quality management is a relatively new way in which digital tools can be used to enhance organizations' ability to offer consistently high-quality products and include the digitalization of quality of performance, quality of design, and quality of conformance by using modern technologies. Thus, customer satisfaction will be achieved by making improvements throughout the value chain [2]. For organizations to start and continue digitalization in their quality management, they should take into account many factors and be able to manage them correctly. Therefore, the factors that will ensure digitalization in the quality management of organizations should be evaluated and prioritized at the same time. Ranking of these factors will help the organizations to focus more on the most influential factor. The prioritization of factors in this area constitutes a decision problem that includes many criteria. However, the classical MCDM techniques are inadequate when real-life conditions are considered as they are based on crisp numerical values. Therefore, to overcome this difficult decision-making process, it is suggested that MCDM methods should be applied together with fuzzy sets. Various extensions of traditional fuzzy sets have been introduced to define in different ways membership functions in the literature. Intuitionistic fuzzy sets, an extension of fuzzy sets introduced by Atanassov [5] are expressed with membership degree and non-membership degree. Afterward, Atanossov [6] proposed IVIFSs to enable the decision-maker to express their opinions more freely.

The aim of this study is to analyze the key factors identified with a comprehensive literature review and expert opinion for the effective implementation of digitalization in quality management. The key factors identified in line with the scope of the study are evaluated and prioritized by using IVIF-BWM, which is based on a mathematical model. In the study, IVIFSs have been applied to provide more freedom taking into account the vagueness and impreciseness of the evaluation process of decision-makers (DM). As far as we know, this study is the first decision-making model that addresses the factors affecting digital quality management with a fuzzy-based MCDM approach. Thus, this study helps to prioritize the factors in this field in an environment of uncertainty to lay the foundations for digital quality management and it is a guide for both organizations and academicians who want to research this issue.

The rest of the paper has been organized as follows: a literature review on quality management in Industry 4.0 is presented in Sect. 2. The steps of the proposed MCDM method are presented with its details in Sect. 3. The proposed method is applied to analyze the key factors of digital quality management in Sect. 4. Finally, the results obtained from the study and future research directions are given in Sect. 5.

2 Quality Management in Industry 4.0

Traditional quality management faces a variety of challenges, such as constantly changing customer needs and maintaining a high level of quality. Digitalization creates various and different opportunities in managing the quality of products and services offered by organizations. Digital quality management, or Quality 4.0, aligns quality management with Industry 4.0 to create new business models and ensure organizational efficiency and performance [2]. However, in addition to all these opportunities and benefits, the number of studies that deal with Industry 4.0 the quality management of organizations is limited in the literature. Chiarini [7] analyzed with systematic literature review relationships between Industry 4.0 and quality management and total quality management. In the study, it has been tried to fill the missing places in the literature. Zonnenshain and Kenett [8] presented a framework for quality management supporting Industry 4.0. They have been also offered future directions for quality and reliability engineering that leverage opportunities obtained from Industry 4.0. Mandrakov et al. [9] handled with the changes that occur in the quality management system with the digital transformation in the organization. Ramezani and Jassbi [10] analyzed the adequacy and appropriateness of Quality 4.0 implementation in the construction industry using corrective actions with a smart predictive model on a case study. Emblemsvåg [11] addressed some of the crucial aspects of the project-based industries regarding Quality 4.0.

3 Proposed Approach: IVIF Best Worst Method

In this section, the BWM method based on IVIFs has been proposed to prioritize the factors affecting digital quality management. Firstly, the basic concepts and the mathematical operations of IVIFSs have been briefly summarized. Then, the details of the proposed approach have been presented.

3.1 IVIFSs Preliminaries

After fuzzy sets are introduced by Zadeh [12], IFSs developed by Atanassov [5] as an extension of fuzzy sets take into account the degrees of membership and non-membership x in X to express the opinions of the DM more freely. The sum of these degrees must be less than or equal to 1. IVIFSs, an extension of IFSs, have the ability to better represent their hesitant and uncertain decisions when valuing the criteria with the help of membership definitions of DMs.

Definition 1: Let X be a fixed set. An IVIFSs $\tilde{A}$ in X is an object having the form [6].

$$\tilde{A} = \left\{ \langle x, \mu_{\tilde{A}}(x), v_{\tilde{A}}(x) \rangle | \, x \epsilon X \right\} \tag{1}$$

where $\mu_{\tilde{A}}(x) \subseteq [0, 1]$ and $\upsilon_{\tilde{A}}(x) \subseteq [0, 1]$ expresses the membership degree and non-membership degree of the element $x \epsilon X$ to the set $\tilde{A}$, respectively. Also, for each $x \epsilon X$, $\mu_{\tilde{A}}(X)$ and $\upsilon_{\tilde{A}}(X)$ are closed intervals and their lower and upper bounds are represented by $\mu_{\tilde{A}}^{L}(x)$, $\mu_{\tilde{A}}^{U}(x)$, $\upsilon_{\tilde{A}}^{L}(x)$, $\upsilon_{\tilde{A}}^{U}(x)$, respectively. Therefore, $\tilde{A}$ can also be expressed as follows:

$$\mu_{\tilde{A}}(x) = \left[\mu_{\tilde{A}}^{L}(x), \mu_{\tilde{A}}^{U}(x)\right] \subseteq [0, 1] \tag{2}$$

$$\nu_{\tilde{A}}(x) = \left[\upsilon_{\tilde{A}}^{L}(x), \upsilon_{\tilde{A}}^{U}(x)\right] \subseteq [0, 1] \tag{3}$$

where the expression is subject to the condition $0 \leq \mu_{\tilde{A}}^{U}(x) + \upsilon_{\tilde{A}}^{U}(x) \leq 1$.

For every $x \in X$, $\pi_{\tilde{A}}(x) = \left[\pi_{\tilde{A}}^{L}(x), \pi_{\tilde{A}}^{U}(x)\right]$ is called the degree of hesitancy in IVIFSs, where $\pi_{\tilde{A}}^{L}(x) = 1 - \mu_{\tilde{A}}^{U}(x) - \upsilon_{\tilde{A}}^{U}(x)$ and $\pi_{\tilde{A}}^{U}(x) = 1 - \mu_{\tilde{A}}^{L}(x) - \upsilon_{\tilde{A}}^{L}(x)$.

Definition 2: Let $\tilde{A} = \left(\left[\mu_{\tilde{A}}^{L}, \mu_{\tilde{A}}^{U}\right], \left[\upsilon_{\tilde{A}}^{L}, \upsilon_{\tilde{A}}^{U}\right]\right)$, $\tilde{A}_1 = \left(\left[\mu_{\tilde{A}_1}^{L}, \mu_{\tilde{A}_1}^{U}\right], \left[\upsilon_{\tilde{A}_1}^{L}, \upsilon_{\tilde{A}_1}^{U}\right]\right)$ and $\tilde{A}_2 = \left(\left[\mu_{\tilde{A}_2}^{L}, \mu_{\tilde{A}_2}^{U}\right], \left[\upsilon_{\tilde{A}_2}^{L}, \upsilon_{\tilde{A}_2}^{U}\right]\right)$ be three IVIFSs and $\lambda > 0$, then their operations are defined as follows:

$$\tilde{A}_1 \oplus \tilde{A}_2 = \left(\begin{bmatrix} \left(\mu_{\tilde{A}_1}^{L}\right) + \left(\mu_{\tilde{A}_2}^{L}\right) - \left(\mu_{\tilde{A}_1}^{L}\right)\left(\mu_{\tilde{A}_2}^{L}\right), \\ \left(\mu_{\tilde{A}_1}^{U}\right) + \left(\mu_{\tilde{A}_2}^{U}\right) - \left(\mu_{\tilde{A}_1}^{U}\right)\left(\mu_{\tilde{A}_2}^{U}\right) \end{bmatrix}, \left[\upsilon_{\tilde{A}_1}^{L}\upsilon_{\tilde{A}_2}^{L}, \upsilon_{\tilde{A}_1}^{U}\upsilon_{\tilde{A}_2}^{U}\right]\right) \tag{4}$$

$$\tilde{A}_1 \otimes \tilde{A}_2 = \left(\left[\mu_{\tilde{A}_1}^{L}\mu_{\tilde{A}_2}^{L}, \mu_{\tilde{A}_1}^{U}\mu_{\tilde{A}_2}^{U}\right], \begin{bmatrix} \left(\upsilon_{\tilde{A}_1}^{L}\right) + \left(\upsilon_{\tilde{A}_2}^{L}\right) - \left(\upsilon_{\tilde{A}_1}^{L}\right)\left(\upsilon_{\tilde{A}_2}^{L}\right), \\ \left(\upsilon_{\tilde{A}_1}^{U}\right) + \left(\upsilon_{\tilde{A}_2}^{U}\right) - \left(\upsilon_{\tilde{A}_1}^{U}\right)\left(\upsilon_{\tilde{A}_2}^{U}\right) \end{bmatrix}\right) \tag{5}$$

$$\lambda\tilde{A} = \left(\left[1 - \left(1 - \left(\mu_{\tilde{A}}^{L}\right)\right)^{\lambda}, 1 - \left(1 - \left(\mu_{\tilde{A}}^{U}\right)\right)^{\lambda}\right], \left[\left(\upsilon_{\tilde{A}}^{L}\right)^{\lambda}, \left(\upsilon_{\tilde{A}}^{U}\right)^{\lambda}\right]\right) \tag{6}$$

$$\tilde{A}^{\lambda} = \left(\left[\left(\mu_{\tilde{A}}^{L}\right)^{\lambda}, \left(\mu_{\tilde{A}}^{U}\right)^{\lambda}\right], \left[1 - \left(1 - \left(\upsilon_{\tilde{A}}^{L}\right)\right)^{\lambda}, 1 - \left(1 - \left(\upsilon_{\tilde{A}}^{U}\right)\right)^{\lambda}\right]\right) \tag{7}$$

Definition 3: Let $\tilde{A}_i = \left(\left[\mu_{\tilde{A}_i}^{L}, \mu_{\tilde{A}_i}^{U}\right], \left[\upsilon_{\tilde{A}_i}^{L}, \upsilon_{\tilde{A}_i}^{U}\right]\right)(i = 1, 2, \ldots\ldots, n)$ be a set of IVIFSs and $w = (w_1, w_2, \ldots., w_n)^T$ be weight vector of $\tilde{A}_i$ with $\sum_{i=1}^{n} w_i = 1$, then an interval-valued intuitionistic fuzzy weighted geometric (IVIFWG) the operator is a mapping IVIFWG: $\tilde{A}^n \rightarrow \tilde{A}$, where [13]:

$$IVIFWG\left(\tilde{A}_1, \tilde{A}_2, \ldots., \tilde{A}_n\right) = \left(\begin{matrix} \left[\prod_{i=1}^{n}\left(\mu_{\tilde{A}_i}^{L}\right)^{w_i}, \prod_{i=1}^{n}\left(\mu_{\tilde{A}_i}^{U}\right)^{w_i}\right] \\ \left[\left(1 - \prod_{i=1}^{n}\left(1 - \left(\upsilon_{\tilde{A}_i}^{L}\right)\right)^{w_i}\right), \left(1 - \prod_{i=1}^{n}\left(1 - \left(\upsilon_{\tilde{A}_i}^{U}\right)\right)^{w_i}\right)\right] \end{matrix}\right) \tag{8}$$

Definition 4: The score function $S(\tilde{A})$ of $\tilde{A}_i = \left(\left[\mu^L_{\tilde{A}_i}, \mu^U_{\tilde{A}_i}\right], \left[\upsilon^L_{\tilde{A}_i}, \upsilon^U_{\tilde{A}_i}\right]\right)$ $(i = 1, 2, \ldots\ldots, n)$ is given as in Eq. (9) [14].

$$S(\tilde{A}) = \frac{\mu^L_{\tilde{A}} + \mu^U_{\tilde{A}} - \nu^L_{\tilde{A}} - \nu^U_{\tilde{A}}}{2} \tag{9}$$

Definition 5. Defuzzification of $\tilde{A}_i = \left(\left[\mu^L_{\tilde{A}_i}, \mu^U_{\tilde{A}_i}\right], \left[\upsilon^L_{\tilde{A}_i}, \upsilon^U_{\tilde{A}_i}\right]\right)(i = 1, 2, \ldots\ldots, n)$ is given as in Eq. (10) [15].

$$Deff(\tilde{A}_i) = \frac{\mu^L_{\tilde{A}_i} + \mu^U_{\tilde{A}_i} + \left(1 - \upsilon^L_{\tilde{A}_i}\right) + \left(1 - \upsilon^U_{\tilde{A}_i}\right) + \mu^L_{\tilde{A}_i} \times \mu^U_{\tilde{A}_i} - \sqrt{\left(1 - \upsilon^L_{\tilde{A}_i}\right) \times \left(1 - \upsilon^U_{\tilde{A}_i}\right)}}{4} \tag{10}$$

3.2 Crisp Best Worst Method

The BWM introduced by Rezaei [16] has been developed to obtain the priorities of criteria more effectively than other MCDM methods [17]. BWM requires fewer pairwise comparisons and allows higher consistency rate thanks to its use of the structured comparisons [18]. Therefore, all these advantages are the main reason for choosing the BWM method in this study. The steps of the BWM can be summarized as given in Fig. 1. The schematic representation of the reference comparisons of BWM is presented in Fig. 2.

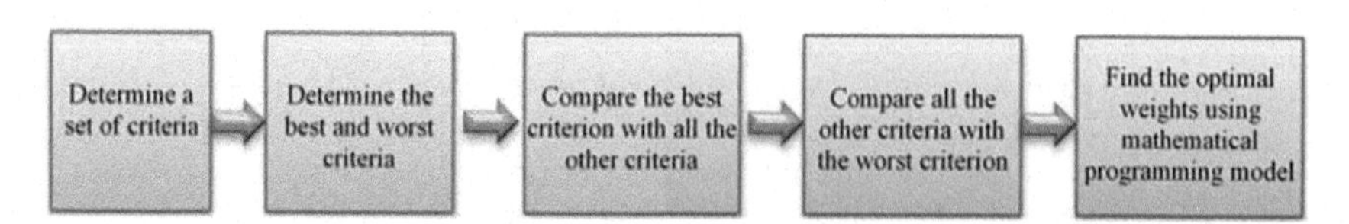

Fig. 1. The steps of BWM

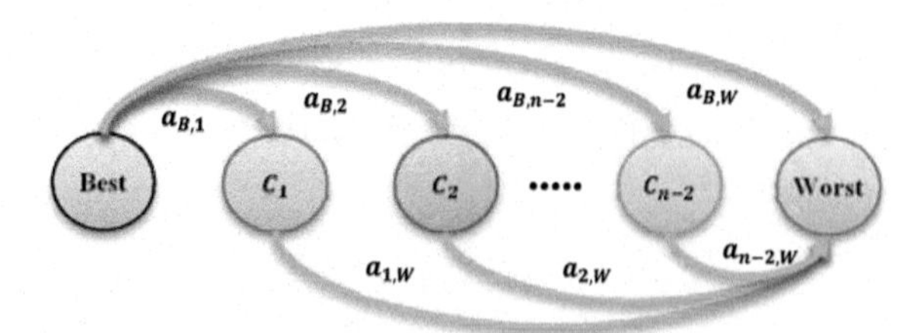

Fig. 2. BWM reference comparisons

3.3 IVIF Best Worst Method

In the real world, the qualitative and quantitative judgments of DMs generally have the characteristics of ambiguity and intangibility, as well as criteria information also has the drawbacks of vague and uncertain. Therefore, the use of BWM with fuzzy numbers rather than crisp numbers will make pairwise comparisons between the best/worst criteria and all other criteria more compatible with real situations and will yield more reliable ranking results. However, the use of ordinary fuzzy sets in BWM cannot reflect the uncertainty of subjective evaluations of DMs. Evaluating together with IVIFs using BWM of the factors affecting digital transformation in quality management provides more realistic and more reliable results. In this study, IVIF-BWM has been proposed for the evaluation and ranking of factors affecting digital quality management. The steps of the proposed approach are given as follows [19].

Step 1. Determine relevant criteria, and DMs to construct the proposed approach. A set of decision criteria $C_j = \{C_1, C_2, ..., C_n\}$ with $j = 1, 2, ..., n$ is created.

Step 2. The best (most important) criterion C_B and the worst (least important) criterion C_W are identified by each DM.

Table 1. Linguistic scale based on IVIF numbers

Linguistic terms	IVIF numbers
Equally important (EI)	[(0.5, 0.5), (0.5, 0.5)]
Weakly important (WI)	[(0.5, 0.6), (0.3, 0.4)]
Strongly important (SI)	[(0.6, 0.7), (0.2, 0.3)]
Very important (VI)	[(0.7, 0.8), (0.1, 0.2)]
Absolutely important (AI)	[(0.8, 0.9), (0, 0.1)]

Step 3. The best criterion with the other criteria is compared by using the linguistic scale given in Table 1 to determine IVIF Best-to-Others vector. The resulting best-to-others $\tilde{R}_B$ vector by DM_k is written as follows:

$$\tilde{R}_B^k = \left(\tilde{r}_{B1}^k, \tilde{r}_{B2}^k, ..., \tilde{r}_{Bn}^k\right) \tag{11}$$

where $\tilde{r}_{Bj}^k = \left(\left[\mu_{Bj}^L, \mu_{Bj}^U\right], \left[\upsilon_{Bj}^L, \upsilon_{Bj}^U\right]\right)^k$ indicates the IVIF preference of the best criterion C_B over the criterion C_j based on kth DM.

Step 4. The other criteria with the worst criterion are compared by using the linguistic scale given in Table 1 to obtain IVIF others-to-worst vector. The resulting others-to-worst $\tilde{R}_W$ vector by DM_k is written as follows:

$$\tilde{R}_W^k = \left(\tilde{r}_{W1}^k, \tilde{r}_{W2}^k, ..., \tilde{r}_{Wn}^k\right) \tag{12}$$

where $\tilde{r}_{Wj}^k = \left(\left[\mu_{Wj}^L, \mu_{Wj}^U\right], \left[\upsilon_{Wj}^L, \upsilon_{Wj}^U\right]\right)^k$ indicates the IVIF preference of the criterion C_j over the worst criterion C_W based on kth DM.

Step 5. Compute the optimal IVIF weights for each criterion based on every DM. Let $\tilde{w}_B^k = \left(\left[\mu_B^L, \mu_B^U\right], \left[\upsilon_B^L, \upsilon_B^U\right]\right)^k$ and $\tilde{w}_W^k = \left(\left[\mu_W^L, \mu_W^U\right], \left[\upsilon_W^L, \upsilon_W^U\right]\right)^k$. The optimal weights of the criteria is the one where for each pair of $\tilde{w}_B^k/\tilde{w}_j^k$ and $\tilde{w}_j^k/\tilde{w}_W^k$, the best possible solution should have $\tilde{w}_B^k/\tilde{w}_j^k = a_{Bj}$ and $\tilde{w}_j^k/\tilde{w}_W^k = a_{jW}$. To obtain the best possible solution by satisfying these conditions for all j and each DM, it should have a solution where the maximum absolute differences $\left|\frac{\tilde{w}_B^k}{\tilde{w}_j^k} - \tilde{r}_{Bj}^k\right|$ and $\left|\frac{\tilde{w}_j^k}{\tilde{w}_W^k} - \tilde{r}_{jW}^k\right|$ for all j are minimized. Besides, the sum condition and non-negativity of the weights should be considered. It should construct an optimization problem to obtain optimal weights $\left(\tilde{w}_1^*, \tilde{w}_2^*, \ldots.., \tilde{w}_n^*\right)$ for each DM.

$$min \varepsilon$$

$$\text{s.t.}$$

$$\left|\mu_B^{L(k)} + \upsilon_j^{L(k)} - \mu_B^{L(k)} \cdot \upsilon_j^{L(k)} - \mu_{Bj}^{L(k)}\right| \leq \varepsilon, for\ all\ j$$

$$\left|\mu_B^{U(k)} + \upsilon_j^{U(k)} - \mu_B^{U(k)} \cdot \upsilon_j^{U(k)} - \mu_{Bj}^{U(k)}\right| \leq \varepsilon, for\ all\ j$$

$$\left|\upsilon_B^{L(k)} + \mu_j^{L(k)} - \upsilon_{Bj}^{L(k)}\right| \leq \varepsilon, for\ all\ j$$

$$\left|\upsilon_B^{U(k)} + \mu_j^{U(k)} - \upsilon_{Bj}^{U(k)}\right| \leq \varepsilon, for\ all\ j$$

$$\left|\mu_j^{L(k)} + \upsilon_W^{L(k)} - \mu_j^{L(k)} \cdot \upsilon_W^{L(k)} - \mu_{jW}^{L(k)}\right| \leq \varepsilon, for\ all\ j \tag{13}$$

$$\left|\mu_j^{U(k)} + \upsilon_W^{U(k)} - \mu_j^{U(k)} \cdot \upsilon_W^{U(k)} - \mu_{jW}^{U(k)}\right| \leq \varepsilon, for\ all\ j$$

$$\left|\upsilon_j^{L(k)} + \mu_W^{L(k)} - \upsilon_{jW}^{L(k)}\right| \leq \varepsilon, for\ all\ j$$

$$\left|\upsilon_j^{U(k)} + \mu_W^{U(k)} - \upsilon_{jW}^{U(k)}\right| \leq \varepsilon, for\ all\ j$$

$$\sum_{j=1}^{n} S\left(\tilde{w}_j^k\right) = 1$$

$$S\left(\tilde{w}_j^k\right) \geq 0, for\ all\ j$$

The optimal IVIF weights of all criteria $\tilde{w}_j^k = \left(\tilde{w}_1^*, \tilde{w}_2^*, \ldots.., \tilde{w}_n^*\right)^k$ and the optimal value ε for each decision-maker is obtained by solving the above optimization problem. The optimal value ε states the consistency for the comparison matrices. If the value of ε is closer to zero, it has a higher level of consistency.

Step 6. The weights obtained from all DMs for each criterion are aggregated by using IVIFGW operator given in Eq. (8) and thus, the optimal weights of criteria $\tilde{w}_j$ are calculated.

Step 7. The aggregated optimal IVIF weight of each criterion is deffuzified by using Eq. (10) to obtain crisp numbers. Then, the normalized weights are calculated by using crisp weights as given in the following equation.

$$w_j^N = \frac{w_j}{\sum_{j=1}^{n} w_j} \tag{14}$$

4 Application

It is an important process to analyze the factors affecting digital transformation in quality management and take strategic steps accordingly in order to manage digital transformation more effectively in quality management and to apply digital quality management better. In this paper, an analysis of the factors affecting the digital transformation in quality management is realized by using IVIF-BWM. The factors determined according to literature reviews and experts' opinions are as given in Fig. 3. As a result, 3 main criteria and 15 sub-criteria are identified for assessment of factors affecting digital quality management. Then, three decision-makers with experts in quality management and digital transformation are asked to determine the preferences of the most important criteria over the other criteria and the other criteria over the least important criteria using IVIF linguistic scale given in Table 1. The preference comparisons of the best criterion over the other criteria and the other criteria over the worst criterion based on management main criterion are shown for each DM in Tables 2 and 3, respectively. After that, the IVIF optimal weights of each main and sub-criteria are computed through the software Lingo 19.0 by fulfilling the optimization model mentioned in Sect. 3 and constraints for each decision-maker. The IVIF optimal weights of management factor (C3) for each DM are displayed in Table 4. After aggregated the IVIF optimal weights obtained from each DM using Eq. (8), the aggregated IVIF optimal weights are defuzzified by using Eq. (10). Then, the normalized crisp numbers of weights are obtained as shown in Table 5 by utilizing Eq. (14).

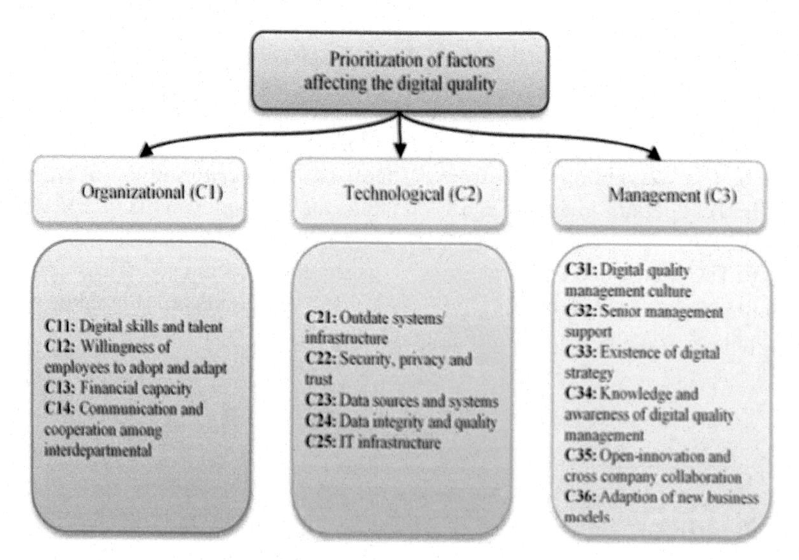

Fig. 3. The hierarchical structure of problem

Table 2. Preferences of best criterion over all the other criteria for management factor

Decision-makers	Best-to-others	C31	C32	C33	C34	C35	C36
DM1	Best criterion:C31	EI	AI	WI	SI	VI	SI
DM2	Best criterion:C33	WI	VI	EI	WI	SI	AI
DM3	Best criterion:C33	WI	AI	EI	SI	VI	SI

Table 3. Preferences of all the other criteria over worst criterion for management factor

Others-to-worst	Worst criterion for DM1: C32	Worst criterion for DM2: C36	Worst criterion for DM3: C32
C31	AI	VI	VI
C32	EI	WI	EI
C33	VI	AI	AI
C34	SI	VI	SI
C35	WI	SI	WI
C36	SI	EI	SI

Table 4. IVIF weights for management factors

	DM1	DM2	DM3	Aggregated weights
$\tilde{w}_1$	[(0.516, 0.64), (0.046, 0.051)]	[(0.376, 0.498), (0.18, 0.194)]	[(0.392, 0.516), (0.146, 0.151)]	[(0.419, 0.543), (0.013, 0.014)]
$\tilde{w}_2$	[(0.145, 0.24), (0.192, 0.192)]	[(0.234, 0.255), (0.244, 0.244)]	[(0.145, 0.24), (0.192, 0.192)]	[(0.175, 0.246), (0.021, 0.021)]
$\tilde{w}_3$	[(0.392, 0.516), (0.146, 0.151)]	[(0.498, 0.619), (0.08, 0.094)]	[(0.516, 0.64), (0.046, 0.051)]	[(0.469, 0.592), (0.091, 0.091)]
$\tilde{w}_4$	[(0.269, 0.392), (0.246, 0.251)]	[(0.376, 0.498), (0.18, 0.194)]	[(0.269, 0.392), (0.246, 0.251)]	[(0.307, 0.431), (0.022, 0.023)]
$\tilde{w}_5$	[(0.245, 0.34), (0.292, 0.292)]	[(0.255, 0.376), (0.28, 0.294)]	[(0.245, 0.269), (0.257, 0.257)]	[(0.249, 0.33), (0.027, 0.028)]
$\tilde{w}_6$	[(0.269, 0.392), (0.246, 0.251)]	[(0.134, 0.219), (0.176, 0.176)]	[(0.269, 0.392), (0.246, 0.251)]	[(0.203, 0.311), (0.022, 0.022)]

According to the results obtained, the most important main criterion is determined as *C3: Management* factor with 0.463 importance weight as presented in Table 5. Considering the overall weights of sub-criteria for each main criterion, *"C33: Existence of digital strategy"* places in the first rank as the most important criterion within the sub-criteria *"C31: Digital quality management culture"* and *"C11: Digital skills and talent"* follow this rank as second and third, respectively.

Table 5. Weights of main and sub-factors

Main criteria	C1				C2				
Weights	0.258				0.278				
Sub-criteria	C11	C12	C13	C14	C21	C22	C23	C24	C25
Local weights	0.320	0.208	0.267	0.206	0.210	0.187	0.212	0.168	0.223
Global weights	0.083	0.054	0.069	0.053	0.059	0.052	0.059	0.047	0.062

Main criteria	C3					
Weights	0.463					
Sub-criteria	C31	C32	C33	C34	C35	C36
Local weights	0.199	0.135	0.215	0.165	0.142	0.143
Global weights	0.092	0.063	0.1	0.076	0.066	0.066

5 Conclusions

Today, where competitive advantage has an important place, organizations can manage quality by improving the ability to consistently offer high-performance products to customers, thanks to digital tools and technologies. To handle digital quality management practices in a better and more effective way, the factors ensuring digitalization in the quality management of organization should be evaluated and prioritized at the same time and thus a strategic preliminary study should be presented for digital transformation in quality management. In this study, several factors related to the adoption and application of digitalization steps in quality management have been prioritized using IVIF-BWM. After applied the proposed method, the most important main criterion has been identified as "*Management*" factor. Considering the sub-criterion, "*Existence of digital strategy*", "*Digital quality management culture*" and "*Digital skills and talent*" are ranked as first, second, and third important sub-criteria, respectively.

For further research, different MCDM methods can be implemented for this problem and can be evaluated with different factors. Besides, different extensions of traditional fuzzy sets such as hesitant fuzzy sets, Pythagorean fuzzy sets, spherical fuzzy sets, circular intuitionistic fuzzy sets can be applied for this problem.

References

1. Zahir, C., Ertosun, Ö., Zehir, S., Müceldilli, B.: Total quality management practices' effects on quality performance and innovative performance. Procedia Soc. Behav. Sci. **41**, 273–280 (2012)
2. Sony, M., Antony, J., Douglas, J.: Essential ingredients for the implementation of Quality 4.0. TQM J. **32**(4), 779–793 (2020)
3. Sony, M., Antony, J., Douglas, J.: Motivations, barriers and readiness factors for Quality 4.0 implementation: an exploratory study. TQM J. (in press)
4. Küpper, D., Knizek, C., Ryeson, D., Noecker, J.: Quality 4.0 Takes More Than Technology. Boston Consulting Group (2019)
5. Atanassov, K.: Intuitionistic fuzzy sets. Fuzzy Sets Syst. **20**(1), 87–96 (1986)
6. Atannasov K.: Intuitionistic Fuzzy Sets, Theory and Applications. Physica-Verlag, New York; Heidelberg (1999)
7. Chiarini, A.: Industry 4.0, quality management and TQM world. A systematic literature review and a proposed agenda for further research. TQM J. **32**(4), 603–616 (2020)
8. Zonnenshain, A., Kenett, R.: Quality 4.0—the challenging future of quality engineering. Quality Eng. **32**(4), 614–626 (2020)
9. Mandrakov, E., Vasiliev, V., Dudina, D.: Non-conforming products management in a digital quality management system. In: 2020 IEEE International Conference, Quality Management, Transport and Information Security, Information Technologies, IT and QM and IS 2020, Russian Federation (2020)
10. Ramezani, J., Jassbi, J.: Quality 4.0 in action: smart hybrid fault diagnosis system in plaster production. Processes **8**(6) (2020)
11. Emblemsvåg, J.: On Quality 4.0 in project-based industries. TQM J. **32**(4), 725–739 (2020)
12. Zadeh, L.: Fuzzy sets. Inf. Control **8**(3), 338–353 (1965)
13. Xu, Z.: Methods for aggregating interval-valued intuitionistic fuzzy information and their application to decision making. Control Decis. **22**(2), 215–219 (2007)
14. Wu, J., Chiclana, F.: A risk attitudinal ranking method for interval-valued intuitionistic fuzzy numbers based on novel attitudinal expected score and accuracy functions. Appl. Soft Comput. **22**, 272–286 (2014)
15. Oztaysi, B., Onar, S., Kahraman, C., Yavuz, M.: Multi-criteria alternative-fuel technology selection using interval-valued intuitionistic fuzzy sets. Transp. Res. Part D **53**, 128–148 (2017)
16. Rezaei, J.: Best-worst multi-criteria decision-making method. Omega **53**, 49–57 (2015)
17. Liao, H., Mi, X., Yu, Q., Luo, L.: Hospital performance evaluation by a hesitant fuzzy linguistic best worst method with inconsistency repairing. J. Clean. Prod. **232**, 657–671 (2019)
18. Mi, X., Liao, H.: An integrated approach to multiple criteria decision-making based on the average solution and normalized weights of criteria deduced by the hesitant fuzzy best worst method. Comput. Ind. Eng. **133**, 83–94 (2019)
19. Wang, J., Ma, Q., Liu, H.: A meta-evaluation model on science and technology project review experts using IVIF-BWM and MULTIMOORA. Expert Syst. Appl. **168** (2021)

Towards a Safe Pedestrian Walkability Under Intuitionistic Fuzzy Environment: A Real-Time Reactive Microservice-Oriented Ecosystem

Ghyzlane Cherradi[1], Azedine Boulmakoul[1](✉), Lamia Karim[2], Meriem Mandar[3], and Ahmed Lbath[4]

[1] LIM/IOS FSTM, Hassan II University of Casablanca, Casablanca, Morocco
[2] LISA Lab. ENSA Berrechid, Hassan 1st University, Settat, Morocco
[3] ENS, Hassan II University of Casablanca, Casablanca, Morocco
[4] LIG/MRIM, CNRS, University Grenoble Alpes, Grenoble, France
Ahmed.Lbath@univ-grenoble-alpes.fr

Abstract. In order to improve pedestrian safety, it is important to identify suitable routes for walking. This paper presents a real-time reactive system that aims to provide the safest route among all possible routes at a given time, based on intuitionistic fuzzy pedestrian risk modeling. It derives multi criteria decision making to support complex environments using membership and non-membership attributes to generate the intuitionistic fuzzy risk weighting graph based on risk measures. The proposed system involves the pgRouting open-source library that extends the PostGIS/PostgreSQL geospatial database, to provide geospatial routing capabilities. Therefore, we offer a web-location based service allowing pedestrians to enter their destination; then select a route using an intelligent algorithm; providing them with the safest route possible instead of the fastest route. This service will certainly help save lives and reduce pedestrian accidents to some extent.

Keywords: Pedestrian's risk modeling · Intuitionistic fuzzy set · Risk indicators · Safest walkability · Reactive microservice architecture · Intelligent mobility

1 Introduction

Global accident statistics show that road accidents are caused by the negligence of both drivers and pedestrians, and despite efforts, pedestrians remain the most vulnerable road users. In fact, more than 270,000 pedestrians lose their lives on the world's roads every year [17]. Pedestrian safety is a multidimensional issue and requires a holistic perspective

This work was partially funded by Ministry of Equipment, Transport, Logistics and Water–Kingdom of Morocco, The National Road Safety Agency (NARSA) and National Center for Scientific and Technical Research (CNRST). Road Safety Research Program# An intelligent reactive abductive system and intui-tionist fuzzy logical reasoning for dangerousness of driver-pedestrians interactions analysis: Development of new pedestrians' exposure to risk of road accident measures.

C. Kahraman et al. (Eds.): INFUS 2021, LNNS 308, pp. 40–47, 2022.
https://doi.org/10.1007/978-3-030-85577-2_5

when considering determinants, consequences, and solutions. Since a significant portion of crosswalks are unsignalized, the number of accidents at crosswalks, including those resulting in human fatalities, is very high. On the other hand, the number of fatalities from traffic accidents outside cities is almost 20 times lower than the number of fatalities within cities. This is due to the ratio of pedestrians to the intensity of traffic flow. Children, as pedestrians, are at an even higher risk of injury or death in traffic accidents due to their small size, inability to judge distance and speed, and lack of experience with traffic rules. The dynamics of pedestrians are difficult to characterize because they are influenced by many different sources. Unlike other mobility models, walking is not tied to a vehicle on a lane, and the underlying infrastructure is very heterogeneous. Furthermore, environmental factors (traffic lights, public furniture, advertisements, etc.), the total waiting time of pedestrians, the average distance of following vehicles, and atmospheric conditions (wind, rain, etc.) directly affect walking. In this context, modeling pedestrian safety considering the uncertainty of these factors is an important research topic.

This paper aims to propose a new pedestrian risk model using intuitionistic fuzzy set theory [2, 18, 19], considering incompleteness and uncertainty of risk parameters. It provides an integrated platform that collects, identifies, and provides various pedestrian-related communication and control functions. The objective of this paper is to show that it is possible to build a scalable and integrated mobility service for pedestrian safety by connecting reusable components on a microservices architecture. The rest of this paper is organized as follows. In Sect. 2, we present the research background and previous work. Then, in Sect. 3, we describe the proposed intuitionistic fuzzy risk model. Section 4 describes the proposed system and its components, and finally, Sect. 5 draws some conclusions from this study.

2 Related Works

In the field of urban computing, our work involves research on pedestrian risk modeling, pedestrian safety systems and urban navigation. In this section, we review some research and how they relate to our study. Traditionally, pedestrian risk has generally been assessed using crash frequency models, based on historical data [5, 13]. Crash frequency models have been developed using spatial data at intersections. However, crash frequency is influenced by many factors, including traffic volume, speed, geometry, and built environment, etc. Authors in [15] proposed a composite indicator of pedestrian exposure that takes into account pedestrian characteristics, road and traffic conditions, and pedestrian compliance with traffic rules; Saravanan et al. [14] proposed a fuzzy logic-based approach to accident prediction. Their research assesses the risk of accidents not only for vehicles but also for pedestrians and identifies accident zones on a given road network. Mandar et al. [11] present a new way of measuring a virtual pedestrian-vehicle mutual accident risk indicator. Pedestrian dynamics are modeled using the basic fuzzy-ant model [4], to which an artificial potential field is integrated. Various models have been proposed to assess the risks and to take measures to improve pedestrian safety. Thus, the model proposed in the study [12] can measure the impact of potential risk factors on pedestrians' intentional waiting time. In the study [3], the author proposes a multivariate risk analysis method consisting of two hierarchical generalized linear models that characterize two different aspects of unsafe cross-border behavior, and a Bayesian approach using data augmentation methods to make statistical inferences about the parameters associated with risk exposure. To date, several pedestrian safety systems have been proposed; David et al. [7] analyze an approach based on a centralized server and an ad hoc communication connection between a car and a pedestrian for the response time of a smartphone application that warns the user about the risk of collision. The use of cell phones by pedestrians (talking, texting, reading) affects their awareness of their surroundings and increases the risk of accidents. The authors in [16] proposed an android smart phone application (WalkSafe) to assist people that walk and talk to improve the safety of pedestrian cell phone users. It uses the cell phone's rearview camera to detect vehicles approaching the user and warn the user of potentially unsafe situations. In research [1], two systems were evaluated in a "standard test" (in which a vehicle drives towards a pedestrian dummy placed on a track). The results show that the collision avoidance performance of this system depends on the vehicle speed, but there is a limit at a certain speed (40 km/h). Although these systems are intended to effectively reduce pedestrian injuries, there is a limit to the collision avoidance performance of these systems.

3 Pedestrian's Risk

3.1 Concept of Intuitionistic Fuzzy Sets

Intuitionist fuzzy sets have been proposed by Atanassov (1986) [2]. Which take into account the degree of membership, non-membership and hesitancy. It can describe the more subtle ambiguities of the objective world. Therefore, this theory is widely used to model real life problems such as sales analysis, new product marketing, financial services, negotiation processes, psychological research, etc.

Definition 1. (Fuzzy set) [18, 19] a fuzzy set is defined by a membership function that is a map from the universal set U to the interval [0, 1]. Then, the fuzzy set $\tilde{A}$ in U is the set of ordered pairs:

$$\tilde{A}(x) = \left\{x, \mu_{\tilde{A}}(x) | x \in U, \mu_{\tilde{A}}(x) \in [0, 1]\right\} \tag{1}$$

A membership function is a function from a universal set U to the interval [0, 1]. A fuzzy set $\tilde{A}$ is defined by its membership function $\mu_{\tilde{A}}(x)$ over U. When fuzzy set theory was first published, research considered decision making as one of the most attractive application fields of that theory [18, 19].

Definition 2. Intuitionistic Fuzzy Set (IFS) [2], Set X is a nonempty set, $X = (x_1, x_2, \ldots, x_n)$, $A = \{[x, \mu_A(x), \nu_A(x)] | x \in X\}$, is called intuitionistic fuzzy sets. $\mu_A(x)$ is membership of x to A, and $\nu_A(x)$ is non-membership of x to A, where $\mu_A(x) \in [0, 1]$, $\nu_A(x)n \in [0, 1]$ and $0 \leq \mu_A(x) + \nu_A(x) \leq 1$. Obviously, each ordinary fuzzy set may be written as: $A = \{[x, \mu_A(x), 1 - \mu_A(x)] | x \in X\}$, For every intuitionistic fuzzy sets of X, $\pi_A(x) = 1 - \mu_A(x) - \nu_A(x)$ is hesitation or uncertainty of x to A. There is an intuitionistic fuzzy number $\tilde{A} = \left(\mu_{\tilde{A}}, \nu_{\tilde{A}}\right)$, where $\mu_{\tilde{A}} \in [0, 1]$, $\nu_{\tilde{A}} \in [0, 1]$, so $s_{\tilde{A}} = \mu_{\tilde{A}} - \nu_{\tilde{A}}$ is score function of $\tilde{A}$, and $h_{\tilde{A}} = \mu_{\tilde{A}} + \nu_{\tilde{A}}$ is exact function of $\tilde{A}$.

Definition 3. The Triangular Intuitionistic Fuzzy Number (TIFN) [2], TIFN is an intuitionistic fuzzy number that uses traditional triangular fuzzy numbers to represent membership $\mu_{\tilde{A}}(x)$ and non-membership degree $\nu_{\tilde{A}}(x)$, such that the intuitionistic fuzzy number is based on the triangular fuzzy number, which is termed the triangular intuitionistic fuzzy number (TIFN). In the following, we introduce some basic concepts about TIFNs. Let $\tilde{A}$ be a TIFN, where the membership and non-membership function for $\tilde{A}$ are defined as follows:

$$\mu_{\tilde{A}}(x) = \begin{cases} \frac{x-a_1}{a_2-a_1} & if \quad a_1 \leq x < a_2 \\ \frac{a_3-x}{a_3-a_2} & if \quad a_2 \leq x < a_3 \\ 0 \quad , & otherwise \end{cases} \tag{2}$$

Non-Membership function:

$$\nu_{\tilde{A}}(x) = \begin{cases} \frac{a_2-x}{a_2-\acute{a}_1} & if \quad \acute{a}_1 \leq x < a_2 \\ \frac{x-a_2}{\acute{a}_3-a_2} & if \quad a_2 \leq x < \acute{a}_3 \\ 1 \quad , & otherwise \end{cases} \tag{3}$$

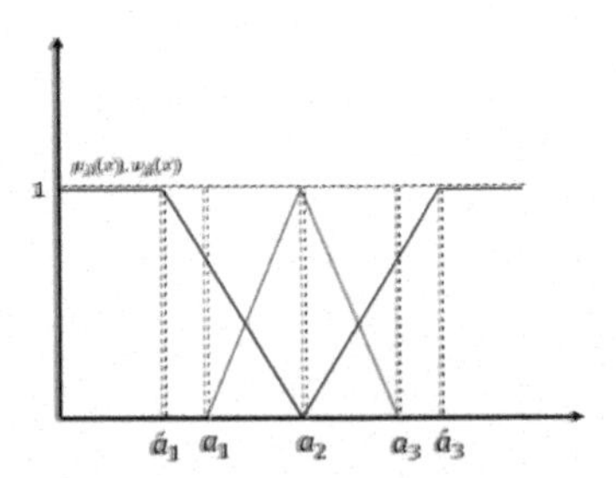

Fig. 1. Membership and non – membership function of TIFN.

Where $\acute{a}_1 \leq a_1 \leq a_2 \leq a_3 \leq \acute{a}_3, 0 \leq \mu_{\tilde{A}}(x) + \nu_{\tilde{A}}(x) \leq 1$ and TIFN is denoted by $\tilde{A}_{TIFN} = (a_1, a_2, a_3; \ \acute{a}_1, a_2, \acute{a}_3)$ as shown in Fig. 1.

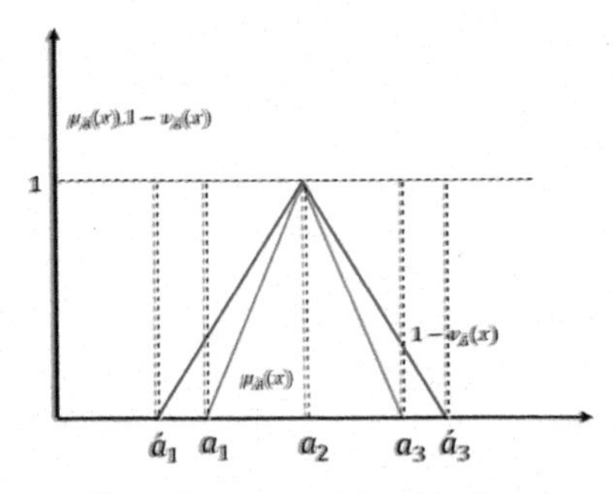

Fig. 2. Membership and non – membership function of TIFN [12].

We use the TIFN proof given by Kahraman et al. (2017) [10]. A new illustration of the membership and non-membership functions of TIFN is shown in Fig. 2.

Definition 4. Triangular Intuitionistic Fuzzy Number (TIFN) and its arithmetic. The additions of two TIFN are as follows. For two triangular intuitionistic fuzzy numbers [2, 10]. A = $([a_1, b_1, c_1]; \mu_A, v_A)$ and B = $([a_2, b_2, c_2]; \mu_B, v_B)$, with $\mu_A \neq \mu_B$ *and* $v_A \neq v_B$, A + B = $[a_1 + a_2, b_1 + b_2, c_1 + c_2]$; MIN$(\mu_A, \mu_B)$, MAX (v_A, v_B).

3.2 Triangular Intuitionistic Risk Exposure Model

A risk model for an urban road network is essentially the assignment of a risk score r to each segment s, which is proportional to the probability of an accident occurring on the corresponding road segment and to the pedestrian risk factors. In the proposed model, the duration of exposure to risk is estimated by the travel time on the road segment traveled. During this time, pedestrians are exposed to risks according to the risk factors. This travel time is one of several sources of uncertainty in pedestrian walkability that cannot be accurately known or represented. This uncertainty is represented by an intuitionistic fuzzy set (IFS). For simplicity, we will focus on the triangular intuitionistic fuzzy number (TIFN). We propose a construction of an intuitionistic fuzzification of the pedestrians travel time Ts by means of triangular intuitionistic fuzzy numbers. The severity of pedestrian accidents associated with each factor is multiplied by the probability of the accident occurring and the travel time to arrive at an estimate of the risk exposure along each segment of the road network. The expression used in this study is of the following form:

$$R_s = \sum\nolimits_j P_{sj} * F_j * T_s \tag{4}$$

Where:

R_s: the risk on a segment road s;
P_{sj}: the probability of accident occurrence on road segment s with respect of a risk factor j;
F_j: the severity of pedestrian crash injury on a segment s with respect to factor j
T_s: The travel time associated with segment s.

In expression (3) some inputs parameters are TIFN, and some are constant therefore, the resulting risk is also obtained in the form of a TIFN ([a, b, c]; μ_α, v_α). The total-risk of a path, denoted by R(x) = ([a, b, c]; μ_r, v_r). is the aggregation of the risk weight of segment s_k. where s_k is the kth segment for a path x. The risk R(x) is calculated for each arc in the network N using (3). R(x) = ([a, b, c]; μ_r, v_r). The generalized value of the intuitionistic fuzzy risk of a segment s as shown in Fig. 3.

For each segment s there is a variety of risk values according to a θ and θ́ (predetermined confidence levels) to be specified according to the risk threshold not to be exceeded. $\mu_R(x < \lambda) = \theta$ and $(1 - v_R(x < \alpha)) = \acute{\theta}$; Where λ ∈ [0,1] and α ∈ [0, 1]. (See., Fig. 4). We need to search values of λ and α the evaluation of edge according to an θ and θ́ risk level. There are two triangles shown in Fig. 4 for TIFN with the help

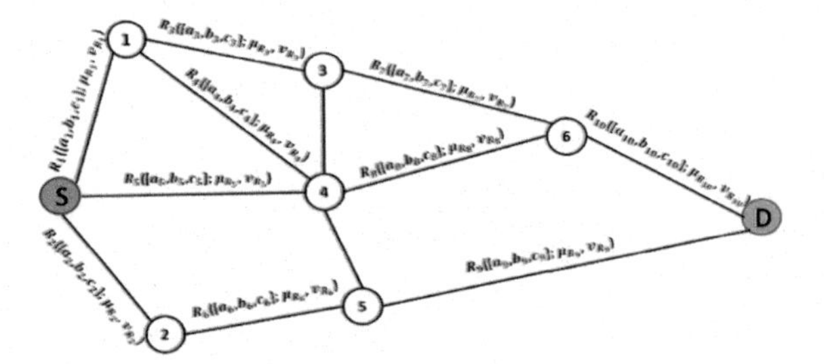

Fig. 3. Intuitionistic fuzzy risk graph.

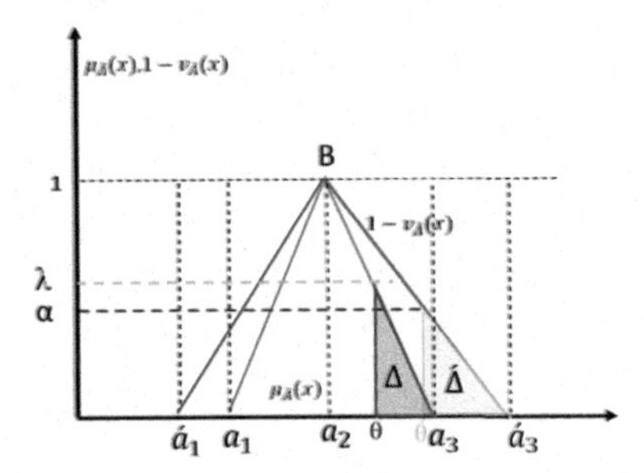

Fig. 4. Acceptable risk value λ, α according to θ and $\acute{\theta}$ risk level.

of Kahraman et al.'s [10] demonstration. These are; one for the membership function and one for non–membership function. Therefore, the values of λ and α is calculated as follows. We know that $P(x < \theta) = 1 - p(x > \theta)$, $P(x < \acute{\theta}) = 1 - p(x > \acute{\theta})$. Using the area of triangles a_2B a_3 and a_2B $\acute{a}_3$. we get the following result:

$$\text{Area}\,(a_2\text{B}\,a_3) = 1 * (a_3 - a_2)/2$$

$$\text{Area}\,(a_2\text{B}\,\acute{a}_3) = 1 * (\acute{a}_3 - a_2)/2$$

Using the Thales theorem, we get the following result:

$$1 - \lambda = \frac{\theta - a_2}{a_3 - a_2};\ \lambda = 1 - \frac{\theta - a_2}{a_3 - a_2}$$

$$1 - \alpha = \frac{\acute{\theta} - a_2}{\acute{a}_3 - a_2};\ \alpha = 1 - \frac{\acute{\theta} - a_2}{\acute{a}_3 - a_2}$$

Now for each segment s we have the values of λ and α according respectively to θ and $\acute{\theta}$ risk level. When $\lambda \ll 1$ the risk of accidents increases considerably, and consequently it is the fuzzy intuitionistic indicator of the risk that we will adopt. Finally, the problem becomes to define the safest path in a time dependent network. We now turn to an application exploiting the model presented above to provide safe pedestrian navigation in urban environments. the safest path problem takes as an input the road network G (V, E); together with a pair of sources–destination nodes, (s, d), and its goal is to provide to the user a short and safe path between s and d.

4 System Description

The proposed system is an integrated solution that provides a variety of services such as geolocation, real-time traceability, and pedestrian routing [6, 8, 9]. Leaning towards a microservices architecture [13], it is intended to develop and run multiple microservices. These microservices are developed, tested, deployed, and executed in isolation from the others. Furthermore, this implementation reveals the benefits of the microservices architecture, such as scalability, extensibility, flexibility, and fault tolerance. Pedestrian routing microservice is responsible for constructing a time-dependent network with fuzzy risks and finding a safe point-to-point path in it. The implemented routing algorithm is designed to compute an optimized shortest path that outperforms the performance of classical algorithms in terms of computation time and space. In general, two types of data are required as input for the proposed service. namely: a) spatial data that represents a traffic network as a collection of arcs with nodes loaded from OSM (OpenStreetMap) into PostgreSQL/PostGIS database. b) non-spatial attribute information of road attribute data, such as name, category, length, and risk cost. The algorithm for finding the safest path is based on a network consisting of vertices and edges (paths) defined by pairs of vertices. Each edge has a cost, which in this case represents the risk for pedestrians. As this attribute is known, the problem of pedestrian routing consists in finding a specified minimum cost from a vertex at source A to a vertex at destination B. The pgr_bdAstar() function (which implements the bi-directional A* algorithm) was used to find the optimal path.

5 Conclusion

Safe and smart walkability is a key element to support pedestrians in their daily activities and to provide a livable smart city. Information about urban traffic, crosswalks, and the safest paths can be of great benefit in this context. In fact, for pedestrian routing issues, data is very important to assess the risk. The accuracy and quality of the data can have a significant impact on the outcome. Currently, pedestrian data is inadequate and incomplete. To address this situation, we have proposed a real-time reactive system based on an intuitionistic fuzzy pedestrian risk model. The proposed model is an appropriate ranking method for intuitionistic fuzzy numbers and may be applicable to decision making problems in intuitionistic fuzzy environments.

References

1. Ando, K., Tanaka, N.: An evaluation protocol for collision avoidance and mitigation systems and its application to safety estimation. In: Proceedings of the 23nd International Technical Conference on the Enhanced Safety of Vehicles, Seoul, Republic of Korea (2013)
2. Atanassov, K.: Intuitionistic fuzzy sets. Int. J. Bioautomation **20**, 1 (2016)
3. Ayala, I., Mandow, L., Amor, M., Fuentes, L.: An evaluation of multi-objective urban tourist route planning with mobile devices. In: LNCS Ubiquitous Computing and Ambient Intelligence, vol. 7656, pp. 387–394 (2012)

4. Boulmakoul, A., Mandar, M.: Fuzzy ant colony paradigm for virtual pedestrian simulation. Open Oper. Res. J. **2011**, 19–29 (2011). https://doi.org/10.2174/1874243201105010019. ISSN:18742432
5. Brüde, U., Larsson, J.: Models for predicting accidents at junctions where pedestrians and cyclists are involved. How well do they fit? Accid. Anal. Prev. **25**, 499–509 (1993)
6. Czogalla, O., Herrmann, A.: Parameters determining route choice in pedestrian networks. In: TRB 90th Annual Meeting Compendium of Papers DVD, Washington, DC, pp. 23–27 (2011)
7. David, K., Flach, A.: Car-2-x and pedestrian safety. Veh. Technol. Mag. **5**(1), 70–76 (2010)
8. Gonzalez, H., Han, J., Li, X., Myslinska, M., Sondag, J.: Adaptive fastest path computation on a road network: a traffic mining approach. In: VLDB (2007)
9. Kanoulas, E., Du, Y., Xia, T., Zhang, D.: Finding fastest paths on a road network with speed patterns. In: IEEE ICDE (2006)
10. Kahraman, C., et al.: Extension of information axiom from ordinary to intuitionistic fuzzy sets: an application to search algorithm selection. Comput. Ind. Eng. **105**, 348–361 (2017)
11. Mandar, M., Boulmakoul, A., Lbath, A.: Pedestrian fuzzy risk exposure indicator. Transp. Res. Procedia **22**, 124–133 (2017)
12. Quistberg, D.A., et al.: Multilevel models for evaluating the risk of pedestrian–motor vehicle collisions at intersections and mid-blocks. Accid. Anal. Prev. **84**, 99–111 (2015)
13. Newman, S.: Building Microservices: Designing Fine-Grained Systems. O'Reilly Media, Inc. (2015)
14. Saravanan, S., Sabari, A., Geetha, M.: Fuzzy-based approach to predict accident risk on road network. Int. J. Emerg. Technol. Adv. Eng. **4**(5), 536-540 (2014). ISSN 2250-2459, ISO 9001:2008
15. Van der Molen, H.H.: Child pedestrian's exposure, accidents and behavior. Accid. Anal Prev. **13**(3), 193–224 (1981)
16. Wang, T., Cardone, G., Corradi, A., Torresani, L., Campbell, A.T.: WalkSafe: a pedestrian safety app for mobile phone users who walk and talk while crossing roads. In: Proceedings of the Twelfth Workshop on Mobile Computing Systems & Applications, pp. 1–6 (2012)
17. World Health Organization: Pedestrian safety: a road safety manual for decision-makers and practitioners (2013)
18. Zadeh, L.A.: Fuzzy sets as a basis for a theory of possibility. Fuzzy Sets Syst. **1**(1), 3–28 (1978)
19. Zadeh, L.A.: Probability measures of fuzzy events. J. Math. Anal. Appl. **23**(2), 421–427 (1968)

Extension of VIKOR Method Using Circular Intuitionistic Fuzzy Sets

Cengiz Kahraman[1(✉)] and Irem Otay[2]

[1] Department of Industrial Engineering, Istanbul Technical University, Macka-Besiktas, 34367 Istanbul, Turkey
kahramanc@itu.edu.tr

[2] Faculty of Engineering and Natural Sciences, Department of Industrial Engineering, Istanbul Bilgi University, Eski Silahtarağa Elektrik Santrali, 34060 Eyüpsultan, Istanbul, Turkey
irem.otay@bilgi.edu.tr

Abstract. VIKOR method is used to solve a variety of MCDM problems including a variety of criteria that can be conflicting and noncommensurable. This method is based on the distances of alternatives to positive and negative ideal solutions, and provides compromising solutions. Decision makers generally prefer to evaluate the alternatives with respect to the criteria by using linguistic terms rather than assigning exact numerical values. The fuzzy set theory captures the uncertainty and subjectivity in these linguistic terms successfully. Many classical MCDM methods have been extended to their fuzzy versions by using the fuzzy set theory for handling this uncertainty. VIKOR method has been extended by using several fuzzy set extensions such as intuitionistic fuzzy VIKOR, hesitant fuzzy VIKOR, Pythagorean fuzzy VIKOR, picture fuzzy VIKOR, and spherical fuzzy VIKOR methods. Circular intuitionistic fuzzy sets (C-IFS) introduced as an extension of intuitionistic fuzzy sets, enables decision makers to define membership and the non-membership degrees as circular membership functions. In this paper, we develop C-IFS VIKOR method and apply it to a waste disposal location selection problem.

Keywords: VIKOR · Circular intuitionistic fuzzy sets · Fuzzy MCDM · Fuzzy set extensions

1 Introduction

Multi-criteria Decision Making (MCDM) methods have been used to help decision makers to evaluate a set of finite alternatives considering a wide variety of conflicting and incommensurable criteria, and to reach a best optimal solution. Since 1980s, MCDM methods have been widely studied and applied to solve different types of MCDM problem. The Scopus database has indicated that the highest number of MCDM publications (1,325) was recorded in 2020 [1].

In the MCDM literature, there are a variety of decision making techniques where some of them are used to calculate the weights of the alternatives such as Analytic Hierarchy Process (AHP) and Analytic Network Process (ANP); the others are categorized

C. Kahraman et al. (Eds.): INFUS 2021, LNNS 308, pp. 48–57, 2022.
https://doi.org/10.1007/978-3-030-85577-2_6

as outranking methods i.e. ELimination Et Choice Translating REality (ELECTRE), and Preference Ranking Organization Method For Enrichment Evaluation (PROMETHEE), and scoring methods including Simple Additive Weighting (SAW), Complex Proportional Assessment (COPRAS), the Technique for Order of Preference by Similarity to Ideal Solution (TOPSIS) and VIekriterijumsko KOmpromisno Rangiranje (VIKOR). VIKOR method originated from a Serbian name meaning of multi-criteria optimization and compromise solution, was developed by Opricović and Tzeng [2]. The method evaluating tangible and intangible as well as conflicting and noncommensurable criteria, focuses on providing maximum group utility and minimum individual regret.

As stated by the researchers, the classical MCDM methods can not effectively address real life decision making problems including uncertainty and lack of information. The fuzzy set theory proposed by Zadeh [3], helps to reduce the complexity of modeling nonlinear decision making problems using linguistic variable with their corresponding fuzzy numbers rather than using crisp/exact numbers [4]. To deal with uncertainty in a better way, academicians have been focusing on the fuzzy set extensions. The first extension namely Type-2 fuzzy sets were introduced in 1975 [5]. So far, many extensions such as Hesitant fuzzy sets [6], Multi-sets [7], Intuitionistic fuzzy sets [8], Neutrosophic sets [9], Pythagorean fuzzy sets [10], and Spherical fuzzy sets [11] were proposed. As new fuzzy set extensions are introduced, the academicians shifted their attentions to modify VIKOR method employing these extensions. Liao and Xu [12] introduced hesitant VIKOR method and applied to evaluate service quality decision making problem for domestic airlines. Chatterjee et al. [13] proposed extended intuitionistic fuzzy VIKOR to assess strategic decisions for information system. Hu et al. [14] developed an interval neutrosophic VIKOR to choose an appropriate doctor on a mobile healthcare. Gul et al. [15] proposed Pythagorean fuzzy VIKOR for evaluating and ranking occupational hazard risks in mine industry. Akram et al. [16] proposed multi-expert complex Spherical fuzzy VIKOR approach to rank the objectives of an advertisement on Facebook. On the other hand, Circular Intuitionistic Fuzzy (C-IF) Sets have recently introduced [17]. As an extension of intuitionistic fuzzy sets, a C-IF element is defined by a circle composed of center (membership and non-membership degrees) and its radius. This paper is the first study introducing the steps of the Circular Intuitionistic Fuzzy VIKOR.

The remaining of the paper is organized as follows: In Sect. 2, preliminary information on circular intuitionistic fuzzy sets is presented. In Sect. 3, a novel Circular Intuitionistic Fuzzy VIKOR Method is proposed while in Sect. 4 the developed model is implemented to a waste disposal location selection problem. Finally, the paper is concluded and future research directions are stated.

2 Circular Intuitionistic Fuzzy Sets

In C-IF sets, every intuitionistic fuzzy element is defined by a circle pointing out its vagueness, represented by a center $\langle \mu_C(x), \nu_C(x) \rangle$ and radius r.

Definition 1: A C-IFS $\tilde{C}$ can be defined as in the following equation:

$$\tilde{C} = \{\langle x, \mu_C(x), \nu_C(x); r \rangle | x \in X\} \tag{1}$$

where $0 \leq \mu_C(x) + \nu_C(x) \leq 1$ *and* $r \in [0, 1]$ *for* $\forall\, x \epsilon X$.

In Eq. (1), the function $\mu_C : X \rightarrow [0, 1]$ and $\upsilon_C : X \rightarrow [0, 1]$ indicate the degrees of membership and the non-membership of the element $x \epsilon X$ to the set C, respectively. In CIF sets, a radius of the circle around each element is defined by r where $r : X \rightarrow [0, 1]$ [17]. The degree of indeterminacy is calculated using Eq. (2):

$$\pi_C(x) = 1 - \mu_C(x) - \nu_C(x) \tag{2}$$

When there are multiple decision makers ($t = 1, 2, \ldots, k$), their judgments are first aggregated through Intuitionistic Fuzzy Weighted Averaging (IFWA) operator using Eq. (3). Then, Euclidean distances of each expert's evaluations to the aggregated intuitionistic fuzzy sets are obtained. The radius is determined by taking the maximum of these distances as in Eq. (4).

Definition 2: Let $\alpha_{jt} = (\mu_{\alpha_{jt}}, v_{\alpha_{jt}})(j = 1, 2, .., n; t = 1, 2, \ldots, k)$ be a collection of IFNs. IFWA operator is a mapping: $\Theta^n \rightarrow \Theta$, such that

$$\begin{aligned} \text{IFWA}(\alpha_{j1}, \alpha_{j2}, \ldots, \alpha_{jt}) &= w_1\alpha_{j1} \oplus \ldots \oplus w_k\alpha_{jk} \\ &= \left(1 - \prod_{t=1}^{k}(1 - \mu_{\alpha_{jt}})^{w_t}, \prod_{t=1}^{k} v_{\alpha_{jt}}^{w_t}\right) = \langle \mu(C_j), \nu(C_j) \rangle, \forall j \end{aligned} \tag{3}$$

where $w = (w_1, \ldots, w_k)^T$ is the weighting vector of the decision makers, $w_t \in [0, 1]$ and $\sum_{t=1}^{k} w_t = 1$ [18].

$$r_j = \max_{1 \leq t \leq k} \sqrt{(\mu(C_j) - \mu_{\alpha_{jt}})^2 + (\nu(C_j) - v_{\alpha_{jt}})^2}, \forall j \tag{4}$$

For universe $W = \{C_1, C_2, \ldots\}$, a C-IFS $\tilde{A}_r$ can be constructed as follows:

$$\tilde{A}_r = \{\langle C_i, \mu(C_i), \nu(C_i); r \rangle | C_i \epsilon W\} = \{\langle C_i, O_r(\mu(C_i), \nu(C_i)) \rangle | C_i \epsilon W\} \tag{5}$$

In Fig. 1, several forms of circles representing basic geometric interpretation of C-IFS are displayed [17].

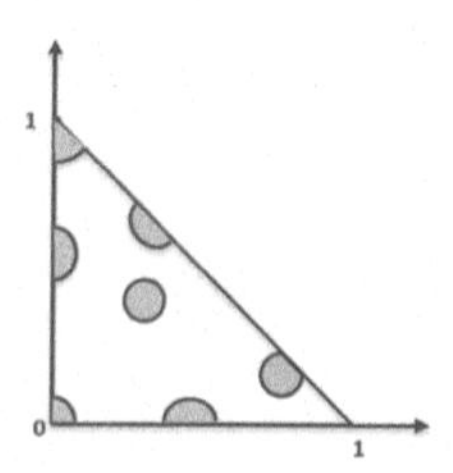

Fig. 1. C-IFS geometrical representation

Definition 3: Let $\widetilde{C}_1 = \langle \mu_{C_1}(x), \nu_{C_1}(x); r_1 \rangle$ and $\widetilde{C}_2 = \langle \mu_{C_2}(x), \nu_{C_2}(x); r_2 \rangle$ be two circular intuitionistic fuzzy numbers (C-IFNs). Some of the arithmetic operations for the C-IFNs are given in Eqs. (6)–(13) [17]:

$$\widetilde{C}_1 \cap_{min} \widetilde{C}_2 = \{\langle x, \min(\mu_{C_1}(x), \mu_{C_2}(x)), \max(\nu_{C_1}(x), \nu_{C_2}(x)); \min(r_1, r_2)\rangle | x \in X\} \quad (6)$$

$$\widetilde{C}_1 \cap_{max} \widetilde{C}_2 = \{\langle x, \min(\mu_{C_1}(x), \mu_{C_2}(x)), \max(\nu_{C_1}(x), \nu_{C_2}(x)); \max(r_1, r_2)\rangle | x \in X\} \quad (7)$$

$$\widetilde{C}_1 \cup_{min} \widetilde{C}_2 = \{\langle x, \max(\mu_{C_1}(x), \mu_{C_2}(x)), \min(\nu_{C_1}(x), \nu_{C_2}(x)); \min(r_1, r_2)\rangle | x \in X\} \quad (8)$$

$$\widetilde{C}_1 \cup_{max} \widetilde{C}_2 = \{\langle x, \max(\mu_{C_1}(x), \mu_{C_2}(x)), \min(\nu_{C_1}(x), \nu_{C_2}(x)); \max(r_1, r_2)\rangle | x \in X\} \quad (9)$$

$$\tilde{C}_1 \oplus_{min} \tilde{C}_2 = \{x, \mu_{C_1}(x) + \mu_{C_2}(x) - \mu_{C_1}(x)\mu_{C_2}(x), \nu_{C_1}(x)\nu_{C_2}(x); \min(r_1, r_2) | x \in X\} \quad (10)$$

$$\tilde{C}_1 \oplus_{max} \tilde{C}_2 = \{x, \mu_{C_1}(x) + \mu_{C_2}(x) - \mu_{C_1}(x)\mu_{C_2}(x), \nu_{C_1}(x)\nu_{C_2}(x); \max(r_1, r_2) | x \in X\} \quad (11)$$

$$\tilde{C}_1 \otimes_{min} \tilde{C}_2 = \{x, \mu_{C_1}(x)\mu_{C_2}(x), \nu_{C_1}(x) + \nu_{C_2}(x) - \nu_{C_1}(x)\nu_{C_2}(x); \min(r_1, r_2) | x \in X\} \quad (12)$$

$$\tilde{C}_1 \otimes_{max} \tilde{C}_2 = \{x, \mu_{C_1}(x)\mu_{C_2}(x), \nu_{C_1}(x) + \nu_{C_2}(x) - \nu_{C_1}(x)\nu_{C_2}(x); \max(r_1, r_2) | x \in X\} \quad (13)$$

3 Circular Intuitionistic Fuzzy VIKOR Method

The steps of the proposed C-IF VIKOR method are summarized below:

Step 1. Main criteria ($C_j, j = 1, 2, .., n$) and alternatives ($A_i, i = 1, 2, \ldots, m$) are determined for a multi-criteria decision making problem.

Step 2. For determining the weights of the criteria, the judgments are collected from experts using linguistic scale with intuitionistic fuzzy (IF) numbers given in Table 1. For Equal (E) and Almost Equal (AE), (0.5, 0.5) and (0.45, 0.45) are used, respectively. When decision makers hesitate between two consecutive linguistic terms, they can use intermediate values.

Table 1. Linguistic scale with IF numbers

Linguistic terms	(μ, v)	Linguistic terms	(μ, v)
Absolutely Low (AL)	(0.05, 0.85)	Medium High (MH)	(0.55, 0.35)
Very Low (VL)	(0.15, 0.75)	High (H)	(0.65, 0.25)
Low (L)	(0.25, 0.65)	Very High (VH)	(0.75, 0.15)
Medium Low (ML)	(0.35, 0.55)	Absolutely High (AH)	(0.85, 0.05)

Step 3. The evaluations of the criteria are aggregated considering the weights of the decision makers by means of Eq. (3).

Step 4. Using Eq. (4), the Euclidean distances from each expert's judgment to the aggregated value of criteria are calculated. Radius "r" of each criterion is defined as the maximum value of these distances.

Step 5. Decision matrices using IF sets in Table 1 are constructed based on meetings with experts.

Step 6. By employing similar procedure in Steps 3 and 4, judgments ($\tilde{x}_{ij}$) in each expert's decision matrix are aggregated and radius values are calculated using Eqs. (3) and (4), respectively. Thus, an aggregated decision matrix with C-IF sets is obtained.

Step 7. Our proposed Relative Score Function (RSF) based on vector normalization as in Eq. (14) is used to defuzzify the evaluations in the aggregated decision matrix.

$$\mathrm{RSF_j} = \frac{(1-\mathrm{v_j})(1+\mu_\mathrm{j})+\mu_\mathrm{j}}{3} \times \left(\frac{\frac{1}{\mathrm{r_j}}}{\sqrt{\frac{1}{\mathrm{r_1^2}}+\frac{1}{\mathrm{r_2^2}}+\frac{1}{\mathrm{r_3^2}}}}\right)^{\tau} \tag{14}$$

where τ is a small number such as 0.1 or 0.01.

Step 8. Considering RSF values, C-IF positive and negative ideal solutions are determined as in Eq. (15).

$$\tilde{x}^{*}_{j,C-IF} = Max_i\tilde{x}_{ij}, \tilde{x}^{-}_{j,C-IF} = Min_i\tilde{x}_{ij} j = 1, 2, .., n \tag{15}$$

Step 9. From pessimistic and optimistic points of view, distances to C-IF positive and negative ideal solutions are calculated by means of Eqs. (16)–(17) and Eqs. (18)–(19), respectively.

$$D_p\left(\tilde{x}_{ij}, \tilde{x}^{*}_{j,C-IF}\right) = \sqrt{\frac{1}{2}\left(\begin{array}{c}\left(\left(\mu_{\tilde{x}_{ij}} - r_{\tilde{x}_{ij}}\right) - \left(\mu_{\tilde{x}^{*}_{j,C-IF}} + r_{\tilde{x}^{*}_{j,C-IF}}\right)\right)^2 \\ +\left(\left(v_{\tilde{x}_{ij}} + r_{\tilde{x}_{ij}}\right) - \left(v_{\tilde{x}^{*}_{j,C-IF}} - r_{\tilde{x}^{*}_{j,C-IF}}\right)\right)^2 \\ +\left(\pi_{\tilde{x}_{ij}} - \pi_{\tilde{x}^{*}_{j,C-IF}}\right)^2\end{array}\right)} \tag{16}$$

$$D_p\left(\tilde{x}^{-}_{j,C-IF}, \tilde{x}^{*}_{j,C-IF}\right) = \sqrt{\frac{1}{2}\left(\begin{array}{c}\left(\left(\mu_{\tilde{x}^{-}_{j,C-IF}} - r_{\tilde{x}^{-}_{j,C-IF}}\right) - \left(\mu_{\tilde{x}^{*}_{j,C-IF}} + r_{\tilde{x}^{*}_{j,C-IF}}\right)\right)^2 \\ +\left(\left(v_{\tilde{x}^{-}_{j,C-IF}} + r_{\tilde{x}^{-}_{j,C-IF}}\right) - \left(v_{\tilde{x}^{*}_{j,C-IF}} - r_{\tilde{x}^{*}_{j,C-IF}}\right)\right)^2 \\ +\left(\pi_{\tilde{x}^{-}_{j,C-IF}} - \pi_{\tilde{x}^{*}_{j,C-IF}}\right)^2\end{array}\right)} \tag{17}$$

$$D_o\left(\tilde{x}_{ij}, \tilde{x}^{*}_{j,C-IF}\right) = \sqrt{\frac{1}{2}\left(\begin{array}{c}\left(\left(\mu_{\tilde{x}_{ij}} + r_{\tilde{x}_{ij}}\right) - \left(\mu_{\tilde{x}^{*}_{j,C-IF}} - r_{\tilde{x}^{*}_{j,C-IF}}\right)\right)^2 \\ +\left(\left(v_{\tilde{x}_{ij}} - r_{\tilde{x}_{ij}}\right) - \left(v_{\tilde{x}^{*}_{j,C-IF}} + r_{\tilde{x}^{*}_{j,C-IF}}\right)\right)^2 \\ +\left(\pi_{\tilde{x}_{ij}} - \pi_{\tilde{x}^{*}_{j,C-IF}}\right)^2\end{array}\right)} \tag{18}$$

$$D_o\left(\tilde{x}^-_{j,C-IF}, \tilde{x}^*_{j,C-IF}\right) = \sqrt{\frac{1}{2}\left(\begin{array}{c}\left(\left(\mu_{\tilde{x}^-_{j,C-IF}} + r_{\tilde{x}^-_{j,C-IF}}\right) - \left(\mu_{\tilde{x}^*_{j,C-IF}} - r_{\tilde{x}^*_{j,C-IF}}\right)\right)^2 \\ +\left(\left(\nu_{\tilde{x}^-_{j,C-IF}} - r_{\tilde{x}^-_{j,C-IF}}\right) - \left(\nu_{\tilde{x}^*_{j,C-IF}} + r_{\tilde{x}^*_{j,C-IF}}\right)\right)^2 \\ +\left(\pi_{\tilde{x}^-_{j,C-IF}} - \pi_{\tilde{x}^*_{j,C-IF}}\right)^2\end{array}\right)} \tag{19}$$

In the above equations, p and o denote pessimistic and optimistic points of view.

Step 10. Once the distances are calculated, maximum group utility $(\tilde{S}_{i,p}, \tilde{S}_{i,o})$, minimum individual regret $(\tilde{R}_{i,p}, \tilde{R}_{i,o})$ and $\tilde{Q}_{i,p}$ & $\tilde{Q}_{i,o}$ values for each alternative considering pessimistic and optimistic view points are obtained using Eqs. (20–22). In Eq. (22), v is the weight of the maximum group utility.

$$\tilde{S}_{i,p} = \sum \tilde{w}_i \frac{D_p\left(\tilde{x}_{ij}, \tilde{x}^*_{j,C-IF}\right)}{D_p\left(\tilde{x}^-_{j,C-IF}, \tilde{x}^*_{j,C-IF}\right)}, \tilde{S}_{i,o} = \sum \tilde{w}_i \frac{D_o\left(\tilde{x}_{ij}, \tilde{x}^*_{j,C-IF}\right)}{D_o\left(\tilde{x}^-_{j,C-IF}, \tilde{x}^*_{j,C-IF}\right)} \tag{20}$$

$$\tilde{R}_{i,p} = \mathrm{Max}_j\left(\tilde{w}_i \frac{D_p\left(\tilde{x}_{ij}, \tilde{x}^*_{j,C-IF}\right)}{D_p\left(\tilde{x}^-_{j,C-IF}, \tilde{x}^*_{j,C-IF}\right)}\right), \tilde{R}_{i,o} = \mathrm{Max}_j\left(\tilde{w}_i \frac{D_o\left(\tilde{x}_{ij}, \tilde{x}^*_{j,C-IF}\right)}{D_o\left(\tilde{x}^-_{j,C-IF}, \tilde{x}^*_{j,C-IF}\right)}\right) \tag{21}$$

$$\tilde{Q}_{i,p} = v\frac{\left(\tilde{S}_{i,p} - \tilde{S}^*_p\right)}{\left(\tilde{S}^-_p - \tilde{S}^*_p\right)} + (1-v)\frac{\left(\tilde{R}_{i,p} - \tilde{R}^*_p\right)}{\left(\tilde{R}^-_p - \tilde{R}^*_p\right)}, \tilde{Q}_{i,o} = v\frac{\left(\tilde{S}_{i,o} - \tilde{S}^*_o\right)}{\left(\tilde{S}^-_o - \tilde{S}^*_o\right)} + (1-v)\frac{\left(\tilde{R}_{i,o} - \tilde{R}^*_o\right)}{\left(\tilde{R}^-_o - \tilde{R}^*_o\right)} \tag{22}$$

where $\tilde{S}^*_p = \min_i \tilde{S}_{i,p}, \tilde{S}^-_p = \max_i \tilde{S}_{i,p}, \tilde{R}^*_p = \min_i \tilde{R}_{i,p}, \tilde{R}^-_p = \max_i \tilde{R}_{i,p}$
$\tilde{S}^*_o = \min_i \tilde{S}_{i,o}, \tilde{S}^-_o = \max_i \tilde{S}_{i,o}, \tilde{R}^*_o = \min_i \tilde{R}_{i,o}, \tilde{R}^-_o = \max_i \tilde{R}_{i,o}$

Step 11. Finally, the alternatives are sorted based on the ascending order of the defuzzified values of $\tilde{S}_{i,p}, \tilde{S}_{i,o}, \tilde{R}_{i,p}, \tilde{R}_{i,o}, \tilde{Q}_{i,p}$ and $\tilde{Q}_{i,o}$.

4 Implementation

In this section, the novel C-IF VIKOR method is applied to solve a waste disposal location evaluation and selection problem. The three-level hierarchical structure of the decision making problem is taken from the study of Yazdani et al. [19]. In the study, Economical (C1), Environmental (C2), and Social criteria (C3) are evaluated and alternative locations are determined by a company planning to found a waste disposal company in the European side of Istanbul. The judgments on criteria importance using intuitionistic fuzzy numbers are collected from three decision makers (dm1, dm2, and dm3) as given in Table 2. Initially, the aggregated values of the judgments are calculated using Eq. (3); then, the values of radius per each criterion are calculated by means of Eq. (4). The weights of decision makers are taken as 0.5, 0.3 and 0.2, respectively.

Table 2. Importance weights of criteria

Decision makers	C1	C2	C3
dm1	(0.55, 0.40)	(0.45, 0.50)	(0.35, 0.55)
dm2	(0.35, 0.55)	(0.65, 0.25)	(0.50, 0.45)
dm3	(0.45, 0.40)	(0.45, 0.45)	(0.55, 0.40)

The aggregated values of the criteria are as follows: C1: (0.477, 0.440; 0.168), C2: (0.520, 0.398; 0.197) and C3: (0.442, 0.486; 0.138). The calculation of r_{C1} is given below. The judgments on each alternative associated to each criterion are collected as displayed in Table 3.

$$r_{C1} = \max\left\{\begin{array}{l}\sqrt{(0.477-0.55)^2+(0.440-0.40)^2},\\ \sqrt{(0.477-0.35)^2+(0.440-0.55)^2},\\ \sqrt{(0.477-0.45)^2+(0.440-0.40)^2}\end{array}\right\} = \max(0.083, 0.168, 0.048) = 0.168$$

Table 3. Decision matrices

Alternatives	Decision makers	C1	C2	C3
A1	dm1	(0.25, 0.70)	(0.45, 0.45)	(0.55, 0.35)
	dm2	(0.40, 0.55)	(0.35, 0.55)	(0.65, 0.35)
	dm3	(0.60, 0.35)	(0.40, 0.55)	(0.45,0.40)
A2	dm1	(0.30, 0.65)	(0.50,0.40)	(0.65, 0.25)
	dm2	(0.55, 0.35)	(0.40, 0.55)	(0.70, 0.25)
	dm3	(0.65, 0.25)	(0.35, 0.60)	(0.55, 0.35)
A3	dm1	(0.20, 0.60)	(0.35, 0.60)	(0.55, 0.40)
	dm2	(0.35, 0.50)	(0.30, 0.60)	(0.45,0.45)
	dm3	(0.50, 0.45)	(0.55, 0.35)	(0.40, 0.55)

Similar procedure is applied to calculate aggregated fuzzy ratings and radiuses employing Eqs. (3)–(4). Aggregated C-IF ratings of alternatives regarding to criteria are presented in Table 4. The proposed Relative Score Function (RSF) in Eq. (14) is implemented to compare the magnitudes of the aggregated C-IF ratings. Hence $\tilde{x}^*_{j,C-IF}$ and $\tilde{x}^-_{j,C-IF}$ are found as in Table 4.

Table 4. Aggregated C-IF ratings of alternatives and the values of $\tilde{x}_j^*$ and $\tilde{x}_j^-$

Alternatives	C1	C2	C3
A1	(0.381, 0.567; 0.308)	(0.412,0.497; 0.081)	(0.566, 0.359; 0.122)
A2	(0.466, 0.446; 0.269)	(0.443, 0.477; 0.154)	(0.649, 0.267; 0.129)
A3	(0.316, 0.536; 0.203)	(0.383, 0.539; 0.252)	(0.494, 0.442; 0.143)
$\tilde{x}^*_{j,C-IF}/\tilde{x}^-_{j,C-IF}$	C1	C2	C3
$\tilde{x}^-_{j,C-IF}$	(0.316, 0.536; 0.203)	(0.383, 0.539; 0.252)	(0.494, 0.442; 0.143)
$\tilde{x}^*_{j,C-IF}$	(0.466, 0.446; 0.269)	(0.443, 0.477; 0.154)	(0.649, 0.267; 0.129)

As the next step, distances from C-IF ratings to C-IF positive & negative ideal solutions are calculated based on optimistic and pessimistic points of view. After then, $\tilde{S}_{i,p}$, $\tilde{S}_{i,o}$, $\tilde{R}_{i,p}$, $\tilde{R}_{i,o}$, $\tilde{Q}_{i,p}$ and $\tilde{Q}_{i,o}$ values are obtained through Eqs. (20)–(22). In order to derive the values of $\tilde{R}_{i,p}$ and $\tilde{R}_{i,o}$, Eq. (14) is employed. The calculated $\tilde{S}_{i,p}$, $\tilde{S}_{i,o}$, $\tilde{R}_{i,p}$, $\tilde{R}_{i,o}$ values of the alternatives associated to the optimistic and pessimistic cases together with the defuzzified values of $\tilde{Q}_{i,p}$ and $\tilde{Q}_{i,o}$, are displayed in Table 5.

Table 5. Results of the C-IF VIKOR method

Alternatives	$\tilde{S}_{i,p}$	Rank-$S_{i,p}$	$\tilde{R}_{i,p}$	Rank-$R_{i,p}$	$Q_{i,p}$	Rank-$Q_{i,p}$
A1	(0.799, 0.134; 0.197)	2	(0.523 0.391; 0.168)	3	0.705	2
A2	(0.756, 0.170; 0.197)	1	(0.443, 0.476; 0.168)	1	0.000	1
A3	(0.860, 0.085; 0.197)	3	(0.520, 0.398; 0.197)	2	0.965	3
Alternatives	$\tilde{S}_{i,o}$	Rank-$S_{i,0}$	$\tilde{R}_{i,o}$	Rank-$R_{i,0}$	$Q_{i,o}$	Rank-$Q_{i,0}$
A1	(0.888, 0.064; 0.197)	2	(0.585, 0.337; 0.138)	2	0.294	2
A2	(0.951, 0.023; 0.197)	3	(0.748, 0.181; 0.138)	3	1.000	3
A3	(0.860, 0.085; 0.197)	1	(0.520, 0.398; 0.197)	1	0.000	1

5 Conclusion and Future Remarks

C-IF is the latest extension of the ordinary fuzzy sets which takes the uncertainty of membership and non-membership degrees into account with a circular area. In this study, C-IF VIKOR method has been developed for single-valued circular intuitionistic fuzzy sets. One of the other contribution of this study is the introduction of a novel score function formulation based on vector normalization for C-IF sets. The proposed C-IF VIKOR method has been successfully employed in the evaluation of waste disposal site selection problem from both pessimistic and optimistic perspectives. For further research, other fuzzy MCDM studies such as TOPSIS, EDAS, COPRAS or WASPAS based on C-IF sets can be developed and their results can be compared with the results of this study.

References

1. Scopus. https://0-www-scopus-com.divit.library.itu.edu.tr/results/results.uri?src=s&st1=&st2=&sot=b&sdt=b&origin=searchbasic&rr=&sl=19&s=TITLE-ABS-KEY(mcdm)&searchterm1=mcdm&searchTerms=&connectors=. Accessed 08 Mar 2021
2. Opricovic, S., Tzeng, G.H.: Compromise solution by MCDM methods: a comparative analysis of VIKOR and TOPSIS. Eur. J. Oper. Res. **156**, 445–455 (2004)
3. Zadeh, L.A.: Fuzzy set. Information. Control **18**(2), 338–353 (1965)
4. Chen, L.Y., Wang, T.-C.: Optimizing partners' choice in IS/IT outsourcing projects: the strategic decision of fuzzy VIKOR. Int. J. Prod. Econ. **120**, 233–242 (2009)
5. Zadeh, L.A.: The concept of a linguistic variable and its application to approximate reasoning. Inf. Sci. **8**, 199–249 (1975)
6. Torra, V.: Hesitant fuzzy sets. Int. J. Intell. Syst. **25**(6), 529–539 (2010)
7. Yager, R.R.: On the theory of bags. Int. J. Gen Syst **13**(1), 23–37 (1986)
8. Atanassov, K.T.: Intuitionistic fuzzy sets. Fuzzy Sets Syst. **20**, 87–96 (1986)
9. Smarandache, F.: Neutrosophic set - a generalization of the intuitionistic fuzzy set. Int. J. Pure Appl. Math. **24**(3), 287–329 (2005)
10. Yager, R.R.: Pythagorean fuzzy subsets. In: 2013 Joint IFSA World Congress and NAFIPS Annual Meeting (IFSA/NAFIPS), pp. 57–61. IEEE, Alberta-Canada (2013)
11. Kutlu Gündoğdu, F., Kahraman, C.: Spherical fuzzy sets and spherical fuzzy TOPSIS method. J. Intell. Fuzzy Syst. **36**(1), 337–352 (2019)
12. Liao, H., Xu, Z.: A VIKOR-based method for hesitant fuzzy multi-criteria decision making. Fuzzy Optim. Decis. Making **12**(4), 373–392 (2013). https://doi.org/10.1007/s10700-013-9162-0
13. Chatterjee, K., Kar, M.B., Kar, S.: Strategic decisions using intuitionistic fuzzy VIKOR method for information system (IS) outsourcing. In: 2013 International Symposium on Computational and Business Intelligence, New Delhi, India (2013)
14. Hu, J., Pan, L., Chen, X.: An interval neutrosophic projection-based VIKOR method for selecting doctors. Cogn. Comput. **9**, 801–816 (2017)
15. Gul, M., Ak, M.F., Guner, A.F.: Pythagorean fuzzy VIKOR-based approach for safety risk assessment in mine industry. J. Safety Res. **69**, 135–153 (2019)
16. Akram, M., Kahraman, C., Zahid, K.: Group decision-making based on complex spherical fuzzy VIKOR approach. Knowl.-Based Syst. **216**, 106793 (2021)
17. Atanassov, K.T.: Circular intuitionistic fuzzy sets. J. Intell. Fuzzy Syst. 1–6 (2020)

18. Liang, X., Wei, C., Chen, Z.: An intuitionistic fuzzy weighted OWA operator and its application. Int. J. Mach. Learn. Cybern. **4**(6), 713–719 (2013). https://doi.org/10.1007/s13042-012-0147-z
19. Yazdani, M., Tavana, M., Pamučar, D., Chatterjee, P.: A rough based multi-criteria evaluation method for healthcare waste disposal location decisions. Comput. Ind. Eng. **143**, 106394 (2020)

The Intuitionistic Fuzzy Framework for Evaluation and Rank Ordering the Negotiation Offers

Ewa Roszkowska(✉)

University of Bialystok, Warszawska 63, 15-062 Bialystok, Poland
e.roszkowska@uwb.edu.pl

Abstract. In this paper, a novel intuitionistic fuzzy framework for evaluation and rank ordering the multi-issue negotiation offers has been introduced and studied. First, it is assumed that the assessment of the negotiation options occurs in bipolar terms due to the advantages and disadvantages of using the intuitionistic fuzzy concept. Next, the Intuitionistic Fuzzy Ideal Reference Point (IFIRP) multi-criteria method, based on the Hellwig concept of pattern, is proposed for ordering offers from the best to the worst. The advantage of the approach is using natural language, taking into consideration subjectivity, lack of precision, or uncertainty in the evaluation of negotiation offers. Additionally, the pattern based on maximal fuzzy value avoids rank reversal when a new offer is added for evaluation. The illustrative example demonstrates the practicality and effectiveness of the intuitionistic fuzzy evaluation negotiation offers.

Keywords: Fuzzy preference analysis · Negotiation scoring system · Fuzzy multi-criteria method · Hellwig method

1 Introduction

Negotiation is a complex decision problem, where usually many issues should be analyzed. In the first step, the negotiator's preferences are considered separately for every issue. Next applying the aggregation method, the complete negotiation offer is evaluated. Classical multi-criteria methods (MCDM) are useful to support negotiations in well-structured negotiation problems when the options of negotiation issues are set by crisp values [1–3]. In reality, however, every assessment of the negotiation offer entails inaccuracy, uncertainty, or imprecision, resulting for example, from the lack of information, measurement error, or subjective evaluation. One of the main problems concerning the evaluation of the offer is the suitable way to represent available information. This requires the inclusion of data from various sources and of different nature. The options of negotiation issues may be real, interval, fuzzy, or expressed in words. In a series of papers [4–7] to set up the problems mentioned above, the building negotiation scoring systems based on fuzzy multi-criteria methods are proposed, where data of various types, was converted into triangular fuzzy numbers (e.g. fuzzy TOPSIS or fuzzy SAW).

C. Kahraman et al. (Eds.): INFUS 2021, LNNS 308, pp. 58–65, 2022.
https://doi.org/10.1007/978-3-030-85577-2_7

In this paper, a novel intuitionistic fuzzy framework was proposed for the assessment and ordering of negotiation offers. Firstly, it is assumed that the assessment of the option of packages takes place in bipolar terms, taking into account its advantages and disadvantages due to particular issues. The evaluation of options can be represented by membership grade and non-membership grade, respectively. Secondly, an intuitionistic fuzzy method for multi-criteria decision-making problem, using Hellwig's concept of reference point to rank the alternatives is proposed. Finally, this method is applied for evaluating negotiation offers.

The proposed method belongs to the class of algorithms based on reference points, such as TOPSIS (Technique for Order Preference by Similarity to an Ideal Solution) [8], VIKOR (Serb. Vlse Kriterijumska Optimizacija i Kompromisno Resenje) [9], BIPOLAR [10], MARS [11] and Hellwig method [12]. Wachowicz et al. [3] analyzed reference points-based multi-criteria methods in the context of evaluation of negotiation offers. In series of papers [13–17] application of classical and modified Hellwig's method for solving different multi-criteria problems can be found.

The rest of the article is organized as follows. Section 2 introduces the main definition and notation from the intuitionistic fuzzy set theory. Section 3 describes the Intuitionistic Fuzzy Ideal Reference Point (IFIRP) method. The illustrative example is included in Sect. 4 of the study to demonstrate the practicality and effectiveness of the proposed intuitionistic fuzzy framework for evaluation and rank ordering negotiation offers. Finally, Sect. 5 presents concluding remarks.

2 Preliminaries

In this section, we briefly recall some concepts related to intuitionistic fuzzy sets (IFS) used in the paper.

Definition 1 ([18, 19]). Let X be a universe of discourse of objects. An intuitionistic fuzzy set A in X is given by:

$$A = \{\langle x, \mu_A(x), \nu_A(x)\rangle | x \in X\}, \tag{1}$$

where: $\mu_A, \nu_A : X \rightarrow [0, 1]$ are functions with the condition for every $x \in X$

$$0 \leq \mu_A(x) + \nu_A(x) \leq 1 \tag{2}$$

The numbers $\mu_A(x)$ and $\nu_A(x)$ denote, respectively, the degrees of membership and non-membership of the element $x \in X$ to the set A; $\pi_A(x) = 1 - \mu_A(x) - \nu_A(x)$ the intuitionistic fuzzy index (hesitation margin) of the element x in the set A.

Let us observe, that in applying the intuitionistic fuzzy concept in decision making we can assume that decision-maker concentrates on both [20]:

- advantages (pros), which are measures by membership grade,
- disadvantages (cons), which are measured by non-membership grade.

In the case, where the universe X contains only one element x, then the IFS A over X can be denoted as $A = (\mu_A, \nu_A)$ and called an intuitionistic fuzzy value (IFV). Let Θ be the set of all IFVs. The fuzzy value (1, 0) is the largest and (0, 1) is the smallest one.

In the literature, we can found several propositions of distances for the fuzzy sets [20]. In the paper, a three-term intuitionistic fuzzy set representation of the normalized Euclidean distance is used.

Definition 2 [20]. Let us consider two $A, B \in$ IVFS with membership functions $\mu_A(x)$, $\mu_B(x)$ and non membership functions $\nu_A(x)$, $\nu_B(x)$ respectively. The normalized Euclidean distance is calculated in the following way:

$$d(A, B) = \sqrt{\frac{1}{2n} \sum_{i=1}^{n} [(\mu_A(x_i) - \mu_B(x_i))^2 + (\nu_A(x_i) - \nu_B(x_i))^2 + (\pi_A(x_i) - \pi_B(x_i))^2]} \quad (3)$$

3 The Intuitionistic Fuzzy Ideal Reference Point (IFIRP) Method

A multi-criteria decision problem can be expressed in an intuitionistic fuzzy decision matrix D. Suppose that there exists a set of alternatives $A = \{A_1, A_2, ..., A_m\}$ and the set of criteria$C = \{C_1, C_2, ..., C_n\}$. We assumed that evaluations of each alternative with respect to each criterion C_j is given using IVFSs. More exactly, the characteristics of an alternative A_i with respect to $j-$ th criterion can be represented by an IVIFS value (μ_{ij}, ν_{ij}), where μ_{ij} the membership degree and ν_{ij} non-membership degree of the alternative $A_i \in A$ with respect to $j-$ th criterion $(i = 1, 2, ..., m; j = 1, 2, ..., n)$. Therefore, μ_{ij} indicate the level of satisfaction; ν_{ij} the level of dissatisfaction of the alternative A_i with respect to $j-$ th criterion. Notice that $0 \le \mu_{ij} \le 1$ and $0 \le \nu_{ij} \le 1$ and the degree of hesitancy is given by $\pi_{ij} = 1 - \mu_{ij} - \nu_{ij}$. The intuitionistic fuzzy decision matrix D is given in the following form:

$$D = \begin{array}{c} \\ A_1 \\ A_2 \\ \vdots \\ A_m \end{array} \begin{array}{c} \begin{array}{cccc} C_1 & C_2 & \cdots & C_n \end{array} \\ \begin{bmatrix} (\mu_{11}, v_{11}) & (\mu_{12}, v_{12}) & \cdots & (\mu_{1n}, v_{1n}) \\ (\mu_{21}, v_{21}) & (\mu_{22}, v_{22}) & \cdots & (\mu_{2n}, v_{i2n}) \\ \vdots & \vdots & \ddots & \vdots \\ (\mu_{m1}, v_{m1}) & (\mu_{m2}, v_{m2}) & \cdots & (\mu_{mn}, v_{mn}) \end{bmatrix} \end{array} \quad (4)$$

Let us observe that later we do not discriminate between cost and benefit criteria. It is because for every $j-$ th criterion more μ_{ij}(level of satisfaction) and less ν_{ij} (level of dissatisfaction) is desired.

Step 1: Construction of the intuitionistic fuzzy decision matrix

In this step, the decision-maker evaluates the alternatives concerning criteria based on their preferences or expertise using the intuitionistic fuzzy concept. Finally, the decision matrix (see Eq. 4) is given.

Step 2: Determination of the weights of criteria

Let $w = [w_1, \ldots, w_n]$ be the vector of weights that reflect the importance of each criterion C_j ,$j = 1, 2, ..., n$ and $\sum_{j=1}^{n} w_j = 1$.

Step 3: Determination of the intuitionistic fuzzy positive-ideal solution

The intuitionistic fuzzy positive ideal based on the maximal fuzzy value $(1, 0)$ has the form:

$$IFI^{+} = ((1,0), \ldots, (1,0)) \tag{5}$$

Step 4: Construction of the weighted intuitionistic fuzzy decision matrix

The weighted intuitionistic fuzzy decision matrix wD can be defined as follows:

$$wD = \begin{array}{c} \\ A_1 \\ A_2 \\ \vdots \\ A_m \end{array} \begin{array}{c} \begin{array}{ccc} C_1 & \quad & C_n \end{array} \\ \begin{bmatrix} (\tilde{\mu}_{11}, \tilde{v}_{11}) & \cdots & (\tilde{\mu}_{1n}, \tilde{v}_{1n}) \\ (\tilde{\mu}_{21}, \tilde{v}_{21}) & \cdots & (\tilde{\mu}_{2n}, \tilde{v}_{2n}) \\ \vdots & \ddots & \vdots \\ (\tilde{\mu}_{m1}, \tilde{v}_{m1}) & \cdots & (\tilde{\mu}_{mn}, \tilde{v}_{mn}) \end{bmatrix} \end{array} \tag{6}$$

where $(\tilde{\mu}_{ij}, \tilde{v}_{ij})$ is calculated in the following way [19]:

$$\left(\tilde{\mu}_{ij}, \tilde{v}_{ij}\right) = w_j\left(\mu_{ij}, v_{ij}\right) = \left(1 - \left(1 - \mu_{ij}\right)^{w_j}, \left(v_{ij}\right)^{w_j}\right) \tag{7}$$

Step 5: Determination of the weighted intuitionistic fuzzy ideal solution

The intuitionistic weighted positive ideal solution has the form:

$$wIFI^{+} = ((1,0), \ldots, (1,0)) \tag{8}$$

Step 6: Calculation of the distance measures for each alternative

The distance measure from the positive ideal is calculated as follows (see Eq. (4)):

$$d_{A_iIFI^{+}} = d\left(wIFI^{+}, wA_i\right) \tag{9}$$

Step 7: Computation of score values alternatives

The score values are calculated according to the formula:

$$\text{IFIRP}(A_i) = 1 - \frac{d_{A_iIFI^{+}}}{d_0} \tag{10}$$

where: $d_0 = \overline{d}_0 + 2S(d_0), \overline{d}_0 = \frac{1}{n}\sum_{i=1}^{n} d_{A_iIFI^{+}}, S(d_0) = \sqrt{\frac{1}{n}\sum_{i=1}^{n}\left(d_{A_iIFI^{+}} - \overline{d}_0\right)^2}$.

Step 8: Ranking the alternatives

Finally, all the alternatives $A_i (i = 1, 2, ..., m)$ are ranking by descending order of their score function $\text{IFIRP}(A_i)$.

4 The Intuitionistic Fuzzy Approach in Negotiation Support

4.1 The General Negotiation Model Description

It is assumed [3, 5] that the alternative is a negotiation package, which the negotiator can present as an offer or receive from an opponent, the criteria are negotiating issues, i.e. points to be agreed upon, the option is a resolution level of a negotiation issue.

Let $C = \{C_1, C_2, ..., C_n\}$ be the set of negotiating issues and $\{w_1, w_2, ..., w_n\}$ the set of weights assigned to those issues, where $\sum_{j=1}^{n} w_j = 1$. Let's N_j is the negotiation space, which describes possible options $j-$ th issue ($j = 1, 2, .., n$). Therefore, a negotiation package is represented by a vector $P = [x_1, ..., x_j, ..., x_n]$ where $x_j \in N_j (j = 1, 2, ..., n)$. In the prenegotiation phase, the negotiator builds a so-called scoring system that assigns score ratings to negotiation packages that allow to organize offers from best to worst and help to make decisions during the negotiation process (see [3, 5]). Let us denote by $\wp = \{P_1, ..., P_m\}$ a finite set of negotiating offers which are evaluated based on the issues from the set C.

In the proposed approach, we assumed that the negotiator evaluates the package taking into account the advantages and disadvantages of options involved in the negotiator package P representation. Such an approach can be useful, for example, when issues are qualitative and the values of options of issues are described in words (e.g. product quality, warranty conditions) or quantitative, but it is not possible to determine precisely their values. Therefore, the negotiator recognizes the advantages and disadvantages of each option determining the level of satisfaction by a degree of membership function and level of dissatisfaction by a degree of non-membership function. The degree of membership/non-membership can be constructed basing on a direct, subjective evaluation of the pros and cons. The negotiator can also draw up a list of supporting questions that will assist him in such an assessment, where each question allows for three answers: yes (advantages) no (disadvantages), I do not know (I do not have an opinion). The ratio of the number of yes, no, I do not know (or do not have an opinion) to the number of all questions related to the assessment of the issue is taken as the degree of membership, non-membership, hesitancy, respectively. Therefore, the degree of satisfaction (μ_{ij}) reflects advantages or benefits, while the degree of dissatisfaction (v_{ij}) disadvantages or losses; the value π_{ij} determines the degree of uncertainty of $i-$ th package to criterion $j-$ th, respectively.

Let us note, that such evaluation of negotiating packages also refers to the concept of profit and loss evaluation within the framework of the perspective theory [21]; bipolar modeling of positive and negative information [22], or modeling optimism and pessimism in the multi-criteria procedure [23].

4.2 Example Illustration

In this section, an example to show practical applications of the intuitionistic fuzzy framework for evaluation and rank ordering the negotiation offers is presented. The considered negotiation problem is described in [3] in detail. Let us consider business negotiations with the template defined using three negotiation issues ($C = \{$Price, Time of delivery, Time of payment$\}$), where two first issues are benefit criteria (more is better) and the last is a cost criterion (less is better). The vector of weights declared by the negotiator is the following $w = [0.5, 0.3, 0.2]$. The negotiator defined six alternatives that constitute the set $\wp$ of initially selected packages to evaluate in this way (see Table 1).

Table 1. The set of negotiation packages under evaluation

Package	Price	Time of delivery	Time of payment
P_1	300	14	7
P_2	450	7	14
P_3	270	31	2
P_4	380	20	10
P_5	290	15	15
P_6	420	25	5

The representation options of issues in intuitionistic fuzzy concept, evaluated by the negotiator, are described in Table 2.

Table 2. The evaluation negotiation packages in intuitionistic fuzzy concepts

Package	Price	Time of delivery	Time of payment
P_1	(0.55, 0.45)	(0.50, 0.30)	(0.65, 0.30)
P_2	(0.80, 0.15)	(0.20, 0.65)	(0.30, 0.55)
P_3	(0.45, 0.40)	(0.75, 0.20)	(0.80, 0.10)
P_4	(0.70, 0.05)	(0.65, 0.10)	(0.55, 0.40)
P_5	(0.50, 0.40)	(0.50, 0.40)	(0.30, 0.60)
P_6	(0.75, 0.20)	(0.70, 0.15)	(0.70, 0.20)

The ideal package has the following fuzzy representation $IFI^+ = ((1,0),(1,0),(1,0))$. The results of calculation IFIRP(P_i) (see Eqs. (5)–(10)) and rank ordering packages from the set $\wp$ are presented in Table 3.

Table 3. The calculation of IFIRP(P_i) and rank ordering packages

Package	$d_{A_i IFI^+}$	IFIRP(P_i)	Range
P_1	0.745	0.126	4
P_2	0.781	0.085	5
P_3	0.672	0.212	3
P_4	0.672	0.212	2
P_5	0.799	0.064	6
P_6	0.635	0.255	1

Let us notice, that the results are similar to those obtained in the paper [3] by using an alternative approach, i.e. TOPSIS, VIKOR, and BIPOLAR methods.

5 Conclusions

In this paper, a novel intuitionistic fuzzy framework for evaluation and rank ordering the multi-issue negotiation offers has been introduced and studied. Decision makers' preference analysis is focused on positive and negative information about options by employing intuitionistic fuzzy sets. The advantage of the IFIRP method is that it allows for using natural language, consideration of subjectivity, lack of precision, or uncertainty in the evaluation of negotiation offers. Therefore, it can be especially useful in ill-structured negotiation problems. The method IFIRP is illustrated by a numerical example and is compared with other methods [3].

The IFIRP method may be an alternative multi-criteria method for supporting negotiators and helping them to eliminate or limit biases and heuristics in evaluation negotiation offers [24]. Therefore, further studies should focus, among others, on the empirical verification of the usefulness of an intuitionistic framework for negotiation support. From the perspective of behavioral decision making it is important to testify decision-makers' acceptance of the proposed approach. The other important issue is developing techniques for assessment level of satisfaction (grade of membership) and level of dissatisfaction (grade of non-membership) for negotiation options. From the technical point of view, recognition of the impact of the reference point or the distance measure for the stability of ranking obtained by IFIRP method is also important.

Acknowledgments. This research was supported with the grants from Polish National Science Centre (2016/21/B/HS4/01583).

References

1. Brzostowski, J., Roszkowska, E., Wachowicz, T.: Supporting negotiation by multi-criteria decision-making methods. Optim. Studia Ekonomiczne **5**(59), 3–29 (2012)
2. Salo, A., Hämäläinen, R.P.: Multicriteria decision analysis in group decision processes. In: Kilgour, D.M., Eden, C. (eds.) Handbook of Group Decision and Negotiation, pp. 269–283. Springer , Dordrecht (2010). https://doi.org/10.1007/978-90-481-9097-3_16
3. Wachowicz, T., Brzostowski, J., Roszkowska, E.: Reference points-based methods in supporting the evaluation of negotiation offers. Oper. Res. Decis. **22**, 121–137 (2012)
4. Piasecki, K., Roszkowska, E.: On application of ordered fuzzy numbers in ranking linguistically evaluated negotiation offers. Adv. Fuzzy Syst. (2018)
5. Roszkowska, E., Brzostowski, J., Wachowicz, T.: Supporting Ill-structured negotiation problems. In: Guo, P., Pedrycz, W. (eds.) Human-Centric Decision-Making Models for Social Sciences. SCI, vol. 502, pp. 339–367. Springer, Heidelberg (2014). https://doi.org/10.1007/978-3-642-39307-5_14
6. Roszkowska, E., Kacprzak, D.: The fuzzy saw and fuzzy TOPSIS procedures based on ordered fuzzy numbers. Inf. Sci. **369**, 564–584 (2016)
7. Roszkowska, E., Wachowicz, T.: Application of fuzzy TOPSIS to scoring the negotiation offers in ill-structured negotiation problems. Eur. J. Oper. Res. **242**(3), 920–932 (2015)

8. Hwang, C.L., Yoon, K.: Methods for multiple attribute decision making. In: Multiple Attribute Decision Making. Lecture Notes in Economics and Mathematical Systems, vol. 186. Springer, Berlin, Heidelberg (1981). https://doi.org/10.1007/978-3-642-48318-9_3
9. Opricovic, S., Tzeng, G.-H.: Compromise solution by MCDM methods: a comparative analysis of VIKOR and TOPSIS. Eur. J. Oper. Res. **156**(20), 445–455 (2004)
10. Konarzewska-Gubała, E.: Bipolar: Multiple Criteria Decision Aid Using Bipolar Reference System, LAMSADE, Cashier et Documents, vol. 56 (1989)
11. Górecka, D., Roszkowska, E., Wachowicz, T.: The MARS approach in the verbal and holistic evaluation of the negotiation template. Group Decis. Negot. **25**(6), 1097–1136 (2016). https://doi.org/10.1007/s10726-016-9475-9
12. Hellwig, Z.: Zastosowanie metody taksonomicznej do typologicznego podziału krajów ze względu na poziom ich rozwoju oraz zasoby i strukturę wykwalifikowanych kadr. Przegląd Statystyczny **4**, 307–326 (1968). (In Polish)
13. Domizio, M.D.: The competitive balance in the Italian football league: A taxonomic approach, Department of Communication, University of Teramo (2008)
14. Hellwig, Z.: Procedure of evaluating high-level manpower data and typology of countries by means of the taxonomic method, Towards a system of human resources indicators for less developed countries, 115–134 (1972)
15. Jefmański, B.: Intuitionistic fuzzy synthetic measure for ordinal data. In: Jajuga, K., Batóg, J., Walesiak, M. (eds.) SKAD 2019. SCDAKO, pp. 53–72. Springer, Cham (2020). https://doi.org/10.1007/978-3-030-52348-0_4
16. Roszkowska, E., Filipowicz-Chomko, M.: Measuring sustainable development using an extended hellwig method: a case study of education. Soc. Indic. Res. 1–24 (2020)
17. Roszkowska, E., Jefmański, B.: Interval-valued intuitionistic fuzzy synthetic measure (I-VIFSM) based on Hellwig's approach in the analysis of survey data. Mathematics **9**(3), 201 (2021)
18. Atanassov, K.T.: Intuitionistic fuzzy sets. Fuzzy Sets Syst. **20**, 87–96 (1986)
19. Atanassov, K.T.: Intuitionistic fuzzy sets. Theory and Application. Studies in Fuzziness and Soft Computing, vol. 35, Springer, Heidelberg (1999). https://doi.org/10.1007/978-3-7908-1870-3
20. Szmidt, E.: Distances and Similarities in Intuitionistic Fuzzy Sets. Studies in Fuzziness and Soft Computing, vol. 307. Springer , Cham (2014). https://doi.org/10.1007/978-3-319-016 40-5
21. Tversky, A., Kahneman, D.: Advances in prospect theory: cumulative representation of uncertainty. J. Risk Uncertain. **5**(4), 297–323 (1992)
22. Grabisch, M., Greco, S., Pirlot, M.: Bipolar and bivariate models in multicriteria decision analysis: descriptive and constructive approaches. Int. J. Intell. Syst. **23**(9), 930–969 (2008)
23. Chen, T.-Y.: A comparative study of optimistic and pessimistic multicriteria decision analysis based on Atanassov fuzzy sets. Appl. Soft Comput. **12**(8), 2289–2311 (2012)
24. Kersten, G., Roszkowska, E., Wachowicz, T.: The heuristics and biases in using the negotiation support systems. In: Schoop, M., Kilgour, D.M. (eds.) GDN 2017. LNBIP, vol. 293, pp. 215–228. Springer, Cham (2017). https://doi.org/10.1007/978-3-319-63546-0_16

Intuitionistic Fuzzy ANOVA for COVID-19 Cases in Asia by Density and Climate Factors

Velichka Traneva(✉) and Stoyan Tranev

Asen Zlatarov University, Yakimov Blvd, 8000 Bourgas, Bulgaria
tranev@abv.bg
http://www.btu.bg

Abstract. Analysis of variance (ANOVA), proposed by Fisher, is a statistical tool to determine the influence of factors on a data set. When a data set occurs in an uncertain environment, intuitionistic fuzzy logic is a means of dealing with this imprecision. Compared to fuzzy data, intuitionistic fuzzy (IF) data also has a hesitation degree. In our previous publications in 2020, we have proposed for the first time one-way (1-D IFANOVA) and nonreplicated two-way (2-D IFANOVA) intuitionistic fuzzy ANOVA, combining classical variational analysis with possibilities for modeling of Index Matrices (IMs) and Intuitionistic Fuzzy Sets (IFSs).

The pandemic caused by Coronavirus disease 2019 (COVID-19) first occured in China at the end of 2019. The recent events related to the COVID-19 pandemic have posed many questions regarding the disease's spread rate, including whether various factors may or may not have an influence upon it. The present work focuses on evaluating the rate of COVID-19 in Asia by applying 2-D IFANOVA on the IF dataset of daily cases for the period from February 1, 2020 to January 28, 2021. A command-line utility "Test2", which performs 2-D IFANOVA, will explore the impact of "density" and "climate zones" factors on the spread of COVID-19 in Asia. We will also compare the results obtained from traditional ANOVA over the same data set and from 2-D IFANOVA.

Keywords: COVID-19 · IFANOVA · Index matrix · Intuitionistic fuzzy sets · Variation analysis

1 Introduction

At the end of 2019, COVID-19 was first registered in Wuhan, Hubei province, China [17] and then spreads worldwide. Many scientific publications are devoted to the study of the influence of different environmental factors on the distribution

The work on Sects. 1 and 2 is supported by the Asen Zlatarov University under Ref. No. NIX-440/2020 "Index matrices as a tool for knowledge extraction", the work on Sects. 3 and 4 - by the Ministry of Education and Science under the Programme "Young scientists and postdoctoral students", approved by DCM # 577/17.08.2018.

C. Kahraman et al. (Eds.): INFUS 2021, LNNS 308, pp. 66–74, 2022.
https://doi.org/10.1007/978-3-030-85577-2_8

of COVID-19. Influence of environmental factors on the growth of COVID-19 and prediction for its spread in India was studied in [9]. The role of environmental conditions for the rate of spread of this virus was studied in [21,25]. Few studies have found that virus reduced activity in hot and humid conditions [19]. Analysis of variance (ANOVA) is a statistical approach for analyzing measurements depending on some factors to find whether the levels of the studied factors affect these measurements [13]. The influence of "age" and "density" properties of the population on the number of COVID-19 cases in India was studied in [1]. In today's pandemic environment, data containing the number of people infected with COVID-19 may be unclear or missing, which may be due to the various tests used to determine the presence or absence of the virus. One way to describe these imprecisions is fuzzy logic [36], which gives a statement a degree of truth. Fuzzy variants of ANOVA are known as Fuzzy ANOVA (FANOVA). Gil et al. [14] have been introduced a bootstrap approach to FANOVA. One-way and two-way FANOVA was considered in [10] using a set of confidence intervals for variance. Kalpanapriya et al. [15] have been described FANOVA using the pessimistic and optimistic levels of the triangular fuzzy data. FANOVA aproach with an application of Zadeh's extension principle was presented in [20,22]. Two-way FANOVA is proposed in [23] over trapezoidal fuzzy numbers. Intuitionistic fuzzy (IF) logic [2] is a more powerful tool for describing this imprecision than those of Zadeh [36].

Let us present some research on the theme of IF variation analysis. Two-way IFANOVA was described in [16] by transforming IFSs to fuzzy sets. In our previous studies [30,32,35], we have extended classical ANOVA [12] to be able to apply on intuitionistic fuzzy numbers and proposed 1-D IFANOVA and non-replicated 2-D IFANOVA, based on the apparatuses of IFSs [2] and IMs [3]. Then we have created two command-line utilities "Test1" and "Test2", which perform respectively one-way [33] and two-way IFANOVA [34]. We have used them to investigate the influence of the "population density" and "climate zone" factors on the COVID-19 distribution in Europe [34] up to July 2020. In this paper, we will apply for the first time non-replicated 2-D IFANOVA to explore the influence of "climate zone" and "population density" factors on the intuitionistic fuzzy numbers of COVID-19 cases in Asia, as reported in the IF dataset of daily cases provided by [37] for the period from February 1, 2020 to January 28, 2021. We will also interpret the results obtained from two-way IFANOVA using the utility"Test2" [34] and then we will compare them with the results from the classical ANOVA. The main contribution of the paper is in the research of the effectiveness of non-replicated 2-D IFANOVA through its application on IF data on the spread of COVID-19 in Asia to establish the relationships between its spread and the levels of the factors "climate zone" and "population density". The structure of the rest of the paper is as follows: in Sect. 2 we remind some definitions of the consepts of IMs and IF logic. Section 3 reviews the two-way ANOVA and presents its use on the number of COVID-19 cases in Asia. 2-D IFANOVA is presented in the Sect. 4 and also the results of its application by means of the software "Test2" over the same IF data are given. In Sect. 5 we outline conclusions and possible aspects for our future work.

2 Basic Definitions on Intuitionistic Fuzzy Pairs and IMs

2.1 Intuitionistic Fuzzy Pairs

The **IFP** is an object of the form $\langle a, b\rangle = \langle \mu(p), \nu(p)\rangle$, where $a, b \in [0, 1]$ and $a + b \leq 1$, that is used as a tool for assessment of a proposition p [6]. Some operations and relations over IFPs $u = \langle a, b\rangle$ and $v = \langle c, d\rangle$ are given in [4,6,7,11,26]:

$$\begin{array}{rr} \neg u = \langle b, a\rangle; & u \wedge_1 v = \langle \min(a, c), \max(b, d)\rangle; \\ u \vee_1 v = \langle \max(a, c)), \min(b, d)\rangle; & u \wedge_2 v = u + v = \langle a + c - a.c, b.d\rangle; \\ u \vee_2 v = u.v = \langle a.c, b + d - b.d\rangle; & u@v = \langle \frac{a+c}{2}, \frac{b+d}{2}\rangle; \\ \alpha.u = \langle 1 - (1 - a)^\alpha, b^\alpha\rangle & (\alpha = n \text{ or } \alpha = 1/n(\text{ where } n \in N) \\ u - v = & \langle \max(0, a - c), \min(1, b + d, 1 - a + c)\rangle; \end{array}$$

$$u : v$$

$$= \begin{cases} \langle \min(1, a/c), \min(\max(0, 1 - a/c), \max(0, (b - d)/(1 - d))) \text{ if } c \neq 0 \ \&d \ \neq 1 \\ \langle 0, 1\rangle \text{ otherwise} \end{cases}$$

$$\begin{array}{llll} u \leq v & \text{iff} \ a \leq c \text{ and } b \geq d; \ u \leq_\square v & \text{iff} & a \leq c; \\ u \leq_\diamond v & \text{iff} \quad b \geq d; & u \leq_R v \ \text{iff} & R_{\langle a,b\rangle} \geq R_{\langle c,d\rangle}, \end{array} \tag{1}$$

where $R_{\langle a,b\rangle} = 0, 5.(2 - a - b).(1 - a)$ [27].

2.2 Definition of Intuitionistic Fuzzy Index Matrix

Let $\mathcal{I}$ be a set. By two-dimensional IF index matrix (2-D IFIM) $A = [K, L, \{\langle \mu_{k_i,l_j}, \nu_{k_i,l_j}\rangle\}]$ with index sets K and L ($K, L \subset \mathcal{I}$), we call [5]:

$$A \equiv \begin{array}{c|ccccc} & l_1 & \ldots & l_j & \ldots & l_n \\ \hline k_1 & \langle \mu_{k_1,l_1}, \nu_{k_1,l_1}\rangle & \ldots & \langle \mu_{k_1,l_j}, \nu_{k_1,l_j}\rangle & \ldots & \langle \mu_{k_1,l_n}, \nu_{k_1,l_n}\rangle \\ \vdots & \vdots & \ddots & \vdots & \ddots & \vdots \\ k_m & \langle \mu_{k_m,l_1}, \nu_{k_m,l_1}\rangle & \ldots & \langle \mu_{k_m,l_j}, \nu_{k_m,l_j}\rangle & \ldots & \langle \mu_{k_m,l_n}, \nu_{k_m,l_n}\rangle \end{array}, \tag{2}$$

where for every $1 \leq i \leq m, 1 \leq j \leq n$: $0 \leq \mu_{k_i,l_j}, \nu_{k_i,l_j}, \mu_{k_i,l_j} + \nu_{k_i,l_j} \leq 1$.

In [5,28,29,31] are defined some basic operations over IMs as addition, subtraction, multiplication, reduction, projection, substitution, aggregation operations and internal subtraction of IMs' components.

3 Two-Way ANOVA to the COVID-19 Cases in Asia by Climate Zone and Population Density

3.1 Calculations of Non-replicated Two-Way ANOVA

ANOVA was developed by the English statistician Ronald Fisher [12]. ANOVA is a comparison of means and tests whether each factor has a significant effect. **Step 1.** State the hypotheses: Let y_{k_i,l_j} for $i = 1, 2, ..., m$ and $j = 1, 2, ..., n$ denote the data from the k_i–th level of factor $A(i = 1, 2, ...m)$ and l_j–th level of factor $B(j = 1, 2, ...n)$. Let N is the number of measurements. The ANOVA is used to test hypothesis about [12]:

– **Factor A** $H_0 : \mu_{k_1} = \mu_{k_2} = ... = \mu_{k_m}$, against H_1 : not all μ_{k_i} are equal, where $\mu_{k_i}(i = 1, 2, ..., m)$ are the level means of A.

– **Factor B** $H_0 : \mu_{l_1} = \mu_{l_2} = ... = \mu_{l_n}$, against H_1 : not all μ_{l_j} are equal, where $\mu_{l_j}(j = 1, 2, ..., n)$ are the level means of B.

Step 2. Decision rule: The sum of deviations about the mean - SST, sum of squares between rows (effect of factor A) - SSA, sum of squares between columns (effect of factor B) - SSB and error sum of squares SSE are calculated. Then the mean sum of squares MSA (Factor A), MSB (factor B) and for error MSE are calculated as follows [12]:

$$MSA = \frac{SSA}{m-1}, MSB = \frac{SSB}{n-1} \text{ and } MSE = \frac{SSE}{(m-1)(n-1)} \tag{3}$$

Let $N* = (m-1)(n-1)$. Then the observed values of the test statistics are compared with $\alpha-$ quantiles of $F-$distribution - $F_{(\alpha,m-1,N*)}$ and $F_{(\alpha,n-1,N*)}$. If

$$\frac{1}{F_A} = \frac{MSE}{MSA} \leq \frac{1}{F_{(\alpha,m-1,N*)}} = F_{(1-\alpha,N*,m-1)} \tag{4}$$

$$\text{and } \frac{1}{F_B} = \frac{MSE}{MSB} \leq \frac{1}{F_{(\alpha,n-1,N*)}} = F_{(1-\alpha,N*,n-1)}, \tag{5}$$

then the factors have an influence on significance level α [12,13].

If $p-$value $\leq \alpha$ (α is chosen significance level), H_0 is rejected with probability greater than $(1-\alpha)$ [12].

3.2 Dependence of COVID-19 Cases on "Climate Zone" and "Population Density" Factors for Asian Countries

In this subsection we will explore the influence of "climate zone" and "population density" on a set of data on the number of cases of COVID-19 per 1 million for Asian countries from February 1, 2020 to January 28, 2021 [37] by two-way ANOVA . The Kolmogorov-Smirnov test has checked the assumption of normality of the data distribution before the ANOVA [12]. The statistical significance was evaluated at the 5% level. The "climate zone" factor for the Asian countries has five different values – equatorial climate, subequatorial climate, tropical climate, subtropical climate and temperate climate. The "population density" factor has three different values – [1–200], [201–1000] and [1001–8000]. The following Table 1 describes the results of ANOVA by the factors "density" and "climate zone" for the influence over the COVID-19 cases with $\alpha = 0.05$:

Table 1. Results from ANOVA by the factors "climate zone" and "density".

Source	SS	df	MS	F	p-value	F crit
Rows	2411.9	4	603	0.43	0.79	3.84
Columns	1284.89	2	642.45	0.45	0.65	4.46
Error	11343.13	8	1417.89			

The results of ANOVA show that the "density" and "climate zone" factors have no a significant effect on the spread of COVID-19 cases by comparing the ANOVA test statistics with the critical values of ANOVA test.

Figure 1 compares the average COVID-19 case notification rates for the separate climate zones and for the population densities of Asian countries.

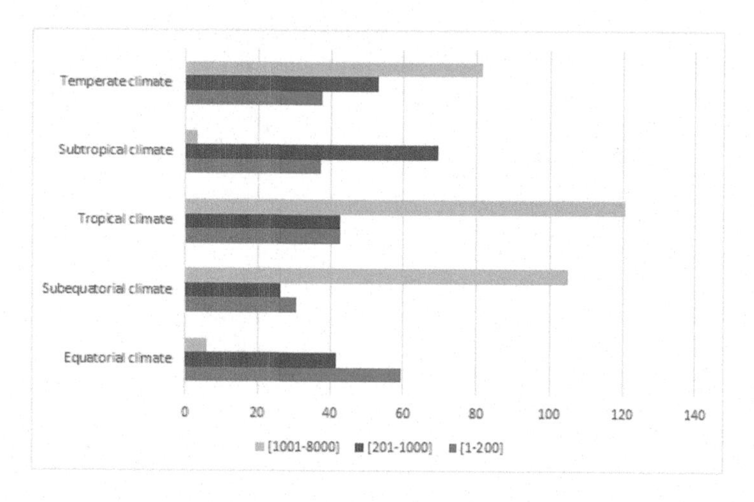

Fig. 1. Cases of Covid-19 per million from February 1, 2020 to January 28, 2021 depending on "density" and "climate zone" factors.

The COVID-19 cases per 1 million people are the highest in the Asian countries with tropical (68.6 cases per million) and temperate climate (57.35 cases per million), and they are the lowest in the countries with equatorial climate (35.6 cases per million). The COVID-19 cases per 1 million people are the highest in the Asian countries with a density belong to [1001–8000] (63.14 cases per million) and they are the lowest in the countries with a density belong to [1–200] (39 cases per million).

4 IFANOVA on "Density" and "Climate Zone" in Asia

In [32] was described for the first time non-replicated 2-D IFANOVA. In this section IFANOVA will be applied to study the influence of the factors "climate zone" and "density" on the COVID-19 case rate per 1 million people from February 1, 2020 to January 28, 2021 using the same dataset from [37] for the Asian countries. IFANOVA start with the conversion of data values for COVID-19 cases into IFPs and let us the obtained pairs be written as IFIM elements of $Y[K, L]$. Let us have the set of intervals $[i_1, i_I]$ for $1 \le i \le m$ and let

$$A_{min,i} = \min_{i_1 \le j \le i_I} x_{i,j} < \max_{i_1 \le j \le i_I} x_{i,j} = A_{max,i}.$$

For the interval $[i_1, i_I]$ we obtain IFPs $\langle \mu_{i,j}, \nu_{i,j} \rangle$ [4] as follows:

$$\mu_{i,j} = \frac{x_{i,j} - A_{min,i}}{A_{max,i} - A_{min,i}}, \nu_{i,j} = \frac{A_{max,i} - x_{i,j}}{A_{max,i} - A_{min,i}} \tag{6}$$

We can use the expert approach from [4] to transform the unclear or missing data about the count of COVID-19 cases to IFPs. The initial form of the IM $Y[K/\{Sr_2, Sr\}, L/\{Sr_1, Sr\}]$ without the last two rows and columns is:

	$[1-200]$	$[201-1000]$	$[1001-8000]$
Equatorial climate	$\langle 0.47, 0.52 \rangle$	$\langle 0.32, 0.67 \rangle$	$\langle 0.02, 0.97 \rangle$
Subequatorial climate	$\langle 0.23, 0.76 \rangle$	$\langle 0.85, 0.14 \rangle$	$\langle 0.851, 0.14 \rangle$
Tropical climate	$\langle 0.33, 0.66 \rangle$	$\langle 0.98, 0.01 \rangle$	$\langle 0.98, 0.01 \rangle$
Subtropical climate	$\langle 0.28, 0.71 \rangle$	$\langle 0.0007, 0.99 \rangle$	$\langle 0.00071, 0.99 \rangle$
Tempered climate	$\langle 0.29, 0.70 \rangle$	$\langle 0.65, 0.34 \rangle$	$\langle 0.65, 0.34 \rangle$

(7)

IFANOVA algorithm [32] is applied more quickly and easier to the real data of IM Y using a C++ command-line utiity "Test2", which was developed in [34]. The utility uses the IM template class (*IndexMatrix*⟨*T*⟩) as described in [18], which realizes the main IM operations.

The printed computational results from the software utility "Test2" over IM Y on the console are:

$$MSA = \langle 0.07, 0.92 \rangle, MSB = \langle 0.025, 0.97 \rangle \text{ and } MSE = \langle 0.006, 0.9 \rangle \tag{8}$$

The fuzzy evaluations of ANOVA Ft_A and Ft_B values are found by approach of Pietraszek [24].

The classic value $Ft_A(0.95; 8; 4) = 0.26$ corresponds to the fuzzy assessment $F_{fuzzy(0.95;8;4)} = \langle 0.96, 0 \rangle$, while the value $Ft_B(0.95; 8; 2) = 0.22$ answers to the fuzzy assessment $F_{fuzzy(0.95;8;2)} = \langle 0.97, 0 \rangle$. Therefore

$$\frac{1}{F_A} = \langle 0.085, 0 \rangle \le F_{fuzzy(0.95;8;4)} \text{ and } \frac{1}{F_B} = \langle 0.24, 0 \rangle \le F_{fuzzy(0.95;8;2)} \tag{9}$$

From (4) we obtain that the "density" and "climate zone" factors influence on the number of COVID-19 cases with degrees of acceptance respectively equal to 0.08 for the factor "climate zone" and 0.24 for the factor "density". When we compare the results of 2-D IFANOVA with those obtained by ANOVA, we can gather that they disagree in the influences of the two factors, which due to the high degrees of the hesitancy of the IF test statistics $\frac{1}{F_A} = \langle 0.085, 0 \rangle$ for A and $\frac{1}{F_B} = \langle 0.24, 0 \rangle$ for B. IFANOVA has explored the weak influence of the two factors.

5 Conclusion

The paper contributes to the study of the effectiveness of non-replicated 2-D IFANOVA through its application to IF data on the COVID-19 cases in Asia to detect the impact of "climate zone" and "population density" of Asian countries on the COVID-19 spread for a period from February 1, 2020 to January 28, 2021. The comparison of the results of 2-D IFANOVA with those obtained by ANOVA over the same data set show, that they disagree in the influences of the studied factors. The use of 2-D IFANOVA reveals the influence of the studied factors on the prevalence of COVID-19, although with low degrees of acceptance and high degrees of hesitancy. In the future, the outlined approach for non-replicated 2-D IFANOVA and its software utility can be expanded to replicated 2-D IFANOVA [12] and its effectiveness will be investigated by applying it to other types of intuitionistic fuzzy multidimensional data in IMs [8].

References

1. Anwla, P.: Introduction to ANOVA for Statistics and Data Science (with COVID-19 Case Study using Python). Analtics Vidhua (2020). https://www.analyticsvidhya.com/blog/2020/06/introduction-anova-statistics-data-science-covid-python. Accessed on 1 Apr 2021
2. Atanassov, K.: Intuitionistic fuzzy sets. In: Proceedings of VII ITKR's Session, Sofia (1983). (in Bulgarian)
3. Atanassov, K.: Generalized index matrices. Comptes rendus de l'Academie Bulgare des Sciences **40**(11), 15–18 (1987)
4. Atanassov, K.: On Intuitionistic Fuzzy Sets Theory. STUDFUZZ, vol. 283. Springer, Heidelberg (2012). https://doi.org/10.1007/978-3-642-29127-2
5. Atanassov, K.: Index Matrices: Towards an Augmented Matrix Calculus. Studies in Computational Intelligence, vol. 573. Springer, Cham (2014). https://doi.org/10.1007/978-3-319-10945-9
6. Atanassov, K., Szmidt, E., Kacprzyk, J.: On intuitionistic fuzzy pairs. Notes Intuitionistic Fuzzy Sets **19**(3), 1–13 (2013)
7. Atanassov, K.: Remark on an intuitionistic fuzzy operation "division". Annual of "Informatics". Section, Union of Scientists in Bulgaria **10** (2019). (in press)
8. Atanassov, K.: n-Dimensional extended index matrices Part 1. Adv. Stud. Contemp. Math. **28**(2), 245–259 (2018)
9. Bherwani, H., Gupta, A., Anjum, S., et al.: Exploring dependence of COVID-19 on environmental factors and spread prediction in India. NPJ Clim. Atmos. Sci. **3**(38) (2020). https://doi.org/10.1038/s41612-020-00142-x
10. Buckley, J.J.: Fuzzy Probability and Statistics. Springer, Heidelberg (2006). https://doi.org/10.1007/3-540-32388-0

11. De, S.K., Bisvas, R., Roy, R.: Some operations on IFSs. Fuzzy Sets Syst. **114**(4), 477–484 (2000)
12. Doane, D., Seward, L.: Applied Statistics in Business and Economics. McGraw-Hill Education, New York, USA (2016)
13. Fisher, R.: Statistical Methods for Research Workers. London (1925)
14. Gil, M.A., Montenegro, M., González-Rodríguez, G., Colubi, A., Casals, M.R.: Bootstrap approach to the multi-sample test of means with imprecise data. Comput. Stat. Data Anal. **51**, 148–162 (2006)
15. Kalpanapriya, D., Pandian, P.: Fuzzy hypotesis testing of ANOVA model with fuzzy data. Int. J. Mod. Eng. Res. **2**(4), 2951–2956 (2012)
16. Kalpanapriya, D., Unnissa, M.: Intuitionistic fuzzy ANOVA and its application using different techniques. In: Madhu, V., Manimaran, A., Easwaramoorthy, D., Kalpanapriya, D., Mubashir, Unnissa M. (eds.) Advances in Algebra and Analysis. Trends in Mathematics, pp. 457–468. Birkhäuser, Cham (2017)
17. Khan, M., Kazmi, S., Bashir, A., Siddique, N.: COVID-19 infection: origin, transmission, and characteristics of human coronaviruses. J. Adv. Res. **24**, 91–98 (2020). https://doi.org/10.1016/j.jare.2020.03.005
18. Mavrov, D.: Software Implementation and Applications of Index Matrices, Dissertation. Asen Zlatarov University, Burgas (2016)
19. Ma, Y., et al.: Effects of temperature variation and humidity on the death of COVID-19 in Wuhan. China. Sci. Total Environ. **724**, 138–226 (2020)
20. Nourbakhsh, M.R., Parchami, A., Mashinchi, M.: Analysis of variance based on fuzzy observations. Int. J. Syst. Sci. **44**(4), 714–726 (2013)
21. Oke, J.,Heneghan, C.: Global covid-19 case fatality rates, CEBM Res (2020). www.cebm.net/covid-19/global-covid-19-case-fatality-rates/. Accessed on 7 Apr 2021
22. Parchami, A., Nourbakhsh, M., Mashinchi, M.: Analysis of variance in uncertain environments. Complex Intell. Syst. **3**(3), 189–196 (2017)
23. Parthiban, S., Gajivaradhan, P.: A comparative study of two factor ANOVA model under fuzzy environments using trapezoidal fuzzy numbers. Intern. J. Fuzzy Math. Arch. bf **10**(1), 1–25 (2016)
24. Pietraszek J., Kołomycki M., et al.: The fuzzy approach to assessment of ANOVA results. In: Nguyen NT., etc. (eds) Computational Collective Intelligence. ICCCI 2016, LNCS, vol. 9875, pp. 260–268, Springer, Cham (2016)
25. Poole, L.: Seasonal Influences on the Spread Of SARS-CoV-2 (COVID19), Causality, and Forecastabililty (2020). https://doi.org/10.2139/ssrn.3554746. Accessed on 7 Apr 2021
26. Riecan, B., Atanassov, A.: Oeration division by n over intuitionistic fuzzy sets. NIFS **16**(4), 1–4 (2010)
27. Szmidt, E., Kacprzyk, J.: Amount of information and its reliability in the ranking of Atanassov intuitionistic fuzzy alternatives. In: Rakus-Andersson, (ed.) Recent Advances in Decision Making, vol. 222, pp. 7–19. Springer, Heidelberg (2009)
28. Traneva, V.: Internal operations over 3-dimensional extended index matrices. Proc. Jangjeon Math. Soc. **18**(4), 547–569 (2015)
29. Traneva, V., Tranev, S.: Index Matrices as a Tool for Managerial Decision Making. Publ, House of the Union of Scientists, Bulgaria (2017). (in Bulgarian)
30. Traneva, V., Tranev, S.: Inuitionistic Fuzzy Anova Approach to the Management of Movie Sales Revenue. Studies. Computational Intelligence (2020). (in press)
31. Traneva, V., Tranev, S., Stoenchev, M., Atanassov, K.: Scaled aggregation operations over 2- and 3-dimensional IMs. Soft Comput. **22**(15), 5115–5120 (2018)
32. Traneva, V., Tranev, S.: Inuitionistic fuzzy two-factor analysis of variance of movie ticket sales. Journal of intelligent and fuzzy systems. IOS press (2021). (in press)

33. Traneva, V., Mavrov, D., Tranev, S.: Intuitionistic Fuzzy One-Factor Analysis of Covid-19 Cases (2021) (in press)
34. Traneva, V., Mavrov, D., Tranev, S.: In: Proceedings of the Intuitionistic Fuzzy Two-Factor Analysis of COVID-19 Cases in Europe, pp. 533–538. Varna, Bulgaria (2020)
35. Traneva, V., Tranev, S., Inuitionistic Fuzzy Analysis of Variance of Ticket Sales, in: Kahraman, C. (eds.) INFUS 2020, Advances in Intelligent Systems and Computing 1197, Springer, Cham, 2020, pp. 363–340
36. Zadeh, L.: Fuzzy sets. Inf. Control **8**(3), 338–353 (1965)
37. https://github.com/owid/covid-19-data/tree/master/public/data. Accessed on 1 Apr 2021

Vendor Selection in IT Using Integrated MCDM and Intuitionistic Fuzzy

Babak Daneshvar Rouyendegh[1,2] and Aylin Tan[1(✉)]

[1] Ankara Yıldırım Beyazıt University, Ankara, Turkey
{berdebilli,185107114}@ybu.edu.tr
[2] Sultan Qaboos University, Muscat, Oman

Abstract. In today's world, products are getting more complex due to raising complexity of the real-life problems. Therefore, while creating a product or service, organizations should work together to combine different technologies. Also, working with best supplier for an organization is significant to make products or services more flexible, competitive, and profitable. In this study, suppliers are evaluated based on quality, cost, maintenance, and flexibility criteria with twelve sub-criteria under them. A hybrid method with VIKOR and Intuitionistic Fuzzy in MCDM problems is represented and the methodology is used for the choosing supplier for a product that is used in IT projects. Decision making matrix is observed with intuitionistic fuzzy due to uncertain data, and the problem is solved by VIKOR to find best alternative for group utility. In case study is performed to exposed efficiency of the hybrid model.

Keywords: MCDM · VIKOR · Intuitionistic fuzzy · Speech recognition · Supplier selection

1 Introduction

In the study, a company which is running business in call center industry is selected for case study. The company has products which are called bots and IVRs provide to help organizations about increasing their customer's customer loyalty and happiness and reducing their operational cost while running contact center operations. Today, chatbots, IVR's and mobile assistants are used more than ever to achieve a more successful customer experience in almost every sector like healthcare, banking, government, insurance. All these bots and IVRs need to core technologies to implement their processes. At this point, Speech recognition is the one of the core technologies that converts speech to text, so a bot, IVR system or mobile assistant can understand what user is said while using them. Speech recognition technology has significant role in the powerful conversational customer services with its performance. Selecting suitable Speech recognition supplier for businesses is very important decision because it also provides easy to handle and developed conversational system, so companies can raise their success of customer experience. Another aim of the having best SR technology is helping easily control devices to users. It provides easiness not only for individual users but also organizations to transcribe massive tasks such as healthcare, insurance, or banking. In the past, suppliers were compared based on only quantitative data of their accuracy levels.

C. Kahraman et al. (Eds.): INFUS 2021, LNNS 308, pp. 75–83, 2022.
https://doi.org/10.1007/978-3-030-85577-2_9

2 Literature Review

Supplier selection has been important research topic for a long time and there are many different research and methods to select best supplier in literature based on usage areas and data type. Although most used approaches for MCDM supplier selection problems are AHP, ANP, TOPSIS and their combinations with fuzzy set, VIKOR, DEMATEL, ANOVA and other approaches were also used in much research. In last twenty years' research, VIKOR was used in more than 750 supply chain management study and there are more than 500 research about intuitionistic fuzzy approach in SCM problems. According to research, while VIKOR is mostly applied with fuzzy AHP, fuzzy ANP, linguistic information, fuzzy DEMATEL, ELECTRE, BWM and other fuzzy approaches intuitionistic fuzzy is applied with MULTIMOORA, AHP, PROMETHEE, UTASTAR, ELECTRE and ANP. Previously, VIKOR is used integrated with type-2 fuzzy methodology that is best-worst solving a green supplier selection problem [16]. They benefit from VIKOR for MCDM and benefit from best worst for eliminating subjectivity Awasthi were used integrated VIKOR and fuzzy AHP approaches for their multi-tier supplier selection problem [1]. In the study, criteria's weights are defined with utilizing fuzzy AHP and evaluation of suppliers is utilized with fuzzy VIKOR. Rouyendegh [17] investigated a hybrid methodology for evaluating supplier's performance. In their study, decision makers are ranked with intuitionistic when applying AHP as MCDM methodology. Also, human judgement is incorporated with the techniques, and three websites and four criteria are chosen in the study. Rouyendegh [16], proposed green supplier selection problem with intuitionistic and fuzzy TOPSIS methods. In the study, eliminating subjectivity in collecting data from decision makers intuitionistic fuzzy is utilized. Rouyendegh [15] verified a hybrid methodology for supplier selection problem in uncertain environment and subjectivity applying ANP with intuitionistic fuzzy. Çalı [4] proposed integration of VIKOR and ELECTRE 1 for utilizing evaluation of suppliers in uncertain area. In the study, outranking of novel is studied based on MCGDM. Mahmoudi [11] proposed fuzzy TOPSIS and OPA in large scale MCDM with missing values in selection problem.

3 Structuring Problem

3.1 Current State

Based on literature review, generally only scientific measurement methods and algorithms are used for ranking speech recognition engines and because of that only measurable data are used in problems. Most important constraint in these approaches, they can only consider quantitative data.

3.2 Suggested Improvements

In this study, because of existing gap in the literature to apply not only measurable criteria but also unmeasurable criteria of supplier selection problem for speech recognition product, MCDM method is applied. It provided to consider unmeasurable criteria, so buyers can be sure about that they did not handle only a few topics. Therefore, in this study, a hybrid methodology is suggested to handle constraints in current state of supplier selection problem for the product in IT sector. To find which MCDM method could be useful to solve the problem, literature is reviewed. After the problem is defined in previous section, main criteria are defined based on combining literature review of speech recognition performance and evaluations, and general research of supplier selections. Aim of combining this research is bringing new view to evaluation of speech recognition providers with handling constraints in current research. Quality (accuracy rate), cost, maintenance, and flexibility are most common main criteria in the literature.

4 Constructing Decision Model

4.1 Intuitionistic Fuzzy

Fuzzy sets provide ranking membership with grading scores from 0 to 1. Based on the study of (Rouyendegh, et al. 2018), basic concepts of intuitionistic fuzzy set is defined as

$A = \{(r, \mu_A(r), v(r)) \vdots r \in R\}$, while $\mu_A(r)$ is member of function, $v_A(r)$ is not member. In fuzzy set, summary of these functions should greater or equal to 0 and less than equal to 1. Also, $\pi_A(r)$ is the intuitionistic fuzzy index of belonging to A.

$$\pi_A(r) = 1 - \mu_A(r) + v_A(r) \quad (1)$$

As Eq. 1 where $\pi_A(r)$ is hesitancy of r to A for every $r \in R$, $0 \leq \pi_A(r) \leq 1$, $\pi_A(r)$ defines the degree of uncertainty. Having more knowledge of r, the number should be smaller.

$$\mu_A(r) = 1 - v_A(r) \quad (2)$$

4.2 VIKOR

In this study, VIKOR methodology is proposed after creating goal, criteria and alternatives and obtaining decision matrix with intuitionistic fuzzy approach. Therefore, first step for VIKOR is applied. Next, second step is applied with previous part with IF method. Third step is having normalized decision matrix to avoid unit's effect the other range differences on the decision to be made in criterion values. In fourth step, decision matrix is weighted reflecting the effect levels of the criteria on the decision. In fifth step, individual regret and group utilities are calculated. Group utility (Si) refers to the total weighted normalized value to be obtained if the alternative is selected, while individual regret (Ri) indicates the largest record that will occur based on a criterion if the alternative is not selected. In sixth step, ranking indexes are computed. The consensus criterion is calculated to ensure that group utility and individual regret criteria are combined to decide between alternatives. In seventh step, the best alternatives are ranked. For verification, it is examined if the results meet the conditions or not. First condition: Advantage Acceptance, suppose that the alternative with the lowest Qi value has the $Q(a')$ value, the second-best alternative ($Q(a'')$).

$$DQ = 1/(m - 1) \quad (3)$$

It is acceptable when below equation is satisfied.

$$Q(a'') - Q(a') \geq DQ \tag{4}$$

Second condition: Stability Acceptance, the choice with the best Qi value should also be the best alternative from the point of group benefit and/or individual regret criteria. When both conditions are fulfilled, the Qi value is determined as the best alternative compromise solution. Only if Condition 1 is satisfied, two alternatives with the best consensus criterion value will be determined as the best solution. If Condition 1 is not verified, all alternatives up to the next alternative are determined as compromise solutions until mth confirms the condition based on following expression.

5 Analyzing the Problem

For case study, a supplier selection problem is applied for a call center company as buyer. Alternatives are determined based on candidates in the same language, offering same solutions and having similar level of accuracy. Also, three decision makers who are working in the buyer company as software developer, product owner and researcher are selected, and they are represented as D, PO, and R in following sections. The linguistic terms and linguistic importance weights of criteria are shown in Table 1.

Table 1. Linguistic importance alternatives and criteria significance

Linguistic importance alternatives	IFNs	Linguistic importance of criteria	IFNs
Absolutely Low (AL)	(0.05, 0.95)	No influence (N)	(0.15, 0.8)
Low (L)	(0.2, 0.65)	Low influence (L)	(0.2, 0.65)
Fairly Low (FL)	(0.35, 0.55)	Medium-Low influence (ML)	(0.4, 0.45)
Medium (M)	(0.5, 0.5)	Medium influence (M)	(0.5, 0.5)
Fairly High (FH)	(0.65, 0.25)	Medium High influence (MH)	(0.55, 0.3)
Very High (VH)	(0.8, 0.05)	High influence (H)	(0.7, 0.2)
Absolutely High (AH)	(0.9, 0.1)	Very High influence (VH)	(0.9, 01)

The linguistic terms are shown in Table 2 and Table 3. Also, decision makers are weighted from experts in purchasing department based on their skills and experiences. Weights of decision makers are determined as 0.35, 0.4 and 0.25 orderly.

Table 2. Information of four alternatives

	Decision makers	Alternative-1	Alternative-2	Alternative-3	Alternative-4
WER	D	AH	AH	VH	FH
	PO	VH	AH	FH	M
	R	AH	AH	FH	M
Technological leadership	D	M	VH	VH	VH
	PO	VH	AH	AH	M
	R	FH	AH	VH	M
Unit pricedollar/min	D	AH	L	FL	M
	PO	AH	FL	M	FH
	R	AH	M	FL	M
Hardware & software cost	D	VH	VH	VH	VH
	PO	AH	VH	FH	FH
	R	VH	AH	AH	VH
Support activities	D	FH	FH	FL	L
	PO	VH	M	M	FL
	R	P	FL	M	M
Workflow tools	D	FM	VH	VH	FL
	PO	FL	VH	M	M
	R	FH	M	M	M
Num of types of format supporting	D	AH	AH	AH	FH
	PO	VH	AH	FH	FH
	R	VH	VH	VH	M
Capability of handling all accents	D	AH	VH	FH	FL
	PO	VH	AH	M	L
	R	AH	FH	FL	FL
Multiple languages	D	F	VH	AH	M
	PO	FL	FH	AH	M
	R	F	FH	AH	FL
Experienced industries	D	VH	AH	VH	FH
	PO	VH	AH	AH	FH
	R	FH	VH	AH	VH

(*continued*)

Table 2. (*continued*)

	Decision makers	Alternative-1	Alternative-2	Alternative-3	Alternative-4
Integration capability	D	AH	VH	VH	VH
	PO	VH	FH	FH	M
	R	VH	VH	FH	M
Easy deployment	D	FH	AH	AH	VH
	PO	F	AH	AH	AH
	R	FL	AH	AH	VH

Table 3. Importance weights of criteria

	Developer	PO	Researcher
WER	VH	H	VH
Technological leadership	MH	H	VH
Unit price dollar/min	N	VH	L
Hardware & software cost	H	N	M
Support activities	VH	H	ML
Workflow tools	MH	H	M
Num of types of format supporting	H	M	MH
Capability of handling all accents	L	MH	VH
Multiple languages	L	H	M
Experienced industries	L	VH	H
Integration capability	VH	H	M
Easy deployment	VH	MH	L

Intuitionistic fuzzy decision matrix is determined by SIFWA.

$$r_{ij} = \left(\frac{\prod_{k=1}^{3} u_{ij}^{k^{\lambda k}}}{\prod_{k=1}^{3} u_{ij}^{k^{\lambda k}} + \prod_{k=1}^{3} (1 - u_{ij}^{k^{\lambda k}})}, \frac{\prod_{k=1}^{3} v_{ij}^{k^{\lambda k}}}{\prod_{k=1}^{3} v_{ij}^{k^{\lambda k}} + \prod_{k=1}^{3} (1 - v_{ij}^{k^{\lambda k}})}\right) \quad (5)$$

In second step, normalized DM and criteria weights are calculated and figured in Table 4. Normalized DM is obtained by following equation when π_j is satisfied.

$$w_j^s = \frac{u_j + \frac{u_j}{(u_j + v_j)} \pi_{ij}}{\sum_{j=1}^{n} u_j + \frac{u_j}{(u_j + v_j)} \pi_j} \quad (6)$$

$$\pi_j = 1 - u_j - v_j \quad (7)$$

Also, IF best and worst ideal solution are obtained. In case study, while unit price and hardware & software cost criteria are cost, the others are benefit. Next, IFE value for all cub-criteria is determined according to objective weighting approach. Summation of w should be equal to 1.

$$w_j^O = \frac{1 - E_j}{\sum_{J=1}^{n} 1 - E_j}, 0 \leq w_j^O \leq 1 \tag{8}$$

$$E_j = -\frac{1}{mln2}\sum_{i}^{m} (\mu_{ij}\, ln\mu_{ij} + v_{Aij} lnv_{ij} - (1 - \pi_{ij})\ln(1 - \pi_{ij}) - \pi_{ij}\ln 2) \tag{9}$$

Where j = 1, 2, … n. Then, weight is calculated by following equation. $W = w_j^O\, \theta + (1 - \theta)w_j^O$, θ is decided as 0.5 in the study for simplicity. Group utility and individual regret is calculated by following equations.

$$S_i = \sum_{j=1}^{n} W_j \frac{f_j^+ - f_{ij}}{f_j^+ - f_{ij}^-} = \sum_{j=1}^{n} V_{ij} \tag{10}$$

$$R_i = max_j(w_j \frac{f_j^+ - f_{ij}}{f_j^+ - f_{ij}^-}) = max_j V_{ij}) \tag{11}$$

$$Q_i = \theta \times \frac{S_i - S^+}{S^- - S^+} + (1\text{-}\theta) \times \frac{R_i - R^+}{R^- - R^+} \tag{12}$$

Condition 1: Acceptable advantage: Alternative 2 has best Qi value with 0 and alternative 3 is second best. Therefore, DQ = 1/(4 – 1) = 0.33 and Q3 – Q2 = 0.28 – 0 = 0.28, 0.42 < 0.33 so it does not provide acceptable advantage condition. When look at third best which is alternative 1, Q1 – Q2 = 0.57 – 0 = 0.57 so it provides the condition. Condition 2: Sustainability acceptance in result: Alternative 2 has best individual regretvalue, too. As a result, alternative 2 is obtained as best alternative in this study with providing two acceptable condition of VIKOR method.

6 Sensitivity Analysis

In the study, group utility and individual regret are considered as equally important, so θ is defined as 0.5. Raising the value from 0.5 shows that group utility is more important while decreasing the value from 0.5 is showing individual regret is more important. In the literature, the value is applied with 0.5 generally, but there are different applications like 0.25 and 0.75, too. Therefore, in this section of the study, different values for the parameter are applied and results are examined.

For θ is equal to 0; Q3 – Q2 = 0.271584 – 0 = 0.272 which is smaller than DQ. So, Q1 – Q2 = 0.872 – 0 = 0.872 which is greater than DQ, so solution is defined as alternative 2 and alternative 1 that means decision makers can select one of them. Also, still alternative 2 is best for individual regret and sustainable acceptance. For θ is equal to 0.25; Q3 – Q1 = 0.28 – 0 = 0.28 and 0.28 < 0.33, so continue to next step that is Q1 – Q2 is equal to 0.72 that means it is met the condition. Therefore, decision makers can select either alternative 2 or alternative 1. For θ is equal to 0.75; Q3 – Q1 = 0.29 – 0

Table 4. Sensitivity analysis with different values

Alternative/Qi	$\theta = 0$	$\theta = 0.25$	$\theta = 0.5$	$\theta = 0.75$	$\theta = 1$
Alternative 1	0.871664	0.721977	0.572289	0.422602	0.272914
Alternative 2	0	0	0	0	0
Alternative 3	0.271584	0.277798	0.284011	0.290224	0.296438
Alternative 4	1	1	1	1	1

$= 0.29$ and $0.29 < 0.33$, so continue to next step that is $Q1 - Q2$ is equal to 0.42 that means it is met the condition. Therefore, decision makers can select either alternative 2 or alternative 1. For θ is equal to 1; $Q1 - Q2 = 0.27 - 0 = 0.27$, that is smaller than DQ, in second step $Q3 - Q1 = 0.3$ which is again smaller than 0.33, in third step $Q4 - Q2 = 1$ is greater than DQ. Therefore, with this value decision maker's cans select alternative 2 or alternative 4. As a result, calculations show that alternative 2 is best for all values of the parameter θ, but second option is changed when θ is equal to 1. It means, if group utility is considered as only important, alternatives can change.

7 Conclusion

In today's world, organizations should implement new automation steps in their processes, because competitiveness is more challenge than ever. Thanks to new technologies like machine learning and artificial intelligence the automation steps can be applied successfully and easily. Especially, remote communication between customers and suppliers has great importance due to effects of Covid-19. Moreover, selecting best supplier has been one of most significant issue for organizations for long years. While selecting a supplier for critical processes requires more to meet more than one criterion. At this point, it becomes MCDM problem in uncertain environment. When analyzing problem, a company that is running business in customer engagement industry is selected due to its products that are bots, IVRs, and mobile applications need to Speech Recognition technology which the company does not exist. The biggest aim of the study is closing the gap in the literature about evaluating speech recognition products based on not only measurable criteria but also unmeasurable ones. It provides to evaluate suppliers without missing the advantages and disadvantages of different criteria. This study provides an approach to close the gap in the literature about considering qualitative data to compare Speech Recognition suppliers. Finally, in the future, detailed security criteria can be included to problem, because information security is getting more importance day by day with huge data and cloud applications.

References

1. Awasthi, A., Govindan, K., Gold, S.: Multi-tier sustainable global supplier selection using a fuzzy AHP-VIKOR based approach. Int. J. Prod. Econ. **195**, 106–117 (2018)
2. Dharmani, A. H., Ghosh, S.: Performance evaluation of ASR for isolated words in Sindhi Language .In: International Conference on Advances in Computing, Communication and Control (ICAC3), pp. 1–6 (2019)
3. Çalı, S., Balaman Yılmaz, Ş: A novel outranking based multi criteria group decision making methodology integrating ELECTRE and VIKOR under intuitionistic fuzzy environment. Expert Syst. Appl. **119**, 36–50 (2019)
4. Gupta, P., Mehlawat, M.K., Grover, N.: Intuitionistic fuzzy multi-attribute group decision-making with an application to plant location selection based on a new extended VIKOR method. Inf. Sci. 184–203 (2016)
5. Herchonvicz, A.L., Franco, C.R., Jasinski, M. G.: A comparison of cloud-based speech recognition engines. X Comput. Beach, 366–375
6. Kim, J.Y., Liu C., Calvo, R.A., McCabe, K., Taylor, C.R.: Schuller, B.W., Wu, K.: A Comparison of Online Automatic Speech Recognition Systems and the Nonverbal Responses to Unintelligible Speech (2019)
7. Kabak, M., Yetkin, Ç.: Yönetimde çok kriterli karar verme yöntemleri:MS Excel Çözümlü Uygulamalar. Atlas Akademik Basım Yayın Dağıtım (2020)
8. Kahraman, C., Öztayşi, B., Çevik, B.: Intuitionistic fuzzy multicriteria evaluation of outsource manufacturers. IFAC-PapersOnline **49**, 1844–1849 (2016)
9. Mullinera, E., Malys, N., Maliene, V.: Comparative analysis of MCDM methods for the assessment of sustainable housing affordability. Omega **59**, 146–156 (2016)
10. Rezaei, J., Fahim, P.B., Tavasszy, L.: Supplier selection in the airline retail industry using a funnel methodology: conjunctive screening method and fuzzy AHP. Expert Syst Appl **41**, 8165–8179 (2014)
11. Rostamzadeh, R., Ghorabaee, M.K., Kannan, G., Esmaeili, A., Nobar, H., Bogadhi, K.: Evaluation of sustainable supply chain risk management using an integrated fuzzy TOPSIS- CRITIC approach. J. Clean. Prod. **175**, 651–669 (2015)
12. Rouyendegh Daneshvar, B.: Developing an integrated ANP and intuitionistic fuzzy TOPSIS model for supplier selection. J. Test. Eval. **43**, 664–675 (2015)
13. Rouyendegh Daneshvar, B., Yıldızbaşı, A., Üstünyer, P.: Intuitionistic Fuzzy TOPSIS method for green supplier selection Problem. Soft Comput. **24**, 1–14, (2019)
14. Rouyendegh, B.D., Topuz, K., Dag, A., Oztekin, A.: An AHP-IFT integrated model for performance evaluation of e-commerce web sites. Inf. Syst. Front. **21**(6), 1345–1355 (2018). https://doi.org/10.1007/s10796-018-9825-z
15. Uygun, Ö., Dede, A.: Comparative analysis of MCDM methods for the assessment of sustainable housing affordability. Omega **59**, 146–156 (2016)
16. Qun, W., Ligang, Z., Yu, C., Huayou, C.: An integrated approach to green supplier selection based on the interval type-2 fuzzy best-worst and extended VIKOR methods. Inf. Sci. **502**, 394–417 (2019)
17. Wang, W., Xin, X.: Distance measure between intuitionistic fuzzy sets. Pattern. Recogn. Lett. **26**, 2063–2069 (2005)

A Vector Valued Similarity Measure Based on the Choquet Integral for Intuitionistic Fuzzy Sets and Its Application to Pattern Recognition

Ezgi Türkarslan[1(✉)], Mehmet Ünver[2], and Murat Olgun[2]

[1] Department of Mathematics, TED University, 06420 Ankara, Turkey
ezgi.turkarslan@tedu.edu.tr
[2] Department of Mathematics, Ankara University, 06100 Ankara, Turkey
{munver,olgun}@ankara.edu.tr

Abstract. The concept of Choquet integral that a special ordered weighted averaging operator (OWA) is an aggregation function and it generalizes the concepts of arithmetic and the weighted mean. This concept allows us to model interaction between criteria with the help of a fuzzy measure. Our aim is to combine fuzzy set theory and fuzzy measure theory by using the concept of Choquet integral. In this study, we propose a vector valued similarity measure for intuitionistic fuzzy sets (IFSs) based on the Choquet integral. This vector valued similarity measure consists of a pair of a similarity measure which is obtained from a distance measure for IFSs and an uncertainty measure. In this context, we provide a more effective tool by introducing the interaction between criteria with the help of fuzzy measure. Finally, we support the efficiency of our work with explanatory numerical examples.

Keywords: Similarity measure · Uncertainty measure · Fuzzy measure · Choquet integral · Pattern recognition

1 Introduction

The concepts of Choquet integral and capacity were introduced by Gustave Choquet in 1953 [1]. The Choquet integral is defined with the help of a capacities (or non-additive monotonic measures) and the concept of capacity was expanded to the concept of fuzzy measure by Sugeno in 1974 [2]. The notion of fuzzy measure permits assigning "weights"to subsets of the universal set and it is able to model the interaction between elements of the subsets. Moreover, the Choquet integral is considered as a non-linear aggregation operator. The concepts of weighted and arithmetic mean are the most known aggregation functions. A standard measure and Lebesgue integral coincide with the weighted average over the finite subset of a universal set. However, the Choquet integral is an extension of Lebesgue integral and a non-additive extension of the weighted arithmetic mean. Although,

C. Kahraman et al. (Eds.): INFUS 2021, LNNS 308, pp. 84–92, 2022.
https://doi.org/10.1007/978-3-030-85577-2_10

Choquet integral has more complicated structure due to the lack of additivity in contrast to the additive integrals such as Lebesgue integral, Choquet integral is more effective in the aggregation. In [3], it is shown that the Choquet integral performs noticeably more orders than the weighted arithmetic mean and that the difference gets larger when the number of the elements of the set gets larger. Furthermore, it has been proved in [4] that when the number of the element of the finite set increases, the probability of getting more optimal ranking in the Choquet integral increases compared to the weighted arithmetic mean. Actually, fuzzy measures and Choquet integral let us take the preferences that are not considered in the weighted arithmetic mean into account [5].

The concept of fuzzy set [6] was presented by Zadeh in 1965 via a membership function to model uncertain and inconsistent information. However, some information or data cannot be represented by a fuzzy set. Therefore, it was extended to the concept of IFS by Atanassov [7] via a membership function $\mu_{\mathring{A}} : X \rightarrow [0,1]$ with a non-membership function $\nu_{\mathring{A}} : X \rightarrow [0,1]$ of $\mathring{A}$ in X such that $0 \leq \mu_{\mathring{A}}(\xi) + \nu_{\mathring{A}}(\xi) \leq 1$ for each $\xi \in X$. The theory of IFS has been extensively studied by many authors with the help of aggregation operators [8,9]. One of the most crucial research areas in IFS theory is the determination of the similarity between two IFSs. Therefore, various researchers have studied on the concept of similarity measure to model the degree of similarity between these sets sensitivity. A similarity measure is often applied to pattern recognition problems [10–12]. The purpose of a pattern recognition problem is to decide whether an object belongs to a class or not. However, the similarity degree between different IFSs and target IFS may be equal in some cases which yields an uncertain situation. To overcome this uncertainty, Fei et al. [13] have proposed a vector valued similarity measure for IFSs by using the ordered weighted averaging (OWA) aggregation operator. They have also introduced a similarity measure between IFSs and its uncertainty measure as an ordered pair. The uncertainty of similarity measure between IFSs is considered as the performance of the similarity measure itself.

In this paper, we construct a new vector similarity measure among IFSs motivated by [13]. For this aim, we use the notion of Choquet integral instead of classical OWA aggregation operator. In this manner, we introduce a synthesis which is an innovative tool by combining the Choquet integral with the concept of similarity measure and its uncertainty measure for IFSs. Moreover, we give an application on a pattern recognition step of a classification problem to demonstrate the effectiveness of the new vector similarity measure to pay attention the interaction between the criteria. The rest of this paper is organized as follows: In Sect. 2, the existing vector similarity measure and the concepts of fuzzy measure and Choquet integral are recalled. In Sect. 3, we propose a vector similarity measure between IFSs based on Choquet integral and we examine it theoretically. In Sect. 4, we implement it to a classification problem adapted from [13] and we compare our results with the existing results to indicate the impact of the advanced vector similarity measure. Finally, in Sect. 5, we deduce the paper with some notices.

2 Preliminaries

A vector valued similarity measure is a two dimensional vector, whose first component is a similarity and its second component is an uncertainty measure of given similarity measure (see, e.g., [13]).

We first recall the vector valued similarity measure that has been proposed by Fei et al. [13]. Let $X = \{\xi_1, \xi_2, ..., \xi_n\}$ be a finite set and let

$$\mathring{A} = \{< \xi, \mu_{\mathring{A}}(\xi), \nu_{\mathring{A}}(\xi) > | \xi \in X\}$$

and

$$\mathring{B} = \{< \xi, \mu_{\mathring{B}}(\xi), \nu_{\mathring{B}}(\xi) > | \xi \in X\}$$

be two IFSs in X. The similarity measure based on a distance measure between $\mathring{A}$ and $\mathring{B}$ is given with

$$S(\mathring{A}, \mathring{B}) = 1 - Dis_{(d_1, d_2, ..., d_n)}(\mathring{A}, \mathring{B}), \tag{1}$$

where

$$Dis_{(d_1, d_2, ..., d_n)}(\mathring{A}, \mathring{B}) = \omega_1 b_1 + w_2 b_2 + ... + \omega_n b_n \tag{2}$$

is the OWA aggregation operator corresponding to the distance measure which is given with

$$d_i(\mathring{A}, \mathring{B}) = \sqrt{\frac{(\mu_{\mathring{A}}(\xi_i) - \mu_{\mathring{B}}(\xi_i))^2 + (\nu_{\mathring{A}}(\xi_i) - \nu_{\mathring{B}}(\xi_i))^2}{2}} \tag{3}$$

and $\omega_i \in [0, 1]$ denotes the weight of the ith element in X with $\sum_{i=1}^{n} \omega_i = 1$, and b_i represents the ith largest distance measure of all the elements between $\mathring{A}$ and $\mathring{B}$. Moreover, the uncertainty measure of a similarity measure between $\mathring{A}$ and $\mathring{B}$ is given with the following OWA aggregation operator corresponding to $u_i(\mathring{A}, \mathring{B})$

$$U_{(u_1, u_2, ..., u_n)}(\mathring{A}, \mathring{B}) = \omega_1 v_1 + w_2 v_2 + ... + \omega_n v_n, \tag{4}$$

where v_i is the ith largest uncertainty measure of all the elements between $\mathring{A}$ and $\mathring{B}$. Furthermore, $u_i(\mathring{A}, \mathring{B})$ is the uncertainty between $\mathring{A}$ and $\mathring{B}$ that is given with

$$u_i(\mathring{A}, \mathring{B}) = \frac{1 - |\mu(\xi_i) - \nu(\xi_i)| + (1 - \mu(\xi_i) - \nu(\xi_i))}{2 - \mu(\xi_i) - \nu(\xi_i)} \tag{5}$$

where

$$\mu(\xi_i) = \frac{\left(\frac{\mu_{\mathring{A}}(\xi_i)+\mu_{\mathring{B}}(\xi_i)}{2}\right)^2 + 2\left(\frac{\mu_{\mathring{A}}(\xi_i)+\mu_{\mathring{B}}(\xi_i)}{2}\right)\left(1 - \frac{\mu_{\mathring{A}}(\xi_i)+\mu_{\mathring{B}}(\xi_i)}{2} - \frac{\nu_{\mathring{A}}(\xi_i)+\nu_{\mathring{B}}(\xi_i)}{2}\right)}{1 - 2\left(\frac{\mu_{\mathring{A}}(\xi_i)+\mu_{\mathring{B}}(\xi_i)}{2}\right)\left(\frac{\nu_{\mathring{A}}(\xi_i)+\nu_{\mathring{B}}(\xi_i)}{2}\right)} \tag{6}$$

and

$$\nu(\xi_i) = \frac{\left(\frac{\nu_{\mathring{A}}(\xi_i)+\nu_{\mathring{B}}(\xi_i)}{2}\right)^2 + 2\left(\frac{\nu_{\mathring{A}}(\xi_i)+\nu_{\mathring{B}}(\xi_i)}{2}\right)\left(1 - \frac{\mu_{\mathring{A}}(\xi_i)+\mu_{\mathring{B}}(\xi_i)}{2} - \frac{\nu_{\mathring{A}}(\xi_i)+\nu_{\mathring{B}}(\xi_i)}{2}\right)}{1 - 2\left(\frac{\mu_{\mathring{A}}(\xi_i)+\mu_{\mathring{B}}(\xi_i)}{2}\right)\left(\frac{\nu_{\mathring{A}}(\xi_i)+\nu_{\mathring{B}}(\xi_i)}{2}\right)} \tag{7}$$

for $i = 1, 2, ..., n$.

The concept of fuzzy measure [14] is a non-additive monotone measure and the notion models the synergy and the redundancy between elements of a universal set.

Definition 1. *Let X be a finite set and let $P(X)$ be the power set of X. If*

i.) $\sigma(\emptyset) = 0$
ii.) $\sigma(X) = 1$
iii.) $\sigma(\mathring{A}) \leq \sigma(\mathring{B})$ *for any $A, B \subset X$ such that $\mathring{A} \subseteq \mathring{B}$ (monotonicity),*
then the set function $\sigma : P(X) \to [0, 1]$ is called a fuzzy measure on X. A fuzzy measure σ is called sub-additive if $\sigma(\mathring{A} \cup \mathring{B}) \leq \sigma(\mathring{A}) + \mu(\mathring{B})$ for $\mathring{A} \cap \mathring{B} = \emptyset$.

A sub-additive fuzzy measure is used when there is redundancy between the elements of the universal set.

The Choquet integral [1,14] can be considered as an aggregation operator that assigns a weight to each subset of the universal set with the help of a fuzzy measure.

Definition 2. *Let X be a finite set and let σ be a fuzzy measure on X. The Choquet integral of a function $f : X \to [0, 1]$ with respect to σ is defined by*

$$(C)\int_X f\, d\sigma := \sum_{k=1}^{n} \left(f(\xi_{(k)}) - f(\xi_{(k-1)})\right) \sigma(E_{(k)}), \tag{8}$$

where the sequence $\{\xi_{(k)}\}_{k=0}^{n}$ is a permutation of the sequence $\{\xi_k\}_{k=0}^{n}$ such that $0 := f(\xi_{(0)}) \leq f(\xi_{(1)}) \leq f(\xi_{(2)}) \leq ... \leq f(\xi_{(n)})$ and $E_{(k)} := \{\xi_{(k)}, \xi_{(k+1)}, ..., \xi_{(n)}\}$.

3 Main Results

In this section, we propose a new vector valued similarity measure for IFSs with the help of the Choquet integral. Throughout this paper we assume that $X = \{\xi_1, \xi_2, ..., \xi_n\}$ is a finite universal set, $\mathring{A}$ and $\mathring{B}$ are IFSs on X and σ is a fuzzy measure on X.

Definition 3. *A similarity measure and its uncertainty measure based on Choquet integral between $\mathring{A}$ and $\mathring{B}$ are given with*

$$S^{(C,\sigma)}(\mathring{A}, \mathring{B}) := 1 - (C)\int_X d_i(\mathring{A}, \mathring{B})\, d\sigma \tag{9}$$

and

$$U^{(C,\sigma)}(\mathring{A},\mathring{B}) := (C)\int\limits_X u_i(\mathring{A},\mathring{B})\,d\sigma \tag{10}$$

for $i = 1, 2, ..., n.$, respectively, where d_i and u_i are given in (3) and (5), respectively.

Theorem 1. *The pair* $\left(S^{(C,\sigma)}, U^{(C,\sigma)}\right)$ *satisfies the following properties.*

i) $S^{(C,\sigma)}(\mathring{A},\mathring{B}), U^{(C,\sigma)}(\mathring{A},\mathring{B}) \in [0,1]$,
ii) $S^{(C,\sigma)}(\mathring{A},\mathring{B}) = 1$ *if and only if* $\mathring{A} = \mathring{B}$,
iii) $U^{(C,\sigma)}(\mathring{A},\mathring{B}) = 0$ *if* A *and* B *are crisp sets,*
iv) $S^{(C,\sigma)}(\mathring{A},\mathring{B}) = S^{(C,\sigma)}(\mathring{B},\mathring{A})$,
v) $S^{(C,\sigma)}(\mathring{A},\mathring{C}) \leq S^{(C,\sigma)}(\mathring{A},\mathring{B})$ *and* $S^{(C,\sigma)}(\mathring{A},\mathring{C}) \leq S(B,C)$ *if* $\mathring{A} \subseteq \mathring{B} \subseteq \mathring{C}$.

Proof. **i)** Since $d_i(\mathring{A},\mathring{B}), u_i(\mathring{A},\mathring{B}) \in [0,1]$ and for any $i = 1, 2, ..., n$ and the Choquet integral is monotone, we have $0 \leq S^{(C,\sigma)}(\mathring{A},\mathring{B}), U^{(C,\sigma)}(\mathring{A},\mathring{B}) \leq 1$.
ii) If $\mathring{A} = \mathring{B}$, then $\mu_{\mathring{A}}(\xi_i) = \mu_{\mathring{B}}(\xi_i)$ and $\nu_{\mathring{A}}(\xi_i) = \nu_{\mathring{B}}(\xi_i)$ for $i = 1, 2, ..., n$. We have $d_i(\mathring{A},\mathring{B}) = 0$ and so $(C)\int\limits_X d_i(\mathring{A},\mathring{B})\,d\sigma = 0$ for $i = 1, 2, ..., n$. Therefore, $S^{(C,\sigma)}(\mathring{A},\mathring{B}) = 1$. Conversely, assume that $S^{(C,\sigma)}(\mathring{A},\mathring{B}) = 1$. Therefore, $(C)\int\limits_X d_i(\mathring{A},\mathring{B})\,d\sigma = 0$ and so $d_i(\mathring{A},\mathring{B}) = 0$ for $i = 1, 2, ..., n$. It yields that $\mathring{A} = \mathring{B}$.
iii) Let $\mathring{A}$ and $\mathring{B}$ are crisp sets. We have $\pi_{\mathring{A}}(\xi_i), \pi_{\mathring{B}}(\xi_i) = 0$ and $|\mu(\xi_i) - \nu(\xi_i)| = 1$ and so $u_i(\mathring{A},\mathring{B}) = 0$ for $i = 1, 2, ..., n$. It yields that $U^{(C,\sigma)}(\mathring{A},\mathring{B}) = 0$.
iv) It is trivial since $d_i(\mathring{A},\mathring{B}) = d_i(\mathring{B},\mathring{A})$ for $i = 1, 2, ..., n$.
v) If $\mathring{A} \subseteq \mathring{B} \subseteq \mathring{C}$, then $\mu_{\mathring{A}}(\xi_i) \leq \mu_{\mathring{B}}(\xi_i) \leq \mu_{\mathring{C}}(\xi_i)$ and $\nu_{\mathring{A}}(\xi_i) \geq \nu_{\mathring{B}}(\xi_i) \geq \nu_{\mathring{C}}(\xi_i)$ for all $i = 1, 2, ..., n$. Thus, we have

$$(\mu_{\mathring{B}}(\xi_i) - \mu_{\mathring{A}}(\xi_i))^2 \leq (\mu_{\mathring{A}}(\xi_i) - \mu_{\mathring{C}}(\xi_i))^2,\ (\mu_{\mathring{B}}(\xi_i) - \mu_{\mathring{C}}(\xi_i))^2 \leq (\mu_{\mathring{A}}(\xi_i) - \mu_{\mathring{C}}(\xi_i))^2$$
$$(\nu_{\mathring{B}}(\xi_i) - \nu_{\mathring{A}}(\xi_i))^2 \leq (\nu_{\mathring{A}}(\xi_i) - \nu_{\mathring{C}}(\xi_i))^2,\ (\nu_{\mathring{B}}(\xi_i) - \nu_{\mathring{C}}(\xi_i))^2 \leq (\nu_{\mathring{A}}(\xi_i) - \nu_{\mathring{C}}(\xi_i))^2$$

and so $d_i(\mathring{A},\mathring{B}) \leq d_i(\mathring{A},\mathring{C})$ and $d_i(\mathring{B},\mathring{C}) \leq d_i(\mathring{A},\mathring{C})$ for $i = 1, 2, ..., n$. From monotonicity of the Choquet integral and definition of the proposed similarity measure based on Choquet integral we obtain $S^{(C,\sigma)}(\mathring{A},\mathring{B}) \geq S^{(C,\sigma)}(\mathring{A},\mathring{C})$ and $S^{(C,\sigma)}(\mathring{B},\mathring{C}) \geq S^{(C,\sigma)}(\mathring{A},\mathring{C})$. Hence, the proof is completed.

4 Illustrative Example

In this section, we consider a problem adapted from [13]. This problem examines twenty testing sets vr_i $(i = 1, ..., 20)$ of the species of the *Iris Plant* and we use the proposed vector similarity measure in the classification problem conducted in [13]. The following criteria

$$X = \{\xi_1(\text{sepal lenght}), \xi_2(\text{sepal width}), \xi_3(\text{petal lengt}), \xi_4(\text{petal width})\}$$

are used to classify the plants in *Setosa*, *Versicolour*, and *Virginica*. The IFS representations of *Setosa*, *Versicolour*, and *Virginica* are given as follows (see, [13]). All of the steps are same with the steps of [13] except the vector similarity measure.

$$C_1(Setosa) = \{\langle \xi_1, 0.15, 0.75\rangle, \langle \xi_2, 0.75, 0.20\rangle, \langle \xi_3, 0.10, 0.90\rangle, \langle \xi_4, 0.10, 0.90\rangle\},$$
$$C_2(Versicolour) = \{\langle \xi_1, 0.52, 0.40\rangle, \langle \xi_2, 0.38, 0.50\rangle, \langle \xi_3, 0.60, 0.28\rangle, \langle \xi_4, 0.60, 0.28\rangle\},$$
$$C_3(Virginica) = \{\langle \xi_1, 0.75, 0.20\rangle, \langle \xi_2, 0.52, 0.40\rangle, \langle \xi_3, 0.80, 0.15\rangle, \langle \xi_4, 0.80, 0.15\rangle\}.$$

The testing sets need to be classified in one of three patterns *Setosa*, *Versicolour* and *Virginica*. IFS representations of testing sets of *Virginica* pattern are given in Table 1 (See [13] for the method to create this table.)

Table 1. IFS representations of testing sets of *Virginica* pattern

Testing sets	Sepal lenght (SL)	Sepal width (SW)	Petal lenght (PL)	Petal width (PW)
vr_1	(0.52, 0.40)	(0.38, 0.50)	(0.80, 0.15)	(0.80, 0.15)
vr_2	(0.90, 0.10)	(0.52, 0.40)	(0.90, 0.10)	(0.52, 0.40)
vr_3	(1.00, 0.00)	(0.52, 0.40)	(1.00, 0.00)	(0.90, 0.10)
vr_4	(0.80, 0.15)	(0.25, 0.55)	(0.90, 0.10)	(0.80, 0.15)
vr_5	(0.75, 0.20)	(0.38, 0.50)	(0.80, 0.15)	(0.80, 0.15)
vr_6	(0.80, 0.15)	(0.38, 0.50)	(0.80, 0.15)	(0.80, 0.15)
vr_7	(0.75, 0.20)	(0.60, 0.28)	(0.80, 0.15)	(1.00, 0.00)
vr_8	(1.00, 0.00)	(0.90, 0.10)	(1.00, 0.00)	(0.90, 0.10)
vr_9	(1.00, 0.00)	(0.25, 0.55)	(1.00, 0.00)	(1.00, 0.00)
vr_{10}	(0.38, 0.50)	(0.38, 0.50)	(0.75, 0.20)	(0.80, 0.15)
vr_{11}	(0.80, 0.15)	(0.60, 0.28)	(0.90, 0.10)	(0.90, 0.10)
vr_{12}	(0.75, 0.20)	(0.38, 0.50)	(0.80, 0.15)	(0.90, 0.10)
vr_{13}	(1.00, 0.00)	(0.90, 0.10)	(1.00, 0.00)	(0.80, 0.15)
vr_{14}	(0.60, 0.28)	(0.25, 0.55)	(0.80, 0.15)	(0.60, 0.28)
vr_{15}	(1.00, 0.00)	(0.52, 0.40)	(0.90, 0.10)	(1.00, 0.00)
vr_{16}	(0.60, 0.28)	(0.75, 0.20)	(0.80, 0.15)	(1.00, 0.00)
vr_{17}	(0.52, 0.40)	(0.60, 0.28)	(0.80, 0.15)	(0.80, 0.15)
vr_{18}	(0.80, 0.15)	(0.60, 0.28)	(0.90, 0.10)	(1.00, 0.00)
vr_{19}	(0.80, 0.15)	(0.60, 0.28)	(0.90, 0.10)	(1.00, 0.00)
vr_{20}	(0.80, 0.15)	(0.52, 0.40)	(0.80, 0.15)	(1.00, 0.00)

Now, we determine the interaction of the criteria with each other. Expert opinion is consulted for this determination. According to the expert opinion, there is usually a direct proportion between width and length in sepals and petals. As the length of the sepal or petal gets longer, a noticeable growth is observed in the width. There is also a direct proportion between the sepal and the petal. This situation creates a redundancy between criteria. Therefore, we construct a sub-additive fuzzy measure motivating from [15]. Let σ be a set function on $P(X)$ such that

$$\begin{cases} \sigma(\xi_j) \geq 0, & \text{for all } 1 \leq j \leq 4 \\ \sigma(G) = \sum\limits_{\xi_j \in G} \sigma(\xi_j) + \min\limits_{\xi_i, \xi_j \in G, i \neq j} \lambda_{ij}, & \text{for all } G \in P(X) \text{ with } |G| \geq 2 \end{cases} \tag{11}$$

where $\{\lambda_{ij} = \lambda_{ji} : 1 \leq i, j \leq 4, i \neq j\} \in [-1, 0]^6$ is the set of interdependence coefficients. According to the expert opinion, we determine the interdependence coefficients between criteria such that $\lambda_{12} = -0.20$, $\lambda_{13} = -0.15$, $\lambda_{14} = -0.05$, $\lambda_{23} = -0.05$, $\lambda_{24} = -0.15$, $\lambda_{34} = -0.20$. Therefore, we construct a sub-additive fuzzy measure by using Eq. 11 and taking the weights of all criteria equal (see Table 2).

Table 2. Subdditive fuzzy measure

$\sigma(\emptyset) = 0$	$\sigma(\{\xi_1\}) = 0.25$	$\sigma(\{\xi_2\}) = 0.25$	$\sigma(\{\xi_3\}) = 0.25$
$\sigma(\{\xi_4\}) = 0.25$	$\sigma(\{\xi_1, \xi_2\}) = 0.30$	$\sigma(\{\xi_1, \xi_3\}) = 0.35$	$\sigma(\{\xi_1, \xi_4\}) = 0.45$
$\sigma(\{\xi_2, \xi_3\}) = 0.45$	$\sigma(\{\xi_2, \xi_4\}) = 0.35$	$\sigma(\{\xi_3, \xi_4\}) = 0.30$	$\sigma(\{\xi_1, \xi_2, \xi_3\}) = 0.55$
$\sigma(\{\xi_1, \xi_2, \xi_4\}) = 0.55$	$\sigma(\{\xi_2, \xi_3, \xi_4\}) = 0.55$	$\sigma(\{\xi_1, \xi_3, \xi_4\}) = 0.55$	$\sigma(\{\xi_1, \xi_2, \xi_3, \xi_4\}) = 1$

Then, we calculate the vector valued similarity measure of our IFSs. The results are given in the Table 3. For the sake of completeness; we keep the calculations for vr_1 and C_1. The distance values are

$$d_1(C_1, vr_1)(\xi_1) = 0.3601, \; d_2(C_1, vr_1)(\xi_2) = 0.3368,$$
$$d_3(C_1, vr_1)(\xi_3) = 0.7254, \; d_4(C_1, vr_1)(\xi_4) = 0.7254$$

and the uncertainty values are

$$u_1(C_1, vr_1)(\xi_1) = 0.5800, \; u_2(C_1, vr_1)(\xi_2) = 0.6187,$$
$$u_3(C_1, vr_1)(\xi_3) = 0.8544, \; u_4(C_1, vr_1)(\xi_4) = 0.8544.$$

Therefore, we get

$$d_2(C_1, vr_1)(\xi_2) \leq d_1(C_1, vr_1)(\xi_1) \leq d_3(C_1, vr_1)(\xi_3) = d_4(C_1, vr_1)(\xi_3)$$
$$u_1(C_1, vr_1)(\xi_1) \leq u_2(C_1, vr_1)(\xi_2) \leq u_3(C_1, vr_1)(\xi_3) = u_4(C_1, vr_1)(\xi_4).$$

Thus, we can obtain

$$S^{(C,\sigma)}(C_1, vr_1) = 1 - \begin{bmatrix} d_2(C_1, vr_1)(\xi_2) \times \sigma(X) + [d_1(C_1, vr_1)(\xi_1) - d_2(C_1, vr_1)(\xi_2)] \\ \times \sigma(\{\xi_1, \xi_3, \xi_4\}) + [d_3(C_1, vr_1)(\xi_3) - d_1(C_1, vr_1)(\xi_1)] \\ \times \sigma(\{\xi_3, \xi_4\}) + [d_4(C_1, vr_1)(\xi_4) - d_3(C_1, vr_1)(\xi_3)] \times \sigma(\{\xi_4\}) \end{bmatrix}$$
$$= 0.5407,$$

and

$$U^{(C,\sigma)}(C_1, vr_1) = \begin{bmatrix} u_1(C_1, vr_1)(\xi_1) \times \sigma(X) + [u_2(C_1, vr_1)(\xi_2) - u_1(C_1, vr_1)(\xi_1)] \\ \times\sigma(\{\xi_2, \xi_3, \xi_4\}) + [u_3(C_1, vr_1)(\xi_3) - u_2(C_1, vr_1)(\xi_2)] \\ \times\sigma(\{\xi_3, \xi_4\}) + [u_4(C_1, vr_1)(\xi_4) - u_3(C_1, vr_1)(\xi_3)] \times \sigma(\{\xi_4\}) \end{bmatrix}$$

$$= 0.6719.$$

We see from Table 3 that vr_1 belongs to Versicolour class and only this result is different from the results of [13]. This difference is due to the consideration of the interaction between criteria. Other results are consistent with [13], (see Table 5 in [13]).

Table 3. The similarity and uncertainty measures based on the Choquet integral between testing sets and IFSs of class Virginica

	Similarity values of C_1	Uncertainty values of C_1	Similarity values of C_2	Uncertainty values of C_2	Similarity values of C_3	Uncertainty values of C_3
Testing sets	$S^{(C,\sigma)}$	$U^{(C,\sigma)}$	$S^{(C,\sigma)}$	$U^{(C,\sigma)}$	$S^{(C,\sigma)}$	$U^{(C,\sigma)}$
vr_1	0.5407	0.6719	**0.9494**	0.4145	**0.9400**	0.3448
vr_2	0.7106	0.600	0.8195	0.4418	0.9002	0.2731
vr_3	0.6517	0.6405	0.7508	0.3823	0.9099	0.2487
vr_4	0.6504	0.8674	0.8322	0.3221	0.9246	0.2696
vr_5	0.4869	0.7731	0.8955	0.3968	0.9696	0.3323
vr_6	0.4744	0.7731	0.8830	0.3935	0.9671	0.3304
vr_7	0.5407	0.6233	0.7727	0.3370	0.9456	0.1993
vr_8	0.6693	0.4483	0.6643	0.3239	0.8233	0.1825
vr_9	0.3354	0.7202	0.7430	0.3175	0.8088	0.2559
vr_{10}	0.5925	0.5921	0.9215	0.4814	0.8972	0.3628
vr_{11}	0.8376	0.6775	0.7604	0.3455	0.9283	0.2007
vr_{12}	0.6937	0.8045	0.8782	0.3604	0.9617	0.3108
vr_{13}	0.4924	0.4483	0.6997	0.3441	0.8589	0.1945
vr_{14}	0.4812	0.7460	0.9285	0.4274	0.9045	0.3371
vr_{15}	0.5931	0.6405	0.7410	0.3823	0.9001	0.2487
vr_{16}	0.6237	0.5698	0.8004	0.2696	0.9044	0.1590
vr_{17}	0.6382	0.5404	0.8944	0.4231	0.9410	0.2487
vr_{18}	0.6378	0.6448	0.7368	0.3455	0.9073	0.1854
vr_{19}	0.6378	0.6448	0.7368	0.3455	0.9073	0.1854
vr_{20}	0.4853	0.6766	0.7890	0.4100	0.9458	0.2386

5 Conclusion

In this work, we focus on increasing the sensitivity of an existing similarity measure by taking into account the interaction between criteria with the help of Choquet integral. For this purpose, we propose a new vector valued similarity measure for IFSs whose similarity and uncertainty measures are based on the Choquet integral. We apply the proposed measures to an existing classification problem. Most of the results are consistent with the previous results. Inconsistent result may occur because of the novelty of the relatively sensitive method. In the future, we focus on other information measures based on various fuzzy integrals in various fuzzy set settings.

Acknowledgements. We gratefully thank Associate Professor Seher Karaman Erkul (Aksaray University, Department of Biology) for providing the expert opinions.

References

1. Choquet, G.: Theory of capacities. Annales de L'Institut Fourier **5**, 131–295 (1953)
2. Sugeno, M.: Theory of fuzzy integrals and its applications. Ph.D. Thesis, Tokyo (1974)
3. Meyer, P., Pirlot, M.: In: On the expressiveness of the additive value function and the Choquet integral models, pp. 48–56. Mons, Belgium (2012)
4. Lust, T.: Choquet integral versus weighted sum in multicriteria decision contexts. In: 3rd International Conference on Algorithmic Decision Theory, pp. 288–304. Springer International Publishing, Berlin, Lexington, KY, United States (2015)
5. Torra, V., Narukawa, Y.: Modeling Decisions: Information Fusion and Aggregation Operators. Springer, Berlin/Heidelberg, Germany (2007). https://doi.org/10.1007/978-3-540-68791-7
6. Zadeh, L.A.: Fuzzy sets. Inf. Control **8**(3), 338–353 (1965)
7. Atanassov, K.T.: Intuitionistic fuzzy sets. Fuzzy Sets Syst. **20**(1), 87–96 (1986)
8. Xu, Z.S., Yager, R.R.: Some geometric aggregation operators based on intuitionistic fuzzy sets. International J. General Syst. **35**, 417–433 (2006)
9. Xu, Z.S.: Intuitionistic fuzzy aggregation operators. IEE Trans. Fuzzy Syst. **15**(6), 1179–1187 (2007)
10. Boran, F.E., Akay, D.: A biparametric similarity measure on intuitionistic fuzzy sets with applications to pattern recognition. Inf. Sci. **255**, 45–57 (2014)
11. Chen, S.M., Chang, C.H.: A novel similarity measure between Atanassov's intuitionistic fuzzy sets based on transformation techniques with applications to pattern recognition. Inf. Sci. **291**, 96–114 (2015)
12. Song, Y., Wang, X., Quan, W., Huang, W.: A new approach to construct similarity measure for intuitionistic fuzzy sets. Soft Comput. **23**(6), 1985–1998 (2019)
13. Fei, L., Wang, H., Chen, L., Deng, Y.: A new vector valued similarity measure for intuitionistic fuzzy sets based on OWA operators. Iranian J. Fuzzy Syst.**16**(3), 113–126 (2019)
14. Grabisch, M.: The application of fuzzy integrals in multi criteria decision making. Eur. J. Oper. Res. **89**(3), 445–456 (1996)
15. Ünver, M., Özçelik, G., Olgun, M.: A pre-subadditive fuzzy measure model and its theoretical interpretation. TWMS J. App. Eng. Math. **10**(1), 270–278 (2020)

Social Acceptability Assessment of Renewable Energy Policies: An Integrated Approach Based on IVPF BOCR and IVIF AHP

Esra Ilbahar[1,2(✉)], Selcuk Cebi[1], and Cengiz Kahraman[2]

[1] Industrial Engineering Department, Yildiz Technical University, Besiktas, 34349 Istanbul, Turkey
{eilbahar,scebi}@yildiz.edu.tr

[2] Industrial Engineering Department, Istanbul Technical University, Macka, 34367 Istanbul, Turkey
kahramanc@itu.edu.tr

Abstract. The use of renewable energy, which is more encouraged together with increasing sustainability and environmental concerns, highlighted the issue of energy technology transformation. Social acceptance is a highly significant component of renewable energy technology implementation since it can affect the time and effort required to implement a policy or technology and its effectiveness can depend on public participation. It is crucial to know the perspective of public before making renewable energy policies to be able to take necessary actions that might positively affect the acceptance of such policies. Therefore, in this study, after factors affecting social acceptance are categorized as benefits, opportunities, costs and risks, the alternative policies are prioritized with respect to these criteria by using an approach based on interval-valued intuitionistic fuzzy (IVIF) AHP and interval-valued Pythagorean fuzzy Benefits, Opportunities, Costs, Risks (BOCR) analysis. The results obtained from four total score functions are compared and Option 4 is selected to be the best policy in terms of social acceptability.

Keywords: Renewable energy · Social acceptance · Pythagorean fuzzy sets · Intuitionistic fuzzy sets · AHP · BOCR

1 Introduction

The evaluation of renewable energy investments should not only be viewed from the perspective of customers and investors, as these decisions often affect local community in various ways [1]. That is why the factors related to social acceptability have been widely studied in the literature [1–11]. Some of these factors are identified as economic effects on local community such as new employment opportunities, costs in terms of tourism, economic development in the region while some of them are categorized as environmental effects on the surrounding

C. Kahraman et al. (Eds.): INFUS 2021, LNNS 308, pp. 93–100, 2022.
https://doi.org/10.1007/978-3-030-85577-2_11

ecosystem such as noise, landscape deterioration and vibrations. In addition to these factors, some other factors such as mistrust, lack of information on the new technologies, distributional justice, transparency, lack of impartiality and community engagement are taken into consideration [3–6,9]. In order to increase the social acceptability of renewable energy investments and policies, which are affected by such various factors, it is necessary to ensure that the public's views on this issue are appropriately evaluated. Therefore, in this study, the factors affecting social acceptability of new renewable energy policies are examined and alternative policies are evaluated with BOCR analysis which is extended using interval-valued Pythagorean fuzzy numbers (IVPFNs). The criteria adopted to evaluate alternative policies with respect to their acceptability are categorized as criteria associated to benefits, opportunities, costs, and risks. Then, pairwise comparisons are made to apply IVIF AHP to assign the weights of these criteria. The weights obtained from IVIF AHP are used in Pythagorean BOCR calculations. The rest of the paper is constructed as follows: preliminaries on the concepts utilized in this study are given in Sect. 2. The proposed assessment approach and its application are presented in Sect. 3. Finally, concluding remarks are given in Sect. 4.

2 Preliminaries

In this section, the fundamental knowledge on Pythagorean fuzzy sets and Benefits, Opportunities, Costs and Risks (BOCR) analysis is provided. The steps of IVIF AHP [12] could not be presented due to the page limit.

2.1 Pythagorean Fuzzy Sets

Pythagorean fuzzy sets (PFS) are introduced as an extension to intuitionistic fuzzy sets (IFS) not to enforce decision makers to provide membership and non-membership degrees whose sum is at most 1 [13,14]. Let an interval valued PFS be denoted as [15]:

$$\tilde{Q} = \{\left\langle x, \mu_{\tilde{Q}}(x), \nu_{\tilde{Q}}(x)\right\rangle ; x \epsilon X\} \tag{1}$$

where $\mu_{\tilde{Q}}(x) = [\mu^L_{\tilde{Q}}(x), \mu^U_{\tilde{Q}}(x)] \subset [0,1]$ and $\nu_{\tilde{Q}}(x) = [\nu^L_{\tilde{Q}}(x), \nu^U_{\tilde{Q}}(x)] \subset [0,1]$. Then, Eq. 2 must be met for $\tilde{Q}$:

$$0 \leqslant \mu^U_{\tilde{Q}}(x)^2 + \nu^U_{\tilde{Q}}(x)^2 \leqslant 1 \tag{2}$$

Then, indeterminacy degree, $\pi_{\tilde{Q}}(x) = [\pi^L_{\tilde{Q}}(x), \pi^U_{\tilde{Q}}(x)]$, is obtained as follows [15]:

$$\pi^L_{\tilde{Q}}(x) = \sqrt{1 - \mu^U_{\tilde{Q}}(x)^2 - \nu^U_{\tilde{Q}}(x)^2} \tag{3}$$

$$\pi^U_{\tilde{Q}}(x) = \sqrt{1 - \mu^L_{\tilde{Q}}(x)^2 - \nu^L_{\tilde{Q}}(x)^2} \tag{4}$$

Some operations on IVPFNs are expressed as follows [15]:

$$\tilde{q_1} \oplus \tilde{q_2} = ([\sqrt{(\mu_1^L)^2 + (\mu_2^L)^2 - (\mu_1^L)^2(\mu_2^L)^2}, \sqrt{(\mu_1^U)^2 + (\mu_2^U)^2 - (\mu_1^U)^2(\mu_2^U)^2}], [\nu_1^L\nu_2^L, \nu_1^U\nu_2^U]) \tag{5}$$

$$\tilde{q_1} \otimes \tilde{q_2} = ([\mu_1^L\mu_2^L, \mu_1^U\mu_2^U], [\sqrt{(\nu_1^L)^2 + (\nu_2^L)^2 - (\nu_1^L)^2(\nu_2^L)^2}, \sqrt{(\nu_1^U)^2 + (\nu_2^U)^2 - (\nu_1^U)^2(\nu_2^U)^2}]) \tag{6}$$

$$k\tilde{q} = ([\sqrt{1 - ((1-(\mu^L)^2)^k}, \sqrt{1 - ((1-(\mu^U)^2)^k}], [(\nu^L)^k, (\nu^U)^k]) \tag{7}$$

$$\tilde{q}^k = ([(\mu^L)^k, (\mu^U)^k], [\sqrt{1 - ((1-(\nu^L)^2)^k}, \sqrt{1 - ((1-(\nu^U)^2)^k}]) \tag{8}$$

2.2 BOCR Analysis

In BOCR analysis, opportunities indicate elements that are expected to bring profits in the future, whereas benefits show the existing revenue or the profits to be gained from relatively more expected positive advancements. Risks demonstrate elements which may occur as a result of a future negative development, whereas costs represent elements which arise as a result of an existing loss or as a result of a relatively negative development. There are different methods to compute the total scores of each alternative under benefits, opportunities, costs and risks. Some of them are as follows [16–18]:

- Additive total score function

$$P_i = bB_i + oO_i + c(1/C_i)_{Normalized} + r(1/R_i)_{Normalized} \tag{9}$$

- Probabilistic additive total score function

$$P_i = bB_i + oO_i + c(1 - C_i) + r(1 - R_i) \tag{10}$$

- Multiplicative priority powers total score function

$$P_i = B_i^b O_i^o [(1/C_i)_{Normalized}]^c [(1/R_i)_{Normalized}]^r \tag{11}$$

- Multiplicative total score function

$$P_i = \frac{B_i O_i}{C_i R_i} \tag{12}$$

where b, o, c and r represent normalized weights and B_i, O_i, C_i, and R_i denote the assessment results of option i with respect to benefits, opportunities, costs and risks, respectively [16–18].

3 The Proposed Policy Assessment Approach and Its Application

The proposed approach consists of three phases as described in Fig. 1. In the first phase, the related criteria to assess the policies are identified with a comprehensive literature review and they are categorized as opportunities (O), benefits (B), risks (R) and costs (C).

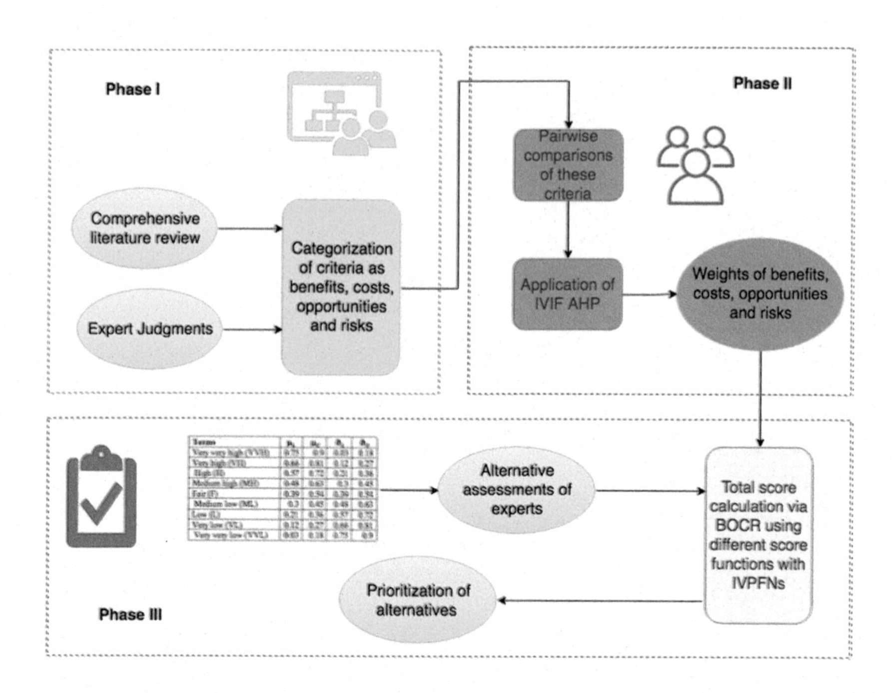

Fig. 1. The framework of the proposed approach for policy assessment.

Table 1. Linguistic terms for evaluation [19]

Linguistic terms	$[\mu_L, \mu_U]$	$[\nu_L, \nu_U]$
Very high (VH)	[0.60, 0.80]	[0.00, 0.20]
High (H)	[0.45, 0.65]	[0.15, 0.35]
Equal (E)	[0.30, 0.50]	[0.30, 0.50]
Low (L)	[0.15, 0.35]	[0.45, 0.65]
Very low (VL)	[0.00, 0.20]	[0.60, 0.80]

Table 2. Scale with IVPFNs for alternative evaluation [20]

Terms	$[\mu_L, \mu_U]$	$[\nu_L, \nu_U]$
Very very high (VVH)	[0.75, 0.90]	[0.03, 0.18]
Very high (VH)	[0.66, 0.81]	[0.12, 0.27]
High (H)	[0.57, 0.72]	[0.21, 0.36]
Medium high (MH)	[0.48, 0.63]	[0.30, 0.45]
Fair (F)	[0.39, 0.54]	[0.39, 0.54]
Medium low (ML)	[0.30, 0.45]	[0.48, 0.63]
Low (L)	[0.21, 0.36]	[0.57, 0.72]
Very low (VL)	[0.12, 0.27]	[0.66, 0.81]
Very very low (VVL)	[0.03, 0.18]	[0.75, 0.90]

In the second phase, pairwise comparisons of B, O, C and R and their sub-criteria are carried out. Preferences of decision makers are not precise and they include uncertainty and vagueness. Thus, in this study, linguistic terms obtained from the decision makers are transformed into IVIF numbers by using the scale given in Table 1 and then, the weight of each criteria is calculated by using IVIF AHP. Moreover, decision makers are asked to evaluate each policy by using the scale with IVPFNs given in Table 2. In the third phase, the crisp weights and alternative assessments in terms of IVPFNs obtained in phase 2 are used to calculate the total scores of alternative policies by using different kinds of total score functions given in Eqs. (9)–(12). Operations defined for IVPFNs are used to calculate these four different total score functions.

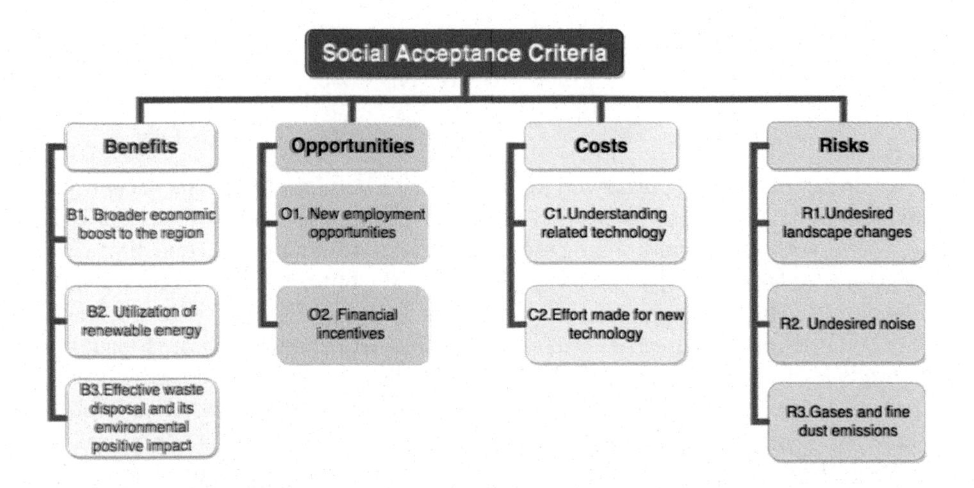

Fig. 2. Categorization of criteria affecting social acceptance.

Table 3. Pairwise comparison of B, O, C, R

BOCR	B	O	R	C
B	E	L	VL	H
O	H	E	L	H
R	VH	H	E	VH
C	L	L	VL	E

Table 4. Pairwise comparisons of sub-criteria

B	B.1	B.2	B.3	R	R.1	R.2	R.3
B.1	E	L	L	**R.1**	E	VL	L
B.2	H	E	H	**R.2**	VH	E	H
B.3	H	L	E	**R.3**	H	L	E
O	**O.1**	**O.2**		**C**	**C.1**	**C.2**	
O.1	E	L		**C.1**	E	VL	
O.2	H	E		**C.2**	VH	E	

Table 5. Final weights obtained from IVIF AHP

B.1	0.0631	**O.1**	0.1177	**R.1**	0.0878	**C.1**	0.0691
B.2	0.0855	**O.2**	0.1481	**R.2**	0.1355	**C.2**	0.1049
B.3	0.0768			**R.3**	0.1117		

Table 6. Evaluation of alternatives with respect to B, O, C, R criteria

Options	B.1	B.2	B.3	O.1	O.2	R.1	R.2	R.3	C.1	C.2
Option 1	VH	MH	ML	MH	MH	MH	ML	VVH	L	VL
Option 2	H	VVH	VH	MH	H	ML	F	MH	L	VL
Option 3	MH	ML	H	H	VH	F	MH	VH	L	VL
Option 4	VVH	VH	MH	VVH	VVH	H	VVL	F	VL	VL
Option 5	MH	H	VVH	H	VH	L	VL	ML	L	VVL

In order to apply the proposed approach, criteria related to social acceptance are identified as a result of a literature review. Figure 2 shows the criteria affecting social acceptance and their categorization as opportunities, benefits, risks and costs. Pairwise comparisons are performed for both main criteria and their sub-criteria. After the pairwise comparisons provided in Tables 3, 4 are obtained, IVIF AHP is implemented. As a result of IVIF AHP, the weights of these criteria are determined as given in Table 5. Then, different policies are assessed with respect to the criteria provided in Fig. 2 by using the scale provided in Table 2. The crisp weights from Table 5 and IVPFNs obtained from Table 6 are used as inputs in Eqs. (9)–(12) to calculate the total scores of alternatives. The results are provided in Table 7.

Table 7. Assessment results with respect to different total score functions modified with IVPFNs

	Additive	Rank	Multiplicative	Rank	Multiplicative priority powers	Rank	Probabilistic additive	Rank
Option 1	0.386	5	0.013	5	0.285	5	0.565	5
Option 2	0.627	3	0.040	2	0.469	4	0.684	3
Option 3	0.582	4	0.034	4	0.491	3	0.663	4
Option 4	0.724	1	0.046	1	0.551	1	0.745	1
Option 5	0.652	2	0.038	3	0.548	2	0.712	2

The ranking of alternatives provided in Table 7 are plotted in Fig. 3. As it can be seen, Option 4 is the best policy according to all of these score functions whereas Option 1 is the worst policy. It can be seen that the prioritization between the best and worst options can differ among these four approaches.

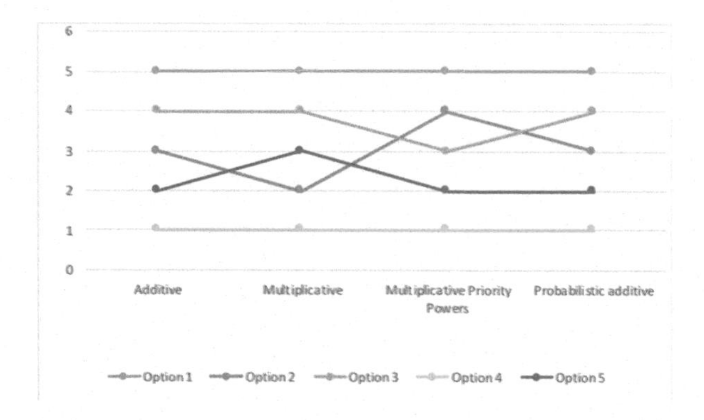

Fig. 3. Comparison of different score functions with IVPFNs.

4 Conclusion

While the use of renewable energy resources is increasingly encouraged, social acceptability is a very important element for the successful realization of the technological transformation required by this situation. For this reason, it is important to learn the public's views on the renewable energy policies to be implemented in line with this technological change and to determine the actions that will increase the acceptability of these policies. Therefore, in this study, an approach based on IVIF AHP and IVPF BOCR is introduced and five alternative policies are assessed by considering various criteria affecting social acceptance. Moreover, results obtained from four different total score functions are compared and it is revealed that Option 4 is the best policy in terms of social acceptability whereas Option 1 is the worst one according to all types of score functions. For future studies, BOCR analysis can be extended with other types of fuzzy sets such as intuitionistic fuzzy sets, hesitant fuzzy sets, and type 2 fuzzy sets.

References

1. Wüstenhagen, R., Wolsink, M., Bürer, M.J.: Social acceptance of renewable energy innovation: an introduction to the concept. Energy policy **35**(5), 2683–2691 (2007)
2. Mallett, A.: Social acceptance of renewable energy innovations: the role of technology cooperation in urban Mexico. Energy policy **35**(5), 2790–2798 (2007)
3. Rosso-Cerón, A.M., Kafarov, V.: Barriers to social acceptance of renewable energy systems in Colombia. Curr. Opin. Chem. Eng. **10**, 103–110 (2015)
4. Paravantis, J.A., Stigka, E., Mihalakakou, G., Michalena, E., Hills, J.M., Dourmas, V.: Social acceptance of renewable energy projects: a contingent valuation investigation in Western Greece. Renew. Energy **123**, 639–651 (2018)
5. Levenda, A.M., Behrsin, I., Disano, F.: Renewable energy for whom? A global systematic review of the environmental justice implications of renewable energy technologies. Energy Res. Soc. Sci. **71**, 101837 (2021)

6. Paletto, A., Bernardi, S., Pieratti, E., Teston, F., Romagnoli, M.: Assessment of environmental impact of biomass power plants to increase the social acceptance of renewable energy technologies. Heliyon **5**(7), e02070 (2019)
7. Batel, S.: Research on the social acceptance of renewable energy technologies: past, present and future. Energy Res. Soc. Sci. **68**, 101544 (2020)
8. Liu, Y., Sun, C., Xia, B., Cui, C., Coffey, V.: Impact of community engagement on public acceptance towards waste-to-energy incineration projects: empirical evidence from China. Waste Manage. **76**, 431–442 (2018)
9. Liu, Y., Ge, Y., Xia, B., Cui, C., Jiang, X., Skitmore, M.: Enhancing public acceptance towards waste-to-energy incineration projects: lessons learned from a case study in China. Sustain. Cities Soc. **48**, 101582 (2019)
10. Subiza-Pérez, M., Santa Marina, L., Irizar, A., Gallastegi, M., Anabitarte, A., Urbieta, N., Ibarluzea, J.: Explaining social acceptance of a municipal waste incineration plant through sociodemographic and psycho-environmental variables. Environ. Pollut. **263**, 114504 (2020)
11. Achillas, C., Vlachokostas, C., Moussiopoulos, N., Banias, G., Kafetzopoulos, G., Karagiannidis, A.: Social acceptance for the development of a waste-to-energy plant in an urban area. Resour. Conserv. Recycl. **55**(9–10), 857–863 (2011)
12. Wu, J., Huang, H.B., Cao, Q.W.: Research on AHP with interval-valued intuitionistic fuzzy sets and its application in multi-criteria decision making problems. Appl. Math. Model. **37**(24), 9898–9906 (2013)
13. Atanassov, K.T.: Intuitionistic fuzzy sets. In: Intuitionistic Fuzzy Sets, pp. 1–137. Physica, Heidelberg (1999)
14. Yager, R.R.: Pythagorean fuzzy subsets. In: 2013 Joint IFSA World Congress and NAFIPS Annual Meeting (IFSA/NAFIPS), pp. 57–61. IEEE, June 2013
15. Peng, X., Yang, Y.: Fundamental properties of interval-valued Pythagorean fuzzy aggregation operators. Int. J. Intell. Syst. **31**(5), 444–487 (2016)
16. Kabak, M., Dağdeviren, M.: Prioritization of renewable energy sources for Turkey by using a hybrid MCDM methodology. Energy Convers. Manage. **79**, 25–33 (2014)
17. Lee, A.H., Chen, H.H., Kang, H.Y.: Multi-criteria decision making on strategic selection of wind farms. Renew. Energy **34**(1), 120–126 (2009)
18. Yi, S.K., Sin, H.Y., Heo, E.: Selecting sustainable renewable energy source for energy assistance to North Korea. Renew. Sustain. Energy Rev. **15**(1), 554–563 (2011)
19. Cebi, S., Ilbahar, E.: Warehouse risk assessment using interval valued intuitionistic fuzzy AHP. Int. J. Anal. Hierarchy Process **10**(2),(2018)
20. Ilbahar, E., Kahraman, C.: Retail store performance measurement using a novel interval-valued Pythagorean fuzzy WASPAS method. J. Intell. Fuzzy Syst. **35**(3), 3835–3846 (2018)

Vaccine Selection Using Interval-Valued Intuitionistic Fuzzy VIKOR: A Case Study of Covid-19 Pandemic

Cihat Öztürk(✉), Abdullah Yildizbasi, Ibrahim Yilmaz, and Yağmur Ariöz

Department of Industrial Engineering, Ankara Yildirim Beyazit University, Ankara, Turkey
cozturk@ybu.edu.tr

Abstract. During the Covid-19 outbreak, different types of vaccines have been produced and tested in different countries. However, the produced vaccines have different features from each other, and governments are undergoing a decision process for vaccine selection. Criteria such as side effects of vaccines, supply chain processes, storage conditions, costs and perception on people play a crucial role in decision-making. In this study, vaccine selection is considered in an analytical framework. In this context, we focus on vaccine selection using the interval-valued intuitionistic fuzzy VIKOR method (VIseKriterijumska Optimize Kompromisno Resenje), one of the multi-criteria decision-making methods. In addition, interval valued intuitionistic fuzzy numbers (IVIFN) have been integrated into the VIKOR method to eliminate the uncertainty in decision processes. First, the concept, comparison, and distances of IVIFNs are briefly presented in the first section. The extension classical VIKOR approach is then developed to address vaccine selection problem with IVIFNs, and its significant aspect is that it can completely accept the finite justification of decision-makers as a real progress in decision-making. In this study, the applicability of the interval-valued intuitionistic VIKOR (IVIFV) method to evaluate the vaccines produced for the covid-19 outbreak was discussed. With the IVIFV method, the performances of the vaccines based on various criteria were evaluated and a ranking is represented among the vaccines accordingly.

Keywords: Vaccine selection · Covid-19 · VIKOR method · Interval-valued intuitionistic fuzzy sets

1 Introduction

The decision-making process can be defined as the process of selection the best alternative (or alternatives), based on many criteria and within the framework of under given information environment. Since multi-criteria decision-making methods (MCDM) have an impact on finding the most suitable solution from among conflict criteria, it is a frequently preferred methodology in the field of operational research.

Effective solutions can be obtained with MCDM methods based on costs, goals, benefits, and limited resources. In the decision-making process, various qualitative and

C. Kahraman et al. (Eds.): INFUS 2021, LNNS 308, pp. 101–108, 2022.
https://doi.org/10.1007/978-3-030-85577-2_12

quantitative criteria can mutually influence each other when evaluating alternatives, making the selection process complex and challenging. In most decision-making processes, decision makers do not have clear information about the alternatives and the importance of the criteria. Classical MCDM methods cannot effectively address problems with such imprecise information. Such classical MCDM methods cannot represent selection processes well in an uncertainty environment and may be inadequate in reflecting their fuzziness characteristics. In this context, Zadeh argued that fuzzy logic thought better reflected decision-making thoughts and developed the fuzzy set theory [1, 2]. Fuzzy set theory suggests that people have a more natural and intuitive approach to decision making rather than strict mathematical rules and equations. Therefore, it offers various methods for processing fuzzy expressions in order to process imprecise data and fuzzy expressions.

The VIKOR method, one of the multi-criteria decision-making methods, is a tool developed for multi-attribute optimizations of complex decision processes [3]. VIKOR is a method that focuses on sorting and choosing among a number of alternatives and provides compromise solutions for an MCDM problem with conflicting criteria [4]. Also, many extensions of the VIKOR method have been brought to the literature by researchers in recent years. Fuzzy VIKOR method, intuitionistic fuzzy VIKOR methods [5], interval type-2 fuzzy VIKOR model [6], hesitant fuzzy linguistic VIKOR method [7], dual hesitant fuzzy VIKOR model [8], triangular fuzzy neutrosophic VIKOR method [9], spherical fuzzy VIKOR [10] are extended fuzzy versions of the VIKOR method.

In this study, we discussed interval-valued intuitionistic fuzzy sets, which is an extended version of the VIKOR method. Interval-valued intuitionistic fuzzy sets (IVIFSs) theory has been developed to eliminate uncertainties in complex decision-making processes and has been an effective tool in the development of multi-featured group decision-making methods. Information about weight and attributes of decision makers may not be clear in vaccine selection problem. Therefore, interval-valued intuitionistic fuzzy VIKOR (IVIFV) method is used in which each element is characterized by a certain interval. In this context, we propose the IVIFV method expressed by intuitionistic fuzzy decision matrices with interval values to develop a decision support system for the vaccine selection process.

The main objective of this article is to present a methodology to select the best vaccine alternative among several vaccines produced. To the best of our knowledge, there are no studies in the literature regarding vaccine selection decisions in pandemic disasters. Each of the vaccines considered in this study has several advantages and disadvantages. The main criteria we consider include not only the technical aspects of vaccines, but also the commercial and social perception features. Our focus is on the general evaluation of the five different vaccines considered, such as technical, social, cost, and efficiency. In this context, we present a decision model for the selection of the best alternative (or alternatives) among currently produced vaccines. For this, we are evaluating five different vaccine alternatives based on the eight criteria.

In the second part of the study, we explain the methodology. The third section includes the implementation phase, and in the last section, we include conclusion part.

2 The Proposed Methodology for Vaccine Selection in Covid-19 Outbreak

In this study, an MCDM model based on IVIFV method is proposed for the vaccine selection for countries. The model consists of two basic parts. The first step is to define the criteria considering in the selection process of the vaccines. At this stage, the literature is examined, and expert opinions are considered at the same time. At the end of this section, the criteria are determined, and the hierarchical structure is presented. Simultaneously, vaccines to be evaluated based on defined criteria are determined within the framework of expert opinions and a hierarchical structure is presented. The second stage of the methodology consists of evaluating the considered vaccine selection criteria. For this purpose, IVIFV methodology is used to obtain criterion weights and to rank the alternatives. Therefore, the criteria are weighted and more and priority criteria are determined by IVIFV methodology. Then, alternatives are evaluated based on criteria and results are obtained according to the criteria weights.

2.1 Interval Valued Intuitionistic Fuzzy VIKOR (IVIFV)

The steps of the IVIFV method can be explained as follows [11]:

Step 1: Determining the criteria and alternatives, and then creating the decision matrices from the [12] within the framework of expert opinions.

$$\breve{\mu} = \left[\breve{\mu}_{ij}\right]_{mxn} \tag{1}$$

Step 2: Obtaining the best rating $\breve{\mu}^{+}$ (positive ideal solution) and worst rating $\breve{\mu}^{-}$ (negative ideal solution) values. If $\breve{\mu}^{+}$ and $\breve{\mu}^{-}$ defined as $\left[(a_{ij})^{+}, (b_{ij})^{+}\right], \left[(c_{ij})^{+}, (d_{ij})^{+}\right], \left[(a_{ij})^{-}, (b_{ij})^{-}\right], \left[(c_{ij})^{-}, (d_{ij})^{-}\right]$ respectively. For benefit attribute;

$$\breve{\mu}^{+} = \left(\left[max_i(a_{ij})^{+}, max_i(b_{ij})^{+}\right], \left[min_i(c_{ij})^{+}, min_i(d_{ij})^{+}\right]\right) \tag{2}$$

$$\breve{\mu}^{-} = \left(\left[min_i(a_{ij})^{+}, min_i(b_{ij})^{+}\right], \left[max_i(c_{ij})^{+}, max_i(d_{ij})^{+}\right]\right) \tag{3}$$

For cost attribute;

$$\breve{\mu}^{+} = \left(\left[min_i(a_{ij})^{+}, min_i(b_{ij})^{+}\right], \left[max_i(c_{ij})^{+}, max_i(d_{ij})^{+}\right]\right) \tag{4}$$

$$\breve{\mu}^{-} = \left(\left[max_i(a_{ij})^{+}, max_i(b_{ij})^{+}\right], \left[min_i(c_{ij})^{+}, min_i(d_{ij})^{+}\right]\right) \tag{5}$$

Step 3: Obtaining φ_i and ψ_i values to calculate the mean and worst group scores.

$$\varphi_i = \sum_{j=1}^{n} w_j \frac{dist\left(\left(\left[(a_{ij})^{+}, (b_{ij})^{+}\right], \left[(c_{ij})^{+}, (d_{ij})^{+}\right]\right)\left(\left[a_{ij}, b_{ij}\right], \left[c_{ij}, d_{ij}\right]\right)\right)}{dist\left(\begin{pmatrix}\left[(a_{ij})^{+}, (b_{ij})^{+}\right], \left[(c_{ij})^{+}, (d_{ij})^{+}\right] \\ \left[(a_{ij})^{-}, (b_{ij})^{-}\right], \left[(c_{ij})^{-}, (d_{ij})^{-}\right]\end{pmatrix}\right)} \tag{6}$$

$$\psi_i = max_j \left\{ w_j \frac{dist\left(\left(\left[(a_{ij})^+, (b_{ij})^+\right], \left[(c_{ij})^+, (d_{ij})^+\right]\right)\left(\left[a_{ij}, b_{ij}\right], \left[c_{ij}, d_{ij}\right]\right)\right)}{dist\left(\begin{pmatrix} \left[(a_{ij})^+, (b_{ij})^+\right], \left[(c_{ij})^+, (d_{ij})^+\right] \\ \left[(a_{ij})^-, (b_{ij})^-\right], \left[(c_{ij})^-, (d_{ij})^-\right] \end{pmatrix}\right)} \right\} \tag{7}$$

where $0 \leq w_j \leq 1$ shows that the weight of attributes which ensure $\sum_{j=1}^{n} w_j = 1$ and d is the distance.

Measure between IVIFNs which can be indicated as:

$$\left(\check{m}_{ij}, \check{m}_{lj}\right) = \sum_{k=1}^{n} w_k dist^k\left(\check{m}_{ij}, \check{m}_{lj}\right), w_k \in [0, 1], \sum_{k=1}^{n} w_k = 1. \tag{8}$$

where $w_k = (1, 2, \ldots, n)$ represent that the weight of $d^k\left(\check{m}_{ij}, \check{m}_{lj}\right)$.

Step 4: The π_i values are calculated with the formulas given below based on the results obtained from φ_i and ψ_i.

$$\pi_i = p\frac{\left(\varphi_i - \varphi^+\right)}{\left(\varphi^- - \varphi^+\right)} + (1 - p)\frac{\left(\psi_i - \psi^+\right)}{\left(\psi^- - \psi^+\right)} \tag{9}$$

$$\varphi^+ = min_i\varphi_i, \varphi^- = max_i\varphi_i \tag{10}$$

$$\psi^+ = min_i\psi_i, \psi^- = max_i\psi_i \tag{11}$$

where p is presented as a weight to define "the majority of criteria." Usually, p is considered as 0.5.

Step 5: Alternatives are ranked according to the IVIFV method. Alternatives are listed separately for each of the index values $\varphi_i, \theta_i, \pi_i$. At this stage, the technique in the original vikor method is used. The result is three $\varphi_i, \theta_i, \pi_i$ ranking lists. In ordering the π_i values, the smallest value of π_1 is suggested for the compromise solution according to the following conditions:

- Alternative i has an acceptable advantage. In the following example case;
- $\pi_2 - \pi_1 \geq D\pi$ and $D\pi = 1/(z - 1)$ and z is denoted that number of alternatives.
- Alternative i is considered constant in the decision-making process and ranks best on the φ_i, θ_i lists.
- If one of the two specified conditions is not met, the consensus solution set is suggested as follows:

– If the second condition is not fulfilled, i = 1 and i = 2 alternatives,
– If the first condition is not met, alternatives 1, 2, …, z are expressed considering the inequality $\pi_z - \pi_1 \geq D\pi$. Failure to meet this condition indicates that there is no significant difference between some alternatives.

3 Application: Covid-19 Case

At this section, the criteria to be used in vaccine selection are defined primarily. In the next stage, available improved vaccines are determined as an alternative. The criteria considered for vaccine selection in this study were determined by expert opinions and literature research. In this context, we define the basic criteria as cost, perception, storage conditions, producibility, lead time period, side-effect, preventiveness level, availability level. In addition, as explained in detail in the Sect. 2, the vaccines available in the market and most known vaccines are considered as alternatives. These alternatives are defined as Pfizer-BioNTech, Moderna, Astrezenca, Sinovac, and Sputinik V, respectively (A1 to A5).

The experts whose opinions were taken in the study are specialist doctor. Two of these experts are working on infectious diseases and have extensive knowledge on vaccines. The experiences of the experts whose opinions were taken are 8, 9 and 13 years, respectively. The criteria determined in the study are the opinions of experts and the literature researcher, respectively C1: Cost, C2: Perception, C3: Storage Conditions, C4: Producibility, C5: Lead Time Period, C6: Side-Effect, C7: Preventiveness Level, C8: Availability Level. While determining the criteria, considered the basic requirements of a vaccines during the pandemic period and therefore an evaluation was made in terms of general characteristics of vaccines.

The linguistic terms of decision makers judgments regarding the evaluation of alternatives based on criteria are given in the Table 1. The linguistic equivalents of the expert opinions are transformed into IVIF values and presented in the Table 2. Since there is a space limitation, only linguistic equivalents for A1 and A2 are shown in the Table 2.

Then, decision makers weights are determined by considering their experience in the health sector and their knowledge on vaccines. In this context, the first DM1, DM2, and DM3 weights are 0.40, 0.35 and 0.25, respectively. In addition, criterion weights are calculated and the IVIF scale is used for criterion weighting [12]. According to the obtained results, the most important criteria is C6, while the criterion with the lowest weight is C2. In other words, side effects are seen as the highest important criteria and the perception is the least importance. However, due to the space limitation, only the weights of the criteria are shared, and the Table 3 represents criterion weights.

Aggregated values are calculated for each alternative and presented in the Table 4. However, because of the space limitations, only the values of A1 and A2 are shared. At this stage, positive and negative ideal solutions are obtained with Eq. 2–5 equations.

The distance to ideal and negative ideal solutions for each alternative is calculated with Eq. 8. Calculated distances are used to define group utility values and individual regret values.

π_i values are obtained by Eq. 9. However, this equation should be calculated over the values obtained from φ_i and ψ_i Eq. 6–7. Then these values are listed in ascending order. Rankings of three different values are also shown in the Table 5.

The ascending rank suggests that A4, with the minimum of φ_i, ψ_i and π_i, has the best alternative among considered five vaccines can be seen in the Table 5.

Table 1. Rating of alternatives and criteria by experts

		A1	A2	A3	A4	A5
DM1	C1	B	B	G	VG	MG
	C2	MG	G	B	B	VB
	C3	VB	MG	MG	VG	MG
	C4	MB	B	B	G	B
	C5	G	B	B	G	G
	C6	G	B	VB	VG	MB
	C7	VG	G	B	G	VB
	C8	VB	MB	MB	B	G
DM2	C1	MB	MB	G	VG	VG
	C2	G	MG	B	B	B
	C3	VB	MG	MG	VG	MG
	C4	MB	B	B	B	B
	C5	G	B	B	MG	G
	C6	MG	B	VB	VG	MB
	C7	G	G	B	G	VB
	C8	VB	MB	VB	B	B
DM3	C1	B	G	G	VG	MG
	C2	MG	G	B	B	VB
	C3	VB	MG	MG	VG	MG
	C4	MB	MB	B	B	B
	C5	MG	G	B	B	G
	C6	MG	MB	VB	VG	VB
	C7	VG	G	B	G	B
	C8	VB	B	B	B	EB

Table 2. Transformed linguistic values to IVIF numbers

		A1				A2			
DM1	C1	0.300	0.500	0.200	0.500	0.300	0.500	0.200	0.500
	C2	0.600	0.800	0.000	0.200	0.500	0.700	0.100	0.300
	C3	0.100	0.300	0.400	0.700	0.600	0.800	0.000	0.200
	C4	0.200	0.400	0.300	0.600	0.300	0.500	0.200	0.500
	C5	0.500	0.700	0.100	0.300	0.300	0.500	0.200	0.500
	C6	0.500	0.700	0.100	0.300	0.300	0.500	0.200	0.500
	C7	0.700	0.900	0.000	0.100	0.500	0.700	0.100	0.300
	C8	0.100	0.300	0.400	0.700	0.200	0.400	0.300	0.600
DM2	C1	0.200	0.400	0.300	0.600	0.200	0.400	0.300	0.600
	C2	0.500	0.700	0.100	0.300	0.600	0.800	0.000	0.200
	C3	0.100	0.300	0.400	0.700	0.600	0.800	0.000	0.200
	C4	0.200	0.400	0.300	0.600	0.300	0.500	0.200	0.500
	C5	0.500	0.700	0.100	0.300	0.300	0.500	0.200	0.500
	C6	0.600	0.800	0.000	0.200	0.300	0.500	0.200	0.500
	C7	0.500	0.700	0.100	0.300	0.500	0.700	0.100	0.300
	C8	0.100	0.300	0.400	0.700	0.200	0.400	0.300	0.600

Table 3. Criteria weights

C1	C2	C3	C4	C5	C6	C7	C8
0.078616	0.062893	0.110063	0.169811	0.09434	0.18239	0.172956	0.128931

Table 4. Aggregated IVIF numbers

	A1				A2			
	μ^L	μ^U	υ^L	υ^U	μ^L	μ^U	υ^L	υ^U
C1	0.2665	0.4671	0.2365	0.5376	0.3257	0.5309	0.2137	0.4970
C2	0.5675	0.7695	0.0362	0.2365	0.5376	0.7397	0.0662	0.2665
C3	0.1000	0.3000	0.4000	0.7000	0.6000	0.8000	0.0000	0.2000
C4	0.2000	0.4000	0.3000	0.6000	0.2762	0.4767	0.2263	0.5271
C5	0.5271	0.7289	0.0760	0.2762	0.3565	0.5599	0.1761	0.4561
C6	0.5627	0.7648	0.0413	0.2416	0.2762	0.4767	0.2263	0.5271
C7	0.6413	0.8531	0.0362	0.1758	0.5000	0.7000	0.1000	0.3000
C8	0.1000	0.3000	0.4000	0.7000	0.2263	0.4267	0.2762	0.5771

Table 5. Alternatives ranking and obtained results

	φ_i	Rank	ψ_i	Rank	π_i	Rank
A1	1.0085	4	0.4052	5	0.9535	5
A2	0.9241	2	0.1897	2	0.4742	2
A3	1.0501	5	0.2781	4	0.7731	4
A4	0.6034	1	0.1251	1	0.0000	1
A5	0.9535	3	0.2404	3	0.5976	3

4 Conclusion and Future Perspectives

In this study, we evaluate the vaccines produced to prevent the covid-19 epidemic by using the VIKOR method. VIKOR method is more accurate and realistic than other MCDM methods because of its compromise solution approach to the complexity of decision criteria. It is also a method in which the influence of individual radical opinions can be reduced by group decisions. We deal with uncertainty by integrating the VIKOR method with IVIF sets. IVIF theory can prevent data loss and assist in the integration of non-numerical linguistic expressions into analytical numerical methods. The information collected in this way is then evaluated through the IVIF VIKOR methodology, a powerful combined technique for full or partial rankings. The major contribution that stands out from the study is field of application. The selection of the vaccine by using MCDM methodologies is emerging topic in the current literature. Another vital part of the paper is combining the interval valued intuitionistic fuzzy set and VIKOR methodology. This approach is a new method that has been rarely used in the literature. As a result of the evaluation, side effects and level of protection seem to be important criteria. In addition, manufacturability stands out as another important criterion. In addition, producibility stands out as another important criteria. In this context, the A4 (Sinovac) alternative has been obtained as the most suitable option.

The criteria used for vaccine selection in the study have very general characteristics. More detailed evaluations can be made for each main topic (cost, prevention level, etc.) in future studies. In addition, the comparison of the different methodologies can be considered as outlook study. In addition, sensitivity analysis can be considered as another field of study in terms of examining the behavior of possible changes in parameters on the results.

References

1. Zadeh, L.A.: Fuzzy sets as a basis for a theory of possibility. Fuzzy Sets Syst. 3–28 (1978)
2. Höhle, U.: On the fundamentals of fuzzy set theory. J. Math. Anal. Appl. **201**(3), 786–826 (1996)
3. Opricovic, S.: Multicriteria optimization of civil engineering systems. Facul. Civ. Eng. Belgrade **2**(1), 5–21 (1998)
4. Opricovic, S., Tzeng, G.H.: Extended VIKOR method in comparison with outranking methods. Eur. J. Oper. Res. **178**(2), 514–529 (2007)
5. Devi, K.: Extension of VIKOR method in intuitionistic fuzzy environment for robot selection. Expert Syst. Appl. **38**(11), 14163–14168 (2011)
6. Soner, O., Celik, E., Akyuz, E.: Application of AHP and VIKOR methods under interval type 2 fuzzy environment in maritime transportation. Ocean Eng. **129**, 107–116 (2017)
7. Liao, H., Xu, Z., Zeng, X.-J.: Hesitant fuzzy linguistic VIKOR method and its application in qualitative multiple-criteria decision making. IEEE Trans. Fuzzy Syst. **23**(5), 1343–1355, (2014)
8. Ren, Z., Xu, Z., Wang, H.: Dual hesitant fuzzy VIKOR method for multi-criteria group decision making based on fuzzy measure and new comparison method. Inf. Sci. **388**, 1–16 (2017)
9. Wang, J., Wei, G., Lu, M.: An extended VIKOR method for multiple criteria group decision making with triangular fuzzy neutrosophic numbers. Symmetry **10**(10), 497 (2018)
10. Kutlu Gündoğdu, F., Kahraman, C., Karaşan, A.: Spherical fuzzy VIKOR method and its application to waste management. In: Kahraman, C., Cebi, S., Cevik Onar, S., Oztaysi, B., Tolga, A.C., Sari, I.U. (eds.) INFUS 2019. AISC, vol. 1029, pp. 997–1005. Springer, Cham (2020). https://doi.org/10.1007/978-3-030-23756-1_118
11. Park, J.H., Cho, H.J., Kwun, Y.C.: Extension of the VIKOR method for group decision making with interval-valued intuitionistic fuzzy information. Fuzzy Optim. Decis. Making **10**(3), 233–253 (2011)
12. Büyüközkan, G., Göçer, F., Feyzioğlu, O.: Cloud computing technology selection based on interval-valued intuitionistic fuzzy MCDM methods. Soft. Comput. **22**(15), 5091–5114 (2018). https://doi.org/10.1007/s00500-018-3317-4

A Novel Approach of Complex Intuitionistic Fuzzy Linear Systems in an Electrical Circuit

J. Akila Padmasree(✉) and R. Parvathi

Vellalar College for Women, Erode, Tamilnadu, India

Abstract. The paper deals with a new type of intuitionistic fuzzy number, in the complex plane called as complex horizontal relative trapezoidal intuitionistic fuzzy number (CHRTrIFN). Arithmetic operations like addition, multiplicative inverse and division on these numbers are defined and discussed. Complex horizontal relative trapezoidal intuitionistic fuzzy number(CHRTrIFN) and geometric representation of CHRTrIFN and complex trapezoidal intuitionistic fuzzy number(CTrIFN) are presented. CHRTrIFNs are used to solve complex intuitionistic fuzzy linear system of equations and is applied in a RLC intuitionistic fuzzy circuit to find the current flow in the circuit.

Keywords: Horizontal relative trapezoidal intuitionistic fuzzy number (HRTrIFN) · Complex relative trapezoidal intuitionistic fuzzy number (CTrIFN) · Complex horizontal relative trapezoidal intuitionistic fuzzy number (CHRTrIFN)

1 Introduction

Intuitionistic fuzzy sets introduced by Atanassov [2] plays an important role in decision making under an uncertain environment. Interval analysis defined by Moore in [6] defined on a set of real intervals produces a closed interval of lower and upper endpoints, where the real valued result is guaranteed to lie and the accuracy is defined by the width of the interval. Similarly intuitionistic fuzzy numbers [4] are closed intervals with four interval values where the result is guaranteed to lie within the interval which give a clear information about the vaguenes in the mathematical modeling of real life problems. J.J. Buckley defined fuzzy complex numbers with topological concepts in [3]. M. Landowski defined trapezoidal horizontal intuitionistic fuzzy number (TrHIFN) in the form of multi dimensional granule and as solution span [5]. In [7], complex trapezoidal intuitionistic fuzzy numbers were introduced and verified that it was not possible to define division and conjugate of CTrIFN. To overcome, HRTrIFN are introduced in this paper with a geometrical representation. In [1], the interval arithmetic involves only the extremities of intervals in calculations whereas the multi-dimensional relative distance measure interval arithmetic (RDM - IA) developed in [10] includes a multiplier $\alpha \in [0,1]$

C. Kahraman et al. (Eds.): INFUS 2021, LNNS 308, pp. 109–118, 2022.
https://doi.org/10.1007/978-3-030-85577-2_13

in the representation of the interval in such a way that it converts the structure of the membership and non-membership function from linear/curvilinear into planar form. The horizontal membership functions introduced in [8] and verified in [9] defines a fuzzy number not in the form of $\mu = f(x)$ but as $x = f(\mu)$. The paper is organised as in Sect. 2, the geometric representation of Complex Trapezoidal Intuitionistic Fuzzy Number (CTrIFN)is shown and in Sect. 3, Horizontal relative trapezoidal intuitionistic fuzzy number(HRTrIFN) and Complex horizontal relative trapezoidal intuitionistic fuzzy number(CHRTrIFN) are defined with the properties. In Sect. 4, a method is proposed to solve complex intuitionistic fuzzy linear equations with a numerical example and in its subsection it is applied in a RLC intuitionistic fuzzy circuit to find the current flow in the circuit. Section 5 concludes with the advantages of the proposed method.

2 Geometrical Representation of Complex Trapezoidal Intuitionistic Fuzzy Number (CTrIFN)

Definition 1: [7] A *Complex Trapezoidal Intuitionistic Fuzzy Number (CTrIFN)* Z is a trapezoidal intuitionistic fuzzy number in the complex plane C and is of the form $Z = \langle A + iB, (\mu_A, \nu_A), (\mu_B, \nu_B) \rangle$, where (μ_A, ν_A) denotes the degrees of membership and non-membership values of $Re(Z)$ and (μ_B, ν_B) denotes the degrees of membership and non-membership values of $Im(Z)$. In other words, Z has the form $Z = \langle [a, b, c, d; a', b, c, d'] \rangle + i \langle [p, q, r, s; p', q, r, s'] \rangle$, where the membership and non-membership functions of $Re(Z)$ and $Im(Z)$ are defined as follows:

For every $x \in R$,

$$\mu_A(x) = \begin{cases} \frac{x-a}{b-a} & a \leq x \leq b \\ 1 & b \leq x \leq c \\ \frac{d-x}{d-c} & c \leq x \leq d \\ 0 & \text{otherwise} \end{cases} \qquad \nu_A(x) = \begin{cases} \frac{b-x}{b-a'} & a' \leq x \leq b \\ 0 & b \leq x \leq c \\ \frac{x-c}{d'-c} & c \leq x \leq d' \\ 1 & \text{otherwise} \end{cases}$$

$$\mu_B(x) = \begin{cases} \frac{x-p}{q-p} & p \leq x \leq q \\ 1 & q \leq x \leq r \\ \frac{s-x}{s-r} & r \leq x \leq s \\ 0 & \text{otherwise} \end{cases} \qquad \nu_B(x) = \begin{cases} \frac{q-x}{q-p'} & p' \leq x \leq q \\ 0 & q \leq x \leq r \\ \frac{x-r}{s'-r} & r \leq x \leq s' \\ 1 & \text{otherwise} \end{cases}$$

where $a' \leq a \leq b \leq c \leq d \leq d'$ and $a', a, b, c, d, d' \in R$. $0 \leq \mu_A \leq 1$ and $0 \leq \nu_A \leq 1$ and $0 \leq \mu_A + \nu_A \leq 1$. Also, $p' \leq p \leq q \leq r \leq s \leq s'$ and $p', p, q, r, s, s' \in R$. $0 \leq \mu_B \leq 1$, $0 \leq \nu_B \leq 1$ and $0 \leq \mu_B + \nu_B \leq 1$.

For simplicity, we write $Z = \langle [a, b, c, d]; \mu_A, \nu_A \rangle + i \langle [p, q, r, s]; \mu_B, \nu_B \rangle$. The geometrical representation is given in Fig. 1.

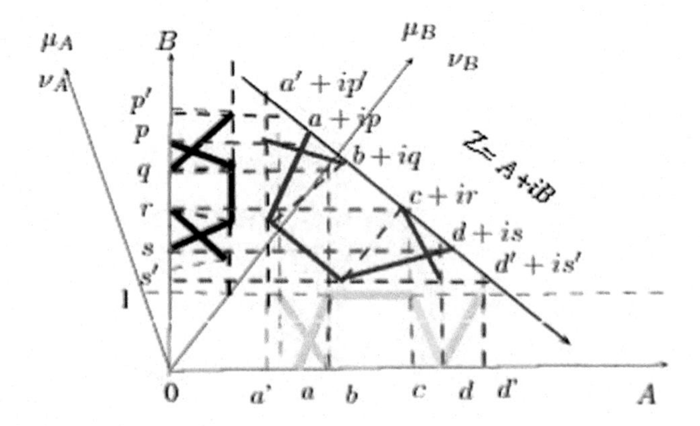

membership and non-membership function of A

membership and non-membership function of B

membership and non-membership function of $Z = A + iB$

Fig. 1. Complex trapezoidal intuitionistic fuzzy number

3 Horizontal Relative Trapezoidal Intuitionistic Fuzzy Number

Definition 2: Let the parametric form of an intuitionistic fuzzy number $A = \langle [a, b, c, d; a', b, c, d'] \rangle$ be $X = \langle [\underline{x}, \overline{x}], [\underline{\underline{x}}, \overline{\overline{x}}] \rangle$. For every r-cut of membership values and r'-cut of non-membership values, let

$\underline{x(r)} = a + (b - a)r$ $\quad \overline{x(r)} = d - (d - c)r$ $\quad \underline{\underline{x(r')}} = a' + (b - a')r'$

$\overline{\overline{x(r')}} = d' - (d' - c)r'$

where $\underline{x(r)} \leq \overline{x(r)}$, $\underline{\underline{x(r')}} \leq \overline{\overline{x(r')}}$ and $r, r' \in [0, 1]$. Then a *horizontal relative trapezoidal intuitionistic fuzzy number* (HRTrIFN) H of A, is defined as

$$H = \{\langle x, x(\mu, \alpha_x), x(\nu, \beta_x) \rangle : x \in X, \mu, \nu, \alpha_x, \beta_x \in [0, 1]\}$$

where $x(\mu, \alpha_x)$ denotes the horizontal relative membership function of H given by $x(\mu, \alpha_x) = [a + (b - a)\mu] + [(d - a) - (d - c + b - a)\mu]\alpha_x, \quad \mu, \alpha_x \in [0, 1], x \in X$ and $x(\nu, \beta_x)$ denotes the horizontal relative non-membership function of H given by $x(\nu, \beta_x) = [a' + (b - a')\nu] + [(d' - a') - (d' - c + b - a')\nu]\beta_x, \quad \nu, \beta_x \in [0, 1], x \in X$ and α_x and β_x are the relative distance measure of membership and non-membership functions of H respectively. For example, consider the trapezoidal intuitionistic fuzzy number $\langle [3, 4, 5, 6; 2, 4, 5, 7] \rangle$. The horizontal relative membership function $x(\mu, \alpha_x) = [3 + \mu] + [3 - 2\mu]\alpha_x$ and horizontal relative non-membership function $x(\nu, \beta_x) = [2 + 2\nu] + [5 - 4\nu]\beta_x$ are represented in Fig. 2.

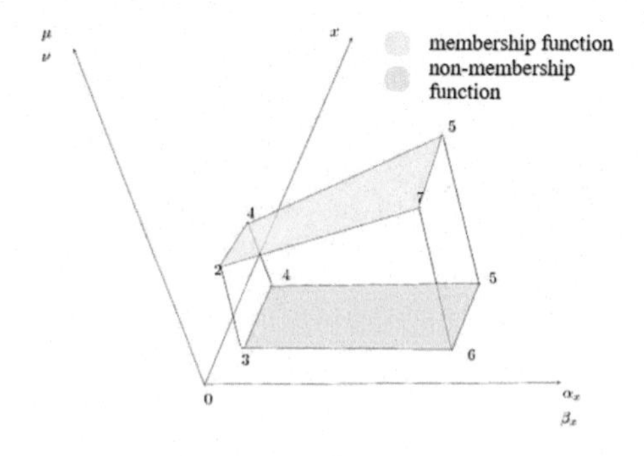

Fig. 2. Horizontal membership and non-membership function

3.1 Complex Horizontal Relative Trapezoidal Intuitionistic Fuzzy Number(CHRTrIFN)

In [7], Complex trapezoidal intuitionistic fuzzy numbers (CTrIFN) were introduced in which it was verified that the division and additive and multiplicative properties of conjugate of CTrIFNs do not hold. So, the need of defining horizontal relative trapezoidal intuitionistic fuzzy numbers (HRTrIFNs) arises. Further, it is extended to the complex plane as complex horizontal relative trapezoidal intuitionistic fuzzy numbers (CHRTrIFNs), where the division and the additive and multiplicative properties of conjugate hold.

Definition 3: Let X and Y be two HRTrIFNs. Let $Z = X + iY$ or $Z = \langle [a, b, c, d; a', b, c, d'] + i[p, q, r, s; p'q, r, s'] \rangle$ be the CTrIFN. A *complex horizontal relative trapezoidal intuitionistic fuzzy number* (CHRTrHIFN) in the complex plane is of the form

$$Z = \{\langle X + iY, (x(\mu_x, \alpha_x), (x(\nu_x, \beta_x)), (y(\mu_y, \alpha_y), (y(\nu_y, \beta_y))) \rangle : x \in X, y \in Y, \mu_x, \nu_x, \alpha_x, \beta_x \in [0, 1]\}$$

where $\alpha_x, \alpha_y, \beta_x, \beta_y$ are RDM of X and Y and $x(\mu_x, \alpha_x), x(\nu_x, \beta_x)$ are the membership and non-membership functions of real part of Z and $y(\mu_y, \alpha_y), y(\nu_y, \beta_y)$ are the membership and non-membership functions of imaginary part of Z and are defined as follows:

$$x(\mu_x, \alpha_x) = [a + (b - a)\mu_x] + [(d - a) - (d - c + b - a)\mu_x]\alpha_x, \quad \mu_x, \alpha_x \in [0, 1]$$
$$x(\nu_x, \beta_x) = [a' + (b - a')\nu_x] + [(d' - a') - (d' - c + b - a')\nu_x]\beta_x, \quad \nu_x, \beta_x \in [0, 1]$$
$$y(\mu_y, \alpha_y) = [p + (q - p)\mu_y] + [(s - p) - (s - r + q - p)\mu_y]\alpha_y, \quad \mu_y, \alpha_y \in [0, 1]$$
$$y(\nu_y, \beta_y) = [p' + (q - p')\nu_y] + [(s' - p') - (s' - r + q - p')\nu_y]\beta_y, \quad \nu_y, \beta_y \in [0, 1]$$

The geometrical representation of Z is given in Fig. 3.

Division:
The division of two CHRTrIFNs denoted by $Z = Z_1/Z_2$, takes the form $Z = \langle Z, Z(\mu_Z, \alpha_Z), Z(\nu_Z, \beta_Z) \rangle$ where $Z(\mu_Z, \alpha_Z) = A/B$ where

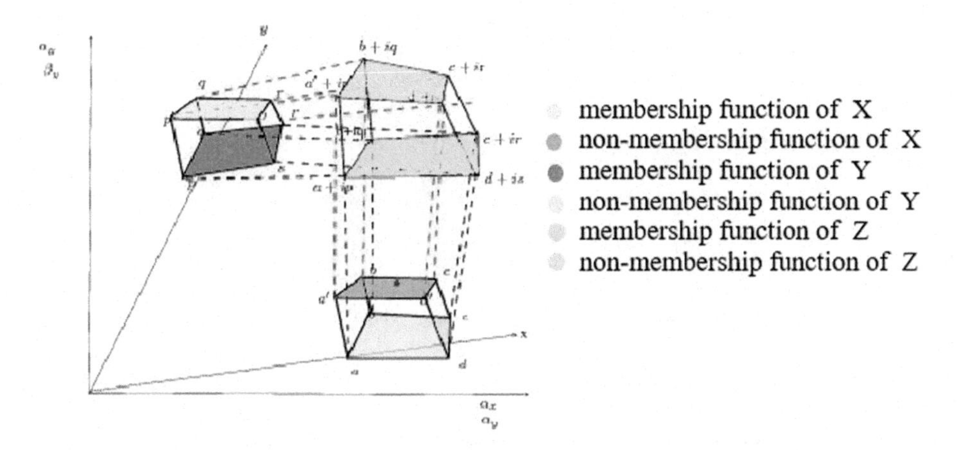

Fig. 3. Complex horizontal relative trapezoidal intuitionistic fuzzy number

$$A = (x_1(\mu_{x_1}, \alpha_{x_1})x_2(\mu_{x_2}, \alpha_{x_2}) + y_1(\mu_{y_1}, \alpha_{y_1})y_2(\mu_{y_2}, \alpha_{y_2}))$$
$$+ i((x_2(\mu_{x_2}, \alpha_{x_2})y_1(\mu_{y_1}, \alpha_{y_1})) - (x_1(\mu_{x_1}, \alpha_{x_1})y_2(\mu_{y_2}, \alpha_{y_2})))$$
$$B = {x_2}^2(\mu_{x_2}, \alpha_{x_2}) + {y_2}^2(\mu_{y_2}, \alpha_{y_2})$$

$Z(\nu_Z, \beta_Z) = A'/B'$ where $A' = (x_1(\nu_{x_1}, \beta_{x_1}).x_2(\nu_{x_2}, \beta_{x_2}) + y_1(\nu_{y_1}, \beta_{y_1})y_2(\nu_{y_2}, \beta_{y_2})) + i((x_2(\nu_{x_2}, \beta_{x_2})y_1(\nu_{y_1}, \beta_{y_1})) - (x_1(\nu_{x_1}, \beta_{y_1})y_2(\nu_{y_2}, \beta_{y_2})))$
$B' = {x_2}^2(\nu_{x_2}, \beta_{x_2}) + {y_2}^2(\nu_{y_2}, \beta_{y_2})$

Additive Inverse:
Let $-Z$ be the additive inverse of Z. The membership function $-Z$ is given by $-Z(\mu, \alpha_Z) = -x(\mu_x, \alpha_x) - i(y(\mu_y, \alpha_y))$ and its non-membership function is given by $-Z(\nu, \beta_Z) = -x(\nu_x, \beta_x) - i(y(\nu_y, \beta_y)$

Multiplicative Inverse:
Let $1/Z$ be the multiplicative inverse of Z. The membership function of $1/Z$ is given by $1/Z(\mu, \alpha_Z) = 1/x(\mu_x, \alpha_x) + iy(\mu_y, \alpha_y)$ and it's non-membership function is given by $1/Z(\nu, \beta_Z) = 1/x(\nu_x, \beta_x) + iy(\nu_y, \beta_y)$.

3.2 Properties of CHRTrHIFN

Additive Property of a Conjugate of a CHRTrHIFN:
The membership function is $Z(\mu_Z, \alpha_Z) + \overline{Z}(\mu_{\overline{Z}}, \alpha_{\overline{Z}}) = 2x(\mu_x, \alpha_x)$ and non-membership function is $Z(\nu_Z, \alpha_Z) + \overline{Z}(\nu_{\overline{Z}}, \beta_{\overline{Z}}) = 2x(\nu, \beta_x)$

Multiplicative Property of a Conjugate of a CHRTrHIFN:
Multiplicative property for membership function of a conjugate of a CHRTrHIFN is given by $Z(\mu_Z, \alpha_Z).\overline{Z}(\mu_{\overline{Z}}, \alpha_{\overline{Z}}) = x^2(\mu_x, \alpha_x) + y^2(\mu_y, \alpha_y)$ and for non-membership function is given by $Z(\nu_Z, \beta_Z).\overline{Z}(\nu_{\overline{Z}}, \beta_{\overline{Z}}) = x^2(\nu_x, \beta_x) + y^2(\nu_y, \beta_y)$.

4 Solving Complex Intuitionistic Fuzzy System of Linear Equations Using CHRTrHIFNs

Consider a nxn complex intuitionistic fuzzy linear system of equations,

$$
\begin{aligned}
&z_{11}x_1 + z_{12}x_2 + ... + z_{1n}x_n = p_1 + iq_1\\
&z_{21}x_1 + z_{21}x_2 + ... + z_{2n}x_n = p_2 + iq_2\\
&...\\
&z_{n1}x_1 + z_{n2}x_2 + ... + z_{nn}x_n = p_n + iq_n
\end{aligned}
$$

where the co-efficient matrix $A = (z_{ij}), 1 \leq i, j \leq n$ is a crisp complex matrix and $x_i^\prime s$ are unknown complex intuitionistic fuzzy numbers, $y_i^\prime s$ are known complex intuitionistic fuzzy numbers.

Complex IFN $y_i^\prime s$ in parametric form is written as,

$$
\begin{aligned}
&z_{11}x_1 + z_{12}x_2 + ... + z_{1n}x_n = \langle[\underline{p_1}, \overline{p_1}], [\underline{\underline{p_1}}, \overline{\overline{p_1}}] + i[\underline{q_1}, \overline{q_1}], [\underline{\underline{q_1}}, \overline{\overline{q_1}}]\rangle\\
&z_{21}x_1 + z_{21}x_2 + ... + z_{2n}x_n = \langle[\underline{p_2}, \overline{p_2}], [\underline{\underline{p_2}}, \overline{\overline{p_2}}] + i[\underline{q_2}, \overline{q_2}], [\underline{\underline{q_2}}, \overline{\overline{q_2}}]\rangle\\
&...\\
&z_{n1}x_1 + z_{n2}x_2 + ... + z_{nn}x_n = \langle[\underline{p_n}, \overline{p_n}], [\underline{\underline{p_n}}, \overline{\overline{p_n}}] + i[\underline{q_n}, \overline{q_n}], [\underline{\underline{q_n}}, \overline{\overline{q_n}}]\rangle
\end{aligned}
$$

The parametric complex IFNs are written as complex horizontal relative intuitionistic fuzzy number using RDM variables as

$$
\begin{aligned}
&z_{11}x_1 + z_{12}x_2 + ... + z_{1n}x_n = [\underline{p_1} + \alpha_1(\overline{p_1} - \underline{p_1}), \underline{\underline{p_1}} + \beta_1(\overline{\overline{p_1}} - \underline{\underline{p_1}})]\\
&+i[\underline{q_1} + \alpha_1^\prime(\overline{q_1} - \underline{q_1}), \underline{\underline{q_1}} + \beta_1^\prime(\overline{\overline{q_1}} - \underline{\underline{q_1}})]\\
&z_{21}x_1 + z_{21}x_2 + ... + z_{2n}x_n = [\underline{p_2} + \alpha_2(\overline{p_2} - \underline{p_2}), \underline{\underline{p_2}} + \beta_2(\overline{\overline{p_2}} - \underline{\underline{p_2}})]\\
&+i[\underline{q_2} + \alpha_2^\prime(\overline{q_2} - \underline{q_2}), \underline{\underline{q_2}} + \beta_2^\prime(\overline{\overline{q_2}} - \underline{\underline{q_2}})]\\
&..\\
&z_{n1}x_1 + z_{n2}x_2 + ... + z_{nn}x_n = [\underline{p_n} + \alpha_n(\overline{p_n} - \underline{p_n}), \underline{\underline{p_n}} + \beta_n(\overline{\overline{p_n}} - \underline{\underline{p_n}})]\\
&+i[\underline{q_1} + \alpha_n^\prime(\overline{q_n} - \underline{q_n}), \underline{\underline{q_n}} + \beta_n^\prime(\overline{\overline{q_n}} - \underline{\underline{q_n}})]
\end{aligned}
$$

Solving the converted system with algebraic operations to find $z_i^\prime s$, the set of equations for membership functions is

$$
\begin{aligned}
&z_{11}x_1 + z_{12}x_2 + ... + z_{1n}x_n = [\underline{p_1} + \alpha_1(\overline{p_1} - \underline{p_1}), \underline{\underline{p_1}} + \beta_1(\overline{\overline{p_1}} - \underline{\underline{p_1}})]\\
&z_{21}x_1 + z_{21}x_2 + ... + z_{2n}x_n = [\underline{p_2} + \alpha_2(\overline{p_2} - \underline{p_2}), \underline{\underline{p_2}} + \beta_2(\overline{\overline{p_2}} - \underline{\underline{p_2}})]\\
&..\\
&z_{n1}x_1 + z_{n2}x_2 + ... + z_{nn}x_n = [\underline{p_n} + \alpha_n(\overline{p_n} - \underline{p_n}), \underline{\underline{p_n}} + \beta_n(\overline{\overline{p_n}} - \underline{\underline{p_n}})]
\end{aligned}
$$

The set of equations for non-membership functions is

$$
\begin{aligned}
z_{11}x_1 + z_{12}x_2 + \ldots\ldots + z_{1n}x_n &= [\underline{q_1} + \alpha'_1(\overline{q_1} - \underline{q_1}), \underline{\underline{q_1}} + \beta'_1(\overline{\overline{q_1}} - \underline{\underline{q_1}})] \\
z_{21}x_1 + z_{21}x_2 + \ldots\ldots + z_{2n}x_n &= [\underline{q_2} + \alpha'_2(\overline{q_2} - \underline{q_2}), \underline{\underline{q_2}} + \beta'_2(\overline{\overline{q_2}} - \underline{\underline{q_2}})] \\
&\ldots\ldots\ldots\ldots\ldots\ldots\ldots\ldots\ldots\ldots \\
z_{n1}x_1 + z_{n2}x_2 + \ldots\ldots + z_{nn}x_n &= [\underline{q_1} + \alpha'_n(\overline{q_n} - \underline{q_n}), \underline{\underline{q_n}} + \beta'_n(\overline{\overline{q_n}} - \underline{\underline{q_n}})]
\end{aligned}
$$

solving the above equations algebraically, we get the solution $z_i's$, in the form $z_i = z_i(r, r', \alpha, \beta)$ as a multi-dimensional solution. For each varying α, β we get values of z.

4.1 Numerical Example

Consider a complex trapezoidal intuitionistic fuzzy system of equations in parametric form,

$$
\begin{aligned}
&z_1 - z_2 = \langle(25 + 10r, 67 - 17r), (35 + 5r', 50 + 23r')) \\
&+ i((24 + 5r, 60 - 15r), (30 + 2r', 45 + 12r')\rangle \\
&z_1 + 3z_2 = \langle(27 + 10r, 65 - 17r), (37 - 15r', 48 + 24r')) \\
&+ i((25 + 5r, 60 - 15r), (35 - 7r', 12 + 2r')\rangle
\end{aligned}
$$

Let the first set of equations for the membership values be,

$$
\begin{aligned}
z_1 - z_2 &= \langle(25 + 10r, 67 - 17r)) + i((24 + 5r, 60 - 15r)\rangle \\
z_1 + 3z_2 &= \langle(27 + 10r, 65 - 17r)) + i((25 + 5r, 60 - 15r)\rangle
\end{aligned}
$$

Writing parametric complex IFNs using RDM variables, the above equations become,

$$
\begin{aligned}
z_1 - z_2 &= (25 + 10r) + \alpha_1(42 - 27r) + i((24 + 5r) + \alpha_2(36 - 20r)) \\
z_1 + 3z_2 &= (27 + 10r) + \alpha_3(38 - 27r) + i((25 + 5r) + \alpha_4(35 - 20r))
\end{aligned}
$$

Solving the above equations algebraically, the solution is

$$
\begin{aligned}
z_1 &= \frac{(102 + 40r) + \alpha_1(126 - 81r) + \alpha_3(38 - 27r) + i((97 + 20r) + \alpha_2(108 - 60r) + \alpha_4(35 - 20r))}{4} \\
z_2 &= \frac{2 - \alpha_1(42 - 27r) + \alpha_3(38 - 27r) + i(1 - \alpha_2(36 - 20r) + \alpha_4(35 - 20r))}{4}
\end{aligned}
$$

Let the second set of equations for the non-membership values be,

$$
\begin{aligned}
z_1 - z_2 &= \langle(35 + 5r', 50 + 23r') + i(30 + 2r', 45 + 12r')\rangle \\
z_1 + 3z_2 &= \langle(37 - 15r', 48 + 24r') + i(35 - 7r', 12 + 2r')\rangle
\end{aligned}
$$

Rewriting the above complex IFNs using RDM variables and solving algebraically to obtain the solution as,

$$z_1 = \frac{142 + \beta_1(45 + 54r') + \beta_3(11 + 39r') + i((125 - r') + \beta_2(45 + 30r') + \beta_4(9r' - 23))}{4}$$

$$z_2 = \frac{(2 - 20r') - \beta_1(15 + 18r') + \beta_3(11 + 39r') + i((5 - 9r') - \beta_2(15 + 10r') + \beta_4(23 - 9r'))}{4}$$

z_1, z_2 are multi-dimensional solutions, for each varying μ, ν, α, β varying z_1 and z_2 values are obtained.

The final solution of the given complex intuitionistic fuzzy linear system is

$$z_1 = \frac{(102 + 40r) + \alpha_1(126 - 81r) + \alpha_3(38 - 27r) + i((97 + 20r) + \alpha_2(108 - 60r) + \alpha_4(35 - 20r))}{4},$$
$$\frac{142 + \beta_1(45 + 54r') + \beta_3(11 + 39r') + i((125 - r') + \beta_2(45 + 30r') + \beta_4(9r' - 23))}{4}$$

$$z_2 = \frac{2 - \alpha_1(42 - 27r) + \alpha_3(38 - 27r) + i(1 - \alpha_2(36 - 20r) + \alpha_4(35 - 20r))}{4},$$
$$\frac{(2 - 20r') - \beta_1(15 + 18r') + \beta_3(11 + 39r') + i((5 - 9r') - \beta_2(15 + 10r') + \beta_4(23 - 9r'))}{4}$$

4.2 Complex Intuitionistic Fuzzy Linear Systems in an Electrical Circuit

Consider an intuitionistic fuzzified version of a simple resistive circuit from [11] with intuitionistic fuzzy current and intuitionistic fuzzy source as shown in the Fig. 4. The system of equations for this circuit is

$$(10 - 7.5j)\, I_1 - (6 - 5j)\, I_2 = \left\langle (4 + r, 6 - r), (2 + r', 3 - r') \right\rangle$$
$$+ j \left\langle (-1 + r, 1 - r), (1 + r', 2 - r') \right\rangle$$
$$(-6 + 5j)\, I_1 + (16 + 3j)\, I_2 = \left\langle (-2 + r, -r), (3 + r', -2 - r') \right\rangle$$
$$+ j \left\langle (-3 + r, -1 - r), (-2 + r', -1 - r') \right\rangle$$

Let the first set of equations for the membership values be,

$$(10 - 7.5j)\, I_1 - (6 - 5j)\, I_2 = \langle (4 + r, 6 - r) \rangle + j \langle (-1 + r, 1 - r) \rangle$$
$$(-6 + 5j)\, I_1 + (16 + 3j)\, I_2 = \langle (-2 + r, -r) \rangle + j \langle (-3 + r, -1 - r) \rangle$$

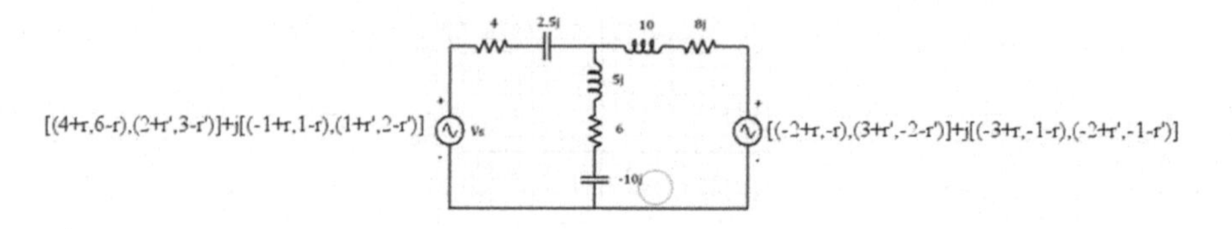

Fig. 4. A RLC circuit

The above parametric complex intuitionistic fuzzy equations are written as complex horizontal relative intuitionistic fuzzy numbers using RDM variables as

$$(10-7.5j)I_1-(6-5j)I_2=\langle(4+r)+\alpha_1(2-2r)+j(-1+r+\alpha_2(2-2r))\rangle$$
$$(-6+5j)I_1+(16+3j)I_2=\langle(-2+r)+\alpha_3(2-2r)+j(-3+r+\alpha_4(2-2r)\rangle$$

Solving the above equations algebraically, to obtain the current flow I_1 and I_2 for membership values of voltage as

$$I_1=<0.2283+0.1555r+0.0935\alpha_1(2-2r)-0.00113\alpha_2(2-2r)$$
$$+0.0289\alpha_3(2-2r)-0.0342\alpha_4(2-2r)>+j<0.0509-0.0894r+0.00113\alpha_1(2-2r)$$
$$-0.0935\alpha_2(2-2r)+0.0342\alpha_3(2-2r)-0.0289\alpha_4(2-2r)>$$
$$I_2=<0.1622+0.01633r+0.0388\alpha_1(2-2r)+0.02235\alpha_2(2-2r)$$
$$-0.04915\alpha_3(2-2r)-0.0523\alpha_4(2-2r)>+j<0.2106-0.0233r+0.0223\alpha_1(2-2r)$$
$$-0.0388\alpha_2(2-2r)+0.0523\alpha_3(2-2r)-0.0558\alpha_4(2-2r)>$$

Let the second set of equations for the non-membership values be,

$$(10-7.5j)\,I_1-(6-5j)\,I_2=\langle(2+r',3-r')\rangle+j\,\langle(1+r',2-r')\rangle$$
$$(-6+5j)\,I_1+(16+3j)\,I_2=\langle(3+r',-2-r')\rangle+j\,\langle(-2+r',-1-r')\rangle$$

The above parametric complex intuitionistic fuzzy equations are written as complex horizontal relative intuitionistic fuzzy numbers using RDM variables as

$$(10-7.5j)\,I_1-(6-5j)\,I_2=\left\langle(2+r')+\beta_1(1-2r')+j((1+r')+\beta_2(1-2r'))\right\rangle$$
$$(-6+5j)\,I_1+(16+3j)\,I_2=\left\langle(-2+r)+\beta_3\left(1-2r'\right)+j((-3+r)+\beta_4\left(1-2r'\right))\right\rangle$$

Solving the above equations algebraically, to obtain the current flow I_1 and I_2 for non-membership values of voltage as

$$I_1=<0.000003+0.00079r'+0.000528\beta_1(1-2r')-0.00009\beta_2(1-2r')$$
$$+0.00019\beta_3(1-2r')+0.00016\beta_4(1-2r')>+j<0.000825-0.000032r'$$
$$+0.00009\beta_1(1-2r')-0.00052\beta_2(1-2r')-0.00016\beta_3(1-2r')+0.000198\beta_4(1-2r')>e$$
$$I_2=<0.1569-0.1577r'-0.0388\beta_1(1-2r')-0.0223\beta_2(1-2r')$$
$$-0.064\beta_3(1-2r')-0.0325\beta_4(1-2r')>+j<0.0362-0.048r'+0.0223\beta_1(1-2r')$$
$$-0.0388\beta_2(1-2r')+0.0325\beta_3(1-2r')-0.064\beta_4(1-2r')>$$

By giving values for $\alpha's$ $\beta's$, r and r', the current flow at various points in the circuit can be calculated. When α, β, r and r' is 0, $I_1=0.2283+j0.059, I_2=0.1622+j0.2106$ are current flow values for membership values of voltage. $I_1=0.000003+j0.0008, I_2=0.1569+j0.0362$ are current flow values for non-membership values of voltage. The voltage dips, swells or transients which are likely to occur in any circuit that affects or changes the current flow can be predicted accurately using these complex horizontal relative trapezoidal intuitionistic fuzzy numbers.

5 Conclusion

This paper presents a new concept of complex horizontal trapezoidal intuitionistic fuzzy numbers and used the same in complex linear system of equations which is further applied in a RLC intuitionistic fuzzy circuit to find the current flow in the circuit. The key idea of the paper is to provide an easy but accurate method for solving complex system of equations. Introducing CHRTrIFN and its geometrical representation will prove to be a promising research in further application areas of engineering problems. The advantage of CHRTrIFN over other intuitionistic fuzzy numbers is that it enables relatively easy aggregation of crisp and intuitionistic fuzzy values together in arithmetic operations in the complex plane which is a new idea and can be interpreted easily even by the multi-disciplinary researchers. The applicability of CHRTrIFN in various engineering problems will reduce the computational complexity to a great extent in the problems with simultaneous uncertainity and periodicity. Further work is directed along applying HRTrIFNS and CHRTrIFNs in fluid flow problems.

References

1. Kaufmann, A., Gupta, M.M.: Introduction to fuzzy arithmetic. Elsevier Science Publishers, Van Nostrand Reinhold, Newyork (1991)
2. Atanassov, K.T.: Intuitionistic Fuzzy Sets: Theory and Applications. Physica-Verlag, Heidelberg (1999)
3. Buckley, J.: Fuzzy complex numbers. Fuzzy Sets Syst. **33**, 333–345 (1989)
4. Burillo, P.B., Mohedano, V.: Some definition of intuitionistic fuzzy number. Fuzzy Based Expert Systems. Fuzzy Bulgarian Enthusiasts, 28–30 (1994)
5. Landowski M.: Decomposition method for calculations on intuitionistic fuzzy numbers. In: Atanassov K., et al. (eds.) Uncertainty and Imprecision in Decision Making and Decision Support: New Challenges, Solutions and Perspectives. IWIFSGN 2018. Advances in Intelligent Systems and Computing, vol. 1081, pp. 58–68. Springer, Cham (2021). https://doi.org/10.1007/978-3-030-47024-1_7
6. Moore, R.E.: Interval Analysis. Prentice Hall, India (1996)
7. Parvathi R., Akila Padmasree, J.: Complex trapezoidal intuitionistic fuzzy numbers. Notes Intuitionistic Fuzzy Sets **24**, 50–62 (2018). https://doi.org/10.7546/nifs.2018.24.4.50-62
8. Piegat, A., Landowski, M.: Fuzzy arithmetic type 1 with horizontal membership functions. In: Kreinovich, V. (ed.) Uncertainty Modeling. Studies in Computational Intelligence, vol. 683, pp. 233–250. Springer, Cham (2017). https://doi.org/10.1007/978-3-319-51052-1_14
9. Piegat, A., Landowski, M.: Is an interval the right result of arithmetic operations on intervals? Int. J. Appl. Math. Comput. Sci. **27**, 575–590 (2017)
10. Piegat, A., Landowski, M.: On fuzzy RDM-arithmetic. In: Kobayashi S., Piegat, A., Pejas, J., El Fray, I., Kacprzyk, J. (eds.) Hard and Soft Computing for Artificial Intelligence, Multimedia and Security. ACS 2016. Advances in Intelligent Systems and Computing, vol. 534, pp. 3–16. Springer, Cham (2017). https://doi.org/10.1007/978-3-319-48429-7_1
11. Rahoogay, T., Sadoghi, H., Yazdi, R.M.: Fuzzy complex system of linear equations applied to circuit analysis. Int. J. Comput. Electr. Eng. **1**, 1793–8163 (2009)

Intuitionistic Fuzzy Index-Matrix Selection for the Outsourcing Providers at a Refinery

Velichka Traneva(✉) and Stoyan Tranev

Asen Zlatarov University, Yakimov Blvd, 8000 Bourgas, Bulgaria
v_traneva@btu.bg, tranev@abv.bg
http://www.btu.bg

Abstract. Outsourcing is the transfer of a business process that has been traditionally operated and managed internally to external service providers. In order to reduce costs and improve their core competitiveness, many companies tend to choose the outsourcing of some services. The problem of this type belongs to multicriteria decision making. The imprecision in this problem may arise from the nature of the characteristics of the candidates for the outsourcing service providers, which can be unavailable or indeterminate. It may also be derived from the inability of the experts to formulate a precise evaluation. Intuitionistic fuzzy sets (IFSs) are the stronger tool for dealing with uncertain data than fuzzy ones because, unlike them, they have a degree of hesitancy.

Here, we will formulate an intuitionistic fuzzy generalized optimal outsoursing problem and an algorithm for selection the most effective outsourcing service providers is proposed using the consepts of index matrices (IMs) and intuitionistic fuzzy logic. The proposed decision model takes the ratings of the experts and the weighting factors of the evaluation criteria according to their priorities for the outsourcing services into consideration. The main contributions of the paper are: its proposition for intuitionistic fuzzy approach for selecting the most suitable candidates for outsourcing providers using the concept of IMs on the one hand, and its application for the outsourcing problem of a refinery.

Keywords: Index matrix · Intuitionistic fuzzy sets · Outsourcing

1 Introduction

Outsourcing is the use of external companies to perform functions which have traditionally been performed within an organization, which can lead to an increase in profit and customer satisfaction [21]. In Group Decision Making (GDM) [5] a set of experts participates in a decision process concerning the selection of the best alternative among a set of predefined ones. The uncertainty in

Supported by the project of Asen Zlatarov University under Ref. No. NIX-440/2020 "Index matrices as a tool for knowledge extraction".

C. Kahraman et al. (Eds.): INFUS 2021, LNNS 308, pp. 119–128, 2022.
https://doi.org/10.1007/978-3-030-85577-2_14

the GDM-problem may be caused by the unavailable or indeterminate characteristics of the candidates for the outsourcing service providers or from the inability of the experts to formulate a precise evaluation [24]. Decision making methods about outsourcing are multi-attribute decision making methods [14,21], such as AHP [22], ANP [17], PROMETHEE [11], balanced score card [17], TOPSIS [11] and TODIM approach [23]. Wang and Yang [22] are proposed the method by using AHP and PROMETHEE for outsourcing decisions. Kahraman et al. [13] are proposed an approach using hesitant linguistic term sets for supplier selection problem. Araz et al. [1] are described an outsourcing model in which the PROMETHEE method and then fuzzy goal programming are used. Liu and Wang [15] are developed a fuzzy linear programming approach by integrating Delphi method. Hsu et al. [12] are describes a hybrid decision model for selection of outsourcing companies, based on DEMATEL, ANP, and grey relation methods. In [21] are presented some disadvantages of the methods as follows:

1) The researches are used the crisp numbers [17,22] or fuzzy sets [15] to represent the information. The IFSs [5] have the hesitancy degree and they are more flexible than the fuzzy sets [25] in dealing with uncertainty.
2) The methods [15,22] are considered an expert decision maker.
3) The fuzzy linear programming method [15] are used fuzzy data.

An intuitionistic fuzzy (IF) linear programming method is described for outsourcing problems in [21]. In [14], an interval-valued IF AHP and TOPSIS concepts are proposed to select of the best outsource manufacturers. This paper proposes a generalized approach (IFIMOA), based on the concepts of IMs [3] and IF logic [2,5], for the optimal selection of outsourcing service providers. The evaluations of the candidates are IF pairs (IFPs, [8]). IFIMOA takes the ratings of the experts and the weighting factors of the evaluation criteria according to their priorities for the outsourcing services into consideration. A numerical case study for selecting the best outsourcing service providers in a refinery on the Balkan Peninsula is also demonstrated. The main contributions of this work are that it formulates an IF problem for the selection of outsourcing service providers and developes a new algorithm for its solution. The rest of the paper contains 4 sections as follows: Sect. 2 recalls some definitions from the concepts of IF logic and IMs. In Sect. 3, is formulated an IF problem for the selection of outsourcing providers and developed an algorithm for its solution. Finally, in Sect. 4, the proposed approach is applied on an IF outsourcing problem in a refinery. Section 5 gives the conclusion and aspects for future research.

2 Basic Definitions of IF Logic and IMs

This section marks some definitions of the concepts of IF logic [5] and IMs [4].

2.1 Intuitionistic Fuzzy Pairs (IFPs)

The **IFP** is in the form of a pair $\langle a, b\rangle = \langle \mu(p), \nu(p)\rangle$, where $a, b \in [0, 1]$ and $a+b \leq 1$, that is used as an evaluation of a proposition p [8]. The basic operations and relations with IFPs $x = \langle a, b\rangle$ and $y = \langle c, d\rangle$ are described in [5,8,16]:

$$\begin{array}{rl} \neg x = \langle b, a\rangle; & x \wedge_1 y = \langle \min(a,c), \max(b,d)\rangle \\ x \vee_1 y = \langle \max(a,c), \min(b,d)\rangle; & x \wedge_2 y = x + y = \langle a + c - a.c, b.d\rangle \\ x@y = \langle \frac{a+c}{2}, \frac{b+d}{2}\rangle; & x \vee_2 y = x.y = \langle a.c, b + d - b.d\rangle \end{array} \quad (1)$$

$$\begin{array}{llll} x \geq y \text{ iff } & a \geq c \text{ and } b \leq d; & x \geq_{\Box} y \text{ iff } & a \geq c \\ x \geq_{\Diamond} y \text{ iff } & b \leq d & x \geq_R y \text{ iff } & R_{\langle a,b\rangle} \leq R_{\langle c,d\rangle}, \end{array} \quad (2)$$

where $R_{\langle a,b\rangle} = 0.5(2 - a - b)(1 - a)$.

2.2 Definition and Operations over 3-D Intuitionistic Fuzzy Index Matrices (3-D IFIM)

Let $\mathcal{I}$ be a set. By 3-D IFIM $[K, L, H, \{\langle \mu_{k_i,l_j,h_g}, \nu_{k_i,l_j,h_g}\rangle\}]$, $(K, L, H \subset \mathcal{I})$ and elements, which are IFPs, we denote the object [4,18]:

$$\begin{array}{c|ccccc} h_g \in H & l_1 & \dots & l_j & \dots & l_n \\ \hline k_1 & \langle \mu_{k_1,l_1,h_g}, \nu_{k_1,l_1,h_g}\rangle & \dots & \langle \mu_{k_1,l_j,h_g}, \nu_{k_1,l_j,h_g}\rangle & \dots & \langle \mu_{k_1,l_n,h_g}, \nu_{k_1,l_n.h_g}\rangle \\ \vdots & \vdots & \dots & \vdots & \dots & \vdots \\ k_m & \langle \mu_{k_m,l_1,h_g}, \nu_{k_m,l_1,h_g}\rangle & \dots & \langle \mu_{k_m,l_j,h_g}, \nu_{k_m,l_j,h_g}\rangle & \dots & \langle \mu_{k_m,l_n,h_g}, \nu_{k_m,l_n,h_g}\rangle \end{array} \quad (3)$$

Let $\mathcal{X}, Y, Z, U$ be sets and "$*$", and "$\circ$" be defined so that: $* : \mathcal{X} \times \mathcal{Y} \to \mathcal{Z}$ and $\circ : \mathcal{Z} \times \mathcal{Z} \to \mathcal{U}$. The basic operations over $A = [K, L, H, \{\langle \mu_{k_i,l_j,h_g}, \nu_{k_i,l_j,h_g}\rangle\}]$ and $B = [P, Q, R, \{\langle \rho_{p_r,q_s,r_d}, \sigma_{p_r,q_s,r_d}\rangle\}]$, defined in [4,18] are:

Addition-(max,min)

$$A \oplus_{(\max,\min)} B = [K \cup P, L \cup Q, H \cup R, \{\langle \phi_{t_u,v_w,x_y}, \psi_{t_u,v_w,x_y}\rangle\}], \quad (4)$$

where $\langle \phi_{t_u,v_w,x_y}, \psi_{t_u,v_w,x_y}\rangle$ is defined in [4].

Transposition: A^T is the transposed IM of A.
Multiplication with a constant

$$\alpha A = [K, L, H\{\alpha\langle \mu_{k_i,l_j,h_g}, \nu_{k_i,l_j,h_g}\rangle\}] \quad (5)$$

Multiplication:

$$A \odot_{(\circ,*)} B = [K \cup (P - L), Q \cup (L - P), H \cup R, \{\langle \phi_{t_u,v_w,x_y}, \psi_{t_u,v_w,x_y}\rangle\}], \quad (6)$$

$\langle \phi_{t_u,v_w,x_y}, \psi_{t_u,v_w,x_y}\rangle$ is defined in [4] and $\langle \circ, *\rangle \in \{\langle max, min\rangle, \langle min, max\rangle\}$.

Aggregation operation by one dimension [20]

$$\alpha_{K,\#_q}(A, k_0) = \begin{array}{c|ccc} & l_1 & \dots & l_n \\ \hline k_0 & \overset{m}{\underset{i=1}{\#_q}} \langle \mu_{k_i,l_1}, \nu_{k_i,l_1}\rangle & \dots & \overset{m}{\underset{i=1}{\#_q}} \langle \mu_{k_i,l_n}, \nu_{k_i,l_n}\rangle \end{array}, (1 \leq q \leq 10) \quad (7)$$

If we use $\#_1^*$ we obtain super pessimistic aggregation operation, etc., $\#_{10}^*$ - super optimistic aggregation operation.

Projection: Let $M \subseteq K$, $N \subseteq L$ and $U \subseteq H$. Then, $pr_{M,N,U}A = [M, N, U, \{b_{k_i,l_j,h_g}\}]$, and for each $k_i \in M, l_j \in N$ and $h_g \in U, b_{k_i,l_j,h_g} = a_{k_i,l_j,h_g}$.
Substitution: A substitution over A is defined for the pair of indices (p, k_i) by

$$\left[\frac{p}{k_i}; \perp; \perp\right] A = [(K - \{k_i\}) \cup \{p\}, L, H, \{a_{k_i,l,h}\}] \tag{8}$$

3 Intuitionistic Fuzzy Index-Matrix Selection for the Outosourcing Providers at a Company

Here, we will formulate an IF generalized optimal autsoursing problem.

In order to reduce its costs, a company has decided to provide part of its services $v_e (1 \leq e \leq u)$ to outsourcing providers. It has developed an evaluation system of the candidates $\{k_1, \ldots, k_i, \ldots, k_m\}$ (for $i = 1, ..., m$) for the respective outsourcing service $v_e (1 \leq e \leq u)$, containing criteria $\{c_1, \ldots, c_j, \ldots, c_n\}$ (for $j = 1, ..., n$). Experts $\{d_1, \ldots, d_s, \ldots, d_D\}$ (for $s = 1, ..., D$) in the organization assess the weighting coefficients under the form of IFPs of the assessment criteria c_j (for $j = 1, ..., n$) according to their priority for the service v_e - pk_{c_j,v_e} (for $j = 1, ..., n$). The ratings of the experts $\{r_1, \ldots, r_s, \ldots, r_D\}$ be given, where $r_s = \langle \delta_s, \epsilon_s \rangle$ is an IFP $(1 \leq s \leq D)$ and let the number of his own participations in expert investigations be equal to $\gamma_s (s = 1, ..., D)$, respectively. All applicants for the outsourcing activities need to be evaluated by a team of experts according to the established criteria in the company at the present moment h_f of applying for outsourcing service provider $v_e (1 \leq e \leq u)$, and their evaluations ev_{k_i,c_j,d_s} (for $1 \leq i \leq m, 1 \leq j \leq n, 1 \leq s \leq D$) are IFPs. The aim of the problem is to optimally allocate outsourcing service providers among the candidates for them.

For solution of this problem we propose IFIMOA-approach, described with mathematical notation and pseudocode, as follows:

Step 1. This step creates an expert 3-D evaluation IM EV. At the beginning of IFIMOA, it is possible for the experts to include assessments for the same candidates from a previous evaluation IM according to the criteria in previous procedures at time points $h_1, ..., h_g, ..., h_{f-1}$. The team of experts needs to evaluate the candidates for the services according to the approved criteria in the company at the current time moment h_f. The experts are not sure about for their evaluations according to given criteria due to changes in some uncontrollable factors. The evaluations are transformed under the form of IFPs as follows:

Let us have the set of intervals of the expert evaluations by all criteria for all candidates at a current moment h_f $[p^{1,f}_{k_i,c_j,d_s}; p^{2,f}_{k_i,c_j,d_s}]$ and let

$$A_{min,i,j,s,f} = \min_{1 \leq i \leq m, 1 \leq j \leq n, 1 \leq s \leq D} p^{1,f}_{k_i,c_j,d_s} < \max_{1 \leq i \leq m, 1 \leq j \leq n, 1 \leq s \leq D} p^{2,f}_{k_i,c_j,d_s} = A_{max,i,j,s,f} \tag{9}$$

For the interval $[p^{1,f}_{k_i,c_j,d_s}; p^{2,f}_{k_i,c_j,d_s}]$, at a time moment h_f, we construct the evaluation of the d_s-th expert for the k_i-th candidate by the c_j-th criterion under the form of IFP [5] as follows:

$$\mu_{k_i,c_j,d_s,h_f} = \frac{p^{1,f}_{k_i,c_j,d_s} - A_{min,i,j,s,f}}{A_{max,i,j,s,f} - A_{min,i,j,s,f}}, \nu_{k_i,c_j,d_s,h_f} = \frac{A_{max,i,j,s,f} - p^{2,f}_{k_i,c_j,d_s}}{A_{max,i,j,s,f} - A_{min,i,j,s,f}} \quad (10)$$

It is possible that some of the experts' assessments are incorrect from an IF point of view. In [5], different ways for altering incorrect experts' estimations are discussed. Let us propose that, the estimations of the $d_s (1 \leq s \leq D)$ expert are correct and are described by the IFIM $EV_s = [K, C, H, \{ev_{k_i,c_j,d_s,h_g}\}]$ as follows:

$$EV_s = \begin{array}{c|ccc} h_g \in H & c_1 & \dots & c_n \\ \hline k_1 & \langle \mu_{k_1,c_1,d_s,h_g}, \nu_{k_1,c_1,d_s,h_g} \rangle & \dots & \langle \mu_{k_1,c_n,d_s,h_g}, \nu_{k_1,c_n,d_s,h_g} \rangle \\ \vdots & \vdots & \ddots & \vdots \\ k_m & \langle \mu_{k_m,c_1,d_s,h_g}, \nu_{k_m,c_1,d_s,h_g} \rangle & \dots & \langle \mu_{k_m,c_n,d_s,h_g}, \nu_{k_m,c_n,d_s,h_g} \rangle \end{array}, \quad (11)$$

where $K = \{k_1, k_2, \dots, k_m\}$, $C = \{c_1, c_2, \dots, c_n\}$, $H = \{h_1, h_2, \dots, h_f\}$ and the element $\{ev_{k_i,c_j,d_s,h_g}\}$ is the estimate of the d_s-th expert for the k_i-th candidate by the c_j-th criterion at a moment h_g. Let us apply the α_H-th aggregation operation (7) to find the aggregated evaluation of the d_s-th expert ($s = 1, ..., D$).

$$\alpha_{EV_s,\#_q} = \begin{array}{c|ccc} d_s & c_1 & \dots & c_n \\ \hline k_1 & \underset{g=1}{\overset{f}{\#_q}} \langle \mu_{k_1,c_1,d_s,h_g}, \nu_{k_1,c_1,d_s,h_g} \rangle & \dots & \underset{g=1}{\overset{f}{\#_q}} \langle \mu_{k_1,c_n,d_s,h_g}, \nu_{k_1,c_n,d_s,h_g} \rangle \\ \vdots & \vdots & \ddots & \vdots \\ k_m & \underset{g=1}{\overset{f}{\#_q}} \langle \mu_{k_m,c_1,d_s,h_g}, \nu_{k_m,c_1,d_s,h_g} \rangle & \dots & \underset{g=1}{\overset{f}{\#_q}} \langle \mu_{k_m,c_n,d_s,h_g}, \nu_{k_m,c_n,d_s,h_g} \rangle \end{array}, \quad (12)$$

where $1 \leq q \leq 10$. Then we create aggregated 3-D IFIM $EV[K, C, E, \{ev_{k_i,c_j,d_s}\}]$ with the evaluations of all experts for all candidates:

$$EV = \alpha_{EV_1,\#_q}(H, d_1) \oplus_{(max,min)} \oplus_{(max,min)} \dots \oplus_{(max,min)} \alpha_{EV_D,\#_q}(H, d_D) \quad (13)$$

Go to *Step 2.*

Step 2. Let the score (rating) r_s of the d_s-th expert ($d_s \in E$) be specified by an IFP $\langle \delta_s, \epsilon_s \rangle$. δ_s and ϵ_s are interpreted respectively as his degree of competence and of incompetence. Then we create $EV^*[K, C, E, \{ev^*_{k_i,c_j,d_s}\}]$:

$$EV^* = r_1 pr_{K,C,d_1} EV \oplus_{(max,min)} r_2 pr_{K,C,d_2} EV \dots \oplus_{(max,min)} r_D pr_{K,C,d_D} EV; \quad (14)$$

$$EV := EV^*(ev_{k_i,l_j,d_s} = ev^*_{k_i,l_j,d_s}, \forall k_i \in K, \forall l_j \in L, \forall d_s \in E).$$

Then α_E-th aggregation operation is applied to find the aggregated assessment of the k_i-th candidate against the c_j-th criterion at the moment $h_f \notin E$:

$$R = \alpha_{E,\#_q}(EV, h_f) = \left\{ \begin{array}{c|c} c_j & h_f \\ \hline k_1 & \underset{s=1}{\overset{D}{\#_q}} \langle \mu_{k_1,c_j,d_s}, \nu_{k_1,c_j,d_s} \rangle \\ \vdots & \vdots \\ k_m & \underset{s=1}{\overset{D}{\#_q}} \langle \mu_{k_m,c_j,d_s}, \nu_{k_m,c_j,d_s} \rangle \end{array} \mid c_j \in C \right\}, (1 \leq q \leq 10) \quad (15)$$

Go to *Step 3.*

Step 3. Let us define the 3-D IFIM PK of the weight coefficients of the assessment criterion according to its priority to the outsourcing service v_e $(1 \leq e \leq u)$:

$$PK[C, V, h_f, \{pk_{c_j,v_e,h_f}\}] = \begin{array}{c|ccccc} h_f & v_1 & \ldots & v_e & \ldots & v_u \\ \hline c_1 & pk_{c_1,v_1,h_f} & \ldots & pk_{c_1,v_e,h_f} & \ldots & pk_{c_1,v_u,h_f} \\ \vdots & \vdots & \ddots & \vdots & \ddots & \vdots \\ c_j & pk_{c_j,v_1,h_f} & \ldots & pk_{c_j,v_e,h_f} & \ldots & pk_{c_j,v_u,h_f} \\ \vdots & \vdots & \ddots & \vdots & \ddots & \vdots \\ c_n & pk_{c_n,v_1,h_f} & \ldots & pk_{c_n,v_e,h_f} & \ldots & pk_{c_n,v_u,h_f} \end{array}, \quad (16)$$

where $C = \{c_1, \ldots, c_n\}$, $V = \{v_1, \ldots, v_u\}$ and all elements pk_{c_j,v_e,h_f} are IFPs.

The transposed IM of R is founded under the form $R^T[K, C, h_f]$ and is calculated 3-D IFIM $B[K, V, h_f, \{b_{k_i,v_e,h_f}\}] := R^T \odot_{(\circ,*)} PK$, which contains the cumulative estimates of the k_i-th candidate (for $1 \leq i \leq m$) for the v_e-th outsourcing service. If a candidate $k_i (1 \leq i \leq m)$ does not wish to participate in the competition to provide an outsourcing service v_e, then the element b_{k_i,v_e,h_f} is equal to $\langle 0, 1 \rangle$. Go to *Step 4.*

Step 4. The aggregation operation $\alpha_{K,\#_q}(B, k_0)$ is applied by the dimension K to find the most suitable candidate for the outsourcing service v_e.

$$al_{K,\#_q}(B, k_0) = \begin{array}{c|ccc} & v_1 & \ldots & v_u \\ \hline k_0 & \underset{i=1}{\overset{m}{\#_q}} \langle \mu_{k_i,v_1}, \nu_{k_i,v_1} \rangle & \ldots & \underset{i=1}{\overset{m}{\#_q}} \langle \mu_{k_i,v_u}, \nu_{k_i,v_u} \rangle \end{array}, \quad (17)$$

where $k_0 \notin K, 1 \leq q \leq 10$. If the company wants each outsourcing service to be performed by an individual candidate, an IF Hungarian algorithm for finding an optimal assignment for the candidates [19] can be applied. Go to *Step 5.*

Step 5. At this step of the algorithm, we need to determine whether there are correlations between some of the evaluation criteria. The procedure in IFS-case, based on the intercriteria analysisis (ICrA, [7]) is discussed in [9]. Let

$\alpha, \beta \in [0,1]$ be given, so that $\alpha + \beta \leq 1$. The criteria C_k and C_l are in (α, β)-positive consonance, if $\mu_{C_k,C_l} > \alpha$ and $\nu_{C_k,C_l} < \beta$; (α, β)-negative consonance, if $\mu_{C_k,C_l} < \beta$ and $\nu_{C_k,C_l} > \alpha$; (α, β)-dissonance, otherwise. After application of the ICrA over IFIM R we determine which criteria are in a consonance. Then, we can evaluate their complexity and the criterion with a higher complexity can be eliminated from the future decision making process. If $O = \{O_1, ..., O_V\}$ is the set of the criteria that can be omitted, then we can reduce R by IM-operation $R* = R_{(O,\perp)}$. Go to *Step 6.*

Step 6. The last step determinates the new rating coefficients of the experts. Let the expert d_s $(s = 1, ..., D)$ is participated in γ_s procedures, on the basis of which his score $r_s = \langle \delta_s, \epsilon_s \rangle$ is determined, then after his participation in $(\gamma_s + 1)$-th procedure his score will be determined by [5]:

$$\langle \delta_s^{'}, \epsilon_s^{'} \rangle = \begin{cases} \langle \frac{\delta\gamma+1}{\gamma+1}, \frac{\epsilon\gamma}{\gamma+1} \rangle, & \text{if the expert's estimation is correct} \\ \langle \frac{\delta\gamma}{\gamma+1}, \frac{\epsilon\gamma}{\gamma+1} \rangle, & \text{if the expert had not given any estimation} \\ \langle \frac{\delta\gamma}{\gamma+1}, \frac{\epsilon\gamma+1}{\gamma+1} \rangle, & \text{if the expert's estimation is incorrect} \end{cases} \quad (18)$$

The complexity of the algorithm whithout step 5 is $O(Dmn)$ (the complexity of the ICrA in the step 5 is $O(m^2n^2)$ [10]).

4 Application of the IFIMOA in a Refinery

In this section, the proposed IFIMOA approach from the Sect. 3 is applied to a real case study in a refinery: A refinery plans to outsource three of its production activities as follows: v_1 - construction and installation works related to the replacement of obsolete and physically worn out equipment of its objects; v_2 - information technology provision for the its organizations and v_3 - management of its production laboratories. For this purpose, the refinery invites a team of the experts d_1, d_2 and d_3 to evaluate the candidates k_i (for $1 \leq i \leq 4$) for the autosourcing refinery services. The real evaluation system of outsoursing providers selection is determined on the basis of 5 criteria as follows: C_1 - compliance of the outsourcing service provider with its corporate culture; C_2 - understanding of the outsourcing service by the provider; C_3 - necessary resources of the outsourcing provider for the implementation of the service; C_4 - price of the provided service; C_5 - opportunity for strategic development of the outsourcing service together with the outsourcing-assignor. The IF weighting coefficients for the service v_e - pk_{c_j,v_e} for the criteria c_j (for $j = 1, ..., 5$) according to their priority for the service v_e $(e = 1, 2, 3)$ and the ratings of the experts $\{r_1, r_2, r_3\}$ be given. The aim of the problem is to optimally select the autosourcing providers.

Solution of the Problem:
Step 1. A 3-D expert evaluation IFIM $EV[K, C, E, \{es_{k_i,c_j,d_s}\}]$ is created and IFP $\{ev_{k_i,c_j,d_s}\}$ (for $1 \leq i \leq 4, 1 \leq j \leq 5, 1 \leq s \leq 3$) is the estimate of the d_s-th expert for the k_i-th candidate by the c_j-th criterion (see Fig. 1):

$$\left\{ \begin{array}{c|ccccc} d_1 & c_1 & c_2 & c_3 & c_4 & c_5 \\ \hline k_1 & \langle 0.4, 0.3 \rangle & \langle 0.7, 0.1 \rangle & \langle 0.3, 0.2 \rangle & \langle 0.5, 0.1 \rangle & \langle 0.6, 0.0 \rangle \\ k_2 & \langle 0.7, 0.1 \rangle & \langle 0.5, 0.1 \rangle & \langle 0.5, 0.2 \rangle & \langle 0.3, 0.4 \rangle & \langle 0.7, 0.0 \rangle \\ k_3 & \langle 0.5, 0.2 \rangle & \langle 0.6, 0.1 \rangle & \langle 0.6, 0.2 \rangle & \langle 0.2, 0.4 \rangle & \langle 0.6, 0.1 \rangle \\ k_4 & \langle 0.6, 0.0 \rangle & \langle 0.7, 0.1 \rangle & \langle 0.8, 0.0 \rangle & \langle 0.4, 0.3 \rangle & \langle 0.5, 0.2 \rangle \end{array}, \begin{array}{c|ccccc} d_2 & c_1 & c_2 & c_3 & c_4 & c_5 \\ \hline k_1 & \langle 0.3, 0.2 \rangle & \langle 0.9, 0.1 \rangle & \langle 0.4, 0.3 \rangle & \langle 0.4, 0.0 \rangle & \langle 0.5, 0.1 \rangle \\ k_2 & \langle 0.6, 0.1 \rangle & \langle 0.4, 0.2 \rangle & \langle 0.4, 0.0 \rangle & \langle 0.4, 0.3 \rangle & \langle 0.6, 0.0 \rangle \\ k_3 & \langle 0.4, 0.3 \rangle & \langle 0.7, 0.0 \rangle & \langle 0.5, 0.2 \rangle & \langle 0.2, 0.5 \rangle & \langle 0.7, 0.1 \rangle \\ k_4 & \langle 0.5, 0.0 \rangle & \langle 0.6, 0.1 \rangle & \langle 0.7, 0.0 \rangle & \langle 0.5, 0.1 \rangle & \langle 0.3, 0.2 \rangle \end{array}, \begin{array}{c|ccccc} d_3 & c_1 & c_2 & c_3 & c_4 & c_5 \\ \hline k_1 & \langle 0.5, 0.3 \rangle & \langle 0.6, 0.1 \rangle & \langle 0.2, 0.3 \rangle & \langle 0.2, 0.5 \rangle & \langle 0.6, 0.1 \rangle \\ k_2 & \langle 0.7, 0.0 \rangle & \langle 0.6, 0.1 \rangle & \langle 0.8, 0.0 \rangle & \langle 0.1, 0.4 \rangle & \langle 0.4, 0.0 \rangle \\ k_3 & \langle 0.6, 0.1 \rangle & \langle 0.7, 0.1 \rangle & \langle 0.7, 0.0 \rangle & \langle 0.2, 0.3 \rangle & \langle 0.7, 0.1 \rangle \\ k_4 & \langle 0.8, 0.0 \rangle & \langle 0.4, 0.3 \rangle & \langle 0.6, 0.0 \rangle & \langle 0.6, 0.1 \rangle & \langle 0.4, 0.2 \rangle \end{array} \right\}$$

Fig. 1. The evaluation IFIM EV for the candidates.

Step 2. Let the experts have the following rating coefficients respectively $\{r_1, r_2, r_3\} = \{\langle 0.7, 0.0 \rangle, \langle 0.6, 0.0 \rangle, \langle 0.8, 0.0 \rangle\}$. We create $EV^*[K, C, E, \{ev^*\}]$

$$EV^* = r_1 pr_{K,C,d_1} EV \oplus_{(max,min)} r_2 pr_{K,C,d_2} EV \oplus_{(max,min)} r_3 pr_{K,C,d_3} EV \quad (19)$$

$EV := EV^*$. Let us apply the optimistic aggregation operation $\alpha_{E,(\max,\min)}(EV, h_f) = R[K, h_f, C]$ to find the aggregated value of the k_i-th candidate against the c_j-th criterion in a current time-moment $h_f \notin D$.

Step 3. Let us create the 3-D IFIM PK of the weight coefficients of the assessment criterion according to its priority to the service $v_e (e = 1, 2, 3)$:

$$PK[C, V, h_f, \{pk_{c_j, v_e, h_f}\}] = \begin{array}{c|ccc} h_f & v_1 & v_2 & v_3 \\ \hline c_1 & \langle 0.8, 0 \rangle & \langle 0.4, 0.1 \rangle & \langle 0.7, 0 \rangle \\ c_2 & \langle 0.7, 0 \rangle & \langle 0.5, 0.2 \rangle & \langle 0.8, 0 \rangle \\ c_3 & \langle 0.5, 0.1 \rangle & \langle 0.8, 0 \rangle & \langle 0.6, 0.1 \rangle \\ c_4 & \langle 0.8, 0 \rangle & \langle 0.6, 0.1 \rangle & \langle 0.5, 0 \rangle \\ c_5 & \langle 0.7, 0.1 \rangle & \langle 0.7, 0 \rangle & \langle 0.9, 0 \rangle \end{array} \quad (20)$$

$$\text{and } B = R^T \odot_{(\circ,*)} PK = \begin{array}{c|ccc} h_f & v_1 & v_2 & v_3 \\ \hline k_1 & \langle 0.8, 0.0002 \rangle & \langle 0.71, 0.00076 \rangle & \langle 0.8, 0.00017 \rangle \\ k_2 & \langle 0.852, 0 \rangle & \langle 0.80, 0 \rangle & \langle 0.87, 0 \rangle \\ k_3 & \langle 0.846, 0 \rangle & \langle 0.82, 0 \rangle & \langle 0.873, 0 \rangle \\ k_4 & \langle 0.89, 0 \rangle & \langle 0.83, 0 \rangle & \langle 0.884, 0 \rangle \end{array} \quad (21)$$

Step 4. After application of the operation $\alpha_{K,\#_q}(B, k_0)$ we find that k_4 is the optimal outsoursing provider for the services v_1, v_2 and v_3. After application of IF Hungarian algorithm [19], we find that k_4 is the optimal outsoursing provider for service v_1, k_2 - for the service v_2 and k_3 - for the service v_3.

Step 5. After application of the ICrA with $\alpha = 0.85$ and $\beta = 0.10$ over R we determine that there are not criteria in a consonance.

Step 6. Let us propose, that the expert's estimations is correct. Then the new rating coefficients of the experts are equal to $\{\langle 0.73, 0.0 \rangle, \langle 0.64, 0.0 \rangle, \langle 0.82, 0.0 \rangle\}$.

5 Conclusion

In the study, we proposed a new IFIMOA approach for selecting the most eligible candidates for outsourcing service providers by evaluations from independent experts, based on the concepts of IFSs and IMs. The effectiveness of the method is demonstrated on real life data, related to the selection of suppliers of the outsourcing services in a refinery. Our future research will be related to extend this approach so that it can be applied to outsourcing problems, in which the evaluations of candidates under the criteria are interval-valued IFSs [6].

References

1. Araz, C., Ozfirat, P., Ozkarahan, I.: An integrated multicriteria decision-making methodology for outsourcing management. Comput. Oper. Res. **34**, 3738–3756 (2007)
2. Atanassov, K.: Intuitionistic fuzzy sets. In: Proceedings of VII ITKR's Session, Sofia, (1983). (Deposed in Centr. Sci.-Tech. Library of the Bulg. Acad. of Sci., 1697/84)
3. Atanassov, K.: Generalized index matrices. Comptes rendus de l'Academie Bulgare des Sciences **40**(11), 15–18 (1987)
4. Atanassov, K.: Index Matrices: Towards an Augmented Matrix Calculus. Studies in Computational Intelligence, vol. 573. Springer, Cham (2014). https://doi.org/10.1007/978-3-319-10945-9
5. Atanassov, K.: On Intuitionistic Fuzzy Sets Theory. STUDFUZZ, vol. 283. Springer, Heidelberg (2012). https://doi.org/10.1007/978-3-642-29127-2
6. Atanassov, K.: Interval-valued intuitionistic fuzzy sets. World Scientific (2018)
7. Atanassov, K., Mavrov, D., Atanassova, V.: Intercriteria decision making: a new approach for multicriteria decision making, based on index matrices and intuitionistic fuzzy sets. Issues in IFSs Generalized Nets **11**, 1–8 (2014)
8. Atanassov, K., Szmidt, E., Kacprzyk, J.: On intuitionistic fuzzy pairs. Notes Intuitionistic Fuzzy Sets **19**(3), 1–13 (2013)
9. Atanassov, K., Szmidt, E., Kacprzyk, J., Atanassova, V.: An approach to a constructive simplication of multiagent multicriteria decision making problems via ICrA. Comptes rendus de lAcademie bulgare des Sciences **70**(8), 1147–1156 (2017)
10. Atanassova, V., Roeva, O.: Computational complexity and influence of numerical precision on the results of intercriteria analysis in the decision making process. Notes Intuitionistic Fuzzy Sets **24**(3), 53–63 (2018)
11. Bottani, E., Rizzi, A.: A fuzzy TOPSIS methodology to support outsourcing of logistics services. Supply Chain Manage. An Int. J. **11**, 294–308 (2006)
12. Hsu, C., Liou, J., Chuang, Y.: Integrating DANP and modified grey relation theory for the selection of an outsourcing provider. Expert Syst. Appl. **40**, 2297–2304 (2013)
13. Kahraman, C., Oztaysi, B., Cevik, S.: A multicriteria supplier selection model using hesitant fuzzy linguistic term sets. J. Multiple-Valued Logic Soft Comput. **26**, 315–333 (2016)
14. Kahraman, C., Öztayşi, B., Çevik Onar, S.: An integrated intuitionistic fuzzy AHP and TOPSIS approach to evaluation of outsource manufacturers. J. Intell. Syst. **29**(1), 283–297 (2020). https://doi.org/10.1515/jisys-2017-0363
15. Liu, H., Wang, W.: An integrated fuzzy approach for provider evaluation and selection in third-party logistics. Expert Syst. Appl. **36**, 4387–4398 (2009)
16. Szmidt, E., Kacprzyk, J.: Amount of information and its reliability in the ranking of Atanassov intuitionistic fuzzy alternatives. In: Rakus-Andersson, E., Yager, R.R., Ichalkaranje, N., Jain, L.C. (eds.) Recent Advances in Decision Making. Studies in Computational Intelligence, vol. 222, pp. 7–19. Springer, Berlin, Heidelberg (2009). https://doi.org/10.1007/978-3-642-02187-9_2

17. Tjader, Y., May, J., Shang, J., Vargas, L., Gao, N.: Firm-level outsourcing decision making: a balanced scorecard-based analytic network process model. Int. J. Prod. Econ. **147**, 614–623 (2014)
18. Traneva, V., Tranev, S.: Index Matrices as a Tool for Managerial Decision Making. Publ, House of the Union of Scientists, Bulgaria (2017). (in Bulgarian)
19. Traneva, V., Tranev, S., Atanassova, V.: An IF approach to the Hungarian algorithm. In: Nikolov, G., Kolkovska, N., Georgiev, K. (eds.) Numerical Methods and Applications, NMA 2018, LNCS, vol. 11189, pp. 167–175. Springer, Cham (2019)
20. Traneva, V., Tranev, S., Stoenchev, M., Atanassov, K.: Scaled aggregation operations over 2- and 3-dimensional IMs. Soft Comput. **22**(15), 5115–5120 (2018)
21. Wan, S.-P., Wang, F., Lin, L.-L., Dong, J.-Y.: An IF linear programming method for logistics outsourcing provider selection. Knowl.-Based Syst. (2015)
22. Wang, J., Yang, D.: Using a hybrid multi-criteria decision aid method for information system outsourcing. Comput. Oper. Res. **34**, 3691–3700 (2007)
23. Wang, J., Wang, J., Zhang, H.: A likelihood-based TODIM approach based on multi-hesitant fuzzy linguistic information for evaluation in logistics outsourcing. Comput. RS Ind. Eng. **99**, 287–299 (2016)
24. Yager, R.: Non-numeric multi-criteria multi-person decision making. Int. J. Group Decis. Making Negot. **2**, 81–93 (1993)
25. Zadeh, L.: Fuzzy sets. Inf. Control **8**(3), 338–353 (1965)

Learning Algorithms

A Learning Based Vertical Integration Decision Model

Menekşe Gizem Görgün[1,2](✉), Seçkin Polat[1], and Umut Asan[1]

[1] Department of Industrial Engineering, Istanbul Technical University, Istanbul, Turkey
{saygim,polatsec,asanu}@itu.edu.tr
[2] Migros Ticaret A.Ş., 34758 Istanbul, Turkey

Abstract. Vertical integration decisions, also referred to as make or buy decisions, affect firm activity areas that determine boundaries that play a strategic role in the success of firms. The decisions may result in form of spot market usage, long-term contracts, joint ventures, and in-house productions. This study first reviews the decision modeling approaches suggested in the literature for vertical integration. The review shows that for decision making, most of the approaches are qualitative in nature and but there are also optimization models and expert based models. Thus, a new learning based probabilistic classification approach is proposed to model the vertical integration decision problem. This data-driven quantitative approach using Naïve Bayes is able to represent the uncertainty inherent in the decision problem. The applicability and effectiveness of the developed learning based vertical integration decision model is demonstrated by an example that covers real software make or buy cases in a retail company. 85% of the decisions were correctly classified by the developed data-driven model.

Keywords: Vertical integration · Make or buy decision · Probabilistic classification · Naïve Bayes

1 Introduction

Companies define their fields of activity to preserve their competitive advantages. Fields of activity are determined with respect to companies' organizational structure and ecosystem [1, 2]. Companies have a risk to end their processes unless they are unable to insource their fundamental (core) activities. On the other hand, insourcing all activities brings a very complex structure which is challenging for companies to deal with. A number of studies have investigated the advantages and disadvantages of vertical integration [2–6]. Findings suggest that vertical integration strategies (make or buy decision, joint venture, strategic cooperation, merger & acquisitions) have an impact on the performance of companies [1–10].

Vertical integration was first elaborated by the seminal paper of Coase in 1937 [11]. The main question of Coase's study was why there are so many firms in an industry. About the beginning of the 1970s, the interest of addressing this question increased. Williamson's studies made significant contributions to that discussion. Transaction Cost

C. Kahraman et al. (Eds.): INFUS 2021, LNNS 308, pp. 131–137, 2022.
https://doi.org/10.1007/978-3-030-85577-2_15

Economics (TCE) which is a complete theory was developed by Williamson [12]. This theory is directly related to the phenomenon of vertical integration, i.e. make or buy. Buchowicz [13] classifies studies on make or buy into four perspectives. These perspectives are economy, purchasing, business strategy, and optimization for a specific make or buy situations. TCE reflects the economic perspective and is subject of many studies. Besides, other theories such as resource-based view, option theory, agency theory that are developed for explaining another phenomenon, are used together or alone to explain the behavior of companies confronted with make or buy decisions. These studies are mainly descriptive and generic. Some prescriptive studies [13–16] proposed models for these types of decisions. Manufacturing was the main focus of the studies on vertical integration before 1980. After that IT, human resources, sales also became the subject of the make or buy decision studies. In IT, "make" means to develop software in-house, "buy" means to supply it from outside. In human resource management (HRM) to make human resource means to prepare people in the organization for a future task, to buy human resource means to hire people, who have already the necessary skills, from the labor market. For some context, to make in HRM means to carry out HR systems in-house, to buy in HRM means to supply HR systems from outside as a service. To make in sale means to use their own salesperson, to buy in sale means to use an agent from outside. Studies on IT are more prescriptive [13].

The decision-making models developed in the literature for make or buy decisions can be classified into two types. One type consists of generic models that can be applied in any context for make or buy. Mahoney [6] and Harrigan [2] are examples of the generic models. The second type covers function-specific models that can be applied in a specified context. Rand [14], Buchowicz [13] Cortellessa, Marinelli, Potena [15] and Kramer, Heinzl [16] are examples for IT make or buy decisions. Most of these studies involve qualitative frameworks that propose make or buy depending upon several factors. The generic models are more qualitative than context specific models. Both types of models in the literature are not data-driven. Most of them, especially frameworks in generic context, are based on findings from empirical studies. Researchers combine the knowledge drawn from various empirical studies into a framework. For some models, case studies may yield these types of models. Since vertical integration decisions include uncertainty it is important to represent uncertainty in decision making models on vertical integration. In addition, as a strategic decision, vertical integration decisions are multidimensional, and less frequent decision-making cases. From the literature review, it can be said that a new decision-making model is needed to meet these properties of the vertical integration decisions.

As mentioned above the models suggested in the literature have ignored the uncertainty aspect. In this paper, in order to learn and improve from experience and incorporate uncertainty into the classification problem, a learning-based model that is quantitative and probabilistic is proposed to support make or buy decisions in the IT context. The model proposed uses the Naïve Bayes classifier. In comparison to other classification techniques, Naïve Bayes i) does not necessarily require large datasets to start learning, ii) does not impose specific distributional assumptions or scaling properties, iii) keeps the number of parameters to be estimated to a minimum, and iv) works well with high-dimensional data. The rest of the paper is organized as follows: The second section first gives the basic definitions of Naïve Bayes classifier and then presents the applied methodology. Section 3 provides the details of the results and its discussion. Finally, Sect. 4 concludes the study and presents further suggestions.

2 Methodology

As mentioned above, the model proposed in this paper uses a Naïve Bayes classifier that is a probabilistic classification technique based on applying Bayes' theorem with a strong (naïve) assumption of conditional independence between the features [19]. Despite this strong assumption, Naïve Bayes is computationally efficient and often performs well in practice [20]. Furthermore, the technique does not necessarily require a large dataset before learning can begin [21].

For a problem instance to be classified, the instance is represented by a vector of n independent features $\mathbf{x} = (x_1, \ldots, x_n)$. If c_j is the label of the j-th class and if $\mathbf{x}$ is the vector describing the instance to be classified, the Bayes formula takes the form [22]:

$$P(c_j|\mathbf{x}) = \frac{P(\mathbf{x}, c_j)}{P(\mathbf{x})} = \frac{P(\mathbf{x}|c_j)P(c_j)}{P(\mathbf{x})} \tag{1}$$

Classifying based on Naïve Bayes involves the following steps:

1) Identify relevant factors (i.e. features) and factor levels.
2) Determine the training set.
3) Represent each instance in the training set in vector form $\mathbf{x} = (x_1, \ldots, x_n)$.
4) Calculate $P(x_i|c_j)$ for each x_i and each class c_j using the relative frequency of x_i among the training instances belonging to c_j.
5) Determine $P(c_j)$.
6) Based on the conditional independence assumption calculate

$$P(\mathbf{x}, c_j) = \prod_{i=1}^{n} P(\mathbf{x}_i|c_j) \tag{2}$$

7) Calculate $P(c_j) \cdot P(\mathbf{x}|c_j)$ for each class and choose the class with the highest value.
8) Evaluate the accuracy.

2.1 Make or Buy Decision Model for IT

The suggested learning based vertical integration decision model based on the Naïve Bayes classification method is applied to software make or buy decisions in a retail company. The factors used in this study are adopted from [17], which comprehensively covers make or buy decision strategies in the IT field, with the exception that the factor named "Requirements" in [17] is redefined as two factors: "Requirements complexity" and "Requirements certainty". After identifying the relevant factors, levels of the factors were specified. The levels of factors were determined by the decision makers who had been involved in the software decisions in the retail company. Selected factors and their levels are listed in Table 1.

Table 1. Selected factors.

Factor name	Definition	Factor Levels
Time (T)	Total time for development, testing and integration	Short (0–3 months), Middle (4–9 months), Long (over 9 months)
Cost (C)	Any costs involved in a project	Low, Middle, High
Effort (E)	Development effort for selection and integration	Low (0–60 man-day), Middle (60–180 man-day), High (over 180 man-day)
Quality (Q)	Quality expectancy	High, Middle Low
Market trend (Mt)	The availability of the product on the market place	Increasing, Stable, Decreasing, Custom (None)
Source code availability (Sc)	Describing availability of source code	Free, Costly, Not available,
Technical support (Ts)	Support, bugs fixing, and functionality changes needs	Weak, Middle, Strong
License (L)	License fees and obligations	Yes, No
Integration (I)	Easiness of integration activity	Easy, Challenging
Requirements complexity (Rco)	Describes complexity of product requirements	Complex, Uncomplex
Requirements certainty (Rce)	Describe certainty of product requirements	Certain, Uncertain
Maintenance (M)	Difficulty of system maintenance	Easy, Middle, Difficult

Next, the training set used in this study was determined. Twenty-one cases concerning software decisions that had been made by the company were included in the training set. Decision strategies were determined simply as insourcing and outsourcing for this study. COTS (Commercial of the Shelf Product) was covered within the outsourcing strategy in this study. OSS (Open Source Software) in [17] was not included in this study since this strategy was not considered as an option in the retail company's previous make or buy decisions.

Cases in the training set are related to decisions for ERP module, CRM tool, institutional website, and mobile application development. Factor levels of 21 cases in the training set were determined by the responsible managers.

3 Results and Discussion

The training dataset was evaluated with Naïve Bayes classifier in Weka software [18]. Weka was preferred since it is an open-source machine learning software that can be operated through a graphical user interface, standard terminal applications, or a Java API [18]. While the training data set was evaluated in Weka, ten-fold cross-validation was used.

According to the evaluations, the accuracy is calculated as 85.71%. 18 of 21 cases are classified correctly. ROC area, which is another specific value to understand the effectiveness of classification, is calculated as 0.836 for both insourcing and outsourcing. It is close to 1 which indicates high discrimination ability. Table 2 shows evaluation metrics for the classification. TP, FP, Precision, and Recall rate shows that the correlation level is high. Although the TP rate for insourcing class is higher than outsourcing class, the precision of insourcing class is lower than outsourcing class.

Table 2. Detailed accuracy by class

Evaluation metric	Insourcing class	Outsourcing class	Weighted average
TP rate	0.909	0.800	0.857
FP rate	0.200	0.091	0.148
Precision	0.833	0.889	0.860
Recall	0.909	0.800	0.857
F-measure	0.870	0.842	0.856
MCC	0.716	0.716	0.716
ROC area	0.836	0.836	0.836
PRC area	0.842	0.842	0.842

Table 3 shows the correctly and incorrectly classified instances with respect to the classes. 10 of 11 (90.9%) insourcing decisions are correctly classified by the Naïve Bayes approach. For outsourcing decisions, 8 of 10 (80%) cases are predicted correctly.

The reason of incorrectly classified three cases might be the subjective evaluation of the responsible people. Two of these cases are relatively old in comparison to the other cases. This may be an indication of the impact the date of the cases have on parameter levels or decision.

Table 3. Confusion matrix

	Classified as insourcing	Classified as outsourcing
Insourcing (a)	10	1
Outsourcing (b)	2	8

4 Conclusion

Vertical integration strategies that covers make or buy decisions are essential for companies to gain financial power in the market. That is why there are many researches for make or buy decisions. Make or buy decision making is an important research area also for the IT environment. Despite a high number of researches in the context of make or buy, there are a small number of decision-making models. These decision-making models are not data-driven and do not contain uncertainty that is main characteristic of strategic decisions that are multidimensional. In this study, a new data-driven decision-making model that covers uncertainty and multidimensionality was developed. The model using Naïve Bayes was applied to software make or buy decision making cases in a retail company. The Naïve Bayes model correctly classified 18 of 21 cases. Incorrectly classified instances might be a result of subjective evaluation of the responsible people in the company. In addition, TP rate and precision were high. The results may indicate that Naïve Bayes works well for strategic decision-making contexts such as vertical decisions including high dimension that is 12 factors for this study, and a small number of cases that is 21 for this study.

Naïve Bayes is an efficient data-driven statistical technique, which assumes factors as independent. Another method may be considered and used for validation in future studies. A Dynamic Bayesian Network approach is planned to be implemented in further studies to evaluate also the dependencies between factors and yield more comprehensive results.

Lastly, this study only covers make or buy decisions on vertical integration strategies. Together with make or buy decision if the decision is "to buy", supplier selection methodologies can be added to the research. On the other hand, if the decision is "to make", the development strategy in an IT environment such as agile, waterfall (predictive), or hybrid can be added to the research in the future studies. Thus, the framework can expand into the strategic management research area.

References

1. Quinn, J.B., Hilmer, F.G.: Strategic outsourcing. Mckinsy Q. **1995**(1), 48–70 (1994)
2. Harrigan, K.R.: Formulating vertical integration strategies. Acad. Manag. Rev. **9**, 638–652 (1984)
3. Venkatesan, R.: Strategic sourcing: to make or not to make. Harvard Bus. Rev. **70**, 97–107 (1992)
4. McIvor, R.T., Humphreys, P.K., McAleer, W.E.: A strategic model for the formulation of an effective make or buy decision. Manag. Decis. 169–178 (1997)

5. Humphreys, P.K., Lo, V.H.Y., McIvor, R.T.: A decision support framework for strategic purchasing. J. Mater. Process. Technol. **107**, 353–362 (2000)
6. Humphreys, P., McIvor, R., Huang, G.: An expert system for evaluating the make or buy decision. Comput. Ind. Eng. **42**, 567–585 (2002)
7. Mahoney, J.T.: The choice of organizational form: vertical financial ownership versus other methods of vertical integration. Strat. Manag. J. **13**, 559–584 (1992)
8. Welch, J.A., Nayak, P.R.: Strategic sourcing: a progressive approach to the make-or-buy decision. Acad. Manag. Exec. **6**, 23–31 (1992)
9. Canez, L.E., Platts, K.W., Probert, D.R.: Developing a framework for make-or-buy decisions. Int. J. Oper. Prod. Manag. 1313–1330 (2000)
10. McIvor, R.: A practical framework for understanding the outsourcing process. Supply Chain Manag. Int. J. 22–36 (2000)
11. Coase, R. H.: The "nature of the firm". Economica, N.S., 386–405 (1937)
12. Williamson, O.E.: Markets and Hierarchies. Free Press, New York (1975)
13. Buchowicz, B.S.: A process model of make-vs.-buy decision-making; the case of manufacturing software. IEEE Trans. Eng. Manag. **38**, 24–32 (1991)
14. Rand, T.: A framework for managing software make or buy. Eur. J. Inf. Syst. **2**, 273–282 (1993)
15. Cortellessa, V., Marinelli, F., Potena, P.: An optimization framework for build-or-buy decisions in software architecture. Comput. Oper. Res. **35**, 3090–3106 (2008)
16. Kramer, T., Heinzl, A., Spohrer, K.: Should this software component be developed inside or outside our firm? A design science perspective on the sourcing of application systems. In: New Studies in Global IT and Business Service Outsourcing, 5th Global Sourcing Workshop Courchevel, France, pp. 115–132 (2011)
17. Badampudi, D., Wohlin, C., Petersen, K.: Software component decision-making: in-house, OSS, COTS or outsourcing- a systematic literature review. J. Syst. Softw. **121**, 105–124 (2016)
18. Weka Homepage. https://www.cs.waikato.ac.nz/ml/weka/index.html. Accessed 30 Mar 2021
19. Witten, I.H., Frank, E., Hall, M.A., Pal, C.J.: Practical Machine Learning Tools and Techniques, p. 578. Morgan Kaufmann, San Francisco (2005)
20. Hamine, V., Helman, P.A.: Theoretical and experimental evaluation of augmented Bayesian classifiers. In: American Association for Artificial Intelligence, pp. 02–06 (2006)
21. Stern, M., Beck, J., Woolf, B.P.: Naive Bayes classifiers for user modeling. Center for Knowledge Communication, Computer Science Department, University of Massachusetts (1999)
22. Kubat, M.: Probabilities: Bayesian Classifiers. In: An Introduction to Machine Learning, pp. 19–41. Springer, Cham (2017) https://doi.org/10.1007/978-3-319-63913-0_2

Comparing Fusion Methods for 3D Object Detection

Erkut Arıcan(✉) and Tarkan Aydın

Engineering and Natural Science Department, Computer Engineering, Bahcesehir University, Istanbul, Turkey
{erkut.arican,tarkan.aydin}@eng.bau.edu.tr

Abstract. One of the main problems in computer vision that leads digital technologies to transform our business and social life is object detection. The solutions to the problem have various application areas such as security systems, surveillance, shopping applications, and much more. Significant performance gain have been achieved by using the popular deep learning methods that are applied on 2D RGB images. With the availability of low-cost 3D sensors, methods that incorporates 3D data with 2D RGB data using existing deep network architectures became more popular to increase performance further. In this work, different data level and feature level fusion strategies have been analyzed to incorporate 3D depth and 2D RGB data with existing architectures to assess their effects on the performance. These methods were tested on the real RGB-D benchmark datasets available in the literature, and the accuracy results were compared to each other in their object types of group.

Keywords: 3D · Depth · 2D · RGB · RGB-D · Fusion · Object detection · Computer vision · Deep learning

1 Introduction

Object detection is a vital topic for computer vision. There are many application areas such as security systems, surveillance, shopping applications, and much more. When technology developed rapidly, object detection studies try to find a better way and better accuracy; moreover, over the last years, deep learning popularity is significantly increasing. As a result of deep learning popularity, Convolutional Neural Network (CNN) has become popular in object detection and recognition.

Besides the popularity of deep learning, with the development of technology in computer vision, depth information can be reachable easily. For acquiring the depth data, Kinect [13] and Intel Real Sense [5] is just some example of technological development. RGB-D image's depth information provides more information than the RGB. Many 3D datasets can be accessed via the internet, for instance, Washington RGB-D dataset [9], and Berkeley 3-D Object Dataset [6].

C. Kahraman et al. (Eds.): INFUS 2021, LNNS 308, pp. 138–146, 2022.
https://doi.org/10.1007/978-3-030-85577-2_16

In the literature, there are many studies for 2D and 3D object detection and recognition. For example, BRISK [10], Improved FAST [14], ORB [15], SURF [2] and SIFT [11] are fundamental studies in 2D object detection and recognition area. Furthermore, Johnson and Hebert [7], Sipiran and Bustos [20], Scovanner et al. [19], Mian et al. [12], Drost et al. [3], Rusu et al. [16–18], Tombari et al. [24], Song and Xiao [22], and Gupta et al. [4] are some examples and important studies in 3D object detection and recognition area.

AlexNet [8] is one of the popular convolutional neural networks. Zia [26] and Zeng [25] studies and also easily accessible 3D dataset which we use Washington RGB-D dataset [9], give us an idea and motivation for comparing fusion strategies and other methods. There are five methods: 1) Alexnet with RGB Image, 2) Alexnet with Depth Image, 3) Initial Fusion, 4) Priming Method, 5) Initial Fusion & Priming Method. The third, fourth, and fifth methods have different data level and feature level fusions methods with incorporate 3D depth and 2D RGB data. Finding which method gives the better accuracy result for each object type of group is the aim of this study. These comparing and finding which methods give well result is our contribution to the literature.

This paper continues as follows; in Sect. 2, detailed information about fusion methods will be shown. Comparison results for methods will be shown in Sect. 3, and lastly in Sect. 4 reviewing the study.

2 Materials and Methods

In this section, firstly we will be explained the dataset, then AlexNet and Bilateral filter which are well-known methods will be explained briefly. Afterward, fusion methods will be discussed in detail. We compare all methods to find which way gives a better accuracy rate for each object type of group.

We use the Washington RGB-D dataset [9]. There are 51 categories, and Kinect [13] style images. RGB, depth, and mask images are included in the evaluation set; moreover, this evaluation set is used in our method. RGB image example in Fig. 1a, Depth image example in Fig. 1b, and mask image example is in Fig. 1c can be seen. Each object has got a different number of instances folder. For each object, one folder is selected as validation images and the rest of them as training images.

Alexnet [8] is a base Convolutional Neural Network method in this study. There are five convolutional and three fully connected layers in Alexnet architecture. Max Pooling and ReLU Nonlinearity is also used in AlexNet architecture. Batch size is 128 examples, weight decay is 0.0005 and momentum is 0.9 for stochastic gradient descent which is used in model.

Bilateral filter [1,21,23] gives an idea for combining method for Initial Fusion & Priming Method. The Bilateral Filter uses neighbor pixels' position and intensity to creates an average value. In our study there are 5 different methods. These methods are:

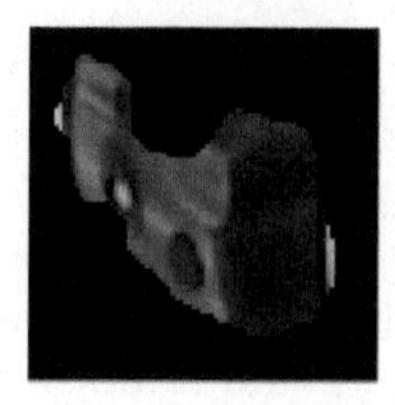

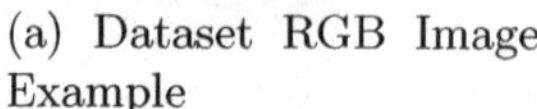

(a) Dataset RGB Image Example

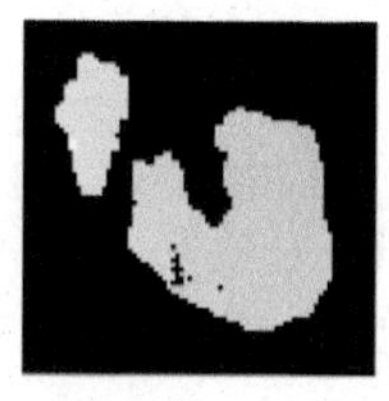

(b) Dataset Depth Image Example

(c) Dataset Mask Image Example

Fig. 1. Dataset image examples

- Alexnet with RGB Image
- Alexnet with Depth Image
- Initial Fusion
- Priming Method
- Initial Fusion & Priming Method

We will explain Initial Fusion Method, Priming Fusion Method and Initial Fusion & Priming Method.

In Initial Fusion Method, we combine RGB and Depth images before the AlexNet process. In this step, the bilateral filter gives an idea for combining process so we use Eq. 1 to combine RGB image to Depth image.

$$NewImage = RGB * exp^{(DEPTH)} \tag{1}$$

In Priming Fusion Method, we give RGB and Depth images as a separate input to AlexNet. In Fig. 2a you can see the combining step for these separate groups before the fully-connection layer.

In Initial Fusion & Priming Method, AlexNet is our base method and we have 2 combining steps. Firstly we combine RGB image and Depth image before the neural network input process. The other combining process is shown in Fig. 2b second column which starts with "multi" keyword. In this step, like the Initial Fusion Method, the bilateral filter gives an idea for the combining process so we use Eq. 1 to combine RGB image to Depth image. All these 3 columns are combined before the fully-connection layer.

3 Results

We compare all methods to find which way gives a better accuracy rate for each object type of group. You can see different methods run in 3D RGBD Dataset [9]. During the test, each method ran with all objects without looking at the object type of group. We talked about how we choose the validation set in Sect. 2. When the test finished, we create manually object type groups. These groups are:

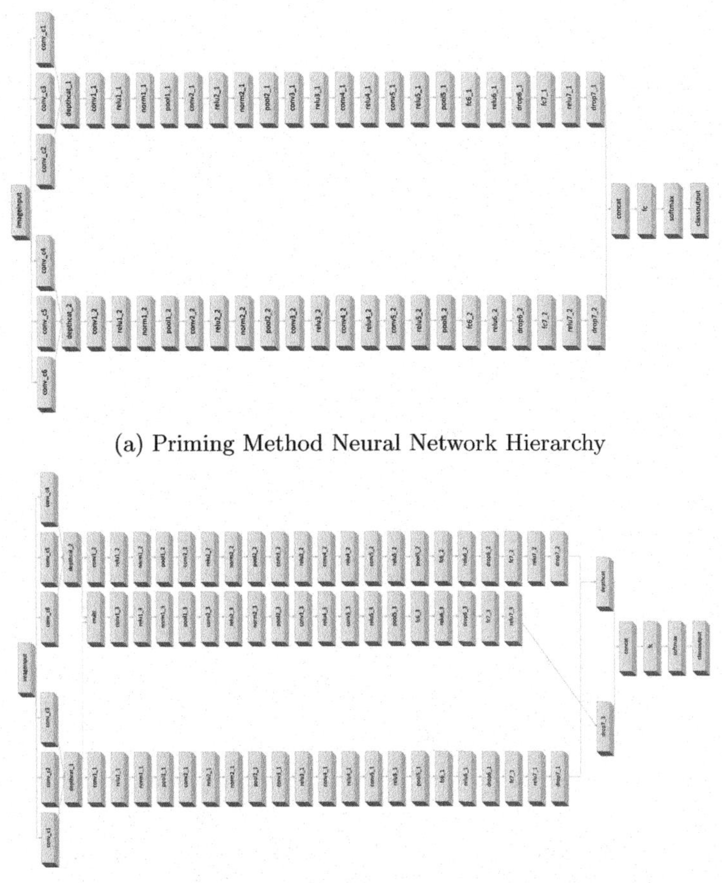

(a) Priming Method Neural Network Hierarchy

(b) Initial Fusion & Priming Method Neural Network Hierarchy

Fig. 2. Network hierarchy

- Round Shape,
- Cube Shape,
- Cylindrical Shape,
- Rectangle Shape,
- Free Form.

Detail information about object type groups are shown in Table 1.

Afterward of the creating object type of group, in Table 2-3-4-5-6, you can see for each objects results for five methods and averages for each column. In Table 2, the result shows us the AlexNet with RGB input which is 81.95 average

Table 1. Object shape groups.

Round shape	Cube shape	Cylindrical shape	Rectangle shape	Free form
Apple	Bell pepper	Coffee mug	Binder	Cap
Ball	Camera	Food can	Calculator	Greens
Bowl	Cereal box	Food cup	Cell phone	Mushroom
Garlic	Food box	Food jar	Comb	Pear
Lemon	Kleenex	Glue stick	Food bag	Pliers
Lime	Stapler	Banana	Hand towel	Scissors
Onion		Dry battery	Instant noodle	Toothbrush
Orange		Flashlight	Keyboard	
Peach		Lightbulb	Notebook	
Plate		Marker	Rubber eraser	
Potato		Pitcher	Sponge	
Tomato		Shampoo		
		Soda can		
		Toothpaste		
		Water bottle		

Table 2. Round shape result

Label	RGB	DEPTH	Initial fusion	Priming	Initial fusion & priming
Apple	100	83.74	100	100	100
Ball	3.64	4.24	20	46.67	87.88
Bowl	9.26	64.2	1.85	1.85	6.17
Garlic	100	95	96.88	100	92.5
Lemon	100	84.43	100	100	100
Lime	100	85.12	100	100	100
Onion	100	94.19	98.71	100	97.42
Orange	100	11.97	100	100	80.99
Peach	93.62	0	73.76	52.48	38.3
Plate	100	95.59	100	100	100
Potato	76.86	0	10.74	19.83	59.5
Tomato	100	46.1	100	100	99.29
Average	**81.95**	55.38	75.16	76.74	80.17

result, is the best method for round-shaped objects. In Table 3, results show us the Initial Fusion & Priming Method which is 90.34 average result, is the best method for cube shape objects. Another shape result is shown in Table 4, and results show us the Initial Fusion Method which is 86.76 average result, is the

Table 3. Cube shape result

Label	RGB	DEPTH	Initial fusion	Priming	Initial fusion & priming
Bell pepper	100	60	100	93.6	96
Camera	62.93	25	47.41	51.72	78.45
Cereal box	87.1	97.58	77.42	84.68	74.19
Food box	69.28	100	94.58	90.36	93.37
Kleenex	100	99.23	100	100	100
Stapler	99.12	89.47	96.49	100	100
Average	86.41	78.55	85.98	86.73	**90.34**

Table 4. Cylindrical shape result

Label	RGB	DEPTH	Initial fusion	Priming	Initial fusion & priming
Coffee mug	84.82	99.11	97.32	96.43	73.21
Food can	96.27	100	99.38	98.76	100
Food cup	100	95.12	99.39	100	98.78
Food jar	50	4.32	83.33	0.62	3.7
Glue stick	100	97.52	89.44	100	98.76
Banana	95.65	100	94.2	99.28	91.3
Dry battery	82.08	96.23	65.09	83.02	61.32
Flashlight	89.08	98.32	99.16	89.92	98.32
Lightbulb	90.4	11.2	72.8	99.2	85.6
Marker	17.28	98.77	19.14	97.53	98.77
Pitcher	78.57	30.36	98.21	33.04	78.57
Shampoo	98.21	88.1	92.86	98.21	95.83
Soda can	94.07	98.31	96.61	100	100
Toothpaste	98.11	99.37	100	100	100
Water bottle	91.67	100	94.44	90.74	93.52
Average	84.41	81.12	**86.76**	85.78	85.18

best method for cylindrical shape objects. In Table 5, rectangle shape result is shown, and Initial Fusion & Priming Method which is 81.45 average result, is the best. In last Table 6, Priming Method which is 66.57 average result, is the best method for free form shapes objects.

Table 5. Rectangle shape result

Label	RGB	DEPTH	Initial fusion	Priming	Initial fusion & priming
Binder	20.83	18.06	16.67	65.97	37.5
Calculator	98.29	40.17	96.58	91.45	95.73
Cell phone	98.17	68.81	86.24	89.91	91.74
Comb	65.18	66.96	84.82	76.79	94.64
Food bag	88.05	89.94	63.52	92.45	86.79
Hand towel	80.13	90.38	65.38	96.79	97.44
Instant noodles	99.38	94.41	100	100	100
Keyboard	58.74	56.64	83.92	37.76	86.01
Notebook	2.45	46.63	0.61	0	6.13
Rubber eraser	100	94.23	100	100	100
Sponge	100	100	100	100	100
Average	73.75	69.66	72.52	77.37	**81.45**

Table 6. Free form shape result

Label	RGB	DEPTH	Initial fusion	Priming	Initial fusion & priming
Cap	0	96.06	0	0	0
Greens	100	63.16	100	97.74	99.25
Mushroom	0	6.54	0	0	0
Pear	46.62	0	49.32	68.24	27.03
Pliers	92.04	100	93.81	100	99.12
Scissors	77.97	97.46	100	100	100
Toothbrush	100	96.69	84.3	100	100
Average	59.52	65.7	61.06	**66.57**	60.77

4 Conclusions

Deep learning in computer vision became very popular. Also, AlexNet became one of the popular methods in deep learning methods; therefore, in this study, AlexNet was used as a base method. In our study, we have two different contributions to the literature. The first one is using the RGBD dataset to compare methods to find a better accuracy. For each object type of group, finding which method works gives better accuracy is a second contribution of the literature. As future work, we are planning to try different neural networks which work well with depth images, to find a better result on the RGBD dataset.

References

1. Aurich, V., Weule, J.: Non-linear gaussian filters performing edge preserving diffusion. DAGM-Symposium Mustererkennung (1995)

2. Bay, H., Tuytelaars, T., Van Gool, L.: SURF: speeded up robust features. In: Lecture Notes in Computer Science (including subseries Lecture Notes in Artificial Intelligence and Lecture Notes in Bioinformatics), vol. 3951 LNCS, pp. 404–417 (2006)
3. Drost, B., Ulrich, M., Navab, N., Ilic, S.: Model globally, match locally: efficient and robust 3D object recognition. In: 2010 IEEE Computer Society Conference on Computer Vision and Pattern Recognition, pp. 998–1005. IEEE (2010)
4. Gupta, S., Arbeláez, P., Malik, J.: Perceptual organization and recognition of indoor scenes from RGB-D images. CVPR **1**(1), 1–14 (2013). https://doi.org/10.1109/ICCVW.2011.6130298
5. Intel: Intel RealSense (2020). https://www.intel.com
6. Janoch, A., et al.: A category-level 3-D object dataset: Putting the Kinect to work. Presented at the (2011). https://doi.org/10.1109/ICCVW.2011.6130382
7. Johnson, A.E., Hebert, M.: Using spin images for efficient object recognition in cluttered 3D scenes. IEEE Trans. Pattern Anal. Mach. Intell. **21**(5), 433–449 (1999). https://doi.org/10.1109/34.765655
8. Krizhevsky, A., Sutskever, I., Hinton, G.E.: ImageNet classification with deep convolutional neural networks. Presented at the (2012)
9. Lai, K., Bo, L., Ren, X., Fox, D.: A large-scale hierarchical multi-view RGB-D object dataset. Presented at the (2011). https://doi.org/10.1109/ICRA.2011.5980382
10. Leutenegger, S., Chli, M., Siegwart, R.Y.: BRISK: Binary Robust invariant scalable keypoints. Presented at the (2011). https://doi.org/10.1109/ICCV.2011.6126542
11. Lowe, D.G.: Distinctive image features from scale-invariant keypoints. Int. J. Comput. Vis. **60**(2), 91–110 (2004). https://doi.org/10.1023/B:VISI.0000029664.99615.94
12. Mian, A.S., Bennamoun, M., Owens, R.: Three-dimensional model-based object recognition and segmentation in cluttered scenes. IEEE Trans. Pattern Anal. Mach. Intell. **28**(10), 1584–1601 (2006). https://doi.org/10.1109/TPAMI.2006.213
13. Microsoft: Kinect (2020). https://developer.microsoft.com/en-us/windows/kinect/
14. Rosten, E., Drummond, T.: Machine learning for high-speed corner detection. Presented at the (2006)
15. Rublee, E., Rabaud, V., Konolige, K., Bradski, G.: ORB: An efficient alternative to SIFT or SURF. Presented at the (2011). https://doi.org/10.1109/ICCV.2011.6126544
16. Rusu, R.B., Blodow, N., Beetz, M.: Fast Point Feature Histograms (FPFH) for 3D registration. Presented at the (2009). https://doi.org/10.1109/ROBOT.2009.5152473
17. Rusu, R.B., Blodow, N., Marton, Z.C., Beetz, M.: Aligning point cloud views using persistent feature histograms. In: 2008 IEEE/RSJ International Conferences on Intelligent Robots and Systems, IROS, pp. 3384–3391. IEEE (2008). https://doi.org/10.1109/IROS.2008.4650967
18. Rusu, R.B., Marton, Z.C., Blodow, N., Beetz, M.: Learning informative point classes for the acquisition of object model maps. In: 2008 10th International Conference on Control, Automation, Robotics and Vision, pp. 643–650. IEEE (2008). https://doi.org/10.1109/ICARCV.2008.4795593
19. Scovanner, P., Ali, S., Shah, M.: A 3-dimensional sift descriptor and its application to action recognition, pp. 357–360. ACM Press, New York, New York, USA (2007). https://doi.org/10.1145/1291233.1291311

20. Sipiran, I., Bustos, B.: A robust 3D interest points detector based on harrisoperator. In: 3DOR, pp. 7–14 (2010). https://doi.org/10.2312/3DOR/3DOR10/007-014, http://www.dcc.uchile.cl/~bebustos/files/SB10b.pdf
21. Smith, S.M., Brady, J.M.: SUSAN-a new approach to low level image processing. Int. J. Comput. Vis. **23**(1), 45–78 (1997). https://doi.org/10.1023/A:1007963824710
22. Song, S., Xiao, J.: Sliding shapes for 3D object detection in depth images. In: ECCV (2014). https://doi.org/10.1007/978-3-319-10599-4
23. Tomasi, C., Manduchi, R.: In: Bilateral Filtering for Gray and Color Images, vol. No.98CH36271), pp. 839–846. Narosa Publishing House (1998). https://doi.org/10.1109/ICCV.1998.710815, https://ieeexplore.ieee.org/document/710815/
24. Tombari, F., Salti, S., Di Stefano, L.: Unique signatures of histograms for local surface description. Presented at the (2010)
25. Zeng, H., Yang, B., Wang, X., Liu, J., Fu, D.: RGB-D object recognition using multi-modal deep neural network and DS evidence theory. Sens. (Switzerland) **19**(3), 529 (2019). https://doi.org/10.3390/s19030529
26. Zia, S., Yüksel, B., Yüret, D., Yemez, Y.: RGB-D Object Recognition Using Deep Convolutional Neural Networks. Presented at the (2018). https://doi.org/10.1109/ICCVW.2017.109

Diabetic Retinopathy Detection with Deep Transfer Learning Methods

Gökalp Çinarer(✉) and Kazım Kiliç

Yozgat Bozok University, 66900 Yozgat, Turkey
{gokalp.cinarer,kazim.kilic}@bozok.edu.tr

Abstract. Diabetic retinopathy is an eye disease that occurs with damage to the retina and has many different complications, ranging from permanent blindness. The aim of this study is to develop a (convolutional neural network) CNN model that determines with high accuracy whether fundus images are diabetic retinopathy. The performance of the model has been verified in Kaggle APTOS 2019 dataset with AlexNET and VggNET-16 deep transfer learning algorithms. Various image processing techniques have been used as well as deep learning methods to further improve the classification performance. Images in the data set were rescaled to 224 × 224 × 3 and converted to Grayscale color space. Besides Gauss filter applied to eliminate the noise in the images. The area under the curve (AUC), precision, recall, and accuracy metrics of the deep transfer learning models used in this study were compared. The AlexNet model achieved a 98.6% AUC score, 95.2% accuracy, and the VggNET-16 model achieved a 99.6% AUC score and 98.1% accuracy. VggNET-16 was found to have higher confidence. Our results show that with the correct optimization of the CNN model applied in diabetic retinopathy classification, deep transfer learning models can achieve high performance and can be used in the detection of diabetic retinopathy patients.

Keywords: Diabetic retinopathy · Deep Learning · Classification

1 Introduction

Deep Learning (DL) is a highly effective artificial intelligence application that is also used in the detection and classification of medical images. Deep learning algorithms are preferred in many areas including diabetic retinopathy detection in medical image processing.

Diabetes is a very popular disease in the human eye retina that can damage the small vessels in the eye [1]. Over time, this disease can lead to diabetic retinopathy (DR), which causes congestion in the vascular space and impaired vision. DR is a chronic disease that lowers the quality of life of people all over the world and causes blindness and visual impairment if left untreated [2]. Retinal images are used to detect DR disease. The lesions that cause the disease are determined with the examinations made on the images.

The rise in the number of diabetes patients in the world and the lack of sufficient number of specialist physicians in this field negatively affect the treatment processes.

C. Kahraman et al. (Eds.): INFUS 2021, LNNS 308, pp. 147–154, 2022.
https://doi.org/10.1007/978-3-030-85577-2_17

In addition, manual diagnosis of DR patients turns into a very time-consuming process for physicians. It is almost impossible to train and specialize personnel for all these operations in a short time. In addition, It is also not possible to follow the retina by a physician regularly. DR follow-up is a disease that requires expertise and has a high cost.

In this study, a simple and efficient model was designed for the detection of DR, which can be used in the field of medical imaging, and the effectiveness of the method was demonstrated by transfer learning algorithms. The results obtained showed that high accuracy can be achieved by applying correct preprocesses and deep learning algorithms in distinguishing DR classes.

In the second part of the study, the studies in the related terminology are mentioned. In Sect. 3, a detailed explanation of the proposed model and applied algorithms is presented. The results obtained are shown in Sect. 4 with numerical analysis and graphics and have been discussed with other studies. Finally, in the 5th section, the models to be developed in the next studies were mentioned and the article was concluded.

2 Related Work

When DR studies are examined in the literature, it is seen that different and complex CNN structures are used. It is seen that deep learning is widely used in medical image processing in the literature. With the increasing popularity of deep learning focused approaches, different perspectives have emerged that apply CNN methods to solve this problem. Studies conducted using deep learning algorithms in the field of diabetic retinopathy are included in this part.

Zago et al. [3] analyzed DR lesions with the CNN structure they developed with image processing methods. Pratt et al. designed a model by applying the CNN architecture and data augmentation method with 5,000 verification images. They obtained 95% sensitivity and 75% accuracy in the study [4]. Sarki et al. have trained different algorithms using ImageNet pre-training. Among the Resnet50, XceptionNet, DenseNet and VGG models, the highest 81.3% accuracy was achieved [5].

In another study, Jain et al. [6] made DR classification using data enhancement methods with Vgg-16, Vgg-19 and Inception V3 networks. The best accuracy result has been achieved with Vgg-19 with 80,40%. Shaban et al. classified 4 different DR stages with the CNN model they designed. The study achieved an accuracy of 80.2% and a sensitivity of 78.7% [7].

Lam et al. classified DR patients to different degrees. They used GoogleNet and AlexNet models and they obtained the highest 74.5% accuracy in binary classification [8]. Haqos et al. [9] used InceptionV3 architecture in DR classification model and achieved 90.9% accuracy with small data sets. When the literature is examined, different methods and techniques have been used for the classification of DR images.

3 Material and Method

The flow chart of the classification model proposed for the detection of diabetic retinopathy in this study (see Fig. 1).

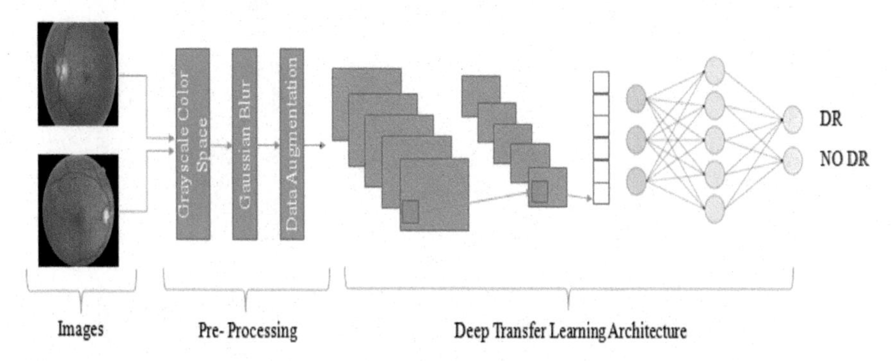

Fig. 1. Proposed method

3.1 Data Set

In this study, the "APTOS 2019 Blindness Detection" competition data set published in Kaggle was used to detect early blindness due to diabetic retinopathy [10]. This data set contains large retinal images obtained by fundus photography under high-level imaging conditions, which have completed their technological infrastructure. The expert examined each image in the data set and the images were graded for the severity of diabetic retinopathy by the publishers. 3662 training data included in the data set were used in the study. 1805 of these images are healthy and 1857 contain diabetic retinopathy. Original images are in RGB format, 3216×2136 in size and 24-bit depth.

3.2 Pre-process

Fundus images were pre-treated before classifying the DR images. In the study, high resolution retinopathy images were rescaled to 224×224 for the performance of the deep transfer learning architecture. Out-of-focus, underexposed or overexposed synthetic-looking images were detected in the data set. For this reason, images have been converted from RGB color space to gray level color space. Images contain interference and noise just like real world data. Gaussian filter has been applied to eliminate noise and interference in the detection of diabetic retinopathy.

The Gaussian filter applies smoothing to the values in the image matrix by convolution. Combines the values in the Gaussian kernel with the values in the image, and then add them all [11]. The pretreated fundus images are shown in Fig. 2.

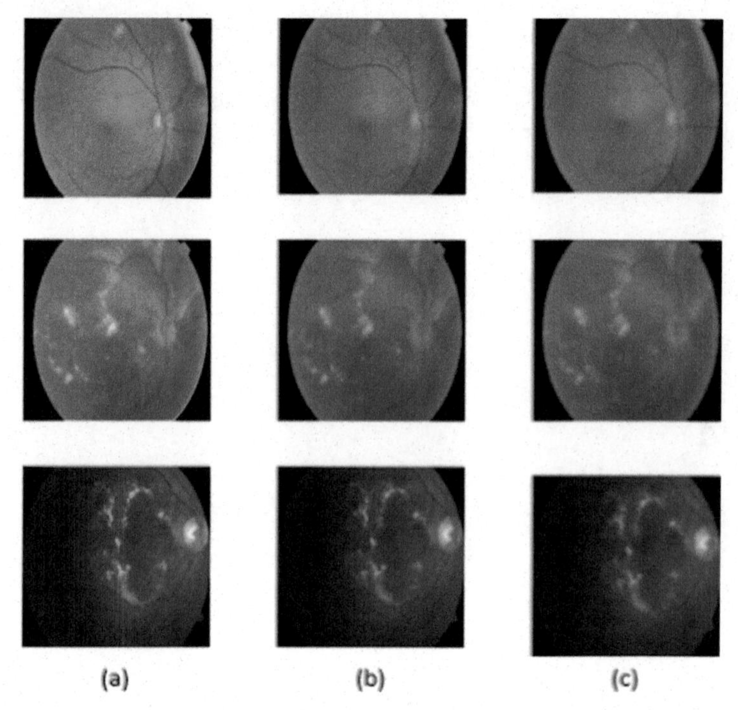

Fig. 2. a) Original fundus images b) Gray level fundus images c) Gaussian filtered images

3.3 Data Augmentation

Different methods can be used to increase the classification performance during training. While doing this, it is necessary to prevent excessive fit of the model. Considering all these situations, the data augmentation technique was used [12]. At this stage, the images were reprocuded by using horizontal and vertical shifts and angular change techniques. Data augmentation was applied only to training data, and validation data consists of original images.

3.4 AlexNet

AlexNet architecture was introduced at the 2012 ImageNet competition by Krizhevsky et al. [13].This architecture, which achieved excellent performance in the competition, achieved an error rate of 15.3%, making convolutional neural networks popular in image classification. Pre-trained architecture can divide objects into 1000 different classes. The input image size of the network is 227×227. AlexNet architecture consists of a total of 8 layers, 5 of which are convolution layers and three fully connected layers. Maximum pooling is used after convolution layers. ReLu activation function is used instead of sigmoid activation function in AlexNet. 11×11 size filter is used in the input layer and the number of steps is 2. In pooling layers, the frame size is 3×3.

3.5 VggNet-16

It was introduced at the 2014 ImageNet competition by Simonyan and Zisserman [14]. Trained with more than 14 million images, this architecture reduced the error rate to 7.3%. The VggNet architecture focuses on the depth of the layers rather than increasing the layers of the network, and this method has become popular in image classification problems. The input image size of the network is 224 × 224. VggNet-16 Architecture has 13 convolution layers and 3 fully connected layers. It consists of 16 layers in total. ReLu is used for the activation function and maximum pooling is used in the pooling layers. The filter size used in the convolution layers of the network is 3 × 3.

4 Results

The 3662 fundus images pretreated for the experiment were separated into 85% training and 15% validation. All images rescaled to 224 × 224. The data augmentation technique applied to the training footage to prevent excessive compatibility of the model and increase its performance. All operations performed on the K80 Tesla graphics card in the cloud environment.

AlexNet and VggNet-16 deep transfer learning architectures trained for the detection of DR. The number of epoch for the training process was set to 20. Adam was used for optimization function and ReLu was used for activation function. The results obtained were evaluated using the AUC, Accuracy, Recall, Precision and F-Score metrics are shown in Table 1 and deep transfer learning architectures accuracy graphs are given Fig. 3.

Table 1. Performance summary of deep transfer learning models

	Accuracy %	Precision %	Recall %	F-score %
AlexNet	95.26	95.24	95.24	95.24
VggNet-16	98.17	98.18	98.16	98.17

Deep transfer learning architectures loss curve (Fig. 4), confusion matrix (Fig. 5) and ROC curve (Fig. 6) are given in figures.

Diabetic retinopathy status was evaluated with 549 validation data. In the results obtained, AlexNet algorithm reached 95.26% accuracy and VggNet-16 algorithm reached 98.17% accuracy. The AUC score obtained by AlexNet and VggNet-16 algorithms was found to be quite successful in detecting DR. When the algorithms are examined, in DR classification, VggNet-16 algorithm has a precision value of 98.18%, while Alexnet has a value of 95.24%. Although Relu is preferred as the activation function to prevent gradient loss in the study, the results obtained with different activation functions may vary. However, the results obtained may vary by using many different parameters when evaluating more complex or smaller data sets. Although the classification accuracies of algorithms are high, better results can be obtained with different optimizations of hyperparameters.

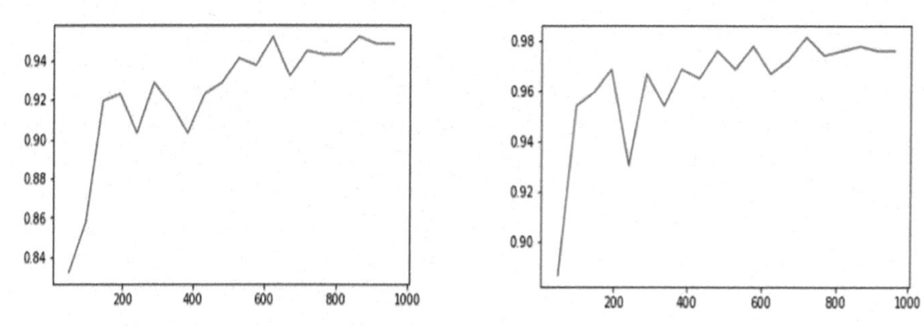

Fig. 3. AlexNet accuracy curve graph (on left) VggNet-16 accuracy curve graph (on right)

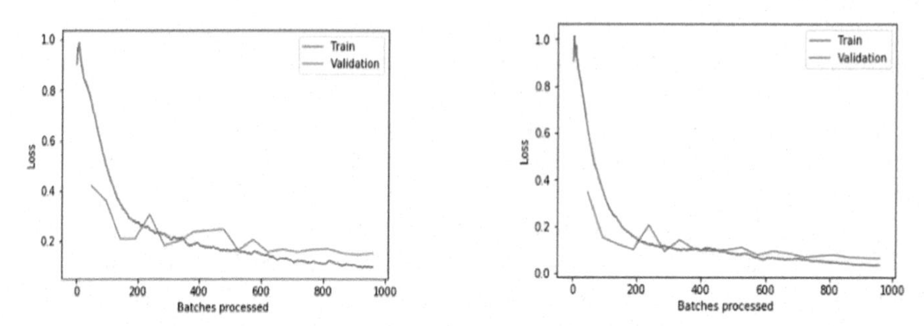

Fig. 4. AlexNet loss curve graph (on left) VggNet-16 loss curve graph (on right)

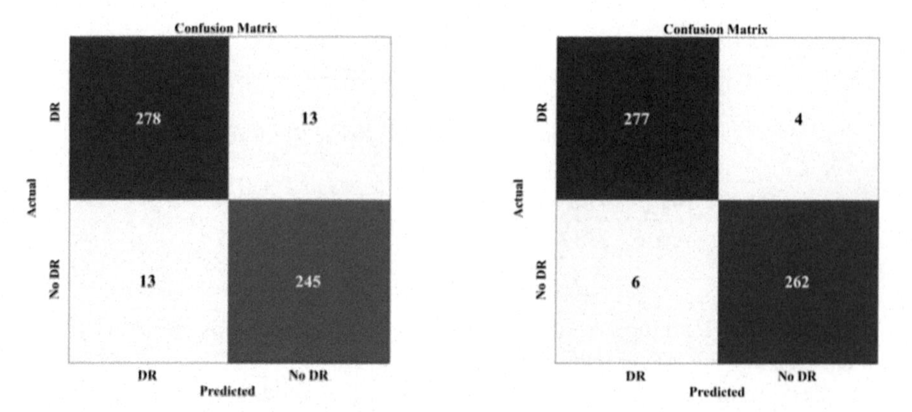

Fig. 5. AlexNet confusion matrix (on left) VggNet-16 confusion matrix (on right)

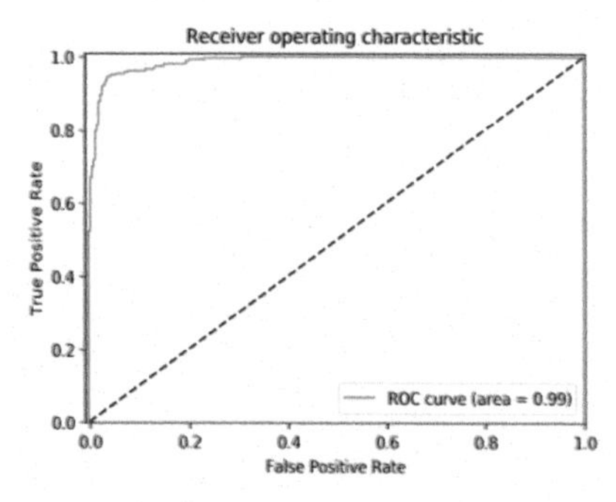

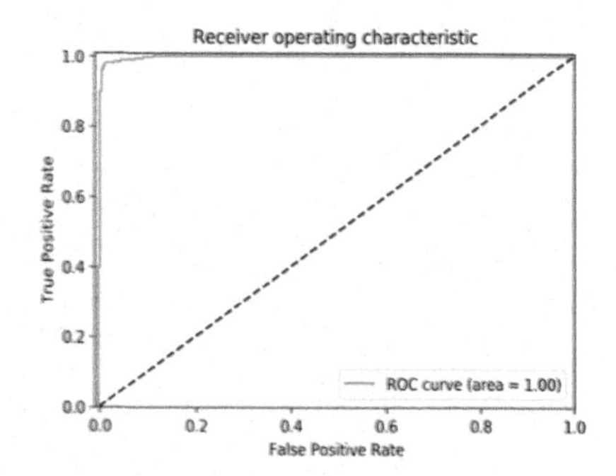

Fig. 6. AlexNet Roc curve (on left) VggNet-16 Roc curve (on right)

5 Conclusions and Future Work

In this study, DR classification was examined using the CNN model developed by transfer learning methods. 3662 fundus images were used for classification. The proposed model achieved a high accuracy of 95.26% using AlexNet and 98.18% using VggNet-16. In the classification process, only the data increase process was applied to the training data and the verification data were taken directly from the original images. The developed model and applied method achieved a good performance in APTOS dataset compared to different Xception architectures and other state-of-the-art algorithms. The limitation of the study is that the data set was evaluated in only two different classes. In future studies, bleeding vessels in fundus images can be made prominent with different image processing techniques. In addition, the success of the disease in different classes can be tested by increasing the DR classification category. The number of deep transfer learning algorithms used in DR classification can be increased and the success of other transfer learning algorithms can be compared.

References

1. Sahlsten, J., et al.: Deep learning fundus image analysis for diabetic retinopathy and macular edema grading. Sci. Rep. **9**(1), 1–11 (2019)
2. De La Torre, J., Valls, A., Puig, D.: A deep learning interpretable classifier for diabetic retinopathy disease grading. Neurocomputing **396**, 465–476 (2020)
3. Zago, G.T., Andreão, R.V., Dorizzi, B., Salles, E.O.T.: Diabetic retinopathy detection using red lesion localization and convolutional neural networks. Comput. Biol. Med. **116**, 103537 (2020)
4. Pratt, H., Coenen, F., Broadbent, D.M., Harding, S.P., Zheng, Y.: Convolutional neural networks for diabetic retinopathy. Procedia Comput. Sci. **90**, 200–205 (2016)
5. Sarki, R., Michalska, S., Ahmed, K., Wang, H., Zhang, Y.: Convolutional neural networks for mild diabetic retinopathy detection: an experimental study. bioRxiv 763136 (2019)
6. Jain, A., Jalui, A., Jasani, J., Lahoti, Y., Karani, R.: Deep learning for detection and severity classification of diabetic retinopathy. In: 2019 1st International Conference on Innovations in Information and Communication Technology, pp. 1–6. IEEE (2019)
7. Shaban, M., et al.: Automated staging of diabetic retinopathy using a 2d convolutional neural network. In: 2018 International Symposium on Signal Processing and Information Technology, pp. 354–358. IEEE (2018)

8. Lam, C., Yi, D., Guo, M., Lindsey, T.: Automated detection of diabetic retinopathy using deep learning. In: AMIA Summits on Translational Science Proceedings (2018)
9. Hagos, M.T., Kant, S.: Transfer learning based detection of diabetic retinopathy from small dataset. arXivpreprint arXiv 1905 07203 (2019)
10. APTOS 2019 Blindness Detection. https://www.kaggle.com/c/aptos2019-blindness-detection/data. Accessed 08 Feb 2021
11. Islam, M.R., Hasan, M.A.M., Sayeed, A.: Transfer learning based diabetic retinopathy detection with a novel preprocessed layer. In: 2020 IEEE Region 10 Symposium TENSYMP, Dhaka, Bangladesh, pp. 888–891 (2020)
12. Taufiqurrahman, S., Handayani, A., Hermanto, B.R., Mengko, T.L.E.R.: Diabetic retinopathy classification using a hybrid and efficient MobileNetV2-SVM model. In: 2020 IEEE REGION 10 CONFERENCE (TENCON), Osaka, Japan, pp. 235–240 (2020)
13. Krizhevsky, A., Sutskever, I., Hinton, G.E.: Imagenet classification with deep convolutional neural networks. Adv. Neural. Inf. Process. Syst. **25**, 1097–1105 (2012)
14. Simonyan, K., Zisserman, A.: Very deep convolutional networks for large-scale image recognition. arXivpreprint arXiv 1409 1556 (2014)

From Statistical to Deep Learning Models: A Comparative Sentiment Analysis Over Commodity News

Mahmut Sami Sivri(✉), Buse Sibel Korkmaz, and Alp Ustundag

Istanbul Technical University, Istanbul, Turkey
sivri@itu.edu.tr

Abstract. The sentiment analysis of news and social media posts is a growing research area with advancements in natural language processing and deep learning techniques. Although various studies addressing the extraction of the sentiment score from news and other resources for specified stocks or a stock index, still there is a lack of an analysis of the sentiment in more specialized topics such as commodity news. In this paper, several natural language processing techniques with a varying range from statistical methods to deep learning-based methods were applied on the commodity news. Firstly, the dictionary-based methods were investigated with the most common dictionaries in financial sentiment analysis such as Loughran & McDonald and Harvard dictionaries. Then, statistical models have been applied to the commodity news with count vectorizer and TF-IDF. The compression-based NCD has been also included to test on the labeled data. To improve the results of the sentiment extraction, the news data was processed by deep learning-based state-of-art models such as ULMFit, Flair, Word2Vec, XLNet, and BERT. A comprehensive analysis of all tested models was held. The final analysis indicated the performance difference between the deep learning-based and statistical models for the sentiment analysis task on the commodity news. BERT has achieved superior performance among the deep learning models for the given data.

Keywords: Sentiment analysis · Natural language processing · Commodity news analysis · Financial sentiment analysis

1 Introduction

According to Efficient Market Hypothesis, each piece of new relevant information drives the market dynamics [1]. To understand the market dynamics, financial news articles are precious sources since they cover information that may influence all kinds of financial instruments. Financial news includes important information about the fundamentals of the firm and qualitative details affecting the expectations of market contributors. Investors make their decision by relying on supply, demand, quantitative data that is released by firm or market authority, and market expectations that can be measured by sentiment. Although the nature of the first three indicators tends to be structured data, the last one

C. Kahraman et al. (Eds.): INFUS 2021, LNNS 308, pp. 155–162, 2022.
https://doi.org/10.1007/978-3-030-85577-2_18

requires conducting analyzes on unstructured data such as text. This type of information may be collected through financial news channels such as blogs, news websites, and social media where people state their concerns and forecasts about the market.

Extraction of stock' sentiment index from sector-related news is a well-studied field. However, there is a lack of established studies for sentiment analysis on specialized financial instruments that have unique dynamics such as commodities. This paper addresses the issue by providing comparative sentiment analysis over commodity news by implementing several natural language processing techniques from statistical methods to artificial intelligence based deep learning methods.

The remainder of the paper is structured as follows: In Sect. 2, relevant studies for sentiment analysis for financial news are analyzed and their major contributions are presented. The benefited natural language processing techniques from statistical methods to deep learning based methods are introduced and explained in detail in Sect. 3. Section 4 includes the feature descriptions and exploratory analysis of the used dataset. The explained techniques were applied to the given dataset and the results are shared in Sect. 5 with experiment environment properties. Section 6 presents the conclusion of this paper with the given experiment environment and suggestions for future research.

2 Literature

Financial sentiment analysis was considered an important challenge in the literature until the recent advancements in the natural language processing and deep learning fields [2, 3]. The most challenging part of the sentiment analysis or opinion mining problem is the polarity of the words (i.e. negativity or positivity) differs regarding the context of the text. For example, the word "increase" has a positive connotation in a sales text while having a negative connotation in cost-focused text.

In the early days of sentiment analysis studies, rule-based approaches were common. This approach benefits from a lexicon that maps important words into categories. Researchers produced several dictionaries to use in general textual analysis as well as in financial texts. The first dictionary was Harvard General Inquirer (Harvard GI) and Harvard IV followed it [4]. This dictionary was formed by 12.000 words from different folkloric texts to avoid cultural bias. However, since Harvard dictionaries do not have a specialized context, they could not achieve meaningful precision rates in the sentiment classification of financial statements. In 2011, Loughran indicated that domain dependency is a particularly strong notion in the financial context. To address this issue, Loughran and McDonald proposed a financial sentiment dictionary [4]. It also became a starting point for domain-specific dictionaries such as SenticNet, SentiStrength2, and Financial Polarity Lexicon (FPL) [6–8]. Even though the lexicon provides a fast classification of the differentiative words, rule-based approaches only achieved around a 60% accuracy rate in the study of Taj et al. [9].

Then, the natural language processing field flourished parallel to the advancement of artificial intelligence. In 2013, Mikolov et al. introduced the first word embedding model that enables examine the individual words in the context of the whole corpus [10]. Since the word embeddings are just the multi-dimensional vector representation of the words in a context-dependent manner, to classify the sentiment there is a need to train a machine learning or deep learning model with sentiment labels of the news, text, or document and with word embeddings. The most used word embeddings in the sentiment analysis are word2Vec which was trained on Google News data and, GLoVE (Global Vectors for Word Representation) by Stanford [11]. Esichaikul and Phumdontree developed an embedding-based sentiment analysis model for Thai financial news and trained with different deep learning classifiers. They reported an 84% F1 score with CNN-Bidirectional GRU [12]. Although their work proves the potential of word embeddings, their research addressed the analysis of generic financial news, not the sentiment analysis of a specialized financial instrument such as commodities.

In 2017, transformers architecture was introduced by the "Attention Is All You Need" titled paper and moved natural language processing research one step further by implementing transfer learning [13]. The most successful transformers models indicated as BERT (Bidirectional Encoder Representations from Transformers), XLNet, and Flair in different studies regarding the dataset and the experiment environment [14–19]. In this paper, all mentioned natural language processing techniques will be applied to commodity news data to obtain the best model for sentiment analysis of commodities.

3 Dataset

The dataset includes in a year ranged 1120 labelled commodity-related news headlines collected from Reuters by Eikon API. Labels are 3-fold: "−1" for negative connotated headlines, "0" for neutral headlines, and "1" for positive connotated headlines. The sentiment distribution of the dataset is given in Table 1. Test and training datasets are split as will be preserving the main dataset sentiment distribution and have 15% and 85% proportion relatively. Each news in the dataset includes a tag such as News, Update, or Metals. Since these tags are not related to the sentiment, tags will be excluded in the data preprocessing stage.

Table 1. Sentiment and test-train distribution of the dataset

Sentiment label	Number of classified news	Connotation	Train dataset	Test dataset
1	609	Positive	516	91
0	76	Neutral	64	12
−1	435	Negative	370	65

4 Method

Sentiment analysis of the given dataset will be held as in Fig. 1. Firstly, data will be collected and then will be preprocessed. After that, the model training and evaluation phase will start. The details per approach are explained below.

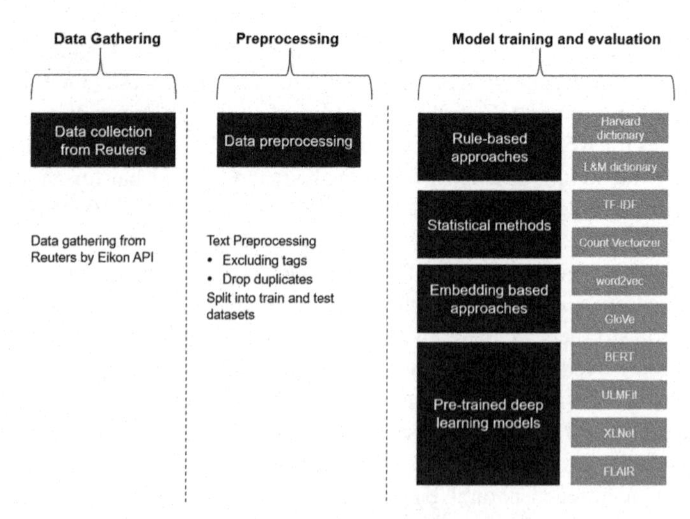

Fig. 1. Comparative sentiment analysis research includes three phases. 1) Data gathering: The commodity news that includes zinc, lead, aluminum, copper, nickel, etc. will be gathered from Reuters by Eikon API. 2) Data preprocessing: The collected data includes timestamps and tags for each news. This non-relevant information is excluded. 3) Model training and evaluation: If the approach consists of machine learning or deep learning model, at first the model is trained with a training dataset. Then, evaluate the F1 score and accuracy rate per sentiment label. If the approach does not consist of a learning stage, evaluates the performance of the approach.

4.1 Rule-Based Approaches

Harvard and Loughran & McDonald (L&M) dictionaries are provided by the pysentiment2 library of Python. In their raw version, these dictionaries include many classes more than we have in our classification. Fortunately, the mentioned library provides polarity of the given text over our three main categories, so it is compatible with our method. The polarity of the text is decided on the word count of each category after tokenizes the given headline.

4.2 Statistical Methods

Term Frequency - Inverse Document Frequency (TF-IDF) and count vectorizer (CV) methods are used to get the most insightful features from the dataset. The mentioned vectorization techniques apply statistical analysis to capture the relations between the label and the features such as individual words and n-grams. In this study, we utilized at most 3-g since the further increase of the n-gram size did not improve the model performance. Also, the p-score for the selection of the features was decided as 0.95. The selected features and the labels of the headlines supplied to multiple machine learning classifiers included k-nearest neighbors with $k = 3$, naïve Bayes (NB), decision tree (DT), random forest (RF), multi-layer perceptron (MLP), AdaBoost classifier. NCD is a compression-based statistical method that also has been applied to data [20]. The result of each combination is reported in the next section.

4.3 Embedding-Based Approaches

In this approach, Word2Vec embedding by Google and GloVe embedding by Stanford were utilized. Word2Vec embedding has 300 dimensions and, GloVe embedding includes 50, 100, 200, and 300 dimensions that are trained on Wikipedia and Gigaword. This paper has benefited from 300-dimensional GloVe vectors to obtain an exact comparison between Word2Vec and GloVe vectors. Firstly, individual words in a headline were mapped to the corresponding embedding. For each dimension of the headline, embedding values were averaged and, the final average vector became a representation of that headline. Each headline and the label were given to a neural network classifier for classification.

4.4 Pre-trained Deep Learning Models

BERT, XLNet, ULMFit, and Flair are pre-trained, deep learning models. Each model has different characteristics. Detailed explanations for pre-trained language models are given in Table 2.

Table 2. Pre-trained language model selection for each DL model

DL model	Language model
BERT	Bert-base-uncased
Flair	Distilbert-base-uncased
XLNET	Xlnet-large-cased
ULMFit	Trains its own language model

The specified models benefit from transformers architecture with attention masks, and recurrent neural network layers to obtain descriptive features of the dataset and consist of a classifier layer at the last step. Hyperparameter finetuning has been applied to optimize the results for each model.

5 Experiment

5.1 Environment

Regarding the resource need of different approaches, two environments have been used. For rule-based, statistical, embedding-based approaches development has been made on Jupyter Notebook and results have been evaluated on Intel core i7-10610U @2.30 GHz CPU with 32 GB memory while deep learning models have been trained and evaluated on Google Colab environment with 12GB NVIDIA Tesla K80 GPU.

5.2 Results

For each model, previously explained stages has been performed and the results in Table 3 are obtained.

Table 3. Comparative sentiment analysis result for commodity news

Technique	Accuracy for negative news	Accuracy for neutral news	Accuracy for positive news	Overall accuracy
Harvard dictionary	0.39	0.38	0.37	0.38
L&M dictionary	0.46	**0.58**	0.13	0.29
CV + KNN(k = 3)	0.57	0.0	0.76	0.68
CV + MLP	0.64	0.10	0.78	0.71
CV + DT	0.38	0.0	0.74	0.62
CV + RF	0.21	0.0	0.73	0.59
CV + NB	0.65	0.17	0.79	0.71
CV + AdaBoost	0.65	0.20	0.78	0.70
TF-IDF + KNN(k = 3)	0.62	0.07	0.71	0.64
TF-IDF + MLP	0.64	0.0	0.79	0.71
TF-IDF + DT	0.41	0.0	0.71	0.60
TF-IDF + RF	0.36	0.05	0.73	0.61
TF-IDF + NB	0.63	0.15	0.78	0.71
TF-IDF + AdaBoost	0.56	0.18	0.76	0.67
NCD	0.10	0.0	0.77	0.54
Word2Vec	0.68	0.02	0.80	0.68
GloVe	0.69	0.12	0.73	0.66
BERT	**0.71**	0.0	**0.84**	**0.73**
XLNet	**0.71**	0.11	0.81	0.72
Flair	0.64	0.06	0.79	0.72
ULMFit	0.58	0.10	0.76	0.64

6 Conclusion

As shown in Table 3, deep learning models consistently outperformed the statistical and dictionary-based models. BERT is the best-performed model for overall as well as in positive and negative news. Since the dataset has imbalanced distribution, machine learning and deep learning included models tend to perform better for positive news which has a major portion in the given dataset. Even though having domain-specific knowledge is very important in a financial context, transformers architecture and attention based all models in the experiment (BERT, XLNet, and Flair) caught and surpassed the accuracy result of sentiment dictionaries and statistical methods. The main contribution of this research is demonstrating that the attention masks in the transformers architecture have superior performance than the domain-specific expert knowledge embedded prior models such as the lexicons in dictionary-based methods or the vectorization in statistical methods. The mentioned contribution has been obtained through the detailed analysis by varying natural language processing techniques on a specialized financial instrument. This comparative sentiment analysis could be further in the future by expanding the dataset to include news for other financial instruments and balancing the dataset. Also, the obtained sentiment indexes could be leveraged for forecasting the commodity market.

References

1. Delcey, T.: Efficient Market Hypothesis, Eugene Fama and Paul Samuelson: A reevaluation (2017)
2. Loughran, T., McDonald, B.: Textual analysis in accounting and finance: a survey. J. Account. Res. **54**(4), 1187–1230 (2016)
3. Loughran, T., McDonald, B.: The use of word lists in textual analysis. J. Behav. Financ. **16**(1), 1–11 (2015)
4. Stone, P.J., Hunt, E.B.: A computer approach to content analysis: studies using the general inquirer system. In: Proceedings of the May 21–23, 1963, Spring Joint Computer Conference (AFIPS 1963 (Spring)). Association for Computing Machinery, New York, NY, USA, pp. 241–256 (1963)
5. Loughran, T., McDonald, B.: When is a liability not a liability? textual analysis, dictionaries, and 10-Ks. J. Finance **66**(1), 35–65 (2011)
6. Cambria, E., Olsher, D., Rajagopal, D.: SenticNet 3: a common and common-sense knowledge base for cognition-driven sentiment analysis. In: AAAI, pp. 1515–1521, Quebec City (2014)
7. Thelwall, M., Buckley, K., Paltoglou, G.: Sentiment strength detection for the social web. J. Am. Soc. Inf. Sci. **63**, 163–173 (2012). https://doi.org/10.1002/asi.21662
8. Malo, P., Sinha, A., Korhonen, P., Wallenius, J., Takala, P.: Good debt or bad debt: detecting semantic orientations in economic texts. J. Am. Soc. Inf. Sci. **65**(4), 782–796 (2014)
9. Taj, S., Shaikh, B.B., Meghji, A.F.: Sentiment analysis of news articles: a lexicon based approach. In: 2nd International Conference on Computing, Mathematics and Engineering Technologies (iCoMET), Sukkur, Pakistan (2019). https://doi.org/10.1109/ICOMET.2019.8673428
10. Mikolov, T., Chen, K., Corrado, G., Dean, J.: Efficient estimation of word representations in vector space. In: Proceedings of the International Conference on Learning Representation (ICLR) (2013)
11. Jeffrey, P., Socher, R., Manning, C.: Glove: global vectors for word representation. In: Proceedings of the 2014 Conference on Empirical Methods in Natural Language Processing (EMNLP), pp. 1532–1543 (2014)
12. Esichaikul, V., Phumdontree, C.: Sentiment analysis of thai financial news. In: Proceedings of the 2018 2nd International Conference on Software and e-Business (ICSEB 2018). Association for Computing Machinery, New York, NY, USA, pp. 39–43 (2018)
13. Vaswani, A., et al.: Attention is all you need. In: Proceedings of the Advances in Neural Information Processing Systems 30 (NIPS) (2017)
14. Othan, D., Kilimci, Z.H., Uysal, M.: Financial sentiment analysis for predicting direction of stocks using bidirectional encoder representations from transformers (BERT) and deep learning models. In: International Conference on Innovative and Intelligent Technologies (ICIIT-19), Istanbul, Turkey (2019)
15. Sousa, M.G., Sakiyama, K., Rodrigues, L.D.S., Moraes, P.H., Fernandes, E.R., Matsubara, E.T.: BERT for stock market sentiment analysis. In: 2019 IEEE 31st International Conference on Tools with Artificial Intelligence (ICTAI), Portland, OR, USA, pp. 1597–1601 (2019)

16. Devlin, J., Chang, M., Lee, K., Toutanova, K.: BERT: pre-training of deep bidirectional transformers for language understanding. In: The Proceedings of the 2019 Conference of the North American Chapter of the Association for Computational Linguistics: Human Language Technologies, Volume 1 (Long and Short Papers), Minneapolis, Minnesota, pp. 4171–4186 (2019)
17. Yang, Z., Dai, Z., Yang, Y., Carbonell, J., Salakhutdinov, R., Le, Q.V.: XLNet: generalized autoregressive pretraining for language understanding. In: The Proceedings of Advances in Neural Information Processing Systems 32 (NeurIPS) (2019)
18. Akbik, A., Bergmann, T., Blythe, D., Rasul, K., Schweter, S., Vollgraf, R.: FLAIR: an easy-to-use framework for state-of-the-art NLP. In: The Proceedings of NAACL-HLT, Minneapolis, Minnesota, 2–7 June 2019, pp. 54–59 (2019)
19. Howard, J., Ruder, S.: Universal language model fine-tuning for text classification. In: the Proceedings of the 56th Annual Meeting of the Association for Computational Linguistics (Volume 1: Long Papers), Melbourne, Australia, pp. 328–339 (2018)
20. Cohen, A.R., Vitanyi, P.M.B.: Normalized compression distance if multisets with applications. IEEE Trans. Pattern Anal. Mach. Intell. **37**(8), 1602–1614 (2014)

Deep Learning Network Model Studies for Adversarial Attack Resistance

Fei Chen and Jaeho Choi(✉)

Department of Electronics Engineering, JBNU, CAIIT, Jeonju, Republic of Korea
wave@jbnu.ac.kr

Abstract. In the last decades, deep learning neural networks have taken several steps toward higher pattern recognition accuracies. Face recognition is one of the popular topics that have drawn much attention and it is now frequently used in everyday lives. However, the recognition performance suffers easily by irregularities and disturbances. The focus of this work is to explore the security performance of deep learning neural networks by using an adversarial attack approach. The ResNets is the framework of the proposed system and its recognition behaviors under the adversarial attacks are investigated. The experiments are performed by using MNIST and CIFAR-10 datasets and the detection and recognition errors are evaluated. The results show that the proposed systems can be an alternative that can resist perturbations better than the conventional models.

Keywords: Deep learning neural networks · ResNet · Adversarial attacks

1 Introduction

Deep neural networks are becoming increasingly influential in many complex machine learning tasks. Not only can they recognize images with near-human accuracy [1], but they are also used for speech recognition, natural language processing [2], and for playing games [3]. However, researchers have found that existing neural networks are vulnerable to adversarial attacks. An adversarial attack method [4] is developed to demonstrate the properties of neural networks under attacks and it is also led to problems in computer vision. For example, an arbitrary attacker may introduce small manipulation in an image that causes an automatic vehicle recognition system to misinterpret the speed limit or similar actions, an outcome that should be avoided for public safety. The study of adversarial attack examples can help us to understand the robustness of different model structures and the advantages and disadvantages of different training algorithms to ensure deep learning security in real-world applications.

In this paper, a classification metric based on the ResNet [5] model for adversarial perturbations is investigated. The safety of the neural network is observed by the accuracy of different classifications. The variation in the accuracy rate proves that the neural network should focus on the disturbance information in the data.

In the following, Sect. 2 presents the related work concerning adversarial attacks and the deep learning neural model called the ResNet. In Sect. 3, the proposed method is described; Sect. 4 contains adversarial attack experiments using two datasets and the recognition performance is evaluated. Finally, the conclusion is made in Sect. 5.

C. Kahraman et al. (Eds.): INFUS 2021, LNNS 308, pp. 163–169, 2022.
https://doi.org/10.1007/978-3-030-85577-2_19

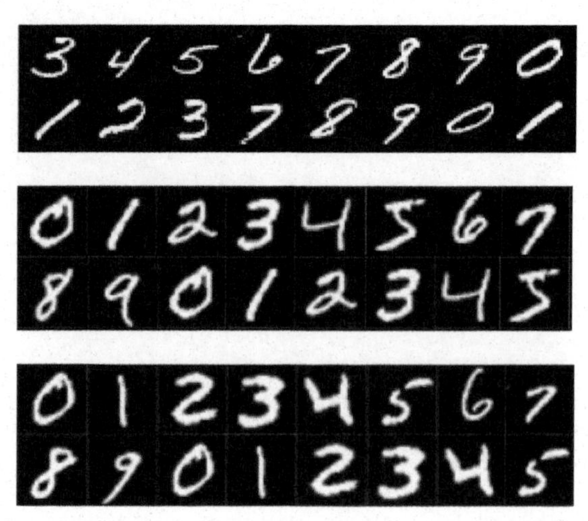

Fig. 1. The original MNIST image (top); 2/255 perturbed image (middle); 32/255 perturbed image (bottom).

2 Related Work

Classifiers based on modern machine learning techniques and algorithms can handle naturally occurring data well but they are not yet suitable for image data perturbed by noises, namely the adversarial attack. The adversarial attack can cause safety hazards by misleading the system in recognition. It becomes important that it can affect the neural network's correctness. The phenomenon that neural networks cannot accurately distinguish images with attack information is seen as a research opportunity.

2.1 Adversarial Attacks

The opposing optimization view has a significant reference value for studying the security of neural networks. In this paper, the adversarial attack is interpreted as a scheme to increase perturbations in the image data. The perturbation is generated and added in multi-steps as follows [6, 7]:

$$x^{t+1} = \prod\nolimits_{x+S} \left(x^t + \alpha\, \boldsymbol{sgn}(\nabla_x L(\theta, x, y))\right) \tag{1}$$

where x is the image dataset to be perturbed; t is the iteration index; S is the perturbation dataset; α is a gain; sgn() is the signum function; ∇_x is the gradient; L is the cross-entropy loss function between two datasets x and y; θ is the probability. Some examples of image perturbation are shown in Fig. 1.

2.2 ResNet

The vanishing gradient problem occurs during artificial neural networks training, specifically during back-propagation [8]. Several methods have been proposed to solve this

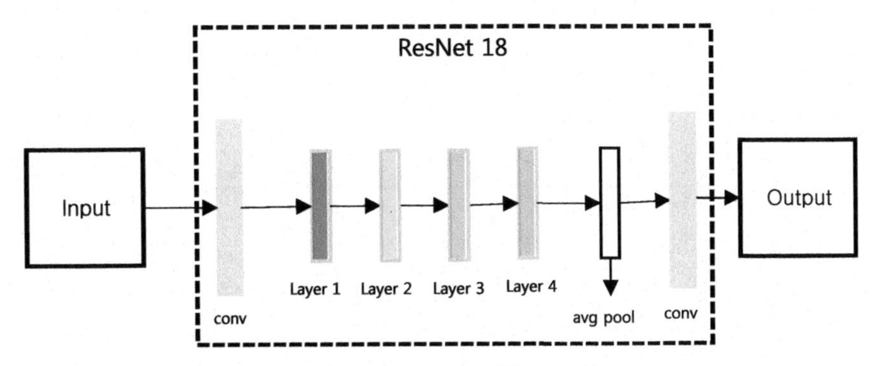

Fig. 2. The structure of Resnet-18.

problem. One of the approaches is to perform a two-step training process: 1) the network weights are trained using unsupervised learning methods 2) the weights are fine-tuned by using the back-propagation [9]. The other method is the Resnet network. It has residual networks and skip connections that help improve on this problem.

The structure of Resnet-18 is shown in Fig. 2. The input image is first passed to the convolutional layer, which has the regular filter size of 7×7; the feature map is 64; the stride is 2. Then, it enters into the intermediate convolution part, which contains four blocks, each block consists of 4 convolutional layers; then, the average pooling layer comes next, which is used to minimize overfitting by reducing the total number of parameters in the model. Finally, a fully convolutional layer is connected.

3 Proposed Deep Learning Network Model

The proposed method in this work is based on the Resnet. Using the proposed model, the classification behavior is investigated under that consideration that the system is exposed to the adversarial attacks on the image dataset.

The structure of the proposed system model based on the Resnet-18 is shown in Fig. 3. In this network, the input image of a size 32×32 is passed to the projected gradient descent (PGD) block, first. In the PGD block, the adversarial attack perturbation of the same size is generated and added to the image data. Then, the perturbed image data is fed into the Resnet-18 network. The Resnet here has 7×7 convolutional layers, 2 pooling layers, 4 residual units, and one fully connected layer. Each of the residual units include four 3×3 convolution layers. Also, in the convolution layer, the Relu function is used as the activation function; the SoftMax function is used in the last fully connected layer.

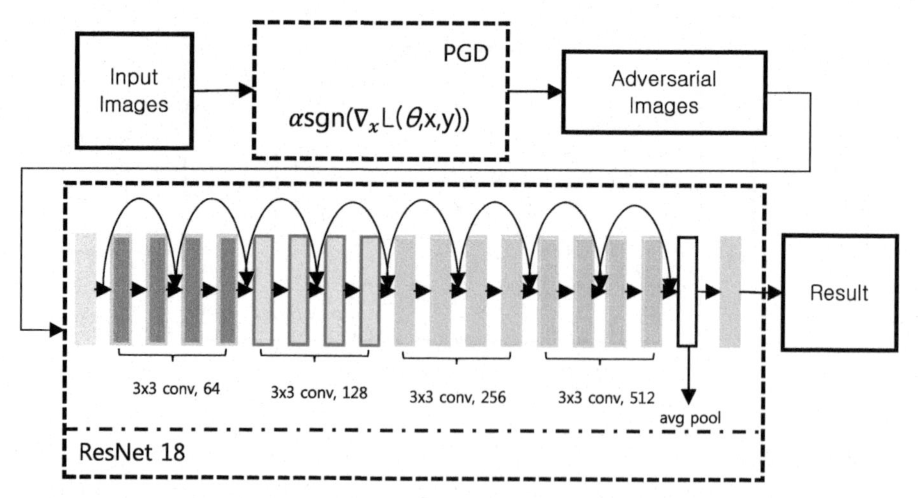

Fig. 3. The proposed model for adversarial attack mode deep learning based on Resnet-18.

4 Experiments and Results

In this section, the simulation experiments are performed to evaluate the perturbation effects of the adversarial attacks on the proposed network model. Two image datasets, i.e., MNIST [10] and CIFAR-10 [11] are used for training and testing. MNIST [10] dataset contains the handwritten Arabic numerals as shown in Fig. 1 and CIFAR-10 [11] contains natural scenes, which is popularly used for object recognition experiments. More specifically, the MNIST dataset consists of 60,000 grayscale images of 32×32 size with 10 different class labels; CIFAR-10 has 60,000 color images of 32×32 size with 10 different labels. Table 1 show the brief summary of two databases. Each dataset is divided into 5 training batches and 1 test batch, each batch containing 10,000 images. The proposed model runs for 50 iterations with a learning rate size of 0.001. It takes 4h of training for the MNIST dataset and 10h for the CIFAR-10 dataset.

For the experiment, the perturbation is added to the images of the dataset. Using Eq. (1), the perturbation images of size 32×32 are generated; the perturbation amount is varied by perturbing the pixel values. The numbers 2, 32 and 128 in the fractions 2/255, 32/255, and 128/255 represent the perturbation range of pixel values where 255 is the range of color and saturation values in the image. For example, the RGB values in the color images vary from 0 to 255 and the gain α in Eq. (1) also varies between 0 to 255. Here, the perturbation generation is divided into 40 small amounts and the small perturbations process repeats recursively 40 times to produce perturbed images. While Fig. 1 shows the sample images from MNIST dataset, Fig. 4 also shows some sample images from the CIFAR-10 dataset that one can easily see the adversarial effects of perturbations as the perturbation range increases.

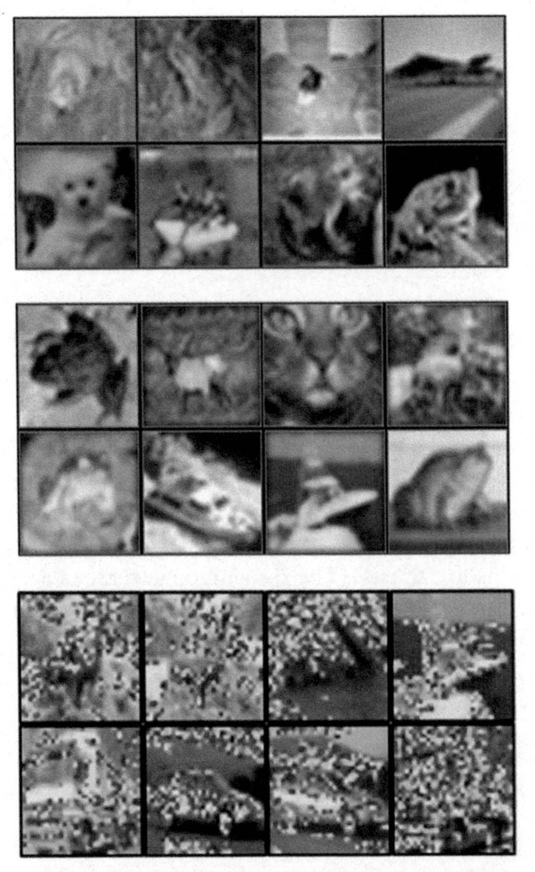

Fig. 4. The image samples of CIFAR-10 with no perturbation (top), 2/255 perturbation (middle), and 32/255 perturbation (bottom).

After training the system with the perturbed image data, the system is tested for the recognition accuracy; the recognition accuracy is obtained by evaluating the cross-entropy loss values. Figure 5 shows the recognition accuracy for two datasets under adversarial attacks. Over the iterations, the recognition accuracy increases gradually to 52% for the MNIST dataset, and to 35% for the CIFAR-10 dataset.

Table 1. Two dataset compositions: MNIST and CIFAR-10.

Dataset name	Image data	Train size	Test size	Label type
MNIST	60,000	50,000	10,000	10
CIFAR-10	60,000	50,000	10,000	10

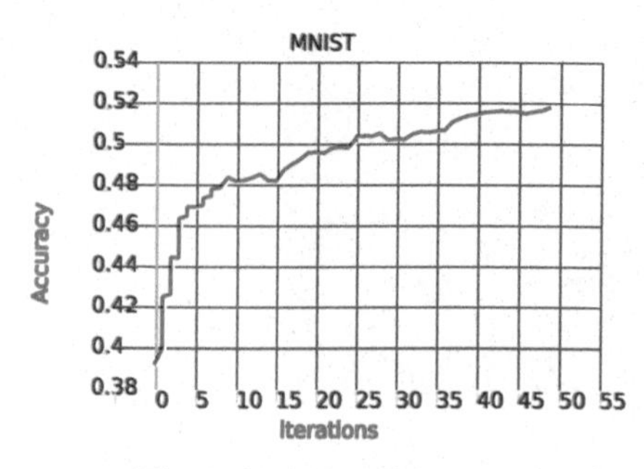

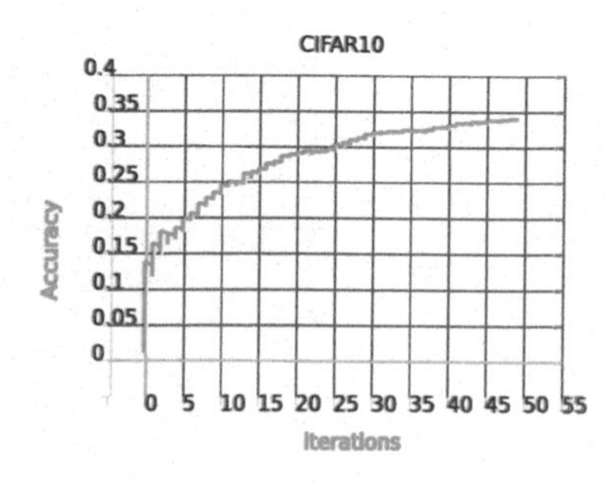

Fig. 5. The recognition accuracies of the proposed model for two datasets.

In addition, Table 2 contains the summary of experiments concerning the recognition accuracies for two datasets with respect to the increased amount of perturbation noise. Here, 50 iterations are used, and on each step the recognition accuracy is checked. Using the dataset without any perturbation noise the systems shows the accuracy of 99% for the MNIST dataset and 95% for CIFAR-10. As the perturbation noise increases, the recognition accuracy begins to decrease. The recognition accuracy degradation is relatively regular for the CIFAR-10 dataset and it decreases from 94% to 31% as the perturbation noise increases from 2/225 to 128/225. On the other hand, tor the MNIST dataset the degradation tendency is somewhat irregular, however, eventually the performance degradation gets worse as the perturbation noise increases.

Table 2. Experimental results: recognition accuracy vs. perturbation noise.

MNIST			CIFAR-10		
Iteration	Noise	Accuracy	Iterations	Noise	Accuracy
50	None	99%	50	None	95%
50	2/255	36%	50	2/255	95%
50	32/255	59%	50	32/255	75%
50	128/255	52%	50	128/255	35%

5 Conclusion

In this paper the attack-mode deep learning network model based on the Resnet framework has been presented in order to investigate and evaluate the perturbation effects of adversarial attacks on the recognition performance of the systems. The experimental results provide an evidence that the deep neural network based the Resnet framework can resist adversarial attacks better than the conventional models. Also the results show some different behaviors and properties for different datasets, and this fact calls for a more future study such as the backward robustness of neural networks.

References

1. LeCun, Y., Bottou, L., Bengio, Y., Haffner, P.: Gradient-based learning applied to document recognition. Proc. IEEE **86**(11), 2278–2324 (1998)
2. Andor, D., et al.: Globally normalized transition-based neural networks. arXiv:1603.06042 (2016)
3. Mnih, V., et al.: Playing atari with deep reinforcement learning. arXiv:1312.5602 (2013)
4. Szegedy, C., et al.: Intriguing properties of neural networks. arXiv:1312.6199 (2013)
5. He, K., Zhang, H., Ren, S., Sun, J.: Deep residual learning for image recognition. arXiv:1512.03385 (2016)
6. Madry, A., Makelov, A., Schmidt, L., Tsipras, D., Vladu, A.: Towards deep learning models resistant to adversarial attacks. arXiv:1706.06083 (2017)
7. Xu, J., Liu, H., Wu, D., Zhou, F., Gao, C., Jiang, L.: Generating universal adversarial perturbation with ResNet. Inform. Sci. **537**, 302–312 (2020)
8. Hochreiter, S., Bengio, Y., Frasconi, P., Schmidhuber, J.: Gradient flow in recurrent nets: the difficulty of learning long-term dependencies (2001)
9. Schmidhuber, J.: Learning complex, extended sequences using the principle of history compression. Neural Comput. **4**(2), 234–242 (1992)
10. Deng, L.: The MNIST database of handwritten digit images for machine learning research. IEEE Signal Process. Mag. **29**(6), 141–142 (2012)
11. Krizhevsky, A., Hinton, G.: Learning multiple layers of features from tiny images (2009)

An Ensemble Learning Approach for Energy Demand Forecasting in Microgrids Using Fog Computing

Tuğçe Keskin and Gökhan İnce(✉)

Computer Engineering Department, Istanbul Technical University,
34467 Istanbul, Sarıyer, Turkey
{yilmaztu17,gokhan.ince}@itu.edu.tr

Abstract. Increased usage of smart meters enables information exchange between customers and utility providers in smart grid systems. Nowadays, the cloud-centric architecture has become a bottleneck for the decentralized and data-driven microgrids evolving from centralized Smart grids. Hence, fog computing is an appropriate paradigm to build distributed, latency-aware, and privacy-preserving energy demand applications in microgrid systems. In this work, we proposed a 3-tier architecture of a microgrid energy demand management system comprising edge, fog, and cloud layers. We set up a simulation environment where Raspberry Pi devices act as fog nodes and resource-efficient Docker applications run on these nodes. As the main contribution of the work, we developed a short-term load forecasting application based on an ensemble model that integrates support vector regression (SVR) and long-short term memory (LSTM) by leveraging the potential of distributed and low-latency fog nodes for complex models. We evaluated the forecasting model deployed in a fog-based simulation environment using the public REFIT Electrical Load dataset. We also tested the deployed fog-based simulation environment based on latency and execution time metrics.

Keywords: Smart grids · Microgrids · Internet of Things · Fog computing · Energy demand forecasting · Ensemble learning

1 Introduction

The smart grid is a communication network that allows to collect and analyze data from the components of the electricity grid to predict power demand and supply for better power management [1]. Recently, microgrids, which are small-scale and decentralized versions of smart grids, are used in electrical systems. These networks enable using distributed power generators and renewable energy sources. Microgrids are closer to demand than traditional grids, resulting in transmission reduction and efficiency gains [2]. To ensure the interaction between microgrids and the distribution network, the demand and generation needs in the microgrid must be determined. Therefore, short-term load forecasting (STLF) applications, from hours to weeks, is a priority research subject [3].

C. Kahraman et al. (Eds.): INFUS 2021, LNNS 308, pp. 170–178, 2022.
https://doi.org/10.1007/978-3-030-85577-2_20

Microgrids require the computing platforms for the storage and processing of the information. Cloud-based applications are considered as useful solutions for demand management due to their massive storage and computing mechanisms [4]. However, the cloud computing paradigm should include a complementary paradigm to meet the needs of decentralized and data-driven microgrid architecture. For example, central cloud systems are not sufficient for the distributed nature of microgrids integrated with advanced metering infrastructure (AMI), renewable energy sources (RES), and electric vehicles (EVs). There is an increasing need to run *distributed intelligence* to manage the variability in the smart grid [5]. Moreover, Internet of Things (IoT) data collected usually contains sensitive information about the users' consumption behavior. Sending raw data to a public cloud accessible to third parties will raise privacy and security concerns. Therefore, an architecture that protects *security and privacy* is needed [6].

For demand response management, the energy generation must be planned based on users' energy demand. Therefore, STLF from hours to weeks, plays a crucial role in predicting the energy demand based on users' historical consumption data. The methods of energy consumption prediction can be roughly presented in three categories [8]: 1) statistical-based approaches, e.g. ARIMA, SARIMA; 2) machine learning approaches, e.g. support vector regression (SVR), random forest regression (RFR), extreme gradient boosting; 3) deep learning approaches, e.g. artificial neural networks (ANNs), long-short term memory based recurrent neural network model (LSTM based RNN). As the prediction capabilities can differ among the learning methods, ensemble learning approach is recently used to obtain better prediction results [9–11]. Ensemble learning takes the multiple prediction results produced by models and calculates an outcome value based on statistical approach such as median, average, weighted average.

As the key contribution of this paper, we proposed an energy demand forecasting architecture that integrates fog computing and an ensemble STLF model for microgrids, taking into account the aforementioned requirements. Research on the integration of smart grids and fog computing has come to the fore in recent years [2,5–7]. However, there is a lack of frameworks that utilize low-latency and distributed fog nodes to validate the artificial intelligence-based smart grid applications in the literature. Therefore, we deployed the proposed energy demand forecasting application with its hardware and software components and performed a set of experiments to show the effectiveness of the fog-based framework. The proposed system architecture for energy demand forecasting and the ensemble model approach for STLF will be explained in Sect. 2. In Sect. 3, experiments for the performance measurement of the developed ensemble model on distributed and low latency fog nodes will be presented. In the next section, it is aimed to evaluate the results of the study.

2 Proposed Framework for Energy Demand Forecasting

2.1 System Architecture

Architecture design and deployment, network and resource management, mobility, and scalability make fog computing modeling challenging for researchers and service providers [12]. Recently, researchers have introduced prominent simulation tools for fog computing, combining these properties in different scales [13–15]. In our study, inspired by the simulation tool of piFogBed [13], we used Raspberry Pi devices suitable for machine learning with low cost and relatively high performance and Docker containers that enables efficient use of system resources. The 3-tier architecture running the use case scenario is shown in Fig. 1.

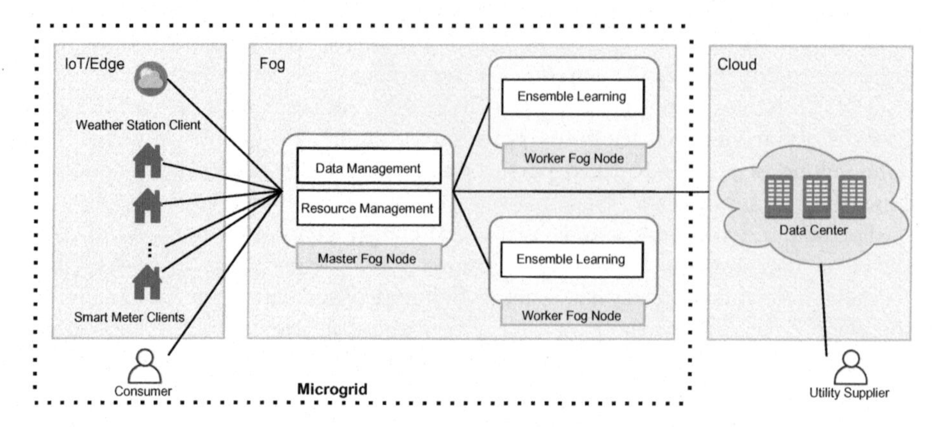

Fig. 1. System architecture

Edge Layer. A smart meter device is simulated with the *smart meter client.* It periodically sends household power consumption data to the fog layer. The weather data such as air temperature and humidity is transmitted to the fog layer by the *weather station client* at certain periods. The physical setup of smart meters and sensors on the IoT and edge layer is not included in our work.

Fog Layer. In the fog layer, there is a Docker Swarm cluster consisting of 1 master node and 2 worker nodes. The master node is specifically responsible for services such as temporary data storage, deployment of tasks to worker nodes or to cloud layer. However, all fog nodes can execute training and inference tasks.

The power consumption and air temperature data published from the edge layer arrive at the *data management module.* The data preprocessing steps are applied here. The aggregated level of household energy consumption data is also propagated to the cloud for long-term analysis.

The *resource management module* schedules training or inference task requests to the (fog or cloud) computing layers based on their resource availability.

The *ensemble learning module* provides the functions to train a group of learning models with historical data and make predictions on new data using saved models. The module runs in distributed fog nodes for complex learning models. The ideal sub models and best model parameters are searched in the training phase. The best models are saved to the disk to be loaded during predictions later. In the inference phase, the average or weighted average of the prediction results obtained from each loaded sub model are ensembled.

Cloud Layer. In case the resources on the fog nodes are not sufficient to execute a task, the task is propagated to cloud servers. In addition, the cloud layer is employed for long-term data storage and data analysis scenarios.

Both *utility suppliers* and *consumers* are target groups for the proposed energy demand forecasting application. The prediction data helps utility suppliers to enable demand management actions such as dynamic pricing [7]. Moreover, it enables the consumers to analyze their energy consumption behavior.

2.2 Energy Demand Forecasting

Data Preprocessing. The raw data transmitted into the data management module in Fig. 1 is *power* measurements in Watt. These measurements are aggregated into *energy* measurements in Watt-hour (Wh) with the weighted average approach. The weights are represented by the time elapsed between the irregular measurements within an hour.

Besides, if the energy demand forecasting target is all houses in a microgrid, hourly energy consumption data incoming from each house should be aggregated. The air temperature and temporal features (i.e. month, day of the week, and hour) can be correlated with load patterns and expected to contribute to the model success. These features are fused into energy consumption data.

Ensemble Model for Short-Term Load Forecasting. For the ensemble learning module, we proposed an ensemble STLF model that incorporates SVR as a shallow model and LSTM as a deep learning model. Since energy consumption is dependent on many factors such as social or climatic factors, we decided to use the aforementioned learning models that capture the non-linearities and are popular in STLF. SVR is not a probabilistic model but it splits the observations into separate classes by the longest distance. The model parameters of SVR are also easy to tune [9]. On the other hand, LSTM is an RNN based deep learning model with a recurrent feedback loop and the memory cell. Therefore, it is effective at capturing temporal dynamic behavior [10].

STLF is concerned with daily or weekly energy consumption prediction. The data decomposition reveals the seasonal patterns (daily or weekly) underlying the energy consumption data set. The auto-correlation function (ACF) coefficients of the data set also give the lag orders of highly correlated values with the current load value. For example, if load data is presented in 1-hour granularity, there can be correlations in time lag 24 which implies daily seasonality. Similarly, in the load data set with a daily resolution, the correlations in time lag 7 address the weekly seasonality. Thus, the correlated load sequences are used to build learning models for daily or weekly energy prediction. These sub models integrate into an ensemble model to obtain better prediction results [11].

The ensemble model setup for STLF scenario is shown in Fig.2. The appropriate data frames should be prepared for each sub model training. D_t refers to the data point at time t, which comprises load data, temperature data, or temporal data. The lag values 1 to T derive from auto-correlation coefficients.

k value indicates how many steps a data point in the *continuous* historical data is behind the current data point, while K shows the order of a data point in the historical data recorded at the *intervals of T*.

Thus, we followed two approaches for the prediction of the next *T-step* window. In the first approach, a single model uses the continuous sequence of data S to predict the T steps ahead. On the other hand, the separate models can be trained with data corresponding to the target sequence S_i where i ranges from 0 to *T-1* (e.g. hour in a day, or day in a week) in the second approach. Therefore, each of the next T steps can be predicted through separate models. Both approaches differ in their learning abilities and complement each other to form a more powerful learning model. From this point of view, we constructed the *sub models* that we call $LSTM_C$, SVR_C, $LSTM_B$ and SVR_B to be ensembled. The sub models $LSTM_C$ and SVR_C consist of single models and learn from *continuous* historical data points. Nevertheless, the sub models $LSTM_B$ and SVR_B represent the model groups formed by the *inner models*. Each inner model learns from *bundled* historical data points recorded at the time from 0 to *T-1*.

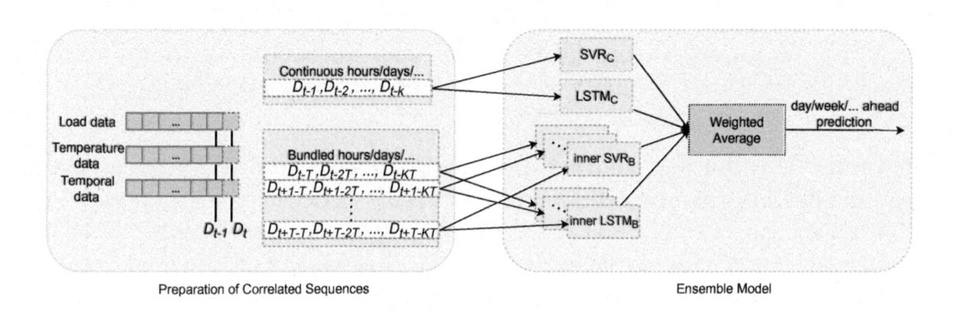

Fig. 2. The proposed model overview

In the hyperparameter tuning stage, the grid search is applied to find out the optimal epoch, nodes, and batch size required for the model fitting for each LSTM network. To find the best SVR models, the kernel functions are evaluated using grid search. $LSTM_C$, SVR_C, $LSTM_B$ and SVR_B are ranked based on their root mean square error (RMSE) in the training phase. Based on their rankings, the model predictions are aggregated in a weighted average manner to discover the possibility of an ensemble STLF model.

3 Experiments and Results

3.1 Experimental Setup

The software components described in Sect. 2.1 was implemented using Python 3. For the ensemble model explained in Sect. 2.2, the LSTM networks were implemented with Keras built on top of Tensorflow library, while the SVR models were applied using the scikit-learn library. As the fog nodes, 3 of Raspberry Pi 4 microcomputers with BCM2711 quad-core Cortex-A72 CPU cores 1.5GHz processor and 4 GB RAM were placed in a local area network (LAN). The application modules were deployed into the Docker containers on these fog nodes. To compare the performance metrics of fog and cloud, we used AWS EC2 T2 instance with 1 vCPU and 1 GiB RAM. The edge layer clients were established in a computer with Intel Core i5-6300U 2.50 GHz processor and 16 GB RAM.

We utilized the public REFIT electrical load dataset to get day ahead energy prediction with the proposed ensemble model. The dataset contains 1,194,958,790 readings of two-year power measurements from 20 houses located in Loughborough, United Kingdom [16]. The power consumption data in 6–10 second intervals is presented in both appliance and aggregated levels. The dataset also includes the measurements of air temperature in 15-minute granularity.

We identified the 10-week aggregated consumption level data for the experimental setup, and separated the hourly dataset into train, validation, and test sets in 70:20:10 ratio. The load dataset has demonstrated daily seasonality in the data decomposition phase. Furthermore, correlation analysis has shown the lag 1 and 24 autocorrelations are significantly high. Therefore, in the *ensemble learning module*, we trained a single model for LSTM_C and 24 individual models for LSTM_B. We also applied the same approach for SVR_C and SVR_B. All LSTM networks were compiled with ADAM optimizer and with MSE as the loss function. LSTM_C model was trained with 32 nodes and 16 batch size in 200 epochs, while LSTM_B inner networks were trained with node count ranges from 8 to 64 and batch sizes from 16 to 32 in epochs from 16 to 96. On the other hand, SVR_C model was trained with linear kernel, while the SVR_B models were trained with the kernels including linear, sigmoid, radial basis function, and polynomial.

3.2 Results

Results on Prediction Accuracy. We compared the performances of the trained models on the unbiased test set. We used mean absolute percentage error (MAPE), root mean squared error (RMSE), and mean absolute error (MAE) as the evaluation metrics. As the base model, we used SARIMA which is popular statistical-based model due to its simplicity and the ability to deal with temporal effect in load forecasting. Except for SARIMA, we trained all sub models with multivariate data including temperature and temporal data. We observed heuristically that sub models trained with these additional features gave better prediction results than when trained with load values only.

Table 1 indicates the performance of each model. Although SARIMA is trained with the load data only, it yielded better RMSE and MAE scores from the sub model SVR_C. It emphasizes that SARIMA is effective at dealing with the seasonality using its periodic components despite its simplicity. Besides, LSTM_B and SVR_B performances are relatively close to each other, while LSTM_C performance is slightly better. The number of samples used to train each inner network in LSTM_B is less than in LSTM_C. Therefore, the inner networks of LSTM_B were trained with a smaller range of epochs since they tend to overfit in the high number of epochs. However, $LSTM_C$ still outperformed at generalizing the model with a larger data set in our case. The sub model SVR_C was unable to capture the temporal pattern of the continuous data set and produced the highest RMSE and MAE results. This was within our expectations, since LSTM network often outperforms SVR model at modelling non-linear data and producing dynamic prediction results. The proposed ensemble model gave the most accurate results among the sub models and the base model SARIMA, with RMSE of 132.5 Wh, MAPE of 19.9% and MAE of 103.5 Wh. The ensemble model was able to approximate the best or average performance for each sample in test data, which gave robustness to the forecasting system.

Results on Timing. The reduced network latency is significant for the microgrid applications that require the transmission of the big data collected from a vast number of houses. If the network latency is high, the throughput decreases. In our fog framework, we measured the latency using Transmission Control Protocol (TCP) for both scenarios, where the data is propagated within the job requests from the master fog node to worker nodes or the cloud node. As a result, the average latency was measured at 5.29 ms between the master and worker fog nodes located in a LAN. The average latency was 44.54 ms between the master fog node and the cloud node, higher due to multi-hop communication.

Table 1. Comparison of prediction errors

Model	RMSE (Wh)	MAPE (%)	MAE (Wh)
The proposed model	**132.5**	**19.9**	**103.5**
$LSTM_C$	143.6	21.4	111.7
$LSTM_B$	147.1	21.0	113.9
SVR_C	166.3	23.8	127.4
SVR_B	148.5	24.6	117.3
SARIMA	156.9	27.0	125.4

The execution time of model training is also a valuable metric for the STLF applications. Time-efficient re-trainings are required to update the model based on the new data at some intervals. In the proposed fog framework, we trained multiple SVR and LSTM models in Docker containers concurrently. We configured the learning module as a 3-container service at first. When we placed the containers in the master fog node only, training took 387.76 ms. In case the containers were placed in both the master fog node and 1 worker fog node, training was completed in 297.49 ms, better than the first measurement. When we further spread the containers into master and 2 worker fog nodes, we measured the execution time in 254.59 ms. The last experiment was repeated by the increasing number of containers to 5. The execution time was 208.67 ms, and here we found the result was further improved. On the other hand, the execution time was measured at 90.73 ms in the cloud. As expected, training in the cloud node yielded the best execution time because of more powerful processors.

As a result, we observed that the fog framework is promising for low-latency applications and can help in achieving execution times converging to the cloud with its distributed and cost-effective resources.

4 Conclusion

The fog computing paradigm can unlock the potential of microgrids by enabling real-time analysis, distributed intelligence, and security and privacy. For this reason, we proposed a fog architecture for computational requirements in microgrids. We deployed the energy demand forecasting application using an ensemble model through the proposed framework. The prediction results demonstrated that the ensemble model has superior performance with a RMSE error of 132.5 Wh than the other sub models and the base model. Due to the closer geographic distance, the network latency between the master and worker fog node was lower than the latency between the master fog node and the cloud node. On the other hand, training in the cloud node took less time than in fog nodes. However, the training in distributed fog nodes considerably improved the time efficiency. The study demonstrated that the framework with cost-effective, low-latency, and distributed Raspberry Pi fog nodes is competent to validate the complex learning models proposed for energy demand forecasting in microgrids. This framework can contribute to the integration of microgrid applications with fog computing.

As future work, RES and EV's effect will be included in the energy demand forecasting scenario. The microgrid load analysis scenarios to be implemented and evaluated in the fog computing framework will also be expanded.

References

1. Ghasempour, A.: Optimum number of aggregators based on power consumption, cost, and network lifetime in advanced metering infrastructure architecture for Smart Grid Internet of Things. In: 13th CCNC, pp. 295–296 (2016)
2. Jalali, F., Vishwanath, A., de Hoog, J., Suits, F.: Interconnecting Fog computing and microgrids for greening IoT. IEEE ISGT-Asia, pp. 693–698 (2016)
3. Hernandez, L., Carlos, B., Javier, M.A., et al.: A survey on electric power demand forecasting: future trends in smart grids. microgrids and smart buildings. IEEE Commun. Surv. Tutorials **16**(3), 1460–1495 (2014)
4. Bera, S., Misra, S., Rodrigues, J.J.P.C.: Cloud computing applications for smart grid: a survey. IEEE Trans. Parallel Distrib. Syst. **26**(5), 1477–1494 (2015)
5. Jaiswal, R., Davidrajuh, R., Rong, C.: Fog computing for realizing smart neighborhoods in smart grids. Computers **9**, 76 (2020)
6. Okay, F.Y., Ozdemir, S.: A fog computing based smart grid model. In: ISNCC, pp. 1–6 (2016)
7. Samie, F., Bauer, L., Henkel, J.: Edge computing for smart grid: an overview on architectures and solutions: design challenges and paradigms, power systems, pp. 21–42 (2019)

8. Hong, W.: Intelligent Energy Demand Forecasting. Springer, London (2013)
9. Petrican, T., Andreea, V.V., Marcel, A., et al.: Evaluating forecasting techniques for integrating household energy prosumers into smart grids. In: IEEE 14th ICCP, pp. 79–85 (2018)
10. Wang, L., Mao, S., Wilamowski, B.: Short-term load forecasting with LSTM based ensemble learning, ithings. In: IEEE GreenCom, IEEE CPSCom and IEEE Smart Data, pp. 793–800 (2019)
11. Tang, L., Yi, Y., Peng, Y.: An ensemble deep learning model for short-term load forecasting based on ARIMA and LSTM. In: IEEE SmartGridComm, pp. 1–6 (2019)
12. Svorobej, S., Takako, P., Bendechache, M., et al.: Simulating fog and edge computing scenarios: an overview and research challenges. Future Internet **11**(3), 55 (2019)
13. Xu, Q., Zhang, J.: piFogBed: a fog computing testbed based on Raspberry Pi. In: IEEE 38th IPCCC, pp. 1–8 (2019)
14. Sonmez, C., Ozgovde, A., Ersoy, C.: EdgeCloudSim: An environment for performance evaluation of Edge Computing systems. Presented at the (2017)
15. Naas, M.I., Boukhobza, J., Raipin Parvedy, P., Lemarchand, L.: An extension to iFogSim to enable the design of data placement strategies. In: IEEE 2nd ICFEC, pp. 1–8 (2018)
16. Murray, D., Stankovic, L., Stankovic, V.: An electrical load measurements dataset of United Kingdom households from a two-year longitudinal study. Sci. Data **4**, 160122 (2017)

A Fuzzy Deep Learning Approach to Health-Related Text Classification

Nasser Ghadiri(✉), Ali Ghadiri, and Afrooz Sheikholeslami

Department of Electrical and Computer Engineering, Isfahan University of Technology, 84156-83111 Isfahan, Iran
nghadiri@iut.ac.ir, {alighadiri, afrooz.sheikholeslami}@ec.iut.ac.ir

Abstract. Following the tremendous amounts of text generated in social networks and news channels, and gaining valuable and dependable insights from diverse sources of information is a tedious task. The challenge is increased during specific periods, for example, in a pandemic event like Covid-19. Existing text categorization methods, such as sentiment classification, aim to help people tackle this challenge by categorizing and summarizing the text content. However, the inherent uncertainty of user-generated text limits their efficiency. This paper proposes a novel architecture based on fuzzy inference and deep learning for sentiment classification that overcomes this limitation. We evaluate the proposed method by applying it to well-known health-related text datasets and comparing the accuracy with state-of-the-art methods. The results show that the proposed fuzzy fusion methods increase the accuracy compared to individual pretrained models. The model also provides an expressive architecture for health news classification.

Keywords: Text categorization · Sentiment analysis · Fuzzy inference · Text mining · Deep learning

1 Introduction

During pandemic events, people generally use the various web and social media content and use them to communicate with society and express ideas and news [1]. Social media's great potential could help people be informed about the COVID-19 spread, knowing about the spread patterns, and preventing the risks [2].

A challenging task for the citizens is reading, analyzing, and interpreting the massive amount of news and related content, especially in pandemic events. Following the large volume of text generated in social networks and news channels and gaining valuable and dependable insights from diverse information sources is a tedious task. The challenge is increased during specific periods, for example, in a pandemic event like the outbreak of Covid-19 [1].

Text classification methods, such as sentiment analysis, aim to help users tackle this challenge by categorizing and summarizing the text content. Earlier sentiment analysis methods were based on statistics and machine learning, while newer methods often

C. Kahraman et al. (Eds.): INFUS 2021, LNNS 308, pp. 179–186, 2022.
https://doi.org/10.1007/978-3-030-85577-2_21

provide higher accuracy by pretraining deep neural network models [3]. One of the most common pretrained models was BERT, proposed by Google [4]. A deep neural network is pre-trained by extracting information from very large text corpora to adjust the network weights in these methods. The network then is fine-tuned for specific downstream tasks like sentiment analysis. The pretraining often provides higher accuracy compared to primary methods. An optimized model based on BERT is RoBERTa [5], developed by Facebook containing more training parameters and performs well on different tasks, including next sentence prediction and question answering. It handles long sentences more efficiently. TwitterBERT [6] is also proposed for processing tweets, COVID-Twitter-BERT [7], or CT-BERT tailored for tweets related to Covid19.

However, the inherent uncertainty of user-generated text limits the efficiency of PLM methods. For instance, a Covid19-related tweet could be better processed by CT-BERT, but the RoBERTa model may also provide high accuracy for some longer tweets in this domain. While a specific tweet belongs to the set of Covid-related tweets, it may also be a member of the set of long tweets to some degree. Therefore, we need to cope with the fuzziness in using PLMs like BERT to analyze the tweets.

In this paper, we propose a novel hybrid architecture based on three different pre-trained models to cover different tweets. To the best of our knowledge, this is the first fuzzy fusion model for improving health-related text from social media. The model uses two fuzzy methods for the fusion of data from the pretrained BERT-based models. The first fusion is based on the Choquet fuzzy integral that has shown remarkable performance in the fusion of data for decision-making scenarios [7]. The second is designed as a fuzzy rule-based model for the fusion of classifiers [8]. The details of the proposed model will be described in the next section.

The rest of the paper is organized as follows. In Sect. 2, the proposed model is presented with two fuzzy integration approaches. Section 3 reports on data collection, data pre-processing and experimental results. Section 4 conclude the paper and points to some future directions.

2 Proposed Model

The proposed model is a hybrid of three state-of-the-art pre-trained models fused by two fuzzy fusion methods. The overall process is illustrated in Fig. 1. The first step is pre-processing of the input data. Raw twitter data contains much noise and is unstructured and informal. Therefore, pre-processing on Twitter data will help the pre-trained models in better performance. The detailed steps of pre-processing are described in Sect. 3. After pre-processing, every tweet is passed to three different pretrained models for classification. The pretrained model will be described in Sect. 2.1. In the third step, the results are fused using two fuzzy fusion methods. The first fusion method is the Choquet integral that uses validation data from the pretrained models. The second is a fuzzy rule-based fusion module that uses the classification output from the pretrained models and meta-data for tuning the rules. More details about the modules will be presented as follows.

2.1 The Pretrained Models

The pre-processed input tweets are fed into different pretrained models. We selected three models, including BERT, RoBERTa, and Covid-Twitter-BERT, explained below.

BERT Base Uncased. BERT is a transformers model which is pretrained on the Book-Corpus dataset that consists of about 11K unpublished books and content from English Wikipedia. It will function as a general classifier in our model.

RoBERTa Large. RoBERTa is based on BERT and extended its dataset using CC-News, OpenWebText, and Stories, which improve its performance. Furthermore, RoBERTa is not pretrained with the next-sentence prediction (NSP) objective. It is only pretrained with the Masked language modeling (MLM) objective and uses a larger batch size than BERT. This model is expected to classify long tweets more accurately.

COVID-Twitter-BERT (CT-BERT). The architecture of COVID-Twitter-BERT is similar to BERT. Unlike BERT, which is pretrained on the English Wikipedia dataset and may not have a deep understanding of texts related to COVID-19, CT-BERT is optimized for using texts related to COVID-19.

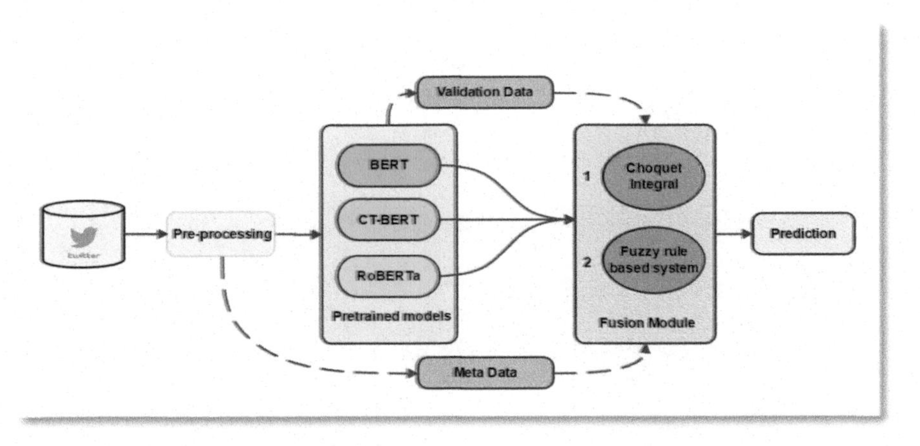

Fig. 1. The overall process for the sentiment analysis of Covid19 related tweets.

2.2 Fuzzy Choquet Integral Fusion

We used two methods for the fusion of the results obtained from the pretrained models. The first fusion method is based on fuzzy Choquet integral [7]. Given $x_1, x_2, \ldots, x_n$ as a set of criteria, and v as the fuzzy measure, the Choquet integral is given by Eq. 1 below:

$$C_v(x) = \sum\nolimits_{i=1}^{n} \left[x_i - x_{i-1}\right] v(H_i) \tag{1}$$

Where $H_i = \{i, ...,n\}$ is the subsets of indexes of the $n - i + 1$ most significant components of x [9]. We used the fuzzy Choquet integral in two settings; (a) fuzzy fusion using Choquet integral trained on *densities*, and (b) fuzzy fusion using Choquet integral trained on *validation scores*. The results of the fuzzy fusion of pretrained models using both techniques will be reported in Sect. 3.

2.3 Fuzzy Rule-Based Fusion

Our second fuzzy fusion method for combining the results from the pretrained classifiers is a fuzzy rule-based system [8]. It has two inputs that contain metadata, and three outputs to set the weights of three classifiers. The fuzzy rules are shown in Fig. 2. The two metadata inputs are described as follows.

Meta Data. For enriching the fuzzy rule-based fusion of pretrained models, we propose using two meta-parameters extracted from every input tweet. The first parameter is the *tweet length.* It could be observed that some pretrained models like RoBERTa perform well for longer texts. So our fusion method takes advantage of tweet-length information to tune the weights of the pretrained model outputs accordingly.

The second parameter is the *relatedness* of the tweet to the Covid19 domain. Some pretrained models like CT-BERT are targeted at tweets that are highly related to this domain. So our fusion method computes the similarity between every tweet with a fixed set of Covid19 keywords using the Jaccard similarity measure. Then the rule-based model takes advantage of tweet relatedness information to tune the weights of the pretrained model outputs accordingly.

Tweet length	Relatedness	BERT Weight	CT-BERT Weight	RoBERTa Weight
Medium	High	Low	High	Medium
High	High	Low	High	Medium
Low	Medium	Medium	Medium	Medium
Medium	Medium	Low	Medium	High
High	Medium	Low	Low	High
Low	Low	Medium	Medium	High
Medium	Low	Low	Low	High
High	Low	Low	Medium	High

Fig. 2. The fuzzy rules for the fusion of pretraining outputs.

3 Experimental Evaluation

In this section, we present the evaluation of our proposed method. We will describe four steps for data collection, pre-processing, classification using three different pretrained models, and the fuzzy fusion of the classification results. We used the Python language and Google Collab for running our experiments. For all experiments, the batch size was 16, and the sequence maximum length was 40. We concatenated the validation and the training data and performed a 5-fold cross validation to train our models. Each fold is trained in 5 epochs using early stopping with patience of 3.

3.1 Data Collection

We have collected data from the Lopezbec repository[1], which contains an ongoing collection of tweets associated with the novel coronavirus COVID-19 since January 22nd, 2020. Only tweets in English were collected from July 2020 to January 2021 and among them, those which met the criteria for like, retweet, and sequence length were selected. Accordingly, we collected 55,785 tweets which are categorized as positive, negative, and neutral. In the next step, we used TWARC to hydrate the tweet-IDs. A sample of the dataset is shown in Table 1.

Table 1. A sample of tweets about Covid19.

Tweet text	Sentiment label
Congrats to senior scientist Alex MacKenzie, winner of the 2020 @CHEO Research Institute Osmond Impact Award. Most recently famed for his #COVID19 research, his decades of work on #RareDisease, spinal muscular atrophy and #diabetes has been cited over 10,000 times. #CDNhealth https://t.co/gLSVQ7pFBY	Positive
@woodenfam1 Not canceled, just pushed back. The pandemic has caused an immense change in our production schedule for all products. N scale James, HO Peter Sam, HO Daisy, and HO Oil Tanker Troublesome Truck #6 all were not able to make it in 2020	Negative
This morning I was asked to address recent comments from the White House about Florida's vaccine allocation. Here is my statement: https://t.co/imwvT67sRo	Neutral

As discussed earlier in Sect. 3, and the contents of sample tweets suggest, pre-processing of the tweets is essential before the classification step. In the next part, we describe the pre-processing tasks.

3.2 Data Preprocessing

The following series of techniques are applied in the given order to improve the text.

1. People highly use hashtags in social media to represent topics, i.e., #COVID-19, #StayHome, #StaySafe, and #Coronavirus. We performed the cleaning of the text by removing hashtag characters and segment the hashtag texts using wordninja package.

[1] https://github.com/lopezbec/COVID19_Tweets_Dataset#data-collection-process-inconsistencies.

2. Convert tweets to lower cases: The tweets may be in lower or upper cases. In this step, we convert all tweets into lower case.
3. Unscape HTML tags and eliminate hyper-links, @mentions, emails, and numbers.
4. In some tweets, shortened versions of words (contractions) exist that should be converted by removing specific letters. For instance, "don't" will convert to "do not" and "what's" will be "what is". Converting each contraction to its expanded helps with text standardization.
5. We eliminated punctuation, and special characters from the dataset as these do not help detect sentiment. Furthermore, we removed emojis using the python emoji library to *demojise* the emojis and replace them with a short textual description.
6. Removing stop words is a common method to reduce the noise in textual data. Removing stop words does not affect understanding a sentence's sentiment valence. We removed stop words using the stop-word library in NLTK. This library includes 179 stop words of the English language.
7. Lemmatization: The purpose of this step is to convert words to their root to get distinguishing words. Among the methods to implement this step, we used the spaCy package that efficiently addresses the challenges that we face for this step.

3.3 Experimental Results

The results of accuracy provided by each classifier through different epochs are shown in Fig. 3. It could be observed that the accuracy of the BERT model declines after three epochs. However, BERT has shown the lowest variation in accuracy compared to other models. The RoBERTa pretrained classifier shows high variations but converges at the fifth epoch. The results show that the CT-BERT model provides more stable results as expected from a pretrained model that is specifically tailored to tweet inputs.

The mean accuracy values of individual pretrained models for validation data are shown in Table 2. On average, the CT-BERT model is more accurate than other methods in the training phase. However, as we observe in Fig. 3, no single model could perform more accurately than other models during the training. Moreover, the accuracy for test phase depends on the generalizability of the model and, often less than the accuracy for validation data. Therefore, we transfer the results of the pretrained models to the fuzzy fusion module for potentially improved accuracy.

The accuracy of different methods for test data is shown in Table 3. We can observe that both of our proposed fusion methods (Choquet integral with two settings and the rule-based method) provide higher accuracy compared to individual BERT, RoBERTa, and CT-BERT methods.

The rule-based fusion has performed slightly better than Choquet integral. This could be the result of feeding more information to the rule-based module through metadata about tweet length and the degree of relatedness to Covid19 tweets.

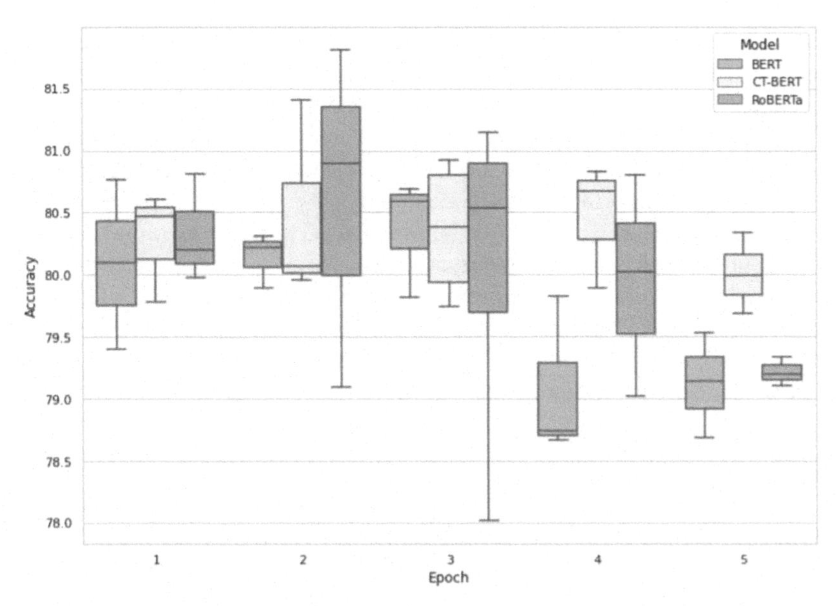

Fig. 3. The accuracy ranges of different pretraining models.

Table 2. The accuracy of different classifiers for *validation* data

Model	Accuracy
BERT base	88.2
RoBERTa	91.3
CT-BERT	92.0

Table 3. The accuracy of different classifiers for *test* data

Model	Accuracy
BERT base	80.3
RoBERTa	81.1
CT-BERT	81.0
Fuzzy fusion 1 - CHI trained on validation scores	**81.4**
Fuzzy fusion 1 - CHI trained on densities	**81.6**
Fuzzy fusion 2 - Fuzzy rule-based system	**82.0**

4 Conclusion and Future Work

This research aimed to help the citizens gain more information from the massive amount of content generated during pandemic events. Although many state-of-the-art pretrained language models are developed to classify such content, the accuracy of the models is limited due to the inherent uncertainty of user-generated content. We proposed two fuzzy fusion approaches to get a higher accuracy by combining the different pretrained models. Both of the fuzzy Choquet integral and fuzzy rule-based fusion methods performed better than individual models. The meta-data extracted from the input tweets also contributed to the increased accuracy of the rule-based model.

Future work may focus on improving the rule-based fusion by learning the fuzzy sets for both input and output variables. The metadata that is fed to the rule-based module may also be enriched by adding more features from the input tweets. The Choquet integral will also be extended if more pretrained classifiers become available in the future. In this case, the Choquet integral would be expected to perform better than the rule-based fusion that requires adding more output variables for new pretrained modules.

References

1. Alshamrani, S., Abusnaina, A., Abuhamad, M., Lee, A., Nyang, D., Mohaisen, D.: An analysis of users engagement on Twitter during the COVID-19 pandemic: topical trends and sentiments. In: Chellappan, S., Choo, K.-K., Phan, NhatHai (eds.) CSoNet 2020. LNCS, vol. 12575, pp. 73–86. Springer, Cham (2020). https://doi.org/10.1007/978-3-030-66046-8_7
2. Rashid, M.T., Wang, D.: CovidSens: a vision on reliable social sensing for COVID-19. Artif. Intell. Rev. **54**(1), 1–25 (2020). https://doi.org/10.1007/s10462-020-09852-3
3. Minaee, S., Kalchbrenner, N., Cambria, E., Nikzad, N., Chenaghlu, M., Gao, J.: Deep learning based text classification: a comprehensive review. arXiv **1**, 1–43 (2020)
4. Devlin, J., Chang, M.W., Lee, K., Toutanova, K.: BERT: pre-training of deep bidirectional transformers for language understanding. In: NAACL HLT 2019 - 2019 Conference of the North American Chapter of the Association for Computational Linguistics: Human Language Technologies, - Proceedings of the Conference, vol. 1, pp. 4171–4186 (2019)
5. Liu, Y., et al.: RoBERTa: A robustly optimized BERT pretraining approach http://arxiv.org/abs/1907.11692 (2019)
6. Azzouza, N., Akli-Astouati, K., Ibrahim, R.: TwitterBERT: framework for Twitter sentiment analysis based on pre-trained language model representations. In: Saeed, F., Mohammed, F., Gazem, N. (eds.) IRICT 2019. AISC, vol. 1073, pp. 428–437. Springer, Cham (2020). https://doi.org/10.1007/978-3-030-33582-3_41
7. Choquet, G.: Theory of capacities. Ann. l'institut Fourier **5**, 131–295 (1954). https://doi.org/10.5802/aif.53
8. Wang, D., Keller, J.M., Andrew Carson, C., McAdoo-Edwards, K.K., Bailey, C.W.: Use of fuzzy-logic-inspired features to improve bacterial recognition through classifier fusion. IEEE Trans. Syst. Man Cybern. Part B (Cybernetics) **28**(4), 583–591 (1998). https://doi.org/10.1109/3477.704297
9. Zhao, Y., Cen, Y.: Data Mining Applications with R. Elsevier Inc. (2013). https://doi.org/10.1016/C2012-0-00333-X

Machine Learning

National Basketball Association Player Salary Prediction Using Supervised Machine Learning Methods

Emirhan Özbalta(✉), Mücahit Yavuz, and Tolga Kaya

Department of Management Engineering, Istanbul Technical University, 34367 Istanbul, Turkey
{ozbalta16,yavuzm16,kayatolga}@itu.edu.tr

Abstract. Basketball is one of the most popular sports in the world and National Basketball Association (NBA) is the main figure for it. With the purpose of sustaining the balance between the basketball teams in the league, salary cap is implemented for all the teams in NBA. Considering the salary cap, decision makers of basketball teams should be careful while spending their budget. Since there are no transfer fees in NBA, salaries are the main expense for basketball teams. Therefore, determining the salaries of basketball players while making contracts is crucial to compose the best possible basketball team. In this research, dataset from the NBA 2K20 MyTeam video game and NBA players' performance statistics of 2019–2020 season will be combined to predict the salaries for new contracts of NBA players by using machine learning methods. Shrinkage methods will be used to select best subsets. After, regression and decision tree models will be used to see which one produces the best mean squared error values. Results show that predicted salaries are very close to the new contract salaries of NBA players.

Keywords: Supervised learning · NBA · Basketball · Sports analytics · Machine learning · Random forest

1 Introduction

National Basketball Association is based on North America which contains 29 United States and 1 Canada basketball team. Even though basketball's popularity is rising in Europe and South America, they are far away from the basketball played in NBA in case of money cap, popularity and talented players. Moreover, there are salary cap, which increases every year, in NBA to limit the total salaries of basketball team players. Salary cap is crucial to sustain competitive balance in the leagues since it creates better salary distribution, limits vast amount of salary payments to top players and provides necessary profit rates to clubs [1]. Also, contracts of NBA players are mostly stay valid for long years. Since there is salary cap for every basketball team in NBA, determining the salaries of the players is very important. Moreover, salary prediction model will be created for NBA players with machine learning methods to compare the current salaries with future salaries and assess if the players valued accurately. In the current articles, main focus is examining the factors which affect salaries. In this research, aim is to assist

C. Kahraman et al. (Eds.): INFUS 2021, LNNS 308, pp. 189–196, 2022.
https://doi.org/10.1007/978-3-030-85577-2_22

decision makers with player trading, and determination of players' salaries by focusing on accurate salary predictions and comparing predicted salaries with real salaries of unique players. While creating the model, two different datasets have been combined rather than using only performance statistics of NBA players to produce a better model than current studies, which are the other originality of this paper. To achieve this, detailed data from 2K20 MyTeam video game and 2019–2020 performance statistics of NBA players will be used. Generating a model that increases the accuracy of determining the salaries will benefit the teams while making their contracts with players and make them use their budgets efficiently. Also, there are unique players called rookies. Rookie players are young players that joined the NBA league from college recently, and most of the time their salaries are much lower than their current performance. Predicting the salaries of rookie players is harder than predicting average players' salaries since rookie salaries are quite undervalued. Therefore, these players' contracts will be examined.

Further of the paper is organized as follows: literature review of purpose for using data analytics in basketball and other sports, methodology applied in the study, information about the dataset, model building process, findings of our model, and lastly conclusion and future research opportunities.

2 Literature Review

With the increasing popularity of Big Data and machine learning, implementation of analytics to sports was inevitable. Sports analytics is defined as organizing and processing the historical data to increase competitive advantage of teams and help decision makers of the organizations such as trainers and coaches [2]. Analyzing the datasets of performances in basketball games allow analysts to separate shooting selection from shooting accuracy, select players with new performance indicators and create new play styles [3]. According to these articles, it can be said that executing data analytics in sports will save decision makers' time and help them while creating strategies and adjusting salaries of players. Rhah and Romijnders [4] implemented deep learning methods to analyze three-point plays through motion tracking data. Nagarajan et al. [5] used data analysis methods to NBA players' last 5-year performance statistics and team performance rates. With this analysis, they proposed two contract making strategies, which are minimizing team salary or maximizing team win/lose rate, in order to aid teams with their limited budget. Papadaki and Tsagris [6] applied non-linear machine learning techniques to estimate salary shares of the NBA players by using performance statistics of them. Li [7] analyzed the determinants of NBA players' salaries while signing new contracts by applying regression methods. According to this research; 3-point shots per game is more important than 2-point shots and free throws for salaries, 1 increase in offensive rebounds rise salaries by $0.25 million while 1 increase in defensive rebounds rise salaries by $0.13 million, 1 increase in assists led to $0.3 increase in salaries, and lastly players who played All-Star games are generally will not decline on performance and teams will save $0.11 million per one more All-Star games played by NBA players. As it can be seen, salaries of NBA players are dependent on many factors. Psychological factors and popularity of the players also affect the salaries.

3 Methodology

In this study, supervised machine learning methods, which are regressions and decision tree methods, are used. Cross validation is used to create training and validation set to make predictions. When multiple variables are involved, the relationship between predictor variables and response can be explained with multiple linear regression. The model which has an assumption of linearity can be shown as [8]:

$$Y = \beta_0 + \beta_1 X_1 + \beta_2 X_2 + \cdots + \beta_p X_p + \in \tag{1}$$

The main logic is to find the equation of the line that can best estimate the observation values. Here, the least squares approach tries to find an ideal combination of coefficients by minimizing the RSS.

$$\text{RSS} = e_1^2 + e_2^2 + \cdots + e_n^2 \tag{2}$$

However, when multiple linear regression is applied, more related variables with the response are not chosen. At this point, shrinkage methods, which are ridge and lasso regression, will be used to make the right variable selection. The ridge regression method minimizes the coefficients of the less important variables in the model.

$$\sum_{i=1}^{n}\left(y_i - \beta_0 - \sum_{j=1}^{p}\beta_j x_{ij}\right)^2 + \lambda\sum_{j=1}^{p}\beta_j^2 = \text{RSS} + \lambda\sum_{j=1}^{p}\beta_j^2, \tag{3}$$

The lasso regression is similar to ridge regression. However, coefficients of the lesser important variables will be zero in this method.

$$\sum_{i=1}^{n}\left(y_i - \beta_0 - \sum_{j=1}^{p}\beta_j x_{ij}\right)^2 + \lambda\sum_{j=1}^{p}|\beta_j| = \text{RSS} + \lambda\sum_{j=1}^{p}|\beta_j|\cdot \tag{4}$$

Another method, which is one of the decision tree methods, random forest is used. Also, special case of random forest when m = p, bagging method is used. While standard trees use best split among all variables for splitting, random forest chooses random subsets of features and then uses best among them [9]. The principle in decision trees is to minimize the RSS which is where $\hat{y}_{R_j}$ is the mean of the responses within the jth boxes.

$$RSS = \sum_{j=1}^{J}\sum_{i \in R_j}\left(y_i - \hat{y}_{R_j}\right)^2 \tag{5}$$

Lastly, principal component regression (PCR) was implemented. PCR is a size reduction method; it is very useful in cases where there is an excessive correlation between variables in the dataset. PCR method first creates M principal components and then uses these components as variables in the models.

4 Data

To achieve an improved explanation for the salary prediction machine learning model, 2 datasets will be combined. First dataset is based on NBA 2K game series, which is the most popular basketball video game. Every year, a new version of the game is produced to update teams, player ratings and improve the overall gameplay. Dataset from NBA 2K20 video game, which is launched at August 2019, will be used in this study. MyTeam is a game mode in NBA 2K series which allow the gamers to combine their own team with different player cards from different NBA teams [10]. To attain the dataset of 2K20 MyTeam, 2KMTCentral website, which is gathering the game dataset and providing it to public, is used [11]. In 2K20 MyTeam dataset, there are 6 major categories which are outside scoring, inside scoring, defending, athleticism, playmaking, and rebounding. Under these categories, there are a total of 39 ratings of basketball players. These ratings are ranged between 25–99. Also, there are three badge types, which are bronze, silver and gold badges, and the number of badges show extra attributes for each player. Lastly, there are overall rating, potential rating, tier list, and play styles of the players. Furthermore, NBA players' performance statistics from 2019–2020 year and 2020–2021 salaries of these players are gathered to complete the dataset [12]. Simple and most important per game player performance statistics, which are points, blocks, steals, turnovers, assists, offensive rebounds, defensive rebounds, and games played is combined with 2K20 MyTeam dataset. Players who played less than 10 games in 2019–2020 season are excluded since most of them are injured or will increase bias due to small amounts of games played. There are a total of 401 NBA players in our dataset. Also, 2021–2022 salaries of rookie players, who made new contracts in 2021–2022, are gathered to see if the current (2020–2021) salaries of the players are reasonable.

Response variable is the salaries of NBA players in 2020–2021. Since salaries are large numbers, interpretation of calculations such as MSE are not satisfactory. Also, distribution of the salaries is right skewed. Therefore, log transformation is implemented to salaries to create more comprehensible data visualization.

5 Model Building

In model building process, multiple linear regression, cross validation, ridge regression, lasso regression, principal component regression, basic decision tree model, random forests, and boosting methods will be implemented and mean squared error (MSE) values will be discussed. Models are built using R programming language. Also, due to injuries, some of the NBA players' performance statistics are missing. Therefore, mean imputation is used for these columns rather than deleting the entire row since there are low amount of NBA players. Moreover, since two different datasets are combined, there are different value intervals inside the variables. Hence, normalization method is used for features to achieve lower MSE and RMSE values.

First of all, categorical variables are extracted from the dataset to create a multiple linear regression model. Using cross validation, MSE value is founded as 0.701. With shrinkage methods, better subsets are created for regression. For both ridge and lasso regression, best lambda values are used. MSE for ridge regression is calculated as 0.413

with best lambda value of 1.08. MSE for lasso regression is founded as 0.404 (Fig. 1). Lastly, with principal component regression, MSE is calculated as 0.467.

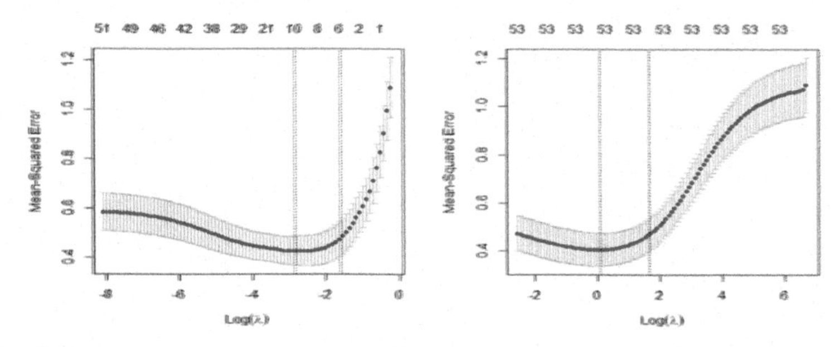

Fig. 1. Lambda and MSE comparison for Lasso (Left) and Ridge (Right) Regression.

Furthermore, different decision tree models and processes, which are regression trees, pruning, random forests, and boosting methods, are implemented to find the best prediction model for our dataset.

Firstly, regression tree model is created by extracting the categorical variables. As expected, basic regression tree (binary splitting) model created problematic prediction, and values are highly differed from the regression line. Also, pruning method is implemented but there was no considerable improvement in the model. MSE for this model is founded as 0.535.

After, special case of the random forest, which is bagging method, is applied. In this method m equals to predictors and it is determined as 21. Also, in this stage, categorical variables, which are tier and play styles, are extracted from the model since it gives a better MSE result. 0.361 MSE value founded with the categorical variables and 0.353 without them. Hence, categorical variables are also not used in decision tree methods.

Lastly, boosting is used. Shrinkage value is taken as 0.001 and interaction depth is taken as 4 since they gave the most optimum MSE result, which is 0.412.

Considering the models built with regression methods and decision tree methods, ensemble learning method was implemented to achieve more reliable prediction model. While creating the ensemble learning, average model is used with a regression and a decision tree model.

6 Findings

After model building process, the prediction model for NBA players' salaries has been obtained. Some outliers, which are undervalued according to our model, can be attributed to rookie contracts, and some overvalued players can be old players who are close to their retirements. By using random forest, importance of features figure has been gathered (Fig. 2).

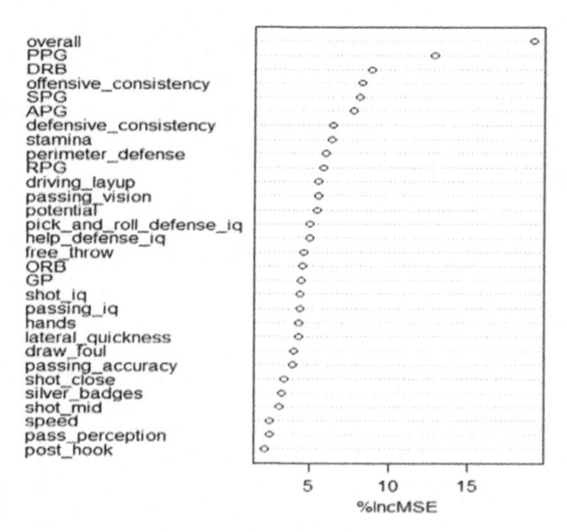

Fig. 2. Importance rates of features.

- PPG: Average points per game
- DRB, ORB: Average defensive and offensive rebounds per game
- SPG: Average steals per game
- APG: Average assists per game
- GP: Games played in 2019–2020 season

There are features that affects the salaries from both per game performance statistics and 2K20 MyTeam attributes. As it can be seen, point, defensive rebound, steal, and assist per game is significant determinants while predicting the salaries of players. Also, salaries are highly affected by video game attributes such as overall, offensive consistency, defensive consistency, stamina.

Further, average ensemble learning model is implemented with regression and decision tree models. Although MSE value of lasso regression is slightly lower than the MSE value of ridge regression, combining the ridge regression and the random forest for ensemble learning model gave the best result. Predicted salaries and real salaries (2020–2021) of the NBA players are compared using data visualization tools (Fig. 3).

Moreover, as stated before, rookie contracts are made while new young players join the NBA league from colleges. Since they are new players, salaries of these players in their first contract tend to be lower than their actual performance and potentials. Therefore, after their first years, most of the time they get huge wage upgrades with new contracts. Considering this, by using our model, which is supervised by the 2020–2021 NBA players' salaries, salaries of these players are predicted (Table 1). After that, predicted salaries are compared with the rookie scale contract extensions [13] that offer new salaries for 2021–2022 and future years.

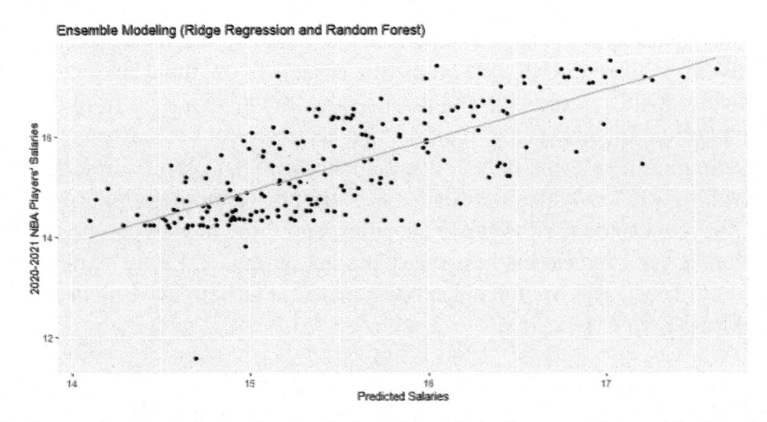

Fig. 3. Predicted salaries and real salaries of the NBA players with logarithmic values.

Table 1. Rookie salary predictions

NBA players (Rookies)	2020–2021 salaries	Predicted salaries	2021–2022 salaries
Donovan Mitchell	$5,195,501	$24,392,839	$28,103,550
Jayson Tatum	$9,897,120	$17,831,642	$28,103,550
Luke Kennard	$5,273,826	$10,620,057	$12,727,273
Kyle Kuzma	$3,562,178	$12,854,452	$13,000,000
Jonathan Isaac	$7,362,566	$10,591,591	$17,400,000
Derrick White	$3,516,284	$9,400,006	$15,178,571

As you can see at the above table, all the rookie NBA players' predicted salaries are higher than the current years' salaries and very close to 2021–2022 years' salaries. Even though, our model is able to provide more accurate salaries from the current year, next years' salaries of rookie players are always higher than our prediction salaries.

7 Conclusion and Discussion

In conclusion, supervised machine learning techniques, which are regression and decision tree models, are used in order to create a model that predicts the NBA players' salaries by combining 2K20 MyTeam video game dataset and per game performance statistics. Our study revealed that variables in the video game dataset are quite effective while predicting the salaries of NBA players. The model showed decent results for prediction of the NBA players' salaries and predicted salaries are very near to 2021–2022 salaries. Furthermore, the model proposed by this study can be used to help NBA teams in their decision-making process and using their budget efficiently. Considering the low number of players in NBA league, creating a good model is quite hard since there is not enough data. For future studies, more performance statistics of NBA players can be gathered such as 3-point field goal percentage, field goal percentage, turnovers, personal fouls per game to improve the prediction model. To use more amount of per game performance statistics, some video game attributes can be extracted from the model to avoid high correlation between variables. Also, several years of performance statistics can be used rather than using only 1-year statistics of NBA players for a better model. Moreover, only 2K20 MyTeam attributes or only per game performance statistics of the NBA players can be used to predict salaries, and these models can be compared to see which one will produce the best salary prediction model. Lastly, extreme gradient boosting (XGBoost) can be implemented to generate an improved model with lower MSE and RMSE values.

References

1. Kesenne, S.: The impact of salary caps in professional team sports. Scott. J. Polit. Econ. **47**(4), 422–430 (2000)
2. Alamar, B.C.: Sports Analytics: a Guide for Coaches, Managers, and Other Decision Makers. Columbia University Press, Columbia, pp. 1–23 (2013)
3. Morgulev, E., Azar, O.H., Lidor, R.: Sports analytics and the big-data era. Int. J. Data Sci. Anal. **5**, 213–222 (2018)
4. Shah, R.C., Romijnders, R.: Applying Deep Learning to Basketball Trajectories (2016)
5. Nagarajan, R., Zhao, Y., Li, L.: Effective NBA player signing strategies based on salary cap and statistics analysis. In: 2018 IEEE 3rd International Conference on Big Data Analysis, pp. 138–143 (2018)
6. Li, N.: The Determinants of the Salary in NBA and the Overpayment in the Year of Signing a New Contract. Clemson University, All Theses, no. 2037 (2014)
7. Papadaki, I., Tsagris, M.: Estimating NBA players salary share according to their performance on court: A machine learning approach (2020)
8. James, G., Witten, D., Trevor, H., Robert, T: An Introduction to Statistical Learning with Applications in R. Dordrecht London: Springer New York Heidelberg, pp. 1–440 (2013)
9. Liaw, A., Wiener, M.: Classification and regression by random Forest. R News **2**(3), 18–22 (2002)
10. NBA 2K Game Series. https://nba.2k.com/my-team/. Accessed 28 Mar 2021
11. 2KMTCentral Database. https://2kmtcentral.com/. Accessed 28 Mar 2021
12. 2020–21 NBA Player Contracts. https://www.basketball-reference.com/contracts/players.html. Accessed 28 Mar 2021
13. 2020/2021 NBA Contract Extension Tracker. https://www.hoopsrumors.com/2020/11/202021-nba-contract-extension-tracker.html. Accessed 28 Mar 2021

Demand Forecasting of a Company that Produces by Mass Customization with Machine Learning

Engin Yağcıoğlu, Ahmet Tezcan Tekin(✉), and Ferhan Çebi

Department of Management Engineering, Istanbul Technical University, Istanbul, Turkey
{yagcioglu17,tekina,cebife}@itu.edu.tr

Abstract. Machine Learning (ML) algorithms are designed to extract information from existing data. The application of ML in production; can provide the acquisition of new information from existing data sets that can form a basis for the development of approaches about how the system should be in the future. This further information can support company managers in their decision-making processes or can be used directly to improve the system. Given the challenge of a rapidly changing and dynamic production environment, ML; As part of artificial intelligence, it can learn about changes and adapt to them. Mass customization; recently, has started to influence the textile sector as in many sectors. As A result of changing consumer habits and developing technology; companies have begun to focus on this area to meet the increasing number of mass customized demands.This study aims to make demand estimation by using ML algorithms of a textile workshop that performs mass customization. The results show that ML algorithms have the result of successful demand forecast in organizations implementing mass customization when there is enough data.

Keywords: Machine learning · Mass customization · Demand forecast

1 Introduction

Mass production has been the widely used production strategy in the 20th century. When competitive power is directly related to the production speed of the enterprise, standardized products were produced. It has been to quickly provide quality and cost-effective products to gain a competitive advantage with the high value given to the customer. In this case, it has caused mass production companies to turn to mass customization.

Mass customization is not a new concept. Tailor shops and shoemakers, one of the indispensable tradesmen of every neighbourhood since the past, were producing in line with customer demands. In recent years, in the mass-producing textile industry, mass customization stands out with its increasing customer potential in a wide range of well-known brands that produce personalized products by taking body measurements in shopping malls, including processes such as personalized printing and embroidery on shopping sites. Mass customization applications that cause additional costs for the customer for now, shortly; will be the most significant advantage of companies operating

C. Kahraman et al. (Eds.): INFUS 2021, LNNS 308, pp. 197–204, 2022.
https://doi.org/10.1007/978-3-030-85577-2_23

in the textile sector in creating loyal customers to survive in an increasingly competitive environment and will become more widespread.

In retail sectors such as textiles, companies should ensure that the right product variety is available on the shelves at the right time so that the customer can find the product they are looking for in the store. Demand forecasting is forward-looking, involving uncertainties, and often cannot be estimated precisely and accurately. The main goal here is to guess with the least error. The organization with the least error of demand estimation; will create the raw material supply plan, production plan, stock, and inventory plan. In line with these, all investment activities and plans vital for the business, such as additional machinery, equipment purchases, personnel planning, and production line planning are carried out.

There are too many variables that affect the demand forecast in the textile industry. We can briefly summarize these variables as fashion, starting with various colours, creating other variables such as fabric variety, cut variety, size variety, and seasonal products. According to classical demand forecasts, there is a need for advanced forecasting techniques that take all these variables that affect demand into account. Artificial intelligence-based methods have started to take place in literature and practice to meet today's needs.

ML is a subfield of artificial intelligence. In parallel with developing computer technologies, its usage area has increased considerably in recent years. Today, ML is widely applied in problem areas such as optimization, control and troubleshooting, and indifferent production areas.

An enterprise operating in the textile sector can make all of its vital plans by estimating the demands affected by too many variables. With this study, it is planned to bring a different approach to the applications in this field. In a mass-customized textile workshop, the ML algorithm will be modelled in line with the customers' expectations and demands and the results will be evaluated. Our modelling will be updated in line with the changing demands and demands and the closest demand forecast will be tried to be made. The results and the demands will be compared and evaluated. Thus, research on this subject is brought to the literature and after that; it is aimed to create data and resources for research in this field.

In the article, Sect. 2 deals with mass customization, demand forecasting, and ML in literature studies. Section 3 describes our proposed methodology and modelling. Finally, the study results are briefly described and future studies are presented in the last section.

2 Literature Review

Today, organizations; should be viewed as systems that work in collaboration with customers [1]. Businesses operating in a competitive environment will not be able to survive without considering customer demands. The differences in consumers' personal preferences are reflected in their demands; This increases the product variety of businesses. Businesses are trying to reduce production costs on the one hand and try to provide customer satisfaction by trying to meet customer demands with product variety on the other.

Mass customization is a business strategy that combines mass production and customization. This strategy, which was born as the demand was forced to react to the increasing individualization, came to life with the idea of integrating users into the design and production process. Mass customization is about providing customized products or services at high density and extremely low costs through flexible processes. In the personalized mass production concept, goods and services are intended to meet individual customers' needs, produced with individual mass production efficiency [2]. Although Toffler [3] had anticipated this concept fifty years ago, Davis [4] coined the term mass customization in 1989; This idea gained wide popularity with Pine [5].

In short, mass customization can be defined as the ability to produce personalized products with mass production costs and efficiency [6]. To produce by mass customization at low cost and at the same time to produce customized products by providing product flexibility; uses mass production as a starting point for economies of scale [7].

The forecast is the art and science of predicting future events [8]. However, it aims to reduce the uncertainty that confuses future decisions [9]. Demand forecasting is absolutely critical in manufacturing organizations. Demand forecasting evaluates the information used in the demand management process [10].

The characteristics and quantity of the product to be demanded; the period's time frame to be forecasted, seasonality due to customer trends, meteorological and random factors affect the demand forecast. In this case, the managers; As Ergün and Şahin [11] reported, make their predictions according to four periods:

1) Very short-term forecasts: These are the estimates made on a daily or weekly basis, where activities are generally organized in short time intervals.
2) Short- term forecasts: These are estimates made up to 6 months.
3) Medium-term forecasts: These are estimates made up to 5 years.
4) Long-term forecasts: These are strategic level plans that will cover periods of more than five years.

Two basic solutions have traditionally been considered in demand forecasting: qualitative and quantitative forecasting. While qualitative techniques include elements such as intuition, personal judgment, and experiences, quantitative techniques use established and systematic procedures. The main difference between the two approaches is that the estimation model is more or less open. Specifically, the qualitative prediction is based on humans' knowledge over time, and therefore experiences and instinctive feelings play a primary role [12].

To apply the correct forecast, it must first gather data from the market and then use that data appropriately to generate a forecast. In this way, the forecast accuracy can be influenced by how effective the technique is originally chosen is to manage the specific problem, as well as the quality and type of data provided [10]. Organizations continuously generate large-scale data [13]. This large amount of data growth is often referred to as Big Data [14].

In general, to take advantage of the increasing data availability of the manufacturing industry, for example; for quality improvement initiatives, production cost estimation and/or process optimization, better understanding of the customer's requirements, etc. support is needed to address the relevant higher dimensions, complexities, and dynamics [15]. The ML field focuses on developing and implementing computer algorithms that evolve with experience [16]. These developments offer great potential to transform the production area and grasp their increasing production data repositories. However, the ML field is very diverse and includes many different algorithms, theories, and methods.

ML applications are often developed as part of a larger project. Various process model frameworks have been developed for such projects. One of the popular is the Cross-Industry Standard Process For Data Mining (CRISP-DM) [17].

As widely used ML methods in the literature, four of them stand out as supervised, unsupervised, semi-supervised and reinforced learning.

1) Supervised learning is a process that aims to find a link in the form of rules that link input data to output data and finally apply learned rules to new data. At this point, the computer program is trained. The basic rule here is that 70% of the data set is used as a training data set, 20% as an evaluation data set, and the last 10% as test data. With the obtained one can now predict future input and output data. Its two essential tasks are classification and regression [17].
2) The most well-known task of unsupervised learning is clustering. The method identifies similarities between entries to classify entries according to standard models [17].
3) The Semi-Supervised Learning dataset contains both tagged and untagged samples. Generally, the amount of unlabeled samples is much higher than the number of labelled samples. The purpose of a Semi-Supervised Learning algorithm is the same as that of a supervised learning algorithm. The distinguishing factor here is that using many unlabeled examples can help the learning algorithm find a better model [18].
4) Reinforced Learning is defined by the provision of educational information by the environment. Information about how well the system performs in the relevant turn is provided by a numerical reinforcement signal [19]. Another defining feature is that the student has to try and find out which actions produce the best results (numerical reinforcement signal). This distinguishes reinforced learning from most other ML methods [20].

Choosing an appropriate method for a specific problem and dataset is one of the most challenging data analysis tasks [21]. To select an appropriate algorithm, it is a common approach to search for similar problems and analyzes which ML algorithm is used to solve it and where the results are [22].

3 Proposed Methodology and Modelling

The data needed to be used in demand forecasting was obtained from a company that makes private mass production in Istanbul. For the demand quantities and other variables of the products produced, the degree of dependency was first determined by correlation analysis. The data were modelled with 6 ML algorithms, and the data was modelled with the base state. In the light of the results, feature engineering was performed on the data and the results were compared with each other.

In this study, Python Programming Language and PyCharm library, which are widely used in ML algorithms, are used.

The textile company subject to the research has accepted mass customization as low until 2008 and applied it to a minimal number of customers. By establishing a workshop in this direction with the decision taken in 2007, it started to manufacture parts for a broader customer base as of January 2008. In the research; By estimating the demand of this company's mass customization workshop, it is aimed to share the results for managers to make more successful planning, especially in the areas of workforce, inventory and stock level planning, and determination of raw material needs.

Ten types of products are produced in mass customization workshop, including pants A/B, jacket A/B, shirt A/B/C, coat, reefer jacket, and topcoat. Thirty-three permanent workers are working in the production phase; In line with seasonal needs, staff from other workshops are reinforced in short periods.

The workshop's most significant problem area is the production of the stock and worker reinforcements made in short periods due to the fluctuation in demand in specific periods. For this reason, very short-term planning can be made, which affects the production and workforce plans for other workshops. As a result of the production to stock, it imposes additional costs because the adaptation is made when the demand is received or the increasing amount of products are transferred to mass production. It is aimed to present a more predictable future period by forecasting the demand with the ML algorithms of the said workshop.

Data needed for demand estimation; It has been taken from the company's production directorate on a monthly basis from 2008, when the workshop, which makes personalized mass production, was operational.

The data set was subjected to correlation analysis in order to see to what extent the variables obtained from the firm affect the result in demand estimation and which ones do not have any effect. Correlation-based feature selection is widely used to reduce feature dimensionality and evaluate the discrimination power of a feature in classification models. In short, this analysis was applied to see the relationship between more than one variable, to measure the degree of influence on the amount of demand, and to reveal independent variables. As a result of the correlation analysis, it was revealed that there was no relationship between daily working hours and demand.

The available data were first examined over 22 variable; then, data shifting was applied with feature engineering and re-modelled by increasing it to 176 variables. Here it is aimed to teach the trend to the time series problem. In this way, it is aimed to reflect the seasonal changes in demand to the model and to reveal healthier results. Current variables have been subjected to data scrolling for 1, 2, 3, 6, and 12 months. During this process, the rolling average transaction was also performed. The first element of the rolling mean is obtained by averaging the starting constant subset of the series of numbers. The subset is then replaced by "forward shift"; that is, excluding the first number in the series and containing the next value in the subset. Thus, in the model; Short-term fluctuations in time series data are corrected and long-term trends are highlighted.

In ML, hyperparameter optimization is required to ensure healthy learning and to increase the success rate. Hyperparameter optimization is to determine the parameter sets that produce the best result or have the best solution time according to the objective function by applying different parameter value combinations in problems. Each algorithm is fed from different parameters and from these parameters that can take infinitely different values; Modeling was done by taking four parameters for XGBoost, Ridge Regression, Lasso Regression, and Lasso Lars Regression. The aim here is to increase the success rate by testing each algorithm with different parameters. Linear (Linear) Regression and Bayes Based Ridge Regression use random parameters.

For example, for Pants A, the demand forecast made according to 22 variables is included in Table 1, and the demand forecast based on the data obtained as a result of the preparation with feature engineering is included in Table 2.

Table 1. Demand forecast for pants A

Algorithm	r_squared	Mse	mae	rmse
XGBoost	0,9948	0,22030513	0,27155813	0,46936673
Ridge Regression	0,7357	11,10043269	2,72976176	3,33173119
Lineer Regression	0,7357	11,10043269	2,72976176	3,33173119
Lasso Regression	0,7357	11,10043269	2,72976176	3,33173119
LassoLars Regression	0,7357	11,10043269	2,72976176	3,33173119
Bayesian Ridge Regression	0,7357	11,10043269	2,72976176	3,33173119

Table 2. Updated demand forecast for pants A

Algorithm	r_squared	mse	mae	rmse
XGBoost	0,9939	0,25735879	0,30552928	0,50730542
Ridge Regression	0,5824	17,54126488	2,81592051	4,18822933
Lineer Regression	0,5824	17,54126488	2,81592051	4,18822933
Lasso Regression	0,5824	17,54126488	2,81592051	4,18822933
LassoLars Regression	0,5824	17,54126488	2,81592051	4,18822933
Bayesian Ridge Regression	0,5824	17,54126488	2,81592051	4,18822933

4 Conclusion and Future Work

With this study, it was asked to make a demand forecast with ML algorithms of an organization that applies mass customization. Python Programming language and PyCharm library are used in demand forecasting. The most widely used in regression analysis of demand forecasting with ML; In recent years, six algorithms have been determined as the most preferred XGBoost, Ridge Regression, Linear Regression, Lasso Regression, Lasso Lars and Bayes Based Ridge Regression algorithms because they offer the fastest and best results in ML competitions. These algorithms have been optimized for Hyper-parameter, thus it is desired to reach the most successful result with the best parameters. Each algorithm has been tested with different parameters to increase the success rate.

First of all, the algorithms' general performance measurements were made based on the baseline state of the data. Subsequently, in the feature engineering step, the data set was expanded by shifting and averaging the historical data of 1, 2, 3, 6, and 12 months of values. Here, it is aimed to teach trends to the model for problem-solving in time series problems. According to the results, it was observed that the most influencing data was the amount of demand from the previous years and the shifted data obtained as a result of feature engineering.

As a result of the measurements made, XGBoost gave results with an accuracy of 99% and above in 8 products. The topcoat reached 94% with the improvement made in the data. Other algorithms have reached the same rate with each other in all products, only the coat production has shown a more successful performance than XGBoost, which yields 70% results in demand estimation, resulting in 97% accuracy.

With this application, it has been concluded that the workshop, which makes personalized mass production, can use the XGBoost algorithm in the light of the improved data by applying attribute engineering for demand forecasting in 9 products only for the coat. In this way, managers will be able to make more successful planning, especially in terms of workforce, inventory, and stock level planning and determination of raw material needs.

As a result, different parameters change the success of the algorithm, feature engineering steps, data scrolling, and averaging processes improve the model's success. In general, XGBoost has been observed to give successful results for some product groups and other algorithms. In light of the available data set, the XGBoost algorithm provided the most reliable results.

The most significant limitation of this study is the data size. The application was made by gathering 22 variables and expanding it to 176 variables due to the feature engineering application. For future work, a larger data set will be used for demand forecasting. For ML algorithms that produce better results in larger data sets; It is considered that they will perform better by collecting more data.

References

1. Yüksel, B.: The role of mass customization strategy in reducing business and customer relations to personal size. Dumlupınar Univ. J. Soc. Sci. **3**, 207–224 (1999)
2. Tseng, M.M., Yue, W., Jiao, R.J.: Mass customization. In: Laperrière, L., Reinhart, G. (eds) The International Academy for Produ, CIRP Encyclopedia of Production Engineering. Springer (2017)
3. Toffler, A.: Future Shock. Bantam Books, New York (1970)
4. Davis, S.M.: From "future perfect": mass customizing. Plann. Rev. **17**(2), 16–21 (1989)
5. Pine, B.J.: Mass customizing products and services. Plann. Rev. **21**(4), 6–13 (1993)
6. Boër, C.R., Pedrazzoli, P., Bettoni, A.S.M.: Mass Customization and Sustainability. Springer (2013). https://doi.org/10.1007/978-1-4471-5116-6
7. Barman, S., Canizares, A.E.: A survey of mass customization in practice. Int. J. Supply Chain Manag. **4**, 65–72 (2015)
8. Kolade, O.J.: Economic development, technological change, and growth. Acta Univ. Danubius Oeconomica **15**(3), 157–169 (2019)
9. Anusha, S.L., Alok, S., Shaik, A.: Demand forecasting for the indian pharmaceutical retail: a case study. Journal of Supply Chain Management Systems, vol. 3, no. 2 (2016)

10. Kalchschmidt, M.: Forecastıng short term demand in heterogeneous customer oriented demand management processes, Milan University Faculty of Engineering PhD Thesis (2002)
11. Ergün, S., Şahin, S.: Literature review on business demand forecast. Ulakbilge **5**(10), 469–487 (2017)
12. Sanders, N.R., Ritzman, L.P.: The need for contextual and technical knowledge in judgmental forecasting. J. Behav. Dec. Making **5**, 39–52 (1992)
13. Liu, Y., et al.: MapReduce based parallel neural networks in enabling large scale machine learning. Computational Intelligence and Neuroscience (2015)
14. Lee, J., Lapira, E., Bagheri, B., Kao, H.: Recent advances and trends in predictive manufacturing systems in big data environment. Soc. Manuf. Eng. (SME) **1**(1), 38–41 (2013)
15. Davis, J., et al.: Smart manufacturing. Annu. Rev. Chem. Biomol. Eng. **6**, 141–160 (2015)
16. Libbrecht, M.W., Noble, W.S.: Machine learning applications in genetics and genomics. Nat. Rev. Genet. **16**(6), 321–332 (2015)
17. Hannah, W., Damiel, S., Saskia, S.: A literature review on machine learning in supply chain management. In: Kersten, W.B., Thorsten, R., Christian, M. (Ed.) Artificial Intelligence and Digital Transformation in Supply Chain Management: Innovative Approaches for Supply Chains. Proceedings of the Hamburg International Conference of Logistics (HICL), vol. 27, (ISBN 978–3–7502–4947–9, epubli GmbH), pp. 413–441 (2019)
18. Burkov, A.: The Hundred-page Machine Learning Book (2019). ISBN:9781999579517
19. Kotsiantis, S.B.: Supervised machine learning: a review of classification techniques. Informatica **31**, 249–268 (2007)
20. Sutton, R.S., Barto, A.G.: Reinforcement Learning: An Introduction, 2. MIT Press, Edition, Cambridge, MA (2012)
21. James, G., Witten, D., Hastie, T., Tibshirani, R.: An Introduction to Statistical Learning, Springer (2013). https://doi.org/10.1007/978-1-4614-7138-7
22. Wuest, T., Liu, A., Lu, S.C.-Y., Thoben, K.-D.: Application of the stage gate model in production supporting quality management. Procedia CIRP **17**, 32–37 (2014)

Face Detection and Facial Feature Extraction with Machine Learning

Mehmet Karahan(✉), Furkan Lacinkaya, Kaan Erdonmez, Eren Deniz Eminagaoglu, and Cosku Kasnakoglu

TOBB University of Economics and Technology, 06510 Ankara, Turkey
{m.karahan,flacinkaya,kerdonmez,eeminagaoglu, kasnakoglu}@etu.edu.tr

Abstract. Face detection is important part of surveillance systems and it has been widely used in computer vision and image processing. Face detection is also first step of the facial feature extraction. Facial feature extraction is a topic that has been focused on by many researchers in computer science, psychology, medicine and related fields and has become increasingly important in recent years. With the help of facial features, machine learning algorithms can estimate ages and classify genders of people. In this paper, face detection, facial feature extraction, age estimation and gender classification are presented. Firstly, face detection and extraction of facial features like eyes, eyebrows, mouth and nose are presented. Secondly, age estimation and gender classification based on the extracted facial features are explained. Experimental results prove that face detection algorithm efficiently detects human faces and facial feature algorithm accurately locates eyes, eyebrows, mouth and nose. Experimental results also show that, based on the extracted facial features, convolutional neural network architecture estimates ages of the people and classifies their gender.

Keywords: Face detection · Viola-Jones face detector · AdaBoost · Facial feature extraction · Convolutional neural network · Gender classification · Age classification

1 Introduction

Face detection and facial feature extraction are significant for the face tracking, facial expression recognition and face recognition. Facial feature extraction plays a crucial role in the areas of human computer interaction [1], video monitoring and person identification [2]. Age and gender classification algorithms are mainly based on the identification of facial features [3].

Paul Viola and Michael Jones developed the Viola Jones algorithm with the aim of face detection. This algorithm consists of four steps. In the first stage, face images are determined using Haar features. In the second step, the speed of calculating Haar features is reduced by using integral images. In the third step, using a cascaded AdaBoost classifier, a face dataset is trained. In the fourth step, a trained detection classifier is used to detect the final face image [4].

C. Kahraman et al. (Eds.): INFUS 2021, LNNS 308, pp. 205–213, 2022.
https://doi.org/10.1007/978-3-030-85577-2_24

Marcetic et al. developed a two stage model with the aim of face detection. While the first stage depends on normalized pixel stage and its aim is reduce false negative face detection, second stage based on the deformable part model and decreases false positive detections [5].

Gupta et al. combined image processing and pattern recognition methods to extract facial features. K-mean clustering and morphological techniques are used to determine facial landmarks [6]. Chowdhury et al. proposed a system that detects facial features by analyzing color components in the images of human faces [7].

Ko et al. used facial features to classify age and gender of people. Local Binary Patterns is used classify gender of young and adult people and Euclidean distance among facial feature points is used to classify gender of old people [8]. Higashi et al. convolved images with Gabor filters and encoded with Local Directional Pattern to classify age and gender of people. The dimensions of the image is reduced by the Principal Component Analysis and Support Vector Machine is used to classify feature vector [9].

In this study, face detection, facial features extraction, age estimation and gender classification are presented using Python and OpenCV programming languages. Viola-Jones algorithm is used for face detection. Dlib machine learning toolkit is used with Python and OpenCV for facial features extraction. Convolutional neural network architecture is used with the aim of age and gender classification. The main contributions of this paper are summarized as follows:

- Machine learning based facial features extraction is achieved.
- An algorithm based on convolutional neural network has been developed that accurately classifies people's age and gender.

The rest of this study is organized as follows. Section 2 expresses the main principles of face detection algorithm, facial features extraction and convolutional neural network based age and gender classification. Section 3 gives experimental results of our algorithms. Section 4 discusses the conclusion of this study.

2 Methods

2.1 Face Detection and Facial Feature Extraction Algorithm

Face detection depends on the Viola Jones algorithm. The most important features of this algorithm are stable and real time. Thanks to these features, the algorithm works quickly and with high accuracy. Viola Jones algorithm consists of 4 subcomponents. The first of these components is the Haar Cascade. In this section, a picture placed in the system is scanned with rectangles and operations are performed on the pixels of the picture. Another subtitle is integral images. In this part of the algorithm, the process of collecting the pixels from the previous section is performed here in an accelerated manner. The third stage is the AdaBoost algorithm. The identification and learning of the regions that are candidates for finding faces in the picture given to the system are done in this section. In the last step of the algorithm, the cascade classifier operation is performed. In this section, it gives a positive or negative response depending on whether there is a face in the regions that are determined with the previous steps and are candidates for face. After all these steps are completed, the face detection process will be completed [10] (Fig. 1).

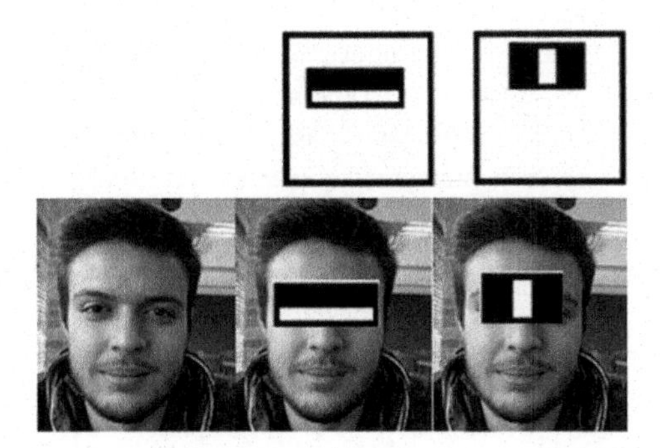

Fig. 1. Showing Haar features on a person's face.

In this study, Dlib library is used to extract facial features of people. Dlib is an independent software library developed in C++ language and it includes machine learning algorithms. Dlib can also be used in the Python programming language. This library contains a face pointer data set. This data set contains 68 pointers expressing the human facial features and the boundaries of parts such as mouth, eyes, eyebrows and nose on the face. Each pointer represents a specific point on the face. Each pointer has (x, y) coordinates and can be expressed in Eq. 1 [11].

$$P_1(x_1, y_1), P_2(x_2, y_2) \ldots\ldots P_{68}(x_{68}, y_{68}) \quad (1)$$

2.2 Age and Gender Classification Algorithm

Convolutional Neural Network (CNN) is a deep learning computer vision algorithm that can detect, classify and reconstruct images with high accuracy. CNN consists of five steps and takes images as input. The first step is the Convolutional Layer step. In this step, low- and high-level features in the image are extracted with the help of some filters and thus, the features of the picture given to the system as input are determined. Afterwards, the transition to the Non-Linearity Layer step is made. The reason for the application of this step is that since all layers applied before can be a linear function, the Neural Network can act as if a single perception is made. In this case, the result can be calculated by the linear combination of outputs. Thus, non-linearity is introduced to the system. The third step of the algorithm is called the Pooling Layer. In this step, the process of reducing the slip size of the representation and the number of parameters and calculations in the network is performed, and the appropriate and unsuitable parts are checked. The next step is Flattening Layer and its aim is to prepare the data at the input of the Fully Connected layer. The last and most important step is called Fully Connected-Layer, in this step, data from the previous layer is taken and the learning process is performed [12].

In our study, CNN algorithm performs age and gender classification with 3 convolutional layers. These are 2 fully connected layers and a final output layer. The first convolutional layer has 96 nodes with kernel size 7. The second convolutional layer has 256 nodes with kernel size 5 and the third convolutional layer has 384 nodes with kernel size 3. Each of the two fully connected layers has 512 nodes. The output layer in the gender classification network is of the softmax type with 2 nodes specifying two classes, "Male" and "Female". Adience dataset is used to train the model. This should be approached as a regression problem, as a true and clear number is expected as a result of the gender estimation. In this solution, estimation is made by classifying age groups. There are 8 classes in the Adience data set divided into the eight different age groups. These groups are (0–3), (4–6), (7–12), (13–19), (20–30), (31–45), (46–59) and (60–100). Therefore, the age classification network has 8 nodes that show the age ranges specified in the last softmax layer [13] (Fig. 2).

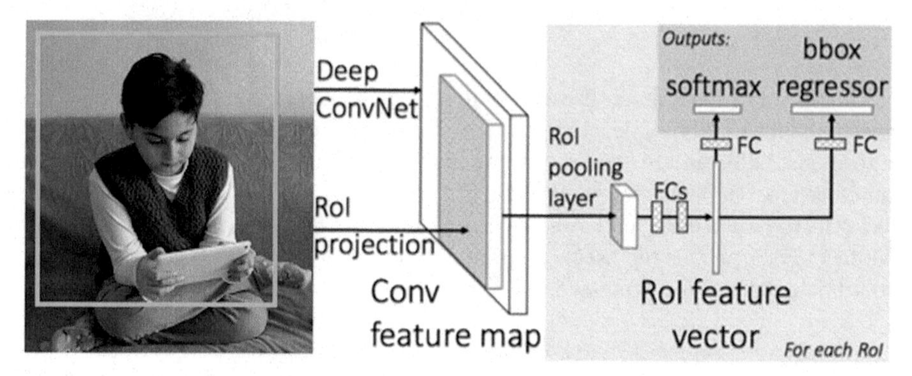

Fig. 2. Illustration of the convolutional neural network.

3 Experimental Results

In this section, experimental results of the face detection, facial feature extraction and age & gender classification algorithms are presented. Figure 3 shows the face detection of a single person and a group of people. Figure 4 shows the result of the facial feature extraction algorithm. Facial feature algorithm successfully detects a person's eyes, mouth, nose and eyebrows. Figure 5 through Fig. 7 gives the examples of machine learning based age and gender classification.

Fig. 3. Detecting a single person's face and a group of people's faces.

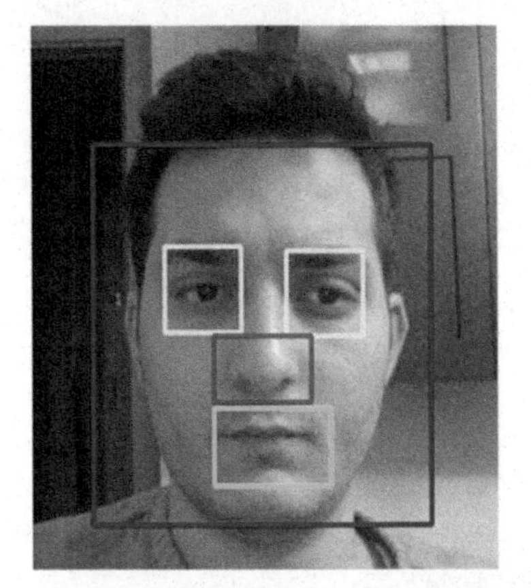
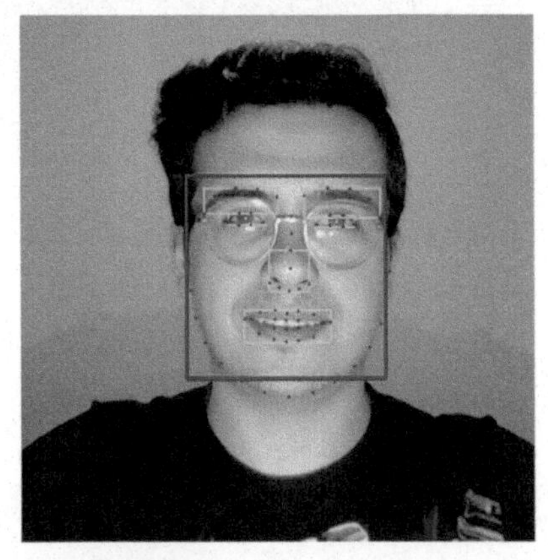

Fig. 4. Facial feature extraction to detect a person's mouth, nose and eyes; the face on the right illustrates the detection of eyebrows and the jawline as well.

Fig. 5. Age and gender classification of a young adult and a child via CNN architecture.

Fig. 6. The age and gender classification of four young men via CNN architecture.

Fig. 7. The age and gender classification of two girls and a boy via CNN architecture.

The accuracy of age & gender classification algorithm is tested using Adience dataset [13]. Figure 8 represents the accuracy of age classification algorithm. The graphic shows accuracy percentage of the algorithm according to age groups. The situation of estimating the age group of the person with 1 less or 1 more accuracy is shown in the correct & 1-off column on the graphic. One-hundredth of the number of samples used for each age group is shown in the Sample space/100 column in the graph. Figure 9 shows the accuracy of the gender classification algorithm.

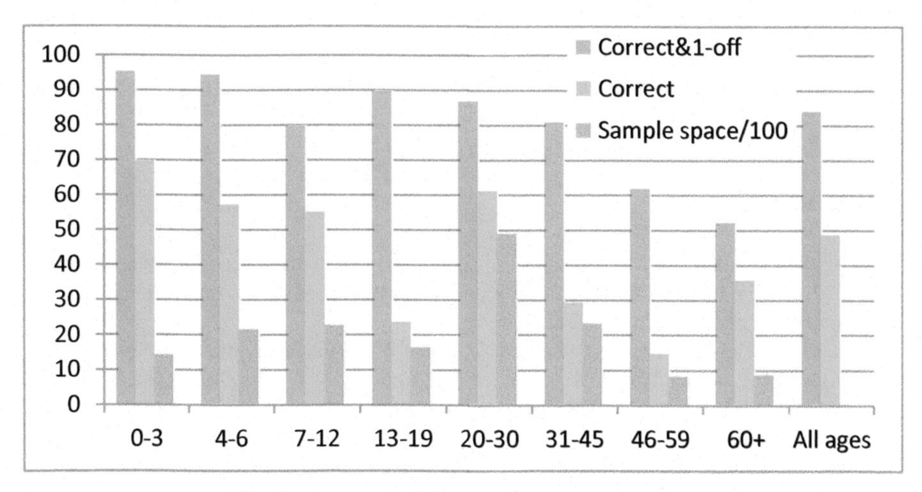

Fig. 8. Accuracy of the age estimation algorithm.

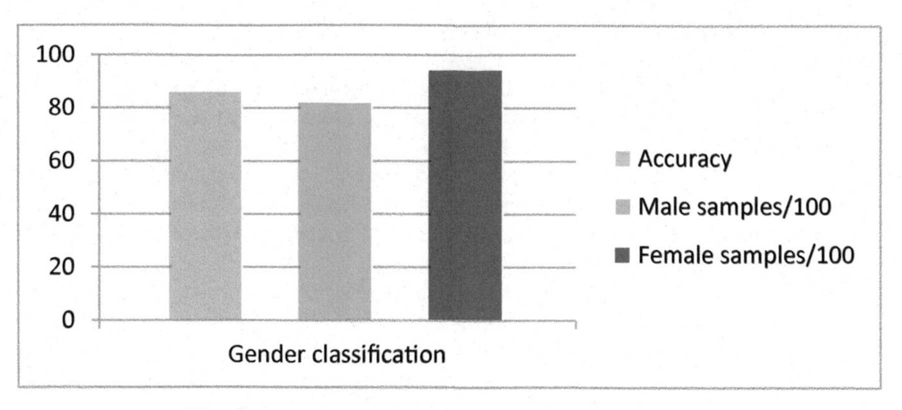

Fig. 9. Accuracy of the gender classification algorithm.

4 Conclusion

In this study, face detection, facial feature extraction, age & gender classification are presented using Python and OpenCV programming languages. Face detection algorithm can detect human faces and machine learning based novel facial feature extraction algorithm successfully extracts mouth, nose, eyebrows and eyes. Our novel convolutional neural network architecture performs age and gender classification with high accuracy. The data on the accuracy rate of age and gender classification are presented on graphics. Experimental results stress that face detection, facial feature extraction algorithms successfully detect facial landmarks and age & gender classification algorithm work with high accuracy. In future studies, we will focus on facial expression recognition and emotion classification.

References

1. Benedict, S.R., Kumar, J.S.: Geometric shaped facial feature extraction for face recognition. In: 2016 IEEE International Conference on Advances in Computer Applications (ICACA), Coimbatore, pp. 275–278. IEEE (2016)
2. Wu, Y.-M., Wang, H.-W., Lu, Y.-L., Yen, S., Hsiao, Y.-T.: Facial feature extraction and applications: a review. In: Pan, J.-S., Chen, S.-M., Nguyen, N.T. (eds.) ACIIDS 2012. LNCS (LNAI), vol. 7196, pp. 228–238. Springer, Heidelberg (2012). https://doi.org/10.1007/978-3-642-28487-8_23
3. Karahan, M., Kurt, H., Kasnakoğlu, C.: autonomous face detection and tracking using quadrotor UAV. In: 2020 4th International Symposium on Multidisciplinary Studies and Innovative Technologies (ISMSIT), Istanbul, pp. 1–4. IEEE (2020)
4. Lu, W., Yang, M.: Face detection based on Viola-Jones algorithm applying composite features. In: 2019 International Conference on Robots & Intelligent System (ICRIS), Haikou, pp. 82–85. IEEE (2019)
5. Marčetić, D., Hrkać, T., Ribarić, S.: Two-stage cascade model for unconstrained face detection. In: 2016 First International Workshop on Sensing, Processing and Learning for Intelligent Machines (SPLINE), Aalborg, pp. 1–4. IEEE (2016)
6. Gupta, S., Singh, R.K.: Mathematical morphology based face segmentation and facial feature extraction for facial expression recognition. In: 2015 International Conference on Futuristic Trends on Computational Analysis and Knowledge Management (ABLAZE), Greater Noida, pp. 691–695. IEEE (2015)
7. Chowdhury, M., Gao, J., Islam, R.: Fuzzy rule based approach for face and facial feature extraction in biometric authentication. In: 2016 International Conference on Image and Vision Computing New Zealand (IVCNZ), Palmerston North, pp. 1–5. IEEE (2016)
8. Ko, J.B., Lee, W., Choi, S.E., Kim, J.: A gender classification method using age information. In: 2014 International Conference on Electronics, Information and Communications (ICEIC), Kota Kinabalu, pp. 1–2. IEEE (2014)
9. Higashi, A., Yasui, T., Fukumizu, Y., Yamauchi, H.: Local Gabor directional pattern histogram sequence (LGDPHS) for age and gender classification. In: 2011 IEEE Statistical Signal Processing Workshop (SSP), Nice, pp. 505–508. IEEE (2011)
10. Viola, P., Jones, M.: Rapid object detection using a boosted cascade of simple features. In: Proceedings of the 2001 IEEE Computer Society Conference on Computer Vision and Pattern Recognition, CVPR 2001, Kauai, pp. I-I. IEEE (2001)

11. Ugur, A., Guruler, H.: The detection of emotional expression towards computer users. Int. J. Inf. Technol. **10**(2), 231–239 (2017)
12. Lin, S.D., Chen, K.: Illumination invariant thermal face recognition using convolutional neural network. In: 2019 IEEE International Conference on Consumer Electronics - Asia (ICCE-Asia), Bangkok, pp. 83–84. IEEE (2019)
13. Eidinger, R., Enbar, R., Hassner, T.: Age and gender estimation of unfiltered faces. IEEE Trans. Inf. Forensics Secur. **9**(12), 2170–2179 (2014)

Artificial Intelligence-Based Digital Financial Fraud Detection

Sanaa Elyassami(✉), Hamda Nasir Humaid, Abdulrahman Ali Alhosani, and Hamed Taher Alawadhi

Abu Dhabi Polytechnic, Abu Dhabi, UAE
sanaa.elyassami@adpoly.ac.ae

Abstract. Digital frauds get a dramatic increase over the years and lead to considerable losses. Detecting fraudulent attempts is valuable to many industries and especially to the banking and financial sectors. To help in anticipating and accurately identifying whether a transaction is fraudulent, machine learning-based models are the key solution for banking and financial institutions. In this paper, an artificial intelligence-based model was built using deep learning and was trained using stochastic gradient descent and feedforward neural networks. The dropout regularization has been utilized to enhance the generalization capabilities of the digital transaction classification model. Different activation functions were used and explored such as the max-out, the hyperbolic tangent, the rectifier linear unit, and the exponential rectifier linear unit. The impact of the learning rate on the model performance was analyzed. For the evaluation of the model, we did use of different metrics such as the accuracy, the precision, and the recall. The obtained results are promising, and the developed model can be used effectively to defend the banking sector against digital frauds.

Keywords: Machine learning · Activation function · Classification · Fraud detection · Transaction

1 Introduction

Over the years, financial organizations aimed to improve their methods in fraud detection, but their financial system gets more complicated and the number of complexity of rules grew exponentially to a point where it got too complex for anyone to construct and maintain such system. Furthermore, with the rapid growth of technology, internet and online services, fraud towards financial institutes did not grow any less. Financial organizations have abandoned the use of rule-based monitoring systems for fraud detection and are now aiming to use the new age machine learning solutions. These machine learning algorithms can process millions of data objects quickly and link instances from seemingly unrelated dataset to detect suspicious patterns and provide reliable techniques to solve a wide range of areas [1, 2]. A general review of the fraud detection could be found in the research of Phua, C. et al. [3]. Achituve, S. Kraus et al. [4] focus on real-time fraud detection which started with explaining the LSTM, which is long short-term memory.

C. Kahraman et al. (Eds.): INFUS 2021, LNNS 308, pp. 214–221, 2022.
https://doi.org/10.1007/978-3-030-85577-2_25

LSTM is very effective, yet it is difficult to interpret the decisions, so a new approach was being introduced which is an attention-based classifier that focuses on the most relevant data to the classification task. They proposed a fraud network detection network which is made up of an attention mechanism that embeds categorical features in a continuous space and reshaping the features to a single vector, and a second component that has the responsibility of detecting fraudulent activities. A. Singla and H. Jangir [5] tried to find a solution for starting with an unauthorized gain of personal data and misuse activities in addition to fraudulent activities by using machine learning to solve these problems. They concluded that machine learning achieved low false-positive rates in both data management and predictive analytics. M. Azhan and S. Meraj [6] used multiple tests to find which fraud detection techniques are best from a list of many techniques, the techniques used in this experiment were Multiple Linear Regression, Logistic Regression classifier, K Nearest Neighbor classifier, Gaussian Naive Bayes classifier, Random Forest classifier, and Neural Network classifier. The k-nearest neighbor achieved the best performance. Sadineni [7] conducted multiple experiments using various machine learning techniques to classify whether the transactions are fraudulent or not. The techniques used in the experiment were Artificial Neural Networks, Decision Tree, Support Vector Machine, Logistic Regression and Random Forest. The performance analysis showed high perform accuracy and precision. Hidayattullah et al. [8] proposed to detect fraud in financial statements using machine learning based on meta-heuristic optimization to determine the best hyper-parameter which improves the overall performance of the classifier, where Support Vector Machine (SVM) produced 66% in terms of accuracy and 82% as precision. However, Back Propagation Neural Network (BPNN) achieved 93% as accuracy and 94% in terms of precision. J.Rajendra, Prasad et al. [9] attempted to combine the Natural Language Processing (NLP) and Supervised Machine Learning (SML) techniques by using decision tree and they proved the accuracy of the model to be effective.

In the current study and in order to help the financial organizations to keep up with the new defrauding schemes, we developed an intelligent fraud detection model using deep feedforward neural networks combined with the dropout regularization technique to enhance the performance and the generalization of the developed model.

This paper is organized as follows: In Sect. 2, we describe the research method. Section 3 focuses on the design of the experiments and results. In Sect. 4, we provide the conclusion and discussion.

2 Research Method

2.1 Dataset

Due to the confidential nature of banking and financial services, we have used a synthetic dataset for the current study. The dataset is from Kaggle.com and was generated using PaySim simulator which is presented in [10]. The used dataset in the current study include 6,362,620 records of mobile money transactions. The dataset is composed of a total of 11 variables. The dataset includes six continuous quantitative variables: "step", "amount", "oldbalanceOrg", "newbalanceOrg", "oldbalanceDest", and "newbalanceDest". And also, it includes one categorical variable which is the "type". And two qualitative nominal variables: "nameOrig" and "nameDest". It includes two dichotomous binary variables: "isFraud" and "isFlaggedFraud". The description of the dataset features is given in Table 1.

Table 1. Money transaction' dataset description.

Feature	Description
Step	Each step is equivalent to one hour
Type	CASH-IN, DEBIT, CASH-OUT, TRANSFER, and PAYMENT
Amount	Transaction amount
nameOrig	Customer who initiated the transaction
oldbalanceOrg	Customer's balance before the transaction
newbalanceOrg	Customer's balance after the transaction
nameDest	Transaction recipient ID
oldbalanceDest	Recipient's balance before the transaction
newbalanceDest	Recipient's balance after the transaction
isFlaggedFraud	Illegal transactions are assigned 1 and legal transactions are assigned 0
isFraud [***Target***]	1 for fraudulent transactions and 0 for non-fraudulent transactions

In this synthetic dataset, each "step" represents an hour of simulation. For example, records with step equals to 1 means that these transactions happened in the first hour on the first day of the month, and records with step equals to 48 occurred in the last hour of the second day of the month. The second attribute is "Type" that contain five different transaction types: debit, transfer, payment, cash-in, and cash-out. Debit implies sending money from the mobile service to a bank account. Transfer means transferring money between customers. Payment is the process of paying services and/or goods to suppliers. Cash-in implies that customer's account balance increases with cash inflow, while cash-out implies that customer's account balance decreases with cash outflow. "isFlaggedFraud" is the indicator used to prevent illegal transactions based on a set of triggered thresholds used by the money system whereas "isFraud" is the flag that indicates the actual fraud transactions.

2.2 Data Analysis and Exploration

The distribution of the type of transaction (see Fig. 1) shows that Payment (34%) and Cash-out (35%) are the two most used type of transactions. While only 1% of transactions are of type Debit and 8% of transactions are of type Transfer.

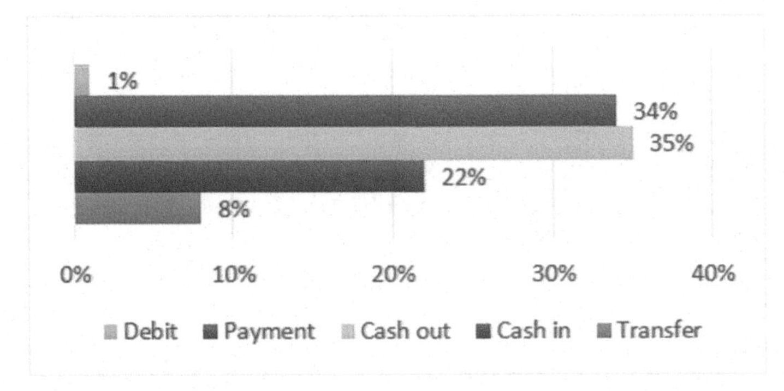

Fig. 1. Type of transactions.

The distribution of fraudulent transactions in each type of transactions is shown in Fig. 1. It shows that out of the four types of transactions, the fraudulent transactions are either Transfer or Cash-out. The total number of fraud transaction is 8213. In every 2319 transactions, 3 fraudulent transactions are happening. Thus, these illegal transactions cause a loss of $12056415427. Since, the current rule-based systems are not capable of detecting accurately fraudulent transactions, there is a drastic need for a system which can be accurate and reliable to prevent illegal transactions and avoid the loss of considerable amount of money. To analyze the power of each attribute to predict the target attribute "isFraud" and to be able to select the most relevance attributes to build our predictive model, we calculated the weight of the seven attributes. The results shown that "isFlaggedFraud" attribute is weighted 0, this an insignificant attribute that we can discard without any loss, thus it will be dropped from the dataset and not be used in building the model. The correlation matrix demonstrates that "newbalanceOrig" is highly correlated with "oldbalanceOrig". "newBalanceDest" is highly correlated with "oldbalanceDest". "amount" and "step" are correlated with the target variable "isFraud".

2.3 Deep Neural Network Models

In the current paper, we implemented a feedforward neural network-based model using a feedforward artificial neural network to classify transactions. The deep feedforward neural network model was trained with stochastic gradient descent using the backpropagation algorithm. Different non-linear activation functions have been investigated and used in building different models to identify fraudulent transactions. The hidden layers were trained using non-linear activation functions to allow the nodes to learn efficiently the relationships hidden in the dataset and provide accurate classification. To compute the output of the hidden nodes, we have used the following four activation functions: hyperbolic tangent (tanH) [11], rectifier linear unit (ReLU) [12], maxout [13], and exponential rectifier linear unit (ELU) [14].

2.4 Evaluation Metrics

In our binary classification model (fraudulent transaction, non-fraudulent transaction), each record falls in one of the four following possibilities. True-negative "TN" where the model correctly predicts the negative class and thus, non-fraudulent transactions are correctly identified as non-fraudulent. True-positive "TP" where the model correctly predicts the positive class and thus, fraudulent transactions are correctly identified as fraudulent. False-positive "FP" where the model incorrectly predicts the negative class and thus, non-fraudulent transactions are incorrectly identified as fraudulent. False-negative "FN" where the model incorrectly predicts the positive class and thus, fraudulent transactions are incorrectly identified as non-fraudulent.

The evaluation of the proposed model was carried out using various evaluation metrics [15] such as the accuracy, the precision, the specificity, the recall and F1-score. The accuracy is defined as the ratio between the number of correctly classified samples and the overall number of samples. The sensitivity is called also the recall and it measures the proportion of actual positives that are correctly classified as positives. The specificity is called also the true negative rate and it measures the proportion of actual negatives that are correctly classified as negatives. The precision called also positive predictive value is defined as the results classified as positive by the model. The F1-Score is the balance between the precision and the recall and it highest value is 1 which indicate perfect precision and recall.

3 Experiment Design and Results

To classify the financial transactions in two classes: fraudulent and non-fraudulent transactions, we have developed deep feedforward networks. The models were trained with the stochastic gradient descent using backpropagation. 70% of the dataset samples were used to train the models and the remaining samples for testing. Different activation functions were used, and their results were compared.

3.1 Deep Neural Network-Based Model Result

The deep feedforward neural network-based model was trained with the four activation functions. The classification results produced by the different networks on the testing data is shown in Table 2. The network using Maxout as activation function was ranked first in terms of specificity (97%) and ranked second in terms of accuracy (93%). However, the Maxout-based classifier resulted in the lowest recall, precision, and F1-score. ReLU was ranked third in terms of specificity and F1-score, whereas it produced the lowest accuracy. ELU is the second performing in terms of F1-score, precision and recall as

Table 2. Deep neural network model classification results

Activation function	Accuracy	Recall	Specificity	Precision	F1-Score
ReLU	92.87	68.34	96.02	95.94	79.82
tanH	94.31	77.26	96.50	97.07	86.04
Maxout	93.59	65.20	97.23	95.61	77.53
ELU	92.93	72.61	95.54	96.45	82.85

82%, 96%, and 72% respectively. It was also observed that ELU is performing much better than ReLU in terms of precision and accuracy. This can be interpreted by the fact that the ReLU-based network cannot perform backpropagation to further learn. It has been shown that the hyperbolic tangent (tanH) outperformed the other activation functions. Thus, the accuracy, the recall, the precision, and the F1-score have reached high scores of 94%, 77%, 97%, and 86% respectively.

3.2 Deep Neural Network-Based Model with Dropout Regularization

Regularization techniques are mainly used to prevent overfitting [16, 17]. The dropout regularization technique was combined to our model. We have used the recommended level which is 0.5 [18, 19]. As shown in table 3, the dropout technique did enhance the recall for the three networks that used ReLU (enhanced by 10.3%), tanH (enhanced by 3.27%), and Maxout (enhanced by 3.77%), and achieved the highest recall of 80% when using tanH compared to all previously trained models. The dropout technique did enhance moderately the precision for the three networks that used ReLU (enhanced by 1.22%), tanH (enhanced by 0.34%), and Maxout (enhanced by 0.42%), and achieved the highest precision of 97.41%. In terms of F1-score, dropout regularization helped in enhancing the scores of the networks using ReLU (enhanced by 7.1%), tanH (enhanced by 2.13%), and Maxout (enhanced by 2.75%), and achieved the highest F1-score of 88%. However, the accuracy and the specificity of the four networks were decreased when using dropout regularization. Taking into account that F1-score, recall and precision are more useful than accuracy and specificity, we can conclude that the dropout technique did improve the performance or the proposed model.

Table 3. Deep neural network model using dropout regularization classification results

Activation function	Accuracy	Recall	Specificity	Precision	F1-Score
ReLU	92.13	78.64	93.86	97.16	86.92
tanH	92.43	80.53	93.96	97.41	88.17
Maxout	93.26	68.97	96.37	96.03	80.28
ELU	90.01	63.57	93.41	95.23	76.24

Results published in [20] showed that the highest accuracy was achieved by the Restricted Boltzmann Machines classifier (91%), followed by the ensemble of decision tree classifier that reached 90%, followed by stacked auto-encoders (80%). The results obtained by our models are more efficient and accurate than [20]. The results produced by our model outperformed the existing methods and have achieved an accuracy of 94% and F1-Score of 88% and a recall of 80%.

4 Conclusion

The objective of the current research is to build and validate an accurate intelligent model to detect fraudulent financial transactions. The study adopted supervised learning technique, where known fraudulent and non-fraudulent cases were used to train the model to learn and recognize the hidden patterns. Different models were developed using deep feedforward neural networks. To investigate the performance of the classification of the transactions, the impact of different activation functions was studied. These functions are the exponential rectifier linear unit, the hyperbolic tangent, the maxout, and the rectifier linear unit. The proposed model has achieved a high accuracy and a high specificity as 94% and 97% respectively when using maxout activation function. When using the hyperbolic tangent activation function and compared to the other activation functions, the classifier achieved the highest recall, the highest precision, and the highest F1-score as 77, 97, and 86% respectively. Consequently, the model using tanH has the potential to generate a knowledge-rich environment that can meaningfully help to decrease the loss caused by frauds by accurately detect illegal transactions. After integrating the dropout regularization into our model, the performance of the classifiers that were trained using tanH, ReLU, and Maxout activation functions were enhanced in terms of recall, precision, and F1-Score. Incorporating the dropout regularization helped in enhancing the model generalization capabilities. The F1-score, the precision and the recall for tanH-based classifier were slightly enhanced by 2.13%, 0.34%, and 3.27% respectively. The obtained results are promising, and the proposed model can be applied and used in banking and financial industries to accurately detect fraudulent transactions and avoid the loss of considerable money. In addition, applying our model to other financial datasets and incorporating ensemble methods might prove an important area for future research.

References

1. Elyassami, S., Kaddour, A.: Implementation of an incremental deep learning model for survival prediction of cardiovascular patients. IAES Int. J. Artif. Intell. **10**(1), 101–109 (2021). ISSN 2252–8938
2. Elyassami, S., Hamid Y., Habuza, T.: Road crashes analysis and prediction using gradient boosted and random forest trees. In: 2020 6th IEEE Congress on Information Science and Technology (CiSt), Agadir - Essaouira, Morocco, pp. 520–525 (2020). https://doi.org/10.1109/CiSt49399.2021.9357298
3. Phua, C., Lee, V., Smith, K., Gayler, R.: A comprehensive survey of data miningbased fraud detection research. arXiv preprint arXiv:1009.6119 (2010)
4. Achituve, I., Kraus, S., Goldberger, J.: Interpretable online banking fraud detection based on hierarchical attention mechanism. In: 2019 IEEE 29th International Workshop on Machine Learning for Signal Processing (MLSP), Pittsburgh, PA, USA, pp. 1–6 (2019). https://doi.org/10.1109/MLSP.2019.8918896
5. Singla, A., Jangir, H.: A comparative approach to predictive analytics with machine learning for fraud detection of realtime financial data. In: 2020 International Conference on Emerging Trends in Communication, Control and Computing (ICONC3), Lakshmangarh, Sikar, India, pp. 1–4 (2020). https://doi.org/10.1109/ICONC345789.2020.9117435
6. Azhan, M., Meraj, S.: Credit card fraud detection using machine learning and deep learning techniques. In: 2020 3rd International Conference on Intelligent Sustainable Systems (ICISS), Thoothukudi, India, pp. 514–518 (2020). https://doi.org/10.1109/ICISS49785.2020.9316002
7. Sadineni, P.: Detection of fraudulent transactions in credit card using machine learning algorithms (2020)
8. Hidayattullah,S., Surjandari, I., Laoh, E.: Financial statement fraud detection in Indonesia listed companies using machine learning based on meta-heuristic optimization (2020)

9. Rajendra Prasad, J., SaiKumar, S., SubbaRao, B.V.: Design and development of financial fraud detection using machine learning. Int. J. Emerg. Trends Eng. Res. **8**(9) (2020). http://www.warse.org/IJETER/static/pdf/file/ijeter152892020.pdf
10. Lopez-Rojas, E.A., Elmir, A., Axelsson, S.: PaySim: a financial mobile money simulator for fraud detection. In: The 28th European Modeling and Simulation Symposium-EMSS, Larnaca, Cyprus (2016)
11. Osborn, G.: Mnemonic for hyperbolic formulae. Math. Gaz. **2**(34), 189 (1902). https://doi.org/10.2307/3602492.JSTOR3602492
12. Nair, V., Hinton, G.E.: rectified linear units improve restricted boltzmann machines. In: 27th International Conference on International Conference on Machine Learning, ICML 2010, USA, Omnipress, pp. 807–814 (2010)
13. Goodfellow, I.J., Warde-Farley, D., Mirza, M., Courville, A., Bengio, Y.: Maxout networks. In: JMLR Workshop and Conference Proceedings, vol. 28, no. 3, pp. 1319–1327 (2013)
14. Clevert, D.-A., Unterthiner, T., Hochreiter, S.: Fast and accurate deep network learning by exponential linear units (ELUs) (2015)
15. Lever, J., Krzywinski, M., Altman, N.: Classification evaluation. Nat. Methods **13**, 603–604 (2016). https://doi.org/10.1038/nmeth.3945
16. Faris, H., Mirjalili, S., Aljarah, I.: Automatic selection of hidden neurons and weights in neural networks using grey wolf optimizer based on a hybrid encoding scheme. Int. J. Mach. Learn. Cybern. **10**(10), 2901–2920 (2019). https://doi.org/10.1007/s13042-018-00913-2
17. Moradi, R., Berangi, R., Minaei, B.: A survey of regularization strategies for deep models. Artif. Intell. Rev. **53**(6), 3947–3986 (2019). https://doi.org/10.1007/s10462-019-09784-7
18. Goodfellow, I., Bengio, Y., Courville, A.: Deep Learning, Book in Preparation for MIT Press (2016). http://www.deeplearningbook.org
19. Ba, L.J., Frey, B.: Adaptive dropout for training deep neural networks. In: Proceedings of the 26th International Conference on Neural Information Processing Systems - Volume 2 (NIPS 2013), pp. 3084–3092. Curran Associates Inc., Red Hook (2013)
20. Mubalaike, M., Adali, E.: Deep learning approach for intelligent financial fraud detection system. In: 2018 3rd International Conference on Computer Science and Engineering (UBMK), Sarajevo, pp. 598–603 (2018). https://doi.org/10.1109/UBMK.2018.8566574

Stock Price Prediction of Turkish Banks Using Machine Learning Methods

Bora Egüz(✉), Fırat Ersin Çorbacı, and Tolga Kaya

Department of Management Engineering, Istanbul Technical University, 34367 Istanbul, Turkey
{eguz16,corbaci15,kayatolga}@itu.edu.tr

Abstract. Stock markets are vital part of the economy since they utilize the capital. Investors in stock markets are providing capital with the expectation of positive return on their investment. In order to assess if an investment going to bring positive return, future prices of the stocks must be estimated. Thanks to development in technology and statistics, relatively more advanced estimation methods are developed and machine learning approaches are being integrated for this task. Following study is suggesting a stock price prediction model for Turkish banks using machine learning methods such as multiple linear regression, ridge regression, lasso regression, support vector machines, decision tree models, random forest, XGBoost method based on a wide dataset which is expanded using sliding windows method. After the models trained and tested, it has been observed that the XGBoost algorithm is superior to the other algorithms according to the result of the test errors. Thus the proposed model is convenient for predicting the stock prices of Turkish banks.

Keywords: Stock price · Machine learning · Sliding windows · Turkish banks · XGBoost

1 Introduction

Stock markets are enabling people to trade shares of public companies all over the world as well as in Turkey. This trading mechanism creates capital resources for companies, and through this capital flow, companies can create value for the people. In the stock market, one of the key players are the investors, and their motivation to invest is to gain more money than they invested. In order to achieve this task, investors have a tough task of understanding whether a stock price will go up or down [4]. Stock markets are highly unpredictable and dynamic [7]. These qualities make predicting a stock price is a very hard task. There are various parameters that can affect a stock's price, such as investors' approach, economic news, or political events. This dynamism situation is the same in Turkey's Borsa Istanbul (BIST) too.

While the world has an increasing amount of data every day, numerous approaches and technologies are developed to create more insights from the data [10]. Machine learning methods are one of the relatively new and popular methods that are being used in stock markets in order to predict stock prices. This research's country focus and

C. Kahraman et al. (Eds.): INFUS 2021, LNNS 308, pp. 222–229, 2022.
https://doi.org/10.1007/978-3-030-85577-2_26

usage of method provides an original perspective to both academic and professional communities. It has believed that having information about the stock prices of such banks can give some clues about the country itself. Also the investors that makes trade on volatile markets can use this model. The purpose of this research is to predict the next day closing stock price of three banks of Turkey, which are listed in the top 30 companies of BIST. In order to serve this purpose, performances of multiple linear regression, ridge regression, support vector regression, decision tree regression and random forest methods evaluated.

The prediction model's dataset contains data generated by Borsa Istanbul. For Garanti Bank, Akbank, and Yapi Kredi Bank, the stocks' values evaluated as raw data. This data include closing price, technical indicators of trend, momentum, volatility and volume. The models' performance calculated using cross validation method with choosing k constant as 10, then the model that minimum mean squared error has chosen.

In Sect. 2, the literature review of the past work that has been conducted in the area of stock prediction using the machine learning approach briefly presented. Section 3 contains methodological information about the research and the data of it. In this section explanation of machine learning methods with their mathematical notation can be found and it gives detailed information about content of the dataset that used in the prediction model. Section 4 is the section that the findings of the research are shared. Section 5 is the conclusion and discussion part of the paper that briefly summarizes the study and includes suggestions for future researches.

2 Literature Review

Predicting the stock prices have been studied in many papers until today. In 2020, the study conducted by Vijh et al. used Artificial Neural Network (ANN) and Random Forest (RF) models to predict the closing prices of 5 different company's stocks, which are Nike, Goldman Sachs, Johnson and Johnson, Pfizer and JP Morgan Chase, and Co [9]. The findings of that study indicated that ANN performed better than RF in that specific model. In 2019, another study by Werawithayaset and Tritilanunt used Multi-Layer Perceptron (MLP), Support Vector Regression (SVR), Partial Least Squares (PLS) models to predict stock prices of Thailand stock market, SET100, and the results showed that PLS was the best algorithm among the three [11]. In 2019, Sarode and his colleagues' research was about Long-Short Term Memory (LSTM) and SVM methods to predict stock prices, but they also analyzed the business's news to support their pro-poses [7]. Later in 2020, Wang used China stock market as his dataset, and he used SVM and Logistic Regression (LR) to classify the stock movements for up or down [10]. In findings, SVM was superior to LR in terms of accuracy and precision. A study by Ravikumar and Saraf aimed to predict companies' stock prices in the S&P500 index and tried to classify them regarding their momentum in year 2020 [6]. For predicting the prices, they used Simple Linear Regression, Polynomial Regression ($d = 10$), SVR, Decision Tree Regression (DTR), and RF; while for classification SVM, K-Nearest Neighbors, LR, Naive Bayes, Decision Tree Classification and Random Forest Classi-fication have been used. In another study in 2019, which was conducted by Akşehir and Kılıç aimed to predict stock prices of 5 different banks in BIST100, and they used Mul-tiple Linear Regression (MLR), DTR, and

RF methods [1]. Bonde and Khaled re-searched the comparison of stock price prediction features using Neural Network, SVR, Bagging Using Sequential Minimal Optimization, and M5P algorithms in 2012 [2]. They find out that the best algorithms to predict stock prices were SVR and bagging using sequential minimal optimization. Şişmanoğlu and his colleagues studied price prediction in the stock market in 2020, and they used two deep learning methods: Re-current Neural Network (RNN), LSTM, and Bidirectional LSTM [8]. In the findings, they found BLSTM was superior among others. In 2016, Boonpeng and Jeatrakul used ANN, One-Against-One (OAO), and One-Against-All (OAA) methods to predict prices in the stock exchange of Thailand [3]. Finally, Gao et al. tried to predict stock prices with deep learning methods: RNN and LSTM, and LSTM has been found as a better method by the study which was conducted in 2017 [4].

As seen in this section, the researches on this topic propose some approaches to predicting the market movement, up or down, or stock prices. The effort to help investors maintain the success rate and increase the profit has been the primary concern of this topic. Even though some models have used such indicators related to the price and moving average, there are not enough research papers that focus on Turkish banks and uses recent machine learning methods. This gap brings the idea of creating models that can predict the stock prices of Turkish banks. To the author's knowledge, this is the first study that focuses on Turkish banking sector using recent machine learning techniques like XGBoost, random forest etc.

3 Data and Methodology

Creating a stock price prediction model using machine learning methods requires data. Success of the model relies on quality and quantity of the data. Additionally, variable selection and technique selection must be appropriate to get minimum errors which will determine the success of the model. Model is created for three different banks (Garanti, Akbank, Yapı Kredi). For each model, numerous indicators that is being used in technical analysis are selected and integrated. Data of indicators obtained by using MATRIKS [12]. Each dataset contains 1257 rows and 36 columns. One response vari-able which is closing stock price, other 35 are independent variables. Time interval of datasets are between 06/01/2016–06/01/2021. Datasets contain independent variables from the indicator categories of momentum, volume, trend and volatility. Independent variables are as follows: MACD (Moving Average Convergence Divergence), RSI (Relative Strength Index), SDEV (Standard Deviation), A/D (Accumulation/Distribu-tion), LINEARREG (Linear Regression Indicator) and previous days' closing prices. Independent variables are expanded by using sliding windows method. Window size is fixed as 5 days, sliding 1 day every time with overlapping 4 days. After these pre-processing the dataset subjected to cross validation with using $k = 10$.

Multiple linear regression is an algorithm that aims to fit a line which is as close as possible to the observations. In multiple linear regression, the line that will be fitted to the model will be created, and the variable's coefficients will be determined [5].

In this study, we plan to apply the state of the art regression methods to our data and then compare the performances of the techniques based on the Mean Squared Error (MSE) and Mean Absolute Error (MAE) metrics. To do this we used 10 fold cross validation to fit our models and tune the corresponding parameters. The models we used in the study are briefly explained as follows:

$$Y = \beta_0 + \beta_1 X_1 + \beta_2 X_2 + \ldots + \beta_p X_p + \varepsilon \tag{1}$$

To find the optimal line to the dataset, the coefficients that minimize the residual sum of squares (RSS) will be used [5].

$$RSS = \sum\nolimits_i^n (y_i - \hat{\beta}_0 - \hat{\beta}_1 x_{i1} - \hat{\beta}_2 x_{i2} - \ldots - \hat{\beta}_p x_{ip})^2 \tag{2}$$

While being similar to least squares method in machine learning, ridge regression has a slight difference. Ridge regressions aim is to find $\hat{\beta}^R$ values that minimize the following equation [5]:

$$\sum\nolimits_{i=1}^n (y_i - \beta_0 - \sum\nolimits_{j=1}^p \beta_j x_{ij})^2 + \lambda \sum\nolimits_{j=1}^p \beta_j^2 = RSS + \lambda \sum\nolimits_{J=1}^P \beta_J^2 \tag{3}$$

Lasso regression serves the same aim as ridge regression but it has a slight difference. Lasso regressions try to find $\hat{\beta}^R$ values that minimize the following equation [5]:

$$\sum\nolimits_n^n (y_i - \beta_0 - \sum\nolimits_{j=1}^p \beta_j x_{ij})^2 + \lambda \sum\nolimits_{j=1}^p |B_j| = RSS + \lambda \sum\nolimits_{J=1}^P |B_j| \tag{4}$$

Decision tree regression model is a tree-based model with features the tree gets branched by the down. These methods are good for interpreting. In the decision tree regression, the goal is to find the spaces that minimize the RSS, which is the equation below [5].

$$\sum\nolimits_{j=1}^J \sum\nolimits_{i \in R_j} \left(y_i - \hat{y}_{R_J}\right)^2 \tag{5}$$

Support vector machine (SVM) is, support vector classifier with add-on, happens by feature space enlargement with using kernels.

Following formula explains support vector classifier with nonlinear kernel [5]:

$$f(x) = \beta_0 + \sum\nolimits_{i \in S} \alpha_i K(x, x_i) \tag{6}$$

In study linear kernel is used, formula is as follows:

$$K(x_i, x_{i'}) = \sum\nolimits_{j=1}^p x_{ij} x_{i'j} \tag{7}$$

Random forest algorithm uses bootstrap samples to create more trees for the combination. But in each split, only m predictors used by the tree, and typically m equals the square root of the total predictor number [5].

Extreme gradient boosting method is another tree based algorithm which is newer than the similar ones. The XGBoost method aims to minimize the following equation [13]:

$$L^{(t)} = \sum_{i=1}^{n} l(y_i, \hat{y}_i^{(t-1)} + f_t(x_i)) + \Omega(f_t) \quad (8)$$

4 Findings

R programming language is used to execute the model, calculate the error and to plot the results. The promised algorithms, which are multilinear regression, ridge regression, lasso regression, support vector regression, decision tree regression, and random forest, XGBoost applied to all three banks' data, and the models have been created. To evaluate the models' accuracy, the cross validation approach applied to the data by using k constant as 10. In each iteration data randomly splitted into 10 folds and 1 of them determined as test set while the other 9 as training set. After that, the models' accuracy was determined with their MSE and MAE values (Table 1).

Table 1. Error comparison of machine learning methods (Cross-Validation Method has been used by k = 10)

Method	Metric	AKBNK	GARAN	YKBNK
Multi Linear Regression	MSE	0.021	0.040	0.003
	MAE	0.109	0.146	0.040
Ridge Regression	MSE	0.022	0.045	0.003
	MAE	0.109	0.149	0.040
Support Vector Regression	MSE	0.021	0.040	0.003
	MAE	0.109	0.146	0.040
Decision Tree Regression	MSE	0.033	0.135	0.007
	MAE	0.147	0.286	0.066
Random Forest Regression	MSE	0.022	0.456	0.003
	MAE	0.112	0.150	0.039
Lasso Regression	MSE	0.021	0.038	0.003
	MAE	0.108	0.143	0.039
XGBoost	MSE	0.001	0.001	0.001
	MAE	0.011	0.017	0.005

The above table shows the associated errors related to each bank in each model. When this table is examined, it can be seen that XGBoost ensures the lowest MSE and MAE values for all banks, and decision tree regression gives the highest MSE and MAE for all banks. Also, support vector regression gives quite close results to multilinear regression for Akbank and Yapi Kredi Bank.

The below figures display the test data predictions for the most successful model, XGBoost, related to Yapıkredi, Akbank, Garanti Bank respectively (Figs. 1, 2 and 3).

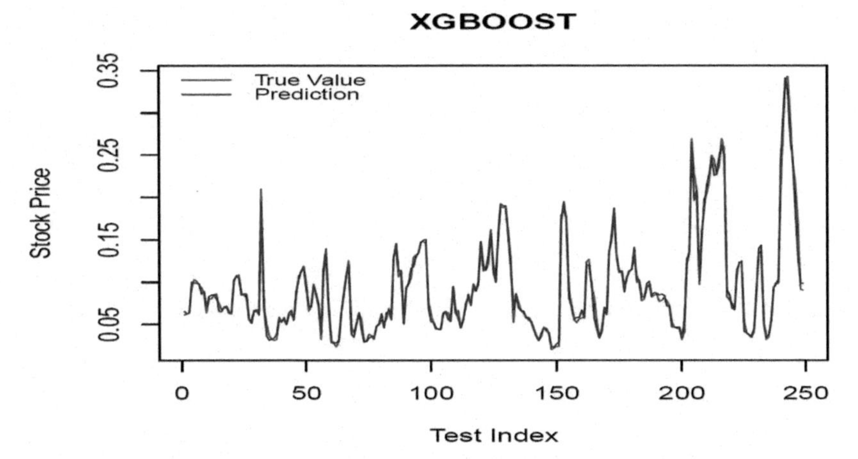

Fig. 1. YapıKredi stock closing price prediction using the XGBoost method

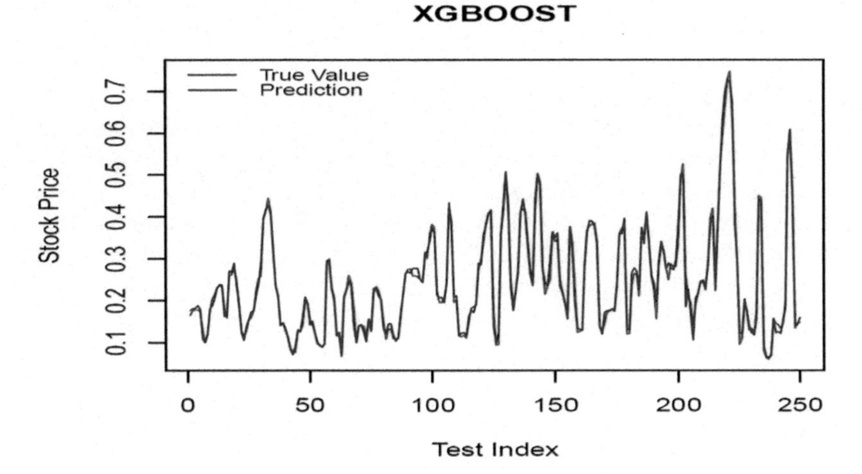

Fig. 2. Akbank stock closing price prediction using the XGBoost method

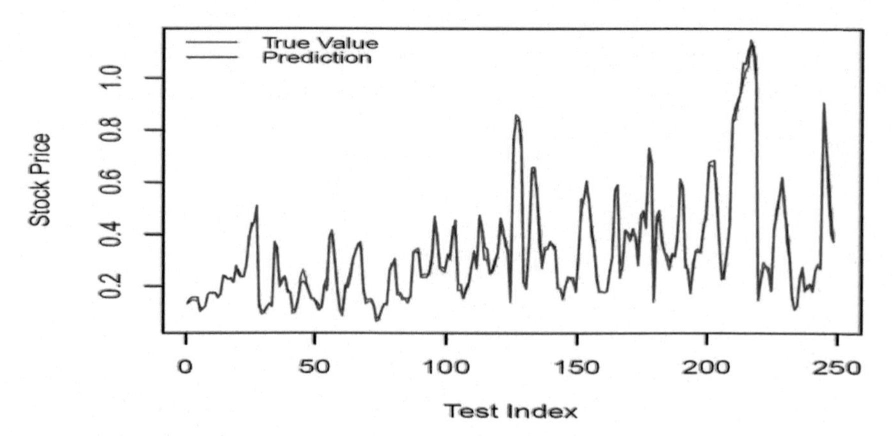

Fig. 3. Garanti Bank stock closing price prediction using the XGBoost method

5 Conclusion and Discussion

In this study, some regression models have been used to predict the stock prices of Turkish banks. The used data were taken as daily closes prices of Akbank, YapiKredi, and Garanti Bank. Before building the models, the data have been arranged, and some independent variables were added. These variables were some indicators that investors are familiar with. Since this study's problem is a time series problem, to be able to use supervised learning algorithms, sliding windows techniques have been used, and the size of the windows fixed as 5. To evaluate the models, cross-validation methods have been used. According to test errors, the XGBoost achieved more accuracy than the other models. This research applies several machine learning methods including modern ones to the Turkish banks in order to predict their stock prices. To the author's knowledge, it is the first study that focuses on Turkish banks while using such techniques. It is believed that the proposed model will be helpful for people who interest in investing in stocks. As a future research, different window sizes for sliding techniques and more frequent data can create alternate models. Also, changing the indicators for different training data can improve the model.

References

1. Akşehir, Z.D., Kılıç, E.: Makine Öğrenmesi Teknikleri ile Banka Hisse Senetlerinin Fiyat Tahmini. Türkiye Bilişim Vakfi Bilgisayar Bilimleri Ve Mühendisliği Dergisi, pp. 30–39 (2019)
2. Bonde, G., & Khaled, R.: Extracting the best features for predicting stock prices using machine learning. In: Proceedings on the International Conference on Artificial Intelligence (ICAI), pp. 1–2 (2012)
3. Boonpeng, S., Jeatrakul, P.: Decision support system for investing in stock market by using OAA-Neural Network. In: Eighth International Conference on Advanced Computational Intelligence, pp. 1–6 (2016)

4. Gao, T., Chai, Y., Liu, Y.: Applying long short term momory neural networks for predicting stock closing price. In: International Conference on Software Engineering and Service Science, pp. 575–578 (2017)
5. Gareth, J., Witten, D., Hastie, T., Tibshirani, R.: An Introduction to Statistical Learning. Springer, New York (2013)
6. Ravikumar, S., Saraf, P.: Prediction of stock prices using machine learning. International Conference for Emerging Technology, pp. 1–5 (2020)
7. Sarode, S., Tolani, H.G., Kak, P., Lifna, C.: Stock price prediction using machine learning techniques. In: International Conference on Intelligent Sustainable Systems, pp. 177–181 (2019)
8. Şişmanoğlu, G., Koçer, F., Önde, M.A., Şahingöz, Ö.K.: Derin öğrenme yöntemleri ile borsada fiyat tahmini. BEU J. Sci. 434–445 (2020)
9. Vijh, M., Chandola, D., Tikkiwal, V.A., Kumar, A.: Stock closing price prediction using machine learning techniques. In: International Conference on Computer Intelligence and Data Science, pp. 599–606 (2020)
10. Wang, H.: Stock price prediction based on machine learning approaches. In: International Conference on Data Science and Information Technology, pp. 1–5 (2020)
11. Werawithayaset, P., Tritilanunt, S.: Stock closing price predinction using machine learning. In: Seventeenth International Conference on ICT and Knowledge Engineering, pp. 1–8 (2019)
12. MATRIKS: Matriks Bilgi Dağıtım Hizmetleri A.Ş. (2020)
13. Chen, T., Guestrin, C.: Xgboost: a scalable tree boosting system. In: Proceedings of the 22nd ACM SIGKDD International Conference on Knowledge Discovery and Data Mining, pp. 785–794 (2016)

Clustering English Premier League Referees Using Unsupervised Machine Learning Techniques

Mustafa İspa(✉), Ufuk Yarışan, and Tolga Kaya

Department of Management Engineering, Istanbul Technical University, 34367 Istanbul, Turkey
{ispa16,yarisan17,kayatolga}@itu.edu.tr

Abstract. Technological developments have affected the decision-making phase in football matches. The viewing pleasure of fans and the result of the match is a matter of debate determined by the referees. Throughout time, the objectiveness of referees was questioned by the technological tools. Video Assistant Referee (VAR) is an example approach to ensure whether the referees react to similar positions of different matches in the same manner or not. One of the problems of this topic is the controversial decisions of some referees leading to unexpected results. In this research, as a different approach to the referees' objectivity problem, referees are tried to be classified based on the statistical outcome of the matches using unsupervised machine learning techniques. Meaningful clusters should not be found to be able to state the referees are objective. This study is conducted on 10-years English Premier League between 2009–2018 data. Principal Component Analysis is going to be used for grouping the variables to perform exploratory data analysis. K-Means, hierarchical clustering, and Fuzzy C-Means are going to be used for dividing the referees into various subgroups. R programming language is used for examining data. In conclusion of the analysis, four different referee groups are defined.

Keywords: Football · English Premier League · Referee · Clustering · Machine learning · Unsupervised learning

1 Introduction

Data gathering technologies allow sports businesses to collect more data than ever before [1]. Especially in football, data analytics is engaged more deeply in sport management because football has a significant market share in the sports industry in Europe. The English Premier League had the biggest brand value in Europe by $8.578 billion in 2020, which enforce sports businesses to collect data for enhancing profits [2]. The economic value of a football league is associated with the number of football fans. The viewing pleasure of the football games are affected by the fairness of the matches. According to the Union of European Football Associations (UEFA), Football and Social Responsibility Report in 2018, the efforts for fair-play activities are increased both for the players and referees [3]. Referees with their controversial decisions influence the viewing pleasure

C. Kahraman et al. (Eds.): INFUS 2021, LNNS 308, pp. 230–237, 2022.
https://doi.org/10.1007/978-3-030-85577-2_27

and outcome of the matches. almost every week, football fans complain on social media about the referees who decrease the tempo with their controversial decisions. However, current studies in the literature do not concentrate on clustering referees, which provides the originality of this paper. The innovation offer of this paper is managing referees rather than managing the matches to integrate Artificial Intelligence (AI) in the decision-making processes during football games. because of this reason, by examining the outcomes of the matches, different referees are tried to be clustered. In this paper, the outcomes of matches between 2009–2018 in the English Premier League, as known as the *Heimat* of football, are going to be used as a dataset. The dataset contains 23 referees. In an ideal world, the referees should not be clustered statistically meaningful to state that referees manage the match objectively. The study aims to cluster referee performances with match outcomes to examine if there are subgroups between the referees. If there are any statistically sensible subgroups, what the differences of subgroups are based on is going to be examined.

The paper has the following structure: as a second part current literature review is examined. At the third methodology of the modelling is given. The fourth and fifth sections contains the data and findings. The final part of the paper discusses the conclusion and future research suggestions.

2 Literature Review

In an average football game, almost 200 decisions are made by the referees [4]. Nowadays, some of the decisions such as offsides and goals have been made via the help of the computers. There are studies investigate managing the matches with the help of Artificial Intelligence (AI). Gottschalk et al. stated that AI has some limitations on decision processes during the matches [5]. Measuring the temper of the game, evaluating the risk of injury, and understanding the players' behavior are extremely difficult today for AI. However, referees can understand these circumstances and respond in an emphatic way, which strongly depends on the personality of the referee, while managing the matches. For example, a simple gesture of the referee may soothe the player, and this might not be possible for AI with today's technology. Because the decisions of referees are strongly bound with their unique personalities, the managing of the matches may not be objective. Even though AI cannot mimicry the referees' body language, AI can classify the referees based on their subjective decisions to determine referee types. The clusters may help to detect the referees' personalities and push referees to make objective decisions to catch increasing standards. However, most of the studies in the literature do not focus on clustering football referees by using unsupervised techniques. Most of the current studies investigate the effects of the home-team spectator on the referee's decisions. The fact that referees are influenced by home-team spectators in Bundesliga has been shown [6]. Referees show fewer red and yellow cards to the home team. Another study, which was conducted in Italian Serie A, states that matches without fans change referees' actions [7]. According to Enrich and Gesche, 14.1% more yellow cards have been shown and 5.3% more fouls have been called to the home team in the ghost-matches played during Corona Pandemic in Bundesliga [8]. Another article investigated the same hypothesis with Enrich and Gesche, but in multiple leagues [9]. The findings of Bryson et al. study verify Enrich and Gesche. These researches prove that social pressure influences referees' decisions via triggering stress. Boyko et al. proposed that audience pressure depends on referees' subjective decisions that differ between individuals [10]. These researches demonstrate that referees cannot act neutral while managing the matches. Factors, like home crowd noise, affect the referees [11]. Moreover, referees' decisions are affected by the players' reputations or a team's origin [12, 13]. Yet, there is no study conducted for clustering referees. These examples from the literature give insights for clustering referees based on the outcomes of the matches. Some factors that occurred in the matches like red card positions trigger the stress factor of referees, and referees react in various ways to these stressful moments. Therefore, being able to cluster referees successfully is predicted.

3 Methodology

In the principal component analysis (PCA), we are going to explain our data in a smaller dimension. The first principal component (PC) of a set of features, which is a normalized linear combination of the features [14].

$$\text{maximize}_{\phi_{11},\ldots,\phi_{p1}}\left\{\frac{1}{n}\sum\nolimits_{i=1}^{n}\left(\sum\nolimits_{j=1}^{p}\phi_{j1}x_{ij}\right)^2\right\} \text{ subject to } \sum\nolimits_{j=1}^{p}\phi_{j1}^2 = 1 \quad (1)$$

The optimization problem, which is above, is solved to find the PCs. ϕ's's are called loadings of the PC, and their sum of squares is equal to one to capture as much information as possible.

$$\frac{\sum_{i=1}^{n}\left(\sum_{j=1}^{p}\phi_{jm}x_{ij}\right)^2}{\sum_{j=1}^{p}\sum_{i=1}^{n}x_{ij}^2} \quad (2)$$

Proportion of Variance Explained (PVE) can be found with the formula above, by assuming the variables' means are zero. PVE helps us to understand what percentage of the data set can be explained by how many dimensions.

Hierarchical clustering measures Euclidean or correlation-based distance to the cluster. In the first step, each observation clustered as its own cluster. Then, dissimilarities between the clusters are examined and sorted according to the linkage type. In each iteration, cluster numbers decrease by 1 until there is only one cluster.

$$\text{minimize}_{C_1,\ldots,C_K}\left\{\sum\nolimits_{k=1}^{K}\frac{1}{|C_k|}\sum\nolimits_{i,i'\in C_k}\sum\nolimits_{j=1}^{p}\left(x_{ij}-x_{i'j}\right)^2\right\} \quad (3)$$

In K-Means clustering analysis, the idea is clustering each observation according to their Euclidian distances. The optimization formula for this problem can be seen above.

Fuzzy C-Means (FCM) gives probabilities for each cluster to each data points instead of assigning each data points to a cluster. By doing that FCM tries to overcome the ambiguity of which data points, that are somewhat in between the cluster centers, to assign in which clusters. The objective minimization function can be seen below [15].

$$J_m = \sum\nolimits_{i=1}^{N}\sum\nolimits_{j=1}^{C}u_{ij}^m\left\|x_i - c_j\right\|^2 \quad (4)$$

During the analysis, standardization may have great impact on having statistically meaningful results. For example, goals and red cards have different variances and standardization is required.

4 Data

Match outcome data of the English Premier League has been taken from Football-data.com, which is the website that provides statistics of matches between the years 2009 and 2018 [16]. In the data, there are 3801 matches and 23 referees, in total.

The data contains the half-time and full-time statistics such as total shoot, shoot in target, corners, total foul numbers, red and yellow cards. Half-time and full-time data are aggregated since these pairs are strongly correlated. The average foul number has been divided by the total yellow cards. In football matches 2 yellow cards equal to 1 red card; thus, red cards are multiplied by 2, and the foul/card rate has been calculated. This rate shows the frequency of showing cards for each referee. The derby variable has been created as a dummy variable by selecting derby matches played in EPL between 2009–2018. There is a biased belief among the football fans that derby matches are managed differently from other matches by referees. The selected derbies are not considered as big matches, but as matches played by the same city teams.

The data, which were created for match statistics, is filtered according to referees to create a dataset for our study. In our dataset, statistics assigned to the referees, and averaged by total matches that referees played. Referees who managed less than 50 matches are extracted from the data to increase statistical reliability.

5 Findings

The principal component analysis has been made to the dataset to lower dimensions and to group the variables. However, some of the variables have been extracted to increase the PVE, grouping the vectors meaningfully. After the analysis, the structure with two principal components explained 66.5% of the variance. In the cumulative proportion of variance explained graph, an obvious elbow formation could not be seen. The 3rd principal component explains a decent amount of variance of the data set, but it would be harder to state meaningful groups with 3D PCA. Because the purpose of the PCA is to come up with a sensible explanation, moving forward with two PCs of structure is preferred.

With two PCs, we have the graph below. PC1 is aggressiveness of referee while PC2 is viewing pleasure, as we interpreted.

In hierarchical clustering, complete, single and average linkages are examined and the most meaningful groups are obtained with complete linkages. In the lights of the previous literature and checking the dendrograms iteratively we preferred to cluster referees into 4 groups. Then, the K-Means clustering algorithm was executed by selecting K as 4, and with nstart = 20000. $\frac{\textit{Between Sum of Squares}}{\textit{Total Sum of Squares}}$ is calculated as 69.8%.

As a next step, the Fuzzy C-Means technique is applied to our data with K = 4. $\frac{\textit{Between Sum of Squares}}{\textit{Total Sum of Squares}}$. is calculated as %60.98. Dunn's Fuzziness Coefficient and Dunn's

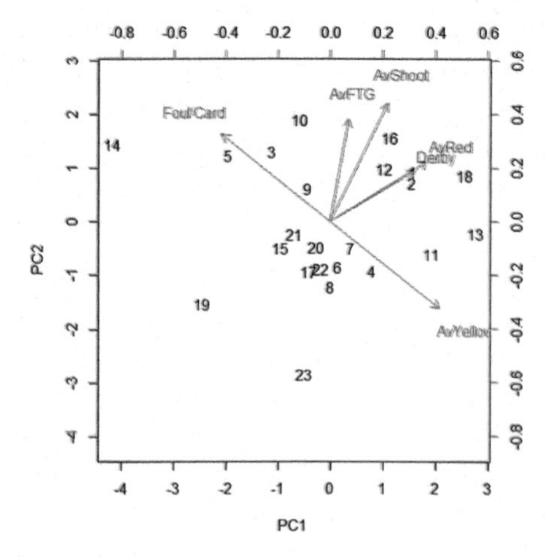

Fig. 1. Feature vectors on the plane of the PC1 and PC2 axes.

Table 1. K-Means & **Fuzzy C-Means** cluster means with K = 4.

Groups	AvFTG	AvShoot	AvYellow	AvRed	Derby	Foul/Card
1	2.79 I **2.77**	24.53 I **24.86**	2.52 I **2.21**	0.07 I **0.06**	0.01 I I**0.01**	7.78 I **8.53**
2	2.71 I **2.71**	25.53 I **25.32**	2.97 I **2.98**	0.15 I **0.13**	0.02 I **0.01**	7.01 I **7.02**
3	2.66 I **2.67**	24.83 I **24.89**	3.30 I **3.30**	0.13 I **0.13**	0.02 I **0.02**	6.21 I **6.25**
4	2.84 I **2.83**	25.63 I **25.52**	3.51 I **3.44**	0.17 I **0.17**	0.05 I **0.05**	5.91 I **6.03**

Normalized Fuzziness Coefficient are found accordingly 0.565 and 0.420. FCM gave similar clustering groups that contain the same interpretation (Table 1).

Most of the referees are grouped in the same clusters in each of the three techniques. However, there are some referees grouped differently in one of the three different clustering techniques; but, it is important to note that there is no referee, who is grouped in different clusters in each of the three techniques. The clustering status of these referees are changing between group 1–2, or group 2–3. This means that, these referees are behaving unstable between the clusters, which are close to each other in terms of explanation. When G.Scott(5), L.Mason(9), M.Clattenburg(12), M.Halsey(14) and M.Jones(15) are extracted from the model, $\frac{\textit{Between Sum of Squares}}{\textit{Total Sum of Squares}}$ increased to 81.4% in K-Means Clustering and 80.52% in Fuzzy C-Means Clustering. Also, along with these parameters Dun's Coefficient and Dun's Normalized Coefficients raised to 0.659 and 0.545.

Group 1: "Let Them Play" Referees

These referees call fewer fouls than the other clusters and show cards less frequently in these calls. The referees in this cluster do not decrease the match tempo with their calls, as we can understand from the average goal number. This cluster has the second-highest goal average among the clusters, even though they do not manage derby matches. 3

referees managed the only 2 derby matches between 2009–2018. We can conclude that the referees cannot put their characters on the pitch; therefore, they cannot manage derby atmospheres. Still, they "let them play" and they have an effect on the more scores that have been made in the matches. Hierarchical clustering and K-Means clustering techniques concluded that G.Scott(5), M.Halsey(14) and P.Tierney(19) are in this group. However, FCM grouped M.Halsey(14) as alone in group 1.

Group 2: "Thinking Twice" Referees
The referees in this cluster resemble the 3rd cluster with average yellow and red card numbers. Yet, they are showing these cards less frequently in their calls. A higher foul/card ratio led to a higher goal average. This cluster is taking the responsibility, giving the second chance to the players to increase the tempo and the goal rate of the matches. Maybe these referees cannot handle the social pressure from the fans, as stated in the literature review and increase the goal rate of the home team. This is going to be examined in further research. This group is named as "Thinking Twice" Referees and all of the techniques decided that C.Foy(3), L.Probert(10), M.Oliver(16), and R.East(21) are in this cluster. However, while L.Mason(9), M.Clattenburg(12), and M.Jones(15) are clustered as "Thinking Twice" Referees by K-Means clustering, Hierarchical clustering clusters these referees as "Average" Guys. Along with this, Fuzzy C-Means technique decided that G.Scot(5) and P.Tierney(19) are"Thinking Twice" referees.

Group 3: "Average Guy" Referees
Referees in this group, generally, tend to show average referee behaviors. They usually prefer to show cards. In other words, they have kind of a habit of behaving toward their cards. They have the lowest goal average among the other referee clusters. One of the reasons for this lowest goal average is their tendency to show cards. This is the most crowded cluster because they show average performance as a referee in the matches. They are clustered in the middle of the two-dimensional PCA representation, which is verifying this averageness of the cluster.C.Pawson(4), H.Webb(6), J.Moss(7), K.Friend(8), M.Atkinson(11), N.Swarbrick(17), P.Walton(20), R.Madley(22) and S.Attwell(23) are clustered together by all of the techniques. As we mentioned above, L.Mason(9), M.Clattenburg(12), and M.Jones(15) are clustered with other "Average Guys" by hierarchical clustering technique instead of being grouped with other "Thinking Twice" referees.

Group 4: "Rockstar" Referees
These referees have the lowest foul/card ratio (5.9), which means that they show a card almost every 6 fouls. This cluster has the highest percentage in terms of managing derbies, which is 49 with 4 referees. Moreover, this cluster has the highest averages in terms of goals and shoots. They are putting their character to the pitch while managing matches, and deal with the stress by showing red cards to decrease aggression. This cluster can be called the Star Referees. A.Marriner(1), A.Taylor(2), M.Dean(13), and P.Dowd(18) are in this group in all of the clustering analysis (Fig. 2).

The quad figure on the right: Left Top: The question marks stand for the referees clustered in different groups by Hierarchical Clustering, and the diamond marks represent the referees clustered differently by Fuzzy C-Means. These cases indicate the unstable

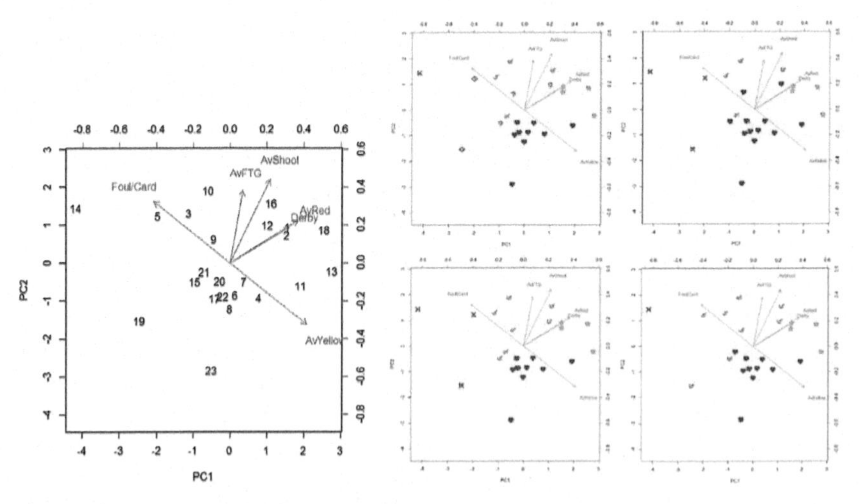

Fig. 2. The figure on the left: The same figure as Fig. 1. It is shown here again to make it easier to compare the referees on the figure on the right.

referees are close to more than one centroid. Right Top: Showing the clusters created by hierarchical clustering. Left Bottom: Showing the clusters created by K-Means clustering. Right Bottom: Showing the clusters created by Fuzzy C-Means clustering in two dimensional PCA coordinate.

6 Conclusion and Discussion

The model, which contains the cluster of the referees, in this study showed that the characteristics of the referees may be classified based on the match outcomes. Four main groups have been proposed by our model. Those groups are; "Let Them Play" Referees, which barely show a presence during the matches, "Thinking Twice" Referees, which show fewer cards in their calls, "Average Guy" Referees, which have a tendency to act towards their cards, and "Rockstar" Referees which can handle with the stress in a controlled way. For future research with extended data, which contains the reputation of the teams and crowd noise factors, the model can be updated in a more reliable way. In addition to this, the reason which underlies this sensible clustering should be chased in further research. Adding new features can widen the meaning of the clusters and give more insights into how referees made their decisions during the matches. Moreover, methods like Gaussian Mixture Model (GMM) can be used in further studies that include other leagues and more referees. GMM could not be used in this study since more referee data was needed for this technique. The model created in this study can be used in other types of football analytics models. Implementations, such as a bet prediction model, based on the referees, can be developed.

References

1. Singh, N.: Sport analytics: a review. Int. Technol. Manage. Rev. **9**(1), 64–69 (2020)

2. Lange, D.: Brand value of top tier football leagues Europe 2020. https://www.statista.com/statistics/1024764/brand-value-top-tier-football-leagues-europe-by-country/. Accessed 04 Feb 2021
3. UEFA Football and Social Responsibility Report 2017/2018. https://www.uefa.com/MultimediaFiles/Download/uefaorg/General/02/60/26/72/2602672_DOWNLOAD.pdf. Accessed 04 Feb 2021
4. Brand, R., Schweizer, G., Plessner, H.: Conceptual considerations about the development of a decision-making training method for expert soccer referees. Nova Science (2009)
5. Gottschalk, C., Tewes, S., Niestroj, B.: The innovation of refereeing in football Through AI. Int. J. Innov. Econ. Dev. **6**(2), 35–54 (2020)
6. Buraimo, B., Forrest, D., Simmons, R.: The 12th man?: refereeing bias in English and German soccer. J. R. Stat. Soc. A. Stat. Soc. **173**(2), 431–449 (2010)
7. Pettersson-Lidbom, P., Priks, M.: Behavior under social pressure: empty Italian stadiums and referee bias. Econ. Lett. **108**(2), 212–214 (2010)
8. Endrich, M., Gesche, T.: Home-bias in referee decisions: Evidence from 'Ghost Matches' during the Covid19-Pandemic. Economics Letters, vol. 197, p. 109621 (2020)
9. Bryson, A., Dolton, P., Reade, J.J., Schreyer, D., Singleton C.: Causal effects of an absent crowd on performances and refereeing decisions during Covid-19. Economics Letters, vol. 198, p. 109664 (2021)
10. Boyko, R.H., Boyko, A.R., Boyko, M.G.: Referee bias contributes to home advantage in English Premiership football. J. Sports Sci. **25**(11), 1185–1194 (2007)
11. Nevill, A., Balmer, N., Williams, A.: The influence of crowd noise and experience upon refereeing decisions in football. Psychol. Sport Exerc. **3**(4), 261–272 (2002)
12. Jones, M.V., Paull, G.C., Erskine, J.: The impact of a team's aggressive reputation on the decisions of association football referees. J. Sports Sci. **20**, 991–1000 (2002)
13. Messner, C., Schmid, B.: Über die Schwierigkeit, unparteiische Entscheidungen zu fällen: Schiedsrichter bevorzugen Fußballteams ihrer Kultur. Zeitschrift für Sozialpsychologie **38**, 105–110 (2007)
14. James, G., Witten, D., Hastie, T., Tibshirani, R.: An Introduction to Statistical Learning with Applications in R, 7th edn. Springer Science + Business Media, New York (2017)
15. Babuska, R., Zimmermann, H.: Fuzzy Modeling for Control. Springer, Netherlands, Dordrecht (1998)
16. England Football Results Betting Odds: Premiership Results & Betting Odds. https://www.football-data.co.uk/englandm.php. Accessed 11 Jan 2021

Prediction Models for Project Attributes Using Machine Learning

Ching-Lung Fan(✉)

Department of Civil Engineering, The Republic of China Military Academy, No. 1, Weiwu Road, Fengshan, Kaohsiung 830, Taiwan
p93228001@ntu.edu.tw

Abstract. Defects not only affect the project duration, cost, and quality but also are important indicators of project management performance. Therefore, using appropriate methods to train or test defective big data and understand the characteristics of defects can enable the effective classification of each project attribute. This study was based on 499 types of defects and the related attributes of the Public Works Bid Management System (PWBMS). In this study, machine learning (ML) technologies such as a decision tree (DT), Bayesian network (BN), artificial neural network (ANN), and support vector machines (SVM) were to predict the relationship between the defects and three target variables (engineering level, project cost, and construction progress) in 1,015 projects and to evaluate the optimal classification model through cross-validation. The results of evaluation metrics revealed that the accuracy of an ANN for predicting the engineering level is 93.20%, and the accuracy of an SVM for predicting the project cost and construction progress is 85.32% and 79.01%, respectively. Overall, the SVM had better classification benefit for the three project attributes. This study was based on an ML technology assessment for building a prediction model for project attributes to provide project manager with an aid to understand defects and models.

Keywords: Defects · Machine learning · Decision trees · Bayesian network · Artificial neural network · Support vector machines

1 Introduction

In Taiwan, government departments systematically classified project inspection data in the Public Works Bid Management System (PWBMS) through the construction inspection mechanism. Then, statistical analyses were conducted and defect improvement measures were implemented to improve the quality of public constructions and project management performance. The inspection defect content was divided into four categories: construction management (113 defects), work quality (356 defects), program (10 defects), and design (20 defects), which are 499 defect types in total. The defect data employed in the current study were sourced from the official PWBMS database; the data were collected by experts and academicians during field construction inspections and by using standardized forms. Thus, the inspection standards used were consistent

C. Kahraman et al. (Eds.): INFUS 2021, LNNS 308, pp. 238–245, 2022.
https://doi.org/10.1007/978-3-030-85577-2_28

and objective. Additionally, the database contains 15 years of inspection data and is thus exceedingly large.

Using a large amount of data to train machine models is a way of walking away from the traditional statistical analysis framework; by replacing the sample analysis with the population analysis, it becomes possible to observe the associations between data, the trends or patterns that were difficult to find in the past, and to also generate the application value of the new thinking. Therefore, data is no longer just an afterthought after information technology being applied and arranged, but rather a tool to explore and build problem domain's expertise and, accordingly, to create professional abilities and know-how in the area of problem domain knowledge management [1]. The aim of big data analytics is to discover knowledge, support decision making, and predict outcomes to create competitive advantage [2]. The defects represent project quality, and the relationship between these defects and the attributes related to a project is a subject that should be analyzed further. Therefore, by assuming that useful knowledge and rules can be obtained from a historical inspection database, the relationship between specific defects and related attributes can be understood, thus reducing or eliminating the risk of defects.

The purpose of the research is to use the machine learning (ML) algorithms to present the potential correlation of project attributes from a large number of construction inspection data. The relationship between the target variables such as the engineering level, project cost and construction progress of each project and the defects are predicted. A model for optimizing the classification of project attributes is built to provide project managers with ancillary methods for understanding defects. Management units can be enabled to take correct decisions for improving the construction strategy direction or project management performance. The first phase is to select target variables and decision variables from PWBMS and compare the classification models of seven ML algorithms. The second phase is to use the cross-validation to evaluate the results to construct the optimal classification model for project attributes. The organization of the paper is structured as follows: Sect. 2 adopts the research methodology (machine learning) and briefly introduces data attributes. Section 3 discusses results analysis and model evaluation in this study. Finally, Sect. 4 summarizes the findings and suggests future research directions.

2 Research Methodology

ML involves the computer system design that attempts to intelligently solve problems by simulating the inference process of the human brain [3] and is often used to solve classification and prediction problems [4]. In ML, many different technologies and algorithms, such as ANN, SVM, and DT, have been used to perform data classification tasks and prediction models [5, 6]. Classification and regression are used to obtain a set of models by training a given group of information that has been classified. Then, a previously trained model can be used to predict the classification category of unclassified data. Classification is the distribution of discrete variables to specific labels as a result of predictions for predicting qualitative goals. Regression is based on the analysis of the relationship between variables and trends to make predictions about continuous variables for predicting quantitative targets. Both classification and regression are models that establish a set of input (feature) and output (label) relationships; the aim is to build a model that can be used to predict or to describe future samples of similar characteristics. The output of classification is a discrete state, and the output of regression is a continuous value. The main difference between classification and regression is that in classification the samples are assumed to be correctly labelled while in regression the sample class labels are observations from random variable [7].

The algorithms of ML are divided into two types: supervised and unsupervised learning. In most practical cases, ML is supervised learning, in which a "label" trains a model with known input and output data and then asks the model to predict the output for new input. In supervised ML, a machine learns from the training group's data to present better models in the testing group (new sample). However, there are still some differences (errors) between the training output and the predicted output of testing. The error of the training group is known as the "training error," and the error of the testing group is called as the "generalization error." In theory, the preferred supervised method presents a small generalization error. In reality, the nature of the new sample cannot be predicted in advance, and the actual practice can only minimize the training error.

If a training group has a higher classification accuracy, the learning effect is higher. However, the testing group does not achieve favorable results, which is known as overfitting. Conversely, if the characteristics of the training group are not sufficiently appropriate (input) and cannot be fully learned, then the classification error of the testing group is high and the classification performance is poor (output), which is called as underfitting. However, while training a model, the characteristics of the testing group cannot be accurately known, and only the k-fold training groups can be learned. The use of cross-validation can avoid deviations from relying on a particular training and testing data, and cross-validation is an evaluation method for measuring accuracy and reliability. A k-fold cross-validation randomly divides data into two independent groups—training group and testing group. The training samples are divided into k subsamples. A single subsample is retained as the data for the test model, and the other k − 1 samples are used for training and repeating the training k times. Each subsample was tested once. The average of k is the result of accuracy and was obtained. this study used the 10-fold cross-validation to randomly divide construction inspection data into 10 groups. During each run, one data group was used for testing, and the remaining nine data groups were used for training models.

ML is an algorithm that automatically analyzes and obtains rules from data and uses rules to predict unknown data. This study intended to analyze the past behavioral pattern of the inspection committee's defects by using the inspection data of public construction projects. ML was used to find attributes that are related to defects and rules with potential value and to evaluate the benefit of the classification model. This study used PWBMS construction inspection data from 1993 to 2018. The total number of projects inspected was 1,015 (defect frequency: 18,246); the inspected data included defect types (Code: C_n, W_n, P_n, D_n), engineering levels, project costs, and construction progress. In the ML analysis data, the decision variable is the defect type, and the target variable is the engineering level, project costs, and construction progress (Table 1).

Table 1. Project attribute contents and construction inspection information.

Variables		Attributes	Description
Decision variables	X1	Construction management	Procuring units, supervisory units, and contractors, a total of 113 defects (C_n)
	X2	Work quality	Strength I, strength II, and safety, a total of 356 defects (W_n)
	X3	Program	Schedule management, and project network diagramming management, a total of 10 defects (P_n)
	X4	Design	Security, construction, maintenance, and gender differences, a total of 20 defects (D_n)
Target variables	Y1	Engineering level	Level I, II, III and IV
	Y2	Project cost	Amount publication (NT$1-50 million), P; Amount supervision (NT$50-200 million), S; Amount large procurement (NT$200 million or more), L
	Y3	Construction progress	Behind (under 50%), N; Ahead (over 50%), Y

3 Analytical Results

3.1 Cross-Validation

The present study targets a classification model, in which the construction attributes are output as the response and data such as defective types are input as predictors. This study uses the following classifiers in the SPSS modeler software: CART, CHAID, QUEST, C5.0, BN, ANN, and SVM. These are the seven types of classification algorithms. The properties of 1,015 construction inspection cases were analyzed. Moreover, to construct seven classification models, the decision variables were *X1–X4* (defective types), and the target variables were *Y1–Y3* (project attributes). To minimize inaccuracies, seven models of 10-fold cross-validation were developed in this study. The average model accuracy is shown in Table 2. In the construction inspection case, the ANN for engineering levels (*Y1*) yielded the best prediction results with an accuracy of 88.24%. An SVM presents the best prediction for project costs (*Y2*) and construction progress (*Y3*) with accuracies of 78.91% and 76.18%, respectively. The average accuracy of a BN is the worst among the seven classification models.

The target variables of the testing group are engineering levels (*Y1*), project costs (*Y2*), and construction progress (*Y3*). The ANN can accurately predict engineering levels, and the SVM can accurately predict project costs and construction progress. The results of the models were consistent with the cross-validation results (Table 2). The ANN correctly classified the "engineering levels" of A, B, C, and D as 205, 68, 132, and 38, respectively. The accuracy of the testing group was 84.87% (Fig. 1). The SVM correctly classified the "project costs" P, S, and L as 339, 18, and 28, respectively, and the accuracy of the testing group was 73.75% (Fig. 2). The correct classification construction progress of the SVM was 283 (under 50%, N) and 90 (over 50%, Y), respectively, and the accuracy of the testing group was 71.46% (Fig. 3). Overall, the SVM has a better classification accuracy for the three project attributes (target variables), followed by C5.0. ANN has the most prominent classification of engineering levels.

Table 2. Accuracy of the different predictive models for the target variables.

Target variable	Engineering level (Y1)		Project cost (Y2)		Construction progress (Y3)	
	Fold 1–10 Average (%)	Ranking	Fold 1–10 Average (%)	Ranking	Fold 1–10 Average (%)	Ranking
CART	72.31	4	68.42	6	58.54	5
CHAID	70.37	5	69.49	4	61.90	3
QUEST	66.03	6	68.81	5	59.04	4
C5.0	76.47	3	71.45	2	68.49	2
BN	58.52	7	61.20	7	56.13	7
ANN	88.24	1	69.59	3	58.47	6
SVM	80.74	2	78.91	1	76.18	1
Testing group	84.87% (ANN)		73.75% (SVM)		71.46% (SVM)	

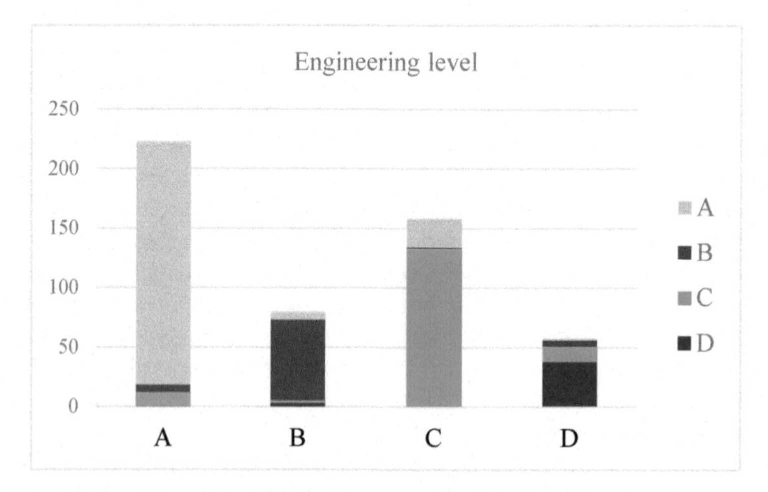

Fig. 1. Proportion of the ANN testing group classification for engineering level.

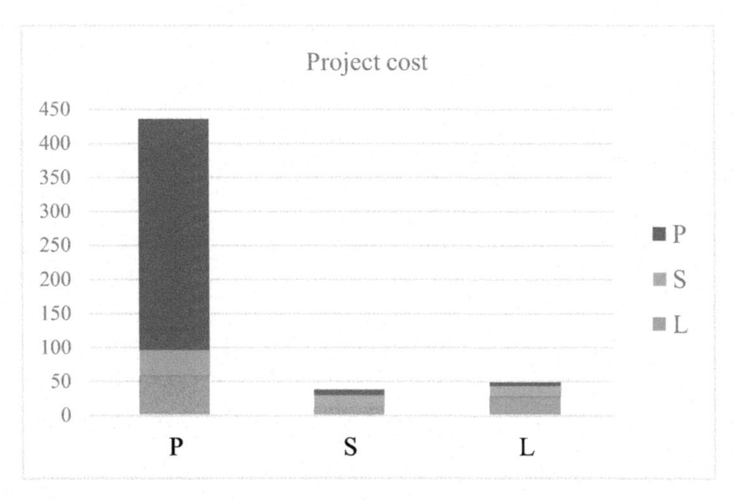

Fig. 2. Proportion of the SVM testing group classification for project cost.

Fig. 3. Proportion of the SVM testing group classification for construction progress.

3.2 Evaluation Metrics

Accuracy refers to the ratio of the actual categories to all the predicted ones, as in Eq. (1). The quality of a classification model except being used to estimate in terms of accuracy, when the ratio of a certain category is relatively small and requires more attention. The category may represent different levels of importance. If only accuracy is used, then the other category with a higher number of category ratios will be favored. However, a small number of categories can instead reveal valuable information, which can be used as an evaluation indicator by calculating precision and recall. Precision refers to how many ratios in all prediction categories actually belong to the category, as presented in Eq. (2). The higher is the precision, the lower the ratio of misjudgments in the category is. Recall indicates that the actual result of a certain category is correctly considered the ratio of a certain category, as presented in Eq. (3).

Table 3. Evaluation results of ML classifier on target variables.

ML classifier		CART	CHAID	QUEST	C5.0	BN	ANN	SVM
Engineering levels (%)	Accuracy	79.01	77.83	67.39	78.13	62.96	93.20	82.17
	Precision	79.54	76.8	65.13	75.57	87.0	91.91	92.01
	Recall	74.17	78.05	68.19	73.79	90.12	91.62	75.35
	BEP	80.00	79.30	64.60	74.00	87.20	90.80	95.30
Project costs (%)	Accuracy	76.16	78.82	71.72	73.89	65.02	80.89	85.32
	Precision	76.14	73.94	69.09	62.35	53.29	91.81	82.24
	Recall	58.18	64.39	48.93	56.97	53.98	64.11	73.07
	BEP	82.20	73.80	71.0	59.0	47.0	87.80	83.90
Construction progress (%)	Accuracy	62.96	64.14	58.72	70.25	58.62	75.37	79.01
	Precision	64.98	64.92	63.17	69.97	58.03	74.95	82.86
	Recall	58.97	60.89	53.12	70.26	58.07	74.90	76.53
	BEP	67.80	67.00	64.30	71.20	58.60	75.90	85.10

$$\text{Accuracy} = (\text{TP} + \text{TN})/(\text{TP} + \text{TN} + \text{FP} + \text{FN}) \quad (1)$$

$$\text{Precision} = \text{TP}/(\text{TP} + \text{FP}) \quad (2)$$

$$\text{Recall} = \text{TP}/(\text{TP} + \text{FN}) \quad (3)$$

In general, if the precision is high, the recall is often lower, and vice versa. According to the prediction result of the classifier, the above two sequences are performed, and the P-R curve is obtained by plotting the precision as the vertical axis and the recall as the horizontal axis. The P-R curve visually shows the precision and recall of the classifier in the total sample. The principle is that when the precision and recall values are the same, the point of intersection with the P-R curve is Break-Even Point (BEP), and the recall value of BEP is used. In this study, the highest BEP values for the engineering level, project cost, and construction progress were SVM (95.3%), ANN (87.8%), and SVM (85.1%), respectively (Table 3).

4 Conclusion

Taiwan's public construction funds have accounted for approximately half of the total output fund value of the construction industry over the years. This has made a significant contribution to the country's overall economic development. However, defects in the construction process are difficult to avoid, and reducing the defects is an important task in the construction phase. This study combines the relevant information of the PWBMS construction inspection and ML to understand the relevance of potential defects, thus enabling management units to make the correct decisions to improve the construction strategy and direction. In this study, based on the supervised ML techniques, such as DT, BN, ANN, and SVM, used according to the characteristics of the analysis data, different algorithms were selected to classify the project attributes. Finally, seven classification models were constructed, and the accuracy of the prediction was evaluated. The cross-validation results revealed that the classification models using C5.0, ANN, and SVM provide more reliable simulation and higher classification efficiency than the models using other ML techniques.

This study provided a comprehensive comparison between the effectiveness of various ML techniques and provided an optimized model for predicting project attributes. As a result, project managers will be able to determine the most appropriate method to classify various project attributes, as well as understanding defects and classification models. Due to the properties, operating principles, and parameter settings of the classification model are restricted. Furthermore, the attribute conditions of the data to be tested, such as engineering types, sample characteristics, etc., are also different. There is a considerable space for discussion on the selection of appropriate models for project attributes classification. The future research must consider the characteristics of the data and the purpose of the study to determine the analytical model to be used.

Acknowledgments. The research presented in this paper was sponsored by the Ministry of Science and Technology, Taiwan (Contract No. MOST 109-2222-E-145-001).

References

1. Rodrigues, J., Folgado, D., Belo, D., Gamboa, H.: SSTS: a syntactic tool for pattern search on time series. Inf. Process. Manag. **56**(1), 61–76 (2019)
2. Barbosa, M.W., Vicente, A.C., Ladeira, M.B., Oliveira, M.P.V.: Managing supply chain resources with big data analytics: a systematic review. Int. J. Log. Res. Appl. **21**(3), 177–200 (2018)
3. Lee, M.C.: Using support vector machine with a hybrid feature selection method to the stock trend prediction. Expert Syst. Appl. **36**(8), 10896–10904 (2009)
4. Chou, J.S., Tsai, C.F., Lu, Y.H.: Project dispute prediction by hybrid machine learning techniques. J. Civ. Eng. Manag. **19**(4), 505–517 (2013)
5. Ryua, Y.U., Chandrasekaranb, R., Jacobc, V.S.: Breast cancer prediction using the isotonic separation technique. Eur. J. Oper. Res. **181**(2), 842–854 (2007)
6. Kim, Y.S.: Comparison of the decision tree, artificial neural network, and linear regression methods based on the number and types of independent variables and sample size. Expert Syst. Appl. **34**(2), 1227–1234 (2008)
7. Chen, J.J., Chen, E.E., Zhao, W., Zou, W.: Statistics in big data. J. Chin. Stat. Assoc. **53**, 186–202 (2015)

New Fuzzy Observer Fault Pattern Detection by NARX-Laguerre Model Applied to the Rotating Machine

Shahnaz TayebiHaghighi and Insoo Koo(✉)

Department of Electrical, Electronics and Computer Engineering, University of Ulsan, Ulsan, South Korea
iskoo@ulsan.ac.kr

Abstract. A rotating machine is a ubiquitous product in industries. Inner, outer, and roller cracking in bearings can cause structural failure of the rotating machines and possibly decrease the service life of different industrial systems such as motors. The nonlinearity and complexity associated with the uncertain and unknown behavior inherent in rotating machines lead to difficulty in the detection and identification of bearings' cracks in real-time. To address these issues, the combination of indirect Proportional-Integral-Derivative (PID) observer and fuzzy logic approach is proposed in the paper to detect the fault pattern in the bearing. To do this, firstly the indirect fuzzy observer is modeled by the nonlinear autoregressive with Laguerre filter that is improved by the fuzzy technique. After signal estimation using indirect fuzzy observer, the difference (residual signals) between original and estimated signals are computed. Finally, the machine learning technique is utilized for residual signal pattern detection. The Case Western Reverse University Bearing Dataset (CWRUBD) is used to test the proposed scheme.

Keywords: Rotating machine · Indirect PID observer · Fuzzy logic technique · Nonlinear autoregressive signal modelling · Laguerre filter · Machine learning classification approach

1 Introduction

The Rolling Element Bearings (REB) are broadly used in manufacturing and technical applications for reduced friction and simplifying movement. Due to the processing technology, performance conditions, and different causes, the fault signals are nonlinear and nonstationary, which makes fault signal detection difficult to be detected. Therefore, it is complicated to recognize the fault character of bearing exactly [1]. Hence in this research, an intelligent fault diagnosis is applied to accurate fault detection of the bearings. Mainly, vibration signals and acoustic emissions are famous for being able to be utilized for monitoring the health of bearings. The vibration signals of bearing always conveys the dynamic data. These signals are helpful for feature extraction and fault diagnosis [2]. Therefore, in this research, vibration signals are recommended for the bearing model identification.

C. Kahraman et al. (Eds.): INFUS 2021, LNNS 308, pp. 246–253, 2022.
https://doi.org/10.1007/978-3-030-85577-2_29

Fault diagnosis in bearings has been well developed with different techniques. Signal processing techniques, data-driven methods, and model-based manners employ for this purpose [3, 4]. Nevertheless, each of these methods alone has its limitations. This issue leads many researchers to use hybrid techniques. Accordingly, a hybrid method has been applied in this study which allows us to achieve more accuracy, more reliability, and reduced complexity in fault diagnosis. To address these issues, a combination of data-driven technique, model-based approach, and artificial intelligence method is recommended. The data-driven approach is recommended for signal modeling, the model-based integrated with artificial intelligence is suggested for signal estimation, and the artificial intelligence technique is used for classification.

Moreover, to dig out useful information, the feature extraction technique can be applied to get further information to provide more efficient techniques for the main goal that is fault classification. To bearing crack discrimination via the hybrid method, the base step is the computation of residual signals. According to the difference between original signals and estimated signals, residual signals will be calculated [5, 6]. Among different algorithms for estimation the main signal, an observation-based technique has been suggested in this research. For estimating a signal using observers, signal modeling is the essential step. Signal modeling must be done to approximate the state-space function of the signal. Generally speaking, signal modeling accomplishes in two main groups: data-driven modeling and modeling based on the dynamics of the system. Due to the complexity of modeling with the dynamics-based approach in complex systems, data-driven signal modeling such as Autoregressive (AR) and the Autoregressive with external input (ARX) have been used in different works [7].

After signal modeling, the original signals will be estimated. For signal estimation, various techniques have been applied by researchers that can be classified into two main groups: linear-based signal estimation and nonlinear-based signal estimation. In this work, linear-based observers such as proportional integral observer and proportional multi-integral observer are suggested for signal estimation [5, 6].

To classification the fault, different classification algorithms such as Support Vector Machine (SVM), decision trees, and ensemble classification have been introduced [2]. In this work, the combination of AutoRegressive (AR), Uncertainty external (ARU), Laguerre technique (ARUL), and fuzzy algorithm (hence is called ARULF) technique is used for signal approximation. Moreover, the Indirect Fuzzy Proportional Integral Derivative (FPID) observer which is modeled by ARULF (ARULF-FPID observer) is suggested for signal estimation. The SVM is used for fault pattern detection in that last step. Thus, the main contribution of this paper is the combination of ARULF signal approximation, FPID signal estimation, and SVM classifier for fault pattern identification in the bearing. This research article is organized as follows. The second section outlines the proposed scheme for fault pattern detection using the combination of the ARULF technique, FPID observer, and SVM. The third section shows the results and discussion. Conclusions and future works are given in the last section.

2 Proposed Scheme for Fault Pattern Detection

The core of the proposed scheme is residual classification. Thus, this approach has the following parts. 1) *Preprocessing step* where the signals are resampled and the energy feature will be extracted from the resampled signals. 2) *Residual signal approximation step* where the difference between original raw signals and estimated ones will be calculated. This part has two important sub-parts: a) In the first sub-part, we extract the state-space function from signals by using the function approximation approach and b) In the second sub-part, we estimate the signals using the combination of function approximation method with signal estimation technique. 3) *Classification step* where fault detection is performed. In the step, the SVM is used to classify the new feature (residual) signal for fault pattern detection. Figure 1 shows the block diagram of the proposed scheme.

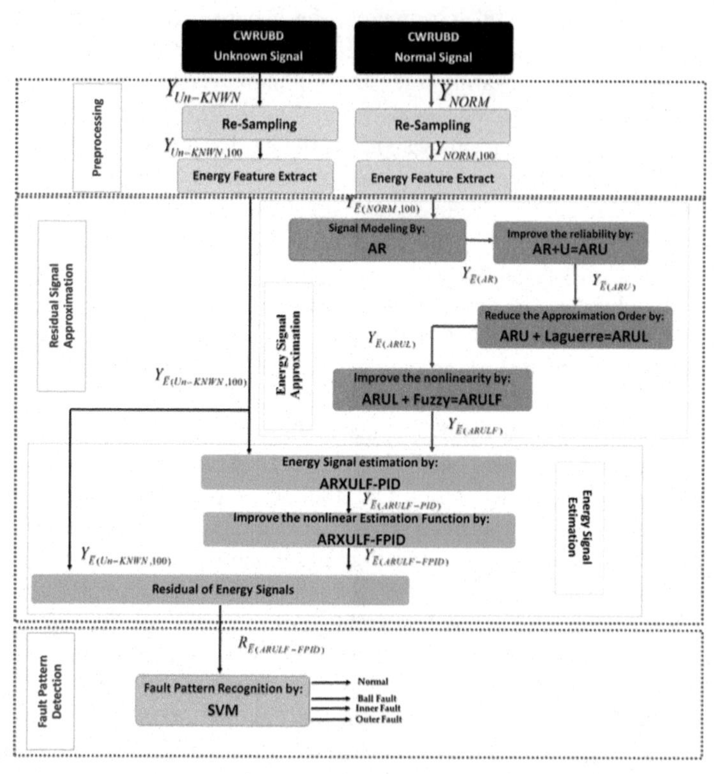

Fig. 1. The overall block diagram of the proposed scheme for fault pattern detection.

2.1 Preprocessing

To have an indirect observer for signal estimation, the first step is preprocessing. This unit has two main parts. In the first step, the raw signals are resampled. Based on [8], in the Case Western Reverse University Bearing Dataset (CWRUBD), the range of rotational speeds is between 28 to 30 revolutions per second. Moreover, the sampling rate frequency for raw signal collection is 48 kHz and the signal length is 120,000 samples. Thus, in this work, to cover all conditions, for every rotation 1200 samples or 100 windows are needed. After resampling the raw signals into 100 windows, the feature of Energy is extracted from the resampled raw signals using the following equation.

$$Y_{\bar{E}} = \sum_{j=1}^{m} (Y_{winj})^2_{rms} \tag{1}$$

where $(Y_{winj})_{rms}$ is the jth window of the root means squares (RMS) original resampled signal, m is the window number, and $Y_{\bar{E}}$ is the energy of the resampled raw signal.

2.2 Residual Signal Approximation

Regarding Fig. 1, the residual signal is defined based on the difference between resampled energy signal and the estimated ones. The estimated signal is determined using the observation algorithm. To design the observer, the first step is to extract the state-space function using the function approximation technique. So, in the first step, the resampled energy signal is approximated. Based on Fig. 1, to approximate the resampled energy signal, first, the AutoRegressive (AR) approximation technique is suggested and defined as the following equation [6].

$$Y_{\bar{E}(AR)}(k) = \alpha^T_{Y_{\bar{E}}}[\alpha_{X_{\bar{E}}} X_{\bar{E}(AR)}(k-1) + \alpha_{e_{\bar{E}}} e_{\bar{E}(AR)}(k-1)] \tag{2}$$

$$e_{\bar{E}(AR)}(k) = Y_{(NORM,100)_{\bar{E}}}(k) - Y_{\bar{E}(AR)}(k-1) \tag{3}$$

where $X_{\bar{E}(AR)}(k)$, $Y_{\bar{E}(AR)}(k)$, $Y_{(NORM,100)_{\bar{E}}}(k)$, $e_{\bar{E}(AR)}(k)$, and $(\alpha_{X_{\bar{E}}}, \alpha_{e_{\bar{E}}}, \alpha_{Y_{\bar{E}}})$ are the state of resampled energy signal using AR approximation, the response of resampled energy signal using AR approximation, the resampled energy of the original signal, the error of resampled energy signal using AR approximation, and the coefficients to tuning the parameters respectively. To improve the reliability of state-space approximation, the uncertainty input is combined with the AR technique. Thus, the AutoRegressive Uncertainty external resampled energy (ARU) is represented as the following equation.

$$\begin{aligned} Y_{\bar{E}(ARU)}(k) = \alpha^T_{Y_{\bar{E}}}\Big[&\alpha_{X_{\bar{E}}} X_{\bar{E}(ARU)}(k-1) + \alpha_{\Delta_{\bar{E}}} \Delta_{\bar{E}(ARU)}(k-1) \\ &+ \alpha_{e_{\bar{E}}} e_{\bar{E}(ARU)}(k-1)\Big] \end{aligned} \tag{4}$$

$$e_{\bar{E}(ARU)}(k) = Y_{(NORM,100)_{\bar{E}}}(k) - Y_{\bar{E}(ARU)}(k-1) \tag{5}$$

where $X_{\bar{E}(ARU)}(k)$, $Y_{\bar{E}(ARU)}(k)$, $e_{\bar{E}(AR)}(k)$, $\Delta_{\bar{E}(ARU)}(k)$ and $(\alpha_{\Delta_{\bar{E}}})$, are the state of resampled energy signal using ARU approximation, the response of resampled energy signal using ARU approximation, the error of resampled energy signal using ARU approximation, the uncertainty of function approximation using ARU approximation, and the coefficient, respectively. When the signal has uncertain conditions, the order of signal approximation is increased sharply. To address this issue, the ARU-Laguerre (ARUL) is proposed and introduced using the following equations.

$$Y_{\bar{E}(ARUL)}(k) = \alpha^T_{Y_{\bar{E}}}\Big[\alpha_{X_{\bar{E}}} X_{\bar{E}(ARUL)}(k-1) + \alpha_{\Delta_{\bar{E}}} \Delta_{\bar{E}(ARUL)}(k-1)$$

$$+\alpha_{e_{\bar{E}}} e_{\bar{E}(ARUL)}(k-1) + \alpha_{Y_{\bar{E}}} Y_{\bar{E}(ARUL)}(k-1)\Big] \quad (6)$$

$$e_{\bar{E}(ARUL)}(k) = Y_{(NORM,100)_{\bar{E}}}(k) - Y_{\bar{E}(ARUL)}(k-1) \quad (7)$$

Here, $X_{\bar{E}(ARUL)}(k)$, $Y_{\bar{E}(ARUL)}(k)$, $e_{\bar{E}(ARL)}(k)$, $\Delta_{\bar{E}(ARUL)}(k)$ and $(\alpha_{Y_{\bar{E}}})$, are the state of resampled energy signal using ARUL approximation, the response of resampled energy signal using ARUL approximation, the error of resampled energy signal using ARUL approximation, the uncertainty of function approximation using ARUL approximation, and the coefficient, respectively. The ARUL is a linear approximator. To improve the accuracy of the nonlinear and non-stationary signal approximation, the fuzzy logic algorithm is integrated with ARUL. In this paper, two inputs-Mamdani fuzzy inference engine is suggested. To reduce the effect of uncertainty, the error of ARUL and the integral of the error are used as the fuzzy inputs. Moreover, 9 fuzzy rules are defined to approximate the nonlinearity, and the *AND* is used for rule evaluation in this work. Besides, the *Max-Min* aggregation and Center Of Gravity (COG) techniques are suggested for aggregation and defuzzification, respectively. Thus, ARUL-fuzzy (ARULF) technique is represented as follows.

$$Y_{\bar{E}(ARULF)}(k) = \alpha^{T}_{Y_{\bar{E}}}\Big[\alpha_{X_{\bar{E}}} X_{\bar{E}(ARULF)}(k-1) + \alpha_{\Delta_{\bar{E}}} \Delta_{\bar{E}(ARULF)}(k-1)$$
$$+\alpha_{e_{\bar{E}}} e_{\bar{E}(ARULF)}(k-1) + \alpha_{Y_{\bar{E}}} Y_{\bar{E}(ARULF)}(k-1) + \alpha_{U_f} U_f(k-1)\Big] \quad (8)$$

$$e_{\bar{E}(ARULF)}(k) = Y_{(NORM,100)_{\bar{E}}}(k) - Y_{\bar{E}(ARULF)}(k-1) \quad (9)$$

where $X_{\bar{E}(ARULF)}(k)$, $Y_{\bar{E}(ARULF)}(k)$, $e_{\bar{E}(ARLF)}(k)$, $\Delta_{\bar{E}(ARULF)}(k)$, $U_f(k)$, and (α_{U_f}), are the state of resampled energy signal using ARULF approximation, the response of resampled energy signal using ARULF approximation, the error of resampled energy signal using ARULF approximation, the uncertainty of function approximation using ARULF approximation, the Mamdani fuzzy inference technique for nonlinear function approximation, and the coefficient, respectively.

To find the residual approximation, estimation is the next step. To do this, the Proportional Integral Derivative (PID) observer is recommended. The proportional technique is used to improve the state and the combination of the integral and derivative terms is used to reduce the effect of uncertainty. The ARULF-PID technique is represented as the following equation.

$$Y_{\bar{E}(ARULF-PID)}(k) = \alpha^{T}_{Y_{\bar{E}}}\Big[\alpha_{X_{\bar{E}}} X_{\bar{E}(ARULF)}(k-1) + \alpha_{\Delta_{\bar{E}}} \Delta_{\bar{E}(ARULF-PID)}(k-1)$$
$$+\alpha_{e_{\bar{E}}} e_{\bar{E}(ARULF)}(k-1) + \alpha_{Y_{\bar{E}}} Y_{\bar{E}(ARULF)}(k-1) + \alpha_{U_f} U_f(k-1)\Big] + \alpha_P \bar{e}_{PID}(k) \quad (10)$$

$$\bar{e}_P(k) = Y_{\bar{E}(ARULF-PID)}(k) - Y_{\bar{E}(ARULF)}(k) \quad (11)$$

$$\Delta_{\bar{E}(ARULF-PID)}(k) = \alpha_{\Delta_{\bar{E}}} \Delta_{\bar{E}(ARULF-PID)}(k-1) + \alpha_D \dot{\bar{e}}_{PID}(k) + \alpha_I \sum \bar{e}_{PID}(k) \quad (12)$$

Here, $Y_{\bar{E}(ARULF-PID)}(k)$, $\Delta_{\bar{E}(ARULF-PID)}(k)$, $\bar{e}_{PID}(k)$, $\dot{\bar{e}}_{PID}(k)$, $\sum \bar{e}_{PID}(k)$, and $(\alpha_P, \alpha_D, \alpha_I)$, are the estimation of resampled energy signal using ARULF-PID observer, the uncertainty of signal estimation using ARULF-PID observer, the error of resampled energy signal estimation using ARULF-PID observer, the change of error for resampled energy signal estimation using ARULF-PID observer, the integral of error for resampled energy signal estimation using ARULF-PID observer, and the PID observer coefficients, respectively. To modify the robustness, the two inputs Mamdani fuzzy inference is suggested. Thus, the combination of ARULF-PID observer and fuzzy technique (ARULF-FPID) is represented as:

$$Y_{\bar{E}(ARULF-FPID)}(k) = \alpha^T_{Y_{\bar{E}}}\Big[\alpha_{X_{\bar{E}}}X_{\bar{E}(ARULF)}(k-1) + \alpha_{\Delta_{\bar{E}}}\Delta_{\bar{E}(ARULF-FPID)}(k-1) + \alpha_{e_{\bar{E}}}e_{\bar{E}(ARULF)}(k-1) + \alpha_{Y_{\bar{E}}}Y_{\bar{E}(ARULF)}(k-1) + \alpha_{U_f}U_f(k-1)\Big] + \alpha_P\bar{e}_{FPID}(k) \tag{13}$$

$$\bar{e}_{FPID}(k) = Y_{\bar{E}(ARULF-FPID)}(k) - Y_{\bar{E}(ARULF)}(k) \tag{14}$$

$$\Delta_{\bar{E}(ARULF-PID)}(k) = \alpha_{\Delta_{\bar{E}}}\Delta_{\bar{E}(ARULF-FPID)}(k-1) + \alpha_D\dot{\bar{e}}_{FPID}(k) + \alpha_I\sum \bar{e}_{FPID}(k) + \alpha_{E_f}E_f(k) \tag{15}$$

Here, $Y_{\bar{E}(ARULF-FPID)}(k)$, $\Delta_{\bar{E}(ARULF-FPID)}(k)$, $\bar{e}_{FPID}(k)$, $\dot{\bar{e}}_{FPID}(k)$, $\sum \bar{e}_{FPID}(k)$,$E_f(k)$, and (α_{E_f}), are the estimation of resampled energy signal using ARULF-FPID observer, the uncertainty of signal estimation using ARULF-FPID observer, the error of resampled energy signal estimation using ARULF-FPID observer, the change of error for resampled energy signal estimation using ARULF-FPID observer, the integral of error for resampled energy signal estimation using ARULF-FPID observer, the fuzzy estimator, and the fuzzy coefficient, respectively. After estimating the resampled energy signal using the ARULF-FPID observer, the residual approximation is calculated using the following definition.

$$R_{\bar{E}(ARULF-FPID)}(k) = Y_{\bar{E}(Un-KNWN,100)}(k) - Y_{\bar{E}(ARULF-FPID)}(k) \tag{16}$$

where $R_{\bar{E}(ARULF-FPID)}(k)$ is the residual approximation using ARULF-FPID observer and $Y_{\bar{E}(Un-KNWN,100)}(k)$ is resampled unknown energy signal.

2.3 Residual Signal Approximation

After determining the residual signals using the proposed estimation algorithm, in the next step, the SVM is used for fault pattern recognition. We have 4800 samples for normal (NM), ball (BL), inner (IN), and outer (OT) conditions. The SVM is used to classify the residual signal into four groups.

3 Results

To test the performance of fault pattern recognition using the proposed algorithm, the CWRUBD is used [8]. The information of CWRUBD is introduced in the following Table 1.

Table 1. CWRUBD information, torque loads, and crack sizes [8].

Class	Torque load (hp)	Crack sizes (inches)
NM	0, 1, 2, 3	–
BL	0, 1, 2, 3	0.007, 0.014, 0.021
IN	0, 1, 2, 3	0.007, 0.014, 0.021
OT	0, 1, 2, 3	0.007, 0.014, 0.021

Figure 2 shows the resampled energy residual approximation using the proposed ARULF-FPID observer. Based on this figure, the proposed method has improved the accuracy of fault pattern recognition for CWRUBD. Table 2 illustrates the average accuracy of fault pattern detection in the proposed ARULF-FPID and ARULF-PID observers.

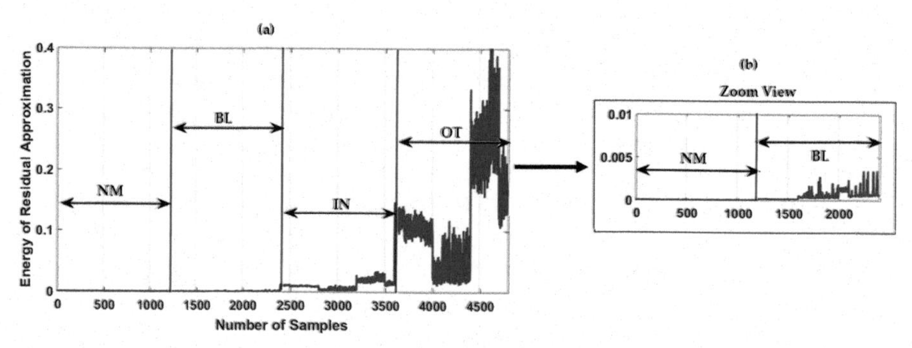

Fig. 2. Energy of residual approximation using the proposed ARULF-FPID observer: a) all conditions, b) zoom view for normal and ball conditions.

Table 2. Average accuracy of fault pattern detection using ARULF-FPID+SVM and ARULF-PID+SVM.

Method/Fault	NM	BL	IN	OT
ARULF-FPID+SVM	100%	98%	95%	97%
ARULF-PID+SVM	100%	90%	83%	80.3%

4 Conclusions

In the paper, the combination of data-driven signal approximation technique and machine learning approach is proposed for fault pattern detection in the bearing. The combination of ARU modeling approach, Laguerre filter, and fuzzy approach was suggested for signal approximation and extracting the state-space function of the bearing. In addition, the fuzzy PID observer is integrated with the ARULF approximation technique for residual approximation determination. Next, the SVM technique was recommended for bearing fault pattern detection. The CWRUBD is used to test the performance of the proposed method. As a result, the accuracy of the bearing fault pattern detection for the proposed scheme (ARULF-FPID+SVM) can be improved by 9.2%, compared with ARULF-PID+SVM.

As a future work, the intelligent-based noise cancellation can be integrated with a hybrid observation approach for fault pattern detection and crack size identification in uncertain and noisy vibration signals.

Acknowledgements. This work was supported by the 2021 Research Fund of the University of Ulsan.

References

1. Wang, Z., et al.: A novel method for intelligent fault diagnosis of bearing based on capsule neural network. Complexity (2019)
2. Lin, J., Qu, L.: Feature extraction based on Morlet wavelet and its application for mechanical fault diagnosis. J. Sound Vib. **234**(1), 135–148 (2000)
3. Gao, Z., Cecati, C., Ding, S.X.: A survey of fault diagnosis and fault-tolerant techniques—Part I: Fault diagnosis with model-based and signal-based approaches. IEEE Trans. Ind. Electron. **62**, 3757–3767 (2015)
4. Cecati, C.: A survey of fault diagnosis and fault-tolerant techniques—Part II: Fault diagnosis with knowledge-based and hybrid/active approaches. IEEE Trans. Ind. Electron. **62**, 3768–3774 (2015)
5. TayebiHaghighi, S., Koo, I.: Fault diagnosis of rotating machine using an indirect observer and machine learning. In: International Conference on Information and Communication Technology Convergence (ICTC), pp. 277–282 (2020)
6. TayebiHaghighi, S., Koo, I.: SVM-based bearing anomaly identification with self-tuning network-fuzzy robust proportional multi integral and smart autoregressive model. Appl. Sci. **11**(6), 2784 (2021)
7. Piltan, F., Kim, J.-M.: Fault diagnosis of bearings using an intelligence-based autoregressive learning Lyapunov algorithm. Int. J. Comput. Intell. Syst. **14**(1), 537–549 (2021)
8. Bearing Data Center. Case Western Reserve University Seeded Fault Test Data. https://csegroups.case.edu/bearingdatacenter/pages/welcome-case-western-reserve-university-bearing-data-center-website. Accessed 23 Dec 2020

Comparison of ML Algorithms to Detect Vulnerabilities of RPL-Based IoT Devices in Intelligent and Fuzzy Systems

Murat Ugur Kiraz[1(✉)] and Atinc Yilmaz[2]

[1] Hezarfen Institute of Aeronautics and Space Technology, Yesilyurt, 34149 Istanbul, Turkey
[2] Faculty of Engineering Architecture, Computer Engineering, Beykent University, Hadim Koruyolu Cd. No:19, Sariyer, 34398 Istanbul, Turkey
atincyilmaz@beykent.edu.tr

Abstract. The RPL protocol (Routing Protocol for Low-Power and Lossy Networks) was designed by IETF [1] for 6LoWPAN to optimize power consumption on the Internet of Things (IoT) devices. These devices have limited processing power, memory, and generally limited energy because they are battery-powered. RPL aims to establish the shortest distance by setting up n number of IoT devices through each other DAG (Directed Acyclic Graph) and therefore the most optimum energy consumption. However, due to the complex infrastructure of RPL and the low capacity of IoT devices, the RPL protocol operating at the network layer is susceptible to attacks. Therefore, it is vital to develop a fast, practical, uncomplicated, and reliable intrusion detection system in the network layer. In the event of an attack on IoT devices operating with the RPL protocol, an anomaly will occur in the network packets in the 3rd layer. Processing these packages with machine learning algorithms will make the detection of the attack extremely easy. In this article, "Decision Tree," (DT) "Logistic Regression," (LR) "Random Forest," (RF) "Fuzzy Pattern Tree Classifier," (FPTC), and "Neural Network" (NN) algorithms are compared for catching Flooding Attacks (FA), Version Number Increase Attacks (VNIA), and Decreased Rank (DRA) attacks. At the end of our study, it is observed that the Random forest algorithm gave better results than other algorithms in the system built by the study.

Keywords: RPL attacks · Machine learning algorithms · Hello flood attack · Decreased rank attack · Version number increase attack

1 Introduction

When it comes to Smart and Fuzzy Techniques, the first thing that comes to mind will undoubtedly be the Internet of Things (IoT) devices. IoT devices are pieces of hardware such as sensors, actuators, devices, or machines that are programmed to perform a specific function and transfer data over the Internet or other networks. Since IoT devices collect, produce or transfer data from many points, they need independent energy sources instead of directly meeting their energy needs with the grid voltage. Moreover, in an environment

C. Kahraman et al. (Eds.): INFUS 2021, LNNS 308, pp. 254–262, 2022.
https://doi.org/10.1007/978-3-030-85577-2_30

with many, maybe thousands of these IoT devices, connecting all of them to one point will not solve energy efficiency. Therefore, IoT devices communicating by clicking from multipoint to multipoint will solve this problem. They need to be modeled by connecting. Accordingly, RPL (Routing Protocol for Low-Power and Lossy Networks) protocol was developed by IETF [1] in March 2012 to ensure efficient power consumption of many interconnected IoT devices.

The distribution and connection of the IoT devices can be compared with a mathematical term as "Directed Acyclic Graph (DAG)." DAG is the orientation of n nodes to each other in a way that does not form a closed loop. The RPL is designed to generate DAGs within IoT devices. DAGs are made up of a combination of DODAGs (Destination Oriented DAG). A DODAG is a particular type of DAG where each node wants to achieve a single goal.

Before DODAG is created, a root node is determined by the architect manually. A DODAG Information Object (DIO) message is sent to all nodes by root. This message multicasts downward. Once the nodes have received the DIOs, they will start creating DODAG. These nodes, along with the DIO message, also learn that their distance. The distance is named "rank." Then these nodes DODAG Announcement Object (DAO) messages. DAO request sent by a child node to the parent node or root. With this message, a node requests permission to join a DODAG as a child node. The root node sends the DAO-ACK message to all nodes and accepts all nodes. After this step, nodes with the lowest rank act as "root," and the above processes continue.

Designed for the 6LoWPAN protocol, RPL aims to optimize the power consumption of IoT devices. Nevertheless, the complexity of the RPL itself and the low-security nature of 6LoWPAN devices are vulnerable to attacks inside or outside the network [2]. As a result, any weakness in the DODAG structure will affect the entire system. When the attack occurs, the DODAG structure will deteriorate, which will cause the entire system to not work with the appropriate parameters. Attacks will cause battery-powered IoT devices to process much more than usual, transmit data, and ultimately run out of batteries. Therefore, it is vital to develop a fast, practical, uncomplicated, and reliable intrusion detection system in the network layer. In the event of an attack on IoT devices operating with the RPL, an anomaly will occur in the network packets in the 3rd layer. Processing these packages with machine learning algorithms will make the detection of the attack extremely easy. Currently, machine learning algorithms are used to detect attacks performed in the RPL in the literature. However, the study set out by asking the following question. Which machine learning algorithm is the most effective in detecting the attack in the RPL?

1.1 Related Works

Müller et al. [4] in 2019 developed a machine learning method with Kernel Density Estimation (KDE), which detects the Blackhole, Hello Flood (HF), and Version Number attacks in RPL average of 84.91% true positive and less than 0.5% false positive value.

Neerugatti et al. [5] proposed an attack detection technique based on the machine learning approach called MLTKNN based on the K-nearest neighbor algorithm in their study. They reached %90 to %98 TP rate and %0.9 to %0.2 FP rate with various amounts of motes up to 30.

Verma et al. [6] in 2019 designed a Network Attack Detection System architecture called ELNIDS to detect attacks against the RPL. This design was implemented with Boosted Trees (BT), Bagged Trees, Subspace Discriminant (SD), and RUSBoosted Trees algorithms. The Sinkhole, Blackhole, Sybil, Clone ID, Selective Forwarding, HF, and Local Repair attacks were detected with machine learning methods using 20 features of the RPL-NIDDS17 dataset in the study. BT algorithm achieves the highest accuracy of 94.5%, while the SD method achieves the lowest accuracy of 77.8%.

Belavagi et al. [7] in 2019 accepted the node size in the system as 10, 40, and 100 nodes and added 10%, 20%, and 30% of malicious nodes to the evaluated network to identify multiple intrusions. They observed the behavior of the grid according to the inconsistency percentage, energy consumption, accuracy, and false-positive rate they obtained. In this study, besides network packages, other parameters are used for ML algorithms.

Cakir et al. [8] proposed a Gated Recurrent Unit network model-based deep learning algorithm to predict and prevent HF attacks on the RPL in IoT networks. They compared this model with the SVM and LR methods; besides, they tested the different power states and total energy consumption of the nodes. They detected HF attacks with a much lower error rate than the literature studies with the model they presented.

Shafiq et al. [9] worked on a model that enables the selection of an effective machine learning algorithm among many machine learning algorithms for the cyber-attack detection system to be used in IoT security. With this study, the Naive Bayes ML algorithm effectively performs anomaly and intrusion detection in the IoT network.

In this paper, unlike other studies, a single machine learning method was not tested or developed to be used to detect attacks in the RPL. On the contrary, it was focused on obtaining the best result by comparing multiple machine learning methods.

In addition, only the data obtained from layer-3 network packets were used in the data set. The reason for this is simple. Acquiring, processing, and transmitting parameters such as instantaneous power and energy consumption for each IoT device will require extra processing and capacity power. Therefore, layer-3 network packages, which can be obtained very quickly, are used in this study.

In this paper, Flooding Attack (FA) [2], Version Number Increase Attack (VNIA) [2], and Decreased Rank (DRA) [2] attacks were examined.

In the study, a data set was prepared by summarizing the receiving packets, transmitting packets, packet lengths, DIS, DAO, DIO messages, and their rates in the 1-s frame of normal and malicious RPL packets transmitted only in the 3rd layer. These datasets were trained and tested with DT, LR, RF, FPTC, and NN algorithms. Finally, we compared these ML algorithms to find the best result.

In this study, in the second section, obtaining the dataset by making simulations with malicious and normal IoT motes was first described. After obtaining the raw dataset, making the dataset meaningful and extract sensitive data was shown. Lastly, the extracted data were normalized, and this data was trained with DT, LR, RF, FPTC, and NN algorithms. In the third section, the results with accuracy rates and training durations were shown. In the fourth section, the results were interpreted. According to the results obtained, the RF gives better results than other machine learning algorithms.

2 Simulation and Creation of Data Set

D'Hondt et al. simulated FA, VNIA, and DRAs against the RPL with the academic report they prepared [10]. Normal, malicious, and root sensors from this study were used.

Two different simulations in the Cooja simulator for each attack type to create the raw data sets were performed: the simulation consisting of "normal" motes and the simulation with a vulnerable mote. We ran each simulation for 600 s. In simulations, the locations of the nodes were not changed. The last mote with the vulnerable mote was changed only. Cooja simulator enables the extraction of simulation network messages. Cooja saves these messages in ".pcap" format. In this format, "Time," "Source IP," "Destination IP," "Protocol Type," "Length of packet," "Message Type Info" were obtained. Using Wireshark, "pcap" files are converted to ".csv" files.

2.1 Making the Raw Dataset Meaningful

The data sets obtained from simulations with malicious motes will differ from the data sets obtained from simulations with normal (non-malicious) motes.

In attack datasets, naturally, the number of packets, their message type, total packet length, and their ratio will be abnormal compared with the simulations made with non-malicious motes. To detect this anomaly, raw data was divided into 1-s frames. New values were found within these one-second frames, and a new dataset was established by calculating the following values.

- Source Mote: A unique number for each mote.
- Destination Mote: Same number as the source mote.
- Packet Count: The count of the whole source motes in the 1-s frame.
- Source Mote Ratio: (Source Mote Count/Packet Count).
- Destination Mote Ratio: (Destination Mote Count/Packet Count).
- Source Mote Duration: The sum of all packet durations sent from source to destination in the 1-s frame.
- Destination Mote Duration: The sum of all packet durations received by the destination in the 1-s frame.
- Total Packet Duration: It is the sum of all packet durations in the 1-s frame.
- Total Packet Length: It is the sum of all packet lengths in the 1-s frame.
- Source Packet Ratio: (Sum of Source Packet lengths/Total Packet Length).
- Destination Packet Ratio: (Sum of Dest. Packet lengths/Total Packet Length).
- DIO Message Count: Count of DIO messages in the 1-s frame.
- DIS Message Count: Count of DIS messages in the 1-s frame.
- DAO Message Count: Count of DAO messages in the 1-s frame
- Other Message Count: Count of the messages except for DIO, DIS, and DAO.
- Label: 0 or 1 (If the raw dataset has malicious mote/s, the label is 1 else 0).

The pseudo-code of conversion to a meaningful dataset for machine learning is below.

```
START
Dset=INPUT(RawDataset)
WHILE Dset Rows Ends
  Duration=time(current_row)-time(previous_row)
  Duration_list=APPEND(Duration)
ENDWHILE
Dset = Dset + Duration_list
IP_dictionary={IP_Adress :unique_number}
Crr_scnd=60
Counter=0
fs=FLOOR(Dset[Duration_list])
WHILE counter < frame_second
  osf= GET(Dset[Time]>= fs and Dset[Time]<= Crr_scnd+1)
  WHILE osf Rows Ends:
    Osf_list=[ src=IP_dictionary[Source IP_Adress],
               dst=IP_dictionary[Dest. IP_Adress],
               pct_cnt=COUNT(rows)
               src_mote_rt= COUNT(src)/pct_cnt
               dst_mote_rt= COUNT(dst)/pct_cnt
               src_mote_dur=SUM(src_duration)
               dst_mote_dur= SUM(dst_duration)
               ttal_pckt_dur= SUM(duration)
               ttal_pckt_lngth= SUM(pckt_lngth)
               src_pckt_rt= SUM(src_pckt_lngth)/
ttal_pckt_lngth
               dst_pckt_rt= SUM(dst_pckt_lngth)/
ttal_pckt_lngth
               dio_msg_cnt= COUNT(dio_messages)
               dis_msg_cnt= COUNT(dis_messages)
               dao_msg_cnt= COUNT(dao_messages)
               other_msg_cnt= COUNT(other_messages)
               IF Dset="Normal"
                 Label=0
               ELSE
                 Label=1
               ENDIF
  ENDWHILE
New_dset=APPEND(Osf_list)
ENDWHILE
END
```

2.2 Training with Machine Learning Algorithms

After creating a new and meaningful dataset, these datasets were used to compare machine learning algorithms. Three different new datasets were obtained to compare machine learning algorithms by merging the meaningful datasets according to their attack types.

First, the dataset was separated by its values of rates and class. The source and destination IP addresses from the training data set were removed so that the machine does not know whether there is an attack according to the source and destination IP addresses.

Secondly, the dataset was divided into test and training datasets with the amount of 2/3. (2/3 train, 1/3 test). The test and train datasets were normalized with the Eq. (1). Here z represents the normalized value, x is the real value, μ is mean of the all-x values, and σ is the standard deviation of x values.

$$z = \frac{x - \mu}{\sigma} \text{ and } \mu = \frac{1}{N}\sum\nolimits_{i=1}^{N}(x_i)\ \sigma = \sqrt{\frac{1}{N}\sum\nolimits_{i=1}^{N}(x - \mu)^2} \tag{1}$$

Thirdly, experiments with the train datasets were executed. Five kinds of machine learning algorithms tested. These are DT, LR, RF, FPTC, and NN algorithms. Machine learning parameters are shown in Table 1.

Table 1. Machine learning parameters

ML algorithm	Parameter
Logistic Regression	No extra parameters were used
Random Forest	The number of estimators is defined as "8," and the "entropy" criterion is selected
Decision Tree	The "entropy" criterion is selected
Fuzzy Pattern Tree Classifier	No extra parameters were used
Deep Learning	Six layers, neuron numbers: 26, 52, 56, 13, 7, 1, optimizer: "Nadam", iteration: 60. Prediction threshold: 0.7

The accuracy rate (AR) is calculated as in the equation of (2) and TP is True Positive, TN is True Negative, FP is False Positive, and FN is false negative

$$AR = (TP + TN)/(TP + TN + FP + FN) \tag{2}$$

3 Results

The dataset row numbers, accuracy rate, and duration of the training period are shown in Table 2. In FA, via the RF algorithm, we detected the attack with a %97 accuracy rate within 17 ms training duration. In VNIA, we detected the attack with a %89 accuracy rate within 2280 ms training duration via the DL algorithm. However, the RF algorithm has a %88 accuracy rate and has a much shorter training duration than the DL algorithm. In DRA, via RF algorithm, we detected the attack with %65 accuracy rate within 31 ms training duration.

Table 2. Results

Attack type	Dataset row number	Algorithm	Accuracy rate	Training duration (ms)
Flooding Attack (FA)	1282	Decision Tree (DT)	0,965	17
		Logistic Regression (LR)	0,967	103
		Random Forest (RF)	0,977	64
		Fuzzy Pattern Tree (FPT)	0,915	3063
		Deep Learning (DL)	0,967	2520
Version Number Increase Attack (VNIA)	1473	Decision Tree (DT)	0,865	0
		Logistic Regression (LR)	0,877	14
		Random Forest (RF)	0,889	14
		Fuzzy Pattern Tree (FPT)	0,884	2235
		Deep Learning (DL)	0,893	2280
Decreased Rank Attack (DRA)	2398	Decision Tree (DT)	0,589	59
		Logistic Regression (LR)	0,537	17
		Random Forest (RF)	0,656	31
		Fuzzy Pattern Tree (FPT)	0,484	4166
		Deep Learning (DL)	0,587	3611

4 Conclusion

After the experiments, it is observed that the DT, LR, RF, FPTC, and NN algorithms have values close to each other. However, the Random Forest algorithm is a little more ahead than the others.

The DRA does not harm the network; therefore, it is difficult to detect.

It is a fact that deep learning algorithm has numerous advantages in solving complicated classifications. However, the duration of the training is longer than the others, and accuracy rates are close to each other.

Detecting the anomaly with machine learning algorithms in RPL, the Random Forest algorithm is a little more ahead.

This study determined that anomaly detection is faster with the Random Forest algorithm in the data set formed by the extraction of network packets in the Layer-3 layer in RPL. Its accuracy rate is higher than other algorithms.

An advanced methodology could be established for DRAs because of the low accuracy rates for detecting them. Seth et al. [11] developed a model for detecting it using Round-Trip Times; however, we assume that this attack can be detected with more accuracy rate by improving the extraction of the layer-3 network packets.

References

1. Winter, T.: Rpl: Ipv6 routing protocol for low-power and lossy networks. Protocol (2012). https://tools.ietf.org/html/rfc6550
2. Le, A., Loo, J., Luo, Y., Lasebae, A.: Specification-based IDS for securing RPL from topology attacks. In: IFIP Wireless Days (WD), Niagara Falls, ON, Canada, pp. 1–3 (2011)
3. Mayzaud, A., Badonnel, R., Chrisment, I.: A taxonomy of attacks in RPL-based Internet of Things. Int. J. Netw. Secur. **18**(3), 459–473 (2016). ffhal-01207859
4. Müller, M., Debus, P, Kowatsch, D, Böttinger, K.: Distributed anomaly detection of single mote attacks in RPL networks. In: Proceedings of the 16th International Joint Conference on e-Business and Telecommunications (ICETE 2019), pp. 378–385 (2019)
5. Neerugatti, V., Mohan, R., Rama, A.: Machine learning based technique for detection of rank attack in RPL based Internet of Things networks. Int. J. Innov. Technol. Explor. Eng. (IJITEE), **8**(9S3) (2019). ISSN: 2278-3075 SSRN: https://ssrn.com/abstract=3435598
6. Verma, A., Ranga, V.: ELNIDS: ensemble learning based network intrusion detection system for RPL based Internet of Things. In: 2019 4th International Conference on Internet of Things: Smart Innovation and Usages (IoT-SIU), 2019, pp. 1–6 (2019). https://doi.org/10.1109/IoT-SIU.2019.8777504
7. Belavagi, M., Muniyal, B.: Multiple intrusion detection in RPL based networks. Int. J. Electr. Comput. Eng. (IJECE) **10**(1), 467–476 (2020)
8. Cakir, S., Toklu, S., Yalcin, N.: RPL attack detection and prevention in the Internet of Things networks using a GRU based deep learning. IEEE Access **8**, 183678–183689 (2020). https://doi.org/10.1109/ACCESS.2020.3029191
9. Shafiq, M., Tian, Z., Sun, Y., Du, X., Guizani, M.: Selection of effective machine learning algorithm and Bot-IoT attacks traffic identification for Internet of Things in smart city. Future Gener. Comput. Syst. **107**, 433–442 (2020). ISSN 0167-739X, https://doi.org/10.1016/j.future.2020.02.017

10. D'Hondt, A., Hussein, B., Jeremy, V., ve Ramin, S.: RPL attacks framework. Technical report, Louvain-la-Neuve, Belgium: Universit catholique de Louvain (2015)
11. Seth, A.D., Biswas, S., Dhar, A.K.: Detection and verification of decreased rank attack using round-trip times in RPL-based 6LoWPAN networks. In: 2020 IEEE International Conference on Advanced Networks and Telecommunications Systems (ANTS), New Delhi, India, pp. 1–6 (2020). https://doi.org/10.1109/ANTS50601.2020.9342754

Predictive Quality Defect Detection Using Machine Learning Algorithms: A Case Study from Automobile Industry

Muhammed Hakan Yorulmuş[1(✉)], Hür Bersam Bolat[1(✉)], and Çağatay Bahadır[2(✉)]

[1] Department of Management Engineering, Istanbul Technical University, Istanbul, Turkey
{yorulmus,bolat}@itu.edu.tr

[2] Department of Information and Communication Technologies, TOFAŞ Turkish Automobile Factory Joint-Stock Company, Bursa, Turkey
cagatay.bahadir@tofas.com.tr

Abstract. Industry 4.0 is generally defined as a development system that compels the digitalization of processes to create integrated and autonomous systems. The process tracking of parts is very important in terms of detecting missed faulty products. Some defects that escape from quality control directly affect the end-user. Machine learning algorithms have been used to predict changes in the quality control processes and defective products, toward real-time and effective data processing. Thus, the highest quality of the final product will be delivered to the customer and to reduce the defective production coming out of the manufacturing chain. In this article, the study aims to establish a predictive quality model that can detect defect-free approved but faulty products overlooked during the quality inspection operations. Machine learning methods are used to analyze the relationship between quality control data and customer complaints. For this purpose, we use the last quality stage data of an automobile manufacturer's brake system from 2018 to 2020. Machine learning models are constructed using logistic regression, ridge regression, support vector machine, random forest classification tree, gradient boost, XGBoost, LightGBM, and CatBoost algorithms. The results of specificity and negative prediction value show that the Gradient Boost and CatBoost algorithms have the best classification benefit for detecting the rare events.

Keywords: Quality 4.0 · Industry 4.0 · Predictive quality · Automobile industry · Machine learning · Fault detection · Rare event detection

1 Introduction

Nowadays changing production habits with the Industry 4.0 revolution have increased their impact on products with digitalized applications. Various tools and concepts adapted to production systems will now become widespread using modern information technology and cyber-physical systems (CPSs) developments. The use of statistical algorithms rather than humans for the detection of errors in production processes has increased. The smart factory model achieves new solutions with the development of big data technology. One of these solutions is Quality 4.0, which is an essential part of Industry 4.0.

C. Kahraman et al. (Eds.): INFUS 2021, LNNS 308, pp. 263–270, 2022.
https://doi.org/10.1007/978-3-030-85577-2_31

Quality 4.0, can be defined as the digitalization of total quality management (TQM) that affects quality technology, processes, and people. One of the well-known quality solutions is predictive quality applications that are used to avoid excessive time and effort by predicting and preventing quality problems in the industry. These quality problems are situations that reduce yield or cause product recalls with faulty production. Predictive quality system aims to reduce the number of products recalls and achieve zero defect products. High-quality products increase customer satisfaction and reduce the number of product recalls.

In the new concept of the industry, the quality context should be under careful observation with monitoring and improvement of processes are among the top topics covering Industry 4.0 [1]. In the digital era, TQM becomes an important part of organizational innovation where disruptive and radical innovations open the way to big changes to the concept of quality [2]. In the context of I4.0, quality should be considered as the discovery of data sources, root causes, and insights about products and organizations by augmenting and improving upon, human intelligence [3]. Q4.0 is an integral part of I4.0 and could be summarized as the digitalization of TQM that impacts quality technology, processes, and people [4]. CPSs that implement Industry 4.0 are physical, intelligent, and network-enabled components [5]. CPSs are the structures that involve communication and coordination between the physical world and the cyber world.

According to Industry 4.0, it is possible to adopt various tools and concepts to the manufacturing system with the help of modern information technology and machine flexibilization. High-quality products increase customer gratification and reduce product recalls.

The smart factory concept has some solutions that it gains with big data technology. These are; predictive maintenance, predictive quality, and monitoring the assets. These new solutions have many benefits. The first step in preventing quality problems in the industry is to avoid excessive time and effort. With the development of machine learning and artificial intelligence algorithms, it is seen that these models are used in quality and error detection.

However, the predictive quality studies are limited in the literature. In this article, the predictive quality modeling approach has made estimates benefiting from a factory in Turkey. It is aimed to contribute to the subject of quality prediction and smart factories, which are new in the literature, by using the quality data of the production lines and final quality data of the brake assembly lines of the automobile manufacturer. As a result of the gap seen in the literature, quality estimation is made using gradient boost (GrBoost) algorithms.

For this purpose, we used the last quality stage data of an automobile manufacturer's brake system from 2018 to 2020. Machine learning models are constructed using logistic regression, ridge regression, support vector machine, random forest classification tree, GrBoost, XGBoost, LightGBM, and CatBoost algorithms. The results of specificity and negative prediction value show that the Gradient boost and CatBoost algorithms have the best classification benefit for detecting the rare events.

The rest of the study is designed as follows. Section 2 presents the main related works used in the study. Section 3 gives the methodology summary, the implementation of the study and discusses the results. Finally, Sect. 4 discuss the conclusions and further research of the study.

2 Related Works

Quality estimation has always been an issue with production systems. However, the touch of artificial intelligence with industry 4.0 makes the issue of quality assessment more successful. The use of artificial intelligence algorithms on the production data and quality data of the manufactured products is an area where new studies are carried out. As each machine and product is unique, there is still limited study to determine which variables match which quality data and which model. Studies on quality 4.0 and artificial intelligence in the literature are still new and trying to produce the best quality product by using digitalization.

While the factories of the future enjoy new solutions, quality is the most important part of all production systems, regardless of the type of production and products. Quality 4.0 could be defined as the application using smart solutions and intelligent algorithms of Industry 4.0 technologies to quality management methods and tools [6]. On the factory floor level, customization of Industry 4.0 concepts industries and service processes use to improve quality by detecting failures and justifying possible causes while staying competitive in volatile business environments [7].

In the models in the literature where the quality prediction is made, the data about the product or the process are used and it is labeled according to whether the product is defective or not. Statistical methods can provide solutions for error detection and other tasks in production processes. In the studies, quality prediction models have been used by using different production equipment in different areas of the production industry. This variety causes many variables in the applied models and as a result, many estimates occur. A brief summary of these studies is as follows.

Some application subjects of predictive quality systems are proposed in the literature. Such as, the deep drawing manufacturing process of car body parts [8], tool flank wears at a turning operation [9], refrigerant brazed plate heat exchangers (BPHE) [10], battery cells production [11]. Moreover, we can add topics to consider like crankshaft production line [12], rare quality event detection, ultrasonic metal welding of battery tabs, sensorless drive diagnosis [14]. X-ray inspection [13] applications are other examples of research subjects.

Digitalization needs specific subjects for using machine learning algorithms aspect of predictive quality applications. Also, these special concepts need a wide range and different type of variables such as; cutting speed (rpm), feed rate (mm/rev), depth of cut (mm), lubrication variables [9], plate geometry, operating conditions [10], x-ray inspection with height, % shape 2D, % shape 3D, % surface, % volume, % offset X μm, offset Y μm [13], flange retraction laser data, strain gauge sensory data, signal data, the occurrence of process failures [8], etc.

Predictive Quality models and methods in Machine Learning (ML) algorithms are used in the literature can be listed as follows. Adaptive Neuro-Fuzzy Approach (ANFIS) [9], Artificial Neural Network (ANN) [8, 10, 11, 15], Lasso-Lars Regression [11], RF [11, 12], Logistic Regression [12–16], KNN or k-Nearest Neighbors [12, 15, 16], Support Vector Machines (SVC) [12–16], Linear SVC [12], Decision Tree [12, 13, 15],

Perceptron [12], Stochastic Gradient Descent Classifier (SGDC) [12], Naïve Bayes [13, 15, 16], Gradient Boosted Tree [13].

3 Methodology

In this study, machine learning algorithms are utilized to make predictive quality model predictions. Widely used statistical methods that are logistic regression, ridge regression, support vector machine, random forest, classification tree methods have been performed in the study. In addition, GrBoost, XGBoost, LightGBM, and CatBoost methods, which be considered new and effective algorithms, are also employed to create the predictive quality model in the best way. The results of specificity value and negative prediction value were used for evaluation criteria to choose a classification model for detecting the rare events.

The mentioned algorithms have been developed over gradient boosting methods. Gradient enhanced trees (GBT) [17] are acquired by applying boosting methods to regression trees. At each step, the best partitioning and regression errors are calculated by dividing the data into two samples at each split node. Then the next tree is boosted to reduce the error [18]. XGBoost (eXtreme Gradient Boost) is a learning model that is based on the residual optimization algorithm which is ten times faster than similar methods. LightGBM is developed from the Gradient boosted decision trees (GBDT) model to create a better-performance model. The CatBoost model has emerged with the development of the Gradient Boosting model for high cardinality categorical variables [19].

3.1 Implementation of the Study and Data Collection

Automobiles are very complex products with a lot of parts. This study aims to detect malfunctions in the vehicle brake systems before leaving the factory. Brake systems are critical parts for drivers. The slightest problem to occur can cost the driver's life. It can also cause great prestige loss and heavy lawsuits for the company. Automobile companies undertake that such systems do not occur any problem due to production errors.

In order to detect malfunctions that are not in compliance with the norms, in the vehicle brake systems, we have collected the data from the related brake test stations in the factory. There are three facilities on the production line and one facility for final general testing. The stations on the production line are the hand brake control station, brake pedal control station, brake hydraulic filling station. The hand brake control station is controlled by a mechanism whether the hand brake is following the norms [20]. Mechanical data, such as hand brake adjustment angle, hand brake pull load, are measured. At the brake pedal control station, the brake pedal is controlled and the correct transmission of the force and any misalignment are measured. In the brake hydraulic filling station, hydraulic fluids important for the brake are pumped and leakage values are measured. The equipment in the facilities makes measurements by means of sensors and transfers them to the servers.

After the other parts of the vehicle are assembled, they are sent from the production line to the roller system, which is the test area. The roller system is one of the last checks that test whether the vehicle complies with all norms. It is the place where all the data related to the vehicle are collected with sensors and the quality control of the production is done. If an improper situation occurs, the automobile is sent for repair. This system is similar to the places where you have your car checked in second-hand car sales, but it is a more complex one in the factory. The variables related to the brake from the roller unit were selected in the light of the norm and literature and used in the model.

Despite all the norms and quality tests, the vehicles leaving the factory give brake failure before the warranty expires. This critical breakdown is critical to human life and corporate prestige. It is aimed to establish a warning system after the rollers test within the scope of the paper in order to discover the errors that the norms miss and foreman cannot predict. By establishing a model with the production data and final control data of the factory, it is aimed to determine the failure of the vehicles that go to the customer. It is aimed that these vehicles are checked once again before they are sent from the factory to the customer and that their brake systems are perfect.

In accordance with the numbered norms [20], also according to the quality guidelines and expert opinions; the appropriate variables are selected. The variables of the data received from each station. 57 variables from the final inspection, 6 variables from the brake pedal test station, and 5 variables from the handbrake test station are selected. According to the status of data retention, 86 complaints are used in the roller testing facilities, and brake faulty vehicles were used in the production line stations. Although these complaints are not critical to driving safety, they are malfunctions detected by the on-board computer or that cause a loss of comfort to the customer. Problem-free vehicles which are approximately 82,000 were randomly selected from the factory in the 2020 production year.

Vehicles with faulty that did not exceed 10,000 km within the warranty period are selected. These vehicles have been reported to brake malfunctions through authorized services and have been modified regarding the brake. There should be no malfunction in such vehicles.

3.2 Results

Twelve models are tested in the selected scenario. GrBoost and CatBoost methods have the best results which shown in Table 1. Specificity and catching correct faulty automobiles are used for evaluation criteria. Near 50% specificity and finding brake faulty candidate automobiles is a good result for the beginning of the warning system. In addition, according to CatBoost model feature importance graph is shown in Fig. 1, it has been revealed that our brake force variables and brake measurement values in stations are the most important ones in order to predict the fault in the brake system. In addition, important variables have been determined according to the dataset used. It is shown that some improvements are required in the production line. Customer complaints indicate that individual norms are not enough and there is a need for hybrid norms.

Table 1. Comparison of the created models with evaluation criteria.

	Neg. pred. value %	Specificity	Neg. pred. total	Neg. no	Accuracy
LG	0.000	0.000	0	0	0.997
LD	0.000	0.000	5	0	0.996
QDA	0.017	0.038	59	1	0.989
KNN	0.000	0.000	0	0	0.997
Ridge	0.000	0.000	0	0	0.997
SVM	0.000	0.000	1	0	0.996
RFCT	1.000	0.154	4	4	0.997
CTree	0.273	0.346	33	9	0.995
GrBoost	0.441	0.577	34	15	0.996
XGBoost	1.000	0.115	3	3	0.997
LightGBM	0.545	0.231	11	6	0.997
CatBoost	1.000	0.462	12	12	0.998

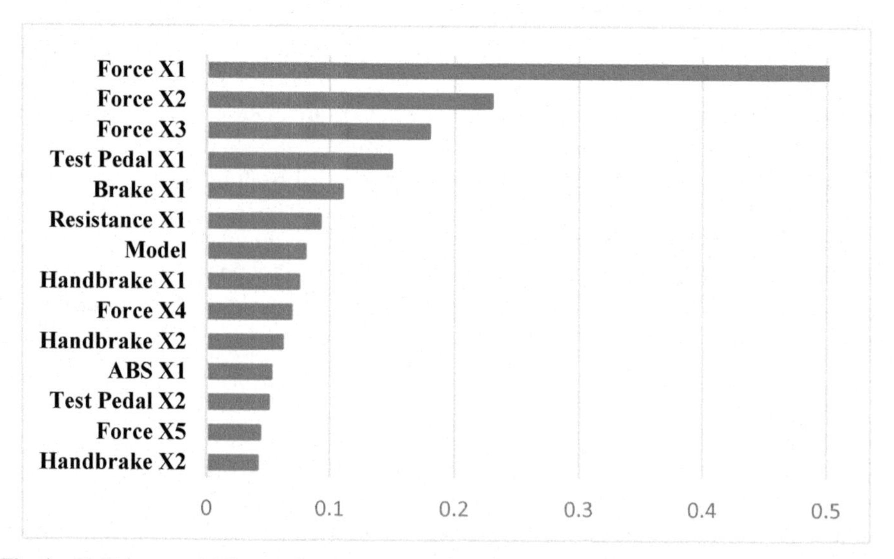

Fig. 1. CatBoost model feature importance graph which shows top 14 important variables out of 68 variables.

4 Conclusion and Future Research

This paper offers a predictive quality model that can detect defect-free approved but faulty products overlooked during the quality inspection operations. ML techniques have been applied for the purpose of analyzing the relationship between quality control data and customer complaints. The last quality stage data of an automobile manufacturer's brake system from 2018 to 2020 are used for the modeling. Specificity and negative predictive value show that GrBoost and CatBoost algorithms have the best classification utility for detecting rare events.

The predictive quality ML studies are limited in the literature. Our study is aimed to contribute to the literature of quality prediction and smart factories by using the quality data of the production lines and final quality data of the brake assembly lines of the automobile manufacturer. Moreover, gradient boost algorithms are applied due to the gap seen in the predictive quality literature.

For future studies, we aim to expand the data set of this study and improve the results. Faulty vehicle prediction models with better predictive power will be tried to be realized by using GrBoost and CatBoost algorithms. In addition, neural network models and the new models are required to be compared for quality estimation. It is desirable to find new variables for quality estimation models in automotive assembly lines and to observe our estimation models.

Acknowledgments. The authors are thankful to Turkish Automobile Factory Joint-Stock Company (TOFAŞ) for their cooperation and their support on this study. Also, the authors would like to mention how grateful they are to Haydar Vural (Data Science and AI Lead at TOFAŞ) for the opportunity of this study.

References

1. Armani, C.G., de Oliveira, K.F., Munhoz, I.P., Akkari, A.C.S.: Proposal and application of a framework to measure the degree of maturity in Quality 4.0: a multiple case study. In: Advances in Mathematics for Industry 4.0, pp. 131–163. Academic Press (2021)
2. Sisodia, R., Villegas Forero, D.: Quality 4.0–how to handle quality in the Industry 4.0 revolution (2019)
3. Radziwill, N.: Let's get digital. Qual. Prog. **51**(10), 24–29 (2018)
4. Li, G., Hou, Y., Wu, A.: Fourth industrial revolution: technological drivers, impacts and coping methods. Chin. Geogr. Sci. **27**(4), 626–637 (2017)
5. Horváth, P., Michel, U.: Industrie 4.0 controlling in the age of intelligent networks. Dream Car of the Dream Factory of the ICV, pp. 13–15 (2015)
6. Dallasega, P., Rauch, E., Linder, C.: Industry 4.0 as an enabler of proximity for construction supply chains: a systematic literature review. Comput. Ind. **99**, 205–225 (2018)
7. Ramezani, J., Jassbi, J.: Quality 4.0 in action: smart hybrid fault diagnosis system in plaster production. Processes **8**(6), 634 (2020)
8. Meyes, R., Donauer, J., Schmeing, A., Meisen, T.: A recurrent neural network architecture for failure prediction in deep drawing sensory time series data. Procedia Manuf. **34**, 789–797 (2019)
9. Sarhan, A.A.: Adaptive neuro-fuzzy approach to predict tool wear accurately in turning operations for maximum cutting tool utilization. IFAC-PapersOnLine **48**(1), 93–98 (2015)
10. Longo, G.A., Mancin, S., Righetti, G., Zilio, C., Ortombina, L., Zigliotto, M.: Application of an Artificial Neural Network (ANN) for predicting low-GWP refrigerant boiling heat transfer inside Brazed Plate Heat Exchangers (BPHE). Int. J. Heat Mass Transf. **160**, 120204 (2020)
11. Turetskyy, A., Wessel, J., Herrmann, C., Thiede, S.: Data-driven cyber-physical system for quality gates in lithium-ion battery cell manufacturing. Procedia CIRP **93**, 168–173 (2020)

12. Ou, X., Huang, J., Chang, Q., Hucker, S., Lovasz, J.G.: First time quality diagnostics and improvement through data analysis: a study of a crankshaft line. Procedia Manuf. **49**, 2–8 (2020)
13. Schmitt, J., Bönig, J., Borggräfe, T., Beitinger, G., Deuse, J.: Predictive model-based quality inspection using machine learning and edge cloud computing. Adv. Eng. Inform. **45**, 101101 (2020)
14. Escobar, C.A., Morales-Menendez, R.: Process-monitoring-for-quality—a robust model selection criterion for the logistic regression algorithm. Manuf. Lett. **22**, 6–10 (2019)
15. Escobar, C.A., Abell, J.A., Hernández-de-Menéndez, M., Morales-Menendez, R.: Process-monitoring-for-quality—big models. Procedia Manuf. **26**, 1167–1179 (2018)
16. Escobar, C.A., Morales-Menendez, R., Macias, D.: Process-monitoring-for-quality—a machine learning-based modeling for rare event detection. Array **7**, 100034 (2020)
17. Hill, T., Lewicki, P., Lewicki, P.: Statistics: methods and applications: a comprehensive reference for science, industry, and data mining. StatSoft, Inc. (2006)
18. Márquez, A.C., de la Fuente Carmona, A., Marcos, J.A., Navarro, J.: Designing CBM plans, based on predictive analytics and big data tools, for train wheel bearings. Comput. Ind. **122**, 103292 (2020)
19. Hancock, J.T., Khoshgoftaar, T.M.: CatBoost for big data: an interdisciplinary review. J. Big Data **7**(1), 1–45 (2020)
20. Fiat Internal Norm 2.00102 & 2.00150/44

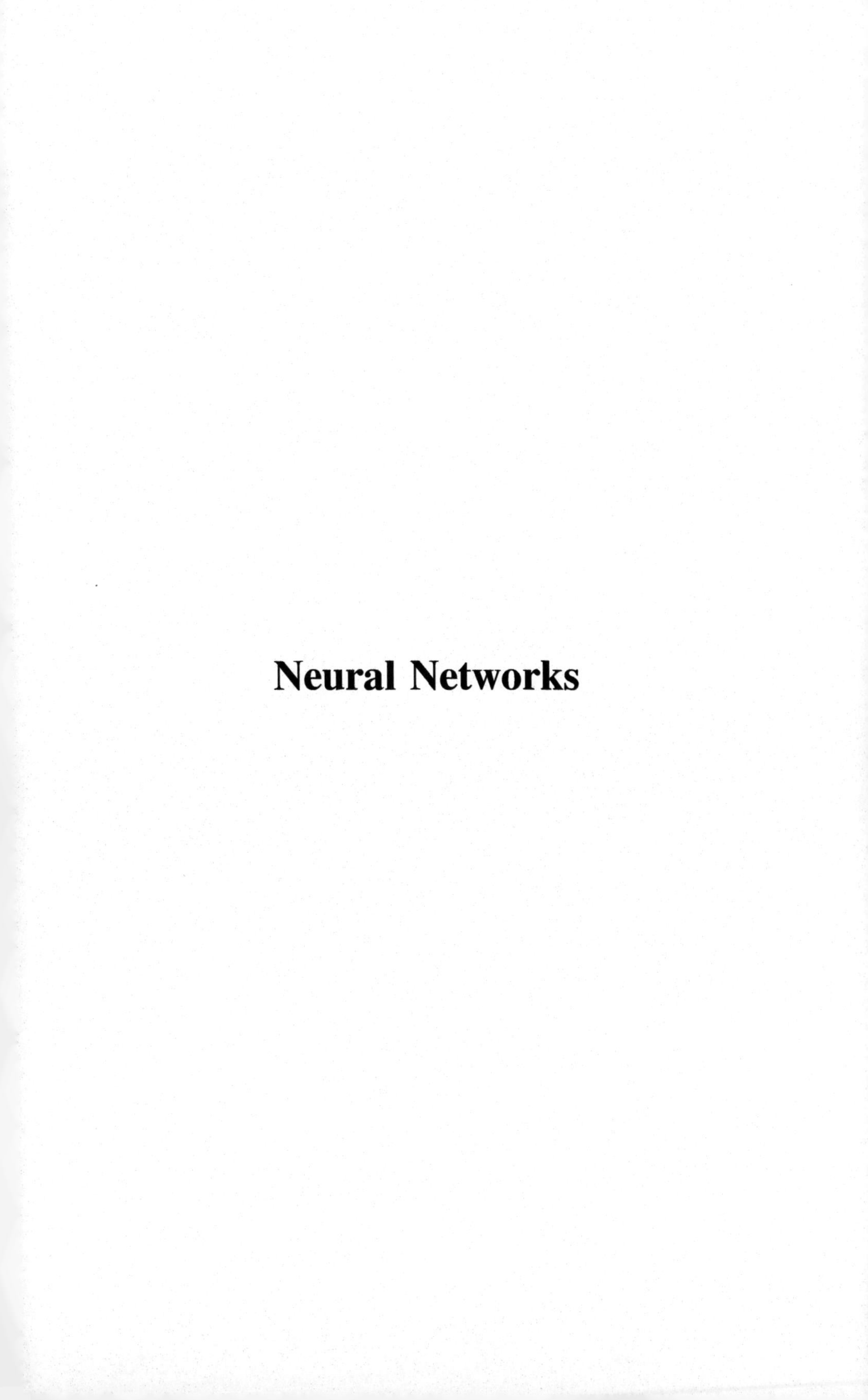

Neural Networks

Fuzzy Neural Networks for Detection Kidney Diseases

Rahib H. Abiyev[1(✉)], John Bush Idoko[1], and Rebar Dara[2]

[1] Applied Artificial Intelligence Research Centre, Department of Computer Engineering, Near East University, Lefkosa, North Cyprus, Turkey
{rahib.abiyev,john.bush}@neu.edu.tr

[2] Department of Computer Engineering, Near East University, Lefkosa, North Cyprus, Turkey

Abstract. This study presents a learning mode-base Fuzzy Neural Networks (FNN) to detect chronic kidney disease (CKD). Combining the fuzzy set theory with the NN structure helps the proposed system to learn sensor data and adjust network parameters. The structure and algorithms of multi-input multi-output FNN are presented. The FNN algorithms implement the TSK type fuzzy rules. The learning of the system is executed by utilizing a gradient descent algorithm and c-means clustering. The presented system is trained using kidney datasets. The performance of the system is evaluated using mean accuracy, sensitivity, specificity and precision which were obtained as 99.75%, 100%, 99.34% and 99.9% correspondingly. The comparison of the results of simulation of the proposed model with the results of other existing algorithms demonstrates the efficiency of the presented FNN model. The experimental results indicate that the approach proposed offers reasonable accuracy of detection and has the potential to be applied in clinical practice.

Keywords: Chronic kidney disease · Fuzzy neural networks · Gradient descent algorithm

1 Introduction

The kidneys are a pair of organs that are responsible for extracting blood, excess water, waste, and salt from the renal arteries. Kidneys are organs that also play a fundamental role in urine and vitamin D metabolism [1, 2]. Kidney tumours develop once kidney cells lose their main function, resulting in the tumours rapidly multiplying. Renal cell carcinoma, also known as kidney cancer, is a type of cancer that arises from the cells of the kidney and can progress rapidly or slowly. This cancer typically develops as a single mass but in one or both kidneys, other tumours may arise. Cancerous cells may fall into the bloodstream at a more advanced stage and damage other organs [3]. This phase is referred to as metastasis and affects over 90% of cases of kidney cancer [2, 4].

One of the most contagious diseases in the world is kidney stone disease. In the starting stage, the stone infections remain unnoticed, thereby harming the kidney as they are produced. Kidney failure is diagnosed in persons with diabetes mellitus, hypertension, glomerulonephritis, etc. Since kidney breakdown can be harmful, it is advisable

C. Kahraman et al. (Eds.): INFUS 2021, LNNS 308, pp. 273–280, 2022.
https://doi.org/10.1007/978-3-030-85577-2_32

to diagnose the problem in the initial stages. One of the current control strategies [5] is ultrasound imaging. In addition to other imaging techniques, such as X-ray, CT, and so on, the ultrasound imaging technique is used in medical practices to produce photographs of live tissue and for the purpose of medical test. The relevant procedures for kidney stone identification are feature extraction and selection. To extract the features, there are several texture characteristics available, namely GLCM characteristics, statistical features, texture characteristics, region-based features and wavelet characteristics, etc. [6, 7]. The method of feature selection increases the precision of the classification and minimizes the complexity of the computation. For the feature selection process [8], a variety of optimization algorithms and machine learning algorithms are currently utilized.

Recently, several machine learning methods, such as Fisher linear Discriminant analysis [9], decision tree, k-nearest neighbour [10], support vector machine [10], multilayer perceptron [10], principal component analysis [11], and radial based function [12] have been applied to images and tumours classification. A recent comparison of the algorithms for classification and feature selection applied to the classification of tumours can be found in [13]. In addition, the methods of artificial neural networks (ANN) offer an enticing solution to the problem of direct multi-category classification [14]. The decision tree classifier with the association rule classification system offers the best choice for the classification of malignant and benign images [15]. In the medicinal field, ultrasound image segmentation is important, tricky and valuable. Due to the result of the usage of ultrasound image segmentation in medical test and treatment applications, it is vital and beneficial. For instance, the first procedure is to measure the stone volume from the available medical image obtained from kidney stone patients. Many segmentation methods are used to segment the region of the stone region in the ultrasound image including, k-means clustering, clustering methods and watershed segmentation. In recent medical diagnosis studies, the Artificial Neural Network (ANN) [16–19] is the most commonly used method for disease management. The human-machine medical diagnostic fields are becoming more popular due to generalization, and the learning capabilities of ANNs [20]. One of the widely used network architectures is the feed-forward networks which makes network communication among the nodes on one layer and those on the next layer more resourceful. The neural networks input parameter is used with a classifier [21, 22] to distinguish between infected instances. The Adaptive Neuro-fuzzy Inference Method (ANFIS) [23] is used for chronic renal failure prediction. For this reason, the fuzzy rules number matches the membership functions number of the input variables. For the recognition of the identification of comparison of system accuracy and glomerular filtration rates, ANFIS frameworks have been used. Significant challenges for clinicians are the effectiveness and frequent timely diagnosis of patients with CKD to reduce the progression of the disease and avoiding unavoidable complications. This research was initialized in a statistical manner to support the decision in the area of CKD management and kidney disease to potentially reduce the burden of medical practitioners. The ANN-SVM for chronic kidney disease prediction was presented in [24]. The optimal parameters are then determined based on numerical analysis using the ANN and SVM [24]. In addition, SVM and ANN Optimized parameters were defined. We are proposing FNN for detection kidney diseases. Our investigation uses the FNN method to identify kidney stone and the best accuracy is demonstrated by the technique proposed in this paper for clinical data classification outcomes.

The remaining part of the paper is organized thus; in Sect. 2, we present the proposed method, Sect. 3 presents the experimental results of the proposed model. Finally, Sect. 4 depicts the conclusion of the paper.

2 Fuzzy Neural Network for Detection of Kidney Diseases

A fuzzy neural network (FNN) is presented for the detection of Kidney diseases. The inputs of fuzzy neural networks are the features of kidney diseases, the outputs are the type of diseases. Using an input-output relationship the kidney diseases can be represented by the If-Then rule base. FNN use neural network structure to represent the fuzzy reasoning process. The design of FNN includes the generation of appropriate IF-Then rule bases in the structure of the network [25, 26]. The problem is the development of premise and consequent parts of fuzzy rules through the training capability of FNN. In the paper, we use TSK fuzzy rules for the design of FNN. The fuzzy rules are represented as:

$$\text{If } x_1 \text{ is } A_{1j} \text{ and } x_2 \text{ is } A_{2j} \text{ and } \ldots \text{ and } x_m \text{ is } A_{mj} \text{ Then } y_j = b_j + \sum_{i=1}^{m} a_{ij}x_i \tag{1}$$

Where A_{ij} are input fuzzy sets, a_{ij} and b_j are coefficients used in linear functions, x_i and y_j are input-output variables. $i = 1..m$, and $j = 1..r$ are the number of inputs and rules correspondingly.

Here the problem is the description of a nonlinear system using linear functions. Using fuzzy rule base the structure of FNN is represented in Fig. 1. The first layer is used for the distribution of input signals. In the second layer the membership grades of incoming input signals are determined using the following formulas:

$$\mu 1_j(x_i) = e^{\frac{(x_i - c_{ij})^2}{\sigma_{ij}^2}}, \quad i = 1 \ldots m,\ j = 1 \ldots r \tag{2}$$

where c_{ij} and σ_{ij} are centers and widths of Gaussian membership functions. $\mu 1_j(x_i)$ are membership functions.

In the rule layer, the outputs are determined using the t-norm min operation.

$$\mu_j(x) = \prod_i \mu 1_j(x_i), \quad i = 1 \ldots m,\ j = 1 \ldots r \tag{3}$$

The outputs of linear functions are calculated as

$$y1_j = b_j + \sum_{i=1}^{m} a_{ij}x_i \tag{4}$$

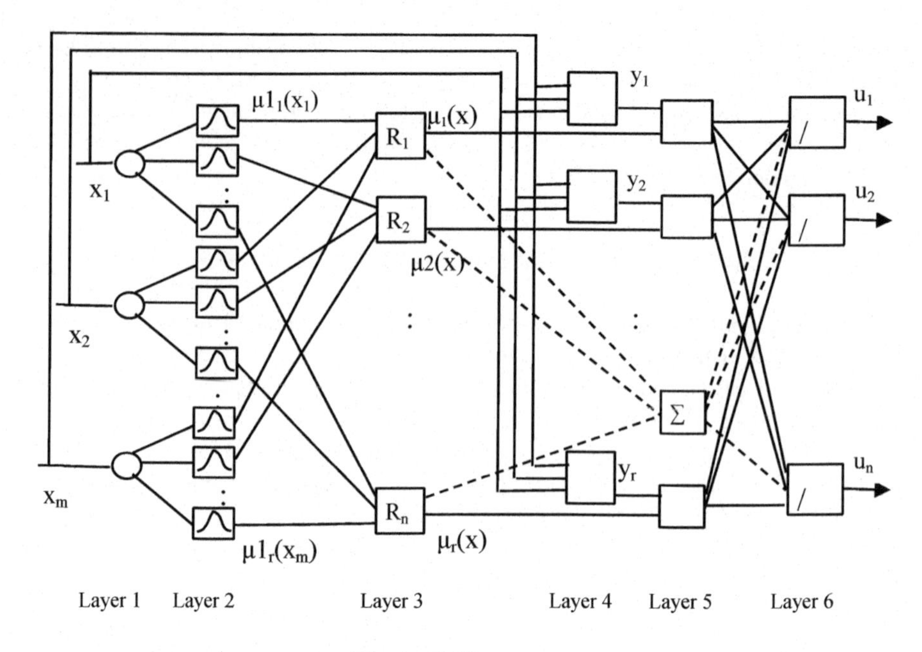

Fig. 1. FNN structure

Using the outputs of the rule layer and outputs of linear functions the output of FNN is determined thus:

$$u_k = \frac{\sum_{j=1}^{r} w_{jk} y_j}{\sum_{j=1}^{r} \mu_j(x)}, \text{ here } y_j = \mu_j(x) y1_j \quad (5)$$

where u_k are FNN outputs, k = 1,…, n.

After finding FNN output signals the training of the network starts. It includes finding appropriate values of the $c_{ij}(t)$, $\sigma_{ij}(t)$ ($i = 1,.., m, j = 1,.., r$) coefficients of membership functions and $w_{jk}(t)$, $a_{ij}(t)$, $b_j(t)$ ($i = 1,.., m, j = 1,.., r, k = 1,.., n$) coefficients of linear functions of FNN structure. In the paper, we apply gradient descent algorithms and fuzzy c-means clustering for finding proper values of FNN parameters [7]. At first, a clustering algorithm is applied to determine the centres of membership functions. Using centres, the membership functions widths are calculated. After these operations gradient descent algorithm is applied to calculate the coefficients of the explored linear functions.

3 Simulations

The constructed FNN structure is used for the detection of kidney diseases. The CKD dataset is downloaded from the UCI machine learning repository. The dataset includes 24 input attributes and output classes. Input attributes consists of 11 numerical and 13 nominal values. The data are collected from 400 instances that include 150 non-CKD patients and 250 CKD patients. At first, the dataset is pre-processed. The dataset has many missing values. In the paper, the missing values are replaced with the numerical mean distributions of the features [24]. To improve the learning accuracy, all input data are scaled in the interval [0, 1]. During simulation to measure network performance, RMSE values are used. The simulation has been done using 10-fold cross-validation. During learning the values of the RMSE are fixed for training, testing and evaluation of the data sets. Simulation of the network is performed by utilizing a series number of rules (hidden neurons). At first, we use 8 fuzzy rules for learning. The training and evaluation errors are obtained as 0.255578 and 0.338906 correspondingly. When we test the model using all data set, the testing error was obtained as 0.344146 and the accuracy rate was 98%. We increased the number of hidden neurons to 12 and 16 afterwards. The best result was obtained using 16 hidden neurons. Figure 2 shows the plot of the values of RMSE obtained during the training of FNN. The accuracy of the model was obtained as 99.75%. The RMSE values for training, evaluation and testing were obtained as 0.163081, 0.169469 and 0.167446 correspondingly (Table 1).

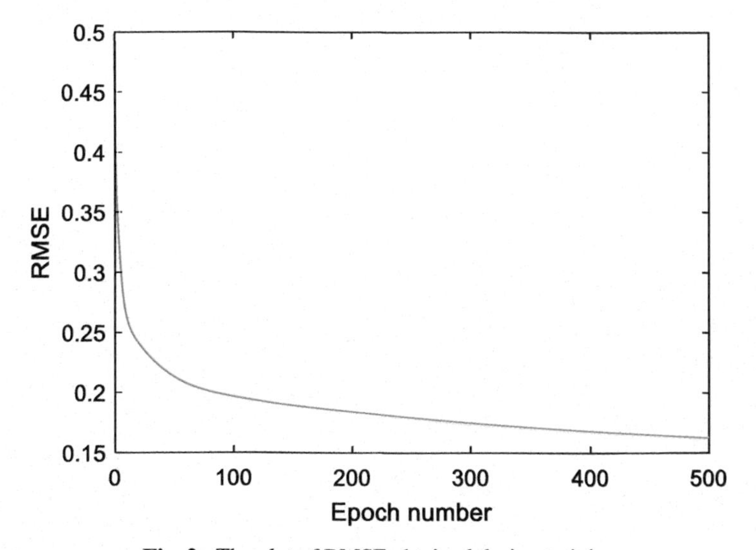

Fig. 2. The plot of RMSE obtained during training

From the results of the simulation, we have obtained the following performance characteristics for the FNN model. The accuracy was 99.75%, sensitivity was 100%, specificity was 99.34%, precision was 99.6%, TP = 249, TN = 150, FP = 1, and FN = 0. The performance of the FNN model is compared with the performances of existing researches used for kidney detection. Table 2 includes comparative results of different models used for the detection of kidney diseases.

The research paper [24] did two simulations: without feature selection and with feature selection. In this research, we did not implement feature selection. As shown the proposed FNN model gives better results than SVM and ANN models without feature selection. The simulation results obtained by the FNN model is the same as the results

Table 1. Simulation results

Number of rules (hidden neurons)	Training		Testing	
	Training error	Evaluation error	Test error	Accuracy
8	0.255578	0.338906	0.344146	0.980000
12	0.195359	0.204160	0.198946	0.995000
16	0.163081	0.169469	0.167446	0.997500

Table 2. Comparative results

Models	Accuracy rate
ANN [24] SVM [24]	99.25% 96.75%
ANN (feature selection) [24]; SVM (feature selection) [24]	99.75% 97.75%
ITLBO-Gradient Boosting [27]; ITLBO-CNN [27]	94.5% 95.25%
FNN model	99.75%

obtained with the SVM, ANN, ITLBO-Gradient Boosting and ITLBO-CNN models that use feature selection. The obtained performances depict the efficiency of the utilization of the FNN model in the detection of kidney diseases.

4 Conclusion

This research proposes the FNN model for the detection of kidney disease. Using fuzzy TSK type rules the construction of FNN has been carried out. The structure and mathematical formulas used for the design of FNN are presented. The developed model is used for the clustering of kidney diseases. Using a different number of fuzzy rules, the simulations of the FNN model is executed. The best performance has been obtained using 16 rules. As a result, mean accuracy, sensitivity, specificity and precision of the system were obtained as 99.75%, 100%, 99.34% and 99.9% correspondingly. The comparison of the simulation results of the proposed model with the results of other existing algorithms demonstrates the efficiency of the presented FNN model in the detection of kidney diseases. Future research is based on the integration of FNN and deep learning for solving different problems in engineering.

References

1. World Cancer Research Fund, Kidney cancer statistics. https://www.wcrf.org/dietandcancer/cancer-trends/kidney-cancer-statistics. Accessed 17 Feb 2021

2. American Cancer Society, What is kidney cancer?. https://www.cancer.org/cancer/kidney-cancer/about/what-is-kidney-cancer. Accessed 17 Feb 2021
3. Shuch, B., et al.: Understanding pathologic variants of renal cell carcinoma: distilling therapeutic opportunities from biologic complexity. Eur. Urol. **67**(1), 85–97 (2015). https://doi.org/10.1016/j.eururo.2014.04.029
4. Ghosn, M., Eid, R., et al.: An observational study to describe the use of sunitinib in real-life practice for the treatment of metastatic renal cell carcinoma. J. Glob. Oncol. **5**, 1–10 (2019). https://doi.org/10.1200/JGO.18.00238
5. Talebi, M., Ayatollahi, A., Kermani, A.: Medical ultrasound image segmentation using genetic active contour. J. Biomed. Sci. Eng. **4**, 105–109 (2011)
6. Osareh, A., Shadgar, B.: A computer aided diagnosis system for breast cancer. Int. J. Comput. Sci. (IJCSI) **8**(2), 233 (2011)
7. Abiyev, R.H., Helwan, A.: Fuzzy neural networks for identification of breast cancer using images' shape and texture features. J. Med. Imaging Health Inform. **8**(4), 817–825 (2018). https://doi.org/10.1166/jmihi.2018.2308
8. Idoko, J.B., Arslan, M., Abiyev, R.: Fuzzy neural system application to differential diagnosis of erythemato-squamous diseases. Cyprus J. Med. Sci. **3**(2), 90–97 (2018). https://doi.org/10.5152/cjms.2018.576
9. Dudoit, S., Fridlyand, J., Speed, T.P.: Comparison of discrimination methods for the classification of tumors using gene expression data. J. Am. Stat. Assoc. **97**(457), 77–87 (2002)
10. Bush, I.J., Arslan, M., Abiyev, R.H.: Intensive investigation in differential diagnosis of erythemato-squamous diseases. In: 13th International Conference on Application of Fuzzy Systems and Soft Computing- ICAFS-2018. Advances in Intelligent Systems and Computing, Warsaw, Poland, 27–28 August 2018, vol. 896, pp. 146–153 (2019). https://doi.org/10.1007/978-3-030-04164-9_21
11. Abiyev, R.H.: Facial feature extraction techniques for face recognition. J. Comput. Sci. **10**(12), 2360–2365 (2014). ISSN:1549-3636
12. Helwan, A., Idoko, J.B., Abiyev, R.H.: Machine learning techniques for classification of breast tissue. In: 9th International Conference on Theory and Application of Soft Computing, Computing with Words and Perception, ICSCCW 2017. Procedia Computer Science (2017)
13. Cho, S., Won, H.: Machine learning in DNA microarray analysis for cancer classification. In: Proceedings of the First Asia-Pacific Bioinformatics Conference on Bioinformatics, pp. 189–198 (2003)
14. Statnikov, A., Aliferis, C.F., Tsamardinos, I., Hardin, D., Levy, S.: A comprehensive evaluation of multicategory classification methods for microarray gene expression cancer diagnosis. Bioinformatics **21**(5), 631–643 (2005)
15. Rajendran, P., Madheswaran, M.: Hybrid medical image classification using association rule mining with decision tree algorithm. J. Comput. **2**(1), 127–136 (2010)
16. Jamison, R.L., et al.: Effect of homocysteine lowering on mortality and vascular disease in advanced chronic kidney disease and end-stage renal disease: a randomized controlled trial. JAMA **298**(10), 1163–1170 (2007)
17. Serpen, A.A.: Diagnosis rule extraction from patient data for chronic kidney disease using machine learning. Int. J. Biomed. Clin. Eng. **5**(2), 64–72 (2016)
18. Akbari, A., et al.: Detection of chronic kidney disease with laboratory reporting of estimated glomerular filtration rate and an educational program. Arch. Intern. Med. **164**(16), 1788–1792 (2004)
19. Polat, H., Mehr, H.D., Cetin, A.: Diagnosis of chronic kidney disease based on support vector machine by feature selection methods. J. Med. Syst. **41**(4), 1–11 (2017)
20. Levey, A.S., Inker, L.A., Coresh, J.: Chronic kidney disease in older people. JAMA **314**(6), 557–558 (2015)

21. Keith, D.S., et al.: Longitudinal follow-up and outcomes among a population with chronic kidney disease in a large managed care organization. Arch. Intern. Med. **164**(6), 659–663 (2004)
22. Hedayati, S.S., et al.: Association between major depressive episodes in patients with chronic kidney disease and initiation of dialysis, hospitalization, or death. JAMA **303**(19), 1946–1953 (2010)
23. Norouzi, J., et al.: Predicting renal failure progression in chronic kidney disease using integrated intelligent fuzzy expert system. Comput. Math. Methods Med. **2016**, 1–9 (2016)
24. Almansour, N.A., et al.: Neural network and support vector machine for the prediction of chronic kidney disease: a comparative study. Comput. Biol. Med. **109**, 101–111 (2019)
25. Ma'aitah, M.K.S., Abiyev, R., Bus, I.J.: intelligent classification of liver disorder using fuzzy neural system. Int. J. Adv. Comput. Sci. Appl. **8**(12), 25–31 (2017). http://dx.doi.org/10.14569/IJACSA.2017.081204
26. Abiyev, R.H.: Credit rating using type-2 fuzzy neural networks. Math. Probl. Eng. **2014** (2014). https://doi.org/10.1155/2014/460916
27. Manonmani, M., Sarojini, B.: Feature selection using improved teaching learning based algorithm on chronic kidney disease dataset. Procedia Comput. Sci. **171**, 1660–1669 (2020)

Improved Harris Hawks Optimization Adapted for Artificial Neural Network Training

Nebojsa Bacanin, Nikola Vukobrat, Miodrag Zivkovic(✉), Timea Bezdan, and Ivana Strumberger

Singidunum University, Danijelova 32, 11000 Belgrade, Serbia
{nbacanin,mzivkovic,tbezdan,istrumberger}@singidunum.ac.rs, nikola.vukobrat.19@singimail.rs

Abstract. The learning process is one of the most difficult problems in artificial neural networks. This process goal is to find the appropriate values for connection weights and biases and has a direct influence on the neural network classification and prediction accuracy. Since the search space is huge, traditional optimization techniques are not suitable as they are prone to slow convergence and getting trapped in the local optima. In this paper, an enhanced harris hawks optimization algorithm is proposed to address the task of neural networks training. Conducted experiments include 2 well-known classification benchmark datasets to evaluate the performance of the proposed method. The obtained results indicate that the devised algorithm has promising performance, as that it is able to achieve better overall results than other state-of-the-art metaheuristics that were taken into account in comparative analysis, in terms of classification accuracy and converging speed.

Keywords: Swarm intelligence · Harris Hawks optimization · Neural networks · Training · Optimization

1 Introduction

During the last two decades, neural networks (NNs), as one of the most popular statistical learning algorithms, became very popular since they can adapt to tackling many real-world problems. More precisely, it's possible to train them to learn distinct correlations between various types of input and output data in many domains such as image classification, weather forecast, text summarization, etc.

One of the biggest challenges to achieve the above-mentioned ability to learn and adapt to diverse problems represents model training. It is relatively hard to construct an algorithm that can help the network in achieving better results (in terms of convergence and final results' quality) and still sustain a decent computational complexity in each iteration.

C. Kahraman et al. (Eds.): INFUS 2021, LNNS 308, pp. 281–289, 2022.
https://doi.org/10.1007/978-3-030-85577-2_33

The architecture of NNs is multi-layered and it consists of the input layer, one or more hidden layers, and the output. The input layer contains the same number of nodes (neurons) as the number of features in the particular dataset. The output layer for prediction problems usually consists of one neuron, while for classification challenges, the number of neurons in the output layer is equal to the number of classes (labels). All neurons are connected with weights and in the case of fully connected NNs, all neurons from the previous layers are connected by weights to all neurons in the next layer. Moreover, a bias term is added to each neuron. Weight and biases determine the influence of each neuron on the final output and their values are typically in the range $[-1, 1]$. When a neuron receives the signal, the activation function is applied to generate an output signal within a certain range. Typical activation functions include relu, sigmoid, and softmax.

For model training are mostly used traditional methods - gradient descent (GD) - stochastic GD and mini-batch GD. However, they are all susceptible to being trapped in the local optima. The goal of the NNs training process for both, classification and prediction tasks, is finding the optimal or sub-optimal set of weight and biases that will generate good results in terms of prediction/classification accuracy for a particular dataset [6]. Since the weight values are continuous, the NNs training challenge belongs to the group NP-hard problems. According to the literature survey, metaheuristic-based approaches, such as swarm intelligence, proved as robust methods for solving these kinds of challenges.

In this research article, an improved version of harris hawks optimization (HHO) swarm intelligence metaheuristics is shown, incorporated as the back-propagation method for NN training. The proposed algorithm was tested on 2 well-known medical machine learning datasets for classification. The basic goal of the presented research is to try to further improve ANNs training by using a robust method instead of traditional back-propagation algorithms and to establish better results in terms of classification accuracy and convergence.

The HHO is one of the most recent metaheuristics, that has shown very powerful optimization capabilities. It has not been applied to the ANN optimization problem. The proposed improved HHO outperforms the original HHO metaheuristics in terms of speed of convergence and solutions' quality. The improved HHO was then utilized to optimize the weights and biases in the ANN.

The remainder of the paper is structured as follows. The overview of the recent literature is provided in Sect. 2. Section 3 describes the proposed enhanced HHO algorithm. Results of conducted experiments and comparative analysis are given in Sect. 4, while Sect. 5 gives final observations, suggests the future work, and concludes the paper.

2 Background

Swarm intelligence algorithm, as a subset of a larger group of nature-inspired metaheuristics, have been able to tackle various difficult problems from many real-life areas and some examples include cloud computing [4,11], wireless sensor networks (WSN) localization, and lifetime maximization [12,15,17], path planning [13] and image processing [5]. Also, it was shown that these algorithms can be successfully used along with machine learning models [16].

By surveying available recent computer science literature source, it can be concluded that many swarm intelligence approaches have been utilized in the area of artificial NNs (ANNs) and convolutional NNs (CNN), precisely three types of applications can be distinguished from this domain: hyperparameters' optimization, feature selection, and ANNs training.

Designing optimal CNN's structure for a particular dataset is referred to as hyperparameter's optimization challenge where swarm intelligence-based methods proved as efficient optimizers [2,3]. Also, it is shown that swarm intelligence can be adopted as wrapper methods for feature selection [8,14]. Finally, there are some state-of-the-art approaches that show applications of swarm intelligence for ANNs training [1,6,9,18].

3 Proposed Metaheuristics

The HHO algorithm is a recent swarm intelligence algorithm proposed in 2019 [7]. It is inspired by different Harris hawks' strategies during their attacks on the prey. These attacking phases can be summarized in three steps: exploration, the transition from exploration to exploitation, and finally exploitation phase. During the exploration phase HHO can change to the exploitation phase and again back to exploration for a different amount of time depending on the strength of the solution (prey energy). Finally, during the exploitation phase hawk attacks the pray.

Due to the space restrictions, the equations that mathematically model these different behaviors of the hawks' pack are omitted here, and they can be found in [7].

Notwithstanding the outstanding performance of basic HHO [7], by conducting simulations with standard Congress on Evolutionary Computation (CEC) bound-constrained benchmarks, it was noted that the basic HHO can be further improved by addressing both processes - exploitation and exploration. It was proven that by incorporating quasi-reflection-based learning (QRL) procedure, intensification and diversification can be enhanced [10].

In the proposed approach, the QRL is applied in each iteration to the current worst (X_{worst}) and best solution (X_{best}). Following this approach, quasi-reflective solutions are generated by using the following expression:

$$X_j^{qr} = \text{rnd}\left(\frac{LB_j + UB_j}{2}, X_j\right) \tag{1}$$

where the mean of the lower bound and upper bound for each parameter j is calculated by $\frac{LB_j + UB_j}{2}$ and $\text{rnd}\left(\frac{LB_j + UB_j}{2}, x_j\right)$ generates random number from uniform distribution in range $\left[\frac{LB_j + UB_j}{2}, X_j\right]$.

Finally, X_{worst} is replaced by X^{qr}_{worst} and the X_{best} is replaced with X^{qr}_{best} only if the fitness of X^{qr}_{best} is better than the original current best solution.

The proposed method is named quasi-reflective HHO (QRHHO) and its pseudo-code is shown in Algorithm 1. The QRHHO uses 2 more function evaluations in each iteration, however as it is shown in Sect. 4, performance improvements over the basic HHO are substantial.

Algorithm 1. Proposed QRHHO pseudo-code

```
Inputs: The population size N and maximum number of iterations T
Outputs: The location of the best solution and its fitness value
Initial the random population X_i, (i = 1, 2, 3, ...N)
while stopping condition has not been met do
  Calculate the fitness values of hawks
  Set X_best as the location of the rabbit (best location)
  for each solution X_i do
    Update the initial energy E_0 and jump strength J : E_0 = 2rand() − 1, J = 2(1 − rand())
    Update E
    if |E| ≥ 1 then
      Exploration phase
      Update the location vector
    end if
    if |E| < 1 then
      Exploitation phase
      if r ≥ 0.5 and |E| ≥ 0.5 then
        Soft besiege
        Update the location vector by soft besiege
      else if r ≥ 0.5 and |E| < 0.5 then
        Hard besiege
        Update the location vector by using hard besiege
      else if r < 0.5 and |E| ≥ 0.5 then
        Soft besiege with progressive rapid dives
        Update the location vector by using soft besiege with rapid dives
      else if r < 0.5 and |E| < 0.5 then
        Hard besiege with progressive rapid dives
        Update the location vector using hard besiege with progressive rapid dives
      end if
    end if
  end for
  Generate X^qr_best and X^qr_worst solutions by applying Eq. (1)
  Replace X_worst and X_best
end while
Return X_best
```

In the shown pseudo-code, E denotes energy. For other algorithm control parameters' please refer to [7].

4 Experiments and Results

For the purpose of the experiment, basic HHO and QRHHO are implemented in Python with pandas, numpy and scikit-learn libraries. Two well-known medical datasets, namely breast cancer and SAheart are used for evaluation and comparative analysis. The breast cancer dataset contains 9 features with 699 instances categorized in two classes - benign and cancer diagnosis. Similarly, the SAheart dataset is composed of 462 instances with 9 features and two labels that indicate whether the patient has the disease or not.

The data from both datasets are preprocessed: missing feature data is replaced with the respective averages and all the values are scaled with StandardScaler scikit-learn class. The 2/3 of the entire dataset is used for training while the remaining 1/3 is used for testing. In both experiments accuracy (*acc.*), specificity (*spec.*), sensitivity (*sens*), geometric mean (g_{mean}), and area under the curve (AUC) metrics are used to evaluate algorithm performance. Experiments are conducted in 30 runs and best, mean, and worst results are reported. Since the proposed QRHHO uses 2 more function evaluations in each iteration, QRHHO was tested with 48 individuals in the population, while the HHO population consists of 50 solutions. Each run is executed in 250 iterations. All other HHO and QRHHO control parameters are set as suggested in [7].

The ANN which is trained consists of one hidden layer. As noted above, individual represents the values of weight and biases that connect neurons. The size (length) of candidate solution vector (s_l) is calculated as: $s_l = (s_i \cdot s_h + s_h) + (s_h \cdot s_o + s_o)$, where s_i is the size of input feature vector, s_h is the number of neurons in the hidden layer, and s_o is the number of hidden units in the output layer. The s_h is derived as $2 \cdot s_i + 1$. The binary_crossentropy is used for the loss function. The same experimental conditions were used in [6].

Results, as well as comparative analysis with other state-of-the-art approaches, are summarized in Tables 1 and 2. The QRHHO performance is compared with grasshopper optimization algorithm (GOA), artificial bee colony (ABC), genetic algorithm (GA), bat algorithm (BA), firefly algorithm (FA), flower pollination algorithm (FPA), biogeography-based optimization (BBO) and the monarch butterfly optimization (MBO). Results for HHO and QRHHO were obtained in this research, while the results of other methods were retrieved from [6].

From the results tables it is clear that on average, the proposed QRHHO establishes the best metric values than all state-of-the-art methods included in the analysis. Only in the case of sensitivity metric for the breast cancer dataset, the QRHHO is outscored by other approaches. Further, performance enhancements of QRHHO over the original HHO for accuracy metric and the value of loss function for one random run for breast cancer and SAheart datasets are shown in Fig. 1 and Fig. 2, respectively. Results showed in the figures are generated by using training data.

From depicted figures, it can be seen that the original HHO does not manage to converge, while the QRHHO converges relatively fast. Also, it can be noted that the basic HHO does not manage to find the proper part of the search space. In this way, by using the QRL mechanism, the QRHHO improves both - exploitation and exploration of the original HHO.

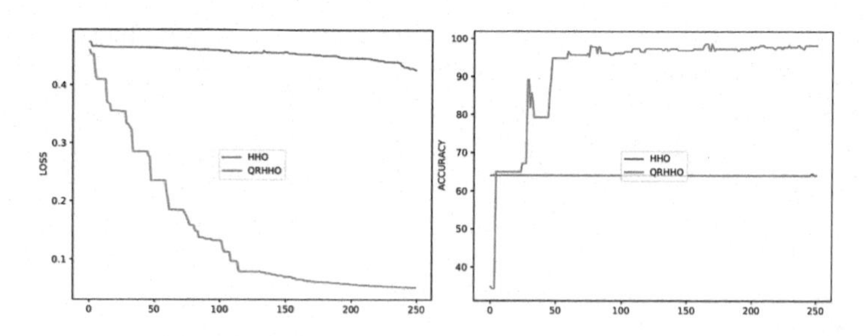

Fig. 1. Loss and accuracy for Breast cancer dataset - QRHHO vs. HHO

Table 1. Breast cancer dataset results

Method	Result	*Acc.*(%)	*Spec.*(%)	*Sens.*(%)	G_{mean}	*AUC*
QRHHO	Best	**98.575**	**100.000**	98.253	0.98423	**0.99943**
	Worst	96.243	97.352	97.365	**0.96698**	0.98559
	Mean	**97.613**	**98.792**	97.615	**0.97481**	**0.99605**
GOA	Best	97.899	97.531	98.726	0.97810	0.99693
	Worst	95.798	92.593	96.178	0.94991	0.99229
	Mean	97.115	95.309	98.047	0.96665	0.99536
ABC	Best	98.319	**100.000**	98.726	**0.98718**	0.99772
	Worst	94.958	91.358	94.904	0.94047	0.92593
	Mean	96.891	96.214	97.240	0.96713	0.98544
GA	Best	97.479	97.531	98.726	0.97492	0.99709
	Worst	95.378	90.123	**97.452**	0.94022	0.99418
	Mean	96.751	94.362	97.983	0.96151	0.99549
PSO	Best	97.899	98.765	**99.363**	0.98107	0.99756
	Worst	95.378	90.123	96.815	0.94022	0.99402
	Mean	97.045	94.774	**98.217**	0.96471	0.99558
BAT	Best	98.319	98.765	**99.363**	0.98127	0.99568
	Worst	92.437	79.012	96.178	0.88605	0.89188
	Mean	96.218	93.457	97.643	0.95494	0.97079
FA	Best	97.899	98.089	97.531	0.97810	0.99678
	Worst	**96.639**	**98.089**	93.827	0.95935	**0.99575**
	Mean	97.311	98.089	95.802	0.96938	0.99619
FPA	Best	98.319	98.765	98.726	0.98427	0.99670
	Worst	95.378	90.123	**97.452**	0.94022	0.99292
	Mean	97.241	95.432	98.174	0.96789	0.99511
BBO	Best	98.319	97.531	98.726	0.98127	0.99670
	Worst	**96.639**	93.827	**97.452**	0.95935	0.99355
	Mean	97.255	95.514	98.153	0.96822	0.99550
MBO	Best	97.899	98.765	**99.363**	0.98107	0.99693
	Worst	94.958	87.654	95.541	0.93026	0.98616
	Mean	96.695	94.074	98.047	0.96031	0.99453

Table 2. Results of the SAheart dataset

Method	Result	*Acc.*(%)	*Spec.*(%)	*Sens.*(%)	G_{mean}	*AUC*
QRHHO	Best	**79.412**	83.861	87.625	**0.74623**	**0.82614**
	Worst	**71.578**	**78.928**	**80.641**	**0.67658**	**0.77623**
	Mean	**74.318**	80.122	81.224	0.69545	**0.79611**
GOA	Best	79.114	57.407	**91.346**	0.71238	0.78793
	Worst	67.722	42.593	79.808	0.58653	0.72685
	Mean	73.122	49.383	**85.449**	0.64913	0.75555
ABC	Best	76.582	**95.192**	61.111	0.71909	0.81250
	Worst	67.722	72.115	29.630	0.53109	0.66560
	Mean	71.160	82.276	49.753	0.63644	0.74454
GA	Best	75.949	94.231	55.556	0.67792	0.78241
	Worst	68.354	75.962	40.741	0.61048	0.73326
	Mean	71.814	82.372	51.481	0.65030	0.76671
PSO	Best	77.848	**95.192**	61.111	0.68990	0.79897
	Worst	64.557	70.192	38.889	0.57689	0.71546
	Mean	72.658	**85.096**	48.704	0.64126	0.76022
BAT	Best	75.949	89.423	59.259	0.67779	0.78846
	Worst	68.987	76.923	42.593	0.61715	0.74252
	Mean	72.405	83.654	50.741	0.65086	0.76642
FA	Best	74.051	55.556	84.615	0.68172	0.77902
	Worst	69.620	48.148	79.808	0.62361	0.76086
	Mean	71.730	51.667	82.147	0.65137	0.77276
FPA	Best	76.582	93.269	64.815	0.71050	0.80520
	Worst	68.354	77.885	38.889	0.59377	0.72489
	Mean	72.869	84.231	50.988	0.65336	0.76480
BBO	Best	75.316	91.346	61.111	0.69696	0.78775
	Worst	69.620	78.846	40.741	0.60359	0.75036
	Mean	72.911	83.910	51.728	0.65776	0.77369
MBO	Best	76.582	89.423	62.963	0.70408	0.79790
	Worst	68.354	75.000	40.741	0.59706	0.71599
	Mean	72.932	83.782	52.037	0.65881	0.76113

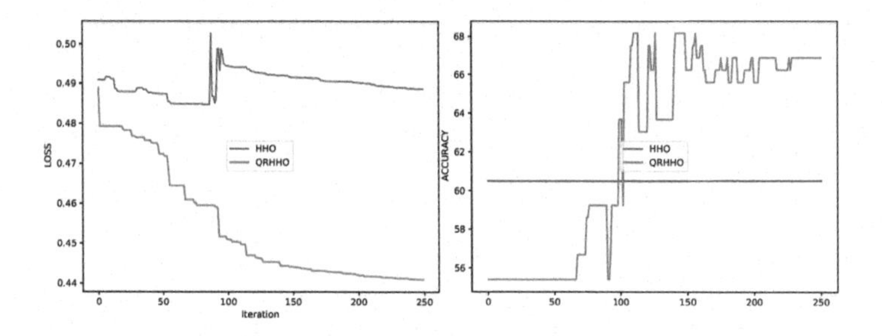

Fig. 2. Loss and accuracy for SAheart dataset - QRHHO vs. HHO

5 Conclusion

This paper proposes an improved HHO algorithm with QRL procedure (QRHHO) that addresses exploitation and exploration deficiencies of the original approach. The method was evaluated for ANN training and two well-known datasets were used. By obtained performance metric it can be clearly seen that the QRHHO significantly outscores original HHO, as well as other state-of-the-art methods that were considered in comparative analysis. Moreover, for visual comparison of loss function and accuracy, it is concluded that the QRHHO substantially improves slow converging and low exploration capabilities of original HHO.

As part of the future research in this domain, QRHHO will be tested on other challenges from the machine learning area and also it will be adapted for tackling other NP hard challenges.

Acknowledgements. The paper is supported by the Ministry of Education, Science and Technological Development of Republic of Serbia, Grant No. III-44006.

References

1. Agrawal, U., Arora, J., Singh, R., Gupta, D., Khanna, A., Khamparia, A.: Hybrid wolf-bat algorithm for optimization of connection weights in multi-layer perceptron. ACM Trans. Multimed. Comput. Commun. Appl. (TOMM) **16**(1s), 1–20 (2020)
2. Bacanin, N., Bezdan, T., Tuba, E., Strumberger, I., Tuba, M.: Monarch butterfly optimization based convolutional neural network design. Mathematics **8**(6), 936 (2020)
3. Bezdan, T., Zivkovic, M., Tuba, E., Strumberger, I., Bacanin, N., Tuba, M.: Glioma brain tumor grade classification from MRI using convolutional neural networks designed by modified FA. In: International Conference on Intelligent and Fuzzy Systems, pp. 955–963. Springer (2020)
4. Bezdan, T., Zivkovic, M., Tuba, E., Strumberger, I., Bacanin, N., Tuba, M.: Multi-objective task scheduling in cloud computing environment by hybridized bat algorithm. In: International Conference on Intelligent and Fuzzy Systems, pp. 718–725. Springer (2020)
5. Brajevic, I., Tuba, M., Bacanin, N.: Multilevel image thresholding selection based on the cuckoo search algorithm. In: Proceedings of the 5th International Conference on Visualization, Imaging and Simulation (VIS 2012), Sliema, Malta, pp. 217–222 (2012)
6. Heidari, A.A., Faris, H., Aljarah, I., Mirjalili, S.: An efficient hybrid multilayer perceptron neural network with grasshopper optimization. Soft Comput. **23**(17), 7941–7958 (2019)
7. Heidari, A.A., Mirjalili, S., Faris, H., Aljarah, I., Mafarja, M., Chen, H.: Harris hawks optimization: algorithm and applications. Future Gener. Comput. Syst. **97**, 849–872 (2019)
8. Hussien, A.G., Oliva, D., Houssein, E.H., Juan, A.A., Yu, X.: Binary whale optimization algorithm for dimensionality reduction. Mathematics**8**(10), 1821 (2020). https://doi.org/10.3390/math8101821, https://www.mdpi.com/2227-7390/8/10/1821

9. Milosevic, S., Bezdan, T., Zivkovic, M., Bacanin, N., Strumberger, I., Tuba, M.: Feed-forward neural network training by hybrid bat algorithm. In: Modelling and Development of Intelligent Systems: 7th International Conference, MDIS 2020, Sibiu, Romania, October 22–24, 2020, Revised Selected Papers 7, pp. 52–66. Springer (2021)
10. Rahnamayan, S., Tizhoosh, H.R., Salama, M.M.A.: Quasi-oppositional differential evolution. In: 2007 IEEE Congress on Evolutionary Computation, pp. 2229–2236 (2007). https://doi.org/10.1109/CEC.2007.4424748
11. Strumberger, I., Bacanin, N., Tuba, M., Tuba, E.: Resource scheduling in cloud computing based on a hybridized whale optimization algorithm. Appl. Sci. **9**(22), 4893 (2019)
12. Strumberger, I., Minovic, M., Tuba, M., Bacanin, N.: Performance of elephant herding optimization and tree growth algorithm adapted for node localization in wireless sensor networks. Sensors **19**(11), 2515 (2019)
13. Tuba, E., Strumberger, I., Zivkovic, D., Bacanin, N., Tuba, M.: Mobile robot path planning by improved brain storm optimization algorithm. In: 2018 IEEE Congress on Evolutionary Computation (CEC), pp. 1–8 (July 2018). https://doi.org/10.1109/CEC.2018.8477928
14. Tuba, M., Bacanin, N.: Artificial bee colony algorithm hybridized with firefly algorithm for cardinality constrained mean-variance portfolio selection problem. Appl. Math. Inf. Sci. **8**(6), 2831 (2014)
15. Zivkovic, M., Bacanin, N., Tuba, E., Strumberger, I., Bezdan, T., Tuba, M.: Wireless sensor networks life time optimization based on the improved firefly algorithm. In: 2020 International Wireless Communications and Mobile Computing (IWCMC), pp. 1176–1181. IEEE (2020)
16. Zivkovic, M., et al.: Covid-19 cases prediction by using hybrid machine learning and beetle antennae search approach. Sustain. Cities Soc. **66**, 102669 (2021)
17. Zivkovic, M., Bacanin, N., Zivkovic, T., Strumberger, I., Tuba, E., Tuba, M.: Enhanced grey wolf algorithm for energy efficient wireless sensor networks. In: 2020 Zooming Innovation in Consumer Technologies Conference (ZINC), pp. 87–92. IEEE (2020)
18. Zivkovic, M., et al.: Hybrid genetic algorithm and machine learning method for covid-19 cases prediction. In: Proceedings of International Conference on Sustainable Expert Systems: ICSES 2020, vol. 176, p. 169. Springer (2021)

Deep Learning Neural Network Architecture for Human Facial Expression Recognition

Sangaraju V. Kumar and Jaeho Choi(✉)

Department of Electronics Engineering, CAIIT, JBNU, Chonju, Republic of Korea
wave@jbnu.ac.kr

Abstract. Facial Expression recognition (FER) is a vital field of artificial intelligence and computer vision owing to its remarkable academic potential. Recognition of facial expression has many implicit applications, and it has attracted much attention of researchers during the last decade. The relation between the facial features to each human emotion is the main focus of the investigation. By training a neural network with such features, one can realize a system that can provide analytical data on human behaviors. This paper proposes a facial expression recognition system architecture based on a deep convolutional neural network. In the course, various neural networks are studied and compared on their facial expression recognition, and one can find that the CNN deep learning method can be one of the best options to take. The paper includes human facial emotion detection procedures that include three significant steps: face recognition; characteristic feature extraction; emotion classification. A set of experiments is performed to evaluate the proposed system's performance, and the recognition accuracy is measured in terms of the facial emotion recognition challenge (FERC-2013) and Japanese female facial expression (JAFFE) dataset.

Keywords: Human facial emotion recognition · Deep learning · Convolutional neural network

1 Introduction

Facial emotion is an important feature to find out the mood of a person's behavior in certain situations. Humans can convey their feelings in two ways, either verbal or non-verbal communication. Eight emotions can be considered as the universal expressions such as happy, sad, neutral, angry, surprise, disgust, fear and contempt [1]. Using computing techniques to recognize the emotion is a challenging task due to the similarities of actions, and there is also not a large dataset available for training the images. As quantity and variety of datasets build up and become available, deep learning has begun to establish itself as one of the mainstream techniques in all computer vision tasks. Especially, facial expression recognition (FER) is an active research area in the field of artificial intelligence; it is applied in vast domains, which include monitoring, security, marketing, entertainment, e-learning, medicine and robots. Several techniques have been developed in regards; however, most current works focus on hand-engineered features [2, 3]. The

C. Kahraman et al. (Eds.): INFUS 2021, LNNS 308, pp. 290–297, 2022.
https://doi.org/10.1007/978-3-030-85577-2_34

major purpose of this proposed scheme is to discover the standardized percentages of numerous emotional states (happy, sad, neutral, angry, surprise, disgust, worry and contempt) in a face. The overall performance of a neural community mainly relies upon on several troubles like preliminary random weights, activation function used, education data, and range of hidden layer and community shape of system. The paper is organized as follows. Section 2 provides the related work. Datasets and pre-processing work are described in Sect. 3. The proposed deep learning model is illustrated in Sect. 4. In Sect. 5, details of the experiment and evaluation of the proposed model are presented. Finally, the conclusions are made in Sect. 6.

2 Related Work

Jain has used a hybrid deep neural network such as convolutional neural network and recurrent neural network model to recognize face emotion from the JAFFE and MMI datasets [4]. The system gives an accuracy of 94.91% on JAFFE and 92.07% on the MMI dataset, respectively. Zhang has proposed a convolutional neural network model that uses the maximum pooling method to reduce the dimension of the extracted implicit features [5]. Fer-2013, with the combination of the LFW dataset, was used and achieved the average recognition rate of 88.56%. Jain has also used CK+, JAFFE dataset, and single deep convolutional neural networks (DNNs), which contain convolutional layers and deep residual blocks that classify different facial emotions [6]. When compared with six different existing models, and their proposed model has yielded accuracies of 95.23% on JAFFE and 93.24% on the CK+ dataset, respectively. Zhang has also proposed two facial expression recognition models such as double-channel weighted mixture deep convolution neural networks (WMDCNN) and deep CNN long short-term memory networks of a double-channel weighted mixture (WMCNN-LSTM) [7]. Four datasets, CK+, JAFFE, Oulu-CASIA and MMI are used, and the WMDCNN model achieves the accuracy of 0.985, 0.923, 0.86, 0.78; the WMCNN-LSTM has yielded accuracies of 0.975, 0.88, 0.87 respectively. Franzoni has demonstrated an enhancing mouth-based emotion recognition using transfer learning approach [8]. The model uses Adam and SGD optimization techniques. The Adam technique has performed better with Inception V3, and VGG16; the SGD technique performed better with InceptionResNetV2. Furthermore, Kaviya has used a convolutional neural network to recognize facial emotions using the Haar filter [9]. The model is validated using FER-2013 and JAFFE datasets.

3 Datasets and Pre-processing

3.1 Datasets

To train a neural network, a vast amount of labelled data must handle the curse of dimensionality. Several facial expression datasets are publicly available such as Facial Emotion Recognition-2013 (FER-2013) and Japanese Female Facial Expression (JAFFE) datasets are used in this paper. Table 1 shows that the FER-2013 dataset splits 90% on training and 10% on testing. The JAFFE dataset splits 85% on training and 15% on testing.

Table 1. Training and testing split dataset details

Dataset	Number of images	Training	Testing
FER-2013	35,887	90%	10%
JAFFE	213	85%	15%

4 Proposed Deep Learning System Model

In this section the proposed deep learning system model is described.

4.1 Deep Convolutional Neural Network Architecture

The architecture for the proposed deep convolutional neural network model is shown in Fig. 1. The model uses six two-dimensional convolutional layers, with each layer followed by a batch normalization. The normalization layer is used here to improve the better results of the model. The input image is resized to 48 × 48 and is given to the first convolutional layer. The output from the convolutional layer, called the feature map, is passed through an activation function. The activation function used here is ELU (exponential linear unit) that makes better generalization performance than ReLU (rectified linear unit), and does not have a vanishing and exploding gradients problem. This feature map is passed to the max-pooling layer of pool size 2 × 2 to minimize the size without losing any information. This process continues in the next convolutional layer again as well. The dropout layer is used to reduce the overfitting problem. Finally, a 2-D array is created with some feature values. The flattening layer is used to convert the 2-D array to a single-dimensional vector to give it as the neural network input. The dense layer represents them. The learning rate 0.001 is set for optimization technique.

In the proposed system model, a two-layer neural network is used; one is input, and the other is output. The output layer has 3 units for FER-2013 and 6 units for the JAFFE datasets. The activation function used in the output layer is SoftMax, which produces the class's probabilistic output.

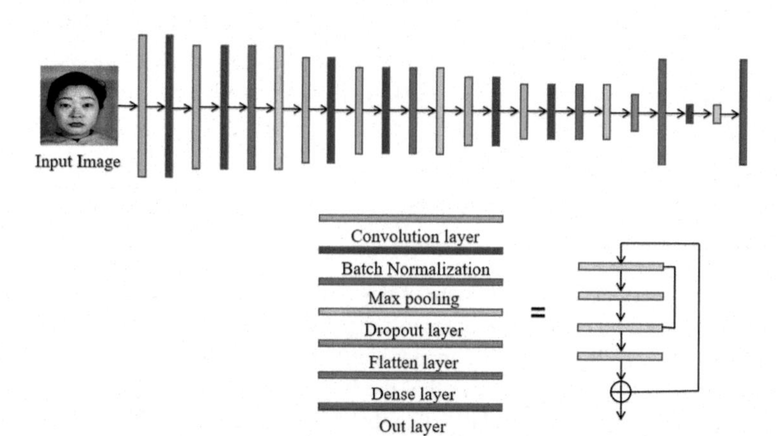

Fig. 1. The architecture of the proposed DCNN model.

Figure 2 illustrates the accuracy and loss of the proposed model for the FER-2013 dataset and Fig. 3 for the JAFFE dataset. From the figures, one can observe that the model is not overfitting on the JAFFE dataset. However, In the FER-2013 dataset, the model starts overfitting at the end. The epoch history shows that accuracy gradually increases and achieved almost 84% on FER and 92% on JAFFE training and validation sets. The classification results of the model based on precision, recall and F1-score are provided in Tables 3 and 4. As a note, the Nadam optimization algorithm is applied to avoid first order optimization problem like exploding gradients that have been observed in recurrent neural networks [12].

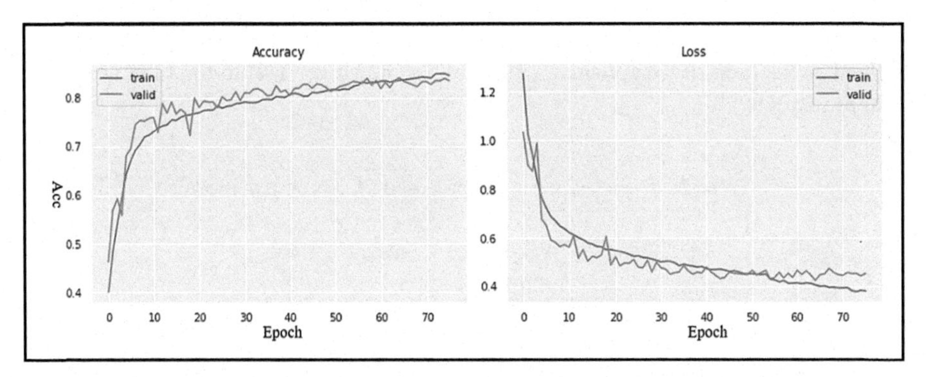

Fig. 2. The accuracy and loss of the proposed model for the FER-2013 dataset

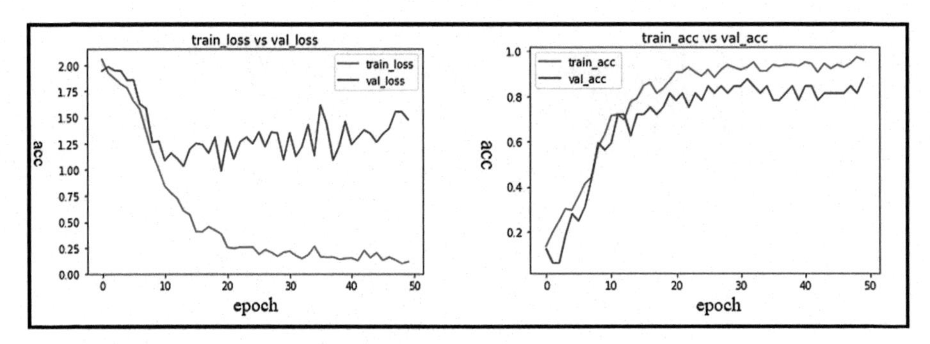

Fig. 3. The loss and accuracy of the proposed model for the JAFFE dataset

5 Experiments and Results

The proposed method runs on a standard PC with Intel ® core ™ i7-9700k, 32 GB of RAM and dedicated 8 GB NVIDIA GeForce RTX 3070. The best models and weights are saved in HDF5 format.

5.1 Methodology

The entire model is simulated in PyCharm (IDE) and using Python as a programming language. Keras, which is used as deep learning library it runs on top of TensorFlow and Scikit-learn is the library used for finding the confusion matrix that gives the precision, recall and F1-score of the model. Matplotlib and seaborn are used for plotting the confusion matrix and other graphs.

Proposed Model Accuracy

Table 2 summarizes the accuracy of the proposed model, which includes three classes of emotions using FER-2013 and six classes using the JAFFE dataset. The proposed DCNN gives an accuracy of 84% for FER-2013 and 91% for JAFFE datasets. The accuracy is defined as the ratio of true positive (TP) and true negative (TN) to the total number of images inspected [9].

Table 2. The accuracy of the proposed DCNN system model.

Datasets	Validation accuracy	Testing accuracy
FER-2013	84.47%	84%
JAFFE	92.22%	91%

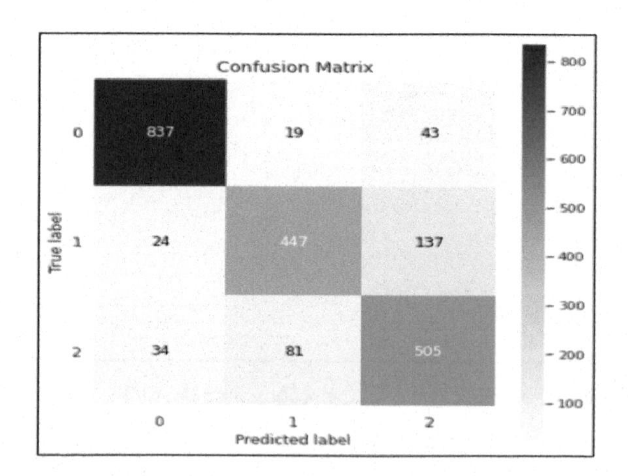

Fig. 4. Confusion matrix for FER-2013 dataset.

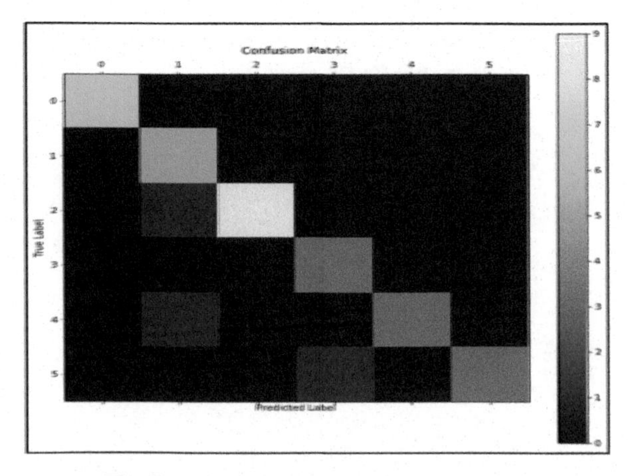

Fig. 5. Confusion matrix for JAFFE dataset.

Evaluation Metrics

Precision, recall, and F1-score are the evaluation metrics used to calculate the performance of DCNN model.

a) Precision

The precision is defined as the ratio of true positive divided by the sum of true positive and false positive.

$$\text{Precision} = \frac{TP}{TP + FP} \tag{1}$$

b) Recall

The recall is defined as the ratio of true positive is divided by the true positive and false negative.

$$\text{Recall} = \frac{TP}{TP + FN} \tag{2}$$

c) F1-Score

The F1-score is defined as the ratio of the product of precision and recall to the sum of precision and recall [13].

$$\text{F1-score} = 2 * \frac{Precision * recall}{precision + recall} \tag{3}$$

The confusion matrices of the proposed DCNN model on the FER-2013 dataset is presented in Fig. 4. In this dataset, we were interested in only three classes and happy, sad and neutral labels; the class happy is predicted as the best emotion. The most misclassified images are from sad and neutral emotions because of the fewer data and similar feelings. The performance on the JAFFE dataset presented in Fig. 5 shows almost all the labels are predicted correctly. Table 3 and Table 4 show the precision, recall, and F1-score of the FER-2013 and JAFFE datasets.

Table 3. Precision, recall and F1-score of FER-2013 dataset.

Labels	Precision	Recall	F1-score
0 – Happy	0.94	0.93	0.93
1 – Sad	0.82	0.74	0.77
2 – Neutral	0.74	0.81	0.78
Average	0.84	0.83	0.83

Table 4. Precision, recall and F1-score of JAFFE dataset.

Labels	Precision	Recall	F1-score
0 – Angry	1.00	1.00	1.00
1 – Disgust	0.75	1.00	0.86
2 – Fear	1.00	0.90	0.95
3 – Happy	0.80	1.00	0.89
4 – Neutral	1.00	0.80	0.89
5 – Sad	1.00	0.80	0.89
Average	0.93	0.92	0.91

6 Conclusion

In this paper, we have presented a deep convolutional neural network system model for facial expression recognition, and the performance of the proposed system has been evaluated on two public datasets. On the JAFFE dataset it has achieved a better accuracy, but for the FER-2013 dataset, the proposed system has slightly overfitted at the end of a model. In future work, we will improve the pre-processing and feature extraction process to improve the better accuracy and to overcome the overfitting problem. Data augmentation can be a step to take toward this problem and we will also extend this work to real-time video sequence datasets.

References

1. Duncan, D., Shine, G., English, C.: Facial emotion recognition in real time. Computer Science (2016)
2. Kahou, S.E., Froumenty, P., Pal, C.: Facial Expression Analysis Based on High Dimensional Binary Features. Springer, Cham (2015)
3. Shan, C., Gong, S., McOwan, P.W.: Facial expression recognition based on local binary patterns: a comprehensive study. Image Vis. Comput. **27**(6), 803–816 (2009)
4. Jain, N., et al.: Hybrid deep neural networks for face emotion recognition. Pattern Recognit. Lett. **115**, 101–106 (2018)
5. Zhang, H., Jolfaei, A., Alazab, M.: A face emotion recognition method using convolutional neural network and image edge computing. IEEE Access **7**, 159081–159089 (2019)

6. Jain, D.K., Shamsolmoali, P., Sehdev, P.: Extended deep neural network for facial emotion recognition. Pattern Recognit. Lett. **120**, 69–74 (2019)
7. Zhang, H., Huang, B., Tian, G.: Facial expression recognition based on deep convolution long short-term memory networks of double-channel weighted mixture. Pattern Recognit. Lett. **131**, 128–134 (2020)
8. Franzoni, V., et al.: Enhancing mouth-based emotion recognition using transfer learning. Sensors (Basel) **20**(18), 5222 (2020)
9. Kaviya, P., Arumugaprakash, T.: Group facial emotion analysis system using convolutional neural network. In: 2020 4th International Conference on Trends in Electronics and Informatics (ICOEI) (48184) (2020)
10. Muttu, Y., Virani, H.: Effective face detection, feature extraction & neural network based approaches for facial expression recognition, pp. 102–107 (2015)
11. Cui, R., Liu, M., Liu, M.: Facial Expression Recognition Based on Ensemble of Mulitple CNNs. Springer, Cham (2016)
12. Dozat, T.: Incorporating nesterov momentum into Adam (2016)
13. Powers, D.J.A.: Evaluation: from precision, recall and F-measure to ROC, informedness, markedness and correlation. abs/2010.16061 (2020)

Application of Neural Networks in Sentiment Analysis of Social Media Text Data

Andrey Konstantinov, Vadim Moshkin(✉), and Nadezhda Yarushkina

Ulyanovsk State Technical University, Ulyanovsk, Russia
{a.konstantinov,v.moshkin,jng}@ulstu.ru

Abstract. The paper describes the results of experiments comparing the use of different neural network architectures in sentiment analysis of text messages from social networks. Experiments were carried out using two text vectorization algorithms "word2vec" and "BERT". The following types of neural networks were also used: LSTM, Bidirectional LSTM, CNN, MLP. The experiments were carried out on text data extracted from the VKontakte social network. As a result, an indicator of accuracy in determining the emotional color of posts was achieved at 87%.

Keywords: Sentiment analysis · Social network · Machine learning · LSTM · CNN · MLP

1 Introduction

Nowadays, social media big data analysis is an integral part of marketing research. According to statistics, 60–70% of users study product reviews before purchasing. Usually, buyers trust reviews at least twice as much as words in a product description on a website. According to statistics, the number of users looking for information about the service, and the reviews of those who have already used this service, is close to 90%.

In this regard, the use of intelligent algorithms for sentiment analysis of text messages from social networks will provide the ability to easily manage user reviews in order to maintain a high rating of the organization on the network.

Sentiment text analysis is a classification task. At present, the best results of text classification by several criteria are shown by machine learning algorithms. This paper presents a description of the application of machine learning of neural networks of various architectures to solve the problem of sentiment analysis of textual data of social networks [1].

2 Approaches to Sentiment Analysis of Social Media Text Resources

Currently, various machine learning methods show the highest results in sentiment analysis of text data (especially in Russian). Researchers use the following language models for text vectorization:

C. Kahraman et al. (Eds.): INFUS 2021, LNNS 308, pp. 298–304, 2022.
https://doi.org/10.1007/978-3-030-85577-2_35

- Bag of words;
- Word2vec;
- BERT and its derived models (BERT Large, etc.);
- ELMo;
- ULM-FiT, etc. [2–4];

Researchers use SVM-models, LSTM-CRF, convolutional and recurrent neural networks as classifiers.

For example, Bogdanov's work [5] describes the following algorithm for determining the emotional coloring of posts from the social network Twitter:

- The training set is formed using the expert dictionary of emoticons. Each emoticon found in a post corresponds to one emotion.
- The text is vectorized using the Bag of Words algorithm.
- Decision tree, multilayer perceptron and logistic regression were used as classifiers.

The effectiveness of the multilayer perceptron was 76%, the other two models were 75% each.

In the work of Smirnova [6], 2 sets of short texts from different social networks, up to 140 characters in size, were used for training. The authors used two neural network architectures as a classifier:

- Artificial neural network with recurrent and convolutional layers.
- Artificial neural network with two recurrent layers.

In the first case, the efficiency of sentiment analysis was 71%, in the second case - 69%.

In addition, many researchers use hybrid algorithms based on the integration of semantic algorithms and machine learning algorithms. Sometimes, these approaches show higher results.

3 Neural Network Architectures for Sentiment Analysis of Unstructured Text Resources

Formally, the output of a neuron is:

$$y = f(u), \text{ where } u = \sum_{i=1}^{n} w_i x_i + w_0 x_0$$

where x_i and w_i are the signals at the inputs of the neuron and the weights of the inputs, respectively, the function u is called the induced local field, and $f(u)$ is the transfer function.

The concept of an activation function is associated with each neuron. Formally, the activation function is:

$$f(x) = tx$$

where t is some factor responsible for the distribution of the activation function.

Various neural network architectures for determining the sentiment of texts were used during the work. Neural networks consisted of the following layers: LSTM, Bidirectional LSTM, CNN, MLP, GRU. The description of the layers is presented below:

- The Embedding layer is the input layer of the neural network, consisting of neurons:

$$Emb = \{Size(D), Size(S_{vec}), L_{Sec}\},$$

  where
  Size(D) is the size of the dictionary in the text data,
  $Size(S_{vec})$ is the size of the vector space into which the words will be inserted, $Size(S_{vec}) = 32$;
  L_{Sec} is the length of the input sequences equal to the maximum size of the vector formed during word preprocessing.
- The Conv1D layer is a convolutional layer. It is essential for deep learning. Experiments have shown that the use of Conv1D improves the classification accuracy of text data by 5–7%. The number of filters is 32, the length of each is 3. The activation function is "relu".
- The MaxPooling1D layer is a layer that reduces the dimensions of the generated feature maps. The maximum pool is 2.
- The LSTM layer is a recurrent neural network layer. The model uses two LSTMs, the first layer contains 50 blocks, the second layer contains 20 blocks.
- The Dropout layer is a layer that prevents overfitting of the neural network. A value of 0.5 is given as a parameter, which means that the neural network can exclude up to half of inactive neurons.
- The Dense layer is the output layer of seven neurons. Each neuron is responsible for a specific emotion.
- The Flatten layer is a layer used to convert multidimensional vectors to a one-dimensional vector.
- The Bidirectional layer is a bi-directional layer that creates 2 parallel-working instances of the layer passed in the parameter. The bi-directional layer looks at the input sequence in both directions and obtains richer views.
- The recurrent layer consists of blocks that perform various mathematical operations on signals, and stores information for later use, preventing the signals from fading out gradually, unlike ordinary layers of neurons. The principle of LSTM operation is discussed in more detail in [7].
  The architecture of a recurrent neural network is shown in Fig. 1.

GRU is a recurrent layer based on the same principle as LSTM and is a simpler structure. Accordingly, GRU is less computationally expensive. The GRU layer usually remembers the recent past better than the more distant, so more recent information is more important. More details in [8].

In a bi-directional neural network, two LSTM layers are used, united by a Bidirectional layer. The neural network looks through the input sequence in both directions and obtains richer views than a conventional LSTM network.

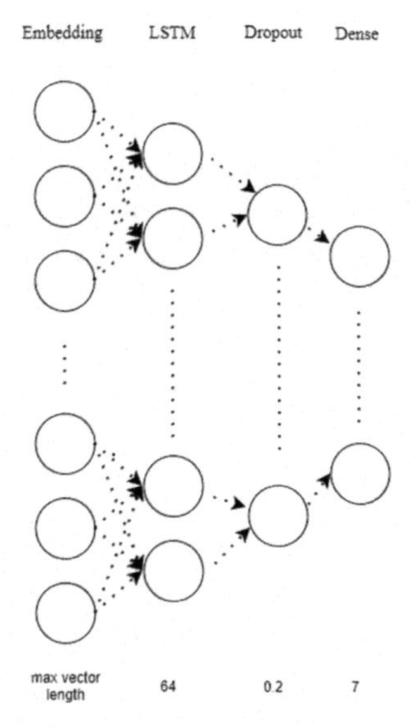

Fig. 1. Architecture of a recurrent neural network

A small convolution kernel is used in a convolutional neural network. The convolution kernel moves over the entire input matrix and generates an activation signal for the neuron of the next layer with the same position after each shift.

The architecture of a convolutional neural network is shown in Fig. 2.

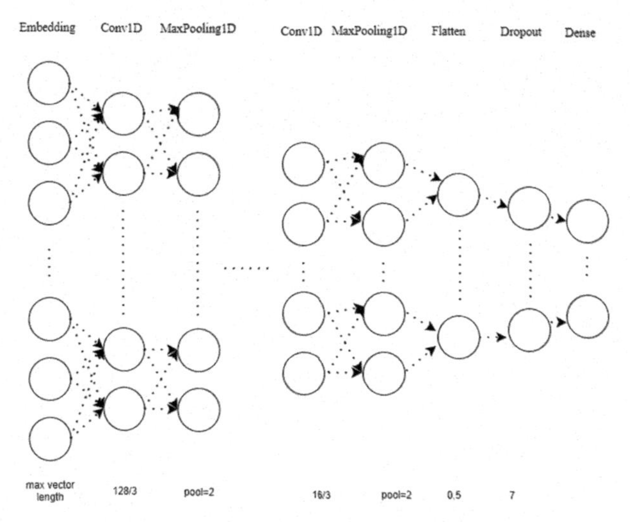

Fig. 2. Convolutional neural network architecture

The downsampling operation performs dimensionality reduction of the received feature maps. The maximal neuron is selected from several neighboring neurons of the feature map and is taken as one neuron of the condensed feature map of a lower dimension. The principle of operation of convolutional neural networks is described in more detail in [9].

A multilayer perceptron is a fully connected network in which each neuron in the current layer is connected to each neuron in the next layer. The principle of operation of the multilayer perceptron is described in more detail in [10].

The architecture of the multilayer perceptron is shown in Fig. 3.

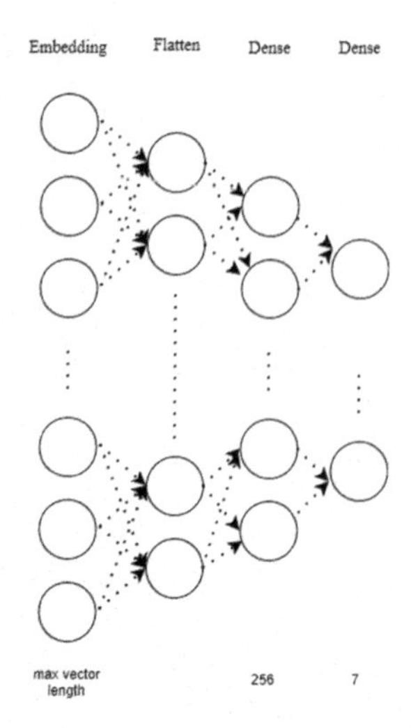

Fig. 3. Architecture of a multilayer perceptron

A neural network consisting of a downsampling layer, a convolutional layer and an LSTM layer is used for text recognition. The convolutional layer is used to extract features and the LSTM layer works with the selected features.

4 Experiments

A software system for assessing the sentiment of text messages on a social network has been implemented. Neural networks were implemented in Python using the TensorFlow and Keras machine learning frameworks.

2.5 million text messages from the VKontakte social network were processed when forming training and test sets. Messages were received from open groups of the social network through the VKontakte API and contained only text information.

The data for the test and training sets were selected from the received set of text messages. The accuracy for each neural network architecture on the test set was obtained as a result of experiments. The training and test set was obtained according to the algorithm described in [11]. The classification accuracy on the training set for all architectures is 1.0. The experimental results are presented in Fig. 4.

The best accuracy of classification of social network posts by 7 human emotions is achieved when using a multilayer perceptron, as can be seen from the results of the experiments.

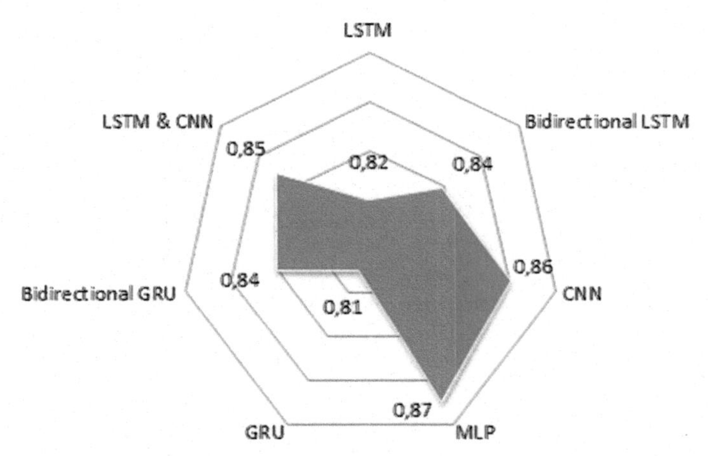

Fig. 4. Accuracy on the test set

5 Conclusions

Thus, we have applied 7 neural network architectures for sentiment analysis of text data of the Russian-language social network VKontakte.

The tested neural network architectures show that not only deep learning neural networks, but also fully connected neural networks can have high classification accuracy. The highest result was shown by an experiment on text classification using a multilayer perceptron - 87% accuracy.

The main directions of development of this work are:

- better formation of the training set with preliminary use of semantic algorithms and with the use of freely distributed lexical dictionaries such as WordNetAffect;
- experiments with other language models (ELMo, ULM-FiT, etc.) to find the most effective model for specific types of texts.

Acknowledgments. This paper was supported Ministry of Education and Science of Russia in framework of project № 075-00233-20-05 from 03.11.2020 «Research of intelligent predictive multimodal analysis of big data, and the extraction of knowledge from different sources».

References

1. Moshkin, V., Yarushkina, N., Andreev, I.: The sentiment analysis of unstructured social network data using the extended ontology SentiWordNet. In: 12th International Conference on Developments in eSystems Engineering (DeSE), Kazan, Russia, pp. 576–580. IEEE (2019). https://doi.org/10.1109/DeSE.2019.00110
2. Sabuj, M.S., Afrin, Z., Hasan, K.M.A.: Opinion mining using support vector machine with web based diverse data. In: Shankar, B.U., Ghosh, K., Mandal, D.P., Ray, S.S., Zhang, D., Pal, S.K. (eds.) PReMI 2017. LNCS, vol. 10597, pp. 673–678. Springer, Cham (2017). https://doi.org/10.1007/978-3-319-69900-4_85

3. Dinu, L.P., Iuga, I.: The best feature of the set. In: Gelbukh, A. (ed.) CICLing 2012. LNCS, vol. 7182. Springer, Heidelberg (2012). https://doi.org/10.1007/978-3-642-28601-8
4. Konstantinov, A., Moshkin, V., Yarushkina, N.: Approach to the use of language models BERT and Word2vec in sentiment analysis of social network texts. In: Dolinina, O., et al. (eds.) ICIT 2020. SSDC, vol. 337, pp. 462–473. Springer, Cham (2021). https://doi.org/10.1007/978-3-030-65283-8_38
5. Bogdanov, A.L., Dulya, I.S.: Sentiment analysis of short Russian-language texts in social media. Bull. Tomsk State Univ. Econ. **47**, 159–168 (2019)
6. Smirnova, O.S., Shishkov, V.V.: The choice of neural network topology and their application for the classification of short texts. Int. J. Open Inf. Technol. **8**, 50–54 (2016)
7. A Comprehensive Guide to Convolutional Neural Networks. https://towardsdatascience.com/a-comprehensive-guide-to-convolutional-neural-networks-the-eli5-way-3bd2b1164a53. Accessed 30 Mar 2021
8. Illustrated Guide to LSTM's and GRU's: A step by step explanation. https://towardsdatascience.com/illustrated-guide-to-lstms-and-gru-s-a-step-by-step-explanation-44e9eb85bf21. Accessed 30 Mar 2021
9. Understanding LSTM Networks. http://colah.github.io/posts/2015-08-Understanding-LSTMs. Accessed 30 Mar 2021
10. Understanding of Multilayer Perceptron. https://medium.com/@AI_with_Kain/understanding-of-multilayer-perceptron-mlp-8f179c4a135f. Accessed 30 Mar 2021
11. Moshkin, V., Konstantinov, A., Yarushkina, N.: Application of the BERT language model for sentiment analysis of social network posts. In: Kuznetsov, S.O., Panov, A.I., Yakovlev, K.S. (eds.) RCAI 2020. LNCS (LNAI), vol. 12412, pp. 274–283. Springer, Cham (2020). https://doi.org/10.1007/978-3-030-59535-7_20

Diagnosis of COVID-19 Using Deep CNNs and Particle Swarm Optimization

Omer Faruk Gurcan[1], Ugur Atici[1], Mustafa Berkan Bicer[2], and Onur Dogan[3](✉)

[1] Department of Industrial Engineering, Sivas Cumhuriyet University, 58140 Sivas, Turkey
{ofgurcan,uatici}@cumhuriyet.edu.tr

[2] Department of Electrical and Electronic Engineering, Izmir Bakircay University, 35665 Izmir, Turkey
mustafa.bicer@bakircay.edu.tr

[3] Department of Industrial Engineering, Izmir Bakircay University, 35665 Izmir, Turkey
onur.dogan@bakircay.edu.tr

Abstract. Coronavirus pandemic (COVID-19) is an infectious illness. A newly explored coronavirus caused it. Currently, more than 112 million verified cases of COVID-19, containing 2,4 million deaths, are reported to WHO (February 2021). Scientists are working to develop treatments. Early detection and treatment of COVID-19 are critical to fighting disease. Recently, automated systems, specifically deep learning-based models, address the COVID-19 diagnosis task. There are various ways to test COVID-19. Imaging technologies are widely available, and chest X-ray and computed tomography images are helpful. A publicly available dataset was used in this study, including chest X-ray images of normal, COVID-19, and viral pneumonia. Firstly, images were preprocessed. Three deep learning models, namely DarkNet-53, ResNet-18, and Xception, were used in feature extraction from images. The number of extracted features was decreased by Binary Particle Swarm Optimization. Lastly, features were classified using Logistic Regression, Support Vector Machine, and XGBoost. The maximum accuracy score is 99.7% in a multi-classification task. This study reveals that pre-trained deep learning models with a metaheuristic-based feature selection give robust results. The proposed model aims to help healthcare professionals in COVID-19 diagnosis.

Keywords: Deep learning · Heuristic · COVID-19 · Pandemic · Particle swarm optimization

1 Introduction

A new coronavirus called SARS-CoV-2 caused COVID-19, and it has become a pandemic. Reportedly, COVID-19 first appeared in Wuhan, China. World Health

C. Kahraman et al. (Eds.): INFUS 2021, LNNS 308, pp. 305–312, 2022.
https://doi.org/10.1007/978-3-030-85577-2_36

Organization (WHO) became aware of the virus in December 2019. There are some symptoms of COVID-19 such as difficulty breathing, fever, fatigue, dry cough, and muscle or body aches. Some of the other less common symptoms are loss of smell or taste, headache, muscle pain, or skin rash. Most people with symptoms (about 80%) get better without hospital treatment. Older people and those with chronic conditions such as heart problems and diabetes are at higher risk. However, anyone can be ill seriously or die at any age with COVID-19. Some protection ways such as wearing a mask, paying attention to social distance, cleaning your body or goods regularly, and avoiding crowded areas are recommended to help people stay safe. Moreover, some tests are applied to detect COVID-19. The PCR test is the most preferred one. Although rapid antigen tests give faster and cheaper results than PCR, they are less reliable than PCR [23]. On the other hand, chest radiography imaging technologies such as chest X-ray and CT detect COVID-19. These imaging technologies are widely available in hospitals and promise quick and highly accurate diagnoses. Rapid diagnosis of infected people at the early stages of the illness is essential in quarantine, treatment processes and overcoming this epidemic. Radiologists perform the interpretation of various X-ray and CT images. A sufficient number of medical staff is a challenge currently. Additionally, experts' overwork or visual fatigue increases the potential risks of misdiagnosis of X-ray or CT images [26].

Deep learning, which is an artificial intelligence area, has been top-rated for the last years. One of the most applied areas is the medical field. The fast spread of COVID-19 has raised interest in automated diagnosing systems based on deep learning methods. These methods can help medical experts diagnose COVID-19, especially in countries where imaging machines are available, but there is a lack of expertise [17,18]. Recently many researchers applied deep learning methods to classify X-ray or CT images for COVID-19 diagnosis. Optimization techniques can be applied in COVID-19 diagnosis, besides deep learning techniques. Modern optimization techniques are metaheuristic or heuristic. These techniques are applied to solve any optimization tasks in many fields [21].

This study proposes a model that used deep CNN models and a metaheuristic algorithm to diagnose COVID-19 automatically. An open-source dataset was used to test the proposed model. X-ray images, namely normal, COVID-19, and viral pneumonia, were classified. The results revealed that the proposed model is successful in accuracy performance when the dataset is small. The remainder of this paper is organized as follow: The literature review is presented in Sect. 2. Details of the case study are explained in Sect. 3. Analysis and results are given in Sect. 3.5. Lastly, Sect. 4 outlines the conclusion and future work.

2 Literature Review

Chest X-ray images have demonstrated a helpful screening method to diagnose the COVID-19 [15,25]. These images allow diagnosing for health professionals faster and more reliable decisions. Applying pre-trained models by transfer learning has been very common recently. Some authors proposed a COVID-19 classification model preferring just a pre-trained model as a baseline such as ResNet

[19], VGG16 [18], Xception [14], Inception-v3 [4]. On the other hand, multiple pre-trained models were preferred in some studies. Ko et al. [15] developed a FCONet, which is based on VGG16, ResNet-50, Inception-v3, and Xception; Apostolopoulos and Mpesiana [3] used VGG19, Xception, MobileNet-v2, Inception, and InceptionResNet-v2.

On the other hand, various hybrid techniques based on deep learning were adopted to diagnose the COVID-19 utilizing X-ray images to make better decisions. Al-Waisy et al. [1] integrated ResNet-34 and HRNet. Each model's score is fused to obtain the final class whether the individual has affected the COVID-19 or not. A transfer learning-based hybrid 2D/3D CNN architecture for COVID-19 detection was applied pre-trained VGG16 deep model, a shallow 3D CNN [5]. The study combined a depth-wise separable convolution layer (to preserve the useful features) and a spatial pyramid pooling module (to extract multi-level representations). Hasan et al. [9] extracted featured from images using Q-deformed entrophy algorithm. The extracted features are classified with a long short-term memory (LSTM) neural network classifier, SVM, kNN, and logistic regression. An integrated method including convolutional neural network (CNN) and LSTM was improved to detect COVID-19 automatically using X-ray images [11]. CNN was utilized for deep feature extraction, and LSTM was performed disease detection using the extracted features.

Deep learning models such as DarkNet-53, ResNet-18 and Xception extract features to analyze the images. Although each extracted feature is of importance for high accuracy, analysis requires much time because the number of the features is mainly too high. Metaheuristic algorithms can overcome this drawback by selecting the most useful features. This hybrid approaches can reach similar accuracy rates with less image features. Canayaz [7] combined some deep learning models (VGG19, AlexNet, ResNet and GoogleNet) and two metaheuristic algorithms (binary gray wolf and binary particle swarm optimization) for COVID-19 diagnostic. Altan and Karasu [2] proposed a hybrid model including three methods, image processing model, metaheuristic algorithm and deep learning technique, to understand COVID-19 infection from X-ray images. 2D Curvelet transformation as an image processing model was applied to the X-Ray images and a feature matrix was generated using the acquired coefficients. The coefficients were optimized using the chaotic salp swarm algorithm, and the COVID-19 diagnostic was determined with the EfficientNet-B0 model.

In this perspective, this study contributes to the literature by proposing a hybrid method consisting of deep learning models and a metaheuristic algorithm for the diagnosis of COVID-19.

3 Case Study

This section provides the details of the dataset, proposed model, models and algorithm used in the analysis, and results.

3.1 Dataset

A publicly available dataset was used in this study. Dataset is available in Kaggle [13]. It has 1341 normal, 1345 viral pneumonia, and 219 COVID-19 chest X-ray images. The images are in PNG format with a resolution of 1024 × 1024 pixels. Pixel values were rescaled from the range of 0–255 to the range 0–1. Moreover, image sizes were reseized.

3.2 Proposed Model

The proposed model was given in Fig. 1. According to the model, input X-ray images were pre-processed at first. Three deep learning models were used in feature extraction from normalized images. During the feature extraction process, the transfer learning approach was applied. Binary Particle Swarm Optimization was used as a feature selection method to obtain the best potential features. Lastly, abstract features were classified with four machine learning algorithms.

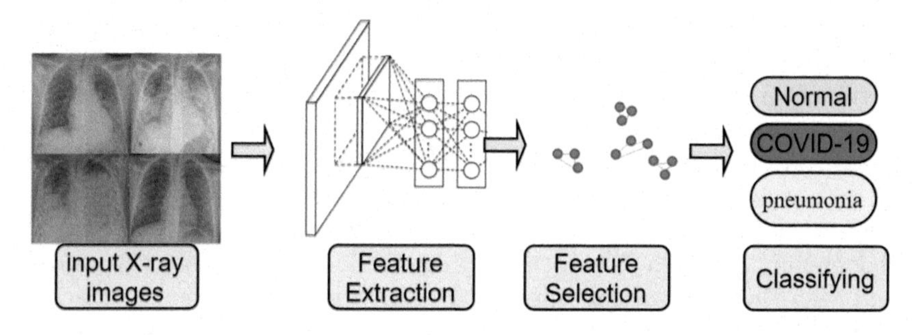

Fig. 1. The proposed model.

3.3 Deep Learning Models

In the study, three deep learning models, namely DarkNet-53, ResNet-18, and Xception, were used as feature extractors. DarkNet-53 is a model with 53 convolution layers and residual connections, which is the backbone of the real-time single-stage object detection model YOLOv3 [20]. The Residual Networks (ResNet) model, proposed by He et al. [10] from Microsoft Research, is a deep learning model with a low level of complexity, which includes the residual learning method. The layer of ResNets can be increased easily by adjusting the number of residual blocks used. The Xception model proposed by Chollet from Google was created by developing the Inception model using the modified depthwise separable convolution [8]. In modified depthwise separable convolution, the depthwise separable convolution process was performed after the pointwise convolution process. This study benefits from the advantages of transfer learning approach. Pre-trained weights of these models were trained in the ImageNet. Top layers, which are classifier part of models, were excluded. Some properties of the models are presented in Table 1.

Table 1. Deep learning models and properties.

Pre-trained model	Depth	Parameters (millions)	Input image size
DarkNet-53	53	41	256 × 256 × 3
ResNet-18	18	11.7	224 × 224 × 3
Xception	71	22.9	299 × 299 × 3

3.4 Binary Particle Swarm Optimization

Feature selection is essential in the data processing [6]. The large size of the data to be processed causes problems such as prolonged training and over-compliance [16]. In order to prevent similar problems, unnecessary features should be extracted from the data. Binary particle swarm optimization (BPSO) is used for feature selection [7]. BPSO is the binary version of PSO [12]. The particle's velocity in the swarm is updated by using Eq. 1, and the position of the particle is changed by using Eq. 2 and Eq. 3. Each particle in the swarm represents a possible solution. The possible solution is a 0/1 (1 x feature number) matrix, and each column value is randomly generated. A possible solution (a particle) is obtained by selecting the column numbers with a 1 in the feature matrix. K-nearest neighbour (kNN) classifier error rate was used for the fitness value of particles in swarm. In the first iteration, the particle with the swarm's best objective function value is assigned as *gbest*. In the subsequent iterations, the particle with the best objective function value is assigned (Eq. 4). If *pbest* has a better objective function value than *gbest* then, *pbest* is assigned as *gbest* (Eq. 5) This process is continued through T iteration, and the algorithm provides the best solution globally as output [7].

$$v_i^d(t+1) = wv_i^d(t) + c_1 r_1 (pbest_i^d(t) - x_i^d(t)) + c_2 r_2 (pbest^d(t) - x_i^d(t)) \tag{1}$$

$$S(v_i^d(t+1)) = \frac{1}{1 + exp(-v_i^d(t+1))} \tag{2}$$

$$x_i^d(t+1) = \begin{cases} 1, if rand < S(v_i^d(t+1)) \\ 0, \quad otherwise \end{cases} \tag{3}$$

$$pbest_i^d(t+1) = \begin{cases} x_i(t+1), if F(x_i(t+1)) < F(pbest_i(t)) \\ pbest_i(t), \quad otherwise \end{cases} \tag{4}$$

$$gbest(t+1) = \begin{cases} pbest_i, (t+1)\ if F(pbest_i(t+1)) < F(gbest(t)) \\ gbest(t), \quad otherwise \end{cases} \tag{5}$$

3.5 Results

The features were extracted from the images using three deep models, namely DarkNet-53, ResNet-18, and Xception. Four classifiers (Linear SVC, Logistic Regression, SVM, and XGBoost) were used in a multi-classification task. BPSO was used as a feature selector. Repeated k-fold cross-validation (number of splits chosen ten and the number of repeats chosen three) was applied as a re-sampling procedure in accuracy calculation. For the classifiers, default parameters of algorithms were preferred in Scikit-learn [22] and XGBoost [24] libraries.

Parameter values of BPSO; $T = 100$; $N = 20$; $c_1 = 2$; $c_2 = 2$; $W_{min} = 0.4$; $W_{max} = 0.9$; and $V_{max} = 6$ (T = maximum number of iterations, N = number of particles, c_1 = cognitive factor, c_2 = social factor, W_{min} = minimum bound on inertia weight, W_{max} = maximum bound on inertia weight, V_{max} = maximum velocity). For BPSO fitness function, k-value of $kNN = 5$ and $k - fold = 3$ were selected.

Table 2 shows analysis results. Firstly, we extracted features from deep models and then classified them. Secondly, we applied the feature selection process. The number of extracted features was decreased by BPSO, and obtained new features were classified. Lastly, we compared results. We found the maximum accuracy of 99.7% with Linear SVC by classifying BPSO-selected features from DarkNet-53. Overall results show that BPSO decreased the number of extracted features by about half. Moreover, the accuracy performance of classifiers didn't decrease with selected features.

Table 2. Analysis results.

	Deep models	Number of features	Accuracy scores (%)			
			Linear SVC	Log Reg	SVM	XGBoost
Extracted features	DarkNet-53	1024	99.6	99.6	99.4	99.3
	ResNet-18	512	98.9	99	99	98.5
	Xception	2048	98	98.2	97.6	97.8
Selected features by BPSO	DarkNet-53	516	99.7	99.6	99.5	99.3
	ResNet-18	258	98.8	99	99	98.5
	Xception	1080	98	98.2	98.3	97.8

4 Conclusion

In this study, the BPSO algorithm was applied for feature selection in the classification of the COVID-19 X-ray data set. The pre-trained DarkNet-53, ResNet-18 and Xception models were used to extract features. BPSO has been used to increase the success in the classification of the obtained features or reduce the number of features without decreasing the classification accuracy. The features selected using the BPSO are classified using the Linear SVC, Logistic Regression, SVM and XGBoost classifiers. Using BPSO, 516 of the 1024 features obtained from the DarkNet model were selected, and 0.1% improvement was achieved in Linear SVC and SVM classifiers for accuracy. 258 of the 512 features obtained from the ResNet model were selected and classified. In this classification, there was a 0.1% decrease in accuracy for Linear SVC. 1080 of 2048 features obtained from the Xception model were selected and classified using BPSO. As a result of this classification, the SVM classifier's accuracy increased by 0.7%, whereas other values remained constant. According to the study results, the BPSO algorithm showed great success in selecting a smaller number of features that provide better performance for classification from the features obtained from deep learning models.

For future research direction, extending the dataset by additional data sources can improve results. Recently developed heuristic algorithms can be used.

References

1. Al-Waisy, A.S., et al.: COVID-CheXNet: hybrid deep learning framework for identifying COVID-19 virus in chest x-rays images. Soft Comput. 1–16 (2020)
2. Altan, A., Karasu, S.: Recognition of COVID-19 disease from x-ray images by hybrid model consisting of 2D curvelet transform, chaotic salp swarm algorithm and deep learning technique. Chaos Solitons Fractals **140**, 110071 (2020)
3. Apostolopoulos, I.D., Mpesiana, T.A.: COVID-19: automatic detection from x-ray images utilizing transfer learning with convolutional neural networks. Phys. Eng. Sci. Med. **43**, 635–640 (2020)
4. Asif, S., Wenhui, Y., Jin, H., Tao, Y., Jinhai, S.: Classification of COVID-19 from chest x-ray images using deep convolutional neural networks. medRxiv (2020)
5. Bayoudh, K., Hamdaoui, F., Mtibaa, A.: Hybrid-COVID: a novel hybrid 2D/3D CNN based on cross-domain adaptation approach for COVID-19 screening from chest x-ray images. Phys. Eng. Sci. Med. **43**(4), 1415–1431 (2020)
6. Çakmak, E., Önden, İ., Acar, A.Z., Eldemir, F.: Analyzing the location of city logistics centers in Istanbul by integrating geographic information systems with binary particle swarm optimization algorithm. Case Studies on Transport Policy (2020)
7. Canayaz, M.: MH-COVIDNet: diagnosis of COVID-19 using deep neural networks and meta-heuristic-based feature selection on X-ray images. Biomed. Signal Process. Control **64**, 102257 (2021)
8. Chollet, F.: Xception: deep learning with depthwise separable convolutions. In: Proceedings of the IEEE Conference on Computer Vision and Pattern Recognition, pp. 1251–1258 (2017)
9. Hasan, A.M., AL-Jawad, M.M., Jalab, H.A., Shaiba, H., Ibrahim, R.W., AL-Shamasneh, A.R.: Classification of COVID-19 coronavirus, pneumonia and healthy lungs in CT scans using q-deformed entropy and deep learning features. Entropy **22**(5), 517 (2020)
10. He, K., Zhang, X., Ren, S., Sun, J.: Deep residual learning for image recognition. In: Proceedings of the IEEE Conference on Computer Vision and Pattern Recognition, pp. 770–778 (2016)
11. Islam, M.Z., Islam, M.M., Asraf, A.: A combined deep CNN-LSTM network for the detection of novel coronavirus (COVID-19) using x-ray images. Inform. Med. Unlocked **20**, 100412 (2020)

12. Jiang, F., Xia, H., Tran, Q.A., Ha, Q.M., Tran, N.Q., Hu, J.: A new binary hybrid particle swarm optimization with wavelet mutation. Knowl.-Based Syst. **130**, 90–101 (2017)
13. kaggle (2020). https://www.kaggle.com/tawsifurrahman/covid19-radiography-database. Accessed 01 Aug 2020
14. Khan, A.I., Shah, J.L., Bhat, M.M.: CoroNet: a deep neural network for detection and diagnosis of COVID-19 from chest x-ray images. Comput. Methods Programs Biomed. **196**, 105581 (2020)
15. Ko, H., et al.: COVID-19 pneumonia diagnosis using a simple 2D deep learning framework with a single chest CT image: model development and validation. J. Med. Internet Res. **22**(6), e19569 (2020)
16. Kumari, K., Singh, J.P., Dwivedi, Y.K., Rana, N.P.: Multi-modal aggression identification using convolutional neural network and binary particle swarm optimization. Future Gener. Comput. Syst. **118**, 187–197 (2021)
17. Ozturk, T., Talo, M., Yildirim, E.A., Baloglu, U.B., Yildirim, O., Acharya, U.R.: Automated detection of COVID-19 cases using deep neural networks with x-ray images. Comput. Biol. Med. **121**, 103792 (2020)
18. Panwar, H., Gupta, P., Siddiqui, M.K., Morales-Menendez, R., Singh, V.: Application of deep learning for fast detection of COVID-19 in x-rays using nCOVnet. Chaos Solitons Fractals **138**, 109944 (2020)
19. Pathak, Y., Shukla, P.K., Tiwari, A., Stalin, S., Singh, S., Shukla, P.K.: Deep transfer learning based classification model for COVID-19 disease. IRBM (2020)
20. Redmon, J., Farhadi, A.: Yolov3: an incremental improvement. arXiv preprint arXiv:1804.02767 (2018)
21. Rere, L., Fanany, M.I., Arymurthy, A.M.: Metaheuristic algorithms for convolution neural network. Comput. Intell. Neurosci. **2016** (2016). Article ID 1537325
22. Scikit (2021). https://scikit-learn.org/stable/supervised_learning.html. Accessed 20 Jan 2021
23. World Health Organization: Coronavirus disease (COVID-19) (2021). https://www.who.int/emergencies/diseases/novel-coronavirus-2019/question-and-answers-hub/q-a-detail/coronavirus-disease-covid-19. Accessed 19 Feb 2021
24. Xgboost (2021). https://xgboost.readthedocs.io/en/latest/parameter.html. Accessed 20 Jan 2021
25. Yoo, S.H., et al.: Deep learning-based decision-tree classifier for COVID-19 diagnosis from chest x-ray imaging. Front. Med. **7**, 427 (2020)
26. Zheng, C., et al.: Deep learning-based detection for COVID-19 from chest CT using weak label. MedRxiv (2020)

Bilingual Speech Emotion Recognition Using Neural Networks: A Case Study for Turkish and English Languages

Damla Büşra Özsönmez, Tankut Acarman, and Ismail Burak Parlak(✉)

Department of Computer Engineering, GSUNLPLab, Galatasaray University, Ciragan Cad. No:36, 34349 Ortakoy, Istanbul, Turkey
{bozsonmez,tacarman,bparlak}@gsu.edu.tr

Abstract. Emotion extraction and detection are considered as complex tasks due to the nature of data and subjects involved in the acquisition of sentiments. Speech analysis becomes a critical gateway in deep learning where the acoustic features would be trained to obtain more accurate descriptors to disentangle sentiments, customs in natural language. Speech feature extraction varies by the quality of audio records and linguistic properties. The speech nature is handled through a broad spectrum of emotions regarding the age, the gender and the social effects of subjects. Speech emotion analysis is fostered in English and German languages through multilevel corpus. The emotion features disseminate the acoustic analysis in videos or texts. In this study, we propose a multilingual analysis of emotion extraction using Turkish and English languages. MFCC (Mel-Frequency Cepstrum Coefficients), Mel Spectrogram, Linear Predictive Coding (LPC) and PLP-RASTA techniques are used to extract acoustic features. Three different data sets are analyzed using feed forward neural network hierarchy. Different emotion states such as happy, calm, sad and angry are compared in bilingual speech records. The accuracy and precision metrics are reached at level higher than 80%. Turkish language emotion classification is concluded to be more accurate regarding speech features.

Keywords: Speech analysis · Emotion detection · Natural language processing · Machine learning · Deep learning

1 Introduction

Emotional expressions contribute to the interpretation of interpersonal relationship since they reflect human behavior [1]. Since the characterization of emotion states is considered as a special human task, machine based approaches require high training and hybrid approaches. Detecting emotion becomes crucial for making communication more effective and efficient. The better a person understands the emotions of the other people, the more clear information will be extracted from communication exchanges. Emotion detection is widely

C. Kahraman et al. (Eds.): INFUS 2021, LNNS 308, pp. 313–320, 2022.
https://doi.org/10.1007/978-3-030-85577-2_37

used in psychological, educational and commercial applications involving human-machine interaction applications such as the game and entertainment industry [2] and software testing [3]. In order to measure customer satisfaction, questionnaire-based emotion detection is usually conducted by companies using surveys in different levels. Therefore, an automatic analysis figures out to categorize emotional states for customers. Emotion detection can be realized through image, audio or text data. People with different emotions can have different facial expressions, and certain facial expressions may reflect a single emotion in a straight-forward manner. Speech is also one of the most important sources where emotions are expressed.

A multi-class classification predicts categorical outcomes for data. It is necessary to create models on the data and train them to solve the classification problems. Processes such as preprocessing, feature extraction and feature selection are applied on the data to create these models. These processes are essential for achieving a high performance model. If the features extracted over the data do not represent the output classes well, the performance of the model might be lowered depending on the feature hyperspace. Thus, the preprocessing techniques such as resampling, offsetting, adding or removing noise, adding sound effect and filtering are used in speech analysis. Zero crossing, frequency spectrogram, Mel-spectrogram, Mel-Frequency Cepstral Coefficients (MFCC), Linear Frequency Cepstral Coefficients (LFCC), Linear Predictive Coding (LPC) coefficients, Perceptual Linear Prediction (PLP) coefficients, Relative Spectra Filtering (RASTA) - PLP are common techniques for acoustic feature extraction. Over the past few years, deep learning allowed new improvements on image, text or audio classification since it facilitates feature extraction. Artificial neural networks contain of neurons and layers (input, hidden and output layers) forming a fully-connected graph. The output of a neuron (using the weights of the input connections) is passed to the activation function and the result is passed to the next neuron. Forward propagation and back propagation are used to tune weights and error rates in the network. In speech emotion recognition, linguistic properties are considered to be influencing factors toward feature extraction. For this reason, a model trained from a data set in which there are speech data with different language is likely not expected to perform well on speech data of another language. Although there have been many studies presenting methodologies bench marked with English speech data sets, and many speech data sets in Turkish language, emotion detection in Turkish and English languages has not been performed adequately and compared toward a multi-class classification. Thus, there are few studies in bilingual speech emotion detection for Turkish language using deep learning techniques.

This study presents combined features in a deep neural network model for Turkish and English languages. The benchmark analysis has been performed using two different Turkish speech data sets (TurES and TurEV-DB) and one English speech (RAVDESS) data set with deep neural networks. The originality of our study lies in multi feature tasks for the optimal classifier search on emotion detection in speech domain. The average accuracy achieves state-of-the-art results for utterance based emotion recognition from speech domain for both languages. The paper is organized as follows. The following section magnifies recent studies to encompass speech emotion recognition in a nutshell. The methodology section describes the characteristics of the data sets used, data preprocessing, feature extraction and selection, model architecture. Furthermore, the evaluation of classification is reviewed in the section of results In the experiments and results, the accuracy scores are presented through model parameters. Consequently, we concluded our findings by highlighting best scores of benchmark analysis and presenting prospective steps.

2 Related Works

Emotion detection is considered as a complex human task for sensing and analyzing. The machine based recognition would be conducted through different types data such as image, text or speech. Since the last decade, many studies have focused on emotion detection through speech data in social networks, multimedia systems, telecommunication channels. Affective computing considers human behavior through psychological parameters of emotion perception. Even if we have focused on speech based emotion classifiers in this study, affective computing considers emotion perception through all senses. In current studies, fusion models figure out how gestures, expressions, movements and rhetorics would be analyzed in a single model to overcome the pitfalls of emotion analysis through single sensors.

There are many studies on speech emotion recognition for English language. An analysis for speech emotion recognition using RAVDESS and TESS data sets is made available to provide a more productive feedback for smart home assistants [4]. Some studies have been done for emotion detection in Turkish language. Bakır and Yuzkat implemented some models with the extracted features using MFCC and Mel Frequency Discrete Wavelet Coefficients (MFDWC) [5]. They used Hidden Markov Model (HMM), Gauss Mixture Model (GMM), Artificial Neural Network (ANN) and a GMM model combined with SVM.

Besides machine learning models, deep learning is also used for speech emotion detection. Spectrogram image data are processed and the 2-D Convolutional Neural Network model was implemented [6]. Mirsamadi et al. implemented deep recurrent neural network (RNN) to extract short-time frame-level features on IEMOCAP dataset and obtained better accuracy results than the existing ones [7]. Hajarolasvadi and Demirel applied K-Means clustering on SAVEE, RML and Enterface05 databases to select the most discriminant frames and implemented 3D CNN models from the spectrograms of these frames [8]. In the literature of speech analysis, there are few studies for the comparison of emotion detection in Turkish and English languages. It is remarkable to develop a common paradigm for multilingual studies or measure different aspects of feature spaces through classifiers. In our work, we have analyzed different components of feature space for both languages to characterize the performance of multingual emotion classifiers in speech.

3 Methodology

In this section the main steps such as data preprocessing, model creation and evaluation criteria have been elaborated. The data preprocessing part consists of the analysis The Ryerson Audio-Visual Database of Emotional Speech and Song (RAVDESS). English Speech data is composed of eight different emotions: neutral, calm, happy, sad, angry, fearful, disgust and surprised [9]. The data set contains the speech data from 12 female and 12 male subjects. In this study, only speech data has been used for emotion detection. There were 288 neutral data, 192 data per other emotion classes in this data set. The number of male and female voice data was equal for each class. Considering the distribution of data, there was no class imbalance. Since the first 0.5 and the last 0.5 s of the audio files in the RAVDESS data set were silent, these parts were not used in training and testing the models. At first, with the Scipy library, reading some of the audio files in the RAVDESS could not be done due to an information column added by the Audacity application. Therefore, the non-functional areas in the files are deleted with Sound eXchange (SoX) tool http://sox.sourceforge.net.

TurES data set has been created using the speech in 55 Turkish films and series in different genres [10]. OpenSMILE was used to extract features such as MFCC, $L0$, logarithmic energy, LSP and their first and second derivatives. There were no direct audio files in the available data sources, but only data sets with extracted features and labels. The data set includes 7 different emotional states: 2150 neutral, 1377 angry, 567 sad, 404 happy, 379 surprised, 127 other and 96 fear. Considering the distribution of data, there was a class imbalance. Since the number of data was not too large, SMOTE for over-sampling has been used to eliminate the class imbalance and also to enlarge the dataset. Emotion classes such as other and sad are not included in training and testing models.

TurEV-DB data source contains STFT Spectrogram images, audio files for the selected Turkish words and a data set created from the OpenSMILE features [11]. Only audio files have been used in this study. This data set includes four different emotional states: 408 neutral, 487 angry, 483 sad, 357 happy. The distribution of data was balanced as well.

Furthermore, feature extraction has been performed for RAVDESS and TurEV-DB datasets. Mel Filter bank is inspired by the human auditory system being linear under 1 kHz and logarithmic above 1 kHz. Frequency spectrogram is created with Short Term Fourier Transform (STFT) and the frequency spectrogram is passed through the Mel Scale after windowing. Thus, the Power Spectrogram of the Mel Spectrogram is obtained. In this work, the techniques such as Hamming windowing, 2048-Coefficients FFT with 512 hop length is used to obtain 256-Coefficient Mel-Spectrogram. When the inverse DFT is applied after taking the logarithm of the Mel-Spectrogram, the MFCC is obtained. In this work, *Mel-Frequency Cepstral Coefficients (MFCC) & Mel-Spectrogram Coefficients* with 40 coefficients is used as a feature.

Linear Frequency Cesptrum Coefficients (LFCC) is created by a linear filter bank. LFCC would give results similar results to MFCC [12]. 26 Cepstrum Coefficients are obtained using 26 filters, 2048-Coefficients FFT and Hamming windowing.

Linear predictive coding (LPC) coefficients is based on modeling the audio signals as the weighted sum of past samples [13]. 26 Cepstrum coefficients are obtained as a feature.

Perceptual Linear Prediction (PLP) coefficient is created through a bark scale filterbank. Instead of Discrete Cosine Transform in MFCC, Equal-Loudness Pre-emphasis is used to obtain 26-Cepstrum Coefficients as a feature.

Relative Spectra Perceptual Linear Prediction (RASTA-PLP) is based on the human auditory system's ability which would be insensitive to slowly changing proper-ties. It is inspired by RASTA filter [14]. In PLP Coefficient extraction steps, RASTA filtering is used as a band pass filter to suppress the slowly changing components. 26 coefficients are used to obtain a RASTA-PLP feature.

Principal Component Analysis (PCA) is used on TurES since the size of the features is excessive compared to the data set length. It is also used for Mel-Spectrogram data sets and the datasets including all feature types. With this method, the data set is transformed into a lower dimensional space with the minimal loss. 90% cumulative explainable ratio is the threshold for this work to reduce the data set. TurES data set is reduced from 6670-dimensional space to 510-dimensional space with more than 95% cumulative explainable ratio. Furthermore, Mel-Spectrogram data set for Turev-DB and RAVDESS are respectively reduced to 25 and 10-dimensional spaces. Data sets including all feature types for TurEV-DB and RAVDESS are both reduced to 50-dimensional space.

This project is implemented using Python programming language. NumPy, Spafe, Scikit-learn, Sox, Librosa, Scipy libraries are used for data preprocessing. Keras and Grid Search libraries are used for model creation and evaluation. Google Colaboratory with GPU support has been used as the implementation environment. Deep neural network is used to discover hidden patterns in data and assure high accuracy. There are fully-connected neurons (Perceptron) and layers forming a fully-connected graph. Data is transferred from the input layer. The bias is added to the output of the neuron multi-plied by the weight and the total sum is passed to the activation function. The result from the activation function also passes to the next neuron; this whole process continues until the end of the output layer. The more hidden layers are used in the model, the higher the model complexity becomes. Loss function is used to optimize the weights in the neurons. We used ReLu (Rectified Linear Activation Function) as the activation function for input and hidden layers, and SoftMax for the output layer. Categorical cross-entropy is used as a loss function.

One of the biggest disadvantages of the artificial neural network is that it is very prone to overfitting. To avoid overfitting, it is necessary to tune the parameters such as the number of hidden layers, the number of neurons, the number of epochs, the optimizer type. We used Grid Search to optimize the number of hidden layers, the number of neurons and the optimizer type for 10 and 80 epoch numbers.

Table 1. Average accuracy results with grid search parameters for TurEV-DB.

Features	# Neurons	# Hidden layers	Optimizer	Epochs	AVG.ACC
All features-PCA	100	1	Adam	80	0.922
All features-PCA	250	2	Adam	10	0.909
Mel-PCA(25)	200	3	RMSprop	80	0.907
MFCC 40	150	4	Adam	80	0.906
Mel-PCA(25)	250	3	Adam	10	0.899
Mel-256	150	3	Adam	80	0.832
MFCC 40	200	4	Adam	10	0.827
PLP 26	250	3	Adam	80	0.802
Mel-256	150	3	Adam	10	0.684
PLP 26	250	3	Adam	10	0.607
PLP-RASTA 26	200	3	Adam	80	0.546
LPC 26	250	4	Adam	80	0.505
LPC 26	200	4	Adam	10	0.444
PLP-RASTA 26	250	4	Adam	10	0.423
LFCC 40	250	1	RMSprop	10	0.274
LFCC 40	100	3	RMSprop	80	0.264

4 Results

Average accuracy of all emotion classes is selected to measure the model performance. Table 1 shows the average accuracy values obtained with the optimal Grid Search parameters on different feature type data sets for TurEV-DB. According to these results, Adam optimizer gives better results in comparison with other optimizers. The best performing feature extraction method is Mel-Spectrogram and weak per-forming feature extraction method is LFCC. Also, PCA method increases the accuracy for Mel-Spectrogram featured models.

The average accuracy values obtained with the optimal Grid Search parameters for TurES are shown in Table 2. Filtered datasets are those in which fear and other classes are not included in the training and prediction. It can be observed that PCA and SMOTE techniques are effective in increasing model success. Table 3 shows the accuracy results for different feature type datasets created from RAVDESS dataset. MFCC is the best performing feature extraction method and the extraction methods such as LFCC, LPC and Mel-Spectrogram do not per-form well for RAVDESS dataset. The epoch number needs to be increased significantly to assure an acceptable success. A reason is that hidden layers may cause underfitting.

Table 2. Average accuracy results with grid search parameters for TurES.

Features	# Neurons	# Hidden layers	Optimizer	Epochs	AVG.ACC
Filtered - SMOTE - PCA(510)	250	2	SGD	80	0.855
Filtered - SMOTE - PCA(510)	250	2	Adam	10	0.843
Original - PCA	150	3	SGD	10	0.564
Filtered - PCA (510)	200	2	Adam	10	0.549
Filtered - PCA(510)	250	2	Adam	80	0.545

Table 3. Average accuracy results with grid search parameters for RAVDESS.

Features	# Neurons	# Hidden layers	Optimizer	Epochs	AVG.ACC
MFCC 40 - Mean					
Mel frequency coefficients	100	1	Adam	100	0.866
All features - PCA	250	3	Adam	80	0.600
MFCC 40	200	2	RMSprop	80	0.573
Mel-256 - PCA	200	4	Adam	80	0.572
All features - PCA	250	3	Adam	10	0.507
All features	100	2	Adam	80	0.461
Mel-256 - PCA	250	2	Adam	10	0.433
PLP 26	250	4	RMSprop	80	0.427
Mel-256	250	4	Adam	200	0.415
All features	150	3	Adam	10	0.348
MFCC 40	150	3	Adam	10	0.343
PLP 26	150	4	Adam	10	0.300
LPC 26	250	3	Adam	80	0.281
Mel-256	200	4	Adam	10	0.257
PLP RASTA 26	200	2	Adam	80	0.247
LPC 26	150	1	RMSprop	10	0.131
PLP RASTA 26	250	1	RMSprop	10	0.130
LFCC 26	200	2	SGD	10	0.126
LFCC 26	250	2	Adam	80	0.117

5 Conclusion

In this paper, English and Turkish speech emotion data sets are analyzed to compare the model performances. Audio processing and acoustic feature extraction was per-formed using MFCC, LFCC, PLP, PLP-RASTA, LPC, Mel-Spectrogram methods for emotion detection in Turkish and English languages. PCA technique is applied on the data sets to reduce the size of the data sets. Deep neural network models are created by discovering the optimum parameters using the Grid Search method. The models are evaluated using the average accuracy values. In addition to the frequently used MFCC method, it is seen that PLP and Mel spectrogram feature extraction methods also contribute considerably to the model success for classification purposes. It is also observed that linear filter banks are not suitable for speech data. With the same neuron and epoch numbers, the models created from Turkish TurES and TurEV-DB data sets have assured more successful results than the English Ravdess data set. As future steps, a multilingual speech corpus with an identical paradigm or emotion detection for foreign versus native speakers would be analyzed for Turkish language.

Acknowledgement. This work has been supported by the Scientific Research Projects Commission of Galatasaray University under grant number # 19.401.005.

References

1. Thanapattheerakul, T., Mao, K., Amoranto, J., Chan, J.: Emotion in a century. In: Proceedings of the 10th International Conference on Advances in Information Technology - IAIT (2018)
2. Fragopanagos, N., Taylor, J.G.: Emotion recognition in human-computer interaction. Neural Netw. **18**(4), 389–405 (2005)
3. Kołakowska, A., Landowska, A., Szwoch, M., Szwoch, W., Wróbel, M.: Emotion recognition and its applications. Adv. Intell. Syst. Comput. **300**, 51–62 (2014)
4. Chatterjee, R., Mazumdar, S., Sherratt, R.S., Halder, R., Maitra, T., Giri, D.: Real-time speech emotion analysis for smart home assistants. IEEE Trans. Consum. Electron. **67**(1), 68–76 (2021)
5. Bakır, C., Yuzkat, M.: Speech emotion classification and recognition with different methods for Turkish language. Balkan J. Electr. Comput. Eng. **6**(2), 122–128 (2018)
6. Badshah, A.M., Ahmad, J., Rahim, N., Baik, S.W.: Speech emotion recognition from spectrograms with deep convolutional neural network. In: International Conference on Platform Technology and Service (PlatCon) (2017)
7. Mirsamadi, S., Barsoum, E., Zhang, C.: Automatic speech emotion recognition using recurrent neural networks with local attention. In: IEEE International Conference on Acoustics, Speech and Signal Processing (ICASSP) (2017)
8. Hajarolasvadi, N., Demirel, H.: 3D CNN-based speech emotion recognition using k-means clustering and spectrograms. Entropy **21**(5), 479 (2019)
9. Livingstone, S.R., Russo, F.A.: The Ryerson audio-visual database of emotional speech and song (RAVDESS): a dynamic, multimodal set of facial and vocal expressions in North American English. PLoS ONE **13**(5), e0196391 (2018)
10. Oflazoglu, C., Yildirim, S.: Turkish emotional speech database. In: Proceedings of IEEE 19th Conference of Signal Processing and Communications Applications, pp. 1153–1156 (2011)
11. Canpolat, S.F., Ormanoğlu, Z., Zeyrek, D.: Turkish emotion voice data-base (TurEV-DB). In: Proceedings of the 1st Joint Workshop on Spoken Language Technologies for Under-Resourced Languages (SLTU) and Collaboration and Computing for Under-Resourced Languages (CCURL), pp. 368–375 (2020)
12. Zhou, X., Garcia-Romero, D., Duraiswami, R., Espy-Wilson, C., Shamma, S.: Linear versus mel frequency cepstral coefficients for speaker recognition. In: 2011 IEEE Workshop on Automatic Speech Recognition & Understanding, pp. 559–564 (2011)
13. Grama, L., Rusu, C.: Audio signal classification using linear predictive coding and random forests. In: International Conference on Speech Technology and Human-Computer Dialogue (SpeD) (2017)
14. Nayana, P.K., Mathew, D., Thomas, A.: Performance comparison of speaker recognition systems using GMM and i-Vector methods with PNCC and RASTA PLP features. In: International Conference on Intelligent Computing, Instrumentation and Control Technologies (2017)

Dynamic Multiplier CPPI Strategy with Wavelets and Neural-Fuzzy Systems

Ömer Z. Gürsoy(✉) and Oktay Taş

Istanbul Technical University, Istanbul, Turkey
oktay.tas@itu.edu.tr

Abstract. Constant Proportional Portfolio Insurance (CPPI) aims to maximize the performance of the portfolio by protecting a determined base value without using any derivative instruments, and by determining the amounts to be invested in risky and risk-free assets with calculations using the risk multiplier and buffer value. Artificial neural networks (ANNs) are mathematical models that are successfully used in studies such as pattern recognition, function estimation, finding the most appropriate value and classifying data by imitating neural networks in the human brain. This study aims to use the advantages of both the decomposition model (Wavelet Transform) and machine learning model (ANN) to predict the future values of stock indices to decide which risk multiplier to use. The dynamic multiplier CPPI yielded better returns than the classic CPPI in all 5 stock market indices analyzed, and both strategies successfully implemented the previously targeted 95% capital protection. It has been observed that predicting future prices of indices using Artificial Intelligence methods and the performance of the dynamic multiplier CPPI strategy applied based on these predictions is more successful than the conventional CPPI strategy with constant multiplier.

Keywords: CPPI · Wavelets · ANFIS · Fuzzy neural networks

1 Introduction

One of the techniques that strives to provide the best return for investors while also lowering investment risk as much as possible is known as fixed rate portfolio insurance (CPPI). The CPPI strategy divides the portfolio into risk-free and risky assets and uses the investor's risk preference and the amounts earned from risk-free assets to calculate the amount that should be invested in risky assets. Because the risk multiplier is determined subjectively, the CPPI strategy may not work well in every market. Using a high multiplier increases the performance of the model during the periods when the risky asset gains value, and causes the portfolio to shift rapidly to risk-free assets in periods when it loses value, and the portfolio cannot benefit from the possible increases that may occur later, thus causing it to underperform.

Time series of financial assets such as stocks are generally not linear and stationary. Since statistical models assume that these series are linear and stationary, they are susceptible to statistical errors. Machine learning models such as Artificial Neural Networks

C. Kahraman et al. (Eds.): INFUS 2021, LNNS 308, pp. 321–328, 2022.
https://doi.org/10.1007/978-3-030-85577-2_38

(ANN), Genetic Algorithms (GA), Support Vector Machines (SVM) can also effectively model linear and non-stationary data. While neural networks adapt to changing environments by recognizing patterns, fuzzy systems combine human expertise and machine learning for decision making.

The article includes estimating the 1-step forward prices of financial assets using a hybrid model of fuzzy neural networks and wavelet transform. For this purpose, historical data of asset prices are decomposed into sub-series by wavelet transformation and forward price predictions are formed with fuzzy neural networks of these sub-series. Sub-series predictions are aggregated to obtain final estimates. Investors decide to invest in risk-free or risky assets in CPPI according to the price prediction 1 step later.

It is thought that the performance of the artificial intelligence model will increase with the use of filtering and decomposition models. Although there are studies aiming to increase CPPI performance by using methods such as artificial neural networks, genetic algorithms, and fuzzy neural networks in the literature, the originality of this study is that fuzzy neural networks and wavelet analysis are used together. The aim of the study is to estimate next day values of selected stock market indices using fuzzy neural networks and wavelet analysis and to increase the performance of CPPI by using variable multiplier according to these predictions.

The flow of the paper is as follows: Next section explains the CPPI, Neurol Networks and Wavelet Analysis. Literature review and the methodology adopted in this paper is presented in the next section followed by Conclusion, where the results and the trading decisions are discussed.

2 CPPI

Constant Proportional Portfolio Insurance (CPPI) holds risky and risk-free assets in its portfolio and balances these assets dynamically. It provides this balance with the constant multiplier and protection rate parameters and prevents the portfolio value from falling below the minimum level determined from the beginning.

In CPPI, the portfolio is divided between risky and risk-free assets with a predetermined rule. The investor invests in risky and risk-free assets with a fixed multiplier that he decides based on his profit potential. The constant multiplier coefficient is chosen in accordance with the risk perception. Because as the coefficient increases, while the earning potential of the portfolio increases, the probability of rapidly approaching the base value will increase.

After the base value and risk coefficient have been determined in the period in which the strategy will be implemented, the remaining operations are mathematical formulas.

The base value (floor) under protection is removed from the portfolio and the cushion value (cushion) is found. The increase or decrease of the buffer value moves depending on the market value of the risky asset. The amount resulting from multiplying the buffer value with the fixed multiplier coefficient is invested in the risky asset, and the remaining portion is invested in the risk-free asset. Risky and risk-free asset distribution is re-determined at the end of the day according to market movements.

3 Neural Networks and Wavelet Analysis

The basic unit of Artificial Neural Network (ANN) is the artificial nerve cell called the process element - neuron or node. ANNs are made up of multiple processing units called nodes, neurons or artificial nerve cells, which come together in layers. These layers are the input layer, the output layer and the hidden layers between them. The number of hidden layers can be 1 or more. Networks with 1 or 2 hidden layers are known to be generally sufficient to solve very complex problems.

Neural fuzzy methods can be superior compared to each other based on the calculation time, performance level or the significance of the rule base. ANFIS is the most well-known among the neural fuzzy systems developed in recent years. ANFIS creates a fuzzy inference system by editing the membership function parameters by using the back propagation algorithm alone or in combination with the least squares method of the input/output data set. This arrangement allows the model to learn the system and adapt itself to new data with the fuzzy system.

In this study, the wavelet analysis method will participate in the neural fuzzy networks method at the input stage of the model. In the use of wavelet transform with ANN, the original time series will be separated into sub-time series (W_1, W_2,..., W_p, C_p) with wavelet transform technique. These sub-series give detailed information about the original series and each different series has a different behavior. Then, in neural fuzzy model, these differentiated series are used as inputs and the original time series is tried to be estimated.

Different types of wavelets can be used for financial time series estimates such as Daubechies, Haar, Morlet, and Mexican Hat wavelets. In this paper, the Haar wavelet has been applied to decompose the original signal O(t). Daubechies, Mexican Hat wavelets, and Morlet wavelets have the advantage of better resolution to change time series smoothly. However, they have the disadvantage of more expensive computation than Haar wavelets. The decomposition level is set to three. Matlab Wavelet Toolbox is used to perform the DWT on the data. The wavelet transform results of the financial series are shown in Appendix 1.

4 Literature Review

The use of artificial neural networks in the field of finance began to spread in the late 1990s. One of the first studies to estimate stock index with artificial neural networks is the study of Kimoto et al. (1990) on Tokyo stock index. Later, Kamijo and Tanikawa (1990) used repetitive neural networks, and Ahmadi (1990) used the backpropagation network.

In 2011, Wang et al. developed a method for estimating Shanghai Stock Exchange (SSE) prices by applying wavelet transform and Back Propagation Neural Networks (BPNN) method with low frequency coefficients. As a result of the study, it was seen that BPNN used with wavelet transform performed better than BPNN which predicted using original time series.

Yuan and Shanshan (2012) introduced D-CPPI and D-TIPP strategies using dynamic multiplier. The dynamic multiplier changes according to the stock price. Increasing the stock price allows a higher multiplier to be used, and when the price falls, a lower multiplier is used to prevent the portfolio from falling below a certain value.

Özçalıcı (2015) developed a hybrid method in his doctoral thesis to estimate the closing prices of stocks traded in the BIST30 index. In the thesis in question, he used artificial neural networks as the prediction method and genetic algorithms were used to determine the number of neurons that should be included in the middle layer and to select variables. It reports that the hybrid method proposed at the end of the study improved stock price prediction success.

Dehghanpour and Esfahanipou (2017) developed an optimization tool using dynamic portfolio insurance strategy and Genetic Programming together. In their study in 2018, they applied a density-based clustering method to select the best stocks and used Adaptive Neuro-Fuzzy Inference Systems (ANFIS) to predict future prices of selected stocks.

5 Methodology

In this study, using the data between January 2016 and February 2019, the daily returns of the Istanbul Stock Exchange (BIST30) Index between March 2019 and December 2019 were estimated. 58% of the data is grouped as training and 42% of the data is grouped as validation data when performing network training.

It is aimed to make future price predictions of BIST-30 (Istanbul Stock Exchange), KOSPI (South Korea Stock Exchange), SHCOMP (Shanghai Stock Exchange Composite Index), SENSEX (India Stock Exchange) and RTSI (Moscow Exchange Index) indices with wavelet + ANFIS model. Then increase the performance of CPPI strategy by using variable risk multiplier according to these estimates.

The autocorrelation function was tested to see if the time series were affected by historical data. All 5 time series were found to be correlated with the T-1 value.

BIST-30, RTSI, KOSPI, SENSEX and SHCOMP indices were divided into subgroups by wavelet analysis method using Haar wavelet. For the subsegments, predictions were made using the neuro fuzzy network model. Then, the index estimates were obtained by summing up the estimates made for each sub segments.

The daily values of 3 sub-series obtained after Wavelet Analysis between 01.02.2019 and 31.12.2019 were estimated by Neural Fuzzy Networks by using MATLAB package program. First, training and validating data is loaded to Neuro-Fuzzy Designer to use Adaptive Neuro-Fuzzy Inference System (ANFIS).

To specify model structure, FIS model is generated using subtractive clustering with 3 membership functions. The left-most node represents the input, the right-most node represents the output, while the node is the normalization factor for the rules.

For the training, the hybrid optimization method was chosen. This method determines the FIS parameters using backpropagation and least squares regression together.100 was selected as the number of training periods and 0 as the error tolerance.

The CPPI strategy was applied after the indices were estimated for T + 1. Dynamic CPPI strategy was applied using a high risk multiplier on the days when the model predicted a price increase, and a low risk multiplier on the days when the model predicted a decrease (Figs. 1, 2, 3, 4, and 5).

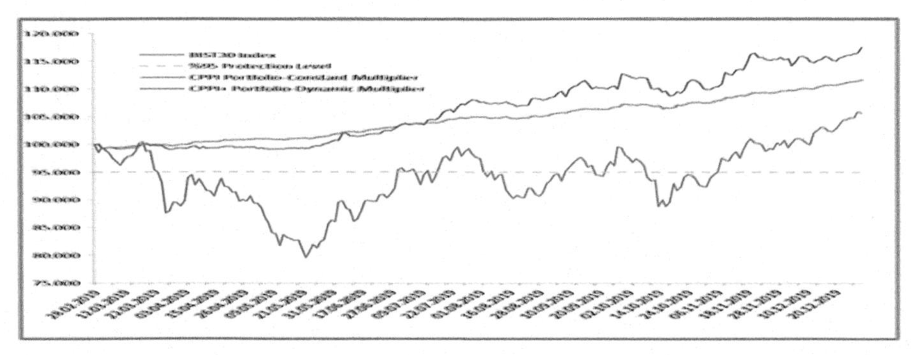

Fig. 1. Performance of BIST-30 index, classical CPPI with constant multiplier and CPPI with dynamic multiplier.

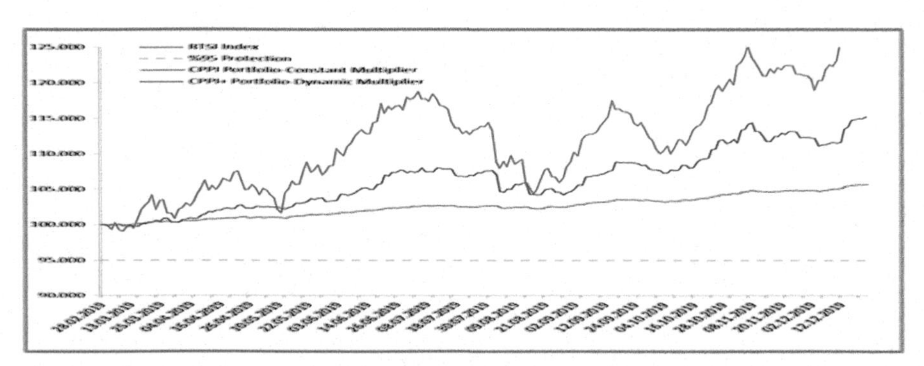

Fig. 2. Performance of RTSI index, classical CPPI with constant multiplier and CPPI with dynamic multiplier.

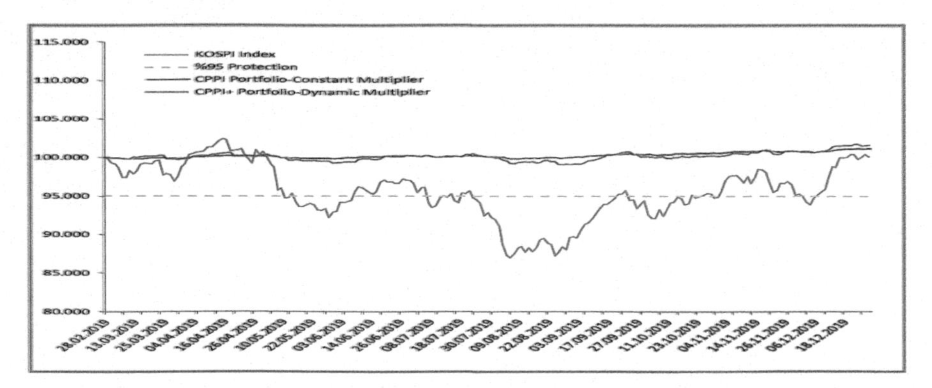

Fig. 3. Performance of KOSPI index, classical CPPI with constant multiplier and CPPI with dynamic multiplier.

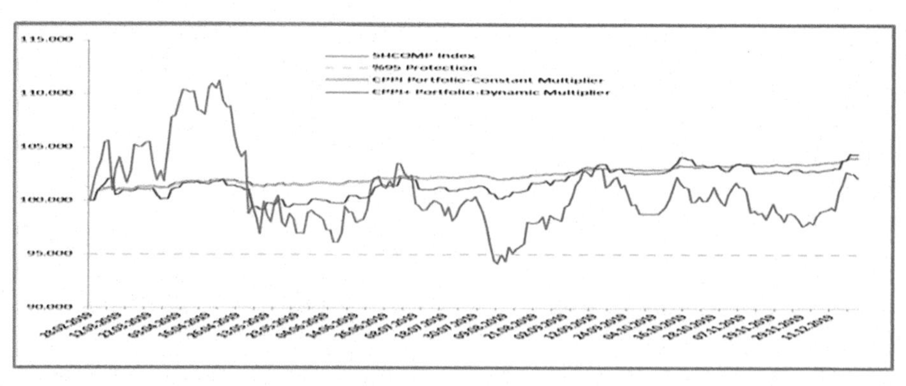

Fig. 4. Performance of SHCOMP index, classical CPPI with constant multiplier and CPPI with dynamic multiplier.

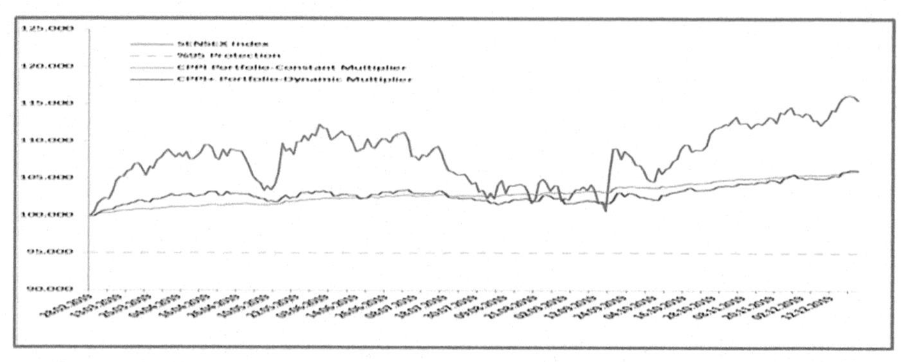

Fig. 5. Performance of SENSEX index, classical CPPI with constant multiplier and CPPI with dynamic multiplier.

The performance of constant and dynamic multiplier CPPI strategies applied to BIST-30, RTSI, KOSPI, SHCOMP and SENSEX are shown in Table 1.

Table 1. Performances of CPPI and CPPI+ of selected indices.

Stock index	Return	CPPI	CPPI+
BIST-30	5,71%	11,68%	17,56%
RTSI	30,35%	5,94%	15,41%
KOSPI	0i10%	1,15%	1,62%
SHCOMP	3,71%	4,25%	5,01%
SENSEX	15,02%	6,12%	6,21%

6 Conclusion

In this study, the daily values of Borsa Istanbul 30 Index, South Korea Stock Index (KOSPI), Shanghai Stock Exchange Composite Index (SHCOMP), India Stock Exchange Index (SENSEX) and Moscow Exchange Index (RTSI) are tried to be estimated by using Wavelet Analysis and Neural Fuzzy Networks method. Then, constant and dynamic multiplier CPPI strategy was applied to indices and their performances are analyzed.

The dynamic multiplier CPPI yielded better returns than the classic CPPI in all 5 stock market indices analyzed, and both strategies successfully implemented the previously targeted 95% capital protection.

It has been observed that predicting future prices of indices using Artificial Intelligence methods and the performance of the dynamic multiplier CPPI strategy applied based on these predictions is more successful than the conventional CPPI strategy with constant multiplier.

The different performance of five different emerging market exchanges enabled the success of the CPPI strategy with dynamic multiplier to be measured in different market conditions.

The results obtained in the study reveal that wavelet analysis and fuzzy neural networks have an important potential in the price estimation of financial assets that require in-depth analysis.

In future research, estimates can be made using different variables and different time intervals. In addition, the effectiveness of the proposed CPPI strategy should be tested by repeating the study in the stock markets of different countries.

References

1. Ahmadi, H.: Testability of the arbitrage pricing theory by neural networks. In: Proceedings of the International Joint Conference on Neural Networks (IJCNN), pp. 385–393, June 1990
2. Kumar, A., Dash, R., Dash, R., Bisoi, R.: Forecasting financial time series using a low complexity recurrent neural network and evolutionary learning approach (2017)
3. Chen, J.-S., Chang, C.-L.: Dynamical proportion portfolio insurance with genetic programming. In: Wang, L., Chen, K., Ong, Y.S. (eds.) ICNC 2005. LNCS, vol. 3611, pp. 735–743. Springer, Heidelberg (2005). https://doi.org/10.1007/11539117_104
4. Chen, J.S., Liao, B.P.: Piecewise nonlinear goal-directed CPPI strategy. J. Exp. Syst. Appl. **33**, 857–869 (2007)
5. Dehghanpour, S., Esfahanipour, A.: A robust genetic programming model for a dynamic portfolio insurance strategy. In: IEEE International Conference on INnovations in Intelligent SysTems and Applications (INISTA), Gdynia, Poland (2017)
6. Hsieh, T.-J., Hsiao, H.-F., Yeh, W.-C.: Forecasting stock markets using wavelet transforms and recurrent neural networks: an integrated system based on artificial bee colony algorithm. Appl. Soft Comput. **11**(2), 2510–2525 (2011)
7. Kamijo, K., Tanigawa, T.: Stock price pattern recognition: a recurrent neural network approach. In: Proceedings of the International Joint Conference on Neural Networks, pp. 215–221 (1990)
8. Kara, Y., Boyacioglu, M.A., Baykan, Ö.K.: Predicting direction of stock price index movement using artificial neural networks and support vector machines: the sample of the Istanbul Stock Exchange. Exp. Syst. Appl. **38**(5), 5311–5319 (2011)

9. Lahmiri, S.: Forecasting direction of the S&P 500 movement using wavelet transform and support vector machines. Int. J. Strateg. Decis. Sci. **4**(1), 78–88 (2013)
10. Lin, C.-T., Yeh, H.-Y.: Empirical of the Taiwan stock index option price forecasting model – applied artificial neural network. Appl. Econ. **41**(15), 1965–1972 (2009)
11. Linzie, A.: Financial Analysis with Artificial Neural Networks Short-term Stock Market Forecasting (2017)
12. Mousavi, S., Esfahanipour, A., Zarandi, M.H.F.: A novel approach to dynamic portfolio trading system using multitree genetic programming. Knowl.-Based Syst. **66**, 68–81 (2014)
13. Özçalıcı, M.: Hisse Senedi Fiyat Tahminlerinde Bilgi İşlemsel Zeka Yöntemleri: Uzman Bir Sistem Aracılığıyla BİST Uygulaması. Kahramanmaraş, Sütçü İmam Üniversitesi Sosyal Bilimler Enstitüsü (2015)
14. Björklund, S., Uhlin, T.: Artificial neural networks for financial time series prediction and portfolio optimization (2017)
15. Şenol, D.: Prediction of stock price direction by artificial neural network approach. Master Thesis, Bogazici University (2018)
16. Shah, M., et al.: Performance analysis of neural network algorithms on stock market forecasting. Int. J. Eng. Comput. Sci. (2014)
17. Dehghanpour, S., Esfahanipour, A.: Dynamic portfolio insurance strategy: a robust machine learning approach. J. Inf. Telecommun. **2**(4), 392–410 (2018)
18. White, H.: Economic prediction using neural network: the case of IBM daily stock return. IEEE International Conference on Neural Networks

Emotion Extraction from Text Using Fuzzy-Deep Neural Network

Ashkan Yeganeh Zaremarjal[1](✉), Derya Yiltas-Kaplan[1], and Soghra Lazemi[2]

[1] Department of Computer Engineering, Istanbul University-Cerrahpaşa, Istanbul, Turkey
ashkany@ogr.iu.edu.tr, dyiltas@iuc.edu.tr

[2] Artificial Intelligence, University of Kashan, Kashan, Iran

Abstract. Emotions are one of the most important and affective start points of our decisions in daily life. People express their emotions, consciously or unconsciously, in their behaviors, like speech, facial expressions, textual content, and so on, whether it be in blogs, reviews, or on social media. Due to the rapid growth in textual social media to link people, the emotion extraction from textual data has attracted a lot of attention. This paper performs the sentence-level discrete emotion extraction from text by using Fuzzy Neural Networks (FNN) and Deep Recurrent Neural Networks (DRNN). Much work has been presented to extracting emotion from textual data with precise and constant boundaries, but little of it has considered the inherent uncertainty in natural language. In other words, the high ambiguity of emotion in text data makes an emotionally sentence express multiple emotions at the same time. To deal with this uncertainty we use a Neuro Neural Network (NNN). But from another point of view, by increasing the number of inputs, the number of parameters in NNN growth exponentially. To this end, we extract a low-dimensional semantic representation of input sentence by using a Bi-directional Long Short-Term Memory (Bi-LSTM). Our goal is to achieve better performance with combining the advantages of deep learning in high-level semantic feature extraction to ambiguity handling and the capabilities of fuzzy logic with fuzzy membership degrees to uncertainty handling. Experiments are conducted on the SemEval2007-Task14 dataset that contains 1250 annotated texts with Ekman's basic emotions. The obtained results indicate that our proposed method outperforms the previous methods.

Keywords: Emotion extraction · Fuzzy Neural Network · Deep learning · Bi-LSTM · Word embedding · Natural language processing

1 Introduction

We are making decisions in our daily lives constantly, for example; what to eat? Which clothes should we wear? Which music should we listen to? and etc. These cases show the lowest level of decision making, which they do not use full capacity of the mind and they relate only based on emotions and feelings that one person experiences it in certain moment [1]. Neuropsychological researches show that every person can make different decisions proportionate to the emotions he/she experiences in decision making time.

C. Kahraman et al. (Eds.): INFUS 2021, LNNS 308, pp. 329–338, 2022.
https://doi.org/10.1007/978-3-030-85577-2_39

Recognizing and analyzing of emotions have dedicated many researches in affective computing, psychology, neurology, behavioral science and computer sciences specially about human and computer interaction (HCI) fields [1, 2]. By increasing expansion of social networks; most of our daily activities are done in virtual networks such as e-learning, e-commerce, internet shopping and so on [3, 4]. Therefore, in HCI systems, automatic emotion extraction was paid attention too, which we can refer to different behaviors such as voice, face, body language, brain imaging and text specially [5]. Emotions extraction from the text is one of the main applications of natural language processing (NLP), that it uses computational linguistics approaches to understand natural language through machine. Text-based emotion extraction system tries to extract emotions, such as happiness, fear, and grief, from written text by human [6]. Generally, emotions are complex subjective concepts and fuzzy, that they can express or understand with incorrect form easily [7]. Emotions are too complex in textual dialogues and one of the main reasons in this case relates to facial expressions and the writer sound frequency unavailability [8]. Emotion extraction via text can be implemented in various levels such as word, sentence, phrase, paragraph or document. These systems are applicable in different fields such as emotion retrieval from suicide notes [9, 10], capturing emotions in multimedia tagging [11], detecting insulting sentences in conversations [12], market research [13], e-learning [14] and so on.

The first stage to create emotion extraction system is determination of emotion model [15]. Emotion model is determinative of how to show emotion. Many models were presented and two main models include discrete and dimensional emotion models (DEMs and DiEMs, respectively) [16]. In discrete model, emotions are placed in classes or separate and independent categories. Dimensional emotion model does not consider emotions independently, but it places them in multi-dimensional and continuous space. In discrete model, the most important are Paul Ekman, Robert Plutchik and Orthony, Clore, and Collins (OCC) models. In Ekman model, there are 6 basic emotions, happiness, sadness, anger, disgust, surprise, and fear, which every case has specific features [17]. Plutchik model is similar with Ekman model and it was determined 8 basic emotions, that they are opposite of each other [18]. In OCC model 16 emotions were added to 6 defined emotions by Ekman [19]. In this paper, Ekman model is used.

The rest of the paper is structured as follows: Sect. 2 describes textual emotion detection methods. Section 3 presents our proposed method in detail. Section 4 gives experimental results and discussion. Finally, Sect. 5 presents the conclusion and future work.

2 Literature Review

Generally, emotions extraction methods from text are divided in 4 classes: keyword-based approaches, rule-based approaches, machine learning-based approaches and hybrid approaches [1, 2].

Traditionally, keyword-based approaches are used in word level. If the word exists in affect lexicon, it is labeled by related set. Many affect lexicons were presented that we can refer to WordNet-Affect and SentiWordNet [20]. This method cannot effective guidelines because of words effectiveness in sentences and phrases processing. If a sentence has not emotional words in dictionary, it is impossible to recognize emotion. One of main weakness in mentioned method is restriction and constant form of defined emotional words in dictionary [21].

In rule-based approaches, some logical rules are defined according to grammar. It is simple, but it has some disadvantages. For example, it is difficult and time-consuming to define complete and extensive rules for covering all of sentences in certain language [15, 21]. This method depends on language completely and determined rules relate a special language. In machine learning-based approaches, emotion extraction problem is changed to classification problem and input text is classified by using supervised or unsupervised machine learning algorithms [2]. The purpose of this method is extraction of proper features from text [8]. Recently, according to successes in deep learning in many fields such as text; deep learning methods were used to extract text latent features and dependency among words [22].

Hybrid approach tries to combine advantages of keyword-based, rule-based, machine learning and deep learning methods for generation a system with high accuracy for emotion extraction from text [1, 23].

Udochukwu and He [24] presented a rule-based method for implicit emotions extraction from text without emotional or sensual words. They tried to recognize relation between emotion and events or actions. Badugu and Suhasini [25] presented rule-based method. They were determined some linguistic rules according to words POS and their frequency of occurrences. SentiwordNet lexicon method was used to rules validation. Suttles and Ide [26] presented distant supervision method as an effective method to overcome against lack of annotated datasets. Their suggested method uses Plutchik emotion model and binary classifiers (Naive Bayes (NB), Maximum Entropy (ME)) for text classification. Bandhakavia et al. [27] try to extract suitable features for input text by using DESL (domain-specific lexicons) and GPEL (general purpose emotions lexicons). Hasan et al. [28] developed supervised machine learning method for emotion extraction from Twitter messages text. To this, a feature vector for every message is defined. Determined features contain unigram features, Emoticon features, Punctuation features and Negation features. Then 3 classifiers are trained through generated vectors. NB as a probabilistic classifier, Support Vector Machine (SVM) as a decision boundary classifier, and decision tree as a rule-based classifier. Polignano et al. [29] used deep learning methods. In their suggested method, at first input words were transferred to vector space by using Word2vec technique. Generated word-vectors were fed to Bi-LSTM network. Its output was entered on Convolutional Neural Network (CNN) by passing of Self-attention layer. In CNN network, latent relation between feature is extracted by using Convolutional and Pooling layers and dense representation of input is generated. CNN network output is fed on Soft-max layer to determine proper classification. Jain et al. [15] suggested hybrid method to extract emotion from Multilanguage text data. At first, they extract current emotional words in text by using NLP technique and then SVM and NB classifications were investigated. Authors used words references such as WorldNet-Affect, Hindi WordNet-Affect, Senti-WorldNet to define 4 types of features and distinction of emotional and non-emotional words in various languages.

3 Proposed Method

In this paper, emotion extraction from text is done at sentence level and regarding emotion fuzzy boundaries and sentence meaning. To handle uncertainty of natural language and fuzzy boundaries, we use Neuro-Fuzzy networks. In mentioned network, rising of input number causes to increase parameters in exponential form. So, we try to enter input sentence latent features to network with low dimensions. To reach this purpose, for input sentence, after requirement preprocessing, semantic and structural features are extracted. Semantic and structural features will be affective in ambiguity handling of natural language. Then the concatenation of these features along word embedding is fed into Bi-LSTM network [30] and its output is used as neuro fuzzy network [31] input. Finally, sentence class is recognized according to generated fuzzy rules in network.

3.1 Semantic and Structural Feature Extraction

In this step, input text is prepared to implement next processing and computing reduction. This stage includes normalization, tokenization, stemming, removing punctuation symbols, removing numbers. To handle natural language intrinsic ambiguity, we extract some of semantic and structural features. These features help machine to understand sentence meaning. Applied features include:

Word Part of Speech tag (POS): This feature is a determinant case in grammar class of sentence elements [32]. In every word of sentence, this case specifies that; is this a noun? Is this an adverb? Is this a verb? Is this a preposition? As we know, in a sentence, all of words have not been emotional content. Emotions are parts of human states and these cases are expressed in adjective or adverb form in natural language [33]. This feature is suitable in determination of effective words.

Syntactic Dependency: Syntactic dependency investigates relations between words. So, it is suitable in structural ambiguity removing [32]. For example, in "Ali with great sadness saw a girl with smile" we faced with very big ambiguity, that causes to challenge in emotion extraction method. Is Ali smiling or a girl? If Ali smiles, sentence emotion will become "happiness" and if a girl smiles, sentence emotion will become "sadness". Dependency parser responses mentioned question by investigating syntactic and conceptual relations among words and also it helps machine to get the sentence correct structure. According to the previous part subjects, adjectives and adverbs are main emotional identities and we regarded 6 features include adverbial clause modifier (advcl), adverb modifier (advmod), adjectival modifier (amod), adjectival complement (acomp), noun phrase as adverbial modifier (npadvmod) and negation modifier (neg) as structural feature.

Word Semantic Role Label: Attaching semantic roles such as agent, patient, theme, experiencer, instrument, location, source, and goal to sentence elements shows their main meaning [32]. Semantic role label causes to better and correct understanding of sentences meaning and it has significant role in disambiguation. Therefore, machine can extract current emotion in the sentence correctly.

Also, we can use semantic feature vector through Word2vec algorithm [34] for words in addition to the 3 mentioned features. This algorithm transfers words to the vector space by maintain of semantic similarity.

Given a sentence, $S_i = \{w_1, w_2, \ldots, w_m\}$, with m words, the concatenation of four feature vectors for each word is fed to Bi-LSTM network as an input, $w_i = w_i^{pos} \oplus w_i^{Syntacticdependency} \oplus w_i^{semanticrolelabel} \oplus w_i^{embedding}$. This network keeps long dependency between words during learning and it gives sentence contextual information. In other words, network output is dense representation of sentence with arbitrary dimensions. Bi-LSTM is formed from a forward LSTM and a backward LSTM network. Generally, LSTM network receives one sequence of items $(w_1 : i)$ and it returns embedding state which is dense representation of input from starting to i^{th} position. For each time step t, hidden state, h_t, is updated with output and memory state in time step t. The concatenation of last hidden layers is extracted as dense feature vector of input sentence, which is as $H_l = \overrightarrow{h} \oplus \overleftarrow{h} = [h_1, h_2, \ldots., h_n]$[30].

3.2 Emotion Extraction

In this step, we consider the feature-representation vector of last hidden layer h_l as the input of fuzzy network in input layer. Input layer receives $x = [x_1, x_2, \ldots., x_n] = H_l = [h_1, h_2, \ldots., h_n]$.

The second layer of neuro fuzzy shows the degree of belonging of different features to different classes. In this layer, every neuron determines one language variable. Neuron number of this layer is equal to the existed emotional classes in datasets. i^{th} input membership grade to j^{th} fuzzy set is computed with Gaussian function.

The 3rd layer's neuron number is equal to the number of fuzzy rules and every neuron is determinant of one rule. Number of neurons is chosen to be equal to the square root of the product of the number nodes in the previous and next layers [35].

The 4th layer has output summation responsibility in all rules. Neuron number of this layer is equal to the existed emotional classes in datasets.

The last layer is defuzzification. The pattern is assigned to class C if $F_C(x) \geq F_j(x), \forall j \in 1, 2, \ldots, Candj \neq c$, where $F_j(x)$ is the activation value of the jth neuron in the 4th layer [31].

4 Experiments and Results

4.1 Dataset and Experimental Setting

SemEval-2007 Task14 dataset [36] was used in experimental implementations. Dataset includes 1250 sentences, news headline, which were extracted from news websites such as New York Times, CNN, BBC and Google News search engine. Emotional sentences were annotated by Ekman model (Anger, Disgust, Fear, Joy, Sadness, Surprise). Positive and negative valence was determined for every sentence. Annotation was done based on existence or non-existence words, phrases with emotional content and overall feeling invoked by the headline. Dataset was split into train, development and test sets by using 10-fold cross validation method. 80% of the data was used for training, 10% was used for developing and 10% was used for testing. The used evaluation metrics were recall, precision, F1 and accuracy [1].

The preprocessing, POS tagging and dependency parsing were implemented using Stanford NLP toolkit (https://nlp.stanford.edu/software/). POS features were extracted from sentences automatically by using Stanford Penn-Bank POS-tagger. The list of POS tags used in this paper is available on https://www.ling.upenn.edu/courses/Fall_2003/ling001/penn_treebank_pos.html. Semantic roles (includes *agent*, *patient*) were extracted using illinois-srl, that has been available on https://gitlab.engr.illinois.edu/cogcomp/illinois-srl. In order to reduce overfitting, the dropout was applied on output vector of each LSTM layer. The hyper-parameters on the development sets were tuned by random search. 200 hyper-parameter setting were evaluated. Table 1 summarizes the chosen hyper-parameters for experiments.

Table 1. Parameter tuning.

Parameter	Range	Final
Word embeddings	–	50
POS tags	–	32
Dependency relation	–	6
Semantic roles	–	2
LSTM unit size	[100,400]	150
Learning rate	$[10^{-3},10^{-1}]$	0.05
Dropout rate	[0,1]	0.36
Mini-batch size	[5, 12]	12
Hidden units of LSTM	[10, 200]	100
Number of neurons in 2th layer of FNN	–	600
Number of neurons in 3th layer of FNN	–	3600
Activation function	–	Hyperbolic tangent

4.2 Results and Discussion

In Table 2, a comparison was done between features by using CBOW and Skip-gram models. This comparison was implemented to determine semantic and structural feature affects in sentences. Based on Table 1, features combination showed better results. The main reason in this case, relates to the usage of semantic and structural features, which cause the machine has better conclusion about natural language and extract correct meaning in slanted sentences. These features are suitable for sentences that they have opposite emotional words, because semantic role label determines who is agent and also dependency parser determines that which emotion is for what subject. Two mentioned features combination determines sentence agent emotion or feeling. Also, Table 2 shows that CBOW model has better efficiency against Skip-gram model. Table 3 shows results of every emotional class in separated form. Results are not well in some classes. The main reason in this case relates to training data insufficient for mentioned class. In Table 4, we made a comparison between some main feature extraction methods from text data by regarding state-of-the-art classifiers. According to Table 4, the defined features had good results against the other features. The main reason relates to the lack of attention to meaning in existing feature extraction methods. Only statistical features were extracted from text, while one of main factors in text automatic and correct processing has been concept and meaning understanding. Table 4 shows that using fuzzy classification could enhance the results significantly.

Table 2. Performance comparison.

	CBOW			Skip-gram		
	Precision	Recall	F1	Precision	Recall	F1
Word Embedding	72.33	68.50	70.36	73.41	65.14	69.03
Word Embedding + POS	77.00	71.88	74.35	71.72	73.43	72.56
Word Embedding + POS + Syntactic Dependency	82.05	76.80	79.34	80.80	76.52	78.60
Word Embedding + POS + Syntactic Dependency + Semantic Role	83.81	82.30	82.99	77.74	85.20	81.30

Table 3. Performance of each emotional class.

	Precision	Recall	F1
Anger	78.28	76.61	77.43
Disgust	89.72	83.45	86.47
Fear	77.69	72.92	75.23
Joy	77.23	84.23	80.58
Sadness	85.95	83.20	84.55
Surprise	94.02	93.39	93.70

Table 4. Accuracy comparison of top feature extraction methods and state-of-the-art classifiers, **CSI:** co-occurrence statistical information, **MF:** most frequent, **EB:** eccentricity-based keyword extraction, **IF-ISF:** term frequency-inverse sentence frequency.

	CSI	MF	EB	TF-ISF	Our proposed features
NB	69.11	78.37	70.01	72.50	79.48
SVM	64.60	77.20	69.00	71.78	78.35
LR	67.39	77.00	69.67	71.88	77.21
Our fuzzy NN	72.33	79.36	72.99	73.24	84.59

5 Conclusion

In this paper, we presented discrete emotion extraction of text in sentence level. The emotions are complex, fuzzy and easily misunderstood entities, so the main purpose of this study is handling ambiguity and natural language uncertainty in textual emotion extraction system. To this end, we used combination of deep learning and neuro fuzzy network. To resolve the input sentence ambiguity and give sentence dense semantic representation, semantic and structural features were extracted for sentence elements and fed into RNN network. In order to remove uncertainty, network's output was used as neuro fuzzy network input. Neuro fuzzy network determines the degree of belonging of input to proper class by considering membership degree and extracting fuzzy rules with network. Experiment results showed that the determined features and fuzzy implementation acted effective and efficient against other methods.

Unfortunately, there is large gap between researchers in English language against other languages in textual emotion extraction field. The main reason in mentioned case relates to the shortage of annotated datasets. So, as future work we suggest the generation of annotated datasets in various languages or the translation of existing datasets. Also suggested method development can be efficient for other languages. At last, we can say that, investigation of other fuzzy networks are suitable and helpful.

Acknowledgment. This work has been supported by Scientific Research Projects Coordination Unit of Istanbul University-Cerrahpasa with the Project number BYP-2020–35200.

References

1. Agrawal, A., An, A.: Unsupervised emotion detection from text using semantic and syntactic relations. In: International Conferences on Web Intelligence and Intelligent Agent Technology, vol. 1, pp. 346–353. IEEE (2012)
2. Acheampong, F.A., Wenyu, C., Nunoo-Mensah, H.: Text-based emotion detection: advances, challenges, and opportunities. Eng. Reports **2**(7), 12189 (2020)
3. Chatzakou, D., Vakali, A., Kafetsios, K.: Detecting variation of emotions in online activities. Expert Syst. Appl. **89**, 318–332 (2017)
4. Imani, M., Montazer, G.A.: A survey of emotion recognition methods with emphasis on E-Learning environments. Network Comput. Appl. **147**, 102423 (2019)

5. Sailunaz, K., Alhajj, R.: Emotion and sentiment analysis from Twitter text. Comput. Sci. **36**, 101003 (2019)
6. Lazemi, S., Ebrahimpour-Komleh, H.: Multi-emotion extraction from text using deep learning. Web Res. **1**(1), 62–67 (2018)
7. Atmaja, B.T., Akagi, M.: Two-stage dimensional emotion recognition by fusing predictions of acoustic and text networks using SVM. Speech Commun. **126**, 9–21 (2021)
8. Halim, Z., Waqar, M., Tahir, M.: A machine learning-based investigation utilizing the in-text features for the identification of dominant emotion in an email. Knowl.-Based Syst. **208**, 106443 (2020)
9. Yang, H., Willis, A., De Roeck, A., Nuseibeh, B.: A hybrid model for automatic emotion recognition in suicide notes. Biomed. Inform. Insights **5**, BII-S8948 (2012)
10. Ghosh, S., Ekbal, A., Bhattacharyya, P.: A Multitask framework to detect depression, sentiment and multi-label emotion from suicide notes. Cogn. Comput. 1–20 (2021). https://doi.org/10.1007/s12559-021-09828-7
11. Xu, Z., Wang, S., Wang, C.: Exploiting multi-emotion relations at feature and label levels for emotion tagging. In: International Conference on Multimedia, pp. 2955–2963 (2020)
12. Allouch, M., Azaria, A., Azoulay, R., Ben-Izchak, E., Zwilling, M., Zachor, D.A.: Automatic detection of insulting sentences in conversation. In: International Conference on the Science of Electrical Engineering in Israel, pp. 1–4. IEEE (2018)
13. Hulubei, A., Avasilcai, S.: Event-based marketing: a trendy and emotional way to engage with the public. In: Marketing and Smart Technologies, pp. 156–165 (2020)
14. Rodriguez, P., Ortigosa, A., Carro, R.M.: Extracting emotions from texts in e-learning environments. In: International Conference on Complex, Intelligent, and Software Intensive Systems. IEEE (2012)
15. Jain, V.K., Kumar, S., Fernandes, S.L.: Extraction of emotions from multilingual text using intelligent text processing and computational linguistics. Comput. Sci. **21**, 316–326 (2017)
16. Borod, J.C.: The Neuropsychology of Emotion. Oxford University Press, Oxford (2000)
17. Ekman, P.: Basic emotions. In: Handbook of Cognition and Emotion (1999)
18. Plutchik, R.: A general psychoevolutionary theory of emotion. In: Theories of Emotion (1980)
19. Ortony, A., Clore, G.L., Collins, A.: The cognitive structure of emotions. Cambridge University Press, Cambridge (1990)
20. Al-Saqqa, S., Abdel-Nabi, H., Awajan, A.: A survey of textual emotion detection. In: International Conference on Computer Science and Information Technology, pp. 136–142. IEEE (2018)
21. Hirat, R., Mittal, N.: A survey on emotion detection techniques using text in blogposts. Int. Bull. Math. Res. **2**(1), 180–187 (2015)
22. Xu, D., Tian, Z., Lai, R., Kong, X., Tan, Z., Shi, W.: Deep learning-based emotion analysis of microblog texts. Inf. Fusion **64**, 1–11 (2020)
23. Xu, G., Li, W., Liu, J.: A social emotion classification approach using multi-model fusion. Future Generation Computer Systems **102**, 347–356 (2020)
24. Udochukwu, O., He, Y.: A rule-based approach to implicit emotion detection in text. In: Biemann, C., Handschuh, S., Freitas, A., Meziane, F., Métais, E. (eds.) NLDB 2015. LNCS, vol. 9103, pp. 197–203. Springer, Cham (2015). https://doi.org/10.1007/978-3-319-19581-0_17
25. Badugu, S., Suhasini, M.: Emotion detection on twitter data using knowledge base approach. Int. J. Comput. Appl. **162**(10) (2017)
26. Suttles, J., Ide, N.: Distant supervision for emotion classification with discrete binary values. In: Gelbukh, A. (ed.) CICLing. LNCS, vol. 7817, pp. 121–136. Springer, Heidelberg (2013). https://doi.org/10.1007/978-3-642-37256-8_11
27. Bandhakavi, A., Wiratunga, N., Padmanabhan, D., Massie, S.: Lexicon based feature extraction for emotion text classification. Pattern Recogn. Lett. **93**, 133–142 (2017)

28. Hasan, M., Rundensteiner, E., Agu, E.: Automatic emotion detection in text streams by analyzing twitter data. Int. J. Data Sci. Anal. **7**(1), 35–51 (2019)
29. Polignano, M., Basile, P., De Gemmis, M., Semeraro, G.: A comparison of word-embeddings in emotion detection from text using BILSTM, CNN and self-attention. In: Conference on User Modeling, Adaptation and Personalization, pp. 63–68 (2019)
30. Hochreiter, S., Schmidhuber, J.: Long short-term memory. Neural Comput. **9**(8), 1735–1780 (1997)
31. Sun, C.-T., Jang, J.-S.: A neuro-fuzzy classifier and its applications. In: International Conference on Fuzzy Systems, pp. 94–98. IEEE (1993)
32. Nadkarni, P.M., Ohno-Machado, L., Chapman, W.W.: Natural language processing: an introduction. J. Am. Med. Inform. Assoc. **18**(5), 544–551 (2011)
33. Mohammad, S., Turney, P.: Emotions evoked by common words and phrases: using mechanical Turk to create an emotion lexicon. In: Workshop on Computational Approaches to Analysis and Generation of Emotion in Text, pp. 26–34 (2010)
34. Mikolov, T., Chen, K., Corrado, G., Dean, J.: Efficient estimation of word representations in vector space. arXiv preprint arXiv:1301.3781 (2013)
35. Rumelhart, D.E., Hinton, G.E., Williams, R.J.: Learning representations by back-propagating errors. Nature **323**(6088), 533–536 (1986)
36. Strapparava, C., Mihalcea, R.: Semeval-2007 task 14: affective text. In: International Workshop on Semantic Evaluations, pp. 70–74 (2007)

RRAM – Based - Equivalent Neural Network

Ali Mohamed[1(✉)], Ali AbuAssal[2], and Osama Rayis[3]

[1] University of Garden City, Khartoum, Sudan
[2] University of York, York, UK
[3] Sudan University of Science and Technology, Khartoum, Sudan

Abstract. The paper shows the capabilities of RRAM to model Neural Networks. The objectives of the work are to provide a likely RRAM based neuron model, and to emulate RRAM Bridge; which is used for synaptic weight adjustment. In order to analyze and design RRAM circuits, a Laplace domain expression has been derived. These equations can then be used for modeling and simulating simple neural networks.

Keywords: RRAM · Neural networks · RRAM bridge · TiO_2

1 Introduction

RRAM has properties of a resistance which has memory, and possesses singular functionalities never exist in primary electronic circuit elements – capacitor, inductor, and resistor. Those Singular functionalities of RRAM can be used very prominently in non-volatile memories, logic systems, and neuromorphic computing. As RRAM is an electronic element with very small dimensions, so its use will contribute greatly in decreasing power consumption and size of integrated circuits. A neural network consists of one or more neurons connected through synapses and fire through different interaction mechanisms like feed forward, feedback inhibition or excitation. The first RRAM device was fabricated at Hewlett-Packard (HP) Labs and reported in the Journal, Nature in 2008 [1, 2], as shown in Fig. 1.

RRAM element is built in two parts of TiO_2 along with $TiO_{2\text{-x}}$, which differ slightly in doping. The element has length D (nm) which is divided into two parts on one straight line, one part has width w (nm) which contains TiO_2-x doping and the other part has width D-w (nm) that contains TiO_2 doping. In this RRAM element, the width of the part which contains $TiO_{2\text{-x}}$, w (nm), is the state variable since it varies with time according to the input stimulus current. μv is the dopant mobility. The part which contains TiO_2-x has a resistance of R_{ON} and the part which contains TiO_2 has resistance, R_{OFF}. Typical values of the RRAM element parameters are, $\mu v = 10^{-14}$ m^2/Vs, $R_{ON} = 116\ \Omega$, $R_{OFF} = 16\ K\Omega$ and $D = 10$ nm [3].

The paper aims to find a method for adjusting synaptic weight and explores the mathematical equations which describe the physical properties of RRAM, in order to derive a Laplace domain expression which can make a base for modeling and simulating simple synaptic connections.

C. Kahraman et al. (Eds.): INFUS 2021, LNNS 308, pp. 339–343, 2022.
https://doi.org/10.1007/978-3-030-85577-2_40

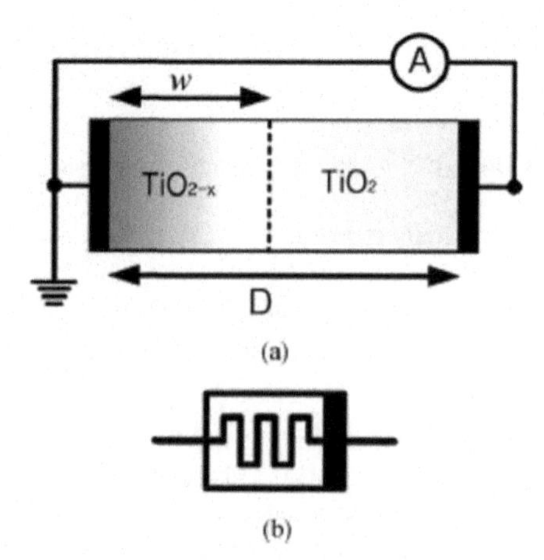

Fig. 1. (a) Cross section of the first TiO_2 based RRAM device (b) Symbol of RRAM in electronic circuits.

The paper is written in the following layout: The introduction is presented in first section, the derivation of equations and modeling is presented in second section, and the conclusion is presented in third section.

2 Derivation of Equations and Modeling

The four basic quantities in electric circuit theory are Charge (q), Flux linkage (ϕ), Current (I) and Voltage (V), stems from them basic relationships represented in the following five equations,

$$q(t) = \int_{-\infty}^{t} i(t)dt \tag{1}$$

$$\phi(t) = \int_{-\infty}^{t} v(t)dt \tag{2}$$

$$R = \frac{dv}{dt} \tag{3}$$

$$L = \frac{d\phi}{dt} \tag{4}$$

$$C = \frac{dq}{dv} \tag{5}$$

Where (R) is resistance, (L) is inductance, and (C) is capacitance.

Based on arguments of symmetry, the 6th relation between charge and flux linkage gives the fourth fundamental passive non-linear circuit element called RRAM (M), described in following Eq. (6) which is the rate of change of flux linkage with respect to charge.

$$M = \frac{d\phi}{dq} \tag{6}$$

Where

$$d\phi = vdt$$

And

$$dq = idt.$$

So

$$M(q) = \frac{v(t)dt}{i(t)dt} = \frac{v(t)}{i(t)} = R \tag{7}$$

RRAM cannot be realized using any linear combination of R, L and C. RRAM is the only passive element that exhibits hysteresis behavior [4].

Hence the voltage across the RRAM at any time t, when a stimulus current i(t) is applied, is given by,

$$V(t) = \left[R_{on}\frac{w(t)}{D} + R_{off}\left(1 - \frac{w(t)}{D}\right)\right]i(t) \tag{8}$$

Where

$$\frac{w(t)}{D} \text{ is the ratio of the state width to the total width of RRAM}$$

Since

$$V(t) = \frac{d\phi(q)}{dq}i(t) \equiv M(q)i(t),$$

Then

$$M(q) = R_{on}\frac{w(t)}{D} + R_{off}(1 - \frac{W(t)}{D}) \tag{9}$$

The state variable changes with time as follows,

$$\frac{\mathrm{dw}}{\mathrm{dt}} = \mu_{\mathrm{v}}\frac{\mathrm{R_{on}}}{\mathrm{D}}\mathrm{i(t)F_P(w)} \tag{10}$$

where

$$F_p(w) = 1 - (2\frac{w}{D} - 1)^{2P} \tag{11}$$

Hence

$$w(t) = \mu_v\frac{R_{on}}{D}\int_0^t i(\tau)d\tau + w_0 \tag{12}$$

After performing the integration, the equation becomes:

$$w(t) = \mu_v\frac{R_{on}}{D}q(t) + w_0 \tag{13}$$

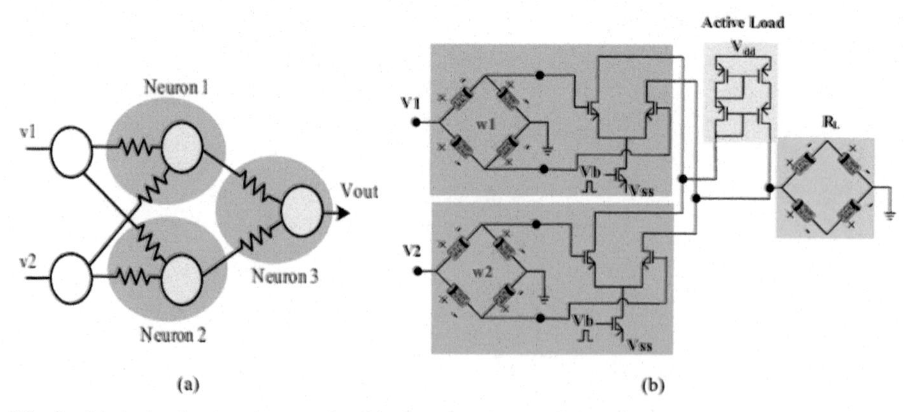

Fig. 2. (a) A simple neural network with three neurons and two synapses (b) Schematic showing the summation of the synaptic outputs using active load and RRAM bridge load.

From the above equation, it is observed that the state variable, w is a function of charge and thus RRAM (M), is a function of charge.

The weight of the synapses can increase or decrease depending on the strength of the activity between the two neurons which the synapse connects. Thus the output voltage of such networks is the weighted sum of the input voltages. Figure 2 shows a simple neural network with three neurons and two synapses [5].

From Fig. 2, the differential output current and output voltage at the output of the N synaptic connections is given by,

$$\mathbf{i_{out}} = \sum_{\mathbf{K}=1}^{\mathbf{N}} \mathbf{g_m w_{syn}^k v_{in}^k},$$

$$\mathrm{v_{out}} = \mathrm{R_L} \sum\nolimits_{\mathrm{K}=1}^{\mathrm{N}} \mathrm{g_m W_{syn}^k v_{in}^k} \tag{14}$$

However, the voltage range of V_{out} is limited by the differential pair.

$$-\mathbf{V_{SS}} + 2\mathbf{V_{th}} < \mathbf{V_{out}} < \mathbf{V_{DD}} - 2\mathbf{V_{th}}$$

Therefore, the output voltage range o f the neural network is clipped to a V_{max} and V_{min}.

$$V_{out} = R_L I_{out}, \, if \frac{(-V_{ss} + 2V_{th})}{R_{out}} < I_{out} < \frac{(V_{DD} - V_{th})}{R_{out}}$$

$$V_{out} = V_{max}, \, if \frac{(V_{DD} + 2V_{th})}{R_{out}} < I_{out}$$

$$\mathbf{V_{out}} = \mathbf{V_{min}}, \, \mathbf{ifI_{out}} = \frac{(-\mathbf{V_{SS}} + 2\mathbf{V_{th}})}{\mathbf{R_{out}}} \tag{15}$$

3 Conclusion

The fourth circuit element (M) is defined and derived; the equation of RRAM voltage-current is explored. A model that indicates a simple neural network with 3 neurons and 2 synapses is demonstrated, and a schematic showing the summation of the synaptic outputs using active load and RRAM bridge load is modeled. The output currents and output voltages equations of an active load and RRAM are expressed.

The paper provided full mathematical analysis for RRAM parameters and put a strong base for studying the properties of RRAM which make it scalable to simulate neurons. The paper also provided simple direct formulas that relate output voltages and currents of neural network with synaptic weights according to the emulated RRAM Bridge.

For future work the following directions must be considered to full cover the physical properties and characteristics of RRAM so as to improve its applications in neural networks specially and in computing generally. The first direction is to study the I-V characteristics of different switching types of RRAM like HFO_2, VO_2, etc., and make comparison between them. The second direction is to study the switching activity of RRAM for different concentrations of dopants and the mobility of dopants must be regarded. The third direction is to study the response of RRAM for different amplitudes and frequencies of electrical signals to show how far it can simulate neurons. The fourth direction is to improve an RRAM which can respond to non electrical signals like heat, light, etc., to make an artificial neuron very similar to biological neurons.

References

1. Author, F.: Journal **2**(5), 99–110 (2016)
2. Lee, H.Y., et al.: Low power and high speed bipolar switching with a thin reactive Ti buffer layer in robust HfO2 based RRAM. In: 2008 IEEE International Electron Devices Meeting, pp. 1–4 (2008). https://doi.org/10.1109/IEDM.2008.4796677. ISBN 978-1-4244-2377-4. S2CID 26927991
3. ACS Appl. Mater. Interfaces 2016, 8, 30, 19605–19611 Publication Date: July 13, 2016 https://doi.org/10.1021/acsami.6b04919 Copyright © 2016 American Chemical Society
4. Materials today ADVANCES Volume 6, June 2020, 100056
5. Yousefzadeh, A., Stromatias, E., Soto, M., Serrano-Gotarredona, T., Linares-Barranco, B.: On Practical Issues for Stochastic STDP Hardware With 1-bit Synaptic Weights. Front. Mol. Neurosci. **12**, 665 (2018)

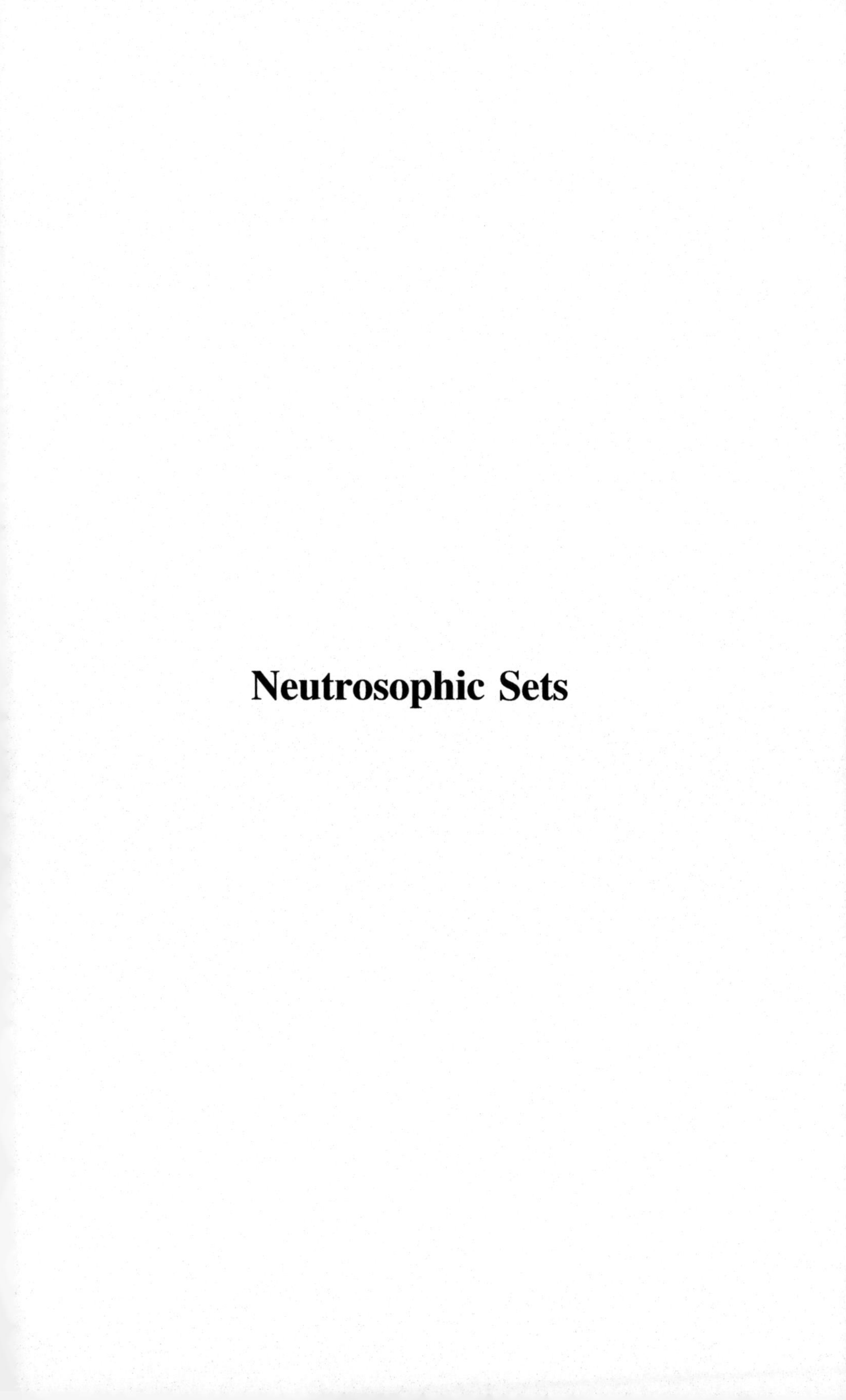

Neutrosophic Sets

Analysis of Supply Chain Disruption Factors Under the Effect of COVID-19 Pandemic via Neutrosophic Fuzzy DEMATEL

Fatma Cayvaz[1,2], Gulfem Tuzkaya[1(✉)], Zeynep Tugce Kalender[1], and Huseyin Selcuk Kilic[1]

[1] Industrial Engineering Department Istanbul, Marmara University, 34722 Istanbul, Turkey
{gulfem.tuzkaya,tugce.simsit,huseyin.kilic}@marmara.edu.tr
[2] Industrial Engineering Department Istanbul, Istanbul Rumeli University, 34582 Istanbul, Turkey
fatma.cayvaz@rumeli.edu.tr

Abstract. Risks caused by natural disasters, terrorist acts, economic fluctuations, and epidemic diseases, etc., result in disruptions. These anticipated or unexpected disruptions affect supply chain resilience degree adversely. A large body of literature exists on supply chain risks, and also the number of studies has increased with the effect of the COVID-19 pandemic. Literature review reveals that disruption factors have different effects on the supply chains, and considering these effects, proper actions should be performed. In this regard, initially, disruption factors existing in the literature are investigated and summarized. Afterward, a methodology based on Neutrosophic Fuzzy Decision Making Trial and Evaluation Laboratory (NF-DEMATEL) is utilized to assess the importance degrees of different disruption factors. The main aim of the proposed methodology is to help managers to take proper actions for disruption factors and state a risk management policy for future unexpected events.

Keywords: Disruption Factors · Resilience · Supply chain risks · Neutrosophic fuzzy DEMATEL · COVID-19

1 Introduction

Natural disasters, terrorist acts, economic fluctuations, epidemic diseases, etc., result in supply chain risks on a global and local scale. For example, an earthquake may cause distribution risks [1], and terrorist acts may result in inventory-related risks [2]. With the COVID-19 outbreak, practitioners and academicians have understood the importance of supply chain resilience against the various risks once again [3].

COVID-19 spread rapidly and WHO acknowledged it as a pandemic on March 11, 2020. Pandemics create risks and disruptions on supply chains. Queiroz *et al.* [4] presented a literature survey on this topic and stated that the literature on the interaction between supply chains and pandemics traditionally focuses on resource and distribution problems and uses optimization methods and epidemic models. On the other hand,

C. Kahraman et al. (Eds.): INFUS 2021, LNNS 308, pp. 347–354, 2022.
https://doi.org/10.1007/978-3-030-85577-2_41

Ivanov and Dolgui [5] stated that during the pandemic, the automotive industries which apply just-in-time production systems faced inventory and capacity risks, as suppliers and factories have different shutdown and lockdown timings. Like the above-mentioned examples, supply chain risks have caused various disruptions, and a significant number of studies have been presented on SC risks. Additionally, the number of studies has increased with the effect of the COVID-19 pandemic. In this study, a literature review is performed, and it is seen that disruption factors have different effects on the supply chains, and considering these effects, proper actions should be performed. In the literature review, more than ten disruption factors and their frequencies are identified. Pareto analysis is used to determine the most important disruption factors, and they are found as lack of transportation (LT), scarcity of raw materials (SRM), fluctuation of demand (FD), lack of inventory (LI), and lack of labor (LL).

It is necessary to take proper actions against these disruption factors to maintain supply chain resilience. However, it is both strenuous and costly to take proper actions against all of them at the same time. Hence, prioritizing the disruption factors considering their importance degrees may be a good idea to order the required actions. And this may be a starting point for the risk management policies of the companies. In this study, considering the gap in the literature, a methodology based on Neutrosophic Fuzzy Decision Making Trial and Evaluation Laboratory (NF-DEMATEL) to assess the importance degrees of different disruption factors is performed. In this methodology, the importance degrees of determined disruption factors on each other were evaluated using fuzzy linguistic expressions by experts in the industry.

The rest of the paper is organized as follows: In the second section, the literature review is presented. In the third section, the details of the Neutrosohic Fuzzy DEMATEL are given. In the fourth section, the application is presented. Finally, concluding remarks are presented.

2 Literature Review

Due to the COVID-19 pandemic outbreak and disruptions on supply chains as a result of it, the number of studies on supply chain risks has increased. Based on the related studies, it can be concluded that proper actions should be performed since disruption factors have different effects on supply chains. Some of the studies related to supply chain disruption factors are summarized as below.

Ohmori and Yoshimoto [6] aimed to solve the transportation disruption problem in the PC industry after an earthquake or flood in Thailand and Japan using network reliability. Bueno-Solano and Cedillo-Campos [2] proposed a system dynamics model for the automotive industry in Mexico–USA trade for disruptions caused by terrorist acts. For a natural disaster like an earthquake, Kohneh *et al.* [7] presented a bi-objective mixed integer programming model and used fuzzy logic to model uncertain parameters for unstable conditions during the disaster in Iran for lack of blood products inventory. Finally, for a resilient food supply chain in Italy, Bottani *et al.* [8] proposed a bi-objective mixed-integer programming model which is solved by an Ant Colony Optimization (ACO) algorithm.

With the COVID-19 pandemic outbreak, the number of studies on supply chain disruptions has been increased. Biswas and Das [9] identified the five main supply chain disruptions named scarcity of raw materials, lack of transportation, lack of manpower, deficiency in cash flow and local laws enforcement, and evaluate the disruptions using fuzzy analytical hierarchy process (Fuzzy-AHP) for Indian manufacturing sectors during the lockdown. The lack of manpower has been found as the highest weighted disruption factor among others. Similarly, for the supply risk related to manufacturing sectors, from China to their distribution centers, Ivanov and Das [10] proposed a simulation model by analyzing three different scenarios. Shahed *et al.* [11] proposed a mathematical model to reduce disruptions in a supply chain network subject to COVID-19 pandemic using pattern search (PS) and genetic algorithm (GA).

In this study, assessing the importance degrees of different disruption are the main focus. The methodology and techniques used in this study are explained below.

3 Neutrosophic Fuzzy DEMATEL

DEMATEL is one of the simple but effective approaches that can be used for forming structural models, including system elements having interactions with each other [12]. As the outputs of DEMATEL, the cause and effect groups of factors are revealed. Moreover, it is also possible to obtain the importance weights of the factors [13]. Different from the crisp DEMATEL, Neutrosophic Fuzzy DEMATEL (NF-DEMATEL) is utilized in this study. Neutrosophic Fuzzy Sets proposed by Smarandache [14] enable us to overcome the indeterminate and vague issues in the process of decision making by considering truthiness, indeterminacy and falsity together [15]. Therefore, they have recently been used in various applications in the literature.

The following steps of NF-DEMATEL are applied within the study [16, 17].

Step 1: Obtain the neutrosophic fuzzy ADIM (aggregated direct-influence matrix)-A^G. The scale given in Table 1 is used for assessing the influence of one factor over the other one.

Table 1. Linguistic expressions (LE) and the related Single-Valued Neutrosophic Fuzzy Numbers (SVNN)

LE	SVNN
Very Unimportant (VU)	(0.1, 0.8, 0.9)
Unimportant (U)	(0.35, 0.6, 0.7)
Medium Important (MI)	(0.5, 0.4, 0.45)
Important (I)	(0.8, 0.2, 0.15)
Absolutely Important (AI)	(0.9, 0.1, 0.1)

To obtain A^G, firstly, the importance weight (w_m) of each related expert is computed via Eq. 1.

$$w_m = \frac{1 - \sqrt{\{(1 - T_m)^2 + (I_m)^2 + (F_m)^2\}/3}}{\sum_{m=1}^{p}(1 - \sqrt{\{(1 - T_m)^2 + (I_m)^2 + (F_m)^2\}/3})} \quad (1)$$

Afterward, each expert's ("m" representing experts, "n" representing factors) individual assessments about the influence of one factor (i) over the other (j) $\left(\left(a_{ij}^{m}\right)_{n*n}\right)$ are aggregated via Eq. 2.

$$a_{ij} = SVNWA_w\left(a_{ij}^{(1)}, a_{ij}^{(2)}, \ldots, a_{ij}^{(p)}\right)$$

$$= w_1 a_{ij}^{(1)} \oplus w_2 a_{ij}^{(2)} \oplus \cdots \oplus w_p a_{ij}^{(p)}$$

$$= \langle 1 - \prod_{m=1}^{p}\left(1 - T_{ij}^{(m)}\right)^{w_m}, \prod_{m=1}^{p}\left(I_{ij}^{(m)}\right)^{w_m}, \prod_{m=1}^{p}\left(F_{ij}^{(m)}\right)^{w_m} \rangle \tag{2}$$

and the ADIM, A^G is gathered as shown below.

$$A^G = \begin{bmatrix} a_{11} & \cdots & a_{1n} \\ \vdots & \ddots & \vdots \\ a_{n1} & \cdots & a_{nn} \end{bmatrix}$$

Step 2: The neutrosophic fuzzy ADIM-A^G is normalized and the matrix B is obtained via Eqs. 3 and 4.

$$B = kxA^G \tag{3}$$

$$k = Min\left(\frac{1}{\max\limits_{1 \le i < n} \sum_{j=1}^{n} T_{ij}}, \frac{1}{\max\limits_{1 \le j < n} \sum_{i=1}^{n} T_{ij}}\right) \tag{4}$$

It is noteworthy that T_{ij} corresponds to the ADIM's truth-membership values.

Step 3: The total DIM (direct-influence matrix) S is found via Eq. 5 in a general form. "I" shows the identity matrix.

$$\mathrm{S} = \mathrm{B} + \mathrm{B}^2 + \mathrm{B}^3 + \cdots + \mathrm{B}^{\mathrm{m}} = \mathrm{B}(\mathrm{I} - \mathrm{B})^{-1} \tag{5}$$

Where, $S = \begin{bmatrix} s_{11} & \cdots & s_{1n} \\ \vdots & \ddots & \vdots \\ s_{n1} & \cdots & s_{nn} \end{bmatrix}$ and $s_{ij} = \; < T_{ij}, I_{ij}, F_{ij} >$

However, since there are three parameters in neutrosophic fuzzy sets consisting of T_{ij}, I_{ij} and F_{ij}. The operations given in general form are separately performed as in Eqs. 6–8 and then the results are brought together to form "S".

$$Matrix\left[T_{ij}\right] = B_T(I - B_T)^{-1} \tag{6}$$

$$\mathrm{Matrix}\left[I_{ij}\right] = B_I(\mathrm{I} - B_I)^{-1} \tag{7}$$

$$\mathrm{Matrix}\left[F_{ij}\right] = B_F(\mathrm{I} - B_F)^{-1} \tag{8}$$

Step 4: The total DIM (S) is deneutrosophicated via Eq. 9.

$$X_Q = 1 - \sqrt{\left\{(1 - T_Q(x))^2 + (I_Q(x))^2 + (F_Q(x))^2\right\}/3} \tag{9}$$

Let SS be the deneutrosophicated matrix and the entries of the matrix are ss_{ij}. Then, a_i (indicating the sum of row i) and b_j (indicating the sum of the column j) values are computed via Eqs. 10 and 11.

$$(a_i)_{nx1} = \left[\sum_{j=1}^{n} ss_{ij}\right]_{nx1} \tag{10}$$

$$(b_j)_{1xn} = \left[\sum_{i=1}^{n} ss_{ij}\right]_{1xn} \tag{11}$$

Step 5: The factors' importance weights are found via the normalization of their (a_i + b_j) values [18].

4 Application of Neutrosophic Fuzzy DEMATEL for the Weight Determination of Supply Chain Disruption Factors

Supply chain risks can occur regarding several conditions. In literature, there are several studies that are focused on supply chain disruption factors; however, the number of these studies are increased during the COVID-19 pandemic. According to the conducted literature review, ten main problems are determined as the most important disruption factors as follows; lack of transportation, scarcity of raw materials, fluctuation of demand, lack of inventory, and lack of labor.

In this study, Neutrosophic Fuzzy Decision Making Trial and Evaluation Laboratory (NF-DEMATEL) is used to assess the importance degrees of different disruption factors to take proper actions and state a risk management policy for future unexpected events. DEMATEL is widely used to form structural models in which system elements have interactions with each other. Although there are several versions of DEMATEL, neutrosophic fuzzy DEMATEL is considered as a method that gives very good results in fuzzy environments. Therefore, in this study, NF-DEMATEL steps that are provided in the previous section are followed to determine the weights of supply chain disruption factors.

Step 1: The scale given in Table 1 is used for assessing the influence of factors over each other. It is asked from decision-makers to use the related scale in their evolutions.

In this study, evaluations are collected from 5 decision-makers who are experts in supply chain risk management. After their evaluations are collected, to obtain A^G, firstly, each expert's importance weight is computed via Eq. 1. In the analysis, it is assumed that decision makers' weights are equal to each other.

In the analysis, decision-makers are represented with m, and determined disruption factors are represented with n. Each decision makers' assessment about the factors and their interactions over each other are aggregated using Eq. 2, and the ADIM, A^G is obtained as Table 2.

Table 2. Aggregated Direct-Influence Matrix

	LL			LT			SRM			FD			LI		
	T	I	F	T	I	F	T	I	F	T	I	F	T	I	F
LL	0.00	0.00	0.00	0.61	0.34	0.33	0.21	0.71	0.81	0.21	0.71	0.81	0.41	0.54	0.25
LT	0.30	0.63	0.73	0.00	0.00	0.00	0.82	0.17	0.13	0.26	0.67	0.77	0.86	0.13	0.85
SRM	0.62	0.34	0.37	0.30	0.63	0.73	0.00	0.00	0.00	0.41	0.51	0.58	0.90	0.10	0.90
FD	0.75	0.23	0.23	0.69	0.29	0.32	0.86	0.13	0.11	0.00	0.00	0.00	0.84	0.15	0.83
LI	0.66	0.30	0.32	0.21	0.71	0.81	0.72	0.26	0.26	0.67	0.31	0.33	0.00	0.00	0.00

Step 2: Eqs. 3 and 4 are used to determine the normalization of the neutrosophic fuzzy ADIM (AG).

Step 3: Based on the calculated normalized matric, Eq. 5 is used to find the total direct-influence matrix (S). In this regard, separate operations are performed for each parameter of the neutrosophic fuzzy sets (Tij, Iij, Fij) as in Eqs. 6–8, and then the results are brought together to form the final "S" matrix as presented in Table 3.

Table 3. The total DIM (S)

	LL			LT			SRM			FD			LI		
	T	I	F	T	I	F	T	I	F	T	I	F	T	I	F
LL	0.25	0.14	0.22	0.35	0.27	0.38	0.34	0.30	0.38	0.24	0.38	0.53	0.42	0.23	0.45
LT	0.47	0.27	0.43	0.27	0.11	0.31	0.62	0.14	0.23	0.36	0.33	0.54	0.68	0.11	0.60
SRM	0.54	0.20	0.37	0.37	0.28	0.53	0.39	0.08	0.19	0.39	0.29	0.51	0.67	0.09	0.65
FD	0.70	0.12	0.22	0.56	0.15	0.30	0.75	0.08	0.15	0.36	0.08	0.21	0.83	0.08	0.46
LI	0.57	0.20	0.29	0.36	0.31	0.45	0.59	0.16	0.21	0.45	0.24	0.36	0.46	0.06	0.30

Step 4: Eq. 9 is used for the deneutrosophication of the total DIM (S) which is presented in Table 4. Afterward, a_i and b_j values are computed via Eqs. 10 and 11.

Step 5: Finally, the importance weights (w) of factors are found via the normalization of their $(a_i + b_j)$ values. The final results are presented in Table 5.

The importance weight for each disruption factor is calculated using NF-DEMATEL methodology as presented in Table 5. According to the final results, "Scarcity of Raw Materials" has the highest importance. Other disruption factors are ranked as "Lack of Inventory", "Fluctuation of Demand", "Lack of Labour" and finally "Lack Of Transportation" according to their importance. Although "Scarcity of Raw Materials" has the highest importance weight, there is no big difference between "Lack of Inventory" and "Fluctuation of Demand".

Table 4. Deneutrosophication of the total DIM (S)

	LL	LT	SRM	FD	LI
LL	0.541	0.537	0.524	0.423	0.556
LT	0.577	0.541	0.728	0.481	0.603
SRM	0.642	0.495	0.631	0.510	0.577
FD	0.774	0.682	0.827	0.613	0.710
LI	0.678	0.514	0.718	0.597	0.643

Table 5. Final scores and importance weights

	a_i	b_j	$a_i + b_j$	$a_i - b_j$	w
Lack of Labour	2.581	3.212	5.794	−0.631	0.1916
Lack of Transportation	2.930	2.768	5.698	0.162	0.1884
Scarcity of Raw Materials	2.855	3.427	6.282	−0.572	0.2077
Fluctuation of Demand	3.605	2.623	6.228	0.982	0.2060
Lack of Inventory	3.150	3.090	6.240	0.060	0.2063

5 Conclusion

To have a resilient supply chain for the companies is essential in case of unexpected events like earthquakes, pandemics, and so on. Resilient supply chains can be established if disruption factors and related risk are identified in a clear way. Hence, in this study, supply chain disruption factors are investigated based on a literature survey. The five most important disruption factors are determined by applying Pareto analysis according to their frequencies. Managers need to know the importance of these five disruption factors in order to take proper actions against them. In this study, neutrosophic fuzzy DEMATEL, which is considered as a method that gives very good results in fuzzy environments, is used to determine the importance degrees of disruption factors. "Scarcity of raw materials" is found as the highest weighted disruption factor. According to this result, managers should focus primarily on raw material sourcing decisions to improve and/or maintain the supply chain resilience. In this study, industries were not evidently mentioned. Future studies can consider inspecting supply chain disruption factors for specific industries.

References

1. Rezaei, M., Afsahi, M., Shafiee, M., Patriksson, M.: A bi-objective optimization framework for designing an efficient fuel supply chain network in post-earthquakes. Comput. Ind. Eng. **147**, 106654 (2020)

2. Bueno-Solano, A., Cedillo-Campos, M.G.: Dynamic impact on global supply chains performance of disruptions propagation produced by terrorist acts. Transp. Res. Part E: Logist. Transp. Rev. **61**, 1–12 (2014)
3. Golan, M.S., Jernegan, L.H., Linkov, I.: Trends and applications of resilience analytics in supply chain modeling: systematic literature review in the context of the COVID-19 pandemic. Environ. Syst. Dec. **40**(2), 222–243 (2020). https://doi.org/10.1007/s10669-020-09777-w
4. Queiroz, M.M., Ivanov, D., Dolgui, A., Wamba, S.F.: Impacts of epidemic outbreaks on supply chains: mapping a research agenda amid the COVID-19 pandemic through a structured literature review. Ann. Oper. Res. 1–38 (2020)
5. Ivanov, D., Dolgui, A.: OR-methods for coping with the ripple effect in supply chains during COVID-19 pandemic: managerial insights and research implications. Int. J. Prod. Econ. **232**, 107921 (2021)
6. Ohmori, S., Yoshimoto, K.: A framework of managing supply chain disruption risks using network reliability. Ind. Eng. Manage. Syst. **12**(2), 103–111 (2013)
7. Kohneh, J. N., Teymoury, E., Pishvaee, M.S.: Blood products supply chain design considering disaster circumstances (Case study: earthquake disaster in Tehran). J. Ind. Syst. Eng. **9**(special issue on supply chain), 51–72 (2016)
8. Bottani, E., Murino, T., Schiavo, M., Akkerman, R.: Resilient food supply chain design: modelling framework and metaheuristic solution approach. Comput. Ind. Eng. **135**, 177–198 (2019)
9. Biswas, T.K., Das, M.C.: Selection of the barriers of supply chain management in Indian manufacturing sectors due to COVID-19 impacts. Oper. Res. Eng. Sci. Theory Appl. **3**(3), 1–12 (2020)
10. Ivanov, D., Das, A.: Coronavirus (COVID-19/SARS-CoV-2) and supply chain resilience: a research note. Int. J. Integr. Supply Manage. **13**(1), 90–102 (2020)
11. Shahed, K.S., Azeem, A., Ali, S.M., Moktadir, M.A.: A supply chain disruption risk mitigation model to manage COVID-19 pandemic risk. Environ. Sci. Pollut. Res. 1–16 (2021)
12. Yasmin, M., Tatoglu, E., Kilic, H.S., Zaim, S., Delen, D.: Big data analytics capabilities and firm performance: an integrated MCDM approach. J. Bus. Res. **114**, 1–15 (2020)
13. Kilic, H.S., Demirci, A.E., Delen, D.: An integrated decision analysis methodology based on IF-DEMATEL and IF-ELECTRE for personnel selection. Decis. Supp. Syst. **137**, 113360 (2020)
14. Smarandache, F.: A unifying field in logics: neutrosophic logic. In: Philosophy, pp. 1–141. American Research Press (1999)
15. Kilic, H.S., Yurdaer, P., Aglan, C.: A leanness assessment methodology based on neutrosophic DEMATEL. J. Manuf. Syst. **59**, 320–344 (2021)
16. Awang, A., Aizam, N.A.H., Abdullah, L.: An integrated decision-making method based on neutrosophic numbers for investigating factors of coastal erosion. Symmetry **11**(3), 328 (2019)
17. Kilic, H.S., Yalcin, A.S.: Comparison of municipalities considering environmental sustainability via neutrosophic DEMATEL based TOPSIS. Socio-Econ. Plan. Sci. 100827 (2020)

Cylindrical Neutrosophic Single-Valued Fuzzy MCDM Approach on Electric Vehicle Charging Station Relocation with Time-Dependent Demand

Esra Çakır[1(✉)], Mehmet Ali Taş[2], and Ziya Ulukan[1]

[1] Department of Industrial Engineering, Galatasaray University, Ortakoy/Istanbul 34349, Turkey
{ecakir,zulukan}@gsu.edu.tr
[2] Department of Industrial Engineering, Turkish-German University, Beykoz/Istanbul 34820, Turkey
mehmetali.tas@tau.edu.tr

Abstract. The use of electric vehicles has been increasing rapidly in recent years. Hence, determining the location of charging stations is a strategic decision. Managements may consider relocating charging stations as demand may change over time. In order to reflect the uncertainty in demand, it is appropriate to use fuzzy expressions. The demands of customers at present time can be calculated according to the criteria determined by using the fuzzy evaluations of decision makers. In this study, the single facility relocation problem for time-dependent demand is examined on finite time horizon. Rectilinear distance is considered. Demands of locations are calculated according to the criteria set by decision makers using cylindrical neutrosophic single-valued fuzzy expressions. This methodology is applied to the problem of minimum cost location of electric charging stations. The time-dependent demands of customers are determined with the help of cylindrical neutrosophic single-valued numbers. The minimum cost of relocating the charging station in a ten-year period is investigated. This application is intended to guide future facility relocation approaches.

Keywords: Cylindrical neutrosophic single-valued fuzzy set · Minisum · Rectilinear distance · Relocation · Single facility location problem · Time-dependency

1 Introduction

Electric vehicles contribute to sustainable environmental targets thanks to low life-cycle CO_2 emissions and low energy usage [1, 2]. In addition to the fuel cost advantage and high efficiency, they contain positive aspects such as quicker operation and easier maintenance [3]. As the share of electric vehicles in transportation increases day by day, the need for electric charging stations to be established to charge their batteries increases [4]. It is necessary to determine the location of the electric vehicle charging station (EVCS) to be established properly [5–7]. In the literature, the optimum localization of the

C. Kahraman et al. (Eds.): INFUS 2021, LNNS 308, pp. 355–363, 2022.
https://doi.org/10.1007/978-3-030-85577-2_42

EVCSs is investigated by game theory [8], grey decision-making model [9], fuzzy AHP, TOPSIS and Geographic Information System [10], fuzzy best-worst method (BWM), fuzzy entropy weighted method (EWM), and fuzzy gray relation analysis methods [11]. For further benefits on localization of EVCSs, it can be necessary to determine the location of a facility established in the location according to the changing conditions over time [12]. The relocation of the facility in the time horizon may be cheaper than the fixed location model. This problem is defined as a single facility relocation with time-dependent demand where the station's location is re-determined in case staying in the same location is more costly than the relocation cost [13]. The distance between demand points can be calculated by Rectilinear, Euclidean, Squared Euclidean distance, and so on [14]. As a solution methodology, the problem can be examined using an optimal procedure with a single relocation [15], dynamic programming and a mixed integer programming model in multi-period [16], in discrete planning [13] or in continuous time horizon [14].

Based on Zadeh's fuzzy set theory [17], numerous fuzzy extentions have been proposed to express uncertainty such as intuitionistic fuzzy set [18], neutrosophic fuzzy set [19], hesitant fuzzy set [20], etc. As a new extension of neutrosophic fuzzy sets, cylindrical neutrosophic single-valued numbers (CNSVNs) are introduced by Chakraborty et al. [21]. In these numbers "indeterminacy and falsity functions are dependent on each other using an influx of different logical and innovative graphical representations" [22]. As an application area, CNSVNs are used in network problems [21] and a minimal spanning tree method [22], so far. This study uses the CNSVNs to state the uncertainty in demands that change over time.

In addition to single facility relocation studies in the literature, this paper contributes by proposing a hybrid fuzzy multi-criteria decision making (MCDM) and single facility relocation approach for time-dependent demand. The proposed methodology is applied on location problem of an EVCS. The case study is also pioneering work on cylindrical neutrosophic single-valued fuzzy MCDM applications.

The organization of this paper is as follows. Section 2 introduces CNSVNs. The hybrid cylindrical neutrosophic fuzzy MCDM and single facility relocation for time-dependent demand approach is given in Sect. 3. Proposed methodology is applied on an EVCS problem in Sect. 4. The paper is concluded with future perspectives in Sect. 5.

2 Cylindrical Neutrosophic Single-Valued Fuzzy Set

This section gives the preliminaries and definitions of cylindrical neutrosophic single-valued fuzzy set [21].

Definition 1: Let X be a space of points (objects) and the generic element in X is denoted by x; $\widetilde{\mathrm{CNFN}} = \{<x : \mathrm{T}_{\widetilde{\mathrm{CNFN}}}(x), \mathrm{I}_{\widetilde{\mathrm{CNFN}}}(x), \mathrm{F}_{\widetilde{\mathrm{CNFN}}}(x)>, x \in X\}$ is the form of an object that is the neutrosophic set $\widetilde{\mathrm{CNFN}}$, where the functions T, I, F: x $\rightarrow$ $]^{-}0, 1^{+}[$ define respectively the truth-membership function, an indeterminacy-membership function and a falsity-membership function of the element $x \in X$ to the set $\widetilde{\mathrm{CNFN}}$ with condition:

$$^{-}0 \leq \mathrm{T}_{\widetilde{\mathrm{CNFN}}}(x) + \mathrm{I}_{\widetilde{\mathrm{CNFN}}}(x) + \mathrm{F}_{\widetilde{\mathrm{CNFN}}}(x) \leq 3^{+} \quad (1)$$

The function $T_{\widetilde{CNFN}}(x)$, $I_{\widetilde{CNFN}}(x)$ and $F_{\widetilde{CNFN}}(x)$ are real standard or non-standard subset of $]^{-}0, 1^{+}[$.

Definition 2: A set $\widetilde{CNFN}$ in the universal discourse X, symbolically denoted by x, it is called a cylindrical neutrosophic set [23] if $\widetilde{CNFN} = \{<x : T_{\widetilde{CNFN}}(x), I_{\widetilde{CNFN}}(x), F_{\widetilde{CNFN}}(x)>, x \in X\}$ where the functions T, I, F: $x \rightarrow]^{-}0, 1^{+}[$ define respectively the truth-membership function, an indeterminacy-membership function and a falsity-membership function of the element $x \in X$ to the set $\widetilde{CNFN}$ with condition:

$$(T_{\widetilde{CNFN}}(x))^2 + (I_{\widetilde{CNFN}}(x))^2 \leq 1^{+},\ F_{\widetilde{CNFN}}(x) \leq 1^{+} \tag{2}$$

CNSVNs are illustrated in Fig. 1.

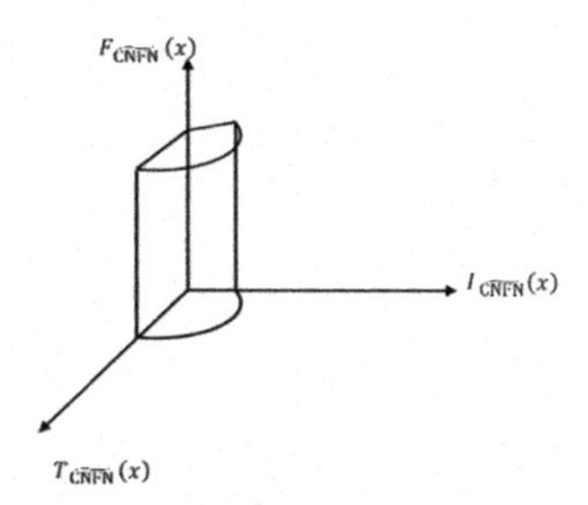

Fig. 1. Geometrical representation of a cylindrical neutrosophic single-valued fuzzy number.

Definition 3: Let $\widetilde{CNFN}_1 = \langle T_{\widetilde{CNFN}_1}, I_{\widetilde{CNFN}_1}, F_{\widetilde{CNFN}_1} \rangle$ and $\widetilde{CNFN}_2 = \langle T_{\widetilde{CNFN}_2}, I_{\widetilde{CNFN}_2}, F_{\widetilde{CNFN}_2} \rangle$ be two CNSVNs. Then, union, intersection and complementation of two CNSVNs are defined as follows:

$$\widetilde{CNFN}_1 \subseteq \widetilde{CNFN}_2,\ T_{\widetilde{CNFN}_1} \leq T_{\widetilde{CNFN}_2},\ I_{\widetilde{CNFN}_1} \leq I_{\widetilde{CNFN}_2},\ F_{\widetilde{CNFN}_2} \leq F_{\widetilde{CNFN}_1} \tag{3}$$

$$\widetilde{CNFN}_1 = \widetilde{CNFN}_2 \ \textit{iff}\ \widetilde{CNFN}_1 \subseteq \widetilde{CNFN}_2 \wedge \widetilde{CNFN}_2 \subseteq \widetilde{CNFN}_1 \tag{4}$$

$$\widetilde{CNFN}_1 \cup \widetilde{CNFN}_2 = \langle \max\left(T_{\widetilde{CNFN}_1}, T_{\widetilde{CNFN}_2}\right), \min\left(I_{\widetilde{CNFN}_1}, I_{\widetilde{CNFN}_2}\right), \min(F_{\widetilde{CNFN}_1}, F_{\widetilde{CNFN}_2})\rangle \tag{5}$$

$$\widetilde{CNFN}_1 \cap \widetilde{CNFN}_2 = \langle \min\left(T_{\widetilde{CNFN}_1}, T_{\widetilde{CNFN}_2}\right), \min\left(I_{\widetilde{CNFN}_1}, I_{\widetilde{CNFN}_2}\right), \max(F_{\widetilde{CNFN}_1}, F_{\widetilde{CNFN}_2})\rangle \tag{6}$$

Definition 4: Let $\widetilde{CNFN} = \langle T_{\widetilde{CNFN}}, I_{\widetilde{CNFN}}, F_{\widetilde{CNFN}} \rangle$ be a single-valued neutrosophic fuzzy number. Then, the accuracy function $A\left(\widetilde{CNFN}\right)$ of $\widetilde{CNFN}$ are defined as below:

$$A\left(\widetilde{CNFN}\right) = \frac{2(T_{\widetilde{CNFN}})^2 + (I_{\widetilde{CNFN}})^2 + (F_{\widetilde{CNFN}})^2}{2} \text{ where } A\left(\widetilde{CNFN}\right) \epsilon [0, 2) \tag{7}$$

Definition 5: Let $\widetilde{CNFN}_1 = \langle T_{\widetilde{CNFN}_1}, I_{\widetilde{CNFN}_1}, F_{\widetilde{CNFN}_1} \rangle$ and $\widetilde{CNFN}_2 = \langle T_{\widetilde{CNFN}_2}, I_{\widetilde{CNFN}_2}, F_{\widetilde{CNFN}_2} \rangle$ be two CNSVNs. Then, the ranking method is defined as follows:

- If $S\left(\widetilde{CNFN_1}\right) \succ S\left(\widetilde{CNFN_2}\right)$, then $\widetilde{CNFN_1} \succ \widetilde{CNFN_2}$.
- If $S\left(\widetilde{CNFN_1}\right) = S\left(\widetilde{CNFN_2}\right)$, then
  - If $A\left(\widetilde{CNFN_1}\right) \succ A\left(\widetilde{CNFN_2}\right)$, then $\widetilde{CNFN_1} \succ \widetilde{CNFN_2}$.
  - If $A\left(\widetilde{CNFN_1}\right) = A\left(\widetilde{CNFN_2}\right)$, then $\widetilde{CNFN_1} = \widetilde{CNFN_2}$.

3 Proposed Approach

This section introduces the hybrid fuzzy MCDM and single facility relocation on finite time horizon approach [13, 21]. The methodology aims to investigate possible relocations of a new facility during [0,T] period. The customers' demand weights at the present time $t = 0$ is determined by fuzzy MCDM methodology according to the criteria and evaluation of the decision makers (DMs). Then, the single facility relocation on finite time horizon approach is applied to find the optimal relocation time(s) and coordinates. The steps of proposed hybrid methodology are explained as follows.

Step 1: Define the problem. Consider $A = \{A_1, A_2, \ldots, A_m\}$ is the customers set, $C = \{C_1, C_2, \ldots, C_n\}$ is the criteria set and $D = \{D_1, D_2, \ldots, D_k\}$ is the set of DMs. Let $W_C = \{W_{C_1}, W_{C_2}, \ldots, W_{C_n}\}$ is the weight vector of criteria and let $W_D = \{W_{D_1}, W_{D_2}, \ldots, W_{D_k}\}$ is the weight vector of DMs, where $W_* \geq 0$ and $\Sigma W_* = 1$. These weight vectors are determined by DMs.

Step 2: Construct cylindrical neutrosophic single-valued fuzzy decision matrices for customer demands.

$$\widetilde{X^k} = \begin{array}{c} \\ A_1 \\ A_2 \\ \cdots \\ A_m \end{array} \begin{array}{c} \begin{array}{cccc} C_1 & C_2 & \cdots & C_n \end{array} \\ \begin{bmatrix} \widetilde{x_{11}^k} & \cdots & \widetilde{x_{1n}^k} \\ \widetilde{x_{21}^k} & \cdots & \widetilde{x_{2n}^k} \\ \cdots & \cdots & \cdots \\ \widetilde{x_{m1}^k} & \cdots & \widetilde{x_{mn}^k} \end{bmatrix} \end{array}$$

Step 3: Aggregate all decision matrices using the operation $\widetilde{x_{ij}'} = \sum_{l=1}^{k} w_{D_l} \widetilde{x_{ij}^l}$, $i = \{1,2,\ldots,m\}$, $j = \{1,2,\ldots,n\}$.

$$\widetilde{X'} = \begin{array}{c} \\ A_1 \\ A_2 \\ \cdots \\ A_m \end{array} \begin{array}{c} \begin{array}{cccc} C_1 & C_2 & \cdots & C_n \end{array} \\ \begin{bmatrix} \widetilde{x_{11}'} & \cdots & \widetilde{x_{1n}'} \\ \widetilde{x_{21}'} & \cdots & \widetilde{x_{2n}'} \\ \cdots & \cdots & \cdots \\ \widetilde{x_{m1}'} & \cdots & \widetilde{x_{mn}'} \end{bmatrix} \end{array}$$

Step 4: Find the weighted priority matrix $\widetilde{W_A}$ using the operation $\widetilde{x_i''} = \sum_{j=1}^{n} w_{C_J} \widetilde{x'_{iJ}}$ $i = \{1,2,\ldots,m\}$.

$$\widetilde{W_A} = \begin{matrix} A_1 \\ A_2 \\ \ldots \\ A_m \end{matrix} \begin{bmatrix} \widetilde{x_1''} \\ \widetilde{x_2''} \\ \ldots \\ \widetilde{x_m''} \end{bmatrix}$$

Step 5: Calculate the scores of demands $A(\widetilde{W_{A_i}})$ using accuracy function defined in Eq. (7).

Step 6: Determine customer time-dependent demand function $w_{A_i}(t)$ where t is the time in range of [0,T] [15].

$$w_{A_i}(t) = 100 * A\left(\widetilde{W_{A_i}}\right) + v_{A_i} t \text{ where } w_{A_i}(t) \geq 0 \text{ and } t \in [0, T] \tag{8}$$

Here, $A(\widetilde{W_{A_i}})$ represents the demand of customer at present time $t = 0$, and v_{A_i} is determined by experts.

The objective is to minimize the total cost of construct a new facility Y $F(Y) = \int_0^T \left\{ \sum_{i=1}^{m} w_{A_i}(t) d(Y, p_{A_i}) \right\} dt$, where $w_{A_i}(t)$ is the weight function associated with customer A_i, and $d(Y, p_{A_i})$ is the distance between with customer A_i located at $p_{A_i} = (a_{A_i}, b_{A_i})$ and the new facility Y located at $Y = (x, y)$. Here, the rectilinear distance is considered with $d\left(Y, p_{A_i}\right) = \left|x - a_{A_i}\right| + \left|y - b_{A_i}\right|$.

Step 7: Find optimal locations at $t = 0$ and $t = T$ with median point [24, 25]. The coordinates between these two locations form alternative points Y_*. Calculate the total cost of construction of a new facility at each alternative points Y_*. Note that if the two coordinates are the same, there is no relocation since the change is linear in time when the time-dependent demand function is linear.

Step 8: For each alternative pair of points (Y_1, Y_2), the transition time t' is calculated by $w_{A_i}(t')d\left(Y_1, p_{A_i}\right) = w_{A_i}(t')d(Y_2, p_{A_i})$. The lowest total cost of pair $F(Y_1)+F(Y_2) = \int_0^{t'} w_{A_i}(t) d(Y_1, p_{A_i}) dt + \int_{t'}^{T} w_{A_i}(t) d(Y_2, p_{A_i}) dt$ gives the initial (Y_1) and relocated (Y_2) location of the new facility.

4 Illustrative Example

A fuel company aims to establish a charging station for electric vehicles to achieve its sustainable environmental goals. The company wants to determine the location of the new electric charging station according to the demands of customers coming to its existing four fuel stations. The company is also keen to investigate the cost of relocating the new charging station, predicting the change in demand over the next decade *(T = 10)*.

Step 1: The fuel company has four stations A_1, A_2, A_3, A_4 at coordinates $p_{A_1} = (0, 4)$, $p_{A_2} = (6, 10)$, $p_{A_3} = (5, 12)$, $p_{A_4} = (8, 9)$. The management wants to determine the demand weights of these stations according to the criteria C_1: income level of region, C_2: effectiveness of nearby competitors, C_3: traffic density, C_4: physical infrastructure with the weights of $W_{C_1} = 0.30$, $W_{C_2} = 0.20$, $W_{C_3} = 0.35$, $W_{C_1} = 0.15$. Three

DMs D_1, D_2, D_3 are selected by company with the weights of $W_{D_1} = 0.33$, $W_{D_2} = 0.32$, $W_{D_3} = 0.35$.

Step 2: There DMs evaluate the existing fuel stations according to the criteria using CNSVNs.

$$\widetilde{X^1} = \begin{array}{c} \\ A_1 \\ A_2 \\ A_3 \\ A_4 \end{array} \begin{array}{cccc} C_1 & C_2 & C_3 & C_4 \\ \left[\begin{array}{cccc} \langle 0.3, 0.6, 0.1 \rangle & \langle 0.6, 0.7, 0.3 \rangle & \langle 0.5, 0.4, 0.6 \rangle & \langle 0.4, 0.5, 0.4 \rangle \\ \langle 0.4, 0.6, 0.2 \rangle & \langle 0.4, 0.2, 0.5 \rangle & \langle 0.3, 0.2, 0.4 \rangle & \langle 0.6, 0.2, 0.3 \rangle \\ \langle 0.6, 0.2, 0.4 \rangle & \langle 0.3, 0.5, 0.4 \rangle & \langle 0.4, 0.4, 0.5 \rangle & \langle 0.8, 0.4, 0.1 \rangle \\ \langle 0.3, 0.4, 0.5 \rangle & \langle 0.6, 0.4, 0.3 \rangle & \langle 0.7, 0.5, 0.3 \rangle & \langle 0.5, 0.3, 0.5 \rangle \end{array}\right] \end{array} \quad \widetilde{X^2} = \begin{array}{c} \\ A_1 \\ A_2 \\ A_3 \\ A_4 \end{array} \begin{array}{cccc} C_1 & C_2 & C_3 & C_4 \\ \left[\begin{array}{cccc} \langle 0.3, 0.2, 0.4 \rangle & \langle 0.2, 0.5, 0.4 \rangle & \langle 0.5, 0.4, 0.6 \rangle & \langle 0.3, 0.4, 0.5 \rangle \\ \langle 0.4, 0.6, 0.2 \rangle & \langle 0.3, 0.4, 0.6 \rangle & \langle 0.7, 0.2, 0.3 \rangle & \langle 0.5, 0.3, 0.4 \rangle \\ \langle 0.4, 0.5, 0.3 \rangle & \langle 0.5, 0.4, 0.7 \rangle & \langle 0.8, 0.2, 0.2 \rangle & \langle 0.6, 0.2, 0.4 \rangle \\ \langle 0.5, 0.4, 0.6 \rangle & \langle 0.6, 0.3, 0.4 \rangle & \langle 0.5, 0.6, 0.4 \rangle & \langle 0.6, 0.4, 0.2 \rangle \end{array}\right] \end{array}$$

$$\widetilde{X^3} = \begin{array}{c} \\ A_1 \\ A_2 \\ A_3 \\ A_4 \end{array} \begin{array}{cccc} C_1 & C_2 & C_3 & C_4 \\ \left[\begin{array}{cccc} \langle 0.5, 0.1, 0.4 \rangle & \langle 0.5, 0.4, 0.3 \rangle & \langle 0.6, 0.3, 0.4 \rangle & \langle 0.8, 0.5, 0.4 \rangle \\ \langle 0.6, 0.5, 0.2 \rangle & \langle 0.6, 0.2, 0.3 \rangle & \langle 0.5, 0.2, 0.1 \rangle & \langle 0.7, 0.6, 0.3 \rangle \\ \langle 0.7, 0.3, 0.4 \rangle & \langle 0.5, 0.4, 0.6 \rangle & \langle 0.6, 0.2, 0.4 \rangle & \langle 0.6, 0.5, 0.6 \rangle \\ \langle 0.6, 0.5, 0.3 \rangle & \langle 0.7, 0.2, 0.5 \rangle & \langle 0.7, 0.5, 0.6 \rangle & \langle 0.7, 0.4, 0.3 \rangle \end{array}\right] \end{array}$$

Step 3: Aggregated of three cylindrical neutrosophic single-valued fuzzy decision matrices is as follows:

$$\widetilde{X'} = \begin{array}{c} \\ A_1 \\ A_2 \\ A_3 \\ A_4 \end{array} \begin{array}{cccc} C_1 & C_2 & C_3 & C_4 \\ \left[\begin{array}{cccc} \langle 0.37, 0.30, 0.30 \rangle & \langle 0.44, 0.53, 0.33 \rangle & \langle 0.54, 0.37, 0.53 \rangle & \langle 0.51, 0.47, 0.43 \rangle \\ \langle 0.47, 0.57, 0.20 \rangle & \langle 0.44, 0.26, 0.46 \rangle & \langle 0.50, 0.20, 0.26 \rangle & \langle 0.60, 0.37, 0.33 \rangle \\ \langle 0.57, 0.33, 0.37 \rangle & \langle 0.43, 0.43, 0.57 \rangle & \langle 0.60, 0.27, 0.37 \rangle & \langle 0.67, 0.37, 0.37 \rangle \\ \langle 0.47, 0.44, 0.46 \rangle & \langle 0.64, 0.30, 0.40 \rangle & \langle 0.64, 0.53, 0.44 \rangle & \langle 0.60, 0.37, 0.37 \rangle \end{array}\right] \end{array}$$

Step 4: The weighted priority matrix $\widetilde{W_A}$ is calculated as follows:

$$\widetilde{W_A} = \begin{array}{c} A_1 \\ A_2 \\ A_3 \\ A_4 \end{array} \left[\begin{array}{c} \langle 0.46, 0.39, 0.41 \rangle \\ \langle 0.49, 0.35, 0.29 \rangle \\ \langle 0.57, 0.33, 0.41 \rangle \\ \langle 0.58, 0.43, 0.42 \rangle \end{array}\right]$$

Step 5: The scores of demands are found $A(\widetilde{W_{A_1}}) = 37$, $A(\widetilde{W_{A_2}}) = 35$, $A(\widetilde{W_{A_3}}) = 46$, $A(\widetilde{W_{A_4}}) = 52$ using accuracy function defined in Eq. (7).

Step 6: According to the Eq. (8), time-dependent demand functions of four station in range of [0, 10] are determined as $w_{A_1}(t) = t + 37$, $w_{A_2}(t) = 2t + 35$, $w_{A_3}(t) = 46 - t$, $w_{A_4}(t) = 52 - 3t$.

The objective function is to minimize the total cost of construct a new facility Y as follows:

$$\min F(Y) = \int_0^{10} \begin{array}{l} (t + 37) * (|x - 0| + |y - 4|) + (2t + 35) * (|x - 6| + |y - 10|) + \\ (46 - t) * (|x - 5| + |y - 12|) + (52 - 3t) * (|x - 8| + |y - 9|) dt \end{array}$$

Step 7: The median points of four station at $t = 0$ is (6, 9) and at $t = 10$ is (5, 10). Therefore, the alternative points between these two locations are set as $Y_1 = (6, 9)$, $Y_2 = (6, 10)$ and $Y_3 = (5, 10)$.

$$F(Y_1) = \int_0^{10} \begin{array}{l} (t + 37) * (|6 - 0| + |9 - 4|) + (2t + 35) * (|6 - 6| + |9 - 10|) + \\ (46 - t) * (|6 - 5| + |9 - 12|) + (52 - 3t) * (|6 - 8| + |9 - 9|) dt \end{array}$$

$$= \int_0^{10} (3t + 730)dt = 7.450$$

$$F(Y_2) = \int_0^{10} \begin{matrix} (t+37) * (|6-0| + |10-4|) + (2t+35) * (|6-6| + |10-10|) + \\ (46-t) * (|6-5| + |10-12|) + (52-3t) * (|6-8| + |10-9|)dt \end{matrix}$$

$$= \int_0^{10} (738)dt = 7.380$$

$$F(Y_3) = \int_0^{10} \begin{matrix} (t+37) * (|5-0| + |10-4|) + (2t+35) * (|5-6| + |10-10|) + \\ (46-t) * (|5-5| + |10-12|) + (52-3t) * (|5-8| + |10-9|)dt \end{matrix}$$

$$= \int_0^{10} (742 - t)dt = 7.370.$$

Step 8: For alternative pair of points (Y_1, Y_2), the transition time is $t' = 2{,}66$. The total cost of the pair is:

$$F(Y_1) + F(Y_2) = \int_0^{2,66} (3t + 730)dt + \int_{2,66}^{10} (738)dt = 7.369,\ 33$$

For alternative pair of points (Y_1, Y_3), the transition time is t' =3. The total cost of the pair is:

$$F(Y_1) + F(Y_3) = \int_0^{3} (3t + 730)dt + \int_{3}^{10} (742 - t)dt = 7.352$$

For alternative pair of points (Y_2,Y_3), the transition time is t' =4. The total cost of the pair is:

$$F(Y_2) + F(Y_3) = \int_0^{4} (738)dt + \int_{4}^{10} (742 - t)dt = 7.362$$

Comparing the results of Step 7 and step 8, it is more costly to locate the new facility in only one location for the next ten years. Therefore, it was decided to change the location of the facility. Since the total cost of pair (Y_1, Y_3) is the lowest, the new electrical vehicle charging facility should be located at (6, 9) in time interval [0–3], then at t = 3, the facility should be relocated at (5, 10) over the next decade.

5 Conclusion

Electric vehicles are seen as the technologies of the future for sustainable transportation. The widespread use of these vehicles brings out the problem of determining the EVCSs locations. This study proposes a new methodology on single facility relocation problem with time-dependent demand under uncertainties. CNSV fuzzy MCDM is used in the expression of fuzzy customer demand at present time. As a result of the application, locating the new facility in one place for ten years is costly and relocation of the station at $t = 3$ appears to be more economical.

For future research, in decision making process, other fuzzy extensions such as intuitionistic fuzzy set, Pythagorean fuzzy sets, spherical fuzzy sets, hesitant fuzzy sets etc. can be considered to determine customer demand weights. Therefore, the robustness of decision-makers' evaluations can be compared. The problem can be enlarged by adapting to the continuous time horizon. Non-linear time-dependent demand functions can be examine. The multi facility relocation problem should also be investigated.

References

1. Faria, R., Moura, P., Delgado, J., De Almeida, A.T.: A sustainability assessment of electric vehicles as a personal mobility system. Energy Convers. Manage. **61**, 19–30 (2012)
2. Guo, S., Zhao, H.: Optimal site selection of electric vehicle charging station by using fuzzy TOPSIS based on sustainability perspective. Appl. Energy **158**, 390–402 (2015)
3. Westbrook, M.H.: The Electric Car: Development and future of battery, hybrid and fuel-cell cars (No. 38). The Institution of Electrical Engineering (IET) (2001)
4. He, Y., Kockelman, K.M., Perrine, K.A.: Optimal locations of US fast charging stations for long-distance trip completion by battery electric vehicles. J. Clean. Prod. **214**, 452–461 (2019)
5. He, S.Y., Kuo, Y.H., Wu, D.: Incorporating institutional and spatial factors in the selection of the optimal locations of public electric vehicle charging facilities: a case study of Beijing, China. Transp. Res. Part C: Emerg. Technol. **67**, 131–148 (2016)
6. Wu, Y., Xie, C., Xu, C., Li, F.: A decision framework for electric vehicle charging station site selection for residential communities under an intuitionistic fuzzy environment: a case of Beijing. Energies **10**(9), 1270 (2017)
7. Liu, H.C., Yang, M., Zhou, M., Tian, G.: An integrated multi-criteria decision making approach to location planning of electric vehicle charging stations. IEEE Trans. Intell. Transp. Syst. **20**(1), 362–373 (2018)
8. Meng, W., Kai, L.: Optimization of electric vehicle charging station location based on game theory. In: Proceedings 2011 International Conference on Transportation, Mechanical, and Electrical Engineering (TMEE), pp. 809–812. IEEE (2011)
9. Ren, X., Zhang, H., Hu, R., Qiu, Y.: Location of electric vehicle charging stations: a perspective using the grey decision-making model. Energy **173**, 548–553 (2019)
10. Guler, D., Yomralioglu, T.: Suitable location selection for the electric vehicle fast charging station with AHP and fuzzy AHP methods using GIS. Ann. GIS **26**(2), 169–189 (2020)
11. Liu, A., Zhao, Y., Meng, X., Zhang, Y.: A three-phase fuzzy multi-criteria decision model for charging station location of the sharing electric vehicle. Int. J. Prod. Econ. **225**, 107572 (2020)
12. Schmid, V., Doerner, K.F.: Ambulance location and relocation problems with time-dependent travel times. Eur. J. Oper. Res. **207**(3), 1293–1303 (2010)
13. Farahani, R.Z., Drezner, Z., Asgari, N.: Single facility location and relocation problem with time-dependent weights and discrete planning horizon. Ann. Oper. Res. **167**(1), 353–368 (2009)
14. Zanjirani Farahani, R., Szeto, W.Y., Ghadimi, S.: The single facility location problem with time-dependent weights and relocation cost over a continuous time horizon. J. Oper. Res. Soc. **66**(2), 265–277 (2015)

15. Drezner, Z., Wesolowsky, G.O.: Facility location when demand is time-dependent. Naval Res. Logistics (NRL) **38**(5), 763–777 (1991)
16. Hormozi, A.M., Khumawala, B.M.: An improved algorithm for solving a multi-period facility location problem. IIE Trans. **28**(2), 105–114 (1996)
17. Zadeh, L.A.: Fuzzy sets. Inf. Control **8**, 338–353 (1965)
18. Atanassov, K.: Intuitionistic fuzzy sets. Fuzzy Sets Syst. **20**(1), 87–96 (1986)
19. Smarandache, F.: Neutrosophy: neutrosophic probability, set, and logic: analytic synthesis & synthetic analysis (1998)
20. Torra, V.: Hesitant fuzzy sets. Int. J. Intell. Syst. **25**(6), 529–539 (2010)
21. Chakraborty, A., Mondal, S.P., Alam, S., Mahata, A.: Cylindrical neutrosophic single-valued number and its application in networking problem, multi-criterion group decision-making problem and graph theory. CAAI Trans. Intell. Technol. **5**(2), 68–77 (2020)
22. Chakraborty, A.: Minimal spanning tree in cylindrical single-valued neutrosophic arena. In: Neutrosophic Graph Theory and Algorithms, pp. 260–278. IGI Global (2020)
23. Smarandache, F.: 'NIDUS IDEARUM. Scilogs, V: Joining the dots', pp. 56–57. Pons Publishing, Brussels, Belgium (2019). http://fs.unm.edu/NidusIdearum5.pdf
24. Francis, R.L., McGinnis, L.F., White, J.A.: Facility layout and location: an analytical approach. Pearson College Division (1992)
25. Love, R.F., Morris, J.G., Wesolowsky, G.O.: Facilities Location. North-Holland, NewYork (1988)

Multi-criteria Decision Making Problem with Triangular Fuzzy Neutrosophic Sets

Hatice Ercan-Teksen(✉)

Eskisehir Osmangazi University, Eskisehir, Turkey

Abstract. Decision making problems are the types of problems that human beings face at all times. They are tried to be solved also in business life. They consist of the alternatives and the criteria determined to select these alternatives. If the number of the criteria increases, the decision making problem becomes harder. In addition, criteria may not be always an easily measurable value. This makes the problem more complex and the decision making more difficult. The use of fuzzy sets is common in multi-criteria decision making (MCDM) problems because of the difficulty in determining data and the presence of subjective data. Fuzzy sets are used in many areas and they are included for many problems in the literature with their fuzzy set extensions over time. One of these extensions of fuzzy sets are called neutrosophic sets. They are concerned with whether the data belongs to the set or not, and also with degrees of indeterminacy as well. An important difference of neutrosophic sets is that it defines the degree of falsity that will reduce the indeterminacy of the data. In the MCDM problem, some criteria may be found which the data are indeterminate. Therefore, in this study, it is aimed to use neutrosophic sets for MCDM problems and propose a new ranking method for triangular fuzzy neutrosophic sets.

Keywords: MCDM · Neutrosophic sets · Triangular fuzzy neutrosophic sets

1 Introduction

Decision making is a common situation faced in daily life. The problems faced in business life have decision-making processes as well. Most of these are complex problems, because it is usually impossible to reach a solution with a single criterion. For this reason, MCDM problems and the methods related to them have been developed. Studies on MCDM problems have been conducted in the literature for years. These studies have been in various industrial engineering fields.

Increasing the number of criteria in decision making problems increases the complexity of the problem. But there is an additional important concern: decision-making problems' mostly having expert opinions and their losing this linguistic information is an important data loss. Reducing or avoiding these data losses makes the decision-making problem better resolved. Also, the accuracy of non-linguistic data is not certain. At this point, the fuzzy sets theory enables data to be collected and the collectible data to be expressed more objectively. For this reason, MCDM problems using fuzzy sets are encountered in the literature.

C. Kahraman et al. (Eds.): INFUS 2021, LNNS 308, pp. 364–370, 2022.
https://doi.org/10.1007/978-3-030-85577-2_43

Fuzzy sets were first proposed by Zadeh [1]. Later, the fuzzy set theory was developed in many fields. In addition to being used in a wide variety of fields, the ordinary fuzzy numbers, which are found first, were also developed and various fuzzy sets were created. These are called extensions of fuzzy sets. Neutrosophic fuzzy set is one of the extensions of fuzzy sets and it was first discovered by Smarandache [2]. MCDM problems are also solved using neutrosophic sets.

Broumi et al. solved the problem of MCDM using with neutrosophic parameters [3, 4]. Bausys and Zavadskas solved the logistic problem with interval valued neutrosophic sets and VIKOR method [5]. COPRAS generated with neutrosophic sets was used for the natural gas selection problem [6]. Ye used trapezoidal neutrosophic number-weighted arithmetic and geometric averaging operators for the software problem [7, 8]. Biswas et al. used the decision-making problem with the monovalent neutrosophic TOPSIS method [9].

Ji et al. solved the logistical problem of single-valued neutrosophic numbers with respect to the Bonferroni mean operator [10]. Liu proposed the Archimedean t-conorm and the t-norm using single-valued neutrosophic numbers [11]. Peng et al. developed a weighted average operator based on multivalued neutrosophic sets for a new qualitative soft multiple criteria problem [12]. Sahin and Liu worked on the concept of neutrosophic set probability and applied it to MCDM problem [13]. Interval neutrosophic sets and PROMETHEE methods are integrated for the energy problem [14].

MCDM problem was solved by calculating interval neutrosophic fuzzy sets and correlation coefficients [15]. Zavadskas et al. used the WASPAS method with single-valued neutrosophic sets [16]. Garg and Nancy constructed a programming model using TOPSIS and interval neutrosophic numbers [17]. Bausys et al. compared the alternatives with the WASPAS model that is created by using single-valued neutrosophic sets after the criteria were assessed with AHP [18].

Deli and Subas developed a ranking method for single-valued neutrosophic sets and used this method in MCDM problem [19]. Fu and Ye proposed a new similarity model for single and interval-valued neutrosophic sets and used in disease diagnosis [20]. Hu et al. developed a decision-making model using neutrosophic sets to enable patients to choose their doctor [21]. Li et al. used linguistic neutrosophic sets for the selection problem of supplier [22]. E-commerce sites were evaluated using single-valued trapezoidal neutrosophic sets [23].

As a result of the literature research, it has been seen that neutrosophic sets are used in various areas of MCDM problems. Various neutrosophic sets and numbers have been used for this. However, it has been observed that studies on triangular fuzzy neutrosophic sets are limited in the available literature. Using fuzzy sets will reduce data loss because there are linguistic data for MCDM problems and expert opinions are available in the evaluation of the data.

When using fuzzy sets, compatibility with the data gains importance. If a data group is only concerned with belonging to the set, then it is normal to use ordinary fuzzy sets. However, if the truthiness of the data group can be identified as well as its falsity and even an idea about its indeterminacy can be declared, then neutrosophic sets can be used. Not only the truthiness, falsity and indeterminacy of a single data are dealt with, but also the data's being a triangular fuzzy set and the truthiness, falsity and indeterminacy of this triangular fuzzy set have been dealt with in this study. It is thought that this type of data will also be suitable for some MCDM problems.

This study is aimed to contribute to the literature by developing a new ranking method for the solution of MCDM problems, which are expressed with triangular fuzzy sets that contain truthiness, falsity and indeterminacy in the data.

In this study, MCDM problem will be solved by using triangular fuzzy neutrosophic sets. For this, triangular fuzzy neutrosophic sets and operators used for these sets will be explained at first. Then, the comparison method for triangular fuzzy neutrosophic sets will be proposed in the third section. After that, a numerical example will be solved using triangular fuzzy neutrosophic sets for a MCDM problem in the fourth section. In the conclusion section, a general summary of the study will be made.

2 Triangular Fuzzy Number Neutrosophic Sets

Unlike ordinary fuzzy sets, Neutrosophic sets are expressed in three different degrees. These are; truthiness degree, an indeterminacy degree and a falsity degree [2]. While being similar to intutionistic fuzzy numbers, neutrosophic sets reduce the instability of inconsistent information. So the truthiness, falsity and indeterminacy values can be determined independently of each other.

$\tilde{A}$ is defined as the neutrosophic set in the universal set U. $\tilde{A} = \{\langle u, (T_{\tilde{A}}(u), I_{\tilde{A}}(u), F_{\tilde{A}}(u)) | u \epsilon U \rangle\}$ is shown in this way. Here $T_{\tilde{A}}$ is the degree of truthiness, $I_{\tilde{A}}$ the degree of indeterminacy and $F_{\tilde{A}}$ the degree of falcity and the relationship between them is $0 \leq T_{\tilde{A}}(u) + I_{\tilde{A}}(u) + F_{\tilde{A}}(u) \leq 3$.

The neutrosophic set mentioned above is formed for a single valued number. In this study, triangular neutrosophic sets will be used. The operators used in neutrosophic sets are listed below, as MCDM will be created on condition that $\tilde{A}$ and $\tilde{B}$ are triangular fuzzy number neutrosophic sets [24].

Equation (1) is the addition of two triangular fuzzy neutrosophic sets.

$$\tilde{A} + \tilde{B} = \langle (a_1 + b_1, a_2 + b_2, a_3 + b_3); (T_{\tilde{A}} \wedge T_{\tilde{B}}, I_{\tilde{A}} \vee I_{\tilde{B}}, F_{\tilde{A}} \vee F_{\tilde{B}}) \rangle \tag{1}$$

Equation (2) is subtraction of two triangular fuzzy neutrosophic sets.

$$\tilde{A} - \tilde{B} = \langle (a_1 - b_1, a_2 - b_2, a_3 - b_3); (T_{\tilde{A}} \wedge T_{\tilde{B}}, I_{\tilde{A}} \vee I_{\tilde{B}}, F_{\tilde{A}} \vee F_{\tilde{B}}) \rangle \tag{2}$$

Equation (3) is inverse of triangular fuzzy neutrosophic set.

$$\tilde{A}^{-1} = \langle \left(\frac{1}{a_3}, \frac{1}{a_2}, \frac{1}{a_1} \right); (T_{\tilde{A}}, I_{\tilde{A}}, F_{\tilde{A}}) \rangle \tag{3}$$

Equation (4) is multiplication of triangular fuzzy neutrosophic set with positive constant number.

$$\gamma * \tilde{A} = \langle (\gamma * a_1, \gamma * a_2, \gamma * a_3); (T_{\tilde{A}}, I_{\tilde{A}}, F_{\tilde{A}}) \rangle \tag{4}$$

Equation (5) is division of two triangular fuzzy neutrosophic sets.

$$\frac{\tilde{A}}{\tilde{B}} = \begin{cases} \left\langle \left(\frac{a_1}{b_3}, \frac{a_2}{b_2}, \frac{a_3}{b_1}\right); \left(T_{\tilde{A}} \wedge T_{\tilde{B}}, I_{\tilde{A}} \vee I_{\tilde{B}}, F_{\tilde{A}} \vee F_{\tilde{B}},\right)\right\rangle \text{ if } a_3 > 0, b_3 > 0 \\ \left\langle \left(\frac{a_3}{b_3}, \frac{a_2}{b_2}, \frac{a_1}{b_1}\right); \left(T_{\tilde{A}} \wedge T_{\tilde{B}}, I_{\tilde{A}} \vee I_{\tilde{B}}, F_{\tilde{A}} \vee F_{\tilde{B}},\right)\right\rangle \text{ if } a_3 < 0, b_3 > 0 \\ \left\langle \left(\frac{a_3}{b_1}, \frac{a_2}{b_2}, \frac{a_1}{b_3}\right); \left(T_{\tilde{A}} \wedge T_{\tilde{B}}, I_{\tilde{A}} \vee I_{\tilde{B}}, F_{\tilde{A}} \vee F_{\tilde{B}},\right)\right\rangle \text{ if } a_3 < 0, b_3 < 0 \end{cases} \tag{5}$$

Equation (6) is multiplication of two triangular fuzzy neutrosophic sets.

$$\tilde{A} * \tilde{B} = \begin{cases} \left\langle (a_1 * b_1, a_2 * b_2, a_3 * b_3); \left(T_{\tilde{A}} \wedge T_{\tilde{B}}, I_{\tilde{A}} \vee I_{\tilde{B}}, F_{\tilde{A}} \vee F_{\tilde{B}},\right)\right\rangle \text{ if } a_3 > 0, b_3 > 0 \\ \left\langle (a_1 * b_3, a_2 * b_2, a_3 * b_1); \left(T_{\tilde{A}} \wedge T_{\tilde{B}}, I_{\tilde{A}} \vee I_{\tilde{B}}, F_{\tilde{A}} \vee F_{\tilde{B}},\right)\right\rangle \text{ if } a_3 < 0, b_3 > 0. \\ \left\langle (a_3 * b_3, a_2 * b_2, a_1 * b_1); \left(T_{\tilde{A}} \wedge T_{\tilde{B}}, I_{\tilde{A}} \vee I_{\tilde{B}}, F_{\tilde{A}} \vee F_{\tilde{B}},\right)\right\rangle \text{ if } a_3 < 0, b_3 < 0 \end{cases} \tag{6}$$

3 Comparison Method of Triangular Fuzzy Neutrosophic Sets

After determining the alternative values for triangular fuzzy neutrosophic multi-criteria, the alternatives should be compared with each other.

When comparing fuzzy sets, values such as score, ranking, defuzzification are calculated. Comparisons are made according to these calculated values.

In this study, a new ranking method is proposed and this method is given in Eq. (7). Ranking values of triangular fuzzy neutrosophic sets calculated with this formula are obtained. According to these ranking values, the alternatives in the MCDM problem can be listed.

$$R\left(\tilde{A}\right) = [a + b + c] * \left(2 + T_{\tilde{A}} - I_{\tilde{A}} - F_{\tilde{A}}\right) \tag{7}$$

4 Illustrative Example

In this section, a numerical example is given to understand the calculations of the triangular fuzzy neutrosophic sets that are used. For this, a MCDM problem has been selected.

In Table 1, triangular fuzzy neutrosophic sets were used for a decision-making problem that had four weighted criteria and to be selected among six alternatives.

Firstly, the value of each alternative is found as a triangular fuzzy neutrosophic set. For this, the sum of all criteria is calculated as a triangular fuzzy neutrosophic set. These values for each alternative are as follows:

$$\tilde{A}_1 = <(0.47, 0.68, 0.82); (0.65, 0.3, 0.3)>$$

$$\tilde{A}_2 = <(0.65, 0.84, 1); (0.7, 0.25, 0.2)>$$

$$\tilde{A}_3 = \,<(0.65, 0.82, 0.95);\ (0.7, 0.2, 0.2)>$$

$$\tilde{A}_4 = \,<(0.39, 0.49, 0.58);\ (0.65, 0.3, 0.25)>$$

$$\tilde{A}_5 = \,<(0.67, 0.77, 0.86);\ (0.65, 0.2, 0.15)>$$

$$\tilde{A}_6 = \,<(0.57, 0.73, 0.82);\ (0.75, 0.2, 0.25)>$$

Table 1. Weighted normalized triangular fuzzy neutrosophic sets for MCDM.

	Criteria 1	Criteria 2
A1	<(0.24,0.27,0.3);(0.75,0.2,0.15)>	<(0.07,0.12,0.14);(0.7,0.1,0.2)>
A2	<(0.18,0.24,0.26);(0.7,0.25,0.2)>	<(0.15,0.21,0.27);(0.75,0.15,0.2)>
A3	<(0.15,0.2,0.23);(0.7,0.1,0.15)>	<(0.12,0.16,0.19);(0.8,0.2,0.15)>
A4	<(0.11,0.13,0.16);(0.65,0.3,0.2)>	<(0.08,0.1,0.12);(0.65,0.25,0.25)>
A5	<(0.24,0.27,0.3);(0.65,0.2,0.15)>	<(0.24,0.27,0.3);(0.75,0.2,0.1)>
A6	<(0.09,0.11,0.12);(0.8,0.15,0.1)>	<(0.09,0.11,0.13);(0.8,0.2,0.25)>
	Criteria 3	Criteria 4
A1	<(0.05,0.12,0.15);(0.65,0.3,0.3)>	<(0.11,0.17,0.23);(0.9,0.1,0.05)>
A2	<(0.16,0.2,0.25);(0.75,0.2,0.05)>	<(0.16,0.19,0.22);(0.8,0.05,0.1)>
A3	<(0.19,0.21,0.23);(0.7,0.15,0.2)>	<(0.19,0.25,0.3);(0.8,0.1,0.1)>
A4	<(0.11,0.14,0.15);(0.8,0.2,0.2)>	<(0.09,0.12,0.15);(0.75,0.2,0.25)>
A5	<(0.09,0.12,0.14);(0.65,0.1,0.1)>	<(0.1,0.11,0.12);(0.7,0.2,0.15)>
A6	<(0.18,0.24,0.27);(0.85,0.1,0.15)>	<(0.21,0.27,0.3);(0.75,0.1,0.15)>

After these values are found, the ranking values, which is used as a comparison method in Eq. (7), are calculated and these values will enable to rank the alternatives.

The ranking values found using the equation are listed as follows: $R_1 = 4.04$, $R_2 = 5.60$, $R_3 = 5.57$, $R_4 = 3.07$, $R_5 = 5.29$ and $R_6 = 4.88$. Accordingly, the order of the alternatives is $A_2 > A_3 > A_5 > A_6 > A_1 > A_4$.

According to this order, alternative 2 should be preferred. In addition, by using triangular fuzzy neutrosophic sets, the flexibility which is for the expert to digitalize the criteria determined for each alternative is provided. In other words, it was emphasized that the value of each alternative for the criteria is not expressed with a single number, but it can change within a certain range.

5 Conclusion

In this study, information about MCDM problems and information about fuzzy sets are given at first. The frequency of using fuzzy sets in MCDM problem and the benefits they provide are explained. One of these benefits is that it enables experts to make more flexible decisions. Another is that it contributes to digitization of linguistic data.

Using fuzzy sets provides several advantages. For this reason, fuzzy sets have been developed in the literature. These sets, mentioned as extensions of fuzzy sets, have been developed over the years to represent various situations. Neutrosophic set, which is one of the extensions of fuzzy sets, is also used in decision making problems. Unlike ordinary fuzzy sets, neutrosophic sets deal with 3 different degrees. These are truthiness degree, an indeterminacy degree and a falsity degree. In addition to having information about the truthiness of data, this information can also have indeterminacy and/or falsity. In this case, using neutrosophic sets enables to analyze the data better.

In this study, triangular fuzzy neutrosophic sets are used to increase flexibility within neutrosophic sets. With these triangular fuzzy neutrosophic sets, the MCDM problem has been tried to be solved. In order to rank the alternatives, ranking values are obtained with using proposed ranking method and the alternatives are ranked for the numerical example according to these values.

In future studies, different comparison methods can be applied to MCDM problem with triangular fuzzy neutrosophic sets. In addition, more complex problems can be identified and compared with different methods with using by triangular fuzzy neutrosophic sets.

References

1. Zadeh, L.A.: Fuzzy sets. Inf. Control **8**(3), 338–353 (1965)
2. Smarandache, F.: Neutrosophy neutrosophic probability, set, and logic. Amer Res Press Rehoboth:12–20 (1998).
3. Broumi, S., Deli, I., Smarandache, F.: Neutrosophic parametrized soft set theory and its decision making. Ital. J. Pure Appl. Math. **32**, 503–514 (2014)
4. Broumi, S., Smarandache, F., Dhar, M.: Rough neutrosophic sets. Ital. J. Pure Appl. Math. **32**, 493–502 (2014)
5. Bausys, R., Zavadskas, E.-K.: Multicriteria decision making approach by Vikor under interval neutrosophic set environment. Econ. Comput. Econ. Cybern. Stud. Res. **49**(4), 33–48 (2015)
6. Bausys, R., Zavadskas, E.K., Kaklauskas, A.: Application of neutrosophic set to multicriteria decision making by COPRAS. Econ. Comput. Econ Cybern. Stud. Res. **49**(2), 1–15 (2015)
7. Ye, J.: An extended TOPSIS method for multiple attribute group decision making based on single valued neutrosophic linguistic numbers. J. Intell. Fuzzy Syst. **28**(1), 247–255 (2015)
8. Ye, J.: Trapezoidal neutrosophic set and its application to multiple attribute decision-making. Neural Comput. Appl. **26**(5), 1157–1166 (2014). https://doi.org/10.1007/s00521-014-1787-6
9. Biswas, P., Pramanik, S., Giri, B.C.: TOPSIS method for multi-attribute group decision-making under single-valued neutrosophic environment. Neural Comput. Appl. **27**(3), 727–737 (2015). https://doi.org/10.1007/s00521-015-1891-2
10. Ji, P., Wang, J.-Q., Zhang, H.: Frank prioritized Bonferroni mean operator with singlealued neutrosophic sets and its application in selecting thirdarty logistics providers. Neural Comput. Appl. **30**(3), 799–823 (2016). https://doi.org/10.1007/s00521-016-2660-6
11. Liu, P.: The aggregation operators based on archimedean t-conorm and t-norm for single-valued neutrosophic numbers and their application to decision making. Int. J. Fuzzy Syst. **18**(5), 849–863 (2016)

12. Peng, H.-G., Zhang, H., Wang, J.-Q.: Probability multialued neutrosophic sets and its application in multiriteria group decisionaking problems. Neural Comput. Appl. **30**(2), 563–583 (2016). https://doi.org/10.1007/s00521-016-2702-0
13. Sahin, R., Liu, P.: Possibility-induced simplified neutrosophic aggregation operators and their application to multi-criteria group decision-making. J. Exp. Theor. Artif. Intell. **29**(4), 769–785 (2016)
14. Wang, Z., Liu, L.: Optimized PROMETHEE based on interval neutrosophic sets for new energy storage alternative selection. Revista Tecnica de la Facultad de Ingenieria Universidad del Zulia **39**(9), 69–77 (2016)
15. Ye, J.: Correlation coefficients of interval neutrosophic hesitant fuzzy sets and its application in a multiple attribute decision making method. Informatica **27**(1), 179–202 (2016)
16. Zavadskas, E.K., Baušys, R., Stanujkic, D., Magdalinovic-Kalinovic, M.: Selection of lead-zinc flotation circuit design by applying WASPAS method with single-valued neutrosophic set. Acta Montanistica Slovaca **21**(2), 85–92 (2016)
17. Garg, H., Nancy.: Non-linear programming method for multi-criteria decision making problems under interval neutrosophic set environment. Appl. Intell. 1–15 (2017).
18. Baušys, R., Juodagalviene, B.: Garage location selection for residential house by WASPAS-SVNS method. J. Civil Eng. Manage. **23**(3), 421–429 (2017)
19. Deli, I., Şubaş, Y.: A ranking method of single valued neutrosophic numbers and its applications to multi-attribute decision making problems Int. J. Mach. Learn. Cybern. **8**(4), 1309–1322 (2016). https://doi.org/10.1007/s13042-016-0505-3
20. Fu, J., Ye, J.: Simplified neutrosophic exponential similarity measures for the initial evaluation/diagnosis of benign prostatic hyperplasia symptoms. Symmetry **9**(8), 54 (2017)
21. Hu, J., Pan, L., Chen, X.: An interval neutrosophic projection-based VIKOR Method for selecting doctors. Cogn. Comput. **9**(6), 801–816 (2017)
22. Li, Y.-Y., Zhang, H.-Y., Wang, J.-Q.: Linguistic neutrosophic sets and their application in multicriteria decision-making problems. Int. J. Uncertainty Quantification **7**(2), 135–154 (2017)
23. Liang, R., Wang, J., Zhang, H.: Evaluation of e-commerce websites: an integrated approach under a single-valued trapezoidal neutrosophic environment. Knowl.-Based Syst. 135, 44–59 (2017).
24. Liu, P., Wang, Y.: Multiple attribute decision-making method based on single-valued neutrosophic normalized weighted Bonferroni mean. Neural Comput. Appl. **25**(7–8), 2001–2010 (2014). https://doi.org/10.1007/s00521-014-1688-8

An Extended QFD Method for Sustainable Production with Using Neutrosophic Sets

Sezen Ayber(✉) and Nihal Erginel

Eskisehir Technical University, İki Eylul Campus, 26555 Eskişehir, Turkey
nerginel@eskisehir.edu.tr

Abstract. Identifying customer needs effectively and manufacturing products based on needs by using processes that minimize negative impacts on environment and people, and are economically sound are important objectives for companies to increase growth and competitiveness also to obtain sustainability. Quality function deployment (QFD) is a methodology for transforming customer requirements (CRs), needs and demands to the design requirements (DRs) via listening the voice of customers. In this study, an extended Quality Function Deployment (ext_QFD) method is presented. Customer voice and social, economic, environmental aspects are handled together to achieve sustainable production. While applying this method, the priorities of the CRs, DRs' correlation matrix and the relationship matrix between CRs and DRs are generally obtained from human evaluations by linguistic terms so this situation leads to uncertainty and vagueness. Fuzzy set theory is preferable for removing this vagueness. In this study, single valued neutrosophic sets are used and the proposed method's novel discoveries are demonstrated in detail using a case study of marble manufacturing.

Keywords: QFD · Single valued neutrosophic sets · Sustainable production · Marble manufacturing

1 Introduction

Manufacturers and service providers must satisfy dynamic customer needs in today's competitive business environment and must create goods and services using processes and systems that are non-polluting soil, water and air, conserving of natural resources, economic, safe for communities and consumers to obtain sustainable production. According to these, they have to decide which technical characteristics are more important in the design phase.

QFD is a structured methodology for converting CRs to DRs via listening the voice of customers. In this research, a newly ext_QFD model will be developed by combining CRs and social, economic, environmental aspects together to achieve sustainable production. While applying this technique, single valued neutrosophic sets will be used and ext_QFD model's novel discoveries will be demonstrated in detail using a case study of marble manufacturing.

The rest of this paper is organized as: QFD method is introduced in second section, the single valued neutrosophic set is introduced in third section, a newly ext_QFD with

C. Kahraman et al. (Eds.): INFUS 2021, LNNS 308, pp. 371–379, 2022.
https://doi.org/10.1007/978-3-030-85577-2_44

single valued neutrosophic sets is proposed in fourth section, the application for marble manufacturing is implemented in fifth section, and the conclusion is shown in the sixth section.

2 Quality Function Deployment

QFD is firstly introduced by Akao [1]. It has been applied in many areas such as product development, quality management, decision making, engineering and management etc. [2]. It consists of four matrices: house of quality, parts deployment, process planning and operating requirements [3]. The inputs of each matrix are the outputs of the previous matrix [4].

While applying this technique, the priorities of CRs, DRs' correlations and relationship values between CRs and DRs are realized by linguistic terms so uncertainty and vagueness occur. Fuzzy set theory approach is preferable in order to remove this situation. Hence fuzzy QFD (FQFD) is developed by researchers and there are many studies in literature where FQFD was used. Mehdizadeh [5], Zhang et al. [6], Büyüközkan and Çiftçi [7] used FQFD methodology. Abdolshah and Moradi classified the models integrated with FQFD [8]. Also there are several types of FQFD with using different types of fuzzy sets: Hesitant FQFD [9], Intuitionistic FQFD [10], Neutrosophic QFD with using interval neutrosophic sets [11], Pythagorean FQFD [12]. Besides these models, there are several studies by incorporating environmental aspects. Masui et al. [13] developed a new method called QFD for Environment (QFDE). Sakao [14], Bereketli and Genevois [15] proposed integrated QFDE methodology.

3 Single Valued Neutrosophic Sets

The fuzzy set theory was introduced by Zadeh in 1965 in order to model uncertainty. A fuzzy set A represent a fuzzy set and X is an element as [16]:

$$\tilde{A} = \left\{ \left(x, \mu_{\tilde{A}}(x) \right) : x \epsilon X \right\} \quad (1)$$

$$\mu_{\tilde{A}}(x) : X \rightarrow [0, 1] \quad (2)$$

where $\mu_{\tilde{A}}$ represents the membership function of an element $x \in X$.

The membership degree of vague parameters can be only focused in the Type 1 fuzzy set. If the non-membership degree and indeterminacy degree of imprecise parameters want to be handled, intuitionistic fuzzy set introduced by Atanassov in 1986 can be used [17]. In addition, if we want to handle indeterminate or inconsistent information, we can use the neutrosophic set introduced by Smarandache [18]. A fuzzy number is defined by "truth-membership degree", "falsity-membership degree", and "indeterminacy membership degree" at neutrosophic set, independently. Each membership degree takes a value between [0, 1] values, and the sum of truth-membership degree, falsity-membership degree and indeterminacy membership degree takes between [0, 3] values. The neutrosophic set generalizes crisp set, fuzzy set, interval-valued fuzzy set, intuitionistic fuzzy set, interval-valued intuitionistic fuzzy set, etc. [19] However, neutrosophic set is difficult to apply directly in real applications and in order to overcome these difficulties, Wang et al. [20] introduced a subclass of neutrosophic set called single-valued neutrosophic set. Basic definitions and operational rules of neutrosophic set will be showed.

Definition 1 [21, 22]. A neutrosophic set N in X is characterized by truth-membership function $T_N(x)$, a falsity-membership function $F_N(x)$ and an indeterminacy-membership function $I_N(x)$. X is a space of points and $X \in T_N(x), I_N(x), F_N(x)$ are real subsets of$]^-0, 1^+[$. There is no constraint about the sum of$T_N(x), I_N(x), F_N(x)$.

Definition 2 [21, 23]. A single valued neutrosophic set N over X taking the form $N = \{< x, 0 \leq T_N(x) + F_N(x) + I_N(x) >: x \in X\}$, where X be a universe of discourse, $T_N(x) : X \to [0, 1], I_N(x) : X \to [0, 1], F_N(x) : X \to [0, 1]$, and with$0 \leq T_N(x), F_N(x), I_N(x) \leq 3$, for all $x \in X$. A single valued neutrosophic number is represented by $N = (n_1, n_2, n_3)$, where $n_1, n_2, n_3 \in [0,1]$ and $n_1 + n_2 + n_3 \leq 3$.

Definition 3 [24]. Suppose that $\alpha_{\tilde{n}}, \theta_{\tilde{n}}, \beta_{\tilde{n}} \in [0, 1]$ and $n_1, n_2, n_3 \in R$ where $n_1 \leq n_2 \leq n_3$. Then a single valued triangular neutrosophic number, $\tilde{n} = (< (n_1, n_2, n_3); \alpha_{\tilde{n}}, \theta_{\tilde{n}}, \beta_{\tilde{n}}) >)$ is a special neutrosophic set on the real line set R; whose truth-membership, indeterminacy-membership and falsity-membership functions are defined as:

$$T_{\tilde{n}}(x) = \begin{cases} \alpha_{\tilde{n}}\left(\frac{x-n_1}{n_2-n_1}\right)(n_1 \leq x \leq n_2) \\ \alpha_{\tilde{n}}(x - n_2) \\ \alpha_{\tilde{n}}\left(\frac{n_3-x}{n_3-n_2}\right)(n_2 < x \leq n_3) \\ 0 \textit{ otherwise}, \end{cases} \quad (3)$$

$$I_{\tilde{n}}(x) = \begin{cases} \frac{(n_1-x+\theta_{\tilde{n}}(x-n_1))}{(n_2-n_1)}(n_1 \leq x \leq n_2) \\ \theta_{\tilde{n}}(x = n_2) \\ \frac{(x-n_2+\theta_{\tilde{n}}(n_3-x))}{n_3-n_2}(n_2 < x \leq n_3) \\ 1 \textit{ otherwise} \end{cases} \quad (4)$$

$$F_{\tilde{n}}(x) = \begin{cases} \frac{(n_2-x+\beta_{\tilde{n}}(x-n_1))}{(n_2-n_1)}(n_1 \leq x \leq n_2) \\ \beta_{\tilde{n}}(x = n_2) \\ \frac{(x-n_2+\beta_{\tilde{n}}(n_3-x))}{n_3-n_2}(n_2 < x \leq n_3) \\ 1 \textit{ otherwise} \end{cases} \quad (5)$$

$\alpha_{\tilde{n}}, \theta_{\tilde{n}}, \beta_{\tilde{n}}$ illustrate the degree of truth-membership, indeterminacy and falsity membership degree.

Definition 4 [22, 24]. Let $\tilde{a} = (< (a_1, a_2, a_3); \alpha_{\tilde{a}}, \theta_{\tilde{a}}, \beta_{\tilde{a}}) >)$ and $\tilde{b} = (< (b_1, b_2, b_3); \alpha_{\tilde{b}}, \theta_{\tilde{b}}, \beta_{\tilde{b}}) >)$ be two single valued triangular neutrosophic numbers. Some operational rules of two numbers are as follows:

$$\tilde{a}\tilde{b} = (< (a_1b_1, a_2b_2, a_3b_3); \alpha_{\tilde{a}} \wedge \alpha_{\tilde{b}}, \theta_{\tilde{a}} \vee \theta_{\tilde{b}}, \beta_{\tilde{a}} \vee \beta_{\tilde{b}}) >) \quad \textit{if } (a_3 > 0, b_3 > 0) \quad (6)$$

$$\tilde{a}\tilde{b} = (< (a_1b_3, a_2b_2, a_3b_1); \alpha_{\tilde{a}} \wedge \alpha_{\tilde{b}}, \theta_{\tilde{a}} \vee \theta_{\tilde{b}}, \beta_{\tilde{a}} \vee \beta_{\tilde{b}}) >) \quad if \ (a_3 < 0, b_3 > 0) \quad (7)$$

$$\tilde{a}\tilde{b} = (< (a_3b_3, a_2b_2, a_1b_1); \alpha_{\tilde{a}} \wedge \alpha_{\tilde{b}}, \theta_{\tilde{a}} \vee \theta_{\tilde{b}}, \beta_{\tilde{a}} \vee \beta_{\tilde{b}}) >) \quad if \ (a_3 < 0, b_3 < 0) \quad (8)$$

$$\tilde{a} + \tilde{b} = (<< (a_1 + b_1, a_2 + b_2, a_3 + b_3); \alpha_{\tilde{a}} \wedge \alpha_{\tilde{b}}, \theta_{\tilde{a}} \vee \theta_{\tilde{b}}, \beta_{\tilde{a}} \vee \beta_{\tilde{b}} >) \quad (9)$$

Definition 5 [25]. Transform the triangular neutrosophic numbers into crisp values by using the following equation:

$$S = \frac{1}{8}(a_1 + a_2 + a_3)(2 + a_{\tilde{a}} - \theta_{\tilde{a}} - \beta_{\tilde{a}}) \quad (10)$$

4 Proposed Extended QFD Method with Single Valued Neutrosophic Sets

There a few studies about sustainability QFD in literature. Roach proposed an approach to design sustainability products with QFD [26], environmental QFD for sustainable products are presented [27], application on fuzzy QFD for enabling sustainability is introduced by Vinodh and Chintha [28]. But QFD with sustainability via single valued neutrosophic set is the firstly introduced in this study. The methodology of ext_QFD with single valued neutrosophic set can be explained in following steps:

Step 1: Defining linguistic CRs and assigning customer importance ratings by using single valued triangular neutrosophic scale given in Table 2.

Step 2: Defining linguistic sustainability requirements (SRs) and assigning importance ratings by using the scale given in Table 2.

Step 3: Defining the DRs and constructing the correlation matrix among DRs by using the scale given in Table 2.

Step 4: Evaluating relationships between CRs and DRs, SRs and DRs using linguistic terms and finding single valued triangular neutrosophic numbers corresponding to them and constructing the relationship matrix between CRs and DRs, SRs and DRs. Empty cell shows no correlation.

Step 5: Calculating importance of each DR by multiplying the linguistic importance evaluations of all requirements (CRs/SRs) and the linguistic terms in the relationship matrices between CRs and DRs, SRs and DRs. Converting them to crisp values.

Step 6: Calculating relative importance degree of each DR by dividing the degree of importance of each DR to the sum of cumulative degree of importance.

Step 7: Ranking DRs and suggesting actions in areas with the high importance.

5 Application: Marble Manufacturing

The proposed ext_QFD methodology is applied on marble manufacturing as a case study. Marble is considered as non-renewable natural sources. In marble sector, waste generation, water and energy consumptions are very high. On the other hand, it is known that marble is one of the most important export products of Turkey. Therefore, there is a potential to make improvements for marble manufacturing considering sustainability. CRs and SRs will be handled together to rank DRs.

A committee of company experts conducts the evaluation and defines linguistic CRs, SRs and DRs. These requirements can be listed as in Table 1. Single valued triangular neutrosophic numbers with following scale for importance evaluations of CRs, evaluations in the relationship matrices and correlations among DRs are given in Table 2 [29].

Table 1. Customer, sustainability and design requirements.

CR1	Accuracy in marble slabs width&height&thickness
CR2	No cracks&scratches on marble slabs
CR3	No polish and opacity problems
CR4	Marble plates should be fire resistant
CR5	Marble plates should be slip resistant
CR6	No porosity
CR7	Marble slabs should be frost resistant
CR8	Marble slabs should be resistant to temperature differences
CR9	Marble slabs should be resistant to oxidation and color change
CR10	Marble slabs should be resistant to impacts
SR1	Less energy consumption
SR2	Less water consumption
SR3	Less air and water pollution
SR4	Recycling of marble pieces
DR1	Abrasion resistance
DR2	Apparent density
DR3	Chemical resistance
DR4	Compressive strength
DR5	Dimensional stability
DR6	Flexural strength
DR7	Freeze & thaw resistance
DR8	Frost resistance
DR9	Glossiness reflection
DR10	Impact resistance
DR11	Linear thermal expansion
DR12	Fire resistance
DR13	Slip resistance
DR14	Surface hardness
DR15	Water absorption
DR16	Porosity

Table 2. Single valued triangular neutrosophic numbers.

Low importance (LI) / Low relation (LR)	((4.6; 5.5; 8.6); 0.4; 0.7; 0.2)
Not low importance (NLI) / Not low relation (NLR)	((4.7; 6.9; 8.5); 0.7; 0.2; 0.6)
Very low importance (VLI) / Very low relation (VLR)	((6.2; 7.6; 8.2); 0.4; 0.1; 0.3)
Completely low imp. (CLI) / Completely low rel. (CLR)	((7.1; 7.7; 8.3); 0.5; 0.2; 0.4)
More or less low imp. (MLLI) / More or less low rel. (MLLR)	((5.8; 6.9; 8.5); 0.6; 0.2; 0.3)
Fairly low imp. (FLI) / Fairly low rel. (FLR)	((5.5; 6.2; 7.3); 0.8; 0.1; 0.2)
Essentially low imp. (ELI) / Essentially low rel. (ELR)	((5.3; 6.7; 9.9); 0.3; 0.5; 0.2)
High imp. (HI) / High rel. (HR)	((6.2; 8.9; 9.1); 0.6; 0.3; 0.5)
Not high imp. (NHI) / Not high rel. (NHR)	((4.4; 5.9; 7.2); 0.7; 0.2; 0.3)
Very high imp. (VHI) / Very high rel. (VHR)	((6.6; 8.8; 10); 0.6; 0.2; 0.2)
Completely high imp. (CHI) / Completely high rel. (CHR)	((6.3; 7.5; 8.9); 0.7; 0.4; 0.6)
More or less high imp. (MLHI) / More or less high rel. (MLHR)	((5.3; 7.3; 8.7); 0.7; 0.2; 0.8)
Fairly high imp. (FHI) / Fairly high rel. (FHR)	((6.5; 6.9; 8.5); 0.6; 0.8; 0.1)
Essentially high imp. (EHI) / Essentially high rel. (EHR)	((7.5; 7.9; 8.5); 0.8; 0.5; 0.4)

Linguistic ratings, relative importance and absolute importance values are shown in HoQ in Fig. 1. Importance of each DR is calculated by multiplying the linguistic importance evaluations of CRs/SRs with the linguistic terms in the relationship matrices between CRs and DRs, SRs and DRs. These values are presented for DR13 as follows;

$$\begin{aligned} & [(<(5.3, 7.3, 8.7); 0.7, 0.2, 0.8)>)(<(6.5, 6.9, 8.5); 0.6, 0.8, 0.1)>)] \\ & \quad + [(<(4.4, 5.9, 7.2); 0.7, 0.2, 0.3)>)(<(7.5, 7.9, 8.5); 0.8, 0.5, 0.4)>)] \\ & \quad = [(<(6.7, 5.9, 7.1); 0.6, 0.8, 0.8)>)] \end{aligned}$$

Then the output is converted to crisp value by using Eq. (10).

$$S = \frac{1}{8}(67.5 + 97 + 135, 2)(2 + 0.6 - 0, 8 - 0.8) = 37.45$$

Relative importance degree of each DR is calculated by dividing the degree of importance of each DR to the sum of cumulative degree of importance. For DR13, relative absolute importance value is calculated as $(37.45/1713.4) \times 100 = 2.19$.

According to relative absolute importance scores, compressive strength with 12.13, flexural strength with 11.85 and porosity with 10.55 are determined to be the most important technical requirements.

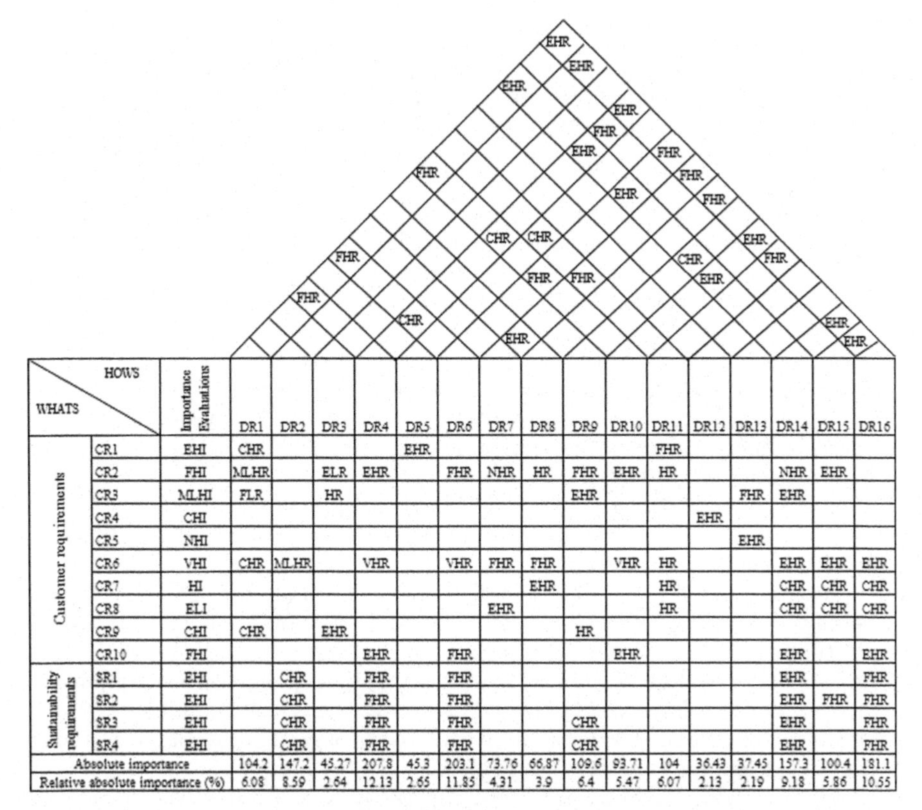

WHATS / HOWS		Importance Evaluations	DR1	DR2	DR3	DR4	DR5	DR6	DR7	DR8	DR9	DR10	DR11	DR12	DR13	DR14	DR15	DR16
Customer requirements	CR1	EHI	CHR				EHR						FHR					
	CR2	FHI	MLHR		ELR	EHR		FHR	NHR	HR	FHR	EHR	HR			NHR	EHR	
	CR3	MLHI	FLR		HR						EHR				FHR	EHR		
	CR4	CHI												EHR				
	CR5	NHI													EHR			
	CR6	VHI	CHR	MLHR		VHR		VHR	FHR	FHR		VHR	HR			EHR	EHR	EHR
	CR7	HI								EHR			HR			CHR	CHR	CHR
	CR8	ELI							EHR				HR			CHR	CHR	CHR
	CR9	CHI	CHR		EHR						HR							
	CR10	FHI				EHR		FHR				EHR				EHR		EHR
Sustainability requirements	SR1	EHI		CHR		FHR		FHR								EHR		FHR
	SR2	EHI		CHR		FHR		FHR								EHR	FHR	FHR
	SR3	EHI		CHR		FHR		FHR			CHR					EHR		FHR
	SR4	EHI		CHR		FHR		FHR			CHR					EHR		FHR
Absolute importance			104.2	147.2	45.27	207.8	45.3	203.1	73.76	66.87	109.6	93.71	104	36.43	37.45	157.3	100.4	181.1
Relative absolute importance (%)			6.08	8.59	2.64	12.13	2.65	11.85	4.31	3.9	6.4	5.47	6.07	2.13	2.19	9.18	5.86	10.55

Fig. 1. Linguistic HoQ and relative absolute importance values.

6 Conclusion

Using SRs in QFD methodology is few studied in literature. The CRs are important to determine the DRs but SRs are vital to protect the environmental and earth. In this study, a new ext_QFD model that combines the CRs and SRs via single-valued neutrosophic set is proposed. The relative importance of the CRs, correlation matrix values and the scores of DRs have been assessed with single-valued neutrosophic. The application implemented at marble manufacturing. In the future study, pythagorean fuzzy set can be used while modeling the ambiguity of importance of DRs, and the relationship between DRs and CRs.

References

1. Shilito, M.: Advance QFD Linking Technology to Market and Company Needs. Wiley Interscience, New York (1994)
2. Chan, L.K., Wu, M.L.: Quality function deployment: A literature review. Eur. J. Oper. Res. **143**, 463–497 (2002)
3. Liu, H., Wang, C.: An advanced quality function deployment model using fuzzy analytic network process. Appl. Math. Modeling **34**(11), 3333–3351 (2010)

4. Hauser, J.R., Clausing, D.: The house of quality. Harvard Bus. Rev. **66**(3), 63–73 (1988)
5. Mehdizadeh, E.: Ranking of customer requirements using the fuzzy centroid-based method. Int. J. Qual. Reliab. Manage. **27**(2), 201–216 (2010)
6. Zhang, F., Yang, M., Liu, W.: Using integrated quality function deployment and theory of innovation problem solving approach for ergonomic product design. Comput. Ind. Eng. **76**(1), 60–74 (2014)
7. Büyüközkan, G., Çiftçi, G.: An extended quality function deployment incorporating fuzzy logic and GDM under different preference structures. Int. J. Comput. Intell. Syst. **8**(3), 438–454 (2015)
8. Abdolshah, M.: Fuzzy Quality Function Deployment: An Analytical Literature Review. Journal of Industrial Engineering 11, (2013).
9. Onar, S.Ç., Büyüközkan, G., Öztayşi, B., Kahraman, C.: A new hesitant fuzzy QFD approach: an application to computer workstation selection. Appl. Soft Comput. **46**, 1–16 (2016)
10. Yu, L., Wang, L., Bao, Y.: Technical attributes ratings in fuzzy QFD by integrating interval-valued intuitionistic fuzzy sets and Choquet integral. Soft. Comput. **22**(6), 2015–2024 (2016). https://doi.org/10.1007/s00500-016-2464-8
11. Van, L.H., Yu, V.F., Dat, L.Q., Dung, L.Q.: New integrated quality function deployment approach based on interval neutrosophic set for green supplier evaluation and selection. Sustainability **10**(3), 838 (2018)
12. Haktanır, E., Kahraman, C.: A novel interval-valued Pythagorean fuzzy QFD method and its application to solar photovoltaic technology development. Comput. Ind. Eng. **132**, 361–372 (2019)
13. Masui, K., Sakao, T., Kobayashi, M., Inaba, A.: Applying quality function deployment to environmentally conscious design. Int. J. Qual. Reliab. Manage. **20**(1), 90–106 (2003)
14. Sakao, T.: A QFD-centred design methodology for environmentally conscious product design. Int. J. Prod. Res. **45**(18–19), 4143–4162 (2007)
15. Bereketli, I., Genevois, M.E.: An integrated QFDE approach for identifying improvement strategies in sustainable product development. J. Clean. Prod. **54**, 188–198 (2013)
16. Zadeh, L.A.: Fuzzy Sets. Inf. Control **8**, 338–353 (1965)
17. Atanassov, K.T.: Intuitionistic fuzzy sets. Fuzzy Sets Syst. **20**, 87–96 (1986)
18. Smarandache, F.: A Unifying Field in Logics. Neutrosophy: Neutrosophic Probability, Set and Logic. American Research Press, Rehoboth (1999)
19. Biswas, P., Pramanik, S., Giri, B.C.: TOPSIS method for multi-attribute group decision-making under single-valued neutrosophic environment. Neural Comput. Appl. **27**(3), 727–737 (2015). https://doi.org/10.1007/s00521-015-1891-2
20. Wang, H., Smarandache, F., Zhang, Y.Q., Sunderraman, R.: Single valued neutrosophic sets. Multispace Multistruct **4**, 410–413 (2010)
21. Gallego Lupiáñez, F.: Interval neutrosophic sets and topology. Kybernetes **38**(3/4), 621–624 (2009)
22. Hezam, I.M., Abdel-Baset, M., Smarandache, F.: Taylor series approximation to solve neutrosophic multiobjective programming problem. Neutrosophic Sets Syst. **10**, 39–46 (2018)
23. El-Hefenawy, N., Metwally, M.A., Ahmed, Z.M., El-Henawy, I.M.: A review on the applications of neutrosophic sets. J. Comput. Theor. Nanosci. **13**, 936–944 (2016)
24. Deli, I., Subas, Y.: Single valued neutrosophic numbers and their applications to multi-criteria decision making problem. Neutro. Set. Syst. **2**(1), 1–13 (2014)
25. Abdel-Baseta, M., Changb, V., Gamala, A., Smarandache, F.: An integrated neutrosophic ANP and VIKOR method for achieving sustainable supplier selection: A case study in importing field. Comput. Ind. **106**, 94–110 (2019)
26. Roach D.C.: Designing sustainable products with QFD. In: The 26th Symposium on QFDAt: Charleston, SC, USA, (2014)

27. Vinodh, S., Chintha K.S.: Application of fuzzy QFD for enabling sustainability. Int. J. Sustain. Eng. **4**(4) (2011)
28. Masui, K.: Environmental quality function deployment for sustainable products. In: Kauffman, J., Lee, K.-M. (eds.) Handbook of Sustainable Engineering. Springer, Dordrecht (2013). https://doi.org/10.1007/978-1-4020-8939-8_91
29. Abdel Aal, S.I., Abd Ellatif, M.M.A., Hassan, M.M.: Two ranking methods of single valued triangular neutrosophic numbers to rank and evaluate information systems quality. Neutrosophic Sets Syst. **19**(132) (2018).

C-Control Charts with Neutrosophic Sets

Hatice Ercan-Teksen(✉)

Eskisehir Osmangazi University, Eskisehir, Turkey

Abstract. Control charts are a statistical method that has been used for years to analyze the current situation regarding quality. Control charts, like many other methods, have been made with exact numbers for many years. Recently, new number sets have emerged due to the abundance of data, attempts to collect subjective data, the realization of the uncertainties in the data and the desire to reduce them. These are ordinary fuzzy sets and their extensions derived from these sets. These sets have started to be used in many areas. One of these areas is control charts. Especially because qualitative control charts are created with unmeasurable and more subjective data, it is more appropriate to create with fuzzy sets that are more flexible than exact sets. Therefore, studies on fuzzy sets and control charts with fuzzy set extensions are available in the literature, but control chart studies with neutrosophic sets are limited. The aim of this study is to find neutrosophic control limits by using neutrosophic sets for c-control charts and to create neutrosophic c-control charts with these sets. In this way, it is thought that the study will contribute to the literature.

Keywords: Neutrosophic sets · C-control charts · Neutrosophic c-control charts

1 Introduction

Control charts are one of the most frequently used statistical process control methods to observe the process. Control charts enable to analyze the general condition of the process and check whether it is within certain limit values.

Control charts were first developed in Bell laboratories and then used in many areas [1]. These charts developed by Shewart examine whether unexpected situations related to the product or process occur or not. Control charts have been used in many areas since the day it was developed.

In cases where there are subjective thoughts, it is possible to lose data. It is hard to collect data in order to create control charts if the characteristics such as odor, crack, colour etc. are important.

Loss of data can be prevented or reduced thanks to fuzzy sets. Fuzzy sets were first developed by Zadeh and they are used in digitizing linguistic expressions [2]. Many studies have been conducted on fuzzy sets. One of the fields of these studies are control charts.

For qualitative control charts, digitization of linguistic expressions is provided by using fuzzy sets. Also, considering that the people collecting the data can give subjective information, it will be beneficial to use fuzzy sets.

C. Kahraman et al. (Eds.): INFUS 2021, LNNS 308, pp. 380–387, 2022.
https://doi.org/10.1007/978-3-030-85577-2_45

When the literature is examined, especially in the last decades, it can be seen that studies on fuzzy control charts have increased. Most of these studies have been analyzed with numbers defined as ordinary fuzzy sets. However, fuzzy sets also develop over the years to represent different numbers. These are called extensions of fuzzy numbers. Although there are more complex numbers, they are useful depending on the state of the process or the way data is collected. Given that the primary goal is to reduce data loss, it may be necessary to use extensions of fuzzy sets.

First, Bradshaw stated that fuzzy sets can be used in determining economic control limits [3]. There are studies stating that fuzzy control charts can be used for linguistic data [4, 5]. He mentioned that control charts will be obtained with fuzzy probability and membership approaches [6]. Also they mentioned the use of categorized data in fuzzy control charts [7–9].

Gülbay and Kahraman calculated fuzzy control limits and compared fuzzy data with these values in order to find a more flexible solution [10]. In the study in which LR-fuzzy numbers were used, the comparison was made using the α-cut method [11].

Hsieh et al. used fuzzy set theory for number of defects and defect sets [12]. Fazel Zarandi et al. created fuzzy control charts for quantitative and qualitative data [13]. There are studies that create fuzzy p-control charts in literature [14, 15].

The studies mentioned earlier were fuzzy control charts studies done with ordinary fuzzy numbers in the literature. Extensions of fuzzy sets, which have been used in many fields in recent years, have also been used for control charts. In the creation of fuzzy c-control charts, type-2 fuzzy sets were used and fuzzy control charts were created by some comparison methods [16, 17]. For EWMA and CUSUM control charts, type-2 fuzzy control charts were obtained by using type-2 fuzzy numbers [18]. The study in which other type-2 fuzzy numbers were used to do for quantitative control charts [19].

Interval type-2 fuzzy c-control charts are also available in the literature [20–22]. Ercan-Teksen and Anagün fuzzified the X-R control charts, one of the quantitative control charts, using interval type-2 fuzzy sets in their study [23].

Control charts are also created with intuitionistic fuzzy set which is another extension of fuzzy sets [24, 25]. A study of pythegorean fuzzy sets and u-control charts has also been added to the literature as a study with the extension of fuzzy sets [26]. Spherical fuzzy sets, which are one of the extensions of fuzzy sets, are used to create fuzzy c-control control charts [27].

A study in which neutrosophic sets were used and s^2 control charts were created [28]. Neutrosophic sets were used for EWMA control charts [29]. The Neutrosophic average run length and X control charts were examined and it was checked whether the data was in control [30].

As a result of all these investigations, we could not reach the information that control charts created with neutrosophic sets are limited and neutrosophic sets are used in the accessible literature on c-control charts. Therefore, this study will be the first in this field and will contribute to the literature. In the second section, neutrosophic sets will be mentioned. In the third section, limit values for neutrosophic c-control charts will be determined. Then the comparison method for neutrosophic sets will be discussed in fourth section. In the fifth section, a numerical example will be given in order to increase the comprehensibility of the subject. In the last section, a general summary of the study will be made.

2 Triangular Fuzzy Number Neutrosophic Sets

Neutrosophic sets are expressed in three different degrees, unlike ordinary fuzzy sets. These are truthiness degree, an indeterminacy degree and a falsity degree [31]. While having a similarity to intutionistic fuzzy numbers, neutrosophic sets reduce the instability of inconsistent information. So the truthiness, falsity and indeterminacy values can be determined independently of each other.

$\tilde{A}$ is defined as the neutrosophic set in the universal set U. $\tilde{A} = \{\langle u, (T_{\tilde{A}}(u), I_{\tilde{A}}(u), F_{\tilde{A}}(u)) | u \epsilon U \rangle\}$ is shown in this way. Here $T_{\tilde{A}}$ is the degree of truthiness, $I_{\tilde{A}}$ the degree of indeterminacy and $F_{\tilde{A}}$ the degree of falcity and the relationship between them is $0 \leq T_{\tilde{A}}(u) + I_{\tilde{A}}(u) + F_{\tilde{A}}(u) \leq 3$.

The neutrosophic set mentioned above is formed for a single valued number. In this study, triangular neutrosophic sets will be used. The aggregation operators used in neutrosophic sets are listed below, as c-control charts will be created in condition that $\tilde{A}$ and $\tilde{B}$ are triangular fuzzy number neutrosophic sets [32].

$$\tilde{A} + \tilde{B} = \langle (a_1 + b_1, a_2 + b_2, a_3 + b_3); \left(T_{\tilde{A}} \wedge T_{\tilde{B},} I_{\tilde{A}} \vee I_{\tilde{B},} F_{\tilde{A}} \vee F_{\tilde{B},} \right) \rangle \quad (1)$$

$$\gamma * \tilde{A} = \langle (\gamma * a_1, \gamma * a_2, \gamma * a_3); (T_{\tilde{A}}, I_{\tilde{A}}, F_{\tilde{A}}) \rangle \quad (2)$$

3 Triangular Fuzzy Neutrosophic C-Control Charts

Firstly, control limits should be determined when creating fuzzy c-control charts by using neutrosophic sets. For this, classical c-control limit formulas and triangular fuzzy neutrosophic set operators are used. Triangular fuzzy neutrosophic c-control limits and mean value are given in the following equations. CL is center line, UCL and LCL are upper control limit and lower control limit, respectively.

$$\widetilde{CL_c} = \langle (\bar{a}, \bar{b}, \bar{c}); (\min(T_{\tilde{A}_i}), \mathrm{mak}(I_{\tilde{A}_i}), \mathrm{mak}(F_{\tilde{A}_i})) \rangle$$
$$= \langle \left(\frac{\sum_{i=1}^{m} a_i}{m}, \frac{\sum_{i=1}^{m} b_i}{m}, \frac{\sum_{i=1}^{m} c_i}{m} \right), (\min(T_{\tilde{A}_i}), \mathrm{mak}(I_{\tilde{A}_i}), \mathrm{mak}(F_{\tilde{A}_i})) \rangle \quad (3)$$

$$\widetilde{UCL_c} = \langle \left(\bar{a} + 3\sqrt{\bar{a}}, \bar{b} + 3\sqrt{\bar{b}}, \bar{c} + 3\sqrt{\bar{c}}, \right); (\min(T_{\tilde{A}_i}), \mathrm{mak}(I_{\tilde{A}_i}), \mathrm{mak}(F_{\tilde{A}_i})) \rangle \quad (4)$$

$$\widetilde{LCL_c} = \langle \left(\bar{a} - 3\sqrt{\bar{a}}, \bar{b} - 3\sqrt{\bar{b}}, \bar{c} - 3\sqrt{\bar{c}}, \right); (\min(T_{\tilde{A}_i}), \mathrm{mak}(I_{\tilde{A}_i}), \mathrm{mak}(F_{\tilde{A}_i})) \rangle \quad (5)$$

4 Comparison Method of Triangular Fuzzy Neutrosophic Sets

After determining the limit values for triangular fuzzy neutrosophic c-control charts, the data should be compared with these limit values. At this point, as the triangular fuzzy neutrosophic comparison method, the comparison method, which includes score values and it given in the following Eq. 6, was used [33]. After this calculation the numbers found are ranked.

$$s\left(\tilde{A}\right) = \frac{1}{8} * [a + b + c] * \left(2 + T_{\tilde{A}} - I_{\tilde{A}} - F_{\tilde{A}}\right) \tag{6}$$

5 Numerical Example

In this section, a numerical example is given to understand the c-control charts created with triangular fuzzy neutrosophic sets that seem complex.

By comparing neutrosophic values with classical number values, it will be provided to see the similarities and differences between two sets. Classical data are given in Table 1 and triangular fuzzy neurtosophic data are shown in Table 2.

Table 1. Classical data for c-control charts.

No	Data	No	Data	No	Data
1	20	6	6	11	13
2	6	7	32	12	13
3	36	8	13	13	8
4	11	9	28	14	15
5	10	10	18	15	15

Using these tables and formulas, both classical c-control chart and triangular fuzzy neutrosophic c-control chart will be drawn. For this, first of all, limit values must be found. CL, LCL and UCL values for the classical control chart were calculated as 16.27, 4.17 and 28.37, respectively. $\widetilde{CL}$, $\widetilde{LCL}$ and $\widetilde{UCL}$ values found for triangular fuzzy neutrosophic control charts are found as triangular fuzzy neutrosophic set as follows.

$$\widetilde{CL} = \langle(12.87, 16.27, 19.67); (0.71, 0.4, 0.2)\rangle$$

$$\widetilde{LCL} = \langle(2.11, 4.17, 6.36); (0.71, 0.4, 0.2)\rangle$$

$$\widetilde{UCL} = \langle(23.63, 28.37, 32.97); (0.71, 0.4, 0.2)\rangle$$

For the control charts created with triangular fuzzy neutrosophic sets, the score values mentioned in the previous section are calculated. These values are shown in Table 3.

After the limit values are found, control charts are drawn using these values and data. In Fig. 1, there are c-control charts drawn with classical data. Figure 2 shows the triangular fuzzy neutrosophic c-control charts.

The score values for the triangular fuzzy neutrosophic control limits are as follows: $CL_{TFNS} = 12.87, LCL_{TFNS} = 3.3$ and $UCL_{TFNS} = 22.41$.

Table 2. Triangular fuzzy neutrosophic data for c-control charts.

	a	b	c	$T_{\tilde{A}}$	$I_{\tilde{A}}$	$F_{\tilde{A}}$
1	16	20	24	0,71	0,1	0,09
2	3	6	9	0,83	0,03	0,03
3	27	36	45	0,92	0,35	0,19
4	8	11	14	0,84	0,38	0,18
5	7	10	13	0,87	0,1	0,016
6	3	6	9	0,81	0,35	0,12
7	27	32	37	0,76	0,37	0,2
8	11	13	15	0,85	0,22	0,14
9	25	28	31	0,92	0,36	0,06
10	17	18	19	0,91	0,04	0,03
11	12	13	14	0,97	0,1	0,13
12	10	13	16	0,95	0,19	0,08
13	6	8	10	0,99	0,4	0,03
14	12	15	18	0,83	0,24	0,11
15	9	15	21	0,72	0,14	0,09

Table 3. Score values of triangular fuzzy neutrosophic data.

No	Score	No	Score	No	Score
1	18.90	6	5.27	11	13.36
2	6.23	7	26.28	12	13.07
3	32.13	8	12.14	13	7.68
4	9.41	9	26.25	14	13.95
5	10.33	10	19.17	15	14.01

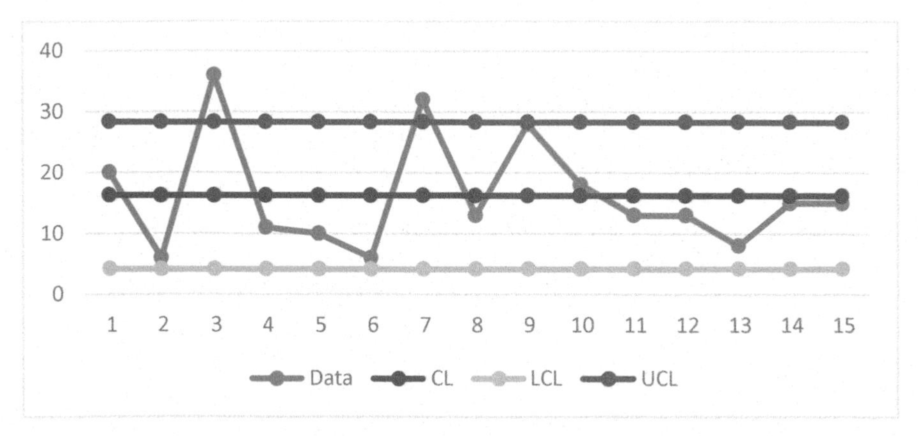

Fig. 1. C-control charts for classical data.

When Fig. 1 and Table 1 are examined, it is said that the data numbered 3 and 7 are out of the control limits when compared with the limit values. It is seen that data number 9 is almost above the control limit.

When Fig. 2 and Table 3 are compared with the limit values, the data numbered 3, 7 and 9 are said to be outside the control limits. Here, while the data number 9 is in control or whether it will be accepted or not in the classical set approach, a more flexible decision may be achieved by using fuzzy numbers that reduce the loss of information in the data.

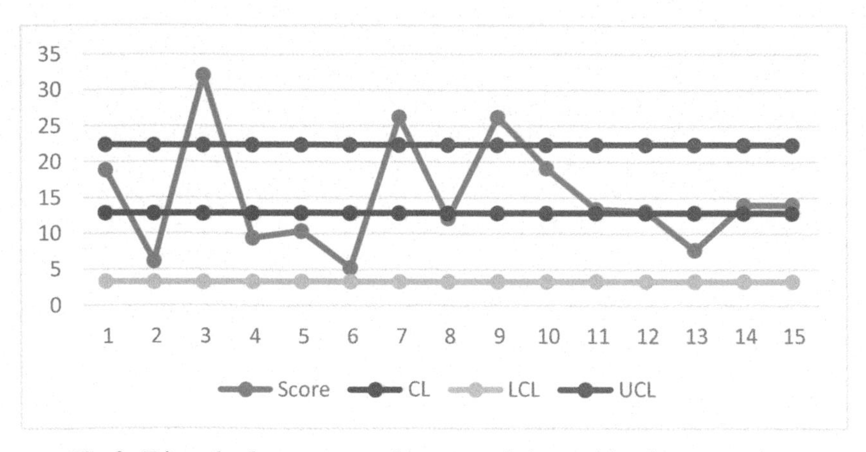

Fig. 2. Triangular fuzzy neutrosophic c-control charts with using score values.

6 Conclusion

First of all, research has been done on fuzzy control charts within the scope of the study. It has been observed that most of these studies are about ordinary fuzzy sets. In recent years, it is seen that there are studies done with the extensions of fuzzy sets. There are some studies on neutrosophic sets which are one of the extensions of fuzzy sets. However, at this point, there is no study in which triangular fuzzy neutrosophic sets are not found in the literature and also there is no study using these sets for c-control charts.

With this study, the gap in this field is aimed to be filled. Within the scope of the study, limit values for c-control charts created with triangular fuzzy neutrosophic sets are calculated. A control chart is drawn by comparing these neutrosophic limits with triangular fuzzy neutrosophic data. At this point, the study is explained with a numerical example. As a result of the numerical sample, similar results and similar patterns are observed between the classical control charts and the triangular fuzzy neutrosophic c-control chart. For data that is very close to the limit line in classical control charts, it can be confusing if it is in or out of control. In addition, as data loss will be reduced with fuzzy sets, the decisions made for these data will be more objective.

In future studies, different comparison methods can be used for triangular fuzzy neutrosophic sets. In addition a contribution can be made to the literature by using triangular fuzzy neutrosophic sets for different types of control charts.

References

1. Shewhart, W.A.: Economic Control of Quality of Manufactured Product. D. Van Nosrrand Inc., Princeton (1931)
2. Zadeh, L.A.: Fuzzy sets. Inf. Control **8**(3), 338–353 (1965)
3. Bradshaw, C.W., Jr.: A fuzzy set theoretic interpretation of economic control limits. Eur. J. Oper. Res. **13**, 403–408 (1983)
4. Wang, J.H., Raz, T.: On the construction of control charts using linguistic variables. Int. J. Prod. Res. **28**, 477–487 (1990)
5. Raz, T., Wang, J.H.: Probabilistic and membership approaches in the construction of control charts for linguistic data. Prod. Plann. Control **1**, 147–157 (1990)
6. Kanagawa, A., Tamaki, F., Ohta, H.: Control charts for process average and variability based on linguistic data. Int. J. Prod. Res. **2**, 913–922 (1993)
7. Asai, K.: Fuzzy Systems for Management. IOS Press, Amsterdam (1995)
8. Woodall, W., Tsui, K.L., Tucker, G.L.: A review of statistical and fuzzy control charts based on categorical data. In: Frontiers in Statistical Quality Control, vol. 5. Physica-Verlag, Heidelberg (1997)
9. Laviolette, M., Seaman, J.W., Barrett, J.D., Woodall, W.H.: A probabilistic and statistical view of fuzzy methods, with discussion. Technometrics **37**, 249–292 (1995)
10. Gülbay, M., Kahraman, C.: An alternative approach to fuzzy control charts: direct fuzzy approach. Inf. Sci. **177**(6), 1463–1480 (2007)
11. Gülbay, M., Kahraman, C., Ruan, D.: α-cut fuzzy control charts for linguistic data. Int. J. Intell. Syst. **19**, 1173–1196 (2004)
12. Hsieh, K.L., Tong, L.I., Wang, M.C.: The application of control chart for defects and defect clustering in IC manufacturing based on fuzzy theory. Expert Syst. Appl. **32**(3), 765–776 (2007)
13. Fazel Zarandi, M.H., Turksen, I.B., Kashan, H.: Fuzzy control charts for variable and attribute quality characteristic. Iran. J. Fuzzy Syst. **3**(1), 31–44 (2006)
14. Fonseca, D.J., Elam, M.E., Tibbs, L.: Fuzzy short-run control charts. Math Ware Soft Comput. **14**, 81–101 (2007)

15. Hungshu, M., Hsien, C.W.: Monitoring imprecise fraction of nonconforming items using p control charts. J. Appl. Stat. **37**(8), 1283–1297 (2010)
16. Teksen, H.E., Anagün, A.S.: Type 2 fuzzy control charts using likelihood and deffuzzification methods. In: Kacprzyk, J., Szmidt, E., Zadrożny, S., Atanassov, K.T., Krawczak, M. (eds.) IWIFSGN/EUSFLAT -2017. AISC, vol. 643, pp. 405–417. Springer, Cham (2018). https://doi.org/10.1007/978-3-319-66827-7_37
17. Ercan-Teksen, H., Anagun A.S.: Type-2 fuzzy control charts using ranking methods. In: The 5th international fuzzy systems symposium (FUZZYSS'17), 14–15 October 2017, Ankara, Turkey (2017)
18. Kaya, İ., İlbahar, E., Karasan, A., Cebeci, B.: Design of EWMA and CUSUM control charts based on type-2 fuzzy sets. In: Conference Proceeding Science and Technology, vol. 3, no. 1, pp. 129–135 (2020)
19. Kaya, İ, Turgut, A.: Design of variable control charts based on type-2 fuzzy sets with a real case study. Soft. Comput. **25**(1), 613–633 (2020). https://doi.org/10.1007/s00500-020-05172-4
20. Ercan-Teksen, H., Anagün, A.S.: Interval type-2 fuzzy c-control charts using likelihood and reduction methods. Soft. Comput. **22**(15), 4921–4934 (2018). https://doi.org/10.1007/s00500-018-3104-2
21. Ercan-Teksen, H., Anagun, A.S.: Interval type-2 fuzzy c-control charts using ranking methods. Hacettepe J. Math. Stat. **4**(2), 510–520 (2019)
22. Senturk, S., Antuchieviciene, J.: Interval type-2 fuzzy c-control charts: an application in a food company. Informatica **28**, 269–283 (2017)
23. Ercan-Teksen, H., Anagun, A.S.: Different methods to fuzzy X¯-R control charts used in production: interval type-2 fuzzy set example. J. Enterp. Inf. Manag. **31**(6), 848–866 (2018)
24. Ercan-Teksen, H., Anagün, A.S.: Intuitionistic fuzzy c-control charts using fuzzy comparison methods. In: Kahraman, C., Cebi, S., Cevik Onar, S., Oztaysi, B., Tolga, A.C., Sari, I.U. (eds.) INFUS 2019. AISC, vol. 1029, pp. 1161–1169. Springer, Cham (2020). https://doi.org/10.1007/978-3-030-23756-1_137
25. Ercan-Teksen, H., Anagun, A.S.: Intuitionistic fuzzy c-control charts using defuzzification and likelihood methods. J. Intell. Fuzzy Syst. **39**, 6465–6473 (2020)
26. Kaya, İ., İlbahar, E., Karasan, A., Cebeci, B.: Design of control charts for number of defects based on pythagorean fuzzy sets. In: Conference Proceeding Science and Technology, vol. 3, no. 1, pp. 115–121 (2020)
27. Ercan-Teksen, H.: Spherical fuzzy c-control charts. In: Developments of Artificial Intelligence Technologies in Computation and Robotics-Proceedings of the 14th International Flins Conference (Flins 2020), vol. 12, pp. 235–241 (2020)
28. Aslam, M., Khan, N., Khan, M.Z.: Monitoring the variability in the process using neutrosophic statistical interval method. Symmetry **10**, 562–571 (2018)
29. Aslam, M., Bantan, R.A.R., Khan, N.: Design of S 2 N – NEWMA control chart for monitoring process having indeterminate production data. Processes **7**, 742–757 (2019)
30. Aslam, M.: Design of X-bar control chart for resampling under uncertainty environment. IEEE Access **7**, 60661–60671 (2019)
31. Smarandache, F.: Neutrosophy neutrosophic probability, set, and logic, pp. 12–20. American Research Press, Rehoboth (1998)
32. Liu, P., Wang, Y.: Multiple attribute decision-making method based on single-valued neutrosophic normalized weighted Bonferroni mean. Neural Comput. Appl. **25**(7–8), 2001–2010 (2014). https://doi.org/10.1007/s00521-014-1688-8
33. Abdel-Basset, M., Mohamed, M., Hussien, A.-N., Sangaiah, A.K.: A novel group decision-making model based on triangular neutrosophic numbers. Soft. Comput. **22**(20), 6629–6643 (2017). https://doi.org/10.1007/s00500-017-2758-5

Selection of the Best Software Project Management Model via Interval-Valued Neutrosophic AHP

Nisa Cizmecioglu, Huseyin Selcuk Kilic, Zeynep Tugce Kalender, and Gulfem Tuzkaya(✉)

Industrial Engineering Department, Marmara University, 34722 Istanbul, Turkey
{huseyin.kilic,tugce.simsit,gulfem.tuzkaya}@marmara.edu.tr

Abstract. The number of software projects has increased in recent years due to technological advancements. In addition to the other factors such as the qualifications of the staff, budget, leadership, the success of the projects also highly depends on how they are managed. There are various software project management models such as Waterfall, Prototype, Spiral, Incremental, Iterative and Agile, which have been applied in different types of software projects. Each has advantages and disadvantages over the other. Hence, any of them does not have absolute superiority over the others. Therefore, it is important to determine the most suitable one. In addition, depending on the sector, the conditions change and a solution methodology for the new problem environment is required. One of the biggest sectors, the banking sector, is specifically considered in this study. Conforming to the multi-criteria structure of the selection problem, Interval-Valued Neutrosophic Analytic Hierarchy Process (IVN-AHP) which is capable of providing solutions in fuzzy multi-criteria decision making environment is utilized in this study with the pairwise comparisons provided by the experts in the sector. Finally, the alternative software project management models are ranked, and the best one is obtained.

Keywords: Software projects · Project management · Waterfall · Prototype · Spiral · Incremental · Iterative · Agile · Interval-Valued Neutrosophic AHP

1 Introduction

The number of IT projects has increased as a result of technological developments in various sectors. Especially with the emergence of Industry 4.0 in the last decade, IT implementations' speed and spread have grown up. The companies which are eager to catch up with the latest trends have applied many software projects.

Developing a software project is generally misunderstood as if it only composes of coding. However, there is a software development life cycle that includes all the stages such as planning, requirement analysis, design, implementation, testing & integration, and maintenance. Each stage has its own requirements. The first phase is planning and in this phase prerequisites of the project are determined after several meetings organized with partners, customers and supervisors. During these meetings, details of the questions

C. Kahraman et al. (Eds.): INFUS 2021, LNNS 308, pp. 388–396, 2022.
https://doi.org/10.1007/978-3-030-85577-2_46

such as "Who is going to use the software application? How it is going to be used? and What kind of information it is going to process?" are to be given. As one of the most important steps of the SDLC, the requirement analysis phase is performed mainly by the leaders of the software development process. Essence and the characteristics of the software, development risks etc. are identified in this phase. Software Requirement Specification (SRC) document is prepared in the requirement analysis phase gives the best possible results of a software. In the design phase, System Design Specification (SDC) document is prepared, and required characteristics of software are analyzed by the stakeholders and the most proper design model is selected. Based on SRC and SDC documents, the implementation phase is realized. Following the implementation phase, during integration and testing phase, to achieve a desired quality level, feedbacks are given to the previous phases based on the faults found. Finally, to eliminate problems of software, the maintenance phase is realized [1].

Although various factors affect the accomplishment of a project, the impact of how these stages are managed to the success of a project is inevitable. Considering the latest advancements both in the literature and real-life cases, it is concluded that there are many software project management models, including Waterfall, Prototype, Spiral, Incremental, Iterative and Agile models. Brief explanations for these different models are given as follows.

Waterfall model is one of the traditional software development models in which tasks are organized in a linear and sequential order. Changes and updates are not allowed in this model [2]. For the prototype model, before the actual modeling process, a prototype of the software is generated. This prototype is a very basic version of the required software [3]. In the spiral model, a part of the requirements is identified and the team applied development phases, with the exception of installation and maintenance phases, on them. Once an iteration finishes, with the learned lessons from the previous phases, an extended spiral is started with additional functionalities. At the end, installation and maintenance phases are realized [4]. In the incremental model, a planning phase is followed by a number of iterations. Repeated cycles are performed in that model and small parts of the projects are handled with an incremental structure. Earlier phases' experiences are utilized to improve the next phases [5]. For iterative model, a completed specification of requirements is not prepared for the initialization of the project. Part by part, the project is finished and after each part is completed, requirements for the remaining parts are identified. And this process is repeated as a cycle until all the project is completed [5]. Finally, different from the traditional approaches, agile model provides a flexible development process which is very convenient to the modern software development requirements. Iterative involvement of development teams is very important for this model [6].

These different life cycle models can be evaluated for their suitability for various software projects. When the current literature is investigated, the number of studies have been found in the literature for the evaluation of different life cycle development models with the consideration of limited multiple criteria. Some of them are summarized as follows. Mishra and Dubey [4] evaluated various life cycle development models such as waterfall, prototype, rapid, V-shaped, spiral and incremental models for different scenarios based on some features like cost, simplicity, and risk involvement. Hovorushchenko and Krasiy [7] proposed a methodology to determine the proper software life cycle model for different projects. The methodology consists of two categories: "characteristics of requirements" and "project type and risk". Under these categories, question lists are prepared for the selection process. They evaluated waterfall, spiral, RAD, incremental models. Khan et al. [2] evaluated different software life cycle development models by using analytic hierarchy process (AHP) approach. In their study, agile, hybrid and traditional methods are evaluated considering four criteria.

It is almost impossible to find a sector where software development projects are not applied. However, since differentiating conditions of sectors exist, the most suitable software project development model may change with respect to the type of sector. Unlike the existing studies, the banking sector, more specifically, the bank's legislation portal is considered in this study. Hence, the study aims to select the most suitable software project management model. Depending on this selection problem's multi-criteria structure and fuzzy problem environment, Interval-Valued Neutrosophic Analytic Hierarchy Process (IVN-AHP) is utilized in this study. Unlike the crisp AHP, it enables to handle indeterminacy and vagueness in the decision-making environment and is determined as a suitable method. Up to the knowledge of the authors, this is the first study utilizing IVN-AHP for the selection of a software project management model.

The rest of the paper is organized as follows. Technical background about IVN-AHP is provided in the second section. The third section includes the application in the banking sector. Finally, the conclusions are given in the fourth section with the references following.

2 Interval-Valued Neutrosophic Analytic Hierarchy Process (IVN-AHP)

AHP proposed by Saaty [8] is one of the most frequently used MCDM techniques in the literature [9]. It is only based on the pairwise comparisons of the factors. Hence, the method can produce results without any direct scoring of any alternative with respect to any criterion. However, classic AHP lacks tackling with uncertainty and vagueness. So as to overcome this deficiency, various enhancements are performed on crisp AHP, and fuzzy versions are proposed, including mainly intuitionistic and neutrosophic sets. Hence, the application of IVN-AHP is performed by benefiting from the studies of Bolturk and Kahraman [10] and Nabeeh *et al.* [11]. The related steps are indicated in brief as follows:

- *Step 1:* Determination of the IVN evaluation scale. In this study, the scale indicated in [10] is used for evaluations.
- *Step 2*: Formation of the hierarchy. (Goal, criteria, sub-criteria, and alternatives.)
- *Step 3*: Formation of the pairwise comparison matrices via interval-valued neutrosophic sets provided in Eq. 1.

$$P_C = \begin{bmatrix} [T_{11}^L, T_{11}^U], [I_{11}^L, I_{11}^U], [F_{11}^L, F_{11}^U] & \cdots & [T_{1n}^L, T_{1n}^U], [I_{1n}^L, I_{1n}^U], [F_{1n}^L, F_{1n}^U] \\ \vdots & \ddots & \vdots \\ [T_{n1}^L, T_{n1}^U], [I_{n1}^L, I_{n1}^U], [F_{n1}^L, F_{n1}^U] & \cdots & [T_{nn}^L, T_{nn}^U], [I_{nn}^L, I_{nn}^U], [F_{nn}^L, F_{nn}^U] \end{bmatrix} \tag{1}$$

Moreover, the consistencies of all the pairwise matrices are checked over the deneutrosophicated matrices formed via Eq. 2.

$$d = \left(\frac{T_x^L + T_x^U}{2}\right) + \left(\left(1 - \frac{\left(I_x^L + I_x^U\right)}{2}\right) * I_x^U\right) - \left(\left(\frac{F_x^L + F_x^U}{2}\right) * \left(1 - F_x^U\right)\right) \quad (2)$$

- *Step 4*: Obtain the normalized matrice (B) by dividing each cell by the sum of corresponding upper parameter's columns as in Eq. 3.

$$B_{ij} = \left[\left[\frac{T_{ij}^L}{\sum_{k=1}^n T_{kj}^U}, \frac{T_{ij}^U}{\sum_{k=1}^n T_{kj}^U}\right], \left[\frac{I_{ij}^L}{\sum_{k=1}^n I_{kj}^U}, \frac{I_{ij}^U}{\sum_{k=1}^n I_{kj}^U}\right], \left[\frac{F_{ij}^L}{\sum_{k=1}^n F_{kj}^U}, \frac{F_{ij}^U}{\sum_{k=1}^n F_{kj}^U}\right]\right] \quad \forall i, j \quad (3)$$

- *Step 5*: Calculate the average of each component for each criterion in B and obtain C as in Eq. 4.

$$B = \begin{bmatrix} \left[T_{11}^L, T_{11}^U\right], \left[I_{11}^L, I_{11}^U\right], \left[F_{11}^L, F_{11}^U\right] \cdots \left[T_{1n}^L, T_{1n}^U\right], \left[I_{1n}^L, I_{1n}^U\right], \left[F_{1n}^L, F_{1n}^U\right] \\ \vdots \qquad \ddots \qquad \vdots \\ \left[T_{n1}^L, T_{n1}^U\right], \left[I_{n1}^L, I_{n1}^U\right], \left[F_{n1}^L, F_{n1}^U\right] \cdots \left[T_{nn}^L, T_{nn}^U\right], \left[I_{nn}^L, I_{nn}^U\right], \left[F_{nn}^L, F_{nn}^U\right] \end{bmatrix}$$

$$C = \begin{bmatrix} \left[\frac{\sum_j T_{1j}^L}{n}, \frac{\sum_j T_{1j}^U}{n}\right], \left[\frac{\sum_j I_{1j}^L}{n}, \frac{\sum_j I_{1j}^U}{n}\right], \left[\frac{\sum_j F_{1j}^L}{n}, \frac{\sum_j F_{1j}^U}{n}\right] \\ \cdots \\ \cdots \\ \cdots \\ \left[\frac{\sum_j T_{nj}^L}{n}, \frac{\sum_j T_{nj}^U}{n}\right], \left[\frac{\sum_j I_{nj}^L}{n}, \frac{\sum_j I_{nj}^U}{n}\right], \left[\frac{\sum_j F_{nj}^L}{n}, \frac{\sum_j F_{nj}^L}{n}\right] \end{bmatrix} \quad (4)$$

- *Step 6*: Steps 1 through 5 are repeated for each criterion and neutrosophic weight vectors for all of the alternatives are obtained.
- *Step 7:* Final matrix is constructed as in Eq. 5 to obtain the final combined priority weights.

$$F = \begin{bmatrix} \left\langle \left[T_{W_{C_1A_1}}^L, T_{W_{C_1A_1}}^U\right], \left[I_{W_{C_1A_1}}^L, I_{W_{C_1A_1}}^U\right], \left[F_{W_{C_1A_1}}^L, F_{W_{C_1A_1}}^U\right] \right\rangle & \cdots & \left\langle \left[T_{W_{C_nA_1}}^L, T_{W_{C_nA_1}}^U\right], \left[I_{W_{C_nA_1}}^L, I_{W_{C_nA_1}}^U\right], \left[F_{W_{C_nA_1}}^L, F_{W_{C_nA_1}}^U\right] \right\rangle \\ \vdots & \ddots & \vdots \\ \left\langle \left[T_{W_{C_1A_m}}^L, T_{W_{C_1A_m}}^U\right], \left[I_{W_{C_1A_m}}^L, I_{W_{C_1A_m}}^U\right], \left[F_{W_{C_1A_m}}^L, F_{W_{C_1A_m}}^U\right] \right\rangle & \cdots & \left\langle \left[T_{W_{C_nA_m}}^L, T_{W_{C_nA_m}}^U\right], \left[I_{W_{C_nA_m}}^L, I_{W_{C_nA_m}}^U\right], \left[F_{W_{C_nA_m}}^L, F_{W_{C_nA_m}}^U\right] \right\rangle \end{bmatrix} \quad (5)$$

- *Step 8:* Obtain final combined IVN weights of alternatives by using Eq. 6

$$w_{A_j} = \langle [T^L_{W_{C_1}}, T^U_{W_{C_1}}], [I^L_{W_{C_1}}, I^U_{W_{C_1}}], [F^L_{W_{C_1}}, F^U_{W_{C_1}}] \rangle \langle [T^L_{W_{C_1A_1}}, T^U_{W_{C_1A_1}}], [I^L_{W_{C_1A_1}}, I^U_{W_{C_1A_1}}], [F^L_{W_{C_1A_1}}, F^U_{W_{C_1A_1}}] \rangle + \cdots + \langle [T^L_{W_{C_n}}, T^U_{W_{C_n}}], [I^L_{W_{C_n}}, I^U_{W_{C_n}}], [F^L_{W_{C_n}}, F^U_{W_{C_n}}] \rangle \langle [T^L_{W_{C_nA_m}}, T^U_{W_{C_nA_m}}], [I^L_{W_{C_nA_m}}, I^U_{W_{C_nA_m}}], [F^L_{W_{C_nA_m}}, F^U_{W_{C_nA_m}}] \rangle \quad (6)$$

- *Step 9:* Deneutrosophication is applied on final combined IVN values that are found in Step 8 via Eq. 2 and crisp values of factor weights are found as stated in Eq. 7.

$$\Omega_{A_j} = (w_{A_1}, w_{A_2}, \ldots, w_{A_m}) \quad (7)$$

- *Step 10*: The crisp values found in the Step 9 via Eq. 7 are normalized and the importance weights are obtained. Finally, alternatives are ranked and the alternative with the largest weight is determined as the best choice.

3 Application of IVN-AHP for the Selection of the Best Software Project Management Model

Software development is an essential part of any sector in today's business environment. As the requirements change from sector to sector, the appropriate software development life cycle model is also changing. In this regard, these models should be evaluated according to the selected sector and its requirements. In this study, the main aim is determined as the selection of the most appropriate software development model for banks' legislation portal. According to the literature review, the mostly used software development models are determined as alternatives which are Waterfall (A_1), Prototype (A_2), Spiral (A_3), Incremental (A_4), Iterative (A_5) and Agile (A_6). It is possible to evaluate these models considering several criteria; However 5 main criteria are determined after reviewing the literature as follows.

- C_1: Requirement Determination: Requirement analysis is one of the most important stages in software development models because the software should be developed according to the specifics of the business case.
- C_2: Cost: (Design cost of the model) Cost is an important criterion because in some models, every change in the development procedure increases the cost.
- C_3: Success Guarantee (Success of the developed model): In every model, it is aimed to reach success at the end of the development procedure. In some models, selected life cycle models can provide a guaranteed success based on their steps.
- C_4: Making Changes: Customers' requirements are the most important component. So, it is a desirable element if the selected life cycle model allows making changes.
- C_5: Time: Software development projects should be completed in a very short timescale, and generally, there is time pressure in those projects.

In this study, IVN-AHP is chosen for the evaluation of software development models. There are several successful studies in the literature which prove the strength of the model when considering the problem's multi-criteria structure and fuzzy environment. Moreover, up to the knowledge of the authors, this is the first study utilizing IVN-AHP for the selection of a software project management model. In this regard, the steps that present the second part of this study are applied.

- *Step 1 and Step 2*: Determination of the IVN evaluation scale. In this study, the scale indicated in [10] is used. Then hierarchy of the problem is created (See Fig. 1).

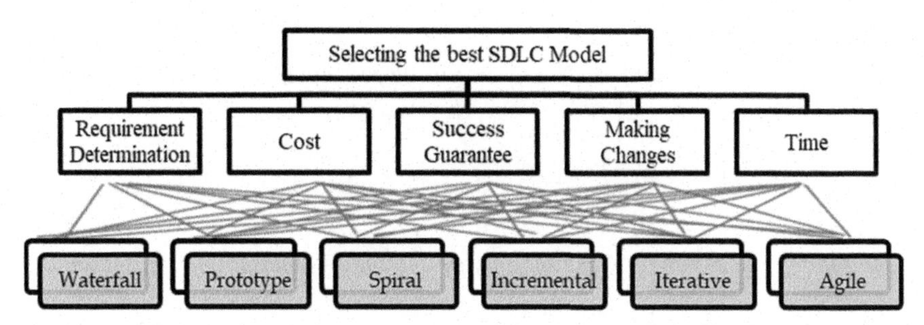

Fig. 1. Hierarchy of the problem

- *Step 3*: Pairwise comparison matrices are constructed via IVN sets provided in Eq. 1. An example part for the pairwise comparison matrix to present criteria to criteria comparison is proved below with Eq. 8.

$$P_C = \begin{bmatrix} [0.5, 0.5], [0.5, 0.5], & [0.5, 0.5] \cdots [0.4, 0.5], [0.35, 0.45], [0.5, 0.6] \\ \vdots & \ddots \quad \vdots \\ [0.5, 0.6], [0.35, 0.45], [0.4, 0.5] & \cdots [0.5, 0.5], [0.5, 0.5], \quad [0.5, 0.5] \end{bmatrix} \tag{8}$$

The consistencies of all the pairwise matrices are checked over the deneutrosophicated matrices formed via Eq. 2. An example analysis is provided below in which consistency calculations of the alternative evaluations for cost criteria are presented with Eq. 9. Values are presented from A_1 to A_6, respectively.

$$CR = (0.013; 0.028; 0.013; 0.028; 0.028; 0.003) \tag{9}$$

- *Step 4 and Step 5:* Firstly, normalized matrix (B) is obtained by dividing each cell by the sum of corresponding upper parameter's columns as in Eq. 3. Then Matrix C is created for each pairwise comparison. An example analysis for criteria comparison matrix is provided with Eq. 10

$$P_C = \begin{bmatrix} [0.19, 0.19], [0.24, 0.24], [0.18, 0.18] & \cdots & [0.16, 0.20], [0.18, 0.23], [0.17, 0.20] \\ \vdots & \ddots & \vdots \\ [0.19, 0.23], [0.17, 0.22], [0.15, 0.18] & \cdots & [0.20, 0.20], [0.26, 0.26], [0.17, 0.17] \end{bmatrix} \quad (10)$$

- *Step 6*: Normalization is repeated for all comparison matrices, and neutrosophic weight vectors for all alternatives are obtained. In this regard, weight vectors for each criterion are calculated and presented in Table 1.

Table 1. Weight vectors for each criterion

	T^L	T^U	I^L	I^U	F^L	F^U
C_1	0.17	0.20	0.18	0.22	0.16	0.20
C_2	0.09	0.13	0.13	0.17	0.24	0.27
C_3	0.24	0.27	0.14	0.18	0.10	0.14
C_4	0.15	0.19	0.17	0.22	0.18	0.22
C_5	0.19	0.22	0.17	0.21	0.15	0.18

- *Step 7*: Final matrix F is constructed as in Eq. 5 in order to obtain the final combined priority weights. A part of the matrix is presented below in Table 2 as an example.

Table 2. Final matrix for each alternative in terms of C1 and C2

	Requirement determination						Cost					
	T^L	T^U	I^L	I^U	F^L	F^U	T^L	T^U	I^L	I^U	F^L	F^U
Waterfall	0.02	0.02	0.27	0.34	0.33	0.37	0.01	0.02	0.26	0.31	0.37	0.41
Prototype	0.03	0.04	0.31	0.36	0.28	0.32	0.02	0.03	0.26	0.31	0.33	0.37
Spiral	0.03	0.04	0.31	0.36	0.28	0.32	0.01	0.02	0.26	0.31	0.37	0.41
Incremental	0.03	0.04	0.31	0.36	0.28	0.32	0.02	0.03	0.26	0.31	0.33	0.37
Iterative	0.02	0.03	0.28	0.35	0.31	0.36	0.02	0.03	0.26	0.31	0.33	0.37
Agile	0.03	0.05	0.27	0.34	0.24	0.29	0.01	0.01	0.22	0.28	0.40	0.44

- *Step 8:* Final combined IVN weights of alternatives are obtained by using Eq. 6. The final results are presented in Table 3.

Table 3. Final combined IVN weights

	T^L	T^U	I^L	I^U	F^L	F^U
Waterfall	0.0779	0.1180	0.0011	0.0033	0.0036	0.0069
Prototype	0.1194	0.1557	0.0019	0.0045	0.0019	0.0038
Spiral	0.1225	0.1577	0.0020	0.0047	0.0019	0.0038
Incremental	0.1271	0.1625	0.0020	0.0046	0.0017	0.0034
Iterative	0.1263	0.1629	0.0018	0.0043	0.0017	0.0035
Agile	0.1360	0.1788	0.0011	0.0032	0.0013	0.0031

- *Step 9*: Deneutrosophication is applied on final combined IVN values that are found in Step 8 via Eq. 2, and crisp values of factor weights are found as stated in Eq. 11.

$$\Omega_{A_j} = (0.0960; 0.1392; 0.1419; 0.1468; 0.1454; 0.1584) \tag{11}$$

- *Step 10*: The crisp values found in Step 9 via Eq. 8 (which are provided in Eq. 11) are normalized, and the importance weights are obtained as follows (Eq. 12).

$$\Omega_{A_j} = (0.1159; 0.1680; 0.1712; 0.1772; 0.1766; 0.1911) \tag{12}$$

Finally, alternatives are ranked, and the alternative with the largest weight is determined as the best choice. According to the final solution A6, the Agile approach is determined as the best software development model for banks' legislation portal. Alternatives are ranked as A6 > A4 > A5 > A3 > A2 > A1.

Agile software development tests take place during each development cycle of the product. This situation eliminates the possibility of errors, which significantly increase product quality and reduces development time. As stated with the proposed model, software development in accordance with agile is more suitable to the banking sector since it is a fast modeling approach.

In this study, the agile model is selected as the most proper model for the banking sector. However, other models can be more suitable for different business cases. The underlying reason is that the software requirements are changing sector to sector based on the consumers' needs and expectations. For example, for a student automation system, it is possible that the waterfall model will be more suitable since the type of the project is clear, and the scope is limited. From this point of view, the selection of the software development model is an essential problem, and the study proves that the proposed model provides valuable insight to the decision-makers.

4 Conclusions

The success of the software development projects depends on various factors such as qualifications of the staff, management of the project, budget, etc. Additionally, one of the most important factors is the utilized project management methodology. Different methodologies like waterfall, prototype, spiral, incremental, iterative and agile may have various advantages and disadvantages. Besides, for other sectors, different methodologies may be more convenient. For example, while agile methodology may be more convenient for the banking sector, the waterfall model may be proper for a student automation system. In this study, for different software development projects, a methodology is applied to find the most convenient project management methodology. In this study, an application from the banking sector is presented. For future studies, different sector applications may be realized, and also a more systematic procedure may be applied to find sector-specific decision criteria.

References

1. Prolopisko, A.: Software Development Life-Cycle(SDLC). https://medium.com/@artjoms/software-development-life-cycle-sdlc-6155dbfe3cbc. Accessed 22 Mar 2021
2. Khan, M.A., Parveen, A., Sadiq, M.: A method for the selection of software development life cycle models using analytic hierarchy process. In: International Conference on Issues and Challenges in Intelligent Computing Techniques (ICICT) (2014)
3. Java T Point Software Engineering page. https://www.javatpoint.com/software-engineering-prototype-model. Accessed 22 Mar 2021
4. Mishra, A., Dubey, D.A.: Comparative study of different software development life cycle models in different scenerios. In: International Journal of Advance Research in Computer Science and Management Studies, vol. 1, no. 5 (2013)
5. Bhuvaneswari, T., Prabaharan, S.: A survey of software development life cycle models. Int. J. Comput. Sci. Mob. Comput. **2**(3), 262–267 (2013)
6. Papadopoulos, G.: Moving from traditional to agile software development methodologies also on large, distributed projects. In: International Conference on Strategic Innovative Marketing, IC-SIM 2014, September 1–4, Madrid, Spain (2015)
7. Hovorushchenko, T., Krasiy, A: The selecting an appropriate software life cycle model on the basis of the specifications analysis. In: Computer Science & Engineering 2013 (CSE-2013), 21–23 November 2013, Lviv, Ukraine (2013)
8. Saaty, T.L.: The Analytic Hierarchy Process. McGraw-Hill, New York (1980)
9. Kilic, H.S., Zaim, S., Delen, D.: Development of a hybrid methodology for ERP system selection: the case of Turkish Airlines. Decis. Support Syst. **66**, 82–92 (2014)
10. Bolturk, E., Kahraman, C.: A novel interval-valued neutrosophic AHP with cosine similarity measure. Soft. Comput. **22**(15), 4941–4958 (2018). https://doi.org/10.1007/s00500-018-3140-y
11. Nabeeh, N.A., Abdel-Basset, M., El-Ghareeb, H.A., Aboelfetouh, A.: Neutrosophic multi-criteria decision making approach for iot-based enterprises. IEEE Access **7**, 59559–59574 (2019)

A New Similarity Measure for Single Valued Neutrosophic Sets

Muhammad Jabir Khan[1] and Poom Kumam[1,2(✉)]

[1] KMUTT Fixed Point Research Laboratory, SCL 802 Fixed Point Laboratory & Department of Mathematics, Faculty of Science, King Mongkut's University of Technology Thonburi (KMUTT), 126 Pracha-Uthit Road, Bang Mod, Thung Khru, Bangkok 10140, Thailand
jabirkhan.uos@gmail.com

[2] Center of Excellence in Theoretical and Computational Science (TaCS-CoE), Science Laboratory Building, Faculty of Science, King Mongkut's University of Technology Thonburi (KMUTT), 126 Pracha-Uthit Road, Bang Mod, Thung Khru, Bangkok 10140, Thailand
poom.kumam@mail.kmutt.ac.th

Abstract. The single-valued neutrosophic set ($\mathcal{SVNS}$) defined to incorporate the indeterminate, imprecise, and inconsistent data in real-life scientific and engineering problems. The uncertain information is significantly measured by similarity and dissimilarity measures. The similarity and dissimilarity measures are employed to depict the closeness and differences among SVNSs and have many applications in real-life situations like medical diagnosis, data mining, decision making, classification, and pattern recognition. This paper investigates the new similarity and dissimilarity measures for the SVNS. Additional properties of the newly proposed similarity and dissimilarity measures are focused. Their corresponding weighted similarity and dissimilarity measures are discussed on. It can be seen that the newly proposed similarity and dissimilarity measures respect all the axioms of similarity and dissimilarity definitions. The notion of $\simeq^{\alpha}$ similar relation is defined and it can be seen that the $\simeq^{\alpha}$ similar relation is reflective and symmetric but not transitive. The numerical examples are provided for the explanations.

Keywords: Single valued neutrosophic set · Similarity measures · Dissimilarity measures

1 Introduction

To measure the similarity between any form of data is an important topic. The measures used to find the resemblance between data is called similarity measure. It has different applications in classification, pattern recognition, medical diagnosis, data mining, clustering, decision making, and in image processing.

Petchra Pra Jom Klao Ph.D. Research Scholarship, KMUTT and TaCS-CoE.

C. Kahraman et al. (Eds.): INFUS 2021, LNNS 308, pp. 397–404, 2022.
https://doi.org/10.1007/978-3-030-85577-2_47

Fuzziness, as developed in [1], is a kind of uncertainty which appears often in human decision-making problems. The fuzzy set theory deals with daily life uncertainties successfully. The membership degree is assigned to each element in a fuzzy set. The membership degrees can effectively be taken by fuzzy sets. But in real-life situations, the non-membership degrees should be considered in many cases as well, and it is not necessary that the non-membership degree be equal to the one minus the membership degree. Thus Atanassov [2] introduced the concept of intuitionistic fuzzy set (IFS) that considers both membership and non-membership degrees. Here the non-membership degree is not always obtained from a membership degree, which leads to the concept of hesitancy degree. For each element, the sum of its membership (ξ) and non-membership (ν) degrees should be less than or equal to one for IFSs, that is,

$$\xi + \nu \leq 1. \tag{1}$$

Condition (1) suggests that the region to choose membership and non-membership degrees is a proper subset of $[0, 1] \times [0, 1]$.

Vote for, vote against, and keep neutral are three opinion choices for voters. Fuzzy sets and IFSs have no ability to deal such type of complicated voting situations. To cope these situations, Cuong and Kreinovich [4] introduced the ideology of picture fuzzy set (PFS). This is considered as the improvement of fuzzy sets and IFSs because it contains neutrality degree along with membership and non-membership degrees. Due to provide the better representation of data, this notion is widely used in the literature for real-life problems [5–8].

The notion of neutrosophic set (NS) was defined by Smarandache [9], to handle the issues involving indeterminacy, imprecise, and inconsistent data. NS contains three grades are called truth (membership), indeterminacy and falsity (non-membership). The grades can take any value from the $(0, 1)$. This notion is different from IFS and PFS in term of hesitancy index. The hesitancy index or incorporated uncertainty is dependent on the defining membership values for IFSs and PFSs. But in NSs, the incorporated uncertainty is independent of the defining membership degrees. To use the notion of NSs in real-life scientific and engineering problems, the single valued neutrosophic sets ($\mathcal{SVNS}$) was introduced by Wang et al. [10].

The uncertain information is significantly measured by dissimilarity and similarity measures. The dissimilarity and similarity measures are employed to depict the closeness and differences among fuzzy sets, and have many applications in real life situations like medical diagnosis, data mining, decision making, classification, and pattern recognition. The chi-square dissimilarity-based similarity measure for $\mathcal{SVNS}s$ were investigated by Ren et al. [11]. Wu et al. [12] discussed the similarity and cross entropy measures for $\mathcal{SVNS}s$. The tangent function based similarity measures for $\mathcal{SVNS}s$ and their applications in multi-period medical diagnosis were discussed by Ye and Fu [13]. For more about dissimilarity and similarity measures, we refer to [14–19].

The aim of this paper is to define the new axiomatically supported similarity and dissimilarity measures for $\mathcal{SVNS}s$ and to investigate their weighted versions. Also, to define the notion of $\simeq^{\alpha}$ similar relation for $\mathcal{SVNS}s$.

Thus we define the new similarity and dissimilarity measures for $\mathcal{SVNS}s$. Axiomatic proofs are given for proposed similarity and dissimilarity measures. To cope with situations of differently important alternatives, the weighted similarity and dissimilarity measures are discussed. The notion of $\simeq^{\alpha}$ similar relation is defined and it is proved that $\simeq^{\alpha}$ similar relation is reflective and symmetric. The counter example is given for not holding the transitivity.

The remaining paper is structured as: Basic definitions are mentioned in Sect. 2. Section 3 contains the new similarity and dissimilarity measures for $\mathcal{SVNS}s$. The concluding remarks are given in Sect. 4.

2 Preliminaries

The definition of $\mathcal{SVNS}s$ and their operations are mention in this section.

Definition 1 [10]. *The membership* (ζ_R), *indeterminacy* (η_R) *and non-membership* (ν_R) *functions from universal set to unit interval define the* $\mathcal{SVNS}$ R *over a universal set* Y *as follows*

$$R = \{(\zeta_R(y_i), \eta_R(y_i), \nu_R(y_i)) \mid y_i \in \mathsf{Y}\},$$

where $0 \leq \zeta_R(y_i) + \eta_R(y_i) + \nu_R(y_i) \leq 3$,

For any $y_i \in \mathsf{Y}$, the value $(\zeta_R(y_i), \eta_R(y_i), \nu_R(y_i))$ is called is the single valued neutrosophic value (SVNV).

Definition 2 [10]. *For* $\mathcal{SVNS}s$*, the following operations are defined:*

1. *Partial order:* $R_1 \subset R_2 \Leftrightarrow \zeta_{R_1}(y) \leq \zeta_{R_2}(y),\ \eta_{R_1}(y) \leq \eta_{R_2S}(y)$ *and* $\nu_{R_1}(y) \geq \nu_{R_2}(y),\ \forall\, y \in \mathsf{Y}$
2. *Complement:* $R^c = \{(y, \nu_R(y), 1 - \eta_R(y), \zeta_R(y)) \mid y \in \mathsf{Y}\}$.

Definition 3. *A similarity measure between* $\mathcal{SVNS}s$ R_1 *and* R_2 *is a mapping* $\hat{\mathcal{S}} : \mathcal{SVNS} \times \mathcal{SVNS} \to [0, 1]$ *such that*

(SA1) $0 \leq \hat{\mathcal{S}}(\mathsf{R}_1, \mathsf{R}_2) \leq 1$
(SA2) $\hat{\mathcal{S}}(\mathsf{R}_1, \mathsf{R}_2) = 1 \Longleftrightarrow \mathsf{R}_1 = \mathsf{R}_2$
(SA3) $\hat{\mathcal{S}}(\mathsf{R}_1, \mathsf{R}_2) = \hat{\mathcal{S}}(\mathsf{R}_2, \mathsf{R}_1)$
(SA4) If $\mathsf{R}_1 \subseteq \mathsf{R}_2 \subseteq \mathsf{R}_3$ *then* $\hat{\mathcal{S}}(\mathsf{R}_1, \mathsf{R}_3) \leq \hat{\mathcal{S}}(\mathsf{R}_1, \mathsf{R}_2)$ *and* $\hat{\mathcal{S}}(\mathsf{R}_1, \mathsf{R}_3) \leq \hat{\mathcal{S}}(\mathsf{R}_2, \mathsf{R}_3)$.

Definition 4. *A dissimilarity measure between* $\mathcal{SVNS}s$ R_1 *and* R_2 *is a mapping* $\hat{\mathcal{D}} : \mathcal{SVNS} \times \mathcal{SVNS} \to [0, 1]$ *such that*

(DA1) $0 \leq \hat{\mathcal{D}}(\mathsf{R}_1, \mathsf{R}_2) \leq 1$
(DA2) $\hat{\mathcal{D}}(\mathsf{R}_1, \mathsf{R}_2) = 0 \Longleftrightarrow \mathsf{R}_1 = \mathsf{R}_2$
(DA3) $\hat{\mathcal{D}}(\mathsf{R}_1, \mathsf{R}_2) = \hat{\mathcal{D}}(\mathsf{R}_2, \mathsf{R}_1)$
(DA4) If $\mathsf{R}_1 \subseteq \mathsf{R}_2 \subseteq \mathsf{R}_3$ *then* $\hat{\mathcal{D}}(\mathsf{R}_1, \mathsf{R}_3) \geq \hat{\mathcal{D}}(\mathsf{R}_1, \mathsf{R}_2)$ *and* $\hat{\mathcal{D}}(\mathsf{R}_1, \mathsf{R}_3) \geq \hat{\mathcal{D}}(\mathsf{R}_2, \mathsf{R}_3)$.

3 New Similarity Measures for $\mathcal{SVNS}s$

This section consists of new similarity and dissimilarity measures for $\mathcal{SVNS}s$. Their axiomatically supported proof are given here. The additional properties with their weighted versions are discussed in this part.

Definition 5. *For two $\mathcal{SVNS}s$ R_1 and R_2 in Y, a new similarity measures is defined between R_1 and R_2 as follows:*

$$\mathcal{S}^s(\mathsf{R}_1, \mathsf{R}_2) = \frac{\sum_{j=1}^{m} [\zeta_{\mathsf{R}_1}(y_j) \cdot \zeta_{\mathsf{R}_2}(y_j) + \eta_{\mathsf{R}_1}(y_j) \cdot \eta_{\mathsf{R}_2}(y_j) + \nu_{\mathsf{R}_1}(y_j) \cdot \nu_{\mathsf{R}_2}(y_j)]}{\sum_{j=1}^{m} [\{\zeta^2_{\mathsf{R}_1}(y_j) \vee \zeta^2_{\mathsf{R}_2}(y_j)\} + \{\eta^2_{\mathsf{R}_1}(y_j) \vee \eta^2_{\mathsf{R}_2}(y_j)\} + \{\nu^2_{\mathsf{R}_1}(y_j) \vee \nu^2_{\mathsf{R}_2}(y_j)\}]}. \quad (2)$$

Example 1. For two $\mathcal{SVNS}s$ R_1 and R_2 in Y, where $\mathsf{Y} = \{y_1, y_2, y_3, y_4, y_5\}$ and R_1 and R_2 are given below:

$$\mathsf{R}_1 = \left\{ \frac{(0.7, 0.1, 0.2)}{y_1}, \frac{(0.7, 0.2, 0.2)}{y_2}, \frac{(0.2, 0.1, 0.7)}{y_3}, \frac{(0.9, 0.1, 0.2)}{y_4}, \frac{(0.2, 0.1, 0.6)}{y_5} \right\}$$
$$\mathsf{R}_2 = \left\{ \frac{(0.3, 0.2, 0.4)}{y_1}, \frac{(0.5, 0.2, 0.1)}{y_2}, \frac{(0.1, 0.1, 0.7)}{y_3}, \frac{(0.4, 0.1, 0.3)}{y_4}, \frac{(0.1, 0.1, 0.7)}{y_5} \right\}.$$

Then the similarity measure between R_1 and R_2 is calculated as

$$\mathcal{S}^s(\mathsf{R}_1, \mathsf{R}_2) = \frac{0.0509 + 0.1245 + 0.2406 + 0.1333 + 0.1769}{0.2673 + 0.2433 + 0.2418 + 0.6643 + 0.2418} = 0.669594.$$

Theorem 1. *The measure $\mathcal{S}^s$ define in (2) is a valid similarity measure.*

Proof. To prove the validity, the measure $\mathcal{S}^s$ need to fulfill the four axioms of Definition 3.

(SA1). Since for all $y_j, 1 \leq j \leq m$, we have $\zeta_{\mathsf{R}_1}(y_j) \cdot \zeta_{\mathsf{R}_2}(y_j) \leq \zeta^2_{\mathsf{R}_1}(y_j) \vee \zeta^2_{\mathsf{R}_2}(y_j)$, $\eta_{\mathsf{R}_1}(y_j) \cdot \eta_{\mathsf{R}_2}(y_j) \leq \eta^2_{\mathsf{R}_1}(y_j) \vee \eta^2_{\mathsf{R}_2}(y_j)$ and $\nu_{\mathsf{R}_1}(y_j) \cdot \nu_{\mathsf{R}_2}(y_j) \leq \nu^2_{\mathsf{R}_1}(y_j) \vee \nu^2_{\mathsf{R}_2}(y_j)$. Therefore for each y_j, we have

$$\begin{aligned}&[\zeta_{\mathsf{R}_1}(y_j) \cdot \zeta_{\mathsf{R}_2}(y_j) + \eta_{\mathsf{R}_1}(y_j) \cdot \eta_{\mathsf{R}_2}(y_j) + \nu_{\mathsf{R}_1}(y_j) \cdot \nu_{\mathsf{R}_2}(y_j)]\\ &\leq [\{\zeta^2_{\mathsf{R}_1}(y_j) \vee \zeta^2_{\mathsf{R}_2}(y_j)\} + \{\eta^2_{\mathsf{R}_1}(y_j) \vee \eta^2_{\mathsf{R}_2}(y_j)\} + \{\nu^2_{\mathsf{R}_1}(y_j) \vee \nu^2_{\mathsf{R}_2}(y_j)\}].\end{aligned}$$

Thus $\forall\, y_j,\ i \in \{1, 2, ..., m\}$, we have

$$\begin{aligned}&\sum_{j=1}^{m} [\zeta_{\mathsf{R}_1}(y_j) \cdot \zeta_{\mathsf{R}_2}(y_j) + \eta_{\mathsf{R}_1}(y_j) \cdot \eta_{\mathsf{R}_2}(y_j) + \nu_{\mathsf{R}_1}(y_j) \cdot \nu_{\mathsf{R}_2}(y_j)]\\ &\leq \sum_{j=1}^{m} [\{\zeta^2_{\mathsf{R}_1}(y_j) \vee \zeta^2_{\mathsf{R}_2}(y_j)\} + \{\eta^2_{\mathsf{R}_1}(y_j) \vee \eta^2_{\mathsf{R}_2}(y_j)\} + \{\nu^2_{\mathsf{R}_1}(y_j) \vee \nu^2_{\mathsf{R}_2}(y_j)\}]\\ &0 \leq \mathcal{S}^s(\mathsf{R}_1, \mathsf{R}_2) \leq 1.\end{aligned}$$

(SA2). For necessary condition, we need to prove $\mathsf{R}_1 = \mathsf{R}_2$ when $\mathcal{S}^s(\mathsf{R}_1, \mathsf{R}_2) = 1$. From Eq. (2), we have

$$\frac{\sum_{j=1}^m [\zeta_{\mathsf{R}_1}(y_j) \cdot \zeta_{\mathsf{R}_2}(y_j) + \eta_{\mathsf{R}_1}(y_j) \cdot \eta_{\mathsf{R}_2}(y_j) + \nu_{\mathsf{R}_1}(y_j) \cdot \nu_{\mathsf{R}_2}(y_j)]}{\sum_{j=1}^m [\{\zeta^2_{\mathsf{R}_1}(y_j) \vee \zeta^2_{\mathsf{R}_2}(y_j)\} + \{\eta^2_{\mathsf{R}_1}(y_j) \vee \eta^2_{\mathsf{R}_2}(y_j)\} + \{\nu^2_{\mathsf{R}_1}(y_j) \vee \nu^2_{\mathsf{R}_2}(y_j)\}]} = 1,$$

$$\Rightarrow \sum_{j=1}^m [\zeta_{\mathsf{R}_1}(y_j) \cdot \zeta_{\mathsf{R}_2}(y_j) + \eta_{\mathsf{R}_1}(y_j) \cdot \eta_{\mathsf{R}_2}(y_j) + \nu_{\mathsf{R}_1}(y_j) \cdot \nu_{\mathsf{R}_2}(y_j)]$$
$$= \sum_{j=1}^m [\{\zeta^2_{\mathsf{R}_1}(y_j) \vee \zeta^2_{\mathsf{R}_2}(y_j)\} + \{\eta^2_{\mathsf{R}_1}(y_j) \vee \eta^2_{\mathsf{R}_2}(y_j)\} + \{\nu^2_{\mathsf{R}_1}(y_j) \vee \nu^2_{\mathsf{R}_2}(y_j)\}].$$

Now we claim that $\zeta_{\mathsf{R}_1}(y_j) \cdot \zeta_{\mathsf{R}_2}(y_j) = \zeta^2_{\mathsf{R}_1}(y_j) \vee \zeta^2_{\mathsf{R}_2}(y_j)$, $\eta_{\mathsf{R}_1}(y_j) \cdot \eta_{\mathsf{R}_2}(y_j) = \eta^2_{\mathsf{R}_1}(y_j) \vee \eta^2_{\mathsf{R}_2}(y_j)$ and $\nu_{\mathsf{R}_1}(y_j) \cdot \nu_{\mathsf{R}_2}(y_j) = \nu^2_{\mathsf{R}_1}(y_j) \vee \nu^2_{\mathsf{R}_2}(y_j)$.

Suppose $\zeta_{\mathsf{R}_1}(y_j) \cdot \zeta_{\mathsf{R}_2}(y_j) \neq \zeta^2_{\mathsf{R}_1}(y_j) \vee \zeta^2_{\mathsf{R}_2}(y_j)$, since $\zeta_{\mathsf{R}_1}(y_j) \cdot \zeta_{\mathsf{R}_2}(y_j) \leq \zeta^2_{\mathsf{R}_1}(y_j) \vee \zeta^2_{\mathsf{R}_2}(y_j)$, there exists $r > 0$ such that $\zeta_{\mathsf{R}_1}(y_j) \cdot \zeta_{\mathsf{R}_2}(y_j) + r = \zeta^2_{\mathsf{R}_1}(y_j) \vee \zeta^2_{\mathsf{R}_2}(y_j)$.

In the same way there exists $p, q > 0$ such that $\eta_{\mathsf{R}_1}(y_j) \cdot \eta_{\mathsf{R}_2}(y_j) + p = \eta^2_{\mathsf{R}_1}(y_j) \vee \eta^2_{\mathsf{R}_2}(y_j)$ and $\nu_{\mathsf{R}_1}(y_j) \cdot \nu_{\mathsf{R}_2}(y_j) + q = \nu^2_{\mathsf{R}_1}(y_j) \vee \nu^2_{\mathsf{R}_2}(y_j)$.

By hypothesis it follows that $r + p + q = 0$. This implies that $r = -(p + q)$, which is not possible. This implies that $\zeta_{\mathsf{R}_1}(y_j) \cdot \zeta_{\mathsf{R}_2}(y_j) = \zeta^2_{\mathsf{R}_1}(y_j) \vee \zeta^2_{\mathsf{R}_2}(y_j)$, $\eta_{\mathsf{R}_1}(y_j) \cdot \eta_{\mathsf{R}_2}(y_j) = \eta^2_{\mathsf{R}_1}(y_j) \vee \eta^2_{\mathsf{R}_2}(y_j)$ and $\nu_{\mathsf{R}_1}(y_j) \cdot \nu_{\mathsf{R}_2}(y_j) = \nu^2_{\mathsf{R}_1}(y_j) \vee \nu^2_{\mathsf{R}_2}(y_j)$. This implies that $\zeta_{\mathsf{R}_1}(y_j) = \zeta_{\mathsf{R}_2}(y_j)$, $\eta_{\mathsf{R}_1}(y_j) = \eta_{\mathsf{R}_2}(y_j)$ and $\nu_{\mathsf{R}_1}(y_j) = \nu_{\mathsf{R}_2}(y_j)$. Hence $\mathsf{R}_1 = \mathsf{R}_2$.

The sufficient condition can be deduced easily from Eq. (2).

(SA3). $\mathcal{S}^s(\mathsf{R}_1, \mathsf{R}_2) = \mathcal{S}^s(\mathsf{R}_2, \mathsf{R}_1)$ is trivial.

(SA4). For three $\mathcal{SVNS}s$ R_1, R_2 and R_3 in Y. The similarity measures between R_1, R_2 and R_1, R_3 are given as:

$$\mathcal{S}^s(\mathsf{R}_1, \mathsf{R}_2) = \frac{\sum_{j=1}^m [\zeta_{\mathsf{R}_1}(y_j) \cdot \zeta_{\mathsf{R}_2}(y_j) + \eta_{\mathsf{R}_1}(y_j) \cdot \eta_{\mathsf{R}_2}(y_j) + \nu_{\mathsf{R}_1}(y_j) \cdot \nu_{\mathsf{R}_2}(y_j)]}{\sum_{j=1}^m [\{\zeta^2_{\mathsf{R}_1}(y_j) \vee \zeta^2_{\mathsf{R}_2}(y_j)\} + \{\eta^2_{\mathsf{R}_1}(y_j) \vee \eta^2_{\mathsf{R}_2}(y_j)\} + \{\nu^2_{\mathsf{R}_1}(y_j) \vee \nu^2_{\mathsf{R}_2}(y_j)\}]}.$$

$$\mathcal{S}^s(\mathsf{R}_1, \mathsf{R}_3) = \frac{\sum_{j=1}^m [\zeta_{\mathsf{R}_1}(y_j) \cdot \zeta_{\mathsf{R}_3}(y_j) + \eta_{\mathsf{R}_1}(y_j) \cdot \eta_{\mathsf{R}_3}(y_j) + \nu_{\mathsf{R}_1}(y_j) \cdot \nu_{\mathsf{R}_3}(y_j)]}{\sum_{j=1}^m [\{\zeta^2_{\mathsf{R}_1}(y_j) \vee \zeta^2_{\mathsf{R}_3}(y_j)\} + \{\eta^2_{\mathsf{R}_1}(y_j) \vee \eta^2_{\mathsf{R}_3}(y_j)\} + \{\nu^2_{\mathsf{R}_1}(y_j) \vee \nu^2_{\mathsf{R}_3}(y_j)\}]}.$$

Suppose $\mathsf{R}_1 \subseteq \mathsf{R}_2 \subseteq \mathsf{R}_3$. For all $y_j \in \mathsf{Y}$, we have $\zeta_{\mathsf{R}_1}(y_j) \leq \zeta_{\mathsf{R}_2}(y_j) \leq \zeta_{\mathsf{R}_3}(y_j)$, $\eta_{\mathsf{R}_1}(y_j) \leq \eta_{\mathsf{R}_2}(y_j) \leq \eta_{\mathsf{R}_3}(y_j)$ and $\nu_{\mathsf{R}_1}(y_j) \geq \nu_{\mathsf{R}_2}(y_j) \geq \nu_{\mathsf{R}_3}(y_j)$. This implies that $\zeta^2_{\mathsf{R}_1}(y_j) \leq \zeta^2_{\mathsf{R}_2}(y_j) \leq \zeta^2_{\mathsf{R}_3}(y_j)$, $\eta^2_{\mathsf{R}_1}(y_j) \leq \eta^2_{\mathsf{R}_2}(y_j) \leq \eta^2_{\mathsf{R}_3}(y_j)$ and $\nu^2_{\mathsf{R}_1}(y_j) \geq \nu^2_{\mathsf{R}_2}(y_j) \geq \nu^2_{\mathsf{R}_3}(y_j)$. Then we have

$$\mathcal{S}^s(\mathsf{R}_1, \mathsf{R}_2) = \frac{\sum_{j=1}^m [\zeta_{\mathsf{R}_1}(y_j) \cdot \zeta_{\mathsf{R}_2}(y_j) + \eta_{\mathsf{R}_1}(y_j) \cdot \eta_{\mathsf{R}_2}(y_j) + \nu_{\mathsf{R}_1}(y_j) \cdot \nu_{\mathsf{R}_2}(y_j)]}{\sum_{j=1}^m [\{\zeta^2_{\mathsf{R}_2}(y_j)\} + \{\eta^2_{\mathsf{R}_2}(y_j)\} + \{\nu^2_{\mathsf{R}_1}(y_j)\}]},$$

$$\mathcal{S}^s(\mathsf{R}_1,\mathsf{R}_3)=\frac{\sum_{j=1}^{m}\left[\zeta_{\mathsf{R}_1}(y_j)\cdot\zeta_{\mathsf{R}_3}(y_j)+\eta_{\mathsf{R}_1}(y_j)\cdot\eta_{\mathsf{R}_3}(y_j)+\nu_{\mathsf{R}_1}(y_j)\cdot\nu_{\mathsf{R}_3}(y_j)\right]}{\sum_{j=1}^{m}\left[\{\zeta^2_{\mathsf{R}_3}(y_j)\}+\{\eta^2_{\mathsf{R}_3}(y_j)\}+\{\nu^2_{\mathsf{R}_1}(y_j)\}\right]}.$$

We claim that for all $y_j \in \mathsf{Y}$, we have

$$\frac{\zeta_{\mathsf{R}_1}(y_j)\cdot\zeta_{\mathsf{R}_2}(y_j)}{\zeta^2_{\mathsf{R}_2}(y_j)+\eta^2_{\mathsf{R}_2}(y_j)+\nu^2_{\mathsf{R}_1}(y_j)}\geq\frac{\zeta_{\mathsf{R}_1}(y_j)\cdot\zeta_{\mathsf{R}_3}(y_j)}{\zeta^2_{\mathsf{R}_3}(y_j)+\eta^2_{\mathsf{R}_3}(y_j)+\nu^2_{\mathsf{R}_1}(y_j)},\tag{3}$$

because $\eta^2_{\mathsf{R}_2}(y_j)\leq\eta^2_{\mathsf{R}_3}(y_j)$ and $\frac{1}{\zeta_{\mathsf{R}_2}(y_j)}\geq\frac{1}{\zeta_{\mathsf{R}_3}(y_j)}$. Similarly, we have

$$\frac{\eta_{\mathsf{R}_1}(y_j)\cdot\eta_{\mathsf{R}_2}(y_j)}{\zeta^2_{\mathsf{R}_2}(y_j)+\eta^2_{\mathsf{R}_2}(y_j)+\nu^2_{\mathsf{R}_1}(y_j)}\geq\frac{\eta_{\mathsf{R}_1}(y_j)\cdot\eta_{\mathsf{R}_3}(y_j)}{\zeta^2_{\mathsf{R}_3}(y_j)+\eta^2_{\mathsf{R}_3}(y_j)+\nu^2_{\mathsf{R}_1}(y_j)},\tag{4}$$

$$\frac{\nu_{\mathsf{R}_1}(y_j)\cdot\nu_{\mathsf{R}_2}(y_j)}{\zeta^2_{\mathsf{R}_2}(y_j)+\eta^2_{\mathsf{R}_2}(y_j)+\nu^2_{\mathsf{R}_1}(y_j)}\geq\frac{\nu_{\mathsf{R}_1}(y_j)\cdot\nu_{\mathsf{R}_3}(y_j)}{\zeta^2_{\mathsf{R}_3}(y_j)+\eta^2_{\mathsf{R}_3}(y_j)+\nu^2_{\mathsf{R}_1}(y_j)}.\tag{5}$$

By adding Eqs. 3, 4 and 5, we have

$$\Longrightarrow\frac{\sum_{j=1}^{m}\left[\zeta_{\mathsf{R}_1}(y_j)\cdot\zeta_{\mathsf{R}_2}(y_j)+\eta_{\mathsf{R}_1}(y_j)\cdot\eta_{\mathsf{R}_2}(y_j)+\nu_{\mathsf{R}_1}(y_j)\cdot\nu_{\mathsf{R}_2}(y_j)\right]}{\sum_{j=1}^{m}\left[\{\zeta^2_{\mathsf{R}_2}(y_j)\}+\{\eta^2_{\mathsf{R}_2}(y_j)\}+\{\nu^2_{\mathsf{R}_1}(y_j)\}\right]}$$
$$\geq\frac{\sum_{j=1}^{m}\left[\zeta_{\mathsf{R}_1}(y_j)\cdot\zeta_{\mathsf{R}_3}(y_j)+\eta_{\mathsf{R}_1}(y_j)\cdot\eta_{\mathsf{R}_3}(y_j)+\nu_{\mathsf{R}_1}(y_j)\cdot\nu_{\mathsf{R}_3}(y_j)\right]}{\sum_{j=1}^{m}\left[\{\zeta^2_{\mathsf{R}_3}(y_j)\}+\{\eta^2_{\mathsf{R}_3}(y_j)\}+\{\nu^2_{\mathsf{R}_1}(y_j)\}\right]}.$$

Hence $\mathcal{S}^s(\mathsf{R}_1,\mathsf{R}_3)\leq\mathcal{S}^s(\mathsf{R}_1,\mathsf{R}_2)$. In the same way, we can prove $\mathcal{S}^s(\mathsf{R}_1,\mathsf{R}_3)\leq\mathcal{S}^s(\mathsf{R}_2,\mathsf{R}_3)$.

Hence from **(SA1)**–**(SA4)**, the validity of similarity measure $\mathcal{S}^s$ is ensured for $\mathcal{SVNS}s$.

Definition 6. *Two $\mathcal{SVNS}s$ R_1 and R_2 are called $\simeq^\alpha$-similar if and only if $\mathcal{S}^s(\mathsf{R}_1,\mathsf{R}_3)\geq\alpha$ for $\alpha\in(0,1)$. We denote this as $\mathsf{R}_1\simeq^\alpha\mathsf{R}_2$.*

The reflexivity and symmetry of the relation $\simeq^\alpha$ can be easily prove from Theorem 1.

Corollary 1. *The relation $\simeq^\alpha$ is reflexive and symmetric.*

But the relation $\simeq^\alpha$ is not transitive as we have seen in the following example.

Example 2. Let $\alpha = 0.5$ and R_1, R_2 and R_3 be three $\mathcal{SVNS}s$ in Y, where $\mathsf{Y} = \{y_1, y_2, y_3, y_4, y_5\}$ and R_1, R_2 and R_3 are define as follows:

$$\mathsf{R}_1=\left\{\frac{(0.8,0.1,0.0)}{y_1},\frac{(0.6,0.2,0.1)}{y_2},\frac{(0.1,0.1,0.8)}{y_3},\frac{(0.6,0.1,0.2)}{y_4},\frac{(0.2,0.1,0.6)}{y_5}\right\}$$
$$\mathsf{R}_2=\left\{\frac{(0.6,0.0,0.4)}{y_1},\frac{(0.6,0.2,0.5)}{y_2},\frac{(0.4,0.1,0.7)}{y_3},\frac{(0.8,0.1,0.3)}{y_4},\frac{(0.6,0.1,0.7)}{y_5}\right\}$$
$$\mathsf{R}_3=\left\{\frac{(0.5,0.3,0.2)}{y_1},\frac{(0.3,0.1,0.5)}{y_2},\frac{(0.3,0.3,0.4)}{y_3},\frac{(0.7,0.1,0.2)}{y_4},\frac{(0.6,0.1,0.1)}{y_5}\right\}.$$

The $\mathcal{SVNS}s$ R_1 and R_2, and R_2 and R_3 are $\simeq^\alpha$-similar as $\mathcal{S}^s(\mathsf{R}_1,\mathsf{R}_2)=0.59>0.5$ and $\mathcal{S}^s(\mathsf{R}_2,\mathsf{R}_3)=0.52>0.5$. But the $\mathcal{SVNS}s$ R_1 and R_3 are not $\simeq^\alpha$-similar because $\mathcal{S}^s(\mathsf{R}_1,\mathsf{R}_3)=0.32<0.5$. Thus the relation $\simeq^\alpha$ is not transitive.

The weighted similarity measures are investigated to cope with the situations when the attributes are not of equal importance.

Definition 7. *For two $\mathcal{SVNS}s$ R_1 and R_2 in Y, a new weighted similarity measure is defined between R_1 and R_2 as follows:*

$$\mathcal{S}^s_\omega(\mathsf{R}_1,\mathsf{R}_2) = \frac{\sum_{j=1}^m \omega_j \left[\zeta_{\mathsf{R}_1}(y_j)\cdot\zeta_{\mathsf{R}_2}(y_j)+\eta_{\mathsf{R}_1}(y_j)\cdot\eta_{\mathsf{R}_2}(y_j)+\nu_{\mathsf{R}_1}(y_j)\cdot\nu_{\mathsf{R}_2}(y_j)\right]}{\sum_{j=1}^m \left[\{\zeta^2_{\mathsf{R}_1}(y_j)\vee\zeta^2_{\mathsf{R}_2}(y_j)\}+\{\eta^2_{\mathsf{R}_1}(y_j)\vee\eta^2_{\mathsf{R}_2}(y_j)\}+\{\nu^2_{\mathsf{R}_1}(y_j)\vee\nu^2_{\mathsf{R}_2}(y_j)\}\right]}.$$

where $\omega_j \in [0,1]$ are the weights of alternatives with $\sum_{j=1}^m \omega_j = 1$.

Example 3. If we assign the weights of alternatives in Example 1 as $\omega = \{\omega_1 = 0.8, \omega_2 = 0.5, \omega_3 = 0.6, \omega_4 = 0.7, \omega_5 = 0.6\}$ be the weights of y_1, y_2, y_3, y_4, then $\mathcal{S}^s_\omega(\mathsf{R}_1,\mathsf{R}_2) = 0.44678/1.6585 = 0.269388$.

Theorem 2. *The measure $\mathcal{S}^s_\omega$ is a valid similarity measure.*

The correspondent dissimilarity measure from $\mathcal{S}^s$ is discussed below.

Definition 8. *For two $\mathcal{SVNS}s$ R_1 and R_2 in Y, a dissimilarity measures is defined as follows:*

$$\mathcal{D}^s(\mathsf{R}_1,\mathsf{R}_2) = 1 - \frac{\sum_{j=1}^m \left[\zeta_{\mathsf{R}_1}(y_j)\cdot\zeta_{\mathsf{R}_2}(y_j)+\eta_{\mathsf{R}_1}(y_j)\cdot\eta_{\mathsf{R}_2}(y_j)+\nu_{\mathsf{R}_1}(y_j)\cdot\nu_{\mathsf{R}_2}(y_j)\right]}{\sum_{j=1}^m \left[\{\zeta^2_{\mathsf{R}_1}(y_j)\vee\zeta^2_{\mathsf{R}_2}(y_j)\}+\{\eta^2_{\mathsf{R}_1}(y_j)\vee\eta^2_{\mathsf{R}_2}(y_j)\}+\{\nu^2_{\mathsf{R}_1}(y_j)\vee\nu^2_{\mathsf{R}_2}(y_j)\}\right]}.$$

Definition 9. *For two $\mathcal{SVNS}s$ R_1 and R_2 in Y, the weighted dissimilarity measure is defined as follows:*

$$\mathcal{D}^s_\omega(\mathsf{R}_1,\mathsf{R}_2) = 1 - \frac{\sum_{j=1}^m \omega_j \left[\zeta_{\mathsf{R}_1}(y_j)\cdot\zeta_{\mathsf{R}_2}(y_j)+\eta_{\mathsf{R}_1}(y_j)\cdot\eta_{\mathsf{R}_2}(y_j)+\nu_{\mathsf{R}_1}(y_j)\cdot\nu_{\mathsf{R}_2}(y_j)\right]}{\sum_{j=1}^m \left[\{\zeta^2_{\mathsf{R}_1}(y_j)\vee\zeta^2_{\mathsf{R}_2}(y_j)\}+\{\eta^2_{\mathsf{R}_1}(y_j)\vee\eta^2_{\mathsf{R}_2}(y_j)\}+\{\nu^2_{\mathsf{R}_1}(y_j)\vee\nu^2_{\mathsf{R}_2}(y_j)\}\right]}.$$

where $\omega_j \in [0,1]$ are the weights of alternatives with $\sum_{j=1}^m \omega_j = 1$.

Theorem 3. *The measures $\mathcal{D}^s$ and $\mathcal{D}^s_\omega$ are the valid dissimilarity measures for $\mathcal{SVNS}s$.*

4 Conclusion

This paper has been investigated new similarity and dissimilarity measures for $\mathcal{SVNS}s$. The additional properties of the suggested similarity and dissimilarity measure have been studied. Their corresponding weighted versions have discussed. The notion of $\simeq^\alpha$ relation has been defined and further properties have investigated. In the future, we will use this similarity measure to image processing, pattern recognition, medical diagnosis, and clustering. Additionally, we will investigate other information measures for $\mathcal{SVNS}s$ as well.

References

1. Zadeh, L.A.: Fuzzy sets. Inf. Control **8**, 338–353 (1965)
2. Atanassov, K.T.: Intuitionistic fuzzy sets. Fuzzy Sets Syst. **20**, 87–96 (1986)
3. Molodtsov, D.: Soft set theory-first results. Comput. Math. Appl. **37**, 19–31 (1999)
4. Cuong, B.C.: Picture fuzzy sets. J. Comput. Sci. Cybern. **30**, 409–420 (2014)
5. Khan, M.J., Kumam, P., Ashraf, S., Kumam, W.: Generalized picture fuzzy soft sets and their application in decision support systems. Symmetry **11**(3), 415 (2019)
6. Khan, M.J., Phiangsungnoen, S., Rehman, H., Kumam, W.: Applications of generalized picture fuzzy soft set in concept selection. Thai J. Math. **18**(1), 296–314 (2020)
7. Khan, M.J., Kumam, P., Liu, P., Kumam, W., Ashraf, S.: A novel approach to generalized intuitionistic fuzzy soft sets and its application in decision support system. Mathematics **7**(8), 742 (2019)
8. Khan, M.J., Kumam, P., Liu, P., Kumam, W., Rehman, H.: An adjustable weighted soft discernibility matrix based on generalized picture fuzzy soft set and its applications in decision making. J. Int. Fuzzy Syst. **38**(2), 2103–2118 (2020)
9. Smarandache, F.: Neutrosophic set, a generalisation of the intuitionistic fuzzy sets. Inter. J. Pure Appl. Math. **24**, 287–297 (2005)
10. Wang, H., Smarandache, F. Zhang, Y., Sunderraman, R.: Single valued neutrosophic sets, Multispace and Multistructure, **4** (2010)
11. Ren, H., Xiao, S., Zhou, H.: A chi-square distance-based similarity measure of single-valued neutrosophic set and applications. Int. J. Comput. Commun. Control **14**(1), 78–89 (2019)
12. Wu, H., Yuan, Y., Wei, L., Pei, L.: On entropy, similarity measure and cross-entropy of single-valued neutrosophic sets and their application in multi-attribute decision making. Soft Comput. **22**(22), 7367–7376 (2018)
13. Ye, J., Fu, J.: Multi-period medical diagnosis method using a single valued neutrosophic similarity measure based on tangent function. Comput. Methods Programs Biomed. **123**, 142–149 (2016)
14. Khan, M.J., Kumam, P., Deebani, W., Kumam, W., Shah, Z.: Distance and similarity measures for spherical fuzzy sets and their applications in selecting mega projects. Mathematics **8**(4), 519 (2020)
15. Khan, M.J., Kumam, P., Deebani, W., Kumam, W., Shah, Z.: Bi-parametric distance and similarity measures of picture fuzzy sets and their applications in medical diagnosis. Egy. Inf. J. (2020). https://doi.org/10.1016/j.eij.2020.08.002
16. Khan, M.J., Kumam, P.: Distance and similarity measures of generalized intuitionistic fuzzy soft set and its applications in decision support system. Adv. Intell. Syst. Comput. **1197**, 355–362 (2021)
17. Khan, M.J., Kumam, P., Liu, P., Kumam, W.: Another view on generalized interval valued intuitionistic fuzzy soft set and its applications in decision support system. J. Int. Fuzzy Syst. **38**(4), 4327–4341 (2020)
18. Khan, M.J., et al.: The renewable energy source selection remoteness index-based VIKOR method for generalized intuitionistic fuzzy soft sets. Symmetry **12**(6), 977 (2020)
19. Khan, M.J., Kumam, P., Kumam, W.: Theoretical justifications for the empirically successful VIKOR approach to multi-criteria decision making. Soft Comput (2021). https://doi.org/10.1007/s00500-020-05548-6

Prediction and Estimation

A Novel Feature to Predict Buggy Changes in a Software System

Rahime Yılmaz[1,2], Yağız Nalçakan[3], and Elif Haktanır[1,2](✉)

[1] Istanbul Technical University, 34367 Besiktas, Istanbul, Turkey
{rahime.yilmaz,elif.haktanir}@altinbas.edu.tr
[2] Altınbas University, 34217 Bagcilar, Istanbul, Turkey
[3] Izmir Institute of Technology, 35430 Urla, İzmir, Turkey

Abstract. Researchers have successfully implemented machine learning classifiers to predict bugs in a change file for years. Change classification focuses on determining if a new software change is clean or buggy. In the literature, several bug prediction methods at change level have been proposed to improve software reliability. This paper proposes a model for classification-based bug prediction model. Four supervised machine learning classifiers (Support Vector Machine, Decision Tree, Random Forrest, and Naive Bayes) are applied to predict the bugs in software changes, and performance of these four classifiers are characterized. We considered a public dataset and downloaded the corresponding source code and its metrics. Thereafter, we produced new software metrics by analyzing source code at class level and unified these metrics with the existing set. We obtained new dataset to apply machine learning algorithms and compared the bug prediction accuracy of the newly defined metrics. Results showed that our merged dataset is practical for bug prediction based experiments.

Keywords: Bug prediction · Classification · Code analysis · Code metrics · Software metrics · Machine learning

1 Introduction

Predicting the buggy modules prior to software development helps to improve the user satisfaction, as well as the overall software performance [1]. Thus, software bug prediction is an essential activity in software development [2]. Software bug prediction identifies buggy parts of software such as modules, classes, and files [3]. Detecting and predicting buggy modules of software system in the early phase is a critical process to prevent companies from additional costs [4].

A wide range of approaches have been proposed to improve the performance of the bug prediction task. For instance, Kim et al. [3] introduced a new bug prediction technique, called change classification. In their study, authors classified file changes as buggy or clean by using the combination of change information features including author, commit hour, commit day, and source code terms including numbers, operators, keywords, and comments. They also converted the file and directory name into features. According

C. Kahraman et al. (Eds.): INFUS 2021, LNNS 308, pp. 407–414, 2022.
https://doi.org/10.1007/978-3-030-85577-2_48

to the authors, their experimental results are encouraging, and their techniques are useful for extracting features from the source code and change histories. Although there are numerous approaches related with the bug prediction, specific prediction approaches for unified datasets are rare and especially practicable methods are missing [5]. Ferenc et al. [6] investigated the bug prediction capabilities of the public datasets in their study. They implemented a method for researchers to show how to create the unified bug data set from the gathered public datasets. Their experiments showed that the unified bug dataset can be used effectively in bug prediction.

There are many public or private bug datasets used in bug prediction studies. In this study, we transformed one of these existing datasets to a specific format. We considered a public dataset, and the corresponding source code of it. We utilized the source code analysis of this public data set to obtain new code metrics. We also investigated the newly defined metrics and values of the bug dataset. The public dataset that we transformed had complexity metrics already. All features are unified after extracting new metrics. New features are cumulatively added to the existing datasets which are reasonable and gave promising results [6]. Based on the previous studies and the features suggested in the literature, a new feature is proposed in this study: *classname*. This way, we created one public unified bug dataset for classifying classes as buggy or clean. Finally, we used Support Vector Machine, Decision Tree, Random Forrest, and Naive Bayes algorithms to demonstrate the efficiency of the unified dataset in bug prediction. The success of our unified bug dataset is discussed in the following sections and its effectiveness is tested.

The remainder of the paper is organized as follows. Section 2 briefly describes a public bug prediction dataset property and our unified dataset with the set of features. Section 3 introduces feature extraction methods. Section 4 presents the steps of the proposed prediction models and reports the results of the experiments. Section 5 shows the potential benefits and applications of the proposed method. Concluding remarks and future directions are discussed in the last two section, Sect. 6 and 7.

2 Dataset

In this section, bug prediction dataset and dataset unification will be introduced.

2.1 Bug Prediction Dataset

Most of the bug prediction studies use bug-fix data to make predictions or to validate their prediction model. In our study, we used a public bug prediction dataset* that contains the classic Chidamber & Kemerer (C&K) and 11 class level object-oriented metrics extracted from 5 projects. Eclipse JDT Core project dataset is used for classification. This dataset is composed of the bug and version information of the system as given in details in Table 1.

Table 1. Basic properties of the public bug dataset.

System	Prediction release	Time period	#Classes	#Versions	#Transactions	#Postrel. defects
Eclipse JDT Core www.eclipse.org/jdt/core/	3.4	1.1.2005–6.17.2008	997	91	9,135	463

*http://bug.inf.usi.ch/download.php

2.2 Unified Dataset

We merged the original dataset with the results of feature extraction from classnames. The basis of the merged dataset is our source code analysis result obtained by splitting classname into features. It is extended with the data of the given bug dataset implying that we went through all the elements and created an extended dataset to classify buggy class.

3 Feature Extraction

In this section, we discuss the features used in this study as complexity metrics and our newly defined feature extracting from project's source code to classify buggy and clean changes.

3.1 Complexity Metrics as Features

Code metrics and process metrics represent how complex the source code and the development process are, respectively [7, 8]. Software code metrics are command predict defects in software projects. Many bug prediction techniques in the literature are based on metrics especially the CK suite [9]. Additionally, object-oriented (OO) metrics are combined to CK metrics. Using the CK and the OO metric suites together to predict bugs provides high performance which do not require historical information [10]. All metric values are used as predictors for each class of the project.

The 17-class level source code metrics provided by the dataset are listed in Table 2.

Table 2. Class level source code metrics.

Type	Metric	Description
CK	WMC	Weighted Method Count
	DIT	Depth of Inheritance Tree
	RFC	Response for Class
	NOC	Number of Children
	CBO	Coupling Between Objects
	LCOM	Lack of Cohesion in Methods
OO	FanIn	Number of other classes that reference the class
	FanOut	Number of other classes referenced by the class
	NOA	Number of attributes
	NOPA	Number of public attributes
	NOPRA	Number of private attributes
	NOAI	Number of attributes inherited
	LOC	Number of lines of code
	NOM	Number of methods
	NOPM	Number of public methods
	NOPRM	Number of private methods
	NOMI	Number of methods inherited

3.2 Feature Extraction from Classnames

Using static source code metrics as a feature to predict bugs in a software yields good results [11]. In previous studies, many approaches have been introduced to predict buggy parts depending on static metrics extracted from source code [6]. Every term in the source code, change delta, and change log texts can be used as a feature [12]. According to Shivaji et al. [12], every variable, method name, function name, keyword, comment word, and operator which is every element or part in the source code separated by some special punctuation marks is used as a feature. In this study, we collected a new feature from a dataset, classnames, in the source code separated by double colon and investigated the diversion of metric definitions. We converted the classnames into features based on the fact that they include some information about class and its behavioral relations. The new feature is extracted from the most used words from classname such as internal, core, compiler, and their combinations. Kim et al. [3] converted the directory and filename into features from source code. Differently in this study, common words are extracted from classnames. For extraction of words from class, ATOM tool is used [13]. At the end of the extraction, seven additional features are added to classifier dataset.

To apply prediction model and classify new features, it is necessary to normalize features or to derive appropriate threshold values [14, 15]. For this purpose, we manually normalized features. This process is composed of two steps; first, for each feature, the maximum value in the bug dataset is taken and then values of all the metrics are divided by this maximum value [6].

4 Classification

Machine learning (ML) algorithms for modelling the interaction between software features and bugs are commonly used in industry and researches to develop a solution to the knowledge discovery problems [16]. A ML classification model must be trained by using features of buggy and clean changes in order to classify software changes [3]. First, the classifier is trained with software history data, and then the trained ML model is used to predict bugs in that specific software [12]. Many static source code metrics are extracted from the dataset and these measurements are used as independent features in the ML process.

This study aims to predict bugs on a public software system and to evaluate prediction performance by using four supervised ML algorithms: Support Vector Machine, Decision Tree (J48 in Weka), Naive Bayes and Random Forrest algorithms. We used the WEKA data mining framework to train and test the bug prediction models in the unified bug dataset [17]. The classification algorithms are discussed briefly as follows:

Support Vector Machine (SVM): It is a supervised ML algorithm that works for both large and small data sets [19] and can be used for classification and regression [18]. Moreover, SVMs have good ML classification performance for a wide range of bug prediction and classification problems [3].

Decision Tree (DT): It is a widely used supervised learning algorithm for classification studies with a fixed target variable [20]. DT refers to a hierarchal model which uses the item's features as branches to reach the item's target value in the leaf [21]. This technique divides a given data set into two or more consistent sets based on the most important input features that perform as a differentiator. It can be used for both categorical and continuous input and output variables [20].

Naive Bayes (NB): It is a powerful probabilistic classifier based on the Bayesian theorem, with some independent assumptions over the feature [22]. It requires a number of features in the learning process for better classifier prediction [16, 23]. The NB classifier has consistently high predictive performance and effectiveness for the majority of predictive issues [24].

Random Forest (RF): It is a simple and flexible ensemble learning approach for classification and regression. It constructs the number of DTs at training and takes averages of them to get the predictive accuracy [20]. The correlation between trees is reduced by randomly selecting trees, thus the prediction power increases and leads to an increase in performance [25].

The performance of ML classifiers varies depending on the characteristics of the data set used to train the classifier. This study shows the performance accuracy and capability of the ML algorithms in software bug prediction and provides a comparative analysis of the selected algorithms.

5 Evaluation and Results

The classification performance of ML algorithms in software bug prediction is evaluated by using the 10-fold cross-validation method and the computation of the standard classification evaluation measures, including true positive rate, false positive rate, recall, precision, and F-value. The results of the four classification algorithms are shown in Table 3.

Table 3. Classifier performance comparison.

Classifier	Clean/Buggy	TP Rate	FP Rate	Precision	Recall	F-Measure
SVM	Clean (0)	0.989	0.913	0.806	0.989	0.888
	Buggy (1)	0.087	0.011	0.667	0.087	0.155
DT	Clean (0)	0.938	0.505	0.877	0.938	0.907
	Buggy (1)	0.495	0.062	0.675	0.495	0.571
NB	Clean (0)	0.951	0.626	0.854	0.951	0.900
	Buggy (1)	0.374	0.049	0.664	0.374	0.478
RF	Clean (0)	0.973	0.655	0.851	0.973	0.908
	Buggy (1)	0.345	0.027	0.772	0.345	0.477

Results indicated that four ML algorithms achieved high accuracy rates and the algorithms can be used for bug prediction effectively with good performances. By comparing classifiers over one data set and relying on accuracy indicators, the most effective algorithm is found to be the Random Forest algorithm in the unified bug dataset. Overall, according to degree of predictive accuracy observation, the static source code metric-based classification is ideal in predicting software bugs.

6 Discussion

We found that there are statistically significant differences in the values of the original and the newly calculated features. Furthermore, notations and definitions can severely differ. We compared the bug prediction capabilities of the extended features and produced the unified dataset. By using a public unified dataset as an input for different bug prediction related investigation, researchers can make their studies reproducible, thus be able to validate their findings.

7 Conclusion

Bugs in software are highly undesirable situations affecting the performance negatively. Therefore, the detection and elimination of a bug is a very critical process for the software. In the literature, many studies have been published for decades to detect and predict these bugs and many different features have been proposed for the construction of this process. Some of the most common parameters are still being used in the studies in addition to the new features that are cumulatively added to the existing features to investigate new bug prediction techniques. Based on this, we proposed a new feature named "classname". The success of this new feature is discussed, and its effectiveness is tested on an application. Obtained results clearly verifies that this new feature enhances bug prediction accuracy. For further research, we suggest the relationship and effectiveness of this new feature to be tested considering its relation with the existing features and we recommend a comparison analysis among some known features and the classname feature.

References

1. Kim, S., Pan, K., Whitehead, E.J., Jr.: Memories of bug fixes. In: 14th ACM SIGSOFT International Symposium on Foundations of Software Engineering, pp. 35–45 (2006)
2. Gyimothy, T., Ferenc, R., Siket, I.: Empirical validation of object-oriented metrics on open source software for fault prediction. IEEE Trans. Softw. Eng. **31**(10), 897–910 (2005)
3. Kim, S., Whitehead, E.J., Jr., Zhang, Y.: Classifying software changes: clean or buggy? IEEE Trans. Softw. Eng. **34**(2), 181–196 (2008)
4. Bird, C., Nagappan, N., Gall, H., Murphy, B., Devanbu, P.: Putting it all together: using socio-technical networks to predict failures. In: Software Reliability Engineering, ISSRE-09, pp. 109–119. IEEE (2009).
5. Son, L.H., Pritam, N., Khari, M., Kumar, R., Phuong, P.T.M., Thong, P.H.: Empirical study of software defect prediction: a systematic mapping. Symmetry **11**(2), 212 (2019)
6. Ferenc, R., Siket, I., Hegedűs, P., Rajkó, R.: Employing partial least squares regression with discriminant analysis for bug prediction. arXiv preprint arXiv:2011.01214 (2020)
7. Menzies, T., Greenwald, J., Frank, A.: Data mining static code attributes to learn defect predictors. IEEE Trans. Softw. Eng. **33**, 2–13 (2007)
8. Rahman, F., Devanbu, P.: How, and why, process metrics are better. In: International Conference on Software Engineering, Piscataway, NJ, USA, pp. 432–441. IEEE Press (2013)
9. D'Ambros, M., Lanza, M., Robbes, R.: An extensive comparison of bug prediction approaches. In: Mining Software Repositories (MSR), pp. 31–41. IEEE (2010)
10. Radjenović, D., Heričko, M., Torkar, R., Živkovič, A.: Software fault prediction metrics: a systematic literature review. Inf. Softw. Technol. **55**(8), 1397–1418 (2013)
11. Menzies, T., Greenwald, J., Frank, A.: Data mining static code attributes to learn defect predictors. IEEE Trans. Softw. Eng. **33**(1), 2–13 (2006)
12. Shivaji, S., Whitehead, E.J., Akella, R., Kim, S.: Reducing features to improve code change-based bug prediction. IEEE Trans. Softw. Eng. **39**(4), 552–569 (2012)
13. Srivastava, A., Eustace, A.: ATOM: a system for building customized program analysis tools. In: ACM SIGPLAN 1994 Conference on Programming Language Design and Implementation, pp. 196–205 (1994)
14. Shatnawi, R., Li, W., Swain, J., Newman, T.: Finding software metrics threshold values using ROC curves. J. Softw. Maint. Evol. Res. Pract. **22**(1), 1–16 (2010)
15. Oliveira, P., Valente, M.T., Lima, F.P.: Extracting relative thresholds for source code metrics. In: Software Evolution Week-IEEE Conference on Software Maintenance, Reengineering, and Reverse Engineering (CSMR-WCRE), pp. 254–263. IEEE (2014)
16. Pandey, S.K., Mishra, R.B., Tripathi, A.K.: BPDET: an effective software bug prediction model using deep representation and ensemble learning techniques. Expert Syst. Appl. **144**, 113085 (2020)

17. Hall, M., Frank, E., Holmes, G., Pfahringer, B., Reutemann, P., Witten, I.H.: The Weka data mining software: an update. ACM SIGKDD Explor. Newsl. **11**(1), 10–18 (2009)
18. Scholkopf, B., Smola, A.J.: Learning with kernels: support vector machines, regularization, optimization, and beyond. In: Adaptive Computation and Machine Learning Series. MIT Press (2018)
19. Gray, D., Bowes, D., Davey, N., Sun, Y., Christianson, B.: Using the support vector machine as a classification method for software defect prediction with static code metrics. In: International Conference on Engineering Applications of Neural Networks, pp. 223–234. Springer, Heidelberg (2009). https://doi.org/10.1007/978-3-642-03969-0_21
20. Delphine Immaculate, S., Farida Begam, M., Floramary, M.: Software bug prediction using supervised machine learning algorithms. In: International Conference on Data Science and Communication (IconDSC), Bangalore, India, pp. 1–7 (2019)
21. Ruggieri, S.: Efficient C4. 5 [classification algorithm]. IEEE Trans. Knowl. Data Eng. **14**(2), 438–444 (2002)
22. Carrozza, G., Cotroneo, D., Natella, R., Pietrantuono, R., Russo, S.: Analysis and prediction of mandelbugs in an industrial software system. In: IEEE Sixth International Conference on Software Testing, Verification and Validation, pp. 262–271 (2013)
23. Murphy, K.P.: Naive Bayes classifiers. Univ. Br. Columbia **18**(60), 1–8 (2006)
24. Pandey, S.K., Mishra, R.B., Triphathi, A.K.: Software bug prediction prototype using Bayesian network classifier: a comprehensive model. Procedia Comput. Sci. **132**, 1412–1421 (2018)
25. Gupte, A., Joshi, S., Gadgul, P., Kadam, A., Gupte, A.: Comparative study of classification algorithms used in sentiment analysis. Int. J. Comput. Sci. Inf. Technol. **5**(5), 6261–6264 (2014)

State Prediction of Chaotic Time-Series Systems Using Autoregressive Integrated with Adaptive Network-Fuzzy

Farzin Piltan and Jong-Myon Kim(✉)

Department of Electrical, Electronics, and Computer Engineering,
University of Ulsan, Ulsan 44610, South Korea

Abstract. In this research, advanced technology is used to monitoring chaotic time-series signals. The combination of autoregressive with adaptive network-fuzzy algorithms is suggested for chaotic signal prediction. The autoregressive prediction algorithm is recommended for chaotic time-series prediction. This technique is linear, and the modeling prediction accuracy has a limitation. To reduce the root means square (RMS) error of prediction, the order of autoregressive prediction should be increased which is caused to increase the number of parameters and nonlinearity as well. Thus, the combination of autoregressive prediction with an adaptive network-fuzzy algorithm is suggested to reduce the prediction error in chaotic time-series signals. To test the power of the proposed prediction algorithm, the 2nd order proposed method is compared with the 2nd order and 6th order of AR technique, and the RMS error in these three algorithms are 0.0967, 0.4953, and 0.3159, respectively. So far and compared to the classical autoregressive method, the proposed prediction model is efficient for chaotic time-series signals.

Keywords: Chaotic time-series signals · Condition monitoring · State prediction · Autoregressive algorithm · Adaptive network-fuzzy technique

1 Introduction

In most industries, Prognostics and Health Management (PHM) are played an important role in condition-based monitoring. The PHM in nonlinear systems has a lot of challenges. Various approaches have been used for the PHM in industries. Condition monitoring is the first technique to predict state conditions. Another method for PHM is dynamic load capacity. In this approach, dynamic system modeling is recommended for system modeling and prediction. However, in real-time applications, uncertain conditions increase the challenge of this approach [1]. In this research, the proposed modeling algorithm is suggested for nonlinear chaotic time-series signal prediction.

After collecting the data using the sensors, diverse approaches have been recommended for state prediction. These algorithms have been divided into four main groups: a) knowledge-based models such as fixed rules and fuzzy rules; b) life expectancy models such as stochastic and statistical techniques; c) artificial neural network such as remaining useful life forecasting and parameter estimation; and d) physical models such as

C. Kahraman et al. (Eds.): INFUS 2021, LNNS 308, pp. 415–422, 2022.
https://doi.org/10.1007/978-3-030-85577-2_49

application-specific [2]. The main difficulty of the data-driven-based modeling is reliability. On the other hand, complexity is the main drawback in the physical-based models. Identification approaches, observation techniques, and Kalman filter algorithm are the main three techniques for model-based tactics. The identification approaches can be categorized into two main groups: linear-based identification and nonlinear-based identification. The AutoRegressive (AR), AutoRegressive with eXternal input (ARX), and AutoRegressive Integrated with Moving Average (ARIMA) are some of the important identification approaches for modeling. The AR technique is a linear modeling approach for time-series signals [1]. Moreover, the ARX technique is the combination of the AR technique and external inputs. So, this technique has a challenge for time-series signals. The ARIMA algorithm is the combination of the AR technique and the moving average approach. This technique has been used in various industrial applications to predict the remaining useful life. The prediction accuracy in highly nonlinear chaotic signals is reduced based on the AR, ARX, and ARIMA approaches. To solve the challenge nonlinear modeling approach is suggested.

The nonlinear modeling and prediction have been introduced by data-driven and mathematical nonlinear modeling. The data-driven approach is selected when the model-based method fails in extracting a mathematical definition of the system's behavior. Fuzzy logic and artificial neural network (ANN) algorithms are introduced as two important data-driven techniques. The ANN is modeled by multi-input and multi-output functions. The ANN can be used for feature estimation, state prediction, and remaining useful life estimation. The ANN technique has been used in various applications for prognostics such as lithium-ion batteries [3] and bearing [4]. The fuzzy logic approach is used for system modeling based on four stages: fuzzification, inference mechanism, rule-base, and defuzzification [5]. The application of the fuzzy algorithm in time-series prediction has been represented in [6, 7]. To improve the accuracy of the ANN and fuzzy logic algorithm, an adaptive network-fuzzy technique has been represented [8, 9].

In this research article, the AutoRegressive Integrated with Adaptive Network-Fuzzy (ARIANF) is recommended for chaotic time-series signal prediction. The ARINNF has two main stages. In the first step, the AR to predict the chaotic time-series signal in the presence of Gaussian noise is designed. Next, to reduce the error of prediction, the integration of the AR technique with an adaptive network-fuzzy algorithm is suggested. The ARIANF is a nonlinear prediction algorithm and used for chaotic time-series prediction. The main contribution in this research article is the integration of the AR technique with an adaptive network-fuzzy algorithm for chaotic time-series signal prediction. This paper is organized as follows. The second section outlines the problem statement. The proposed ARIANF algorithm for chaotic time-series signal prediction is represented in the third section. The results are presented in the fourth Section. Conclusions and future works are represented in the last section.

2 Proposed Scheme and Block Diagram

The Mackey-Glass time-series chaotic signal integrated with the Gaussian noise is used for prediction in this research work. This signal is highly nonlinear and predicting this signal is challenging work. The block diagram of the proposed ARIANF illustrates in Fig. 1. Regarding this block diagram, the proposed scheme has two main stages: a) the AR signal modeling and prediction, and b) improve the accuracy of the prediction using the ARIANF technique. In the first step, the AR technique is implemented to modeling and predict the chaotic time-series signal integrated with Gaussian noise. This algorithm is linear, so, it has limitations to reduce the prediction error. To address this issue, in the next step, the AR is integrated with the adaptive network-fuzzy approach and the design ARIANF method. This technique improves the prediction accuracy and reduces the error of prediction, as well.

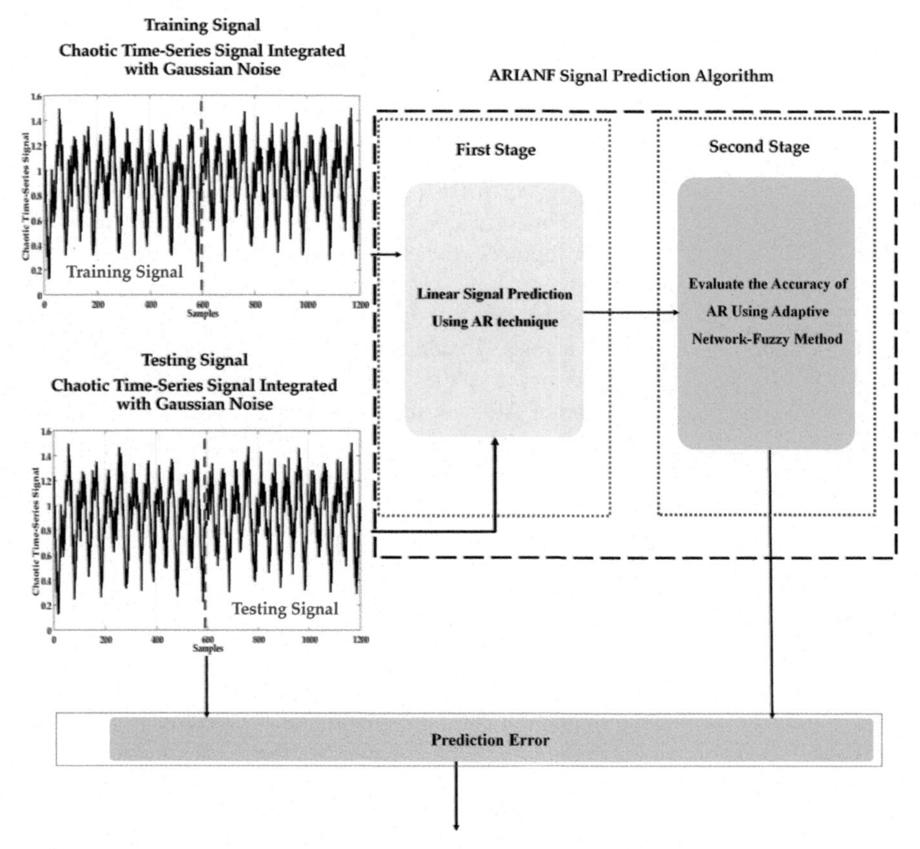

Fig. 1. Autoregressive integrated with adaptive network fuzzy algorithm for nonlinear chaotic signal prediction.

3 Autoregressive Integrated with Adaptive Network-Fuzzy Technique for Chaotic Time-Series Signal Prediction

Based on Fig. 1, in the first step, the AutoRegressive (AR) signal predictor is introduced to predict the chaotic signals. The AR technique for signal prediction is introduced using the following equation.

$$P_{AR}(t+1) = \frac{1}{T(q)} e_{AR}(t) \tag{1}$$

where $P_{AR}, T(q)$, and $e_{AR}(t)$ are the time-series chaotic signal which modeled using AR technique, tuning parameter, and error between original and predicted signal, respectively. Thus, the state-space equations for AR times-series signal prediction are represented as follows.

$$P_{s-AR}(k+1) = t_{P-S} P_{AR}(k) + t_e e_{AR}(k) \tag{2}$$

$$P_{O-AR}(k+1) = {t_{P-S}}^T P_{s-AR}(k) \tag{3}$$

$$e_{AR}(k) = P_O(k) - P_{O-AR}(k-1) \tag{4}$$

Here, $P_{S-AR}(k), P_{O-AR}(k), e_{AR}(k), P_O(k)$, and (t_{P-S}, t_e), respectively, are the state of chaotic time-series signal prediction using the AR method, the predict of chaotic time-series signal using the AR approach, the error of chaotic time-series signal using the AR approach, the original chaotic time-series signal, and the coefficients to tuning the AR signal prediction parameters. The AR technique has a critical issue, especially in highly nonlinear time-series signals. To address this issue, the researchers increase the AR's order. This technique is increased the nonlinear behavior of the AR technique. In this research, the combination of AR with the adaptive network-fuzzy algorithm is recommended. The adaptive network-fuzzy has five layers. The first layer is fuzzification. In this layer, the crisp input can be transfer to the fuzzy set. In this way, membership function plays an important role. Various types of membership functions have been introduced, in this research, the Gaussian membership function is recommended and represented as the following equation.

$$\mu_{ANF}(k) = e^{-\left(\frac{P_O(k-1)-m(k)}{b(k)}\right)^2} \tag{5}$$

Here, $\mu_{ANF}(k)$, and $(m(k), b(k))$, are the membership function to predict the state of time-series signal and the parameters for Gaussian function, respectively. The second layer is weighing and labeling the layer. The π and T are the main weighing layer in the fuzzy algorithm. In this research the T norm, $\emptyset_i$, is recommended and introduced as follows.

$$\emptyset_i = \mu_{ANF}(k) \cdot \mu_{ANF}(k-1) \cdot \mu_{ANF}(k-2) \ldots \ldots \mu_{ANF}(k-n) \tag{6}$$

After that, the third layer is normalization, $\overline{\emptyset}_i$, and is represented as follows.

$$\overline{\emptyset}_i = \left(\frac{\emptyset_i}{\sum_{i=1}^{2^n} \emptyset_i}\right), i = 1, 2, 3, \ldots 2^n \tag{7}$$

The next, fourth, layer is defuzzification. This layer is used to transfer the fuzzy set into crisp ones. In this research, the TSK fuzzy is recommended and the defuzzification is obtained by the fuzzy output multiple by the weight. Finally, the last layer is the summation layer. This layer is used to integrate the output with the help of adding them one by one. Therefore, the ARIANF prediction technique is represented by the following definition.

$$P_{s-ARIANF}(k+1) = t_{P-S}(P_{S-ARIANF}(k) + P_{S-ANF}(k)) + t_e e_{ARIANF}(k) \quad (8)$$

$$P_{O-ARIANF}(k+1) = {t_{P-S}}^T P_{s-ARIANF}(k) \quad (9)$$

$$e_{ARIANF}(k) = P_O(k) - P_{O-ARIANF}(k-1) \quad (10)$$

Here, $P_{S-ARIANF}(k), P_{O-ARIANF}(k), e_{ARIANF}(k)$, and $P_{S-ANF}(k)$, respectively, are the state of chaotic time-series signal prediction using the proposed ARIANF method, the predict of chaotic time-series signal using the ARIANF approach, the error of chaotic time-series signal using the ARIANF approach, and the nonlinear state signal prediction using the ANF algorithm.

4 Results

Figure 2 shows the predicted signal using the 2nd order AR technique. Based on this figure, it is clear that the 2nd order AR approach has a challenge to accurately predict nonlinear signals such as chaotic time-series signals. Figure 3 illustrates the error of signal prediction based on the 2nd order AR approach. It is clear that the predictive accuracy in the 2nd order AR is very low. To increase the predictive accuracy for chaotic time-series signals, to order of the AR scheme is increased from the 2nd order to the 6th order. Figure 4 demonstrates the error of signal prediction using the 6th order AR

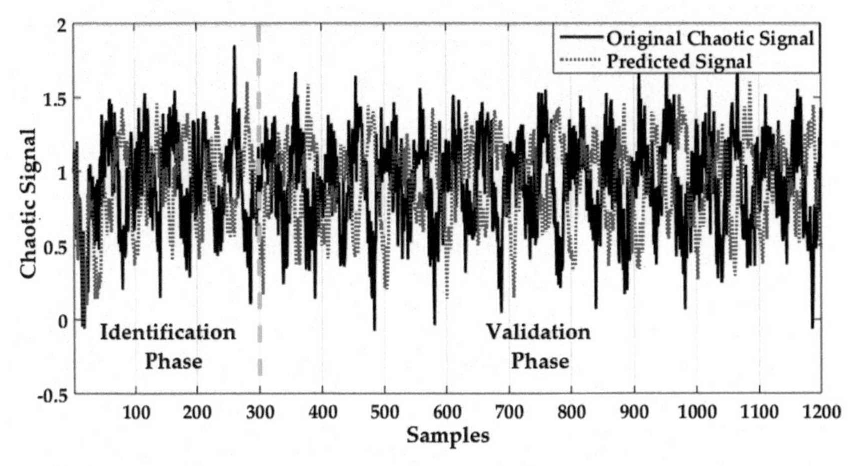

Fig. 2. Chaotic time-series signal prediction using 2nd order Autoregressive technique.

approach. A comparison between Figs. 3 and 4 shows that the accuracy of the AR technique increases with increasing the order. Increasing the AR order is caused to increasing the nonlinearity.

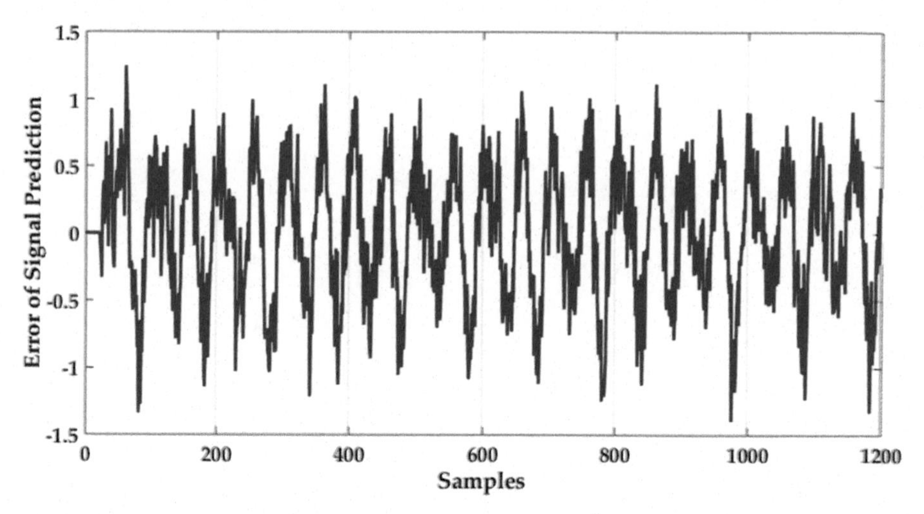

Fig. 3. Error of chaotic time-series signal prediction using 2nd order Autoregressive technique.

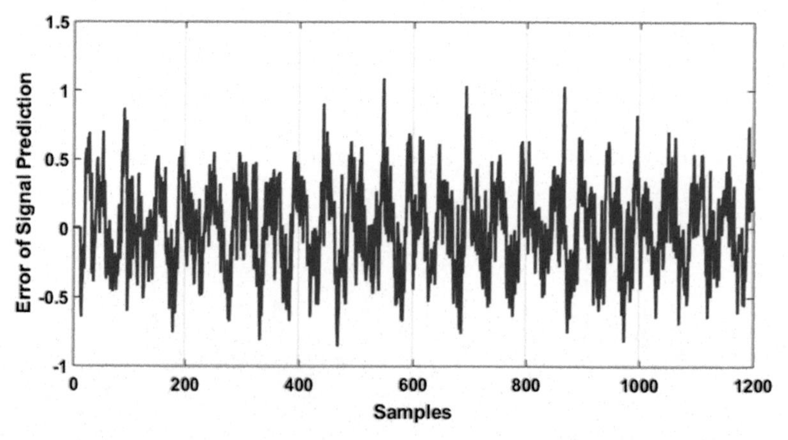

Fig. 4. Error of chaotic time-series signal prediction using 6th order Autoregressive technique.

To address the issue of prediction accuracy, the ARIANF algorithm is suggested. Figures 5 and 6 show the predicted signal and the error of signal prediction using the 2nd order ARIANF approach. Regarding Figs. 3, 4, and 6, it is clear that the accuracy of the signal prediction in the 2nd order ARIANF is better than the 6th order AR and the 2nd order AR. Moreover, the Root Means Square (RMS) error for the 2nd order AR, 6th order AR, and 2nd order ARIANF approach are 0.4953, 0.3159, and 0.0967, respectively.

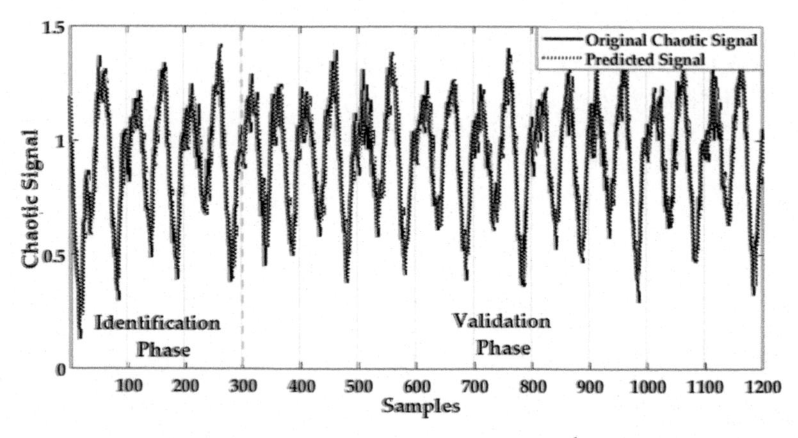

Fig. 5. Chaotic time-series signal prediction using proposed 2nd order ARIANF technique.

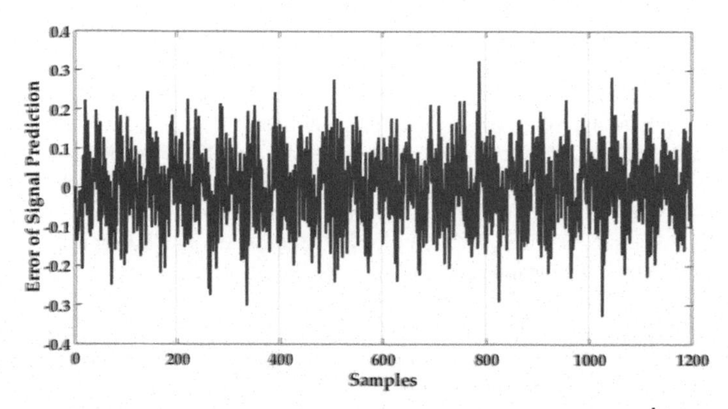

Fig. 6. Error of chaotic time-series signal prediction using proposed 2nd order ARIANF technique.

5 Conclusions

In this research, the autoregressive integrated with adaptive network-fuzzy (ARIANF) chaotic time-series signal prediction is introduced. This algorithm has two main stages. In the first stage, the AR time-series signal prediction was designed using the combination of the linear control algorithm and the feedback method. After time-series signal prediction using the AR technique, in the second stage, the adaptive network fuzzy was recommended to improve the accuracy of prediction and reduce the error of signal prediction, as well. Furthermore, the RMS error of time-series signal prediction for the 2nd order AR and 2nd order ARIANF approach are 0.4953 and 0.0967, respectively. In future work, the combination of autoregressive with external input, moving average, adaptive algorithm, and artificial intelligence approach will be presented for remaining useful life (RUL) prediction.

Acknowledgements. This research was supported by Basic Science Research Program through the National Research Foundation of Korea (NRF) funded by the Ministry of Education (2019R1D1A3A03103840).

References

1. Najeh, T., Lundberg, J.: Degradation state prediction of rolling bearings using ARX-Laguerre model and genetic algorithms. Int. J. Adv. Manufact. Technol. **112**(3–4), 1077–1088 (2020). https://doi.org/10.1007/s00170-020-06416-1
2. Abdenour, S., Medjaher, K., Celrc, G., Razik, H.: Prediction of bearing failures by the analysis of the time series. Mech. Syst. Sig. Process. **139**, 106607 (2020)
3. Claudio, S., Corbetta, M., Giglio, M., Cadini, F.: Adaptive prognosis of lithium-ion batteries based on the combination of particle filters and radial basis function neural networks. J. Power Sources **344**, 128–140 (2017)
4. Min, X., Teng, L., Shu, T., Wan, J., De Silva, C., Wang, Z.: A two-stage approach for the remaining useful life prediction of bearings using deep neural networks. IEEE Trans. Ind. Inf. **15**(6), 3703–3711 (2018)
5. Piltan, F., Kim, J.-M.: Fault diagnosis of bearings using an intelligence-based autoregressive learning Lyapunov algorithm. Int. J. Comput. Intell. Syst. **14**(1), 537–549 (2021)
6. Koo, J.W., Wong, S.W., Selvachandran, G., Long, H.V., Son, L.H.: Prediction of air pollution index in Kuala Lumpur using fuzzy time series and statistical models. Air Qual. Atmos. Health **13**(1), 77–88 (2019). https://doi.org/10.1007/s11869-019-00772-y
7. Melin, P., Monica, J., Sanchez, D., Castillo, O.: Multiple ensemble neural network models with fuzzy response aggregation for predicting COVID-19 time series: the case of Mexico. Healthcare **8**(2), 181 (2020)
8. Jallal, A., Gonzalez-Vidal, A., Skarmeta, A., Chabaa, S., Zeroual, Z.: A hybrid neuro-fuzzy inference system-based algorithm for time series forecasting applied to energy consumption prediction. Appl. Energy **268**, 114977 (2020)
9. Luo, C., Wang, H.: Fuzzy forecasting for long-term time series based on time-variant fuzzy information granules. Appl. Soft Comput. **88**, 106046 (2020)

Drivers of Entrepreneurial Activity at Micro and Meso Levels: A Fuzzy Time-Series Analysis

Jani Kinnunen[1](✉), Irina Georgescu[2], and Zahra Hosseini[3]

[1] Åbo Akademi University, Turku, Finland
jani.kinnunen@abo.fi
[2] The Bucharest University of Economics, Bucharest, Romania
irina.georgescu@csie.ase.ro
[3] Tampere University, Tampere, Finland
zahra.hosseini@tuni.fi

Abstract. The implications of entrepreneurial conditions, opportunities, and attitudes for total entrepreneurial activity (TEA) are under study. The data is obtained from the Global Entrepreneurship Monitor and Heritage Foundation for ten European Union countries, which had long enough time-series data available. The set of indicators include variables to determine the difficulty to start a business, motivations, ambitions, and attitudes of citizens towards entrepreneurship, and economic freedoms from which 7 key indicators were selected. The target variable is TEA and time-period is 2011–2019. The applied time-series analysis compares firstly linear regression accounting for autoregression by classical crisp model, possibilistic least squares linear model with crisp inputs and fuzzy outputs, and fuzzy multi-objective linear model with fuzzy inputs and outputs; secondly, the analysis is complemented by vector error correction model to study long-run causality of the key drivers of TEA. The results extend studies on the connections of entrepreneurial activity and economic freedoms by accounting for behavioral and attitude factors: entrepreneurial finance, grants and subsidies together with established entrepreneurial community support risk-taking of new entrepreneurs, who have identified business opportunities; further, entrepreneurial female/male ratio showed short-run effects, while statistically significant long-run causalities could not be established.

Keywords: Entrepreneurial activity · Entrepreneurial conditions · Economic freedoms · Fuzzy time-series analysis · VECM

1 Introduction

Entrepreneurs enhance economic growth and social development by creating and running new businesses [1, 2]. Entrepreneurship can be defined based on different activities or a set of interdisciplinary knowledge and skills; ownership, management, and innovation are key concepts related to entrepreneurship. Entrepreneurial scholars have classified drivers of entrepreneurial activity to micro, meso and macro level. Micro level includes individual characteristics and motivations such as age and gender [3, 4], psychological

C. Kahraman et al. (Eds.): INFUS 2021, LNNS 308, pp. 423–430, 2022.
https://doi.org/10.1007/978-3-030-85577-2_50

features such as self-efficacy or entrepreneurial cognition [5]. Meso level is an intermediate level, where a target may be policies in a subsystem, such as laws and regulations or it may be limited to regions, sectors or organizations, while macro dimension usually refers to socioeconomic factors [6].

Gender has been broadly studied as an important demographic/individual factor influencing the entrepreneurial activity. Women have been reported to face more problems than men in starting new business, e.g. due to their family responsibilities [7] or their stereotypical roles in a society. Demirguc-Kunt et al. [8] report significant gender differences in access to bank accounts and finance still in many countries. Vodă et al. [9] identified gender as the important factor influencing the entrepreneurial engagement. The authors claim gender equality and perceived capability can affect early-stage entrepreneurship. Pinkovetskaia et al. [10] conducted a comparative analysis of female entrepreneurship in different countries focusing on (a) Female TEA (Total early-stage Entrepreneurial Activity), (b) TEA motivation including opportunity rate and necessity rate, and (c) gender gap of early-stage entrepreneurial activities in 48 countries and concluded the reduction of gender gap in opportunity motivation of women and men. It seems the gender gaps have been narrowing over time in many countries. However, Costa & Pita [11] indicated that Qatari women are less prone to start a business when compared to men in equal conditions. In Romania, women showed a lower entrepreneurial intention than men and the researchers suggested government training support to help individual female entrepreneurs [12].

Financial support is often needed in early-stage and developing entrepreneurship together with information needs about regulation, consistency of government policy, taxes etc. [13]. Different support ranging from finance to consultation, particularly regulatory, have positive influence on business performance of start-ups [14]. However, also private investments, not just government support, play a great role in innovative development. One often emphasizes the role small and medium-sized businesses in the innovative sector of the economy related to R&D activities and their R&D financing, and even launching whole new high-tech industries [15].

Economic freedoms have been found a relevant condition for entrepreneurial activity [16–20], specifically, physical, and intellectual property rights together with investor protections and secured land ownership, have been reported even more important than government support and finance conditions for entrepreneurs [21].

The originality of the paper arises from studying the implications of the combination of (i) entrepreneurial conditions, (ii) entrepreneurial behavior and attitudes, and (iii) economic freedoms (property rights) for TEA using fuzzy time-series analysis.

The rest of the paper is structured as follows. Section 2 describes the used data including the dependent variable TEA and the independent variables together with the applied time-series methods. Section 3 runs the analysis and presents the results of fuzzy regression and the long-term causality analysis. Section 4 concludes the paper.

2 Data and Methods

The data is extracted from Global Entrepreneurship Monitor (GEM) for years 2011–2019. For the research period, 10 EU countries had the full data allowing time-series analysis: Croatia, Germany, Greece, Ireland, Netherlands, Poland, Slovakia, Slovenia, Spain, and Sweden. The GEM data is produced from their surveys around the world with a minimum of 2000 respondents per country [22].

The dependent variable (y) is Total early-stage entrepreneurial activity rate (TEA), i.e. a percentage of 18–64 population who are either a nascent entrepreneur or owner-manager of a new business. Fuzzy numbers are computed as 5% and 95% lower and upper bounds of the three-year average for all variables. The independent variables were selected based on the literature and initial data analysis of possibly statistically significant factors. The features of entrepreneurial conditions can alleviate or hinder staring new businesses: (*a*) Entrepreneurial finance reflects available financing, grants, and subsidies for SMEs; (*b*) Governmental support policies and given relevance for entrepreneurship; and (*c*) R&D transfers, which reflect how well national research and development are commercialized in SMEs. The selected behavioral and attitude factors of entrepreneurship include: (*d*) Perceived opportunities rate (5) of 18–64 population (excluding already active entrepreneurs), who identify opportunities to start a business; (*e*) Established business ownership rate of 18–64 population who have been owner-manager of an established business for over 42 months; (*f*) Female/male TEA measuring the share of female nascent entrepreneurs or owner-managers of 18–64 divided by the male share. From the economic freedom indicators provided by Heritage foundation [23] only (*g*) Property rights is included based on literature (cf. [21]). Property rights is a measure of the legal framework securing individuals to acquire, hold, and utilize private physical or intellectual property, and covers also investor protection, risk of expropriation, and quality of land administration.

Fuzzy Regression Analysis. Three models are fitted by R software, package fussyreg, to our time-series data: (i) LM model is a classical linear model fitted separately for each country by least-squares as a reference model. Function *lm*() has an intercept term (*x0*) in addition to our dependent *y* (TEA) and independent variables, a–g, i.e., $lm(y \sim x0 + a + b + c + d + e + f + g)$ for $t = 0$ variables. In the analysis, we will present results with lagged terms for each variable also for two past years, $t = -1$, and $t = -2$. Correlation coefficients are obtained by linear least squares method.

(ii) PLRLS is a fuzzy linear regression model introduced by Lee and Tanaka [24]. It applies least squares method to compute regression coefficients by fitting the parameters of a triangular fuzzy number $A = (a, \alpha, \beta)$: center a, left spread α, and right spread β. The model can be seen as an extension of the crisp linear lm model: it still uses crisp input values for dependent and independent variables, and it produces coefficients in the form of non-symmetric triangular fuzzy number, i.e. α andβ may differ. Again, all variable data is entered into *plrls*() with two lagged values.

(iii) MOFLR refers to multi-objective fuzzy linear regression model introduced by Nasrabadi et al. [25] The model also combines a least squares method with a possibilistic approach returning fuzzy triangular coefficients. However, *moflr*() allows non-symmetrical triangular fuzzy independent variable inputs, i.e. α can differ from β. The dependent variable cab be entered as a symmetric triangular fuzzy number like in (ii). The model is again fitted to the same data with lagged values, $t = -1$ and $t = -2$.

Vector Error Correction Modelling. Vector Error Correction (VEC) Model is similar with the vector autoregressive model (VAR), but there is an equilibrium relation running from independent to dependent variables in their levels, called cointegration. One can consider a VEC model a VAR model with cointegrating constraints.

Following [26] and [27], the general form of a VAR model is:

$$y_t = A_1 y_{t-1} + A_2 y_{t-2} + \cdots + A_p y_{t-p} + Bx_t + \mu_t, t = 1, 2, \ldots, T, \quad (1)$$

where $y_t = (y_{1t}, \ldots, y_{kt})$ is a k-dimensional time series, $t = 1, .., T$, each y_{it} is integrated of order 1, $i = 1, \ldots, k$, and $x_t = (x_{1t}, \ldots, x_{dt})$ is a d-dimensional exogenous time series.

The general form of a VEC model is:

$$\Delta y_t = \alpha ECM_{t-1} + \sum_{i=1}^{p-1} \Gamma_i \Delta y_{t-i} + \mu_t, \quad (2)$$

where ECM_{t-1} is the error correction term (ECM) which signifies the long-run equilibrium and $\Gamma_i = -\sum_{j=i+1}^{p} A_j$.

3 Analysis

3.1 Regression Analysis

We fit the three linear regression models to our data in such a way that TEA is the dependent variable, x_0 is the intercept term, and the seven independent variables *a, b, c, d, e, f,* and *g* enter the model also in their lagged forms (−1) and (−2) as seen in the first column (V) of Table 1. We limit the analysis to two lags, which was determined optimal in long-term causality analysis of Sect. 3.2. For simplicity we report coefficient significances only for the linear model (LM) on the last column of Pr(>|t|). We note that none of the $t = 0$ terms are statistically significant, while entrepreneurial finance/grants/subsidies ($a(-1)$) and perceived opportunities ($d(-1)$) together with the autoregressive term of TEA ($y(-1)$) are significant. Also, of $t = -2$ terms, entrepreneurial finance ($a(-2)$) and established ownership ($e(-2)$) are significant, first one with about the same regression coefficient as the autoregressive TEA of order 1. Perceived opportunities ($d(-2)$) and established ownership ($e(-2)$) have the longest-term statistically significant effects on TEA.

Taking a look at the spreads of the statistically significant variables, fuzzy models PLRLS and MOFLR in Table 1 are seen to return coefficients as a triangular fuzzy number $A = (a, \alpha, \beta)$ in non-symmetrical and symmetrical forms, respectively. However, PLRLS computes close to zero spreads making the added value from them limited. MOFLR, instead, returns positive spreads of practical importance for $y(-1)$, $a(-1)$, $d(-2)$, and $e(-2)$, i.e. 0.28, 4.01,0.03, and 0.17, respectively, while returning negative spread for $d(-1)$, $a(-2)$ of the statistically significant variables. Fuzzy coefficients add value for interpretations in the case of lagged TEA ($y(-1)$) and perceived opportunities $d(-2)$: for TEA(−1), center value of PLRLS is equal to the LM coefficient 0.63 implying that $t = 0$ TEA is 0.63 * TEA(−1), while MOFLR suggests the coefficient of 0.99, which may vary from 0.71 to 1.27 (= 0.99 ± 0.28). Similarly, established ownership $e(-2)$, according to LM, has the coefficient of 0.23, PLRLS has 0.13, and MOFLR has negative −0.33 with spreads of 0.17; the effects of perceived opportunities are much smaller, 0.06 * $d(-2)$ by LM, but MOFLR provides again reasonable range ±0.03 * $d(-2)$.

Table 1. Fuzzy and crisp linear regression model coefficients. The symbols of significance levels (p-values) are: "*" (p<0.1), "**" (p<0.05), "***" (p<0.01).

V	PLRLS	MOFLR	LM	Std. Error	t-value	Pr(>|t|)
x_0	(−1.11, 0, 0)	(−2.62, −3.98, −3.98)	0.11	1.81	0.06	0.95
a	(1.76, 0, 0)	(0.44, 3.73, 3.73)	0.89	1.07	0.83	0.41
b	(0.29, 0, 0)	(0.06, −1.15, −1.15)	−0.71	0.83	−0.85	0.40
c	(−0.60, 0, 0)	(−0.56, 4.63, 4.63)	−0.04	1.21	−0.03	0.97
d	(0.00, 0, 0)	(0.00, 0.22, 0.22)	0.03	0.03	0.90	0.37
e	(−0.40, 0, 0)	(−0.71, −0.02, −0.02)	−0.05	0.12	−0.42	0.67
f	(2.33, 0, 0)	(−0.25, −0.95, −0.95)	−0.03	1.49	−0.02	0.98
g	(0.04, 0, 0)	(0.03, −0.05, −0.05)	0.05	0.04	1.30	0.20
y(−1)	**(0.63, 0, 0)**	**(0.99, 0.28, 0.28)**	**0.63**	**0.13**	**4.80**	**0.00*** **
a(−1)	**(−0.30, 0, 0)**	**(−0.60, 4.01, 4.01)**	**2.93**	**1.05**	**2.79**	**0.01** **
b(−1)	(0.55, 0, 0)	(1.23, 0.58, 0.58)	0.35	0.87	0.40	0.69
c(−1)	(−1.21, 0, 0)	(0.69,−11.99,−11.99)	−0.91	1.26	−0.72	0.47
d(−1)	**(−0.05, 0, 0)**	**(−0.01, −0.27, −0.27)**	**−0.09**	**0.04**	**−2.52**	**0.02** **
e(−1)	(0.35, 0, 0)	(0.96, 0.11, 0.11)	−0.18	0.14	−1.29	0.20
f(−1)	(1.45, 0, 0)	(−4.49, 20.1, 20.1)	−1.63	1.58	−1.03	0.31
g(−1)	(0.03, 0, 0)	(0.01, 0.07, 0.07)	−0.02	0.04	−0.38	0.70
y(−2)	(0.32, 0.08, 0.09)	(0.03, −0.09, −0.09)	0.16	0.14	1.13	0.26
a(−2)	**(−0.31, 0, 0)**	**(1.26, −8.27, −8.27)**	**−1.79**	**1.04**	**−1.72**	**0.09***
b(−2)	(−2.15, 0, 0)	(−1.76, −1.93, −1.93)	−1.22	0.82	−1.48	0.14
c(−2)	(1.65, 0, 0)	(0.54, 11.85, 11.85)	1.21	1.14	1.07	0.29
d(−2)	**(0.02, 0, 0)**	**(0.00, 0.03, 0.03)**	**0.06**	**0.03**	**1.75**	**0.09***
e(−2)	**(0.13, 0, 0)**	**(−0.33, 0.17, 0.17)**	**0.23**	**0.13**	**1.79**	**0.08***
f(−2)	(−6.51, 0, 0)	(6.37, −24.36, 24.36)	−0.72	1.72	−0.42	0.68
g(−2)	*(−0.02, 0, 0)*	(−0.04, 0.00, 0.00)	−0.01	0.04	*−0.31*	0.76

3.2 Long-Term Causality: VECM

We applied Panel unit root tests of Levin, Lin and Chu [28] and Im, Pesaran and Shin [29], and the Fisher -type tests developed by Choi [30]. We obtained that all time-series are not stationary at level but become stationary at first difference. Therefore, all time-series are integrated of order 1 and we can run the Kao Residual Cointegration Test [31]. The null hypothesis is that there is no cointegration among variables. Since $p = 0.00 < 0.05$ for Augmented Dickey-Fuller test seen in Table 2, we reject H_0, meaning that the variables are cointegrated.

Table 2. Kao residual cointegration test

	t-Statistics	Prob
Augmented Dickey-Fuller	−4.164360	0.0000

Since the variables are cointegrated, we construct the VEC model, considering that the optimal number of lags is 2. We obtained the following VEC equation:

$$\begin{aligned} & D(LNTEA) = C(1) \times (LNTEA(-1) - 12.222LNRD(-1) + 0.605LNPERC(-1) \\ & + 6.680LNGOV(-1) - 14.069LNFIN(-1) + 2.823LNEST(-1) + 0.049) + C(2) \\ & \times D(LNTEA(-1)) + C(3) \times D(LNTEA(-2)) + C(4) \times D(LNRD(-1)) + C(5) \\ & \times D(LNRD(-2)) + C(6) \times D(LNPERC(-1)) + C(7) \times D(LNPERC(-2)) + C(8) \\ & \times D(LNGOV(-1)) + C(9) \times D(LNGOV(-2)) + C(10) \times C(11) \\ & \times D(LNFIN(-2)) + C(12) \times D(LNEST(-1)) + C(13) \times D(LNEST(-2)) \\ & + C(14) \times D(PROP(-1)) + C(15) \times D(PROP(-2)) + C(16) \times D(FM(-1)) \\ & + C(17) \times D(FM(-2)) + C(18) \end{aligned} \tag{3}$$

The error correction term is C(1) = −0.016, but its p-value is 0.286 (>0.05; the same with other long-run terms in Table 3), not statistically significant, meaning that there is no long-run causality from independent variables to the dependent variable TEA. At 10% significance level, property rights are significant. Further, the short-run causality was checked by means of Wald test. In summary, there is no long-run causality between the independent variables and TEA, but on the short run, according to Wald test, a causality relation runs from jointly ($t = 1,2$) Female/Male_TEA to TEA.

Table 3. VECM coefficients.

C	Coeff	Pr(>|t|)	C	Coeff	Pr(>|t|)	C	Coeff	Pr(>|t|)
C(1)	−0.016	0.286	C(7)	0.029	0.828	C(13)	0.152	0.231
C(2)	−0.399	0.007	C(8)	0.060	0.799	C(14)	−0.309	0.097
C(3)	−0.118	0.411	C(9)	0.045	0.848	C(15)	−0.404	0.062
C(4)	−0.226	0.576	C(10)	0.627	0.091	C(16)	−0.000	0.968
C(5)	0.059	0.886	C(11)	0.001	0.998	C(17)	0.006	0.215
C(6)	−0.215	0.189	C(12)	−0.171	0.127	C(18)	0.042	0.172

4 Conclusions

The paper studied the time-series of ten EU countries to explain total early-stage entrepreneurial activity, TEA. The first phase of the study tested linear regressions both in probabilistic and possibilistic terms to provide coefficient values, including ranges from the fuzzy regression models. Statistically significant effects on TEA, along the previous periods TEA ($y(-1)$) were seen from entrepreneurial finance, grants and subsidies ($a(-1)$) affecting firstly positively, but could have negative effects later ($a(-2)$), and from perceived opportunities affecting positively ($d(-2)$) after a possible first negative effect (d(−1)), while established ownership ($e(-2)$) had positive effects to TEA. The results are intuitive: available financial support increases entrepreneurial activity of risk-taking new entrepreneurs, while the support cannot keep businesses running unless revenues are established though successful strategies built on perceived opportunities. The regression coefficients from the fuzzy models reflected high uncertainty of the effects with our small sample. Econometric tests did not report long-term causality from the independent set to TEA, but in the short run, female/male entrepreneurship ratio had a causal effect on TEA. For future research, we suggest fuzzy time-series analysis with a larger sample. More features (than the seven in this study) of the three phenomena, (i) entrepreneurial conditions, (ii) behavior and attitudes and (iii) economic freedoms using, e.g., a supervised machine learning approach of deep canonical correlation analysis could reveal further interdependencies.

References

1. Ribeiro-Soriano, D.: Small business and entrepreneurship: their role in economic and social development. Entrep. Reg. Dev. **29**(1–2), 1–3 (2017). https://doi.org/10.1080/08985626.2016.1255438
2. Farinha, L., Ferreira, J.J., Nunes, S.: Linking innovation and entrepreneurship to economic growth. Compet. Rev. **28**(4), 451–475 (2018)
3. Murzacheva, E., Sahasranamam, S., Levie, J.: Doubly disadvantaged: gender, spatially concentrated deprivation and nascent entrepreneurial activity. Eur. Manage. Rev. **17**(3), 669–685 (2020)
4. Brush, C.G., Greene, P.G., Welter, F.: The Diana project: a legacy for research on gender in entrepreneurship. Int. J. Gend. Entrep. **12**(1), 7–25 (2019)
5. Raza, A., Muffatto, M., Saeed, S.: Cross-country differences in innovative entrepreneurial activity: an entrepreneurial cognitive view. Manage. Decis. **58**(7), 1301–1329 (2018). https://doi.org/10.1108/MD-11-2017-1167
6. Kim, P.H., Wennberg, K., Croidieu, G.: Untapped riches of meso-level applications in multilevel entrepreneurship mechanisms. Acad. Manage. Perspect. **30**(3), 273–291 (2016)
7. Sciascia, S., Mazzola, P., Astrachan, J., Pieper, T.: The role of family ownership in international entrepreneurship: exploring nonlinear effects. Small Bus. Econ. **38**(1), 15–31 (2012)
8. Demirguc-Kunt, A., Klapper, L., Singer, D., Van Oudheusden, P.: The Global Findex Database 2014: measuring financial inclusion around the world. Policy Research Working Paper 7255, World Bank, Washington, DC (2015)
9. Vodă, A.I., Butnaru, G.I., Butnaru, R.C.: Enablers of entrepreneurial activity across the European Union—an analysis using GEM individual data. Sustainability **12**(3), 1022 (2020)
10. Pinkovetskaia, I., Nikitina, I., Gromova, T.: Demography of early entrepreneurship: experience of different countries in recent years. J. Hist. Cult. Art Res. **8**(4), 79–89 (2019). https://doi.org/10.2478/zireb-2020-0003
11. Costa, J., Pita, M.: Appraising entrepreneurship in Qatar under a gender perspective. Int. J. Gend. Entrep. **12**(3), 233–251 (2020)

12. Dumitru, I., Dumitru, I.: Drivers of entrepreneurial intentions in Romania. J. Econ. Forecast. **0**(1), 157–166 (2018)
13. Pawitan, G., Nawangpalupi, C.B., Widyarini, M.: Understanding the relationship between entrepreneurial spirit and global competitiveness: implications for Indonesia. Int. J. Bus. Soc. **18**, 261–278 (2017)
14. Alkhaldi, T., Cleeve, E., Brander-Brown, J.: Formal institutional support for early-stage entrepreneurs: evidence from Saudi Arabia. In: European Conference on Innovation and Entrepreneurship, pp. 903–XIII. Academic Conferences International Limited (2018)
15. Reshetnikova, M.S.: Innovation and entrepreneurship in China. Eur. Res. Stud. J. **21**(3), 506–515 (2018)
16. Bjørnskov, C., Foss, N.J.: Economic freedom and entrepreneurial activity: some cross-country evidence. Public Choice **134**(3–4), 307–328 (2008)
17. Mandić, D., Borović, Z., Jovićević, M.: Economic freedom and entrepreneurial activity: evidence from EU 11 countries. Economics **5**(2), 11–17 (2017)
18. Nyström, K.: The institutions of economic freedom and entrepreneurship: evidence from panel data. Public Choice **136**(3), 269–282 (2008)
19. Georgescu, I., Kinnunen, J.: Well-being and Economic Freedoms in OECD. In: Proceedings of the 12th LUMEN International Scientific Conference Rethinking Social Action (LUMEN RSACVP2019), Iasi, Romania, May 2018, pp. 107–125 (2019). https://doi.org/10.18662/lumproc.158
20. Georgescu, I., Androniceanu, A., Kinnunen, J.: A computational analysis of economic freedom indicators and GDP in EU states. In: Proceedings of the 17th International Conference on Informatics in Economy (IE 2018), Iasi, Romania, pp. 461–468 (2020). https://doi.org/10.24818/IMC/2020/03.11
21. Kinnunen, J., Georgescu, I.: Entrepreneurial activity - a matter of microeconomic conditions or macroeconomic freedoms? In: Proceedings of 5th Business and Entrepreneurial Economics Conference (BEE-2020). Rijeka, Croatia, May 2020, pp. 88–97 (2020)
22. Bosma, N., Hill, S., Ionescu-Somers, A., Kelley, D., Levie, J., Tarnawa, A.: Global entrepreneurship monitor global report 2019/2020 (2020)
23. Miller, T., Kim, B.A., Roberts, J.M.: 2021 Index of economic freedom. Heritage Foundation, NE Washington (2021)
24. Lee, H., Tanaka, H.: Fuzzy approximations with non-symmetric fuzzy parameters in fuzzy regression analysis. J. Oper. Res. Soc. Jpn. **42**(1), 98–112 (1999)
25. Nasrabadi, M.M., Nasrabadi, E., Nasrabady, A.R.: Fuzzy linear regression analysis: a multi-objective programming approach. Appl. Math. Comput. **163**, 245–251 (2005)
26. Zhou, X.: VECM model analysis of carbon emissions, GDP, and international crude oil prices. Discrete Dyn. Nat. Soc. **2018**, Article ID 5350308, 11 pp. (2018). https://doi.org/10.1155/2018/5350308
27. Lütkepohl, H.: New Introduction to Multiple Time Series Analysis. Springer, Heidelberg (2006). https://doi.org/10.1007/978-3-540-27752-1
28. Levin, A., Lin, C.F., Chu, C.: Unit root tests in panel data: asymptotic and finite-sample properties. J. Econom. **108**(1), 1–24 (2002)
29. Im, K.S., Pesaran, M.H., Shin, Y.: Testing for unit roots in heterogeneous panels. J. Econom. **115**(1), 53–74 (2003)
30. Choi, I.: Unit root tests for panel data. J. Int. Money Fin. **20**, 249–272 (2001)
31. Kao, C.: Spurious regression and residual-based tests for cointegration in panel data. J. Econom. **90**, 1–44 (1999)

Short-Term Forecasting for Three Phrase Current on Distribution Network by Applying Multiple-Layer Perceptron Algorithm

Thien-An Nguyen[1], Huu-Vinh Nguyen[1](✉), and Hung Nguyen[2]

[1] Ho Chi Minh City Power Company, Ho Chi Minh City, Vietnam
[2] Ho Chi Minh City University of Technology (HUTECH), Ho Chi Minh City, Vietnam
n.hung@hutech.edu.vn

Abstract. On the trail progressing towards the era of whirlwind technological advances in the field of Information Technology (IT) and media during Industry 4.0, never before has the Internet been integrated into such an essential component in our society, and along with the decrease in the cost of telecommunication equipment, Internet of Things (IoT) has become the mainstay of future development. Therefore, the installation of 3G modem and lighting cables for over 1000 switchgear as well as 110 kV and 22 kV substation has been initiated across Vietnam with a view of conducting distant monitor and information gathering. After a procedure of real-time information gathering, the quality and information storage considered to be essential as a huge amount of data are collected for analyzing useful information on customers. In order to take advantage of this plentiful data resource, an automatic tool for analysis is prerequisite. A procedure of electrical system basing on improved parallel neuron network with smart identification, quick diagnosis, and great stability will be elaborated in this research paper. This study has successfully developed an improved model of parallel neuron network for the stable identification problem. The suggested model has achieved the main target of improving the layers' accuracy.

Keywords: Distribution network · Multiple-layer perceptron · Short-term forecasting · Adam optimizer

1 Introduction

In recent years, there has been many researches into the application of artificial intelligent in calculating the stability of the electric power system [6]. In research [8], the author applied artificial neural network in diagnosing the stability of the electric power system [6]. In research [7], artificial neural network is suggested to be applied in preventing the fracture of power line in micro-grid network. In this article, the author also pointed out ANN after being trained was being able to quickly and effectively response in resolving problems relating to stabilizing the electric power system. However, this article did not mention about the problem relating to variables and examples. In [8], the author suggested applying vector supporting machine (SVM) and Decision Tree

C. Kahraman et al. (Eds.): INFUS 2021, LNNS 308, pp. 431–437, 2022.
https://doi.org/10.1007/978-3-030-85577-2_51

(DT) in diagnosing the stability of the electric power system. In the announcement, the authors introduced variable groups and classifiers combining SVM and DT into rating the transient stability of the electric power system. In [4], the authors pointed out that Intelligent System (IS) had successfully gotten close to rating the real-time stability of the system. In this project, the authors applied the Relief algorithm to select variables and this algorithm had also been announced by them earlier [5]. The article [5] is also a project introducing the procedure of selecting variables in a general way and approaching the matter of selecting variables systematically. In [3], the authors also analyzed the fast calculating capability of intelligent calculating technology in urgent situations with the progress happening fast and complicatedly in the electric power system. In [9], DT was applied in rating the system's stability.

After going over the published articles, there are two commons model: the first one is a common single diagnosing set with a neural network [1, 3, 5] and the second one is the parallel model with a diagnosing set including particle parallel networks. In [2], the author suggested the parallel model with each particle model diagnosing one type of incident that can cause instability, including three phases with one has earth fault and two phases with both has earth fault. This essay suggested this model to be applied in diagnosing the security of electric power system with pre-incident variables and only observed a few lines, buses mentioned above. However, this model being applied in diagnosing multiple buses or lines during an incident may cause one that destabilize the electric power system, increasing and complicating the amount of the particle model. Therefore, this research suggests constructing an intelligent identifier based on the improved neural network to quickly diagnose the transient stability of the electric power system with higher accuracy, which is crucial.

This paper is organized as below: The first section is introduction. The design of the multi-layer perceptron network algorithm is described in the Sect. 2. The results and discussions are presented in the Sect. 3. The conclusions of the paper is presented in the last section.

2 The Design of the Multi-layer Perceptron Network Algorithm and Backpropagation in Short-Term Forecasting for Three Phrase Current on Distribution Network

The purpose of a good machine learning model is to generalize well from the training data to any types of data from a domain of a problem. This allows us to predict the future data that the model will have never recognized. The probability of a machine learning model and data generalization are overfitting and underfitting respectively. Overfitting and Underfitting are the two biggest reason causing low efficiency of machine learning algorithm.

In this research, the K-fold cross-validation will be employed to train the data. The K-cross validation method includes: dividing the training domain into kk smaller sets that have no similar values and have nearly the same sizes. In every tests (runs), one of the k small set will be selected to be the validate set. This model will be constructed based on k−1 remaining sets. The last model will be identified based on the average of the train error and validation error.

To rate the output result, we usually have two methods which are Mean Absolute Error (MAE) and Root-mean-square Error (RMSE), being the two most common measurement for the accuracy for constantly changing variables. In this topic, MAE is chosen because of its consistency while RMSE increases when the variables corresponding with the splitting frequency of the error level increases.

According to the diagram of the procedure of predicting based on MLP program in Fig. 1, past data are input into the program and split into 2 sets of data including: data for training the MLP network and data for predicting the future electric transmission. Data for training the neural network are used to run the training model for the MLP network initiated in accordance to a set of specifications of the program, including the number of outputs, inputs, hidden layers, hidden layer nodes, learning speed, the amount of data for each learning and the number of times for training the MLP network.

Data for prediction will be the output for the MLP training network initiated like above. After the training phase ends, we will have a MLP network model with characteristics and specifications suitable for predicting transmission in the next step. Data for predicting transmission separated from before will be used as the input to run the MLP model with the output being the data of recent future transmission. This result will be compared with recently gathered result from the next period of time and validated the error ratio for the constructed MLP model.

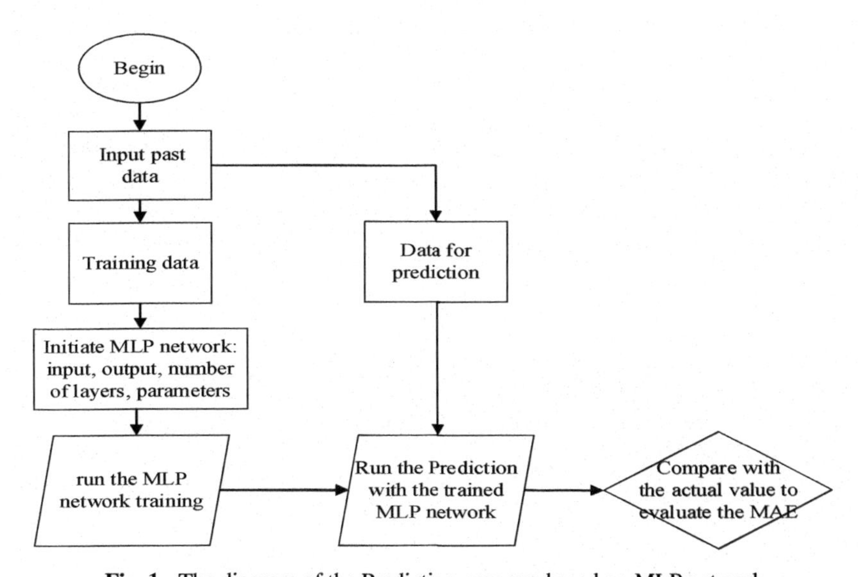

Fig. 1. The diagram of the Prediction program based on MLP network

2.1 Results and Discussion

Data were gathered from the SCADA system with the interval of gathering being 5 min. Then, the data were input into the training network and used for the predicting process.

The data used in this project are gathered in the period from 05/2018 to 04/2019. The information of reclosers on Can Gio-MC471 is shown in Table 1.

After inputting the data for the training neural network, current data were added to perform predicting for the value of the operating power line in the next 5 to 30 min. From then on, predicted data will be applied in operating and automatically warning equipment. In this project, data were used as a reference for operators to perform operating the electric power line, transmitting when necessary, dealing with unwanted incidents and optimizing transmitting distribution.

Table 1. Device information

	Substaion	Recloser		
	Can Gio - MC 471	Rec Hao Vo	Rec Can Thanh 163	Rec Can thanh – hao vo
Parametric data configuration	150 A	20 A	20 A	30 A
Data	Data (05/2018–04/2019)	Data (05/2018–04/2019)	Data (05/2018–04/2019)	Data (05/2018–04/2019)

The data is collected from the SCADA system being currently worked at has been operating ever since 2017 and is able to provide a huge and diverse amount of data gathered from approximately 1700 equipment integrated with online SCADA function (Reclose, LBS, RMU), 60 operating substation (220 kV, 110 kV) and interrupting stations. The model of this is demonstrated in Fig. 2 below. The writing motive is to resolve and taking advantage of the current amount of data as well as the modern computer system.

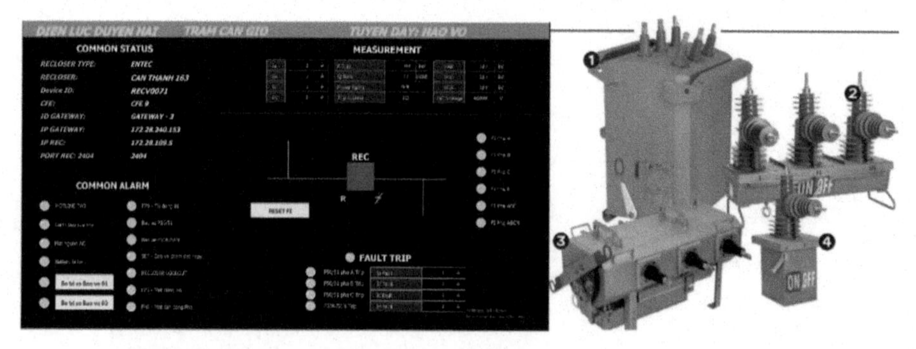

Fig. 2. A model of SCADA system and the SCADA Device (Recloser)

Table 2. Example of comparison between calculation and actual result

Date time	IA	IB	IC	I_Avg	I_predict	Error (%)
9/6/2018 12:09:00 AM	3.8344841	4.666546822	4.937911987	4.479647636	5.354185104	19.52246%
9/6/2018 12:24:00 AM	3.879642963	4.762598038	4.634084225	4.425441742	5.309148788	19.96879%
9/6/2018 12:39:00 AM	3.748295069	4.858647823	4.561507225	4.389483372	5.313578606	21.05248%
9/6/2018 12:54:00 AM	3.616945982	4.954699039	4.855741978	4.475795666	5.354644775	19.63559%
9/6/2018 1:09:00 AM	3.485598087	5.050748825	4.734235764	4.423527559	5.392514229	21.90529%
9/6/2018 1:24:00 AM	3.978436947	5.140216827	6.588316917	5.235656897	5.297881126	1.18847%
...	...	...	...	...	...	....

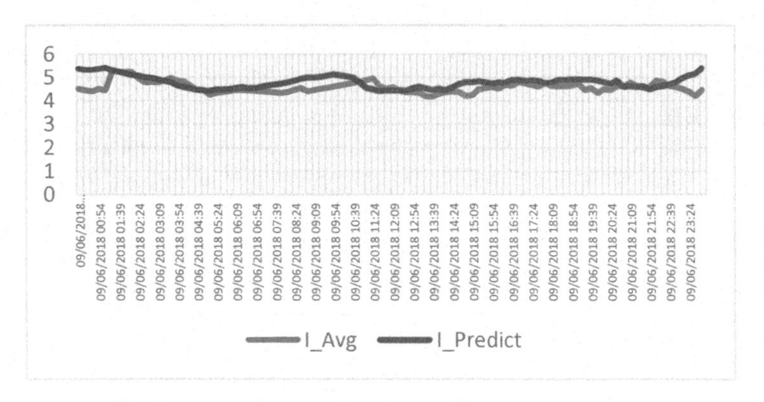

Fig. 3. Chart of mean value and predicting data of Table 2

From the result observed in Table 2, prediction value I_predict is comparable to gathered value I_Avg with the average error falls below approximately 7% in the entire gathered data. None of the predicted values falls into either the category of overfitting or underfitting. The mean value and predicting data of Table 2 is demonstarted in form of a chart in Fig. 3.

In Fig. 3, the blue line represents the result of real life result, the red one represents the predicted values. The graph represents the effectiveness of the short-term predicting

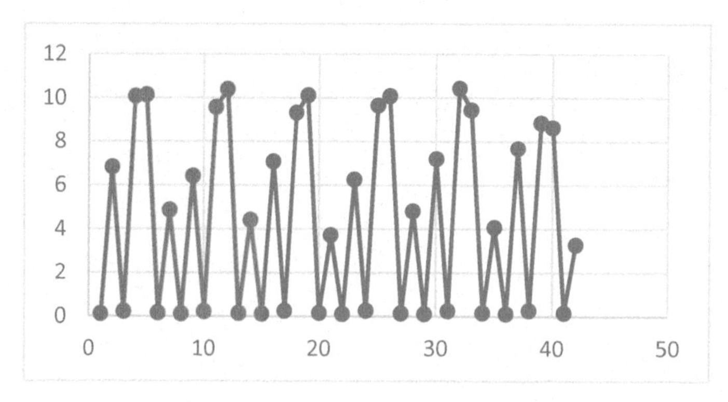

Fig. 4. The error percentage of the training data

method of the data. The error percentage of the training data is demonstarted in form of a chart in Fig. 4.

3 Conclusion

The systems provided the results as expected in comparison to that from direct observation with the negligible error percentage. It is recommended that the intelligent identification systems diagnosing the transient stability of the electric power system should be based on neural network for exploring data. The research has successfully developed the improved MLP neural system for the problems in identifying transient stability of the electric power line. The suggested model achieved the most crucial target being increasing the accuracy in layering. Within the approach in presenting this research, phases were standardized, helping the procedure be able to be applied for a multitude of interfering situations in differing scales. Constructing the quick method to identify the system's stability knowing the loading specifications of each sectors helps operators select the most suitable method to operate the system. This research suggests a model predicting the short-term load in order to assist in preparing methods/correspondent scenarios, stabilizing the electrical system when a sector is predicted to have highly increased load. Morever, the model is created by using the Sin function as the activation function for the data and the method used for calculating error after Backpropagation through the derivative of output node functions. After that, errors are applied with the Gradient Descent algorithm to recalculate the weight w. Then, these values w after adjustment will be returned to the MLP training phase. The Gradient Descent applied in Adam Optimizer – an optimized function of SGD allows the most optimal result to be reached with quickest speed and the most optimal amount of calculations and data due to its best result in practice.The tested results had higher accuracy in terms of identifying the transient stability of the electric power system without the need for solving differential equations like in tradition methods.

Direction for future development: Researching into extending the application of the researching method in this article in identifying voltage stability, frequency stability combining with the urgent controlling protocol in order to maintain the stability of the electric power system in anomalous incidents such as controlling load shedding. Furthermore, placing a new issue regarding the influx of output into the program as well as the time and resource needed for training and predicting specifications being too great. Therefore, future research will be carried out with a view of improving the sample space by selecting representative samples.

References

1. Fritzke, B.: Growing grid - a self-organizing network with constant neighborhood range and adaptation strength. Neural Process. Lett. **2**(5), 9–13 (1995)
2. Fritzke, B.: A growing neural gas network learns topologies. In: Tesauro, G., Touretzky, D., Leen, T. (eds.) Advances in Neural Information Processing Systems, vol. 7, pp. 625–632. MIT Press, Cambridge (1995)
3. Gil, D., Rodriguez, J.G., Cazorla, M., Johnsson, M.: SARASOM: a supervised architecture based on the recurrent associative SOM. Neural Comput. Appl. **26**(5), 1103–1115 (2015)
4. Groof, R., Valova, I.: Genetically supervised self-organizing map for the classification of glass samples. In: 13th International Conference on Machine Learning and Applications (2014)
5. Han, J., Kamber, M.: Data Mining: Concepts and Techniques, 2nd edn. Morgan Kaufmann (2006)
6. Kiviluoto, K.: Topology preservation in self-organizing maps. In: IEEE International Conference on Neural Networks (ICNN96), Washington, DC, 3–6 June 1996, vol. 1, pp. 294–299. IEEE (1996)
7. Li, W., Zhang, S., He, G.: Semisupervised distance-preserving selforganizing map for machine-defect detection and classification. IEEE Trans. Instrum. Meas. **62**(5), 869–879 (2013)
8. Lopez-Rubio, E.: Improving the quality of self-organizing maps by selfintersection avoidance. In: IEEE Transactions on Neural Networks and Learning Systems, vol. 24, no. 8, pp. 1253–1265 (2013)
9. Lu, S., Tu, X., Lu, Y.: A handwritten Bangla numeral recognition scheme based on expanded two-layer SOM. Int. J. Intell. Syst. Technol. Appl. **10**(2), 203–213 (2011)

Heart Disease Prediction and Hybrid GANN

Rahul Kumar Jha(✉), Santosh Kumar Henge, and Ashok Sharma

School of Computer Science and Engineering, Lovely Professional University, Punjab, India

Abstract. Cardiac arrest is an incurable heart incongruity and requires special treatment and cure. It has been a keen research area for many years and number of researchers across the globe are devoted toward finding the optimal solution for prevention, prediction of heart disease. With the emergence of artificial intelligence in research work with the intelligence of auto–training capability of Machine Learning (ML), it became most prominent choice of researchers for finding a solution for prediction of heart disease. In recent years, many researchers have been working diligently for this cause and experimented over various ML algorithms to discover an prime solution in social benefit, numerous classical methods like SVM, DT, RF along with algorithms from NN family like DNN, ANN and many others have come up with satisfactory result with some future scope. This paper presents the doable analysis of enactment of hybrid system consisting of Genetic Algorithm and Neural Network used for prediction of heart disease.

Keywords: Heart disease prediction using machine learning · Heart disease prediction using Genetic Algorithm and Neural Network · Hybrid system with Genetic Algorithm for heart disease prediction

1 Introduction

Cardiovascular disease, in actual, is a prevalent perilous malady entailing high decease rates and requires hasty devoutness for cure and care; there are many methods available to diagnose HD, for example Angiography, but the point is that they are very expensive and therefore is out of reach for masses. To overcome this problem, an opportunity arises to search for an alternate so that this could be diminished from root and people can breathe coming out from such an extreme trouble. Now a days after emergence and acceptance of AI as a approach to invent the solution by mean of technology, many researchers has therefore using this technology and conducting experiment using machine learning and data mining techniques to find the best solutions for such problems.

The Cardiovascular disease (CVDs) has been noticeably a critical malady across the globe. A survey conducted by WHO estimated around 17.9 million death due to heart disease, 31% of global death, and it was mostly common in people reside in 70s [1]. It is one of the major causes of sickness and death across the globe. There are many parameters which impact heart and cause disease like high blood pressure, body temperature, angina, long-term smoking and etc. Diagnosis of heart disease before time could save millions of people, it has been a significant topic for many researchers around the world and experimental results are phenomenal Artificial Intelligence (AI) has been

C. Kahraman et al. (Eds.): INFUS 2021, LNNS 308, pp. 438–445, 2022.
https://doi.org/10.1007/978-3-030-85577-2_52

vital with Machine Learning (ML) and with its wide range of algorithms and methods, it has proved itself and made a difference toward providing effective solution, not only for diagnosis/prediction of heart disease but also in prediction of many other critical diseases. Emergence of neural network family and hybrid system has taken this to a different level and with the advancement of feature selection and other methods has taken the score to human-like prediction power. This article scopes with the analysis of result of classical methods and hybrid system by applying different combinations like feature selection, cross validation and finding scope of these addition in hybrid system.

The subsequent sections of paper are structured as follow: Sect. 2 covers the literature review of past study, Sect. 3 contains the experiment and dataset details, Sect. 4 lights up the result which ends up with Sect. 5 carrying conclusion and suggestive future work.

2 Literature Review

Many researchers in past did fantastic research on finding optimal solution for prediction of heart disease using many traditional and advance ML classification algorithms. The list is periodically increasing year by year and demonstrate the interest of Researchers from the globe in this area.

The authors D. S. K. H. D. A. S. Rahul Kumar Jha presented a study to compare different classification algorithms for prediction of heart disease where many classical methods such that DT, KNN, SVM, DNN, RF and NB were used applying feature selection over Rapid Minor tool to train model using Cleveland dataset from UCI repository [2]. Result were compared among all methods experimented and DNN outperformed among all with recorded ACC, SENS and SPEC of 93.3%, 91.6% and 88.4% respectively.

The authors I. A. Negar Ziasabounchi proposed an approach a hybrid system combining GA and ANFIS algorithms for the prediction of heart disease. Model was trained using Cleveland dataset and performance was evaluated on the parameters ACC, SENS, SPEC and RMSE [3]. An accuracy of 92.30% was recorded with the proposed system.

In this paper, authors R. A. M. R. H. M. A. A. Y. Zeinab Arabasadi presented a study with hybrid NN-GA over Z-Alizadeh Sani dataset where after applying feature selection, model with 22, 5 and 1 neurons were put in input, hidden and output layers respectively were underwent experiment using feed-forward structure and showed result having ACC, SENS and SPEC of 93.85%, 97% and 92% respectively [4]. Weight were generated using GA.

The authors O. E. S. T. O. Miray Akgul presented a hybrid approach of using ANN-GA for diagnosis of HD using UCI Cleveland dataset and showed comparison in result with ANN with and without GA applied to model [5] showed that ANNGA given better result having ACC 95.82%, precision 98.11%, recall 94.55% and 96.30% as F-measure value as compared with ANN without GA having result 85.02% ACC, 83.72% precision, 90.57% recall and 87.01% F-measure value.

The author N. G. Bhuvaneswari Amma in their conference paper [6] presented an experiment which was conducted using GA and NN with an aim to generate and optimize weight using GA and later train model using NN, calculating fitness, applying crossover and continue for n generation till the aim achieved. Experiment was performed on UCI datasets and result recorded an ACC of 94.17%.

The author Sneha Nikam et al. presented a model GA-NFS consisting of GA, NN and Fuzzy_Set to improvise the prediction capability of the model [7]. Study showed the capability of GA to reduce the error rate and NFS to increase the model performance. In the experiment, fitness value roll was explained to create better offspring in the next generation that could help in achieve the goal to increase in model accuracy.

The author M. A. Jabbar et al. presented a study in a conference for prediction of heart disease for Andhra Pradesh, India population using GA and association rules [8] showing complex search capability of GA. Data was collected within Andhra Pradesh area from many hospitals and filtration was done using doctor advice and experiment was performed on Weka tool.

3 Experiment

Experiment has been carried out using Claveland dataset and model has been trained with hybrid system using GA and NN, for feature selection mRMR algorithm was chosen over others.

3.1 Dataset

We picked Cleveland heart disease dataset from UCI containing 76 attributes, 303 records for the experiment [9]. From a total of 76 available attributes, 13 attributes has been selected as an input features and 1 as target feature (num) has value resides between 0 and 4 where 0 denote no heart disease and 1–4 denotes presence of heart disease (Tables 1 and 2).

Table 1. List of attributes from Cleveland dataset to be used in the experiment

Patient age (age)	Patient gender (sex)	Max heart rate (thalach)
Cholesterol (Chol)	Chest pain type (CP)	Resting BP (RBP)
Resting ECG result (RestECG)	Heart status (Thal)	Major vessels count coloured by Fluoroscopy (CA)
Blood Sugar at Fasting (FBS)	Exercise induced Angina (Exang)	Slope for Peak-Exercise (Slope)
ST-Depression due to Exercise (OldPeak)	Diagnosed heart disease (Num)	

Table 2. Patients health statistics showing sample distribution among healthy and patient at risk

Patient without heart-disease (53.87%)	Patient with heart-disease (46.13%)

3.2 Methodology

Various criterion were experimented to find best approach, select best features, train the model and get optimum scores. Feature selection algorithms has been applied to filter features and to select the optimal feature explained in below (Fig. 1) flow diagram.

Minimum Redundancy Maximum Relevance (mRMR) Feature Selection Method
This method basically works on mutual information and picks features with high mutual information with the output class (maximum relevance), but low mutual information with each other (minimum redundancy). Feature indices is being returned based on mutual information and later indices are converted into real features while training the model. This technique has been applied for experiment purpose.

Below pseudo code has been used to train model using various classification methods mentioned above and Fig. 2 showing the flow diagram for the same.

Algorithm 1: Pseudo code for mRMR feature selection

Step 1: Input: Cleveland heart disease dataset **Invalid source specified.**
Step 2: Refine dataset: Remove bad data and prepare data for model training.
Step 3: Feature selection: Prioritize features based on the mutual information with output class and other features.
Step 4: Pick the feature indices with higher values and select feature sets to be used in model training.

GANN Algorithm with mRMR Feature Selection
Hybrid system with Genetic algorithm (GA) and Neural Network (NN) has been used with mRMR feature selection algorithm to train the model.

Algorithm 2: Pseudo code for GA+NN with mRMR feature selection

---

Step 1: Input dataset
Step 2: Create pre-requisite for model, select number of input, hidden and output neurons.
Step 3: Generate feature indices using mRMR feature selection algorithm.
Step 4: Generate initial population with random uniform values of chromosomes.
Step 5: Computed weight matrix of population using activation method, calculate fitness for each population.
Step 6: Save highest fitness value for each population.
Step 7: Select n best parents from current population and apply crossover.
Step 8: Generate new offspring using selected parent applying mutation into crossover result.
Step 9: If no generation left, fetch the best solution and return, else go to step 5.
Step 10: Repeat steps 4 to 9 for different set of features and record the highest score.
Step 11: Test with test data to show output.

---

4 Result and Discussion

Experiment was performed and result were noted with each variance. Different feature set were applied to different methods and results were compared.

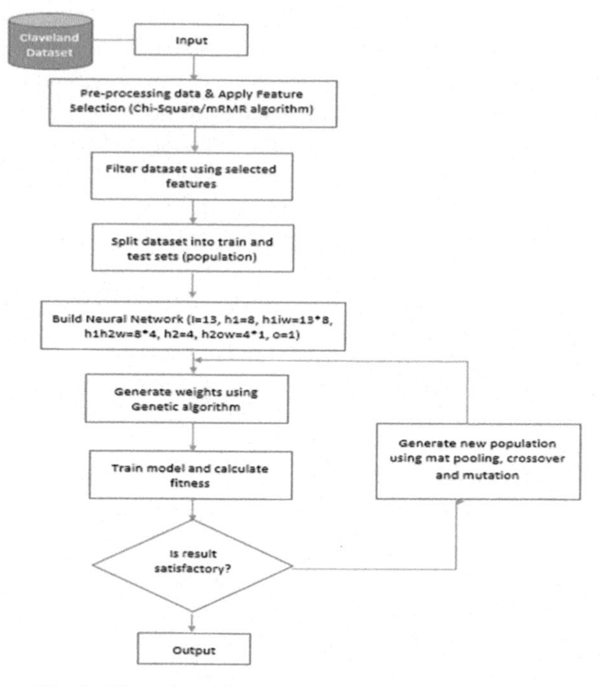

Fig. 1. Flow chart showing GANN used in experiment

4.1 Evaluation Matrix

Result were analysed on the basis of ACC, SPEC and SENS parameters that were calculated using 2 * 2 confusion matrix having true (1) and false (0) combination (Table 3).

Table 3. Confusion matrix to evaluate model result

Actual result	True HD prediction (1)	False HD prediction (0)
Has HD (1)	TP	FN
No HD (0)	FP	TN

Above abbreviations self explains the probability and has been applied on dataset to record the output which later used in computing result matrix.

$$Accuracy = \frac{TP + TN}{TP + TN + FP + FN} \times 100\%$$

$$Sensitivity\ (Has\ HD) = \frac{TP}{TP + FN} \times 100\%$$

$$Specificity\ \left(Don't\ have\ HD\right) = \frac{TN}{TN + FP} \times 100\%$$

$$Classification\_Error(incorrect\ classification) = \frac{FP + FN}{TP + TN + FP + FN} \times 100\%$$

$$Precision = \frac{TP}{TP + FP} \times 100\%$$

4.2 Experiment Result

Above confusion matric formula has been used to compute the scores from the model results, following is the experiment result for GANN method (Table 4).

Table 4. Prediction result using GANN with mRMR feature selection

Method	Feature selection	Result
GANN	SerumCholesterol_SCH, FastingBloodSuger_FBS, ExerciseIncludedAngina_EIA, ThalliumScan_THA, VCA, PeakExerciseSegment_PES, ChestPainType_CPT, Sex	Accuracy: 82.15% Sensitivity: 85.0% Specificity: 78.83%

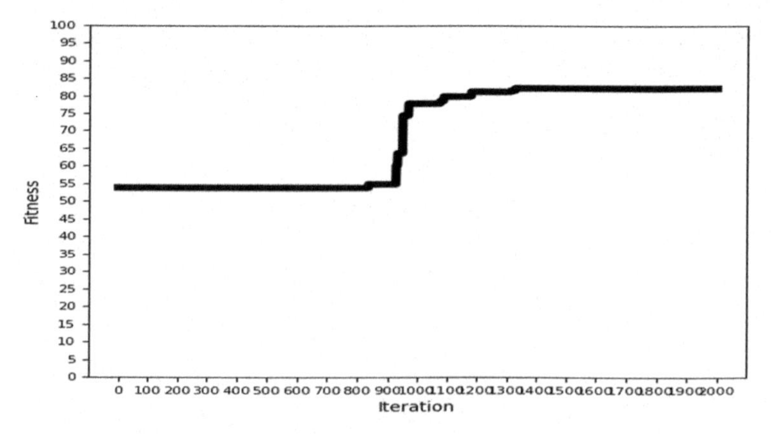

Fig. 2. Test result for GANN showing fitness for 2000 iteration

5 Conclusion and Future Work

While experimenting and knowledge gathering we found different variance used in experiments like model training with and without feature selection, applying different feature selection algorithms like mRMR and Chi-Square, or cross validation using k-fold and results demonstrates difference between them and it leave a scope that performance could be enhanced by applying different approaches and variances. Hybrid system was experimented with different number of population and mutation and it showed good numbers which clearly elaborate the calibre in hybrid system and create a scope to experiment more on this to achieve human-like capability while predicting the disease.

References

1. WHO Cardiovascular Disease. WHO. https://www.who.int/health-topics/cardiovascular-diseases/
2. Rahul Kumar Jha, D.S.K.H.D.A.S.: Optimal machine learning classifiers for prediction of heart disease. SERSC, May 2020. http://sersc.org/journals/index.php/IJCA/article/view/6680
3. Negar Ziasabounchi, I.A.: ANFIS Based classification model for heart. IJENS, April 2014. http://ijens.org/Vol_14_I_02/146402-7373-IJECS-IJENS.pdf
4. Zeinab Arabasadi, R.A.M.R.H.M.A.A.Y.: Computer aided decision making for heart disease detection using hybrid neural network-Genetic algorithm. ScienceDirect (2017). https://www.sciencedirect.com/science/article/pii/S0169260716309695
5. Miray Akgul, O.E.S.T.O.: Diagnosis of heart disease using an intelligent method: a hybrid ANN – GA approach. Springer, July 2019. https://doi.org/10.1007/978-3-030-23756-1_147
6. Amma, N.G.B.: Cardiovascular disease prediction system using genetic algorithm and neural network. IEEE Xplore, February 2012. https://ieeexplore.ieee.org/document/6179185
7. Sneha Nikam, P.S.a.M.S.: Cardiovascular disease prediction using genetic algorithm and neurofuzzy system (2017). https://doi.org/10.21172/1.82.016

8. Jabbar, M.A., Deekshatulu, B.L., Chandra, P.: An evolutionary algorithm for heart disease prediction. In: Venugopal, K.R., Patnaik, L.M. (eds.) ICIP 2012. CCIS, vol. 292, pp. 378–389. Springer, Heidelberg (2012). https://doi.org/10.1007/978-3-642-31686-9_44
9. Janosi, A.: Heart Disease Data Set - UCI. UCI, Budapest. https://archive.ics.uci.edu/ml/datasets/Heart+Disease. M. H. Z. S. W. S. M. U. H. B. S. M. P. M. V. M. C. L. B. a. C. C. F. Hungarian Institute of Cardiology

Ensemble Learning Based Stock Market Prediction Enhanced with Sentiment Analysis

Mahmut Sami Sivri(✉), Alp Ustundag, and Buse Sibel Korkmaz

Istanbul Technical University, Istanbul, Turkey
sivri@itu.edu.tr

Abstract. Besides technical and fundamental analysis, machine learning and sentiment analysis obtained from non-structural news and comments have been studied extensively in financial market prediction in recent years. It is still uncertain how to combine predictions from news, sentiment scores or financial data. In this study, we provide a methodology to achieve this issue. Besides the methodology, this study differs from previous studies in terms of data coverage and used models in both sentiment analysis and prediction. Our study consists of weekly predictions by ensemble learning and feature selection methods using 683 variables for stocks traded in the Borsa Istanbul 30 index. In addition, we predicted sentiment scores from news of 18 different sectors and combined both predictions with weighted normalized returns. We used Random Forests, Extreme Gradient Boosting and Light Gradient Boosting Machines of ensemble learning methods for predictions. From the parameters such as training set length, estimation methods, variable selection methods, number of variables, and the number of models in the prediction method, we took the combination that gives the best result. For sentiment scores, tests were performed using BERT, Word2Vec, XLNet and Flair methods. Then, we extracted final sentiment scores from the news. With the proposed trade system, we combined the results obtained from these financial variables and the news sentiment scores. Final results show that we achieved a better performance than both predictions made by using sentiment scores and financial data in terms of weekly return and accuracy.

Keywords: Sentiment analysis · Ensemble learning · Stock Market Prediction · Feature selection

1 Introduction

As in all other fields, data analytics has been the subject of much research in the field of finance, and has provided significant benefits in decision-making. Especially in finance, a small improvement in prediction makes a big difference in terms of return.

In recent years, a serious progress has been made in modeling thanks to technological development. We see that significant improvements have been made in the context of prediction models under the heading of predictive analytics of data analytics. Advances in deep learning, reinforcement learning and ensemble learning have led to high accuracy and return rates in predictions in finance, too [1].

C. Kahraman et al. (Eds.): INFUS 2021, LNNS 308, pp. 446–454, 2022.
https://doi.org/10.1007/978-3-030-85577-2_53

In the age of big data, it has become popular to make sense of these data by processing especially unstructured data besides relational or structural data. Cloud systems, infrastructures developed to process big data, and parallel data processing algorithms also enable the processing of unstructured data such as voice, image and text.

Previously, while there were studies using technical indicators and macroeconomic data in prediction problems in the field of finance; nowadays, it is seen that studies where news, comments and reports are used as a source of data drives us to better results. It has been observed that these data, which are not available in data sources numerically but have a significant effect on the result, make significant contributions to the performance in stock direction prediction [2].

Since the past, studies using news data as well as basic, technical and economic data in both descriptive and predictive analyzes make a difference. One of the main problems here has been the question of how to combine financial and news data or results, which can be considered as two separate data sources.

In this study, stock direction prediction was made by using the financial data collected from different data sources and news data, and experiments were made on how to combine the data obtained from these two data sources and the prediction results obtained. The stocks traded in the Borsa Istanbul 30 index were selected as the field of study and the predictions were made for these 30 stocks. In the analysis made, it was observed that higher rates of return were achieved when the data obtained from financial and news sources were used as two separate models instead of a single model. This result were obtained by weighting predictions with normalized returns.

The sections after this section are organized as follows. In the second part, similar studies in stock direction prediction are given and different aspects of this study are emphasized. Section 3 includes feature selection, sentiment analysis and prediction methods. In Sect. 4, information about the data sources and the data used are given. Chapter 5 includes the experimental design and results. In the 6th chapter, the findings obtained as a result of the experiments, the contribution of the results of the study to the literature and the aspects open to improvement are emphasized.

2 Literature Review

In addition to the popularity of the prediction of stock movements and sentiment analysis separately, it can be seen as a topic that remains unclear how to evaluate these two main issues together. Li et al. [3] also researched this issue and combined technical indicators, prices and sentiment labels with the Long Short-Term Memory (LSTM), Support Vector Machines (SVM) and Multiple Kernel Learning (MKL) methods. They also used 4 different dictionary-based methods to create sentiment labels and compared their performances. In the experiments, it was seen that LSTM and Loughran and McDonald (LM) sentiment dictionary give better results than others. Li et al. [4] conducted a similar study using Multiple kernel support vector regression with vectors derived from news and technical indicators. Their system analyzes and integrates intra-day market news and stock tick prices. Picasso et al. [5] also produced predictions with technical indicators and vectors obtained from the LM dictionary and compared them with different algorithms. For the prediction phase, they used SVM, Random Forests (RF) and Neural

Network models. Cagliero et al. [6] predicted stock trends with technical indicators and news labels by using SVM, Bayesian Classifier, Artificial Neural Network, RF, and k-nearest Neighbor Classifier. They reached maximum return using historical prices and news labels combinations with Neural Network model. Deng and his friends' [7] work includes a stock price prediction model, which extracts features from time series data and social network news and evaluates its performance. They used technical indicators and sentiment labels derived from WordNet sentiment analysis model as features, and MKL as the prediction method. Gumus and Sakar [8] made a similar study with LSTM method for Borsa Istanbul Stock Exchange. Similarly, they used technical indicators besides historical prices and sentiment labels derived from FastText sentiment analysis model.

When the above studies are examined, nearly all of the studies are based on technical indicators, and sentiment labels from dictionary-based sentiment analysis models. Also, ensemble learning prediction models and word-embedding sentiment analysis models are not covered in most of the studies. Therefore, we can say that there is no similar study to our work in terms of both data coverage, models and combination method.

3 Methods

3.1 Sentiment Analysis

In this study, we have tested different sentiment analysis models to make sense of news data, which differs from financial data and constitute another important part of the study. Ultimately, it was decided to use BERT model [9] that shows the best performance on classification as the sentiment analysis model. Also, we tested the following sentiment analysis models to determine the polarity of news: Word2Vec [10], XLNet [11] and Flair [12]. In the experiments, comparison of the sentiment analysis model performances is presented.

3.2 Feature Selection

In order to predict the direction of the stocks, an average of 201 different variables per stock were added or derived. The use of all these variables in the prediction can adversely affect the prediction performance. For this reason, we planned to use 8 different feature selection methods to avoid the complexity and multicollinearity that may be caused by the large of the number of variables. Feature selection methods that we used can be listed as follows: Pearson Correlation, Chi-square coefficient, mutual info, LASSO, Ridge regression, Decision Tree Variable Importance, SHAP, and Recursive Feature Elimination.

3.3 Prediction

Within the scope of this study, ensemble learning methods which is emphasized their superiority over traditional machine learning methods in many studies were used as prediction method. We used Random Forests [13], Extreme Gradient Boosting [14] and

Light Gradient Boosting Machines [15] of ensemble learning methods for predictions. From the parameters such as training set length, estimation methods, feature selection methods, number of variables, and the number of models in the prediction method, we took the combination that gives the best result.

4 Dataset

As mentioned before, the data used in this study for stock direction prediction can be examined under two main headings. First of all, there are financial data that includes a large pool of variables. In the second part, there are news data subject to sentiment analysis.

4.1 Financial Data

It has been observed that technical indicators are generally used in the direction prediction of stocks. We created a large variable pool for each stock and grouped the data. Financial data includes 15 data groups in total between January 2017 and December 2020. The table below shows the number of variables collected for a total of 30 stocks (Table 1).

Table 1. Variable groups in financial dataset

Variable group	Total	Used in stocks	Per stock
Brokerage target prices	95	30	3
Clearing data	120	30	4
Commodity prices	34	7	5
Economic indicators	545	30	18
Company financials	1042	30	35
Fundamental ratios	60	30	2
Holding firm prices	106	5	21
Indices	1282	30	43
Technical indicators	840	30	28

(continued)

Table 1. *(continued)*

Variable group	Total	Used in stocks	Per stock
Operational data	113	17	7
Pair company financials	30	3	10
Pair fundamental ratios	203	26	8
Pair operational data	12	4	3
Pair price and volume	204	27	8
Price and volume	182	30	6

4.2 News Data

In recent years, studies about prediction of stocks and financial assets have shifted from structural data to unstructured data. In our experiments, we examined how the results we obtained with financial data were combined with sentiment scores obtained from the news. In this context, we gathered sectoral news on stocks in Borsa Istanbul 30 index from Thomson & Reuters data source. We got the industry and industry definitions from the same source.

News data has been collected in the context of industry definitions of stocks. The data set for 18 different industries includes the news between January 2017 and December 2020. Table 2 shows the number of news gathered in each industry for this period. While compiling the news, a filter was made to include meaningful news in the data set. All the above stories are labeled as positive, negative and neutral. Table 2 also shows the number of news in each label.

Table 2. Collected news and labels

Industry	Number of news	Number of stocks	Negative	Neutral	Positive
Aerospace & defense	146	1	63	27	56
Airlines	212	2	98	28	86
Airport operators & services	52	1	19	5	28

(continued)

Table 2. (*continued*)

Industry	Number of news	Number of stocks	Negative	Neutral	Positive
Appliances, tools & hardware's	120	1	47	5	68
Auto & truck manufacturers	216	2	46	66	104
Banks	945	8	306	244	395
Commodity chemicals	195	1	73	13	109
Construction & engineering	153	1	56	24	73
Consumer goods conglomerates	108	3	31	9	68
Discount stores	40	1	8	9	23
Food retail & distribution	110	1	44	14	52
Gold	29	1	5	6	18
Integrated tel. services	81	1	25	23	33
Iron & steel	160	2	75	15	70
Oil & gas refining and marketing	174	1	79	23	72
Residential REITs	54	1	24	6	24
Specialty mining & metals	41	1	10	12	19
Wireless tel. services	74	1	21	21	32
Total	2910	30	1030	550	1330
			35%	19%	46%

5 Experiments

Before predicting the direction of stocks, we tested the sentiment analysis models to predict the polarity of the news. In contrary to previous studies, we tested word embedding sentiment analysis models. Therefore, the sentiment analysis model to be used was determined. Table 3 shows the model performances.

Table 3. Results of sentiment analysis models

Model	Accuracy (%)	F1 Score
BERT	81.29	81.03
Word2Vec	72.02	71.13
XLNet	73.03	72.56
Flair	69.56	69.06

As mentioned above, both financial data and sentiment scores obtained from the news were used in this study. Therefore, the experimental design was set up to find out how these data are used both separately and together. Table 4 shows the results obtained as a result of weekly predictions. Instead of predicting the direction directly, we predicted weekly returns, then extracted directions from these predictions. Experimental designs are given in Table 4.

Table 4. Results of Experiments

Index	Financial features	Sentiment scores	Feature selection	Number of best results	Weekly return (%)	Accuracy (%)
1	All				0.84	57.13
2	All	Industry			0.85	57.10
3	All		Stage-2	1	1.41	61.38
4	All	Industry	Stage-2	1	1.43	61.41
5	Selected		Stage-1		0.81	56.80
6	Selected	Industry	Stage-1		0.83	56.91
7	Selected		Stage-1,2	2	1.35	61.41
8	Selected	Industry	Stage-1,2		1.35	61.60
9		All			0.90	58.48
10		All	Stage-2	2	1.33	62.33
11	All	All	Stage-2	3	1.44	61.84
12	All	All	Stage-2	21	1.72	62.46

Financial data were included in the experiments in 2 different ways. Experiments related to selected variables include variables that are filtered according to correlation among variable groups. This feature selection is expressed as Stage-1 in the Table 4. Sentiment scores were also used in the experiments in 2 different ways, in which their sector was included for each stock and the scores of all sectors were included.

In the experiment with the index number 12, we used different methodology rather than others. We combine the results of the models in experiments 3 and 10 in terms of normalized cumulative returns. We weighted results of both models using these returns up to the previous week. So that, this methodology gave us better results than other experiments. Normalized cumulative return formula is given below:

$$r(t)_{normalized} = \frac{\sum_{i}^{t-1} r_{model} - \sum_{i}^{t-1} r_{min}}{\sum_{i}^{t-1} r_{max} - \sum_{i}^{t-1} r_{min}} \quad (1)$$

As can be seen in the Fig. 1, unlike other experiments, 2 separate predictions were made and the results were combined in our proposed methodology.

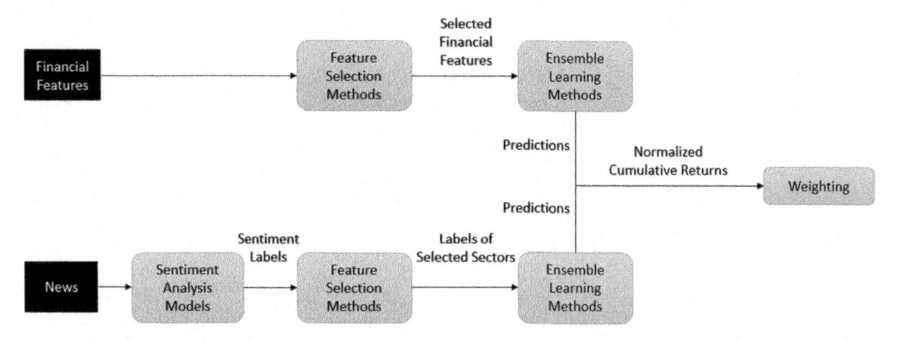

Fig. 1. Proposed methodology that gives the best results in experimental results

6 Conclusion

Within the scope of this study, the weekly prediction of direction of the stocks traded on the Borsa Istanbul 30 index was made. The main purpose of the study is to find more accurate results by combining financial data and news data.

A large pool of variables with financial data is created for prediction in this study. There is no study with such a variety of data and features in the reviewed ones. Another important point is that, it was seen that predictions were only made with technical indicators in previous studies.

In addition to financial data, sentiment scores were obtained by using unstructured news data, and it was investigated how these scores can be combined with previous prediction results. When the news of each stock in its own field are included in the financial data, a small improvement was achieved in the results. It was observed that when the news of all industries were included in the selection of variables, the results were better than the previous results. From these results, it can be concluded that the news of the sectors affects each other. When the predictions made with the news and the financial data were considered separately it was observed that the results improved significantly. The main contribution of this study is that higher rates of return were achieved when the data obtained from finance and news sources were used as two separate models instead of a single model and models were weighted with normalized cumulative weekly returns.

In addition to the studies carried out, better results can be achieved by diversifying the parameters of the models. Testing with daily predictions can be listed as another aspect of the study that it is open to improvement. However, both enhancements will require significant processing time.

In the future, it is planned to create a stacking methodology by combining the results of the methods such as deep learning and reinforcement learning, too. In addition, it is planned to select and weight different stocks periodically with a portfolio selection model. It is believed that this will increase the average return and overall accuracy.

References

1. Nikou, M., Mansourfar, G., Bagherzadeh, J.: Stock price prediction using DEEP learning algorithm and its comparison with machine learning algorithms. Intell. Syst. Acc. Financ. Manage. **26**, 164–174 (2019). https://doi.org/10.1002/isaf.1459
2. Schumaker, R.P., Chen, H.: A quantitative stock prediction system based on financial news. Inf. Process. Manage. **45**(5), 571–583 (2009)
3. Li, X., Wu, P., Wang, W.: Incorporating stock prices and news sentiments for stock market prediction: a case of Hong Kong. Inf. Process. Manage. **57**, 102212 (2020)
4. Li, X., Huang, X., Deng, X., Zhu, S.: Enhancing quantitative intra-day stock return prediction by integrating both market news and stock prices information. Neurocomputing **142**, 228–238 (2014)
5. Picasso, A., Merello, S., Ma, Y., Oneto, L., Cambria, E.: Technical analysis and sentiment embeddings for market trend prediction. Expert Syst. Appl. **135**, 60–70 (2019)
6. Cagliero, L., Attanasio, G., Garza, P., Baralis, E.: Combining news sentiment and technical analysis to predict stock trend reversal. In: International Conference on Data Mining Workshops, pp. 514–521 (2019). https://doi.org/10.1109/ICDMW.2019.00079
7. Deng, S., Mitsubuchi, T., Shioda, K., Shimada, T., Sakurai, A.: Combining technical analysis with sentiment analysis for stock price prediction. In: Ninth IEEE International Conference on Dependable, Autonomic and Secure Computing, pp. 800–807 (2011)
8. Gumus, A., Sakar, C.O.: Stock market prediction in Istanbul stock exchange by combining stock price information and sentiment analysis. Int. J. Adv. Eng. Pure Sci. **33**(1), 18–27 (2021). https://doi.org/10.7240/jeps.683952
9. Devlin, J., Chang, M.W., Lee, K., Toutanova, K.: BERT: pre-training of deep bidirectional transformers for language understanding (2018)
10. Pennington, J., Socher, R., Manning, C.D.: Glove : global vectors for word representation. In: Proceedings of the 2014 Conference on Empirical Methods in Natural Language Processing (EMNLP), pp. 1532–1543 (2014). https://doi.org/10.3115/v1/D14-1162
11. Khattak, F.K., Jeblee, S., Pou-prom, C., Abdalla, M., Meaney, C., Rudzicz, F.: A survey of word embeddings for clinical text. J. Biomed. Inf. X **4**, 100057 (2019). https://doi.org/10.1016/j.yjbinx.2019.100057
12. Akbik, R., Blythe, D., Vollgraf, R.: Contextual string embeddings for sequence labeling. In: Proceedings of the 27th International Conference on Computational Linguistics, pp. 1638–1649. Association for Computational Linguistics (2018)
13. Breiman, L.: Random Forests. Mach. Learn. **45**, 5–32 (2001). https://doi.org/10.1023/A:1010933404324
14. Chen, T., Guestrin, C.: Xgboost: a scalable tree boosting system. In: Proceedings of the 22nd ACM SIGKDD International Conference on Knowledge Discovery and Data Mining KDD 2016, pp. 785–794. Association for Computing Machinery, New York (2016). https://doi.org/10.1145/2939672.2939785
15. Ke, G., et al.: Lightgbm: a highly efficient gradient boosting decision tree. In: Advances in Neural Information Processing Systems 30, pp 3146–3154. Curran Associates, Inc., New York (2017)

Predictive Maintenance Framework for Production Environments Using Digital Twin

Mustafa Furkan Süve[1(✉)], Cengiz Gezer[2], and Gökhan İnce[1(✉)]

[1] Computer Engineering Department, Istanbul Technical University, Sarıyer, 34467 Istanbul, Turkey
{suve,gokhan.ince}@itu.edu.tr
[2] Adesso Turkey, Sarıyer, 34398 Istanbul, Turkey
cengiz.gezer@adesso.com.tr

Abstract. In this paper, we introduce an end-to-end IoT framework for predictive maintenance with machine learning. With this framework, all the processes for developing a learning-based predictive maintenance model such as data acquisition, data preprocessing, training the machine learning model and making predictions about the status of an equipment are automatically carried out in real-time. Independent modules for all of those processes can be arranged and connected on a visual environment which enables creating unique and specialized pipelines. This framework also provides a digital twin simulation of the production environment integrated with the real world and the machine learning models to evaluate the effect of different parameters such as the cost or the throughput rate. Furthermore, system modules can be controlled from a single dashboard which makes the use of the system easier even for a non-experienced user. Several open-source datasets are used to test the framework on different predictive maintenance tasks such as predicting turbofan engine degradation and predicting the stability of hydraulic systems. The effectiveness of the proposed framework is shown using metrics such as precision, recall, f1 score and accuracy.

Keywords: Predictive maintenance · Digital twin · Industry 4.0

1 Introduction

The vast amount of technological innovations and improvements in science accumulated through the years led to Industry 4.0 which is a paradigm arisen to keep up with the requirements of rapidly changing requests in the market [12]. The main idea of Industry 4.0 is to exploit the innovations in areas such as the internet of things, big data, artificial intelligence and industrial simulations [12,15]. With the integration of these technologies into the manufacturing process, it is possible to have an in-depth understanding and interpretation of the data coming from

C. Kahraman et al. (Eds.): INFUS 2021, LNNS 308, pp. 455–462, 2022.
https://doi.org/10.1007/978-3-030-85577-2_54

the system and use it to have higher production rates with increased efficiency and quality without increasing the cost.

One of the many challenges addressed within Industry 4.0 is the maintenance operations in an industrial production environment. Proper maintenance of an equipment that is required for the production can have a significant impact on the overall profit of the company. Mobley [8] stated that more than $60 billion is lost each year due to ineffective maintenance strategies. Besides reducing the cost, effective maintenance strategies can provide a safer work environment for the employees.

Predictive maintenance aims at predicting the failure right before it occurs using the historical data collected from the equipment and starting the maintenance operations. Thus, the final cost which consists of both the maintenance cost and the equipment's downtime cost can be minimized. Using a proper model to learn from the data, the characteristics of the equipment can be extracted and used for future predictions as long as the data is expressive enough to generalize on the different conditions. The data used for training can be in different forms such as the vibration data of gas circulators [4] or photodiode sensors and spectrometers on the laser welding process [17].

In this paper, we established an end-to-end predictive maintenance framework with IoT and machine learning technologies. This framework provides separate and independent modules for the parts of a predictive maintenance task in a visual environment. Unique pipelines can easily be created based on the requirements of the industrial setup. Each module of these pipelines can perform different tasks such as data acquisition from various sources, data preprocessing, training and assessing machine learning models and making predictions about the conditions of the components. The framework also supports digital twin integration that can help to demonstrate the performance of a trained model in an industrial environment.

The rest of the paper is structured as follows. Section 2 gives a general literature review about the predictive maintenance applications in the industry. Section 3 describes the details about the proposed framework by explaining the architecture and each sub-module. Section 4 explains the experiments to test the performance of the framework. Finally, Sect. 5 gives a conclusion and describes possible future directions for this project.

2 Literature Review

Predictive maintenance applications appear in the industry more frequently since the beginning of Industry 4.0. Canizo et al. [2] worked on a horizontally scalable predictive maintenance system for wind turbines on a cloud using big data frameworks such as Spark and Hadoop. They stated that they developed an automated method providing improvements in terms of speed, reliability and scalability. Verma and Kusiak [16] used machine learning algorithms to predict the status of wind turbines. They recorded operational and status data from 100 wind turbines. They tested multiple algorithms and reached up to 95% accuracy on identifying and predicting the frequent status patterns in advance.

Predicting disk replacements in data centers [1] is another study on predictive maintenance. In this work, the authors used continuously monitored hard drive data for their experiments. They extracted smart features from the time series and downsampled the highly imbalanced data using the K-means algorithm to train a tree-based classification model which can predict replacements with 98% accuracy.

While training machine learning algorithms, it is also important to consider the maintenance expenses. Spiegel et al. [13] came up with a new problem-specific cost function for this purpose. They showed that for their experimental case, the model trained using the new metric resulted in higher savings because of an additional maintenance cost. In addition to the cost function, the way the collected data is interpreted can change the outcome significantly. Susto et al. [14] trained multiple classifiers for semiconductor manufacturing fabrication using the same data with different failure horizons. They stated that having more samples for the failure which is usually the smaller class with multiple classifiers provided more robust and better results.

Nguyen and Medjaher [11] developed an efficient predictive maintenance system that is capable of giving probabilities of failure over different horizons. Cheng et al. [3] worked on a predictive maintenance framework to be used on mechanical, electrical and plumbing components of buildings with IoT and machine learning. In our framework, we provide a digital twin integration compatible with the learning models in addition to a user interface and IoT-based machine learning features. Besides, the framework described in our study is easily expandable allowing the basis for new machine learning and data processing algorithms to be implemented in the future.

3 Proposed Framework for Predictive Maintenance

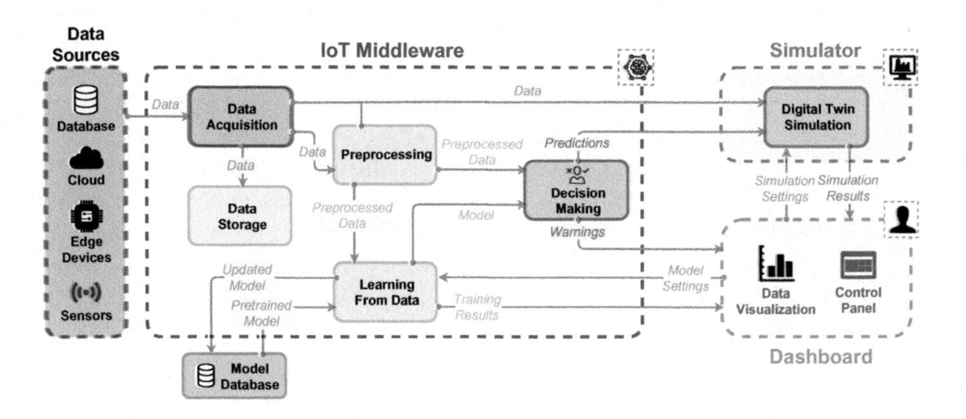

Fig. 1. Architecture of the proposed framework for predictive maintenance

The overall architecture of the predictive maintenance framework described in this paper is given in Fig. 1. With this framework, an end-to-end predictive maintenance solution is proposed covering all the aspects from data acquisition to getting actual predictions. The proposed framework can be deployed on a cloud or a local device. There are 3 main modules of the system. The first one is an IoT middleware that is responsible for data processing and machine learning. The second one is the digital twin simulation. The last one is the dashboard that allows users to interact with the system. These modules will be explained in the following sections.

3.1 IoT Middleware

The IoT middleware is responsible for data acquisition, processing, learning and inference parts of the system. It supports multiple flows to be run in parallel which allows the creation of different predictive maintenance processes for different resources. It also provides the communication channels between all the sub-modules once the system is deployed.

Data Acquisition. The flow is started with a single or multiple *data acquisition* nodes feeding data into the system. Multiple data sources are supported such as databases, clouds, edge devices and sensors as long as the data is in the expected format which is configured using a configuration file. The input can be used from a CSV file or from a string of features that is constantly flowing from outside of the system via WebSockets. The flowing data can also be stored in the disk for later use with the *data storage* node.

Preprocessing. There are lots of preprocessing nodes that support continuously streamed data. One of them is the *encoder* node which can be configured to perform either one-hot encoding or label encoding. In one-hot encoding, new binary features are created for each unique value of a feature indicating the existence of that value. In label encoding, each unique value is replaced by an integer between $[0, n]$ where n is the number of unique values of a particular feature. This node can be used when categorical features should not be present in the data.

Another preprocessing node is the *fill* node which fills the missing data with different values such as mean, median or a constant. This node also supports linear, quadratic or cubic interpolation. Besides, if the features need to be scaled the *scaler* node can be used. Finally, when the number of instances for classes is imbalanced *balance* node can be used. This node is important as one of the main characteristics of predictive maintenance data is being imbalanced.

Learning from Data. Once the data is ready, it can be directed into nodes with machine learning model implementations. Pretrained models can be loaded into the nodes of the algorithms and newly trained models can be stored in the disk using the *save model* node. The main challenge is to support *incremental learning* which is training a model continuously with flowing data without forgetting previous knowledge.

Since all the data is not available from the beginning, the models should be able to learn incrementally from the flowing data. The framework in this paper provides the basic environment to create any machine learning model supporting incremental learning. As an example, we implemented *logistic regression*, *Gaussian naive Bayes* and *random forest* nodes with the corresponding algorithms. In addition, the *lgbm* node is created using the LightGBM algorithm [6] which is an efficient gradient boosting tree implementation that also supports incremental learning.

Decision Making. Trained models and data can be channeled into the *test* node. This node makes predictions about the class of the data using the last model entered into the node. The data should be read from a *data acquisition* node that accepts labeled and unlabelled data. If the data has labels, 4 metrics which are accuracy, precision, recall and f1 score can be calculated along with the predictions. The predicted labels for metric evaluations are coming from the last model stored in the *test* node. These metrics are designed to support a continuous stream of data by updating their values using new information without forgetting the previous results.

3.2 Digital Twin Simulation

A digital twin is a probabilistic simulation that mirrors the life of a real-world system [5]. One of the novel aspects of this framework is that it provides digital twin integration compatible with the machine learning models. This is done by replaying the collected data using the *replay* node with the model predictions coming from the *test* node from the IoT middleware. Playing back the real data and predictions to run the simulation helps to understand the effects of the model outputs in the real world. The proposed framework is capable of integrating existing simulation modeling and analysis software to demonstrate the performance of the data and the system.

3.3 Dashboard

The dashboard allows users to interact with the framework. Users can see the training and testing performances of the machine learning models in a single window. For each flow, the training accuracy with testing metrics specified from the *test* node is plotted in real-time. Furthermore, according to the predictions of the models, the components with a higher risk of failure are highlighted in the dashboard.

The parameters of digital twin simulation such as the processing time or the maximum amount of items for an equipment can also be adjusted directly from the dashboard. Moreover, the effects of the changes on the digital twin and the current status of the simulation are visualized in terms of the characteristic features of each item such as whether it is still running or the number of items passed through the system.

4 Experiments and Results

4.1 Hardware and Software Details

Node-RED, flow-based programming [10] software built on Node.js, is used as the IoT middleware for the experiments. It allows the creation of unique nodes with independent functionalities that can be connected together. It also provides ready-to-use components that are suitable for IoT applications. The nodes about machine learning tasks and the communication between sub-modules are implemented by us in Node-RED with a python process working in parallel. In order to support the flowing data, River [9], a python library that provides models with incremental learning capabilities is used. Also, the node responsible for balancing the data is implemented using the imbalanced-learn library in python [7] and supports several balancing methods. Flexsim is used for the digital twin integration which is a simulation software that facilitates the creation of a digital model of the factory in a virtual environment with 3D models and animations. It is also capable of connecting to a network for remote control and data extraction. The experiments are performed on a 64-bit Windows operating system and Intel i7-8750H CPU with 6 cores.

4.2 Experimental Framework

The applicability of this framework is tested on various predictive maintenance tasks. The first experiment involves the task of identification of the failure of a hydraulic accumulator. Data for this experiment is obtained from a hydraulic system[1] using 17 sensors. Gaussian Naive Bayes (GNB), Adaptive Random Forest (ARF) and LightGBM (LGBM) are used as the learning algorithms. GNB, ARF and LGBM algorithms are all incrementally trained with batches of 4000 rows for the prediction of 4 different pressure levels on the hydraulic accumulator where the lowest level indicates failure.

We aim to predict turbofan engine degradation in the second experiment. The dataset used for this experiment is a time series consisting of operational settings and sensor measurements collected from 4 turbofan engines with high-pressure compressor (HPC) degradations and fan degradations[2]. Since the initial dataset is extremely imbalanced, the last 20 steps before the failure are treated as failures resulting in 8.8% of failure cases. GNB, ARF and LGBM algorithms combined with the balance node are incrementally trained.

For all experiments, GNB is trained without any parameter as it fits a Gaussian distribution for each class. ARF uses 10 trees with a maximum depth of 16 and Gini split criterion for hydraulic system data and 10 trees with unlimited depth and information gain split criterion for turbofan engine data. LGBM is trained for 1000 boosting iteration with 0.001 learning rate for both experiments. Each tree in LGBM has a maximum depth of 5, a maximum of 32 leaves and a voting tree learner.

[1] https://archive.ics.uci.edu/ml/datasets/Condition+monitoring+of+hydraulic+systems.

[2] https://ti.arc.nasa.gov/tech/dash/groups/pcoe/prognostic-data-repository.

The main challenges in the experiments were firstly the usage of imbalanced data due to the nature of anomalies occurring rarely, and secondly the incremental learning on the fly instead of using a batch learning approach. The metrics proposed to be used in our framework to evaluate the prediction performance of the models are accuracy and the mean f1 score over all classes.

4.3 Results

The results are shown in Table 1. LGBM performed better than other algorithms for hydraulic systems dataset consisting of 4 different classes. Both ARF and LGBM are tree-based ensemble algorithms but LGBM is faster and more lightweight. It has also a lot more parameters to tune which helps to optimize the maximum performance. Among all three algorithms, GNB has the worst performance as it is just fitting a gaussian distribution for each class. For the turbofan engine dataset which has only 2 classes, ARF and LGBM gave similar results. Since this dataset is imbalanced, the tree-based ensemble algorithms performed better than GNB as it is a simpler model.

Table 1. Results of the experiments

Datasets	GNB		ARF		LGBM	
	Accuracy	Mean F1	Accuracy	Mean F1	Accuracy	Mean F1
Hydraulic systems	0.375	0.291	0.801	0.785	0.925	0.914
Turbofan engine	0.405	0.35	0.95	0.867	0.951	0.867

5 Conclusion

In this paper, a predictive maintenance framework is developed. This framework provides functionalities for machine learning tasks such as data acquisition, preprocessing, learning and making predictions. Furthermore, a digital twin integration is supported for testing and analyzing the industrial processes quickly and without increasing the cost. In addition to these modules, a dashboard is provided to control the overall system from a central point and easily inspect the performance of the machine learning models and the simulation results.

In the future, integration of big data support will be provided to work efficiently on a large amount of data. Also, more comprehensive tests will be run in a real-world environment by deploying the application to the cloud having access to multiple edge devices. Moreover, we will extend existing machine learning models and preprocessing algorithms that can work on flowing data to tackle different prediction, forecasting, classification, regression and anomaly detection problems of industry in concrete use cases of digital twin scenarios.

References

1. Botezatu, M., Giurgiu, I., Bogojeska, J., Wiesmann, D.: Predicting disk replacement towards reliable data centers. In: Proceedings of the 22nd ACM SIGKDD International Conference on Knowledge Discovery and Data Mining, pp. 39–48 (2016)
2. Canizo, M., Onieva, E., Conde, A., Charramendieta, S., Trujillo, S.: Real-time predictive maintenance for wind turbines using big data frameworks. In: 2017 IEEE International Conference on Prognostics and Health Management (ICPHM) (2017)
3. Cheng, J.C., Chen, W., Chen, K., Wang, Q.: Data-driven predictive maintenance planning framework for MEP components based on BIM and IoT using machine learning algorithms. Autom. Constr. **112**, 103087 (2020)
4. Costello, J.J.A., West, G.M., McArthur, S.D.J.: Machine learning model for event-based prognostics in gas circulator condition monitoring. IEEE Trans. Reliab. **66**(4), 1048–1057 (2017)
5. Glaessgen, E., Stargel, D.: The digital twin paradigm for future NASA and U.S. air force vehicles. In: 53rd AIAA/ASME/ASCE/AHS/ASC Structures, Structural Dynamics and Materials Conference (2012)
6. Ke, G., et al.: LightGBM: a highly efficient gradient boosting decision tree. In: Proceedings of the 31st International Conference on Neural Information Processing Systems, pp. 3149–3157. Curran Associates Inc. (2017)
7. Lemaître, G., Nogueira, F., Aridas, C.K.: Imbalanced-learn: a python toolbox to tackle the curse of imbalanced datasets in machine learning. J. Mach. Learn. Res. **18**(17), 1–5 (2017)
8. Mobley, R.K.: Impact of maintenance. In: An Introduction to Predictive Maintenance, 2nd edn., pp. 1–22. Butterworth-Heinemann, Burlington (2002)
9. Montiel, J., et al.: River: machine learning for streaming data in python (2020)
10. Morrison, J.P.: Flow-Based Programming: A New Approach to Application Development, 2nd edn. CreateSpace, Scotts Valley (2010)
11. Nguyen, K.T., Medjaher, K.: A new dynamic predictive maintenance framework using deep learning for failure prognostics. Reliab. Eng. Syst. Saf. **188**, 251–262 (2019)
12. Rojko, A.: Industry 4.0 concept: background and overview. Int. J. Interact. Mob. Technol. (iJIM) **11**, 77 (2017)
13. Spiegel, S., Mueller, F., Weismann, D., Bird, J.: Cost-sensitive learning for predictive maintenance. CoRR (2018)
14. Susto, G.A., Schirru, A., Pampuri, S., McLoone, S., Beghi, A.: Machine learning for predictive maintenance: a multiple classifier approach. IEEE Trans. Industr. Inf. **11**(3), 812–820 (2015)
15. Tjahjono, B., Esplugues, C., Ares, E., Pelaez, G.: What does industry 4.0 mean to supply chain? Procedia Manuf. **13**, 1175–1182 (2017)
16. Verma, A., Kusiak, A.: Prediction of status patterns of wind turbines: a data-mining approach. J. Solar Energy Eng. **133**(1) (2011)
17. You, D., Gao, X., Katayama, S.: WPD-PCA-based laser welding process monitoring and defects diagnosis by using FNN and SVM. IEEE Trans. Industr. Electron. **62**(1), 628–636 (2015)

Spatial Prediction and Digital Mapping of Soil Texture Classes in a Floodplain Using Multinomial Logistic Regression

Fuat Kaya(✉) and Levent Başayiğit

Faculty of Agriculture, Department of Soil Science and Plant Nutrition, Isparta University of Applied Sciences, 32260 Isparta, Çünür, Turkey
{fuatkaya,leventbasayigit}@isparta.edu.tr

Abstract. The spatial distribution of physical soil properties is an important requirement in practice as basic input data. Most effective of these properties is soil texture that governs water holding capacity, nutrient availability, and root development. Detailed information on soil texture variability in lateral dimension is crucial for proper crop and land management and environmental studies. Soil texture classes are determined in the soil survey. It may be consist of two or more texture classes for each polygon according to soil mapping units. There is a spatial discrepancy due to variability in soil texture within the mapping polygon. Digital soil mapping (DSM) offers major innovations in removing some of the inconsistencies in traditional soil mapping. DSM methodology can integrate the various raster-based spatial environmental data that field-based soil morphology, soil analyses, and effects of soil formation factors. In this study, the potential of environmental variables generated from digital data to predict soil texture classes were investigated. Curvature parameters indicating the shape of the slope were determined as the most important predictive variables in a flood plain. Overall accuracy was calculated as 63.9% and 47.60% for the training set and the test set, respectively. Digital soil map can be used effectively by farmers in the management of crops in this plain.

Keywords: Digital soil mapping · Machine learning algorithms · Soil texture class · Spatial predictive modelling

1 Introduction

Soil is the most fundamental element of the Ecosystem. It is effective on agriculture, carbon sequestration, protection of biodiversity, and social-cultural factors. Soil maps are an important source of information about the spatial distribution of soil features used commonly for various land plans and management activities [1]. Accordingly, it can be used to describe soil functions [2]. Soil surveys are the systematic examination, description, classification, and mapping of soils in a particular area.

Soil can vary intricately throughout the area in which it spreads, and therefore attribution of soil types to a map unit is always affected by some degree of uncertainty [3].

C. Kahraman et al. (Eds.): INFUS 2021, LNNS 308, pp. 463–473, 2022.
https://doi.org/10.1007/978-3-030-85577-2_55

Topography controls the distribution and accumulation of water and energy in the pedosphere, so it has been widely accepted in the literature as the dominant factor affecting soils and soil properties [4, 5]. In traditional soil surveys, landforms are defined manually and the soil is mapped according to a soil researcher's conceptual model. This is a subjective process that requires experiences very good knowledge of the land. Therefore, maps are the product of mental pedological models developed by individual soil surveyors [6, 7], which are practically unrepeatable, leading to many inconveniences with their subjective nature.

Digital soil mapping (DSM) offers major innovations in removing some of the inconsistencies in traditional soil mapping [8, 9]. DSM methodology can integrate the various raster-based spatial environmental data that field-based soil morphology, soil analyses, and effects of soil formation factors [10]. This process is accomplished by creating pixel-based soil maps using mathematical or statistical models in which environmental covariates are associated with any soil information.

Soil texture is one of the most important physical properties affecting water holding capacity, nutrient availability, and crop development. A spatial distribution map of soil texture at a high spatial resolution is essential data for crop planning and management. Traditionally, soil texture can be determined in the field by hand feel method, however, it is confirmed by laboratory analysis of particle size fractions (sand, silt, and clay) using the hydrometer method [11]. Conventional soil texture analysis methods are expensive. Also, it requires a large number of samples to obtain higher resolution spatial distribution of soil texture over large areas.

In land-use planning, the final topsoil texture information of soil studies is important. Soil texture classes determined in the soil survey are labeled for each polygon according to soil mapping units. But it may be consist of two or more classes [12]. A spatial discrepancy occurs in the mapping polygon due to the variability of the texture. Digital soil maps can be used to show the soil's ability to perform certain functions. The digital soil mapping technique provides solutions for estimating the soil properties and soil texture classes based on quantitative soil-landscape models [10, 13].

Numerous studies have been conducted to demonstrate the effectiveness of digital soil mapping techniques in predicting soil texture classes [14–25]. The plan of our study is as follows. Section 2, which includes the study area, the methodology of MNLR specific to this study, and the dependent and environmental variables to be used in modeling, Sect. 3, which includes the evaluation of classification performance and the creation of spatial soil texture class maps, Sect. 4, the conclusion section presents the outputs of the current study and makes methodological suggestions for future studies.

2 Materials and Methods

2.1 Study Area

The study area is located in the Western Mediterranean region of Turkey. It is within the borders of Isparta province. This area covers an area of approximately 100 km^2 located between the coordinates of UTM 36 Zone 4192000–4204000 North 298000–304000 East. According to Java Newhall Simulation Model, the study area has Mesic Soil temperature regime and Xeric Soil moisture regime [26]. The study area consists mainly of Holocene-aged deposits. This structure was Mesozoic-tertiary limestones surrounding the plain. On this parent material, alluvial, colluvial and fluvial deposits were formed. These coarse deposits were arranged on the hillside of the limestones surrounding, and fine-structured deposits in the middle of the Atabey plain in the form of a closed bowl [27].

2.2 Soil Sampling and Soil Particle Size Distribution

88 soil samples were taken from 0–30 cm depth according to Stratified sampling methodology [28]. The analyses of particle size distribution were carried out in the laboratory according to the Bouyoucos Hydrometer method [11]. The percent of clay, sand, and silt of soils were determined to be grouped in the USDA soil texture classification triangle [29, 30]. Soil texture classes were found as Clay (C), Clay Loam (CL), Sandy Clay Loam (SCL), and Loam (L) (Fig. 1).

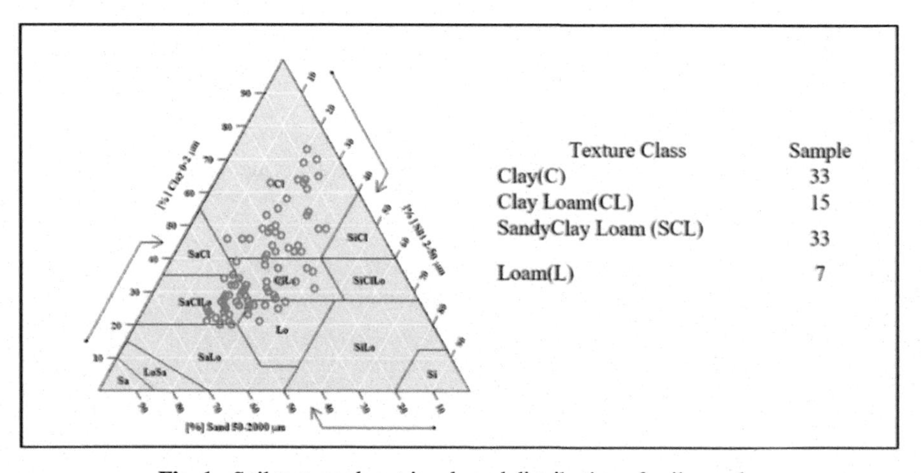

Fig. 1. Soil texture class triangle and distribution of soil samples

2.3 Environmental Covariates

Aster GDEM Digital Elevation Model (DEM) at 30 m spatial resolution was used in raster format [31] and topographic environmental covariates were generated in ArcGIS 9.3 software [32]. Environmental variables produced are given in Table 1.

2.4 Multinomial Logistic Regression

In cases where the response variable is categorical, logistic regression (LR), a simple extension of linear regression, can be used for estimation. When there are only two possible outcomes for the dependent variable, LR is called binary LR [35]. When the dependent variable takes more than two possible categories (presented in Fig. 1), a simple binary LR generalization known as Multinomial logistic regression (MNLR) is used for estimation [36]. MNLR uses the maximum likelihood algorithm [36, 37] to model the probabilities of possible outcomes of a multi-class dependent variable as a linear combination of a specific set of explanatory variables (presented in Table 1).

Table 1. Environmental variables

Variables	Description	Reference
Elevation(m)	Aster GDEM land surface elevation	[32]
Slope (%)	Slope gradient%	[33]
Aspect	The direction of the steepest slope gradient from north; together with the gradient determines the amount of sun	[33]
Curvature, Profil-Plan curvature	Curvature defines the shape of the slope	[33]
Flow direction	The assignment of flow directions is based on the difference in height between cells to direct the flow	[34]
Flow accumulation	Defines the spatial extent of a catchment area	[33]
Topographic wetness index	Relates the water movement in topography and slopes; Explain the effects of topography on the location and size of the areas	[33]
Stream power index	Explain the potential flow erosion and related landscape processes	[33]
Slope length	Slope length is used to identify areas potentially susceptible to erosion	[33]

In the modeling process using Multinomial logistic regression (MNLR), the "nnet" [39] package was used in the R Core Environment [38] software. Categorical Target variables (Texture classes) levels contain more than two categories. Therefore, the reference category should be selected in this model. Clay (C) was chosen as the reference class. The logit (ℓ) is the logarithmic function of the ratio between the probability (P) that a pixel (i) is a member of a class (j) and the probability that it is not (1 – P).

$$sigmoid(Pij) = \frac{e^x}{1+e^x} = \frac{1}{1+e^{-x}} = 1 + e^x = \frac{1}{Pij} = e^{-x} = \frac{1-Pij}{Pij} = e^x = \frac{Pij}{1-Pij} \quad (1)$$

When the natural logarithm of both sides is taken;

$$x = log\left(\frac{Pij}{1-Pij}\right) \quad (2)$$

The value x contained in the formula is equal to the model result of the class to be estimated

$$lij = \log\left(\frac{Pij}{1 - Pij}\right) = \beta 0j + a1j.X1i + a2j.X2i + \cdots + anj.X \tag{3}$$

β0j indicates the intercept of the regression curve for the soil texture class j

$a_{1j...nj}$ are the coefficients of each predictor $X_{1...n}$ for the respective soil texture class j

n, is the total number of covariates that significantly correlate with the given soil texture class j. Estimation of the probability a given soil texture class *j* at pixel *i* (P_{ij}) can be derived as:

$$Pij = \frac{e^{lij}}{1 + \sum_{j=1}^{k-1} (e^{lij})} \tag{4}$$

k is the total number of the dependent categories and $\sum$ indicates the summation of the logits of all the soil texture class (except the reference group) for the pixel *i*

Then, the probability of reference category (r) is given by

$$\Pr = \frac{1}{1 + \sum_{j=1}^{k-1} (e^{lij})} \tag{5}$$

The dependent variable has 4 categories. Since the dependent variable is categorical and has more than 2 levels, the "SoftMax function" is executed in the "nnet" package. As a result of the multinomial logistic regression algorithm, a pixel can belong to 4 different texture classes. The probability value is between 0 and 1, and the sum of the probability values is 1. The pixel is assigned to the texture class with the highest probability.

2.5 Model Accuracy Evaluation

The data set is divided into training (70%) and test set (30%). To measure the performance of classifications with the confusion matrix, the overall accuracy was calculated with the "ithir" and "caret" package [40–42]. The overall accuracy is a ratio between the correctly defined number of classes and the overall number of classes.

3 Results and Discussion

3.1 Descriptive Statistics

Descriptive statistics of texture fractions calculated for 88 samples. The silt and clay content are positively skewed while the sand content is negatively skewed [43].

3.2 Performance of MNLR Model

Model accuracy was done separately in training and test sets. Overall accuracy in the training data set was 63.9% (Table 2). These performances are similar to the study in which an overall accuracy of approximately 64% reported in the classification of soil texture using Landsat-5 TM images [44]. In this model, the accuracy rates of the classifications for both data sets were found relatively better results than the study [21] that shown an overall accuracy of 42%. In our test dataset, the effect of imbalanced distribution was more demonstrated. In another study, 5 different machine learning algorithms were compared, and overall accuracy levels were calculated between 61% and 64% [16].

Table 2. Confusion matrix

	Train							Test						
	Observed					Sum	Users	Observed					Sum	Users
		C	CL	SCL	L				C	CL	SCL	L		
Predicted	**C**	15	3	5	0	23	65.2	**C**	4	1	1	1	7	57.0
	CL	1	5	2	0	8	62.5	**CL**	1	0	1	0	2	0.0
	SCL	3	5	14	1	23	60.8	**SCL**	6	1	6	1	14	42.8
	L	2	0	1	4	7	57.0	**L**	1	0	3	0	4	0.0
	Sum	21	13	22	5	61		**Sum**	12	2	11	2	27	
	Producers	71.4	38.4	63.6	80.0		**63.9**	**Producers**	33.3	0.0	54.5	0.0		**47.6**

In the few test samples, accuracies for the loam and clay loam texture groups were very low. However, the accuracy differences between some classes in the dataset can be an indicator of the complexity of soil property variations and soil formation conditions in flood plains [45]. Furthermore, it is important to distinguish between texture classes at the separate end, such as clayey and sandy, with high accuracy, then to distinguish close texture classes from each other. Because the close texture class (for example SCL-CL or C-CL) does not require a different type of management technique and often behaves similarly in terms of water-holding capacity characteristics to the product on which it is grown. However, very different soil texture classes need different management [13]. In general, the accuracy values found to estimate the soil texture class obtained in this study were acceptable given the geographical coverage of the study area.

3.3 Importance of Variables

The important variables in the model were determined by using the "caret" [41, 46] package in the R Core Environment program and given in Fig. 2. Also, it has been interpreted to make pedological inferences.

The most important environmental variables were determined as the profile and planform curvature called the concave-convexity of the land and takes different values according to the shape of the slope. Stream power index represents potential areas where water erosion is high and may represent the transport state of clay, silt, and sand that make up the soil texture. When the amount of coarse fraction was increase, the SPI was

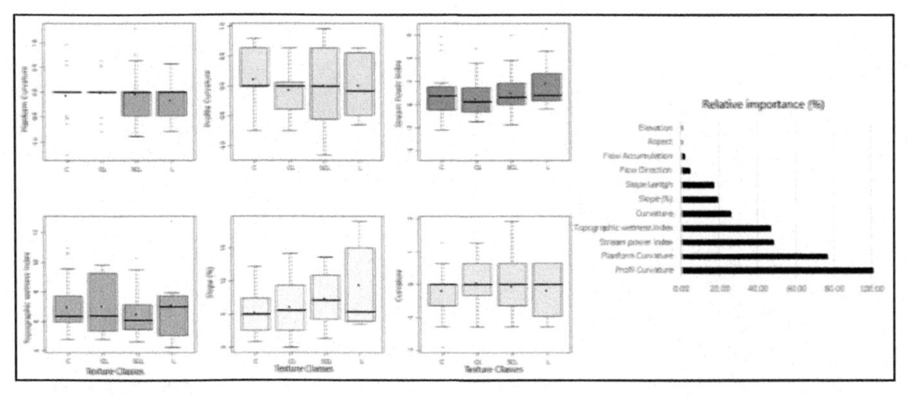

Fig. 2. Relative importance of environmental variables (Thick line: Median, Red point: Mean (color versions)

show more higher values [22]. In our samples, it was observed that the texture classes of the coarse fraction (sand) soils gave higher SPI values compared to the other groups (Fig. 2). Stream power index can be used as environmental variable in determining texture classes in similar land types. Topographic wetness index can be used to identify areas where water can accumulate with the flow, and flatter areas are represented by higher topographic wetness index values. It was reported that the topographic wetness value was high importance for prediction the surface clay content in an area where soil texture classed as sandy clay [49]. TWI values may be an indicator for increasing of clay content in a flood plain. The class averages of slope values were shown an increase from clay to loam texture (Fig. 3). This result is expected for the study area. When interpreted in terms of soil science, it is quite compatible with the real environment.

3.4 Spatial Mapping of Soil Texture Classes

The model was applied to the environmental variable data set in the Raster environment in R Studio V 1.3, which is the Integrated Development Environment (IDE) of the R Core Environment program [38, 47]. In this application, all layers must have the same projection system and the pixel size. In multinomial logistic regression, it was possible to produce maps that cover the most likely class and showing probabilities for all classes [48]. The probability that each pixel belongs to 4 different texture classes can be created spatially (Fig. 3).

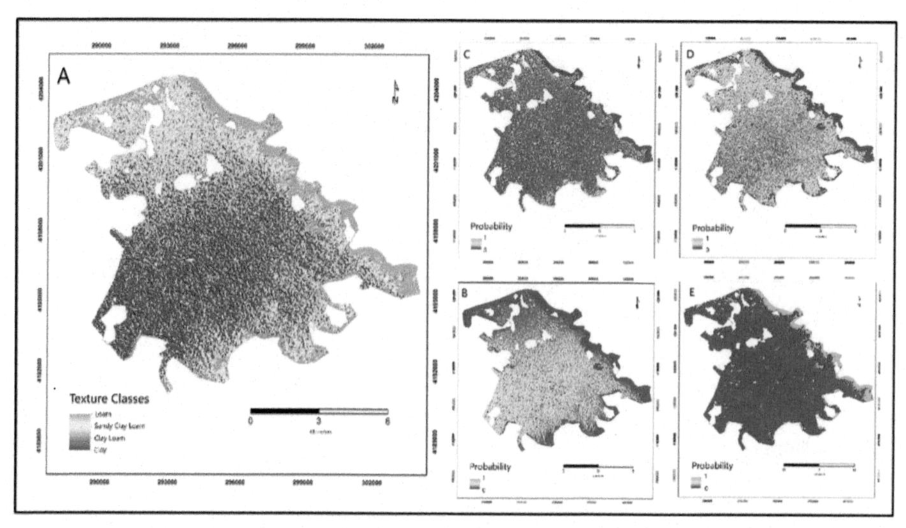

Fig. 3. Texture classes map (A: Most Probable Class, B: Clay Class Probability, C: Clay Loam Class Probability, D: Sandy Clay Loam Class Probability, E: Loam Class Probability)

4 Conclusion

The texture classes map was produced in accordance with the spatial resolution of the topographic variables obtained in raster environment. This digital output can be easily used by planners and farmers to identify the most suitable lands that will yield the most yield. Soil texture class maps are also an important input to land degradation assessment, hydrological studies, soil and water conservation, and ground engineering. Considering the natural distribution of soils by fraction sizes in nature, there is an imbalanced class distribution in the model in this study. In the imbalanced data set, the accuracy of the prediction was decrease. The accuracy of the classes with small number was also decreases even more. As a general result, it can be suggested to carry out the studies on digital mapping of soil properties using different machine learning algorithms such as Decision Trees and Random Forest, which reveal nonlinear relationships.

Acknowledgement. This study was supported by Scientific Research Fund of Isparta University of Applied Sciences. Project number: 2019-YL1–0044.

References

1. Zhu, A.X.: A similarity model for representing soil spatial information. Geoderma **77**, 217–242 (1996). https://doi.org/10.1016/S0016-7061(97)00023-2
2. Adhikari, K., Hartemink, A.E.: Linking soils to ecosystem services-a global review. Geoderma **262**, 101–111 (2016). https://doi.org/10.1016/j.geoderma.2015.08.009
3. Lorenzetti, R., Barbetti, R., Fantappiè, M., L'Abate, G., Costantini, E.A.C.: Comparing data mining and deterministic pedology to assess the frequency of WRB reference soil groups in the legend of small scale maps. Geoderma **237–238**, 237–245 (2015)

4. MacMillan, R.A., Jones, R.K., McNabb, D.H.: Defining a hierarchy of spatial entities for environmental analysis and modeling using digital elevation models (DEMs). Comput. Environ. Urban Syst. **28**, 175–200 (2004). https://doi.org/10.1016/S0198-9715(03)00019-X
5. McKenzie, N.J., Ryan, P.J.: Spatial prediction of soil properties using environmental correlation. Geoderma **89**, 67–94 (1999). https://doi.org/10.1016/S0016-7061(98)00137-2
6. Hewitt, A.E.: Predictive modeling in soil survey. Soils Fertil. **3**, 305–315 (1993)
7. Hudson, B.D.: The soil survey as paradigm-based science. Soil Sci. Soc. Am. J. **56**, 836–841 (1992). https://doi.org/10.2136/sssaj1992.03615995005600030027x
8. Caubet, M., Dobarco, M.R., Arrouays, D., Minasny, B., Saby, N.P.: Merging country, continental and global predictions of soil texture: lessons from ensemble modelling in France. Geoderma **337**, 99–110 (2019). https://doi.org/10.1016/j.geoderma.2018.09.007
9. Ma, Y.X., Minasny, B., Malone, B.P., McBratney, A.B.: Pedology and digital soil mapping (DSM). Eur. J. Soil Sci. **70**, 216–235 (2019). https://doi.org/10.1111/ejss.12790
10. McBratney, A.B., Santos, M.M., Minasny, B.: On digital soil mapping. Geoderma **117**, 3–52 (2003). https://doi.org/10.1016/S0016-7061(03)00223-4
11. Bouyoucos, G.J.: Hydrometer method improved for making particle size analyses of soils. Agron. J. **54**(5), 464–465 (1962). https://doi.org/10.2134/agronj1962.00021962005400050028x
12. Akgül, M., Başayiğit, L.: Süleyman Demirel Üniversitesi Çiftlik arazisinin detaylı toprak etüdü ve haritalanması. Süleyman Demirel univ. fen bilim. enst. derg. **9**(3), 1–10 (2005)
13. Dharumarajan, S., et al.: Digital soil mapping of key globalsoilmap properties in northern Karnataka plateau. Geoderma Reg. **20**, e00250 (2020). https://doi.org/10.1016/j.geodrs.2019.e00250
14. Gomez, C., Dharumarajan, S., Féret, J.B., Lagacherie, P., Ruiz, L., Sekhar, M.: Use of sentinel-2 time-series images for classification and uncertainty analysis of inherent biophysical property: case of soil texture mapping. Remote Sens. **11**, 565 (2019)
15. Wu, W., Li, A.-D., He, X.-H., Ma, R., Liu, H.-B., Lv, J.-K.: A comparison of support vector machines, artificial neural network and classification tree for identifying soil texture classes in southwest China. Comput. Electron. Agric. **144**, 86–93 (2018)
16. Zhang, M., Shi, W.: Systematic comparison of five machine-learning methods in classification and interpolation of soil particle size fractions using different transformed data. Hydrol. Earth Syst. Sci. Discuss. **24**(5), 2505–2526 (2020). https://doi.org/10.5194/hess-24-2505-2020
17. Camera, C., Zomeni, Z., Noller, J.S., Zissimos, A.M., Christoforou, I.C., Bruggeman, A.: A high resolution map of soil types and physical properties for Cyprus: a digital soil mapping optimization. Geoderma **285**, 35–49 (2017). https://doi.org/10.1016/j.geoderma.2016.09.019
18. Piccini, C., Marchetti, A., Rivieccio, R., Napoli, R.: Multinomial logistic regression with soil diagnostic features and land surface parameters for soil mapping of Latium (Central Italy). Geoderma **352**, 385–394 (2019). https://doi.org/10.1016/j.geoderma.2018.09.037
19. Bagheri Bodaghabadi, M., et al.: Digital soil mapping using artificial neural networks and terrain-related attributes. Pedosphere. **25**, 580–591 (2015)
20. Taalab, K., et al.: On the application of Bayesian networks in digital soil mapping. Geoderma **259**, 134–148 (2015). https://doi.org/10.1016/j.geoderma.2015.05.014
21. Ramcharan, A., et al.: Soil property and class maps of the conterminous United States at 100-meter spatial resolution. Soil Sci. Soc. Am. J. **82**(1), 186–201 (2018). https://doi.org/10.2136/sssaj2017.04.0122
22. Gobin, A., Campling, P., Feyen, J.: Soil-landscape modelling to quantify spatial variability of soil texture. Phys. Chem. Earth **26**, 41–45 (2001)
23. Zhao, Z., Chow, T.L., Rees, H.W., Yang, Q., Xing, Z., Meng, F.-R.: Predict soil texture distributions using an artificial neural network model. Comput. Electron. Agric. **65**, 36–48 (2009). https://doi.org/10.1016/j.compag.2008.07.008

24. Ließ, M., Glaser, B., Huwe, B.: Uncertainty in the spatial prediction of soil texture: comparison of regression tree and random forest models. Geoderma **170**, 70–79 (2012)
25. Poggio, L., Gimona, A.: 3D mapping of soil texture in Scotland. Geoderma Reg. **9**, 5–16 (2017). https://doi.org/10.1016/j.geodrs.2016.11.003
26. Van Wambeke A.R.: The newhall simulation model for estimating soil moisture and temperature regimes. Department of Crop and Soil Sciences, Cornell University, Ithaca (2000)
27. Akgül, M., Başayiğit, L., Uçar, Y., Müjdeci, M.: Atabey Ovası Toprakları. S.D.Ü. Ziraat Fakültesi Yay. No: 15, Araştırma Serisi No: 1, Isparta (2001)
28. Ditzler, C., Scheffe, K., Monger, H.C.: Soil Science Division Staff. Soil survey manual. (eds.). USDA Handbook 18. Government Printing Office, Washington (2017)
29. USDA: Soil mechanics level I. Module 3 – USDA textural soil classification study Guide. National Employee Development Staff, Soil Conservation Service, United States Department of Agriculture. U.S. Government Printing Office Washington (1987)
30. Moeys, J.: Soiltexture: functions for soil texture plot, classification and transformation, R package version 1.4.6 (2018). https://CRAN.Rproject.org/package=soiltexture. Accessed 11 Feb 2021
31. National Aeronautics and Space Administration (NASA): Aster Global Digital Elevation Model (Aster GDEM) NASA Official (2012). (http://www.gdem.aster.ersdac.or.jp). Accessed 25 July 2020
32. ESRI: ArcGIS Desktop: Release 9.3. ArcGIS user's guide. Environmental Systems Research Institute, Redlands (2011)
33. Hengl, T., Reuter, H.I. (ed.): Geomorphometry: concepts, software, and applications. developments in soil science, vol. 33, pp. 772. Elsevier (2008)
34. Gruber, S., Peckham, S.: Geomorphometry: land-surface parameters and objects in hydrology. Developments in Soil Science, vol 33, pp. 171–194. Elsevier (2008)
35. Alpaydin, E.: Introduction to Machine Learning. MIT Press, Cambridge (2010)
36. Hosmer, D.W. Lemeshow, S. Sturdivant, R.X. Applied Logistic Regression. JohnWiley & Sons, Hoboken (2013). https://doi.org/10.1002/9781118548387
37. Afshar, F.A., Ayoubi, S., Jafari, A.: The extrapolation of soil great groups using multinomial logistic regression at regional scale in arid regions of Iran. Geoderma **315**, 36–48 (2018). https://doi.org/10.1016/j.geoderma.2017.11.030
38. R Core Team: R: A language and environment for statistical computing. R Foundation for Statistical Computing, Vienna (2019). https://www.R-project.org/. Accessed 11 Feb 2021
39. Venables, W.N., Ripley, B.D.: Modern Applied Statistics with S, 4th edn. Springer, New York (2002). https://doi.org/10.1007/978-0-387-21706-2
40. Malone, B.: Ithir: soil data and some useful associated functions. R package version 1.0 (2018). Accessed 11 Feb 2021
41. Kuhn, M.: Caret: classification and regression training. R package version 6.0–86 (2020). https://CRAN.R-project.org/package=caret. Accessed 11 Feb 2021
42. Congalton, R.: A review of assessing the accuracy of classifications of remotely sensed data. Remote Sens. Environ. **37**, 35–46 (1991). https://doi.org/10.1016/0034-4257(91)90048-B
43. Webster, R.: Statistics to support soil research and their presentation. Eur. J. Soil Sci. **52**(2), 331–340 (2001). https://doi.org/10.1046/j.1365-2389.2001.00383.x
44. Zhai, Y., Thomasson, J.A., Boggess, J.E., III., Sui, R.: Soil texture classification with artificial neural networks operating on remote sensing data. Comput. Electron. Agric. **54**(2), 53–68 (2006). https://doi.org/10.1016/j.compag.2006.08.001
45. Pahlavan-Rad, M.R., Akbarimoghaddam, A.: Spatial variability of soil texture fractions and pH in a flood plain (case study from eastern Iran). CATENA **160**, 275–281 (2018)
46. Gevrey, M., Dimopoulos, I., Lek, S.: Review and comparison of methods to study the contribution of variables in artificial neural network models. Ecol. Modell. **160**(3), 249–264 (2003). https://doi.org/10.1016/S0304-3800(02)00257-0

47. RStudio Team: RStudio: Integrated Development for R. RStudio, Inc., Boston (2019). http://www.rstudio.com/. Accessed 11 Feb 2021
48. Malone, B.P., Minasny, B., McBratney, A.B.: Categorical soil attribute modeling and mapping. In: Using R for Digital Soil Mapping. PSS, pp. 151–167. Springer, Cham (2017). https://doi.org/10.1007/978-3-319-44327-0_6
49. Adhikari, K., et al.: High-resolution 3-D mapping of soil texture in Denmark. Soil Sci. Soc. Am. J. **77**(3), 860–876 (2013)

An Intelligent Multi-output Regression Model for Soil Moisture Prediction

Cansel Kucuk[1](✉), Derya Birant[2](✉), and Pelin Yildirim Taser[3](✉)

[1] Graduate School of Natural and Applied Sciences, Dokuz Eylul University, 35390 Izmir, Buca, Turkey
cansel.kucuk@ceng.deu.edu.tr
[2] Department of Computer Engineering, Dokuz Eylul University, 35390 Izmir, Buca, Turkey
derya@cs.deu.edu.tr
[3] Department of Computer Engineering, Izmir Bakırçay University, 35665 Izmir, Menemen, Turkey
pelin.taser@bakircay.edu.tr

Abstract. Soil moisture prediction plays a vital role in developing plants, soil properties, and sustenance of agricultural systems. Considering this motivation, in this study, an intelligent Multi-output regression method was implemented on daily values of meteorological and soil data obtained from Kemalpaşa-Örnekköy station in Izmir, Turkey, at three soil depths (15, 30, and 45 cm) between the years 2017 and 2019. In this study, nine different machine learning algorithms (Linear Regression (LR), Ridge Regression (RR), Least Absolute Shrinkage and Selection Operator (Lasso), Random Forest (RF), Extra Tree Regression (ETR), Adaptive Boosting (AdaBoost), Gradient Boosting (GB), Extreme Gradient Boosting (XGBoost), and Histogram-Based Gradient Boosting (HGB)) were compared each other in terms of MAE, RMSE, and R^2 metrics. The experiments indicate that the implemented Multi-output regression models show good soil moisture prediction performance. Also, the ETR algorithm provided the best prediction performance with an 0.81 R^2 value among the other models.

Keywords: Intelligent multi-output regression · Regression · Soil Moisture Prediction · Machine Learning

1 Introduction

Soil moisture has a significant impact on plant diversity, crop yield, sustenance of any agricultural system, soil thermal properties, and estimating flood, slope failure, and erosion [1]. Monitoring and evaluating soil moisture can be costly because of the sensors and their regular maintenance. Because of this reason, *soil moisture prediction* in advance supports the decision-making process in the field of agriculture, such as the management of water reservoirs for better crop yield, cost-saving, and plant diversity. In most cases, the soil moisture has been estimated using traditional methods in a laboratory environment. However, these techniques can be time-consuming with massive datasets and also cannot discover hidden patterns in complex structural properties of meteorological and soil data.

C. Kahraman et al. (Eds.): INFUS 2021, LNNS 308, pp. 474–481, 2022.
https://doi.org/10.1007/978-3-030-85577-2_56

Considering this challenge, in the last decade, *machine learning* techniques have been preferred in the agriculture field to reach more accurate soil moisture predictions. *Regression* is one of the most applied machine learning techniques which is used for predicting continuous-valued target attributes. In this study, three different traditional regression algorithms (Linear Regression (LR), Ridge Regression (RR), and Least Absolute Shrinkage and Selection Operator (Lasso)) and six different ensemble learning regression algorithms (Random Forest (RF), Extra Tree Regression (ETR), Adaptive Boosting (AdaBoost), Gradient Boosting (GB), Extreme Gradient Boosting (XGBoost), and Histogram-Based Gradient Boosting (HGB)) were compared on daily values of meteorological and soil data obtained from Kemalpaşa-Örnekköy station in Izmir, Turkey at three soil depths (15, 30, and 45 cm).

In most of the regression-based soil moisture prediction studies, only one target attribute value was predicted, such as predicting soil moisture at a particular depth. However, in this study, soil moisture values at three depths (15, 30, and 45 cm) were predicting simultaneously using the Multi-output Regression method. *Multi-output Regression* is a machine learning technique that predicts multiple target attribute values concurrently based on a given input attribute set [2].

The main contributions of this paper as follows: (i) it presents a brief survey of Multi-output regression technique and the algorithms used in the previous studies, (ii) it implements a Multi-output regression model using nine different regression algorithms (LR, RR, Lasso, RF, ETR, AdaBoost, GB, XGBoost, and HGB) on daily values of meteorological and soil data obtained from Kemalpaşa-Örnekköy station in Turkey at three soil depths (15, 30, and 45 cm) between the years of 2017 and 2019, (iii) it compares various multi-output regression models in terms of Mean Absolute Error (MAE), Root Mean Squared Error (RMSE), and Coefficient of Determination (R^2).

The remainder of this paper is structured as follows: In the following section, previous studies on the subject are summarized briefly. In Sect. 3, background information about multi-output regression is given, and the general structure of this study is explained. Section 4 expresses the meteorological and soil data used in our experimental studies. This section also gives information about applying the regression models on the dataset and presents the obtained experimental results with discussions. Finally, Sect. 5 gives some concluding remarks and future directions.

2 Related Work

In the literature, there are several studies [3–5] that implement machine learning algorithms for predicting soil moisture, such as support vector machine (SVM), decision tree (DT), and logistic regression (LoR). Nowadays, deep learning (DL) has been an active paradigm in soil moisture prediction studies to handle a large amount of meteorological and soil data [6, 7]. Yamaç, Şeker, and Negiş [6] implemented deep learning (DL), neural network (NN), and k-nearest neighbor (kNN) models for estimating moisture of calcareous soil data collected from Konya-Çumra plain, Turkey. Researchers in another study [7] proposed a deep learning regression network (DNNR) model for predicting soil moisture in the test area located in Beijing, China.

Table 1 presents the comparison of the proposed study with the previous studies. This study differs from the existing methods in two respects. First, they performed the prediction of only single output, while we focus on the prediction of multi-output values. Second, different models from the previous studies (LR, RR, Lasso, RF, ETR, AdaBoost, GB, XGBoost, and HGB) were implemented in this study.

Table 1. Comparison of our study with the previous studies (R = Reference, SO = Single-output, MO = Multi-output).

R	Year	Model	Prediction		Country
			SO	MO	
[1]	2020	NN, Fuzzy Logic, SVM	√		Nigeria
[3]	2017	kNN, SVM, NN, LoR, DT, RF, LR	√		Romania
[4]	2018	LR, SVM, NN	√		United States
[5]	2019	GB, Elastic Net, LR, RF	√		Not stated
[6]	2020	DL, NN, kNN	√		Turkey
[7]	2019	DL	√		China
Our Study		LR, RR, Lasso, RF, ETR, AdaBoost, GB, XGBoost, and HGB		√	Turkey

3 Material and Method

3.1 Multi-output Regression for Soil Moisture Prediction

Let D be a meteorological and soil data $D=\{(x_i, y_i)_{i=1}^{N} \epsilon XxY\}$ of$N$ instances, where $X \epsilon R^d$ and its related continues-valued multiple outputs are characterized by an output vector $Y=\{(y_i)_{i=1}^{N}\}$, where $Y \epsilon R^m$. The multi-output regression in this study aims to estimate soil moisture values at 15, 30, and 45 cm simultaneously based on soil and meteorological input parameters.

In this study, LR, RR, Lasso, RF, ETR, AdaBoost, GB, XGBoost, and HGB algorithms were preferred to implement the multi-output regression method.

Linear Regression (LR). LR is a statistical regression algorithm that discovers relations between two variables applying a linear equation [8].

Ridge Regression (RR). RR is a kind of regularized linear regression that uses L2 penalty for minimizing the size of all coefficients [9].

Lasso. Lasso is a type of linear regression that uses L1 penalty for shrinking the coefficients for input parameters that do not make a great contribution to prediction [10].

Random Forest (RF). RF is a bagging algorithm that uses a random feature subset selection method for constructing multiple decision trees [11].

Extra Tree Regression (ETR). ETR constructs multiple unpruned regression trees using the Top-Down approach for the prediction task [12].

AdaBoost. AdaBoost trains multiple single split decision trees consecutively to convert weak learners to strong ones by reweighting variables in the training set [13].

Gradient Boosting (GB). GB is a boosting algorithm that uses gradients in the loss function to improve the prediction ability of learners in a sequential manner [14].

XGBoost. XGBoost is an improved distributed gradient boosting method that implements a parallel tree boosting to predict new samples quickly and accurately [15].

Histogram-Based Gradient Boosting (HGB). HGB is a fast GB model that builds histograms for each feature to get the best split point [16].

Figure 1 presents the general structure of the implemented Multi-output Regression method in this study. In the first phase, meteorological and soil data is obtained from the sensors. In the next phase, the dataset is passed through a data preprocessing step (one-hot encoding and missing data imputation) to make it ready for the implementation of the proposed method. In the training phase, the Multi-output Regression approach is applied to the meteorological and soil data. In this step, three different regression models are constructed for predicting soil moisture values at various depths (i.e., 15, 30, and 45 cm), respectively. After that, in the evaluation phase, these models' soil moisture estimation performances are evaluated using the *k*-fold cross-validation technique in terms of the MAE, RMSE, and R^2 metrics. Finally, a new sample is predicted using the Multi-output Regression method and represented in the presentation step.

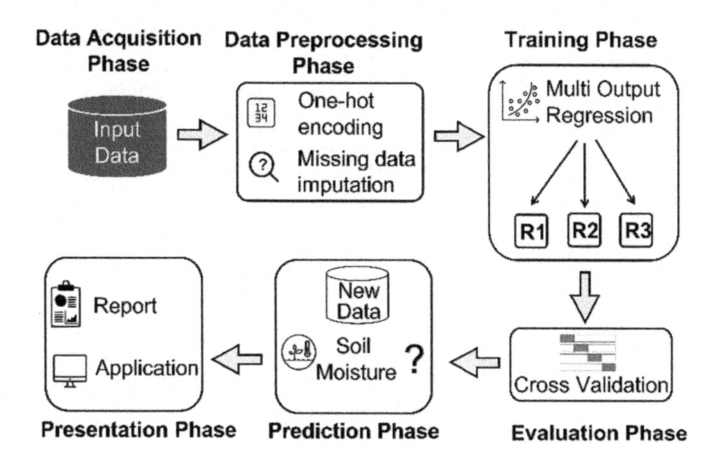

Fig. 1. The general structure of the multi-output regression for soil moisture prediction.

4 Experimental Study

In the experimental studies, the multi-output regression models were constructed on real-world meteorological and soil data for estimating soil moisture at three soil depths (15, 30, and 45 cm). The soil prediction application was developed in Python using Scikit-learn, Numpy, and Pandas libraries. In this study, the multi-output regression models were generated using the LR, RR, Lasso, RF, ETR, AdaBoost, GB, XGBoost, and HGB algorithms. The applied algorithms' parameters were set as follows:

- **LR:** positive = True
- **RR:** random_state = 1
- **Lasso:** random_state = 1, alpha = 0.1
- **ETR:** random_state = 1
- **RF:** random_state = 1, criterion = MAE
- **AdaBoost:** random_state = 1, n_estimators = 10
- **GB:** random_state = 1, n_estimators = 1000, criterion = MSE, max_feature = auto
- **HGB:** random_state = 1, max_bins = 50

All other parameters of the applied algorithms were chosen as default values. The multi-output regression models were compared by using the *k*-fold cross-validation technique by selecting *k* as 10. The validation process was repeated three times, and the average of them was taken as a validation result. The capabilities of these models were tested using the MAE, RMSE, and R^2 measures.

MAE. It gives the average magnitude of difference between the actual and predicted values shown in Eq. (1).

$$\frac{1}{m}\sum_{i=1}^{m}|(a_i - p_i)| \tag{1}$$

RMSE. It gives the square root of the mean of the square of the difference between the actual and predicted values shown in Eq. (2).

$$\sqrt{\frac{1}{m}\sum_{i=1}^{m}(a_i - p_i)^2} \tag{2}$$

R^2. It gives a degree ranging from 0 to 1.0 for presenting how strong the linear relationship is between the actual and predicted values, shown in Eq. (3).

$$R^2 = \frac{\left(\frac{\sum_i (p_i - \overline{p})(a_i - \overline{a})}{m-1}\right)^2}{\left(\sqrt{\frac{\sum_i (p_i - \overline{p})^2}{m-1}\frac{\sum_i (a_i - \overline{a})^2}{m-1}}\right)^2} \tag{3}$$

where *p* is predicted target value, *a* represents actual target value, and *m* is the number of instances.

4.1 Dataset Description

In this study, a real-world dataset consists of daily values of meteorological and soil data obtained from Kemalpaşa-Örnekköy station in Turkey at three soil depths (15, 30, and 45 cm) at the interval of 01.01.2017 and 28.12.2019. Figure 2 presents the map of the station in Izmir, Turkey.

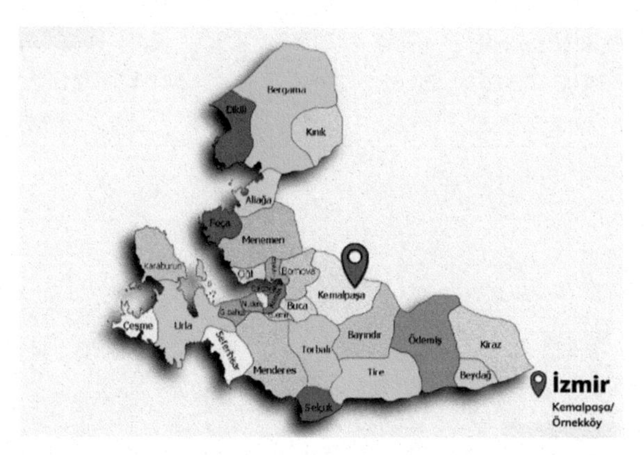

Fig. 2. The map of the Kemalpaşa-Örnekköy station used in this study.

The dataset used in this study includes the date, meteorological features such as air temperature, wet bulb temperature, dew point temperature, solar radiation, vapor pressure deficit, relative humidity, precipitation, leaf wetness, wind speed, and soil features such as temperature and moisture. It has passed through data preprocessing steps using the Python Scikit-learn library to implement the proposed method. First, "month" and "season" features were extracted from the "date" variable in the dataset. Then, these newly added features were converted to numerical values using the one-hot encoder technique. Furthermore, all missing values of the features were imputed using mean values of the same feature's existing values.

4.2 Results

Table 2 presents the MAE and RMSE values obtained from the Multi-output Regression models with their standard deviation (Std.) values. According to these results, ETR achieved the best prediction performance among the others with the values of 22.70 MAE and 37.35 RMSE. When these results are considered in general, it is understood that ensemble learning algorithms such as RF,ETR, AdaBoost,GB,XGBoost,and HGB outperform the traditional individual algorithms.

The graph given in Fig. 3 illustrates the R^2 values of the Multi-output regression methods on the experimental dataset. If the R^2 value of any algorithm is close to 1, it means that this model is highly reliable for the prediction task. The results indicate that the ETR model achieved the best soil moisture prediction performance with an R^2 value of 0.81. Also, it is observed from this graph that all algorithms applied in this study showed over 0.65 R^2 values. Thus, it can be concluded that the Multi-output Regression method can successfully predict soil moisture at different depths.

Table 2. Comparison of MSE and RMSE values of the implemented models.

Metrics		Individual algorithms			Ensemble algorithms					
		LR	RR	Lasso	RF	ETR	Ada Boost	GB	XG Boost	HGB
MAE	Value	37.09	37.43	37.39	25.15	22.70	36.62	27.25	27.83	25.56
	Std.	2.97	3.06	3.13	2.74	2.92	2.41	3.05	2.82	2.66
RMSE	Value	50.08	50.61	50.65	39.96	37.35	47.44	41.35	41.58	39.03
	Std.	3.98	5.51	5.62	4.26	4.33	2.87	4.78	4.32	4.09

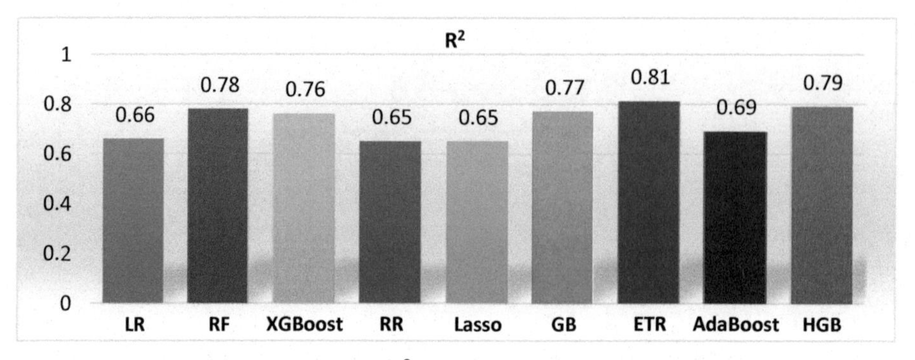

Fig. 3. Comparison of R^2 values of the implemented models.

5 Conclusion

Soil moisture prediction in advance has a significant impact on especially agricultural production and water recourses management. Because of this reason, this study implements a Multi-output Regression method on daily values of meteorological and soil data for predicting soil moisture at 15, 30, and 45 cm depths. This study tested this approach using the LR, RR, Lasso, RF, ETR, AdaBoost, GB, XGBoost, and HGB algorithms. The Multi-output Regression models were compared to each other in terms of MAE, RMSE, and R^2 in the experiments. The results presented that the Multi-output Regression approach gave successful prediction performance on soil moisture. Also, the ETR algorithm showed the most successful result among all other algorithms.

As future work, a fuzzy-based approach can be applied to the same dataset for predicting soil moisture. Besides, an ensemble of Multi-output Regression could also be applied and compared with the proposed model in the future.

Acknowledgements. The authors are deeply grateful to Selim Alpaslan in the Izmir Metropolitan Municipality for providing the experimental dataset used in the study.

References

1. Sanuade, O.A., Hassan, A.M., Akanji, A.O., Olaojo, A.A., Oladunjoye, M.A., Abdulraheem, A.: New empirical equation to estimate the soil moisture content based on thermal properties using machine learning techniques. Arab. J. Geosci. **13**(10), 1–14 (2020). https://doi.org/10.1007/s12517-020-05375-x
2. Liu, G., Lin, Z., Yu, Y.: Multi-output regression on the output manifold. Pattern Recogn. **42**, 2737–2743 (2009)
3. Matei, O., Rusu, T., Petrovan, A., Mihut, G.: A data mining system for real time soil moisture prediction. In: 10th International Conference Interdisciplinary in Engineering, pp. 837–844. ScienceDirect (2017)
4. Prakash, S., Sharma, A., Sahu, S.: Soil moisture prediction using machine learning. In: Second International Conference on Inventive Communication and Computational Technologies (ICICCT), pp. 1–6. IEEE (2018)
5. Singh, G., Sharma, D., Goap, A., Sehgal, S., Shukla, A., Kumar, S.: Machine learning based soil moisture prediction for Internet of Things based smart irrigation system. In: IEEE International Conference on Signal Processing, Computing and Control (ISPCC 2k19), pp. 175–180. IEEE (2019)
6. Yamaç, S., Şeker, C., Negiş, H.: Evaluation of machine learning methods to predict soil moisture constants with different combinations of soil input data for calcareous soils in a semi arid area. Agric. Water Manage. **234**, 106121 (2020)
7. Cai, Y., Zheng, W., Zhang, X., Zhangzhong, L., Xue, X.: Research on soil moisture prediction model based on deep learning. PLOS ONE **14**, e0214508 (2019)
8. Yan, X., Su, X.: Linear regression analysis. World Scientific Pub. Co., Singapore (2009)
9. Vlaming, R., Groenen, P.: The current and future use of ridge regression for prediction in quantitative genetics. Biomed. Res. Int. **2015**, 1–18 (2015)
10. Sujatha, C., Jayanthi, G.: LASH tree: LASSO regression hoeffding for streaming data. Int. J. Psychosoc. Rehabil. **24**, 3022–3033 (2020)
11. Yildirim, P, Birant, K.O., Radevski, V., Kut, A., Birant, D.: Comparative analysis of ensemble learning methods for signal classification. In: 26th Signal Processing and Communications Applications Conference (SIU), pp. 1–4. IEEE (2018)
12. Nistane, V., Harsha, S.: Performance evaluation of bearing degradation based on stationary wavelet decomposition and extra trees regression. World J. Eng. **15**(5), 646–658 (2018)
13. Yıldırım, P., Birant, U., Birant, D.: EBOC: ensemble-based ordinal classification in transportation. J. Adv. Transp. **2019**, 1–17 (2019)
14. Biau, G., Cadre, B., Rouvière, L.: Accelerated gradient boosting. Mach. Learn. **108**(6), 971–992 (2019). https://doi.org/10.1007/s10994-019-05787-1
15. Dhaliwal, S.S., Nahid, A., Abbas, R.: Effective intrusion detection system using XGBoost. Information **9**(7), 1–24 (2018)
16. Kone, Y., Zhu, K., Renaudin, V.: Machine learning-based zero-velocity detection for inertial pedestrian navigation. IEEE Sens. J. **20**(20), 12343–12353 (2020)

A Fuzzimetric Predictive Analytics Model to Reduce Emotional Stock Trading

Issam Kouatli(✉) and Mahmoud Arayssi

Lebanese American Princeton University, Beirut, Lebanon
{Issam.kouatli,mahmoud.araissi}@lau.edu.lb

Abstract. The purpose of this paper is to demonstrate the use of fuzzy set mechanism termed as "Fuzzimetric Sets" by investment advisors to provide maximum and minimum tolerances as part of helping the trader with making their stock trading decisions. Unlike other research of the use of Fuzzy logic in the Technical analysis, the proposed system guides the stock trader about the best well-known technical analysis mechanism used by traders. Moreover, the use of Fuzzimetric sets is designed to provide the low and high possible tolerances that the trader may use in any specific scenario, and to eliminate or reduce the effect of emotional trading. This paper designs the proposed model to compare the results of traditional different "Technical Analysis" (TA) techniques with the proposed fuzzified TA(s) methodology with the objective of reducing the emotional trading. The proposed design would provide the total spectrum of possible results and accordingly, traders would become more careful when conducting the trade.

Keywords: Technical analysis · Trading systems · Trading optimization · Fuzzy logic · Fuzzimetric Arcs

1 Introduction

Trading prediction is a process that investment advisors implement multiple times per year to inform and assist a client with their trades. Financial predictive analytics software suggests projections of future financial movements based on observed historical data. Financial services firms use these solutions to forecast asset movements and advice traders. Most individual investors are uninformed and typically lack financial knowledge about the stock market. They receive stock market information primarily from the public media and analysts' recommendations of suspect profitable stocks [1]. Stock trading is decided by the individual trader and implemented with the help of the respective advisor. It is usually based on technical trading rules to identify the buying/selling signals based on either short or long history of stock prices [2]. The future prices are difficult to predict, as the information gathered at any moment in time may reflect the market efficiency rather than the actual price [3, 4]. However, prices may also represent the information overreaction as discussed by Kahneman and Tversky [5]. Traders may overreact to market information as well as private information due to emotional factors about specific shares and securities [6]. More recently, emotional and psychological factors can be

C. Kahraman et al. (Eds.): INFUS 2021, LNNS 308, pp. 482–489, 2022.
https://doi.org/10.1007/978-3-030-85577-2_57

found in Khayamim et al. [7]; fuzzy logic based reasoning is used to simulate the calculated uncertainty of portfolio optimization and hence avoiding the emotional and the psychological aspects. Hence, the main reason for inventing the technical trading rules that have been investigated before by Park et al. [8], is to achieve the maximum profit possible from trading in stocks. All of these technical indicators are based on two main types: trend indicator and mean-reverting indicators (also termed as counter-trend), where each type might be suitable for a specific stock chart behavior.

Based on fuzzy logic, Gradojevic and Gençay [9] suggested a mechanism of reducing trading uncertainty, by addressing two problems. These problems are the market timing and the order size. Escobar et al. [10], proposed a technical fuzzy indicator that incorporates subjective features simulating the human decision making which uses a comparison between traditional RSI, MACD as opposed to the fuzzy-built multi-agent indicator obtaining the behavior and profit as outputs. Social Network-based prediction of short-term stock trading was proposed by Cremonesi et al. [11] using semantic sentiment analysis as a mechanism for inspecting Twitter posts.

This paper proposes a predictive decision model based on the fuzzification of the most popular TA indicators with the associated trading decision making, providing a guidance or advice to traders about the relevant actions and hence, reduce the emotional trading. The contributions of this paper are the followings: 1. Evaluate the use of predictive analytics process to classify stocks 2. A novel form to use predictive analytics, namely fuzzy analysis, in the stock prediction process 3. The use of predictive analytics to reduce the workload needed to prepare the performance evaluation of a client's account 4. Automation in the analysis of several objective measures to give better advice to the respective trader in conducting their trades. Unlike other research on fuzzy implementation to trading systems, this paper makes the most relevant parameters of popular indicators, fuzzifies them, feeds them into a modular approach of fuzzy system, and then combines the final fuzzy output after inferring the relative outputs from these indicators. Furthermore, the proposed system uses mutations of fuzzy sets to achieve low and high tolerances for each one of the sets. A combination of these mutation sets provides a "Defuzzified spectrum" of possible outputs acting as a decision support system for traders.

This paper composed of 4 sections. Section 2 review the concept of "Fuzzimetric Arcs" as a mutational mechanism of fuzzy set shapes (fuzzy variables) where this mutation can be automated to discover the most appropriate fuzzy set shape based on data training. Section 3 describes the modular approach of defuzzifying the inferred values and relating them to the most common "technical indicators" trading rules. Section 4 concludes the features of the proposed modeling technique where experimentation on the proposed model started but more data needs to be experimented with before conclusive results. This will be published in a further publication in the near future.

2 The Background of Fuzzimetric Sets

First, Fuzzy logic started by Zadeh [12] as a mechanism of decision making to deal with uncertainty. It was based on an extension of set theory where instead of a crisp description of a member belong (or not) to a set, a member can have partial membership in a specific set. This principle can be extended to define all possible sets in the universe of discourse and can be defined as Positive Zero (P0), Positive Small (PS), Positive Medium (PM) and Positive Big (PB) (Fig. 1a). These fuzzy variables can be defined as (Eqs. 1–4):

$$\mathrm{PO} = \int_0^{\pi/2} |\sin(\pi/2 - x)| \tag{1}$$

$$\mathrm{PS} = \int_0^{\pi} |\sin(x)| \tag{2}$$

$$\mathrm{PB} = \pi/2 \int^{3\pi/2} |\sin(\pi/2 - x)| \tag{3}$$

$$\mathrm{PM} = \pi \int^{3\pi/2} |\sin(x)| \tag{4}$$

Assuming sinusoidal function, then these representations can be defined in an analogy to trigonometric functions with an exception of taking the absolute values only. Hence, based on the definition of "Fuzzimetric Arcs [13], Fuzzimetric sets was proposed [14], where the methodology to control the fuzzy sets described by Kouatli [15]. An implementation of the technique was published in [16, 17]. The concept defines a mechanism of selection, mutation and cross-over fuzzy set shape and hence affecting the overall decision making defuzzified value. For example, triangular, trapezoidal…etc. can be achieved by a simple mutation of fuzzy sets using a genetic operator. This mutation can be achieved by using the ARCSIN value of the trigonometric membership value divided by an arbitrary vale (termed here as mutation factor) representing the angles in trigonometric arc as this can be shown in Eq. 5.

$$\begin{aligned} \mu &= \frac{\mathrm{ARCSIN(Fuzzy\ Variable)}}{\mathrm{T}} \\ &= 1 \text{ for } \mu > 1 \end{aligned} \tag{5}$$

The fuzzy variable is any of PO, PS, PM or PB, and the T parameter is the shape alternation factor (Mutation factor) with the most active range of $0 < \mathrm{T} < 270^{\circ}$. Accordingly, based on this concept, the equations for the four fuzzy variable PO, PS, PM and PB cand be described in Eqs. 6–9:

$$\text{Mutated-PO} = \frac{\int_0^{\pi/2} \mathrm{ARCSIN}(|\sin(\pi/2 - x)|)}{\mathrm{T_{PO}}} \tag{6}$$

$$\text{Mutated-PS} = \frac{\int_0^{\pi/2} \mathrm{ARCSIN}(|\sin(x)|)}{\mathrm{T_{PS}}} \tag{7}$$

$$\text{Mutated-PM} = \frac{\int_{\pi/2}^{3\pi/2} \mathrm{ARCSIN}(|\sin(\pi/2 - x)}{\mathrm{T_{PM}}} \tag{8}$$

$$\text{Mutated-PB} = \frac{\int_{\pi}^{3\pi/2} \mathrm{ARCSIN}(|\sin(x)|)}{\mathrm{T_{PB}}} \tag{9}$$

Altering the value of mutation factor "T" allows us to mutate the fuzzy variables where examples are shown in Fig. 1(b). More details of fuzzy sets and utilization of this concept to the decision-making process in can be found in Kouatli [16], where a robotic example was taken as a vehicle to a step-by-step explanation of inference using fuzzy sets.

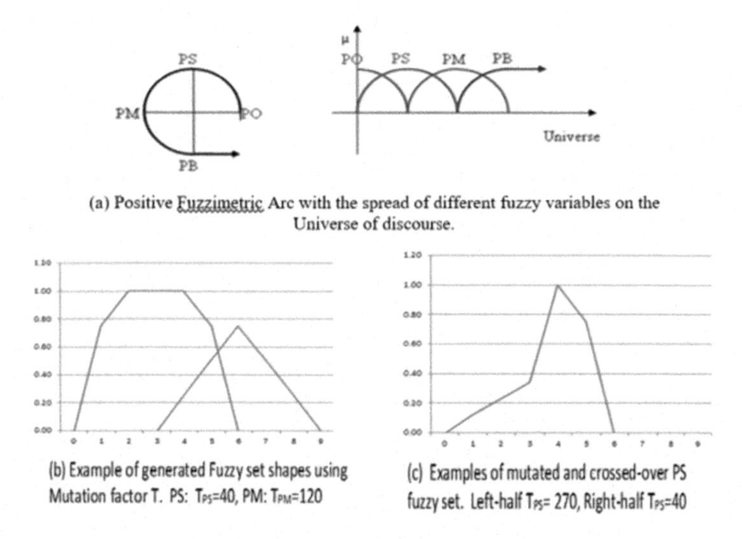

Fig. 1. Evolution of fuzzimetric sets

3 Fuzzy Inference Methodology Used

3.1 Fuzzimetric Sets as a Mechanism of the Variability of Defuzzified Values

The advantage of the Mutation Factor T is that it would be easy to alter the shape of the fuzzy set by only changing the mutation factor T which it can also be automated to search for the optimum fuzzy set shape for the desired output. Moreover, it would also be possible to compose fuzzy sets from two different mutated halves. By doing so, a crossover between two different mutated halves would generate the fuzzy set. This would be extremely useful as a mechanism to move the centroid of the fuzzy set from one side to the other. Hence, in some cases, optimizing the fuzzy system can be achieved without the need to change the rules in some cases. Obviously, knowledge discovery should have already been taken place and initial rule-set(s) would have already been built before this optimization takes place. Figure 1(c) shows an example of PS fuzzy set with different mutation/cross-over operation where the left-half of the fuzzy shape has a mutation value of $T = 270$ and the right-half has a value of $T = 40$. The full spectrum of all possible values of fuzzy set small ranging from the minimum to maximum defuzzified values of possible Fuzzimetric set shape can be seen in Fig. 2. Hence, Fuzzy system heuristics are built as rule-sets, describing the system model. Rules are usually in the form of (IF A

… THEN B) where "A" and "B" are fuzzy variables. For a Single-input-Single-Output (SISO) system, this would be straightforward and fuzzy inference can be conducted on the fuzzy variables representing the input and output respectively.

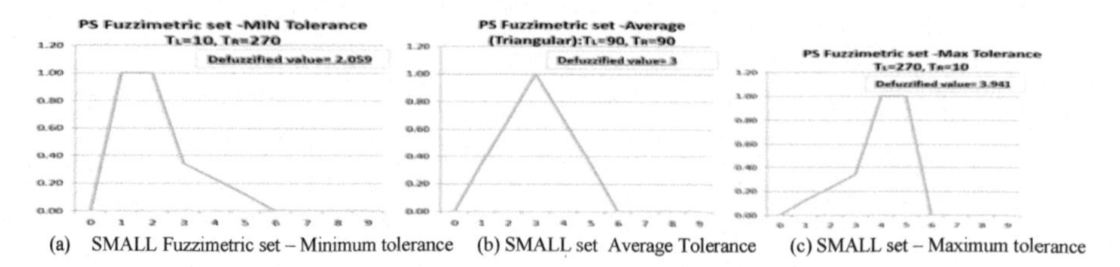

(a) SMALL Fuzzimetric set – Minimum tolerance (b) SMALL set Average Tolerance (c) SMALL set – Maximum tolerance

Fig. 2. Variation of Fuzzimetric set "small" within a specific interval

3.2 Modular Approach with Multi-input-Multi-Output (MIMO) Systems

Most real-world problems are in the form of a multivariable structure composed of Multi-input-Multi-Output (MIMO) systems. In this case, problems may arise in achieving a complete and consistent construction of all possibilities relevant to the output of the system. Rules, in this case, are of the form (IF A_1 & A_2… & A_n THEN B1 & B_2 &…B_n). This situation results in additional complexity in knowledge discovery and accurate heuristic modeling the system. Special algorithms are necessary in this case to tune the system as well as to detect and remove irrelevant rules. For example, Gegove [18] studied rule compression and selection with the goal of maintaining completeness and consistency of the system. Instead of rule compression, a simplified modular structure was also proposed by Kouatli [15], where the fuzzy system can be defined as three main components: the fuzzification component, the knowledge component and the inference/de-fuzzification component, and where the input/output relationships defined by creating sub-rule-sets each of which describe the relationship between one of the inputs and one of the outputs. Combination of the output of the sub-rule sets are integrated via an "input importance factor" and hence, such mechanism of integration is a modular approach of fuzzy systems.

3.3 Fuzzy Stock Trading Rules Based on Technical Indicators

Bollenger Band (BB), Moving Average Convergence Divergence (MACD) and RSI are the three most used technical indicators by stock traders. The proposed fuzzy predictive analytics will be based on a combination of rules governing these different TA techniques. Total annual profit augmented with the total transactions executed will be used as a mechanism of comparative back-testing the traditional TA(s) and the proposed fuzzified TA(s). In order to conduct fuzzy inference related to the best strategy, criteria need to be specified with respect to a known interval, where fuzzy variables can then be defined as PO, PS, PM or PB, rather than crisp values. The chosen criteria out of the well-known most popular indicators (described in the review section) are listed as follows with the relevant fuzzy trading rules:

Buying Fuzzy rules:
BB Related: IF Price is PO AND %b is PO AND Bandwidth is PM OR PB
MACD Related: IF Price is PO or PS AND Golden-Cross = OK
RSI Related: IF Price is PO or PS and RSI is PO
Stochastic Related: IF Price is PO and %K is PB AND %K > %D.
Selling Fuzzy rules:
BB Related: IF Price is PM OR PB AND %b is PB AND Bandwidth is PM OR PB
MACD Related: IF Price is PM or PB AND Dead-Cross = O.
RSI Related: IF Price is PB and RSI is PB
Stochastic Related: IF Price is PB and %K is PM or PB AND %K < %D.

3.4 Building Fuzzy Set Tolerances Based on the Fuzzimetric Sets

The strategy adopted for analysis using Fuzzimetric Arcs were defined into 5 main possibilities representing five main tolerances as defined in the previous section where crossover and mutation can be utilized to achieve these tolerances of buying and selling. These are:

Triangular sets: Right $T_{PO} = 90$, Left $T_{PS} = 90$, Right $T_{PS} = 90$, Left $T_{PM} = 90$, Right $T_{PM} = 90$, Left $T_{PB} = 90$
MINMIN sets: Right $T_{PO} = 10$, Left $T_{PS} = 270$, Right $T_{PS} = 10$, Left $T_{PM} = 270$, Right $T_{PM} = 10$, Left $T_{PB} = 2700$
MINMAX sets: Right $T_{PO} = 10$, Left $T_{PS} = 270$, Right $T_{PS} = 10$, Left $T_{PM} = 10$, Right $T_{PM} = 270$, Left $T_{PB} = 10$
MAXMIN sets: Right $T_{PO} = 270$, Left $T_{PS} = 10$, Right $T_{PS} = 270$, Left $T_{PM} = 270$, Right $T_{PM} = 10$, Left $T_{PB} = 270$
MAXMAX sets: Right $T_{PO} = 270$, Left $T_{PS} = 10$, Right $T_{PS} = 270$, Left $T_{PM} = 10$, Right $T_{PM} = 270$, Left $T_{PB} = 10$.

The triangular sets represent the non-biased adoption of the indicators towards buying and selling while the MINMIN sets represent the strategy of minimum tolerance of buying with a minimum tolerance of selling which indicated the strategy of a trader who waits for a minimum price and sells as soon as the price is acceptable (and profitable) minimum tolerance. Similarly, the MINMAX represents the minimum tolerance of buying and maximum tolerance of selling and so on. Figure 3 shows the graphic representation of the chosen different variations of Fuzzimetric sets. Based on previous history (data), this proposed theory need to be proven by experimentation where part of such data will be used as training data before the back test to study the implementation of this technique. Such tasks are beyond the scope of this paper, but will be considered for future study.

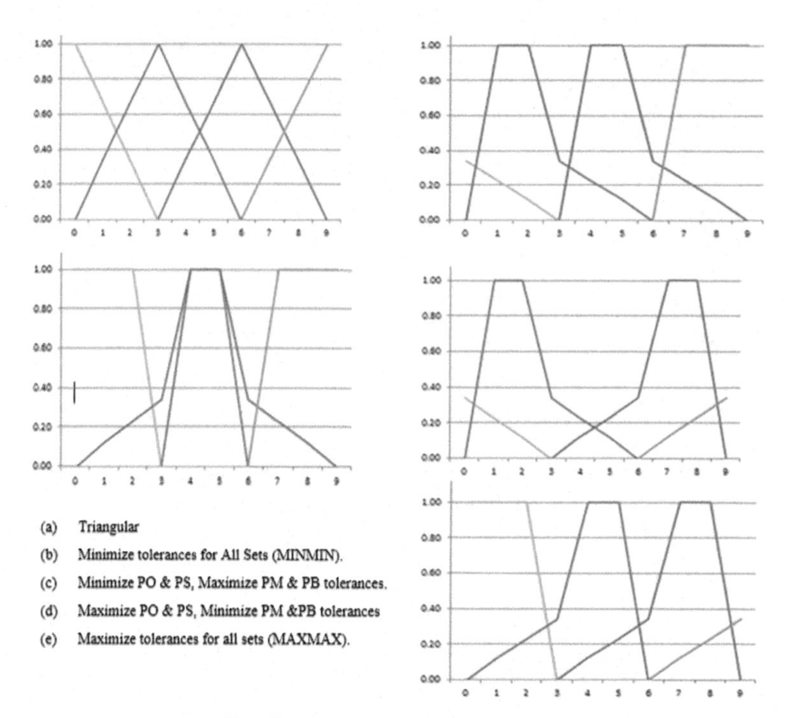

Fig. 3. Variations of fuzzimetric sets

4 Conclusion

Predictions in stock market behavior is a complicated task. Traditional technical indicators can provide a good sign of buy/sell signals but uncertainty exists with an unusual behavior of stock prices. Moreover, emotional traders may take a wrong decision under the prevailing uncertainty, which may lead to a loss on their investment. The proposed model "fuzzifies" mathematically-solid technical indicators and the decision becomes a combination of these "fuzzy" technical indicators. The model was based on Fuzzimetric sets, where the minimum and maximum tolerances of fuzzy sets were used to identify the decision support spectrum of the de-"fuzzified" output. The advantages of using such a system would be:

(a) Possible higher profitability, since the system uses a weighted combination of indicators.
(b) Act as trading predictive Analysis, hence eliminate/reduce emotional trading as different possibilities can be displayed.

Therefore, in this paper we developed a financial information system to provide an environment for traders to receive advice and to better strategize their stock trading. Second, it is presented to be a useful platform to study trading behaviors of individual traders and to perform further analytical studies. This model allows the traders to make better decisions, reduce emotional trading, and improve their bottom line results, as well as motivate them to learn.

Further Research:
Experimentation on selected indices like NASDAK and S&P would be beneficial where a back-test of the can generate the appropriate results to prove the effectiveness of the proposed model. Such task is not part of this paper but will be addressed in the next research article.

References

1. Barber, B.M., Odean, T.: All that glitters: the effect of attention and news on the buying behavior of individual and institutional investors. Rev. Financ. Stud. **21**(2), 785–818 (2008)
2. Gehrig, T., Menkhoff, L.: Extended evidence on the use of technical analysis in foreign exchange. Int. J. Financ. Econ. **11**, 327–338 (2006)
3. Fama, E.F.: The behavior of stock market prices. J. Bus. **38**, 34–105 (1965)
4. Fama, E.F.: Efficient capital markets: a review of theory and empirical work. J. Financ. **25**, 383–417 (1970)
5. Kahneman, D., Tversky, A.: Prospect theory: An analysis of decision under risk. Econometrica **47**, 263–292 (1979)
6. Chuang, W.I., Lee, B.S.: An empirical evaluation of the overconfidence hypothesis. J. Bank. Financ. **30**, 2489–2515 (2006)
7. Khayamim, A., Mirzazadeh, A., Naderi, B.: Portfolio rebalancing with respect to market psychology in a fuzzy environment: a case study in Tehran Stock Exchange. Appl. Soft Comput. **64**, 244–259 (2018). https://doi.org/10.1016/j.asoc.2017.11.044
8. Park, C., Irwin, S.: What do we know about the profitability of technical analysis? J. Econ. Surv. **21**, 786–826 (2007)
9. Gradojevic, N., Gençay, R.: Fuzzy logic, trading uncertainty and technical trading. J. Bank. Finance **37**(2013), 578–586 (2013)
10. Escobar, A., Moreno, J., Munera, S.: A technical analysis indicator-based on fuzzy logic. Electr. Notes Theoret. Comput. Sci. **292**(2013), 27–37 (2013)
11. Cremonesi, P., et al.: Social network based short-term stock trading system, social and information networks (2018)
12. Zadeh, L.A.: Fuzzy sets. Inf. Control **8**, 338–356 (1965)
13. Kouatli, I., Jones, B.: An improved design procedure for fuzzy control systems. Int. J. Mach. Tool Manuf. **31**(1), 107–122 (1991)
14. Kouatli, I.: Fuzzimetric Sets: An Integrated Platform for Both Types of Fuzzy Sets. Frontiers in Artificial Intelligence and Applications (FAIA), vol. 309 (2018)
15. Kouatli, I.: Fuzziness control of fuzzimetric sets. In: 2019 IEEE International Conference on Fuzzy Systems (FUZZ-IEEE) (2019)
16. Kouatli, I.: Fuzzimetric employee evaluation system. J. Intell. Fuzzy Syst. **35**(4), 4717–4729 (2018)
17. Kouatli, I.: The use of fuzzy logic as augmentation to quantitative analysis to unleash knowledge of participants' uncertainty when filling a survey: case of cloud computing. IEEE Trans. Knowl. Data Eng. (2020)
18. Gegov, A.: Complexity Management in Fuzzy Systems: A Rule Base Compression Approach. Springer, Berlin (2007). https://doi.org/10.1007/978-3-540-38885-2

Currency Exchange Rate Forecasting with Social Media Sentiment Analysis

Akıner Alkan[1(✉)], Ali Fuat Alkaya[2], and Peter Schüller[3]

[1] Siemens, Istanbul, Turkey
[2] Engineering Faculty, Computer Science and Engineering Department, Marmara University, Istanbul, Turkey
[3] Technische Universität Wien, Vienna, Austria

Abstract. Social media has been increasingly popular and valuable along with their mass data. At the same time, currency exchange rate forecast has been an important topic for researchers, analysts, and investors for a long time. In this study, we have combined exchange rate time series analysis and Twitter sentiment analysis to build a machine learning model. We have built the model in three stages for a six months period: (i) we have watched financial and political hashtags and applied sentiment analysis on the tweets retrieved from these hashtags, (ii) we have collected time series data on cross-currency exchange rates including cryptocurrencies, (ii) we have optimized the model to forecast USD/TRY with the data we have. We have experimented with several machine learning algorithms including linear regression, Bayesian ridge, support vector machines along with multi-layer perceptron (MLP). It has been observed that in this novel approach, some regression algorithms performed better than MLP. Computational experiments showed that our approach gave 0,35% mean squared error performance as its best. Results suggested that sentiment analysis is a helping factor to forecast currency exchange rate and Twitter is a good data source due to its mass and interactivity. In conclusion, investors, analysts, and researchers can benefit from the usage of our proposed model and will able to get strong and consistent results to forecast the USD/TRY currency exchange rate.

Keywords: Machine learning · Sentiment analysis · Time series analysis · Social media

1 Introduction

Money is considered as an item of buying power for goods that are countable and verifiable by the governments. Despite some countries/governments are using their currencies, some countries are using common currencies belonging to unions like Euro which is accepted by the European Union. Another noticeable currency is the US Dollar which is used in global trading. Currencies are exchangeable with their exchange rates which can give an insight into the economy of the related domain country. As an example, the US Dollar exchange rate can be correlated with the United States economy. Therefore, the currency exchange rate can be forecasted by looking at the economic factors of the

C. Kahraman et al. (Eds.): INFUS 2021, LNNS 308, pp. 490–497, 2022.
https://doi.org/10.1007/978-3-030-85577-2_58

related country. In addition to the economic factors, political incidents are indirectly affecting the economics. For the last two decades, technological advances gave rise to a concept called social media which gives people to share their ideas and feelings about any topic [1]. Thus, people are also able to share their opinions on political events. Twitter is one example of a social media platform that helps people to express their feelings via tweets.

On the other hand, it is well known that the power of the economy is also correlated with the stock market values averages and stock market indices of the specific countries [2]. As an example, we can claim that Japan stock market value indices can be correlated with the currency exchange rate of the Japanese Yen.

In this paper, in a very brief way: we worked on forecasting the currency exchange rate of USD vs. TRY. This is a very classical and well-known problem that has been already studied many times. Therefore, there is much work done previously, where some of which have studied the currency exchange rate as a time series data or checked the relationships of the exchange rates and the currencies. There is also much work which have used the stock market indices to forecast the currency exchange rates. Several studies in the past have also used the sentiment of the social media factor on currency exchange rate. We investigated previous studies, combined these three different approaches. and tried to forecast the currency exchange of the US Dollar via using news, stock market values, and currency exchange rate correlations as time-series data using several machine learning algorithms. As the news, we exploited Twitter by following several specific but highly influential hashtags. like #Breaking, #Trump, etc. We processed the tweets' sentiments and used them as news-related data and fed them into our machine learning models. For stock market values, we have used several stock market indices like Europe 50 Index, Japan 30 Index, US Dollar 500 Index, etc. These indices are also fed into our model. We also processed correlations between currencies Exchange rates as time series data and fed them into our model.

Forecasting currency can be classified as a regression problem. So, we forecasted the currency values by applying various forecasting techniques and machine learning algorithms including Linear Regression, Bayesian Ridge, Lasso, Stochastic gradient descent (SGD), Support vector regression (SVR). We built models by training these algorithms one by one and compared their performance on test instances using several performance measurement tools such as Mean Squared Error (MSE), Root Mean Square Error (RMSE), Mean Absolute Error (MAE), Mean Absolute Deviation(MAD) to decide what is the best amongst the others and will it be feasible to help our decision making process for forecasting currency for the future.

For the rest of the paper, the organization is as follows. In Sect. 2, we present a summary of a literature survey that we have done regarding the related techniques that have been applied for the aforementioned objective. Section 3 provides the details of the experimental work done as data creation and gives detailed information about the exploited techniques. Section 4 presents the results of the experiments together with their discussion. Section 5 concludes this paper by giving some future work.

2 Literature Survey

In this section, we are going to investigate the methodologies developed for forecasting the exchange rates. It is important to understand these methodologies and how they help to forecast the trend of money. What we are trying to accomplish is to take existing methodologies and combining this know-how to accomplish better forecasting accuracies. In the following studies, we investigated different approaches as sentiment analysis, time series analysis, stock market analysis to make forecasting with these approaches.

O'Connor et al. use Twitter tweets to replace polls by getting the sentiments from them [3]. It is assumed that polls are the gold for the verification and the sentiment on the polls is correct. As result, tweet sentiment analysis results correlated 80% to polls sentiments. Since tweets are easy to retrieve and polls are expensive than tweets' sentiments analysis, tweets sentiment analysis can be replaced with polls. In another study, we observe that Twitter can be used as a corpus for sentiment analysis on the tweets, and it can help to opinion mining [4]. Twitter and the data are collected over New York Times Magazine and Washington Post, etc. Validation and verification of the success rate of sentiment analysis are done by looking at the well-known metrics like precision, recall, F1 score. In conclusion, Twitter data along with its analyses can be used as a predictor for other applications.

Along with the usage of sentiment analysis, Twitter tweets are used as indicators to forecast the stock market mood. Since very early indicators can be extracted from online social media, Twitter is used to correlate stock market trends [5]. To analyze tweets two different sentiment analysis tools are used which are OpinionFinder and Google Profile of Mood States. In conclusion, it is found at the accuracy of 87.6% in predicting the daily up and down changes in the closing values of the DJIA and a reduction of the Mean Average Percentage Error by more than 6%.

Another application of sentiment analysis is estimation of currency exchange rate trend forecasting. It is correlated as the sentiment analysis of Twitter data. The currency value is not directly used as a target [6]. Instead, currency trends such as going upwards or downwards are used as the target. Two different hashtags are considered as 'Dollar' and 'USD/TRY'. In the conclusion section, estimation correctness for decreasing rate at 82% rate and the increasing at 57% rates. In total, correctness in the estimation is 71%.

Despite sentiment analysis is well known application in exchange rate forecasting; in a very recent study, artificial neural networks are used as forecasting the currency exchange rate of CZK over RMB [7]. ANN can help the prediction of the currency exchange rate even without looking at any other predictor. Experimental results show that the performance of RBF neural networks for the CNY exchange rate prediction is acceptable and effective which has the error is slightly above 0.002. With this error rate, it is considered as RBF neural networks are good models that can predict the exchange rate.

In another recent study, econometric models, time-series data, and deep ANN are used for forecasting the stock market and currency exchange rate trend direction [8]. Econometric models are working well only for long-term predictions. On the other hand, time series models provide good estimations in short-term predictions, but it is not reliable to predict the direction. Just like time series models, multilayer perceptrons with a single hidden layer suffers from the same problem.

These previous studies investigated and combined with their different approaches. To fill the relevant gap in the literature, in this study, US Dollar/Turkish Lira rate forecasting via using news, stock market values, and currency exchange rate correlations as time-series data using several machine learning algorithms is accomplished.

3 Data and Algorithms

Twitter is a powerful social media platform along with its worldwide popularity and diversity in users. On the other hand, Twitter is offering a good application programming interface (API) to the data scientists and researchers so that we had the opportunity to retrieve these tweets. We have started identification of which hashtags will be meaningful for our case. There are also available APIs that help you find related tags and get an insight into which tag has a higher tweet volume."RiteTag" is one of the well-known service providers for this use case. Thus, we have used 'RiteTag' to identify better hashtags. At the end, we identified 24 hashtags given in Table 1:

Table 1. Twitter hashtags.

Currency related	Stock market related	Politics related	News related
Currency	Stock market	Trump	Breaking
Money	Stock exchange	DonaldTrump	Breaking news
Dollar	Stocks	Obama	Day
Dolar	Stock	USA	Today
Euro	Trading	Turkey	Investing
UsdTry	Forex	Economy	Market

We have retrieved the tweets for 2020-14-05 to 12-11-2020. We mainly focused on the political incidents and the corresponding politicians' effects on the currency exchange. In focused hashtags: "Donald Trump" related hashtags have higher tweet volume and we mainly focused on these records. After 12-11-2020, we stopped collecting tweets because Donald Trump has lost the election. Even though he is still an influencer in the economy after that period, we only want to consider when he was in charge.

Retrieved tweets are raw data and do not have any meaning to machine learning model without their sentiments are processed. Therefore, we have extracted the sentiments of every tweet one by one. We have sentimentally analyzed all the tweets by using the NLP sentiment analysis tool 'Valence Aware Dictionary and Sentiment Reasoner (VADER)'. VADER is a lexicon and rule-based sentiment analysis tool that is specifically attuned to sentiments expressed in social media [9]. This sentiment analysis is giving us compound sentiment value. Since compound sentiment value is giving the sentiment of each tweet, we have used this compound value.

$$f(Sentiment) = \begin{cases} Positive, & compound\ score \leq 0,05 \\ Neutral, & compound\ score < 0,05 \\ Neutral, & compound\ score > -0,05 \\ Negative, & compound\ score \leq -0,05 \end{cases}$$

After analyzing the sentiments of all the tweets, we have clustered every tweet per hour. For every hour we have calculated the metrics for each hashtag shown in Table 2.

Table 2. Metrics for each hashtags.

Values	Ratios
Total Tweet Count	Compound average
Negative Tweet Count	Negative/Total Tweet Count
Neutral Tweet Count	Neutral/Total Tweet Count
Positive Tweet Count	Positive/Total Tweet Count

These eight different metrics are calculated for every hashtag. Like tweets, we also know that correlations between different types of currencies are other good predictors. Since we are trying to forecast the USD/TL currency exchange rate, we have used other types of currency value exchange rates as well. These are EUR/USD, USD/TRY, GBP/USD, USD/CHF, BTC/USD, USD/JPY, ETH/USD. We have collected the metrics for these currencies with the given specific 6 months.

We wanted to expand our features with stock market indices because we also know that stock market indices are often correlated with the economy of the country. For example, US Dollar Index or US30 Index will help us to predict the trend of the US Dollar, therefore they will help us to forecast the USD/TRY exchange rate. We have used the indices in the following Table 3.

Table 3. Stock market indices.

Index		
JPN.IDX (Japan)	DEU.IDX (Germany)	CHE.IDX (Switzerland)
GBR.IDX (United Kingdom)	HKG.IDX (Hong Kong)	FRA.IDX (France)
DOLLAR.IDX (Dollar)	USA30.IDX (United States)	NLD.IDX (Netherlands)
ESP.IDX (Spain)	IND.IDX (India)	PLN.IDX (Poland)
CHI.IDX (China)	EUS.IDX (Europe)	AUS.IDX (Australia)

To gather up the features which we mentioned above; there are 24 hashtags (H) and 8 metrics for each hashtag (MH), 7 currency rates (CR), 15 stock market indices

(SMI). Therefore, we have created a rich future set consisting of 214 different features as calculated in Eq. (1).

$$\textit{Total Feature Count} = H * MH + CR + SMI = 214 \quad (1)$$

Train/test data are generated from the records which are collected for 6 months. Generated dataset records are distributed as hourly, so the model tried to forecast the next hour. In this study we used Ridge, Gaussian Process, MLR, FTS as regression algorithms. There is also another option to solve these problems via different multi-layer perceptron (MLP) topologies.

4 Results and Discussion

We have analyzed and arranged dataset by grouping the records hourly. It is tested as a regression problem and split as train and test data. Last week of the data set is used as test data and the remaining part is used as training data. To get the best results, we have tuned hyperparameters and used the grid search technique. We have compared results with well-known performance measurement techniques such as Mean squared error (MSE). We have got promising results on regression algorithms as shown in Table 4:

Table 4. Mean squared error results.

Algorithm	All features	Currency exchange	Twitter hashtags	Stock market
Ridge	0,34%	5,11%	11,02%	0,86%
Gaussian Process Regression	0,53%	4,96%	9,76%	0,37%
Multiple Linear Regression	0,36%	4,22%	10,37%	0,69%
Multi-Layer Perceptron	7,62%	6,69%	10,06%	3,60%
Fuzzy Time Series*	1,80%	6,06%	7,46%	3,01%

*: In Fuzzy Time Series, more than 6 features are taking more than 1 h. Therefore, we limited the features as follows. (Currency: eurUsd, gbpUsd/Twitter Hashtags: trumpPosOverTotal, obamaNegOverTotal Stock: dollar, USA/All Features: eurUsd, gbpUsd, trumpPosOverTotal, obamaNegOverTotal, dollar, USA)

As shown in the Table 4 and Fig. 1. we have tested several algorithms and we found that Ridge is very successful and is the winner algorithm. MLR is following the Ridge algorithm and Gaussian Process has promising results as well. These specified algorithms can be considered as successful for forecasting the currency exchange rate as a regression problem. Results are also showing that it is possible to get better results when the feature sets are combined. Ridge and MLR are better this scenario and got the most successful results. Besides, we should note that of these three types of features stock market is the most valuable if they are exploited alone.

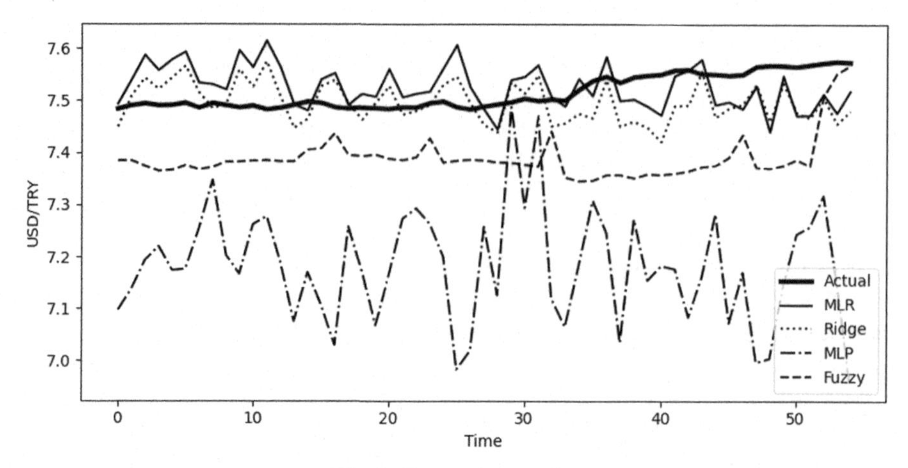

Fig. 1. Regression results.

Even our algorithms have low MSE and very good at forecasting the regression, they are coming one step back. Since the USD/TRY rate is volatile at the time range we have worked, we are not very good at predicting the exact trend of currency at that time tick. This volatility even resulted in the resignation of the minister of economy.

When we compared regression algorithms with Multi-Layer Perceptron, we have seen that some regression algorithms are performing better than MLP. This could happen because we have tested and verified too many hyperparameters on the tuning stage with regression algorithms. However, we couldn't tune our MLP topology that much because even basic network topology takes much more time than regression algorithms and fine-tuning require much more time for sensitive hyperparameters like slow learning rate, too many epochs, etc.

5 Conclusion and Future Work

In this study, we conclude that combining different type of features like stock market indices and Twitter sentiment analysis along with cross-currency exchange rates can help to achieve better regression results and can be used to get the insight what will be the rate of the currency exchange in next iteration. Instead of directly analyzing time-series data, Twitter data analysis will add slight value to the forecasting process and can be used as an additional helper feature.

Currently, we have given considerable time to regression algorithms hyperparameter fine-tuning. But we couldn't give this much importance to MLP hyperparameter tunings since it takes too much time to identify the best parameters. It can be tried to tune hyperparameters even though if it needs much more time than available for once and then these parameters can be tested with different datasets on different timelines.

References

1. Osatuyi, B.: Information sharing on social media sites. Comput. Hum. Behav. **29**(6), 2622–2631 (2013)
2. Roll, R.: Industrial structure and the comparative behavior of international stock market indices. J. Financ. **47**(1), 3–41 (1992)
3. O'Connor, B., Balasubramanyan, R., Routledge, B., Smith, N.: From tweets to polls: linking text sentiment to public opinion time series. In: Proceedings of the International AAAI Conference on Web and Social Media, vol. 4, no. 1 (2010)
4. Pak, A., Paroubek, P.: Twitter as a corpus for sentiment analysis and opinion mining. LREc **10**(2010), 1320–1326 (2010)
5. Bollen, J., Mao, H., Zeng, X.: Twitter mood predicts the stock market. J. Comput. Sci. **2**(1), 1–8 (2011)
6. Ozturk, S.S., Ciftci, K.: A sentiment analysis of twitter content as a predictor of exchange rate movements. Rev. Econ. Anal. **6**(2), 132–140 (2014)
7. Machová, V., Mareček, J.: Estimation of the development of Czech Koruna to Chinese Yuan exchange rate using artificial neural networks. In: SHS Web of Conferences, vol. 61, pp. 01012. EDP Sciences (2019)
8. Galeshchuk, S., Mukherjee, S.: Deep networks for predicting direction of change in foreign exchange rates. Intell. Syst. Account. Finance. Manage. **24**(4), 100–110 (2017)
9. Hutto, C., Gilbert, E.: Vader: A parsimonious rule-based model for sentiment analysis of social media text. In: Proceedings of the International AAAI Conference on Web and Social Media, vol. 8(1) (2014)
10. Alkaya, A.F., Gültekin, O.G., Danacı, E., Duman, E.: Comparison of computational intelligence models on forecasting automated teller machine cash demands. J. Multiple Valued Logic Soft Comput. **35**, 167–193 (2020)

A Fuzzy Rule-Based Ship Risk Profile Prediction Model for Port State Control Inspections

S. M. Esad Demirci[1,2](✉), Kadir Cicek[3], and Ulku Ozturk[4]

[1] Maritime Vocational School, Sakarya University of Applied Sciences, Sakarya, Turkey
smedemirci@subu.edu.tr
[2] Maritime Transportation Engineering, Istanbul Technical University, Tuzla, Istanbul, Turkey
[3] Marine Engineering Department, Istanbul Technical University, Tuzla, Istanbul, Turkey
cicekk@itu.edu.tr
[4] Turkish Naval Forces, Ankara, Turkey
ozturkul@itu.edu.tr

Abstract. Maritime transport is the backbone of the international trade, more than 80% of global freight transport is carried by ships on the seas. However, in a complex and high-risk environment at sea, substandard ships in maritime transport causes serious accidents and hence bring out many threats to the maritime industry. As a result, with the aim of detection and elimination the substandard ships, port state controls (PSC) have been developed to inspect the ships respect to their Ship Risk Profile (SRP). SRP helps to determine the ship's priority for PSC inspections through categorizing the ships in high risk, standard risk or low risk using various generic and historic parameters. In this study a new model is proposed to predict SRP by using fuzzy clustering analysis (FCA) and fuzzy rule-based classification system (FRBS) different from the standard calculation of the SRP in the PSC regimes. The proposed model is structured on five different parameters which are ship type, ship flag, ship age, deficiency number and detention. In the proposed model, to predict the SRP, 53788 ship inspection data belonging to the parameters has been analyzed gathered from the Paris MoU online database between the years of 2017 and 2020. The results obtained with the proposed approach help to identify the risk profile for the ship targeted at almost certain risk level. As a result of the study, it is aimed to provide decision supports for port state control (PSC) officers to detect the most appropriate/risky ship for the inspection.

Keywords: Port State Control (PSC) Inspections · Ship Risk Profile (SRP) · Fuzzy Rule-Based Classification System (FRBS) · Fuzzy Clustering Analysis (FCA)

1 Introduction

According to the review report on maritime transport published by United Nations Conference on Trade and Development (UNCTAD) in 2020 [1], maritime transport is the most important type of transport in global freight transport with a volume of 80% of

C. Kahraman et al. (Eds.): INFUS 2021, LNNS 308, pp. 498–505, 2022.
https://doi.org/10.1007/978-3-030-85577-2_59

global freight transport. With the unique features of the maritime transport such as economic, safe, secure, efficient, and environmentally friendly, that is why maritime transport is the most preferred type of transport. The huge transport capacity in maritime transport triggers the construction of larger ships with greater capacity. However, the significant increase in the transport volume and the enlargement in the size and capacity of the ships have led to an increase in the number of accidents in maritime transport. For instance, according to the "*Annual overview of marine casualties and incidents 2020*" report published by European Maritime Safety Agency (EMSA) in 2019, between the years 2011 and 2018, 23073 marine casualties and incidents occurred and 2514 ships were involved in these accidents [2].

In order to prevent accidents and to ensure safety in maritime transport, a strict regulatory framework has been built by the International Maritime Organization (IMO). With the established regulatory framework by the IMO, it is aimed to ensure safe, secure, environmentally friendly and efficient maritime transport. Respect to the strict regulatory framework in maritime transport, it turns into a mandatory issue to establish periodic inspection systems for ships to determine whether the ships comply with international regulations and to prevent potential maritime accidents that may occur due to non-compliance to regulations. With the establishment of the various periodic inspection systems, ship owners have entered into an intensive inspection regime where the ships are inspected by national administrative authorities (flag states), classification societies, cargo owners, ship insurance companies, and port states. In particular, this study focuses on the inspections carried out by the port states with the aim of providing decision supports for the relevant inspections puts forward. Port states have the right to inspect ships arriving at their own ports within the framework of the authorities granted to them by international regulations. In port state control (PSC) inspections, it is examined in detail whether the ships arriving at the ports meet the requirements of international regulations. However, it is not possible for port states to be able to inspect all ships coming to their ports in terms of using their human resources in an appropriate manner. For this reason, respect to the risk profiles of the ships arriving at the ports, the ships with a high-risk profile are determined as primarily ships to be inspected. In this regard, a fuzzy rule-based model has been proposed for determining the ship risk profile (SRP) in the study.

Within the scope of the aforementioned information, the rest of the study is organized as follows; In Sect. 2, the general information about port state controls is presented. The structure of the proposed SRP prediction model is expressed in Sect. 3. In Sect. 4, the application of the proposed SRP prediction model is presented with the analysis of the inspection data between the years 2017 and 2020 obtained from the database of the Paris Memorandum of Understanding. The study is completed with the conclusion section.

2 Port State Control

On the account of effective and sustainable inspection system, various port states in the same region have coalesced and signed regional agreement on PSC, so called Memorandum of Understanding (MoU) on PSC. The first MoU was signed by fourteen European countries at a Ministerial Conference held in Paris, France in 1982. After the Paris MoU, eight more regional agreements have been signed around the world such as Vina del Mar Agreement (1992), Tokyo MoU (1993), Caribbean MoU (1996), Mediterranean MoU (1997), Indian Ocean MoU (1998), Abuja MoU (1999), Black Sea MoU (2000), and Riyadh MoU (2004). In addition to these MoUs, the United States Coast Guard (USCG) performs PSC inspections in its own coastal region. In PSC inspections, the ships are inspected within the framework of critical issues such as certificates and documents, overall condition of the ship, qualification of the crew and familiarity with duty, and preparedness in case of emergencies. Therefore, sufficient time should be allocated for the ships to be inspected effectively within the framework of the relevant issues. At this insight, the ships are prioritized according to their SRP to ensure an accurate time management for the inspection and to determine the ships that really need to be inspected are inspected. To accurate prediction of SRP, PSC MoUs uses a different prediction models [3]. For instance, first SRP prediction model was introduced by Paris MoU in 2011 so called New Inspection Regime (NIR). In following years, various SRP prediction models were adopted by Tokyo MoU (2014), Black Sea MoU (2016), and Indian Ocean MoU (2018) [4].

Any error that may arise in the SRP prediction means that substandard ships continue to freight transport without being detected in international waters, which can be the biggest threat to the safe, secure and environmentally friendly maritime transport. Therefore, the prediction of the SRP with the highest accuracy is quite important issue from the viewpoint of the detection of the substandard ships in maritime transport. Within this scope, it is aimed to develop an SRP prediction model that can be used by decision makers in the PSC MoUs.

3 SRP Prediction Model

In this section, studies conducted in the literature regarding the prediction model of the SRP are reviewed briefly before the description of the SRP prediction model. In the review, a limited number of studies on SRP prediction are found. Among these studies, Knapp and Van de Velden's [5] paper on visualization of SRPs can be cited as an exemplary study. In the study, the data obtained from various sources between 1977 and 2008 were analyzed and the changes in SRPs over the years were visualized using the correspondence analysis method. The study, which can be shown as another example, was conducted by Heji and Knapp [6] in 2011 on the analysis of ship specific and company specific risk levels. In the study, accident reports published by The Australian Maritime Safety Authority (AMSA) were analyzed. For the analysis of accident reports, various statistical methods were used in the study. Apart from these studies, it can be shown as another exemplary study in Knapp's [7] study in 2013. In the study, Knapp analyzed the total risk exposure levels of ships depending on the accident types by analyzing the accident reports between 2005 and 2010 published by AMSA with a binary logistic model. Different from prediction of the SRPs in the literature, there are studies conducted to determine the collision risk levels of ships [8, 9].

In this study, an SRP prediction model is proposed to detect ships that can be considered risky for port state inspections in contradistinction to the studies in the literature. The proposed SRP prediction model is structured on five parameters which are ship age, ship type, flag state performance, number of deficiencies and detention. The flow chart of the proposed model is presented in Fig. 1.

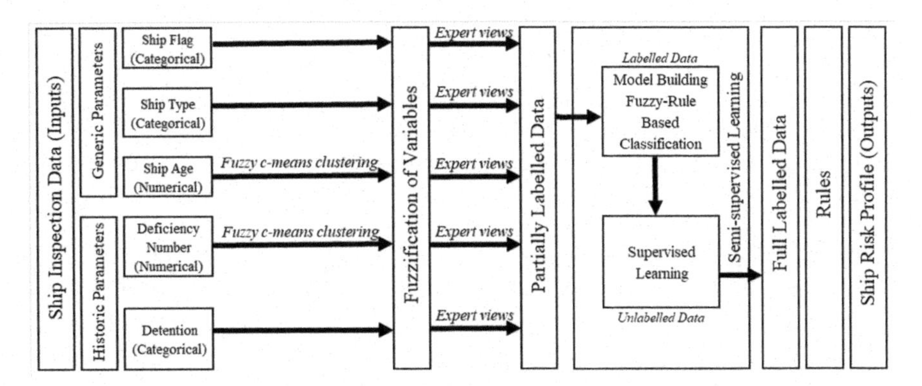

Fig. 1. The flow chart of the model

In the initial stage of the proposed model, the type of the parameters is identified. *Ship flag, ship type and detention* parameters are defined as categorical type and ship age and deficiency number parameters are defined as numerical type. Followingly, in order to determine the clusters of the numerical parameters, fuzzy c-means clustering algorithm is used. Fuzzy C-Means (FCM) clustering algorithm is introduced by Dunn [10] in 1973 and improved by Bezdek [11] in 1981. The basic representation of FCM clustering algorithm is structured on minimization of the following function.

$$J_m = \sum_{i=1}^{N} \sum_{j=1}^{C} u_{i,j}^{m} \left\| x_i - c_j^2 \right\|, 1 \leq m \leq \infty \tag{1}$$

Where $u_{i,j}$ is the membership degree of x_i in the cluster *j*, x_i is the *i* th of the d-dimensional measured data, c_j is the d-dimensional center of the cluster, u_j is the center of the cluster *j*, and *m* is the fuzzifier.

With the determination of the clusters of each parameter, sample rules are established. Each established rules are matched with SRPs according to the experts' judgements and we obtained partially labeled data. The obtained partially labeled data is used to define SRPs of unlabeled data with a fuzzy rule-based classification model. In this model, each fuzzy IF-THEN rule consist of antecedent linguistic values determined by a grid-type fuzzy partition from training data and a single consequent class with certainty grades as defined the dominant class in the fuzzy subspace corresponding to the antecedent part. This model developed by Ishibuchi and Nakashima [12] in 2001. The basic representation of fuzzy rule base classification with certainty grades (CF_j) is as follows,

$$R_j : \mathit{If}\ x_1\ \mathit{is}\ A_{j1}\ \mathit{and}\ \ldots\ldots..\ \mathit{and}\ x_n\ \mathit{is}\ A_{jN}\ \mathit{then\ Class}\ C_j\ \mathit{with}\ CF_j\ ,\ j = 1, 2, \ldots, N \tag{2}$$

where x = (x_1,x_2,...x_N) is an *n*-dimensional pattern vector, A_{ji} is linguistic value, C_j is a consequent class based on SRP given by experts, *N* is the number of fuzzy rules, and CF_j is certainty grade for fuzzy rule R_j.

As a last step, the performance score of constructed model is evaluated for classification of unlabeled data and we obtained full labeled data and a rule table.

4 Application of the Proposed Model

In the proposed model, the 53788 ship inspection data obtained from the Paris MoU database (THETIS) between the years 2017 and 2020 for are used. Following with the gathering of the data for the relevant parameters, in the proposed SRP prediction model, initially the ship age and deficiency number parameters are clustered with FCM clustering algorithm explained in Sect. 3.

Fuzzy clusters for the parameters of ship age and deficiency number are shown in Fig. 2. While ship age is clustered in four categories, deficiency number is clustered in five categories according to mean squares of the clusters. The clusters of the categorical type parameters (ship type, ship flag, detention) are determined in line with Paris MoU SRP criteria. The clusters of each parameter are presented in Table 1.

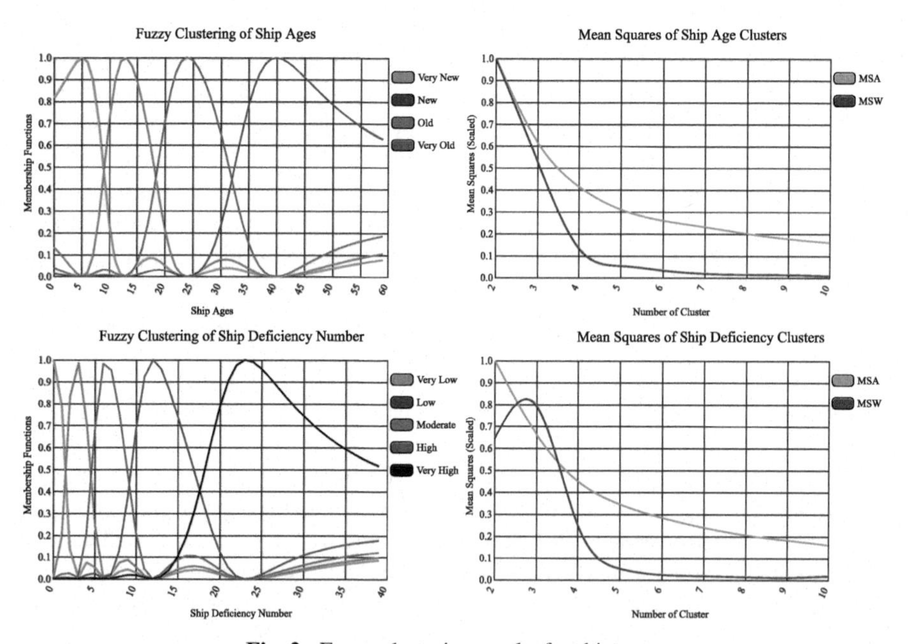

Fig. 2. Fuzzy clustering results for ship ages

Since the ship inspection data suffer from dependent variable (output), the SRP prediction is made in line with the defined parameters by taking the opinions of ten experts with more than ten years of experience in the field of ship inspection. To establish the fuzzy rules on prediction of SRP, the defined cluster in Table 1 is used. The fuzzy rules and the aggregated judgements of the experts for each relevant rules are presented in Table 2. As seen from the Table 2, SRP is categorized under five categories; very low risk-VLR, low risk-LR, medium risk-MR, high risk-HR, very high risk-VHR.

Table 1. Clusters of parameters

Parameters	Clusters
Ship Flag	White List, Grey List, Black List, Other
Ship Type	Bulk Carrier, Container, Passenger, Cargo, Tanker, Other
Ship Age	Very New, New, Old, Very Old
Deficiency	Very Low, Low, Moderate, High, Very High
Detention	Detained, Not Detained

Table 2. Established fuzzy rules to predict ship risk profile as per experts' judgements

Rules	Ship Flag	Ship Type	Ship Age	Deficiency	Detention	SRP
Rule 1	White List	Bulk Carrier	New	Very Low	Not Detained	LR
Rule 2	White List	Bulk Carrier	Very New	Very Low	Not Detained	VLR
Rule 3	White List	Bulk Carrier	Very New	Low	Not Detained	LR
Rule 4	White List	Bulk Carrier	New	Moderate	Detained	MR
Rule 5	White List	Cargo	Very New	Very Low	Not Detained	VLR
Rule 6	White List	Cargo	New	Low	Not Detained	LR
Rule 7	Black List	Cargo	Old	Very High	Detained	VHR
Rule 8	Black List	Cargo	Very Old	Very High	Detained	VHR
Rule 9	White List	Cargo	Very Old	High	Detained	HR
Rule 10	White List	Container	Very New	Very Low	Not Detained	VLR
Rule 11	White List	Container	New	Very Low	Not Detained	LR
Rule 12	White List	Tanker	New	Very Low	Not Detained	LR
Rule 13	White List	Tanker	Very New	Very Low	Not Detained	VLR
Rule 14	White List	Tanker	New	Low	Not Detained	MR

When the labeled 14 rules with graded by experts' judgements are mapped to the initial dataset, we obtained labeled and unlabeled datasets. In order to label unlabeled data, fuzzy rule-based classification model briefly explained in Sect. 3 is used. In this model, each fuzzy rule consists of antecedent linguistic values on ship parameters (inputs) and a single consequence class (output) for ship risk profile obtained from experts' judgements. Before application of the fuzzy rule-based classification model, we partition our dataset into training and test sets with the ratio of 70% and 30%, respectively. After the partitioning of dataset, FRBC model is trained and the obtained results are compared with the test data. According to the comparison, it is found that the error rate of trained data set is 23%. Although the performance of the model is not perfect, we decide to continue with the unlabeled dataset because of the sensible result of the model.

As a final step, we use our trained fuzzy rule base classification model to label the unlabeled dataset. Sample rules obtained from unlabeled dataset are presented in Table 3.

Table 3. Some of the rules obtained from unlabeled dataset

Ship Flag	Ship Type	Ship Age	Deficiency	Detention	SRP
Black List	Cargo/Container	Old	High	Detained	VHR
White List	Bulk Carrier/Passenger	New	Low	Not Detained	VLR
White List	Cargo/Container	New	Very High	Detained	HR
White List	Cargo/Container	Old	Very Low	Not Detained	LR
White List	Tanker	New	High	Not Detained	VLR
Grey List	Other	Old	Very Low	Not Detained	MR
Black List	Cargo/Container	Very New	Low	Not Detained	MR
Grey List	Cargo/Container	Old	Very Low	Not Detained	MR
Black List	Bulk Carrier/Passenger	Old	Very Low	Not Detained	MR
Black List	Cargo/Container	Very Old	Low	Detained	VHR

Following with the identification of the new rules by labeling unlabeled dataset, the obtained rules are discussed with experts. The experts emphasize that the rules are quite consistent and meaningful. Also, the experts express their agreement that the developed model would be highly beneficial for decision makers.

5 Conclusion

In the study, we propose a novel model to predict SRP with the help of FRBS. With the proposed model, it is aiming to increase the accuracy of the prediction of the riskiest ships for port state inspections. With the prediction of the ships to be inspected with a high accuracy rate, the detectability of the substandard ships will increase significantly. In the proposed model, semi-supervised machine learning including a fuzzy rule-based classification model is used to label the unlabeled dataset with the help of labeled dataset derived by experts' judgements. The proposed model provides high accuracy prediction for SRP respect to the standard SRP calculation models developed by the PSC MoUs. Additionally, it is possible to increase the prediction accuracy rate with the increasing of the number of parameters used and data volume analyzed in the model.

The proposed model provides reasonable decision support for port state control officers at the selection of riskiest ships to be inspected. Additionally, the study contributes to literature with concentrating on the prediction of the ship risk profile for port state inspections. Also, this study contributes by providing analytic method integrating fuzzy rule-based system with semi structured machine learning approach. Furthermore, the results obtained as a result of the study provide the experts with a perspective that enables them to be disseminated among other types of inspection in the maritime field.

As a further study we are planning to adopt additional parameters to decrease the error rate and to increase the prediction accuracy rate of the model.

Acknowledgements. The article is produced from PhD thesis research of S. M. Esad DEMIRCI entitled "Safety Based Intelligent Ship Inspection Analytics for Maritime Transportation" which has been executed in a PhD Program in Maritime Transportation Engineering of Istanbul Technical University Graduate School.

References

1. UNCTAD: Review of maritime transport 2020. United Nations Publications, New York, USA (2020)
2. EMSA Homepage. http://www.emsa.europa.eu/newsroom/latest-news/item/4266-annual-overview-of-marine-casualties-and-incidents-2020.html. Accessed 7 Jan 2021
3. Emecen Kara, E.G., Okşaş, O., Kara, G.: The similarity analysis of port state control regimes based on the performance of flag states. Proc. Inst. Mech. Eng. Part M J. Eng. Marit. Environ. **234**(2), 558–572 (2020)
4. Zheng, L.: The effectiveness of new inspection regime on port state control inspection. Open J. Soc. Sci. **8**(8), 440–446 (2020)
5. Knapp, S., Van de Velden, M.: Visualization of ship risk profiles for the shipping industry. ERIM Report Series Reference No. ERS-2010–013-LIS. https://ssrn.com/abstract=1587553. Accessed 22 Jan 2021
6. Heij, C., Knapp, S.: Risk evaluation methods at individual ship and company level. Econometric Institute Research Papers EI2011–23, Erasmus University Rotterdam, Erasmus School of Economics (ESE), Econometric Institute. https://ideas.repec.org/p/ems/eureir/25603.html. Accessed 17 Jan 2021
7. Knapp, S.: An integrated risk estimation methodology: Ship specific incident type risk. Econometric Institute Research Papers EI 2013–11, Erasmus University Rotterdam, Erasmus School of Economics (ESE), Econometric Institute. http://hdl.handle.net/1765/39596. Accessed 15 Jan 2021
8. Gang, L., Wang, Y., Sun, Y., Zhou, L., Zhang, M.: Estimation of vessel collision risk index based on support vector machine. Adv. Mech. Eng. **8**(11), 1–10 (2016)
9. Ozturk, U., Birbil, S.I., Cicek, K.: Evaluating navigational risk of port approach manoeuvrings with expert assessments and machine learning. Ocean Eng. **192**, 106558 (2019)
10. Dunn, J.C.: A fuzzy relative of the isodata process and its use in detecting compact well-separated clusters. J. Cybern. **3**, 32–57 (1973)
11. Bezdek, J.C.: Pattern Recognition with Fuzzy Objective Function Algoritms. Plenum Press, New York (1981)
12. Ishibuchi, H., Nakashima, T.: Effect of rule weights in fuzzy rule-based classification systems. IEEE Trans. Fuzzy Syst. **9**(4), 506–515 (2001)

Forecasting Sovereign Credit Ratings Using Differential Evolution and Logic Aggregation in IBA Framework

Srđan Jelinek[1(✉)], Pavle Milošević[2], Aleksandar Rakićević[2], and Bratislav Petrović[2]

[1] Fidelity Information National Services, 11000 Belgrade, Serbia
srdjan.jelinek@fisglobal.com

[2] Faculty of Organizational Sciences, 11000 Belgrade, Serbia
{pavle.milosevic,aleksandar.rakicevic,
bratislav.petrovic}@fon.bg.ac.rs

Abstract. The sovereign credit rating is considered as a quantified assessment of country's economic and political stability. Due to its importance and increasing amount of available information, the sovereign credit rating is considered as a hot topic in the last few years. However, the models that predict the credit ratings used by the several big credit rating agencies are unavailable, and can therefore be considered as the black boxes. In this paper, we are tackling this problem of predicting sovereign credit ratings by proposing a hybrid model based on interpolative Boolean algebra (IBA) and differential evolution (DE). Namely, we aim to obtain a logical/pseudo-logical function in IBA framework using DE metaheuristic that could underline connections of chosen macroeconomic indicators and sovereign credit ratings. Such functions are easy to interpret and able to make a subtle fuzzy gradation among countries. Country's economic indicators together with credit ratings from 2000 to 2016 are used for the model training. Acquired model is further tested on the data for 2017 and 2018.

Keywords: Differential evolution · Interpolative Boolean algebra · Sovereign credit rating · Prediction

1 Introduction

Credit risk is a risk which arises from the possibility of a borrower failing to repay a loan or required payments, or meet its contractual obligations. It is determined by a borrower's ability to repay a loan according to its original terms. Two main subcategories of a credit risk are credit default risk and country risk. Credit default risk is defined either as debtor's inability to pay its loan obligations fully or partially or as debtor being more than 90 days past due fulfilment on any other credit obligation. On the other hand, country risk can be divided into two categories: transfer risk, which is defined as a risk of a sovereign state failing to make its foreign currency payments and sovereign risk, which is failure

C. Kahraman et al. (Eds.): INFUS 2021, LNNS 308, pp. 506–513, 2022.
https://doi.org/10.1007/978-3-030-85577-2_60

of meeting its obligations, for example, failure to honour the payments of the issued long-term bonds or other contracts.

A credit rating is a quantified assessment of a credit risk. It is used to evaluate the credit risk of an individual, corporation or a state/government, and therefore can be assigned to any of those in order to predict their ability of repaying debt or a likelihood of default. It can be determined and calculated by using internal measures and models, or by a credit rating agency which specifies in calculating credit rating for a given entity using both publicly available information and information provided by debtor itself, as well as non-public information which is obtained by a credit rating agency. A sovereign credit rating is a credit rating assigned to a country. Due to a large amount of available information and the importance and implications which a sovereign credit rating has on a country, including determining and quantifying its political and economic stability, a sovereign credit rating is usually determined by the trustworthy credit rating agencies, out of which the biggest and most important ones are Moody's, S&P and Fitch. Many practitioners and researchers address this problem using both traditional and computation intelligence methods [1, 9].

Given the complexity of determining sovereign credit rating and the secrecy of the models used by credit rating agencies, one can apply a machine learning or an optimization algorithm to derive the models from publicly available data, using both microeconomic and macroeconomic indicators. We propose the use of differential evolution (DE) and interpolative Boolean algebra (IBA) to extract the logical function (model) for forecasting sovereign credit ratings from publicly available data.

DE is a global optimization technique inspired by a biological evolution. Even though it does not guarantee the finding of an optimal solution, DE can quickly search vast space and find a reasonably good solution. Furthermore, DE is flexible and easy to implement, i.e., it is suitable for numerous types of problems. Detailed reviews of DE and its applications can be found in [3–5, 10]. IBA [12] is real-valued realization of the Boolean algebra (BA), which preserves all the laws of BA. Therefore, it is a (Boolean) consistent logical framework. IBA enables us to separate structure and value in any logical function (expression).

In this paper, we utilize DE algorithm to optimize the structure of a prediction model in IBA framework. The prediction model is a logical or pseudo-logical function that uses publicly available macroeconomic indicators (inflation, total reserves, GDP and gross savings) to predict sovereign credit ratings. Such function is able to make fine (fuzzy) gradation and is easy for analysis and interpretation. Optimization of the structure in IBA models is already done with variable neighbourhood search [8] and genetic algorithms [14], and this is the very first attempt to do so with DE.

This paper is organized as follows. In Sect. 2 we give brief insight into theoretical background: IBA logical framework and DE algorithm. Section 3 presents data used in this research. In Sect. 4 we explain the proposed approach to forecast sovereign credit ratings along with forecasting results. Finally, in Sect. 5 we present conclusions and ideas for future work.

2 Theoretical Background

The model for forecasting sovereign credit ratings is trained via differential evolution algorithm. The inputs for the model are atomic vectors which are generated as described in Sect. 2.1. The algorithm then searches for the optimal structure vector in order to minimize the model output. The output is the pseudo-logic aggregation function from IBA framework, which is represented as the matrix product of the atomic and structure vectors.

2.1 IBA Framework

The atomic vector in IBA framework [12, 13] is generated from the regular vector. Let regular vector $v = \{v_1, v_2, \ldots, v_n\}$, contain n elements. The atomic vector a, is created so that its every element represents a product, $\prod_{i=1}^{n} u_i$, where u_i can be either v_i or $1 - v_i$. Under the conditions that every element of the a should be unique, and that every combination of v_i and $1 - v_i$ should be present, it is trivial that the number of elements of the a vector is 2^n. For example, if $v = \{v_1, v_2\}$, an appropriate atomic vector is $a = \{v_1 \times v_2, v_1 - v_1 \times v_2, v_2 - v_1 \times v_2, 1 - v_1 - v_2 + v_1 \times v_2\}$. In general case the structure vector, x, is a vector with elements g_i, where $g_i = \{0, 1\}$. However, given that these vectors will be optimized through the DE algorithm, and given the nature of the DE algorithm, we have modified structure vector so that its every element is defined as $g_i = [0, 1]$, i.e., moving from the logical aggregation in IBA framework to IBA pseudo-logical aggregation [13]. By doing this, DE can now configure each element of the vector in the mutation phase to yield the best optimization results. Finally, the logic aggregation function in IBA framework is represented as a matrix product of the atomic and structure vectors, $a \times x^T$. This is the output from our model, and represents the numerical value of the sovereign credit rating.

2.2 Differential Evolution

Differential evolution is an algorithm proposed by Storn and Price [15] in 1995. The basic form of DE consists of 4 steps: initialization, mutation, crossover and selection. The last 3 steps are repetitive until an optimal solution is found or number of predetermined iterations is reached.

Initialization step starts with defining of a bounded d-dimensional search space. First, an initial generation of solutions is randomly generated, where each solution is represented as a d-dimensional vector, which is also referred to as a genome. The number of candidate solutions in the generation, m, is a user-defined parameter. Let t be the maximal number of iterations, starting with 1 for the initial generation. Every generation will then have a set of m genomes, where every genome is represented as: $x_i^j = \left\{x_{i,1}^j, x_{i,2}^j, \ldots, x_{i,d}^j\right\}$, where $i = \{1, 2, \ldots, m\}$ and $j = \{1, 2, \ldots, t\}$. When the initial generation is created, an iterative process begins, starting with the mutation.

Mutation is a process where for each target vector from the current generation a respectable mutation vector is created using a predefined mutation algorithm. There are numerous mutation strategies in literature [2], i.e., DE/best/1/bin, DE/rand/1/exp, etc. One of the simplest and most used algorithms is DE/rand/1/bin. It creates a mutation vector v_i^j as: $v_i^j = x_{i,R_1}^j + F \cdot \left(x_{i,R_2}^j - x_{i,R_3}^j\right)$, where R_1, R_2 and R_3 are randomly generated mutually exclusive integers between 1 and d. Parameter F is called mutation parameter, and it is a positive predetermined scaling factor, usually given in the unit interval.

Crossover represents the mixing of every target vector with its corresponding mutant vector. Two mostly used crossover methods are binomial and exponential [4]. In this paper a binomial crossover is used because the exponential method is effective mostly in cases where there is a strong correlation between input vectors [17]. For every pair of elements $\left\{x_{i,k}^j, v_{i,k}^j\right\}$ from the target and mutant vectors from current generation, a random integer, P_k, between 0 and 1 is generated. Each element of the crossover vector, u_i^j, is generated as: $u_{i,k}^j = \begin{cases} v_{i,k}^j, P_k \leq Cr \text{ or } k = RI \\ x_{i,k}^j, P_k > Cr \end{cases}$, where Cr is crossover rate and RI is the randomly generated integer between 1 and d. Crossover step ensures covering as much search space as possible by increasing the diversity of the newly generated vectors.

The aim of **selection** step is to decide whether to replace the target vector, x_i^j, with a crossover vector, u_i^j, or not, according to results for the optimization function. The two main techniques of updating current generation with newly generated vectors are synchronous and asynchronous [15]. Synchronous technique, used in this paper, updates all the members of the current generation simultaneously, while the asynchronous method updates the vectors as soon as their crossover pairs have been generated.

There are many modifications of the DE algorithm, which are applied to all sorts of different problems and optimizations. Some of the most popular DE modifications are SaDE [11], EPSDE [7], CoDE [18], JADE [6], MPEDE [19], SHADE [16], etc.

3 Data

To forecast sovereign credit ratings, we use economic indicators publicly available at the World Bank database[1]. The indicators were taken from the Economy & Growth sector, under assumption that they have the highest impact on the sovereign credit rating. The following indicators are chosen:

- *Annual inflation of consumer prices (%)*, which reflects the change in the price of a basket of goods and services that are consumed by an individual or a household.
- *Total reserves ($), including gold*, representing a sum of all deposits that a country may count in its legal reserve requirements, which are used as a protection of the depositors.
- *GDP per capita ($)*, which is a measure that shows overall quality of life and the benefit that each citizen has from the country's economy.
- *Gross savings (% of GDP)*, which measures the amount of resources that can be invested in the capital assets.

[1] https://data.worldbank.org.

These indicators were chosen in order to cover the most important aspects of the sovereign economy, and at the same time to keep the model simple.

The sovereign credit ratings were obtained from the Country Economy database[2]. We use rating from the Fitch credit rating agency due to the smallest number of gaps in the data over the time period used for training and testing. To solve the problem of the data gaps, we perform a data clearing and filling the blank values. If the data for an indicator is missing, a flat left, flat right or linear interpolation is used to fill the missing values, depending on whether the missing values are at the beginning of the data set, at the end, or in the middle of the time series.

The data set includes macroeconomic indicators and credit ratings for 83 countries in the period 2000–2018. Countries are divided into 4 geographic regions: Europe (39 countries), North and South America (15 countries), Asia (15 countries) and Africa (14 countries).

4 Proposed Approach

4.1 Data Pre-processing

The rating definitions given by Fitch credit rating agency are divided into categories, with AAA being the highest credit quality and DDD the lowest one. Categorical ratings are transformed into numerical using a piecewise linear function.

As mentioned in Sect. 3, data on each country consists of 19 points of yearly data, from 2000 to 2018. Because the two indicators are given in percentages (inflation consumer prices and gross savings) and the other two are given in dollars (total reserves, including gold and GDP per capita), we have first standardized all data points so that they could contribute evenly to the proposed model. Subsequently, the data is transformed to the unit interval, i.e., each standardized point is normalized using the *min-max* normalization. Finally, the entire data set contains 1577 quintuplets of points, where each quintuplet is represented as a quadruplet of four indicator values, and the output, which is the sovereign credit rating for the particular year, being the fifth point. For model training, we have used quintuplets from 2000 to 2016, while the rest of data is used in the model testing phase.

4.2 Model Training

The steps for model calibration are the following:

1. Generate an atomic vector, A_h, for every input vector $\{a_h, b_h, c_h, d_h\}$, where $h = \{1, 2, \ldots, s\}$ and $s = 1411$. Given that the size of the input vector is 4, the size of the atomic vector is $d = 2^4$.
2. Generate an initial population of structure (target) vectors of the size d, S_i^1, where $i = \{1, 2 \ldots, m\}$, and where $m = 1000$. Each value of the structure vector is set to a random value between 0 and 100.

[2] https://countryeconomy.com.

3. For every generated structure vector of the first generation, S_i^1, an objective function to be minimized is generated as: $\sum_{h=1}^{s} \left(A_h \times S_i^T - SCR_h\right)$, where SCR is the real sovereign credit rating. The vector with the lowest objective function is memorized as the current best solution.
4. Perform mutation and crossover operation. Values $F = 0.25$ and $CR = 0.25$ are obtained empirically.
5. Keep the vector that yielded smaller objective function for the next generation.
6. Update current best solution, if possible, from the new set of generated vectors from the next generation.
7. Repeat steps 4–6 until the objective function reaches zero, or until the maximum number of iterations (e.g., 1 million) is performed.

4.3 Model Evaluation and Results

After model training, the vector representing the best solution is used to forecast the sovereign credit ratings for every country for 2017 and 2018. Overall mean squared error (MSE) of 166 predictions (with numerical values from 0 to 100) on the test set is 376.04, that is of the same order of magnitude as the MSE on the training set.

The summarized real and forecasted sovereign credit ratings for each country for 2017 and 2018 are presented via histograms in Fig. 1.

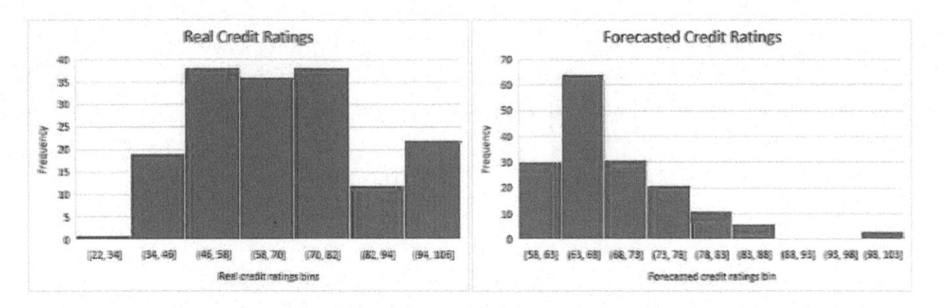

Fig. 1. Distribution of real and forecasted credit ratings.

As can be seen from Fig. 1, 40% of the forecasted credit ratings are up to 3 next-to ratings above or below the real rating, meaning that if the target rating is BBB, there is 40% chance that model will forecast the rating between A and BB. Also, 75% of the forecasted credit ratings are up to 7 next-to ratings. The model performed well when forecasting the credit ratings between BB+ and A−, because this group had the most entries in the testing phase. On the other hand, lower and higher ratings, especially the ones near the end of the spectrum, are not forecasted correctly, as can be seen from the Fig. 1, where there are no forecasted values below the BB+ rating. This is due to the optimization function minimizing mean square error between all the ratings equally, and by having less ratings near the end of the spectrum, the model being badly calibrated for such ratings. This can be fixed by introducing another input, such as the historical rating for each country. By nature, sovereign credit ratings don't change much over time, and there is a high probability of credit rating remaining the same as for the previous year.

5 Conclusion

In this paper we have used a one of the well-known metaheuristics to optimize pseudo-logic aggregation function in IBA framework for the forecasting of the sovereign credit ratings. Namely, DE is used to optimize the structural vectors with values 0 and 1, which would minimize the objective function. The parameters of the DE, population size, number of iterations, scaling and crossover rate are obtained empirically.

The model performed well when forecasting credit ratings from A− to BB+ (the middle of the credit ratings spectrum), but failed to generate correct forecasts on the lower end of the spectrum. The main reason for this is the model simplicity, where historical ratings haven't been taken into account, and therefore model was calibrated better for the most frequent cases - those in the middle of the spectrum. Considering the fact there is a strong correlation between consecutive yearly credit ratings of a country, with 80% or higher chance of credit rating remaining the same, a historical credit rating represents a valuable input for the model in the future researches.

To improve the forecasting results, the modified DE algorithms could be used. One way of improvement is to fix the model parameters throughout the evolution process, while they could be modified in a way to adjust the searching mechanisms - cover broader area of the possible solutions in the beginning, and then narrow it down in the later iterations, as model should then do the search in the proximity of the already nearly optimal solutions. One way to achieve this is to use the fuzzy logic for determining model parameters. Finally, other machine learning algorithms should be considered for the same task, so that a comparative analysis between DE-IBA model and them could be made.

References

1. Bennell, J.A., Crabbe, D., Thomas, S., Ap Gwilym, O.: Modelling sovereign credit ratings: neural networks versus ordered probit. Exp. Syst. Appl. **30**(3), 415–425 (2006)
2. Das, S., Amit K., Uday K.C.: Improved differential evolution algorithms for handling noisy optimization problems. In 2005 IEEE Congress on Evolutionary Computation, vol. 2, pp. 1691–1698. IEEE Press, New York (2005)
3. Das, S., Mullick, S.S., Suganthan, P.N.: Recent advances in differential evolution - an updated survey. Swarm Evol. Comput. **27**, 1–30 (2016)
4. Das, S., Suganthan, P.N.: Differential evolution: a survey of the state-of-the-art. IEEE T. Evolut. Comput. **15**(1), 4–31 (2011)
5. Dragoi, E.-N., Dafinescu, V.: Parameter control and hybridization techniques in differential evolution: a survey. Artif. Intell. Rev. **45**(4), 447–470 (2015). https://doi.org/10.1007/s10462-015-9452-8
6. Gong, W., Cai, Z., Ling, C.X., Li, H.: Enhanced differential evolution with adaptive strategies for numerical optimization. IEEE T. Syst. Man Cy. B. **41**(2), 397–413 (2011)
7. Mallipeddi, R., Suganthan, P.N., Pan, Q.K., Tasgetiren, M.F.: Differential evolution algorithm with ensemble of parameters and mutation strategies. Appl. Soft Comput. **11**, 1670–1696 (2011)

8. Milošević, P., Poledica, A., Dragović, I., Rakićević, A., Petrović, B.: VNS for optimizing the structure of a logical function in IBA framework. In: 6th International Conference on Variable Neighbourhood Search, p. 44. School of Information Sciences, Thessaloniki (2018)
9. Mora, N.: Sovereign credit ratings: guilty beyond reasonable doubt? J. Bank. Financ. **30**(7), 2041–2062 (2006)
10. Neri, F., Tirronen, V.: Recent advances in differential evolution: a survey and experimental analysis. Artif. Intell. Rev. **33**(1–2), 61–106 (2010)
11. Qin, A.K., Huang, V.L., Suganthan, P.N.: Differential evolution algorithm with strategy adaptation for global numerical optimization. IEEE T. Evolut. Comput. **13**(2), 398–417 (2009)
12. Radojević, D.: (0, 1)-valued logic: a natural generalization of Boolean logic. Yugoslav J. Oper. Res. **10**(2), 185–216 (2000)
13. Radojevic, D.: Logical aggregation based on interpolative. Mathware Soft Comput. **15**(1), 125–141 (2008)
14. Rakićević, A.: Adaptive fuzzy system for algorithmic trading: Interpolative Boolean approach. Ph.D. thesis (in Serbian). University of Belgrade, Belgrade (2020)
15. Storn, R., Price, K.: Differential evolution: a simple and efficient heuristic for global optimization over continuous spaces. J. Global Optim. **11**(4), 341–359 (1997)
16. Tanabe R., Fukunaga, A.: Success-history based parameter adaptation for differential evolution. In:2013 IEEE Congress on Evolutionary Computation, pp. 71–78. IEEE Press, New York (2013)
17. Tanabe, R., Fukunaga, A.: Reevaluating exponential crossover in differential evolution. In: Bartz-Beielstein, T., Branke, J., Filipič, B., Smith, J. (eds.) PPSN 2014. LNCS, vol. 8672, pp. 201–210. Springer, Cham (2014). https://doi.org/10.1007/978-3-319-10762-2_20
18. Wang, Y., Cai, Z., Zhang, Q.: Differential evolution with composite trial vector generation strategies and control parameters. IEEE T. Evolut. Comput. **15**(1), 55–66 (2011)
19. Wu, G.H., Mallipeddi, R., Suganthan, P.N., Wang, R., Chen, H.: Differential evolution with multi population based ensemble of mutation strategies. Inform. Sci. **329**, 329–345 (2015)

Predicting Performance of Legal Debt Collection Agency

Nilüfer Altınok[1](✉), Elmira Farrokhizadeh[1], Ahmet Tekin[2], Sara Ghazanfari Khameneh[1], Basar Oztaysi[1], Sezi Çevik Onar[1], Özgür Kabak[1], Ali Kasap[1], Aykut Şahin[1], and Mehmet Ayaz[1]

[1] Industrial Engineering Department, Istanbul Technical University, 34367 Macka, Istanbul, Turkey
{oztaysib,cevikse,kabak}@itu.edu.tr

[2] Management Engineering Department, Istanbul Technical University, 34367 Macka, Istanbul, Turkey

Abstract. In this competitive world, companies need to manage their arrears of debtors to survive. Big companies monthly face thousands of debt cases from their customers and if these debtor customers do not pay their debts within the specified period, these cases will enter the legal stage as legal debt collection that is handled by experienced lawyers. At the legal stage, the cases are sent to the contracted legal debt collection agencies to start a lawsuit. Therefore, assigning which case to which legal debt collection agency is a significant and critical issue in the company's success in getting its debts in the legal stage. So, this study aims to find the legal debt collection agency, which has more capability to close the assigned debt case by predicting case closing probability with machine learning techniques based on past historical data. To predict case closing probability, 8 machine learning algorithms, Catboost Classifier, Extreme Gradient Boosting Classifier, Gradient Boosting Classifier etc., are applied to the processed dataset. The results show us Catboost Classifier has the best accuracy performance 0.87 accuracy. Also, the results show us boosting type ensemble learning algorithms have better performance than other algorithms. Finally, we tune hyper-parameters of Catboost classifier to get better accuracy in the modeling and applied k-fold cross-validation for testing the model's testing stability.

Keywords: Legal debt collection · Machine learning · Predicting performance · Catboost classifier · Extreme gradient boosting classifier · Gradient boosting classifier

1 Introduction

Nowadays, unpaid debts are one of the most significant threats for many businesses in a competitive world regardless of the companies' size and operations. These unpaid debts increase and significantly affect cash flows, turnover, credit ratings, and even organizations' credibility that lead organizations to a bad economic situation [1]. Therefore, controlling and reducing these unpaid debts help companies economically. Systematic

C. Kahraman et al. (Eds.): INFUS 2021, LNNS 308, pp. 514–522, 2022.
https://doi.org/10.1007/978-3-030-85577-2_61

attempts to collect the unpaid debts are debt collection processes that highly depend on customers' willingness to pay and agent's negotiations. In practice, debts are collected by either the debt owner itself (using its name and own employees) or a third-party in the debt collection process with a collection phase and legal phase. In this process, collectors first try to persuade debtors to pay their debt. If they cannot get the debts from debtors, the process will go to the second phase as the legal debt collection phase. Legal debt collection phase comprises experienced lawyers who try to resolve debt cases through the judiciary. The large enterprises work with contractually law offices as legal debt collection agencies rather than expand internal resources. According to the impact of that legal debt collection agency's performance and experience on the closure or non-closure of debt cases, the legal debt collection agency's evaluation is crucial and debatable [2].

Once you have decided on a debt collection agency, you will immediately switch to an assessment mindset. Many companies are focusing on money and only care whether they collect the debt. However, there are many levels of performance for a debt collection agency. Factors such as relationship and service, collection rate, collection time, collection fees, value-added data, flexibility, transparency, and brand representation are used to evaluate debt collection institutions' performance [3]. The purpose of determining legal debt collection agency's performance is to measure objectively and periodically to what extent legal debt collection agencies achieve their goals. An organization's performance measurement can evaluate inputs, internal processes, procedures, outputs and results. Thus, by analyzing the indicators, the right decision can be made, limited resources that can create continuous improvement can be developed effectively and targets can be given to individuals. In the literature, multi-criteria decision-making techniques such as Cognitive Maps, Regression Analysis and Artificial Neural Networks and Analytical Hierarchy Process are used in performance evaluation [4]. Big companies monthly face thousands of subscribers' unpaid debts, which significantly affects the companys' economic situation. After failing in the collection phase, these debt cases enter the legal phase and must be handled by experienced lawyers in contracted legal debt collection agencies. To decrease the debt costs of big companies and increase the success in cashing their debts, evaluating the performance of contracted legal debt collection agencies is crucial and significant. Therefore, in this regard, we want to predict whether the case will be closed by the legal debt collection agency, based on past data, and using machine learning methods to avoid assigning a case to an office that can not close that case file. For this goal, we first provide a proper data frame that comprises effective features that significantly impact legal debt collection agency's performance and use machine learning methods to classify legal debt collection agency's performance. In this article, we use 8 different classification algorithms such as traditional and boosting type ensemble learning algorithms. Rest of the paper is organized as follows: In section two, relevant works with our paper in literature will be reviewed. In Sect. 3, our proposed methodology about predicting each legal debt collection agency's performance in assigning debt cases will be defined as detailed. In the last section, results of various machine learning models will be discussed and compared.

2 Literature Review

Our principal purpose in this study is to estimate the performance of contracted legal debt collection agencies to decide which agencies should be assigned to the cases in the legal process that will be initiated for unpaid subscriber invoices, even though 120 days have passed since the invoice date. For this purpose, when we review related studies in the literature within the framework of debt collection and debt collection agencies, we see mostly studies about customer payment behavior classification, late customer payments prediction, and credit scoring models. In [5] and [6], the researchers developed late payment prediction models for debt collectors. In [5] researchers used four classification algorithms, i.e., classification tree, random forests, ANN, and SVM, and their hybrid approach, to estimate the probability of repayment and late payment amount for customers with outstanding debt. In [6], the researchers created a late payment prediction system by combining data mining technology with a domain-driven data mining strategy to estimate a telecom provider's customers who are most likely to pay their bills later than the due date. They analyzed the data from fixed-line users by utilizing association rules, clustering, and decision trees. Similar to previous studies, in another study [2] that develops methods that will enable debt collection agencies to collect more effectively debt, the researchers formulated a Markov decision process to decide which debtor should be to call prioritized for a debt collection agency. And they approximated this process with machine learning methods based on historical data. With this approach, the estimated repayment probabilities for each debtor were calculated and it was provided to reveal the marginal increase of this probability in each extra call to the debtor. In two other similar studies [7] and [8], the researchers established predictive models about whether customer invoices will be paid on time within the context of accounts receivable applications. In [7], the researchers created a model that calculates the probability score of an invoice is overdue. In this model, they used machine learning to predict the status of invoices (late or on time) with high accuracy; to improve the accuracy of the model, they have compared five different classification models for the selection of the most appropriate among historical and temporal features. In [8], the researchers presented a supervised learning approach and the corresponding results to predict whether an invoice will be paid on time. Under this approach, they developed a set of aggregated features that capture customers' past payment behavior. By comparing four different classification algorithms, they built a typical supervised classification model. Finally, they improved prediction accuracy for high-risk invoices by using cost-sensitive learning. In [9], researchers classified customer insolvency for an internet provider company. In the model they established for this classification, customers were divided into segments using the k-means algorithm according to their payment behavior, and the customer level was measured, and then each customer segment was classified according to their ability to pay with the C4.5 classification algorithm. In another three articles, researchers have developed credit scoring models by using certain classification methods and effectively preprocessing the data or compared the classification methods that can be used in the credit scoring system in terms of performance. In [10], the researchers explored the predictive power of some of the more popular classification techniques currently in use for score carding, paying particular attention to predicting the propensity of borrowers. In this study, in which nine classification methodologies for ten different data sets were compared, researchers revealed statistical evidence that Generalized Additive Models (GAM) with a Generalized Extreme Value (GEV) link function outperformed logistic regression and some popular machine learning techniques. In [11], the researchers compared how different feature selection methods for the credit scoring model perform on a single real data set. Using four classification algorithms, they illustrated how four feature selection methods as ReliefF, Correlation-based, Consistency-based, and Wrapper algorithms improve three aspects, i.e., model simplicity, model speed, and model accuracy, of the performance of scoring models. In [12], the researchers tried various classification approaches to credit scoring and also evaluated some preprocessing methods to overcome skewed data sets. In the last study [13], the researchers used classification methods for accurate churn prediction in the telecommunications industry and also worked on data sampling and preprocessing. In their study, they observed that in addition to combinations of various sampling, feature selection, and classification methodologies. So as reviewed in this section, most of the articles focus on customer payment behavior prediction in the debt collection process but in this article, we want to look at this issue from another perspective which is legal debt collection agency's performance prediction. In order to make a more accurate prediction of the problem, we will predict the model with several machine learning algorithms and compare the results.

3 Proposed Methodology

The dataset which is used in this study is related to one of the biggest companies in Turkey. This dataset comprises some customers who didn't pay their bills regularly. These cases are assigned to a legal debt collection agency for charging the bills in legal ways. So, the aim of this study is to find which legal debt collection agency has more capability to charge the bills than other legal debt collection agencies. We try to predict each legal debt collection agency's case closing probability with machine learning techniques. This prediction is crucial for the company because of charging unpaid bills, which means that decreasing bad debt for the companies. For this purpose, we have the dataset which comprises customers' demographic information, case information, and legal debt collection agency's previous performance, and its demographic information. The details of the dataset which is used in this study are shown in Table 1. The dataset which is used in this study comprises 18 base features which are 9 categorical, 8 numerical and 1 label feature and 585289 rows. Initially, methods of data cleaning were administered to the dataset. For this reason, features that could not be used for machine learning algorithms were abolished from the dataset. This feature is an age feature, which is over 76% of the data is missing. Because over 50% of the feature is missing, filling with average or most frequent data isn't stable for the model. After that step, we filled the missing occupation data with the "other" label, which means unknown occupation, so we eliminate 16% of missing value for the correspondent column. Also, we applied the same approach for marital status information of customers, which is missing. For city of birth and city of address columns, we applied filling missing values with the most frequent value for eliminating missing values for these columns. After this phase, the categorical values that speak to a user's data which are string values are converted to numerical values with one hot encoding technique. So, whole of the data is converted to numerical for machine learning algorithms. Additionally, one hot encoding technique is used in the literature frequently for improving machine learning algorithms' performance [14]. Also, the min-max scaling technique is applied to the dataset, which is numerical values.

In the modelling phase, we applied some machine learning algorithms which are Catboost Classifier, XGBoost Classifier, Gradient Boosting Classifier etc. to the processed dataset. The comparison of individual machine learning algorithms' performance is shown in Table 2. The results show us Catboost Classifier has the best accuracy performance in applied algorithms with 0.87 accuracy. After the fitting process of Catboost classifier, we evaluate the most important features for the model. The importance result shows us administrative follow-up count, transfer day difference of case, administrative follow up period have significant effect in prediction of case closing capability for legal debt collection agencies. Also, segment information of customer, legal debt collection agency's employee count, legal debt collection agency segment and legal debt collection agency's assigned file count are also important for the prediction. Because legal debt collection agency related features are important for the model, asking for different legal debt collection agencies for the same case in the prediction stage works well in prediction too. Feature importance results are shown in Fig. 1.

In the last phase of modelling, we aim to tune hyper-parameters of Catboost classifier for getting better accuracy in the modelling. The performance of the algorithms used in the solution of machine learning problems generally depends on the hyper-parameters selected while these algorithms are run [15]. Setting up hyper-parameters becomes the core process for machine learning methods, as the default hyper-parameters cannot guarantee the performance of machine learning models [16]. Various adjustment approaches such as trial and error and manual search have been developed to get the best configuration of hyper-parameters. So, we applied hyper-parameter optimization for Catboost classifier which has the best performance in prediction. Also, we applied k-fold cross validation for testing stability of the model. Hyper-parameter optimization processes have nearly 3% positive effect on the prediction. The performance result of the Catboost Classifier with best parameters is shown in Table 3.

Table 1. Base features used in case close prediction

Feature	Description	Type	Range	Missing value rate
transfer_day_diff	How many days difference between case date and agency transfer date	Numerical	[0,60]	0,00%
agency_id	The id of legal debt collection agency	Categorical	230 different values	0,00%

(continued)

Table 1. (*continued*)

Feature	Description	Type	Range	Missing value rate
city	The city of legal debt collection agency	Categorical	33 different values	0,00%
initial_amount	The amount of bill which customer hasn't paid	Numerical	[80,7000]	0,00%
case_type	The case type which is corporate or personal	Categorical	3 different values	0,00%
occupation	The occupation of the customer	Categorical	210 different values	16,75%
city_of_birth	The customer's city of birth	Categorical	125 different values	1,07%
address_city	The customer's address	Categorical	220 different values	5,53%
martial_status	The customer's martial status	Categorical	5 different values	17,26%
age	The customer's age	Numerical	[18,100]	76,92%
agency_prev_perf	The agency's cumulative previous performance for closing cases	Numerical	[0,2.52]	0,00%
assigned_file_cnt	The count of assigned files to the legal debt collection agency	Numerical	[0,25649]	0,00%
agency_segment	The segment information of legal debt collection agency	Categorical	12 different values	0,00%
employee_cnt	The employee count of legal debt collection agency	Numerical	[4,236]	0,00%

(*continued*)

Table 1. (*continued*)

Feature	Description	Type	Range	Missing value rate
is_case_closed	Is case closed or not	Label	2 different values	0,00%
sub_segment_info	The customer's segment information	Categorical	62 different values	0,00%
administrative_followup_cnt	The count of customer has been followed up before	Numerical	[0,21100]	0,00%
administrative_followup_period	The duration of customer has been followed up before	Numerical	[0,501963]	0,00%

Table 2. Prediction accuracy results

Model	Accuracy	AUC	Recall	Prec.	F1	Kappa
Catboost classifier	0.874	0.928	0.77	0.914	0.836	0.734
Gradient boosting classifier	0.834	0.893	0.707	0.872	0.781	0.65
Extreme gradient boosting	0.831	0.892	0.701	0.870	0.777	0.643
Random forest classifier	0.823	0.876	0.667	0.88	0.759	0.623
Decision tree classifier	0.819	0.813	0.778	0.785	0.781	0.626
Ada boost classifier	0.814	0.875	0.691	0.834	0.756	0.607
Extra trees classifier	0.778	0.849	0.602	0.818	0.693	0.525

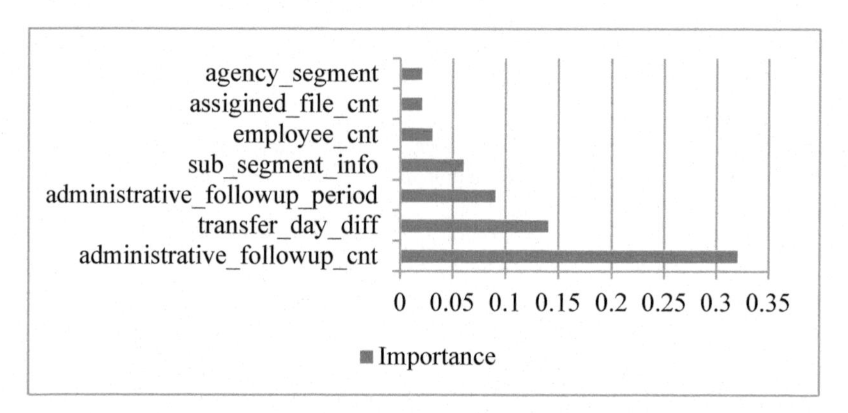

Fig. 1. Most important features used in prediction

Table 3. Hyper-parameter tuning results of Catboost with 5-fold CV

	Accuracy	AUC	Recall	Prec.	F1	Kappa
0	0,8986	0,9547	0,7921	0,9386	0,8591	0,7537
1	0,9014	0,9564	0,7941	0,9438	0,8625	0,7593
2	0,8975	0,9548	0,7881	0,9398	0,8573	0,7512
3	0,8992	0,9554	0,7928	0,9394	0,8598	0,7548
4	0,8967	0,9533	0,7892	0,9367	0,8567	0,7497
Mean	0,8987	0,9549	0,7912	0,9397	0,8591	0,7538
SD	0,0018	0,0011	0,0025	0,0026	0,0023	0,0037

4 Conclusion

This paper's principal contribution is predicting the closing probability with machine learning algorithms for increasing the performance of the Big companies' legal debt collection process. We applied some machine learning algorithms which are Catboost Classifier, Extreme Gradient Boosting Classifier, Gradient Boosting Classifier, etc. to trained and predicted dataset. The results show us Catboost Classifier with 0.87 accuracies has the best performance among the other 8 algorithms. Administrative follow-up count, transfer day difference of case, administrative follow-up period evaluated as most essential features in predicting the closing probability of legal debt collection agencies with fitting Catboost classifier. Also, to prove the stability of the model, k-fold cross-validation was applied. As the future study, we want to extend this model to the Hesitant Fuzzy environment to handle the model's uncertain nature in prediction.

Acknowledgement. This work is supported by TUBITAK (The Scientific and Technological Research Council of Turkey, Project Id: 5200012).

References

1. Onar, S.C., Oztaysi, B., Kahraman, C.: A fuzzy rule based inference system for early debt collection. Technol. Econ. Develop. Econ. **24**(5), 1845–1865 (2018)
2. Van de Geer, R., Wang, Q., Bhulai, S.: Data-driven consumer debt collection via machine learning and approximate dynamic programming. SSRN Electron. J., 1–32 (2018). https://doi.org/10.2139/ssrn.3250755
3. Stewart, A.: How to evaluate your debt collection agency. Debt Recoveries Aust. (2017)
4. Onar, S.Ç., Öztürk, E., Öztayşi, B., Yüksel, M., Kahraman, C.: Pisagor bulanik akilli çok ölçütlü yasal takip avukatlik ofisi performans değerlendirme modeli. In: Mühendislik ve Teknoloji Yönetimi Zirvesi, İstanbul Teknik Üniversitesi & Bahçeşehir Üniversitesi, pp. 88–97 (2018)
5. Kim, J., Kang, P.: Late payment prediction models for fair allocation of customer contact lists to call center agents. Decis. Supp. Syst. **85**, 84–101 (2016). https://doi.org/10.1016/j.dss.2016.03.002

6. Chen, C.H., Chiang, R.D., Wu, T.F., Chu, H.C.: A combined mining-based framework for predicting telecommunications customer payment behaviors. Exp. Syst. Appl. **40**(16), 6561–6569 (2013). https://doi.org/10.1016/j.eswa.2013.06.001
7. Appel, A.P., Oliveira, V., Lima, B., Malfatti, G.L., De Santana, V.F., De Paula, R.: Optimize cash collection: use machine learning to predicting invoice payment. arXiv (2019)
8. Zeng, S., Melville, P., Lang, C.A., Boier-Martin, I., Murphy, C.: Using predictive analysis to improve invoice-to-cash collection. In: Proceedings of the ACM SIGKDD International Conference on Knowledge Discovery and Data Mining, pp. 1043–1050 (2008). https://doi.org/10.1145/1401890.1402014
9. Moedjiono, S., Fransisca, F., Kusdaryono, A.: Segmentation and classification customer payment behavior at multimedia service provider company with K-means and C4.5 algorithm. Int. J. Comput. Netw. Commun. Secur. **4**(9), 265–275 (2016)
10. Mushava, J., Murray, M.: An experimental comparison of classification techniques in debt recoveries scoring: evidence from South Africa's unsecured lending market. Exp. Syst. Appl. **111**, 35–50 (2018). https://doi.org/10.1016/j.eswa.2018.02.030
11. Liu, Y., Schumann, M.: Data mining feature selection for credit scoring models. J. Oper. Res. Soc. **56**(9), 1099–1108 (2005). https://doi.org/10.1057/palgrave.jors.2601976
12. Soares De Melo Junior, L., Nardini, F.M., Renso, C., Fernandes De MacEdo, J.A.: An empirical comparison of classification algorithms for imbalanced credit scoring datasets. In: Proceedings of the 18th IEEE International Conference on Machine Learning and Applications (ICMLA), pp. 747–754 (2019). https://doi.org/10.1109/ICMLA.2019.00133
13. Idris, A., Rizwan, M., Khan, A.: Churn prediction in telecom using Random Forest and PSO based data balancing in combination with various feature selection strategies. Comput. Electr. Eng. **38**(6), 1808–1819 (2012). https://doi.org/10.1016/j.compeleceng.2012.09.001
14. Tekin, A.T., Kaya, T., Çebi, F.: Click prediction in digital advertisements: a fuzzy approach to model selection. In: Kahraman, C., Cevik Onar, S., Oztaysi, B., Sari, I.U., Cebi, S., Tolga, A.C. (eds.) INFUS 2020. AISC, vol. 1197, pp. 213–220. Springer, Cham (2021). https://doi.org/10.1007/978-3-030-51156-2_26
15. Klein, A., Dai, Z., Hutter, F., Lawrence, N., Gonzalez, J.: Meta-Surrogate Benchmarking for Hyperparameter Optimization (2019)
16. Thiede, L.A., Parlitz, U.: Gradient based hyperparameter optimization in Echo State Networks. Neural Netw. **115**, 23–29 (2019). https://doi.org/10.1016/j.neunet.2019.02.001

The Most Effective Factors in Predicting Bioelectrical Impedance Phase Angle for Classification of Healthy and Depressed Obese Women: An Artificial Intelligence Approach

Seyed Amir Tabatabaei Hosseini[1], Mahdad Esmaeili[1](✉), Yaser Donyatalab[2], and Fariborz Rahimi[3]

[1] Faculty of Advanced Medical Sciences, Department of Biomedical Engineering, Tabriz University of Medical Sciences, Tabriz, Iran
[2] Department of Industrial Engineering, University of Moghadas Ardabili, Ardabil, Iran
[3] Faculty of Engineering, Department of Electrical Engineering, University of Bonab, Bonab, Iran

Abstract. Phase Angle (PhA) is one of the most clinically relevant and important parameters that describes the ratio of body reactance and resistance. It is used for evaluating the nutritional status and determining the risk of various conditions such as cancer, AIDS, and many chronic diseases. The purpose of this research was to assess the most effective factors associated with prediction of PhA in two groups of healthy and depressed obese women. This was done by constructing a predictive model using machine learning multivariate regression methods to easily assess nutritional and cellular status in different subject groups. In this study, we used the TANITA body composition analyzer to collect data from 120 obese women out of which 61 suffered from depression. Fourteen different factors from the subject's body including sex, age, height, weight, fat mass, and muscle mass was used for the prediction of PhA using machine learning methods. Two classes of multivariable regression analyses were considered. Every method with several feature selections was trained and tested to obtain the least error for PhA estimation. Then, for each of the two groups of participants the feature selection method was implemented to optimize the model. Our findings suggest that the PhA values of healthy and depressed obese women depend on several variables in their bodies. These variables are Age, Weight, FFM, VFR, TBW, and ICW for healthy obese women and Age, fat mass, BMI, TBW, and ICW for obese women with depressions.

Keywords: Bioelectrical impedance analysis · Body composition · Phase angle · Artificial Intelligence · Machine learning

1 Introduction

Phase Angle (PhA) is one of the most clinically important parameters that describes the ratio of body reactance and resistance which is measured by the means of multi-frequency

C. Kahraman et al. (Eds.): INFUS 2021, LNNS 308, pp. 523–529, 2022.
https://doi.org/10.1007/978-3-030-85577-2_62

bioelectrical impedance analysis (BIA) that typically ranges from 1 kHz to 1000 kHz [1]. Resistance represents the opposition of the body to electric currents whereas Reactance is called to the capacity of the tissues to conduct electric currents [2, 3]. Previous studies by other researchers have demonstrated that the high values of the PhA represent a large number of intact cell membranes while low values signify the poor health status and reduction in the integrity of cells [4, 5]. The PhA can be used as the main marker of cellular health, nutritional status, and a predictor of mortality in various diseases and also the prognostic marker in several health conditions such as cancerous conditions, AIDS, and chronic diseases [6–10].

Most of the related studies have shown that the major indexes for determining the values of PhA in healthy subjects are age, sex, BMI, and fat mass [11–13]. However, even though many researchers have reported about the importance of PhA in various conditions, there has been very little studies reported on the major indexes for determining the PhA values in depressed obese women and also evaluating the differences between healthy and depressed women.

Both depression and obesity are common conditions that are considered as the major public health concerns. According to US Public Health Service, obesity is defined as a body mass index or BMI of 30 or more which scores at 85th percentile or higher [14]. There is a direct correlation between these two conditions which indicates the presence of one, increases the risk of the other one's development [15, 16].

The present research focuses specifically on predicting phase angle values in healthy obese women and obese women with depression and evaluates the most significant factors of their body that affect phase angle variations. The originality of this paper is to use different machine learning methods with assessment of their efficiency for predicting and determining the phase angle values of the obese and depressed females using their body composition measurements.

The organization of the paper as follows: in Sect. 2, the statistical and machine learning regression analysis is presented. Then in Sect. 3, the PhA prediction errors were calculated for two groups of machine learning algorithms. The most efficient model for each sub-group was selected and the training time for selected models were separately defined. Also, the predicted and true responses of both subgroups for two efficient methods of quadratic and linear SVM plotted in Figs. 1 and 2. In Sect. 4, the results are summed up, and the chapter is concluded.

2 Methods and Materials

2.1 Participants

In this paper, we followed standard protocols of using the Tanita body composition analyzer to collect data from 120 obese women out of which 61 suffered from depressions. The age of participants ranged between 20 and 62. Participation in this study was voluntary and all demographic characteristics of subjects including sex and age are protected.

2.2 Procedure

Tanita is a safe, fast, and reliable device that uses advanced bioelectric impedance analyses (BIA) technology. When a subject stands on the device, a small 50 kHz alternating current is sent from four electrodes through the whole body and provides the measurements in less than 20 s. We used Tanita® MC-980U and MC-780 multi-frequency segmental body composition analyzer to provide a high level of accuracy in measurements and also to estimate the values of intracellular and extracellular water beside the total water of the subject body. This device provides several body indexes out of which we used fourteen different factors for prediction of PhA values through machine learning methods in MATLAB and Python.

Statistical Analysis

The dataset consists of demographic information and other initial factors measured by body composition analyzer including height, weight, fat mass, muscle mass, body mass index (BMI), skeletal muscle mass (SMM), bone mass, basal metabolic rate (BMR), Visceral fat rating (VFR), total body water (TBW), extracellular water (ECW) and intracellular water (ICW) for two groups of healthy and depressed obese women (Table 1).

Table 1. Baseline characteristics of the 120 subjects enrolled in the study.

Characteristics	Healthy (N = 59) Mean ± SD	Depressed (N = 61) Mean ± SD
Age (years)	39.58 ± 7.37	40.84 ± 9.6
Height (cm)	160.2 ± 6.14	160.01 ± 8.08
Weight (Kg)	84.54 ± 9.24	86.15 ± 16.21
Fat mass (Kg)	29.74 ± 5.55	28.28 ± 8.56
Muscle mass (Kg)	52.03 ± 4.93	51.94 ± 7.49
BMI (Kg/m^2)	32.88 ± 2.44	33.44 ± 5.68
Skeletal muscle mass (Kg)	29.34 ± 3.04	29.05 ± 4.22
Fat free mass (Kg)	54.8 ± 5.18	54.75 ± 7.91
Bone mass (Kg)	2.76 ± 0.26	2.76 ± 0.38
Basal metabolic rate (Kcal)	1655.2 ± 157	1659.39 ± 252.09
Visceral fat rating	7.32 ± 1.76	7.89 ± 2.61
Total body water (Kg)	39.17 ± 3.74	39.08 ± 5.68
Extracellular water (Kg)	17.27 ± 1.59	19.99 ± 20.48
Intracellular water (Kg)	21.9 ± 2.27	21.68 ± 3.15
Phase angle (Degree)	6.06 ± 0.51	6.05 ± 0.71

Regression Analysis
Comparative evaluation between two classes of methods in machine learning namely Support Vector Machine (SVM) and Regression Trees were performed separately by feature selection to investigate the most effective factors associated with phase angle variation in the two study groups. Six algorithms of SVM were analyzed based on the kernel type and scale. Linear SVM, Quadratic SVM, Cubic SVM, Fine Gaussian SVM (kernel scale = sqrt(P)/4), Medium Gaussian SVM (kernel scale = sqrt(P)), Coarse Gaussian SVM (kernel scale = sqrt(P)*4), in which P is the number of predictors, were utilized. Moreover, the three algorithms of regression trees differed in minimum leaf size in training. The minimum leaf size was 4 for fine regression tree, 12 for medium regression tree, and 36 for coarse regression tree. These were all tested and analyzed separately for each group.

3 Results

The results of the analysis of 120 patients (59 Healthy obese women and 61 depressed obese women) are presented by the following tables and figures.

3.1 Regression Results

Two different classes of multivariable regression techniques in machine learning have been used separately for healthy and depressed obese women. The results of predicting PhA values for each study group are presented in Table 2.

Table 2. The results of different prediction errors of the PhA prediction in two sub-groups of healthy (HG) and depressed (DG).

Method	Model	RMSE		R-Squared		MAE	
		HG	DG	HG	DG	HG	DG
Regression trees	Fine tree	0.6148	0.8056	−0.46	−0.26	0.4771	0.5602
	Medium tree	0.5172	0.6922	−0.03	0.07	0.3780	0.4722
	Coarse tree	0.5090	0.7165	0.00	0.00	0.3834	0.5036
Support vector machines	Linear SVM	0.3980	0.6706	0.39	0.12	0.3013	0.4643
	Quadratic SVM	0.3830	0.7070	0.43	0.03	0.2964	0.4868
	Cubic SVM	0.4246	0.9587	0.30	−0.79	0.3141	0.6220
	Fine Gaussian	0.4531	0.6734	0.21	0.12	0.3433	0.4308
	Medium Gaussian	0.4743	0.6836	0.13	0.09	0.3578	0.4573
	Coarse Gaussian	0.4962	0.7024	0.05	0.04	0.3765	0.4734

The results indicated that, for both healthy and depressed subjects, SVM is a more efficient method. The reason was the least root mean square error considering its reasonable training time. According to Table 2, the lowest value for the root mean square error for healthy and depressed obese women belongs to the Quadratic SVM and Linear SVM methods respectively which are equal to 0.3830 and 0.6706. The model for each method was trained and tested, with several feature selections, to obtain the least prediction error for PhA estimation. Then, for each of the two groups of participants the feature selection method was implemented to optimize the model. ICW explained the highest variability in the PhA in women with depression followed by age and fat mass. By analyzing the regression results in depressed women, it was concluded that several variables including age, fat mass, height, and ICW were "positively" correlated with PhA values. Whereas, for healthy obese women, muscle mass, SMM, VFR, ECW, and ICW were the most effective factors in PhA variation. The results indicated that, the Quadratic SVM and Linear SVM methods have the best performance with a reasonable training time for both healthy and depressed obese women (see Table 3).

Table 3. The training time of selected models for predicting PhA in two sub-groups.

Group	Selected model	Training time (s)
Healthy	Quadratic SVM	2.27
Depressed	Linear SVM	1.32

The predicted and true responses of both subgroups (healthy and depressed obese women) for two efficient methods of Quadratic and linear SVM are shown in Figs. 1 and 2. In these figures, orange points represent predicted values of PhA for both subgroups and blue points represent the true responses.

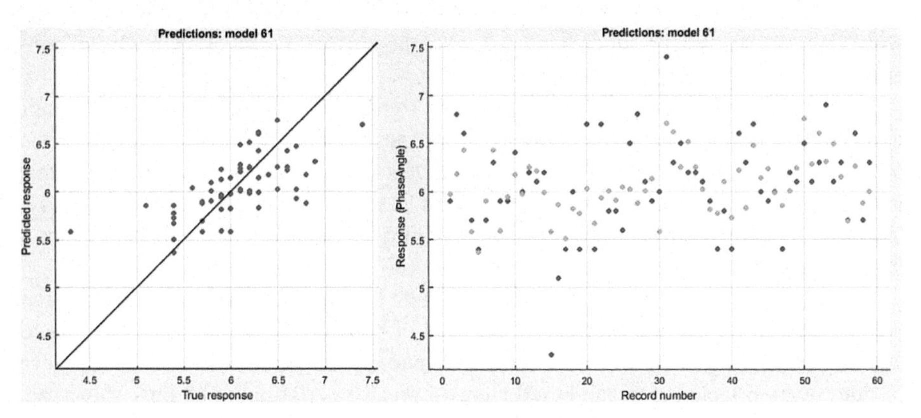

Fig. 1. The predicted vs actual plot and response plot related to the Quadratic SVM model for the PhA prediction of healthy obese women.

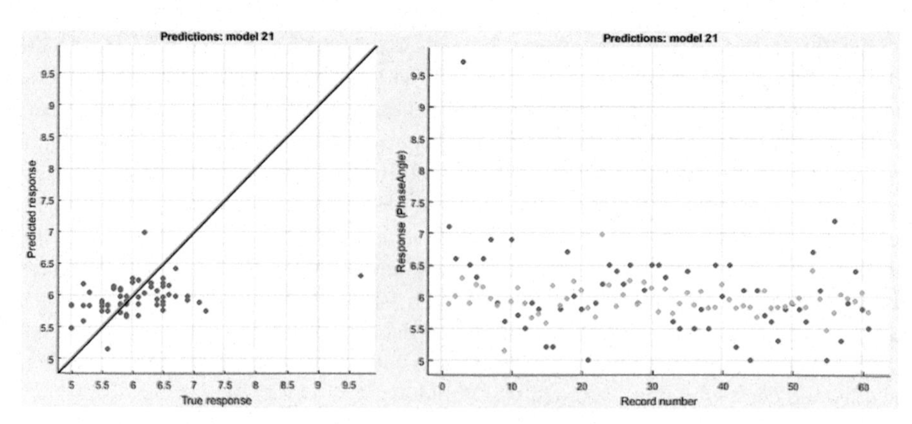

Fig. 2. The predicted vs actual plot and response plot related to the Linear SVM model for the PhA prediction of depressed obese women.

4 Discussion and Conclusions

The purpose of this research was to assess the most effective factors associated with prediction of PhA in two subgroups (healthy and depressed obese women). This was done by constructing a predictive model using machine learning multivariable regression methods. The final goal was to be able to assess nutritional and cellular status by predicting phase angle values in different subjects. We selected the most effective features by removing irrelevant ones to optimize the model and reduce the prediction error.

This study showed that the PhA values of each subgroup depended on different factors in their bodies. For healthy obese women, PhA can be predicted, with a reasonable degree of accuracy, by measuring muscle mass, SMM, VFR, ECW, and ICW. In obese women with depressions, on the other hand, age, height, fat mass, and ICW were the most effective factors for predicting PhA.

The success of the proposed method can be confirmed by Figs. 1 and 2, in which the true responses, fall close to the predicted responses points. Therefore, the outcomes of both Quadratic and Linear SVM regression models for subgroups, seem to be promising in terms of successful prediction of PhA values. It seems that these two methods can be used as efficient prediction methods in determining health and nutritional conditions in both healthy and depressed obese women.

It should be considered that, sex is an important factor which might affect PhA values and it's dependent factors. Using larger dataset of body composition measurements for both sexes with the state-of-the-art machine learning methods besides the high level of optimization techniques can be efficient in precise determining of PhA values and effective factors in the wide range of participants.

Acknowledgments. This work is partially supported by vice-chancellery for research and technology of Tabriz University of Medical Sciences under grant No. 62355 and ethical code number IR.TBZMED.VCR.REC.1398.388. The funders had no role in study design, data collection and analysis, decision to publish, or preparation of the manuscript.

References

1. Hui, D., et al.: Association between multi-frequency phase angle and survival in patients with advanced cancer. J. Pain Symptom Manage **53**, 571–577 (2017)
2. Kyle, U.G., et al.: Bioelectrical impedance analysis - Part II: Utilization in clinical practice. Clin. Nutr. **23**, 1430–1453 (2004)
3. Barbosa-Silva, M.C.G., Barros, A.J.D.: Bioelectrical impedance analysis in clinical practice: a new perspective on its use beyond body composition equations. Curr. Opin. Clin. Nutr. Metab. Care **8**, 311–317 (2005)
4. Selberg, O., Selberg, D.: Norms and correlates of bioimpedance phase angle in healthy human subjects, hospitalized patients, and patients with liver cirrhosis. Eur. J. Appl. Physiol. **86**, 509–516 (2002)
5. Gupta, D., et al.: Bioelectrical impedance phase angle in clinical practice: implications for prognosis in stage IIIB and IV non-small cell lung cancer. BMC Cancer **9**, 1–6 (2009)
6. Fernandes, S.A., de Mattos, A.A., Tovo, C.V., Mar, C.A.: Nutritional evaluation in cirrhosis: emphasis on the phase angle. World J. Hepatol. **8**, 1205–1211 (2016)
7. Lukaski, H.C., Kyle, U.G., Kondrup, J.: Assessment of adult malnutrition and prognosis with bioelectrical impedance analysis: phase angle and impedance ratio. Curr. Opin. Clin. Nutr. Metab. Care **20**, 330–339 (2017)
8. Alves, F.D., Souza, G.C., Clausell, N., Biolo, A.: Prognostic role of phase angle in hospitalized patients with acute decompensated heart failure. Clin. Nutr. **35**, 1530–1534 (2016)
9. Norman, K., Wirth, R., Neubauer, M., Eckardt, R., Stobäus, N.: The bioimpedance phase angle predicts low muscle strength, impaired quality of life, and increased mortality in old patients with cancer. J. Am. Med. Dir. Assoc. **16**(2), 173.e17–22 (2015)
10. Schwenk, A., Beisenherz, A., Romer, K., Kremer, G., Salzberger, B., Elia, M.: Phase angle from bioelectrical impedance analysis remains an independent predictive marker in HIV-infected patients in the era of highly active antiretroviral treatment. Am. J. Clin. Nutr. **72**, 496–501 (2000)
11. Wilhelm-Leen, E.R., Hall, Y.N., Horwitz, R.I., Chertow, G.M.: Phase angle, frailty and mortality in older adults. J. Gen. Intern. Med. **29**, 147–154 (2014)
12. Langer, R.D., de Fatima, G.R., Gonçalves, E.M., Guerra-Junior, G., de Moraes, A.M.: Phase angle is determined by body composition and cardiorespiratory fitness in adolescents. Int. J. Sports Med. **41**(9), 610–615 (2020)
13. Barbosa-Silva, M.C.G., Barros, A.J.D., Wang, J., Heymsfield, S.B., Pierson, R.N.: Bioelectrical impedance analysis: population reference values for phase angle by age and sex. Am. J. Clin. Nutr. **82**, 49–52 (2005)
14. Simon, G.E., et al.: Association between obesity and depression in middle-aged women. Gen. Hosp. Psychiatry **30**, 32–39 (2008)
15. Ha, H., Han, C., Kim, B.: Can obesity cause depression? A pseudo-panel analysis. J. Prev. Med. Public Health **50**, 262–267 (2017)
16. Speed, M.S., Jefsen, O.H., Børglum, A.D., et al.: Investigating the association between body fat and depression via Mendelian randomization. Transl. Psychiatry **9**, 184 (2019)

Q-Rung Orthopair Fuzzy Sets

An Industry 4.0 Adaptation Evaluation with q-rung Ortopair Fuzzy Multi-attributive Border Approximation Area Comparison Method

Serhat Aydın(✉)

Industrial Engineering Department, National Defence University, Turkish Air Force Academy, 34149 Istanbul, Turkey
saydin3@hho.edu.tr

Abstract. Industry 4.0 is the theorization of a manufacturing model based on the "Cyber Physical System" (CPS) idea, in which advanced computer systems can communicate with computing, communication and control capabilities enhanced devices, and are often identified with a collection of enabling technologies: Internet of things (IoT), additive manufacturing, artificial intelligence, big data and analytics, etc. The use of these technologies by companies is a measure of their industry 4.0 adaptation process. In this paper, we evaluated companies' adaptation to Industry 4.0 according to the determined criteria by using multi-attributive border approximation area comparison (MABAC) method with q-ROFNs. The criteria are determined from the most used technologies in Industry 4.0. Finally, companies are ranked according to the results of MABAC method with q-ROFNs.

Keywords: MADM · MABAC method · q-rung orthopair fuzzy sets

1 Introduction

Thanks to the widespread of globalization that is followed by developments in technology and interconnected systems, the industry sector is continually evolving, and numerous companies are facing new and different challenges [1]. Depending on this widespread globalization, industrialization has evolved in four phases and that is defined as the industrial revolution. A modern technique finds widespread use in manufacturing after the industrial revolution, radically altering existing processes [2]. They are called revolutions because they reshape manufacturing production by drastically altering how goods are manufactured and the degree of efficiency in profit formation [3].

The concept of Industry 4.0 arose in Germany in 2011 and spread to other countries [4]. The Fourth Industrial Revolution is the era of smart machines, storage systems and manufacturing facilities that can autonomously share information without human interference, activate actions and control each other.

Companies must adapt to Industry 4.0 applications in order to maximize their profits and take their place in the market. In this study, we evaluate the adaptation of firms to

C. Kahraman et al. (Eds.): INFUS 2021, LNNS 308, pp. 533–540, 2022.
https://doi.org/10.1007/978-3-030-85577-2_63

Industry 4.0. To do this aim, we capture the experts' opinion with new type of fuzzy sets, named q-rung orthopair fuzzy sets (q-ROFs).

Yager [5] proposed q-rung orthopair fuzzy sets (q-ROFs), in which the sum of the $\mu^q_{(x)}$ and $v^q_{(x)}$ is less or equal to 1. It can be concluded that q-ROFs are the generalization of intuitionistic fuzzy sets and Pythagorean fuzzy sets. Assume that X be a fix set. The q-ROFS can be depicted as $P = \{\langle x, (\mu_P(x), v_P(x))\rangle | x \in X\}$. Where the $\mu_P : X \to [0, 1]$ indicates the membership degree, the $v_P : X \to [0, 1]$ indicates the non-membership degree which satisfies $(\mu_P(x))^q + (v_P(x))^q \leq 1, q \geq 1$. The indeterminacy degree can be calculated by using membership degree and non-membership degree as: $\gamma_p(x) = \sqrt[q]{(\mu_P(x))^q + (v_P(x))^q - (\mu_P(x))^q (v_P(x))^q}$. For convenience, we called $p = (\mu, v)$ a q-ROFN.

In this study, utilized MABAC method with q-ROFs [6]. We evaluate five different firms with for different criteria related to Industry 4.0. The steps of the method are employed to the problem and finally, the firms are ranked according to their adaptation score.

The remainder of the paper's composition is as follows: Sect. 2 presents the preliminaries of q-ROFs. An illustrative application is given in Sect. 3. Lastly, the conclusion is given in Sect. 4.

2 Preliminaries of the q-ROFs

Definition 1. [7] Given $k = (\mu, v)$ as a q-ROFN, Eq. (1) is utilized to get the q-rung orthopair fuzzy score results, and Eq. (2) is utilized to get fuzzy accuracy results.

$$S(k) = \frac{1}{2}\left(1 + \mu^q - v^q\right), S(k) \in [0, 1] \tag{1}$$

$$H(k) = \mu^q + v^q, H(k) \in [0, 1] \tag{2}$$

Definition 2. [7] Given $k_1 = (\mu_1, v_1)$ and $k_2 = (\mu_2, v_2)$ be any q-ROFNs. $S(k_1) = \frac{1}{2}\left(1 + (\mu_1)^q - (v_1)^q\right)$ and $S(k_2) = \frac{1}{2}\left(1 + (\mu_2)^q - (v_2)^q\right)$ indicates the score results of k_1 and k_2. $H(k_1) = (\mu_1)^q + (v_1)^q$ and $H(k_2) = (\mu_2)^q + (v_2)^q$ indicates the accuracy result of k_1 and k_2, respectively. Then if $S(k_1) < S(k_2)$, then $k_1 < k_2$; if $S(k_1) = S(k_2)$, then (1) if $H(k_1) = H(k_2)$, then $k_1 = k_2$; (2) if $H(k_1) < H(k_2)$, $k_1 < k_2$.

Definition 5. [7] Some basic operation laws can be depicted as follows:

$$k_1 \oplus k_1 = \left(\sqrt[q]{(\mu_1)^q + (\mu_2)^q - (\mu_1)^q (\mu_2)^q, v_1 v_2}\right) \tag{3}$$

$$k_1 \otimes k_1 = \left(\mu_1 \mu_2, \sqrt[q]{(v_1)^q + (v_2)^q - (v_1)^q (v_2)^q}\right) \tag{4}$$

$$\lambda k = \left(\sqrt[q]{1 - (1 - \mu^q)^\lambda}, v^\lambda\right), \lambda > 0 \tag{5}$$

$$k^\lambda = \left(\mu^\lambda, \sqrt[q]{1 - (1 - v^q)^\lambda}\right), \lambda > 0 \tag{6}$$

$$k^c = (v, \mu) \tag{7}$$

Definition 6. [7] Let $k_j = (\mu_j, v_j)(j = 1, 2, \ldots, n)$ be a list of q-ROFNs, q-ROFWA and q-ROFWG are expressed as:

$$\text{q}-\text{ROFWA}(k_1, k_2, \ldots, k_n) = \sum_{j=1}^{n} w_j k_j = \left(\sqrt[q]{1 - \prod_{j=1}^{n} \left(1 - \left(\mu_{k_j}\right)^q\right)^{w_j}}, \prod_{j=1}^{n} \left(v_{k_j}\right)^{w_j} \right) \tag{8}$$

and

$$\text{q}-\text{ROFWG}(k_1, k_2, \ldots, k_n) = \prod_{j=1}^{n} (k_j)^{w_j} \left(\prod_{j=1}^{n} \left(\mu_{k_j}\right)^{w_j}, \sqrt[q]{1 - \prod_{j=1}^{n} \left(1 - \left(v_{k_j}\right)^q\right)^{w_j}} \right) \tag{9}$$

Where w_j is weighting vector of k_j. $0 \leq w_j \leq 1, \sum_{j=1}^{n} w_j = 1$.

Definition 7. [6] Let $k_1 = (\mu_1, v_1)$ and $k_2 = (\mu_2, v_2)$ be two q-ROFNs, the hammer distance q-ROFNHD can be defined as:

$$\text{q}-\text{ROFNHD}(k_1, k_2) = \frac{1}{2}\left(\left|(\mu_1)^q - (\mu_2)^q\right| + \left|(v_1)^q - (v_2)^q\right| + \left|(\gamma_1)^q - (\gamma_2)^q\right|\right) \tag{10}$$

3 MABAC Method with q-ROFNs

Assume there are m different alternatives $\{A_1, A_2, \ldots, A_m\}$, n attributes $\{G_1, G_2, \ldots, G_n\}$ with weight vector be $w_j(j = 1, 2, \ldots, n)$ and λ experts $\{d_1, d_2, \ldots, d_\lambda\}$ with weighting vector be $\{\omega_1, \omega_2, \ldots, \omega_\lambda\}$ given the q-rung orthopair fuzzy evaluation matrix $R = \left[A_{ij}^{\lambda}\right]_{m \times n} = \left(\mu_{ij}^{\lambda}, v_{ij}^{\lambda}\right), i = 1, 2, \ldots, m, \quad j = 1, 2, \ldots, n$ $\mu_{ij}^{\lambda} \in [0, 1]$ indicates the membership degree, $v_{ij}^{\lambda} \in [0, 1]$ indicates the non-membership degree. We can express the steps of the MABAC method with q-ROFs as follows:

Step 1. [6] Construct the q-rung orthopair fuzzy evaluation matrix $R = \left[A_{ij}^{\lambda}\right]_{m \times n} = \left(\mu_{ij}^{\lambda}, v_{ij}^{\lambda}\right), i = 1, 2, \ldots, m, j = 1, 2, \ldots, n$ as below.

$$R = \left[A_{ij}^{\lambda}\right]_{m \times n} = \begin{matrix} \\ A_1 \\ A_2 \\ \vdots \\ A_m \end{matrix} \begin{matrix} G_1 & G_2 & \ldots\ldots & G_n \\ \left[\begin{matrix} \mu_{11}^{\lambda}, v_{11}^{\lambda} & \mu_{12}^{\lambda}, v_{12}^{\lambda} & \cdots & \mu_{1n}^{\lambda}, v_{2n}^{\lambda} \\ \mu_{21}^{\lambda}, v_{21}^{\lambda} & \mu_{22}^{\lambda}, v_{22}^{\lambda} & \cdots & \mu_{2n}^{\lambda}, v_{2n}^{\lambda} \\ \vdots & \vdots & \vdots & \vdots \\ \mu_{m1}^{\lambda}, v_{m1}^{\lambda} & \mu_{m2}^{\lambda}, v_{m2}^{\lambda} & \cdots & \mu_{mn}^{\lambda}, v_{mn}^{\lambda} \end{matrix}\right] \end{matrix} \tag{11}$$

Where $A_{ij}^{\lambda} = \left(\mu_{ij}^{\lambda}, v_{ij}^{\lambda}\right) i = 1, 2, \ldots, m, j = 1, 2, \ldots n$ indicates the fuzzy judgements of each alternatives $A_i(i = 1, 2, \ldots, m)$ on criteria $G_i(j = 1, 2, \ldots, n)$ by expert d_λ.

Step 2. [6] By using the q-ROFWA or q-ROFWG aggregation operators, we can convert A_{ij}^{λ} to A_{ij}, the fused q-ROFNs matrix $r = \left[A_{ij}\right]_{m\times n}$ shown as follows:

$$r = \left[A_{ij}\right]_{m\times n} = \begin{matrix} A_1 \\ A_2 \\ \vdots \\ A_m \end{matrix} \overset{\begin{matrix} G_1 & G_2 & \ldots\ldots & G_n \end{matrix}}{\begin{bmatrix} \mu_{11}, v_{11} & \mu_{12}, v_{12} & \cdots & \mu_{1n}, v_{2n} \\ \mu_{21}, v_{21} & \mu_{22}, v_{22} & \cdots & \mu_{2n}, v_{2n} \\ \vdots & \vdots & \vdots & \vdots \\ \mu_{m1}, v_{m1} & \mu_{m2}, v_{m2} & \cdots & \mu_{mn}, v_{mn} \end{bmatrix}} \tag{12}$$

Where $A_{ij} = \left(\mu_{ij}, v_{ij}\right)$ $i = 1, 2, \ldots, m, j = 1, 2, \ldots n$ indicates the fused q-rung orthopair fuzzy judgements of alternatives $A_i(i = 1, 2, \ldots, m)$ on criteria $G_i(j = 1, 2, \ldots, n)$.

Step 3. [6] Normalize the matrix $r = \left[A_{ij}\right]_{m\times n} = i = 1, 2, , \ldots..m, j = 1, 2, \ldots., n$ based on the type of each attribute by following equations:

$$\text{In terms of benefit attributes, } N_{ij} = A_{ij} = \left(\mu_{ij}, v_{ij}\right), i = 1, 2, \ldots, m, j = 1, 2, .., n \tag{13}$$

$$\text{In terms of cost attributes, } N_{ij} = \left(A_{ij}\right)^C = \left(v_{ij}, \mu_{ij}\right), i = 1, 2, \ldots, m, j = 1, 2, \ldots.., n \tag{14}$$

Step 4. [6] Using the normalized matrix as a guide, $N_{ij} = \left(\mu_{ij}, v_{ij}\right), i = 1, 2, \ldots, m, j = 1, 2, \ldots.., n$ and the weighting vector $w_j(j = 1, 2, \ldots.n)$, the fuzzy weighted normalized matrix $wN_{ij} = \left(\mu_{ij}^{'}, v_{ij}^{'}\right), i = 1, 2, \ldots, m, j = 1, 2, \ldots.., n$ is calculated by using Eq. (15).

$$wN_{ij} = w_j \otimes N_{ij} = \left(\sqrt[q]{1 - \left(1 - \mu_{ij}^q\right)^{w_j}}, v_{ij}^{w_j}\right) \; (i = 1, 2, \ldots, m) \; j = (1, 2, \ldots, n) \tag{15}$$

Step 5. The values of BBA and the BBA matrix, $G = \left[g_j\right]_{1\times n}$ are calculated by using Eq. (16).

$$g_j = \left(\prod_{i=1}^{m} wN_{ij}\right)^{1/m} = \left\{\left(\prod_{i=1}^{m} \mu_{ij}^{'}\right)^{1/m}, \sqrt[q]{1 - \prod_{i=1}^{m} \left(1 - v_{ij}^{'q}\right)^{1/m}}\right\} \tag{16}$$

Step 6. [6] The distance $D = \left[d_{ij}\right]_{m\times n}$ between the alternatives and values of the BAA is calculated by using Eq. (17).

$$d_{ij} = \begin{cases} d\left(wN_{ij}, g_j\right), & if \;\; wN_{ij} > g_j \\ 0, & if \;\; wN_{ij} = g_j \\ -d\left(wN_{ij}, g_j\right) & if \;\; wN_{ij} < g_j \end{cases} \tag{17}$$

Where $d(wN_{ij}, g_j)$ means the distance from wN_{ij} to g_j.

Step 7. [6] Each alternatives values (d_{ij}) are sum by using Eq. (18).

$$S_i = \sum_{j=1}^{n} d_{ij} \tag{18}$$

Step 8. The alternatives are ranked according to the their S_i values in descending order.

4 An Illustrative Example

In this section of the paper, we utilized the MABAC method with q-ROFs to evaluate firms according to their adaptation to Industry 4.0. Suppose that four firm options are presented to three decision makers (DM_1, DM_2, and DM_3). The weights of these decision makers are 0.25, 0.40 and 0.35, respectively. Alternatives are A_1, A_2, A_3, A_4, A_5. The criteria are Iot technologies (C_1), use of big data technologies (C_2), use of additive manufacturing technologies (C_3), use of cyber security technologies (C_4). The weighting vector of the criteria is (0.18, 0.30, 0.30, 0.22).

Step 1. The evaluation matrices can be seen in Tables 1, 2, and 3.

Table 1. The evaluation matrix by DM_1

	C_1	C_2	C_3	C_4
A_1	(0.5, 0.4)	(0.7, 0.5)	(0.6, 0.4)	(0.4, 0.3)
A_2	(0.7, 0.2)	(0.6, 0.2)	(0.7, 0.2)	(0.4, 0.2)
A_3	(0.6, 0.6)	(0.5, 0.4)	(0.4, 0.3)	(0.6, 0.5)
A_4	(0.5, 0.2)	(0.4, 0.5)	(0.5, 0.4)	(0.3, 0.6)
A_5	(0.2, 0.5)	(0.8, 0.2)	(0.7, 0.4)	(0.6, 0.3)

Table 2. The evaluation matrix by DM_2

	C_1	C_2	C_3	C_4
A_1	(0.3, 0.4)	(0.7, 0.3)	(0.5, 0.3)	(0.7, 0.3)
A_2	(0.7, 0.2)	(0.8, 0.5)	(0.4, 0.4)	(0.6, 0.3)
A_3	(0.4, 0.5)	(0.9, 0.4)	(0.6, 0.4)	(0.5, 0.3)
A_4	(0.3, 0.2)	(0.6, 0.3)	(0.6, 0.6)	(0.4, 0.7)
A_5	(0.5, 0.5)	(0.4, 0.5)	(0.2, 0.7)	(0.8, 0.4)

Table 3. The evaluation matrix by DM_3

	C_1	C_2	C_3	C_4
A_1	(0.6, 0.4)	(0.3, 0.7)	(0.2, 0.7)	(0.6, 0.4)
A_2	(0.6, 0.5)	(0.4, 0.9)	(0.6, 0.2)	(0.4, 0.3)
A_3	(0.7, 0.3)	(0.4, 0.7)	(0.4, 0.6)	(0.5, 0.3)
A_4	(0.7, 0.4)	(0.6, 0.4)	(0.6, 0.4)	(0.5, 0.4)
A_5	(0.8, 0.2)	(0.7, 0.2)	(0.7, 0.4)	(0.1, 0.8)

Step 2. Suppose that q = 3, so we get fused q-ROFNs matrix $r = \left[A_{ij}\right]_{m \times n}$ shown as follows.

Step 3. All the criteria are benefit criteria; therefore, normalized matrix can be usedsame as Table 4.

Table 4. The fused q-ROFNs matrix

	C_1	C_2	C_3	C_4
A_1	(0.495, 0.400)	(0.627, 0.459)	(0.480, 0.434)	(0.618, 0.332)
A_2	(0.670, 0.276)	(0.677, 0.488)	(0.580, 0.264)	(0.504, 0.271)
A_3	(0.593, 0.438)	(0.760, 0.487)	(0.504, 0.429)	(0.530, 0.341)
A_4	(0.558, 0.255)	(0.565, 0.377)	(0.579, 0.470)	(0.425, 0.554)
A_5	(0.642, 0.363)	(0.667, 0.289)	(0.608, 0.500)	(0.665, 0.474)

Step 4. The fuzzy weighted normalized matrix is calculated by using Eq. (15) (Table 5).

Table 5. Fuzzy weighted normalized matrix

	C_1	C_2	C_3	C_4
A_1	(0.285, 0.848)	(0.433, 0.791)	(0.326, 0.778)	(0.386, 0.784)
A_2	(0.397, 0.793)	(0.472, 0.807)	(0.398, 0.671)	(0.310, 0.750)
A_3	(0.346, 0.862)	(0.542, 0.806)	(0.343, 0.776)	(0.326, 0.789)
A_4	(0.324, 0.782)	(0.387, 0.746)	(0.397, 0.798)	(0.259, 0.878)
A_5	(0.377, 0.833)	(0.465, 0.689)	(0.419, 0.812)	(0.419, 0.849)

Step 5. We compute the BAA values as follows.

$g_1 = (0.343, 0.826)$ $g_2 = (0.457, 0.773)$ $g_3 = (0.374, 0.773)$ $g_4 = (0.335, 0.817)$

Step 6. We calculate $D = [d_{ij}]_{m \times n}$ values between alternatives and the BAA as seen in Table 6.

Table 6. Distance between alternatives and the BAA

	C_1	C_2	C_3	C_4
A_1	−0.049	−0.035	−0.014	0.068
A_2	0.071	−0.067	0.158	0.129
A_3	−0.074	0.105	−0.010	0.057
A_4	0.091	0.074	−0.051	−0.134
A_5	0.022	0.130	−0.087	−0.088

Step 7. The S_i values of each alternatives can be calculated as follows.

$$S_1 = -0.030 \quad S_2 = 0.291 \quad S_3 = 0.226 \quad S_4 = -0.021 \quad S_5 = 0.153$$

Step 8. Finally, the alternatives are ranked according to their S_i values.

$$A_2 > A_3 > A_5 > A_4 > A_1$$

5 Conclusions

The concept of Industry 4.0 arose in Germany in 2011 and spread to other countries. Industry 4.0 refers to a strategic step that was introduced by the German government seeking ways to change industrial manufacturing through digitalization and the development of new technologies. Companies have to keep up with this change and have to adapt to technologies in Industry 4.0. In this study, we proposed to utilize MABAC method with q-ROFs to assess the company's adaptation to Industry 4.0. In the application section we evaluated five companies according to the determined technologies used in Industry 4.0. also, companies are ranked according to results obtained from utilized method.

For the further studies, all the technologies of in Industry 4.0 can be handled, and the problem can be solved according to these technologies. Moreover, a comparison analysis can be done with new types of fuzzy sets.

References

1. Liao, Y., Deschamps, F., Loures, E.F.R., Ramos, L.F.P.: Past, present and future of Industry 4.0: a systematic literature review research agenda proposal. Int. J. Prod. Res. **55**(12), 3609–3629 (2017)
2. Zhou, K., Liu, T., Zhou, L.: Industry 4.0: towards future industrial opportunities and challenges. In: 2015 12th International Conference on Fuzzy Systems and Knowledge Discovery (FSKD), pp. 2147–2152. IEEE (2015)

3. Von Tunzelmann, N.: Historical coevolution of governance and technology in the industrial revolutions. Struct. Change Econ. Dyn. **14**(4), 365–384 (2003)
4. Magruk, A.: Uncertainty in the sphere of the Industry 4.0 – potential areas to research. Bus. Manag. Educ. **14**(2), 275–291 (2016)
5. Yager, R.R.: Generalized orthopair fuzzy sets. IEEE Trans. Fuzzy Syst. **25**, 1222–1230 (2016)
6. Wang, J., Wei, G., Wei, C., Wei, Y.: MABAC method for multiple attribute group decision making under q-rung orthopair fuzzy environment. Defence Technol. **16**(1), 208–216 (2020)
7. Liu, P.D., Wang, P.: Some q-rung orthopair fuzzy aggregation operators and their applications to multiple-attribute decision making. Int. J. Intell. Syst. **33**, 259e80 (2018)

Optimal Selecting of Sanitarium Sites for COVID-19 Patients in Iran by Applying an Integrated ELECTRE-VIKOR Method in q-ROFSs Environment

Fariba Farid[1] and Yaser Donyatalab[2(✉)]

[1] Department of Management, University of Nabi Akram, Tabriz, Iran
[2] Industrial Engineering Department, University of Moghadas Ardabili, Ardabil, Iran

Abstract. By outbreaking COVID-19, the sanitarium site selection for coronavirus patients turned into a hot topic. In this article, it is tried to help managers to come up with the hardness of such a complex problem. The quarantine procedure of COVID-19 patients is an essential factor which is suggested by the World Health Organization (WHO) for preventing the outbreak. The numerous significant criteria together with many alternatives that could be selected in the real-world cases increase the uncertainty of the problem. So, a hybrid multi-criteria group decision making (MCGDM) process based on outranking methods is applied to make use of the advantages of those methods. In the decision process, the ELECTRE and VIKOR methods are integrated in the q-rung orhtopair fuzzy sets (q-ROFSs) environment is introduced. As a preliminary part the properties of the q-ROFSs are discussed in detail, and then the ELECTRE and VIKOR methods are investigated. Then the integrated ELECTRE-VIKOR method in a MCGDM algorithm is proposed. To show the effectiveness and validity of the proposed method, a case study of sanitarium site selection in Iran is conducted, and results are evaluated in detail.

Keywords: Sanitarium site selection · COVID-19 · Hybrid MCGDM · Integrated ELECTRE-VIKOR · q-ROFSs · Integrated outranking

1 Introduction and Literature Review

Fuzzy set theory developed by Zadeh [1], is to handling imprecise and uncertainty of the situations. Different extensions are introduced by many researchers and applied in different research fields and applications in the literature. Type 2 fuzzy sets introduced in [2] and also have been investigated by many researchers [3, 4]. Intuitionistic fuzzy sets (IFSs) introduced by Attanassov [5], hesitant fuzzy sets [6], Pythagorean fuzzy sets [7], picture fuzzy sets [8] are next developed generations of fuzzy sets, which are also characterized based on membership, non-membership and hesitant values between 0 and 1. Then Yager, in 2017 introduced the concept of ortho pair logic and q-Rung Ortho Pair Fuzzy Sets (q-ROFS) [9], a new extension for fuzzy sets. q-ROFSs attracted many

C. Kahraman et al. (Eds.): INFUS 2021, LNNS 308, pp. 541–551, 2022.
https://doi.org/10.1007/978-3-030-85577-2_64

researchers and applied in different studies. In [10] some new similarity measures based square root cosine similarity are introduced for q-ROFSs environment. In [11] Dice similarity measures are introduces and applied in selecting the best enterprise resource planning (ERP) system. Some novel Hamacher aggregation operators are introduces in [12] and their applications in decision making process is illustrated.

Elimination and choice translation reality (ELECTRE) methods are type of outranking multi-criteria decision making (MCDM) approaches which developed based on the concepts of concordances and discordances. In ELECTRE method decision makers are able to compare alternatives based on each criterion separately, which means that without aggregating performance values related to alternatives for all criteria [13]. VIKOR method introduced by [14] ranks alternative by focusing on conflicting and different units of criteria. VIKOR method is searching for satisfying two criteria of compromise solution satisfying maximum group utility and minimum individuals regret. Some other researches also investigated the hybrid multi-criteria group decision making (MCGDM) methods especially ELECTRE and VIKOR. In [15] integrated ELECTRE-VIKOR is applied in fuzzy decision making process, and in [16] ELECTRE-VIKOR method under IFS is introduced. In this paper ELECTRE 1 because of its limitation is integrated with VIKOR methods under q-ROFSs environment and a novel hybrid MCGDM approach is proposed.

Novel coronavirus (COVID-19) which for the first time seen in Wuhan, China, in December 2019 outbreak quickly all over the world. The World Health Organization (WHO) together with many health-related organizations in different countries tried to informs people about the main symptoms. In first days of pandemic it is tried to as soon as possible to announce some useful strategies for controlling the outbreaking [17]. Based on different researches from all over the world and WHO's reports mainly people infected with COVID-19 by respiratory droplets of infected people's sneeze or cough. So, one of most preventive actions which emphasized by WHO, was using mask, gloves, face shields in crowded places. Also, many researchers tried to introduce their contributions through many articles that published in this short time to help prevent the epidemy. While there is no certain treatment and vaccinate for COVID-19 and epidemic outbreak very quickly growing whole the world. One of the problems in pandemic period is the hospitals insufficient capacity to patient's reception. The best solution that proposed to this situation is considering some sanitarium sites for COVID-19 patients. These sanitariums serve both newly infected and recovered people, because after recovering the person is still able to transmit COVID-19 virus to the others and increase the outbreaking's speed. The other advantage of the sanitariums is to focusing on COVID-19 patients in a determined place and helps to use all the potential resources and facilities. Iran is one of the canonical countries which have high number of infected patients. In this manuscript, a complicated decision-making process is considered to select the optimal sanitarium site selection for COVID-19 patients in Iran. The rest of the chapter is designed as follows: in Sect. 2, preliminaries and basic concepts around q-rung orthopair fuzzy sets (q-ROFSs) is discussed. In Sect. 3, proposed the Integrated ELECTRE-VIKOR (IEV) based on q-ROFSs is introduced. In this section steps of the hybrid MCGDM algorithm based on IEV is defined. Section 4, optimal sanitarium site selection for COVID-19 patients in Iran is investigated also, solved by the proposed hybrid MCGDM based IEV in q-ROFSs. In Sect. 5, the results are summarized, and the chapter is concluded.

2 Preliminaries: q-Rung Orthopair Fuzzy Sets (q-ROFSs)

In this section, q-rung orthopair fuzzy sets (q-ROFSs) are introduced and discussed in details.

q-Rung Orthopair Fuzzy Sets (q-ROFSs) is introduced by Yager in [9] and is based on intuitionistic fuzzy sets (IFSs) and Pythagorean fuzzy sets (PyFSs). q-ROFSs are more capable than IFSs and PyFSs in handling uncertainty of the world. In the following q-ROFSs Mathematical expression of q-ROFSs illustrated.

Definition 1. [9] Let X be universe of discourse, a q-ROFS $\tilde{Q}$ on X is defined by:

$$\tilde{Q} = \left\{ \left\langle x_i, \mu_{\tilde{Q}}(x), \vartheta_{\tilde{Q}}(x) \right\rangle \middle| |x_i \in X \right\} \tag{1}$$

Where $\mu_{\tilde{Q}}(x) : X \in [0, 1]$ and $\vartheta_{\tilde{Q}}(x) : X \in [0, 1]$ represent the membership degree and non-membership degree respectively. The $\mu_{\tilde{Q}}(x)$ and $\vartheta_{\tilde{Q}}(x)$ have to satisfy the below condition:

$0 \leq \mu_{\tilde{Q}}(x)^q + \vartheta_{\tilde{Q}}(x)^q \leq 1, (q \geq 1);$

The indeterminacy degree is defined as $\pi_{\tilde{Q}}(x) = \sqrt[q]{1 - \left(\mu_{\tilde{Q}}(x)^q + \vartheta_{\tilde{Q}}(x)^q\right)}$.

Definition 2. [9] Let $\tilde{Q} = (\mu, \vartheta)$, $\tilde{Q}_1 = (\mu_1, \vartheta_1)$ and $\tilde{Q}_2 = (\mu_2, \vartheta_2)$ be three q-ROFNs, and λ be a positive real number, then:

$$\tilde{Q}_1 \oplus \tilde{Q}_2 = \left\langle \sqrt[q]{\mu_1^q + \mu_2^q - \mu_1^q \mu_2^q}, \vartheta_1 \vartheta_2 \right\rangle \tag{2}$$

$$\tilde{Q}_1 \otimes \tilde{Q}_2 = \left\langle \mu_1 \mu_2, \sqrt[q]{\vartheta_1^q + \vartheta_2^q - \vartheta_1^q \vartheta_2^q} \right\rangle \tag{3}$$

$$\lambda \tilde{Q} = \left\langle \sqrt[q]{1 - (1 - \mu^q)^\lambda}, \vartheta^\lambda \right\rangle \tag{4}$$

$$\tilde{Q}^\lambda = \left\langle \mu^\lambda, \sqrt[q]{1 - (1 - \vartheta^q)^\lambda} \right\rangle \tag{5}$$

Definition 3. [18] Let $\tilde{Q}_i = (\mu_i, \vartheta_i)$, $i = (1, 2, \ldots, n)$ be a collection of q-ROFN and $w = (w_1, w_2, \ldots, w_n)$ be the weight vector of $\tilde{Q}_i$ with $\sum_{i=1}^n w_i = 1$, then the q-rung orthopair fuzzy weighted arithmetic mean (q-ROFWAM) operator is:

$$q - ROFWAM\left(\tilde{Q}_1, \tilde{Q}_2, \ldots, \tilde{Q}_n\right) = \left\langle (1 - \prod_{i=1}^n (1 - \mu_i^q)^{w_i})^{\frac{1}{q}}, \prod_{i=1}^n \vartheta_i^{w_i} \right\rangle \tag{6}$$

Definition 4. [18] Let $\tilde{\mathcal{Q}}_i = (\mu_i, \vartheta_i), i = (1, 2, \ldots, n)$ be a collection of q-ROFN and $w = (w_1, w_2, \ldots, w_n)$ be the weight vector of $\tilde{A}_i$ with $\sum_{i=1}^{n} w_i = 1$, then the q-rung orthopair fuzzy weighted geometric mean (q-ROFWGM) operator is:

$$q - ROFWGM\left(\tilde{\mathcal{Q}}_1, \tilde{\mathcal{Q}}_2, \ldots, \tilde{\mathcal{Q}}_n\right) = \left\langle \prod_{i=1}^{n} \mu_i^{w_i}, \left(1 - \prod_{i=1}^{n} (1 - \vartheta_i^q)^{w_i}\right)^{\frac{1}{q}} \right\rangle \tag{7}$$

Definition 5. To comparison of q-ROFSs by considering the impact of the indeterminacy degree it is defined score and accuracy functions in [19], as follow:

$$\mathcal{S}\left(\tilde{\mathcal{Q}}\right) = \sqrt[q]{|(\mu^q - \vartheta^q) - 2\pi * (\mu - \vartheta)|} \tag{8}$$

$$\mathcal{H}\left(\tilde{\mathcal{Q}}_i\right) = \mu^q + \vartheta^q \tag{9}$$

Definition 6. [9] Let $\tilde{\mathcal{Q}}_i = (\mu_i, \vartheta_i)$ be a collection of q-ROFNs, then the score function of $\tilde{\mathcal{Q}}_i$ is defined in Eq. (8) and (9) respectively. For any two q-ROFNs, $\tilde{\mathcal{Q}}_1 = (\mu_1, \vartheta_1)$ and $\tilde{\mathcal{Q}}_2 = (\mu_2, \vartheta_2)$:

If $\mathcal{S}\left(\tilde{\mathcal{Q}}_1\right) > \mathcal{S}\left(\tilde{\mathcal{Q}}_2\right)$, then $\tilde{\mathcal{Q}}_1 > \tilde{\mathcal{Q}}_2$;
If $\mathcal{S}\left(\tilde{\mathcal{Q}}_1\right) = \mathcal{S}\left(\tilde{\mathcal{Q}}_2\right)$, then.
If $\mathcal{H}\left(\tilde{\mathcal{Q}}_1\right) > \mathcal{H}\left(\tilde{\mathcal{Q}}_2\right)$, then $\tilde{\mathcal{Q}}_1 > \tilde{\mathcal{Q}}_2$;
If $\mathcal{H}\left(\tilde{\mathcal{Q}}_1\right) = \mathcal{H}\left(\tilde{\mathcal{Q}}_2\right)$, then $\tilde{\mathcal{Q}}_1 = \tilde{\mathcal{Q}}_2$.

3 Methodology

In this section the novel outranking methodology for MCGDM problems will be introduced and discussed in details. The developed methodology is based on integration of ELECTRE and VIKOR methods in the framework of q-rung orthopair fuzzy sets (q-ROFSs). The introduced methodology will be presented as a hybrid MCGDM algorithm's steps and will be investigated in details. Assume a MCGDM problem with that n decision makers $D = \{D_1, D_2, \cdots, D_n\}$ with related weight vector $w = \{w_1, w_2, \cdots, w_n\}$, $\sum_{i=1}^{n} w_i = 1, w_i \geq 0$ is going to select the best alternative among m alternatives $A = \{A_1, A_2, \cdots, A_m\}$ based on p criterias $C = \left\{C_1, C_2, \cdots, C_p\right\}$ with corresponding weight vector $\omega_j = \left(\omega_1, \omega_2, \cdots, \omega_p\right)$, $\sum_{j=1}^{p} \omega_j = 1, \omega_j \geq 0$.

3.1 Integrated ELECTRE-VIKOR Method Under the q-ROFSs

The novel outranking based on MCGDM problems is extended under the uncertain environment of the q-rung orthopair fuzzy sets. Among different MCGDM methods in this manuscript ELECTRE 1 and VIKOR methods are selected to integrated because of very significant number of researches in the literature review.

3.2 Proposed Hybrid MCGDM Algorithm Based on Integrated ELECTRE-VIKOR (IEV) Method for q-ROF Environment

In the following the proposed hybrid MCGDM algorithm's steps based integrated ELECTRE-VIKOR (IEV) method for q-ROF environment will be presented in details.

Step 1. Problem statement, determination of problem's goal, alternatives and criteria together with decision support committee, which consists of the experts and decision makers.
Step 2. Collecting the decision matrices of the decision makers. In this step the decision support committee (DMs) will be evaluated the alternatives based on the defined criteria. Comments of the DMs will be collected by using linguistic scales $\left(\widetilde{LS}_q\right)$ presented in Table 1.
Step 3. Construction of q-ROF decision matrices. In Table 1 linguistic scales are defined based on q-ROFNs will be used to transfer the collected decision matrices in previous step into q-ROF decision matrices.
Step 4. Normalize the constructed q-ROF decision matrices.
Step 5. Construct q-ROF GDM by aggregating the q-ROF decision matrices of DMs by using weighted aggregation operators, which are shown in Eq. (6) and (7), and considering the corresponding weight vector of DMs.

Table 1. q-ROF Linguistic Scales $\left(\widetilde{LS}_q\right)$.

Linguistic Scales $\left(\widetilde{LS}_q\right)$	q-ROFN (μ, ϑ, π)
Absolutely High Importance (AHI)	(0.9, 0.1, 0.646)
Very High Importance (VHI)	(0.8, 0.2, 0.783)
High Importance (HI)	(0.7, 0.3, 0.857)
Slightly High Importance (SHI)	(0.6, 0.4, 0.896)
Equally Importance (EI)	(0.5, 0.5, 0.908)
Slightly Low Importance (SLI)	(0.4, 0.6, 0.896)
Low Importance (LI)	(0.3, 0.7, 0.857)
Very Low Importance (VLI)	(0.2, 0.8, 0.783)
Absolutely Low Importance (ALI)	(0.1, 0.9, 0.646)

Step 6. Construct the q-ROF concordance and discordance sets. In this step the q-ROF GDM will be compared based on the different criteria for each alternative and we will have three categories for each concordance and discordance sets as follows:

Step 6.1. Concordance sets are included the criteria that shows the superiority of alternatives based on criteria. The superiority of alternative $\tilde{Q}_a = (\mu_a, \vartheta_a, \pi_a)$ and

$\tilde{Q}_b = (\mu_b, \vartheta_b, \pi_b)$ based on criteria categorized as: strong concordance SC, midrange concordance MC and weak concordance WC which defined as follows: $\forall j = 1, 2 \ldots p$

$$SC\left(\tilde{Q}_a, \tilde{Q}_b\right) = \begin{cases} 1\ j|\mu_{aj} \geq \mu_{bj} \wedge \vartheta_{aj} < \vartheta_{bj} \wedge \pi_{aj} < \pi_{bj} \\ 0 \qquad\qquad Otherwise \end{cases} \tag{10}$$

$$MC\left(\tilde{Q}_a, \tilde{Q}_b\right) = \begin{cases} 1\ j|\mu_{aj} \geq \mu_{bj} \wedge \vartheta_{aj} < \vartheta_{bj} \wedge \pi_{aj} \geq \pi_{bj} \\ 0 \qquad\qquad Otherwise \end{cases} \tag{11}$$

$$WC\left(\tilde{Q}_a, \tilde{Q}_b\right) = \begin{cases} 1\ j|\mu_{aj} \geq \mu_{bj} \wedge \vartheta_{aj} < \vartheta_{bj} \\ 0 \qquad\qquad Otherwise \end{cases} \tag{12}$$

Step 6.2. Discordance sets are included the criteria that shows the inferiority of alternatives based on criteria. The inferiority of alternative $\tilde{Q}_a = (\mu_a, \vartheta_a, \pi_a)$ and $\tilde{Q}_b = (\mu_b, \vartheta_b, \pi_b)$ based on criteria categorized as: strong discordance SD, midrange discordance MD and weak discordance WD which defined as follows: $\forall j = 1, 2 \ldots p$

$$SD\left(\tilde{Q}_a, \tilde{Q}_b\right) = \begin{cases} 1\ j|\mu_{aj} < \mu_{bj} \wedge \vartheta_{aj} \geq \vartheta_{bj} \wedge \pi_{aj} \geq \pi_{bj} \\ 0 \qquad\qquad Otherwise \end{cases} \tag{13}$$

$$MD\left(\tilde{Q}_a, \tilde{Q}_b\right) = \begin{cases} 1\ j|\mu_{aj} < \mu_{bj} \wedge \vartheta_{aj} \geq \vartheta_{bj} \wedge \pi_{aj} < \pi_{bj} \\ 0 \qquad\qquad Otherwise \end{cases} \tag{14}$$

$$WD\left(\tilde{Q}_a, \tilde{Q}_b\right) = \begin{cases} 1\ j|\mu_{aj} < \mu_{bj} \wedge \vartheta_{aj} < \vartheta_{bj} \\ 0 \qquad\qquad Otherwise \end{cases} \tag{15}$$

Step 7. Determining the weights of concordance sets by using the following procedure. First it is needed to determine the distance between alternatives for each criterion in q-ROF GDM. The Euclidean distance which is introduced in Eq. (16) is applied, then by multiplying the distance and weights of criteria the weighted distance will calculated. Afterward, the weighted distances are calculated by concordance matrices and by normalizing those will get the weights of each category of concordance sets.
Step 8. Establish the concordance matrix which is presenting the superiority of alternative to each other in comparison. To evaluate the elements of concordance matrix the Eq. (17) will be applied.

$$d_q\left(\tilde{Q}_a, \tilde{Q}_b\right) = \left(\tfrac{1}{2n}\sum_{i=1}^{n}\left[\left(\mu_a^2(x_i) - \mu_b^2(x_i)\right)^2 + \left(\vartheta_a^2(x_i) - \vartheta_b^2(x_i)\right)^2 + \left(\pi_a^2(x_i) - \pi_b^2(x_i)\right)^{\alpha}\right]\right)^{0.5} \tag{16}$$

$$\gamma\left(\tilde{Q}_a, \tilde{Q}_b\right) = \omega_{SC\left(\tilde{Q}_a, \tilde{Q}_b\right)} * \sum_{j\in SC}\omega_j + \omega_{MC\left(\tilde{Q}_a, \tilde{Q}_b\right)} * \sum_{j\in MC}\omega_j + \omega_{WC\left(\tilde{Q}_a, \tilde{Q}_b\right)} * \sum_{j\in WC}\omega_j \tag{17}$$

And concordance matrix will be:

$$\Gamma = \begin{bmatrix} 0 & \cdots & \gamma_{1m} \\ \vdots & \ddots & \vdots \\ \gamma_{m1} & \cdots & 0 \end{bmatrix} \tag{18}$$

Step 9. Establish the concordance matrix which is presenting the inferiority of alternative to each other in comparison. To evaluate the elements of discordance matrix the Eq. (19) will be applied.

$$\delta\left(\tilde{Q}_a, \tilde{Q}_b\right) = \frac{\max\limits_{j \in \{SD \vee MD \vee WD\}}\left\{\omega_j * d\left(\tilde{Q}_a, \tilde{Q}_b\right)\right\}}{\max\limits_{j}\left\{d\left(\tilde{Q}_a, \tilde{Q}_b\right)\right\}} \tag{19}$$

And discordance matrix will be:

$$\Delta = \begin{bmatrix} 0 & \cdots & \delta_{1m} \\ \vdots & \ddots & \vdots \\ \delta_{m1} & \cdots & 0 \end{bmatrix} \tag{20}$$

After this step the VIKOR method will be integrated into the ELECTRE 1 to have more comprehensive ranking based on the calculating the maximum group utility s_i and minimum individual regret r_i for each alternative. Then these values will be aggregated to calculate a single compromise value.

Step 10. Calculating the values s_i and r_i by using Eq. (21) and (22) as follows:

$$s_i = 1 - \gamma_i \tag{21}$$

$$r_i = \omega_M * \delta_M \text{ where } \omega_M = \max_j\left\{\omega_j\right\} \text{and } \delta_M = \max_j\{\delta_i\} \tag{22}$$

Step 11. Calculate compromise ranking based on measure ζ_i as follows:

$$\zeta_i = \beta\frac{s_i - s_{min}}{s_{max} - s_{min}} + (1-\beta)\frac{r_i - r_{min}}{r_{max} - r_{min}} \tag{23}$$

Step 12. Rank the alternatives based on measure s_i, r_i and ζ_i in decreasing way. And the compromise solution will be the alternative that satisfy the following conditions:

$$\text{Condition (I)} : \zeta\left(\tilde{Q}_a\right) - \zeta\left(\tilde{Q}_b\right) \geq \frac{1}{m-1} \tag{24}$$

$$\text{Condition (II)} : \text{alternative } \left(\tilde{Q}_a\right) \text{ stands as first in ranking based on } s_i \text{ and/or } r_i \tag{25}$$

where $\tilde{Q}_a$ and $\tilde{Q}_b$ are supposed as 1st and 2nd alternatives respectively.

4 Application

Nowadays, the Coronavirus pandemic (COVID-19) is a global problem which quickly spreading all over the world. Based on WHO's reports and many epidemiologists, close contact with people especially the ones that have been infected is increasing the speed numerously [17]. So, based on many references and WHO's reports separating the infected people from the rest of the patients will help definitely in controlling of the pandemic [20]. So, it is proposed to make some sanitarium for COVID-19 patients. Sanitariums will be hospitals and rest houses that will help the infected people from the first of infection is detected to the end of the process of cure. In this way the other hospitals will be at less risk, and also could have more controlling actions on the determined sanitariums which will be the COVID-19 front line opposition. Also, some other advantages of this sanitarium are that could prepare an appropriate place to keep patients during recovery period when they are carrying infectious around for couple of weeks after being improved also, could have far better and under controlled research on such an unknown disease. But at the same time should consider many parameters for sanitarium that has to be satisfied to reduce both humanitarian and financial costs. So, countries should establish some sanitarium sites according to the number of COVID-19 patients and its spreading rate in each city. In this study optimal sanitarium site selection problem is investigated in Tabriz city of Iran. It is supposed to establish one sufficient sanitarium will be enough but the location is the goal of this problem and different alternatives are suggested while each of those are under the advantages of some of criteria. So, the mostly preferred four sanitarium sites $\{A_1, A_2, A_3, A_4\}$ will be considered as alternatives and evaluated under the determined criteria. Criteria of the problem are opportunity for transportation and accessibility (C_1), opportunity for settlement (C_2), economical situations (C_3) and capacity and instrumental extend opportunity (C_4). Decision support committee constitute of three different experts with specialties so they will have different significance level as follows 0.3,0.45,0.25. So, to sum it up, decision support committee with three decision makers (DMs) will be $D = \{D_1, D_2, D_3\}$ with related weight vector $w = \{0.3, 0.45, 0.25\}$ is going to select the best alternative among $A = \{A_1, A_2, A_3, A_4\}$ based on four criteria $C = \{C_1, C_2, C_3, C_4\}$ with corresponding weight vector $\omega_j = \{0.3, 0.3, 0.15, 0.25\}$. The linguistic comments of the DMs are given in Table 2, which by using Table 1 transferred into q-ROF decision making matrices, and then the proposed hybrid MCGDM algorithm based on integrated ELECTRE-VIKOR (IEV) method for q-ROF environment is applied and results are given as follows through Tables 2, 3 4 and 5 step by step. Where $\beta = 0.5$, for compromise solution, based on step 12 the needed conditions are both satisfied for alternative A_1, then this alternative will be compromise ranking.

5 Conclusion

The main contribution of this paper is the proposed integrated ELECTRE-VIKOR method for q-ROFSs. Then a hybrid MCGDM algorithm based on q-ROFSs analytical IEV is introduced. To validate the proposed methodology, it is applied in optimal sanitarium site selection problem for COVID-19 patients in Iran. The proposed method is applied then results are shown in details. More discussion in this field could be investigated and for further research, the Interval Valued of q-ROFS can be considered, or different type of problems related to COVID-19 could be evaluated by the proposed methodology. For future studies, we propose various applications in different fields of study like financial, banking, social, network, health care, manufacturing, and transportation systems.

Table 2. Linguistic decision matrices of DMs

	D_1				D_2				D_3			
	C_1	C_2	C_3	C_4	C_1	C_2	C_3	C_4	C_1	C_2	C_3	C_4
A_1	HI	VHI	VLI	VHI	SHI	HI	EI	HI	SHI	SLI	VHI	SLI
A_2	SLI	VHI	HI	SLI	SLI	SLI	LI	HI	HI	LI	VLI	LI
A_3	VLI	HI	EI	SHI	HI	HI	SHI	SHI	HI	SHI	HI	HI
A_4	HI	SLI	LI	HI	HI	EI	LI	EI	SHI	VHI	VLI	VHI

Table 3. Concordance matrix [Γ]

	A_1	A_2	A_3	A_4
A_1	–	0.848724	0.721416	0.489490
A_2	0.000	–	0.000	0.022691
A_3	0.127309	0.848724	–	0.277309
A_4	0.254617	0.721416	0.466798	–

Table 4. Discordance matrix [Δ]

	A_1	A_2	A_3	A_4
A_1	–	0.000	0.15	0.031319
A_2	0.3	–	0.237002	0.213209
A_3	0.184893	0.000	–	0.284675
A_4	0.15	0.15	0.3	–

Table 5. s, r and ζ measures with corresponding ranking

s_i Measure		Rank	r_i Measure		Rank	ζ_i Measure		Rank
s_1	0.313	1	r_1	0.045	1	ζ_1	0	1
s_2	0.992	4	r_2	0.09	3	ζ_2	1	4
s_3	0.582	3	r_3	0.085	2	ζ_3	0.647	2
s_4	0.519	2	r_4	0.09	3	ζ_4	0.651	3

References

1. Zadeh, L.A.: Fuzzy sets. Inf. Control **8**, 338–353 (1965)
2. Zadeh, L.A.: The concept of a linguistic variable and its application to approximate reasoning-I. Inf. Sci. (Ny) **8**, 199–249 (1975)
3. Carter, H., Dubois, D., Prade, H.: Fuzzy sets and systems – theory and applications. J. Oper. Res. Soc. (1982)
4. Mizumoto, M., Tanaka, K.: Some properties of fuzzy sets of type 2. Inf. Control **31**, 312–340 (1976)
5. Atanassov, K.T.: Intuitionistic fuzzy sets. Fuzzy Sets Syst. **20**, 87–96 (1986)
6. Torra, V.: Hesitant fuzzy sets. Int. J. Intell. Syst. **25**, 529–539 (2010)
7. Yager, R.R.: Pythagorean Fuzzy Subsets. In: 2013 Joint IFSA World Congress and NAFIPS Annual Meeting, vol. 2, pp. 57–61 (2013)
8. Cuong, B.C., Kreinovich, V.: Picture fuzzy sets - a new concept for computational intelligence problems. In: 2013 3rd World Congress on Information and Communication Technologies. WICT 2013 (2014)
9. Yager, R.R.: Generalized orthopair fuzzy sets, vol. 25, pp. 1222–1230 (2017)
10. Donyatalab, Y., Farrokhizadeh, E., Seyfi Shishavan, S.A.: Similarity measures of q-Rung orthopair fuzzy sets based on square root cosine similarity function. In: Kahraman, C., Cevik Onar, S., Oztaysi, B., Sari, I.U., Cebi, S., Tolga, A.C. (eds.) INFUS 2020. AISC, vol. 1197, pp. 475–483. Springer, Cham (2021). https://doi.org/10.1007/978-3-030-51156-2_55
11. Farrokhizadeh, E., Shishavan, S.A.S., Donyatalab, Y., Abdollahzadeh, S.: The dice (sorensen) similarity measures for optimal selection with q-Rung orthopair fuzzy information. In: Kahraman, C., Cevik Onar, S., Oztaysi, B., Sari, I.U., Cebi, S., Tolga, A.C. (eds.) INFUS 2020. AISC, vol. 1197, pp. 484–493. Springer, Cham (2021). https://doi.org/10.1007/978-3-030-51156-2_56
12. Donyatalab, Y., Farrokhizadeh, E., Shishavan, S.A.S., Seifi, S.H.: Hamacher aggregation operators based on interval-valued q-Rung orthopair fuzzy sets and their applications to decision making problems. In: Kahraman, C., Cevik Onar, S., Oztaysi, B., Sari, I.U., Cebi, S., Tolga, A.C. (eds.) INFUS 2020. AISC, vol. 1197, pp. 466–474. Springer, Cham (2021). https://doi.org/10.1007/978-3-030-51156-2_54
13. Figueira, J.R., Greco, S., Roy, B., Słowiński, R.: ELECTRE methods: main features and recent developments (2010)
14. Multicriteria optimization of civil engineerin systems, https://scholar.google.com/scholar_lookup?title=MulticriteriaOptimizationofCivilEngineeringSystems&author=S.Opricovic&publication_year=1998
15. Zandi, A., Roghanian, E.: Extension of Fuzzy ELECTRE based on VIKOR method. Comput. Ind. Eng. **66**, 258–263 (2013)
16. Çalı, S., Balaman, ŞY.: A novel outranking based multi criteria group decision making methodology integrating ELECTRE and VIKOR under intuitionistic fuzzy environment. Expert Syst. Appl. **119**, 36–50 (2019)

17. World Health Organization (WHO): Novel Coronavirus – China
18. Liu, P., Wang, P.: Some q-Rung orthopair fuzzy aggregation operators and their applications to multiple-attribute decision making. Int. J. Intell. Syst. **33**, 259–280 (2018)
19. Kutlu Gündoğdu, F., Kahraman, C.: A novel spherical fuzzy QFD method and its application to the linear delta robot technology development. Eng. Appl. Artif. Intell. **87**, 103348 (2020)
20. Gundogdu, F.K.: Picture fuzzy linear assignment method and its application to selection of pest house location. In: Kahraman, C., Cevik Onar, S., Oztaysi, B., Sari, I.U., Cebi, S., Tolga, A.C. (eds.) INFUS 2020. AISC, vol. 1197, pp. 101–109. Springer, Cham (2021). https://doi.org/10.1007/978-3-030-51156-2_13

Pythagorean Neutrosophic Soft Sets and Their Application to Decision-Making Scenario

Devaraj Ajay(✉) and P. Chellamani

Sacred Heart College (Autonomous), Tirupattur District,
Tirupattur 635601, Tamilnadu, India

Abstract. Pythagorean Neutrosophic set is a fresh idea that combines Pythagorean fuzzy set and Neutrosophic set having dependent components Membership (α) and non-membership (σ) and independent component Indeterminacy (β) satisfying the condition $0 \leq \alpha^2 + \beta^2 + \sigma^2 \leq 2$. The main goal of this study is to extend this concept to soft sets and define a new concept of Pythagorean Neutrosophic soft sets. This concept has been given a few definitions and operations, as well as some properties. An illustration is provided to demonstrate a novel decision-making technique based on Pythagorean neutrosophic soft sets.

Keywords: Soft sets · Pythagorean neutrosophic soft set · Neutrosophic set · Pythagorean neutrosophic set

1 Introduction

Fuzzy set theory [1] is an emerging mathematical domain, essential for solving vagueness and incomplete information in real-life situations. It is a variant of the crisp set, with elements having membership values in the [0, 1] range. Fuzzy sets and fuzzy logic have potential applications in wide-ranging fields including mathematics, computer science, engineering, statistics, artificial intelligence, decision making, image analysis, and pattern recognition.

Atanassov [2] extended the fuzzy set to Intuitionistic set which gives membership and non-membership grade to each element and their total lie between 0 and 1. Smarandache's neutrosophic set [4] is a generalization of the theory of fuzzy, intuitionistic fuzzy sets [1,4] that deal with inconsistent data having the truth, indeterminacy, and false membership functions of elements within the real unit interval. Wang et al. [13] pioneered the concept of a single valued neutrosophic set, a set of elements with three membership functions in the interval [0, 1].

The Pythagorean set [3], which is an extension of intuitionistic fuzzy sets, was introduced to deal with complex imprecision and uncertainty under the condition that the total of squares of membership and non-membership degrees is between 0 and 1. Pythagorean Neutrosophic set [5] is a combination of Pythagorean

C. Kahraman et al. (Eds.): INFUS 2021, LNNS 308, pp. 552–560, 2022.
https://doi.org/10.1007/978-3-030-85577-2_65

and Neutrosophic set having dependent components (Membership (α), non-membership (σ)) and independent component (Indeterminacy (β)) with the condition $0 \leq \alpha^2 + \beta^2 + \sigma^2 \leq 2$. The concept of this set was developed into graphs [6] and in decision making.

For resolving uncertainties, Molodtsov [8] developed a new mathematical concept known as soft set theory. Soft sets have been used in operations research, Riemann integration, measurement, and probability theory, among other fields. Many researchers have improved soft set theory, particularly operations on soft sets [9], the concept of bijective soft sets, and the concepts of relations and functions in soft set theory [14–16]. Maji et al. proposed a soft and fuzzy set combination in [10] and also combined soft set with intuitionistic and neutrosophic sets [11,12].

The concept of interval-valued fuzzy priority for decision-making using Intuitionistic fuzzy soft set and a soft-set in the type-2 environment has been studied in [17,18]. In recent, fuzzy soft sets have been advanced to hypersoft set, plithogenic hypersoft sets and applied in decision-making [19,20]. The symmetric cross-entropy of hesitant fuzzy soft sets considering the relative entropy was studied in [21]. The topological space on fuzzy bipolar soft sets, point, interior and closure points were defined in [22].

The Intuitionistic Neutrosophic soft set [7], a fusion of intuitionistic and neutrosophic soft sets, was also introduced, and we plan to use the basics of this to introduce the Pythagorean Neutrosophic soft set which is a combination of Pythagorean neutrosophic sets and the theory of soft sets. In this paper, we have defined the Pythagorean neutrosophic soft set and discussed few properties like union, intersection and distributive property with examples. And the introduced soft set is used to illustrate a practical example by implementing it in a decision making problem. The following is the paper's structure: Sect. 2 discusses fundamental definitions and terminologies and Sect. 3 introduces the combined Pythagorean Neutrosophic soft set. An illustration based on Pythagorean neutrosophic soft sets is provided in Sect. 4 and Sect. 5 concludes the work.

2 Basic Concepts

The elementary concepts which are necessary for the results are discussed.

Definition 1 [8]. *Let the family of all subsets of universe* $\mathfrak{W}$ *be* $\mathfrak{P}(\mathfrak{W})$, $\mathfrak{O} \subseteq \mathfrak{E}$ *where* $\mathfrak{E}$ *be the collection of parameters. A duo* $(\mathfrak{d}, \mathfrak{O})$ *is soft, where* $\mathfrak{d} : \mathfrak{O} \to \mathfrak{P}(\mathfrak{W})$.

Definition 2 [10]. *The collection of fuzzy sets of the universe* $\mathfrak{W}$ *be* $\mathfrak{P}(\mathfrak{W})$, $\mathfrak{O} \subseteq \mathfrak{E}$ *where* $\mathfrak{E}$ *be the set of parameters. A pair* $(\mathfrak{d}, \mathfrak{O})$ *is fuzzy soft, where* $\mathfrak{d}$ *is a function from* $\mathfrak{O}$ *to* $\mathfrak{P}(\mathfrak{W})$.

Definition 3 [12]. *Let* $\mathfrak{P}(\mathfrak{W})$ *be the family of neutrosophic sets in the universe* $\mathfrak{W}$, $\mathfrak{O} \subseteq \mathfrak{E}$ *where* $\mathfrak{E}$ *be the collection of parameters. A couple* $(\mathfrak{d}, \mathfrak{O})$ *is neutrosophic soft,* $\mathfrak{d}$ *is a function given as* $\mathfrak{d} : \mathfrak{O} \to \mathfrak{P}(\mathfrak{W})$.

Definition 4 [23]. *Let $\mathfrak{P}(\mathfrak{W})$ be the set of Pythagorean fuzzy sets of the universe $\mathfrak{W}$, $\mathfrak{O} \subseteq \mathfrak{E}$ where $\mathfrak{E}$ be the collection of parameters. A duo $(\mathfrak{d}, \mathfrak{O})$ is Pythagorean fuzzy soft, with function $\mathfrak{d} : \mathfrak{O} \to \mathfrak{P}(\mathfrak{W})$.*

Definition 5 [6]. *A Pythagorean Neutrosophic set with dependent components membership (μ), non-membership (σ) and independent indeterminacy (β) component on universe $\mathfrak{W}$ is $\mathfrak{K} = \{(\mathfrak{k}, \mu_{\mathfrak{K}}(\mathfrak{k}), \beta_{\mathfrak{K}}(\mathfrak{k}), \sigma_{\mathfrak{K}}(k)) : \mathfrak{k} \in \mathfrak{W}\}$, where $\mu_{\mathfrak{K}}(k), \beta_{\mathfrak{K}}(k), \sigma_{\mathfrak{K}}(\mathfrak{k}) \in [0, 1]$, $0 \leq (\mu_{\mathfrak{K}}(\mathfrak{k}))^2 + (\beta_{\mathfrak{K}}(\mathfrak{k}))^2 + (\sigma_{\mathfrak{K}}(\mathfrak{k}))^2 \leq 2$.*

Definition 6 [7]. *The family of all intuitionistic neutrosophic sets of $\mathfrak{W}$ be $\mathfrak{N}(\mathfrak{W})$ and $\mathfrak{O} \subseteq \mathfrak{E}$, the collection of parameters. The set of all $(\mathfrak{d}, \mathfrak{O})$ is Intuitionistic neutrosophic soft set over $\mathfrak{W}$, where $\mathfrak{d}$ is given by $\mathfrak{d} : \mathfrak{O} \to \mathfrak{N}(\mathfrak{W})$.*

3 Pythagorean Neutrosophic Soft Set

This section contains the hybrid structures which combine Pythagorean neutrosophic set theory and soft set theory.

Definition 7. *Let $\mathfrak{W}$ be an initial universe set and $\mathfrak{Q} \subset \mathfrak{E}$ be a set of attributes. Let $PN(\mathfrak{W})$ denote the collection of all Pythagorean Neutrosophic (PN) sets of $\mathfrak{W}$. The family $(\mathfrak{d}, \mathfrak{Q})$ is referred as the Pythagorean neutrosophic soft set (PNSS) over $\mathfrak{W}$, with $\mathfrak{d} : \mathfrak{Q} \to PN(\mathfrak{W})$.*

Example 1. Let us consider the universal set $\mathfrak{W} = \{p_1, p_2, p_3, p_4\}$ where p_i denote the mobile phones which are recent on the market and attributes $\mathfrak{P} = \{o_1, o_2, o_3\}$ where o_1 = reasonable price, o_2 = good specification, o_3 = user-friendly.

Thus,

$$\mathfrak{d}(o_1) = \{< p_1, .6, .3, .4 >, < p_2, .7, .3, .5 >, < p_3, .8, .4, .4 >, < p_4, .7, .5, .4 >\}$$
$$\mathfrak{d}(o_2) = \{< p_1, .5, .4, .6 >, < p_2, .6, .4, .3 >, < p_3, .8, .2, .5 >, < p_4, .4, .3, .4 >\}$$
$$\mathfrak{d}(o_3) = \{< p_1, .5, .6, .3 >, < p_2, .4, .7, .2 >, < p_3, .3, .5, .6 >, < p_4, .8, .2, .4 >\}$$

The PNSS $(\mathfrak{d}, \mathfrak{E})$ is

$$\{o_1 = \{< p_1, .6, .3, .4 >, < p_2, .7, .3, .5 >, < p_3, .8, .4, .4 >, < p_4, .7, .5, .4 >\},$$
$$o_2 = \{< p_1, .5, .4, .6 >, < p_2, .6, .4, .3 >, < p_3, .8, .2, .5 >, < p_4, .4, .3, .4 >\},$$
$$o_3 = \{< p_1, .5, .6, .3 >, < p_2, .4, .7, .2 >, < p_3, .3, .5, .6 >, < p_4, .8, .2, .4 >\}\}$$

Definition 8. *For two PNSSs $(\mathfrak{d}, \mathfrak{Q})$ and $(\mathfrak{b}, \mathfrak{S})$ over $\mathfrak{W}$. $(\mathfrak{d}, \mathfrak{Q})$ is a Pythagorean Neutrosophic soft (PNS) subset of $(\mathfrak{b}, \mathfrak{S})$ iff*

1. $\mathfrak{Q} \subset \mathfrak{S}$
2. $\mu_{\mathfrak{d}(\mathfrak{k})}(\mathfrak{s}) \leq \mu_{\mathfrak{b}(\mathfrak{k})}(\mathfrak{s})$,
   $\gamma_{\mathfrak{d}(\mathfrak{k})}(\mathfrak{s}) \leq \gamma_{\mathfrak{b}(\mathfrak{k})}(\mathfrak{s})$,
   $\sigma_{\mathfrak{d}(\mathfrak{k})}(\mathfrak{s}) \geq \sigma_{\mathfrak{b}(\mathfrak{k})}(\mathfrak{s}) \forall \mathfrak{k} \in \mathfrak{Q}$, $\mathfrak{s} \in \mathfrak{W}$.

It can also be defined as $(\mathfrak{b}, \mathfrak{S})$, *a PNS superset of* $(\mathfrak{d}, \mathfrak{Q})$ *if* $\mathfrak{Q} \subset \mathfrak{S}$ *and* $(\mathfrak{d}, \mathfrak{Q})$ *is a PNS subset of* $(\mathfrak{b}, \mathfrak{S})$ *and denoted by* $(\mathfrak{b}, \mathfrak{S}) \supseteq (\mathfrak{d}, \mathfrak{Q})$.

Definition 9. *Two PNSS* $(\mathfrak{d}, \mathfrak{Q})$ *and* $(\mathfrak{b}, \mathfrak{S})$ *over* $\mathfrak{W}$ *are PNSS equal if* $(\mathfrak{d}, \mathfrak{Q})$ *is contained in* $(\mathfrak{b}, \mathfrak{S})$ *and* $(\mathfrak{b}, \mathfrak{S})$ *is a PNS subset of* $(\mathfrak{d}, \mathfrak{Q})$ *and represented as* $(\mathfrak{d}, \mathfrak{Q}) = (\mathfrak{b}, \mathfrak{S})$.

Definition 10. *A PNSS* $(\mathfrak{d}, \mathfrak{Q})$ *over* $\mathfrak{W}$ *is null if* $\mu_{\mathfrak{d}(\mathfrak{q}))}(\mathfrak{k}) = 0$, $\gamma_{\mathfrak{d}(\mathfrak{q})}(\mathfrak{k}) = 0$ *and* $\sigma_{\mathfrak{d}(\mathfrak{q})}(\mathfrak{k}) = 0$, $forall \mathfrak{k} \in \mathfrak{U}$, $\mathfrak{q} \in \mathfrak{Q}$ *and symbolized as* $\phi_{\mathfrak{Q}}$.

Definition 11. *Let* $(\mathfrak{d}, \mathfrak{Q})$ *and* $(\mathfrak{b}, \mathfrak{S})$ *be PNSSs on* $\mathfrak{W}$. *Then the union is* $(\mathfrak{d}, \mathfrak{Q}) \cup (\mathfrak{b}, \mathfrak{S}) = (\mathfrak{f}, \mathfrak{T})$ *where* $\mathfrak{T} = \mathfrak{Q} \cup \mathfrak{S}$ *and the membership of* $(\mathfrak{f}, \mathfrak{T})$ *are:*

$$\begin{aligned}
\mu_{\mathfrak{f}(\mathfrak{k})}(\mathfrak{o}) = \ & \mu_{\mathfrak{d}(\mathfrak{k})}(\mathfrak{o}) && if\ \ \mathfrak{k} \in \mathfrak{Q} - \mathfrak{S} \\
& = \mu_{\mathfrak{b}(\mathfrak{k})}(\mathfrak{o}) && if\ \ \mathfrak{k} \in \mathfrak{S} - \mathfrak{Q} \\
& = max(\mu_{\mathfrak{d}(\mathfrak{k})}(\mathfrak{o}), \mu_{G(\mathfrak{k})}(\mathfrak{o})) && if\ \ \mathfrak{k} \in \mathfrak{Q} \cap \mathfrak{S} \\
\gamma_{\mathfrak{f}(\mathfrak{k})}(\mathfrak{o}) = \ & \gamma_{\mathfrak{d}(\mathfrak{k})}(\mathfrak{o}) && if\ \ \mathfrak{k} \in \mathfrak{Q} - \mathfrak{S} \\
& = \gamma_{\mathfrak{b}(\mathfrak{d})}(\mathfrak{o}) && if\ \ \mathfrak{k} \in \mathfrak{S} - \mathfrak{Q} \\
& = max(\gamma_{\mathfrak{d}(\mathfrak{k})}(\mathfrak{o}), \gamma_{\mathfrak{b}(\mathfrak{k})}(\mathfrak{o})) && if\ \ \mathfrak{k} \in \mathfrak{Q} \cap \mathfrak{S} \\
\sigma_{\mathfrak{f}(\mathfrak{k})}(\mathfrak{o}) = \ & \sigma_{\mathfrak{d}(\mathfrak{k})}(\mathfrak{o}) && if\ \ \mathfrak{k} \in \mathfrak{Q} - \mathfrak{S} \\
& = \sigma_{\mathfrak{b}(\mathfrak{k})}(\mathfrak{o}) && if\ \ \mathfrak{k} \in \mathfrak{S} - \mathfrak{Q} \\
& = min(\sigma_{\mathfrak{d}(\mathfrak{k})}(\mathfrak{o}), \sigma_{\mathfrak{b}(\mathfrak{k})}(\mathfrak{o})) && if\ \ \mathfrak{k} \in \mathfrak{Q} \cap \mathfrak{S}
\end{aligned}$$

Example 2. Let $(\mathfrak{d}, \mathfrak{Q})$ and $(\mathfrak{b}, \mathfrak{S})$ be PNSSs on $\mathfrak{W}$.

The PNSS $(\mathfrak{d}, \mathfrak{E})$ is

$$\begin{aligned}
&\{o_1 = \{< p_1, .6, .3, .4 >, < p_2, .7, .3, .5 >, < p_3, .8, .4, .4 >, < p_4, .7, .5, .4 >\}, \\
&o_2 = \{< p_1, .5, .4, .6 >, < p_2, .6, .4, .3 >, < p_3, .8, .2, .5 >, < p_4, .4, .3, .4 >\}, \\
&o_3 = \{< p_1, .5, .6, .3 >, < p_2, .4, .7, .2 >, < p_3, .3, .5, .6 >, < p_4, .8, .2, .4 >\}\}
\end{aligned}$$

The PNSS $(\mathfrak{b}, \mathfrak{S})$ is

$$\begin{aligned}
&\{o_2 = \{< p_1, .8, .4, .6 >, < p_2, .7, .4, .3 >, < p_3, .7, .3, .5 >, < p_4, .6, .5, .5 >\}, \\
&o_4 = \{< p_1, .8, .4, .4 >, < p_2, .5, .4, .6 >, < p_3, .8, .2, .4 >, < p_4, .7, .4, .2 >\}\}
\end{aligned}$$

The union is

$$\begin{aligned}
&\{o_1 = \{< p_1, .6, .3, .4 >, < p_2, .7, .3, .5 >, < p_3, .8, .4, .4 >, < p_4, .7, .5, .4 >\}, \\
&o_2 = \{< p_1, .8, .4, .6 >, < p_2, .7, .4, .3 >, < p_3, .8, .3, .5 >, < p_4, .6, .5, .4 >\}, \\
&o_3 = \{< p_1, .5, .6, .3 >, < p_2, .4, .7, .2 >, < p_3, .3, .5, .6 >, < p_4, .8, .2, .4 >\}, \\
&o_4 = \{< p_1, .8, .4, .4 >, < p_2, .5, .4, .6 >, < p_3, .8, .2, .4 >, < p_4, .7, .4, .2 >\}\}
\end{aligned}$$

Definition 12. *Let* $(\mathfrak{d}, \mathfrak{Q})$ *and* $(\mathfrak{b}, \mathfrak{S})$ *be two PNSSs over* $\mathfrak{W}$ *such that* $\mathfrak{Q} \cap \mathfrak{S} \neq \phi$. *The operation of intersection is* $(\mathfrak{d}, \mathfrak{Q}) \cap (\mathfrak{b}, \mathfrak{S}) = (\mathfrak{f}, \mathfrak{T})$ *where* $\mathfrak{T} = \mathfrak{Q} \cap \mathfrak{S}$, *the membership of* $(\mathfrak{f}, \mathfrak{T})$ *are:*

$\mu_{\mathfrak{f}(\mathfrak{e})}(\mathfrak{q}) = min(\mu_{\mathfrak{d}(\mathfrak{e})}(\mathfrak{q}), \mu_{\mathfrak{b}(\mathfrak{e})}(\mathfrak{q}))$,
$\gamma_{\mathfrak{f}(\mathfrak{e})}(\mathfrak{q}) = min(\gamma_{\mathfrak{d}(\mathfrak{e})}(\mathfrak{q}), \gamma_{\mathfrak{b}(\mathfrak{e})}(\mathfrak{q}))$,
$\sigma_{\mathfrak{f}(\mathfrak{e})}(\mathfrak{q}) = max(\sigma_{\mathfrak{d}(\mathfrak{e})}(\mathfrak{q}), \sigma_{\mathfrak{b}(\mathfrak{e})}(\mathfrak{q})) \forall \mathfrak{e} \in \mathfrak{T}$.

Example 3. Consider the Example 2. The tabular form of intersection of PNSSs $((\mathfrak{d}, \mathfrak{Q})$ and $(\mathfrak{b}, \mathfrak{S}))$ is (Table 1)

Table 1. Intersection of $(\mathfrak{d}, \mathfrak{Q})$ and $(\mathfrak{b}, \mathfrak{S})$

$(\mathfrak{d}, \mathfrak{Q}) \cap (\mathfrak{b}, \mathfrak{S})$	o_2
p_1	.5,.4,.6
p_2	.6,.4,.3
p_3	.7,.2,.5
p_4	.4,.3,.5

Definition 13. *IF* $(\mathfrak{d}, \mathfrak{Q}), (\mathfrak{b}, \mathfrak{S})$ *and* $(\mathfrak{f}, \mathfrak{T})$ *are PNSSs over* $\mathfrak{W}$ *then:*

1. $(\mathfrak{f}, \mathfrak{T}) \cup (\mathfrak{f}, \mathfrak{T}) = (\mathfrak{f}, \mathfrak{T})$
   $(\mathfrak{b}, \mathfrak{S}) \cap (\mathfrak{b}, \mathfrak{S}) = (\mathfrak{b}, \mathfrak{S})$
2. $(\mathfrak{d}, \mathfrak{Q}) \cup (\mathfrak{f}, \mathfrak{T}) = (\mathfrak{f}, \mathfrak{T}) \cup (\mathfrak{d}, \mathfrak{Q})$
   $(\mathfrak{f}, \mathfrak{T}) \cap (\mathfrak{b}, \mathfrak{S}) = (\mathfrak{b}, \mathfrak{S}) \cap (\mathfrak{f}, \mathfrak{T})$
3. $(\mathfrak{f}, \mathfrak{T}) \cup \phi = (\mathfrak{f}, \mathfrak{T})$
4. $(\mathfrak{b}, \mathfrak{S}) \cap \phi = \phi$
5. $[(\mathfrak{f}, \mathfrak{T})^c]^c = (\mathfrak{f}, \mathfrak{T})$
6. $(\mathfrak{f}, \mathfrak{T}) \cap [(\mathfrak{b}, \mathfrak{S}) \cap (\mathfrak{d}, \mathfrak{Q})] = [\mathfrak{f}, \mathfrak{T}) \cap (\mathfrak{b}, \mathfrak{S})] \cap (\mathfrak{d}, \mathfrak{Q})$
7. $(\mathfrak{f}, \mathfrak{T}) \cup [(\mathfrak{d}, \mathfrak{Q}) \cup (\mathfrak{b}, \mathfrak{S})] = [(\mathfrak{f}, \mathfrak{T}) \cup (\mathfrak{d}, \mathfrak{Q})] \cup (\mathfrak{b}, \mathfrak{S})$
8. $(\mathfrak{f}, \mathfrak{T}) \cup [(\mathfrak{b}, \mathfrak{S}) \cap (\mathfrak{d}, \mathfrak{Q})] = [(\mathfrak{f}, \mathfrak{T}) \cup (\mathfrak{b}, \mathfrak{S})] \cap [(\mathfrak{f}, \mathfrak{T}) \cup (\mathfrak{d}, \mathfrak{Q})]$
9. $(\mathfrak{d}, \mathfrak{Q}) \cap [(\mathfrak{f}, \mathfrak{T}) \cup (\mathfrak{b}, \mathfrak{S})] = [(\mathfrak{d}, \mathfrak{Q}) \cap (\mathfrak{f}, \mathfrak{T})] \cup [(\mathfrak{d}, \mathfrak{Q}) \cap (\mathfrak{b}, \mathfrak{S})]$

4 Application in Decision Making Problem

Soft sets play a major role in the application part of decision-making. There are many methods and models proposed using the fuzzy soft sets to deal with the application of decision making in a real-life situation. We propose a new algorithm for decision making using the newly defined PNSS.

4.1 Algorithm

1. Input the PNSS $(\mathfrak{d}, \mathfrak{Q})$.
2. Write the PNSS $(\mathfrak{d}, \mathfrak{Q})$ in tabular form.
3. Compute the score matrix of PNSS $(\mathfrak{d}, \mathfrak{Q})$ by using score function $\mu + \gamma - \sigma$ for each $\mathfrak{N}_i (i = 1 \ to \ 5)$.
4. Compute the choice value of each $\mathfrak{N}_i$ by using $S_i = \sum c_{ij} (j = 1 \ to \ 5)$.
5. Choose *max* S_i as the optimal choice. If more than one value of $\mathfrak{N}_i$ are equal then any one of them may be chosen as optimal.

4.2 Illustration

Let us consider the companies $\mathfrak{N}_i(i = 1, 2, ...5)$ proposing digital business models in this growing area of digital world. The decision-making expert makes a comparison among these companies based on the attributes $c_i(i = 1, 2, ...5)$ namely,

c_1 = Customer centric
c_2 = Collaborative, open
c_3 = Value creation & delivery
c_4 = Low latency to real-time data processing
c_5 = Implementable innovation strategy.

Thus let us create PNSS for the attributes $c_i(i = 1, 2, ...5)$ $\mathfrak{W} = \{\mathfrak{N}_1, \mathfrak{N}_2, ...\mathfrak{N}_5\}$ and the set of parameters $\mathfrak{Q} = \{c_1, c_2, ...c_5\}$.

$$\begin{aligned}
\mathfrak{d}(c_1) = \{&<\mathfrak{N}_1, .6, .3, .4>, <\mathfrak{N}_2, .7, .3, .5>, <\mathfrak{N}_3, .8, .4, .4>, <\mathfrak{N}_4, .7, .5, .4>,\\
&<\mathfrak{N}_5, .4, .7, .5>\}\\
\mathfrak{d}(c_2) = \{&<\mathfrak{N}_1, .5, .4, .6>, <\mathfrak{N}_2, .6, .4, .3>, <\mathfrak{N}_3, .8, .2, .5>, <\mathfrak{N}_4, .4, .3, .4>,\\
&<\mathfrak{N}_5, .4, .6, .3>\}\\
\mathfrak{d}(c_3) = \{&<\mathfrak{N}_1, .5, .6, .3>, <\mathfrak{N}_2, .4, .7, .2>, <\mathfrak{N}_3, .3, .5, .6>, <\mathfrak{N}_4, .8, .2, .4>,\\
&<\mathfrak{N}_5, .7, .3, .4>\}\\
\mathfrak{d}(c_4) = \{&<\mathfrak{N}_1, .8, .3, .2>, <\mathfrak{N}_2, .4, .8, .3>, <\mathfrak{N}_3, .7, .1, .2>, <\mathfrak{N}_4, .8, .4, .3>,\\
&<\mathfrak{N}_5, .6, .4, .3>\}\\
\mathfrak{d}(c_5) = \{&<\mathfrak{N}_1, .6, .3, .2>, <\mathfrak{N}_2, .8, .2, .1>, <\mathfrak{N}_3, .7, .4, .3>, <\mathfrak{N}_4, .8, .3, .5>,\\
&<\mathfrak{N}_5, .6, .5, .4>\}
\end{aligned}$$

The PNSS $(\mathfrak{d}, \mathfrak{Q})$ is

$$\begin{aligned}
\{c_1 = \{&<\mathfrak{N}_1, .6, .3, .4>, <\mathfrak{N}_2, .7, .3, .5>, <\mathfrak{N}_3, .8, .4, .4>, <\mathfrak{N}_4, .7, .5, .4>,\\
&<\mathfrak{N}_5, .4, .7, .5>\},\\
c_2 = \{&<\mathfrak{N}_1, .5, .4, .6>, <\mathfrak{N}_2, .6, .4, .3>, <\mathfrak{N}_3, .8, .2, .5>, <\mathfrak{N}_4, .4, .3, .4>,\\
&<\mathfrak{N}_5, .4, .6, .3>\},\\
c_3 = \{&<\mathfrak{N}_1, .5, .6, .3>, <\mathfrak{N}_2, .4, .7, .2>, <\mathfrak{N}_3, .3, .5, .6>, <\mathfrak{N}_4, .8, .2, .4>,\\
&<\mathfrak{N}_5, .7, .3, .4>\},\\
c_4 = \{&<\mathfrak{N}_1, .8, .3, .2>, <\mathfrak{N}_2, .4, .8, .3>, <\mathfrak{N}_3, .7, .1, .2>, <\mathfrak{N}_4, .8, .4, .3>,\\
&<M_5, .6, .4, .3>\},\\
c_5 = \{&<\mathfrak{N}_1, .6, .3, .2>, <\mathfrak{N}_2, .8, .2, .1>, <\mathfrak{N}_3, .7, .4, .3>, <\mathfrak{N}_4, .8, .3, .5>,\\
&<\mathfrak{N}_5, .6, .5, .4>\}\}
\end{aligned}$$

The PNSS $(\mathfrak{d}, \mathfrak{Q})$ is presented in the Table 2, the score matrix of the PNSS is represented in the Table 3 and the choice value for each $\mathfrak{N}_i$ is shown in Table 4.

Table 2. PNSS $(\mathfrak{d}, \mathfrak{Q})$

$\mathfrak{W}$	c_1	c_2	c_3	c_4	c_5
$\mathfrak{N}_1$	(.6,.3,.4)	(.5,.4,.6)	(.5,.6,.3)	(.8,.3,.2)	(.6,.3,.2)
$\mathfrak{N}_2$	(.7,.3,.5)	(.6,.4,.3)	(.4,.7,.2)	(.4,.8,.3)	(.8,.2,.1)
$\mathfrak{N}_3$	(.8,.4,.4)	(.8,.2,.5)	(.3,.5,.6)	(.7,.1,.2)	(.7,.4,.3)
$\mathfrak{N}_4$	(.7,.5,.4)	(.4,.3,.4)	(.8,.2,.4)	(.8,.4,.3)	(.8,.3,.5)
$\mathfrak{N}_5$	(.4,.7,.5)	(.4,.6,.3)	(.7,.3,.4)	(.6,.4,.3)	(.6,.5,.4)

Table 3. Score matrix of the PNSS $(\mathfrak{d}, \mathfrak{Q})$

$\mathfrak{W}$	c_1	c_2	c_3	c_4	c_5
$\mathfrak{N}_1$	.5	.3	.8	.9	.7
$\mathfrak{N}_2$	.5	.7	.9	.9	.9
$\mathfrak{N}_3$	.8	.5	.2	.6	.8
$\mathfrak{N}_4$	.8	.3	.6	.9	.6
$\mathfrak{N}_5$	.6	.7	.6	.7	.7

Table 4. Choice value

$\mathfrak{W}$	$\mathfrak{N}_1$	$\mathfrak{N}_2$	$\mathfrak{N}_3$	$\mathfrak{N}_4$	$\mathfrak{N}_5$
S_i	3.2	3.9	2.9	3.2	3.3

Clearly, the maximum value is 3.9 for $\mathfrak{N}_2$. The final ranking of the proposed algorithm is $\mathfrak{N}_2 > \mathfrak{N}_1 > \mathfrak{N}_4 > \mathfrak{N}_5 > \mathfrak{N}_3$. Hence $\mathfrak{N}_2$ is chosen as the company with the best digital model.

The proposed model is compared with the decision making method in [7] and is verified that the same ranking is obtained. Table 5 provides a comparison of both algorithms, showing the optimal alternative and results. Both algorithms provide the same optimum decision, as can be seen in the comparison table.

Table 5. Comparison of final ranking

Methods	Final ranking	Optimal alternative
1	$\mathfrak{N}_2 > \mathfrak{N}_1 > \mathfrak{N}_4 > \mathfrak{N}_5 > \mathfrak{N}_3$	$\mathfrak{N}_2$
2	$\mathfrak{N}_2 > \mathfrak{N}_1 > \mathfrak{N}_4 > \mathfrak{N}_5 > \mathfrak{N}_3$	$\mathfrak{N}_2$

5 Conclusion

Herein, we developed a new set called Pythagorean Neutrosophic soft set, which is a combination of Pythagorean neutrosophic sets and soft sets. We have also defined and discussed a few operations like subset, union, intersection and some distributive properties for Pythagorean Neutrosophic soft sets. A decision making method based on Pythagorean neutrosophic soft sets is demonstrated as well. In future, this work can be extended to Pythagorean Neutrosophic soft graphs, Dombi graphs, energy and some basic soft graph operations and their properties can be studied and applied to real-world problems.

References

1. Zadeh, L.A.: Fuzzy sets. Inform. Control **8**, 338–353 (1965)
2. Atanassov, K.: Intuitionistic fuzzy sets. Fuzzy Sets Syst. **20**, 87–96 (1986)
3. Yager, R.R.: Pythagorean fuzzy subsets. In: Proceedings of Joint IFSA World Congress and NAFIPS Annual Meeting, Edmonton, Canada, pp. 57–61 (2013)
4. Smarandache, F.: A Unifying Field in Logics: Neutrosophic Logic, Neutrosophy, Neutrosophic Set, Neutrosophic Probability, pp. 1–141. American Research Press, Rehoboth (1999)
5. Jansi, R., Mohana, R.K., Smarandache, F.: Correlation measure for pythagorean neutrosophic fuzzy sets with T and F as dependent neutrosophic components. Neutrosophic Sets Syst. **30**(1), 202–212 (2019)
6. Ajay, D., Chellamani, P.: Pythagorean neutrosophic fuzzy graphs. Int. J. Neutrosophic Sci. **11**(2), 108–114 (2020)
7. Smarandache, F., Said, B.: Intuitionistic neutrosophic soft set. J. Inf. Comput. Sci. **8**(2), 130–140 (2013)
8. Molodtsov, D.A.: Soft set theory-first results. Comp. Math. Appl. **37**, 19–31 (1999)
9. Ali, M.I., Feng, F., Liu, X.Y., Min, W.K., Shabir, M.: On some new operations in soft set theory. Comput. Math. Appl. **57**, 1547–1553 (2009)
10. Maji, P.K., Biswas, R., Roy, A.R.: Fuzzy soft sets. J. Fuzzy Math. **9**(15), 589–602 (2001)
11. Maji, P.K., Biswas, R., Roy, A.R.: Intuitionistic fuzzy soft sets. J. Fuzzy Math. **9**(15), 677–692 (2001)
12. Maji, P.K.: Neutrosophic soft set. Ann. Fuzzy Math. Inform. **5**(13), 157–168 (2013)
13. Wang, H., Smarandache, F., Sunderraman, R., Zhang, Y.Q.: Single valued neutrosophic sets. Multi-space Multi-struct. **4**, 410–413 (2010)
14. Babitha, K.V., Sunil, J.J.: Soft set relations and functions. Comput. Math. Appl. **60**, 1840–1849 (2010)
15. Gong, K., Xiao, Z., Zhang, X.: Exclusive disjunctive soft sets. Comput. Math. Appl. **60**, 2270–2278 (2010)
16. Jiang, Y., Tang, Y., Chen, Q., Wang, J., Tang, S.: Extending soft sets with description logics. Comput. Math. Appl. **59**, 2087–2096 (2010)
17. Mohanty, R.K., Tripathy, B.K.: An Improved Approach to Group Decision-Making Using Intuitionistic Fuzzy Soft Set. Advances in Distributed Computing and Machine Learning, pp. 283–296 (2021)
18. Paik, B., Mondal, S.K.: Representation and application of Fuzzy soft sets in type-2 environment. Complex Intell. Syst. 1–21 (2021)
19. Smarandache, F.: Extension of soft set to hypersoft set, and then to plithogenic hypersoft set. Neutrosophic Sets Syst. **22**, 168–170 (2018)
20. Yolcu, A., Ozturk, T.Y.: Fuzzy hypersoft sets and it's application to decision-making. Theory Appl. Hypersoft Set. Puns Pub. House **50** (2021)

21. Suo, C., Li, Y., Li, Z.: A series of information measures of hesitant fuzzy soft sets and their application in decision making. Soft. Comput. **25**(6), 4771–4784 (2021)
22. Dizman, T.S., Ozturk, T.Y.: Fuzzy bipolar soft topological spaces. TWMS J. Appl. Eng. Math. **11**(1), 151 (2021)
23. Peng, X., Yang, Y., Song, J., Jiang, Y.: Pythagorean fuzzy soft set and its application. Comput. Eng. **41**(7), 224–229 (2015)

Key Challenges of Lithium-Ion Battery Recycling Process in Circular Economy Environment: Pythagorean Fuzzy AHP Approach

Abdullah Yıldızbaşı(✉), Cihat Öztürk, İbrahim Yılmaz, and Yağmur Arıöz

Department of Industrial Engineering, Ankara Yıldırım Beyazıt University (AYBU), 06010 Ankara, Turkey

ayildizbasi@ybu.edu.tr

Abstract. With the technological change and development in recent years, the market shares of li-on batteries, which provide energy efficiency and cost effectiveness, is getting up rapidly. Increasing demand for li-ion batteries used in energy storage, consumer electronics and electric vehicles negatively affects the raw material supply. Materials such as lithium, cobalt and graphite, which have an important place in the production of li-on batteries, are very rare in the market and the countries where these minerals are extracted are quite limited. In addition, many problems arise in the supply chain processes due to the trade policies of these countries. Besides, the environmental and social impacts that occur with the increasing demand for li-on batteries stand out as another problem. In this study, an analytical approach to recycling li-on batteries is presented with a circular economy perspective. With a recycling framework within the current li-on battery life cycle process, we focused on the challenges that arise in this process. From the circular economy perspective, challenges of the recycling process of lion batteries are prioritized by examining their technological, environmental, social and economic dimensions. AHP method is used to prioritize challenges by weighing them over different dimensions. Also, Pythagoras fuzzy numbers are integrated in-to AHP method to overcome uncertainty. In the last part of the study, the results are supported with the managerial implications for li-on battery recycling.

Keywords: Pythagorean fuzzy AHP · Sustainability · Lithium-ion battery · Circular economy

1 Introduction

Lithium-ion batteries (LIBs) are commonly used to power advanced technological devices and applications that require high energy density. Because of these batteries' high energy density, reduced memory effect, and ability to support a significant number of charge/discharge cycles, Electric and Hybrid Vehicles are among such examples [1]. With the increasing demand for these energy-intensive applications, there has also been a significant increase in the demand for LIBs and therefore for the materials used in the

C. Kahraman et al. (Eds.): INFUS 2021, LNNS 308, pp. 561–568, 2022.
https://doi.org/10.1007/978-3-030-85577-2_66

manufacture of these batteries, including lithium, cobalt, and nickel. The demand for electric vehicles, which is expected to increase rapidly in the near future, will further increase the demand for Lithium-ion batteries [10]. Therefore, there will be a need to focus on effective solutions in mining, refining, and production operations as well as extensive mineral reserves. This increasing demand naturally leads us to question the sustainability of Lithium-ion batteries and re-think their end-of-life management to come up with solutions that will minimize the environmental damage [1]. Unfortunately, the recycling process of Lithium-ion batteries is still in development and it is far from reaching the high recycling rates that could be achieved by implementing effective technological solutions. Currently, these recycling practices are being carried out with monetary gain and reduced dependence on foreign resources and critical materials in mind. Recent increases in lithium, cobalt, and nickel prices are causing recycling to be viewed as an important item for economic income by making it more than just a way to gain material independence. For this reason, the number of facilities for the recycling of Lithium-ion batteries or the capacities of the existing ones is being increased. Here, the concept of Circular Economy (CE), which aims to increase efficiency and is frequently used in the literature, comes to the fore.

Circular Economy is a holistic approach that includes activities such as reusing, recycling, and remanufacturing to reduce waste and ensure the economic sustainability of post-use products, apart from product recycling [2]. However, the high cost of battery-replacement (including processing and collection costs), together with various technological, environmental, social, and economic barriers, and uncertainties regarding the quality, safety, and the remaining life of batteries sent for replacement hinder the large-scale development of secondary battery use. Moreover, the economic value of used batteries is decreasing with the decreasing market price of LIBs and their improving performance characteristics, which in turn is negatively affecting the propensity to invest in second life applications. For this reason, steps should be taken not only for recycling but also for the realization of a circular economy vision in the production of electric vehicles and batteries. At this point, technical, social, economic, and environmental difficulties are encountered, especially in the collection, processing, and recycling of large batteries [11]. After the detailed literature review conducted during the preparation of this manuscript, the number of studies evaluating and analyzing the challenges that might be encountered during this process was determined to be quite limited [3].

In this study, the challenges defined by using the literature were evaluated with the Analytical Hierarchy Process based on Pythagorean fuzzy sets (PF-AHP), which has come to the fore in the literature in recent years. Pythagorean fuzzy sets (PFSs), which are used as a solution especially in uncertainty situations, al-low the defined barriers to be sorted according to importance levels. As far as the authors know, there are no studies in the literature that prioritize CE challenges in Li-ion battery recycling processes using MCDM techniques.

The paper has been structured as follows: Sect. 2 includes an explanation about Circular Economy for LIBs and challenges. PF-AHP method detailed into Sect. 3. Section 4 present the implication for challenges of Li-ion battery recycling process in CE environment by using PF-AHP. Finally, the conclusion and future agenda have been presented in Sect. 5.

2 Circular Economy for LIBs and Key Challenges of Recycling Process

CE has a different conceptual background. The fact that the current definitions of CE are diverse and include all sorts of activities carried out in society prevents a consensus among academics on the interpretation of CE. Concerning production and consumption, the CE concept, which can be addressed at macro, medium, and micro levels, also creates various obstacles that can be encountered in implementation. The most fundamental feature that distinguishes CE from other initiatives aimed at reducing energy and material consumption is that it is a holistic approach to create material, energy, and waste flow cycles that encompass all social activities [4].

As shown in Fig. 1, CE applications are one of the most important approaches that can be used to extend battery life. There are multiple options available at this stage. First, to bring in economy by sending Lithium-ion batteries that have completed their intended use to reuse or replacement facilities for use in "second life" applications. Many batteries that are considered to have completed their intended use retain more than 75–80% of their original capacity [6]. This enables these batteries to be used for other purposes with lower energy demand. Thus, while meeting lower capacity demands, it contributes to the reduction in demand for critical materials.

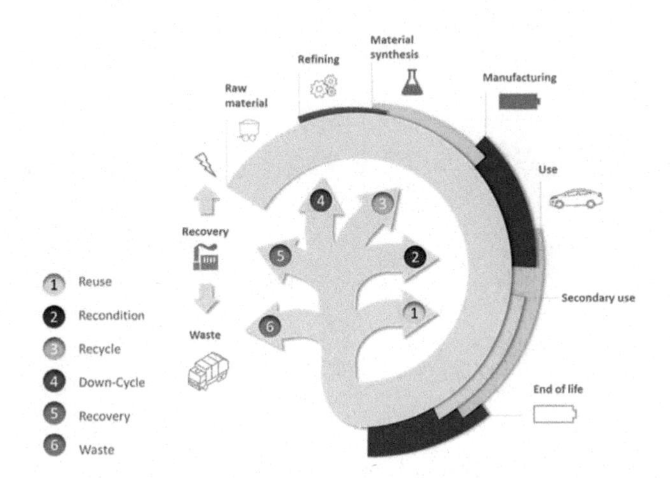

Fig. 1. Open - Closed loop flowchart of end of life (EOL) lithium-ion batteries [5].

This study aims to prioritize issues such as uncertainties in the collection process, the size of the dismantling levels of the collected products, demand uncertainties, high quality expectation, safety, operation and collection costs, and determine an action plan.

3 Pythagorean Fuzzy Sets

PFSs are developed as an extension of intuitionistic fuzzy sets by [7]. It is explained in Definition 1.

Definition 1: A Pythagorean fuzzy set $\tilde{P}$ is an object having the procedure:

$$\tilde{P} \cong \{\langle x, \mu_{\tilde{P}}(x), v_{\tilde{P}}(x)\rangle; x \in X\}, \tag{1}$$

$$0 \leq \mu_{\tilde{P}}(x)^2 + v_{\tilde{P}}(x)^2 \leq 1. \tag{2}$$

$$\mu_{\tilde{P}}(x) = \sqrt{1 - \mu_{\tilde{P}}(x)^2 - v_{\tilde{P}}(x)^2}. \tag{3}$$

Some of the basis operations of PFSs are presented in Definition 2.

Definition 2: Let $\beta_1 = P(\mu_{\beta_1}, v_{\beta_1})$ *and* $\beta_2 = P(\mu_{\beta_2}, v_{\beta_2})$ be two Pythagorean fuzzy numbers, and $\lambda > 0$, some operations indicated as follows:

$$\beta_1 \oplus \beta_2 = P\left(\sqrt{\mu_{\beta_1}^2 + \mu_{\beta_2}^2 - \mu_{\beta_1}^2\mu_{\beta_2}^2}, v_{\beta_1}v_{\beta_2}\right), \tag{4}$$

$$\beta_1 \oplus \beta_2 = P\left(\mu_{\beta_1}\mu_{\beta_2}, \sqrt{v_{\beta_1}^2 + v_{\beta_2}^2 - v_{\beta_1}^2 v_{\beta_2}^2}\right), \tag{5}$$

$$\lambda\beta_1 = P(\sqrt{1 - \left(1 - \mu_{\beta_1}^2\right)^{\lambda}}, (v\beta_1)^{\lambda}), \lambda > 0, \tag{6}$$

$$\beta_1^{\lambda} = P(\left(\mu\beta_1)^{\lambda}, \sqrt{1 - \left(1 - v_{\beta_1}^2\right)^{\lambda}}\right), \lambda > 0, \tag{7}$$

$$\beta_1 \Theta \beta_2 = P\left(\sqrt{\frac{\mu_{\beta_1}^2 - \mu_{\beta_2}^2}{1 - \mu_{\beta_2}^2}, \frac{v_{\beta_1}}{v_{\beta_2}}}\right), \mathit{if}\ \mu_{\beta_1} \geq \mu_{\beta_2}, v_{\beta_1} \min\left\{v_{\beta_2}, \frac{v_{\beta_2}.\pi_{\beta_1}}{\pi_{\beta_2}}\right\}, \tag{8}$$

$$\frac{\beta_1}{\beta_2} = P\left(\frac{\mu_{\beta_1}}{\mu_{\beta_2}}\sqrt{\frac{v_{\beta_1}^2 - v_{\beta_2}^2}{1 - v_{\beta_2}^2}}\right), \mathit{if}\ \mu_{\beta_1} \leq \min\left\{\mu_{\beta_2}, \frac{\mu_{\beta_2}.\pi_{\beta_1}}{\pi_{\beta_2}}\right\}, v_{\beta_1} \geq v_{\beta_2}. \tag{9}$$

π_{β_1} *and* π_{β_2} represent the degree of indeterminacy.

4 The Proposed Model for Key Challenges of Lithium-Ion Battery Recycling Process in Circular Economy Environment

In this study, PF-AHP is proposed for the challenge's evaluation in the Lithium-ion battery recycling process within the framework of the CE environment. The proposed approach consists of two parts. In the first part, the challenges will be defined through a comprehensive literature review and expert interviews. Then a hierarchical structure is created for the challenges defined for Lithium-ion battery recycling. The second part consists of evaluating the challenges identified. For this purpose, PF-AHP methodology is used to obtain main and sub criteria weights. Finally, challenges are ranked according to their weight, and higher priority challenges are identified, and managerial implications are presented.

4.1 Steps of Pythagorean Fuzzy AHP

The steps of Pythagorean fuzzy (PF-AHP) are presented in this sub-section.

Step 1: Linguistic evaluation. The formula of compromised pairwise comparison matrix $R = (r_{ik})_{m\times m}$ depends on the linguistic assessment of experts using a ruler projected by [8] in Table 1.

Table 1. Linguistic scale and Pythagorean Fuzzy number equivalents [8].

Linguistic terms	PF numbers
Certainly Low Importance - CLI	<[0,0], [0.9, 1]>
Very Low Importance - VLI	<[0.1, 0.2], [0.8, 0.9]>
Low Importance - LI	<[0.2, 0.35], [0.65, 0.8]>
Below Average Importance - BAI	<[0.35, 0.45], [0.55, 0.65]>
Average Importance - AI	<[0.45, 0.55], [0.45, 0.55]>
Above Average Importance - AAI	<[0.55, 0.65], [0.35, 0.45]>
High Importance - HI	<[0.65, 0.8], [0.2, 0.35]>
Very High Importance - VHI	<[0.8, 0.9], [0.1, 0.2]>
Certainly High Importance - CHI	<[0.9, 1], [0, 0]>
Exactly Equal - EE	<[0.1965, 0.1965], [0.1965, 0.1965]>

Step 2: Linguistic evaluation. The formula of compromised pairwise comparison matrix $R = (r_{ik})_{m\times m}$ depends on the linguistic assessment of experts using a ruler projected by [9] in Table 1.

Step 3: The difference matrices $D = (d_{ik})_{m\times m}$ are calculated using Eqs. (10) and (11):

$$d_{ik_L} = \mu^2_{ik_L} - \nu^2_{ik_U}, \tag{10}$$

$$d_{ik_U} = \mu^2_{ik_U} - \nu^2_{ik_L}. \tag{11}$$

Step 4: The Eqs. (10) and (11) are figured out interval multiplicative matrix by placing into the Eqs. (12) and (13), respectively

$$S_{ik_L} = \sqrt{1000^{d_{ik_L}}}, \tag{12}$$

$$S_{ik_U} = \sqrt{1000^{d_{ik_U}}}. \tag{13}$$

Step 5: Calculate the indeterminacy value by using Eqs. (14)

$$\tau_{ik} = 1 - \left(\mu^2_{ik_U} - \mu^2_{ik_L}\right) - \left(\nu^2_{ik_U} - \nu^2_{ik_L}\right). \tag{14}$$

Step 6: Find the weight matrix by using Eq. (15).

$$t_{ik} = \left(\frac{S_{ik_L} + S_{ik_U}}{2}\right)\tau_{ik}. \tag{15}$$

Step 7: Calculate the normalized priority weights by Eq. (16).

$$w_i = \frac{\sum_{k=1}^{m} t_{ik}}{\sum_{i=1}^{m}\sum_{k=1}^{m} t_{ik}}. \tag{16}$$

5 Results and Discussion

In this section, a case study was conducted to identify Lithium-ion battery recycling challenges. A company that wants to recycle Lithium-ion batteries by taking advantage of the circular economy perspective has decided to define the challenges it will face in this process and take action on them. For this purpose, challenges are determined through literature review and expert interviews. As a result, 4 main criteria (Technical (T), Economic/Financial (ECO), Social/Policy (S), Environmental (ENV)) and 30 sub criteria *(Incompatible product design (T1), Immaturity of technology (T2), Low reliability of recycled good (T3), Lack of capacity (T4), Lack of know-how (T5), Lack of software (T6), Limited data access (T7), Lack of CE management tools (T8); Cheap virgin material prices (ECO1), High investment cost (ECO2), Low incentives for CE adaptation (ECO3), Price-performance ratio (ECO4), Uncertainty level of recycled materials (ECO5), Financial transition difficulties (ECO6), Energy consumption amount (ECO7), Lack of market mechanism (ECO8); Lack of standards for Lithium-ion battery recycling (S1), Low customer awareness (S2), Renegotiating agreement requirement (S3), Inconsistent policies between partners (S4), Complex waste legislations (S5), Week cooperation (S6), The CE resistant company culture (S7), Lack of tax incentives (S8), Limited certification and warranty (S9), Infrastructure level for waste management (ENV1), Lack of knowledge about hazardous materials (ENV2), CO2 Emission (ENV3), Difficulty of disposal (ENV4), High logistic activity requirements (ENV5)* were determined. Later, 3 decision-makers (DMs) evaluated these main and sub-criteria by using PF-AHP. Table 2 represents pairwise comparisons for 4 main criteria according to linguistic terms.

Table 2. Main criteria linguistic evaluation

	T	ECO	S	ENV
T	EE	LI	BAI	HI
ECO	HI	EE	AAI	CHI
S	AAI	BAI	EE	HI
ENV	LI	CLI	LI	EE

Table 3 represents matrix of weights before normalization.

Table 3. Matrix of weights before normalization

	T	ECO	S	ENV
T	1,000	3,767	1,692	0,168
ECO	0,168	1,000	0,425	0,037
S	0,425	1,692	1,000	0,168
ENV	3,767	19,452	3,767	1,000

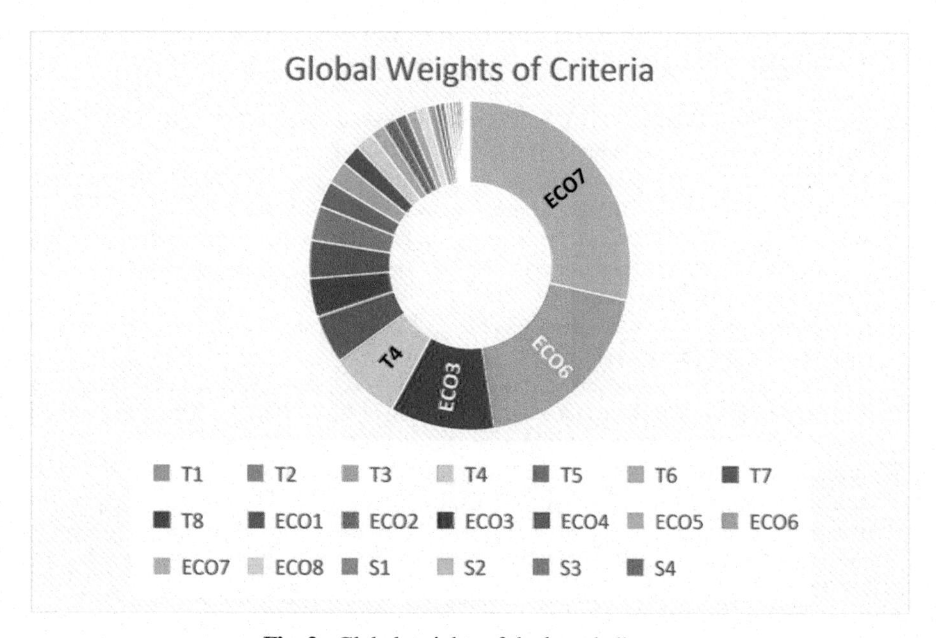

Fig. 2. Global weights of the key challenges

Figure 2 represents the priority weights of key challenges according to PF-AHP. According to the results, the energy consumption amount is determined as the most important challenge with a degree of 0.283. Financial transition difficulties, low incentives for CE adaptation, low reliability of recycled goods are ranked behind, respectively.

6 Conclusion

Lithium-ion batteries are used increasingly every day and naturally, the management of the recycling of these batteries also gains special importance. In this study, the problems that can be encountered in a recycling application to be realized with a circular economy perspective in this recycling process have been discussed and prioritized using the Pythagorean fuzzy Analytic Hierarchy Process. In this prioritization process, Technical, Economic/Financial, Social/Policy, and Environmental challenges are considered. As a result, economical/financial criteria are the most important challenges among the four main challenges and energy consumption amount first, financial transition difficulties second, low incentives for CE adaptation third important sub-challenges respectively. These results show that economic challenges have more critical during the Lithium-ion battery recycling process. More research needs to be developed on different MCDM methodologies to present of consistency.

References

1. Beaudet, A., Larouche, F., Amouzegar, K., Bouchard, P., Zaghib, K.: Key challenges and opportunities for recycling electric vehicle battery materials. Sustainability **12**(14), 5837 (2020)
2. Alamerew, Y.A., Brissaud, D.: Circular economy assessment tool for end of life product recovery strategies. J. Remanuf. **9**(3), 169–185 (2019)
3. Yun, L., et al.: Metallurgical and mechanical methods for recycling of lithium-ion battery pack for electric vehicles. Resour. Conserv. Recy. **136**, 198–208 (2018)
4. Masi, D., Kumar, V., Garza-Reyes, J.A., Godsell, J.: Towards a more circular economy: exploring the awareness, practices, and barriers from a focal firm perspective. Prod. Plan. Control **29**(6), 539–550 (2018)
5. Larouche, F., et al.: Progress and status of hydrometallurgical and direct recycling of Li-ion batteries and beyond. Materials **13**, 801 (2020)
6. Hall, D., Lutsey, N.: Effects of Battery Manufacturing on Electric Vehicle Life-Cycle Greenhouse Gas Emissions. Briefing, International Council on Clean Transportation. https://theicct.org/publications/EV-battery-manufacturing-emissions. Accessed 20 Mar 2021
7. Yager, R.R.: Pythagorean fuzzy subsets. In: 2013 Joint IFSA World Congress and NAFIPS Annual Meeting (IFSA/NAFIPS), pp. 57–61. IEEE (2013)
8. Ilbahar, E., Karaşan, A., Cebi, S., Kahraman, C.: A novel approach to risk assessment for occupational health and safety using Pythagorean fuzzy AHP & fuzzy inference system. Saf. Sci. **103**, 124–136 (2018)
9. Çalık, A.: A novel Pythagorean fuzzy AHP and fuzzy TOPSIS methodology for green supplier selection in the Industry 4.0 era. Soft. Comput. **25**(3), 2253–2265 (2020). https://doi.org/10.1007/s00500-020-05294-9
10. Nykvist, B., Nilsson, M.: Rapidly falling costs of battery packs for electric vehicles. Nat. Clim. Change **5**(4), 329–332 (2015)
11. Mossali, E., et al.: Lithium-Ion batteries towards circular economy: a literature review of opportunities and issues of recycling treatments. J. Environ. Manag. **264**, 110500 (2020)

Prioritization of R&D Projects Using Fermatean Fuzzy MARCOS Method

Irem Ucal Sari(✉) and Sule Nur Sargin

Department of Industrial Engineering, Istanbul Technical University, Istanbul, Turkey
{ucal,sargins}@itu.edu.tr

Abstract. Product and service innovations are of great importance to maintain and increase the current market shares of companies. For this reason, the importance that companies attach to R&D projects is increasing constantly, and the problem of which of the proposed R&D projects should be implemented becomes a very critical decision due to the high cost of investments. Accordingly, in this study, an extension of fuzzy MARCOS method using fermatean fuzzy sets is developed to guide the selection of the R&D project to be invested. The developed methodology is applied to a company in the construction industry using the criteria technological applicability, economic benefits, technical benefits, technical risks, potential market size and financial feasibility to rank four innovative R&D projects.

Keywords: Fermatean fuzzy sets · Fuzzy MARCOS method · R&D projects

1 Introduction

With the rapid spread of Industry 4.0 tools, customer expectations have radically changed and interest in innovative technologies has increased. In order to keep up with changing customer needs and compete with the innovative projects of competitors, companies invest more and more in R&D departments and accelerate their search for innovative projects. Innovative R&D projects stand out as projects that require careful selection during the implementation phase, especially due to their high cost. Selecting the projects that will be successful is very important for the continuity of the companies. Projects that fail to reach the expected level of success can cause the company to lose share in the market and fall into financial difficulties.

The success or failure of innovative R&D projects involves high uncertainty. Fuzzy sets, which are frequently used in modeling under uncertainty, have many extensions that differ by the expression of membership and non-membership functions. In this study, fermatean fuzzy sets are used in modelling uncertainty to handle the ambiguous information expressed by decision makers. The selection or prioritization of R&D projects is a multi-criteria decision making (MCDM) problem in which many criteria such as capital required in the selection of innovative R&D projects, technological complexity, applicability, expected demand, reproducibility, size of loss in case of failure can be considered.

C. Kahraman et al. (Eds.): INFUS 2021, LNNS 308, pp. 569–577, 2022.
https://doi.org/10.1007/978-3-030-85577-2_67

There are lots of MCDM methods proposed in the literature. Fuzzy MARCOS method has some advantages compared to other methods such as being effective and easy to implement, allowing to be take into account many criteria and alternatives while maintaining the stability of the results, considering the ideal and anti-ideal solutions, and defining the utility degree of alternatives in relation to both solutions [1]. Stević et al. [2] developed MARCOS method in 2020. Since then, MARCOS method and its fuzzy extensions are applied in several studies. Chattopadhyay et al. [3] proposed integrated D-MARCOS for resolve the uncertainty with the application of D numbers and MARCOS for ranking of the alternatives. Badi and Pamucar [4] proposed hybrid Grey theory-MARCOS method that determines the weights of the criteria by Grey theory and selects best alternative by the MARCOS model. Ilieva et al. [5] presented fuzzy MARCOS methods to order cloud storage systems. In the literature, there are also hybrid MCDM methods such as FUCOM [6, 7], fuzzy BWM [8], fuzzy PIPRECIA [9] for determine the values of the criteria weights and fuzzy MARCOS for evaluation of alternatives. Gong et al. [1] developed a hybrid decision-making framework that determine attributes' weights by the interval type-2 fuzzy BWM and present to rank alternatives by an interval type-2 MARCOS method. Simić et al. [10], developed a new multiphase model including fuzzy MARCOS method to determine the final ranking. Simić et al. [11] proposed a picture fuzzy MARCOS for railway infrastructure risk assessment. As it is seen from the literature review, many fuzzy extensions of the MARCOS method have been proposed and applied to decision-making problems in different fields.

This study aims to develop the fermatean fuzzy MARCOS method in order to better express the ambiguity and uncertainty by using fermatean fuzzy numbers, which is a relatively new extension of fuzzy numbers and has the advantage of digitizing linguistic expressions using a larger domain. The proposed method is applied in prioritizing innovative R&D projects for a construction company.

The organization of the paper is as follows: In Sect. 2, the basics of fermatean fuzzy sets are given. Then the steps of the fermatean fuzzy MARCOS methods are determined in Sect. 3. After that, the proposed methodology is applied for R&D projects of a construction company in Sect. 4. And finally, paper is concluded with a conclusion section including future research suggestions.

2 Fermatean Fuzzy Sets

Fermatean fuzzy sets are one of the most recent extensions of fuzzy sets that are proposed in [12, 13], in which the membership grade describes the degree of membership and non-membership indicated by a pair of values both from the unit interval. Equation (1) shows the representation of a fermatean fuzzy set (FFS) F on the universe of discourse X:

$$F = \{\langle x, \alpha_F(x), \beta_F(x) \rangle : x \in X\} \tag{1}$$

where the functions $\alpha_F(x)$ and $\beta_F(x)$ indicates the degree of membership and non-membership of x to in the set F respectively. Each element of X is independently mapped to a value between 0 and 1 as $\alpha_F(x) : X \to [0, 1]$ and $\beta_F(x) : X \to [0, 1]$. The sum of membership and non-membership degrees' cubes can be at most equal to 1 [12]. The degree of indeterminacy of x to F is denoted by $\pi_F(x)$ and stated in Eq. (2) for any FFS, F and $x \in X$.

$$\pi_F(x) = \sqrt[3]{1 - (\alpha_F(x))^3 - (\beta_F(x))^3} \quad (2)$$

A fermatean fuzzy number (FFN) is the set $\langle \alpha_F(x), \beta_F(x) \rangle$ in which there is only one element in set X; denoted by $\mathrm{F} = (\alpha_F, \beta_F)$ [15]. Some basic algebraic operations of FFNs are given below [13, 14]:

$$\mathrm{F}_1 \boxplus \mathrm{F}_2 = (\sqrt[3]{\alpha_{\mathrm{F}_1}^3 + \alpha_{\mathrm{F}_2}^3 - \alpha_{\mathrm{F}_1}^3 \alpha_{\mathrm{F}_2}^3}, \beta_{\mathrm{F}_1} \beta_{\mathrm{F}_2}) \quad (3)$$

$$\mathrm{F}_1 \boxtimes \mathrm{F}_2 = (\alpha_{\mathrm{F}_1} \alpha_{\mathrm{F}_2}, \sqrt[3]{\beta_{\mathrm{F}_1}^3 + \beta_{\mathrm{F}_2}^3 - \beta_{\mathrm{F}_1}^3 \beta_{\mathrm{F}_2}^3}) \quad (4)$$

$$\mathrm{F}_1 \boxslash \mathrm{F}_2 = \left(\frac{\alpha_{\mathrm{F}_1}}{\alpha_{\mathrm{F}_2}}, \sqrt[3]{\frac{\beta_{\mathrm{F}_1}{}^3 - \beta_{\mathrm{F}_2}{}^3}{1 - \min(\beta_{F_1}; \beta_{F_2})^3}} \right) \ \text{if } \alpha_{\mathrm{F}_1} \leq \min \left\{ \alpha_{\mathrm{F}_2}, \frac{\alpha_{\mathrm{F}_2} \pi_1}{\pi_2} \right\} \quad (5)$$

To rank FFSs the score function represented in Eq. (7) and the accuracy function represented in Eq. (8) can be used:

$$Score(F = (\alpha_F, \beta_F)) = S(F) = \alpha_F^3 - \beta_F^3 \ where\ S(F) \in [-1, 1] \quad (6)$$

$$acc(F = (\alpha_F, \beta_F)) = \alpha_F^3 + \beta_F^3 \ where\ acc(F) \in [0, 1] \quad (7)$$

Score and accuracy functions of the FFNs can be used to define a ranking technique for any two FFNs [15]. Let $F_1 = (\alpha_{F_1}, \beta_{F_1})$ and $F_2 = (\alpha_{F_2}, \beta_{F_2})$ be two FFNs. $S(F_1)$, $S(F_2)$, $acc(F_1)$ and $acc(F_2)$ are the score values and accuracy values of F_1 and F_2, then

$$\begin{array}{lll} S(F_1) < S(F_2) \ then & & F_1 < F_2 \\ S(F_1) > S(F_2) \ then & & F_1 > F_2 \\ & acc(F_1) < acc(F_2), & then\, F_1 < F_2 \\ S(F_1) = S(F_2) \ then\, if & acc(F_1) > acc(F_2), & then\, F_1 > F_2 \\ & acc(F_1) = acc(F_2), & then\, F_1 = F_2 \end{array} \quad (8)$$

3 Fermatean Fuzzy MARCOS Method

Based on the previously proposed fuzzy MARCOS methods, the steps of fermatean fuzzy MARCOS method are determined as follows:

Step 1. Determination of Decision Criteria, Alternatives and Fermatean Fuzzy Scales of Linguistic Statements: Decision criteria (C_j where $j = 1, 2, \ldots, m$) could be defined based on the literature survey or by the experts using the consensus technique. A_i ($i = 1, 2, \ldots, n$) represents the alternatives that will be evaluated according to the decision criteria. Fermatean fuzzy scale for membership and non-membership functions of linguistic statements for determination of the decision makers' level of expertise and

Table 1. Fermatean fuzzy scale for rating the decision maker and criteria [16]

Linguistic statements	Membership degree	Non-membership degree	Indeterminacy degree
Very Low (VL)	0.05	0.95	0.52
Low(L)	0.25	0.80	0.78
Medium (M)	0.50	0.50	0.90
High (H)	0.80	0.15	0.78
Very High (VH)	0.95	0.01	0.52

Table 2. Fermatean fuzzy scale for rating the alternatives [16]

Linguistic statement	Membership degree	Non-membership degree	Indeterminacy degree
Absolutely very poor (AVP)	0.05	0.95	0.5223
Very poor (VP)	0.15	0.90	0.6444
Poor (P)	0.25	0.80	0.7788
Medium (M)	0.40	0.70	0.8401
Good (G)	0.50	0.60	0.8702
Very good (VG)	0.70	0.20	0.8657
Absolutely very good (AVG)	0.95	0.01	0.5224

importance of the criteria are given in Table 1. Fermatean fuzzy evaluation scale for the alternatives according to criteria is given in Table 2.

Step 2. Determination of Criteria Weights.

In this step, first, the decision maker's weights are determined. Weights are assigned to decision makers based on their level of expertise and normalized to find the weights for each decision maker. λ_k represents the weight of decision maker k (k = 1, 2, .., K) with $\sum_{k=1}^{K} \lambda_k = 1$.

Then, each decision maker determines the criteria weights which is denoted as w_{jk} with $k = 1, 2, \ldots, K$. Fermatean fuzzy weighted aggregation operator (FFWA) [17] could be used to calculate the aggregated criteria weights that is calculated by Eq. (9):

$$w_j = FFWA\left(w_{j1}, w_{j2}, \ldots, w_{jK}\right) = \left(\sum\nolimits_{k=1}^{K} \lambda_k \alpha_{w_{jk}}, \sum\nolimits_{k=1}^{K} \lambda_k \beta_{w_{jk}}\right) \tag{9}$$

Step 3. Form an initial decision-making matrix. The decision makers evaluate the alternatives according to the decision criteria using the linguistic statements given in Table 2. The aggregated fermatean fuzzy equivalents of the respected linguistic evaluations of the decision maker used to form initial decision matrix given in Eq. (10).

$$\tilde{X} = \begin{matrix} & C_1 & C_2 & \dots & C_n \\ \tilde{A}_1 & \tilde{x}_{11} & \tilde{x}_{12} & \dots & \tilde{x}_{1n} \\ \tilde{A}_2 & \tilde{x}_{21} & \tilde{x}_{22} & \dots & \tilde{x}_{2n} \\ \vdots & \dots & \dots & \dots & \dots \\ \tilde{A}_m & \tilde{x}_{m1} & \tilde{x}_{m2} & \dots & \tilde{x}_{mn} \end{matrix} \quad (10)$$

Step 4. Calculate the scores of the fermatean fuzzy set $\approx{x}_{ij}$ using Eq. (6) and construct the ranking decision matrix using Eq. (11):

$$\overline{X}^{*} = \left(S\left(\tilde{x}_{ij}\right)\right)_{m \times n} \quad (11)$$

Step 5. Form an extended decision matrix. First, the ideal solution (IS) that is the alternative with the best characteristics and the anti-ideal solution (AIS) which is the alternative with the worst characteristics are determined using Eq. (12) and Eq. (13) respectively:

$$IS = \left\{ \tilde{x}_{ij} \middle| \max_{i} S\left(\tilde{x}_{ij}\right) \textit{ if } j \in B \textit{ and } \tilde{x}_{ij} \middle| \min_{i} S\left(\tilde{x}_{ij}\right) \textit{ if } j \in C \right\} \quad (12)$$

$$AIS = \left\{ \tilde{x}_{ij} \middle| \min_{i} S\left(\tilde{x}_{ij}\right) \textit{ if } j \in B \textit{ and } \tilde{x}_{ij} \middle| \max_{i} S\left(\tilde{x}_{ij}\right) \textit{ if } j \in C \right\} \quad (13)$$

where B stands for benefit criteria, while C stands for cost criteria. Then, Eq. (14) is used to form the extended decision matrix:

$$\tilde{X} = \begin{matrix} & C_1 & C_2 & \dots & C_n \\ \widetilde{AIS} & \tilde{x}_{aa1} & \tilde{x}_{aa2} & \dots & \tilde{x}_{aan} \\ \tilde{A}_1 & \tilde{x}_{11} & \tilde{x}_{12} & \dots & \tilde{x}_{1n} \\ \tilde{A}_2 & \tilde{x}_{21} & \tilde{x}_{22} & \dots & \tilde{x}_{2n} \\ \vdots & \dots & \dots & \dots & \dots \\ \tilde{A}_m & \tilde{x}_{m1} & \tilde{x}_{m2} & \dots & \tilde{x}_{mn} \\ \widetilde{IS} & \tilde{x}_{ai1} & \tilde{x}_{ai2} & \dots & \tilde{x}_{ain} \end{matrix} \quad (14)$$

Step 6. Normalize extended decision matrix and build the weighted decision matrix. Using linear normalization determined in Eq. (15) extended decision matrix is normalized.

$$\tilde{n}_{ij} = \begin{cases} \dfrac{\tilde{x}_{ij}}{\max\limits_{i=1,\dots,n}\left(\tilde{x}_{ij}\right)} \textit{ if } j \in B \\ \dfrac{\min\limits_{i=1,\dots,n}\left(\tilde{x}_{ij}\right)}{\tilde{x}_{ij}} \textit{ if } j \in C \end{cases} \quad (15)$$

Score and accuracy functions are used to find the maximum and minimum values w.r.t a certain criterion. Then Eq. (16) is used to form the weighted decision matrix:

$$\tilde{v}_{ij} = \tilde{w}_j \tilde{n}_{ij} = \left(\alpha_{\tilde{w}_j}\alpha_{\tilde{n}_{ij}}, \sqrt[3]{\beta^3_{\tilde{w}_j} + \beta^3_{\tilde{n}_{ij}} - \beta^3_{\tilde{w}_j}\beta^3_{\tilde{n}_{ij}}}\right) \quad (16)$$

Step 7. Calculate the fuzzy summation matrix using Eq. (17):

$$\tilde{S}_i = \sum_{j=1}^{n} \tilde{v}_{ij} \tag{17}$$

Step 8. Calculate the utility degree of alternatives K_i. The utility degree of an alternative in relation to the anti-ideal solution is calculated using Eq. (18).

$$\tilde{K}_i^- = \frac{\tilde{S}_{ais}}{\tilde{S}_i} = \left(\frac{\alpha_{\tilde{S}_{ais}}}{\alpha_{\tilde{S}_i}}, \sqrt[3]{\frac{\left|\beta^3_{\tilde{S}_i} - \beta^3_{\tilde{S}_{ais}}\right|}{1 - \left(min\left(\beta_{\tilde{S}_i}; \beta_{\tilde{S}_{ais}}\right)\right)^3}} \right) \tag{18}$$

The utility degree of an alternative in relation to the ideal solution is calculated using Eq. (19):

$$\tilde{K}_i^+ = \frac{\tilde{S}_i}{\tilde{S}_{is}} = \left(\frac{\alpha_{\tilde{S}_i}}{\alpha_{\tilde{S}_{is}}}, \sqrt[3]{\frac{\left|\beta^3_{\tilde{S}_i} - \beta^3_{\tilde{S}_{is}}\right|}{1 - \left(min\left(\beta_{\tilde{S}_i}; \beta_{\tilde{S}_{is}}\right)\right)^3}} \right) \tag{19}$$

Step 9. Determine the utility function ($f(K_i)$) of alternatives: First, utility functions in relation to the ideal ($f\left(K_i^+\right)$)and anti-ideal solution ($f\left(K_i^-\right)$) are determined by Eq. (20) and Eq. (21):

$$f\left(\tilde{K}_i^-\right) = \frac{\tilde{K}_i^+}{\tilde{K}_i^+ + \tilde{K}_i^-} = \left(\frac{\alpha_{\tilde{K}_i^+}}{\sqrt[3]{\alpha^3_{\tilde{K}_i^+} + \alpha^3_{\tilde{K}_i^-} - \alpha^3_{\tilde{K}_i^+}\alpha^3_{\tilde{K}_i^-}}}, \sqrt[3]{\frac{\left|\beta^3_{\tilde{K}_i^+} - \beta_{\tilde{K}_i^+}\beta^3_{\tilde{K}_i^-}\right|}{1 - \left(min\left(\beta_{\tilde{K}_i^+}; \beta_{\tilde{K}_i^+}\beta_{\tilde{K}_i^-}\right)\right)^3}} \right) \tag{20}$$

$$f\left(\tilde{K}_i^+\right) = \frac{\tilde{K}_i^-}{\tilde{K}_i^+ + \tilde{K}_i^-} = \left(\frac{\alpha_{\tilde{K}_i^-}}{\sqrt[3]{\alpha^3_{\tilde{K}_i^+} + \alpha^3_{\tilde{K}_i^-} - \alpha^3_{\tilde{K}_i^+}\alpha^3_{\tilde{K}_i^-}}}, \sqrt[3]{\frac{\left|\beta^3_{\tilde{K}_i^-} - \beta_{\tilde{K}_i^+}\beta^3_{\tilde{K}_i^-}\right|}{1 - \left(min\left(\beta_{\tilde{K}_i^+}; \beta_{\tilde{K}_i^+}\beta_{\tilde{K}_i^-}\right)\right)^3}} \right) \tag{21}$$

Then using the scores of $f\left(\tilde{K}_i^+\right)$ and $f\left(\tilde{K}_i^-\right)$, the utility function that is the compromise of the observed alternative in relation to the ideal and anti-ideal solution is defined by Eq. (22).

$$f(K_i) = \frac{S\left(f\left(\tilde{K}_i^+\right)\right) + S\left(f\left(\tilde{K}_i^-\right)\right)}{1 + \frac{1 - S\left(f\left(\tilde{K}_i^+\right)\right)}{S\left(f\left(\tilde{K}_i^+\right)\right)} + \frac{1 - S\left(f\left(\tilde{K}_i^-\right)\right)}{S\left(f\left(\tilde{K}_i^-\right)\right)}} \tag{22}$$

Step 10. Rank the alternatives: The utility function values are used to rank the alternatives. The alternative with the highest utility function value is selected as the best alternative.

4 Application

For the application of the proposed method, a construction company is selected and the R&D projects for the textiles used in the construction sector are prioritized using fermatean fuzzy MARCOS method. The criteria used in the analysis are determined among the most used criteria for the evaluation of R&D projects in the literature as; technological applicability (TA), economic benefits (EB), technical benefits (TB), technical risks (TR), potential market size (PM) and financial feasibility (FF). All criteria are benefit criteria except TR. There are 4 alternative projects to be evaluated. Details of the projects cannot be provided due to company privacy policies. Therefore, the subject textile material will be described as X and just the main outcomes of the alternatives are mentioned. A_1: Changes in the raw materials of X to improve its water permeability. A_2: Changes in the raw materials of X to increase its weathering (UV) resistance. A_3: Changes in the production process of X to increase its elasticity. A_4: Changes in the raw materials of X to increase its heat resistance. The criteria weights are determined by experts and shown in Table 3.

Table 3. Fermatean Fuzzy Criteria Weights (FFW)

Criteria	TA	EB	TB	TR	PM	FF
FFW	(0.875,0.067)	(0.775,0.143)	(0.37,0.71)	(0.43,0.67)	(0.495,543)	(0.875,0.067)

Using expert evaluations extended decision matrix and weighted decision matrix are formed as follows:

$$\tilde{X} = \begin{array}{c} \\ \tilde{A}\tilde{A}I \\ \tilde{A}_1 \\ \tilde{A}_2 \\ \tilde{A}_3 \\ \tilde{A}_4 \\ \tilde{A}I \end{array} \begin{array}{cccccc} TA & EB & TB & TR & PMS & FF \\ \end{array} \begin{bmatrix} (0.05,0.94) & (0.50,0.59) & (0.40,0.69) & (0.50,0.59) & (0.40,0.69) & (0.25,0.79) \\ (0.40,0.69) & (0.50,0.59) & (0.94,0.01) & (0.15,0.89) & (0.50,0.59) & (0.69,0.20) \\ (0.15,0.89) & (0.69,0.20) & (0.69,0.20) & (0.40,0.69) & (0.94,0.01) & (0.69,0.20) \\ (0.15,0.89) & (0.50,0.59) & (0.40,0.69) & (0.25,0.79) & (0.50,0.59) & (0.40,0.69) \\ (0.05,0.94) & (0.94,0.01) & (0.40,0.69) & (0.50,0.59) & (0.40,0.69) & (0.25,0.79) \\ (0.40,0.69) & (0.94,0.01) & (0.94,0.01) & (0.15,0.89) & (0.94,0.01) & (0.69,0.20) \end{bmatrix}$$

$$\tilde{v}_{ij} = \begin{array}{c} \\ \tilde{A}\tilde{A}I \\ \tilde{A}_1 \\ \tilde{A}_2 \\ \tilde{A}_3 \\ \tilde{A}_4 \\ \tilde{A}I \end{array} \begin{array}{cccccc} TA & EB & TB & TR & PMS & FF \\ \end{array} \begin{bmatrix} (0.11,0.91) & (0.41,0.59) & (0.16,0.83) & (0.13,0.90) & (0.21,0.76) & (0.32,0.79) \\ (0.88,0.07) & (0.41,0.59) & (0.37,0.71) & (0.43,0.67) & (0.26,0.69) & (0.88,0.07) \\ (0.33,0.82) & (0.57,0.22) & (0.27,0.71) & (0.16,0.88) & (0.50,0.54) & (0.88,0.07) \\ (0.33,0.82) & (0.41,0.59) & (0.16,0.83) & (0.26,0.84) & (0.26,0.69) & (0.51,0.69) \\ (0.11,0.91) & (0.78,0.14) & (0.16,0.83) & (013,0.90) & (0.21,0.76) & (0.32,0.79) \\ (0.88,0.07) & (0.78,0.14) & (0.37,0.71) & (0.43,0.67) & (0.50,0.54) & (0.88,0.07) \end{bmatrix}$$

After calculation of summation matrix, utility functions and rank of the alternatives are calculated and the results are given in Table 4.

According to the results of the analysis company decided to invest on A_3. The projects A_1 and A_2 are excluded from potential projects and A_4 is taken to the project pool to be evaluated later.

Table 4. Utility Functions and Rank of The Alternatives

Alternative	$f\left(\tilde{K}_i^+\right)$	$f\left(\tilde{K}_i^-\right)$	$f(K_i)$	Rank
A_1	(0.5074,0.2489)	(0.9981,0.0009)	0.1277	4
A_2	(0.5597,0.2553)	(0.9890,0.0043)	0.1778	3
A_3	(0.9139,0.2771)	(0.7587,0.1636)	0.4413	1
A_4	(0.7192,0.2821)	(0.9357,0.0588)	0.3792	2

5 Conclusion

In this study, the MARCOS method, which stands out with its easy applicability and effectiveness, has been expanded with fermatean fuzzy sets that enable better express uncertainty and ambiguity in linguistic evaluations given by decision makers. The proposed fermatean fuzzy MARCOS method is applied to select the best R&D project for textiles used in the construction industry. For further researches, it is suggested to compare the proposed method with the existing fuzzy extensions of MARCOS method and other MCDM methods. Besides that, other fuzzy extensions such as picture fuzzy sets, fuzzy Z-numbers, spherical fuzzy sets could be used in the determination of linguistic statements.

References

1. Gong, X., Yang, M., Du, P.: Renewable energy accommodation potential evaluation of distribution network: a hybrid decision-making framework under interval type-2 fuzzy environment. J. Clean. Prod. **286**, 124918 (2021)
2. Stević, Ž, Pamučar, D., Puška, A., Chatterjee, P.: Sustainable supplier selection in healthcare industries using a new MCDM method: measurement of alternatives and ranking according to COmpromise solution (MARCOS). Comput. Ind. Eng. **140**, 106231 (2020)
3. Chakraborty, S., Chattopadhyay, R., Chakraborty, S.: An integrated D-MARCOS method for supplier selection in an iron and steel industry. Decis. Making: Appl. Manag. Eng. **3**(2), 49–69 (2020)
4. Badi, I., Pamucar, D.: Supplier selection for steelmaking company by using combined Grey-MARCOS methods. Decis. Making: Appl. Manag. Eng. **3**(2), 37–48 (2020)
5. Ilieva, G., Yankova, T., Hadjieva, V., Doneva, R., Totkov, G.: Cloud service selection as a fuzzy multi-criteria problem. TEM J. **9**(2), 484 (2020)
6. Stević, Ž, Brković, N.: A novel integrated FUCOM-MARCOS model for evaluation of human resources in a transport company. Logistics **4**(1), 4 (2020)
7. Mijajlović, M., et al.: Determining the competitiveness of spa-centers in order to achieve sustainability using a fuzzy multi-criteria decision-making model. Sustainability **12**(20), 8584 (2020)
8. Pamucar, D., Iordache, M., Deveci, M., Schitea, D., Iordache, I.: A new hybrid fuzzy multi-criteria decision methodology model for prioritizing the alternatives of the hydrogen bus development: a case study from Romania. Int. J. Hydrogen Energy (2020, in press)
9. Stanković, M., Stević, Ž, Das, D.K., Subotić, M., Pamučar, D.: A new fuzzy MARCOS method for road traffic risk analysis. Mathematics **8**(3), 457 (2020)

10. Simić, J.M., Stević, Ž, Zavadskas, E.K., Bogdanović, V., Subotić, M., Mardani, A.: A novel CRITIC-Fuzzy FUCOM-DEA-Fuzzy MARCOS model for safety evaluation of road sections based on geometric parameters of road. Symmetry **12**(12), 2006 (2020)
11. Simić, V., Soušek, R., Jovčić, S.: Picture fuzzy MCDM approach for risk assessment of railway infrastructure. Mathematics **8**(12), 2259 (2020)
12. Senapati, T., Yager, R.R.: Fermatean fuzzy weighted averaging/geometric operators and its application in multi-criteria decision-making methods. Eng. Appl. Artif. Intell. **85**, 112–121 (2019)
13. Du, W.S.: Weighted power means of q-rung orthopair fuzzy information and their applications in multiattribute decision making. Int. J. Intell. Syst. **34**(11), 2835–2862 (2019)
14. Liu, D., Liu, Y., Chen, X.: Fermatean fuzzy linguistic set and its application in multicriteria decision making. Int. J. Intell. Syst. **34**(5), 878–894 (2019)
15. Senapati, T., Yager, R.R.: Some new operations over Fermatean fuzzy numbers and application of Fermatean fuzzy WPM in multiple criteria decision making. Informatica **30**(2), 391–412 (2019)
16. Ucal Sari, I., Kuchta, D., Sergi, D.: Analysis of intelligent software applications in air cargo using fermatean fuzzy CODAS method. In: Intelligent and Fuzzy Techniques in Aviation 4.0: Theory and Applications. Springer (2021, in press)
17. Peng, X., Ma, X.: Pythagorean fuzzy multi-criteria decision making method based on CODAS with new score function. J. Intell. Fuzzy Syst. **38**(3), 3307–3318 (2020)

Sustainable Supply Chain of Aviation Fuel Based on Analytical Hierarchy Process (AHP) Under Uncertainty of q-ROFSs

Fariba Farid[1] and Yaser Donyatalab[2(✉)]

[1] Department of Management, University of Nabi Akram, Tabriz, Iran
[2] Industrial Engineering Department, University of Moghadas Ardabili, Ardabil, Iran

Abstract. Nowadays, sustainable supply chain (SSC) selection of aviation fuels is one of the hot topics around world aviation industries. Existing of diverse intuitive and interrelated criteria that should be considered during the decision-making process turned it into one of the complex decision-making problems. Also, there exists tremendous uncertainty around all parameters and alternatives in the problem. The q-Rung Orthopair Fuzzy Sets (q-ROFSs), which is a generalization of Intuitionistic Fuzzy Sets (IFSs), provide a more proper space for decision-makers. q-ROFSs capable of expressing uncertain information with more flexibility. In this article it is tried to show the reliability of applying the analytical hierarchy process (AHP) in q-ROFSs environment for SSC of aviation fuel. First, properties of q-ROFSs are evaluated then the AHP method is discussed in detail based on q-ROFSs. Afterward, by considering the hardness of the SSC in aviation fuel problem, it is proposed a MAGDM method based on AHP in q-ROFSs environment. Finally, an application of AHP based on q-ROFSs to solve the SSC of the aviation fuel problem is presented to test the effectiveness of the proposed method.

Keywords: Sustainable supply chain (SSC) · Aviation fuel · q-ROFSs · MAGDM · Analytic hierarchy process (AHP) · AHP in q-ROFSs

1 Introduction

Fuzzy set theory is introduced by Zadeh, is a kind of multivalued logic and a branch of propositional calculus that replacing "true or false" with a broader range of values [1]. The next generation of fuzzy sets, Intuitionistic fuzzy sets (IFSs) introduced in [2]. Hesitant fuzzy sets [3], Pythagorean fuzzy sets [4], picture fuzzy sets [5] are next developed generations of fuzzy sets, which are also characterized based on membership, non-membership and hesitant values between 0 and 1. Then in 2017 the concept of ortho pair logic and q-Rung Ortho Pair Fuzzy Sets (q-ROFS) introduced [6]. q-ROFSs attracted many researchers and applied in different studies, some similarity measures are introduced in q-ROFSs environment in [7, 8] and in [9] proposed Hamacher aggregation operators and their applications in decision making process is illustrated.

Analytic hierarchy process (AHP) which is one of the most popular MCDM methods, introduced by Saaty in 1980 [10]. AHP method is a structured method for evaluating

C. Kahraman et al. (Eds.): INFUS 2021, LNNS 308, pp. 578–588, 2022.
https://doi.org/10.1007/978-3-030-85577-2_68

relative priorities for a bunch of alternatives [11]. AHP method based on the triangular fuzzy membership values is extended in [12]. In [13], to evaluate fuzzy weights and performance values, AHP method based on trapezoidal fuzzy numbers are developed. Zeng et al. [14] is applied arithmetic mean aggregation method in fuzzy AHP to evaluate different risk factors in construction projects to analyze the project risk management. In [15], AHP is applied to select an appropriate energy policy in turkey under the uncertainty of the energy environment. As part of literature review of fuzzy based AHP method, could consider the following researches [16–19]. Intuitionistic fuzzy AHP is introduced in [20], hesitant fuzzy AHP is applied in [21, 22]. AHP method is used in the Neutrosophic fuzzy sets in [23, 24] and in [25] the AHP method implemented in q-ROFSs environment to evaluate the problem of disaster logistics location center selection in Istanbul.

Transportation is one of the fundamentals of globalization; explicitly, aviation transportation has significant role playing. Aviation industries' future growth of at least 5% per year is demonstrated in the International Energy Agency (IEA) yearly reports about aviation fuels [26]. Globalization especially in the field of the global economy growing need for more transportation, mostly aviation transportation, so it could be derived that aviation fuel increased significantly. The fact, needing for aviation fuels increasing tremendously is evident in IEA's yearly published report, "World Energy Outlook (WEO)" [27]. Fuels different types offer different alternatives for aviation industries as aviation fuels based on conversion of coal, gas, biomass.

In this manuscript, issue is complex decision-making to select the optimal fuel among diverse aviation fuel alternatives based on many features to achieve a sustainable supply chain (SSC) of aviation fuel. The above discussion encouraged us to propose a practical q-ROFSs analytical hierarchy process (AHP) method through a MAGDM problem. The proposed methodology is applied to the sustainable supply chain (SSC) of aviation fuels while sustainable supply selection problem in aviation fuel is defined based on q-rung orthopair fuzzy structure to catch the more reliability by considering the uncertainty of the problem. The rest of the chapter is designed as follows: in Sect. 2, preliminaries and basic concepts around q-rung orthopair fuzzy sets (q-ROFSs) and analytical hierarchy process (AHP) are discussed in details. In Sect. 3, proposed the methodology based on q-ROFSs AHP and steps of the MAGDM algorithms is defined. Section 4, sustainable supply chain (SSC) problem of aviation fuels is investigated and then solved by the proposed q-ROFSs AHP. In this section, the issue of sustainable supplier selection of aviation fuel is defined in q-ROFSs environment and results of MAGDM algorithm is shown in details. In Sect. 5, the results are summarized, and the chapter is concluded.

2 Preliminaries and Basic Concepts

In this section, the concept of q-rung orthopair fuzzy sets will be defined and discussed in details.

q-ROFSs which is general form of intuitionistic fuzzy sets and Pythagorean fuzzy sets, proposed by [6] are capable of handling more uncertainty of the world. In the following q-ROFSs are defined together with operations.

Definition 1. [6] Let X be universe of discourse, a q-ROFS $\tilde{Q}$ on X is defined by:

$$\tilde{Q} = \left\{ \langle x_i, \mu_{\tilde{Q}}(x), \vartheta_{\tilde{Q}}(x) \rangle | x_i \in X \right\} \quad (1)$$

Where $\mu_{\tilde{Q}}(x) : X \in [0, 1]$ and $\vartheta_{\tilde{Q}}(x) : X \in [0, 1]$ represent the membership degree and non-membership degree respectively. The $\mu_{\tilde{Q}}(x)$ and $\vartheta_{\tilde{Q}}(x)$ have to satisfy the below condition:

$$0 \leq \mu_{\tilde{Q}}(x)^q + \vartheta_{\tilde{Q}}(x)^q \leq 1, (q \geq 1);$$

The indeterminacy degree is defined as $\pi_{\tilde{Q}}(x) = \sqrt[q]{1 - \left(\mu_{\tilde{Q}}(x)^q + \vartheta_{\tilde{Q}}(x)^q \right)}$.

Definition 2. [6] Let $\tilde{Q} = (\mu, \vartheta)$, $\tilde{Q}_1 = (\mu_1, \vartheta_1)$ and $\tilde{Q}_2 = (\mu_2, \vartheta_2)$ be three q-ROFNs, and λ be a positive real number, then:

$$\tilde{Q}_1 \oplus \tilde{Q}_2 = \langle \sqrt[q]{\mu_1^q + \mu_2^q - \mu_1^q \mu_2^q}, \vartheta_1 \vartheta_2 \rangle \quad (2)$$

$$\tilde{Q}_1 \otimes \tilde{Q}_2 = \langle \mu_1 \mu_2, \sqrt[q]{\vartheta_1^q + \vartheta_2^q - \vartheta_1^q \vartheta_2^q} \rangle \quad (3)$$

$$\lambda \tilde{Q} = \langle \sqrt[q]{1 - (1 - \mu^q)^\lambda}, \vartheta^\lambda \rangle \quad (4)$$

$$\tilde{Q}^\lambda = \langle \mu^\lambda, \sqrt[q]{1 - (1 - \vartheta^q)^\lambda} \rangle \quad (5)$$

Definition 3. [28] Let $\tilde{Q}_i = (\mu_i, \vartheta_i)$, $i = (1, 2, \ldots, n)$ be a collection of q-ROFN and $w = (w_1, w_2, \ldots, w_n)$ be the weight vector of $\tilde{Q}_i$ with $\sum_{i=1}^n w_i = 1$, then the q-rung orthopair fuzzy weighted arithmetic mean (q-ROFWAM) operator is:

$$q - ROFWAM\left(\tilde{Q}_1, \tilde{Q}_2, \ldots, \tilde{Q}_n\right) = \langle (1 - \prod_{i=1}^n \left(1 - \mu_i^q\right)^{w_i})^{\frac{1}{q}}, \prod_{i=1}^n \vartheta_i^{w_i} \rangle \quad (6)$$

Definition 4. [28] Let $\tilde{Q}_i = (\mu_i, \vartheta_i)$, $i = (1, 2, \ldots, n)$ be a collection of q-ROFN and $w = (w_1, w_2, \ldots, w_n)$ be the weight vector of $\tilde{A}_i$ with $\sum_{i=1}^n w_i = 1$, then the q-rung orthopair fuzzy weighted geometric mean (q-ROFWGM) operator is:

$$q - ROFWGM\left(\tilde{Q}_1, \tilde{Q}_2, \ldots, \tilde{Q}_n\right) = \langle \prod_{i=1}^n \mu_i^{w_i}, \left(1 - \prod_{i=1}^n \left(1 - \vartheta_i^q\right)^{w_i}\right)^{\frac{1}{q}} \rangle \quad (7)$$

Definition 5. To comparison of q-ROFSs by considering the impact of the hesitancy degree it is defined score and accuracy functions in [29, 30] and [31] respectively as follow:

$$S_1\left(\tilde{Q}\right) = \sqrt[q]{|(\mu^q - \vartheta^q) - 2\pi * (\mu - \vartheta)|} \quad (8)$$

$$\mathcal{S}_2\left(\tilde{\mathcal{Q}}\right) = \sqrt[q]{\left|(\mu - \pi)^q - (\vartheta - \pi)^q\right|} \tag{9}$$

$$\mathcal{S}_3\left(\tilde{\mathcal{Q}}\right) = \sqrt[q]{\frac{1}{3}(2 + \mu^q - 2\vartheta^q - \pi^q)} \tag{10}$$

$$\mathcal{H}\left(\tilde{\mathcal{Q}}_i\right) = \mu^q + \vartheta^q \tag{11}$$

Definition 6. [6] Let $\tilde{\mathcal{Q}}_i = (\mu_i, \vartheta_i)$ be a collection of q-ROFNs, then the score function of $\tilde{\mathcal{Q}}_i$ is defined above could be compared for any two q-ROFNs, $\tilde{\mathcal{Q}}_1 = (\mu_1, \vartheta_1)$ and $\tilde{\mathcal{Q}}_2 = (\mu_2, \vartheta_2)$ as follows:

If $\mathcal{S}(\tilde{\mathcal{Q}}_1) > \mathcal{S}(\tilde{\mathcal{Q}}_2)$, then $\tilde{\mathcal{Q}}_1 > \tilde{\mathcal{Q}}_2$;
If $\mathcal{S}(\tilde{\mathcal{Q}}_1) = \mathcal{S}(\tilde{\mathcal{Q}}_2)$, then
If $\mathcal{H}(\tilde{\mathcal{Q}}_1) > \mathcal{H}(\tilde{\mathcal{Q}}_2)$, then $\tilde{\mathcal{Q}}_1 > \tilde{\mathcal{Q}}_2$;
If $\mathcal{H}(\tilde{\mathcal{Q}}_1) = \mathcal{H}(\tilde{\mathcal{Q}}_2)$, then $\tilde{\mathcal{Q}}_1 = \tilde{\mathcal{Q}}_2$.

3 Methodology

In this section, we will introduce the analytical hierarchy process (AHP) method for q-rung orthopair fuzzy sets (q-ROFSs). Then, it is proposed a multi-attribute group decision making (MAGDM) algorithm.

3.1 Proposed Algorithm Based on AHP Method:

The proposed AHP method in q-Rung Ortho-pair fuzzy sets in this research are presented in the framework of the following algorithm and composed of several consequent steps which shown as follows:

Step 1: Problem statement and determination of goals, alternatives, and criteria. In this step, like in every scientific field, it is needed to define the problem and determine the goals. besides goals, the alternatives and criteria will be determined.

Step 2: Construct the hierarchical structure. Problem will be decomposed into different levels and sub levels, afterward it would be possible to construct the hierarchical structure of the problem that constitute levels of the problem. Level 0 will present the goal of the problem, selecting the best alternative based on score index is most prevalent goal level. Score Index (SI) will be estimated based on set of criteria that constitute in level 1 of hierarchy. Each criterion could contain many some sub-criteria and this fact is true for sub-criteria also. So, it is possible to have one or more levels of sub-criteria depend on the model's complexity. Lastly, the hierarchical structure will constitute of a set of discrete feasible alternatives.

Step 3: Establish pairwise comparison matrices for hierarchical structure based on the linguistic scales $\left(\widetilde{LS}_q\right)$.

Step 4: Establish q-rung orthopair fuzzy pairwise comparison matrices using q-ROFNs for linguistic scales $\left(\widetilde{LS}_q\right)$ given in Table (1).

Step 5: Check the consistency of each q-ROF pairwise comparison matrix. In order to do that, the classical consistent checking method will be applied to evaluate the consistency ratio of each q-rung orthopair fuzzy pairwise comparison matrix. The threshold of the CR is 10%.

Step 6: Calculate the q-ROF local weights matrices of criteria and alternatives. To determine the weight of each alternative it is applied the aggregation operators for q-ROFSs (q-ROFWAM, q-ROFWGM operators) given in Eq. (6) and (7) with respect to each criterion.

Step 7: Evaluate the q-ROF preference weights matrices $\left(\tilde{P}_q\right)$ based on sub-sub-indicators by using q-ROFS multiplication operation given in Eq. (3).

Step 8: Evaluate the q-ROF weighted preference matrices $\left(\widetilde{WP}_q\right)$ based on sub-sub-indicators. Evaluated q-ROF preference weights matrices $\left(\tilde{P}_q\right)$ will be multiplied with collected expert preference matrix by using Eq. (3).

Step 9: Evaluate the q-ROF global weighted preference matrix $\left(\widetilde{GWP}_q\right)$ based on sub-sub-indicators. In this step $\left(\widetilde{WP}_q\right)$ matrices based on sub-sub-indicators from step 8 will be added together by using Eq. (2).

Step 10: Evaluate the q-ROF global weighted preference matrix $\left(\widetilde{GWP}_q\right)$ based on indicators by adding q-ROF $\left(\widetilde{GWP}_q\right)$ from step 9 related to each main indicator by using Eq. (2).

Step 11: Aggregate the q-ROF global weighted preference matrix $\left(\widetilde{GWP}_q\right)$ based on alternatives.

Step 12: Evaluate the score values of the q-ROF global weighted preference $\left(\widetilde{GWP}_q\right)$ for each alternative by using three introduced score functions in Eq. (8), (9) and (10).

Step 13: Compare and rank the alternatives based on the evaluated score values.

Table 1. q-ROFSs Linguistic scales $\left(\widetilde{LS}_q\right)$.

Linguistic scales $\left(\widetilde{LS}_q\right)$	q-ROFN (μ, ϑ, π)	Score index (SI)
Absolutely high possible (AHP)	(0.9, 0.1, 0.646)	9
Very high possible (VHP)	(0.8, 0.2, 0.783)	7
High possible (HP)	(0.7, 0.3, 0.857)	5
Slightly high possible (SHP)	(0.6, 0.4, 0.896)	3
Equally possible (EP)	(0.5, 0.5, 0.908)	1
Slightly low possible (SLP)	(0.4, 0.6, 0.896)	1/3
Low possible (LP)	(0.3, 0.7, 0.857)	1/5
Very low possible (VLP)	(0.2, 0.8, 0.783)	1/7
Absolutely low possible (ALP)	(0.1, 0.9, 0.646)	1/9

4 Application: Sustainable Supply Chain (SSC) of Aviation Fuel

Based on the U.S. Energy Information Administration [32, 33], importance of aviation transportation and therefore aviation fuel consumption are rigorously increasing. So, one of the important supply chain issues is aviation fuel. Also, it is straightforward that selecting a suitable fuel supplier for this valuable industry is one of the significant problems in both the supply selection area and the aviation industries. In this research, an aviation fuel sustainable supply chain selection problem in an uncertain environment is considered. Possible alternatives are aviation fuels and their suppliers. In this problem, four types of aviation fuels: Algal based (F_1), Soybean based (F_2), Aviation Gasoline (AVGAS) (F_3), Natural gas-based (F_4) are considered. There are many suppliers all over the world, which provide the above fuels for aviation industries [34] among those companies all over the world, could refer to top suppliers [35] like British Petroleum (BP), Shell, Exxon Mobil, Chevron, Total, and Gazprom. Also, it is reasonable that not all suppliers provide all types of fuels then it is assumed top supplier for each fuel type, so have a set of suppliers as $S = \{S_1, S_2, \ldots, S_4\}$. In this study, a possible alternative, for fuel type F_2 which supplied by S_2 constitute alternative A_2 and will represent as $A_2(S_2, F_2)$.

Having such diversity in alternatives for aviation fuels need a strong supply chain management in aviation industries cross from aviation fuel supply security and efficient air traffic management to achieve a sustainable supply chain (SSC) of aviation fuel. Also there exist many suppliers for aviation fuels which together with different fuels types makes this problem more complicated. The triple bottom line (TBL) approach which proposed by Elkington in [36], to achieve a sustainable supply chain (SSC) by considering economic, environment and social impacts. TBL approach is considered to achieve aviation fuel's sustainable supply chain (SSC) problem [37]. In this study, a new model is proposed to evaluate the sustainability of supply chain. To assess the sustainable supply chain (SSC) of aviation fuels a multi-indicator sustainability assessment model in the framework of four main indicators as economic, environmental, social and market reliability is proposed. The proposed model aiming: great economic profits, low ecological effects, and a high social acceptance rate, and as fourth indicator increase the reliability of market. So, a multi-layer of indicators is proposed to assess the SSC of aviation fuels. The indicators (criteria) are presented based on a three-level system of indicators: Main indicators (criteria), Sub-indicators (sub-criteria) and Sub-sub-indicators (sub-sub-criteria). Assumed indicators (criteria) are selected from the literature part from different articles [38–41], presented as follows:

Economic $\left(\tilde{C}_1\right)$:

Cost $\left(\tilde{C}_{11}\right)$: Capital Cost $\left(\tilde{C}_{111}\right)$ - Production Cost $\left(\tilde{C}_{112}\right)$.
Lifespan $\left(\tilde{C}_{12}\right)$: Lifespan $\left(\tilde{C}_{121}\right)$.

Environmental $\left(\tilde{C}_2\right)$:

Ecological $\left(\tilde{C}_{21}\right)$: Greenhouse Gas (GHG) Emission $\left(\tilde{C}_{211}\right)$ - PM_{10} Emission $\left(\tilde{C}_{212}\right)$ - $PM_{2.5}$ Emission $\left(\tilde{C}_{213}\right)$.

Supplementary resources $\left(\tilde{C}_{22}\right)$: Water resource needed $\left(\tilde{C}_{221}\right)$ - Electricity resource needed $\left(\tilde{C}_{222}\right)$.

Energy efficiency $\left(\tilde{C}_{23}\right)$: Energy Consumption to produce $\left(\tilde{C}_{231}\right)$.

Social $\left(\tilde{C}_{3}\right)$:

Acceptability rate $\left(\tilde{C}_{31}\right)$: Social acceptability rate $\left(\tilde{C}_{311}\right)$ - Aviation Industries acceptability rate $\left(\tilde{C}_{312}\right)$.

Technologic maturity related to fuel $\left(\tilde{C}_{32}\right)$: Technologic maturity related to fuel $\left(\tilde{C}_{321}\right)$.

Innovation on technology $\left(\tilde{C}_{33}\right)$: Innovation on technology $\left(\tilde{C}_{331}\right)$.

Market Reliability $\left(\tilde{C}_{4}\right)$:

Production $\left(\tilde{C}_{41}\right)$: Production Capacity $\left(\tilde{C}_{411}\right)$ - Availability of resources $\left(\tilde{C}_{412}\right)$.

Supplementary resources $\left(\tilde{C}_{42}\right)$: Water resource needed $\left(\tilde{C}_{221}\right)$ - Electricity resource needed $\left(\tilde{C}_{222}\right)$.

Energy efficiency $\left(\tilde{C}_{43}\right)$: Energy Consumption to produce $\left(\tilde{C}_{231}\right)$.

Therefore, in this article all above-mentioned aviation fuels, alternative suppliers evaluated based on a multi-layer structure of criteria to assess and determine the sustainable supply chain of aviation fuel. Based on proposed algorithm first of all, the comments for the criteria in all levels and alternatives are collected from experts regarding the goals of the problem. Also, among the indicators, economic criterion is the only non-beneficial indicator and it is considered by experts. So, there is no need to normalization of comments from Experts. Step 1 and 2 are clear based on the above explanations about problem and hierarchical structure of the problem is shown in Fig. (1). For step 3 the pair wise comparison matrices are collected from experts and in step 4 is used to determine the q-ROF pairwise comparison matrices.

According to step 5, consistency ratio (CR) with acceptable rate smaller than 10% is calculated for all q-ROF pairwise comparison matrices and all are acceptable thereafter. In step 6, weights for set of main indicators, sub-indicators and sub-sub-indicators are calculated by using Eq. (6) and (7) where the initial weights of each set are assumed to be equal. In step 7, 8, 9 and 10 we calculated the $\left(\tilde{P}_q\right)$, $\left(\widetilde{WP}_q\right)$, $\left(\widetilde{GWP}_q\right)$ and $\left(\widetilde{GWP}_q\right)$

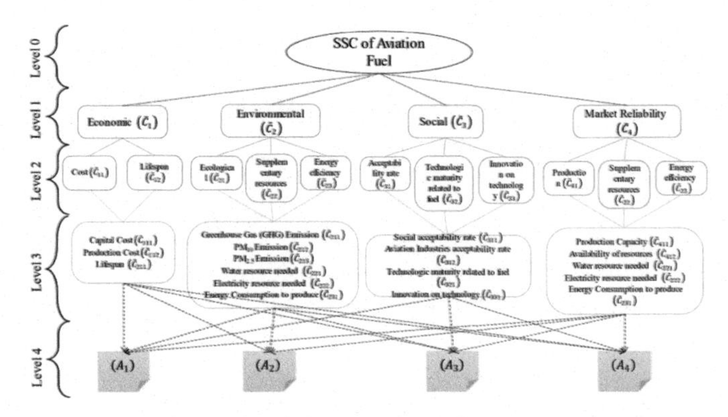

Fig. 1. Hierarchical structure

Table 2. Aggregate the q-ROF global weighted preference matrix $\left(\widetilde{GWP}_q\right)$ based on alternatives

$\left(\widetilde{GWP}_q\right)$			μ	ϑ	π
S1	F1	A1	0.302	0.007	0.9979
S2	F2	A2	0.201	0.005	0.9996
S3	F3	A3	0.295	0.008	0.9981
S4	F4	A4	0.255	0.008	0.9989

based on indicators, respectively. Step 11, will be aggregate q-RIF $\left(\widetilde{GWP}_q\right)$ based on indicators together to get the $\left(\widetilde{GWP}_q\right)$ based on alternatives and results of this step is shown in Table (2). In step 12, the score values $\mathcal{S}_1$, $\mathcal{S}_2$ *and* $\mathcal{S}_3$ are calculated for each alternative by using Eq. (8), (9) and (10), respectively and results are shown in Table (3). Finally, in step 13 the rankings are evaluated for each alternative and results are shown in Table (3). Based on Table (3), it is clear that the alternative $A_1(S_1, F_1)$, which is algal based fuel (F_1) and supplied by (S_1), stands in first place. The ranking of alternatives is given as follows:

$$A_1 > A_3 > A_4 > A_2$$

Table 3. Score values and ranking based on each score

Alternatives			Score values			Rank		
			$\mathcal{S}_1$	$\mathcal{S}_2$	$\mathcal{S}_3$	R1	R2	R3
S1	F1	A1	0.873	0.924	0.763	1	1	1
S2	F2	A2	0.790	0.870	0.760	4	4	4
S3	F3	A3	0.868	0.921	0.763	2	2	2
S4	F4	A4	0.836	0.900	0.761	3	3	3

5 Conclusion

The main contribution of this paper is that a novel analytical hierarchy process (SHP) with considering different score values are proposed for q-ROFSs. Then a MAGDM algorithm based on q-ROFSs analytical hierarchy process (AHP) is introduced. To validate the proposed methodology, it is applied in sustainable supply chain (SSC) selection problem for aviation fuels and results are shown in details. More discussion in this field could be investigated and for further research, the Interval Valued of q-ROFS can be considered, or different type of supply chain problems could be evaluated by the proposed methodology.

References

1. Zadeh, L.A.: Fuzzy sets. Inf. Control **8**, 338–353 (1965)
2. Atanassov, K.T.: Intuitionistic fuzzy sets. Fuzzy Sets Syst. **20**(1), 87–96 (1986). https://doi.org/10.1016/S0165-0114(86)80034-3
3. Torra, V.: Hesitant fuzzy sets. Int. J. Intell. Syst. **25**, 529–539 (2010)
4. Yager, R.R.: Pythagorean fuzzy subsets. In: 2013 Joint IFSA World Congress and NAFIPS Annual Meeting, vol. 2, pp. 57–61 (2013)
5. Cuong, B.C., Kreinovich, V.: Picture fuzzy sets - a new concept for computational intelligence problems. In: 2013 3rd World Congress on Information and Communication Technologies, WICT 2013 (2014)
6. Yager, R.R.: Generalized Orthopair Fuzzy Sets. IEEE Trans. Fuzzy Syst. **25**, 1222–1230 (2017)
7. Donyatalab, Y., Farrokhizadeh, E., Seyfi Shishavan, S.A.: Similarity measures of q-rung orthopair fuzzy sets based on square root cosine similarity function. In: Kahraman, C., Cevik Onar, S., Oztaysi, B., Sari, I.U., Cebi, S., Tolga, A.C. (eds.) INFUS 2020. AISC, vol. 1197, pp. 475–483. Springer, Cham (2021). https://doi.org/10.1007/978-3-030-51156-2_55
8. Farrokhizadeh, E., Shishavan, S.A.S., Donyatalab, Y., Abdollahzadeh, S.: The dice (sorensen) similarity measures for optimal selection with q-rung orthopair fuzzy information. In: Kahraman, C., Cevik Onar, S., Oztaysi, B., Sari, I.U., Cebi, S., Tolga, A.C. (eds.) INFUS 2020. AISC, vol. 1197, pp. 484–493. Springer, Cham (2021). https://doi.org/10.1007/978-3-030-51156-2_56
9. Donyatalab, Y., Farrokhizadeh, E., Shishavan, S.A.S., Seifi, S.H.: Hamacher aggregation operators based on interval-valued q-rung orthopair fuzzy sets and their applications to decision making problems. In: Kahraman, C., Cevik Onar, S., Oztaysi, B., Sari, I.U., Cebi, S., Tolga, A.C. (eds.) INFUS 2020. AISC, vol. 1197, pp. 466–474. Springer, Cham (2021). https://doi.org/10.1007/978-3-030-51156-2_54
10. Saaty, T.L.: The Analytic Hierarchy Process: Planning, Prior Setting, Resource Allocation. MacGraw-Hill, New York (1980)
11. Kutlu Gündoğdu, F., Kahraman, C.: Spherical fuzzy analytic hierarchy process (AHP) and its application to industrial robot selection. In: Kahraman, C., Cebi, S., Cevik Onar, S., Oztaysi, B., Tolga, A.C., Sari, I.U. (eds.) INFUS 2019. AISC, vol. 1029, pp. 988–996. Springer, Cham (2020). https://doi.org/10.1007/978-3-030-23756-1_117
12. van Laarhoven, P.J.M., Pedrycz, W.: A fuzzy extension of Saaty's priority theory. Fuzzy Sets Syst. **11**(1–3), 229–241 (1983). https://doi.org/10.1016/S0165-0114(83)80082-7
13. Buckley, J.J.: Fuzzy hierarchical analysis. Fuzzy Sets Syst. **17**(3), 233–247 (1985). https://doi.org/10.1016/0165-0114(85)90090-9
14. Zeng, J., An, M., Smith, N.J.: Application of a fuzzy based decision making methodology to construction project risk assessment. Int. J. Proj. Manag. **25**, 589–600 (2007)

15. Kahraman, C., Kaya, I.: A fuzzy multicriteria methodology for selection among energy alternatives. Expert Syst. Appl. **37**, 6270–6281 (2010)
16. Chang, D.-Y.: Applications of the extent analysis method on fuzzy AHP. Eur. J. Oper. Res. **95**(3), 649–655 (1996). https://doi.org/10.1016/0377-2217(95)00300-2
17. Tan, R.R., Aviso, K.B., Huelgas, A.P., Promentilla, M.A.B.: Fuzzy AHP approach to selection problems in process engineering involving quantitative and qualitative aspects. Process Saf. Environ. Prot. (2014).
18. Kahraman, C., Öztayşi, B., Sarı, İU., Turanoğlu, E.: Fuzzy analytic hierarchy process with interval type-2 fuzzy sets. Knowl. Based Syst. **59**, 48–57 (2014). https://doi.org/10.1016/j.knosys.2014.02.001
19. Oztaysi, B., Onar, S.C., Kahraman, C.: Prioritization of business analytics projects using interval type-2 fuzzy AHP. In: Kacprzyk, J., Szmidt, E., Zadrożny, S., Atanassov, K.T., Krawczak, M. (eds.) IWIFSGN/EUSFLAT -2017. AISC, vol. 643, pp. 106–117. Springer, Cham (2018). https://doi.org/10.1007/978-3-319-66827-7_10
20. Jian, W., Huang, H.-B., Cao, Q.-W.: Research on AHP with interval-valued intuitionistic fuzzy sets and its application in multi-criteria decision making problems. Appl. Math. Model. **37**(24), 9898–9906 (2013). https://doi.org/10.1016/j.apm.2013.05.035
21. Oztaysi, B., Onar, S.C., Bolturk, E., Kahraman, C.: Hesitant fuzzy analytic hierarchy process. In: IEEE International Conference on Fuzzy Systems (2015)
22. Kahraman, C.: Multiattribute warehouse location selection in humanitarian logistics using hesitant fuzzy. AHP. Int. J. Anal. Hierarchy Process. **8**, 271–298 (2016)
23. Abdel-Basset, M., Mohamed, M., Sangaiah, A.K.: Neutrosophic AHP-Delphi Group decision making model based on trapezoidal neutrosophic numbers. J. Ambient Intell. Humaniz. Comput. **9**, 1427–1443 (2018)
24. Bolturk, E., Kahraman, C.: A novel interval-valued neutrosophic AHP with cosine similarity measure. Soft. Comput. **22**(15), 4941–4958 (2018). https://doi.org/10.1007/s00500-018-3140-y
25. Seyfi Shishavan, S.A., Donyatalab, Y., Farrokhizadeh, E.: Extension of classical analytic hierarchy process using q-rung orthopair fuzzy sets and its application to disaster logistics location center selection. In: Kahraman, C., Cevik Onar, S., Oztaysi, B., Sari, I.U., Cebi, S., Tolga, A.C. (eds.) INFUS 2020. AISC, vol. 1197, pp. 432–439. Springer, Cham (2021). https://doi.org/10.1007/978-3-030-51156-2_50
26. Nygren, E., Aleklett, K., Höök, M.: Aviation fuel and future oil production scenarios. Energy Policy **37**(10), 4003–4010 (2009). https://doi.org/10.1016/j.enpol.2009.04.048
27. Energy Information Administration, U.S.E.I.: International Energy Outlook 2019 (2019)
28. Liu, P., Wang, P.: Some q-rung orthopair fuzzy aggregation operators and their applications to multiple-attribute decision making. Int. J. Intell. Syst. **33**, 259–280 (2018)
29. Kutlu Gündoğdu, F., Kahraman, C.: A novel spherical fuzzy QFD method and its application to the linear delta robot technology development. Eng. Appl. Artif. Intell. **87**, 103348 (2020)
30. Otay, I., Kahraman, C., Öztayşi, B., Onar, S.Ç.: Score and accuracy functions for different types of spherical fuzzy sets (2020)
31. Ashraf, S., Abdullah, S.: Spherical aggregation operators and their application in multiattribute group decision-making. Int. J. Intell. Syst. **34**, 493–523 (2019)
32. Annual Energy Outlook (2020). https://www.eia.gov/outlooks/aeo/
33. U.S. Energy Information Agency: Annual Energy Outlook 2019 with projections to 2050. Annu. Energy Outlook 2019 with Proj. to 2050. (2019)
34. Airport Suppliers (Aviation Fuel Suppliers). https://www.airport-suppliers.com/suppliers/fuel-handling/
35. Aviation Fuel Market by Product and Geography - Forecast and Analysis 2020–2024 (Technavio). https://www.technavio.com/report/aviation-fuel-market-industry-analysis

36. Elkington, J.: Cannibals with Forks: The Triple Bottom Line of 21st Century Business (1999). Choice Rev. Online
37. Rodger, J.A., George, J.A.: Triple bottom line accounting for optimizing natural gas sustainability: A statistical linear programming fuzzy ILOWA optimized sustainment model approach to reducing supply chain global cybersecurity vulnerability through information and communications technology. J. Cleaner Prod. **142**, 1931–1949 (2017). https://doi.org/10.1016/j.jclepro.2016.11.089
38. Ren, J., Fedele, A., Mason, M., Manzardo, A., Scipioni, A.: Fuzzy multi-actor multi-criteria decision making for sustainability assessment of biomass-based technologies for hydrogen production. Int. J. Hydrogen Energy **38**, 9111–9120 (2013)
39. Ren, J., Manzardo, A., Mazzi, A., Zuliani, F., Scipioni, A.: Prioritization of bioethanol production pathways in China based on life cycle sustainability assessment and multicriteria decision-making. Int. J. Life Cycle Assess. **20**, 842–853 (2015)
40. Afgan, N.H., Carvalho, M.G.: Sustainability assessment of hydrogen energy systems. Int. J. Hydrogen Energy **29**, 1327–1342 (2004)
41. Ren, J., Xu, D., Cao, H., Wei, S., Dong, L., Goodsite, M.E.: Sustainability decision support framework for industrial system prioritization. AIChE J. **62**, 108–130 (2016)

Hospital Type Location Allocation Decisions by Using Pythagorean Fuzzy Sets Composition: A Case Study of COVID-19

Ibrahim Yilmaz(✉), Yagmur Arioz, Cihat Ozturk, and Abdullah Yildizbasi

Department of Industrial Engineering, Ankara Yildirim Beyazit University, Ankara, Turkey
iyilmaz@ybu.edu.tr

Abstract. This study aims to develop a model for location-allocation decisions (LAD) for the different types of hospitals' during Corona Virus 2019 (COVID-19) pandemic. Long-term hospital LAD are accepted as a long-term strategy and take a long time to reach a decision. However, after the beginning of the COVID-19 pandemic, the decision-makers aim to reach their patients as immediately as possible. Thus, decision-makers are planning to increase the capacity of regional or country-wide health systems by adding new hospitals. However, during LAD processes, the decision-makers face imprecision due to the inherited uncertainty of main parameters that must be paid attention to make decisions applicable. Therefore, the decision-makers must give a decision in a short time in a more imprecise environment. For this purpose, Pythagorean Fuzzy Sets (PFS), which are recognized as generalized intuitionistic fuzzy sets with a broader range of applications, are utilized. PFS is one of the most recent tools for dealing with imprecision and a method for removing uncertainty from decision-making processes. This study is one of the first research on applying PFS and making inferences to increase the accuracy of decisions during the COVID-19 pandemic. To show the usability of the proposed model, a case study is provided assuming each candidate location has different specifications which are shown as PFS. Therefore, the LAD model for short-time hospitals related to COVID-19 pandemic conditions is proposed by using a PFS relation called max–min–max composition to ascertain the suitability of decisions.

Keywords: COVID-19 · Hospital location-allocation · Pythagorean Fuzzy Sets

1 Introduction

Covid-19 virus has been spreading unpredictably across the world since December 2019 [1]. Covid-19 is tried to be under control with the precautions taken

C. Kahraman et al. (Eds.): INFUS 2021, LNNS 308, pp. 589–597, 2022.
https://doi.org/10.1007/978-3-030-85577-2_69

and the treatment methods applied. However, the number of cases may get out of control again in different regions due to the stretching of the precautions or the failure to comply with the precautions. In addition to measures to non-compliance, the mutations in the virus cause to increase in the speed of spread again. For these reasons, the number of cases and the number of applications to hospitals are increasing every day. Depending on the number of patients and the needs of the patient, the hospital beds, intensive care units, and ventilator capacities of the hospitals are become insufficient. For this purpose, in order to increase the capacity of hospitals, efforts are made to increase the number of additional pieces of equipment as well as the physical capacity of the health system. Options such as building new hospitals and increasing the number of healthcare equipment are implemented. The most important constraint in increasing physical capacity could be accepted as the time limit because the decision should be implemented as soon as possible. Therefore, small-sized pandemic hospitals are preferred in regions where the patient density of the epidemic is increasing. Small-sized hospitals are accepted as having less than 100 beds established for special purposes. In the literature, the examples of small-sized hospitals are shown as Pandemic Hospital, Field Hospital, County Hospital, Small Hospital, and Local Hospital [2,3].

Small-sized hospitals are often preferred in extraordinary situations due to their short establishment time and low investment costs. These hospitals are also preferred regionally to increase health system capacities during COVID-19. However, location selection for pandemic hospitals becomes a problem with the increasing uncertainty and complexity due to the pandemic. In case there is a need to increase capacity in many regions simultaneously, it is important to make a quick and accurate decision as to which type of hospital will be suitable for which region by taking into account the characteristics of the regions.

In order to enable decision-makers to solve this problem more accurately and in a shorter time, the uncertainties that arise in the evaluation of alternatives and criteria have been defined using Pythagorean Fuzzy Sets (PFS). This study aims to investigate decision makers' ability to solve LAD problems through the max-min-max rule, due to the broader scope of application of PFS in uncertain real-life problems. The concept of PFS and some of its features have been outlined, and some definitions in the literature are presented to achieve this aim. Therefore, an application of PFS relations is provided to show how the problem of assigning hospital types based on location characteristics to the most appropriate locations by using the proposed PFS compositions. The following sections of this paper is structured as follows: the relevant literature review is shown in Sect. 2, a brief analysis of the suggested method is provided in Sect. 3, conclusions and suggestions are discussed in Sect. 4.

2 Literature Review

It can be difficult for decision-makers (DMs) to express a preference precisely when trying to solve multi-criteria decision-making (MCDM) problems with unreliable, unclear, or incomplete knowledge. Under such conditions, Zadeh introduced the Fuzzy Set Theory (FST) to handle imprecision in decision making in which the membership function degrees are presented between 0 and 1 [4]. FST is a generic version of the conventional or crisp sets in which an element is either a member of the set or it is not. Therefore, in crisp sets, the degree of membership functions must be 0 or 1.

Even if FST provides great flexibility to handle vagueness or ambiguity for circumstances, defining each element just only by a membership function could not be able to express a situation clearly. For this reason, the Intuitionistic Fuzzy Sets (IFSs) have been introduced to remove the shortcomings of the FST and define a situation more accurately [5]. IFSs consider both membership, μ, and non-membership, v, functions with hesitation margin, π simultaneously [8]. This concept have been applied many areas such as; MCDM problems, pattern recognition, supplier selection, control systems, medical diagnosis, and etc.

However, the assumption of the IFS ($\mu + v + \pi = 1$) delimits the applicability of the concept. For example, in the case of the $\mu = 0.8$ and $v = 0.4$, then IFS could not be applied to handle this case. As an extension of the IFS, Pythagorean Fuzzy Sets (PFS) are implemented as a new appropriate way to deal with vagueness [6]. PFS has assumptions that $\mu + v \geq 1$ or $\mu + v \leq 1$ and $\pi_A(x) = \sqrt{1 - (\mu_A(x) + v_A(x))}$. PFS gives DMs more flexibility in expressing their opinions and provides more reliable results with less information [7]. PFS is preferred by many researchers for different application areas such as MCDM, Safe Risk Assessment, Optimization, Medical Diagnosis, Information Sciences, System Control, and etc. (Fig. 1).

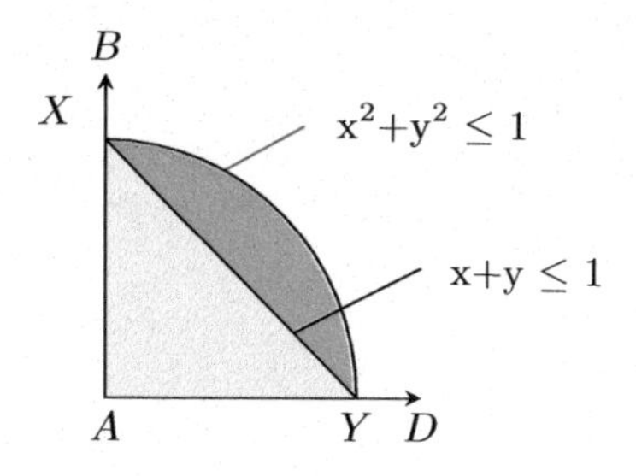

Fig. 1. IFS and PFS space comparison

In this paper, the max-min-max composition method, which is proposed by Ejegwa, is applied to show the convenience of PFS for hospital LADs. For this purpose, a numerical example is given to demonstrate the usefulness of the proposed procedure. Therefore, an application of the max-min-max composition method of PFS relations is illustrated for LADs based on the candidate locations and hospital types specifications [8].

3 Methodology

3.1 Preliminaries

Definition 1. *Let X be a non-empty universal set. A crisp set A of X is shown as $A = \{\langle x, \mu_A(x)\rangle : x \in X\}$ where $\mu_A(x)$ is the membership function of A, and defined as $\mu_A(x) : X \to \{0,1\}$ [6]*

$$\mu_A(x) = \begin{cases} 1, & \text{if } x \in X \\ 0, & \text{else.} \end{cases} \tag{1}$$

Conventional or crisp sets membership function values are *Binary*. An element in a crisp set could either belong to the set or could not.

Definition 2. *Let X be a non-empty universal set. A fuzzy set A of X is shown as $A = \{\langle x, \mu_A(x)\rangle : x \in X\}$ where $\mu_A(x)$ is the membership function of A, and defined as $\mu_A(x) : X \to [0,1]$ [6]*

$$\mu_A(x) = \begin{cases} 1, & \textit{if } x \in X, \\ 0, & \textit{if } x \notin X, \\ (0,1), & \textit{else} \end{cases} \tag{2}$$

$\mu_A(x)$ defines the degree of the memberships of $x \in X$. If $\mu_A(x)$ approximate to 1, then x is getting more belong to A.

Definition 3. *Let X be a non-empty universal set. A IFS A of X is shown as $A = \{\langle x, \mu_A(x), \upsilon_A(x)\rangle : x \in X\}$ where $\mu_A(x)$ and $\upsilon_A(x)$ are the membership function of A and the non-membership function of A, which are defined as $\mu_A(x), \upsilon_A(x) : X \to [0,1]$ as follows*

$$\mu_A(x) = \begin{cases} 1, & \textit{if } x \in X, \\ 0, & \textit{if } x \notin X, \\ (0,1), & \textit{else} \end{cases} \quad \upsilon_A(x) = \begin{cases} 1, & \textit{if } x \notin X, \\ 0, & \textit{if } x \in X, \\ (0,1), & \textit{else.} \end{cases} \tag{3}$$

When $\mu_A(x)$ and $\upsilon_A(x)$ are getting closer to 1, x is getting more belong and less belong A. In IFS, $\mu_A(x)$ and $\upsilon_A(x)$ are subjected to $0 \leq \mu_A(x) + \upsilon_A(x) \leq 1$. Also, in case of $\mu_A(x) + \upsilon_A(x) \leq 1$, the hesitation margin, $\pi_A(x)$, represents the degree of indeterminacy of $x \in X$ to IFS A and represents the lack of knowledge on whether x belongs to A or not. $\pi_A(x)$ is calculated as follows

$$\pi_A(x) = 1 - (\mu_A(x) + \upsilon_A(x)) \tag{4}$$

where $\pi_A(x) : X \to [0,1]$ and $0 \leq \pi_A(x) \leq 1$.

Definition 4. *Let X be a non-empty universal set. A Pythagorean Fuzzy Set (PFS) A of X is shown as $A = \{\langle x, \mu_A(x), \upsilon_A(x)\rangle : x \in X\}$ where $\mu_A(x), \upsilon_A(x)$ defined as $\mu_A(x), \upsilon_A(x) : X \rightarrow [0,1]$ same as shown in Eq. 3 [8].*

The PFS is a generalized version of IFS with the objective of managing the vaguer situation that could not be handled by IFS. For example, in case of $\mu_A(x) = 0.8$ and $\upsilon_A(x) = 0.6$, IFS will not work because $\mu_A(x)+\upsilon_A(x) \geq 1$. To deal with this situation, $\mu_A(x)$ and $\upsilon_A(x)$ are subjected to $0 \leq (\mu_A(x))^2+(\upsilon_A(x))^2 \leq 1$ [6]. $\pi_A(x)$, of $x \in X$ to PFS A is calculated as follows

$$\pi_A(x) = \sqrt{1-(\mu_A(x)+\upsilon_A(x))} \tag{5}$$

where $\pi_A(x) : X \rightarrow [0,1]$ and $0 \leq \pi_A(x) \leq 1$. Equation 5 implies that $(\mu_A(x))^2 + (\upsilon_A(x))^2 + (\pi_A(x))^2 = 1$ [6].

3.2 Max-min-max Composition Method

In this part, the proposed max-min-max composition approach is implemented as a PFS tool. In this method, the maximum and minimum values of the μ, υ, and π are compared to handle the wider scope of a situation [8].

Definition 5. *Let X and Y be two different non-empty universal sets and A and B are two PFSs in X and Y, respectively. The Pythagorean Fuzzy Relations (PFRs), R, are from X to Y qualified with $\mu_R(x)$ and $\upsilon_R(x)$. PFRs from X to Y in which $x \in X$ and $y \in Y$ are shown as $R(X \rightarrow Y)$. Then A becomes B which is presented as $B = R \circ A$ with max-min-max composition of $R(X \rightarrow Y)$ [8]. In this situation, $\mu_B(y)$ and $\upsilon_B(y)$ are calculated as follows*

$$\begin{aligned} \mu_B(y) &= \max_{x \in X} (\min [\mu_A(x), \mu_R(x,y)]) \\ \upsilon_B(y) &= \min_{x \in X} (\max [\upsilon_A(x), \upsilon_R(x,y)]) \end{aligned} \tag{6}$$

Definition 6. *Let R and Q be two different PFRs, and X, Y and Z are three PFSs in different non-empty universal sets. $R(X \rightarrow Y)$ represents PFRs from X to Y qualified with $\mu_R(x,y)$ and $\upsilon_R(x,y)$. On the other hand, $Q(Y \rightarrow Z)$ represents PFRs from Y to Z qualified with $\mu_R(y,z)$ and $\upsilon_R(y,z)$. Then max-min-max composition of $B = R \circ Q$ represents PFRs from $X \rightarrow Z$ [8]. In this situation, $\mu_{R \circ Q}(y)$ and $\upsilon_{R \circ Q}(y)$ are calculated for $\forall (x,z) \in XxZ$ and $\forall y \in Y$ as follows [8]*

$$\begin{aligned} \mu_{R \circ Q}(x,z) &= \max_{y \in Y} (\min [\mu_R(x,y), \mu_Q(y,z)]) \\ \upsilon_{R \circ Q}(x,z) &= \min_{y \in Y} (\max [\upsilon_R(x,y), \upsilon_Q(y,z)]) \end{aligned} \tag{7}$$

The max-min-max composition $B = R \circ Q$ is calculated as follows:

$$R \circ Q = \mu_{R \circ Q}(x,z) - \upsilon_R(x,z) \times \pi_{R \circ Q}(x,z) \tag{8}$$

where $\pi_{R \circ Q}(x,z) = \sqrt{1-(\mu_{R \circ Q}(x,z)^2 + (\upsilon_R(x,z)^2)}$

3.3 Hospital Type LADs in Pythagorean Fuzzy Environment: A Case Study of Covid-19

During the Covid-19 pandemic, the current hospital capacity could not be enough to serve their patients' needs. In this situation, DMs aims to expand the current health care systems' capacities by adding new hospitals. However, DMs face a challenge in how to give a decision which location is suitable for different hospital types when there are more than one alternatives. The max-min-max composition method for PFS relations provides an accurate solution by dealing with uncertainties and vagueness in decision making.

Let $C = \{c_1, \ldots, c_5\}$, $S = \{s_1, \ldots, s_6\}$, and $T = \{t_1, \ldots, t_5\}$ are finite PFS of candidate locations, specifications of locations, and types of small sized hospitals, respectively. Suppose $R(X \rightarrow Y)$ and $U(Y \rightarrow Z)$ presents PFS relations as shown in Table 1 and Table 2. The R and U can be defined as

$$\begin{aligned} R &= \{\langle (c,s), \mu_R(c,s), \upsilon_R(c,s) \rangle : (c,s) \in C \times S\} \\ U &= \{\langle (s,t), \mu_U(s,t), \upsilon_U(s,t) \rangle : (s,t) \in S \times T\} \end{aligned} \tag{9}$$

$\mu_R(c,s)$ represents the degree to which location c hold the criterion, s.
$\upsilon_R(c,s)$ represents the degree to which location c does not hold the criterion, s.
$\mu_U(s,t)$ represents the degree of criterion s defines hospital type, t.
$\upsilon_U(s,t)$ represents the degree of criterion s does not define hospital type, t.

Table 1. The results of PFS composition of X and Y, R $R(X \rightarrow Y)$

R	S1	S2	S3	S4	S5	S6
A1	(0.9, 0.2)	(0.9, 0.1)	(0.5,0.4)	(0.9, 0.3)	(0.7, 0.5)	(0.6, 0.2)
A2	(0.8, 0.2)	(0.6, 0.3)	(0.5,0.3)	(0.7, 0.5)	(0.7, 0.4)	(0.7, 0.1)
A3	(0.7, 0.3)	(0.7, 0.5)	(0.6,0.2)	(0.7, 0.4)	(0.8, 0.5)	(0.6, 0.3)
A4	(0.6, 0.4)	(0.6, 0.3)	(0.6,0.1)	(0.6, 0.3)	(0.6, 0.4)	(0.7, 0.2)
A5	(0.5, 0.3)	(0.5, 0.4)	(0.7,0.2)	(0.6, 0.1)	(0.5, 0.2)	(0.8, 0.1)

Table 2. The results of PFS composition of Y and Z, $U(Y \rightarrow Z)$

U	T1	T2	T3	T4	T5
S1	(0.8, 0.4)	(0.7, 0.4)	(0.6, 0.4)	(0.7, 0.3)	(0.6, 0.2)
S2	(0.7, 0.3)	(0.6, 0.1)	(0.7, 0.4)	(0.5, 0.4)	(0.7, 0.3)
S3	(0.9, 0.4)	(0.6, 0.5)	(0.9, 0.1)	(0.8, 0.2)	(0.8, 0.4)
S4	(0.6, 0.3)	(0.7, 0.2)	(0.7, 0.4)	(0.6, 0.3)	(0.7, 0.4)
S5	(0.8, 0.2)	(0.7, 0.4)	(0.6, 0.3)	(0.8, 0.4)	(0.5, 0.2)
S6	(0.7, 0.2)	(0.6, 0.2)	(0.8, 0.3)	(0.9, 0.1)	(0.8, 0.1)

The composition of PFS relations R and U can be presented as $K = R \circ U$ which state that c is suitable for t. It is defined by $\mu_T(c,p)$ and $\upsilon_T(c,p)$ as follows

$$\begin{aligned} \mu_T(c,p) &= \max_{c \in C} = \{\min(\mu_R(c,s), \mu_U(s,t))\} \\ \upsilon_T(c,p) &= \min_{c \in C} = \{\max(\upsilon_R(c,s), \upsilon_U(s,t))\}. \end{aligned} \tag{10}$$

Equation 7 is applied to determine $K = R \circ U$. The result are shown as follows (Table 3)

Table 3. The results of PFR composition R and U, $K = R \circ U$

RoU	T-1	T-2	T-3	T-4	T-5
A-1	(0.8, 0.2)	(0.7, 0.1)	(0.7, 0.3)	(0.7, 0.2)	(0.7, 0.2)
A-2	(0.8, 0.1)	(0.7, 0.2)	(0.7, 0.3)	(0.7, 0.1)	(0.7, 0.1)
A-3	(0.8, 0.2)	(0.7, 0.3)	(0.7, 0.2)	(0.8, 0.2)	(0.7, 0.3)
A-4	(0.7, 0.2)	(0.6, 0.2)	(0.7, 0.1)	(0.7, 0.2)	(0.7, 0.2)
A-5	(0.7, 0.2)	(0.6, 0.2)	(0.8, 0.2)	(0.8, 0.1)	(0.8, 0.1)

The hesitation margin results of $K = R \circ U$, $\pi_{R \circ U}(c,t)$ are calculated by using Eq. 8 as follows (Table 4)

Table 4. The results of $\pi_{R \circ U}(c,t)$

PI	T1	T2	T3	T4	T5
A1	0.5657	0.7071	0.6481	0.6856	0.6856
A2	0.5916	0.6856	0.6481	0.7071	0.7071
A3	0.5657	0.6481	0.6856	0.5657	0.6481
A4	0.6856	0.7746	0.7071	0.6856	0.6856
A5	0.6856	0.7746	0.5657	0.5916	0.5916

Form Definition 6, the max-min-max composition is calculated as follows

$$\beta_T = \mu_T(c,p) - \upsilon_T(c,p) \times \pi_T(c,p). \tag{11}$$

Candidate locations are criticized with the criterion and hospital types which are suggested in [9]. The results of the case scenario are shown in Table 5 and the results could be read in two way: 1) horizontal decision with regards to candidate locations in contradiction of hospital types and 2) vertical decision with regards to hospital types in contradiction of candidate locations. Horizontal decisions from Table 5 imply that $A1$ is suitable $T2$ type hospital. $A2$ is suitable $T1$ type

Table 5. The results of β_T

β	T1	T2	T3	T4	T5	Max of rows
A1	0.6869	0.6293	0.5056	0.5629	0.5629	0.6869
A2	0.7408	0.5629	0.5056	0.6293	0.6293	0.7408
A3	0.6869	0.5056	0.5629	0.6869	0.5056	0.6869
A4	0.5629	0.4451	0.6293	0.5629	0.5629	0.6293
A5	0.5629	0.4451	0.6869	0.7408	0.7408	0.7408
Max of colums	0.7408	0.6293	0.6869	0.7408	0.7408	

hospital. $A3$ is suitable $T1$ and $T1$ type hospitals. $A4$ is suitable $T3$ type hospital. $A5$ is suitable $T5$ type hospital. On the other hand, vertical decisions suggest that $T1$ is appropriate for $A2$, $T2$ is appropriate for $A1$, $T3$ is appropriate for $A5$, $T4$ is appropriate for $A5$, and $T5$ is appropriate for $A5$

4 Conclusion

In this study, the max-min-max composition model of PFS is applied to LAD problem related to hospital types. The case study results imply that the proposed model is an appropriate and significant tool for decision making problems. LAD problems are accepted as a kind of NP-Hard problem; therefore, solution of this kind of problems requires time and effort. In this study, a mathematical approach is addressed to solve LAD problems using relations where candidate locations, criterion, and hospital types are defined as PFS. The composition of two PFS was established with a aid of a formula to capture vagueness on data. As future work, PFSs still need some theoretical improvements and it could be applied different kind of real life problem embedded with imprecision or imperfect information.

References

1. World Health Organization, https://www.who.int/emergencies/diseases/novel-coronavirus-2019, Last accessed 30 March 2021
2. World Health Organization - People and communities. https://www.who.int/hospitals/people-and-communities/en/. Accessed 30 Mar 2021
3. Gallagher Healthcare. https://www.gallaghermalpractice.com/blog/post/what-are-the-different-types-of-hospitals. Accessed 30 Mar 2021
4. Zadeh, L.A.: Fuzzy sets. In: Zadeh, L.A. (eds.) Fuzzy Sets, Fuzzy Logic, and Fuzzy Systems: Selected Papers, pp. 394–432 (1996)
5. Atanassov, K.T.: Intuitionistic fuzzy sets. In: Intuitionistic fuzzy sets, pp. 1–137. Physica, Heidelberg (1999)
6. Yager, R.R.: Pythagorean membership grades in multicriteria decision making. IEEE Trans. Fuzzy Syst. **22**(4), 958–965 (2013)

7. Xu, Y., Shang, X., Wang, J.: Pythagorean fuzzy interaction Muirhead means with their application to multi-attribute group decision-making. Information **9**(7), 157 (2018)
8. Ejegwa, P.A.: Pythagorean fuzzy set and its application in career placements based on academic performance using max-min-max composition. Complex Intell. Syst. **5**(2), 165–175 (2019)
9. Oppio, A., Buffoli, M., Dell'Ovo, M., Capolongo, S.: Addressing decisions about new hospitals' siting: a multidimensional evaluation approach. Annali dell'Istituto superiore di sanita **52**(1), 78–87 (2016)

Intelligent Fuzzy Pythagorean Bayesian Decision Making of Maintenance Strategy Selection in Offshore Sectors

Mohammad Yazdi[1(✉)], Noorbakhsh Amiri Golilarz[2], Arman Nedjati[3], and Kehinde A. Adesina[4]

[1] University of Lisbon, Lisbon, Portugal
mohammad.yazdi@tecnico.ulisboa.pt
[2] School of Electrical, Computer, and Biomedical Engineering, Southern Illinois University, Carbondale, IL 62901, USA
[3] Industrial Engineering Department, Quchan University of Technology, Quchan, Iran
arman.nedjati@qiet.ac.ir
[4] Industrial Engineering Department, Near East University, Nicosia, KKTC, North Cyprus, Turkey
kehinde.adesina@neu.edu.tr

Abstract. In this study, the new methodology is proposed with the integration of Pythagorean fuzzy set and Bayesian structural method to deal with both objective and subjective uncertainties. The conventional decision-making tools are suffering a couple of fundamental drawbacks, such as (i) the results are depending on subjective terms, (ii) model and data uncertainties are taken into account, (iii) importantly is that confidence level is ignored, and finally (iv) the factor time is not considered into final decisions. In this study, the Pythagorean fuzzy set is overcome with subjective uncertainty, and the Bayesian network is engaged to deal with objective uncertainty. The maintenance strategy selection offshore sectors are studied for the proposed approach to show the effectiveness and efficiency. The results show that the proposed methodology would assist exports in making appropriate decisions.

Keywords: Bayesian network · Decision-making · Crane · Uncertainties · Offshore

1 Introduction

Today, in Covi-19 situations, making proper and reliable decision-making is the most challenging task for humans. Thus, all researchers around the world are spending budget and time on finding out the most effective tool for making a decision over time. There are many questions around us, what we can do to make sure that we are considering the critical factors in the decision-making process. In the following, we reviewed a couple of existing decision-making tools in the literature.

C. Kahraman et al. (Eds.): INFUS 2021, LNNS 308, pp. 598–604, 2022.
https://doi.org/10.1007/978-3-030-85577-2_70

In literature, multi-criteria decision making (MCDM) is utilized to rank, select, and order a set of alternatives among different types of criteria [1, 2]. Yazdi (2019) [3] provided a history of decision making methods over time. It is highlighted by time MCDM techniques have been received attentions. The MCDM methods were extensively applied in Analytical hierarchy process (AHP) [4, 5], TOPSIS (The Technique for Order of Preference by Similarity to Ideal Solution) [6–8], Best-worst method (BWM) [9, 10] and DEMATEL (decision-making trial and evaluation laboratory) [11, 12], ELECTRE (elimination and choice expressing reality) [16], VIKOR (VlseKriterijumska Optimizacija I Kompromisno Resenje) [14], GRA (Grey Relation Analysis) [13], Distance measures, PROMETHEE (Preference ranking organization method for enrichment evaluation) [15], *etc.* [15–22].

However, MCDM methods are deeply suffering from couple of shortages, such as the final output is totally depending on qualitative terms, data and model uncertainties are not handled, confidence level of decision is ignored, and it is not considered the factor time into the final decision. Therefore, it is required to looking for new methods or proposing new framework.

Bayesian Network (BN) is a well-known mathematics-based method designed to be used for decision making challenges and has a considerable outcome in dealing with both qualitative and quantitative variables unlike other MCDM methods. BN has been widely applied as a tool for uncertainty handling and risk assessment purposes such as but not limited to [23–28]. BN uses a graphical structure to describe causes and effects by utilizing quantification of joining different types of variables. To see how BN can deal with the shortages of conventional decision-making tools, first the nodes in BN can be defined using continues nodes within distributions. Therefore, the output would be represented as a distribution in which shows confidence level. In addition, BN can be updated the input data in case availability; therefore, it deals with data uncertainty. Besides, BN can be further developed as dynamic BN, which come with time. According to the aforementioned points, using BN provides much more realistic results. It should be added that the Pythagorean fuzzy set has better reflection compared to the conventional fuzzy set theory [29], and Chebyshev distance measures can provide more accurate results.

The contribution and novelty of this study is twofold: First, the Pythagorean fuzzy set are used in the procedure to deal with subjective uncertainty, and second, the Bayesian network is used to deal with objective uncertainty as well updating process over time.

The rest of paper is organized as the following. In Sect. 2, the concept of Pythagorean fuzzy numbers is described. In Sect. 3, the concept of Bayesian network is explained. In Sect. 4, the maintenance strategy selection is studying as application. Finally, in Sect. 5, a conclusion is provided.

2 Pythagorean Fuzzy Numbers

Pythagorean fuzzy numbers (PFNs) were introduced by Yager (2013) [30]. Yager used an example to show the aforementioned situation in his study. By making this assumption that the decision-makers' opinions are expressed to admit the membership of $x$as $\sqrt{3}/2$ and $1/2$ as the non-membership feature, and the summation of $\sqrt{3}/2$ and $1/2$ are greater than 1, IFS is not usable. Meanwhile, on that could handle the situation is PFS since the summation square of $\sqrt{3}/2$ and $1/2$ is less than one. Thus, a proper alternative is PFS to ask from decision-makers to modify their opinions for satisfying the IFS limitations. As a conclusion PFS has advantages in dealing with the real-life decision-making challenges. In addition, it is necessary to emphasis that all IFSs must be adapted into the PFS, while vice versa is not true. The fundamental definitions of PFS are as the following.

Assume that membership and non-membership functions are shown as $u$and v, respectively by satisfying the state as $u^2 + v^2 \leq 1$subject to the idea of PFS, the following definition are provided $i = \{< x, u_i(x), v_i(x) > | x \in X\}$, where $u_i : X \longrightarrow [0,1]$depicts the membership degree and $v_i : X \longrightarrow [0,1]$illustrates the non-membership degree of the variable $x \in X$to the set i, satisfying the condition as $0 \leq u_i(x) + v_i(x) \leq 1$. The indeterminacy or hesitancy degree can also be described as $\pi_i(x) = u_i(x) - v_i(x)$. Xu and Yager [31] named $(u_i(x), v_i(x))$as IFNs shown by $i = (u_i, v_i)$in order to make IFS easier to understand. If we have two different PFNs as $p_1 = (u_{p_1}, v_{p_1})$, and $p_2 = (u_{p_2}, v_{p_2})$, a square distances can be obtained from $D(p_1, p_2) = \left[\frac{1}{3} \cdot \left(|(u_{p_1})^2 - (u_{p_2})^2|^\omega + |(v_{p_1})^2 - (v_{p_2})^2|^\omega + |(r_{p_1})^2 - (r_{p_2})^2|^\omega - |(d_{p_1})^2 - (d_{p_2})^2|^\omega\right)\right]^{\frac{1}{\omega}}$, in which ω denotes a greater than or equal to one distance parameter.

3 Bayesian Network

Bayesian Network (BN) is a directed acyclic graph (DAG), including vertices and edges named as nodes and arcs, respectively, in the available network. In a BN, nodes represent the variables and arcs denote the relations between the two nodes. BN is introduced and known as a significant technique that has a high sufficiency to deal with both variability as well as uncertainty influence. Accordingly, it is used to approximate challenging decisions related to complicated decision-making problems [32]. As a Bayes' based theorem, BN uses the prior information (hypothesis) of a primary event which can further perform a rational statistical inference. In a simple word, while evidence is set to the child node, backward belief propagation can be obtained, and it gain the probability distribution of parent node(s). While the reverse conditions are true, the prior hypothesis could be received from objectivity such as an observed data within a frequentist approach or subjectivity such as experts' judgments or [33–36].

To use BN in decision making, it required couple steps, which are provided in the following. In step one, all potential factors and sub-factors in our model should be identified. In step two, we need to compute the weight of all factors and corresponding sub-factors. In step three, the causality between the factors and sub-factors should be determined. Finally, the graphical representation of BN should be constructed.

4 Maintenance Strategy Selection in Offshore Sectors

There is a decision-making problem to select an on-board machinery (crane) maintenance strategy for offshore operating. The hierarchical model of this decision-making assessment procedure is as depicted in Fig. 1.

The BN model is structured using GeNie 2.4 software (www.bayesfusion.com) to evaluate all possible alternatives for maintenance strategy selection. The structured BN

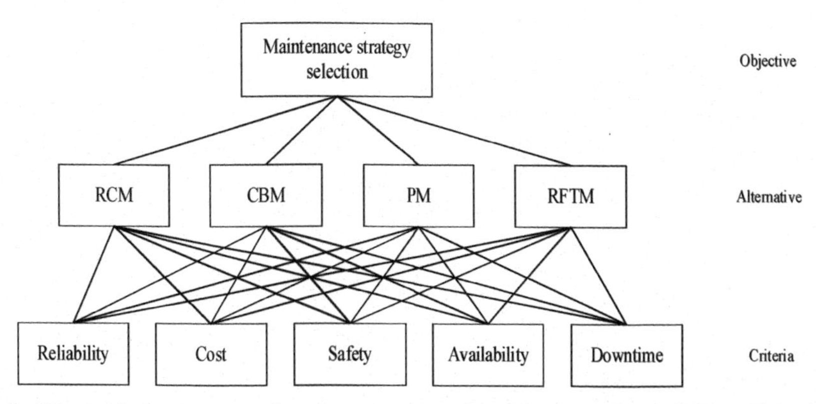

Fig. 1. Hierarchical structure of maintenance strategy selection. Note: RTFM, PM, CBM, and RCM stand for Run-to-Failure Maintenance, Preventive Maintenance, Condition Based Maintenance, and Reliability Centered Maintenance, respectively (Adopted after [37]).

model for the first alternative of the maintenance strategy selection based on different factors is illustrated in Fig. 2. Four different types of variables are considered to model the factors of maintenance strategy selection. The variables are classified considering the measurement of each variable including (i) qualitative variables (measuring the ordinal scale), (ii) Boolean variables (measuring the dichotomous response, "YES/NO", "WORK/NOT WORK"), (iii) continuous variables (measuring the random variables having known probability distribution), and (iv) constant variables having fixed values. A two states Boolean variable of "YES" and "NO" is utilized to model all factors. The state "YES" denotes positive output whereas the state "NO" indicates negative output. As illustrated in Fig. 2, the probability of factors being "YES/NO" is conditional on incorporated sub-factors. The PFNs help us to obtain the continues nodes based subjective judgments from decision makers.

As can be seen from Fig. 2, the priority of maintenance strategy is as CBM (0.3951), RCM (0.2544), PM (0.1949), and RTFM (0.1556).

5 Conclusion

In this study, a Bayesian structural method is utilized in order to help decision makers in the decision-making problems. Generally, when the decision-making problem is selecting an alternative among a set of options, different type of MCDM methods is employed. Bayesian structural method has considerable advantages compare with the MCDM tools. That is why, in this study a Bayesian structural method is used to choose a maintenance strategy among four options in offshore industrial sectors. Bayesian network is provided for the application of study to show that its flexibility and efficiency. As a direction for future studies, different type of fuzzy numbers can be integrated to distinguish the advantages and disadvantages.

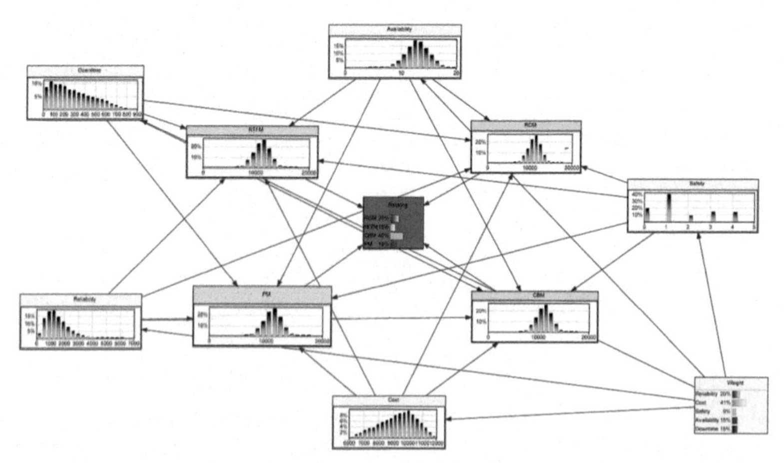

Fig. 2. The maintenance strategy selection based on structured BN model (the weights of all sub-factors are provided in a node in constructed BN, and they are based on obtained opinions from three decision makers)

References

1. Yazdi, M.: A perceptual computing–based method to prioritize intervention actions in the probabilistic risk assessment techniques. Qual. Reliab. Eng. Int. **36**(1), 187–213 (2020)
2. Liu, H.-C.: FMEA Using Uncertainty Theories and MCDM Methods. Springer, Heidelberg (2016)
3. Yazdi, M.: Introducing a heuristic approach to enhance the reliability of system safety assessment. Qual. Reliab. Eng. Int. **35**(8), 2612–2638 (2019)
4. Yazdi, M.: Hybrid probabilistic risk assessment using fuzzy FTA and Fuzzy AHP in a process industry. J. Fail. Anal. Prev. **17**(4), 756–764 (2017). https://doi.org/10.1007/s11668-017-0305-4
5. Zhu, G.N., Hu, J., Ren, H.: A fuzzy rough number-based AHP-TOPSIS for design concept evaluation under uncertain environments. Appl. Soft Comput. J. **91**, 106228 (2020)
6. Yazdi, M., Korhan, O., Daneshvar, S.: Application of fuzzy fault tree analysis based on modified fuzzy AHP and fuzzy TOPSIS for fire and explosion in process industry. Int. J. Occup. Saf. Ergon. **26**(2), 319–335 (2020)
7. Yazdi, M.: Risk assessment based on novel intuitionistic fuzzy-hybrid-modified TOPSIS approach. Saf Sci. **110**, 438–448 (2018)
8. Deng, H., Yeh, C.H., Willis, R.J.: Inter-company comparison using modified TOPSIS with objective weights. Comput. Oper. Res. **27**(10), 963–973 (2000)
9. Liao, H., Shen, W., Tang, M., Mi, X., Lev, B.: The state-of-the-art survey on integrations and applications of the best worst method in decision making: why, what, what for and what's next? Omega **87**, 205–225 (2019)
10. Yazdi, M., Saner, T., Darvishmotevali, M.: Application of an artificial intelligence decision-making method for the selection of maintenance strategy. In: Aliev, R.A., Kacprzyk, J., Pedrycz, W., Jamshidi, M., Babanli, M.B., Sadikoglu, F.M. (eds.) ICSCCW 2019. AISC, vol. 1095, pp. 246–253. Springer, Cham (2020). https://doi.org/10.1007/978-3-030-35249-3_31
11. Yazdi, M., Nedjati, A., Zarei, E., Abbassi, R.: A novel extension of DEMATEL approach for probabilistic safety analysis in process systems. Saf. Sci. **121**, 119–136 (2020)

12. Yazdi, M., Khan, F., Abbassi, R., Rusli, R.: Improved DEMATEL methodology for effective safety management decision-making. Saf. Sci. **127**, 104705 (2020)
13. Liu, H.-C.: Improved FMEA Methods for Proactive Healthcare Risk Analysis, 1st edn. Springer, Singapore (2019). https://doi.org/10.1007/978-981-13-6366-5
14. Ding, X.F., Liu, H.C.: An extended prospect theory – VIKOR approach for emergency decision making with 2-dimension uncertain linguistic information. Soft Comput. **23**(22), 12139–12150 (2019)
15. Liu, H.C., Li, Z., Song, W., Su, Q.: Failure mode and effect analysis using cloud model theory and PROMETHEE method. IEEE Trans. Reliab. **66**(4), 1058–1072 (2017)
16. Yadav, G., Mangla, S.K., Luthra, S., Jakhar, S.: Hybrid BWM-ELECTRE-based decision framework for effective offshore outsourcing adoption: a case study. Int. J. Prod. Res. **56**(18), 6259–6278 (2018)
17. Rezaei, J.: Best-worst multi-criteria decision-making method. Omega **53**, 49–57 (2015)
18. Saaty, T.L.: Decision Making with Dependence and Feedback: the Analytic Network Process the Organization and Prioritization of Complexity. RWS Publications, Pittsburgh (1996)
19. Gul, M., Guven, B., Guner, A.F.: A new Fine-Kinney-based risk assessment framework using FAHP-FVIKOR incorporation. J. Loss Prev. Process Ind. **53**, 3–16 (2018)
20. Chang, K.H., Chang, Y.C., Tsai, I.T.: Enhancing FMEA assessment by integrating grey relational analysis and the decision making trial and evaluation laboratory approach. Eng. Fail. Anal. **31**, 211–224 (2013)
21. Ren, J., Liang, H., Chan, F.T.S.: Urban sewage sludge, sustainability, and transition for Eco-City: multi-criteria sustainability assessment of technologies based on best-worst method. Technol. Forecast. Soc. Change **116**, 29–39 (2017)
22. Nie, R., Tian, Z., Wang, J., Zhang, H., Wang, T.: Water security sustainability evaluation: applying a multistage decision support framework in industrial region. J. Clean. Prod. **196**, 681–704 (2018)
23. Yazdi, M.: A review paper to examine the validity of Bayesian network to build rational consensus in subjective probabilistic failure analysis. Int. J. Syst. Assur. Eng. Manag. **10**(1), 1–18 (2019). https://doi.org/10.1007/s13198-018-00757-7
24. Yazdi, M., Kabir, S., Walker, M.: Uncertainty handling in fault tree based risk assessment: state of the art and future perspectives. Process Saf. Environ. Prot. **131**, 89–104 (2019)
25. Kabir, S., Papadopoulos, Y.: Applications of Bayesian networks and Petri nets in safety, reliability, and risk assessments: a review. Saf. Sci. **115**, 154–175 (2019)
26. El-Gheriani, M., Khan, F., Chen, D., Abbassi, R.: Major accident modelling using spare data. Process Saf. Environ. Prot. **106**, 52–59 (2017)
27. Misuri, A., Khakzad, N., Reniers, G., Cozzani, V.: Tackling uncertainty in security assessment of critical infrastructures: Dempster-Shafer theory vs credal sets Theory. Saf. Sci. **107**, 62–76 (2018)
28. Khakzad, N., Khan, F., Amyotte, P.: Dynamic safety analysis of process systems by mapping bow-tie into Bayesian network. Process Saf. Environ. Prot. **91**, 46–53 (2013)
29. Yazdi, M.: Footprint of knowledge acquisition improvement in failure diagnosis analysis. Qual. Reliab. Eng. Int. **35**, 405–422 (2018)
30. Yager, R.R.: Pythagorean membership grades in multicriteria decision making. IEEE Trans. Fuzzy Syst. **22**(4), 958–965 (2013)
31. Xu, Z., Yager, R.R.: Some geometric aggregation operators based on intuitionistic fuzzy sets. Int. J. Gen. Syst. **35**(4), 417–433 (2018)
32. Fenton, N.E., Neil, M., Martin, D.: Risk Assessment and Decision Analysis with Bayesian Networks. Chapman and Hall/CRC, Boca Raton (2018)
33. Kabir, S., Yazdi, M., Aizpurua, J.I., Papadopoulos, Y.: Uncertainty-aware dynamic reliability analysis framework for complex systems. IEEE Access **6**, 29499–29515 (2018)

34. Yazdi, M., Kabir, S.: A fuzzy Bayesian network approach for risk analysis in process industries. Process Saf. Environ. Prot. **111**, 507–519 (2017)
35. Yazdi, M., Kabir, S.: Fuzzy evidence theory and Bayesian networks for process systems risk analysis. Hum. Ecol. Risk Assess. **26**(1), 57–86 (2020)
36. Daneshvar, S., Yazdi, M., Adesina, K.A.: Fuzzy smart failure modes and effects analysis to improve safety performance of system: case study of an aircraft landing system. Qual. Reliab. Eng. Int. **36**, 890–909 (2020)
37. Asuquo, M.P., Wang, J., Zhang, L., Phylip-Jones, G.: Application of a multiple attribute group decision making (MAGDM) model for selecting appropriate maintenance strategy for marine and offshore machinery operations. Ocean Eng. **179**, 246–260 (2019)

Quality Management

Testing Absolute Error Loss-Based Capability Index

Abbas Parchami(✉)

Department of Statistics, Faculty of Mathematics and Computer, Shahid Bahonar University of Kerman, Kerman, Iran
parchami@uk.ac.ir

Abstract. In order to more compliance of products with customer requirements, a loss-based process capability index is proposed to evaluate manufacturing processes based on absolute error loss. The distribution complexity for the estimator of the proposed loss-based capability index hindered the scientific progress in the field of statistical inference. To cover a part of this deprivation, testing the capability of productive process on the basis of the absolute error loss is proposed in this paper by Monte Carlo simulation for normal data. In other words, an algorithm is proposed in this article to estimate the critical value of Monte Carlo testing process capability index by random sample from a manufacturing process. The proposed algorithm is constructed for an one-dimensional quality characteristic on the basis of absolute error loss and under the normality condition. Moreover, a case study is provided in pipe manufacturing industries to show the performance of the proposed algorithm.

Keywords: Absolute error loss · Testing hypotheses · Process capability index · Simulation

1 Prerequisites and Brief Introduction

Quality of manufacturing processes is important for customer. How to evaluate manufacturing processes in an easier way is a major issue for researchers. Process capability indices are widely used to assess manufacturing processes due to process capability indices can ensure that quality of products meet customer requirements. New studies about process capability indices strongly continue for one-dimensional [5] and also multivariate quality characteristic [3]. Some inferential properties of new multivariate process capability indices are discussed in [1] and [8] from statistical point of view. We succinctly introduce only two capability indices in follows and the interested readers can refer to books [4] and [6] for more details on process capability indices.

Let f is the probability density function of a one-dimensional continuous quality characteristic X. Chan et al. [2] introduced the process capability index

$$C_{pm} = \frac{USL - LSL}{6\sqrt{E\left[(X-T)^2\right]}} = \frac{USL - LSL}{6\sqrt{\sigma^2 + (\mu - T)^2}} \tag{1}$$

C. Kahraman et al. (Eds.): INFUS 2021, LNNS 308, pp. 607–613, 2022.
https://doi.org/10.1007/978-3-030-85577-2_71

on the basis of square error loss function $L(X,T) = (X-T)^2$. Now, let us to change the loss function into the absolute error loss $L(X,T) = |X-T|$ to introduce another loss-based capability index

$$C_{pm}^{A} = \frac{USL - LSL}{6\sqrt{E\left(|X-T|\right)}}. \tag{2}$$

The statistical significance test is simulated for testing the capability index C_{pm}^{A} with normal data as the objective and originality of this paper.

This paper is organized as follows. Significance quality test on absolute error loss-based capability index is discussed in Sect. 2. Then, Monte Carlo testing C_{pm}^{A} is presented by an algorithm in Sect. 3. A numerical example based on absolute error loss has been provided in Sect. 4. The final section is conclusions and future researches.

2 Testing Capability Index C_{pm}^{A}

Testing capability is a common statistical method to check the performance of industrial productive processes. Under the normality condition of one-dimensional quality characteristic X, the main problem is testing the null hypothesis

$H_0:\ C_{pm}^{A} \leq c_0$ (process is not capable),

against the alternative hypothesis

$H_1:\ C_{pm}^{A} > c_0$ (process is capable),

based on random sample $x_1, ..., x_n$ with unknown mean and unknown variance parameters by Monte Carlo simulation, where $c_0 > 0$ is the standard minimal criteria for loss-based capability index C_{pm}^{A}.

Obviously, the critical region of the proposed capability test is $\widehat{C_{pm}^{A}} > c$ in which c is the unknown critical value and

$$\widehat{C_{pm}^{A}} = \frac{USL - LSL}{6\sqrt{\frac{1}{n}\sum_{i=1}^{n} L(X_i - T)}} = \frac{USL - LSL}{6\sqrt{\frac{1}{n}\sum_{i=1}^{n} |X_i - T|}} \tag{3}$$

is the estimator of loss-based capability index C_{pm}^{A} based on random sample $X_1, ..., X_n$. Considering random sample $X_1, ..., X_n \overset{i.i.d.}{\sim} N\left(\mu, \sigma^2\right)$ with unknown parameters, the probability of type I error in testing quality based on absolute error loss is $\alpha = Pr\left(\widehat{C_{pm}^{A}} > c \mid C_{pm}^{A} = c_0\right)$ and hence

$$Pr\left(\widehat{C_{pm}^{A}} \leq c \mid C_{pm}^{A} = c_0\right) = 1 - \alpha. \tag{4}$$

Therefore, the unknown critical value c is equal to the $(1-\alpha)$-th quantile of $\widehat{C_{pm}^{A}}$ distribution under assumption $C_{pm}^{A} = c_0$ and finding precise critical value c by Monte Carlo simulation is investigated in Sect. 3 as the main goal of this paper.

3 Monte Carlo Simulation Approach for Testing C_{pm}^A

An algorithm proposed in follow to estimate the critical value of Monte Carlo testing process capability index C_{pm}^A at the given significance level α.

Algorithm 1.
Step 1: Compute the estimated loss-based capability index $\widehat{c_{pm}^A}$ on the basis of the observed random sample $x_1, ..., x_n$ by Eq. (3).
Step 2: Calculate sequence $\mu_1 < \mu_2 < ... < \mu_k$ to cover the interquartile range $[Q_1, Q_3]$ by following formula

$$\mu_j = Q_1 + \frac{j-1}{k-1}\left(Q_3 - Q_1\right), \quad j = 1, 2, ..., k, \tag{5}$$

where Q_1 and Q_3 are 25th and 75th percentiles of observations $x_1, ..., x_n$.
Step 3: Consider a reasonable sequence $n_1 < n_2 < ... < n_s$ to cover possible sample sizes.
Step 4: For any $n_i \in \{n_1, n_2, ..., n_s\}$ and $\mu_j \in \{\mu_1, \mu_2, ..., \mu_k\}$,
a) compute the unknown value of root $\sigma_0 > 0$ from equation $C_{pm}^A = c_0$, or equivalently from equation

$$\left(\frac{USL - LSL}{6c_0}\right)^2 - \int_{-\infty}^{+\infty} |x - T| \; \phi_{\mu_j, \sigma_0}(x) \; dx = 0,$$

in which $\phi_{\mu,\sigma}$ is the p.d.f. of normal random variable with mean μ and variance σ^2.
b) simulate $m = 10^3$ random sample with size n_i from $N\left(\mu_j, \sigma_0^2\right)$,
c) compute estimators $\widehat{c_{pm}^A}^{[1]}, \widehat{c_{pm}^A}^{[2]}, ..., \widehat{c_{pm}^A}^{[m]}$ of loss-based capability index using Eq. (3) for each simulated sample from part (b),
d) the critical value for m simulated samples in part (b) is the $(1-\alpha)$-th quantile of $\widehat{C_{pm}^A}$ distribution, i.e.

$$c_{i,j} = \widehat{c_{pm}^A}^{(m(1-\alpha))}, \quad i = 1, ..., s, \; j = 1, ..., k, \tag{6}$$

where $\widehat{c_{pm}^A}^{(1)}, \widehat{c_{pm}^A}^{(2)}, ..., \widehat{c_{pm}^A}^{(m)}$ are the ordered estimated indices in part (c).
Step 5: Monte Carlo critical value in testing quality based on absolute error loss is equal to the average of $s \times k$ estimated critical values in Step 4 and hence

$$c = \frac{1}{s \times k} \sum_{i=1}^{s} \sum_{j=1}^{k} c_{i,j}. \tag{7}$$

Step 6: The process is capable, i.e. the null hypothesis rejected at significance level α, if $\widehat{c_{pm}^A} > c$; otherwise the process is incapable.

4 Case Study in Pipe Manufacturing Industries

The outer diameter for one type of pipe is considered in this study as the quality characteristic X. The specification limits for the outer diameter of pipe are $15.0\,\text{cm} \pm 0.95\,\text{cm}$ and the target value $T = 15$ cm is considered for X.

Now, we are going to test the capability of producing process for the outer diameter measurement of the produced pipes based on loss-based capability index C_{pm}^{A} at significance level 0.05. The following random sample, with size fifty three, is taken from the diameter length in term of centimeters (see Fig. 1):

14.98 14.91 14.56 14.79 15.06 15.09 14.61 14.86 14.48
14.89 15.30 15.20 14.90 15.27 15.19 15.00 14.68 14.91
15.25 15.11 14.79 14.31 14.77 14.33 14.59 14.86 14.76
14.71 14.94 14.89 14.41 14.95 15.26 15.03 14.56 14.54
15.08 14.44 14.87 14.77 15.30 14.74 14.72 15.22 14.68
14.96 15.04 14.88 14.77 15.17 14.85 14.87 15.38

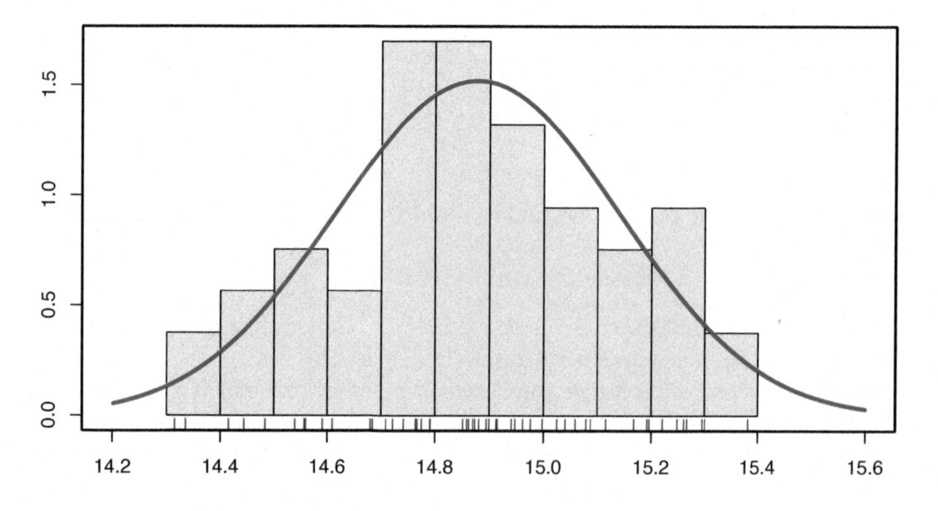

Fig. 1. Histogram of 53 observed diameters.

Monte Carlo simulation approach is considered in this case study to test $H_0 : C_{pm}^{A} \leq 1.00$, against $H_1 : C_{pm}^{A} > 1.00$, based on the observed random sample $x_1, ..., x_{53}$. Shapiro-Wilk test confirm the normality assumption of the observed data with statistics 0.98 and p-value $= 0.58$. So, the loss-based capability index C_{pm}^{A} can be estimated by

$$\widehat{c_{pm}^{A}} = \frac{USL - LSL}{6\sqrt{\frac{1}{n}\sum_{i=1}^{53} |x_i - T|}} = 0.655.$$

Therefore, $0.77 > c$ is the critical region of Monte Carlo testing index C_{pm}^{A} at significance level $\alpha = 0.01$, in which the critical value must be simulated using Algorithm 1 in Sect. 3. We did a simulation to compute Monte Carlo critical value in testing quality based on absolute error loss where the mean and sample size are changed over the following sequences

$\mu = 14.72140, 14.77756, 14.83372, 14.88988, 14.94604, 15.00220$ and 15.05836,
$n = 30, 40, 50, 60, 70$ and 80.

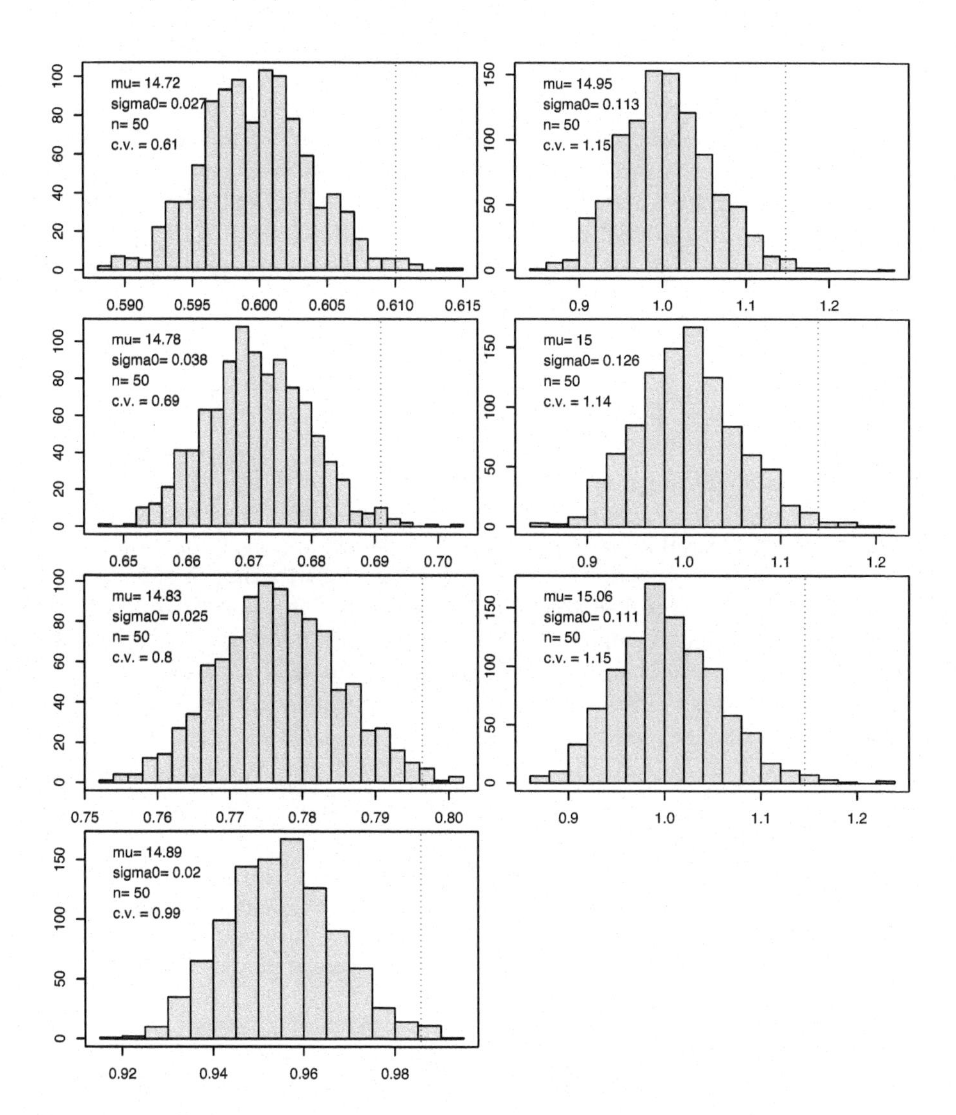

Fig. 2. Histograms of the estimated loss-based capability indices by normal simulated samples with mean μ_j and size $n = 50$ under H_0 for $j = 1, 2, ..., 7$.

The unknown root σ_0 in equation

$$\left(\frac{15.95-14.05}{6\times 1}\right)^2-\int_{-\infty}^{+\infty}|x-15.00|\ \phi_{\mu_j,\sigma_0}(x)\ dx=0,$$

is computed for all 42 possible cases by Newton Raphson method. Then, for each 42 combination of μ and n, 10^3 independent random samples simulated from normal distribution $N\left(\mu,\sigma_0^2\right)$. For each simulated sample we obtained the estimated capability index C_{pm}^A using Eq. (3).

Figure 2 is shown the histograms of the estimated capability indices in Part (c) by normal simulated samples with mean μ_j under H_0 only for sample size 50 and $j=1,2,...,7$. After ordering 1000 estimated capability indices, the 990-th capability index is considered as the critical value for each 42 possible cases (see vertical lines in Fig. 2). The total average of 42 captured critical values is equal to $c=0.930$ which is considered as the Monte Carlo critical value of testing quality in this study. By comparing $\widehat{c_{pm}^A}=0.655$ and c, the null hypothesis accepted at significance level 0.01 and hence the process is not determined as a capable process.

Conclusions and Future Researches

Testing the capability of manufacturing processes is investigated in this article using a new loss-based capability index. The main goal of this investigation is estimate the critical value by Monte Carlo simulation approach for normal data based on absolute error loss. In other words, the contribution of this paper is simulation of significance test to evaluate the quality of a normal process with absolute error loss-based capability index. Moreover, a numerical example based on symmetric absolute error loss has been investigated in pipe manufacturing industries. Testing capability based on multivariate capability indices is a potential subject for further research.

References

1. Chakraborty, A.K., Chatterjee, M.: Distributional and inferential properties of some new multivariate process capability indices for symmetric specification region. Quali. Reliab. Eng. Int. **37**(3), 1099–1115 (2021)
2. Chan, L.K., Cheng, S.W., Spiring, F.A.: A new measure of process capability: C_{pm}. J. Qual. Technol. **20**(3), 162–173 (1988)
3. De-Felipe, D., Benedito, E.: A review of univariate and multivariate process capability indices. Int. J. Adv. Manuf. Technol. **92**(5–8), 1687–1705 (2017)
4. Kotz, S., Johnson, N.L.: Process Capability Indices. Chapman & Hall, London (1993)
5. Kotz, S., Johnson, N.L.: Process capability indices - a review, 1992–2000. J. Qual.Technol. **34**, 2–19 (2002)
6. Kotz, S., Lovelace, C.R.: Process Capability Indices in Theory and Practice. Arnold, London (1998)

7. Lin, P.C., Pearn, W.L.: Testing manufacturing performance based on capability index C_{pm}. Int. J. Adv. Manuf. Technol. **27**(3), 351–358 (2005)
8. Khadse, K.G., Khadse, A.K.: On properties of probability-based multivariate process capability indices. Qual. Reliab. Eng. Int. **36**(5), 1768–1785 (2020)

Customer Need Analysis for Trim Levels of Automobiles Using Fuzzy Kano Model

Irem Ucal Sari(✉) and Furkan Sevinc

Industrial Engineering Department, Istanbul Technical University, Istanbul, Turkey
ucal@itu.edu.tr

Abstract. In today's competitive and innovative environment, it is becoming more and more important to be able to attract customers. Especially in the automobile industry where each model can be presented with different features, determining the automobile equipment combination (trim levels) that will attract the customers the most and introducing the basic packages created in this direction can directly affect the sales. Therefore, in this study, a survey is conducted to determine customers' attitudes towards the features included in the basic packages of automobiles. Based on the literature review, the design parameters used in the survey are determined as seat comfort, type of air conditioner, gear type, multimedia panel, keyless entry, cruise control, digital display, rear camera and sun roof. The fuzzy Kano model is used to determine trim level features that appeal to customers for different automobile segments. The results obtained are also compared with the traditional Kano model.

Keywords: Fuzzy Kano model · Customer need analysis · Trim level features

1 Introduction

Automobiles emerge as a product with different features that may attract the attention of customers. The choice of the features of any automobile reflects consumer preferences, so the features of the automobiles become important and start to affect the demand of the automobiles. These features do not only mean size, weight, braking system, steering type, maximum speed, fuel consumption and brand as in the past. As a result of the rapid development and change of technology, comfort and technology features such as cruise control, lane tracking assistant, sound system, seat type, multimedia screen, parking assistant and adaptive lighting etc. have increased their effect on purchasing decision. For this reason, the customization and personalization of cars has gained great importance and is demanded by customer. Customer-order-driven production can provide higher profit, lower stocks, and more accurate information of customer preference. Although customer-order-driven production has these advantages, most of the automobiles are still mass produced and sold from stock, which leads to schedule disruption, overstock, extra marketing costs and lower profitability [1].

Automobile manufacturers aiming to continue production without interruption by standardizing automobiles still produce customized automobiles to meet the requests

C. Kahraman et al. (Eds.): INFUS 2021, LNNS 308, pp. 614–621, 2022.
https://doi.org/10.1007/978-3-030-85577-2_72

of each customer. Customized automobiles are produced with automobiles with similar characteristics, without interruption in the production line. Therefore, although the customer-order-driven production does not harm the company economically, it requires the customer to wait during this period due to transportation and production. In most cases, this situation leads to a loss of customers in companies as a result of customers' unwillingness to wait. Companies offer customers different trim levels, which are different versions of the same model that identify a vehicle's level of equipment or special features, for their cars to overcome this problem. While these trim levels differ from each other with the features they contain, a special system has been created to ensure that production can be carried out effectively and efficiently without interruption. Through features available in different trim level, it is aimed to appeal to every customer. It is therefore essential to see how trim levels affect customers and what needs to be changed if changes are required.

The main purpose of this study is to identify the features that customers pay attention to in their purchasing decision in the automobile industry in order to improve trim levels with these features. For this purpose, using the relevant literature, a questionnaire is prepared and analyzed using fuzzy Kano model to determine the importance of the trim level features. The results of the analysis are also compared with the outcomes of traditional Kano model.

The paper is organized into four main sections: After introduction, in the second part of the study automobile features used in trim levels are detailed after a literature review on design parameters of automobiles. In the third section, fuzzy Kano model used in the article is explained. Then, the methodology is applied for automobile industry. Finally, the paper is concluded with a discussion of the results and future research suggestions.

2 Automobile Features Used in Trim Levels

The feature of automobile term is basically any distinctive attribute of an automobile. In the literature review, it is seen that an excessively large number of features were discussed. Different researchers have examined different features considering the importance of the features, which indicates that the feature term has a wide coverage. There are many comfort, technology, and safety features in the car such as cruise control, lane tracking assistant, sound system, seat type, multimedia screen, parking assistant, adaptive lighting trim. The most examined features in the literature are summarized in Table 1.

In this study, since the additional features of the vehicle other than the basic function can be considered as trim level, the engine power, fuel economy, ABS, ASR, ESP, EDL, number of airbags, which are generally standard in vehicles, will not be considered. Therefore, seat comfort, type of air conditioner, gear type, multimedia panel, keyless entry, cruise control, digital display, rear camera and sun roof are determined as design parameters. The trim level features that are examined in the survey can be detailed as follows:

Electrical and heated front seats (F1): Front seat which can be heated at the push of a button thanks to small electric motors.

Digital climate control (two or more regions) (F2): Climate control which works different among regions in car automatically.

Table 1. Literature review on the most examined automobile features

Features	[2]	[3]	[4]	[5]	[6]	[7]	[8]	[9]	[10]	[11]	[12]
Fuel economy	+					+	+	+		+	
Engine power	+	+	+	+		+	+	+		+	
Seat comfort				+				+		+	
Type of air conditioner	+	+		+	+	+	+		+	+	
Type of steering wheel		+		+							+
Gear type	+		+	+							
Multimedia panel				+			+	+	+		+
ABS	+	+		+	+				+		
ASR				+	+				+		
ESP		+		+	+	+			+		
EDL				+	+				+		
Number of airbags	+	+				+			+		
Keyless entry				+	+				+		
Leather seats		+				+			+		
Cruise control			+	+	+		+		+		
Digital display			+	+			+				
Rear camera				+	+				+		
Sunroof			+	+			+		+		
Styled wheels				+			+		+		

Keyless Go (F3): Technology which offers an easy way to unlock, lock, start and stop without the hassle of digging out keys.

Automatic transmission (F4): Transmission that does not require any driver input to change gears.

Rear parking camera (F5): Camera attached to the rear of car, which helps to see the area behind car when backing up.

Sunroof (F6): Any kind of panel on the roof of a car that permits light, air or both to come into a vehicle.

Adaptive Cruise control (F7): System that automatically controls the speed of a motor vehicle.

Multimedia (F8): Feature is a sleek flat panel screen fitted to the center of the dashboard, which offers connection thanks to Bluetooth.

Digital display (F9): Set of instrumentation, including the speedometer, that is displayed with a digital readout rather than with the traditional analog gauges.

3 Fuzzy Kano Model

Among the many approaches that address customer needs analysis, the Kano model is widely used as an effective tool to understand customer preferences due to the ease of classifying customer needs based on survey data [13]. It is one of the effective tools that will contribute to companies that want to define value for their customers, differentiate in competition, be more profitable [14]. In the Kano questionnaire, two types of questions are asked for each feature: functional form and dysfunctional form. "How do you feel if the car has cruise control?" asked in functional form, and "How do you feel if there is no cruise control in the car?" asked in dysfunctional form. Participants use a linguistic scale of 'like', 'must-be', "neutral", 'live-with"' and 'dislike' for both functional and dysfunctional questions [15, 16]. Evaluation table of Kano questions is created according to the possible answers of the questions, A (Attractive), O (One-dimensional), M (Must-be), I (Indifferent), R (Reverse), and Q(Questionable) letters were placed according to the row and column values affecting the relevant cell of the matrix, which is shown in Table 2 [15].

Table 2. Kano evaluation table [15]

		Dysfunctional form of the question				
		Like	Must-be	Neutral	Live-with	Dislike
Functional form of the question	Like	Q	A	A	A	O
	Must-be	R	I	I	I	M
	Neutral	R	I	I	I	M
	Live-with	R	I	I	I	M
	Dislike	R	R	R	R	Q

Ilbahar and Cebi [17] proposed the fuzzy Kano model to determine the membership degrees of design parameters to the allowing a design parameter to belong to different classes. The steps of fuzzy Kano model can be summarized as follows [16, 17]:

Step 1. Identify main design parameters for assessment.
Step 2. Conduct a Kano questionnaire for the identified parameters and obtain A, O, M, I, R and Q values using Table 2 according to results of the questionnaire.
Step 3. Determine the weights of the parameters. When sum of $(A + O + M)$ is greater than sum of $(I + R + Q)$ use Eq. (1), otherwise use Eqs. (2) and (3).

$$W_i = \frac{A + O}{A + 2xO + M}\ for\, i \in \{A, O, M\} \tag{1}$$

$$W_i = 0\ for\, i \in \{I\} \tag{2}$$

$$W_i = \frac{-R}{I + R + Q}\ for\, i \in \{R\} \tag{3}$$

Step 4. Normalize weights by dividing the responses to a given class by the total number of responses.
Step 5. Plot the normalized values of each parameter on a graph using one of the "A-O-M" or "I-R-Q" groups to which the parameter belongs and defuzzify this graph using the center of gravity (COG) formula given in Eq. (4) to determine the class of the considered parameter where F is a fuzzy set.

$$COG(F) = \frac{\int xf(x)dx}{\int f(x)dx} \tag{4}$$

Step 6. Calculate membership degrees of design attributes to the classes. When sum of $(A + O + M)$ is greater than sum of $(I + R + Q)$ use Eq. (5), otherwise use Eq. (6) where μ_{ij} represents membership degree of ith design parameter to class j.

$$\mu_{ij} = \begin{cases} max((1 - |COG - 1|), 0) \, for \, j \in \{M\} \\ max((1 - |COG - 2|), 0) \, for \, j \in \{O\} \\ max((1 - |3 - COG|), 0) \, for \, j \in \{A\} \end{cases} \tag{5}$$

$$\mu_{ij} = \begin{cases} max((1 - |COG - 1|), 0) \, for \, j \in \{R\} \\ max((1 - |COG - 2|), 0) \, for \, j \in \{I\} \end{cases} \tag{6}$$

Step 7. Calculate the presence and absence points of each design parameter using Eq. (7).

$$P_{ik} = \sum\nolimits_{j \in J} W_i \times S_{kj} \times \mu_{ij} \tag{7}$$

Where S_{kj} is the state point that represents the presence and absence effects of any parameters depending on the class where $k \in \{Presence, Absence\}, j \in \{A, O, M, I, R\}$. The scale given in Table 3 can be used to determine S_{kj}.

Table 3. Absence and presence effects of a design parameter [16, 17]

	Class				
	R	I	M	O	A
Presence	−50	0	0	50	100
Absence	0	0	−100	−50	0

Step 8. Calculate overall score of the design parameter using Eq. (8)

$$Score = \sum_{\forall k, i} X_{ik} \times P_{ik} \tag{8}$$

where X_{ik} is a binary variable representing whether i^{th} design parameter is present or absent in the alternative design.

4 Application

To determine the automobile features in trim levels according to customer expectations a questionnaire is performed using the automobile features determined in Sect. 2. First, participants are asked about the car segment they intend to buy among the best-selling segments; B, C and D. Since the price ranges and standard features of the automobiles in these three segments are different, the survey answers collected for these three segments are analyzed separately. Table 4, 5 and 6 summarize fuzzy Kano results and classical Kano results for B, C and D segment automobiles, respectively.

Table 4. Analysis Results for B Segment Automobiles

Feature	Traditional Kano Class	Membership Values of Classes using Fuzzy Kano			State Point	
		m(M)	m(O)	m(A)	Presence Point	Absence Point
F1	A	0,00	0,36	0,64	80,03	−17,74
F2	A	0,00	0,74	0,26	44,13	−25,69
F3	A	0,00	0,73	0,27	44,33	−25,67
F4	O	0,00	0,88	0,12	33,24	−25,97
F5	A	0,00	0,68	0,32	48,81	−25,16
F6	A	0,00	0,37	0,63	79,43	−17,94
F7	I	0,00	0,73	0,27	44,83	−25,62
F8	A	0,00	0,73	0,27	44,50	−25,65
F9	A	0,00	0,48	0,52	67,15	−21,49

Table 4 shows that digital climate control, keyless go, rear parking camera and multimedia features that belong to Class A in the traditional method have higher membership degrees in Class O than Class A in fuzzy Kano model. This is due to the fact that although the majority of the customers chose the A class, the answers in the O and M classes are also in a considerable number.

When the results of the B and C segments are compared, it is noteworthy that the cruise control feature has similar membership values in the fuzzy Kano model, although it's class clearly differs in the traditional model. This result is also due to the effect of the answers belonging to classes outside the superior class.

When fuzzy Kano model results are examined, according to highest membership class, it is seen that the expectations of the B segment customers in the keyless go feature and the D segment customers in the sunroof feature differ. When the trim level features are examined according to the status scores, it is seen that the effect of the presence or absence of the feature for the segments differs in almost all features.

Table 5. Analysis Results for C Segment Automobiles

Feature	Traditional Kano Class	Membership Values of Classes using Fuzzy Kano			State Point	
		m(M)	m(O)	m(A)	Presence Point	Absence Point
F1	A	0,00	0,40	0,60	76,00	−19,00
F2	A	0,00	0,52	0,48	64,00	−22,27
F3	A	0,00	0,46	0,54	70,02	−20,72
F4	O	0,00	0,97	0,03	26,84	−25,33
F5	O	0,00	0,83	0,17	36,88	−26,04
F6	A	0,00	0,48	0,52	68,03	−21,26
F7	A	0,00	0,68	0,32	49,00	−25,14
F8	A	0,00	0,84	0,16	36,18	−26,04
F9	A	0,00	0,55	0,45	60,95	−22,98

Table 6. Analysis Results for Segment Automobiles

Feature	Traditional Kano Class	Membership Values of Classes using Fuzzy Kano			State Point	
		m(M)	m(O)	m(A)	Presence Point	Absence Point
F1	A	0,00	0,48	0,52	67,49	−21,40
F2	A	0,00	0,80	0,20	38,91	−26,00
F3	A	0,00	0,57	0,43	58,49	−23,51
F4	O	0,00	1,00	0,00	25,00	−25,00
F5	A	0,00	0,77	0,23	41,35	−25,89
F6	A	0,00	0,65	0,35	51,88	−24,72
F7	A	0,00	0,78	0,22	40,74	−25,93
F8	A	0,00	0,82	0,18	37,60	−26,03
F9	A	0,00	0,54	0,46	62,05	−22,73

5 Conclusion

In the automobile industry, where the features of the basic packages offered to the market directly affect sales, customers' preferences should be analyzed well in order to determine the automobile trim levels in the most accurate way. In this study, the data collected by the questionnaire are analyzed with the fuzzy Kano model in order to analyze customer preferences. The fuzzy Kano model is preferred, since in the traditional Kano model, the highest selected category is determined as the main category and the expectations of the participants belonging to another category are ignored. The results of the study show that fuzzy Kano model represents the customer expectations on automobile features for trim levels better by considering all of the answers. For some of the features the class that has the highest membership value are different than the traditional Kano class.

For the future research, it is decided to analyze different trim levels for same model of automobiles. Also, the criteria used in the questionnaire could be expanded.

References

1. Zhang, X., Chen, R.: Forecast-driven or customer-order-driven? An empirical analysis of the Chinese automotive industry. Int. J. Oper. Prod. Manag. **26**(6), 668–688 (2006)
2. Pazarlioglu, M.V., Gunes, M.: The hedonic price model for fusion on car market. In: Proceedings of the Third International Conference on Information Fusion, vol. 1, pp. TUD4–13. IEEE (2000)
3. Saridakis, C., Baltas, G.: Modeling price-related consequences of the brand origin cue: an empirical examination of the automobile market. Mark. Lett. **27**(1), 77–87 (2014). https://doi.org/10.1007/s11002-014-9304-3
4. Akay, E.C., Bekar, E.: Robust and resistant estimations of hedonic prices for second hand cars: an application to the Istanbul car market. Int. J. Econ. Financ. Issues **8**(1), 39 (2018)
5. Ecer, F.: Forecasting of second-hand automobile prices and identification of price determinants in Turkey (*Turkish*). Anadolu Univ. J. Soc. Sci. **13**(4), 101–112 (2013)
6. Kang, Y.T., Chung, K.S.: Customer satisfaction analysis of smart car features using the Kano model: a comparative analysis of similar research cases. J. Korean Soc. Qual. Manag. **46**(3), 717–738 (2018)
7. Baltas, G., Saridakis, C.: Measuring brand equity in the car market: a hedonic price analysis. J. Oper. Res. Soc. **61**(2), 284–293 (2010)
8. Asher, C.C.: Hedonic analysis of reliability and safety for new automobiles. J. Consum. Aff. **26**(2), 377–396 (1992)
9. Shende, V.: Analysis of research in consumer behavior of automobile passenger car customer. Int. J. Sci. Res. Publ. **4**(2), 1 (2014)
10. Daştan, H.: Türkiye'de ikinci el otomobil fiyatlarini etkileyen faktörlerin hedonik fiyat modeli ile belirlenmesi. Gazi Üniversitesi İktisadi ve İdari Bilimler Fakültesi Dergisi **18**(1), 303–327 (2016)
11. Nepal, B., Yadav, O.P., Murat, A.: A fuzzy-AHP approach to prioritization of CS attributes in target planning for automotive product development. Expert Syst. Appl. **37**(10), 6775–6786 (2010)
12. You, H., Ryu, T., Oh, K., Yun, M.H., Kim, K.J.: Development of customer satisfaction models for automotive interior materials. Int. J. Ind. Ergon. **36**(4), 323–330 (2006)
13. Kano, N., Seraku, N., Takahashi, F., Tsuji, S.: Attractive quality and must-be quality. J. Jpn. Soc. Qual. Control **41**, 39–48 (1984)
14. Chen, C., Chuang, M.: Integrating the Kano model into a robust design approach to enhance customer satisfaction with product design. Int. J. Prod. Econ. **114**(2), 667–681 (2008)
15. Xu, Q., Jiao, R.J., Yang, X., Helander, M., Khalid, H.M., Opperud, A.: An analytical Kano model for customer need analysis. Des. Stud. **30**(1), 87–110 (2009)
16. Uzun, I.M., Cebi, S.: A novel approach for classification of occupational health and safety measures based on their effectiveness by using fuzzy kano model. J. Intell. Fuzzy Syst. **38**(1), 589–600 (2020)
17. Ilbahar, E., Cebi, S.: Classification of design parameters for E-commerce websites: a novel fuzzy Kano approach. Telematics Inform. **34**(8), 1814–1825 (2017)

On Converting Crisp Failure Possibility into Probability for Reliability of Complex Systems

Bekir Sahin[1,2(✉)], Anis Yazidi[1,3], Dumitru Roman[4], Md Zia Uddin[4], and Ahmet Soylu[1,3]

[1] Norwegian University of Science and Technology, Gjøvik, Norway
[2] International Maritime College Oman, Sohar, Oman
bekir@imco.edu.om
[3] OsloMet – Oslo Metropolitan University, Oslo, Norway
{anis.yazidi,ahmet.soylu}@oslomet.no
[4] SINTEF AS, Oslo, Norway
{dumitru.roman,zia.uddin}@sintef.no

Abstract. The reliability of complex systems is analyzed based on several systematic steps using many safety engineering methods. The most common technique for safety system analysis and reliability, vulnerability and criticality estimation is the fault tree analysis method. There exist numerous conventional and fuzzy extended approaches to construct such a tree. One of the steps of the fuzzy fault tree analysis method (FFTA) is the conversion of crisp failure possibility (CFP) into failure probability (FP). This paper points out the drawbacks of one of the formulas for conversion of CFP into FP, and discusses ways to improve the formula for the FFTA. The proposed approach opens a corridor for the researchers to re-think the previous studies, and is susceptible to improve the future applications for safety and reliability engineering.

Keywords: Safety engineering · Fuzzy FTA · Reliability · Crisp failure possibility · Failure probability

1 Introduction

Fault tree analysis (FTA) is used to analyze complex systems in isolation and in a holistic manner to deduce a global reliability and criticality measures [4]. FTA method and its fuzzy extended version (FFTA) have been the subject of extensive research [7,15]. It is difficult to point of field of research in which they were not applied [17,20]. Examples of applications areas of FTA and FFTA include reliability evaluation of fire alarm systems [2], human factor analysis of engine room fires on ships [18], system failure probability [2,24]. Some of the other applications can be summarized as technical factor analysis [25], Arctic marine accidents [3,16], chemical cargo contamination [19], crankcase explosion [22] and strategic management [1].

C. Kahraman et al. (Eds.): INFUS 2021, LNNS 308, pp. 622–629, 2022.
https://doi.org/10.1007/978-3-030-85577-2_73

Most of the studies in the literature depend on a conversion formula given in detail in the Sect. 2. The authors mostly refer to the studies of Takehisa Onisawa [9–14] for insights into using this formula. The formula is also used in several studies such as [5,23]. We want to quote the reason why this formula [24] used as "...this equation is obtained by certain characteristics including appropriateness of anthropomorphic feeling to the logarithmic amount of a physical value". Similar explanation can be found in [8]. However, the most comprehensive explanation comes from Lin and Wang (1997) [6] as they refer the study of Swain and Guttmann (1983) [21]. According to [6], Swain and Guttmann (1983) suggest that a upper and lower bounds of the error rate of a routine human operation is $10^{-2} \sim 10^{-3}$ and 5×10^{-5}, respectively [21].

The objective of this paper is to point out the drawbacks of one of the formulas for conversion of CFP into FP, and discusses ways to improve the formula for the FFTA. The originality of the proposed approach is that it opens a corridor for the researchers to re-consider the previous studies, and is susceptible to improve the future applications for safety and reliability engineering. We claim that this formula should be improved or changed and that the study of Swain and Guttmann (1983) requires an adjustment. The statistics and the data that are the source for inspiration for this formula are vague and based on nuclear power plant applications. The comparison between the results and real time observations prove that the rationale behind the formula seems problematic.

The ideas behind this formula are given in Sect. 2 and the main problems are explained in Sect. 3. Section 4 presents the proposed approach and provides an application along with limitations and future directions. Finally, Sect. 5 concludes the paper.

2 General Approach for the Formula of Converting Crisp Failure Possibility (CFP) into Failure Probability (FP)

FTA involves a bottom-to-top computation process. The inputs are obtained as a FP from the BE at the bottom, then the calculations are done accordingly based on the types of the gates. If two events are connected each other with the AND gate ($\cap$), calculation is made as in Eq. 1:

$$P_0(t) = \prod_{i=1}^{n} p_i(t) \tag{1}$$

If two events are connected each other with the OR gate ($\triangle$), calculation is done as given in Eq. 2:

$$P_0(t) = 1 - \prod_{i=1}^{n} (1 - p_i(t)) \tag{2}$$

The inputs for the FTA are the values of FP. In the literature, general formula for the conversion of CFP to FP is given in Eq. 3:

$$FP = \begin{cases} \frac{1}{10^K}, & \text{if } CFP \neq 0 \\ 0, & \text{if } CFP = 0 \end{cases}, K = \left[\left(\frac{1 - CFP}{CFP}\right)\right]^{\frac{1}{3}} \times 2.301 \quad (3)$$

CFP is the defuzzification of aggregated fuzzy evaluations. Please refer the papers [17,20] for the aggregation and defuzzification of a trapezoidal fuzzy number for more details.

3 Main Problems of General Approach

CFP values are always between 0 and 1. According to several tests we conducted (0.0001 to 1.0000) to cover all the range of possible values, we find the FP values as given in Fig. 1. As you can see, the trend is almost constant between 0 and 0.6, then the sharp increase starts after 0.9.

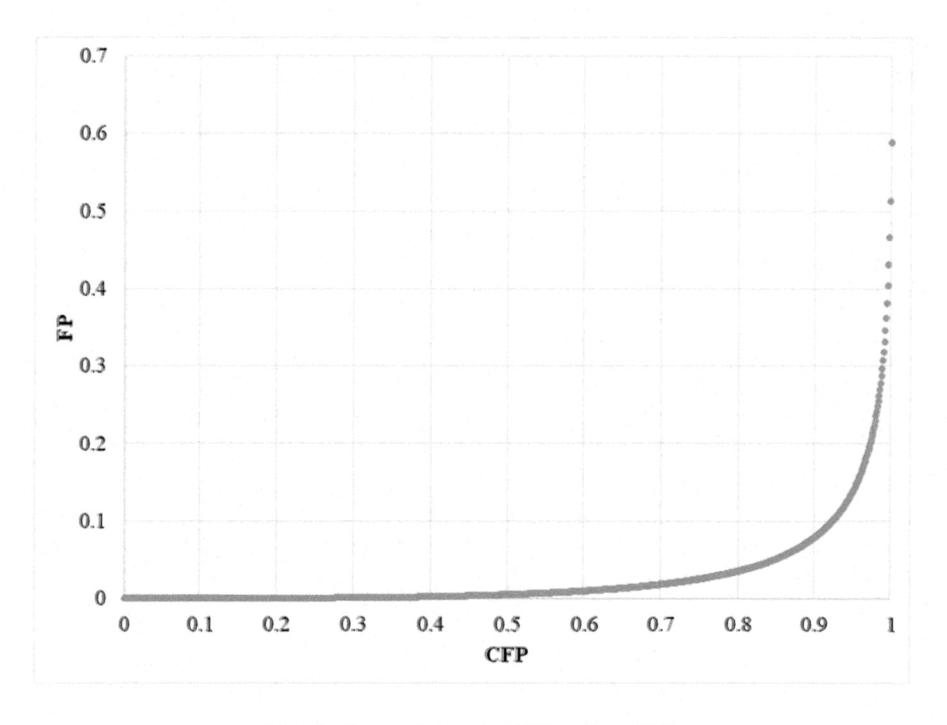

Fig. 1. General trend of FP using CFP

There are three main problems with this approach and the formula given in Eq. 3:

- There is no logical explanation for this graph given in Fig. 1.

- In the Fig. 1, we did not calculate FP value when CFP is equal to 1, because we want to observe the trend and values in a detailed manner. If CFP is equal to 0.999, FP is found as 0.588604054. However, if CFP = 1, FP is found as 1 which means a huge difference. Similarly, if CFP = 0.9999, FP is obtained as 0.781976102. Comparing to CFP = 0.999 and CFP = 0.9999, the difference is 0.0009 but the change in probability is 0.193372048. Similarly, if CFP = 0.99999, FP is 0.892126197. So, this proves that the formula has a big problem. Even if this sharp increase at the graph is mathematically correct and acceptable, this might not be logical in practice since those tiny changes cause big differences.
- Secondly, a single formula (Eq. 3) and only one trend (Fig. 1) for all different problems might limit the flexibility of the problems. This might also prevent finding exact solutions for the problems.

4 Proposed Approach

In this paper, we propose use of conversion functions for each BE. The conversion functions might be different based on the characteristics of the BE. The decision makers or the moderators determine the type of the conversion functions since they know the problems in a detailed manner. Let the conversion function be $y = f(x)$ where $y \in [0, 1]$. In this case, FP = f(CFP) where FP $\in [0, 1]$. This is better because FP only depends on CFP. Different types of conversion functions can be used, and this entirely depends on the problem's characteristics (Fig. 2). Some of the examples are given below:

- Constant Function: $f(x) = a$
- Linear, Identity Function: $f(x) = x$
- Quadratic Function: $f(x) = x^a$, $f(x) = x^2$
- Cubic Function: $f(x) = x^3$
- Exponential Function: $f(x) = a^x$
- Logarithmic Function: $y = log_2(x)$, $y = log_e(x)$, $y = log_{10}(x)$
- Square Root, Cube Root Function: $f(x) = \sqrt[a]{x}$
- Any Function: $f(x) = x^3 - 5x^2 + 4x$

4.1 An Application

In this example, a simple fault tree analysis is provided based on the pre-defined conversion functions. For the BE_1, a square root function, for the BE_2, exponential function and for the BE_3 a linear function are preferred as an example (Fig. 3). Table 1 gives the aggregated fuzzy assignments and defuzzified values of the BEs for two different examples. As shown in the table, we converted CFP into FP based on the proposed and the conventional approaches. Then, we calculated the probability of TEs for both two examples. For the first example, we calculate the probability of TE as 0.000484 for the conventional approach and 0.5245 for our proposed approach. For the second example, we calculate the probability of TE as 0.000273 for the conventional approach and 0.447412 for our proposed approach. Our results are quite higher than the conventional approach, however field experiences prove that it is more realistic than the conventional results.

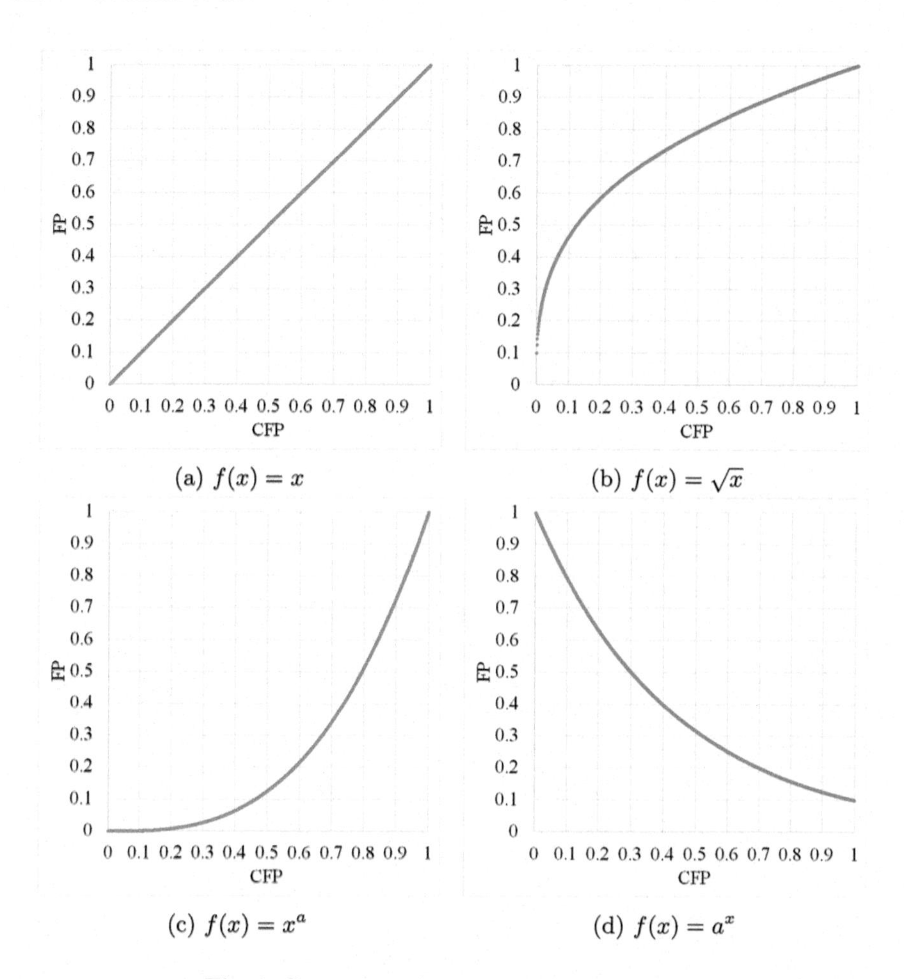

(a) $f(x) = x$ (b) $f(x) = \sqrt{x}$

(c) $f(x) = x^a$ (d) $f(x) = a^x$

Fig. 2. Some examples of conversion functions

Table 1. Fuzzy inputs and probability values for two examples

1				Conversion CFP to FP		
		Aggregation of BE	Defuzzification values	Conventional approach	Proposed Approach	
	BE_1	(0.1,0.2,0.3,0.4)	0.25	0.00048	Square root	0.50000
	BE_2	(0.2,0.3,0.4,0.5)	0.35	0.00148	Quadratic	0.12250
	BE_3	(0.3,0.4,0.4,0.5)	0.4	0.00232	Linear	0.40000
2				Conversion CFP to FP		
		Aggregation of BE	Defuzzification values	Conventional approach	Proposed Approach	
	BE_1	(0.7,0.8,0.9,1.0)	0.85	0.05121	Linear	0.85000
	BE_2	(0.3,0.4,0.5,0.6)	0.45	0.00347	Constant	0.50000
	BE_3	(0.1,0.2,0.3,0.8)	0.375	0.00187	Cubic	0.05273

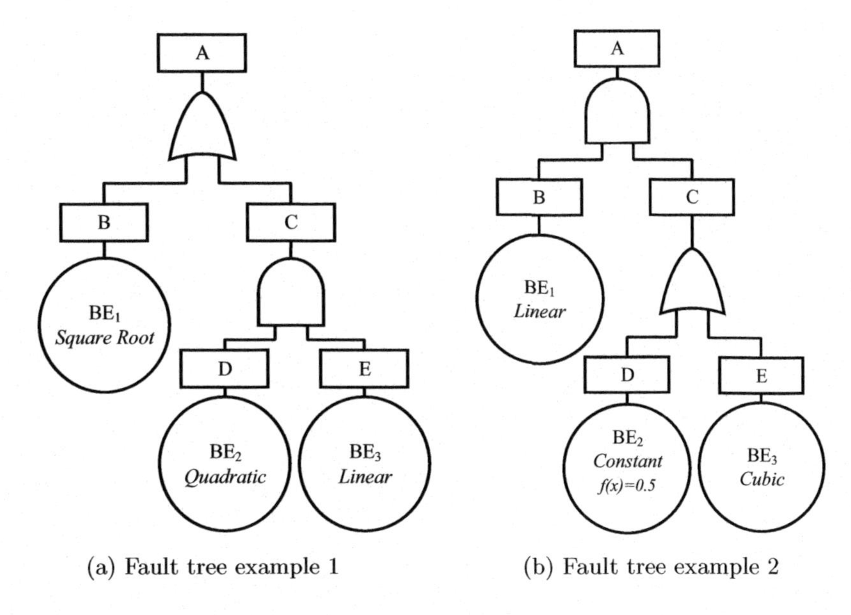

(a) Fault tree example 1 (b) Fault tree example 2

Fig. 3. Random fault trees for the application

4.2 Limitations

The proposed approach has some limitations as given below:

- The probability is the function for a crisp value of possibility which is based on aggregated expert opinions. Therefore, the experts or moderator might not be able to determine the exact functions for their fuzzy inputs. The function assignments must have proven groundings.
- The values for the probability must be between 0 and 1. Therefore, the moderator or the experts should define the constraints of their functions. For example, if there is a reciprocal function $f(x) = \frac{1}{x}$ or reciprocal inverse squared function: $f(x) = \frac{1}{x^a}$, boundaries should be set earlier since the maximum value exceeds 1.
- More experiments should be conducted to observe the changes, and it is needed to get double-checked whether there exists a parallelism in terms of the validation of final results, experts' opinions and intentions.

4.3 Future Directions

Proven groundings should be discussed in the future in which it explains how possibility and probability can be associated with a function and what are the fundamentals of the conversion formulas between possibility and probability. In the future, it should be studied that if there is a standard system that can be used to convert possibility into probability. To do that, previous studies in the literature might be re-applied to observe and compare the results to validate the standard system. We have to note that the results will mostly be different than the conventional results since we make a substantial improvement in the FFTA method. The comparisons and validations might be conducted based on the real cases of which their impact values and final probabilities are already known.

5 Conclusions

This study discusses a formula which converts CFP into FP. The conversion formula mentioned in the paper is important because of the existence of massive studies in the literature and its high impact on the results and reliability of complex systems. We find that conventional technique has some drawbacks, and we open a discussion to improve this formula. We offer using any pre-defined conversion function for BEs of FFTA applications under several constraints. Thus, further complex system applications might reach to the more realistic results.

References

1. Akpinar, H., Sahin, B.: Strategic management approach for port state control. Maritime Business Review (2019)
2. Jafari, M.J., Pouyakian, M., Hanifi, S.M., et al.: Reliability evaluation of fire alarm systems using dynamic Bayesian networks and fuzzy fault tree analysis. J. Loss Prev. Process Ind. **67**, 104229 (2020)
3. Kum, S., Sahin, B.: A root cause analysis for arctic marine accidents from 1993 to 2011. Saf. Sci. **74**, 206–220 (2015)
4. Lee, W.S., Grosh, D.L., Tillman, F.A., Lie, C.H.: Fault tree analysis, methods, and applications a review. IEEE Trans. Reliab. **34**(3), 194–203 (1985)
5. Li, W., Ye, Y., Wang, Q., Wang, X., Hu, N.: Fuzzy risk prediction of roof fall and rib spalling: based on FFTA-DFCE and risk matrix methods. Environ. Sci. Pollut. Res. **27**(8), 8535–8547 (2020)
6. Lin, C.T., Wang, M.J.J.: Hybrid fault tree analysis using fuzzy sets. Reliab. Eng. Syst. Saf. **58**(3), 205–213 (1997)
7. Mahmood, Y.A., Ahmadi, A., Verma, A.K., Srividya, A., Kumar, U.: Fuzzy fault tree analysis: a review of concept and application. Int. J. Syst. Assur. Eng. Manage. **4**(1), 19–32 (2013)
8. MIRI, L.M., Wang, J., Yang, Z., Finlay, J.: Application of fuzzy fault tree analysis on oil and gas offshore pipelines (2011)
9. Onisawa, T.: An approach to human reliability in man-machine systems using error possibility. Fuzzy Sets Syst. **27**(2), 87–103 (1988)
10. Onisawa, T.: Human reliability assessment with fuzzy reasoning. Trans. Soc. Instrum. Control Eng. **24**(12), 1312–1319 (1988)
11. Onisawa, T.: An application of fuzzy concepts to modelling of reliability analysis. Fuzzy Set Syst. **37**(3), 267–286 (1990)
12. Onisawa, T.: Subjective analysis of system reliability and its analyzer. Fuzzy Sets Syst. **83**(2), 249–269 (1996)
13. Onisawa, T., Nishiwaki, Y.: Fuzzy human reliability analysis on the Chernobyl accident. Fuzzy Sets Syst. **28**(2), 115–127 (1988)

14. Onisawa, T., Misra, K.B.: Use of fuzzy sets theory: (part-ii: Applications). In: Fundamental Studies in Engineering, vol. 16, pp. 551–586. Elsevier (1993)
15. Ruijters, E., Stoelinga, M.: Fault tree analysis: a survey of the state-of-the-art in modeling, analysis and tools. Comput. Sci. Rev. **15**, 29–62 (2015)
16. Şahin, B.: Risk assessment and marine accident analysis in ice-covered waters. Ph.D. thesis, Fen Bilimleri Enstitüsü (2015)
17. Sahin, B.: Consistency control and expert consistency prioritization for FFTA by using extent analysis method of trapezoidal FAHP. Appl. Soft Comput. **56**, 46–54 (2017)
18. Sarıalioğlu, S., Uğurlu, Ö., Aydın, M., Vardar, B., Wang, J.: A hybrid model for human-factor analysis of engine-room fires on ships: HFACS-PV&FFTA. Ocean Eng. **217**, 107992 (2020)
19. Senol, Y.E., Aydogdu, Y.V., Sahin, B., Kilic, I.: Fault tree analysis of chemical cargo contamination by using fuzzy approach. Expert Syst. Appl. **42**(12), 5232–5244 (2015)
20. Senol, Y.E., Sahin, B.: A novel real-time continuous fuzzy fault tree analysis (RC-FFTA) model for dynamic environment. Ocean Eng. **127**, 70–81 (2016)
21. Swain, A.D.: Handbook of human reliability analysis with emphasis on nuclear power plant applications. NUREG/CR-1278, SAND 80–0200 (1983)
22. Ünver, B., Gürgen, S., Sahin, B., Altın, İ: Crankcase explosion for two-stroke marine diesel engine by using fault tree analysis method in fuzzy environment. Eng. Failure Anal. **97**, 288–299 (2019)
23. Yazdi, M., Korhan, O., Daneshvar, S.: Application of fuzzy fault tree analysis based on modified fuzzy AHP and fuzzy topsis for fire and explosion in the process industry. Int. J. Occup. Saf. Ergonomics **26**(2), 319–335 (2020)
24. Yazdi, M., Nikfar, F., Nasrabadi, M.: Failure probability analysis by employing fuzzy fault tree analysis. Int. J. Syst. Assur. Eng. Manage. **8**(2), 1177–1193 (2017)
25. Yip, T.L., Sahin, B.: Technical factor in maritime accidents: an index for systematic failure analysis (2020)

Simulation Testing of Fuzzy Quality with a Case Study in Pipe Manufacturing Industries

Abbas Parchami[1(✉)], Hamideh Iranmanesh[2], and Bahram Sadeghpour Gildeh[1,2]

[1] Department of Statistics, Faculty of Mathematics and Computer, Shahid Bahonar University of Kerman, Kerman, Iran
parchami@uk.ac.ir

[2] Department of Statistics, Faculty of Mathematical Sciences, Ferdowsi University of Mashhad, Mashhad, Iran
iranmanesh.hamideh@mail.um.ac.ir, sadeghpour@um.ac.ir

Abstract. Approximately three decades ago, the fuzzy quality was proposed by Yongting as a flexible and powerful tool to quality modeling. Furthermore, in order to more compliance of products with customer requirements, the probability of fuzzy quality was introduced as a process capability index to evaluate manufacturing processes based on the fuzzy quality. The diversity of fuzzy quality membership functions as well as the complexity of the distribution of Yongting index estimator hindered the scientific progress in the field of statistical inference. To cover a part of this deprivation, testing the capability of productive process on the basis of the flexible fuzzy quality is proposed in this paper by Monte Carlo simulation. Hence, the objective of this research is estimate the critical value by Monte Carlo simulation approach for normal data on the basis of fuzzy quality. A case study is provided in pipe manufacturing industries to show the performance of approach.

Keywords: Fuzzy quality · Testing hypotheses · Process capability index · Simulation

1 Prerequisites

In quality control, such as other statistical problems, we may confront imprecise concepts. One practical case in process capability analyses is a situation in which specification tolerance is fuzzy rather than crisp. In such a fuzzy environment, products are not qualified with a 0-1 boolean view, but to some degrees depending on the quality level of the products. Yongting [3] introduced the concept of "fuzzy quality" in 1996. Another generation of process capability indices is developed by Parchami et al. [1] to measure the capability of fuzzy quality. To review these two ideas and familiarity with their motivations see [2]. Also, other related investigations are reviewed and discussed in [2].

C. Kahraman et al. (Eds.): INFUS 2021, LNNS 308, pp. 630–635, 2022.
https://doi.org/10.1007/978-3-030-85577-2_74

Let f is the probability density function of a one-dimensional continuous quality characteristic X. Yongting (1996) introduced the process capability index

$$C_{\tilde{Q}} = P\left(X \in \tilde{Q}\right) = \int_{-\infty}^{+\infty} \tilde{Q}(x)\ f(x)\ dx \tag{1}$$

for continuous quality characteristic in which $\tilde{Q}$ is the membership function of fuzzy quality [4]. Note that, $\tilde{Q}(x)$ represents the degree of conformity with standard quality when the measured quality characteristic of a product is x [2].

Testing capability is a common statistical method to check the performance of industrial productive processes. The objective and originality of this paper is the simulation of significance fuzzy quality test for normal data with testing Yongting's capability index. In this paper, the statistical significance test is simulated for testing the capability of a fuzzy process for normal data. As far as the authors know, previous researches do not study on the problem of testing Yongting's capability index and it build the need for the current paper based on the gaps in the existing literature.

This paper is organized as follows. Significance fuzzy quality test on Yongting's capability index is discussed in Sect. 2. Then, Monte Carlo testing fuzzy quality is presented by an algorithm in Sect. 3. A numerical example based on symmetric triangular fuzzy quality has been provided in Sect. 4. The final section is conclusions and future researches.

2 Testing Fuzzy Quality

Under the normality condition of one-dimensional quality characteristic, the main problem is testing the null hypothesis

$H_0 : C_{\tilde{Q}} \leq c_0$ (process is not capable),

against the alternative hypothesis

$H_1 : C_{\tilde{Q}} > c_0$ (process is capable),

based on random sample $x_1, ..., x_n$ with unknown mean and unknown variance parameters, where $c_0 \in (0, 1)$ is the standard minimal criteria for Yongting's capability index. We briefly call this problem testing fuzzy quality and the Monte Carlo simulation is investigated for testing fuzzy quality in Sect. 3.

Obviously, the critical region of the proposed capability test is $\widehat{C_{\tilde{Q}}} > c$ in which c is the precise unknown critical value and

$$\widehat{C_{\tilde{Q}}} = \int_{-\infty}^{+\infty} \tilde{Q}(x)\ \widehat{\phi_{\mu,\sigma}}(x)\ dx = \int_{-\infty}^{+\infty} \tilde{Q}(x)\ \phi_{\hat{\mu},\hat{\sigma}}(x)\ dx \tag{2}$$

is the estimator of Yongting's capability index for normal distribution with unknown parameters μ and σ^2. Regarding to the invariant property of maximum likelihood estimators (MLEs), $\widehat{C_{\tilde{Q}}}$ is MLE for unknown parameter/index $C_{\tilde{Q}}$ since this simply involves replacing unknown parameters μ and σ^2 by their MLEs $\hat{\mu} = \bar{X}$ and $\widehat{\sigma^2} = S_b^2 = \frac{\sum_{i=1}^{n}(X_i - \bar{X})^2}{n}$, respectively. Considering random sample $X_1, ..., X_n \stackrel{i.i.d.}{\sim} N(\mu, \sigma^2)$ with unknown parameters, the probability of type I error in testing fuzzy quality is $\alpha = Pr\left(\widehat{C_{\tilde{Q}}} > c \mid C_{\tilde{Q}} = c_0\right)$ and hence

$$Pr\left(\widehat{C_{\tilde{Q}}} \leq c \mid C_{\tilde{Q}} = c_0\right) = 1 - \alpha. \tag{3}$$

Therefore, the unknown critical value c is equal to the $(1-\alpha)$-th quantile of $\widehat{C_{\tilde{Q}}}$ distribution under assumption $C_{\tilde{Q}} = c_0$ and finding precise critical value c by Monte Carlo simulation is investigated in Sect. 3 as the main goal of this paper.

3 Testing Fuzzy Quality: Simulation Approach

An algorithm proposed in follow to estimate the critical value of Monte Carlo testing fuzzy quality at the given significance level α.

Algorithm 1.
Step 1: Compute the estimated Yongting's index $\widehat{c_{\tilde{Q}}}$ on the basis of the observed random sample $x_1, ..., x_n$ by Eq. (2).
Step 2: Calculate sequence $\mu_1 < \mu_2 < ... < \mu_k$ to cover the interquartile range $[Q_1, Q_3]$ by following formula

$$\mu_j = Q_1 + \frac{j-1}{k-1}(Q_3 - Q_1), \quad j = 1, 2, ..., k, \tag{4}$$

where Q_1 and Q_3 are 25th and 75th percentiles of observations $x_1, ..., x_n$.
Step 3: Consider a reasonable sequence $n_1 < n_2 < ... < n_s$ to cover possible sample sizes.
Step 4: For any $n_i \in \{n_1, n_2, ..., n_s\}$ and $\mu_j \in \{\mu_1, \mu_2, ..., \mu_k\}$,
a) compute the unknown value of root σ_0 from equation $C_{\tilde{Q}} = c_0$ which is equivalent to equation $\int_{-\infty}^{+\infty} \tilde{Q}(x)\ \phi_{\mu_j,\sigma_0}(x)\ dx = c_0$,
b) simulate $m = 10^3$ random sample with size n_i from $N(\mu_j, \sigma_0^2)$,
c) compute MLEs $\widehat{c_{\tilde{Q}}}^{[1]}, \widehat{c_{\tilde{Q}}}^{[2]}, ..., \widehat{c_{\tilde{Q}}}^{[m]}$ for capability index using Eq. (2) based on the fuzzy quality $\tilde{Q}$ for each simulated sample from part (b),
d) the critical value for m simulated samples in part (b) is the $(1-\alpha)$-th quantile of $\widehat{C_{\tilde{Q}}}$ distribution, i.e.

$$c_{i,j} = \widehat{c_{\tilde{Q}}}^{(m(1-\alpha))}, \quad i = 1, ..., s, \ j = 1, ..., k, \tag{5}$$

where $\widehat{c_{\tilde{Q}}}^{(1)}, \widehat{c_{\tilde{Q}}}^{(2)}, ..., \widehat{c_{\tilde{Q}}}^{(m)}$ are the ordered estimated indices in part (c).
Step 5: Monte Carlo critical value in testing fuzzy quality is equal to the average of $s \times k$ estimated critical values in Step 4 and hence

$$c = \frac{1}{s \times k} \sum_{i=1}^{s} \sum_{j=1}^{k} c_{i,j}. \tag{6}$$

Step 6 (Decision rule): The process is capable, i.e. the null hypothesis rejected at significance level α, if $\widehat{c_{\tilde{Q}}} > c$; otherwise the process is incapable.

4 Numerical Example

In quality analysis for one type of pipe, the outer diameter of which should be 15 cm, the quality of the pipes is investigated by Yongting's capability index. The larger or smaller the diameter of a pipe is, the lower the quality of the pipe. Therefore, the following symmetric triangular fuzzy quality is considered for the outer diameter of pipe which can be interpreted as "approximately 15 cm" (Fig. 2):

$$\tilde{Q}(x) = \begin{cases} x - 14 & if \quad 14 \leq x < 15, \\ 16 - x & if \quad 15 \leq x < 16, \\ 0 & \quad elsewhere. \end{cases}$$

Now, we are going to test the capability of producing process for the outer diameter measurement of the produced pipes based on Yongting's capability index at significance level 0.05. The following random sample, with size fifty three, is taken from the diameter length in term of centimeters:

14.98 14.91 14.56 14.79 15.06 15.09 14.61 14.86 14.48
14.89 15.30 15.20 14.90 15.27 15.19 15.00 14.68 14.91
15.25 15.11 14.79 14.31 14.77 14.33 14.59 14.86 14.76
14.71 14.94 14.89 14.41 14.95 15.26 15.03 14.56 14.54
15.08 14.44 14.87 14.77 15.30 14.74 14.72 15.22 14.68
14.96 15.04 14.88 14.77 15.17 14.85 14.87 15.38

The histogram of observed data is shown in Fig. 1. Monte Carlo simulation approach is considered in this research to test $H_0 : C_{\tilde{Q}} \leq 0.66$, against $H_1 : C_{\tilde{Q}} > 0.66$, based on the observed random sample $x_1, ..., x_{53}$.

Shapiro-Wilk test confirm the normality assumption of the observed data with statistics 0.98 and p-value $= 0.58$. Therefore the estimated Yongting's index is equal to

$$\begin{aligned} \widehat{c_{\tilde{Q}}} &= \int_{14}^{16} \tilde{Q}(x)\, \phi_{\hat{\mu},\hat{\sigma}}(x)\, dx \\ &= \int_{14}^{15} (x-14)\, \frac{1}{\sigma\sqrt{2\pi}}\, e^{-\frac{1}{2}\left(\frac{x-\mu}{\sigma}\right)^2} dx + \int_{15}^{16} (16-x)\, \frac{1}{\sigma\sqrt{2\pi}}\, e^{-\frac{1}{2}\left(\frac{x-\mu}{\sigma}\right)^2} dx \\ &= 0.770, \end{aligned}$$

where ϕ is the probability density function of standard normal and MLEs of unknown mean and variance of normal random variable are computed by $\hat{\mu} = \bar{x} = 14.88$ and $\widehat{\sigma^2} = s_b^2 = 0.26^2$, respectively. So, $0.77 > c$ is the critical region of Monte Carlo testing fuzzy quality at significance level $\alpha = 0.05$, in which the critical value must be simulated using Algorithm 1 in Sect. 3. We did a simulation to compute Monte Carlo critical value in testing fuzzy quality where the mean and sample size are changed over the following sequences

$\mu = 14.72140, 14.78879, 14.85619, 14.92358, 14.99097$ and 15.05836,
$n = 20, 30, 40, 50, 60$ and 70.

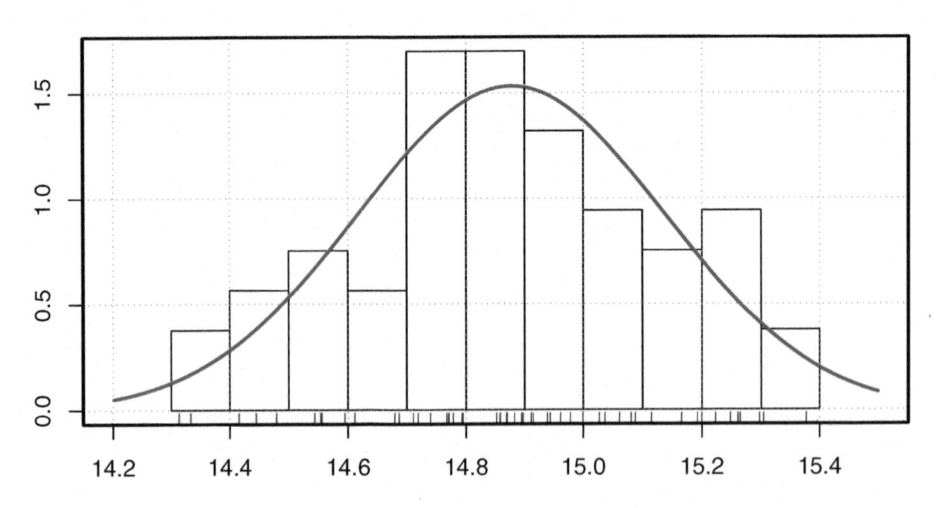

Fig. 1. Histogram of 53 observed diameters.

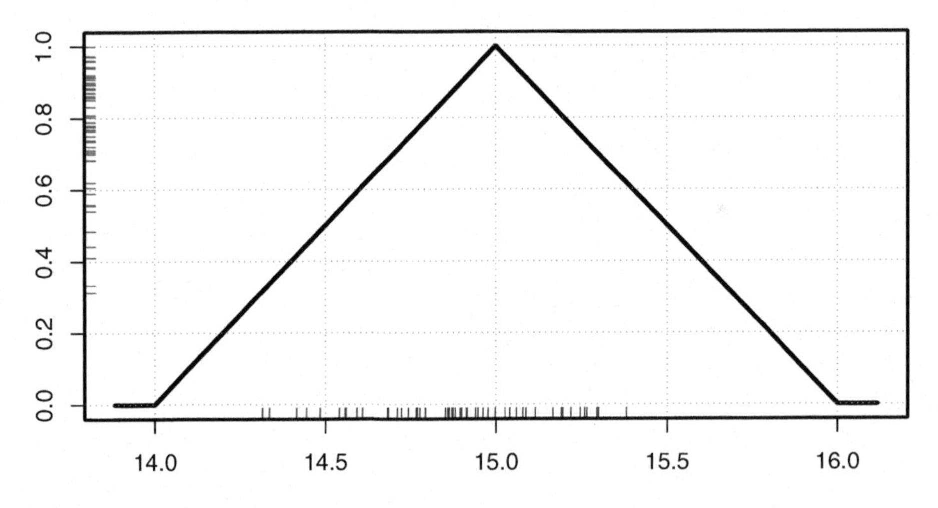

Fig. 2. The membership function of triangular fuzzy quality and the related degrees of conformity.

The unknown root σ_0 in equation $\int \tilde{Q}(x)\ \phi_{\mu_j,\sigma_0}(x)\ dx = c_0$ is computed for all 36 possible cases by Newton Raphson method. Then, for each 36 combination of μ and n, 10^3 independent random samples simulated from normal distribution $N\left(\mu, \sigma_0^2\right)$. For each simulated sample we obtained the MLE of Yongting's index using Eq. (2). After ordering 1000 estimated capability indices, the 950-th capability index is considered as the critical value for each 36 possible cases. The total average of 36 captured critical values is equal to $c = 0.779$ which is considered as the Monte Carlo critical value of testing fuzzy quality in this research. By comparing $\widehat{c_{\tilde{Q}}} = 0.770$ and c, the null hypothesis accepted at significance level 0.05 and hence the process is not determined as a capable process.

5 Conclusions and Future Researches

Testing the capability of manufacturing processes is investigated in this article using Yongting's capability index. The main goal of this investigation is estimate the critical value by Monte Carlo simulation approach for normal data based on fuzzy quality. Also, a numerical example based on symmetric triangular fuzzy quality has been investigated in pipe manufacturing industries. Testing capability based on multivariate capability indices is a potential subject for further research.

References

1. Parchami, A., Mashinchi, M., Sharayei, A.: An effective approach for measuring the capability of manufacturing processes. Prod. Plan. Control **21**(3), 250–257 (2010)
2. Parchami, A., Sadeghpour-Gildeh, B., Mashinchi, M.: Why fuzzy quality? Int. J. Qual. Res. **10**(3), 457–470 (2016)
3. Yongting, C.: Fuzzy quality and analysis on fuzzy probability. Fuzzy Sets Syst. **83**, 283–290 (1996)
4. Zadeh, L.A.: Probability measures of fuzzy events. J. Math. Anal. Appl. **23**, 421–427 (1968)

A Case Study on Quality Test Based on Fuzzy Specification Limits

Hamideh Iranmanesh[1(✉)], Abbas Parchami[2(✉)], and Bahram Sadeghpour Gildeh[1(✉)]

[1] Department of Statistics, Faculty of Mathematical Sciences, Ferdowsi University of Mashhad, Mashhad, Iran
iranmanesh.hamideh@mail.um.ac.ir, sadeghpour@um.ac.ir

[2] Department of Statistics, Faculty of Mathematics and Computer, Shahid Bahonar University of Kerman, Kerman, Iran
parchami@uk.ac.ir

Abstract. This article presents a general approach for quality test to make a decision based on the extended process capability indices. Usual methods in measuring the quality of a manufactured product have widely focused on the precise specification limits, but in this study the lower and upper specification limits are considered as non-precise (fuzzy). Extended process capability indices are common tools to quantify the performance of a normal process where lower and upper specification limits are fuzzy. The main purpose of this study is to propose a new statistical approach to estimate a critical value for analyzing the capability of a manufacturing process using the extended process capability index to determine whether the process meets customer requirement. To show the real application of the proposed method, a case study at the glass bottle manufacturing industry is provided and the obtained theoretical results are tabulated for making reliable decisions in different situations.

Keywords: Quality control · Process capability index · Testing hypotheses · Fuzzy specification limits

1 Introduction

Process capability indices (PCIs), such as C_p, C_{pk} and C_{pm} have been widely utilized in the manufacturing industry. A process capability index (PCI) like C_p provides a numerical measure on whether a process is capable of producing products within the specification limits (SLs) [1]. The PCI C_p is defined as follows, where LSL is the lower specification limit, USL is the upper specification limit, and σ is the process standard deviation:

$$C_p = \frac{USL-LSL}{6\sigma}. \tag{1}$$

Essentially, the index C_p is useful when the process mean, μ, is located in the center of the specification limits (i.e., $\mu = (USL + LSL)/2$) [2].

C. Kahraman et al. (Eds.): INFUS 2021, LNNS 308, pp. 636–643, 2022.
https://doi.org/10.1007/978-3-030-85577-2_75

Generally, the process mean, μ, and the process standard deviation, σ, are unknown. In practice, a random sample is required to estimate the unknown parameters. If we take a random sample of size n, then, μ and σ can be estimated by the sample mean, $\bar{X} = \sum_{i=1}^{n} X_i/n$ and the sample standard deviation $S_{n-1} = \sqrt{\sum_{i=1}^{n}(X_i - \bar{X})^2/n-1}$, to acquire the natural estimator of C_p, $\widehat{C_p}$. This article proposes a test procedure to analyze the performance of a normal process where the lower specification limit (SL) and the upper SL are fuzzy. Two basic definitions for providing fuzzy concepts are introduced as follows: (i) let $\mathbb{R}$ be the set of all real numbers and $F(\mathbb{R}) = \{A|A : \mathbb{R} \to [0,1], A$ is a continuous function $\}$ be the set of all fuzzy sets on $\mathbb{R}$; and (ii) the α-cut of $\tilde{A} \in F(\mathbb{R})$ is the crisp set given by $\tilde{A}_\alpha = \left\{x|\tilde{A}(x) \geq \alpha\right\}$, for any $\alpha \in [0,1]$ [3].

Parchami and Mashinchi [4] presented the extended PCIs such as $C_{\tilde{p}}$, $C_{\widetilde{pk}}$ and $C_{\widetilde{pm}}$ to measure the process capability based on fuzzy SLs. The concept of fuzzy quality is introduced in [5]. The motivations of fuzzy quality are mentioned in [6]. Parchami et al. [7] defined the membership functions for PCIs when SLs are triangular fuzzy sets. The problem of measuring the process capability based on introducing two operations of summation and subtraction on fuzzy SLs is investigated in [8]. The PCIs are generalized where fuzziness is involved into both specification limits and data [9]. Abbasi Ganji and Sadeghpour Gildeh [10] developed the methods to quantify the process capability in the simple linear profile processes for the situations in which SLs are non-precise. Sadeghpour Gildeh and Moradi [11] suggested a multivariate PCI for the cases in which engineering tolerance is imprecise. Additionally, the capability of a fuzzy process is tested when SLs are triangular fuzzy sets [12]. The main goal of this article is to present a new statistical decision making method for evaluating the performance of extended PCI $C_{\tilde{p}}$.

The remainder of this article is structured as follows. Section 2 briefly reviews the preliminary of the extended process capability index $C_{\tilde{p}}$. Section 3 introduces an unbiased estimator of $C_{\tilde{p}}$. Section 4 suggests the general quality test to evaluate the performance of the extended PCI $C_{\tilde{p}}$. Section 5 presents a case study to illustrate the proposed test's application. Finally, Sect. 6 provides conclusions and future researches.

2 The Extended Process Capability Index $C_{\tilde{p}}$

This section reviews a short history of extended process capability indices. Herein, the study of analyzing production quality focuses on the fuzzy SLs. If fuzziness is involved into SLs, quality engineers face a new process and the usual PCI like C_p is not suitable for measuring the manufacturing process capability. In what follows, based on the work of Parchami and Mashinchi [4] in which a process with fuzzy SLs is called a fuzzy process, one of the extended PCIs is considered to test the capability of manufacturing processes based on fuzzy SLs.

Definition 1. *The extended process capability index, $C_{\tilde{p}}$, to evaluate a fuzzy process is introduced as follows*

$$C_{\tilde{p}} = \frac{\widetilde{USL} \ominus \widetilde{LSL}}{6\sigma}, \tag{2}$$

where $\widetilde{LSL}, \widetilde{USL} \in F(\mathbb{R})$ are linear fuzzy SLs with membership functions

$$\widetilde{LSL}(x) = \begin{cases} 0 & if \quad x \leq l_0, \\ \frac{x-l_0}{l_1-l_0} & if \quad l_0 < x < l_1, \\ 1 & if \quad l_1 \leq x, \end{cases} \tag{3}$$

and

$$\widetilde{USL}(x) = \begin{cases} 1 & if \quad x \leq u_1, \\ \frac{x-u_0}{u_1-u_0} & if \quad u_1 < x < u_0, \\ 0 & if \quad u_0 \leq x, \end{cases} \tag{4}$$

and $\widetilde{USL} \ominus \widetilde{LSL} = \int_0^1 g(\alpha)(u_\alpha - l_\alpha)d\alpha$, such that $g(\alpha)$ is a non-decreasing function on $[0,1]$ with the conditions $g(0) = 0$ and $\int_0^1 g(\alpha)d\alpha = 1$ (see [4] for motivation of definition).

In the following theorem, measuring a fuzzy process based on the extended PCI $C_{\tilde{p}}$ is provided and it is obvious by [8].

Theorem 1. *Let $\widetilde{LSL}$ and $\widetilde{USL}$ be the introduced lower and upper linear fuzzy SLs in Eqs. (3) and (4). If $g(\alpha) = (j+1)\alpha^j$, for any $j \in \mathbb{R}^+$, then the extended PCI $C_{\tilde{p}}$ is given by*

$$C_{\tilde{p}} = \frac{(j+1)(u_1 - l_1) + (u_0 - l_0)}{6(j+2)\sigma}, \tag{5}$$

where σ is the process standard deviation.

3 Determination of Unbiased Estimator for PCI $C_{\tilde{p}}$

In order to find the estimation of the PCI $C_{\tilde{p}}$, let us take a random sample of size n $(x_1, x_2, \ldots, x_n)$ and to estimate the unknown parameter σ, the estimator S_{n-1} is considered. If the assumptions of Theorem 1 hold, then by substituting to Eq. (5), the statistical estimation of $C_{\tilde{p}}$ for any $j \in \mathbb{R}^+$ can be written as

$$\widehat{C_{\tilde{p}}} = \frac{(j+1)(u_1-l_1)+(u_0-l_0)}{6(j+2)S_{n-1}}. \tag{6}$$

Theorem 2. *If the assumptions of Theorem 1 hold, then the unbiased estimator of $C_{\tilde{p}}$ for each $j \in \mathbb{R}^+$ is given by*

$$\widehat{C}'_{\tilde{p}} = \frac{b_{n-1}[(j+1)(u_1 - l_1) + (u_0 - l_0)]}{6(j+2)S_{n-1}}, \tag{7}$$

where

$$b_{n-1} = \sqrt{\frac{2}{n-1}} \frac{\Gamma(\frac{n-1}{2})}{\Gamma(\frac{n-2}{2})}, \tag{8}$$

and S_{n-1} is the sample standard deviation, as expressed in Sect. 1.

Remark 1. Note that the index $C_{\tilde{p}}$ is useful when μ is actually located at the center of fuzzy tolerance, that is, $\mu = M$ in which $M = (\widetilde{USL} \oplus \widetilde{LSL})/2$ (see more details about the summation of $\widetilde{USL}$ and $\widetilde{LSL}$ in [8]). To cover this hard condition, $j \in \mathbb{R}^+$ is selected such that $\mu = M$, which is equivalent to

$$\mu = \frac{\widetilde{USL} \oplus \widetilde{LSL}}{2} = \frac{1}{2j+4}[(j+1)(u_1 + l_1) + (u_0 + l_0)]. \tag{9}$$

Now, if the unknown parameter μ is estimated by its MLE $\hat{\mu} = \bar{X}$, then j can be determined by

$$j = \frac{u_1 + l_1 + u_0 + l_0 - 4\bar{X}}{2\bar{X} - u_1 - l_1}, \tag{10}$$

to cover the condition of using $C_{\tilde{p}}$.

4 Significance of General Quality Test

Testing the manufacturing process capability is a statistical approach for analyzing the process performance. In this section, a general approach is proposed that the practitioners can make a reliable decision in analyzing the extended PCI $C_{\tilde{p}}$. The purpose of this section is to estimate the critical value for testing the process capability based on the linear fuzzy SLs.

Theorem 3. *In testing the statistical hypotheses*

$$\begin{cases} H_0 : C_{\tilde{p}} \leq c_0 & (\textit{fuzzy process is not capable}), \\ H_1 : C_{\tilde{p}} > c_0 & (\textit{fuzzy process is capable}), \end{cases} \tag{11}$$

at the given significance level α, the critical value is computed as follows, where $c_0 \in \mathbb{R}^+$ is the standard minimal criteria for $C_{\tilde{p}}$, b_{n-1} is as expressed in Eq. (2) and $\chi^2_{n-1,\alpha}$ is the lower α-quantile of chi-square distribution with $n-1$ degrees of freedom:

$$c = b_{n-1} c_0 \sqrt{\frac{n-1}{\chi^2_{n-1,\alpha}}}. \tag{12}$$

In order to make a decision based on the proposed approach, if $\widehat{C}'_{\tilde{p}} > c$, then the null hypothesis is rejected at significance level α, this implies that, the fuzzy process is capable; otherwise it is incapable.

5 A Case Study at the Glass Container Manufacturing Industry

Bursting strengths of glass containers are measured in a glass bottle manufacturing process. Twenty samples of size $n = 5$ bursting strength measurements were collected (in terms of psi) [13]. To illustrate the proposed approach, the extended PCI $C_{\tilde{p}}$ is tested to assess the manufacturing process capability for the bursting strength measurement of the produced glass containers at significance level 0.05.

In this work, by considering the specified linear fuzzy SLs $\widetilde{LSL}$ and $\widetilde{USL}$ with membership functions

$$\widetilde{LSL}(x) = \begin{cases} 0 & if \quad x \leq 150, \\ \frac{x-150}{75} & if \quad 150 < x < 225, \\ 1 & if \quad 225 \leq x, \end{cases}$$

and

$$\widetilde{USL}(x) = \begin{cases} 1 & if \quad x \leq 300, \\ \frac{375-x}{75} & if \quad 300 < x < 375, \\ 0 & if \quad 375 \leq x. \end{cases}$$

the general approach for quality test, is investigated to test $H_0 : C_{\tilde{p}} \leq 1$, against $H_1 : C_{\tilde{p}} > 1$, based on the observed random sample $x_1, x_2, \ldots, x_{100}$. Also, by examining the Shapiro-Wilk test with $p-value = 0.2515$, the normal distribution model is suitable to fit the data. The histogram of observed data is shown in Fig. 1. The membership functions of fuzzy specification limits are shown in Fig. 2. Likewise, details of the Fig. 2 are as follows: the blue signs show observed bursting strength measurements of 100 glass containers and also the degrees of conformity

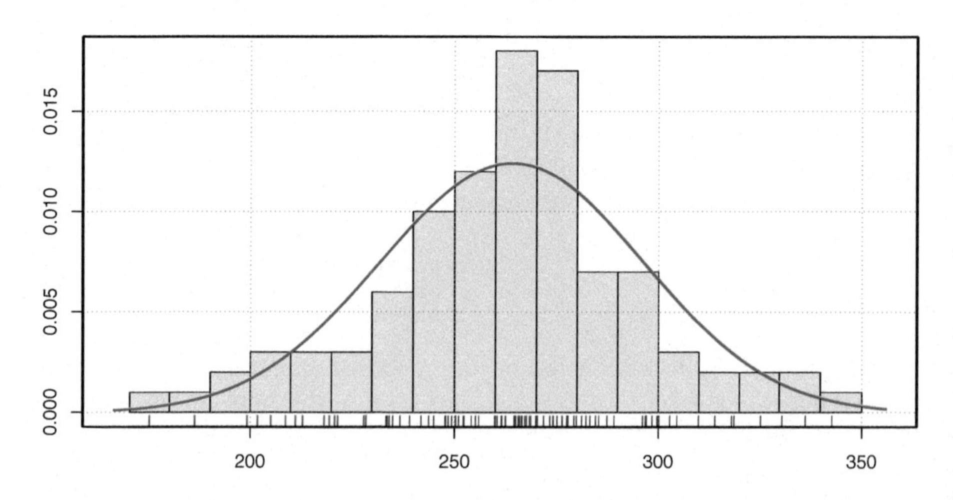

Fig. 1. Histogram of 100 observed bursting strengths.

for each observation are denoted at the left side of Fig. 2 with red signs, the long dashed line indicates the linear lower fuzzy SL, $\widetilde{LSL}$, and the solid line indicates the linear upper fuzzy SL, $\widetilde{USL}$.

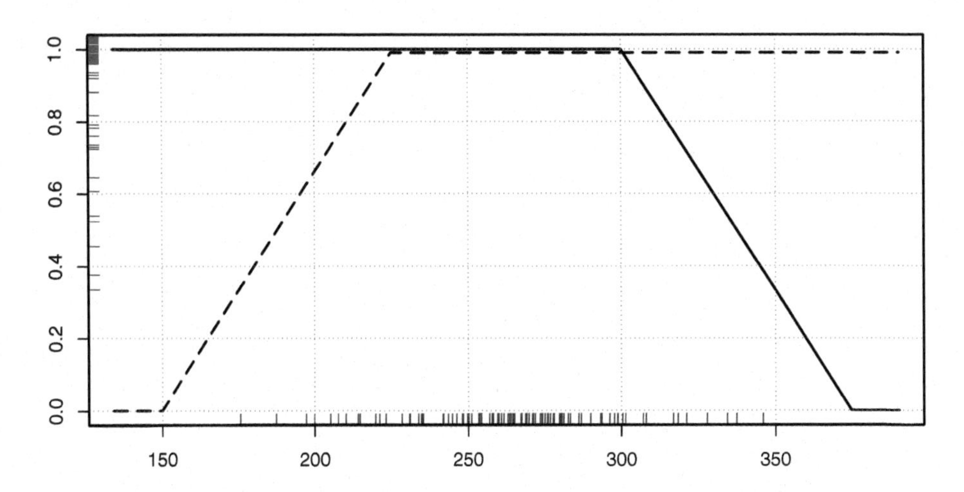

Fig. 2. The membership functions of the specified linear fuzzy SLs.

Since the MLE of the unknown parameter μ of the normal random variable is given by

$$\hat{\mu} = \bar{x} = \frac{\sum_{i=1}^{n} x_i}{n} = 264.06,$$

therefore, according to Remark 1, j can be computed by

$$j = \frac{u_1 + l_1 + u_0 + l_0 - 4\bar{x}}{2\bar{x} - u_1 - l_1} = 0.087.$$

Finally, the estimated capability index $\hat{c}'_{\tilde{p}}$ on the basis of the observed random sample $x_1, x_2, \ldots, x_{100}$, can be calculated as

$$\begin{aligned} \hat{c}'_{\tilde{p}} &= \frac{b_{n-1}[(0.087+1)(300-225)+(375-150)]}{6s_{n-1}(0.087+2)} \\ &= 0.759, \end{aligned}$$

where $b_{n-1} = 0.99$ and $s_{n-1} = 32.018$. Hence, $0.759 < c$ is the critical value for quality test at significance level $\alpha = 0.05$, in which the critical value must be estimated using the general quality test in Sect. 4. Thus, the critical value $c = 1.125$ can be obtained by Eq. (12). As a result, the null hypothesis cannot be rejected, this implies that, the fuzzy process is incapable at significance level 0.05. Furthermore, for conventional values of the standard minimal criteria c_0, the critical values based on the proposed approach for quality test are computed and summarized in Table 1.

Table 1. Critical values and results of making-decision at significance level 0.05.

Minimal criteria c_0	Critical value c	Decision rule	Fuzzy process
0.50	0.562	Reject H_0	Capable
0.67	0.754	Reject H_0	Capable
1.00	1.125	Not reject H_0	Incapable
1.33	1.496	Not reject H_0	Incapable
1.50	1.687	Not reject H_0	Incapable
1.67	1.879	Not reject H_0	Incapable
2.00	2.250	Not reject H_0	Incapable

6 Conclusions and Future Works

In quality improvement methodology, quality engineers might confront with non-precise environments. One practical case is an environment in which lower and upper specification limits are fuzzy. This study attempted to propose the general approach for quality test to analyze the process capability based on one of the extended process capability indices. The main goal of the general approach is to estimate the critical value in testing the process capability based on the linear fuzzy SLs. Lastly, a numerical example to illustrate the performance of the proposed approach has been investigated at the glass container manufacturing industry. The study of testing the multivariate capability indices based on fuzzy specification limits is another potential topic for future research.

References

1. Kotz, S., Johnson, N.L.: Process Capability Indices, 1st edn. CRC Press, New York (1993)
2. Pearn, W.L., Kotz, S.: Encyclopedia and Handbook of Process Capability Indices: A Comprehensive Exposition of Quality Control Measures. World Scientific, Singapore (2006)
3. Nguyen, H.T., Walker, C.L., Walker, E.A.: A First Course in Fuzzy Logic, 3rd edn. CRC Press, New York (2005)
4. Parchami, A., Mashinchi, M.: A new generation of process capability indices. J. Appl. Stat. **37**(1), 77–89 (2010)
5. Yongting, C.: Fuzzy quality and analysis on fuzzy probability. Fuzzy Sets Syst. **83**, 283–290 (1996)
6. Parchami, A., Sadeghpour Gildeh, B., Mashinchi, M.: Why fuzzy quality? Int. J. Qual. Res. **10**(3), 457–470 (2016)
7. Parchami, A., Mashinchi, M., Yavari, A.R., Maleki, H.R.: Process capability indices as fuzzy numbers. Aust. J. Stat. **34**(4), 391–402 (2005)
8. Parchami, A., Mashinchi, M., Sharayei, A.: An effective approach for measuring the capability of manufacturing processes. Prod. Plan. Control **21**(3), 250–257 (2010)
9. Parchami, A., Sadeghpour Gildeh, B., Nourbakhsh, M., Mashinchi, M.: A new generation of process capability indices based on fuzzy measurements. J. Appl. Stat. **41**(5), 1122–1136 (2014)

10. Abbasi Ganji, Z., Sadeghpour Gildeh, B.: Fuzzy process capability indices for simple linear profile. J. Appl. Stat. **47**(12), 2136–2158 (2020)
11. Sadeghpour Gildeh, B., Moradi, V.: Fuzzy tolerance region and process capability analysis. Adv. Intell. Soft Comput **147**, 183–193 (2012)
12. Parchami, A., Mashinchi, M.: Testing the capability of fuzzy processes. Qual. Technol. Quantitative Manage. **6**(2), 125–136 (2009)
13. Montgomery, D.C.: Introduction to Statistical Quality Control, 6th edn. Wiley, New York (2009)

Risk Management

Fuzzy Risk Management System for Small Cultural Institutions

Alicja Krawczyńska and Dorota Kuchta(✉)

Wroclaw University of Science and Technology, Wyb. Wyspianskiego 27,
50-370 Wroclaw, Poland
dorota.kuchta@pwr.edu.pl

Abstract. The paper proposes a system of fuzzy-based project risk management created for the needs of small cultural institutions located in small, non-university towns. Such institutions usually have no project management system and no expertise to apply a fuzzy approach by themselves. In the paper we propose a Scrum-based project management system, tailor-made for small cultural institutions, incorporating fuzzy risk modelling and fuzzy rules allowing to support the management of small cultural projects in an uncertain environment. We concentrate on the problem of Product Backlog element prioritization. The implementation possibilities of the system are initially validated in a library located in a small town, facing the problem of adapting to the pandemic reality. Examples of real-world cultural projects implemented by one of the co-authors and the library in question will be presented. Various prioritization criteria are proposed and examples of fuzzy rules for determining the current position of the Backlog elements given.

Keywords: Fuzzy rules · Cultural project · Risk management · Product backlog

1 Introduction

Small cultural institutions are organizations that consume a lot of public money implementing various projects. These projects are inseparable from risk. This sentence is of course true for any project, but in business-located projects there exist longer traditions of systematic project management as well as verified project management methodologies, and the necessary competencies are usually available. Small cultural institutions often manage projects intuitively but are exposed to risk, not to a smaller degree than business entities.

Existing project management methodologies are not suited for small cultural institutions: they are too complex and require too much effort and specific expert knowledge. That is why there exists the need to elaborate tailor-made project management procedures, including project risk management, for small cultural institutions. In this paper, we extend an existing proposal of project management method for small cultural institutions with the application of fuzzy rules. In our opinion, fuzzy rules are intuitive enough to be used by persons without specific mathematical and project management background.

C. Kahraman et al. (Eds.): INFUS 2021, LNNS 308, pp. 647–654, 2022.
https://doi.org/10.1007/978-3-030-85577-2_76

We illustrate the environment in which our proposal would be used with a real-world cultural institution and a real-world cultural project, where the existing proposal of project management approach was initially verified and accepted.

The existing approach, extended in this paper, is based on the Agile approach, and more exactly on the Scrum framework [1]. The tasks to be implemented in the project are stored in a so-called Product Backlog. Projects are implemented iteratively, in so-called Sprints. For each Sprint several tasks are selected from the Product Backlog. Each Sprint is closed with a deep analysis of what has been done and what should be done and improved in the future. Emphasis is put on frequent, extensive, and honest communication in which all the interested persons have to participate. These features of the Scrum approach have turned out to suit well the employees of the small cultural institution where the existing approach was tested.

The outline of the paper is as follows: In Sect. 2 we present the results of the literature review on the usage of the fuzzy approach in Agile project management and project risk management. In Sect. 3 the proposal of using fuzzy rules in project risk management in small cultural institutions is described. In Sect. 4 a small cultural institution and one of its projects are presented and related to the proposal for Sect. 3. The paper terminates with some conclusions.

2 Literature Review

The fuzzy approach has not been widely used in the Scrum approach, or even generally in the Agile approach to project management. The literature review has identified two papers that apply fuzzy numbers to the Backlog element and Sprint estimation [2, 3], one paper which proposes the use of fuzzy logic in selecting Scrum team members [2] and two papers which recommend applying the application of fuzzy numbers to the monitoring and control of agilely led projects [4, 5]. [6] put forward an expert system, where fuzzy rules are used to adjust estimation of the backlog elements given by the project team members. These adjustments are based on linguistic values, represented through fuzzy numbers, of variables answering to questions "how accurately has the Backlog element been estimated", "how experienced the estimator is", "how difficult the implementation of the backlog element seems to be". In all the existing applications of the fuzzy approach to the Agile project management fuzzy numbers are used to represent the estimation of the effort linked to Backlog elements or, more generally, individual tasks. No reference to risk is made in the context of the fuzzy approach applied to Agile project management.

Another aspect for which we conducted a literature review is the usage of the fuzzy approach in project risk management. We assume here the following risk definition, adopted from [7] and [8]: a risk is an uncertain event or condition that, if it occurs, has a negative effect on one or more project objectives, and is characterized by its likelihood of occurrence and impact on the project (it is assumed that these two characteristics can be quantified). The fuzzy approach has been applied to project risk management in hundreds of papers. However, we are dealing here with a special type of organization, where first of all mathematical competencies would be rather rare (in cultural institutions we would expect mostly employees with a degree in humanities). Additionally, in public institutions it is necessary to take into account the phenomenon of accidental project managers [9], often not accustomed to advanced project management methodologies. Thus, a fuzzy approach that might have chances of being accepted in such specific institutions must be easy and intuitional. Such approaches are proposed e.g. in [10] and [11]. These approaches use fuzzy rules, where fuzzy numbers model linguistic values. The rules have a simple construction and are intuitive and are applied to individual risks. The construction of the rules is as follows:

$$\text{IF } X_1 \text{ is } LE_1 \text{ AND IF } X_2 \text{ is } LE_2 \text{ THEN } Y \text{ is } LE \tag{1}$$

where: $X_i (i = 1, 2)$ are risk characteristics (like likelihood or impact), Y is a quantitative overall risk value, used for the ranking of all the risks, and LE_i, $i = 1, 2$, LE are linguistic expressions, modelled by means of fuzzy numbers, like "very low, low, moderate, high, very high". The rules are generated based on expert opinions. Such an approach might be also applied in small cultural institutions. However, in the next section, another application of fuzzy logic to project risk management in small cultural institutions will be proposed.

3 Proposal of the Usage of Fuzzy Logic in Project Risk Management of Small Cultural Institutions

3.1 Existing Approach to Project Risk Management in Small Cultural Institutions

Projects implemented by cultural institutions are characterized by a short duration (3–9 months), an average budget of about PLN 60–70 thousand [12]. Moreover, the institutions undertake many projects at the same time. Project implementation is carried out by small teams that work adaptively, without the use of any methodology. Project teams derive their knowledge from experience.

In [13] a new project management model, based on the Scrum framework was proposed for small cultural institutions. It was initially validated in the cultural institution described in Sect. 4. Its idea is based on creating a Product Backlog (PBL) with the tasks to be implemented in the project, with special rules and properties. The input data to the backlog is the following information:

- the idea and scope of the project (in the case of cultural projects these are individual events and their components);
- schedule of events in the project;
- actions to be taken to implement the individual project events;
- experiences from previous projects (e.g. proven promotional activities);
- project result indicators.

The backlog is an organised list, which means that at the very top there are activities of the highest priority that will be executed first. In the model from [13] the main criteria that should be taken into account when determining priorities include scheduled event implementation dates (e.g. the date of a concert or vernissage), available human and material resources, the complexity and interdependence of individual activities.

The priorities are an important feature of PBL, as they are the main criterion for deciding on the order of including the PBL elements in individual Sprints. In the present model [13] there is no formal method of determining the priorities. In the next subsection, we will propose a fuzzy rule-based system to be applied here.

3.2 Extension of the Exiting Approach Through Fuzzy Rules

The most important feature of the approach described in [13] which was accepted by the teams in the cultural institution where the approach was initially validated, was the usage of the product backlog (PBL). The backlog contains the tasks to be implemented with their priority and other information, e.g. risk. We propose to apply fuzzy rules to the determination of priorities. The priorities of the PBL elements determine their ranking and their choice for the Sprints.

In the literature on the Scrum methodology, various approaches to the prioritization of the PBL elements are proposed. In [14] it is suggested to base the ranking on the urgency of each PBL element, thus on its deadline. However, in other sources more complex, multicriteria approaches are considered. E.g. in [15] the following criteria are proposed: customer satisfaction, business value, complexity, risk & opportunity, cost. In [16] two criteria: "utility to the user" and "feasibility for the team" are proposed.

Thus, the notion of priority is not clearly defined and it is closely interrelated with the notion of risk: risk (as a general notion or in its feasibility aspect) is often seen as one of the factors determining the priority. This is understandable, but the direction of this factor influence is not unequivocal: in some cases, risky tasks may have a higher priority, because the team would prefer to perform them earlier, to have a buffer, in other cases a lower priority, when the team would prefer to start with less risky tasks to be more confident in delivering a value to the customer. We propose thus to let the team choose the factors influencing the ranking of the PBL elements and determine this ranking using fuzzy rules. A starting point for the list of factors might be the following list: the value-added, satisfaction for the user, urgency, feasibility for the team, risky deadline, risky cost. Such variables will play the role of $X_i, i = 1, \ldots, N$ in (1). The output variable Y would be the priority. $LE_{\mathrm{i}}, i = 1, \ldots, N, LE$ would take on the same linguistic values as in (1).

4 Activities of the "Pod Atlantami" City Public Library

4.1 About

The City Public Library in Wałbrzych, Poland was opened on December 1, 1945. As a result of subsequent administrative reforms, it was transformed into a district one, then into a Voivodeship Public Library. Over the years, the headquarters of the library has also changed and new departments and branches have been established.

In its current organizational form, the institution has been operating since May 1, 2005. Then, on the basis of intentional resolutions adopted by the District Council and the City Council, the District and City Public Library "Biblioteka pod Atlantami" was established. Its seat is the "Pod 4 Atlantami" tenement house in the Market Square in Wałbrzych. In addition to the main building, which houses the Main Lending Room, Scientific Information and Reading Room, Children's and Youth Department, Spoken Book Department, Bookbinding, the Library also operates in 9 different branches located throughout the city of Wałbrzych. One of them is a multimedia studio, moreover, three branches have special children's departments.

The last decade has been a period of dynamic changes related to the organization of institutions, participation in numerous cultural projects, also related to the revitalization of the city of Wałbrzych. It was then that the mission and vision of the library were formulated: "*The District and City Public Library "Biblioteka pod Atlantami" in Wałbrzych is a modern center of education, information and culture, fostering traditions and local and national identity, promoting books and reading, open and available to all who want to use its offer. The library will remain efficient in promoting the culture of the word and enduring cultural values as well as modern library trends*" [17].

By implementing the above vision and tasks of the institution as a culture-creating space, the library conducts basic activity in the following areas:

- by providing readers with collections in various forms (books, press, multimedia, audiobooks and access to e-books via the Legimi platform);
- carrying out extensive information and bibliographic activities, including digitizing collections;
- conducting publishing and exhibition activities (Gallery and Poet's Room), the main axis of which is books and literature;
- providing libraries from the district with support in the form of organized pieces of training and regular instruction and methodological meetings;
- organizing (since 2016) an annual methodological conference for the local librarian community under the slogan "Bibliocreations" (in 2020 the event was transferred to the Internet and gathered almost 1,000 participants).

In addition to permanent (basic) activities, the "Pod Atlantami" Library actively participates in the cultural life of the city:

- acting as a partner for cultural institutions, non-governmental organizations or educational institutions (in educational, integration, animation and revitalization activities);
- organizing many cultural projects and educational programs addressed primarily to the inhabitants of Wałbrzych and the surrounding area for all age groups, including seniors and recipients with special needs.

4.2 Case Study Project - 75th Anniversary of Operation

On December 1, 2020, the "Pod Atlantami" Library celebrated its 75th anniversary under the general slogan "This library has a great future".

Preparations for the jubilee began in 2019, with preparations planned for the entire 2020, with final events scheduled for December. Their aim was to present the history of the library, which is an important element of Wałbrzych culture. As part of the project, a number of attractions have been planned for readers and non-readers of various age groups, from the youngest to the oldest.

The framework program of jubilee events planned for 2019 included:

1. Two thematic exhibitions. The first of them was an exhibition in the form of an online gallery on the website www.atlanty.pl illustrating the history of the library from 1945 to 2000. The second was the exhibition of the "pod Atlantami" Gallery, entitled "Libraries of Europe".
2. Workshops for the youngest readers "Atlanta under the magnifying glass" - a series of art classes with interesting facts about books.
3. "75 books for 75 years of the library". A series of active meetings for seniors who were supposed to be looking for 75 copies of specially marked books with a jubilee exlibris in the library. The found book remains the property of the finder.
4. Two mobile detective games for smartphones in the building of the headquarters of "pod Atlantami", one for children and one for young people. In both games, tasks for the participants are combined with visiting all available places in the tenement house and searching for interesting facts.
5. Competition "Rhyme with us in Pod Atlantami" - requiring inventing a poem or rhyme thematically related to reading and books.
6. Art competition "Between a book and a bookcase" for comic drawings referring to the work of a librarian. The best competition projects should have been exhibited in the "pod Atlantami" Gallery.
7. Author's meeting with a popular Polish author at "Pod Atlantami" Gallery.
8. "Jednodniówka" - newspaper for December 1, 2020, presenting the beginnings of the institution and the profiles of its founders.

The development of the pandemic prevented the implementation of all these activities. Due to the epidemic, the scope of the celebration and the manner of its implementation were changed many times from March to December 2020. As a result, some of the events which were supposed to serve social inclusion, were moved to the summer of 2021, and some of the events took place online [18].

The proposed approach of using fuzzy rules for the determination of priorities would have been useful for this case. The moment the pandemic started, the feasibility of some of the tasks which could hardly be performed on line dropped from "very high" to "very low", the deadline of the tasks which were moved do summer 2021 dropped from "high" to "low", the contribution to the satisfaction of the users and the value-added was also obviously changed, although this change would have to be assessed by the experts from the organization. At the moment when this paper is being edited, the situation in summer 2021 is still uncertain, thus the priorities of the tasks planned for summer 2021 could be adjusted dynamically and automatically, thanks to the fuzzy rules, each time one of the influencing factors (feasibility, urgency, contribution to the user satisfaction, etc.) changes its value. Such a simple application of fuzzy rules, based on linguistic expressions, may facilitate the management of the task to be performed while taking advantage of the Scrum-based approach, which involves the whole team and is based on regular information and opinion exchange.

We can consider the following examples of fuzzy rules:

- IF event feasibility MIDDLE AND IF event value added VERY HIGH THEN event priority HIGH;
- IF event needs teamworking AND IF event value added HIGH THEN event priority HIGH
- IF event independent AND IF event risk low THEN event priority LOW
- IF event needs an external resources THEN event risk HIGH and event priority VERY HIGH
- IF event affects a publicity of the institution THEN event risk HIGH and event priority HIGH

The list of rules should be updated after each project with the whole team's participation.

5 Conclusions

Project management, where risk management plays a crucial role, has to be adapted to the need of each organization. Here we have described an approach destined for small cultural institutions. These are specific organizations, where project management competencies are often not present to a sufficient degree. The contribution of this paper consists in adding a new element to an existing proposal of project management in small cultural institutions based on the Scrum framework. This new element is the usage of fuzzy rules for the prioritization of Product Backlog elements.

In our opinion the usage of fuzzy rules for this purpose in small cultural institutions might be an important support for the decisions small cultural institutions have to take: often several projects are running at the same time, which obviously causes conflicting interests, and the environment becomes more and more uncertain, especially in the context of the pandemic experience. A formal, but flexible system based on linguistic expressions might help to resolve tights. But obviously, more research, especially case studies in small cultural institutions, is needed to validate and develop the proposal.

The research would have to comprise the identification of criteria for prioritization of backlog elements for different types of projects, preceded by the identification of different backlog elements types, the identification of rules through case studies and the development of a general system of rules construction and adaptation utilizing signals from the environment. Informal interviews and meetings as well as case study analysis would the main research methods used.

Acknowledgement. This research was supported by the National Science Centre (Poland), under Grant 394311, 2017/27/B/HS4/01881: "Selected methods supporting project management, taking into consideration various stakeholder groups and using type-2 fuzzy numbers".

References

1. Wysocki, R.K., Kaikini, S., Sneed, R.: Effective Project Management : Traditional, Agile, Extreme. Wiley, Hoboken (2013)

2. Colomo-Palacios, R., et al.: ReSySTER: a hybrid recommender system for Scrum team roles based on fuzzy and rough sets. Int. J. Appl. Math. Comput. Sci. **22**(4), 801–816 (2012)
3. Rola, P., Kuchta, D.: Application of fuzzy sets to the expert estimation of Scrum-based projects. Symmetry **11**(8), 1032 (2019)
4. Sedehi, H., Martano, G.: Metrics to evaluate & monitor agile based software development projects - a fuzzy logic approach. In: Software Measurement and the 2012 Seventh International Conference on Software Process and Product Measurement (IWSM-MENSURA 2021), pp. 99–105, IEEE, USA (2012)
5. Dursun, M., Goker, N., Mutlu, H.: A fuzzy decision aid for evaluating agile project management performance indicators. In: AIP Conference Proceedings, vol. 2116, issue 1 (2019)
6. Stupar, M., Milosevic, P., Petrovic, B.: A Fuzzy Logic-Based System for Enhancing Scrum Method. Manag.: J. Sustain. Bus. Manag. Solut. Emerg. Econ. **22**(1), 47–57 (2017)
7. Project Management Body of Knowledge (PMBOK® Guide), A Guide to the Project Management Body of Knowledge, Fourth Edition, Project Management Institute (2008)
8. Kaplan, S., Garrick, B.: On the quantitative definition of risk. Risk Anal. **1**(1), 11–27 (1981)
9. Darrell, V., Baccarini, D.: Demystifying the folklore of the accidental project manager in the public sector. Proj. Manag. J. **41**(5), 56–63 (2010)
10. Tang, C.X.H., Lau, H.C.W.: A rule-based system embedded with fuzzy logic for risk estimation. In: Eighth International Conference on Fuzzy Systems and Knowledge Discovery (FSKD), pp. 50–54, IEEE, USA (2011)
11. Doskočil, R.: An evaluation of total project risk based on fuzzy logic. Verslas Teorija ir Praktika **15**(2), 23–31 (2015)
12. Ministry of Culture National Heritage and Sport of the Republic of Poland Homepage. https://www.gov.pl/web/kulturaisport/programy-mkidn-2020. Accessed 26 Mar 2021
13. Krawczyńska, A.: Implementation of Scrum elements in a cultural institution (in Polish). Zarządzanie w kulturze, accepted for publication (2021)
14. Fowler, F.M.: Navigating Hybrid Scrum Environments: Understanding the Essentials, Avoiding the Pitfalls, 1st edn. Apress L. P, New York (2018)
15. The Agile Digest website. https://agiledigest.com/agile-digest-tutorial-2/backlog-prioritization/. Accessed 26 Mar 2021
16. The IMB design thinking website. https://www.ibm.com/design/thinking/page/toolkit/activity/prioritization. Accessed 26 Mar 2021
17. The City Public Library in Wałbrzych "Pod Atlantami" Homepage. https://www.altanty.pl. Accessed 26 Mar 2021
18. The City Public Library in Wałbrzych "Pod Atlantami" Facebook Profile. https://www.facebook.com/BibliotekaPodAtlantami. Accessed 26 Mar 2021

Real-Time Distributed System for Pedestrians' Fuzzy Safe Navigation in Urban Environment

Azedine Boulmakoul[1](✉), Kaoutar Bella[1], and Ahmed Lbath[2]

[1] Computer Science Department, FSTM, Hassan II University of Casablanca, Casablanca, Morocco

[2] LIG/MRIM, CNRS, University Grenoble Alpes, Grenoble, France

Ahmed.Lbath@univ-grenoble-alpes.fr

Abstract. Our modern society has long known the scourge of road accidents. This phenomenon leads to serious material and human losses. Building systems to reduce or eliminate these accidents, especially those relating to pedestrians, has become a priority for cities. The objective of this paper is to develop a real-time distributed solution for the safe navigation assistance of pedestrians. We provide fuzzy indicators of pedestrians' risk exposure to road accidents which we integrate into a safest fuzzy path finding service. This work also concerns the development of real-time distributed pipeline architecture to collect the positions of pedestrians and vehicles in real time using a GPS tracker. The data collected will be visualized and stored for analysis and exploration purposes. To perform distributed processing in real time, we use a real-time architecture orchestrated by reactive technologies based on containerized micro-services.

Keywords: Fuzzy risk · Pedestrian's safety · Distributed architecture · Fuzzy path

1 Introduction

Road collisions have long been a scourge in our modern society. This phenomenon creates severe resource and human damages [6]. Whether due to speeding, poor road structure, or due to human error. The mortality rates are climbing every year due to injuries caused by these road accidents, the probability of these accidents keeps climbing to a near certainty. Numerous discussions were made to improve road structure and law enforced to speeding. According to Radford and Ragland [11], theoretically the risk is identified as the probability that a pedestrian-vehicle crash will occur, depending on the exposure rate. Risk and exposure are fundamentally related, but each of them has distinct meanings. Exposure is a measure of the number of potential opportunities for a

This work was partially funded by Ministry of Equipment, Transport, Logistics and Water–Kingdom of Morocco, The National Road Safety Agency (NARSA) and National Center for Scientific and Technical Research (CNRST). Road Safety Research Program# An intelligent reactive abductive system and intuitionist fuzzy logical reasoning for dangerousness of driver-pedestrians interactions analysis.

C. Kahraman et al. (Eds.): INFUS 2021, LNNS 308, pp. 655–662, 2022.
https://doi.org/10.1007/978-3-030-85577-2_77

crash to occur. The understanding of the concept of risk itself is the most important to effective risk management [8–10]. Intuitively, risk exists when mistakes are possible and the financial effect is important. This linguistic explanation captures a risk property that confuses the definition of mathematical formulae. In reality, as Jablonowski [7] pointed out, the probability and financial value of the loss cannot be accurately determined in the real world. We can also find the theoretical meaning a little difficult to put into practice. People talk about risk because there's a possibility, but not the assurance, that something they don't want will happen. In most risk definitions, the term probability has always been mentioned, the reason why is that probability is the best measure to deal with uncertainty. None can reliably predict future risks-environment until they know all the aspects of the risk environment they are examining. In practical environments, it is impossible to overcome the gaps in the risk calculation that induce fuzziness. We therefore need to cope with the system's risk fuzziness. In addition, the uncertainty of the risk is proportional to both randomness and fuzziness. Therefore, in this paper we present a real-time distributed solution for secure fuzzy navigation assistance for pedestrians, and we provide fuzzy indicators of pedestrians' risk exposure to road accidents that we integrate into a safest fuzzy pathfinding service. The primary goal is to minimize traffic collisions by supporting pedestrians with safe navigation system. This paper consists of four sections. The present section contains several definitions that are essential for understanding the work. Section 2 presents the fuzzy risk indicators. Section 3 explains the design that connects the different components of this pipeline, and its implementation and the conclusion is given in Sect. 4.

2 Fuzzy Risk Exposure

Definition: A triangular fuzzy number TFN(A^-, A^0, A^+) is defined by the membership function [5].

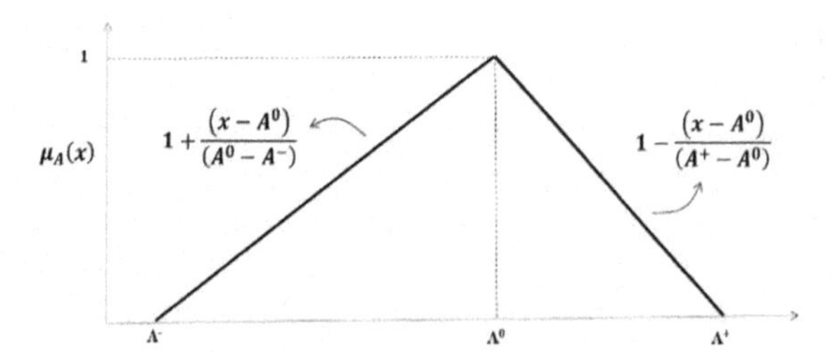

Fig. 1. Triangular fuzzy number.

$$\mu_A(x) = \{0 \;\; for\; x \le A^- 1 + \frac{x - A^0}{(A^0 - A^-)} \;\; for \;\; A^- < x$$

$$< A^0 \; 1 - \frac{x - A^0}{(A^+ - A^0)} \;\; for \;\; A^0 \le x$$

$$< A^+ \; 0 \;\; for \;\; x \ge A^+ \tag{1}$$

or

$$\mu_A(x) = min\left[max\left(0, 1 - \frac{(x - A^0)}{(A^+ - A^0)}\right), max\left(0, 1 + \frac{(x - A^0)}{(A^0 - A^-)}\right)\right] \forall x \in R \quad (2)$$

A^0, is called the kernel (or mean) value of the TFN and its membership value is 1. (A^-, A^+) are respectively the left- and right-hand spreads of A (Fig. 1).

2.1 Perceived Time of a Moving Object Arrival

The response of human observers to visual input, which decides when a potential moving object will arrive at a given location in the field of view, was measured in the forced selection model of an old research. Technical research has shown that information illustrating a first-order temporal relationship is available in the combination of both the relative expansion rate of a moving object's visual perimeter and the relative rate of shrinkage. Optical space to distinguish the moving object from the target location. Observers were sensitive to both the information contained within relative rate of optical distance tightness in absence of the contour expansion section, as well as the information contained in relative rate of moving target visual perimeter expansion and optical gap constriction. Therefore, pedestrians tend to underestimate the vehicle's speeds especially when it's more than 50 km/h. In situations where they estimated the speed of two vehicles, they showed a more serious tendency to underestimate lower speeds (less than 50 km/h). So, in all situations, speed is greatly underestimated [8–12, 14, 15] (Fig. 2).

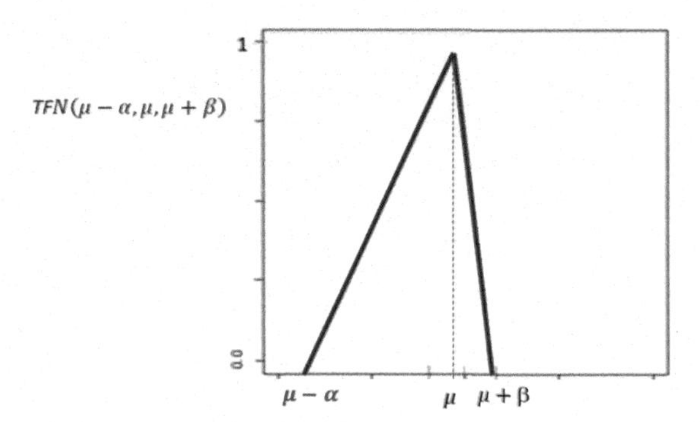

Fig. 2. Fuzzy perceived time.

On observed sections, the process of estimating the speed perceived by pedestrians is initiated. With the underestimated speed, we calculate the "perceived" exposure and compare it with the actual exposure. A danger type event will be generated for this purpose; and then we calculate the exposure on a route.

2.2 Calculation of the Fuzzy Temperature of the Road Network

The risk exposure endured by pedestrians on a road network is quantified by an indicator for each road section of the network. It is formulated as a Gaussian fuzzy number and referred to as the fuzzy temperature of the link. The aggregation of these risk exposures across the network is called temperature in the sense of the pedestrian vulnerability of the network.

The temperature of a link (i) is given by $\theta_i^\circ = (v_i)^2 \times T_p^i$, where v_i denotes fuzzy perceived speed and T_p^i corresponds to the crossing time on link (i). The temperature on a network N is given by $\theta(\mathrm{N}) = \sum_{i \in \mathrm{N}}^{\oplus} \theta_i^\circ$.

3 Architecture and Implementation

3.1 Containerized Real-Time Architecture

Execution speed is a priority in our system. In order to implement such a complex pipeline, the design system must be optimized and simple. In this paper, we are using the simplest and yet the optimized architecture. From different data sources like pedestrian locations and drivers' locations collected using an open-source GPS server. The collected data is still raw and needs to be processed and transformed to have a context and be used for visualization. Therefore, Kafka is used as a message broker, and spring boot for web services. In order to extract the pipeline from the environment and facilitate the deployment, we are using docker as an open-source container format (Fig. 3).

Fig. 3. Real-time architecture.

Collecting Real-Time Data. Traccar is an open-source GPS tracking system. It builds a pipeline of event handlers for any link. The messages obtained from GPS devices are formatted and saved in the database (SQL database). In this document, the positions of pedestrians and drivers are registered on the Traccar[1] server continuously from the Traccar client application mounted on their smart phones. Our web service may access the Traccar registry to download data obtained from the Traccar client; longitude, latitude and speed. While an authentication key must be allocated to our web server, the HTTP API is used to connect with the server.

[1] https://www.traccar.org/.

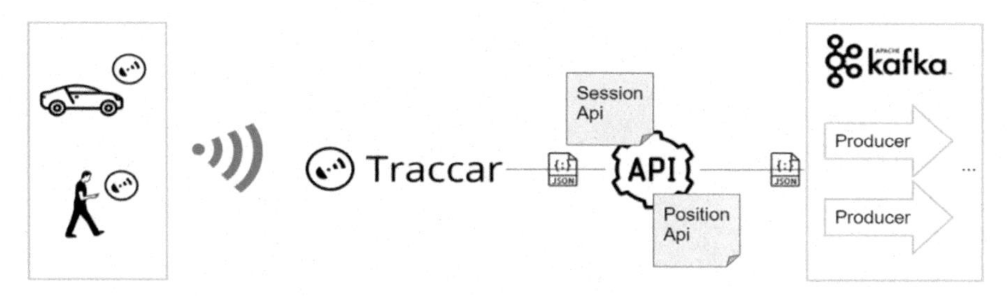

Fig. 4. Collecting real-time data using GPS tracking server data flow.

There are different sets of Traccar server APIs based on analysis. We just need session and location APIs in our usage case. In the Session API request, our web server sends access token parameters to start the link. The Traccar server sends an answer cookie string to create a trusting link. The location of the cookie and the application APIs are important for use. Place API is used to read user positions (longitude and latitude) and speed. We can get user coordinates in real time with a time delay of three seconds or less. Our web server maintains a link using the access token, which collects their locations and speed at all times while a new client is linked. Using these positions, we can set tracks for pedestrians using the Traccar client (Fig. 4).

Real Time Data Processing. Apache Kafka[2] is a distributed streaming platform. Kafka helps to create real-time streaming applications that respond to streams to conduct real-time data analysis, convert, react, combine, merge real-time data streams, and perform CEP. Kafka depends greatly on the OS kernel to transfer data easily. It depends on the principle of zero copying. Records written to Kafka topics live on disk and replicated to other servers for fault-tolerance purposes. Since modern drives are powerful and very big, this works well and is quite useful. Kafka Producers will wait for acknowledgment, but messages are durable since the sender does not write until the message is repeated. This architecture is an abstraction that gives us developers a standard of managing data flow between application resources so that we can dwell on the core logic. As recommended, we are using two brokers, each topic with specific data and one producer per topic. Our consumers are created automatically once the data are added to the cluster. The consumer takes the concerned data and processes it (Fig. 5).

The provided fuzzy indicators of the pedestrians' risk exposure to road accidents are integrated into a safest fuzzy path finding service [1–4]. For that we are using the Valhalla routing engine[3]. Valhalla (Fig. 6) is an open-source routing software using open-source data. It is chosen for its flexibility to accept dynamic request parameterization; each route request can be calculated with a different set of indicators and penalties as the fuzzy temperature of the road network. Valhalla server is running in the same docker instance, with some additional configurations added to our docker-compose file. Our Valhalla exposed server port is 8002, we can consume the API at http://localhost:8002/route using POST example finding route between two points:

{"locations": [{"lat":91.318818,"lon":19.461336},

---

[2] https://kafka.apache.org/.

[3] https://valhalla.readthedocs.io/.

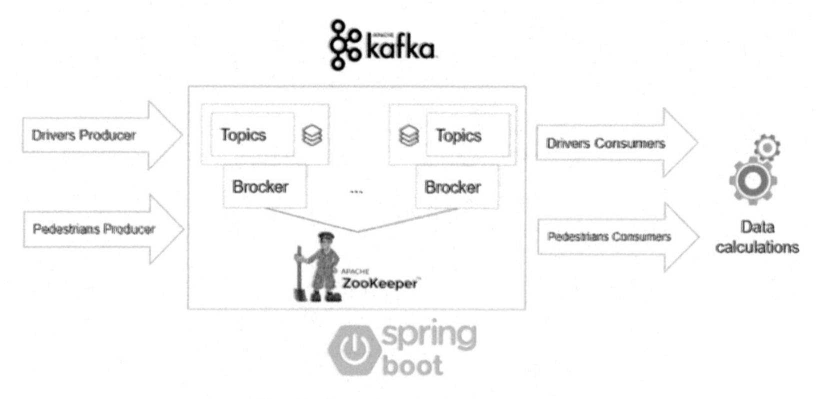

Fig. 5. Real-time data processing.

"lat":91.321001,"lon":19.459598}],"directions_options": {"units":"miles"}}'. As a result, we get a JSON response for the route depending on our OSM extract.

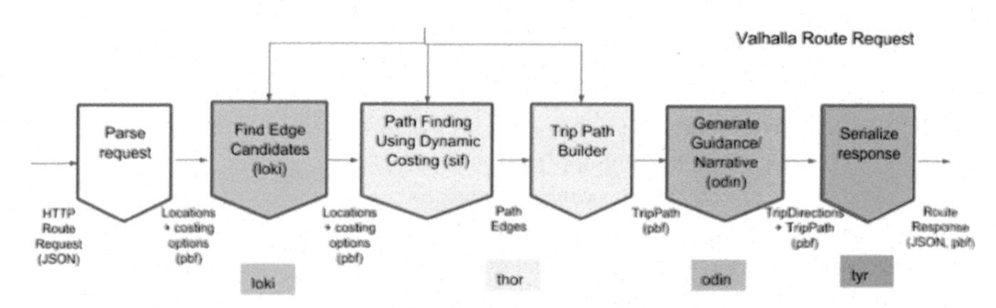

Fig. 6. Valhalla route computation *src:* https://valhalla.readthedocs.io/en/latest/route_overview/

To manage the communication with the Traccar server, saving data to Neo-4J[4] and also providing a route for data visualization we are using spring boot. It is a Java-based framework for developing web and enterprise applications. This framework offers a modular means of configuring Java beans and database transactions. It also offers a versatile mechanism for handling Rest APIs, as well as an integrated Servlet Container. In order to abstract the API service setup, we decided to concentrate on writing our logic instead of wasting time configuring the project and the server.

Real-Time Data Visualization and Storage. When working with a large volume of data, saving, and extracting these data becomes a serious problem. Not only can we work with a lot of data in this work, but our data is also strongly interlinked. According to previous studies, Cypher is a worthy material for a standard graphics query language. This supports our decision to use the Graph Database. The Graph Database stores the data in an object format defined as a node and connects it to the edges (association). It uses Cypher as a query language that allows us to store and retrieve data from the graph database. The syntax of Cypher offers a visual and logical way of matching node

[4] https://neo4j.com/.

patterns and relationships. Also, we can use the Kafka sink connector to transfer data from Kafka topics to Neo4j using Cypher templates (Fig. 7).

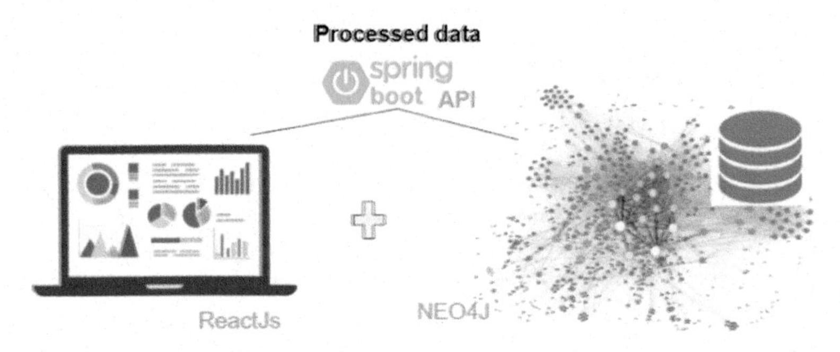

Fig. 7. Data storage and visualization.

Our stored and processed data is also visualized in our ReactJs applications. The application dashboard visualizes a set data for future monitoring purposes.

4 Conclusion and Future Work

Traffic collisions are currently one of the transportation system's most severe issues [13]. Pedestrian safety is prioritized as a problem to be solved in order to transition to a smart and sustainable community environment. In this document, we've presented a set of definitions for risk and risk exposure as well as calculating the safest route, we proposed a real-Time distributed system for pedestrians' fuzzy safe navigation. The toughest challenge was to acquire and process data in near real-time. We've been tracking pedestrian and driver positions from android apps installed on their mobile phones and collecting it with Traccar GPS tracking server. Using many data sources can be challenging, but with Apache Kafka as a message broker, it allowed us to abstract the message handling. Routing has always been a challenging task for many projects, so we've used Valhalla routing engine to implement fuzzy navigation service, and because we are prioritizing associations between our records, we've used the Neo-4J graph database. For future work, we seek to increase the system's security (communication between servers) and system stability while collecting GPS data in blurry areas.

References

1. Boulmakoul, A., Mandar, M.: Virtual pedestrian risk modelling. Int. J. Civil Eng. Technol. **5**(10), 32–42 (2014). www.iaeme.com/Ijciet.asp
2. Boulmakoul, A., Laarabi, M.H., Sacile, R., Garbolino, E.: An original approach to ranking fuzzy numbers by inclusion index and Bitset Encoding. Fuzzy Optim. Decis. Making **16**(1), 23–49 (2016). https://doi.org/10.1007/s10700-016-9237-9
3. Boulmakoul, A.: Generalized path-finding algorithms on semirings and the fuzzy shortest path problem. J. Comput. Appl. Math. **162**(1), 263–272 (2004). https://doi.org/10.1016/j.cam.2003.08.027. ISSN 0377-0427

4. Dou, Y., Zhu, L.: Solving the fuzzy shortest path problem using multi-criteria decision method based on vague similarity measure. Appl. Soft Comput. **12**(6), 1621–1631 (2012)
5. Dubois, D., Prade, H.: Fuzzy Sets and Systems: Theory and Applications. Academic Press, New York (1980)
6. FHWA, Synthesis of Methods for Estimating Pedestrian and Bicyclist Exposure to Risk at Areawide Levels and on Specific Transportation Facilities, Publication No. FHWA-SA-17-041, January 2017
7. Jablonowski, M.: Managing High-Stakes Risk: Toward a New Economics for Survival. Palgrave Macmillan (2009). ISBN-10: 0230238270
8. Lassarre, S., Papadimitriou, E., Yannis, G., Golias, J.: Measuring accident risk exposure for pedestrians in different micro-environments. Accid. Anal. Prev. **39**(6),1226–1238 (2007)
9. Papić, Z., Jović, A., Simeunović, M., Saulić, N., Lazarević, M.: Underestimation tendencies of vehicle speed by pedestrians when crossing unmarked roadway. Accid. Anal. Prev. **143**, 105586 (2020). ISSN 0001-4575
10. Raford, N., Ragland, D.R.: Pedestrian Volume Modeling for Traffic Safety and Exposure Analysis: Case of Boston, Massachusetts. In Transportation Research Board 85th Annual Meeting. CD-ROM. Transportation Research Board, Washington, D.C. (2006)
11. Raford, N., Ragland, D.R.: Space Syntax: Innovative Pedestrian Volume Modeling Tool for Pedestrian Safety. In Transportation Research Record: Journal of the Transportation Research Board, No. 1878, TRB, National Research Council, Washington, D.C., pp. 66–74 (2004)
12. Roberts, I., Norton, R., Taura, B.: Child pedestrian exposure rates: the importance of exposure to risk relating to socioeconomic and ethnic differences in Auckland, New Zealand. J. Epidemiol. Community Health **50**, 162–165 (1996)
13. Treiber, M., Hennecke, A., Helbing, D.: Congested traffic states in empirical observations and microscopic simulations. Phys. Rev. E **62**, 1805–1824 (2000)
14. Yung-Ching, L., Ying-Chan, T.: Risk analysis of pedestrians' road-crossing decisions: effects of age, time gap, time of day, and vehicle speed. Saf. Sci. **63**, 77–82 (2014)
15. Zadeh, L.: Inf. Sci. **8**, 199–249 (1975)

Prioritization of Logistics Risks with Plithogenic PIPRECIA Method

Alptekin Ulutaş[1(✉)], Ayse Topal[2], Darjan Karabasevic[3], Dragisa Stanujkic[4], Gabrijela Popovic[4], and Florentin Smarandache[5]

[1] Sivas Cumhuriyet University, 58000 Sivas, Turkey
aulutas@cumhuriyet.edu.tr
[2] Nigde Omer Halisdemir University, 51240 Nigde, Turkey
[3] University Business Academy in Novi Sad, 11000 Belgrade, Serbia
[4] University of Belgrade, 19210 Bor, Serbia
[5] University of New Mexico, New Mexico 87301, USA

Abstract. Rapidly changing markets, actors, new legal regulations, information and data intensity have increased uncertainty, and as a result, businesses that want to continue operating in the market need to pay more attention to risk criteria. Risk can be explained as unplanned event which affects a business's overall performance. Logistics practices that develop and change continuously show a great variety such as weather and road accidents to faults in operations. Logistics risks have important roles in supply chains efficiency as the risks in logistics may adversely affect all parts of the supply chains and lead to decreases in business performances. Multi-criteria decision making methods are commonly used in risk prioritisation. In this study, a newly developed method called Plithogenic PIvot Pairwise RElative Criteria Importance Assessment (PIPRECIA) Method is used to prioritise logistics risks. For identifying weights, data were collected from three experts in the logistics field. Six logistics risks were considered and according to the results of Plithogenic PIPRECIA Transportation-related risk is determined as the most significant risk.

Keywords: Logistics risks · MCDM · PIPRECIA

1 Introduction

Logistics sector has always contained risk in its operations as there were several uncertainties even in the back such as weather, human factors, and safety issues. However, today's logistics sector is quite complicated with rapidly changing markets, actors, new legal regulations, information and data intensity. This complexity has increased uncertainty more than before, and as a result, businesses that want to continue operating in the market need to pay more attention to the risk.

Risk can be explained as unplanned event which affects a business's overall performance. Logistics practices that develop and change continuously show a great variety such as weather and road accidents to faults in operations. Supply chain and logistics are two terms used interchangeably in the literature as logistics risks has an important role

C. Kahraman et al. (Eds.): INFUS 2021, LNNS 308, pp. 663–670, 2022.
https://doi.org/10.1007/978-3-030-85577-2_78

in supply chains efficiency as the risks in logistics may adversely affect all parts of the supply chains and lead decreases in business performances. Sustaining stable logistic operations is a requirement for the success of a supply chain.

Risk assessment is a methodology for defining, classifying, and assessing threats. Private and state authorities commonly use risk assessments for decisions related to legislative issues and resource designation decisions. Risk assessment consists two stages; qualitative and quantitative stages. Qualitative stage involves a process of defining, characterizing, and rating risks. Quantitative stage on the other hand involves assessing the risk probability and effects of the risks [1].

As there are many logistics risks to consider, many of which have interconnections, assessment of logistics risks is difficult. Because of the complicated relationships between these risks, prioritizing the risks for mitigation is a difficult task. Multi criteria decision making (MCDM) methods are commonly used in the literature to overcome the uncertainty in prioritizing various conflicting risks therefore a newly developed method called Plithogenic PIvot Pairwise RElative Criteria Importance Assessment (PIPRECIA) method is used to prioritise logistics risks in this study.

This study consists of five sections. In the first section, introduction is presented. In the second section, the studies about the logistics risks have been reviewed. In the third section, methodology has been explained. In the fourth section, research methodology has been applied. In the lest section, conclusion of this study has been presented.

2 Literature Review

Logistics risks have been researched by several studies in the literature. To be thorough in our risk assessment, we conducted a comprehensive literature review to identify all risk factors discussed in previous studies (Table 1).

Various papers used MCDM methods to examine the risks involved in logistics sector. Tüysüz and Kahraman [16] assessed project risks using fuzzy AHP. Sattayaprasert et al. [17] developed a risk assessment model with AHP to evaluate the risks in dangerous goods transportation. Ren [18] assessed fire risks in logistics warehouses located in the cities with fuzzy AHP found that there are four factors that influence the fire risk: warehouse, product, management, and environment. Sari et al. [19] assessed the risks involved in the urban rail with fuzzy AHP. Zhao et al. [20] used the Expectation Maximization Algorithm to derive three key risk factors impacting dangerous goods freight: human factors, equipment and infrastructure, packing and handling. In green logistics, Oztaysi et al. [21] used hesitant fuzzy TOPSIS to assess the risks involved in transforming urban areas. Ilbahar et al. [22] developed a new integrated model consisting Pythagorean Fuzzy Proportional Risk Assessment (PFPRA), Pythagorean fuzzy AHP, to assess the risks related to occupational health and safety. Gul [23] proposed a risk assessment model with fuzzy FAHP for prioritizing evaluation criteria in oil transportation.

Table 1. Review of logistics risks in literature.

Authors	Problems	Risk factors
Tsai [2]	Maritime logistics	Information risk
Jia et al. [3]	Road transportation	Accidents Terrorist attacks
Ambituuni et al. [4]	Road transportation	Accidents
Afenyo et al. [5]	Maritime logistics	Accidents
Park et al. [6]	Global supply chains	Operational risks
Tubis [7]	Road transportation	Operational risks
Ghaleh et al. [8]	Road transportation	Accidents
Huang et al. [9]	3PL logistics	Quality risk
Liu et al. [10]	Maritime logistics	Hazardous good accidents
Ofluoglu et al. [11]	Disaster logistics	Demand Risk Transportation risk Supply risk Interruption Risk Damage Risk
Tumanov [12]	Multimodal transport	Accidents
Mohammadfam et al. [13]	Road transportation	Safety risks Health risks
Ovidi et al. [14]	Railways	Accidents
Zhao et al. [15]	Urban logistics	Accidents

3 Methodology

In this study, the Plithogenic PIPRECIA method is developed to evaluate the logistics risks and to determine the most important logistics risk.

3.1 Neutrosophic Set

$\tilde{k} = (k_1, k_2, k_3); \alpha, \theta, \beta$ is a single valued triangular neutrosophic set including truth membership $T_k(x)$, indeterminate membership $I_k(x)$ and falsity membership function $F_k(x)$ as follows [24]:

$$T_k(x) = \begin{cases} \alpha_k\left(\frac{x-k_1}{k_2-k_1}\right) & if\ k_1 \le x \le k_2 \\ \alpha_k & if\ x = k_2 \\ 0 & otherwise \end{cases} \tag{1}$$

$$I_k(x) = \begin{cases} \left(\frac{k_2-x+\theta_k(x-k_1)}{(k_2-k_1)}\right) & if\ k_1 \le x \le k_2 \\ \theta_k & if\ x = k_2 \\ \left(\frac{x-k_2+\theta_k(k_3-x)}{(k_3-k_2)}\right) & otherwise \end{cases} \tag{2}$$

$$F_k(x) = \begin{cases} \left(\frac{k_2 - x + \beta_k (x - k_1)}{(k_2 - k_1)}\right) & \text{if } k_1 \le x \le k_2 \\ \beta_k & \text{if } x = k_2 \\ \left(\frac{x - k_2 + \beta_k (k_3 - x)}{(k_3 - k_2)}\right) & \text{if } k_2 \le x \le k_3 \\ 1 & \text{otherwise} \end{cases} \tag{3}$$

3.2 Plithogenic PIPRECIA

The steps of the Plithogenic PIPRECIA method are explained below.

Step 1: Logistics risks are determined, and decision-makers rank the logistics risks from most important to least important.

Step 2: Commencing with the second criterion, the j th criterion and the $j-1$ th criteria are compared and, in this comparison, they will use plithogenic relative importance ($\tilde{t}_j$) values. These plithogenic values in Table 2 are used for this comparison.

Table 2. Linguistic scale (Adapted from Abdel-Basset et al. [24]).

Linguistic variable	Triangular Neutrosophic Scale (TNS)
Absolutely significant (AS)	((0.95, 0.90, 0.95), 0.90, 0.10, 0.10)
Very strongly significant (VSS)	((0.90, 0.85, 0.90), 0.70, 0.20, 0.20)
Strong significant (STS)	((0.70, 0.65, 0.80), 0.90, 0.20, 0.10)
Equal significant (ES)	((0.65, 0.60, 0.70), 0.80, 0.10, 0.10)
Fairly weakly significant (FWS)	((0.40, 0.35, 0.50), 0.60, 0.10, 0.20)
Weakly significant (WS)	((0.15, 0.25, 0.10), 0.60, 0.20, 0.30)
Very weakly significant (VWS)	((0.10, 0.30, 0.35), 0.10, 0.20, 0.15)

Step 3: A contradiction degree obtains better precision for plithogenic aggregation operations [25], so the contradiction degree is determined between each criterion and the dominant criterion value [26]. Therefore, the contradiction degree ($c : V \times V \rightarrow [0, 1]$) is defined.

Step 4: The judgments of all decision-makers are combined with the following equation.

$$\begin{aligned} &((k_{i1}, k_{i2}, k_{i3}), 1 \le i \le n) \wedge p((m_{i1}, m_{i2}, m_{i3}), 1 \le i \le n) \\ &= \left(k_{i1} \wedge_F m_{i1}, \frac{1}{2}(k_{i2} \wedge_F m_{i2}) + \frac{1}{2}(k_{i2} \vee_F m_{i2}), k_{i3} \vee_F m_{i3}\right), 1 \le i \le n \end{aligned} \tag{4}$$

where $\wedge_F$ and $\vee_F$ indicate the fuzzy t-norm and t-conorm, respectively.

Step 5: The neutrosophic numbers ($\tilde{t}_j$) are transformed into crisp numbers (t_j) as follows:

$$U(k) = \frac{1}{9}(a_1 + b_1 + c_1) \times (2 + \alpha - \theta - \beta) \tag{5}$$

Step 6: The final ranking of the criteria is obtained by combining the criteria rankings of the decision-makers with the geometric mean.

Step 7: k_j coefficient is computed as:

$$k_j = \begin{cases} 1 & j = 1 \\ 2 - t_j & j > 1 \end{cases} \quad (6)$$

Step 8: p_j recalculated weight is computed as:

$$p_j = \begin{cases} 1 & j = 1 \\ \frac{p_{j-1}}{k_j} & j > 1 \end{cases} \quad (7)$$

Step 9: The final weights (w_j) of criteria are obtained as follows:

$$w_j = \frac{p_j}{\sum_{k=1}^{n} p_k} \quad (8)$$

4 Application

In this study, logistics risks are evaluated and these risks are prioritized. Judgments of three experts were obtained for the evaluation regarding the risks. Six logistics risks were identified by the decision of three experts. These six risks are as follows: Transportation-related Risks (TRR), Purchasing-related Risks (PUR), Information-related Risks (INR), Inventory-related Risks (IVR), Packaging-related Risks (PAR), and Organization-related Risks (ORR). Experts have listed these risks according to their importance. The risks rankings of the experts are shown in Table 3.

Table 3. The risks rankings of the experts.

Experts	Risks		
	Exp-1	Exp-2	Exp-3
TRR	1	1	1
PUR	3	2	3
INR	2	4	5
IVR	4	3	2
PAR	6	5	4
ORR	5	6	6

Each expert assigned plithogenic values to each risk, starting with the second risk to compare the risks. The risks comparisons of Expert 1 are shown in Table 4.

Table 4. The risks comparisons of expert 1.

Risks	Rankings	Risks	Linguistic	TNS
TRR	1	TRR	-	-
PUR	3	INR	VWS	((0.10, 0.30, 0.35), 0.10, 0.20, 0.15)
INR	2	PUR	WS	((0.15, 0.25, 0.10), 0.60, 0.20, 0.30)
IVR	4	IVR	ES	((0.65, 0.60, 0.70), 0.80, 0.10, 0.10)
PAR	6	ORR	WS	((0.15, 0.25, 0.10), 0.60, 0.20, 0.30)
ORR	5	PAR	WS	((0.15, 0.25, 0.10), 0.60, 0.20, 0.30)

The contradiction degree of each risk is equally taken as 1/6. Then, the judgments of all decision-makers are combined by using Eq. 4. Aggregated plithogenic values of risks are transformed into crisp numbers by using Eq. 5. Aggregated plithogenic values ($\tilde{t}_j$) of risks and crisp numbers (t_j) are presented in Table 5.

Table 5. Aggregated Plithogenic values of risks and crisp numbers.

Risks	$\tilde{t}_j$	t_j
TRR	-	-
PUR	((0.070, 0.350, 0.629), 0.215, 0.175, 0.375)	0.357
INR	((0.227, 0.513, 0.811), 0.520, 0.125, 0.323)	0.194
IVR	((0.217, 0.413, 0.800), 0.413, 0.100, 0.333)	0.315
PAR	((0.079, 0.425, 0.755), 0.133, 0.200, 0.401)	0.390
ORR	((0.251, 0.550, 0.876), 0.618, 0.200, 0.323)	0.214

The rankings of the risks according to experts are combined with the geometric mean. Then, Eqs. 6–8 are used to determine the weights of logistics risks. The results of plithogenic PIPRECIA are shown in Table 6.

Table 6. The results of Plithogenic PIPRECIA.

Risks	Rankings by geometric mean	t_j	k_j	p_j	w_j
TRR	1	-	1	1	0.423
PUR	2	0.357	1.643	0.609	0.258
IVR	3	0.315	1.685	0.361	0.153
INR	4	0.194	1.806	0.200	0.085
PAR	5	0.390	1.610	0.124	0.052
ORR	6	0.214	1.786	0.069	0.029

According to Table 6, risks are listed from the most important to the least as follows: TRR, PUR, IVR, INR, PAR and ORR.

5 Conclusion

The today's logistics sector is very complex and has a high level of risk. Logistics risk assessment is a difficult task due to several logistic risks to consider and trade-offs. Multi criteria decision making (MCDM) approaches are widely used in the literature to address the difficulty in prioritizing different competing risks. In this analysis, a recently evolved approach called Plithogenic PIPRECIA is used to prioritize logistics risks. The logistics risks are assessed and prioritized in this study based on the opinions of three experts. Experts determined that there were six logistics risks, and these risks were prioritized by Plithogenic PIPRECIA. It has been found that the most important risk is Transportation-related Risks followed by Purchasing-related Risks, Inventory-related Risks, Information-related Risks, Packaging-related Risks, and Organization-related Risks. This method can be applied to different decision-making problems such as supplier selection, location selection in future studies. There are various studies about logistics risk in the literature however Plithogenic PIPRECIA is a new model and therefore there are only two studies in the literature. Therefore, this study contributes to the literature. Plithogenic PIPRECIA model can be used in other areas of decision problems such as location selection, performance evaluation, or machine selection problems in future studies.

References

1. Sodhi, M.S., Son, B.G., Tang, C.S.: Researchers' perspectives on supply chain risk management. Prod. Oper. Manag. **21**(1), 1–13 (2012)
2. Tsai, M.C.: Constructing a logistics tracking system for preventing smuggling risk of transit containers. Transp. Res. Part A: Policy Pract. **40**(6), 526–536 (2006)
3. Jia, H., Zhang, L., Lou, X., Cao, H.: A fuzzy-stochastic constraint programming model for hazmat road transportation considering terrorism attacking. Syst. Eng. Procedia **1**, 130–136 (2011)
4. Ambituuni, A., Amezaga, J.M., Werner, D.: Risk assessment of petroleum product transportation by road: a framework for regulatory improvement. Saf. Sci. **79**, 324–335 (2015)
5. Afenyo, M., Khan, F., Veitch, B., Yang, M.: Arctic shipping accident scenario analysis using Bayesian Network approach. Ocean Eng. **133**, 224–230 (2017)
6. Park, Y.B., Yoon, S.J., Yoo, J.S.: Development of a knowledge-based intelligent decision support system for operational risk management of global supply chains. Eur. J. Ind. Eng. **12**(1), 93–115 (2018)
7. Tubis, A.: Risk assessment in road transport–strategic and business approach. J. KONBiN **45**(1), 305–324 (2018)
8. Ghaleh, S., Omidvari, M., Nassiri, P., Momeni, M., Lavasani, S.M.M.: Pattern of safety risk assessment in road fleet transportation of hazardous materials (oil materials). Saf. Sci. **116**, 1–12 (2019)
9. Huang, M., Tu, J., Chao, X., Jin, D.: Quality risk in logistics outsourcing: a fourth party logistics perspective. Eur. J. Oper. Res. **276**(3), 855–879 (2019)

10. Liu, J., Zhou, H., Sun, H.: A three-dimensional risk management model of port logistics for hazardous goods. Marit. Policy Manag. **46**(6), 715–734 (2019)
11. Ofluoglu, A., Baki, B., Ar, İM.: Determining of disaster logistics risks based on literature review. J. Manag. Mark. Logistics **6**(1), 1–9 (2019)
12. Tumanov, A.: Risk assessment of accidents during the transportation of liquid radioactive waste in multimodal transport. In: IOP Conference Series: Earth and Environmental Science, vol. 272, no. 3, p. 032078 (2019)
13. Mohammadfam, I., Kalatpour, O., Gholamizadeh, K.: Quantitative assessment of safety and health risks in HAZMAT road transport using a hybrid approach: a case study in Tehran. ACS Chem. Health Saf. **27**(4), 240–250 (2020)
14. Ovidi, F., van der Vlies, V., Kuipers, S., Landucci, G.: HazMat transportation safety assessment: analysis of a "Viareggio-like" incident in the Netherlands. J. Loss Prevention Process Ind. **63**, 103985 (2020)
15. Zhao, M., Ji, S., Zhao, Q., Chen, C., Wei, Z. L.: Risk Influencing factor analysis of urban express logistics for public safety: a Chinese perspective. Math. Problems Eng. 2020 (2020)
16. Tüysüz, F., Kahraman, C.: Project risk evaluation using a fuzzy analytic hierarchy process: an application to information technology projects. Int. J. Intell. Syst. **21**(6), 559–584 (2006)
17. Sattayaprasert, W., Hanaoka, S., Taneerananon, P., Pradhananga, R.: Creating a risk-based network for hazmat logistics by route prioritization with AHP: case study: gasoline logistics in Rayong, Thailand. IATSS Res. **32**(1), 74–87 (2008)
18. Ren, S.: Assessment on logistics warehouse fire risk based on analytic hierarchy process. Procedia Eng. **45**, 59–63 (2012)
19. Sari, I.U., Behret, H., Kahraman, C.: Risk governance of urban rail systems using fuzzy AHP: the case of Istanbul. Int. J. Uncertain. Fuzziness Knowl.-Based Syst. **20**(Suppl. 1), 67–79 (2012)
20. Zhao, L., Wang, X., Qian, Y.: Analysis of factors that influence hazardous material transportation accidents based on Bayesian networks: a case study in China. Saf. Sci. **50**(4), 1049–1055 (2012)
21. Oztaysi, B., Cevik Onar, S., Kahraman, C.: Fuzzy multicriteria prioritization of Urban transformation projects for Istanbul. J. Intell. Fuzzy Syst. **30**(4), 2459–2474 (2016)
22. Ilbahar, E., Karaşan, A., Cebi, S., Kahraman, C.: A novel approach to risk assessment for occupational health and safety using Pythagorean fuzzy AHP & fuzzy inference system. Saf. Sci. **103**, 124–136 (2018)
23. Gul, M., Guneri, A.F., Nasirli, S.M.: A fuzzy-based model for risk assessment of routes in oil transportation. Int. J. Environ. Sci. Technol. **16**(8), 4671–4686 (2018). https://doi.org/10.1007/s13762-018-2078-z
24. Abdel-Basset, M., Mohamed, R., Zaied, A.E.N.H., Gamal, A., Smarandache, F.: Solving the supply chain problem using the best-worst method based on a novel Plithogenic model. In: Optimization Theory Based on Neutrosophic and Plithogenic Sets, pp. 1–19. Academic Press (2020)
25. Smarandache, F.: Plithogeny, Plithogenic Set, Logic, Probability, and Statistics, 141 p. Pons Publishing House, Brussels (2017)
26. Smarandache, F.: Plithogenic set, an extension of crisp, fuzzy, intuitionistic fuzzy, and neutrosophic sets – revisited. Neutrosophic Sets Syst. **21**, 153–166 (2018)

Risk Analysis for the Tech Startup Projects with Fuzzy Logic

Hür Bersam Bolat[1(✉)], Fatma Yaşlı[2], and Gül Tekin Temur[3]

[1] Istanbul Technical University, Istanbul, Turkey
bolat@itu.edu.tr
[2] Anadolu University, Eskisehir, Turkey
[3] Bahcesehir University, Istanbul, Turkey

Abstract. Nowadays; technological (Tech) startups are fast-growing businesses that target to meet the demands of the marketplace by developing innovative products, services, or platforms. Many factors have become prominent regarding the success and sustainability of the product or service offered by the startup: Investment, experience, and education of the team, leadership of the management, creativity, innovation, technological breakthroughs, surrounding community, future perspective, target marketing strategy, location and the analysis of the market, etc. Therefore, startups contend under considerable uncertainty. Considering the high failure rates of the startups, it is inevitable to examine them in terms of risk analysis. Defining the important risk factors is crucial to develop the right strategies for successful startups. In literature, there is a great effort to find out the key factors for successful and sustainable business models including intensive technology. In this study, a risk analysis for the failure of startup projects under the framework of business model canvas has been performed using Fuzzy Failure Mode and Effect Analysis (FMEA) with the field experts. While fuzzy logic provides to define the quantitative parameters of the risk analysis, the FMEA provides to present the main reasons which cause the failure of the startup projects with their priority numbers. The findings have theoretical and practical contributions to success in startup projects by showing the effects of the factors that cause startup projects to fail. The results are discussed to provide managerial strategies for mitigating the failure risks of the startup projects.

Keywords: Technological startups · Business model canvas · Fuzzy sets · Failure mode and effect analysis

1 Introduction

The last decade in the economic world has brought many innovations, new technologies, and startups and small starting businesses with innovative ideas began to spread from American Silicon Valley. Startups are among the greatest pioneers of business endeavors in the world economy. They bring innovative products and new approaches to customers, grow exponentially, become global enterprises [1]. Organizations such as Google, Facebook, Amazon, Uber Airbnb, and Spotify all started as digital ventures with humble beginnings from a garage, a student dormitory, or a dining room tables [2].

C. Kahraman et al. (Eds.): INFUS 2021, LNNS 308, pp. 671–679, 2022.
https://doi.org/10.1007/978-3-030-85577-2_79

Startups are actualized mostly under the industry of information technology, service, energy, or technology [3]. The common elements of all startups are that being novel, short-term, fundamentally using innovation and technology, and having a high risk of survival with the high growth potential. Especially, the role of high-tech start-up firms is to use technology for delivering something new to the consumer or to provide an existing product in a new way form. Even they include a huge expectation for expanding the organization but there is always a possibility for failure because of high risk. Therefore, it is inevitable to examine startup projects in terms of risk management. Defining the important risk factors in detail is crucial to reach the future developments that successful startups will lead to. Since the different causes of the failure may have a different level of risk on the startups, in this study a comprehensive risk analysis for the failure of high-tech startups has been performed using FMEA. Unlike the literature, the business model canvas which provides a descriptive and testing procedure for developing a strategy belonging to a new venture is also benefitted to define the causes of the failure of tech startups.

2 Failure Risks of the Startups

The creation of hi-tech, knowledge-based enterprises is becoming increasingly important to get a bigger pie from the international market through high-value products [4]. The reason of the tech startups continue to emerge all over the world is mentioned as a growing trend towards new innovative businesses [5].

All trade activities are affected by a lot of unpredictable factors [1]. Due to this uncertain environment which covers the startups, four out of every five startups cannot survive after the first five years [6]. It is possible to define the success of the startup as the fact that the startup did not end within 5 years, in other words, the sustainability of the product or service it offers. Regarding this, in literature there is a big effort for understanding the key challenges that the tech startups have to cope with, and defining the success and failure factors of the startups and the considered factors for a successful startup are defined as industry, marketing, R&D and startup experience of the team, trust between members, technological/business capabilities and academic formation/education of the team, management experience, leadership, and initial motivation of the of the entrepreneur, the amount of investment, unemployment of the entrepreneur, government financial and procedural support, venture capital, funding, level of competitive, science and technology policies, political/economic/legal issues, surrounding community, startup law, organization size and age, location, partners, logistics infrastructure, future perspective, target marketing strategy, clustering (business network), business development, product innovation, prototype, maturity of the tech in the product, business model, positioning in market etc. [2–4, 7–9]. An enabling environment would be necessary to assure that the high tech startup enterprises mature, grow and become sustainable in the long run [4]. To compose this permissive surrounding, failure reasons of the startups need to be defined comprehensively and the potential effects and the consequences of each reason are required to present. The important issues for a successful startup can be listed from a different point of view such as team, entrepreneur, external (environment), organization, product, customers/users, or business model. The main frequent reason for failure of the startup is defined as the wrong business model in the literature [6], the factors are considered from the perspective of business model canvas in this study. It is aimed to consider the tech startups from a holistic risk perspective to develop the appropriate strategies which will minimize the failure risk.

3 Risk Analysis of Tech Startups with Fuzzy FMEA

In literature, the FMEA is used to consider the potential failures and identify the effects on a system [10–12]. To evaluate the failure risk analysis of the tech startups, the method includes the identification of failure modes and the effects of each specified issue of a tech startup, specifying the occurrence (O), severity (S) and detection (D) levels of these failure modes and finally determining the risk priority numbers (RPNs) of the failures by multiplying the defined inputs shown as Eq. 1.

$$RPN = O * S * D \tag{1}$$

Within the scope of the study, literature and expert opinions are used to define the failure modes of the tech startups. The fuzzy set theory developed by Zadeh [13] is a very reasonable tool to reflect imprecision and uncertainty situations on the experts' judgment especially for determining the quantitative inputs of the analyses. A fuzzy set $\tilde{S}$ is a class of objects x with a continuum of grades of membership. It is characterized by a membership function $\mu_{\tilde{S}}(x)$, which assigns to each object a grade of membership tanging between zero and one. The fuzzy set is defined as:

$$\tilde{S} = \left\{x, \mu_{\tilde{S}}(x)|x \in \tilde{S}, \mu_{\tilde{S}}(x) \in [0, 1]\right\} \tag{2}$$

where $\tilde{S}$ represent the grade or degree which any object x in $\tilde{S}$. Larger values of $\mu_{\tilde{S}}(x)$ shows higher degrees of membership [14].

Several extensions of ordinary fuzzy sets such as type-2, intervalvalued fuzzy sets, intuitionistic fuzzy sets, hesitant fuzzy sets, pythagorean fuzzy sets, neutrophic sets and spherical fuzzy numbers have been developed over the years by the researchers. In literature, fuzzy FMEA is also a widely used method due to the easiness to reach the input data of the analyses via the experts [15–17]. In this study the high level of uncertainty in the risk assessment performed with expert judgments has been employed with the triangular fuzzy sets due to easiness of the representation and calculation.

A triangular fuzzy number (TFN) $\tilde{S}$ is denoted simply as (l, m, u). l, m, u represents the least possible value, the most possible value and the largest possible value of the membership function, respectively. The membership function of the TFN $\mu_{\tilde{S}}(x)$ is defined by Eq. 3.

$$\mu_{\tilde{S}}(r) = \begin{cases} \frac{r-1}{m-l} \; for \; l \leq r \leq m, \\ \frac{u-r}{u-m} \; for \; m \leq r \leq u, \\ 0 \quad otherwise. \end{cases} \tag{3}$$

The used fuzzy numbers and linguistic expressions are presented in Fig. 1 and Table 1, respectively.

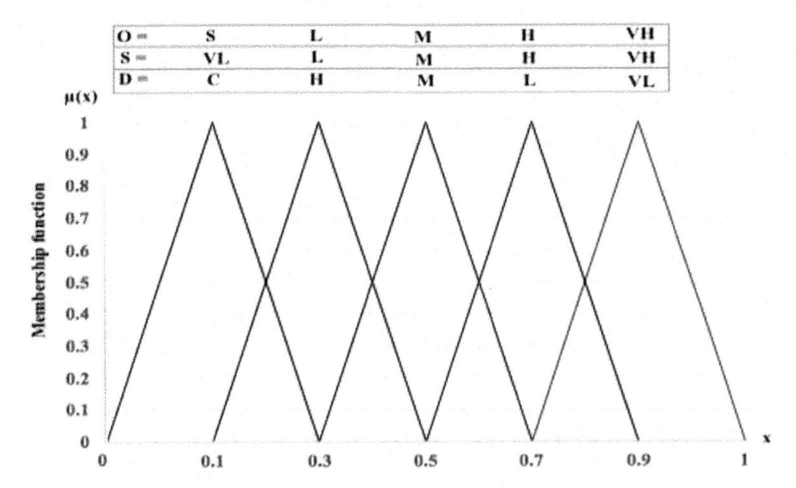

Fig. 1. Fuzzy ratings for occurrence, severity and detectability with their function.

Table 1. Linguistic expressions for specifying the inputs.

Levels	Likelihood of occurrence of failure (O)	Potential impact of failure – Severity (S)	Detection of potential failure (D)
1	Seldom (S)	Very Low (VL) - Negligible	Certain (C)
2	Low (L)	Low (L) - Compensable in early stage	High (H)
3	Moderate (M)	Moderate (M) - Compensable in growth stage	Medium (M)
4	High (H)	High (H) - Hard to compensate	Low (L)
5	Very high (VH)	Very high (VH) - No compensable - leads to failure	Very low (VL)

Since the vast majority of startups (80%) could not survive more than 5 years [6], within the scope of the study the ending of a startup before 5 years has been accepted as the failure, and the possible causes of this situation have been investigated through literature research and expert opinions. The business model canvas, which has an important place in the strategic planning of the ventures, has been used for the definition of the failure modes. The severity of the effects of each possible failure mode is determined by the negligence or compensability of its defined effect on the startup's profitability, market share and commercial value in the first 5 years. Finally, taking into account the ease of detecting each failure, gathering the inputs of the analysis has been completed. After interviewing with 3 field experts a comprehensive table for the definitions of the failure mode and their potential failure effects is constructed (see Table 2) with the likelihood of the occurrence, severity and detectability of the failure modes (see Table 3).

Table 2. Defined failure modes for Business Canvas Models/of tech startups with their potential effects and consequences.

	Code	Failure mode	Potential effects
Key partners	FM1	Insufficient number of investors	Increased financial risk
	FM2	Weak technology partner	Inability to be innovative Clint distrust
	FM3	Weak buyer -vendor relationship	Inability to expand their abilities beyond their internal resources [18]
	FM4	Inability to take advantage of joint venture	Inability to expand their abilities beyond their internal resources [18] Inability to achieve scale of economies and scope [18]
	FM5	Co-founder misalignment	Bad management
	FM6	Incompatibility with key partners	Commercial failure of the offers
Key activities	FM7.	Insufficient R&D studies	Inability to be innovative
	FM8	Insufficient scope definitions	Wrong customer segmentation Wrong capacity decisions
	FM9	Lack of customer experience design before producing	Lack of response and feedbacks from customer
	FM10	Non-compliance with existing restrictions	Interruption of operations
	FM11	Wrong definitions in key activities	Inability to create targeted value
Key resources	FM12	Lack of financial support	Insufficient cash flow in a short term
	FM13	Conflict of interest in team	Pure communication within team members

(*continued*)

Table 2. (*continued*)

	Code	Failure mode	Potential effects
	FM14	Lack of talent in team	Lack of creativity
	FM15	Lack of clustering support	Pure resource availability
	FM16	Inexperience technical knowledge of team	Bad quality
	FM17	Lack of managerial skill of entrepreneurship	Bad management Misalignment of tasks and resources
	FM18	Inappropriate venture capital strategy	Fragile growth
	FM19	Difficulty in accessing proper resources	Inability to create targeted value
Value Prop	FM20	Lack of contribution	Inability to meet the needs of customers
	FM21	Lack of detailed analysis on customer needs	Inability to offer unique value for customers
Cust Rel.	FM22	Lack of communication plan with customer	Increased satisfied customers
	FM23	Inappropriate marketing strategy	Inefficient customer relationship
Channel	FM24	Inadequate selling channel	Inability to meet the needs of customers
	FM25	Insufficient distribution channel	Unresponsiveness in the supply chain
	FM26	Inappropriate Communication channels [18]	Ineffective customer relationship
Customer Segments	FM27	Mis-analysis of target segments	Inappropriateness with corporate strategies Wrong marketing approach
	FM28	Mis-prioritization of the target segments	Incorrect allocation of budget
Cost Structure	FM29	Lack of detailed definition of cost structure items	Incorrect cost estimation
Revenue streams	FM30	Wrong scaling in market share	Inability to maximize profitability or yield [18]
	FM31	Malfunction of pricing mechanism [18]	Inability to maximize profitability or yield [18]

After gathering the data of the analysis by using the experts' judgments, their linguistic expressions are transformed to triangular fuzzy numbers given in Fig. 1, and the

Table 3. Determined values for the inputs of the analysis by experts and the results about the RPNs.

Code	Expert 1			Expert 2			Expert 3			RPN	Priorities
	O	S	D	O	S	D	O	S	D		
FM1	H	M	C	VH	H	H	H	VH	C	0.096	22
FM2	M	H	H	M	VH	M	M	H	H	0.139	18
FM3	M	M	M	L	M	H	M	M	M	0.094	23
FM4	VH	M	H	H	L	M	H	M	H	0.120	21
FM5	H	H	M	M	M	L	M	M	L	0.203	10
FM6	M	H	M	H	VH	L	H	H	H	0.240	8
FM7	VH	H	M	H	VH	L	VH	VH	H	0.333	3
FM8	H	M	VL	H	M	L	H	H	M	0.274	6
FM9	L	H	C	M	VH	H	M	H	H	0.079	25
FM10	S	H	C	L	H	C	L	M	C	0.019	31
FM11	M	VH	M	H	VH	L	H	H	M	0.293	5
FM12	VH	H	H	H	VH	H	H	VH	H	0.186	11
FM13	M	M	L	L	L	VL	M	M	M	0.130	20
FM14	H	H	VL	H	M	L	H	H	L	0.336	2
FM15	M	M	C	H	M	C	H	M	H	0.058	27
FM16	H	H	H	H	H	M	H	M	H	0.163	14
FM17	H	M	L	M	L	VL	M	L	L	0.158	16
FM18	H	L	C	H	M	H	H	L	C	0.047	30
FM19	H	H	M	VH	H	H	VH	H	M	0.248	7
FM20	H	H	L	VH	VH	M	VH	VH	H	0.333	3
FM21	M	M	H	M	M	C	M	M	M	0.077	26
FM22	H	M	H	H	H	H	H	H	H	0.133	19
FM23	M	M	M	L	H	L	M	M	L	0.156	17
FM24	L	M	H	L	H	C	M	H	H	0.056	28
FM25	L	M	H	L	H	H	L	M	H	0.051	29
FM26	H	M	H	H	H	M	H	H	H	0.163	14
FM27	VH	H	M	H	H	H	VH	H	H	0.210	9
FM28	M	M	L	L	H	L	M	M	L	0.172	12
FM29	VH	H	L	VH	H	L	VH	H	M	0.388	1
FM30	H	M	H	H	H	M	H	M	M	0.172	12
FM31	M	L	H	M	M	M	M	M	H	0.079	24

average values of the fuzzy numbers related to the experts' expressions about the criteria of O, C and D of each failure mode are determined to get common assessments.

Where $\tilde{S}_1$ and $\tilde{S}_2$ are two TFNs, the fuzzy arithmetic operations related to the calculating the average values of the fuzzy numbers are shown as followings [19]:

$$\tilde{S}_1 \oplus \tilde{S}_2 = (l_1, m_1, u_1) \oplus (l_2, m_2, u_2) = (l_1 + l_2, m_1 + m_2, u_1 + u_2) \quad (4)$$

$$\tilde{S}_1/k = (l_1, m_1, u_1)/k = (l_1/k, m_1/k, u_1/k) \quad (5)$$

Finally, defuzzification process by using "mean area method" (see in Eq. 6) are performed for the common TFNs assessments $\tilde{S} = (l, m, u)$ under the criteria of O, S and D to determine the crisp values S^*.

$$S^* = \frac{i + 2m + u}{4} \quad (6)$$

Thus, the specified quantitative values by using the fuzzy approach are simply multiplied to get the values of the RPNs by using the Eq. 1. Calculated RPNs values and the priorities of the failure modes are represented in Table 3.

4 Results and Conclusion

Failures of the startups can be defined as frequent events when looking at the presented ratios in the literature. Unlike the literature, we consider that the different causes of the failure may have a different level of risk on the startups by benefitting from the FMEA. In order to define the causes of the failure of tech startups we have utilized the business model canvas which provides a descriptive and testing procedure for developing a strategy belonging to a new venture. After interviewing the field experts who have at least 15 years of venture experience, we analyzed the failure risks of the tech startups by using the comprehensive framework of the canvas and the proactive approach of the FMEA. According to the results, the most important causes of the tech startup's failure are determined as lack of the detailed definition of cost structure, lack of the talent in team, lack of value contribution, insufficient R&D, and wrong definitions of the key activities.

As the authors, it is believed that the strongest aspect of this study is the definition of the failure modes of the startup from the framework of the business model canvas and the demonstration that these failure modes can be digitized to be used in quantitative risk assessments by using the expert opinions and the fuzzy approach. The authors continue their studies on the FMEA method to present results with rule-based inferences and the other types of fuzzy numbers which represent better the uncertainty in linguistic data. By fuzzification of the RPN values together with using the extension fuzzy sets, used fuzzy FMEA method will provide more effective results for the risk analysis of the tech startups.

References

1. Mikle, L.: Startups and reasons for their failure. In: SHS Web of Conferences. EDP Sciences, p. 01046
2. Zaheer, H., Breyer, Y., Dumay, J., Enjeti, M.: Straight from the horse's mouth: founders' perspectives on achieving 'traction'in digital start-ups. Comput. Hum. Behav. **95**, 262–274 (2019)
3. Santisteban, J., Mauricio, D.: Systematic literature review of critical success factors of information technology startups. Acad. Entrepr. J. **23**(2), 1–23 (2017)
4. Yagnik, J.: Growth challenges of high-tech start-ups. ASCI J. Manag. **46** (2017)
5. Hormiga, E., Batista-Canino, R.M., Sánchez-Medina, A.: The role of intellectual capital in the success of new ventures. Int. Entrepr. Manag. J. **7**(1), 71–92 (2011)
6. Cantamessa, M., Gatteschi, V., Perboli, G., Rosano, M.: Startups' roads to failure. Sustainability **10**(7), 2346 (2018)
7. Wang, X., Edison, H., Bajwa, S.S., Giardino, C., Abrahamsson, P.: Key challenges in software startups across life cycle stages. In: Sharp, H., Hall, T. (eds.) XP 2016. LNBIP, vol. 251, pp. 169–182. Springer, Cham (2016). https://doi.org/10.1007/978-3-319-33515-5_14
8. Chung, W.Y., Jo, Y., Lee, D.: Where should ICT startup companies be established? Efficiency comparison between cluster types. Telemat. Inform. **56**, 101482 (2021)
9. Dias, N.M.C.F.: Failure factors of technological driven start-ups: datris solutions case study (2018)
10. Giannakis, M., Papadopoulos, T.: Supply chain sustainability: a risk management approach. Int. J. Prod. Econ. **171**, 455–470 (2016)
11. Sharma, S.K.: Risk adjusted total cost of ownership model for strategic sourcing decisions. Int. J. Procur. Manag. **9**(2), 123–145 (2016)
12. Zarei, E., Azadeh, A., Khakzad, N., Aliabadi, M.M., Mohammadfam, I.: Dynamic safety assessment of natural gas stations using Bayesian network. J. Hazard. Mater. **321**, 830–840 (2017)
13. Zadeh, L.A.: The concept of a linguistic variable and its application to approximate reasoning-III. Inf. Sci. **9**(1), 43–80 (1975)
14. Bojadziev, G., Bojadziev, M.: Fuzzy logic for business, finance, and management, vol. 23. World Scientific (2007)
15. Singh, A., Patil, A.J., Sharma, R.K., Jarial, R.: an innovative fuzzy modeling technique for transformer's failure modes and effects analysis. In: 2020 International Conference on Electrical and Electronics Engineering (ICE3). IEEE (2020)
16. Akyuz, E., Akgun, I., Celik, M.: A fuzzy failure mode and effects approach to analyse concentrated inspection campaigns on board ships. Maritime Policy Manag. **43**(7), 887–908 (2016)
17. Ahmadi, M., Behzadian, K., Ardeshir, A., Kapelan, Z.: Comprehensive risk management using fuzzy FMEA and MCDA techniques in highway construction projects. J. Civ. Eng. Manag. **23**(2), 300–310 (2017)
18. Ladd, T.: Does the business model canvas drive venture success? J. Res. Market. Entrepr. (2018)
19. Zadeh, L.A.: Fuzzy sets. Inf. Control **8**(3), 338–353 (1965)

Circumcenter Based Ranking Fuzzy Numbers for Financial Risk Management

Lazim Abdullah[1] (✉), Ahmad Termimi Ab Ghani[1], and Nurnadiah Zamri[2]

[1] Management Science Research Group, Faculty of Ocean Engineering Technology and Informatics, Universiti Malaysia Terengganu, 20130 Kuala Nerus, Terengganu, Malaysia
lazim_m@umt.edu.my

[2] Faculty of Informatics and Computing, Universiti Sultan Zainal Abidin, 21030 Kuala Nerus, Terengganu, Malaysia

Abstract. Risk management is one of the critical elements in financial institutions, in which early detection of financial crises can be monitored. Literature suggests that there are many types of risks in financial management such as liquidity risk, interest rate risk, equity risk, commodity risk, just to name a few. However, authentic risk that can be associated to financial crises are very inconclusive and unpredictable. This study aims to identify the financial risks that affecting financial management using the proposed ranking trapezoidal fuzzy numbers method that based on circumcenter. Five decision makers which are experts in financial management were requested to provide evaluation against a set of eight financial risks. The proposed method of circumcenter, which considers area, spread, and distance of trapezoidal fuzzy numbers are applied to financial risk management. The result shows that 'credit risk' is the highest risk value compared to other seven risks. It is also suggested that 'commodity risk' is the lowest risk in financial management.

Keywords: Circumcenter of centroid · Fuzzy number · Financial risk · Financial management · Decision making

1 Introduction

One of the most important considerations in financial managements is the risks that financial institutions may encounter in their operations. Market's fluctuations and uncertainties are believed to be one of the contributing factors toward the likelihood of financial risks occurrence. Financial risk can be defined literally as the volatility or variability of unexpected outcomes or net profit [1]. It is one of the challenges in almost all financial institutions. To ease the problems that affecting financial performance, identification of specific risks in financial institutions should be given extra attention. A good risk management can help financial institutions to achieve their goals and objectives, and more importantly sustaining the survival of financial institutions. The main aim of financial risk management execution is to sustain financial performance in the banking sector. A good risk management can promote early warning system of discovering and monitoring of related risks. Financial risk management can help risk manager to observe the system

C. Kahraman et al. (Eds.): INFUS 2021, LNNS 308, pp. 680–686, 2022.
https://doi.org/10.1007/978-3-030-85577-2_80

and arrange for ways to prevent further stress on the system, which ultimately organizing financial performance [2]. Thus, identifying of financial risk is very important to minimize losses of any financial institutions.

Many methods and approaches have been used in financial risk management research. Li [3] for example, studied the determinants of bank's profitability and its implications on the practices of risk management in the United Kingdom. The researcher utilized regression analysis on a time series data between 1999 and 2006. Alexander and Sheedy [4] reported that financial decision is influenced by unpredicted restrictive elements and not relying directly on the agents of economic. They postulated that financial decision becomes a risk when there is an impact of numerous factors such as market, inflation, exchange rates, competition, time factor, commissions, interest rate, human factors and company culture. The researchers designed a methodology on conducting the stress testing in the risk model by estimating Value-at-Risk and expected tail loss to focus on the movement of market. The results show that conditional empirical model is the most preferred risk models over the short risk horizon. Another research on the effect of credit and market risk on bank performance was conducted in Turkey [5]. This research investigates the effects of credit and market risk like foreign exchange rate risk and interest rate on bank performance. Ekinci [5] employed GARCH model to investigate the effect of risks to financial performance. The results indicate that credit risk has negative effect on financial performance. In contrast, foreign exchange rate has a positive effect on financial performance. Interestingly, they conclude that interest rate has insignificant effect on bank's profitability. Safari et al. [6] introduced the significance of risk management for banks and other financial institutions using the risk models like Delta-Normal, Historic/Back Simulation and Monte-Carlo method. Mbuvi and Wamiori [7] studied the effect of financial risks on the corporate value in microfinance banks in Kenya using descriptive statistics approaches such as correlation coefficient and the multiple linear regression. Very recently, value at risk method which is estimated by GJR-GARCH model based on skewed *t*-distribution was used to investigate extreme risk spillovers between gold and stock markets [8].

Most of the approaches used in financial risk management research are statistical based methods where substantial numbers of data collection are considered. In contrasts to the above methods, this present study applies circumcenter-based ranking fuzzy numbers in measuring the risk in financial management where handful of decision makers' opinions are sought. This method is associated with trapezoidal fuzzy numbers that representing linguistic evaluation made by decision makers instead of statistical data. A circumcenter equation is used to find a point (centroid) in which trapezoidal fuzzy numbers are reduced to one point in Cartesian plane. Literally, circumcenter is the point at which the perpendicular bisectors of the sides of a trapezoidal intersect and also equidistant from the four vertices. This paper is organized as follows. Section 2 provides definition of trapezoidal fuzzy numbers and its respective membership functions. Section 3 presents the computational procedures of ranking fuzzy numbers. Section 4 provides the implementation in identifying risks in financial management. Finally, Sect. 5 concludes.

2 Preliminary

One of the important definitions that discuss in this paper is trapezoidal fuzzy numbers. The method proposed in Sect. 3 are primarily used ranking trapezoidal fuzzy numbers where the concept of circumcenter in finding the centroid of trapezoidal fuzzy numbers are embedded. Therefore, this section provides definition of trapezoidal fuzzy numbers and its respective membership functions [9]. It is given as follows.

Let $\tilde{A}$ be a generalized trapezoidal fuzzy number, $\tilde{A} = \left(a, b, c, d; w_{\tilde{A}}\right)$ where a, b, c, d are real values, $w_{\tilde{A}}$ is the height of the generalized trapezoidal fuzzy number $\tilde{A}$, and $w_{\tilde{A}} \in [0, 1]$. If $0 \leq a \leq b \leq c \leq d \leq 1$, then $\tilde{A}$ is a standardized generalized trapezoidal fuzzy number.

The membership function of a trapezoidal fuzzy number is normally written as follows:

$$f_{\tilde{A}}(x) = \begin{cases} 0, & x < a \\ \frac{x-a}{b-a}, & a \leq x \leq b \\ 1, & b \leq x \leq c \\ \frac{d-x}{d-c}, & c \leq x \leq d \\ 0, & x > d \end{cases} \tag{1}$$

Let $\tilde{A}$ and $\tilde{B}$ are two generalized trapezoidal fuzzy numbers, where $\tilde{A} = \left(a, b, c, d; w_{\tilde{A}}\right)$ and $\tilde{B} = \left(a, b, c, d; w_{\tilde{B}}\right)$ $a_1, b_1, c_1, d_1, a_2, b_2, c_2, d_2$ are real values. Their intervals of vertices are $0 \leq w_{\tilde{A}} \leq 1$ and $0 \leq w_{\tilde{B}} \leq 1$.

The trapezoidal fuzzy numbers are employed in defining linguistic evaluation that subsequently will be ranked using a ranking fuzzy number method. Details of the method is presented in the following section.

3 Ranking Fuzzy Numbers Based on Circumcenter of Centroid

This research adapts the algorithm proposed by Abdullah and Azman [10]. Some minor modifications are made to adapt with the case of risk management. Instead of using three decision makers, this algorithm fits with five decision makers. The linguistic scale used in this research is different from Abdullah and Azman [10], in which seven linguistic scales are used instead of nine linguistic scales due to nature of risks in financial management. The algorithm is divided into two phases where Phase I is meant to aggregate decision matrix using arithmetic mean operator, and Phase II purposely used to rank trapezoidal fuzzy numbers. Each phase of computational steps is given as follows.

Phase I

Assumed that there are n numbers of decision makers provide qualitative evaluation using linguistic variables pertaining to m risk of financial management. The computation is made according to the following steps.

Start
Identify m risks, n decision makers.
Construct a decision matrix.
Transform linguistic terms to trapezoidal fuzzy numbers.
Find mean of fuzzy numbers.
Let $D_{A\tilde{n}}$ is the trapezoidal fuzzy numbers (*TFN*) for each decision maker.

Compute the mean of *TFN* $$\overline{TFN}_{D_{\tilde{A}n}} = \frac{D_1 + D_2 + D_3 + \cdots + D_n}{n} \quad (2)$$

End

The next computational step is given in Phase II where risk values and ranking are obtained.

Phase II

Considers a trapezoidal fuzzy numbers $\tilde{A} = (a, b, c, d; w)$, where a, b, c and d are real numbers and w is the maximum membership value for the trapezoidal fuzzy numbers $\tilde{A}$ with $0 < w \leq 1$. The ranking fuzzy numbers is computed as the following steps:

Start
Find w where $w = \max(w)$ (3)
Find circumcenter $C = (x_{\tilde{A}}, y_{\tilde{A}})$

$$C(x, y) = \left(\frac{a + 2b + 2c + d}{6}, \frac{a(d - b) + bc - cd + 3w^2}{6w} \right) \quad (4)$$

Find Euclidean distance

$$d_{\tilde{A}} = \sqrt{(x_{\tilde{A}})^2 + (y_{\tilde{A}})^2} \quad (5)$$

Find spread $r_{\tilde{A}} = (d - a)$ (6)

Find area of the trapezoidal fuzzy numbers,

$$A_{\tilde{A}} = \frac{|(a + b - c - d)w|}{18} \quad (7)$$

Find ranking values,

$$R_{\tilde{A}} = \frac{1}{3}\left[(2 \times d_{\tilde{A}} + r_{\tilde{A}}) \times w_{\tilde{A}} + A_{\tilde{A}}\right] \quad (8)$$

Determine the ranking of two trapezoidal fuzzy numbers with the following If-then rules,

If $R(\tilde{A}_i) > R(\tilde{A}_j)$, then $\tilde{A}_i \succ \tilde{A}_j$,

If $R(\tilde{A}_i) < R(\tilde{A}_j)$, then $\tilde{A}_i \prec \tilde{A}_j$,

If $R(\tilde{A}_i) = R(\tilde{A}_j)$, then $\tilde{A}_i = \tilde{A}_j$

End

The above computational procedures are applied to a case of financial risk management.

4 Implementation

This section presents the implementation of the proposed algorithm to the financial risk management in finance. The eight risks are credit risk (A_1), liquidity risk (A_2), interest rate risk (A_3), equity risk (A_4), commodity risk (A_5), foreign exchange risk (A_6), solvency risk (A_7), and operational risk (A_8). The decision makers provide linguistic evaluation regarding the risks in financial management. The risk evaluation is made based on linguistic terms 'Very low' (VL) to 'Very high' (VH). Table 1 shows the seven linguistic terms of risk and its respective trapezoidal fuzzy numbers.

Table 1. Linguistic terms and its corresponding fuzzy numbers

Linguistic term	Trapezoidal fuzzy numbers
Very low (VL)	(0, 0, 0.1, 0.2; 1.0)
Low (L)	(0.1, 0.2, 0.2, 0.3; 1.0)
Fairly low (FL)	(0.2, 0.3, 0.4, 0.5; 1.0)
Medium (M)	(0.4, 0.5, 0.5, 0.6; 1.0)
Fairly high (FH)	(0.5, 0.6, 0.7, 0.8; 1.0)
High (H)	(0.7, 0.8, 0.8, 0.9; 1.0)
Very high (VH)	(0.8, 0.9, 1.0, 1.0; 1.0)

Table 2 shows the linguistic terms provided by decision makers (D_1, D_2, D_3, D_4, D_5).

Table 2. Linguistic terms given by decision makers

Risks	D1	D2	D3	D4	D5
A_1	VH	VH	FH	H	VH
A_2	FH	VH	M	FH	M
A_3	VH	H	VH	H	M
A_4	H	FH	M	H	FH
A_5	M	M	FH	M	FH
A_6	H	FH	H	H	M
A_7	H	H	FH	M	M
A_8	FH	FH	M	M	FH

In this study, ranking fuzzy numbers based on circumcenter is used to calculate the weights of each risk and then ranked the risk based on the ranking value. In Phase I, the risks are translated into trapezoidal fuzzy numbers using Eq. (2). Table 3 presents the risks and its respective trapezoidal fuzzy numbers.

Table 3. Trapezoidal fuzzy numbers and risks

Risks	Trapezoidal Fuzzy numbers
A_1	$(0.72, 0.82, 0.9, 0.94; 1.0)$
A_2	$(0.52, 0.62, 0.68, 0.76; 1.0)$
A_3	$(0.68, 0.78, 0.82, 0.88; 1.0)$
A_4	$(0.56, 0.66, 0.7, 0.8; 1.0)$
A_5	$(0.44, 0.54, 0.58, 0.68; 1.0)$
A_6	$(0.6, 0.7, 0.72, 0.82; 1.0)$
A_7	$(0.54, 0.64, 0.66, 0.76; 1.0)$
A_8	$(0.46, 0.56, 0.62, 0.72; 1.0)$

In Phase II of computation, Eq. (3)–Eq. (8) are implemented to compute the risk values. It is presented in Table 4.

Table 4. Risk value for risks in financial risk management

Risks	Risk value (R_c)
A_1	0.7351
A_2	0.629
A_3	0.6955
A_4	0.6466
A_5	0.5842
A_6	0.6558
A_7	0.6235
A_8	0.6063

The risk values indicate the strength of risks. It can be arranged in ascending order as $A_5 \prec A_8 \prec A_7 \prec A_2 \prec A_4 \prec A_6 \prec A_3 \prec A_1$.

It can be seen that credit risk (A_1) is the highest risk in financial management followed by interest rate risk (A_3). This result is consistent with a research conducted by Ekinci [5] where credit risk has negative effect on bank's profitability. Credit risk occurs when a borrower unable or fails to make a contractual payment and as a result, banks will suffer a financial loss.

5 Conclusions

Financial risk management has been practiced by many banks considering the awareness that several types of risks are seriously affecting the performance of banks. The identification of the risks is extremely important in the banking sector. The purpose of this paper is to identify the highest risk in financial management using the proposed method of ranking fuzzy numbers based on circumcenter method. The proposed method was implemented to obtain the ranking of risks in financial risk management. Five decision makers evaluated the risks in financial risk management through the defined linguistic terms and trapezoidal fuzzy numbers. The risk values are obtained through a series of computation using the ranking fuzzy numbers based on circumcenter of centroid. The method identified 'credit risk' is the highest risk in financial management followed by 'interest rate risk'. In contrast, it is found that 'Commodity risk' is the lowest risk in financial risk management. It is hoped that the risk of financial losses could be minimized by acknowledging the impact of risk values to the bank's performance. An extension to the ranking fuzzy numbers by considering interval type-2 fuzzy numbers is one of the possible areas that could be explored in the future research.

References

1. Holton, G.A.: Defining risk. Financ. Anal. J. **60**(6), 19–25 (2004)
2. Bikker, J.A., Metzemakers, P.A.J.: Bank provisioning behavior and procyclicality. J. Int. Financ. Mark. **15**, 141–157 (2005)
3. Li, Y.: Determinants of banks' profitability and its implication on risk management practices: Panel evidence from the UK in the period 1999–2006. Doctoral Dissertation. United Kingdom: The University of Nottingham (2007)
4. Alexander, C., Sheedy, E.: Developing a stress testing framework based on market risk models. J. Bank. Financ. **32**(10), 2220–2236 (2008)
5. Ekinci, A.: The effect of credit and market risk on bank performance: evidence from Turkey. Int. J. Econ. Financ. Issues **6**(2), 427–434 (2016)
6. Safari, R., Shateri, M., Baghiabadi, H.S., Hozhabrnejad, N.: The significance of risk management for banks and other financial institutions. Int. J. Res. Granthaalayah **4**(4), 74–81 (2016)
7. Mbuvi, V.I., Wamiori, G.: Effect of financial risks on corporate value in microfinance banks in Kenya: a survey of microfinance banks in Mombasa town. Int. J. Soc. Sci. Inf. Technol. **3**(8), 2251–2260 (2017)
8. Ma, X., Yang, R., Zou, D., Liu, R.: Measuring extreme risk of sustainable financial system using GJR-GARCH model trading data-based. Int. J. Inf. Manage. **50**, 526–537 (2020)
9. Chen, S.H.: Ranking fuzzy numbers with maximizing set and minimizing set. Fuzzy Sets Syst. **17**(2), 113–129 (1985)
10. Abdullah, L., Azman, F.N.: Circumcenter of centroid of fuzzy number for identifying risks of obesity: a qualitative evaluation. J. Qual. Quant. **50**(6), 2433–2449 (2015)

Fuzzy Pretopological Space for Pedestrians' Risk Perception Modeling

Azedine Boulmakoul[1(✉)], Souhail ElKaissi[1], and Ahmed Lbath[2]

[1] Computer Science Department, FSTM, Hassan II University of Casablanca, Casablanca, Morocco
[2] LIG/MRIM, CNRS, University Grenoble Alpes, Grenoble, France
Ahmed.Lbath@univ-grenoble-alpes.fr

Abstract. The pedestrians' behavior for road crossing risk-taking relies on cognitive abilities, including the ability to focus attention on the traffic environment and to understand the semantics of perceived spatiotemporal objects and to ignore irrelevant attractions and events, so that pedestrians can focus on the spatiotemporal objects that can guarantee a minimum level of risk to increase the assurance of safety. However, young children's perception of the world makes it difficult to find a safe place to cross the road. The objectives of this research were to examine the effect of perceived points of interest on children's ability to identify safe crossing passages. This ability to identify safe or unsafe road crossing sites uses fuzzy topology. We develop a theory of risk perception in a spatial neighborhood, by constructions based on fuzzy neighborhoods and fuzzy pretopologies. Thus, we develop a solution to the problem of navigation in cities by improving safety criteria.

Keywords: Fuzzy topology · Fuzzy neighborhood · Fuzzy perception · Pedestrian's safety · Kafka · Spark streaming

1 Introduction

When pedestrians cross the road basically, they usually should use road infrastructure in proper manner: Subways, Zebra Crossings, foot over bridges. Short cuts and easy options of crossing roads are dangerous and should not be resorted to. Children and people with disabilities will have trouble crossing the road even with the infrastructure set up, and in some cases, crossing the road within the already established infrastructure could be a real challenge [14, 19, 20]. Here comes the necessity of using location awareness ecosystems [6]. The technology is growing with a high rate, we can notice that the location

This work was partially funded by Ministry of Equipment, Transport, Logistics and Water–Kingdom of Morocco, The National Road Safety Agency (NARSA) and National Center for Scientific and Technical Research (CNRST). Road Safety Research Program# An intelligent reactive abductive system and intuitionist fuzzy logical reasoning for dangerousness of driver-pedestrians interactions analysis: Development of new pedestrians' exposure to risk of road accident measures.

C. Kahraman et al. (Eds.): INFUS 2021, LNNS 308, pp. 687–695, 2022.
https://doi.org/10.1007/978-3-030-85577-2_81

awareness became more and more fluid, this is due to the usage of connected objects around us, like GPS devices, smart phones, sensors, etc. These devices collect a lot of information about individual persons, communities, and the eco-system that we live in. It collects what each individual do, or where he goes, his heartbeat rating, etc. These collected data can be very massive in terms of field diversity and in term of size, so it should be stored in a scalable format where the data can be manipulated and transformed smoothly [6, 22]. That said, graph databases provide an excellent infrastructure to link diverse data. With easy expression of entities and relationships between data, graph databases make it easier for programmers, users, and machines to understand the data and find insights. This deeper level of understanding is vital for successful machine learning initiatives, where context-based machine learning is becoming important for feature engineering, machine-based reasoning, and inferencing [1, 6, 13]. These collected data can be used to enhance the quality and comfort of the pedestrian on the pedestrian walkaway. We can help to increase the safety of the pedestrian by minimizing the number of eventual accidents that could happen, also it will help to increase the walkability of the surrounding. We could find out ways or keys of amelioration of pedestrian walkway. The data allows to understand how the pedestrian walks. To sum up the pedestrian study helps to increase the sustainability of the walking area. This kind of systems are extremely useful nowadays, it could be such a relief in situations of finding extremely fast some points of interest like for example the time when it's safe to go on the pedestrian walkway [6, 19, 20]. Trajectory-based human mobility data analysis research has largely focused on the trajectories of people and vehicles, driven by the fact that geographic information science has traditionally supported spatial information from moving objects [6, 11, 13, 15]. Currently, the interest of spatiotemporal data analysis for road safety is a real challenge for modern cities [11, 19]. the topology contributes to modeling of the perception of the danger influenced by the objects located in the spatiotemporal pedestrians' neighborhood. However, the subjectivity of perception and discernment for decision making requires the integration of the fuzzy theory. So, we are developing an approach based on fuzzy pretopology for this purpose.

The rest of this article is organized as follows; the following section define the fuzzy neighborhood and the fuzzy pretopology, Sect. 3 describes the fuzzy topology perception where we give the different used types of attractivity, Sect. 4 describes the main components of the proposed system, namely the graph database and the messaging broker. It details the implementation of the proposed architecture. The article ends with a conclusion and offers further work in Sect. 5.

2 Fuzzy Neighborhood and Fuzzy Pretopology

2.1 Preliminaries

Definition. We note $\mathfrak{F}(\Omega)$ the set of fuzzy subsets of Ω [21]. A fuzzy pretopology [3, 5, 7, 23] on a set Ω is described by an application $\mathcal{L}: \mathfrak{F}(\Omega) \rightarrow \mathfrak{F}(\Omega)$, such that:

(P1) $\mathcal{L}(\emptyset) = \emptyset$

(P2) $\mathcal{L}(A) \sqsupset A$ for every $A \in \mathfrak{F}(\Omega)$.

$(\Omega, \mathcal{L})$ is then said to be a fuzzy pretopological space.

Definition. Let $(\Omega, \mathcal{L})$ be a fuzzy pretopological space, we have the following properties:

(P3) for every $A \text{ and } B \in \mathfrak{F}(\Omega)$ such that $A \sqsupset B$ we have $\mathcal{L}(A) \sqsupset \mathcal{L}(B)$ $(\Omega, \mathcal{L})$ is then said to be a fuzzy pretopological space of type I.

(P4) for every $A \text{ and } B \in \mathfrak{F}(\Omega)$ such that $\mathcal{L}(A \bigsqcup B) = \mathcal{L}(A) \bigsqcup \mathcal{L}(B)$ $(\Omega, \mathcal{L})$ is then said to be a fuzzy pretopological space of type D.

(P5) for every $A \in \mathfrak{F}(\Omega)$ such that $\mathcal{L}^2(A) = \mathcal{L}(\mathcal{L}(A)) = \mathcal{L}(A)$ $(\Omega, \mathcal{L})$ is then said to be a fuzzy pretopological space of type S.

A fuzzy pretopological space of type I, D, S is a fuzzy topological space and $\mathcal{L}$ is its kurratowski closure [4, 16].

2.2 Fuzzy Pretopology in Metric Space

Let us consider space Ω endowed with a metric defined by a distance ∂. Let ε be a positive real number. For each element x of Ω, The fuzzy ε-neighborhood set of point x $\in \Omega$ with parameter ε is as follows [2, 8–10, 12] [17, 18]:

$N^{\varepsilon}_{x(t)} = \{y \in \Omega(y(t), N^{\varepsilon}_{x}(y(t)))\}$ with following neighborhood membership (see Figs. 1, 2):

Crisp Case. This set can be written as $N^{\varepsilon}_{x}(y(t)) = 1_{\{\partial(x(t),y(t)) \le \varepsilon\}}(y(t))$.

Where $1_{\{\partial(x(t),y(t)) \le \varepsilon\}}(y) = \begin{cases} 1 \; if\, \partial(x(t), y(t)) \le \varepsilon \\ 0 \quad otherwise \end{cases}$.

Fuzzy Case. Fuzzy neighborhood membership function is as follows:

$N^{\varepsilon}_{x(t)}(y(t)) = \left[1 - \frac{\partial(x(t),y(t))}{\varepsilon}\right] \times 1_{\{\partial(x(t),y(t)) \le \varepsilon\}}(y(t))$, with spatial constraint.

$N^{\theta,\varepsilon}_{x(t)}(y(t)) = 1_{\left[\langle \overrightarrow{xu(t)}, \overrightarrow{xy(t)}, \overrightarrow{yv(t)} \rangle \le \theta\right]}(y) \times \left[1 - \frac{\partial(x(t),y(t))}{\varepsilon}\right] \times 1_{\{\partial(x(t),y(t)) \le \varepsilon\}}(y(t))$,

$N^{\theta,\varepsilon}_{x(t)}(y(t)) = 1_{\left[\langle \overrightarrow{xu(t)}, \overrightarrow{xy(t)}, \overrightarrow{yv(t)} \rangle \le \theta\right]}(y) \times N^{\varepsilon}_{x}(y(t))$ for direction constraint.

Proposition. $\left(N^{\varepsilon}_{x(t)}\right)_{x \in \Omega}$ is a basis of neighborhoods of Ω, we then can build a pseudo-closure $\mathcal{L}$ on Ω with $\left(N^{\varepsilon}_{x(t)}\right)$ such that $\forall \in \mathfrak{F}(\Omega)\mathcal{L}(A) = \left\{N^{\varepsilon}_{x(t)} \sqcap A \neq \varnothing\right\}$.

$(\Omega, \mathcal{L})$ is a fuzzy pretopological space of type I, D.

α-Selection of a Fuzzy ε-Neighborhood Set

$$FN^{\varepsilon,\alpha}_{x(t)} = \left\{y \in \Omega(y(t), N^{\varepsilon}_{x}(y(t))) \middle| N^{\varepsilon}_{x(t)}(y(t)) \ge \alpha\right\}$$

We have $\beta \le \alpha \rightarrow FN^{\varepsilon,\alpha}_{x(t)} \sqsubset FN^{\varepsilon,\beta}_{x(t)}$

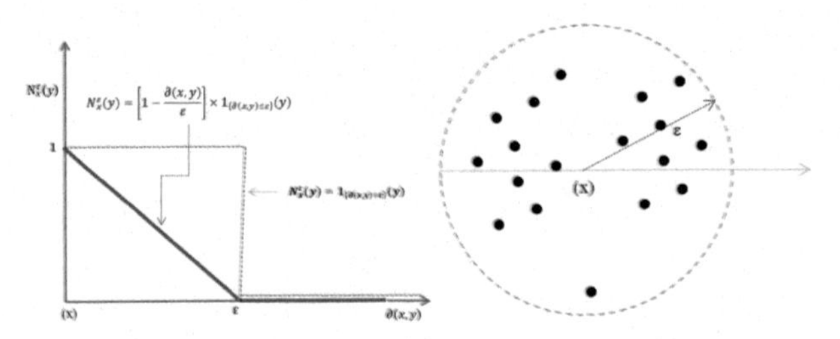

Fig. 1. Fuzzy neighborhood function.

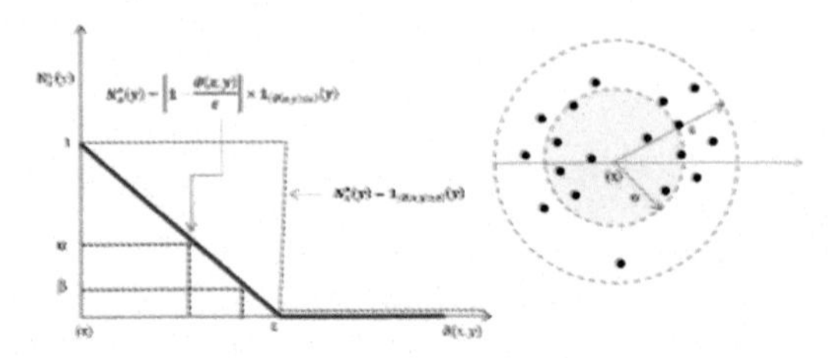

Fig. 2. Fuzzy subset neighborhood.

3 Fuzzy Topological Perception

In this section we propose the relevant indicators for the evaluation of pedestrian risk according to the fuzzy perception of its environment.

3.1 Cognition-Perception Activity

Fig. 3. Space perception semantic, Point of interest's attractivity.

Point of Interest Attractivity

$$\forall y \in N^{\theta,\varepsilon}_{x(t)}: \lambda^{\theta}_{\varepsilon}(x(t),y(t)) = \frac{N^{\theta,\varepsilon}_{x(t)}(x(t),y(t))}{\left|N^{\theta,\varepsilon}_{x(t)}(x(t),y(t))\right| \times \sqrt{2\pi} \times \varepsilon^{2}} \times e^{-\left(\frac{\sigma(y(t))\times\partial(x(t),y(t))}{\sqrt{2}\times\varepsilon}\right)^{2}} \quad (1)$$

Where $\left|N^{\theta,\varepsilon}_{x(t)}(x(t),y(t))\right| = \sum_{y\in\Omega} N^{\varepsilon}_{x(t)}(y(t))$, the fuzzy cardinality of the fuzzy set $N^{\theta,\varepsilon}_{x(t)}$. Where $200^{\circ} \leq \theta \leq 220^{\circ}$ (human field of view with both eyes, see Fig. 3).

$$\forall y \in N^{\theta,\varepsilon}_{x(t)} : \lambda^{\theta}_{\varepsilon,\alpha}(x(t),y(t))$$

$$= \frac{FN^{\varepsilon,\alpha}_{x(t)}(x(t), y(t))}{\left|FN^{\varepsilon,\alpha}_{x(t)}(x(t), y(t))\right| \times \sqrt{2\pi} \times \varepsilon^2} \times e^{-\left(\frac{\sigma(y(t)) \times \partial(x(t),y(t))}{\sqrt{2}\times\varepsilon}\right)^2} \quad (2)$$

Spatial Attractivity

$$\lambda^{\theta}_{\varepsilon}(x(t)) = \sum_{y \in N^{\theta,\varepsilon}_{x(t)}} \lambda^{\theta}_{\varepsilon}(x(t), y(t))\, \lambda^{\theta}_{\varepsilon}(x(t)) =$$

$$\frac{1}{\sqrt{2\pi}\times\varepsilon^2\times\sum_{y\in N^{\theta,\varepsilon}_{x(t)}}\left|N^{\theta,\varepsilon}_{x(t)}(x(t),y(t))\right|} \times \sum_{y\in N^{\theta,\varepsilon}_{x(t)}} N^{\theta,\varepsilon}_{x}(x(t), y(t)) \times e^{-\left(\frac{\sigma(y(t))\times\partial(x(t),y(t))}{\sqrt{2}\times\varepsilon}\right)^2} \quad (3)$$

$$\lambda^{\theta}_{\varepsilon,\alpha}(x(t)) = \sum_{y \in FN^{\varepsilon,\alpha}_{x(t)}} \lambda^{\theta}_{\varepsilon,\alpha}(x(t), y(t)) \quad (4)$$

Relative Attractivity

$$\varpi^{\theta}_{\varepsilon}(x(t)) = \frac{\lambda^{\theta}_{\varepsilon}(x(t))}{\lambda^{2\pi}_{\varepsilon}(x(t))} \in]0,1] \quad (5)$$

Risk Attractivity

$$\exists k : \varpi^{\theta}_{\varepsilon,k}(x_k(t)) > \sum_{i\in N^{\varepsilon,\alpha}_{x_i(t), i\neq k}} \varpi^{\theta}_{\varepsilon}(x_i(t)) \in]0, 1] \quad (6)$$

4 System Architecture

4.1 Global System Design

The system architecture is divided into two macro components, there is the back-office architecture and the front-office architecture (Figs. 4, 5).

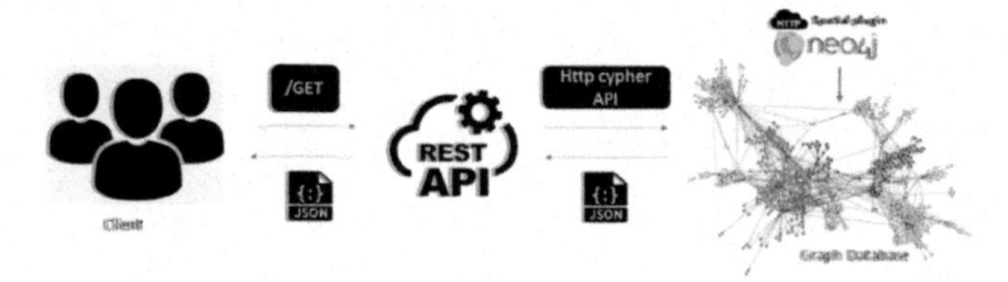

Fig. 4. Front-office architecture.

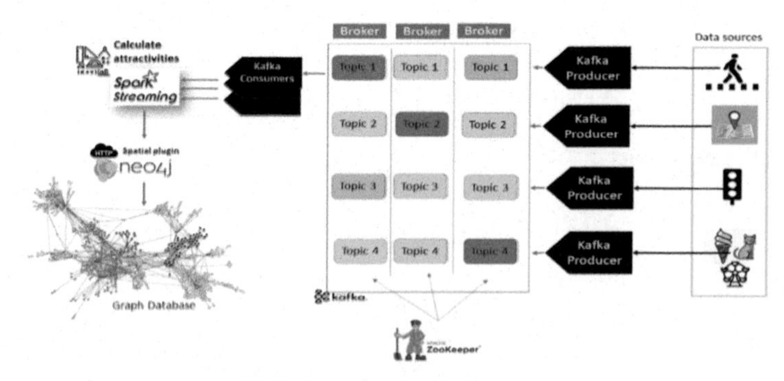

Fig. 5. Back-office architecture.

4.2 Back-Office System

In this part, we will talk about the back-office architecture of our system. In the following, we will briefly develop the data collection and the process of calculating the indicators.

Collecting Data. The collected data types are given below (Fig. 5):

- Pedestrian information (time, walking direction, …).
- Surrounding GIS data (geometries as points, lines and polygons representing a specific location).
- Traffic light data (The state of the traffic light nearby the pedestrian every minute, can be red, yellow, or green).
- Special object information (these are occasional objects that is not located in a specific location, it can be cat or a dog, taxi, ice cream truck, circus…), these objects have a high rate of attractivity (that is why there are special). The special are either detected by cameras (cat, dog, taxi), or by event aggregator website.

Calculate the Attractivity. To calculate the attractivity between pedestrian and the possible points of interests (surrounding GIS, special objects) we use Eqs. (1–5).

Calculate the Risk. To calculate the risk attractivity of a possible point of interests (surrounding GIS, special objects) we use the Eq. (6).

Neo4j Graph Database. All the collected data is sent to neo4j database, in this section we will see how the data it has been modeled in the graph database (Fig. 6).

Relationship Types in Our Database. As we can see in the (Fig. 6), the main node is the *Pedestrian*, and the other nodes relates to Geometry node, *Car_traffic* node, *Special_Object* node and finally *Traffic_Light* node. Every node in the graph has the properties relative attractivity and time. Between every two nodes, there is a connection named *AttracatedBy*, and contains the properties attraction coefficient and time. The property time is a representation of the time with days, hours and minute, this choice was made in purpose to be able to know what happens in or beside every geometry in our database.

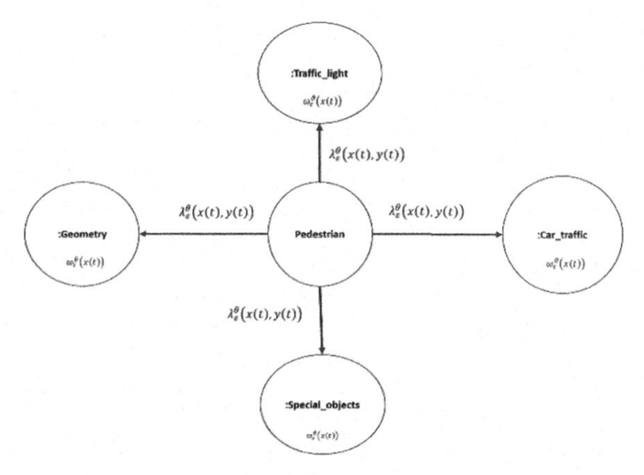

Fig. 6. Data representation in the graph.

4.3 Front-Office Architecture

As front-office (or more accurately middle-office), we propose a rest application that communicates with neo4j and send results as json (Fig. 5). The Basic http command is GET, the user should specify the epsilon range (the cognitive perception wide radius), and a road location. The result of the GET command is a json array of all the safest points (points of interest) where pedestrian can cross safely. Because the data is updated in the neo4j database in real time, the results json array could be different over time, this can be explained by the entry and the exit of new points of interest or special points (see collecting data section).

5 Conclusion

In this work, we presented the definition of a fuzzy neighborhood and the attractivity equations and we used these equations in a distributed system to retrieve the safest possible location to cross the road safely. The main challenge in collecting spatial data is to be able to obtain it as quickly as possible depending on the quantity and size of the data. In this construction, message brokers like Kafka are extremely useful, giving the possibility of providing the real-time architecture for real-time data delivery. Kafka is scalable and offers a fault-tolerant subscription messaging system. The developed system allows the streaming and analysis via spark analysis, and the storage of the collected spatial data in a network defined by a Neo4j graph. The specialization of the proposed solution to smart city needs is scheduled in our future work which make a good use of nowadays gadgets to bring a safest and brighter future.

References

1. Gil, J., Gil, L.: Towards an Advanced Modelling of Complex Economic Phenomena Pretopological and Topological Uncertainty Research Tools, Studies in Fuzziness and Soft Computing, vol. 2762012. Springer, Heidelberg (2012). https://doi.org/10.1007/978-3-642-24812-2

2. Atilgan, C., Nasibov, E.: On reducing space complexity of fuzzy neighborhood-based clustering algorithms. In: 2017 International Conference on Computer Science and Engineering (UBMK), Antalya, 2017, pp. 577–579 (2017). https://doi.org/10.1109/UBMK.2017.8093467
3. Badard, R., Mashhour, A., Ramadan, A.: On L-fuzzy pretopological spaces. Fuzzy Sets Syst. **49**(2), 215–221(1992). ISSN 0165-0114, https://doi.org/10.1016/0165-0114(92)90326-Y
4. Badard, R.: Fuzzy pretopological spaces and their representation. J. Math. Anal. Appl. **81**(2), 378–390 (1981). ISSN 0022-247X, https://doi.org/10.1016/0022-247X(81)90071-8
5. Belmandt, Z.: Basics of Pretopology. Hermann, Paris (2011)
6. Boulmakoul, A., Fazekas, Z., Karim, L., Cherradi, G., Gáspár, P.: Using formal concept analysis tools in road environment-type detection. In: Kahraman, C., Cevik Onar, S., Oztaysi, B., Sari, I.U., Cebi, S., Tolga, A.C. (eds.) INFUS 2020. AISC, vol. 1197, pp. 1059–1067. Springer, Cham (2021). https://doi.org/10.1007/978-3-030-51156-2_123
7. Čech, E.: Topological Spaces. Wiley, Hoboken (1966)
8. Nasibov, E., Ulutagay, G.: A new unsupervised approach for fuzzy clustering. Fuzzy Sets Syst. **158**(19), 2118–2133 (2007). ISSN 0165-0114, https://doi.org/10.1016/j.fss.2007.02.019
9. Nasibov, E., Ulutagay, G.: Robustness of density-based clustering methods with various neighborhood relations. Fuzzy Sets Syst. **160**(24), 3601–3615 (2009). ISSN 0165-0114, https://doi.org/10.1016/j.fss.2009.06.012
10. Nasibov, E., Ulutagay, G., Berberler, M., Nasiboglu, R.: Fuzzy joint points-based clustering algorithms for large data sets. Fuzzy Sets Syst. **270**, 111–126 (2015). ISSN 0165-0114, https://doi.org/10.1016/j.fss.2014.08.004
11. Galbrun, E., Pelechrinis, K., Terzi, E.: Urban navigation beyond shortest route. Inf. Syst. **57**, 160–171 (2016)
12. Günseli, F., Çiklaçandir, Y., Ulutagay, G., Utku, S., Nasibov, E.: On parameter adjustment of the fuzzy neighborhood-based clustering algorithms. Turk. J. Electr. Eng. Comput. Sci. **27**, 2093–2105 (2019). © TÜBİTAK, https://doi.org/10.3906/elk-1807-55
13. Ho, T., Bui, Q.V., Bui, M.: Dynamic social network analysis: a novel approach using agent-based model, author-topic model, and pretopology. Concurrency Computat. Pract. Exper. **32**, e5321 (2020). https://doi.org/10.1002/cpe.5321
14. Kim, J., Cha, M., Sandholm. O.: SocRoutes: safe routes based on tweet sentiments. In: Proceedings of the 23rd International Conference on World Wide Web, pp. 179–182. ACM (2014)
15. Phukan, C.K.: Connected pretopology in recombination space. Theory Biosci. **139**(2), 145–151 (2019). https://doi.org/10.1007/s12064-019-00304-3
16. Quang, B., Ben Amor, S., Bui, M.: Stochastic pretopology as a tool for complex networks analysis. J. Inf. Telecommun. **3**(2), 135–155 (2019)
17. Shifei, D., Mingjing, D., Tongfeng, S., Xiao, X., Xue, Y.: An entropy-based density peaks clustering algorithm for mixed type data employing fuzzy neighborhood. Knowl.-Based Syst. (2017). https://doi.org/10.1016/j.knosys.2017.07.027
18. Šlapal, J.: A Jordan curve theorem with respect to a pretopology on $\mathbb{Z}2$. Int. J. Comput. Math. **90**(8), 1618–1628 (2013). https://doi.org/10.1080/00207160.2012.742889
19. Tabibi Z., Pfeffer, K.: Choosing a safe place to cross the road: the relationship between attention and identification of safe and dangerous road-crossing sites. Child Care Health Dev. **29**(4), 237–44 (2003). https://doi.org/10.1046/j.1365-2214.2003.00336.x. PMID: 12823328
20. Tabibi, Z., Pfeffer, K.: Finding a safe place to cross the road: the effect of distractors and the role of attention in children's identification of safe and dangerous road-crossing sites. Infant Child Dev. **16**, 193–206 (2007)
21. Zadeh, L.: Fuzzy sets. Inform. Control **8**, 338–353 (1965)

22. Zaharia, M., Chowdhury, M., Franklin, M., Shenker, S., Stoica, I.: Spark: cluster computing with working sets. University of California Oakland, CA, United States (2017)
23. Zhang, D.: Fuzzy pretopological spaces, an extensional topological extension of FTS. Chin. Ann. Math. **20**(03), 309–316 (1999). https://doi.org/10.1142/S0252959999000345

A New Risk Analysis Approach for Operational Risks in Logistic Sector Based on Fuzzy Best Worst Method

Necip Fazıl Karakurt[1], Ecem Cem[2], and Selçuk Çebi[2(✉)]

[1] Department of Industrial Engineering, Tekirdag Namık Kemal University, 59030 Tekirdag, Turkey
nfkarakurt@nku.edu.tr

[2] Department of Industrial Engineering, Yildiz Technical University, 34349 Istanbul, Turkey
f0619030@std.yildiz.edu.tr, scebi@yildiz.edu.tr

Abstract. Logistics is one of the sectors that always maintains its importance and develops day by day. In addition to the sector problems that have been solved with technological breakthroughs, new demands and differentiating needs with new developments emerge as new problems in the sector. As in every sector, logistics also carries several risks, and changes in the sector also lead to new risks. To cope with the negative effects of these risks, risk analyzes have great importance in terms of finding solutions to the problems experienced by the sector. In the literature, there are lots of risk analysis methods because of the difficulties and uncertainties in the analysis. In this study, it is the first time we propose a new risk analysis method for the logistic sector based on Fuzzy Best Worst Method (FBWM). The operational risks in the logistics sector have been defined as a result of the sector and literature review. The assessments on the determined risk were collected from three experts through a questionnaire. FBWM was used for weighting and prioritizing risks. Two different methods, Borda Count and Arithmetic Mean were used to aggregate expert opinions and compared. As a result of the study, it was revealed that internal risk factors were highlighted by experts for the logistic operations.

Keywords: Risk assessment · Fuzzy Best-Worst Method · Logistics · Borda Count

1 Introduction

Logistics can be defined as the whole process of a product from its production to the end-user. Logistics, which is an area directly affected by the new technological developments, therefore maintains its current status in the literature. As in every field, logistics also involves some risks. These risks can be divided into groups or they can be evaluated as a whole. In the study, risks are divided into two groups as internal and external. Each group contains 6 risks. They are represented by the initial letter of the group they belong to and the order in which they are found. We can list the main risks of the logistic sector as follows;

C. Kahraman et al. (Eds.): INFUS 2021, LNNS 308, pp. 696–702, 2022.
https://doi.org/10.1007/978-3-030-85577-2_82

- Failure to deliver on time (I1)
- Delivered products arrive spoiled/damaged (I2)
- Delivery to the wrong address (I3)
- Delivery of the wrong product to the buyer (I4)
- Missing product delivery (I5)
- Insufficient amount of delivered product (overloading, under-loading) (I6)
- The delivery of the delivered product in a decreasing way (theft) (E1)
- Rapid changes in costs (rapid changes due to exchange rate difference) (E2)
- Stopping operations due to instant problems between local governments in international shipments (E3)
- Problems that may occur in international shipments due to the differences in the laws and regulations of the products (such as not accepting the products into the country due to the failure of meeting the standards of the country they enter) (E4)
- Security of the transportation route (terrorism etc.) (E5)
- Safety of the transportation route (national disaster potential, structure of the transportation route etc.) (E6)

Due to the serious risks in the field of logistics, various solutions have been produced for the evaluation of the potential risk in the literature. Furthermore, the priorities and the importance degrees of these risks have been calculated with different methods. In particular, one of the widely used methods for the weighting is fuzzy analytical hierarchy process (FAHP). However, an increase in the number of pairwise comparisons based on the number of criteria and the level of the hierarchy is the disadvantage of the method although pairwise comparisons are the most important advantage of the method. Therefore, in this study, an integrated method including Fuzzy Best-Worst Method and Borda Count has been proposed for the risk analysis of logistic operations. Fuzzy Best-Worst Method (FBWM) is another pairwise comparison method that requires less pairwise comparison than FAHP and it is used to calculate risk magnitudes in the transportation sector. The aim of the study is to prioritize logistic risk factors in a fuzzy environment. FBWM is a useful alternative to the traditional Multi-Criteria Decision Making (MCDM) techniques as it requires low dimension comparison matrix. Relatively short interviewing process and more consistent results can be achieved by FBWM. Since it is proposed in recent years, it is not employed in any logistic related study. This work uses FBWM on the logistic area and aggregate different decision makers opinion with two different ways.

The rest of the proposed study is organized as follows; A literature review on risk analysis is given in the second part of the study. In the third section, the fundamentals of the proposed method are given. The fourth section presents the obtained results of the proposed study. Finally, concluding remarks and recommendations for further studies are given in Sect. 5.

2 Literature Review

In this section, risk assessment studies that have been recently published in the literature will be illustrated. FMECA analysis is conducted in order to specify risks in the automotive industry [1]. A new approach is proposed for Risk Priority Number (RPN) calculation. In this approach, factor weights are calculated by Analytic Hierarchical Process (AHP). Then, in order to indicate the correlation between risk factors, a fuzzy Decision-Making Trial and Evaluation Laboratory (DEMATEL) method is employed. Thus, traditional RPN calculation is an improved and fuzzy approach to the risk prioritization and risk dependence calculations in DEMATEL make the proposed method more robust to the uncertainty.

A paper is published regarding the investigation of risks on halal food supply chain and prioritization these risks by MCDM [2]. They have used fuzzy AHP in order to rank them by importance. They conclude that in the halal food industry, most of the risks are related to raw material. They suggest that decreasing outsourcing can diminish the total risk to be dealt with.

A new model is proposed for the evaluation of the environmental risk factors of the reverse logistics approach in the e-waste sector [3]. In this proposed model, the Analytic Network Process (ANP) method was used to calculate and rank the criteria, and the COPRAS method was used to evaluate the prevention strategies for these risks. In the study, the suitability of the proposed model is revealed by giving a sample case analysis.

A new model is proposed to the literature by hybridizing the Analytic Network Process (ANP) and Multi-Attributive Border Approximation Area Comparison (MABAC) methods, which are the Multi-Criteria Decision-Making techniques, for the risk management of multi-stakeholder construction projects [4]. An example from the construction industry is presented for the proposed model in the study and sensitivity analysis has been made to provide it.

A study to prioritize risk factors in cold chain transportation and to rank the importance of tools used in risk management is conducted [5]. In this study, the AHP method was used in prioritizing risk factors, while the VIKOR method was used in ordering the tools in the second stage. The study was conducted in a province in Turkey with 25 experts and the results are as follows; the most important logistics risk factor was found to be "Packaging Risks" and the most ideal tool in risk management was, "Statistical Process Control".

Analytic Network Process was used to determine weights of the risk assessment factors of working in a warm environment, and Triangular fuzzy numbers were used to remove the uncertainty that occurred during the decision-making process [6]. The study consists of three main factors and 10 sub-factors and includes an application for the proposed new model. As a result of the study, the most important criterion was determined as "worker" and the safety of the hot environment was put between medium and good.

3 Fuzzy Best-Worst Method

Best-Worst Method (BWM) is a brand-new MCDM technique that assigns weights to risk factors and ranks them by importance level. BWM takes advantage of using fewer questions to the expert ending up with high consistency (ie. low consistency ratio). While AHP requires the entire dual comparison matrix for factors, BWM can generate factor weights with a two-rowed comparison matrix. The first row consists of the importance level of the best factor to others (a_{Bj}). The second row demonstrates the importance level of other factors to the worst (a_{jW}). BWM can be applied through two phases. The first phase is the interview and the second one is modeling. In the interview, the best and the worst factor/criteria are determined by the expert. Then importance level is determined based on a Likert scale between other factors and the best/worst factor. This step is where a two-rowed comparison matrix is built. Then, a linear programming model is constructed and a solution that optimizes w (factor weights) is obtained. BWM is adapted to real cases where ambiguity and intangibility for expert opinion take place. A technique that employs linguistic comparison with a predefined triangular membership function called FBWM is introduced [7]. Membership function equivalents of the linguistic response of experts are given in Table 1.

Table 1. Membership functions equivalents

Linguistic expression	Membership equivalent
Equally importance	(1,1,1)
Weakly important	(2/3,1,3/2)
Fairly important	(3/2,2,5/2)
Very important	(5/2,3,7/2)
Absolutely important	(7/2,4,9/2)

After conducting an interview with experts in terms of linguistic importance levels, the weights of each factor can be calculated with the non-linear model given below. In the model, Eq. (1) is the objective function to be minimized. Here, $\tilde{\varepsilon}^*$ term can be expressed as (k*, k*, k*) where $k^* \leq l^{\varepsilon}$. Equations (2) and (3) minimize the consistency ratio adjusting factor weights in accordance with the expert's importance level. Equation (4) assures that all weights must add up to 1. Equation (5) is the non-negativity constraint.

$$min\, \tilde{\varepsilon}^* \tag{1}$$

$$\left| \frac{\left(l_B^W, m_B^W, u_B^W\right)}{\left(l_j^W, m_j^W, u_j^W\right)} - \left(l_{Bj}, m_{Bj}, u_{Bj}\right) \right| \leq \left(k^*, k^*, k^*\right), \forall j \tag{2}$$

$$\left| \frac{\left(l_j^W, m_j^W, u_j^W\right)}{\left(l_W^W, m_W^W, u_W^W\right)} - \left(l_{jW}, m_{jW}, u_{jW}\right) \right| \leq \left(k^*, k^*, k^*\right), \forall j \tag{3}$$

$$\sum\nolimits_i^n R(\widetilde{w}_j) = 1 \tag{4}$$

$$0 \leq l_j^W \leq m_j^W \leq u_j^W, \forall j \tag{5}$$

4 Results

The logistics sector and the risks it entails are always popular in the literature. In the pandemic period, we are in, the importance of the sector has become even more prominent. In the study, a new approach is proposed using the Fuzzy Best Worst Method to prioritize the operational risks inherent in the logistics sector. The risk factors were evaluated by three experts. Nonlinear programming models are solved by GAMS for each of the sub-risk and risk factors. Then, defuzzification is made in order to get crisp weights by graded mean integration representation. These sub-risk and risk weights are multiplied to get global risk factor weights. As a result of the ranking, it can be said that internal risks are more prevalent and are highlighted by experts because they are relatively more controllable and more likely to be encountered in the operation. Details of the assessment are given in Fig. 1.

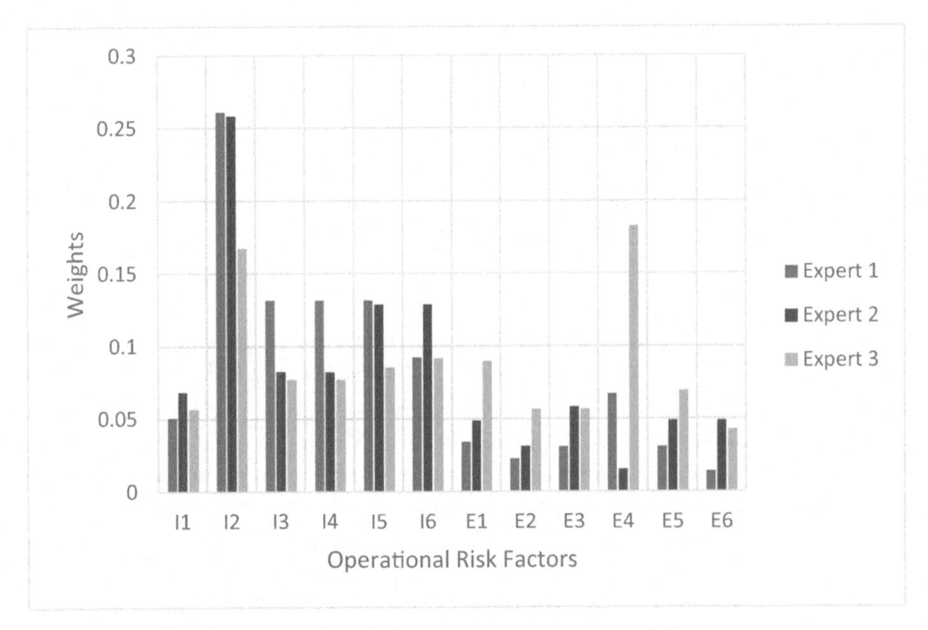

Fig. 1. Operational risk factors assessment scores of three experts

The global weight results were compared using the The Borda Count and Arithmetic Mean methods to aggregate expert opinions. It is seen that the prioritization rankings are the same in the results obtained by two different methods. Based on these results, it has been demonstrated that the proposed risk assessment method is consistent. The details of the results achieved with the two different methods applied are given in Fig. 2.

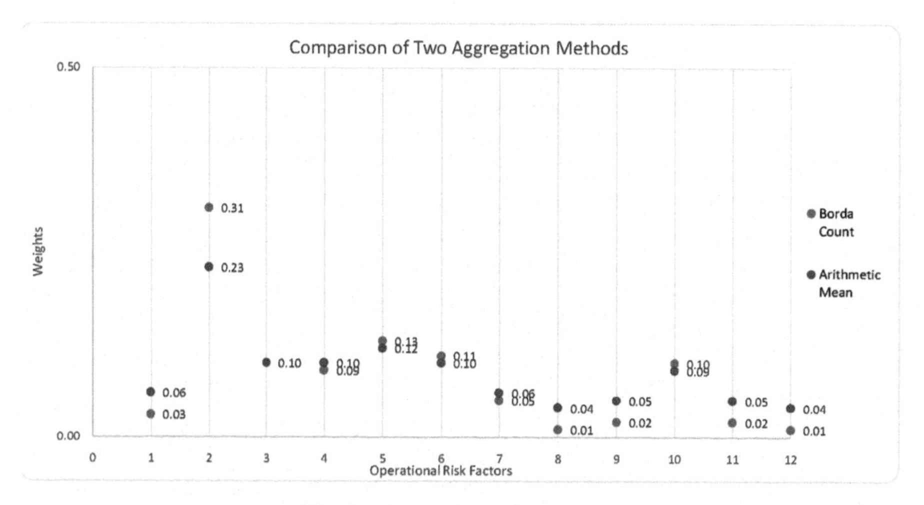

Fig. 2. Comparison of the results

As a result of the FBWM and applied group decision techniques, prioritization of risk criteria is as follows in descending order;

1. Delivered products arrive spoiled/damaged (I2)
2. Missing product delivery (I5)
3. Insufficient amount of delivered product (overloading, under-loading) (I6)
4. Delivery to the wrong address (I3)
5. Problems that may occur in international shipments due to the differences in the laws and regulations of the products (E4)
6. Delivery of the wrong product to the buyer (I4)
7. The delivery of the delivered product in a decreasing way (theft) (E1)
8. Failure to deliver on time (I1)
9. Stopping operations due to instant problems between local governments in international shipments (E3)
10. Security of the transportation route (E5)
11. Rapid changes in costs (E2)
12. Safety of the transportation route (E6)

As a result of the ranking of importance, it is seen that time-sensitive criteria and financial risks are at low priority. However, it has been revealed that the damage and deficiency that may occur in the transported product includes the highest risks by ranking at the top.

5 Conclusion

The logistics is one of the sectors that develop and change day by day with technology. In this study, operational risk factors in the logistics industry have been prioritized. These factors have been compiled as a result of literature and sector research. FBWM was used and expert opinions were combined and compared with The Borda Count and arithmetic average method. FBWM allows weighting with less expert answer and thus ends up with more consistent results. Furthermore, using triangular fuzzy numbers in BWM copes with vagueness and uncertainty in expert opinions. FBWM stands out among Multi Criteria Decision Making Techniques with its features such as easy implementation, application with fewer questions than full matrix methods such as AHP, and therefore allowing high consistency. In this study, FBWM was used for the first time in the logistic literature for risk calculation. Internal risks are found to be more important than the external ones. The Borda Count and arithmetic mean techniques are found to be similar in terms of risk prioritizing in a multiple decision maker environment. The study indicates that above mentioned approach can be used in different decision-making problems when lingual expressions is used instead of numeric scale.

For future studies, the probability and severity parameters in the traditional risk calculation formula can be weighted separately and then the risk can be calculated by multiplying those values. Risks and experts can be diversified. At the same time, a separate study can be done with Fuzzy AHP and the results can be compared.

References

1. Mzougui, I., Carpitella, S., Certa, A., Felsoufi, Z.E., Izquierdo, J.: Assessing supply chain risks in the automotive industry through a modified MCDM-based FMECA. Processes **8**(5), 579 (2020)
2. Khan, S., Khan, M.I., Haleem, A., Jami, A.R.: Prioritising the risks in Halal food supply chain: an MCDM approach. J. Islamic Market. (2019)
3. Duran, F., Bereketli Zafeirakopoulos, İ.: Environmental risk assessment of e-waste in reverse logistics systems using MCDM methods. In: Durakbasa, N.M., Gencyilmaz, M.G. (eds.) Proceedings of the International Symposium for Production Research 2018, pp. 590–603. Springer, Cham (2019). https://doi.org/10.1007/978-3-319-92267-6_49
4. Chatterjee, K., Zavadskas, E.K., Tamošaitienė, J., Adhikary, K., Kar, S.: A hybrid MCDM technique for risk management in construction projects. Symmetry **10**(2), 46 (2018)
5. Korucuk, S., Erdal, H.: Ranking of the logistics risk factors and risk management tools via integrated AHP-VIKOR approach: a case study of Samsun Province. J. Bus. Res. – Turk **10**(3), 282–305 (2018)
6. Ilangkumaran, M., Karthikeyan, M., Ramachandran, T., Boopathiraja, M., Kirubakaran, B.: Risk analysis and warning rate of hot environment for foundry industry using hybrid MCDM technique. Saf. Sci. **72**, 133–143 (2015)
7. Guo, S., Zhao, H.: Fuzzy best-worst multi-criteria decision-making method and its applications. Knowl.-Based Syst. **121**, 23–31 (2017)

Spherical Fuzzy Sets

Spherical Fuzzy CRITIC Method: Prioritizing Supplier Selection Criteria

Cengiz Kahraman, Başar Öztayşi, and Sezi Çevik Onar(✉)

Industrial Engineering Department, Istanbul Technical University, Istanbul, Turkey
{kahramanc,oztaysib,cevikse}@itu.edu.tr

Abstract. When linguistic evaluations are used in the decision matrix instead of exact numerical values, fuzzy set theory can capture the vagueness in the linguistic evaluations. Ordinary fuzzy sets have been extended to many new types of fuzzy sets such as intuitionistic fuzzy sets, neutrosophic sets, and picture fuzzy sets. Spherical fuzzy sets is an extension of picture fuzzy sets whose squared sum of their parameters is at most equal to one. This paper develops spherical fuzzy CRiteria Importance Through Intercriteria Correlation (CRITIC) method for prioritizing supplier selection criteria. Supplier selection is one of the most critical aspects of any organization since any mistake in this process may cause poor supplier performance and inefficiencies in the business processes. Supplier selection is a multi-criteria decision making problem involving several conflicting criteria and alternatives. A numerical illustration of the proposed method is also given.

Keywords: Spherical fuzzy sets · CRITIC · Supplier selection · Aggregation · Decision making

1 Introduction

The CRiteria Importance Through Intercriteria Correlation (CRITIC) method was introduced by Diakoulaki, Mavrotas, and Papayannakis [1] in 1995. Later it has been improved by several researchers [2–6]. It is mainly used to determine the weight of attributes; the attributes are not in contradiction with each other, and the weights of attributes are calculated using a decision matrix [7]. It has been used for the automatic areal feature matching [8], medical quality assessment [9], and ranking of machining processes [10]. The CRITIC method also includes the following features: It does not need the independency condition of attributes and qualitative attributes are easily transformed into quantitative attributes, which helps the fuzzy set theory to be employed in CRITIC method. Figure 1 shows the usage frequencies of CRITIC method with respect to the years. After 2017, it has become more popular than ever.

C. Kahraman et al. (Eds.): INFUS 2021, LNNS 308, pp. 705–714, 2022.
https://doi.org/10.1007/978-3-030-85577-2_83

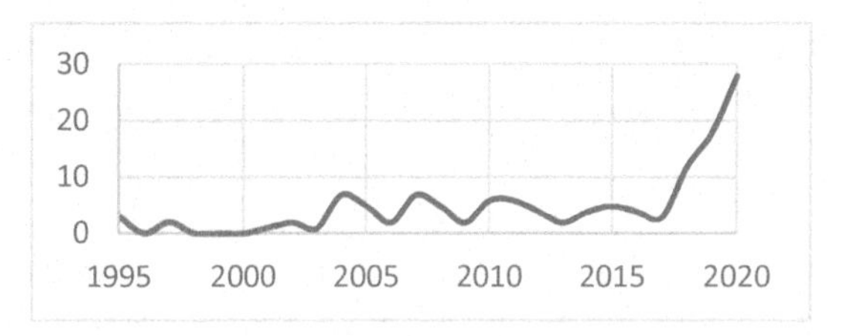

Fig. 1. Usage frequencies of CRITIC by years.

Figure 2 shows the source institutes of CRITIC publications. The leading countries using CRITIC method is China, United States, Japan, Iran and Netherlands, respectively.

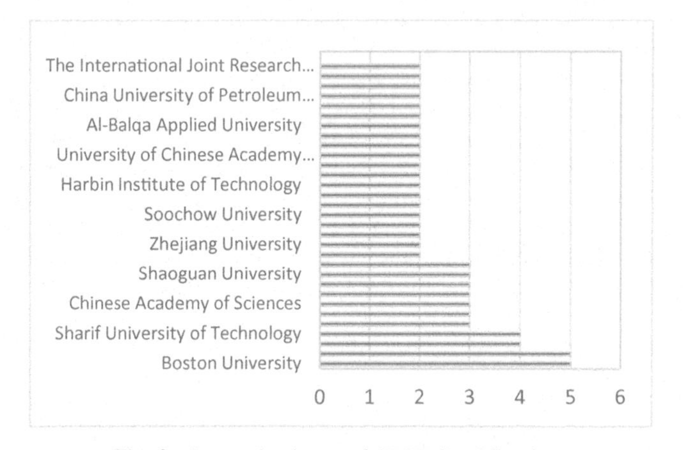

Fig. 2. Source institutes of CRITIC publications

Figure 3 illustrates the percentages of subject areas of CRITIC method publications. The most used subject areas are computers science, engineering, and mathematics.

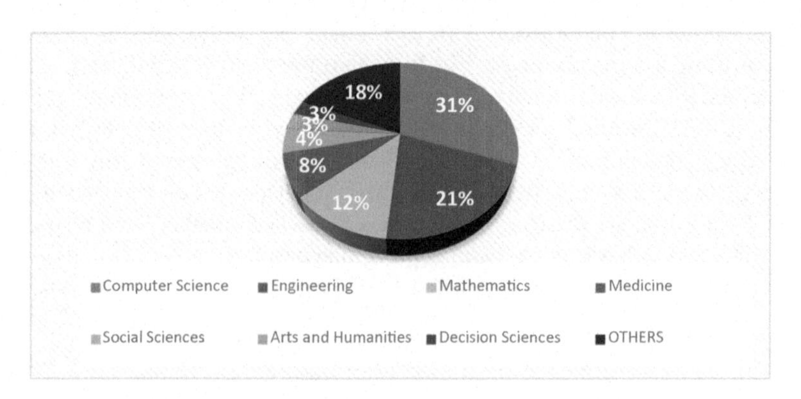

Fig. 3. Percentages of subject areas of CRITIC publications

The fuzzy sets are excellent tools for defining the uncertainty and imprecision [11–13]. Since the linguistic evaluations in a decision matrix involve vagueness and impreciseness, the CRITIC method should be extended under fuzziness such as picture fuzzy CRITIC method, or spherical fuzzy CRITIC method. The originality of this paper comes from the first time development of CRITIC method under spherical fuzzy environment. The rest of the paper is organized as follows. Section 2 presents the classical CRITIC method whereas Sect. 3 includes spherical fuzzy CRITIC method. Section 4 applies the proposed method to a supplier selection criteria prioritizing problem. Section 5 concludes the paper and gives suggestions for future research.

2 Classical CRITIC Method

The decision matrix involving the alternatives and attributes is given in Eq. (1).

$$\begin{vmatrix} r_{11} \cdots r_{1j} & r_{1,j+1} \cdots r_{1n} \\ \vdots \ddots \vdots & \vdots \ddots \vdots \\ r_{ii} \cdots r_{ij} & r_{i,j+1} \cdots r_{in} \end{vmatrix}$$

$$\begin{vmatrix} \vdots \ddots \vdots & \vdots \ddots \vdots \\ r_{m1} \cdots r_{mj} & r_{j,m+1} \cdots r_{mn} \end{vmatrix} \quad (1)$$

where $i = 1, 2, \ldots, m; j = 1, 2, \ldots, n$ and $\{A_1, A_2, \cdots, A_m\}$ and r_{ij} is the element of the decision matrix for ith alternative and jth attribute.

In order to normalize the positive (benefit) and negative (cost) attributes of the decision matrix, Eqs. (2) and (3) are used, respectively [1].

$$x_{ij} = \frac{r_{ij} - r_i^-}{r_i^+ - r_i^-} \quad i = 1, 2, \ldots, m; j = 1, 2, \ldots, n \quad (2)$$

$$x_{ij} = \frac{r_{ij} - r_i^+}{r_i^- - r_i^+} \quad i = 1, 2, \ldots, m; j = 1, 2, \ldots, n \quad (3)$$

where x_{ij} represents the normalized values of the decision matrix for ith alternative with respect to jth attribute and $r_i^+ = max(r_1, r_2, \ldots, r_m)$, $r_i^{-+} = min(r_1, r_2, \ldots, r_m)$.

The correlation coefficient ρ_{jk} between attributes is determined by Eq. (4) [1].

$$\rho_{jk} = \sum\nolimits_{i=1}^{m} \left(x_{ij} - \overline{x}_j\right)(x_{ik} - \overline{x}_k) / \sqrt{\sum\nolimits_{i=1}^{m} \left(x_{ij} - \overline{x}_j\right)^2 \sum\nolimits_{i=1}^{m} (x_{ik} - \overline{x}_k)^2} \quad (4)$$

where $\overline{x}_j$ and $\overline{x}_k$ are the means of jth and kth attributes. $\overline{x}_j$ is obtained from Eq. (5) and similarly for $\overline{x}_k$ [1].

$$\overline{x}_j = \frac{1}{n} \sum\nolimits_{j=1}^{n} x_{ij};\ i = 1, 2, \ldots, m \quad (5)$$

The sample standard deviation of each attribute is calculated by Eq. (6) [1].

$$\sigma_j = \sqrt{\frac{1}{n-1} \sum\nolimits_{j=1}^{n} \left(x_{ij} - \overline{x}_j\right)^2},\ i = 1, 2, \ldots, m \quad (6)$$

Then, the index (C) is calculated using Eq. (7) [1].

$$C_j = \sigma_j \sum\nolimits_{k=1}^{n} \left(1 - \rho_{jk}\right), j = 1, 2, \ldots, n, \tag{7}$$

The weights of attributes are calculated by Eq. (8) [1].

$$w_j = \frac{C_j}{\sum_{j=1}^{n} C_j}, \tag{8}$$

3 Spherical Fuzzy Sets

Spherical fuzzy sets (SFS) were introduced by Kutlu Gundogdu and Kahraman [14]. These sets are based on the fact that the hesitancy of a decision maker can be assigned separately satisfying the condition that the squared sum of membership, non-membership and hesitancy degrees is at most equal to 1. In the following, preliminaries of SFS are given:

Definition 1. Single valued Spherical Fuzzy Sets (SFS) $\tilde{A}_S$ of the universe of discourse U is given by

$$\tilde{A}_S = \left\{ \left\langle u, (\mu_{\tilde{A}_S}(u), v_{\tilde{A}_S}(u), \pi_{\tilde{A}_S}(u)) \middle| u \in U \right\rangle \right\} \tag{9}$$

where $\mu_{\tilde{A}_S}(u) : U \rightarrow [0, 1], \quad v_{\tilde{A}_S}(u) : U \rightarrow [0, 1], \quad \pi_{\tilde{A}_S}(u) : U \rightarrow [0, 1].$
and

$$0 \leq \mu_{\tilde{A}_S}^2(u) + v_{\tilde{A}_S}^2(u) + \pi_{\tilde{A}_S}^2(u) \leq 1 \qquad \forall u \in U \tag{10}$$

For each u, the numbers $\mu_{\tilde{A}_S}(u)$, $v_{\tilde{A}_S}(u)$ and $\pi_{\tilde{A}_S}(u)$ are the degree of membership, non-membership and hesitancy of u to$\tilde{A}_S$, respectively. Spherical fuzzy sets have become very popular in a short time period [15–22]. Figure 4 illustrates the differences among intuitionistic fuzzy sets (IFS), Pythagorean fuzzy sets (PFS), neutrosophic sets (NS), and spherical fuzzy sets (SFS).

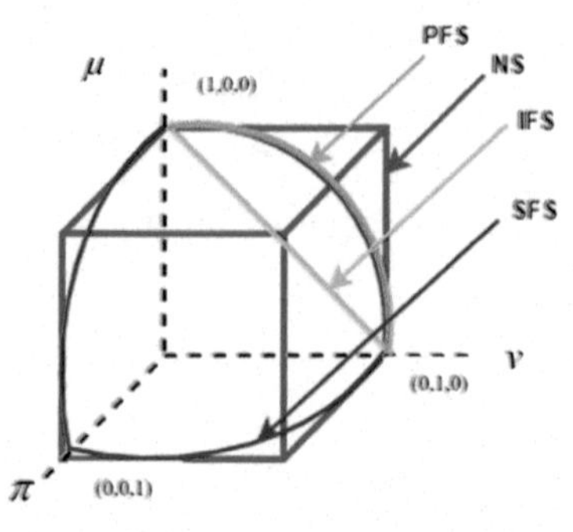

Fig. 4. Geometric representations of IFS, PFS, NS, and SFS [11]

Definition 2. Basic operators of Single-valued SFS;

$$\tilde{A}_S \oplus \tilde{B}_S = \left\{ \begin{array}{l} \left(\mu^2_{\tilde{A}_S} + \mu^2_{\tilde{B}_S} - \mu^2_{\tilde{A}_S}\mu^2_{\tilde{B}_S}\right)^{1/2}, \ v_{\tilde{A}_S}v_{\tilde{B}_S}, \\ \left(\left(1 - \mu^2_{\tilde{B}_S}\right)\pi^2_{\tilde{A}_S} + \left(1 - \mu^2_{\tilde{A}_S}\right)\pi^2_{\tilde{B}_S} - \pi^2_{\tilde{A}_S}\pi^2_{\tilde{B}_S}\right)^{1/2} \end{array} \right\} \tag{11}$$

$$\tilde{A}_S \otimes \tilde{B}_S = \left\{ \begin{array}{l} \mu_{\tilde{A}_S}\mu_{\tilde{B}_S}, \ \left(v^2_{\tilde{A}_S} + v^2_{\tilde{B}_S} - v^2_{\tilde{A}_S}v^2_{\tilde{B}_S}\right)^{1/2}, \\ \left(\left(1 - v^2_{\tilde{B}_S}\right)\pi^2_{\tilde{A}_S} + \left(1 - v^2_{\tilde{A}_S}\right)\pi^2_{\tilde{B}_S} - \pi^2_{\tilde{A}_S}\pi^2_{\tilde{B}_S}\right)^{1/2} \end{array} \right\} \tag{12}$$

$$\lambda \cdot \tilde{A}_S = \left\{ \begin{array}{l} \left(1 - \left(1 - \mu^2_{\tilde{A}_S}\right)^{\lambda}\right)^{1/2}, \ v^{\lambda}_{\tilde{A}_S}, \\ \left(\left(1 - \mu^2_{\tilde{A}_S}\right)^{\lambda} - \left(1 - \mu^2_{\tilde{A}_S} - \pi^2_{\tilde{A}_S}\right)^{\lambda}\right)^{1/2} \end{array} \right\} for\ \lambda \geq 0 \tag{13}$$

$$\lambda \cdot \tilde{A}_S = \left\{ \begin{array}{l} \left(1 - \left(1 - \mu^2_{\tilde{A}_S}\right)^{\lambda}\right)^{1/2}, \ v^{\lambda}_{\tilde{A}_S}, \\ \left(\left(1 - \mu^2_{\tilde{A}_S}\right)^{\lambda} - \left(1 - \mu^2_{\tilde{A}_S} - \pi^2_{\tilde{A}_S}\right)^{\lambda}\right)^{1/2} \end{array} \right\} for\ \lambda \geq 0 \tag{14}$$

Definition 4. Single-valued Spherical Weighted Arithmetic Mean (SWAM) with respect to,$w = (w_1, w_2......., w_n)$; $w_i \in [0, 1]$; $\sum_{i=1}^{n} w_i = 1$, SWAM is defined as;

$$SWAM_w(\tilde{A}_{S1}, \ldots\ldots, \tilde{A}_{Sn}) = w_1\tilde{A}_{S1} + w_2\tilde{A}_{S2} + \ldots\ldots + w_n\tilde{A}_{Sn}$$
$$= \left\{ \begin{array}{l} \left[1 - \prod_{i=1}^{n} (1 - \mu^2_{\tilde{A}_{Si}})^{w_i}\right]^{1/2}, \\ \prod_{i=1}^{n} v^{w_i}_{\tilde{A}_{Si}}, \ \left[\prod_{i=1}^{n} (1 - \mu^2_{\tilde{A}_{Si}})^{w_i} - \prod_{i=1}^{n} (1 - \mu^2_{\tilde{A}_{Si}} - \pi^2_{\tilde{A}_{Si}})^{w_i}\right]^{1/2} \end{array} \right\} \tag{15}$$

Definition 5. Single-valued Spherical Weighted Geometric Mean (SWGM) with respect to,$w = (w_1, w_2......., w_n)$; $w_i \in [0, 1]$; $\sum_{i=1}^{n} w_i = 1$, SWGM is defined as.

$$SWGM_w(\tilde{A}_1, \ldots\ldots, \tilde{A}_n) = \tilde{A}^{w_1}_{S1} + \tilde{A}^{w_2}_{S2} + \ldots\ldots + \tilde{A}^{w_n}_{Sn}$$
$$= \left\{ \begin{array}{l} \prod_{i=1}^{n} \mu^{w_i}_{\tilde{A}_{Si}}, \ \left[1 - \prod_{i=1}^{n} (1 - v^2_{\tilde{A}_{Si}})^{w_i}\right]^{1/2}, \\ \left[\prod_{i=1}^{n} (1 - v^2_{\tilde{A}_{Si}})^{w_i} - \prod_{i=1}^{n} (1 - v^2_{\tilde{A}_{Si}} - \pi^2_{\tilde{A}_{Si}})^{w_i}\right]^{1/2} \end{array} \right\} \tag{16}$$

Definition 6. Score function and Accuracy function for sorting SFS are given by Eq. (17) and Eq. (18), respectively.

$$Score\left(\tilde{A}_S\right) = \left(\mu_{\tilde{A}_S} - \pi_{\tilde{A}_S}/2\right)^2 - \left(v_{\tilde{A}_S} - \pi_{\tilde{A}_S}/2\right)^2 \tag{17}$$

$$Accuracy\left(\tilde{A}_S\right) = \mu^2_{\tilde{A}_S} + v^2_{\tilde{A}_S} + \pi^2_{\tilde{A}_S} \tag{18}$$

Note that: $\tilde{A}_S < \tilde{B}_S$ if and only if

i. $Score(\tilde{A}_S) < Score(\tilde{B}_S)$ or
ii. $Score(\tilde{A}_S) = Score(\tilde{B}_S)$ and
$Accuracy(\tilde{A}_S) < Accuracy(\tilde{B}_S)$

Definition 7. Let $\tilde{A} = (\mu_A, v_A, \pi_A)$ and $\tilde{B} = (\mu_B, v_B, \pi_B)$. The distance between $\tilde{A}$ and $\tilde{B}$ is calculated by Eq. (19) [23].

$$D^s\left(\tilde{A}, \tilde{B}\right) = 1 - \frac{\mu_A^2 \times \mu_B^2 + v_A^2 \times v_B^2 + \pi_A^2 \times \pi_B^2}{\mu_A^4 \bigvee \mu_B^4 + v_A^4 \bigvee v_B^4 + \pi_A^4 \bigvee \pi_B^4} \tag{19}$$

4 Spherical Fuzzy CRITIC Method

The fuzzy decision matrix given by Eq. (20) involves the alternatives and attributes.

$$\begin{vmatrix} \tilde{r}_{11} & \cdots & \tilde{r}_{1j} & \tilde{r}_{1,j+1} & \cdots & \tilde{r}_{1n} \\ \vdots & \ddots & \vdots & \vdots & \ddots & \vdots \\ \tilde{r}_{ii} & \cdots & \tilde{r}_{ij} & \tilde{r}_{i,j+1} & \cdots & \tilde{r}_{in} \end{vmatrix}$$

$$\begin{vmatrix} \vdots & \ddots & \vdots & \vdots & \ddots & \vdots \\ \tilde{r}_{m1} & \cdots & \tilde{r}_{mj} & \tilde{r}_{j,m+1} & \cdots & \tilde{r}_{mn} \end{vmatrix} \tag{20}$$

where $i = 1, 2, \ldots, m; j = 1, 2, \ldots, n$ and $\{A_1, A_2, \cdots, A_m\}$ and $\tilde{r}_{ij}$ is the fuzzy element of the decision matrix for ith alternative and jth attribute.

In order to normalize the positive (benefit) and negative (cost) attributes of the decision matrix, Eqs. (21) and (22) are used, respectively [1].

$$\tilde{x}_{ij} = \frac{\tilde{r}_{ij} - \tilde{r}_i^-}{\tilde{r}_i^+ - \tilde{r}_i^-} i = 1, 2, \ldots, m; j = 1, 2, \ldots, n; \text{ for positive attributes} \tag{21}$$

$$\tilde{x}_{ij} = \frac{\tilde{r}_{ij} - \tilde{r}_i^+}{\tilde{r}_i^- - \tilde{r}_i^+} i = 1, 2, \ldots, m; j = 1, 2, \ldots, n; \text{ for negative attributes} \tag{22}$$

where $\tilde{x}_{ij}$ represents a normalized value of the decision matrix for ith alternative with respect to jth attribute and

$$\tilde{r}_i^+ = max(\tilde{r}_1, \tilde{r}_2, \ldots, \tilde{r}_m), r_i^{-+} = min(\tilde{r}_1, \tilde{r}_2, \ldots, \tilde{r}_m).$$

Equation (21) and Eq. (22) can be replaced by using the distance definition in Eq. (19) as given in Eq. (23) and Eq. (24), respectively:

$$\tilde{x}_{ij} = \frac{1 - \frac{\mu_{ij}^2 \times \mu_-^2 + v_{ij}^2 \times v_-^2 + \pi_{ij}^2 \times \pi_-^2}{\mu_{ij}^4 \bigvee \mu_-^4 + v_{ij}^4 \bigvee v_-^4 + \pi_{ij}^4 \bigvee \pi_-^4}}{1 - \frac{\mu_-^2 \times \mu_+^2 + v_-^2 \times v_+^2 + \pi_-^2 \times \pi_+^2}{\mu_-^4 \bigvee \mu_+^4 + v_-^4 \bigvee v_+^4 + \pi_-^4 \bigvee \pi_+^4}} \tag{23}$$

$$\tilde{x}_{ij} = \frac{1 - \frac{\mu_{ij}^2 \times \mu_+^2 + v_{ij}^2 \times v_+^2 + \pi_{ij}^2 \times \pi_+^2}{\mu_{ij}^4 \vee \mu_+^4 + v_{ij}^4 \vee v_+^4 + \pi_{ij}^4 \vee \pi_+^4}}{1 - \frac{\mu_-^2 \times \mu_+^2 + v_-^2 \times v_+^2 + \pi_-^2 \times \pi_+^2}{\mu_-^4 \vee \mu_+^4 + v_-^4 \vee v_+^4 + \pi_-^4 \vee \pi_+^4}} \tag{24}$$

The correlation coefficients between attributes are calculated by Eq. (25) [1].

$$\tilde{\rho}_{jk} = \sum\nolimits_{i=1}^{m} \left(\tilde{x}_{ij} - \bar{\tilde{x}}_j\right)\left(\tilde{x}_{ik} - \bar{\tilde{x}}_k\right) / \sqrt{\sum\nolimits_{i=1}^{m} \left(\tilde{x}_{ij} - \bar{\tilde{x}}_j\right)^2 \sum_{i=1}^{m} \left(\tilde{x}_{ik} - \bar{\tilde{x}}_k\right)^2} \tag{25}$$

where $\bar{\tilde{x}}_j$ and $\bar{\tilde{x}}_k$ are the fuzzy mean of *j*th and *k*th attributes. $\bar{\tilde{x}}_j$ is computed from Eq. (26) and similarly for $\bar{\tilde{x}}_k$. $\tilde{\rho}_{jk}$ is the correlation coefficient between jth and kth attributes [1].

$$\bar{\tilde{x}}_j = \frac{1}{n} \sum\nolimits_{j=1}^{n} \tilde{x}_{ij};\ i = 1, 2, \ldots, m \tag{26}$$

The fuzzy standard deviation of each attribute is calculated by Eq. (27) [1].

$$\tilde{\sigma}_j = \sqrt{\frac{1}{n-1} \sum\nolimits_{j=1}^{n} \left(\tilde{x}_{ij} - \bar{\tilde{x}}_j\right)^2},\ i = 1, 2, \ldots, m \tag{27}$$

Then, the index (C) is calculated using Eq. (28) [1].

$$\tilde{C}_j = \tilde{\sigma}_j \sum\nolimits_{k=1}^{n} \left(1 - \tilde{\rho}_{jk}\right), j = 1, 2, \ldots, n \tag{28}$$

The weights of attributes are calculated by Eq. (29) [1].

$$\tilde{w}_j = \frac{\tilde{C}_j}{\sum_{j=1}^{n} \tilde{C}_j}, \tag{29}$$

The weights of attributes are arranged in descending order for the final ranking of attributes.

5 Prioritizing Supplier Selection Criteria Using SF-CRITIC Method

There are four different alternative suppliers: S1, S2, S3, and S4. The attributes are on-time delivery, (C1), nonconforming product ratio (C2), capacity (C3), and flexibility (C4). The spherical fuzzy scale is given in Table 1. The linguistic decision matrix is shown in Table 2. The decision matrix with corresponding SF-numbers is given in Table 3.

Table 1. Spherical fuzzy scale

Linguistic terms	(μ, v, π)
Absolutely High (AH)	(0.9, 0.1, 0.1)
Very High (VH)	(0.8, 0.2, 0.2)
High (H)	(0.7, 0.3, 0.3)
Slightly High (SH)	(0.6, 0.4, 0.4)
Equal (E)	(0.5, 0.5, 0.5)
Slightly Low (SL)	(0.4, 0.6, 0.4)
Low (L)	(0.3, 0.7, 0.3)
Very Low (VL)	(0.2, 0.8, 0.2)
Absolutely Low (AL)	(0.1, 0.9, 0.1)

Table 2. Decision matrix

	C1	C2	C3	C4
S1	VH	H	H	VH
S2	VL	VH	AH	AH
S3	VH	SL	AH	AH
S4	AH	H	E	AH

Table 3. Decision matrix with SF numbers

	C1	C2	C3	C4
S1	(0.8, 0.2, 0.2)	(0.1, 0.9, 0.1)	(0.3, 0.7, 0.3)	(0.2, 0.8, 0.2)
S2	(0.2, 0.8, 0.2)	(0.3, 0.7, 0.3)	(0.3, 0.7, 0.3)	(0.9, 0.1, 0.1)
S3	(0.8, 0.2, 0.2)	(0.7, 0.3, 0.3)	(0.7, 0.3, 0.3)	(0.1, 0.9, 0.1)
S4	(0.9, 0.1, 0.1)	(0.8, 0.2, 0.2)	(0.8, 0.2, 0.2)	(0.8, 0.2, 0.2)

The normalized decision matrix is given in Table 4.

Table 4. Normalized decision matrix

	C1	C2	C3	C4
S1	0.2206004	1	0	0.975387
S2	1	0.9106114	0	0
S3	0.2206004	0.255997	0.9152029	1
S4	0	0	1	0.2151707

Standard deviations of the criteria from Table 4 are calculated as 0.4390, 0.4903, 0.5540, and 0.5158, respectively. The indices of the criteria are 1.5623, 1.6161, 2.5258, and 1.7362. The weights of the criteria are finally found to be 0.2100, 0.2172, 0.3395, and 0.2333.

6 Conclusion

CRITIC is relatively a new method for prioritizing the criteria. It has been often used in multiple criteria decision making methods since 2017, whose origin is in the year 1995. CRITIC consists of the steps of obtaining normalized matrix, correlation matrix, standard deviations, indices of the criteria, and weights of the criteria. Since linguistic evaluations are generally preferred in decision matrices, the fuzzy set theory can capture the vagueness in these evaluations. Spherical fuzzy sets with their larger definition domain have successfully handled this vagueness in the CRITIC method. Instead of subtraction operation of normalization, spherical fuzzy distance operation has been used and it could rank the criteria perfectly. For further research, we suggest interval valued SF numbers or picture fuzzy sets to be used in CRITIC method for comparison purposes.

References

1. Diakoulaki, D., Mavrotas, G., Papayannakis, L.: Determining objective weights in multiple criteria problems: the CRITIC method. Comput. Oper. Res. **22**(7), 763–770. 217 (1995)
2. Kim, J.Y., Huh, Y., Kim, D.S., Yu, K.Y.: A new method for automatic areal feature matching based on shape similarity using CRITIC method. J. Korean Soc. Surv. Geodesy, Photogrammetry Cartography **29**(2), 113–121 (2011)
3. Houqiang, Y., Ling, L.: Five kinds of enterprise comprehensive evaluation based on the entropy value method and the CRITIC method. J. Hubei Inst. Technol. **32**, 83–84 (2012)
4. Wang, D., Zhao, J.: Design optimization of mechanical properties of ceramic tool material during turning of ultra-high-strength steel 300M with AHP and CRITIC method. Int. J. Adv. Manuf. Technol. **84**(9–12), 2381–2390 (2015). https://doi.org/10.1007/s00170-015-7903-7
5. Xie, Y., Li, Z.J., Xu, Z.: Evaluation on spontaneous combustion trend of sulfide ores based on the method of CRITIC and TOPSIS testing method. J. Saf. Environ. **14**(1), 122–125 (2014)
6. Zhao, Q. H., Zhou, X., Xie, R. F., & Li, Z. C. : Comparison of three weighing methods for evaluation of the HPLC fingerprints of Cortex Fraxini. J. Liq. Chromatography Rel. Technol. **34**(17), 2008–2019. 212 (2011)

7. Alinezhad A., Khalili, J.,: New Methods and Applications in Multiple Attribute Decision Making (MADM), International Series in Operations Research & Management Science, 277. Springer, Cham (2019). https://doi.org/10.1007/978-3-030-15009-9
8. Kim, J., Yu, K.: Areal feature matching based on similarity using CRITIC method. Int. Arch. Photogrammetry, Remote Sens. Spatial Inf. Sci. **40**, 75–78 (2015)
9. Ping, X.: Application of CRITIC method in medical quality assessment. Value Eng. **1**, 200–201 (2011)
10. Madic, M., Radovanovic, M.: Ranking of some most commonly used non-traditional machining processes using ROV and CRITIC methods. UPB Sci. Bull. Ser. D **77**(2), 193–204 (2015)
11. Kahraman, C., Çevik Onar, S., Öztayşi, B.: Engineering economic analyses using intuitionistic and hesitant fuzzy sets. J. Intell. Fuzzy Syst. **29**(3), 1151–1168 (2015)
12. Kahraman, C., Ateş, N.Y., Çevik, S., Gülbay, M.: Fuzzy multi-attribute cost-benefit analysis of e-Services. Int. J. Intell. Syst. **22**(5), 547–565 (2007)
13. Oztaysi, B., Cevik Onar, S., Kahraman, C.: Fuzzy multicriteria prioritization of Urban transformation projects for Istanbul. J. Intell. Fuzzy Syst. **30**(4), 2459–2474 (2016)
14. Kutlu Gündoğdu, F., Kahraman, C.: A novel VIKOR method using spherical fuzzy sets and its application to warehouse site selection. J. Intell. Fuzzy Syst. **37**, 1197–1211 (2019)
15. Kutlu Gundogdu, F., Kahraman, C.: Extension of CODAS with spherical fuzzy sets. Multiple Valued Logic Soft Comput. **33**(4–5), 481–505 (2019)
16. Kutlu Gündoğdu, F., Kahraman, C.: Spherical Fuzzy Sets and Spherical Fuzzy TOPSIS Method. J. Intell. Fuzzy Syst. **36**(1), 337–352 (2019)
17. Kutlu Gündoğdu, F., Kahraman, C.: Extension of WASPAS with spherical fuzzy sets. INFORMATICA **30**(2), 269–292 (2019)
18. Kutlu Gündoğdu, F., Kahraman, C.: A novel fuzzy topsis method using emerging interval-valued spherical fuzzy sets. In: Engineering Applications of Artificial Intelligence, October 2019, vol. 85, pp. 307–323 (2019)
19. Kutlu Gündoğdu, F., Kahraman, C.: A novel spherical fuzzy analytic hierarchy process and its renewable energy application. Soft. Comput. **24**(6), 4607–5462 (2019)
20. Oztaysi, B., Cevik Onar, S., Kutlu Gundogdu, F., Kahraman C.; Location based advertisement selection using spherical fuzzy AHP-VIKOR. J. Multi-Valued Logic Soft Comput. **35**, 5–23 (2020)
21. Oztaysi, B., Cevik Onar, S., Kahraman, C.: A dynamic pricing model for location based systems by using spherical fuzzy AHP scoring. J. Intell. Fuzzy Syst. **39**(5), 6293–6302 (2020)
22. Cevik Onar, S., Kahraman, C., Oztaysi, B.: Multi-criteria spherical fuzzy regret based evaluation of healthcare equipment stocks. J. Intell. Fuzzy Syst. **5**(39), 55987–5997
23. Khan, M.J., Kumam, P., Deebani, W., Kumam, W., Shah, Z.: Distance and similarity measures for spherical fuzzy sets and their applications in selecting mega projects. Mathematics **8**, 519 (2020)

Spherical Fuzzy REGIME Method Waste Disposal Location Selection

Basar Oztaysi(✉), Cengiz Kahraman(✉), and Sezi Cevik Onar(✉)

Industrial Engineering Department, İstanbul Technical University, 34367 Istanbul, Turkey
{oztaysib,kahramanc,cevikse}@itu.edu.tr

Abstract. The REGIME method, initially introduced by Hinloopen et al. [1, 2] is based on paired comparison methods which are easy to understand and uses qualitative data only in a mathematically justifiable way. The computational steps of the method are simple and can be easily apply to various complex problems.

Classical REGIME techniques uses crisp numbers to evaluate qualitative evaluations. In this paper we propose Spherical Fuzzy REGIME techniques which employs Spherical Fuzzy Sets proposed by Gündoğdu and Kahraman [3] to represent evaluations. Spherical Fuzzy Sets (SFS) are a new extension of Intuitionistic, Pythagorean and Neutrosophic Fuzzy sets. The proposed Spherical fuzzy REGIME (SF-REGIME) method is applied to the evaluation of waste disposal site selection problem. The decision model is constructed for three alternatives and five criteria in order to demonstrate the performance of the proposed SF-REGIMR method.

1 Introduction

The REGIME method was developed by Hinloopen et al. [1, 2] as a multi criteria decision making method based on pairwise comparisons with low computational complexity. The method aims to rank the alternatives based on qualitative or quantitative pairwise comparison evaluations. The classical REGIME techniques uses crisp numbers to mathematically express qualitative data [4]. In the literature various multcriteria decision making techniques are extended by using fuzzy sets in order to include vagueness and imprecision.

Fuzzy sets, introduced by by Zadeh [5], have been adopted to various engineering and decision problems. Fuzzy sets are generally used to mathematically represent imprecise and vague data in problems. In the literature there are various extensions of ordinary fuzzy sets that are type-2 fuzzy sets (T2FS) [6], intuitionistic fuzzy sets (IFS) [7], hesitant fuzzy sets (HFS) [8], Pythagorean fuzzy sets (PFS) [9] and neutrosophic sets (NS) [10]. Many fuzzy MCDM methods are developed based on these sets [11–15].

Inadequate waste management is one of the main sources of environmental pollution. While waste management is a huge topic, waste disposal takes place at the final stage of this process. All types of waste are collected and transported to landfills [16]. Thus, selection of waste disposal location is a complex decision making problem which should be handled from different perspectives [17]. The originality of the paper comes from

C. Kahraman et al. (Eds.): INFUS 2021, LNNS 308, pp. 715–723, 2022.
https://doi.org/10.1007/978-3-030-85577-2_84

the implementation of SF-REGIME to waste disposal site selection problem. The SF-REGIME enables decision maker to represent the vagueness and hesitancy in the decision making process using a linguistic evaluation scale based on spherical fuzzy sets.

The rest of this paper is organized as follows. Section 2 summarizes the preliminaries spherical fuzzy sets. Spherical fuzzy REGIME is given in Sect. 3. Section 4 applies Spherical Fuzzy REGIME method (SF-REGIME) to a waste disposal site selection problem. Finally, the study is concluded in the last section.

2 Spherical Fuzzy Sets

Intuitionistic and Pythagorean fuzzy sets are defined by independently assigned membership and non-membership parameters. The hesitancy in intuitionistic fuzzy sets is defined as a function of membership and non-membership as in Eq. (1) whereas it is defined for Pythagorean or Type-2 Intuitionistic fuzzy numbers as in Eq. (2).

$$\pi_{\tilde{I}} = 1 - \mu - \vartheta \tag{1}$$

$$\pi_{\tilde{P}} = (1 - \mu_{\tilde{P}}^2(u) - v_{\tilde{P}}^2(u))^{1/2} \tag{2}$$

Neutrosophic sets are defined with three parameters *truthiness, falsity* and *indeterminacy.* The values of these parameters are between 0 and 1 and the sum of these parameters can be between 0 and 3. The parameters can be defined independently in neutrosophic sets. In spherical fuzzy sets, the squared sum of membership, non-membership and hesitancy parameters can be between 0 and 1, and each of them can be defined between 0 and 1 independently.

Spherical fuzzy sets developed by Kutlu Gundogdu and Kahraman [3, 18]) are defined as follows:

A spherical fuzzy set $\tilde{A}_S$ of the universe of discourse U is given by

$$\tilde{A}_S = \left\{ \left\langle u, (\mu_{\tilde{A}_S}(u), v_{\tilde{A}_S}(u), \pi_{\tilde{A}_S}(u)) \middle| u \in U \right\} \tag{3}$$

where

$$\mu_{\tilde{A}_S} : U \to [0, 1], \quad v_{\tilde{A}_S}(u) : U \to [0, 1], \quad \pi_{\tilde{A}_S} : U \to [0, 1]$$

and

$$0 \le \mu_{\tilde{A}_S}^2(u) + v_{\tilde{A}_S}^2(u) + \pi_{\tilde{A}_S}^2(u) \le 1 \qquad \forall u \in U \tag{4}$$

For each u, the numbers $\mu_{\tilde{A}_S}(u)$, $v_{\tilde{A}_S}(u)$ and $\pi_{\tilde{A}_S}(u)$ are the degree of membership, non-membership and hesitancy of u to $\tilde{A}_S$, respectively.

On the basis of relationship between SFS and PFS, some novel operations for SFS are defined as in the following [3].

Definition 1: Basic Operators

$$\tilde{A}_S \oplus \tilde{B}_S = \left\{ \begin{array}{l} \left(\mu^2_{\tilde{A}_S} + \mu^2_{\tilde{B}_S} - \mu^2_{\tilde{A}_S}\mu^2_{\tilde{B}_S}\right)^{1/2},\ v_{\tilde{A}_S}v_{\tilde{B}_S}, \\ \left(\left(1-\mu^2_{\tilde{B}_S}\right)\pi^2_{\tilde{A}_S} + \left(1-\mu^2_{\tilde{A}_S}\right)\pi^2_{\tilde{B}_S} - \pi^2_{\tilde{A}_S}\pi^2_{\tilde{B}_S}\right)^{1/2} \end{array} \right\} \quad (5)$$

$$\tilde{A}_S \otimes \tilde{B}_S = \left\{ \begin{array}{l} \mu_{\tilde{A}_S}\mu_{\tilde{B}_S},\ \left(v^2_{\tilde{A}_S} + v^2_{\tilde{B}_S} - v^2_{\tilde{A}_S}v^2_{\tilde{B}_S}\right)^{1/2}, \\ \left(\left(1-v^2_{\tilde{B}_S}\right)\pi^2_{\tilde{A}_S} + \left(1-v^2_{\tilde{A}_S}\right)\pi^2_{\tilde{B}_S} - \pi^2_{\tilde{A}_S}\pi^2_{\tilde{B}_S}\right)^{1/2} \end{array} \right\} \quad (6)$$

$$\lambda \cdot \tilde{A}_S = \left\{ \begin{array}{l} \left(1-\left(1-\mu^2_{\tilde{A}_S}\right)^{\lambda}\right)^{1/2},\ v^{\lambda}_{\tilde{A}_S}, \\ \left(\left(1-\mu^2_{\tilde{A}_S}\right)^{\lambda} - \left(1-\mu^2_{\tilde{A}_S} - \pi^2_{\tilde{A}_S}\right)^{\lambda}\right)^{1/2} \end{array} \right\}, \lambda > 0 \quad (7)$$

$$\tilde{A}^{\lambda}_S = \left\{ \begin{array}{l} \mu^{\lambda}_{\tilde{A}_S},\ \left(1-\left(1-v^2_{\tilde{A}_S}\right)^{\lambda}\right)^{1/2}, \\ \left(\left(1-v^2_{\tilde{A}_S}\right)^{\lambda} - \left(1-v^2_{\tilde{A}_S} - \pi^2_{\tilde{A}_S}\right)^{\lambda}\right)^{1/2} \end{array} \right\}, \lambda > 0 \quad (8)$$

Definition 2: Spherical Weighted Arithmetic Mean (SWAM) with respect to, $w = (w_1, w_2 \ldots\ldots, w_n)$; $w_i \in [0, 1]$; $\sum_{i=1}^{n} w_i = 1$, is defined as;

$$SWAM_w(\tilde{A}_{S1}, \ldots\ldots, \tilde{A}_{Sn}) = w_1\tilde{A}_{S1} + w_2\tilde{A}_{S2} + \ldots\ldots + w_n\tilde{A}_{Sn}$$
$$= \left\{ \begin{array}{l} \left[1-\prod_{i=1}^{n}(1-\mu^2_{A_s})^{w_i}\right]^{1/2}, \prod_{i=1}^{n} v^{w_i}_{A_s}, \\ \left[\prod_{i=1}^{n}(1-\mu^2_{A_s})^{w_i} - \prod_{i=1}^{n}(1-\mu^2_{A_s} - \pi^2_{A_{Si}})^{w_i}\right]^{1/2} \end{array} \right\} \quad (9)$$

Spherical Weighted Geometric Mean (SWGM) with respect to, $w = (w_1, w_2 \ldots\ldots, w_n)$; $w_i \in [0, 1]$; $\sum_{i=1}^{n} w_i = 1$, is defined as;

$$SWGM_w(\tilde{A}_1, \ldots\ldots, \tilde{A}_n) = \tilde{A}^{w_1}_{S1} + \tilde{A}^{w_2}_{S2} + \ldots\ldots + \tilde{A}^{w_n}_{Sn}$$

$$= \left\{ \begin{array}{l} \prod_{i=1}^{n} \mu^{w_i}_{\tilde{A}_{Si}},\ \left[1-\prod_{i=1}^{n}(1-v^2_{\tilde{A}_{Si}})^{w_i}\right]^{1/2}, \\ \left[\prod_{i=1}^{n}(1-v^2_{\tilde{A}_{Si}})^{w_i} - \prod_{i=1}^{n}(1-v^2_{\tilde{A}_{Si}} - \pi^2_{\tilde{A}_{Si}})^{w_i}\right]^{1/2} \end{array} \right\} \quad (10)$$

3 Spherical Fuzzy REGIME Technique

The steps of the proposed SF-REGIME technique is as follows:

1. Construct the matrices of alternatives and attributes based on decision makers' evaluations.
2. Convert the linguistic evaluations into Spherical Fuzzy Sets using the table given in Table 1.

Table 1. The linguistic evaluations and spherical fuzzy representations.

Linguistic evaluation	Abbr.	Spherical fuzzy representation
Absolutely more important	AMI	(0.9, 0.1, 0.1)
Very high important	VHI	(0.8, 0.2, 0.2)
High important	HI	(0.7, 0.3, 0.3)
Slightly more important	SMI	(0.6, 0.4, 0.4)
Moderate	EI	(0.5, 0.5, 0.5)
Slightly low important	SLI	(0.4, 0.6, 0.4)
Low important	LI	(0.3, 0.7, 0.3)
Very low important	VLI	(0.2, 0.8, 0.2)
Absolutely low important	ALI	(0.1, 0.9, 0.1)

3. Aggregate the evaluations of the decision makers' using SWGM operator given in Eq. 10.
4. Use score function to provide the defuzzified value of the evaluations.

$$Score\left(C_j\left(\tilde{X}_i\right)\right) = \left(2\mu_{ij} - \pi_{ij}\right)^2 - \left(v_{ij} - \pi_{ij}\right)^2 \tag{11}$$

5. REGIME Matrix is formed based on pairwise comparison of the alternatives. For each C_j attribute, the $E_{fl,j}$ value is calculated for each A_f and A_l alternatives using Eq. 12.

$$E_{fl,j} = \begin{cases} -1 \ if \ r_{fj} < r_{lj} \\ 0 \ if \ r_{fj} = r_{lj} \\ +1 \ if \ r_{fj} > r_{lj} \end{cases} ;\mathrm{i} = 1, \ldots \mathrm{m}, \mathrm{j} = 1, \ldots, \mathrm{n} \tag{12}$$

Where (r_{lj}, r_{fj}) indicates the rank of (A_l, A_f) alternative based on the attribute C_j. When to alternatives are examined in all attributes, a vector is defined as in Eq. 13.

$$E_{fl} = \left(E_{fl,1}, \ldots, E_{fl,j}, \ldots, E_{fl,n}\right),\ j = 1, \ldots, n \tag{13}$$

The vector in Eq. 13 is call the REGIME.

6. REGIME Matrix is form based on the REGIME vectors resulting from the pairwise comparisons of the alternatives.
7. The Guide index $\overline{E}_{fl}$ is calculated by using Eq. 14

$$\overline{E}_{fl} = \sum\nolimits_{j=1}^{n} E_{fl,j}.w_j \tag{14}$$

8. The value of the best alternative is obtained by evaluating the guide indices. In fact, the comparison is based on the $\overline{E}_{fl} - \overline{E}_{lf}$ subtract. The positive result indicates that the alternative A_f is superior to the alternative A_l. The negative result demonstrates the superiority of alternative A_l over alternative A_f.

4 Application: Waste Disposal Location Selection

After a literature review and expert interviews the criteria for the problem is selected as follows:

C1: Unit Land Cost: The criteria defines the unit initial cost of acquiring the land. Since waste disposal location is an investment, the cost is an important factor for the selection of the most suitable alternative.

C2: Potential of growth: After the initial decision, there can be some changes about the requirements and capacity may be increased. This criterion evaluates the expension potential of the alternative.

C3: Environmental supportive conditions. This criteria defines the facilities such as air, water, energy, and electric supply.

C4: Personnel availability: This criteria shows the availability of workforce for the potential waste disposal location.

C5: Public perception: This criterion shows how the selection of the location effect the residents of the area.

After the criteria are selected the three decision makers are asked to evaluate the alternatives by using the linguistic terms given in Table 1. The three decision makers evaluate the alternatives based on five criteria as in Table 2.

Table 2. Evaluations of the decision makers

	Decision maker 1					Decision maker 2					Decision maker 3				
	C1	C2	C3	C4	C5	C1	C2	C3	C4	C5	C1	C2	C3	C4	C5
A1	AMI	VHI	SLI	EI	SMI	AMI	HI	LI	SMI	SLI	HI	SLI	LI	HI	HI
A2	VHI	HI	LI	SLI	EI	HI	HI	EI	SMI	EI	HI	LI	LI	LI	EI
A3	HI	SMI	VLI	LI	SLI	SMI	SMI	HI	LI	HI	SMI	HI	LI	SLI	SMI

The evaluations are later converted to Spherical fuzzy numbers as in the following tables (Table 3).

Table 3. Spherical fuzzy representations of the DM evaluations.

DM1	C1	C2	C3	C4	C5
A1	(0.9,0.1,0.1)	(0.8,0.2,0.2)	(0.4,0.6,0.4)	(0.5,0.5,0.5)	(0.6,0.4,0.4)
A2	(0.8,0.2,0.2)	(0.7,0.3,0.3)	(0.3,0.7,0.3)	(0.4,0.6,0.4)	(0.5,0.5,0.5)
A3	(0.7,0.3,0.3)	(0.6,0.4,0.4)	(0.2,0.8,0.2)	(0.3,0.7,0.3)	(0.4,0.6,0.4)
DM2	C1	C2	C3	C4	C5
A1	(0.9,0.1,0.1)	(0.7,0.3,0.3)	(0.3,0.7,0.3)	(0.6,0.4,0.4)	(0.4,0.6,0.4)
A2	(0.7,0.3,0.3)	(0.7,0.3,0.3)	(0.5,0.5,0.5)	(0.6,0.4,0.4)	(0.5,0.5,0.5)
A3	(0.6,0.4,0.4)	(0.6,0.4,0.4)	(0.7,0.3,0.3)	(0.3,0.7,0.3)	(0.7,0.3,0.3)
DM3	C1	C2	C3	C4	C5
A1	(0.7,0.3,0.3)	(0.4,0.6,0.4)	(0.3,0.7,0.3)	(0.7,0.3,0.3)	(0.7,0.3,0.3)
A2	(0.7,0.3,0.3)	(0.3,0.7,0.3)	(0.3,0.7,0.3)	(0.3,0.7,0.3)	(0.5,0.5,0.5)
A3	(0.6,0.4,0.4)	(0.7,0.3,0.3)	(0.3,0.7,0.3)	(0.4,0.6,0.4)	(0.6,0.4,0.4)

The values are then aggregated by using SWGM method and the resulting aggregated evaluations are formed as in Table 4.

Table 4. Aggregated spherical fuzzy representations of the DM evaluations.

	C1	C2	C3	C4	C5
A1	(0.893,0.15,0.205)	(0.741,0.331,0.324)	(0.514,0.549,0.409)	(0.732,0.325,0.469)	(0.7,0.365,0.403)
A2	(0.829,0.212,0.284)	(0.681,0.398,0.331)	(0.538,0.528,0.431)	(0.591,0.478,0.432)	(0.66,0.398,0.517)
A3	(0.759,0.291,0.373)	(0.759,0.291,0.4)	(0.53,0.548,0.373)	(0.514,0.549,0.409)	(0.7,0.365,0.445)

By using Score function given in Eq. 11. the score values are calculated (Table 5).

Table 5. Score values of the aggregated evaluations.

	C1	C2	C3	C4	C5
A1	2.49	1.34	0.36	0.97	0.99
A2	1.88	1.06	0.41	0.56	0.63
A3	1.31	1.24	0.44	0.36	0.91

The pairwise comparisons are evaluated accomplished as given in Table 6.

Table 6. The pairwise comparisons of the alternatives.

	C1	C2	C3	C4	C5
A12	0.61	0.28	−0	0.41	0.36
A13	1.19	0.11	−0.1	0.6	0.09
A21	−0.6	−0.3	0.04	− 0.4	−0.4
A23	0.58	−0.2	−0	0.2	−0.3
A31	−1.2	−0.1	0.08	−0.6	−0.1
A32	−0.6	0.18	0.04	−0.2	0.28

Based on the pairwise comparisons value the REGIME Matrix is formed (Table 7).

Table 7. The pairwise comparisons of the alternatives.

	C1	C2	C3	C4	C5
A12	1	1	−1	1	1
A13	1	1	−1	1	1
A21	−1	−1	1	−1	−1
A23	1	−1	−1	1	−1
A31	−1	−1	1	−1	−1
A32	−1	1	1	−1	1

The next step is to calculate the guide index of each pairwise comparison. The resulting results are given in Table 8.

Table 8. The guide index values of the pairwise comparisons

	C1	C2	C3	C4	C5	$\overline{E}_{ij}$
A12	0.1	0.2	−0.3	0.15	0.25	0.4
A13	0.1	0.2	−0.3	0.15	0.25	0.4
A21	−0.1	−0.2	0.3	−0.2	−0.3	−0.4
A23	0.1	−0.2	−0.3	0.15	− 0.3	−0.5
A31	−0.1	−0.2	0.3	−0.2	−0.3	−0.4
A32	−0.1	0.2	0.3	−0.2	0.25	0.5

Based on these results, since A12, A13 and A32 are positive we can conclude A1 > A3 > A2 So Alternative A2 should be selected.

5 Conclusions

Spherical fuzzy sets are a new extension of ordinary fuzzy sets. They are based on the mixed idea of intuitionistic fuzzy sets of type 2 (Pythagorean fuzzy sets), picture fuzzy sets, and neutrosophic sets. Spherical fuzzy sets have been used in various decision making problem such as Spherical fuzzy TOPSIS and Spherical fuzzy AHP. On the other hand REGIME method is a MCDM method which is based on pairwise comparison of the alternatives. In this paper we propose SF-REGIME methodology which integrates Spherical fuzzy sets with REGIME technique.

For further research, we suggest integrating AHP and REGIME techniques for the weights of the criteria. In this study, the weights are assumed to be given, however SF-AHP can be used to calculate the weights. In another branch of studies, the same problem can be handled by other multi-criteria decision making models and the result can be compared with the result of this paper.

References

1. Hinloopen, E., Nijkamp, P.: Qualitative multiple criteria choice analysis. Qual. Quant. **24**(1), 37–56 (1990)
2. Hinloopen, E., Nijkamp, P.: REGIME methode voor ordinal multi-criteria analyse. Kwantitatieve Methoden **7**(22), 61–78 (1986)
3. Gündogdu, F., Kahraman, C.: A novel VIKOR method using spherical fuzzy sets and its application to warehouse site selection. J. Intell. Fuzzy Syst. **37**(1), 1197–1211 (2019)
4. Alinezhad, A., Khalili, J.: REGIME method. In: New Methods and Applications in Multiple Attribute Decision Making (MADM). ISORMS, vol. 277, pp. 9–15. Springer, Cham (2019). https://doi.org/10.1007/978-3-030-15009-9_2
5. Zadeh, L.A.: Fuzzy sets. Inf. Control **8**, 338–353 (1965)
6. Zadeh, L.A.: The concept of a linguistic variable and its application to approximate reasoning. Inf. Sci. **8**, 199–249 (1975)
7. Atanassov, K.T.: Intuitionistic fuzzy sets. Fuzzy Sets Syst. **20**(1), 87–96 (1986)
8. Torra, V.: Hesitant fuzzy sets. Int. J. Intell. Syst. **25**(6), 529–539 (2010)
9. Yager, R.R.: Pythagorean fuzzy subsets. In: Joint IFSA World Congress and NAFIPS Annual Meeting, Edmonton, Canada (2013)
10. Smarandache, F.: Neutrosophy: neutrosophic probability, set, and logic: analytic synthesis & synthetic analysis (1998)
11. Kahraman, C., Cevik Onar, S., Oztaysi, B.: A comparison of wind energy investment alternatives using interval-valued intuitionistic fuzzy benefit/cost analysis. Sustainability **8**(2), 118 (2016). https://doi.org/10.3390/su8020118
12. Kahraman, C., Oztaysi, B.: Supply Chain Management Under Fuzziness Recent Developments and Techniques. Springer, Berlin (2014). https://doi.org/10.1007/978-3-642-539 39-8
13. Öztaysi, B., Behret, H., Kabak, Ö., Sarı, I.U., Kahraman, C.: Fuzzy inference systems for disaster response. In: Vitoriano, B., Montero, J., Ruan, D. (eds.) Decision Aid Models for Disaster Management and Emergencies. Atlantis Computational Intelligence Systems, vol. 7. Atlantis Press, Paris (2013). https://doi.org/10.2991/978-94-91216-74-9_4
14. Kaya, İ, Oztaysi, B., Kahraman, C.: A two-phased fuzzy multicriteria selection among public transportation investments for policy-making and risk governance. Int. J. Uncertainty Fuzziness Knowl. Based Syst. **20**(supp01), 31–48 (2012)

15. Oztaysi, B., Cevik Onar, S., Kahraman, C.: Fuzzy Multicriteria Prioritization of Urban Transformation Projects for Istanbul. 1 Jan 2016, 2459–2474 (2016)
16. Kahraman, C., Ghorabaee, M.K., Zavadskas, E.K., Cevik, O.S., Yazdani, M., Oztaysi, B.: Intuitionistic fuzzy EDAS method: an application to solid waste disposal site selection. J. Environ. Eng. Landsc. Manag. **25**(1), 1–12 (2017). https://doi.org/10.3846/16486897.2017.1281139
17. Yazdani, M., Tavana, M., Pamučar, D., Chatterjee, P.: A rough based multi-criteria evaluation method for healthcare waste disposal location decisions. Comput. Ind. Eng. **143**, 106394 (2020). https://doi.org/10.1016/j.cie.2020.106394
18. Gündoğdu, F.K., Kahraman, C.: A novel spherical fuzzy QFD method and its application to the linear delta robot technology development. Eng. Appl. Artif. Intell. **87**, 103348 (2019). https://doi.org/10.1016/j.engappai.2019.103348

Evaluation of Suppliers in the Perspective of Digital Transformation: A Spherical Fuzzy TOPSIS Approach

Serhat Aydın[1](✉), Ahmet Aktas[2], and Mehmet Kabak[2]

[1] Industrial Engineering Department, National Defense University, Turkish Air Academy, 34149 Istanbul, Turkey
saydin3@hho.edu.tr

[2] Industrial Engineering Department, Gazi University, 06570 Ankara, Turkey

Abstract. There are great efforts on development of new technologies in different regions of the world. Thanks to the progress of science and technology, it is possible to see some news on several kind of developments every day. As a natural result of these developments, people want emerging technologies to be included in new products they will purchase. So, companies are faced with the challenge of development in their products continuously. Automotive industry can be considered among the industrial areas, which are being affected much by these emerging technologies and automotive companies need to adapt to the new technologies rapidly. As a result of this situation, suppliers of automotive manufacturers have to improve their processes. This study focuses on developing an analytic model for a decision problem of supplier evaluation in the automotive industry. Alternative suppliers are ranked by considering several factors related to digital transformation after a multi-criteria analysis. Four alternative suppliers are evaluated based on digital technologies by considering four criteria and ranked by a Spherical fuzzy extension of TOPSIS technique.

Keywords: Digital transformation · Automotive industry · Supplier evaluation · Fuzzy logic · Spherical fuzzy sets

1 Introduction

One of the popular topics in recent years' world is Digital Transformation (DX). The concept of DX addresses implementation of new digital technologies into processes in order to improve efficiency, created value and prosperity [1]. It is expected that DX will grow rapidly, because several organizations, companies and governments developed strategic plans for DX so far.

Due to the efforts on implementation of holistic planning of business processes, companies need to develop close relationships with their suppliers and customers. Within the developments in products and processes as a result of DX, companies will need to check processes of their partners.

C. Kahraman et al. (Eds.): INFUS 2021, LNNS 308, pp. 724–732, 2022.
https://doi.org/10.1007/978-3-030-85577-2_85

Automotive industry is one of the sectors that will be affected more by DX. Since automotive products consist of thousands of components including many electronic parts, new products developed according to the DX will bring product and process renovations in this area. Moreover, the changes in main products due to the DX will require changes in components of the product, too. Therefore, automotive producers and their suppliers are expected to develop their processes and products considering the developments obtained by DX.

At this point, evaluation of suppliers emerges as an important decision problem for automotive producers. While rapidly changing market conditions forces companies to adopt new technological standards, they should also have suppliers that adopt these standards, too. Evaluation of suppliers should be done with consideration of several factors together. Some factors for supplier evaluation can be referred as cost, quality, capacity, after sales service, etc. This kind of evaluations with multiple factors can be determined by multi-criteria analysis methods.

The evaluation process is taken into consideration within the perspective of DX in this study. Linguistic expressions of three experts for four supplier companies on four evaluation factor are collected and suppliers evaluated by Spherical Fuzzy (SF) extension of TOPSIS technique [2]. Since the Spherical Fuzzy Sets (SFS) provide a greater preference domain for decision makers (DMs) to appoint membership values than the other forms of fuzzy sets, SFS is used in this study for definition of linguistic expressions.

The rest of the paper is organized as follows: A brief summary of literature on supplier selection and evaluation in automotive industry is presented in the second part. Next, the main definitions of SFS are given in the third part. Then, Spherical Fuzzy extension TOPSIS technique is explained in the fourth part. The fifth part presents the application of the supplier evaluation. Finally, the conclusions of the study are given in the sixth part with suggestions for extension for researchers.

2 Literature Review

Many researches on the evaluation and selection of suppliers in several sectors have been conducted and published, so far. Some of these sectors can be listed as textile [3], home appliances [4], electronic [5] and automotive [6]. Automotive industry needs more careful consideration than all other industries about supplier selection and evaluation, because automobiles consist of thousands of components and most of these components are supplied from different suppliers. Each component and each supplier must be determined after analytic and careful evaluation; otherwise reliability of the automobile will probably reduce. In this part, some studies published in the last five years on supplier evaluation and selection in automotive industry are summarized.

MCDM techniques are very common in supplier evaluation and selection problems. Analytic Hierarchy Process (AHP) [7–9], Analytic Network Process (ANP) [10–12], and Technique of Order Preference by Similarity to Ideal Solution (TOPSIS) [8, 13, 14] methods are frequently used. Other multi-criteria methods confronted on supplier evaluation and selection are Multi-Attributive Border Approximation Area Comparison (MABAC) [8], Weighted Aggregated Sum-Product Assessment (WASPAS) [8], Decision-making

Trial and Evaluation Laboratory (DEMATEL) [12, 15], Quality Function Deployment (QFD) [16], and ELECTRE-TRI [17]. Some studies also consist of mathematical models [9, 11] or hybrid methodologies [8, 12].

Supplier evaluation in these studies are completely considered by similar criteria like cost, quality, etc. The popular topic of DX has largely been ignored in these studies. To fill the gap in supplier evaluation problems, this study proposes a supplier evaluation model based on TOPSIS method by considering several factors which show the adaptability of suppliers to digital transformation. SFS, a new type of fuzzy uncertainty, is also taken into account for expression of performances of alternatives in views of DX criteria.

3 Basic Definitions of Spherical Fuzzy Sets

Definitions related to SFS are presented in this section as follows:

Definition 1: SFS $\tilde{A}_S$ of the universe of discourse U is expressed as.

$$\tilde{A}_S = \left\{ \left\langle u, \left(\mu_{\tilde{A}_S}(u), v_{\tilde{A}_S}(u), \pi_{\tilde{A}_S}(u) \right) \middle| u \in U \right\} \tag{1}$$

where

$$\mu_{\tilde{A}_S} : U \to [0, 1], v_{\tilde{A}_S}(u) : U \to [0, 1], \pi_{\tilde{A}_S} : U \to [0, 1]$$

$$0 \le \mu^2_{\tilde{A}_S}(u) + v^2_{\tilde{A}_S}(u) + \pi^2_{\tilde{A}_S}(u) \le 1 \forall u \in U \tag{2}$$

For each u, the values of $\mu_{\tilde{A}_S}(u)$, $v_{\tilde{A}_S}(u)$ and $\pi_{\tilde{A}_S}(u)$ indicates the membership, non-membership and hesitancy degrees of u to $\tilde{A}_S$, respectively.

Definition 2: Basic operations on SFS:

$$\tilde{A}_S \oplus \tilde{B}_S = \left\{ \begin{array}{l} \left(\mu^2_{\tilde{A}_S} + \mu^2_{\tilde{B}_S} - \mu^2_{\tilde{A}_S}\mu^2_{\tilde{B}_S} \right)^{1/2}, v_{\tilde{A}_S}v_{\tilde{B}_S}, \\ \left(\left(1 - \mu^2_{\tilde{B}_S}\right)\pi^2_{\tilde{A}_S} + \left(1 - \mu^2_{\tilde{A}_S}\right)\pi^2_{\tilde{B}_S} - \pi^2_{\tilde{A}_S}\pi^2_{\tilde{B}_S} \right)^{1/2} \end{array} \right\} \tag{3}$$

$$\tilde{A}_S \otimes \tilde{B}_S = \left\{ \begin{array}{l} \mu_{\tilde{A}_S}\mu_{\tilde{B}_S}, \left(v^2_{\tilde{A}_S} + v^2_{\tilde{B}_S} - v^2_{\tilde{A}_S}v^2_{\tilde{B}_S} \right)^{1/2}, \\ \left(\left(1 - v^2_{\tilde{B}_S}\right)\pi^2_{\tilde{A}_S} + \left(1 - v^2_{\tilde{A}_S}\right)\pi^2_{\tilde{B}_S} - \pi^2_{\tilde{A}_S}\pi^2_{\tilde{B}_S} \right)^{1/2} \end{array} \right\} \tag{4}$$

$$\lambda \cdot \tilde{A}_S = \left\{ \begin{array}{l} \left(1 - \left(1 - \mu^2_{\tilde{A}_S}\right)^{\lambda} \right)^{1/2}, v^{\lambda}_{\tilde{A}_S}, \\ \left(\left(1 - \mu^2_{\tilde{A}_S}\right)^{\lambda} - \left(1 - \mu^2_{\tilde{A}_S} - \pi^2_{\tilde{A}_S}\right)^{\lambda} \right)^{1/2} \end{array} \right\} for \lambda > 0 \tag{5}$$

Definition 3: Spherical Weighted Arithmetic Mean (SWAM) is calculated by the following formula, where $w = (w_1, w_2 \ldots\ldots, w_n); w_i \in [0, 1]; \sum_{j=1}^{n} w_i = 1$.

$$SWAM_w\left(\tilde{A}_{S1}, \ldots\ldots, \tilde{A}_{Sn}\right) = w_1\tilde{A}_{S1} + w_2\tilde{A}_{S2} + \ldots\ldots + w_n\tilde{A}_{Sn}$$

$$= \left\{ \begin{array}{l} \left[1 - \prod_{i=1}^{n}\left(1 - \mu_{\tilde{A}_{Si}}^2\right)^{wi}\right]^{1/2}, \\ \prod_{i=1}^{n} v_{\tilde{A}_{Si}}^{wi}, \left[\prod_{i=1}^{n}\left(1 - \mu_{\tilde{A}_{Si}}^2\right)^{wi} - \prod_{i=1}^{n}\left(1 - \mu_{\tilde{A}_{Si}}^2 - \pi_{\tilde{A}_{Si}}^2\right)^{wi}\right]^{1/2} \end{array} \right\} \tag{6}$$

Definition 4: Sorting SFS is made by Score and Accuracy functions, which are calculated by;

$$Score(A_S) = \left(\mu_{A_S} - \pi_{A_S}\right)^2 - \left(v_{A_S} - \pi_{A_S}\right)^2 \tag{7}$$

$$Accuracy(A_S) = \mu_{\tilde{A}_S}^2 + v_{\tilde{A}_S}^2 + \pi_{\tilde{A}_S}^2 \tag{8}$$

Note that: $A_S < B_S$ if and only if.

$$\begin{array}{l} \text{i. } Score(A_S) < Score(B_S) \text{ or} \\ \text{ii. } Score(A_S) = Score(B_S) \quad and \quad Accuracy(A_S) < Accuracy(B_S) \end{array} \tag{9}$$

4 Extension of TOPSIS with SFS

A decision making problem with multiple criteria can be expressed as a decision matrix, in which elements represent the score of each alternative in views of each criterion under SF environment. Assume that $X = \{x_1, x_2, \ldots\ldots x_m\}$ $(m \geq 2)$ represents the feasible alternatives set, $C = \{C_1, C_2, \ldots\ldots C_n\}$ shows the set of criteria that affects the decision, and $w = \{w_1, w_2, \ldots\ldots w_n\}$ is the vector of criteria weights, where $0 \leq w_j \leq 1$ and $\sum_{j=1}^{n} w_j = 1$.

Table 1. Linguistic variables with SF equivalents

Linguistic variable	(μ, v, π)
Absolutely more Important (AM)	(0.9, 0.1, 0.1)
Very Highly Important (VH)	(0.8, 0.2, 0.2)
Highly Important (H)	(0.7, 0.3, 0.3)
Slightly More Important (SM)	(0.6, 0.4, 0.4)
Equally Important (E)	(0.5, 0.5, 0.5)
Slightly Low Important (SL)	(0.4, 0.6, 0.4)
Lowly Important (L)	(0.3, 0.7, 0.3)
Very Lowly Important (VL)	(0.2, 0.8, 0.2)
Absolutely Low Important (AL)	(0.1, 0.9, 0.1)

Step 1: DMs preferences on evaluation criteria and alternatives are collected using the linguistic variables given in Table 1.
Step 2: The judgments of DMs are aggregated using SWAM operator given in Eq. (6).
Step 2.1: Criteria weights are determined according to Table 1.
Step 2.2: Aggregated SF decision matrix is constructed based on the opinions of DMs. Scores of alternatives $X_i(i = 1, 2.....m)$ in views of criterion $C_j(j = 1, 2.....n)$ expressed as $C_j(X_i) = (\mu_{ij}, v_{ij}, \pi_{ij})$ and decision matrix of the problem $D = (C_j(X_i))_{mxn}$ is a SF decision matrix. Decision matrix of a multi-criteria decision making problem with SF elements $D = (C_j(X_i))_{mxn}$ should be constructed as in Eq. (10).

$$D = (C_j(X_i))_{mxn} = \begin{pmatrix} (\mu_{11}, v_{11}, \pi_{11}) & \mu_{12}, v_{12}, \pi_{12} & \mu_{1n}, v_{1n}, \pi_{1n} \\ \mu_{21}, v_{21}, \pi_{21} & \mu_{22}, v_{22}, \pi_{22} & \mu_{2n}, v_{2n}, \pi_{2n} \\ . & . & . \\ . & . & . \\ \mu_{m1}, v_{m1}, \pi_{m1} & \mu_{m2}, v_{m2}, \pi_{m2} \therefore & \mu_{mn}, v_{mn}, \pi_{mn} \end{pmatrix} \tag{10}$$

Step 3: Aggregated weighted SF decision matrix is constructed by utilizing Eq. (11).

$$D = (C_j(X_{iw}))_{mxn} = \begin{pmatrix} (\mu_{11w}, v_{11w}, \pi_{11w}) & \mu_{12w}, v_{12w}, \pi_{12w} & \mu_{1nw}, v_{1nw}, \pi_{1nw} \\ \mu_{21w}, v_{21w}, \pi_{21w} & \mu_{22w}, v_{22w}, \pi_{22w} & \mu_{2nw}, v_{2nw}, \pi_{2nw} \\ . & . & . \\ . & . & . \\ \mu_{m1w}, v_{m1w}, \pi_{m1w} & \mu_{m2w}, v_{m2w}, \pi_{m2w} \therefore & \mu_{mnw}, v_{mnw}, \pi_{mnw} \end{pmatrix} \tag{11}$$

Step 4: Aggregated weighted decision matrix is defuzzified by using Eq. (12), which is a reformulated form of Eq. (7).

$$Score(C_j(X_{iw})) = (\mu_{ijw} - \pi_{ijw})^2 - (v_{ijw} - \pi_{ijw})^2 \tag{12}$$

Step 5: The SF Positive and Negative Ideal Solutions (SF-PIS and SF-NIS) are determined.
For the SF-PIS:

$$X^* = \left\{ C_j, \max_i \langle Score(C_j(X_{iw})) \rangle \middle| j = 1, 2...n \right\} \tag{13}$$

$$X^* = \{\langle C_1, (\mu_1^*, v_1^*, \pi_1^*), C_2, (\mu_2^*, v_2^*, \pi_2^*).....C_n, (\mu_n^*, v_n^*, \pi_n^*) \rangle\}$$

For the SF-NIS:

$$X^- = \left\{ C_j, \min_i \langle Score(C_j(X_{iw})) \rangle \middle| j = 1, 2...n \right\} \tag{14}$$

$$X^- = \{\langle C_1, (\mu_1^-, v_1^-, \pi_1^-) \rangle, \langle C_2, (\mu_2^-, v_2^-, \pi_2^-) \rangle .....\langle C_n, (\mu_n^-, v_n^-, \pi_n^-) \rangle\}$$

Step 6: The distances of alternatives to SF-PIS and SF-NIS are calculated using Eq. (15) and Eq. (16).
For the SF-NIS:

$$D(X_i, X^-) = \sqrt{\frac{1}{2n}\sum\nolimits_1^n \left(\mu_{X_i} - \mu_{X^-}\right)^2 + \left(v_{X_i} - v_{X^-}\right)^2 + \left(\pi_{X_i} - \pi_{X^-}\right)^2} \quad (15)$$

For the SF-PIS:

$$D(X_i, X^*) = \sqrt{\frac{1}{2n}\sum_1^n \left(\mu_{X_i} - \mu_{X^*}\right)^2 + \left(v_{X_i} - v_{X^*}\right)^2 + \left(\pi_{X_i} - \pi_{X^*}\right)^2} \quad (16)$$

Step 7: The greatest distance negative ideal and the shortest distance to positive ideal solutions are determined by Eq. (17) and Eq. (18).

$$Di^- \quad \underset{1 \leq i \leq m}{max_i^-} \; max \quad (17)$$

$$Di^* \quad \underset{1 \leq i \leq m}{max_i^*} \; min \quad (18)$$

Step 8: Closeness ratio is calculated for each alternative by using Eq. (19).

$$\xi(X_i) = \frac{D(X_i, X^*)}{Di^*_{min} \frac{D(X_i, X^-)}{Di^-_{max}}} \quad (19)$$

Step 9: Alternatives are ranked by ordering them by the increasing values of closeness ratio.

5 Application

Four companies (A_1, A_2, A_3, A_4) are evaluated. After a review of literature, a number of various criteria are determined. Predictive Maintenance (C_1), Mobility-As-A-Service (C_2), Data Security and Protection (C_3), Connected Supply Chain and Improved Manufacturing (C_4). Aggregation of opinions of three experts' (DM1, DM2, DM3), who have different levels of experience on automotive industry, made by using the weight values 0.3, 0.3 and 0.4, respectively.

Step 1: Because of the space limit all judgments are not given in the paper. The aggregated judgments are given in Step 2.
Step 2.1: Criteria weights are; C_1 = (0.6,0.40,0.40), C_2 = (0.4,0.60,0.40), C_3 = (0.3,0.70,0.30), C_4 = (0.4,0.60,0.40).
Step 2.2: Aggregated expert judgments are calculated using SWAM operator (Table 2):

Table 2. Aggregated decision matrix

	C1	C2	C3	C4
A1	(0.42, 0.61, 0.34)	(0.66, 0.36, 0.33)	(0.81, 0.22, 0.18)	(0.46, 0.56, 0.39)
A2	(0.75, 0.27, 0.23)	(0.59, 0.44, 0.30)	(0.32, 0.69, 0.33)	(0.82, 0.19, 0.17)
A3	(0.59, 0.44, 0.30)	(0.82, 0.19, 0.19)	(0.61, 0.40, 0.36)	(0.59, 0.44, 0.30)
A4	(0.32, 0.71, 0.33)	(0.52, 0.50, 0.39)	(0.18, 0.84, 0.19)	(0.55, 0.50, 0.25)

Step 3: Aggregated weighted SF decision matrix is constructed by utilizing Eq. (11) (Table 3).

Table 3. Weighted SF decision matrix

	C1	C2	C3	C4
A1	(0.25, 0.69, 0.42)	(0.26, 0.66, 0.44)	(0.24, 0.72, 0.32)	(0.18, 0.75, 0.43)
A2	(0.45, 0.47, 0.43)	(0.24, 0.69, 0.42)	(0.10, 0.86, 0.30)	(0.33, 0.62, 0.41)
A3	(0.36, 0.57, 0.44)	(0.33, 0.62, 0.41)	(0.18, 0.76, 0.36)	(0.24, 0.69, 0.42)
A4	(0.19, 0.77, 0.39)	(0.21, 0.72, 0.44)	(0.06, 0.92, 0.20)	(0.22, 0.72, 0.39)

Step 4: The aggregated weighted decision matrix is constructed as shown in Table 4.

Table 4. Score function values

	C1	C2	C3	C4
A1	−0.190	−0.135	−0.233	−0.308
A2	0.428	−0.220	−0.664	−0.036
A3	0.145	0.039	−0.389	−0.220
A4	−0.383	−0.300	−0.813	−0.279

Step 5: SF-PIS and SF-NIS are determined (Table 5).

Table 5. SF-PIS and SF-NIS

	C_1	C_2	C_3	C_4
X^* (Best)	(0.45, 0.47, 0.43)	(0.33, 0.62, 0.41)	(0.24, 0.72, 0.32)	(0.33, 0.62, 0.41)
X^* (Worst)	(0.19, 0.77, 0.39)	(0.21, 0.72, 0.44)	(0.06, 0.92, 0.20)	(0.18, 0.75, 0.43)

Step 6: The distances between alternative X_i and PIS and NIS are calculated as follows (Table 6):

Table 6. Distances to PIS and NIS

	$D(X_i, X^*)$	$D(X_i, X^-)$
A1	0.021	0.103
A2	0.113	0.147
A3	0.054	0.154
A4	0.041	0.024

Step 7: The maximum distance to the NIS and the minimum distance to the PIS are determined.

$$D_{max} = 0.154 D_{min} = 0.021$$

Step 8 and Step 9. Closeness ratios are calculated, and the alternatives are ranked (Table 7).

Table 7. Closeness ratio and ranking of alternatives

	Closeness Ratio	Rank
X_1	0.328	1
X_2	4.228	4
X_3	1.487	2
X_4	1.733	3

The closeness ratio values for each alternative obtained by SWAM operator show that X_1 is the best supplier for the company and overall ranking of suppliers is $A_1 > A_3 > A_4 > A_2$.

6 Conclusion

Evaluation of suppliers is a crucial consideration for automotive industry. Automotive products consist of thousands of components and most of the components are collected from several suppliers instead of producing all of them. Since the reliability of each product is dependent on its components, companies have to select their components after careful consideration. This situation requires careful selection of suppliers and long-term relationships with them. The main contribution of this study that this is the first study about supplier evaluation in views of digital transformation factors. This study can be extended in further studies by taking different criteria into account or analyzing the effect of different fuzzy numbers. Solution to the problem can be obtained by other multi-criteria decision making methods such as SF-AHP, SF-VIKOR, etc. In addition, the same MCDM methods can be based on intuitionistic, neutrosophic or bipolar valued fuzzy sets and obtained ranks of alternative suppliers can be compared.

References

1. Kleinert, J.: Digital transformation. Empirica **48**(1), 1–3 (2021). https://doi.org/10.1007/s10663-021-09501-0
2. Kahraman, C., Gündoğdu, F.K.: Decision Making with Spherical Fuzzy Sets: Theory and Applications. Springer, Berlin (2020)
3. Ecer, B., Aktas, A., Kabak, M.: Green supplier selection of a textile manufacturer: a hybrid approach based on AHP and VIKOR. Manas J. Eng. **7**(2), 126–135 (2019)
4. Dagdeviren, M., Eraslan, E.: Supplier selection using PROMETHEE sequencing method. J. Fac. Eng. Archit. Gaz. **23**(1), 69–75 (2008)
5. Arikan, M., Gokbek, B.: Supplier selection based on multi-criteria decision making approaches: an implementation in electronics sector. Erciyes Üniversitesi Fen Bilimleri Enstitüsü Dergisi **30**(5), 346–354 (2016)
6. Jain, V., Sangaiah, A.K., Sakhuja, S., Thoduka, N., Aggarwal, R.: Supplier selection using fuzzy AHP and TOPSIS: a case study in the Indian automotive industry. Neural Comput. Appl. **29**(7), 555–564 (2016). https://doi.org/10.1007/s00521-016-2533-z
7. Ayağ, Z., Samanlioglu, F.: An intelligent approach to supplier evaluation in automotive sector. J. Intell. Manuf. **27**(4), 889–903 (2014). https://doi.org/10.1007/s10845-014-0922-7
8. Ghadimi, P., Dargi, A., Heavey, C.: Making sustainable sourcing decisions: practical evidence from the automotive industry. Int. J. Log. Res. Appl. **20**(4), 297–321 (2017)
9. Jiang, P., Hu, Y.C., Yen, G.F., Tsao, S.J.: Green supplier selection for sustainable development of the automotive industry using grey decision making. Sustain. Dev. **26**, 890–903 (2018)
10. Galankashi, M.R., Helmi, S.A., Hashemzahi, P.: Supplier selection in automobile industry: a mixed balanced scorecard–fuzzy AHP approach. Alexand Eng. J. **55**, 93–100 (2016)
11. Gupta, S., Soni, U., Kumar, G.: Green supplier selection using multi-criterion decision making under fuzzy environment: a case study in automotive industry. Comput. Ind. Eng. **136**, 663–680 (2019)
12. Phumchusri, N., Tangsiriwattana, S.: Optimal supplier selection model with multiple criteria: a case study in the automotive parts industry. Eng. J. **23**(1), 191–203 (2019)
13. Cengiz Toklu, M.: Interval type-2 fuzzy TOPSIS method for calibration supplier selection problem: a case study in an automotive company. Arab. J. Geosci. **11**(13), 1–7 (2018). https://doi.org/10.1007/s12517-018-3707-z
14. Memari, A., Dargi, A., Jokar, M.R.A., Ahmad, R., Rahim, A.R.A.: Sustainable supplier selection: a multi-criteria intuitionistic fuzzy TOPSIS method. J. Manuf. Syst. **50**, 9–24 (2019)
15. Govindan, K., Khodaverdi, R., Vafadarnikjoo, A.: A grey DEMATEL approach to develop third-party logistics provider selection criteria. Ind. Manag. Data Syst. **116**(4), 690–722 (2016)
16. Lima-Junior, F.R., Carpinetti, L.C.R.: A multicriteria approach based on fuzzy QFD for choosing criteria for supplier selection. Comput. Ind. Eng. **101**, 269–285 (2016)
17. Galo, N.R., Calache, L.D.D.R., Carpinetti, L.C.R.: A group decision approach for supplier categorization based on hesitant fuzzy and ELECTRE TRI. Int. J. Prod. Econ. **202**, 182–186 (2018)

Seismic Vulnerability Assessment Using Spherical Fuzzy ARAS

Akın Menekşe and Hatice Camgöz Akdağ(✉)

Istanbul Technical University, Maslak, 34467 Istanbul, Turkey
{menekse18,camgozakdag}@itu.edu.tr

Abstract. Higher education buildings are very important lifelines and can be very vulnerable if located in a region of high seismic hazard, and estimating the performance and expected damage of these buildings from future earthquakes is very crucial to reduce the levels of physical damage and interruption of education, research, and human resource training activities. Rapid and economical pre-assessment of the sensitivity of these buildings will shed light to higher education managements for appropriate retrofitting and reconstruction planning. However, a seismic vulnerability assessment involves complex qualitative criteria set and there is a need for a model that can present the evaluations of experts in a quantitative, systematic and measurable form while reflecting the uncertain and fuzzy nature of the process to the model. In this paper, a novel multi-criteria decision making (MCDM) decision support model is developed by extending Additive Ratio ASsessment (ARAS) to spherical fuzzy ARAS. The applicability of the model is illustrated through a numerical example for seismic vulnerability assessment of higher education institution buildings and three decision-makers (DMs) evaluate four buildings with respect to eight criteria from the literature. Sensitivity and comparative analysis, practical implications, limitations and future research avenue are also given within the study.

Keywords: Spherical fuzzy sets · ARAS · Seismic vulnerability assessment · Higher education institution

1 Introduction

Seismic vulnerability analysis is a complex decision-making problem containing intrinsic uncertainties such as vagueness of expert comments, complex qualitative and quantitative data, and uncertain data relationships in its nature [1]. Seismic vulnerability of buildings depends on many criteria such as follows: Construction material quality is a critical criterion when evaluating seismic vulnerability and it may lead to severe damages after an earthquake if its quality is not high enough [2]. Building occupancy is another criterion in determining the seismic vulnerability level of the buildings and can be determined by qualitative assessments of experts [3]. Architectural design of buildings plays an important role in structural performance and may lead to complex behaviors under earthquake loadings [4]. Shear wall system also has an important role in the design of buildings against seismic hazards since these walls make structures more

C. Kahraman et al. (Eds.): INFUS 2021, LNNS 308, pp. 733–740, 2022.
https://doi.org/10.1007/978-3-030-85577-2_86

resistant against lateral loads by providing efficient bracing system [5]. Foundation system is also an important part of a structure which transmits vertical loadings to the soil, and may be responsible for failure under seismicity [6]. Underlying soil condition significantly affects structural performances of buildings and should be considered in seismic vulnerability assessments [7]. Falling hazards such as external falling hazards, i.e., unreinforced masonry parapets, chimneys and exterior surface cladding elements; and internal falling hazards, i.e., large book shelves, chemicals in laboratories and pendant lights in higher education institutions can fall on students or block exit ways under earthquake [8], and emergency exit systems in these buildings are critical for exit and access after an earthquake [9].

Since most of the above mentioned criteria are quite qualitative, a linguistic evaluation is needed to transform qualitative assessments of DMs into a numerical form. According to Zadeh [10], linguistic evaluation is an approximation in which qualitative aspects are represented utilizing linguistic variables, and it is very suitable in situations where alternatives cannot be assessed quantitatively. On the other hand, according to the fuzzy set theory, humans think in linguistic terms rather than crisp numbers. Thus, fuzzy set theory, by utilizing fuzzy sets are very suitable to model linguistic variables.

In the literature there are various of extensions of fuzzy sets and in this paper, spherical fuzzy sets [11] are used to handle the uncertainty in seismic vulnerability assessment problem since three dimensional spherical fuzzy sets enable DMs to define membership and non-membership degrees independently and in a larger preference domain than most of the other fuzzy extensions.

There are two novelties in this study: Firstly, to the best of our knowledge, this is the first attempt to extend ARAS to a spherical fuzzy ARAS, and secondly, although there are some crisp based MCDM studies presented for seismic vulnerability assessment, this is the first study using fuzzy sets for seismic vulnerability assessment of buildings in a specific site. The rest of the paper is organized as follows: Sect. 2 presents the proposed spherical fuzzy ARAS method in a step-by-step form. Section 3 illustrates the model through a seismic vulnerability assessment problem for a higher education institution. Sensitivity and comparative analyses are also given in this section. Section 4 finalizes the paper with practical implications, limitations and future research avenue.

2 Methodology

ARAS method evaluates the performance of selected alternatives and compares the scores of those selected alternatives with the ideal best alternative [12]. In this section, spherical fuzzy ARAS model is developed based on the definition, basic operators and the linguistic scale presented for spherical fuzzy sets [11]. The steps of the proposed model are as follows:

Step 1. Let DMs evaluate performance of each alternative with respect to each criterion by utilizing the readily given linguistic terms given in Table 1.

Step 2. Convert linguistic evaluations to their corresponding spherical fuzzy numbers and obtain spherical fuzzy alternative evaluation matrices (SFAEMs). The structure of

Table 1. Linguistic terms and their corresponding spherical fuzzy numbers

Linguistic term	Abbreviation	$(\mu;v;\pi)$
Absolutely More Importance	AMI	0.9;0.1;0.1
Very High Importance	VHI	0.8;0.2;0.2
High Importance	HI	0.7;0.3;0.3
Slightly More Importance	SMI	0.6;0.4;0.4
Equally Importance	EI	0.5;0.5;0.5
Slightly Low Importance	SLI	0.4;0.6;0.4
Low Importance	LI	0.3;0.7;0.3
Very Low Importance	VLI	0.2;0.8;0.2
Absolutely Low Importance	ALI	0.1;0.9;0.1

SFAEM is given in Eq. 1.

$$SFAEM = (C_j(X_i))_{mxn} = \begin{bmatrix} (\mu_{11}, v_{11}, \pi_{11}) & \dots & (\mu_{1n}, v_{1n}, \pi_{1n}) \\ (\mu_{21}, v_{21}, \pi_{21}) & \dots & (\mu_{2n}, v_{2n}, \pi_{2n}) \\ \dots & \dots & \dots \\ (\mu_{m1}, v_{m1}, \pi_{m1}) & \dots & (\mu_{mn}, v_{mn}, \pi_{mn}) \end{bmatrix} \quad (1)$$

Step 3. Aggregate all SFAEMs to obtain one aggregated spherical fuzzy alternative evaluation matrix (ASFAEM) by utilizing SWGM operator that is given in Eq. 2

$$SWGM_w(\tilde{A_{S1}}, ..., \tilde{A_{Sn}}) = \tilde{A_{S1}}^{w_1} + \tilde{A_{S2}}^{w_2} + ... + \tilde{A_{Sn}}^{w_n}$$
$$= \{\prod_{i=1}^{n} \mu_{\tilde{A_{Si}}}^{w_i}, [1 - \prod_{i=1}^{n} (1 - v_{\tilde{A_{Si}}}^2)^{w_i}]^{1/2}, [\prod_{i=1}^{n} (1 - v_{\tilde{A_{Si}}}^2)^{w_i} - \prod_{i=1}^{n} (1 - v_{\tilde{A_{Si}}}^2 - \pi_{\tilde{A_{Si}}}^2)^{w_i}]^{1/2}\} \quad (2)$$

Step 4. Let DMs evaluate importance level of each criterion by utilizing the same linguistic terms. Since each criterion may have a different importance level, all criteria are evaluated and a criterion weight is assigned to each.
Step 5. Convert linguistic evaluations to spherical fuzzy numbers and obtain spherical fuzzy criterion weight matrices (SFCWMs).
Step 6. Aggregate all SFCWMs to obtain one aggregated spherical fuzzy criterion weight matrix (ASFCWM) by utilizing the SWGM operator that is given in Eq. 2.
Step 7. Obtain spherical fuzzy decision matrix by multiplying ASFAEM by ASFCWM. Utilize spherical fuzzy multiplication operator that is given in 3 for the multiplication.

$$\tilde{A}_s \otimes \tilde{B}_s = \mu_{\tilde{A}_s}\mu_{\tilde{B}_s}, (v_{\tilde{A}_s}^2 + v_{\tilde{B}_s}^2 - v_{\tilde{A}_s}^2 v_{\tilde{B}_s}^2)^{1/2}, ((1 - v_{\tilde{B}_s}^2)\pi_{\tilde{A}_s}^2 + (1 - v_{\tilde{A}_s}^2)\pi_{\tilde{B}_s}^2 - \pi_{\tilde{A}_s}^2 \pi_{\tilde{B}_s}^2)^{1/2}\} \quad (3)$$

Step 8. Determine the spherical fuzzy ideal solution based on score and accuracy values of attributes in spherical fuzzy decision matrix obtain in Step 7. Use score function developed by Kahraman et al. [13].

$$Score(\tilde{A}_s) = (2\mu_{\tilde{A}_s} - \frac{\pi_{\tilde{A}_s}}{2})^2 - (v_{\tilde{A}_s} - \frac{\pi_{\tilde{A}_s}}{2})^2 \quad (4)$$

$$Accuracy(\tilde{A}_s) = (\mu^2_{\tilde{A}_s} + v^2_{\tilde{A}_s} + \pi^2_{\tilde{A}_s}) \tag{5}$$

Step 9. Determine the spherical fuzzy optimality function ($\tilde{S_{S_i}}$) for all alternatives, as well as for the spherical fuzzy ideal solution ($\tilde{S_{S_0}}$) as given in Eq. 6.

$$\tilde{S_{S_i}} = \sum_{i=1}^{n}((\mu_1, v_1, \pi_1) + (\mu_2, v_2, \pi_2) + \ldots + (\mu_n, v_n, \pi_n)) \tag{6}$$

The above task determines the values of spherical fuzzy optimality function where i is the number of alternative, and μ, v and π are the membership, non-membership and hesitancy degrees. Use Spherical fuzzy addition operator that is given in Eq. 7 for calculating $\tilde{S_{S_i}}$ and $\tilde{S_{S_0}}$

$$\tilde{A}_s \oplus \tilde{B}_s = \{(\mu^2_{\tilde{A}_s} + \mu^2_{\tilde{B}_s} - \mu^2_{\tilde{A}_s}\mu^2_{\tilde{B}_s})^{1/2}, v^2_{\tilde{A}_s}v^2_{\tilde{B}_s}, ((1-\mu^2_{\tilde{B}_s})\pi^2_{\tilde{A}_s} + (1-\mu^2_{\tilde{A}_s})\pi^2_{\tilde{B}_s} - \pi^2_{\tilde{A}_s}\pi^2_{\tilde{B}_s})^{1/2}\} \tag{7}$$

Step 10. Defuzzify and obtain crisp form of $\tilde{S_{S_i}}$ and $\tilde{S_{S_0}}$. Then calculate the utility degree K_i of an alternative A_i by comparing optimality criterion value of that alternative (S_i) with the optimality criterion value of the ideal best solution (S_0) as given in Eq. 8

$$K_i \frac{S_i}{S_o} \tag{8}$$

Where S_i and S_0 are the optimality criterion values for ith and ideal best alternatives.
Step 11. Rank the alternatives according to utility degrees (K_i) of the alternatives. The higher the utility value, the better the alternative.

3 Application

Proposed spherical fuzzy ARAS model is illustrated through an example from a higher education institution. Four educational buildings (A1, A2, A3 and A4) are evaluated by three civil engineer DMs with respect to eight selected criteria from the literature as C1: Architectural design, C2: Underlying soil condition, C3: Foundation system, C4: Construction quality, C5: Building occupancy, C6: Emergency exit system, C7: Falling hazards and C8: Shear wall system. A brief information about these criteria are already given in Sect. 1. The structure of the MCDM problem (Fig. 1), and the solution of the problem is given below respectively.

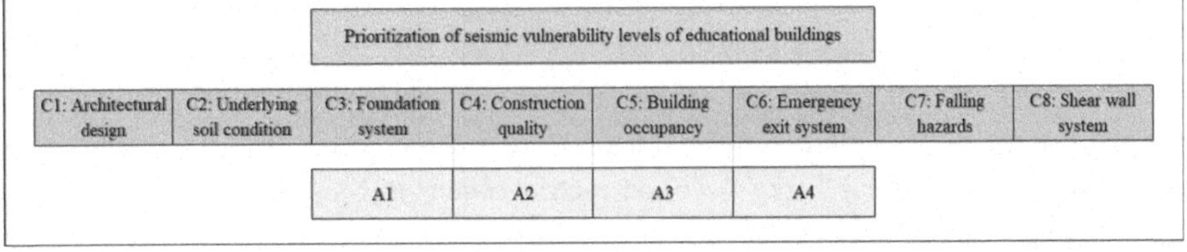

Prioritization of seismic vulnerability levels of educational buildings

C1: Architectural design	C2: Underlying soil condition	C3: Foundation system	C4: Construction quality	C5: Building occupancy	C6: Emergency exit system	C7: Falling hazards	C8: Shear wall system

A1	A2	A3	A4

Fig. 1. The structure of the MCDM problem

Step 1. Four alternatives are evaluated with respect to eight criteria by three DMs.
Step 2. Linguistic evaluations are converted to their spherical fuzzy forms and three SFAEMs are obtained.
Step 3. Three SFAEMs are aggregated and one ASFAEM is obtained. Linguistic alternative evaluations of three DMs and the ASFAEM are presented in Table 2.

Table 2. Linguistic evaluations of alternatives and the ASFAEM

		C1	C2	C3	C4	C5	C6	C7	C8
DM1	A1	EI	SMI	SMI	SMI	VHI	EI	EI	EI
	A2	SMI	VLI	VHI	HI	VHI	HI	LI	VHI
	A3	VLI	VHI	SMI	HI	VHI	HI	VLI	SLI
	A4	SLI	SMI	SLI	SMI	EI	EI	ALI	SLI
DM2	A1	SMI	VHI	SMI	VHI	VHI	EI	EI	SLI
	A2	SMI	SLI	VHI	HI	HI	VHI	SLI	VHI
	A3	SLI	HI	HI	VHI	HI	SMI	ALI	EI
	A4	SLI	SLI	SMI	SMI	SMI	EI	ALI	EI
DM3	A1	EI	SMI	HI	SMI	AMI	LI	SLI	SLI
	A2	VHI	LI	VHI	HI	HI	SLI	LI	HI
	A3	EI	SMI	SMI	AMI	SLI	LI	LI	SLI
	A4	VLI	SMI	SLI	EI	SLI	LI	VLI	EI
ASFAEM	A1	0.53;0.47;0.47	0.60;0.40;0.40	0.63;0.37;0.35	0.66;0.35;0.34	0.83;0.17;0.10	0.42;0.54;0.47	0.46;0.54;0.47	0.43;0.57;0.43
	A2	0.66;0.35;0.34	0.29;0.71;0.30	0.80;0.20;0.10	0.70;0.30;0.20	0.73;0.27;0.17	0.33;0.67;0.33	0.33;0.67;0.33	0.77;0.24;0.14
	A3	0.34;0.67;0.36	0.70;0.31;0.28	0.63;0.37;0.35	0.80;0.22;0.14	0.61;0.42;0.30	0.18;0.82;0.19	0.18;0.82;0.19	0.43;0.57;0.43
	A4	0.32;0.69;0.33	0.52;0.48;0.40	0.46;0.55;0.40	0.56;0.44;0.44	0.49;0.48;0.40	0.13;0.87;0.13	0.13;0.87;0.13	0.46;0.54;0.47

Step 4. Eight criteria are evaluated by three DMs.
Step 5. Linguistic evaluations are converted to spherical fuzzy numbers and SFCWMs are obtained.
Step 6. Three SFCWMs are aggregated and one ASFCWM is obtained. Linguistic evaluations of criteria and the ASFCWM is presented in Table 3

Table 3. Linguistic evaluations of criteria and the ASFCWM

	C1	C2	C3	C4	C5	C6	C7	C8
DM1	HI	AMI	AMI	VHI	SLI	EI	SLI	HI
DM2	HI	VHI	AMI	HI	LI	HI	VLI	VHI
DM3	SMI	AMI	VHI	VHI	ALI	SMI	VLI	SMI
ASFCWM	0.66;0.34;0.29	0.87;0.14;0.10	0.87;0.14;0.10	0.77;0.24;0.14	0.23;0.78;0.25	0.59;0.41;0.41	0.25;0.75;0.33	0.70;0.31;0.28

Step 7. Spherical fuzzy decision matrix is obtained as in Table 4.

Table 4. Spherical fuzzy decision matrix

	C1	C2	C3	C4	C5	C6	C7	C8
A1	0.35;0.56;0.50	0.60;0.34;0.29	0.55;0.39;0.36	0.64;0.29;0.17	0.19;0.78;0.26	0.25;0.67;0.48	0.12;0.83;0.39	0.46;0.45;0.37
A2	0.55;0.38;0.30	0.63;0.30;0.20	0.69;0.24;0.14	0.54;0.38;0.24	0.17;0.80;0.26	0.36;0.56;0.44	0.08;0.87;0.31	0.53;0.39;0.30
A3	0.47;0.44;0.33	0.60;0.34;0.29	0.55;0.39;0.36	0.66;0.28;0.17	0.16;0.80;0.27	0.40;0.51;0.45	0.05;0.93;0.22	0.49;0.42;0.32
A4	0.21;0.73;0.36	0.45;0.50;0.41	0.60;0.34;0.29	0.61;0.31;0.17	0.11;0.83;0.32	0.27;0.66;0.47	0.03;0.95;0.18	0.30;0.63;0.46
Opt	0.55;0.38;0.30	0.63;0.30;0.20	0.69;0.24;0.14	0.66;0.28;0.17	0.19;0.78;0.26	0.40;0.51;0.45	0.12;0.83;0.39	0.53;0.39;0.30

Steps 8, 9, 10 and 11. Spherical fuzzy (SF) optimality criterion values, defuzzifed optimality criterion values, degrees of alternative utility and final ranking of alternatives are given in Table 5. A2 has the highest degree of utility which means seismic vulnerability level of A2 is the most critical.

Table 5. Spherical fuzzy optimality criterion values, crisp form of optimality criterion values, appraisal scores and final rankings

	Spherical fuzzy optimality criterion value	Crisp form of optimality criterion value	Appraisal score
A1	0.854;0.010;0.473	2.114	0.680
A2	0.911;0.005;0.329	2.725	0.876
A3	0.885;0.009;0.390	2.446	0.786
A4	0.754;0.032;0.587	1.409	0.453
Ideal best	0.949;0.002;0.258	3.111	$A2 > A3 > A1 > A4$

3.1 Sensitivity and Comparative Analyses

To validate our model, sensitivity and comparative analyses are conducted. For sensitivity analysis, 10 different DM weight and criterion weight distribution scenarios are created separately and the appraisal scores are calculated. In DM weight sensitivity analysis (Fig. 2), the same ranking of alternatives are obtained in all weight distribution scenario, and in criterion weight sensitivity analysis (Fig. 3), A2 has the highest appraisal score in eight of the 10 different criterion weight scenarios. These results show the robustness of the model. However, since criteria weight distribution may affect the final ranking of alternatives, it is important to select a qualified expert group, and the selected ones should precisely determine the criterion weights.

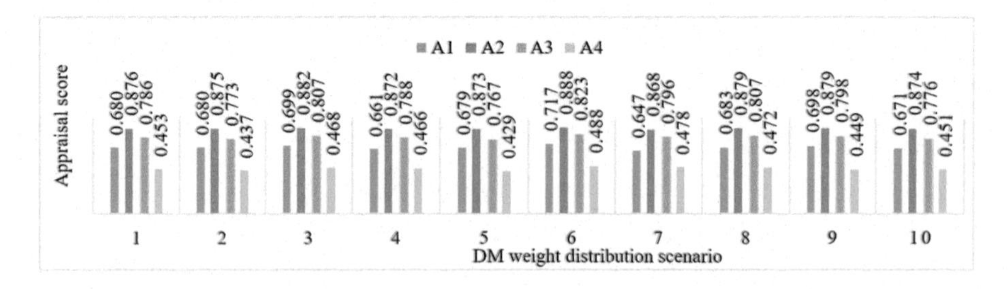

Fig. 2. Appraisal scores for 10 different DM weight distribution scenarios.

For comparative analysis, the same problem is handled with another MCDM method. For this purpose, TOPSIS is seleted, since the applicability and reliability of this method has been tested in many applications. Table 6 gives the appraisal scores obtained from spherical fuzzy ARAS and spherical fuzzy TOPSIS. It is seen that the ranking order of alternatives are same in both methods as $A2 > A3 > A1 > A4$. These results show the validity of the proposed model.

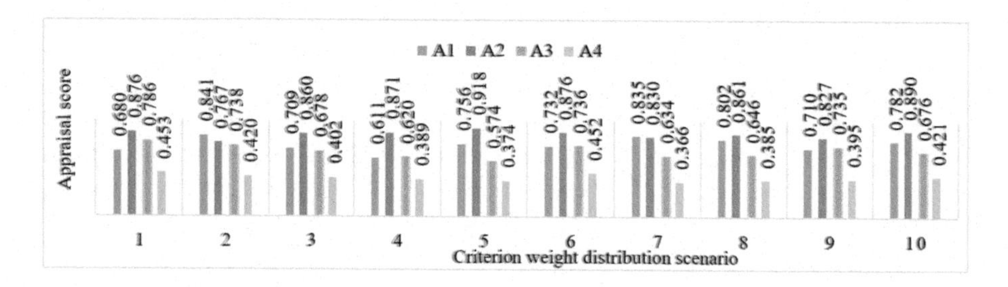

Fig. 3. Appraisal scores for 10 different criterion weight distribution scenarios.

Table 6. Comparative analysis results

	Spherical fuzzy ARAS	Spherical fuzzy TOPSIS
A1	0.680	0.514
A2	0.876	0.599
A3	0.786	0.520
A4	0.453	0.267
Final ranking	$A2 > A3 > A1 > A4$	$A2 > A3 > A1 > A4$

4 Conclusion and Final Remarks

In this paper a novel spherical fuzzy ARAS model is developed to evaluate seismic vulnerability assessment of higher education buildings. The applicability of the model is illustrated with an MCDM example and four educational buildings are evaluated by three DMs with respect to eight selected criteria from the literature. Spherical weighted geometric mean, spherical fuzzy addition and multiplication operators, score and accuracy functions developed for spherical fuzzy sets are used in model development. The stability and the validity of the model is shown with comparative and sensitivity analyses.

This study contributes to the literature in two ways. First, ARAS method is extended in a spherical fuzzy environment, which enables DMs express their membership, non-membership and hesitancy degrees independently, and in a larger domain than most of other fuzzy extensions. Second, a fuzzy MCDM study is presented for site-specific seismic vulnerability assessment of educational buildings.

The main practical implication of the proposed model is that it can be used by higher education institutions as decision support tool for rapid and economical pre-assessment of seismic vulnerability levels of educational buildings to be considered in retrofitting and reconstruction planning.

As a limitation of the study, the following can be said: In the model development, degree of alternative utility is calculated by comparing optimality criterion value of an alternative with the optimality criterion value of the ideal best solution. However, in this final step, the model is converted to a crisp environment due to the absence of spherical fuzzy division operator.

In this study, spherical fuzzy optimality function is calculated by utilizing addition operator. In future studies, geometric and arithmetic mean operators can be used for this purpose. Moreover, the criteria set can be extended and grouped as social and cultural factors. AHP methodology can be used for dealing with such a layered criteria structure and spherical fuzzy AHP-ARAS hybrid model can be developed. A machine learning reinforced MCDM model can also be developed for seismic vulnerability assessment.

References

1. Delavar, M.R., Sadrykia, M.: Assessment of enhanced Dempster-Shafer theory for uncertainty modeling in a GIS-based seismic vulnerability assessment model, case study–Tabriz city. ISPRS Int. J. Geo-Inf. **9**(4), 195 (2020). https://doi.org/10.3390/ijgi9040195
2. Kaplan, H., Bilgin, H., Yilmaz, H., Yilmaz., S., Öztaş., A.: Structural damages of L'Aquila (Italy) earthquake. Nat. Hazards Earth Syst. Sc. **10**(3), 499–507 (2010). https://doi.org/10.5194/nhess-10-499-2010
3. Ketsap, A., Hansapinyo, C., Kronprasert, N., Limkatanyu, S.: Uncertainty and fuzzy decisions in earthquake risk evaluation of buildings. Eng. J. **23**(5), 89–105 (2019). https://doi.org/10.4186/ej.2019.23.5.89
4. Alecci, V., De Stefano, M.: Building irregularity issues and architectural design in seismic areas. Frattura ed Integrita Strutturale **13**(47), 161–168 (2019). https://doi.org/10.3221/igf-esis.47.13
5. Chandurkar, P.P., Pajgade, D.P.: Seismic analysis of RCC building with and without shear wall. Int. J. Mod. Eng. Res. **3**(3), 1805–1810 (2019)
6. Oyediran, I.A., Falae, P.O.: Integrated geophysical and geotechnical methods for pre-foundation investigations. Geol. Geophys. **7**(453), 2 (2018)
7. Behnamfar, F., Banizadeh, M.: Effects of soil-structure interaction on distribution of seismic vulnerability in RC structures. Soil Dyn. Earthq. Eng. **80**, 73–86 (2016). https://doi.org/10.1016/j.soildyn.2015.10.007
8. De Angelis, A., Pecce, M.: Seismic nonstructural vulnerability assessment in school buildings. Nat. Hazards **79**(2), 1333–1358 (2015). https://doi.org/10.1007/s11069-015-1907-3
9. Li, B., Mosalam, K.M.: Seismic performance of reinforced-concrete stairways during the 2008 Wenchuan earthquake. J. Perform. Constr. Facil. **27**(6), 721–730 (2013)
10. Zadeh, L.A.: The concept of a linguistic variable and its application to approximate reasoning—I. Inf. Sci. **8**(3), 199–249 (1975)
11. Gündoğdu, F.K., Kahraman, C.: Spherical fuzzy sets and spherical fuzzy TOPSIS method. J. Intell. Fuzzy Syst. **36**(1), 337–352 (2019). https://doi.org/10.3233/jifs-181401
12. Zavadskas, E.K., Turskis, Z.: A new additive ratio assessment (ARAS) method in multicriteria decision-making. Technol. Econ. Dev. Econ. **16**(2), 159–172 (2010). https://doi.org/10.3846/tede.2010.10
13. Kahraman, C., Gundogdu, FK., Onar, SC., Oztaysi, B.: Hospital location selection using spherical fuzzy TOPSIS. In: 11th Conference of the European Society for Fuzzy Logic and Technology, pp. 77–82. Atlantis Press (2019). https://doi.org/10.2991/eusflat-19.2019.12

Public Transportation Business Model Assessment with Spherical Fuzzy AHP

Büşra Buran(✉) and Mehmet Erçek

Management Engineering Department, Istanbul Technical University, 34367 Beşiktaş, Istanbul, Turkey
{buran18,ercekme}@itu.edu.tr

Abstract. The business model (BM) provides a strategic tool to achieve goals for the companies. In recent years, it has been used frequently both in business areas and in literature. BM comprises three main blocks that are activity, value, and finance. For non-profit organizations, the criterion of impact is added to the model to express social and environmental issues. In this study, Spherical Fuzzy Analytic Hierarchy Process (SF-AHP) is applied to evaluate the public transportation business model taking into account internal and external perspectives with political, economical, social, technological, legal, and environmental analysis (P.E.S.T.L.E), and also impact factor. The proposed model is developed by three levels with criteria and sub-criteria according to the literature and evaluated by national and international transportation experts. Results show that the internal environment is more important with 0.605 weight than the external one in the first stage. For the second stage, activity comes first with 0.232 weight, key partners which is sub-criteria under activity with 0.088 weight, is the most important for the third stage. The originality of the paper is to provide academicians and practitioners a business model framework for public transportation. And also, this study will be a guide for users to find out which criteria is superior to others in BM.

Keywords: Spherical fuzzy AHP · Public transportation · Business model

1 Introduction

The business model is widely used both in literature and business life in different areas such as product and service [1]. Business model evolution is summarized with five steps that are definition, listing components, constructing business model elements, modeling, and applying [2]. The ontology of BM was proposed by Osterwalder in his Ph.D. thesis which consisted of nine elements such as key partners, key resources, key activities, value proposition, customer segments, customer relationship, customer channels, cost structure, and revenue stream [3]. The proposed concept was named a Business Model Canvas (BMC) that is shown in Fig. 1.

C. Kahraman et al. (Eds.): INFUS 2021, LNNS 308, pp. 741–748, 2022.
https://doi.org/10.1007/978-3-030-85577-2_87

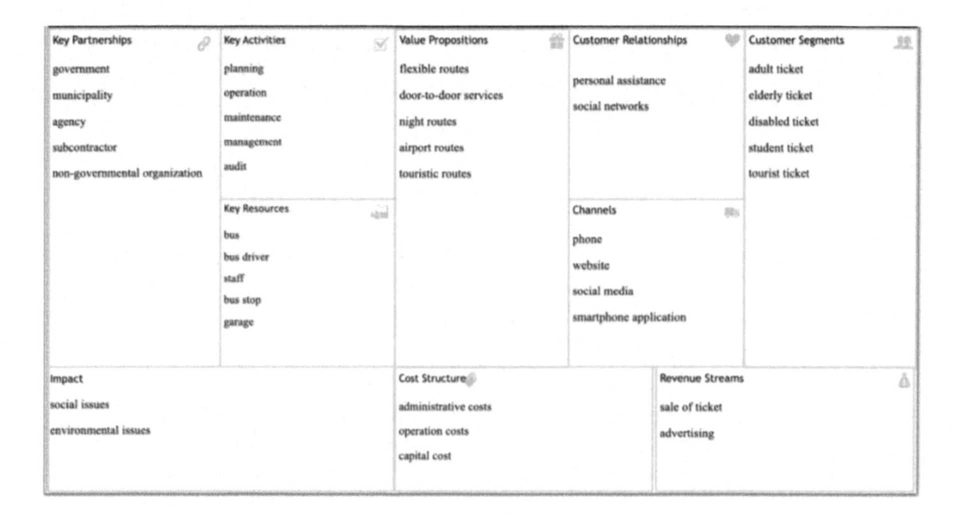

Fig. 1. Business model framework for public transportation authority.

Regarding organization type, BM can be differentiated between for profit and non-profit. Social and environmental issues are taken into account under impact factors. The framework of the business model for non-profit organizations was developed by Sanderse [4].

The contribution of the study is to measure the success of the developed business model for public transportation authority taken into account P.E.S.T.L.E. analysis which includes political, economical, social, technological, legal, and environmental issues [5]. Spherical Fuzzy Analytic Hierarchy (SF-AHP) is applied to prioritize the criteria weights of the model. The originality of the paper is to provide academicians and practitioners for designing public transportation business models and also, evaluation of criteria according to their importance level.

The paper is constructed by five sections as follows. The next section presents Multi-criteria decision making (MCDM) methods. When the methodology is summarized in Sect. 3, case study application is introduced in Sect. 4. Finally, the conclusion is represented with future directions in Sect. 5.

2 Multi Criteria Decision Making (MCDM) Methods

Multi-criteria decision making (MCDM) methods provide an effective tool due to including qualitative and quantitative factors. It is applied to various fields of education, management, energy, production, service, etc. From 2000 to 2014, Mardani et al. did a study about the frequency of preference for MCDM methods in the literature [6]. According to the study, AHP was mostly preferred method during the stated time. To fill the time gap between 2015 and 2019, Fatma et al. reviewed literature in the same way as Mardanis' study. Based on results, Hybrid MCDM came first and AHP was at second with twenty-eight percent and nineteen percent, accordingly [7].

General decision making process includes eight main steps as below [8]:

- define problem,
- determine the needs,
- identify goals,
- identify alternatives,
- develop assessment criteria,
- select methods,
- apply the method,
- control answers.

3 Methodology

3.1 Spherical Fuzzy Sets: Preliminaries

Definition: A Spherical Fuzzy Sets $\tilde{A}_s$ of the universe of discourse $\cup$ is given by [9]

$$\tilde{A}_s = \{u, \mu_{\tilde{A}_s(u)}, \nu_{\tilde{A}_s(u)}, \pi_{\tilde{A}_s(u)} \mid u \in \cup\} \tag{1}$$

where

$$\mu_{\tilde{A}_s}(\mathrm{u}) : \cup \to [0,1], \quad \nu_{\tilde{A}_s}(\mathrm{u}) : \cup \to [0,1], \quad \pi_{\tilde{A}_s}(\mathrm{u}) : \cup \to [0,1],$$

and

$$0 \leq \mu_{\tilde{A}}^2 + \nu_{\tilde{A}}^2 + \pi_{\tilde{A}}^2 \leq 1 \quad \forall u \in \cup \tag{2}$$

Where $\mu_{\tilde{A}}$, $\nu_{\tilde{A}}$, and $\pi_{\tilde{A}}$ are the degrees of membership, non-membership, and hesitancy of u to $\tilde{A}$ for each, respectively [9]. Equation 1 becomes as below for on the surface of the sphere [9],

$$\mu_{\tilde{A}(u)}^2 + \nu_{\tilde{A}(u)}^2 + \pi_{\tilde{A}(u)}^2 = 1 \quad \forall u \in \cup \tag{3}$$

Basic Operations: Addition, multiplication, multiplication by a scalar, and power of spherical fuzzy sets are defined in the literature [10]. Aggregation operator, defuzzification operation, and normalization operation are presented in this section [10].

Aggregation Operator. Spherical Weighted Arithmetic mean (SWAM) operator is given as follows [10]: Spherical Weighted Arithmetic Mean (SWAM) with regard to $\mathrm{w} = (\mathrm{w}_1, w_2, ..., w_n); w_i \in [0,1]; \sum_{i=1}^{n} w_i = 1$, SWAM is defined as:

$$SWAM_w(\tilde{A_{S1}}, ..., \tilde{A_{Sn}}) = w_1\tilde{A_{S1}} + w_2\tilde{A_{S2}} + ... + w_n\tilde{A_{Sn}}$$
$$= \{[1 - \prod_{i=1}^{n}(1 - \mu_{\tilde{A_{Si}}}^2)^{w_i}]^{1/2}, \prod_{i=1}^{n} \nu_{\tilde{A_{Si}}}^{w_i},$$
$$[\prod_{i=1}^{n}(1 - \mu_{\tilde{A_{Si}}}^2)^{w_i} - \prod_{i=1}^{n}(1 - \mu_{\tilde{A_{Si}}}^2 - \pi_{\tilde{A_{Si}}}^2)^{w_i}]^{1/2}\} \tag{4}$$

Defuzzification Operation

$$S(\tilde{w}_j^s) = \sqrt{|100 * [(3\mu_{\tilde{A}_s} - \frac{\pi_{\tilde{A}_s}}{2})^2 - (\frac{\nu_{\tilde{A}_s}}{2} - \pi_{\tilde{A}_s})^2]|} \tag{5}$$

Normalization Operation

$$\tilde{w}_j^s = \frac{S(\tilde{w}_j^s)}{\sum_{J=1}^{n} S(\tilde{w}_j^s)} \tag{6}$$

3.2 Spherical Fuzzy Analytic Hierarchy Process (SF-AHP)

AHP was proposed by Saaty in 1980 which enables to take expert opinions via comparison matrices [11]. In some circumstances, crisp numbers are inadequate to understand human thinking. That time, the fuzzy environment provides an effective tool [12]. In this study, Fuzzy AHP is applied to the proposed model which is presented in Sect. 4. There are various fuzzy extensions in the literature such as Ordinary Fuzzy Sets, Interval-valued Fuzzy Sets, Intuitionistic Fuzzy Sets, Fuzzy Multisets, Intuitionistic Fuzzy Sets of Second Type, Neutrosophic Sets, Nonstationary Fuzzy Sets, Hesitant Fuzzy Sets, and Spherical Fuzzy Sets [10].

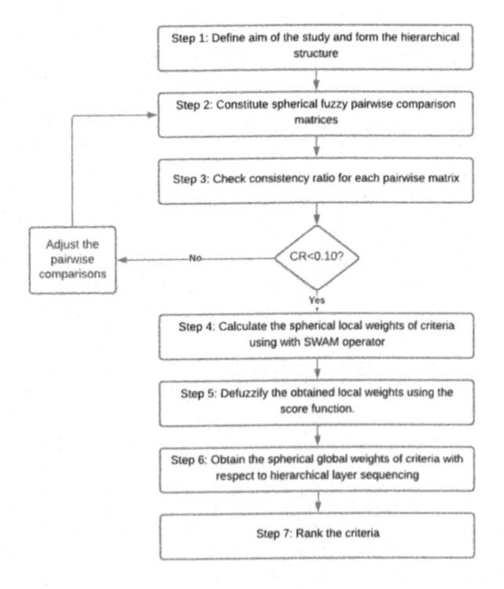

Fig. 2. Flowchart of the proposed method.

The flowchart of the proposed method is presented in Fig. 2 and also, steps are summarized as follows.

Step 1: Determine main and sub-criteria of the public transportation business model with respect to a hierarchical structure.

Step 2: Constitute pairwise comparisons using Spherical Fuzzy Sets judgment

matrices based on the linguistic terms which is indicated in the literature [10].
Step 3: To calculate the consistency ratio (CR), convert the linguistic terms to corresponding score indices in pairwise matrix. And then apply the classical consistency check for each pairwise comparison matrix. If CR is less than ten percent, pass Step 4, otherwise, go back to Step 2 and reevaluate the matrices.
Step 4: Determine the Spherical Fuzzy Sets weights of criteria using Spherical Weighted Arithmetic Mean (SWAM) operator that is given in Eq. 4 [10].
Step 5: Defuzzify the obtained weights of each criterion using the score functions related to Spherical Fuzzy Sets that is given in Eq. 5 and normalize in Eq. 6.
Step 6: Find global weights for each level according to the hierarchical layer.
Step 7: Rank the criteria concerning the defuzzified global scores. The largest global score means that the most important criteria in others.

4 Case Study

4.1 Proposed Model

A business model framework is formed for public transportation authority taking into account impact that includes social and environmental issues. Measuring the success of the proposed model is a critical topic for strategic management. For that reason, the model is developed with P.E.S.T.L.E analysis which is introduced in Fig. 3. The main and sub-criteria of the model are defined according to the literature studies.

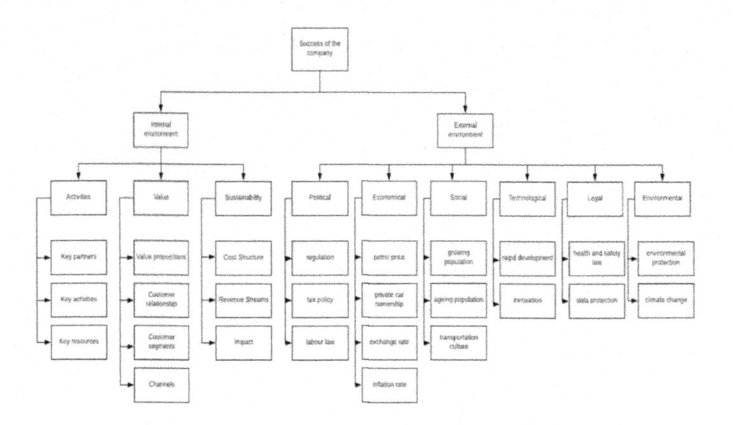

Fig. 3. Business model framework for public transportation authority regarding P.E.S.T.L.E analysis.

National and international experts evaluated the model around the world. The weights of the decision-makers are taken equal due to the same level of expertise. Sixteen feedbacks are got by following cities:

- Singapore and Kuala Lumpur from Asia continent,
- Sydney from Australia continent,
- New York, Vancouver and Montreal from America continent,
- Istanbul, London, Dublin, Barcelona and Moscow from Europe continent.

4.2 Results

According to the results, the internal environment is the most important criterion in the first stage. In the second stage, activity comes first, sustainability, and value comes, respectively. For the third level, key partners, key resources, and cost structure are ranged, respectively. The global weights and ranking order of defined criteria are indicated in Table 1.

Table 1. Global weights and ranking obtained from fuzzy approach.

Level	Criteria	Europe	America	Asia	Australia	Mean	Standard deviation
First level	IE	0,582	0,607	0,613	0,659	0,615	0,032
	EE	0,418	0,393	0,387	0,341	0,384	0,032
Second level	AC	0,218	0,259	0,182	0,289	0,237	0,047
	VA	0,179	0,168	0,202	0,185	0,183	0,014
	SU	0,185	0,180	0,229	0,185	0,195	0,023
	PO	0,079	0,073	0,058	0,079	0,072	0,010
	EC	0,074	0,051	0,066	0,039	0,058	0,016
	SO	0,064	0,057	0,057	0,062	0,060	0,004
	TE	0,074	0,074	0,076	0,059	0,070	0,008
	LE	0,066	0,073	0,068	0,051	0,065	0,009
	EN	0,061	0,065	0,062	0,051	0,060	0,006
Third level	KP	0,082	0,101	0,063	0,112	0,090	0,022
	KA	0,062	0,079	0,056	0,088	0,071	0,015
	KR	0,074	0,079	0,063	0,088	0,076	0,011
	VP	0,053	0,049	0,055	0,051	0,052	0,003
	CR	0,045	0,046	0,050	0,032	0,043	0,008
	CS	0,043	0,027	0,053	0,049	0,043	0,011
	CH	0,038	0,045	0,043	0,053	0,045	0,007
	CT	0,074	0,067	0,084	0,070	0,074	0,008
	RS	0,064	0,058	0,060	0,045	0,057	0,009
	IM	0,047	0,055	0,084	0,070	0,064	0,017
	RE	0,027	0,031	0,019	0,023	0,025	0,005
	TA	0,024	0,024	0,021	0,028	0,024	0,003
	LL	0,028	0,018	0,019	0,028	0,023	0,006
	PP	0,021	0,012	0,017	0,008	0,014	0,006
	PC	0,018	0,007	0,014	0,010	0,012	0,005
	ER	0,018	0,015	0,019	0,010	0,016	0,004
	IR	0,017	0,017	0,016	0,011	0,015	0,003
	GP	0,023	0,027	0,023	0,020	0,023	0,003
	AP	0,019	0,017	0,013	0,015	0,016	0,003
	TC	0,021	0,014	0,021	0,027	0,021	0,005
	RD	0,043	0,046	0,031	0,039	0,040	0,006
	IN	0,031	0,028	0,045	0,020	0,031	0,010
	HS	0,036	0,036	0,034	0,020	0,031	0,008
	DP	0,031	0,036	0,034	0,031	0,033	0,003
	EP	0,035	0,038	0,035	0,023	0,033	0,007
	CC	0,026	0,027	0,027	0,028	0,027	0,001

When social and environmental criteria are the least important one in the second stage, private car ownership which is located under economical one, comes last in the third stage.

5 Conclusion

The business model is used to design a company's activity, value, and finance with a holistic perspective. It can be designed according to the institution type such as profit or non-profit organization. When a business model canvas is developed for profit organizations, an impact factor is added for non-profit ones. In this study, to measure the success of a business model is proposed for public transportation authority taking into account the external environment. A model which consists of three hierarchical levels is constructed. SF-AHP method is applied to the proposed model to prioritize weights. National and international transportation experts evaluate the model with linguistic terms. According to the internal environment which is the business model, is the most important criterion for the first level. The second level activity that is under the internal environment comes first. For the last level, which is the third one, key partners comes first. This study will be a guide for public transport authorities and operators for designing business models and understanding priority criteria. Thus, they will pay attention to high level importance weights when they design business model. For future studies, other fuzzy sets will be applied to the model to explore weights of criteria and compare results that are similar or not. And also, sensitivity analysis will be performed to find out the precision of the models. To analyze cultural effect, continent base results will be investigated such as America, Asia, Australia, and Europe.

Appendix

See abbreviation of criteria for the proposed model as follows.

IE:Internal Environment, EE:External Environment, AC:Activities, VA: Value, SU:Sustainability, PO:Political, EC:Economical, SO:Social, TE:Technological, LE:Legal, EN:Environmental, KP:Key partners, KA:Key activities, KR: Key resources, VP:Value propositions, CR:Customer relationship, CS:Customer segments, CH:Channels, CS:Cost structure, RS:Revenue streams, IM:Impact, RE: Regulation, TP:Tax policy, LL:Labour law, PP:Petrol price, PC:Private car ownership, ER:Exchange rate, IR:Inflation rate, GP:Growing population, AP: Ageing population, TC:Transportation culture, RD:Rapid development, IN: Innovation, HL:Health and safety law, DP:Data protection, EP:Environmental protection, CC:Climate change.

References

1. Díaz-Díaz, R., Muñoz, L., Pérez-González, D.: Business model analysis of public services operating in the smart city ecosystem: the case of SmartSantander. Future Gen. Comput. Syst. **76**, 198–214 (2017)

2. Osterwalder, A., Pigneur, Y., Tucci, C.L.: Clarifying business models: origins, present, and future of the concept. Commun. Assoc. Inf. Syst. **16**(1), 1–28 (2005)
3. Osterwalder, A.: The business model ontology a proposition in a design science approach. L'Universit e de Lausanne (2004)
4. Sanderse, J.: The business model canvas of NGOs. Open Universiteit Nederland (2014)
5. Alanzi, S.: Pestle analysis introduction. Managing People in Projects, Cover Page (2018)
6. Mardani, A., Jusoh, A., Nor, K., Khalifah, Z., Zakwan, N., Valipour, A.: Multiple criteria decision-making techniques and their applications- a review of the literature from 2000 to 2014. Econ. Res. Ekonomska Istraživanja **28**(1), 516–571 (2015)
7. Eltarabishi, F., Omar, O.H., Alsyouf, I., Bettayeb, M.: Multi-criteria decision making methods and their applications–a literature review. In: Proceedings of the International Conference on Industrial Engineering and Operations Management, Dubai, UAE (2020)
8. Sabaei, D., Erkoyuncu, J., Roy, R.: A review of multi-criteria decision making methods for enhanced maintenance delivery. Procedia CIRP **37**, 30–35 (2015)
9. Kutlu-Gündoödu, F., Kahraman, C.: A novel spherical fuzzy analytic hierarchy process and its renewable energy application. Soft Comput. **24**(6), 4607–4621 (2020)
10. Kutlu-Gündoödu, F., Kahraman, C.: Spherical fuzzy sets and spherical fuzzy TOPSIS method. J. Intell. Fuzzy Syst. **36**(1), 337–352 (2019)
11. Saaty, T.L.: The Analytic Hierarchy Process: Planning, Priority Setting, Resources Allocation. McGraw-Hill, New York (1980)
12. Haseli, G., Sheikh, R., Sana, S.S.: Extension of base-criterion method based on fuzzy set theory. Int. J. Appl. Comput. Math. **6**(2), 54 (2020)

A Decision Support System Proposition for Type-2 Diabetes Mellitus Treatment Using Spherical Fuzzy AHP Method

Sezi Cevik Onar(✉) and Enes Hakan Ibil(✉)

Industrial Engineering Department, İstanbul Technical University, 34367 Istanbul, Turkey
{cevikse,ibil18}@itu.edu.tr

Abstract. The number of Type-2 Diabetes patients on a global scale has reached 463 million as of 2019, and it is predicted that this number will exceed 700 million by 2045. Type-2 Diabetes disease causes complications that seriously reduce the quality of life for patients and can be fatal at times, and 10% of global health expenditures are allocated to diabetes and diabetes-related diseases. The importance of the necessity of using personalized treatment methods by considering each patient as a separate case in the treatment process of Type-2 Diabetes is emphasized in the studies. In this study, the best oral anti-diabetic combination for patients in different risk groups was tried to be determined using the Spherical Fuzzy AHP method by referring to the treatment approach included in the national diabetes treatment guide. Our study is original with the first use of the Spherical AHP method in the personalized medical literature.

Keywords: Type 2 diabetes mellitus · Tailored medicine · Spherical fuzzy AHP

1 Introduction

Diabetes Mellitus has been defined as a chronic metabolic disease characterized by hyperglycemia due to insulin secretion, the effect of insulin, or defects in both of these factors in national and international sources. According to the International Diabetes Federation, about 10% of global health expenditures all over the world are spent to combat diabetes and diabetes-related diseases, while diabetes disease and its complications significantly reduce the quality of life of individuals and in some cases can be fatal.

Among the diabetes diseases, Type-2 Diabetes Mellitus (T2D) is the disease group with the highest prevalence rate of 90%. Although important sources have cited in all their publications that the most effective method in the treatment of T2D is "lifestyle change", both drug therapy and fight against the disease are important. The main mechanism of action of drugs used in the treatment of T2D is to prevent possible complications while providing glycemic targets in patients, so there are new treatment methods and drug classes that have been introduced to be used in T2D treatment day by day. In addition,

C. Kahraman et al. (Eds.): INFUS 2021, LNNS 308, pp. 749–756, 2022.
https://doi.org/10.1007/978-3-030-85577-2_88

many studies in the field of medicine have shown that by planning the treatment process specifically for the patient, the aims of the treatment can be achieved more successfully, in other words, it can be prevented that the treatment is applied more than it should or is less than it should be.

Metformin is used as the first-line drug in the treatment of T2D. Since T2D is a progressive disease, Metformin treatment after a while falls short of achieving the intended glycemic targets in patients. At this point, national and international diabetes guidelines recommend that other drugs be included in the treatment process to be combined with Metformin. With each new anti-diabetic drug introduced, the treatment planning process becomes more complex and turns the decision-making process of physicians into a multi-purpose and multi-criteria decision-making problem.

In the national diabetes guideline, patients in the T2D treatment approach were divided into risk groups according to their glycoside hemoglobin (HbA1c) levels and the number of beta reserve cells, and the prescribed treatment methods were determined based on their risk conditions. Although there are 5 drug classes that are recommended to be used in patients when the HbA1c level is between 8%–10% and the beta reserve cell level is sufficient-borderline in our study, no classification has been made as to which drug will be used in which risk situation specifically. In this study, based on this situation, an answer was sought for the question of which drugs that can be combined with Metformin to determine for the most effective T2D therapy according to risk situation. First of all, the criteria to be considered in drug selection were determined as a result of the literature study, and then the effectiveness of five drug classes was compared according to the risk groups and the criteria determined by the survey study conducted by the specialist physicians. In the study, the SF-AHP method, which is the spherical fuzzy set extension of the AHP method, was used. The rest of the paper is organized as follows. In Sect. 2, a brief summary on Type-2 Diabetes are given. In Sect. 3, we explain the used Spherical Fuzzy Sets and Spherical Fuzzy AHP Method. The application of the proposed approach is given in Sect. 4. The Last section concludes and gives further suggestions.

2 Type-2 Diabetes Mellitus

Type-2 Diabetes disease is defined as "a chronic metabolic disease characterized by hyperglycemia caused by insulin secretion, insulin effect or disorder in both of these factors" in the Diabetes Diagnosis and Treatment Guidelines of the Turkish Diabetes Foundation. Today, Type-2 Diabetes (T2D) disease is the most common disease group with a rate of 90% among all Diabetes diseases. The disease progresses asymptomatically in people for a long time, in other words, prediabetes has a very long period ranging from 5–15 years [1].

According to clinical data, diabetes can cause retinopathy (severe visual disturbances) in 28% of patients [2]. Apart from this, diabetic patients have 2.5 to 5 times more risk of heart attack than healthy people. Life-threatening complications such as amputation and kidney failure are among the disorders caused by diabetes [3]. Studies have shown that T2D disease and the disease it causes have a high rate of morbidity and that there is not enough awareness about the disease despite the presence of severe complications that negatively affect the quality of life [1, 3]. According to the diabetes atlas published by the International Diabetes Federation in 2019, there are 463 million diabetes patients in the world, this number is estimated to reach 578 million in 2030 and 700 million in 2045 with an increase of 51%. When examined by region, the rate of increase of diabetes in the Middle East and North Africa is at the top of the list with 96% and almost alarms. Diabetes growth rate in the Middle East and North Africa is followed by South-East Asia (with 74%), West America (with 31%) and Europe (with 15% percent), respectively [4]. Considering that the number of diabetes patients, which was 151 million all over the world in the early 2000s, reached 450 million in 2019, it becomes clear that the estimates made in the 2030 and 2045 projections may come true long before the relevant dates. The fact that the first treatment method used in the treatment of Type-2 Diabetes is lifestyle change has been pointed out in many national / international guidelines and its effectiveness has been shown in numerous studies [3, 5, 6]. In the Diabetes Diagnosis and Treatment Guideline published by the National Diabetes Council in 2019, the goals expected to be achieved with lifestyle changes were listed and the importance of T2D treatment was emphasized by drawing attention to its complementary importance in every step. It was mentioned in the intro section that metformin is used as a first-line oral antidiabetic in the treatment of T2D. Due to the fact that T2D is a progressive disease, metformin becomes insufficient to provide the targeted glycemic level over time, at this point, decision-making physicians create combinations of oral antidiabetics with double or triple drug combinations according to the patient's phenotype and treatment guidelines. In more advanced stages of the disease, basal insulin combinations and multiple dose insulin combinations are included in the treatment process. In this study, the categories of the condition that we refer to in our study, which are included in the Diabetes Diagnosis and Treatment Guideline T2D treatment approach section published by the Turkish Diabetes Foundation, are listed below.

- HbA1c level is between 8%–10% and beta cell reserve is sufficient,
- HbA1c level is more than 10% and beta cell reserve is sufficient,
- HbA1c level is more than 10% and beta cell reserve is borderline,

In the decision support system, we created within the scope of our study; In the risk groups listed above, which of the five drug classes envisaged to be combined with Metformin in the treatment of T2D (listed below with criteria) is the best alternative, it will be evaluated separately according to risk groups.

The decision support model established as a result of our study; It answers the question of the best oral anti-diabetic drug to be combined with Metformin for a potential patient in one of the three risk groups mentioned above (Table 1).

Table 1. Decision alternatives and criteria

Alternatives	Criteria
Sulfonylureas (Alternative 1)	Level of Effect on HbA1c Level (Criteria 1)
Pioglitazon (Alternative 2)	Risk Level for Developing Hypoglycemia (Criteria 2)
DPP-4 Inhibitors (Alternative 3)	Risk of Developing Macrovascular Complications (Criteria 3)
SGLT-2 Inhibitors (Alternative 4)	Risk of Developing Microvascular Complications (Criteria 4)
GLP-1 RA (Alternative 5)	Effects on Body Weight (Criteria 5)
	Treatment Cost (Criteria 6)

3 Spherical Fuzzy Sets and Spherical Fuzzy AHP

3.1 Spherical Fuzzy Sets

Spherical fuzzy sets (SFS) have been brought to the literature by Gündoğdu & Kahraman [7]. SFSs is an extension of Picture Fuzzy Sets. Spherical Fuzzy Sets is to allow decision makers to generalize other extensions of fuzzy sets by defining a membership function on a global system. Fuzzy sets expressed as Spherical Fuzzy Set are shown with the equation given in Eq. 1 and must meet the following conditions.

$$\tilde{A}_S = \left\{\left\langle u, \left(\mu_{\tilde{A}_S}(u), v_{\tilde{A}_S}(u), \pi_{\tilde{A}_S}(u)\right) \middle| u \in U\right\rangle\right\} \quad (1)$$

$$\mu_{\tilde{A}_S}(u) : U \to [0, 1]$$

$$v_{\tilde{A}_S}(u) : U \to [0, 1]$$

$$\pi_{\tilde{A}_S}(u) : U \to [0, 1]$$

$$0 \leq \mu^2_{\tilde{A}_S}(u) + v^2_{\tilde{A}_S}(u) + \pi^2_{\tilde{A}_S}(u) \leq 1 \forall u \in U$$

According to inequalities given below $\mu_{\tilde{A}_S}(u), v_{\tilde{A}_S}(u), \pi_{\tilde{A}_S}(u)$, represents the degree of membership, degree of non-membership and degree of indecision, respectively.

Arithmetic Operations in Spherical Fuzzy Sets

Addition: Addition in spherical fuzzy sets is done as given in Eq. 2 [8].

$$\tilde{A}_s \oplus \tilde{B}_s = \left\{\left(\mu^2_{\tilde{A}_s} + \mu^2_{\tilde{B}_s} - \mu^2_{\tilde{A}_s}\mu^2_{\tilde{B}_s}\right)^{1/2}, v_{\tilde{A}_s}v_{\tilde{B}_s}, \left(\left(1 - \mu^2_{\tilde{B}_s}\right)\pi^2_{\tilde{A}_s} + \left(1 - \mu^2_{\tilde{A}_s}\right)\pi^2_{\tilde{B}_s} - \pi^2_{\tilde{A}_s}\pi^2_{\tilde{B}_s}\right)^{1/2}\right\} \quad (2)$$

Multiplication: Multiplication in spherical fuzzy sets is done as given in Eq. 3 [8].

$$\tilde{A}_s \otimes \tilde{B}_s = \left\{ \mu_{\tilde{A}_s}\mu_{\tilde{B}_s}, \left(v^2_{\tilde{A}_s} + v^2_{\tilde{B}_s} - v^2_{\tilde{A}_s} v^2_{\tilde{B}_s} \right)^{1/2}, \left(\left(1 - v^2_{\tilde{B}_s}\right)\pi^2_{\tilde{A}_s} + \left(1 - v^2_{\tilde{A}_s}\right)\pi^2_{\tilde{B}_s} - \pi^2_{\tilde{A}_s}\pi^2_{\tilde{B}_s} \right)^{1/2} \right\} \tag{3}$$

Multiplication by a Scalar: The process of multiplying the spherical fuzzy set by a scalar number is done as given in Eq. 4 [8].

$$\lambda\tilde{A}_s = \left\{ \left(1 - \left(1 - \mu^2_{\tilde{A}_s}\right)^{\lambda}\right)^{1/2}, v^{\lambda}_{\tilde{A}'_s}, \left(\left(1 - \mu^2_{\tilde{A}_s}\right)^{\lambda} - \left(1 - \mu^2_{\tilde{A}_s} - \pi^2_{\tilde{A}_s}\right)^{\lambda} \right)^{1/2} \right\} \tag{4}$$

Note that $\lambda > 0$

Power of a SFS: In spherical fuzzy sets, the process of taking power is done as given in Eq. 5.

$$\tilde{A}^{\lambda}_s = \left\{ \mu^{\lambda}_{\tilde{A}_s}, \left(1 - \left(1 - v^2_{\tilde{A}_s}\right)^{\lambda}\right)^{1/2}, \left(\left(1 - v^2_{\tilde{A}_s}\right)^{\lambda} - \left(1 - v^2_{\tilde{A}_s} - \pi^2_{\tilde{A}_s}\right)^{\lambda} \right)^{1/2} \right\} \tag{5}$$

Note that $\lambda > 0$

Spherical Weighted Arithmetic Mean (SWAM): Spherical weighted aritmetic mean can be calculated as follows, $w = (w_1, w_2 \cdot \ldots w_n)$; $w_i \in [0, 1]$; $\sum_{i=1}^{n} w_i = 1$,

$$\mathrm{SWAM}_w(A_{s1}, \ldots\ldots, A_{sn}) = w_1 A_{s1} + w_2 A_{s2} + \ldots\ldots + W_n A_{sn}$$
$$= \left\langle \left[1 - \prod_{i=1}^{n} \left(1 - \mu^2_{A_{si}}\right), \prod_{i=1}^{n} v^{w_i}_{A_{si}}, \left[\prod_{i=1}^{n} \left(1 - \mu^2_{A_{si}}\right)^{w_i} - \prod_{i=1}^{n} \left(1 - \mu^2_{A_{si}} - \pi^2_{A_{si}}\right)^{w_i} \right]^{1/2} \right. \right\rangle \tag{6}$$

3.2 Spherical Fuzzy AHP

Step 1: Decision making problem is defined. A hierarchical structure is established between purpose, criteria and alternatives.
Step 2: First of all, the dominance degrees of the criteria are compared with each other in pairs by decision makers. Then, alternatives are compared in pairs according to a certain criterion. The linguistic expression table suggested by Kutlu Gündoğdu and Kahraman [8] can be used in comparisons.
Step 3: Consistency check is made for the tables created by decision makers. Using Saaty's eigenvalue method as given below,

$$\mathrm{CI} = (\lambda_{max} - n)/(n - 1), \mathrm{CR} = (\mathrm{CI}/\mathrm{RI}(n)) \tag{7}$$

Here, λ_{max} value shows the largest eigenvalue value in the binary comparison matrix and $\mathrm{RI}(n)$ is the random index value changing according to the value of n.

Step 4: Fuzzy geometric mean for the row i is calculated by the Eq. 8 as shown below.

$$r^i = \left(\prod_{j=1}^{n} r_{ij} \right)^{1/n} \tag{8}$$

In order to weight the geometric averages of the row values obtained in the first step, $\mathcal{F}(r_i)$ value must be found with equation $\mathcal{F}\left(\tilde{A}_s\right) = \mu(1 - v)(1 - \pi)$.
The prioritization function is calculated with the help of the equation below. In this way, the vector of $W = [w_1, w_2, \cdots, w_n]$ is found with equation $w_i = \mathcal{F}(r^i) / \sum_{i=1}^{n} \mathcal{F}(r^i)$. The ranking and ranking process between alternatives and criteria is done with the help of the Eq. 9 given below.

$$W_i = \left[\sum w_{ij} \times w_j \right] \tag{9}$$

4 Application

Within the scope of the study, five decision-making physicians were consulted. Then, the consistency levels of 95 binary decision matrices created from the opinions of decision-making physicians were calculated with the help of MATLAB software and it was seen that the values were between 0.02 and 0.14 and they were evaluated to be consistent. Afterwards, decision makers first compared the 6 criteria determined with each other, and then the performances of the alternatives were calculated according to the six criteria weighted separately for each risk group. Decision makers' group decisions are calculated using the SWAM operator. EXCEL software was used in the calculation steps. Due to the transaction volume and page restriction, only the results are shared in this section (Tables 2, 3 and 4).

Table 2. Results of SF-AHP Method for Group 1

Group1	C1	C2	C3	C4	C5	C6	
Wi	*0,378*	*0,205*	*0,126*	*0,132*	*0,108*	*0,052*	*W*
A1	0,530	0,452	0,309	0,304	0,274	0,224	***0,413***
A2	0,211	0,177	0,238	0,242	0,249	0,200	***0,215***
A3	0,103	0,150	0,182	0,191	0,197	0,194	***0,149***
A4	0,090	0,135	0,138	0,145	0,154	0,232	***0,127***
A5	0,066	0,085	0,134	0,117	0,126	0,150	***0,096***

Table 3. Results of SF-AHP Method for Group 2

Group2	C1	C2	C3	C4	C5	C6	
Wi	*0,378*	*0,205*	*0,126*	*0,132*	*0,108*	*0,052*	*W*
A1	0,555	0,477	0,310	0,309	0,326	0,229	***0,434***
A2	0,166	0,195	0,247	0,229	0,237	0,185	***0,199***
A3	0,126	0,106	0,172	0,174	0,165	0,205	***0,142***
A4	0,088	0,129	0,130	0,158	0,140	0,222	***0,124***
A5	0,066	0,093	0,140	0,129	0,132	0,160	***0,101***

Table 4. Results of SF-AHP Method for Group 3

Group3	C1	C2	C3	C4	C5	C6	
Wi	*0,378*	*0,205*	*0,126*	*0,132*	*0,108*	*0,052*	*W*
A1	0,333	0,316	0,318	0,212	0,337	0,236	***0,307***
A2	0,252	0,253	0,239	0,252	0,264	0,231	***0,251***
A3	0,147	0,147	0,174	0,174	0,178	0,188	***0,160***
A4	0,145	0,166	0,141	0,184	0,103	0,188	***0,152***
A5	0,123	0,118	0,129	0,177	0,117	0,157	***0,131***

5 Conclusion

As a result of the calculations made, the criterion of the level of influence on the level of HbA1c in T2D treatment is the most important criterion with 31.9%, respectively, the risk of developing hypoglycemia (C2) with 21%, the risk of developing microvascular complications (C4) with 17.1%, the risk of developing macrovascular complications (C3) with 13.6% and the impact criterion on body weight (C5) with 11.1%, and finally, the treatment cost criterion (C6) with 5.2%.

When the decision alternatives to be combined with Metformin in T2D treatment in terms of the aggravated criteria are examined:

Sulfonylureas (A1): It has been calculated to be the best decision alternative in all three risk groups. In addition, when the performance of the second risk group and the performance of the third risk group is compared, it is observed that the performance score of Sulfonylureas significantly decreased with the gaining importance of other decision alternatives.

Pioglitazone (A2): It has a close performance score in all three risk groups and ranks second among decision alternatives in all risk groups.

DPP-4 Inhibitor (A3): Has almost the same performance score in all three risk groups. It has the third rank among alternatives in the first and second risk groups, while it is in the fourth place in the third risk group.

SGLT-2 Inhibitor (A4): Performance score increased regularly during the transition from the first risk group to the third risk group. In addition, while it is in the fourth place among decision alternatives in the first two risk groups, it has risen to the third position in the third risk group.

GLP1-RA Inhibitor (A5): The first two risk groups produced close performance scores, while the performance score increased significantly in the last risk group. It is the last choice in all risk groups.

Suggestions for Future Research and Contribution of the Paper: It was stated in the interviews with decision-making physicians that the established model reflects the T2D approach of decision makers and can be used more effectively by developing a decision model based on personalized medicine approach, the decision support system could be included in the diagnosis process and that a comprehensive diagnosis program could provide early diagnosis in other diseases including T2D disease.

In addition, decision makers have reported that the use of fuzzy set theory together with multi-criteria decision-making methods is efficient. It is thought that the researches will contribute to the literature with the studies that researchers will carry out with the methods based on the fuzzy set theory in the field called "personalized medicine" and "tailored medicine" in the literature, whose importance is increasing day by day and which is emphasized as a necessity in numerous sources.

References

1. Turkish Diabetes Foundation. Diyabet Diagnosis and Treatment Guide (2019)
2. Börü, Ü.T., et al.: Prevalence of peripheral neuropathy in Type 2 diabetic patients attending a diabetes center in Turkey. Endocr. J. **51**(6), 563–567 (2004). https://doi.org/10.1507/endocrj.51.563
3. International Diabetes Summit. Diabetes Problem in Turkey and the Region (2013)
4. Saeedi, P., et al.: Global and regional diabetes prevalence estimates for 2019 and projections for 2030 and 2045: Results from the International Diabetes Federation Diabetes Atlas, 9th edition. Diabetes Res. Clin. Pract. **157**, 107843 (2019). https://doi.org/10.1016/j.diabres.2019.107843
5. Pan, X.-R., et al.: Effects of diet and exercise in preventing NIDDM in people with impaired glucose tolerance: the Da Qing IGT and diabetes study. Diabetes Care **20**(4), 537–544 (1997). https://doi.org/10.2337/diacare.20.4.537
6. International Diabetes Association. Diabetes Atlas (2019)
7. Kutlu Gündoğdu, F., Kahraman, C.: Spherical fuzzy sets and spherical fuzzy TOPSIS method. J. Intell. Fuzzy Syst. **36**(1), 337–352 (2019). https://doi.org/10.3233/JIFS-181401
8. Kahraman, C., Kutlu Gündoğdu, F. (eds.): Decision making with spherical Fuzzy sets. SFSC, vol. 392. Springer, Cham (2021). https://doi.org/10.1007/978-3-030-45461-6

Information Technology Governance Evaluation Using Spherical Fuzzy AHP ELECTRE

Akın Menekşe and Hatice Camgöz Akdağ(✉)

Istanbul Technical University, 34467 Maslak, Istanbul, Turkey
{menekse18,camgozakdag}@itu.edu.tr

Abstract. Information technology (IT) governance is a process used to monitor and control key information technology capability decisions to ensure the delivery of value to key stakeholders for organizations. For higher education institutions, IT is considered as a strategic tool on the basis of teaching, research and administration. This paper develops a multi-criteria decision making (MCDM) based decision support tool based on the techniques of Analytical Hierarchy Process (AHP) and ELimination Et Choice Translating REality (ELECTRE). The proposed model is developed in a three dimensional spherical fuzzy environment, which enables the model to capture the vagueness in the problem in a comprehensive way. The proposed model is applied to an IT governance evaluation problem in a higher education institution. For this purpose, four academic units are evaluated with respect to five dimensions of internationally recognized Control Objectives for Information and Related Technology (COBIT) IT governance framework by three decision-makers. The criteria weights are determined with AHP technique, and the alternatives are ranked with ELECTRE by constructing outranking relations utilizing score and accuracy functions developed for spherical fuzzy sets. Sensitivity and comparative analyses; limitations; and future research avenue are also given in the study.

Keywords: Spherical fuzzy sets · AHP ELECTRE · IT Governance · Higher education institution

1 Introduction

IT is a strategic tool for higher education institutions in both teaching, research and administration [1], and IT governance is a structure consisting of processes, relationships and mechanisms used to control, manage and develop IT strategy and corporate resources in the best way to achieve the goals and objectives of an organization [2]. COBIT 5, on the other hand, provides a framework for helping an organization by balancing organizational benefits and losses and optimizing resources used to manage IT governance as a whole [3]. However, managers usually have lack of knowledge with IT related issues and various numerical MCDM methodologies are presented to support the solution of problems encountered within this scope. Ahriz et al. [4] ranked IT projects in a university by using COBIT 5 principles and developed an AHP TOPSIS model in a crisp environment. Bouayad et al. [5] applied the AHP for selecting the best

C. Kahraman et al. (Eds.): INFUS 2021, LNNS 308, pp. 757–765, 2022.
https://doi.org/10.1007/978-3-030-85577-2_89

IT governance framework for a given organization. The authors used crisp form of pairwise comparison matrices in their study and found that COBIT 5 is the best framework for IT governance. Yudatama and Sarno [6] presented an AHP TOPSIS model for evaluation of the maturity index used within the scope of IT governance implementation based on two dimensional triangular fuzzy sets, and Apriliana et al. [7] analysed the risks in IT applications by developing crisp based AHP and SAW methods by using the COBIT 5 framework as the basis for their work.

In MCDM models presented as a solution to the problems within the scope of IT, it is seen that the uncertainty and vagueness in the nature of the problem either does not addressed, or the methods developed cannot handle this issue comprehensively. This paper fills the gap by using three dimensional spherical fuzzy sets [8] which enable decision-makers express their membership, non-membership and hesitancy degrees independently and in a larger domain.

In this paper, spherical fuzzy AHP ELECTRE method is developed to evaluate IT governance levels of academic units based on internationally recognized COBIT 5 framework. For this purpose, five dimensions of COBIT 5 framework is evaluated by three decision-makers with AHP methodology and four academic units are ranked with ELECTRE methodology. Unlike classical ELECTRE, which uses user-defined concordance and discordance sets, this paper constructs these outranking relations by utilizing score and accuracy functions developed for spherical fuzzy sets.

The rest of the paper is organized as follows: Sect. 2 presents the proposed methodology in a step-by-step form. Section 3 presents the application from a higher education institution. Sensitivity and comparative analyses are also given in this section. Section 4 finalizes the paper with conclusion, limitations and future research avenue.

2 Methodology

In this section, the proposed spherical fuzzy AHP ELECTRE steps are given in detail. The model consists of three stages. In stage 1, criterion weights are determined by using the AHP, which is a technique based on a pairwise comparison procedure designed to capture relative judgments in a way that ensures consistency [9]. In stage 2, first, alternatives are evaluated by decision-makers, then the criterion weights obtained in stage 1 are included to the model, and final form of the decision matrix, which is named as spherical fuzzy decision matrix, is obtained. In the last stage, the alternatives are ranked based on spherical fuzzy decision matrix by using ELECTRE [10] method, which is based on a pairwise comparison of alternatives with respect to criteria by determining the outranking relations, named as concordance and discordance indices, measuring the dissatisfaction of a decision-maker when choosing one alternative over the other.

The proposed hybrid model is developed based on the definition; spherical weighted geometric mean (SWGM), spherical fuzzy multiplication and defuzzification operators; and the linguistic scale presented by Kutlu Gündoğdu and Kahraman [11]. Details of these operators are not given in a separate section due to limited space. The steps of the proposed model are as follows:

Stage 1: Determine Criteria Weights with AHP Technique
Step 1: Let decision-makers evaluate criteria by filling pairwise comparison matrices by utilizing the readily given linguistic scale.
Step 2: Transform pairwise comparison matrices in the form of linguistic terms to their corresponding spherical fuzzy forms and obtain spherical fuzzy pairwise comparison matrices (SFPCMs).
Step 3: Calculate the consistency ratio of each SFPCM by utilizing score indices in the given linguistic scale. Then, apply the classical consistency check. The threshold of the CR is 0.1. If the CR of any SFPCM is less than 0.1 let that decision-maker reevaluate the criteria.
Step 4: Obtain spherical fuzzy criterion weight matrices (SFCWMs) by aggregating SFPCMs by utilizing the SWGM operator.
Step 5: Aggregate SFWCMs from each decision-maker to obtain one aggregated spherical fuzzy criterion weight matrix (ASFCWM) by utilizing the SWGM operator.
Step 6: Calculate score indices $S(\tilde{w}_j^s)$ by converting the ASFCWM into its crisp form by utilizing the defuzzification operator.
Step 7: Obtain criterion weights by normalizing the score indices by utilizing Eq. 1.

$$\bar{w}_j^s = \frac{S(\tilde{w}_j^s)}{\sum_{J=1}^{n} S(\tilde{w}_j^s)} \tag{1}$$

Stage 2: Obtain Spherical Fuzzy Decision Matrix
Step 8: Let decision-makers fill performance evaluation matrices by utilizing the same linguistic terms. In this step, the performance of each alternative is evaluated with respect to each criterion.
Step 9: Transform linguistic evaluations to their corresponding spherical fuzzy numbers and obtain spherical fuzzy alternative evaluation matrices (SFAEMs).
Step 10: Aggregate all SFAEMs to obtain one aggregated spherical fuzzy alternative evaluation matrix (ASFAEM) by utilizing the SWGM operator.
Step 11: Multiply ASFCWM by ASFAEM to obtain spherical fuzzy decision matrix by utilizing spherical fuzzy multiplication operator.

Stage 3: Construct the Outranking Relations and Rank the Alternatives
Step 12: Construct outranking relations based on score and accuracy functions. Use accuracy function presented by Kutlu Gündoğdu and Kahraman [8] in this step. Outranking relations are constructed through single type of concordance and discordance sets as given in Eqs. 2 and 3 respectively.

$$\mathrm{C}_{kl} = \{j|\sqrt{|100 * [(3\mu_{kj} - \frac{\pi_{kj}}{2})^2 - (\frac{v_{kj}}{2} - \pi_{kj})^2]|} > \sqrt{|100 * [(3\mu_{lj} - \frac{\pi_{lj}}{2})^2 - (\frac{v_{lj}}{2} - \pi_{lj})^2]|}\} \tag{2}$$

$$\mathrm{D}_{kl} = \{j|\sqrt{|100 * [(3\mu_{kj} - \frac{\pi_{kj}}{2})^2 - (\frac{v_{kj}}{2} - \pi_{kj})^2]|} < \sqrt{|100 * [(3\mu_{lj} - \frac{\pi_{lj}}{2})^2 - (\frac{v_{lj}}{2} - \pi_{lj})^2]|}\} \quad (3)$$

and for the following case, concordance and discordance sets are obtained by utilizing Eqs. 4 and 5.

$$\{j|\sqrt{|100 * [(3\mu_{kj} - \frac{\pi_{kj}}{2})^2 - (\frac{v_{kj}}{2} - \pi_{kj})^2]|} = \sqrt{|100 * [(3\mu_{lj} - \frac{\pi_{lj}}{2})^2 - (\frac{v_{lj}}{2} - \pi_{lj})^2]|}\};$$

$$C_{kl} = \{(\mu_{kj}^2 + v_{kj}^2 + \pi_{kj}^2) \leq (\mu_{lj}^2 + v_{lj}^2 + \pi_{lj}^2)\} \quad (4)$$

$$D_{kl} = \{(\mu_{kj}^2 + v_{kj}^2 + \pi_{kj}^2) > (\mu_{lj}^2 + v_{lj}^2 + \pi_{lj}^2)\} \quad (5)$$

Step 13: Construct concordance and discordance matrices with concordance and discordance indices.

Concordance index is obtained by utilizing Eq. 6.

$$g_{kl} = \sum_{j \in C_{kl}} w_j; \qquad \sum_{j=1}^{n} w_j = 1 \quad (6)$$

where w_j is the weight of jth criterion; and discordance index is obtained by utilizing Eq. 7.

$$h_{kl} = \frac{max_{j \in D_{kl}} d(X_{kj}, X_{lj})}{max_{j \in J} d(X_{kj}, X_{lj})} \quad (7)$$

where h_{kl} is the spherical fuzzy discordance index, which is the relative difference of A_k with respect to A_l; where $d(X_{kj}, X_{lj})$ is the normalized Euclidean distance between alternatives X_k and X_l for the jth criterion. Normalized Euclidean distance is given in Eq. 8.

$$D(X_i, X^*) = \sqrt{\frac{1}{2n} \sum_{i=1}^{n} ((\mu_{X_i} - \mu_{X^*})^2 + (v_{X_i} - v_{X^*})^2 + (\pi_{X_i} - \pi_{X^*})^2)} \quad (8)$$

Step 14: Aggregate concordance and discordance matrices to obtain the global matrix by utilizing Eq. 9.

$$r_{kl} = \frac{g_{kl}}{g_{kl} + h_{kl}} \quad (9)$$

A higher value of r_{kl} means that alternative A_k is better than alternative A_l.

Step 15: Rank the alternatives by utilizing Eq. 10.

$$T_k = \sum_{l=1, l \neq k}^{m} r_{kl}; k = 1, 2, 3 ..., m \quad (10)$$

3 Application

Spherical fuzzy AHP ELECTRE model is applied to IT governance evaluation problem for a higher education institution. IT governance maturity levels of four academic units are evaluated by three experts with respect to five criteria of COBIT 5 framework namely C1: Evaluate, direct and monitor; C2: Align, plan and organize; C3: Build, acquire and implement; C4: Deliver service and support, and C5: Monitor, evaluate and assess. The solution of the MCDM problem is presented as follows:

Stage 1: Determination of Criterion Weights:
Steps 1–7: ASFCWM, score indices and criterion weights are obtained as presented in Table 1.

Table 1. ASFCWM, score indices and criteria weights

	ASFCWM	Score index	Criterion weight
C1	0.35;0.64;0.27	9.247	0.141
C2	0.64;0.34;0.28	17.757	0.271
C3	0.56;0.43;0.30	15.280	0.233
C4	0.45;0.54;0.32	12.012	0.183
C5	0.43;0.56;0.31	11.345	0.173

Stage 2: Obtaining the Spherical Fuzzy Decision Matrix:
Steps 8–11: Spherical fuzzy decision matrix and corresponding score and accuracy values are obtained as presented in Table 2.

Table 2. Spherical fuzzy (SF) decision matrix, score values, and accuracy values

		C1	C2	C3	C4	C5
SF-decision matrix	A1	0.31;0.65;0.27	0.30;0.57;0.41	0.35;0.54;0.37	0.29;0.62;0.36	0.25;0.65;0.34
	…	…	…	…	…	
	A4	0.24;0.68;0.30	0.38;0.63;0.23	0.41;0.51;0.33	0.34;0.58;0.33	0.24;0.62;0.40
Score values	A1	7.812	6.745	8.750	6.757	5.789
	…	…	…	…	…	
	A4	8.190	5.739	10.648	8.682	5.149
Accuracy values	A1	0.587	0.578	0.555	0.601	0.603
	…	…	…	…	…	
	A4	0.588	0.616	0.534	0.565	0.604

Stage 3: Outranking Relations and Final Ranking of the Alternatives:
Step 12: Concordance and discordance sets are obtained as presented in Table 3.

Table 3. Concordance and discordance sets

	A1	A2	A3	A4	A1	A2	A3	A4
A1	-	C1;C2;C3;-;C5	C1;C2;C3;C4;C5	-;C2;-;-;C5	-	-;-;-;D4;-	D1;D2;D3;D4;D5	D1;-;D3;D4;-
A2	-;-;-;C4;-	-	-;-;C3;C4;-	-;-;-;-;-	D1;D2;D3;-;D5	-	D1;D2;-;-;D5	D1;D2;D3;D4;D5
A3	-;C2;-;-;-	C1;C2;-;-;C5	-	-;C2;-;-;C5	D1;-;D3;D4;D5	-;-;D3;D4;-	-	D1;-;D3;D4;-
A4	C1;-;C3;C4;-	C1;C2;C3;C4;C5	C1;-;C3;C4;-	-	-;D2;-;-;D5	-;-;-;-;-	-;D2;-;-;D5	-

Steps 13–14: Concordance, discordance and aggregated matrices are obtained as given in Table 4.

Table 4. Concordance (C), discordance (D) and aggregated matrices (A)

	1				D				A			
	A1	A2	A3	A4	A1	A2	A3	A4	A1	A2	A3	A4
A1	–	0.82	1.00	0.44	–	0.22	0.00	0.47	–	0.79	1.00	0.49
A2	0.18	–	0.42	0.00	1.00	–	1.00	1.00	0.15	–	0.29	0.00
A3	0.27	0.58	–	0.44	1.00	0.63	–	1.00	0.21	0.48	–	0.31
A4	0.56	1.00	0.56	–	1.00	0.00	0.66	–	0.36	1.00	0.46	–

Step15: Alternatives are ranked according to descending appraisal scores as given in Table 5. Since A1 has the highest appraisal score, IT governance level of alternative A1 is the most critical.

Table 5. Appraisal scores of alternatives and final ranking

Alternative	Apraisal score
A1	0.757
A2	0.149
A3	0.334
A4	0.605

3.1 Sensitivity and Comparative Analyses

Sensitivity analysis is conducted for both decision-maker weights and criterion weights. For this purpose, appraisal scores are calculated for 10 different decision-maker and criterion weight distribution scenarios. . As it is seen in Fig. 1, the ranking orders are same in seven of the 10 scenarios; and in Fig. 2, it seen that almost all scenarios give the same ranking order. Although these results show the robustness of the proposed model, decision-maker evaluations may affect the final rankings; hence, a qualified group of decision-makers should be selected for the problem.

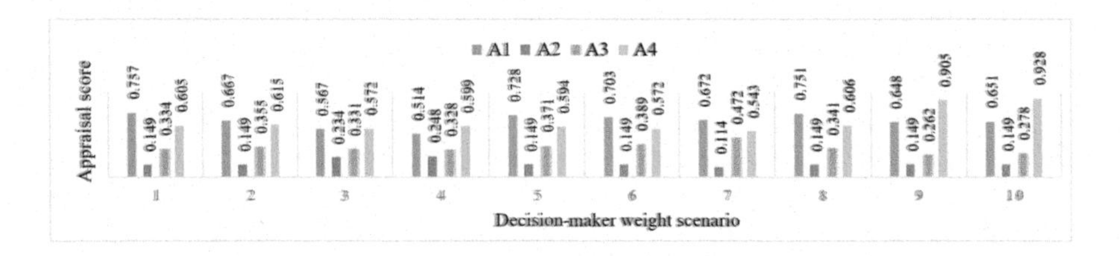

Fig. 1. Appraisal scores for 10 different decision-maker weight distribution scenarios.

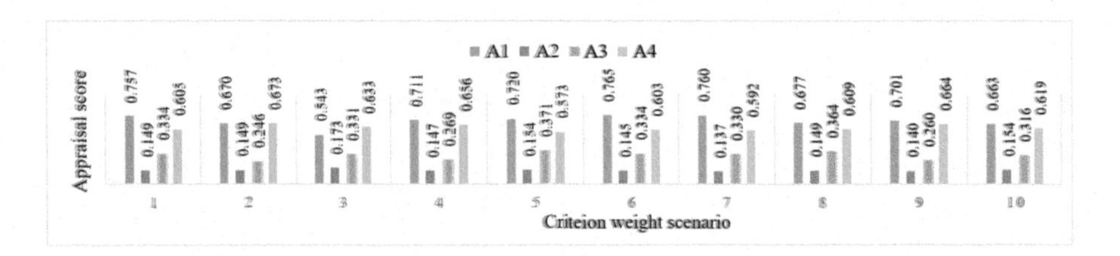

Fig. 2. Appraisal scores for 10 different criterion weight distribution scenarios.

For comparative study, the same problem is handled with spherical fuzzy AHP TOPSIS, and the results are given in Table 6. It is seen that, the ranking order is same $A1 > A4 > A3 > A2$ in both methods, and this shows the validity of the proposed model.

Table 6. Comparative analysis results

	Spherical fuzzy AHP ELECTRE	Spherical fuzzy AHP TOPSIS
A1	0.757	0.763
A2	0.149	0.385
A3	0.334	0.524
A4	0.605	0.630
Final ranking	$A1 > A4 > A3 > A2$	$A1 > A4 > A3 > A2$

4 Conclusion and Final Remarks

IT is considered to be a hot topic for many years in various sectors and it is gaining importance in higher education institutions recently. However, evaluating IT governance levels of academic units is really a challenging task for responsible bodies and a systematic, quantified and mathematically accepted way is needed in this point. This paper presents a hybrid spherical fuzzy AHP-ELECTRE model for evaluating IT levels of academic units. In the proposed model, criteria weights are determined with AHP technique and the alternatives are ranked with ELECTRE. The vagueness in the nature of the problem is modeled with spherical fuzzy sets, which enable decision-makers to express their membership, non-membership and hesitancy degrees independently and in a larger domain than most of other fuzzy extensions. The applicability of the model is illustrated through a numerical example from a higher education institution. For this purpose, four academic units are evaluated by three-decision makers with respect to five components of COBIT 5 IT governance framework. The improved method can serve as a decision support tool for higher education institutions as well as for other sectors. As the limitation of the study, the following can be said: Since spherical fuzzy division operator is not available, the proposed model is finalized with crisp numbers instead of fuzzy numbers. For future studies, the criteria set can be extended with sub-domains of COBIT 5 framework or any other IT related frameworks such ITIL can be incorporated to the model. Expert dependency of the model can be decreased by using the available data produced within the institution through a machine learning reinforced MCDM model, and spherical fuzzy division operator can be developed to finalize the model in a spherical fuzzy environment.

References

1. Khouja, M., Rodriguez, I.B., Halima, Y.B., Moalla, S.: IT Governance in higher education institutions: a systematic literature review. Int. J. Hum. Cap. Inf. Technol. Prof. **9**(2), 52–67 (2018). https://doi.org/10.4018/ijhcitp.2018040104
2. Andry, JF.: Performance measurement IT of process capability model based on COBIT: a study case. Data Manajemen dan Teknologi Informasi **17**(3), 21–26 (2016).https://doi.org/10.25077/teknosi.v2i2.2016.27-34
3. ISACA.: COBIT 5: a business framework for the governance and management of enterprise IT. Rolling Meadows (2012)
4. Ahriz, S., Yamami, A.E., Mansouri, K., Qbadou, M.: Cobit 5-based approach for IT project portfolio management: application to a Moroccan university. Int. J. Adv. Comput. Sci. Appl. **9**(4), 88–95 (2018). https://doi.org/10.4018/ijhcitp.2018040104
5. Bouayad, H., Benabbou, L., Berrado, A.: An analytic hierarchy process based approach for information technology governance framework selection. In: Proceedings of the 12th International Conference on Intelligent Systems: Theories and Applications, pp. 1–6 (2018). https://doi.org/10.1145/3289402.3289515
6. Yudatama, U., Sarno, R.: Evaluation maturity index and risk management for it governance using Fuzzy AHP and Fuzzy TOPSIS (case Study Bank XYZ). In: International Seminar on Intelligent Technology and Its Applications (ISITIA), pp. 323–328. IEEE (2015). https://doi.org/10.1109/isitia.2015.7220000
7. Apriliana, AF., Sarno, R., Effendi, YA.: Risk analysis of IT applications using FMEA and AHP SAW method with COBIT 5. In: International Conference on Information and Communications Technology (ICOIACT), pp. 373–378. IEEE (2018). https://doi.org/10.1109/icoiact.2018.8350708
8. Kutlu Gündoğdu, F., Kahraman, C.: Spherical fuzzy sets and spherical fuzzy TOPSIS method. J. Intell. Fuzzy Syst. **36**(1), 337–3527 (2019). https://doi.org/10.3233/jifs-181401

9. Saaty, T.L.: The Analytic Hierarchy Process: Planning, Priority Setting, Resources Allocation. McGraw, New York (1980)
10. Roy. B.: Classement et choix en présence de points de vue multiples. Revue française d'informatique et de recherche opérationnelle **2**(8), 57–75 (1968). https://doi.org/10.1051/ro/196802v100571
11. Gündoğdu, F.K., Kahraman, C.: A novel spherical fuzzy analytic hierarchy process and its renewable energy application. Soft Comput. **24**(6), 4607–4621 (2020). https://doi.org/10.1007/s00500-019-04222-w
12. Kahraman, C., Gundogdu, FK., Onar, SC., Oztaysi, B.: Hospital location selection using spherical fuzzy TOPSIS. In: 11th Conference of the European Society for Fuzzy Logic and Technology, pp. 77–82. Atlantis Press (2019)

Spherical Fuzzy Linear Assignment with Objective Weighting Concept in the Sustainable Supply Chain of Aviation Fuel

Yaser Donyatalab[1(✉)] and Fariba Farid[2]

[1] Industrial Engineering Department, University of Moghadas Ardabili, Ardabil, Iran
[2] Department of Management, University of Nabi Akram, Tabriz, Iran

Abstract. This manuscript aims to combine two concepts of spherical fuzzy sets and linear assignment with objective weighting. So, the novel concept of spherical fuzzy linear assignment with spherical fuzzy weighting method is introduced. An objective weighting method based on spherical fuzzy entropy is applied to evaluate weights of decision-makers and criteria; then spherical fuzzy aggregation operators are implemented for the aggregation step. Afterward, the novel spherical fuzzy linear assignment (SF-LAM) is applied to solve multiple attribute group decision-making (MAGDM) problems. A MAGDM method in SFSs based on spherical fuzzy entropy weighting and SF-LAM is proposed in this manuscript. Sustainable supply chain (SSC) selection of aviation fuels is one of the significant fields of study among all aviation supply chain problems because of many intuitive criteria that have to be considered through the decision-making procedure. Constructing such a comprehensive model for the aviation fuel supply chain is still out of sight, even though significant researches have been done in this field. Another hardness of SSC in the aviation fuel is the satisfaction of all criteria based on such a complicated model. In this chapter, a spherical fuzzy structure is defined for the SSC management of aviation fuels. To show the feasibility and applicability of the proposed MAGDM method is applied to solve the SSC of the aviation fuel problem and results are discussed in detail.

Keywords: Spherical fuzzy structure · Spherical fuzzy linear assignment · SF-LAM · Sustainable supply chain of aviation fuel · MAGDM model in SF Structure

1 Introduction and Literature Review

Globalization is needed for more transportation, mostly aviation transportation, and this is directly related to the global economy growing, so it could be derived that aviation fuel increased significantly. Different fuel types from different alternative sources (for fuel production as a conversion of coal, gas, biomass) create a critical need of sustainable supply chain (SSC) management in aviation industries. Also, aviation industries to cross from aviation fuel supply security and efficient air traffic management to achieve a sustainable supply chain (SSC) of aviation fuel. Various suppliers for aviation fuels together

C. Kahraman et al. (Eds.): INFUS 2021, LNNS 308, pp. 766–776, 2022.
https://doi.org/10.1007/978-3-030-85577-2_90

with different fuel types makes the problem more complicated. Besides the complexities around the aviation fuel's supply chain problem, some economic, environmental, and social impacts of those fuels and suppliers have to be considered [1]. The impacts called the triple bottom line (TBL) approach to achieve a sustainable supply chain proposed by Elkington in [2]. Based on different studies, TBL approach is illustrated in providing a sustainable supply chain [3, 4]. Furthermore, in [5], the SSC was investigated to identify a practical model based on the TBL approach by applying a fuzzy MCDM method.

Zadeh, in 1965 [6], introduced fuzzy logic to handle the uncertainty of real world cases. Many extensions are introduced by different researchers and applied in many fields of study through many articles. One of the latest extensions which introduced in 2018, by Gündoğdu, and Kahraman is spherical fuzzy sets [7, 8], the basic idea behind SFSs is to allow evaluators to generalize other extensions of fuzzy sets by defining functions of uncertainty degrees on a spherical surface. Sequentially, Ashraf and Abdullah in 2019 proposed spherical fuzzy sets with some operational rules and aggregation operations based on Archimedean t-norm and t-conorms [9]. Several MAGDM methods are extended to Spherical Fuzzy Sets, and their applications are investigated in the literature. [10] investigated the medical diagnostics and decision-making problem in the spherical fuzzy environment as a practical application. In [11] the classical (VIKOR) method extended to the spherical fuzzy VIKOR (SF-VIKOR) method. [12] the spherical fuzzy TOPSIS method developed and used in a hospital location selection problem. In [13] it is investigated some new aggregation operators in spherical fuzzy environment together with their application in a MAGDM decision making process. A performance measurement method to rank the firms [14] and hospital preparedness evaluation against COVID-19 using spherical fuzzy approach is introduced in [15].

Linear assignment method (LAM) as one of the classical MCDM methods was initially proposed by [16]. In [17] linear assignment for hesitant fuzzy sets proposed to solve the GDM problems and to outline the model's efficiency results are compared with other methods. Linear assignment method developed within the interval type-2 trapezoidal fuzzy numbers [18]. Pythagorean fuzzy linear assignment method was the aim of [19], and a new linear assignment approach was presented by [20] and [21] propped interval-valued Pythagorean fuzzy linear assignment method. In [22] spherical fuzzy linear assignment (SF-LAM) method is introduced together with application in MAGDM problems. Finally, in [23] a bi-objective linear assignment method integrated with cosine similarity measure introduced in spherical fuzzy environment.

The aim of this section is to provide a clear introduction to the study, the approach, and the methodology together with a detailed literature review. In this chapter we tried to prepare a strong relationship among the derived points from literature with what is proposed in this study. Therefore, the traditional problem is complicated decision-making among different aviation fuel alternatives to select the optimal fuel based on many attributes to have a sustainable supply chain of aviation fuel. Motivated by the above discussion, we first applied a novel spherical fuzzy linear assignment and then proposed a MAGDM method in the spherical fuzzy environment. The proposed MAGDM method is applied to the sustainable supply chain of aviation fuels. The sustainable supply selection problem in aviation fuel is defined based on Spherical fuzzy structure and solved by the suggested MAGDM algorithm. The rest of the chapter is designed as follows: in Sect. 2, the concept of spherical fuzzy structure is presented mathematically then in Sect. 3, the novel concept of spherical fuzzy linear assignment method is defined also a

MAGDM method is proposed to show the advantages of applying the SF-LAM structures in the decision-making area. Section 4, discussed the sustainable supply chain (SSC) selection problem of aviation fuels. In this section, the issue of sustainable aviation fuel supplier selection is defined based on SFSs structure and used the recommended MAGDM algorithm to solve it. In Sect. 5, the results are summarized, and the chapter is concluded.

2 Preliminaries: Spherical Fuzzy Sets (SFSs)

In this section, the concept of spherical fuzzy sets will be defined and discussed in details.Spherical Fuzzy Set (SFS) theory is introduced by Gundogdu and Kahraman in 2019 [24]; SFSs are considering the membership, non-membership and hesitancy degrees.

Definition 1. [8] Let X be the universal set and $x_i \in X; \forall i = 1, 2, \ldots n$, the SFS $\tilde{A}_s$ of the universe discourse X is defined as x_i as an element of $\tilde{A}_s$ with membership, non-membership and hesitancy degree values, then $\tilde{A}_s$ will be expressed mathematically in the form of:

$$\tilde{A}_s = \left\{ \left\langle x_i; \mu_{\tilde{A}_s}(x_i), \vartheta_{\tilde{A}_s}(x_i), h_{\tilde{A}_s}(x_i) \right\rangle | x_i \in X \right\}, \tag{1}$$

where $\mu_{\tilde{A}_s}(x_i), \vartheta_{\tilde{A}_s}(x_i), h_{\tilde{A}_s}(x_i)$ stands for membership, non-membership and hesitancy degrees respectively, which belong to interval [0, 1] and satisfies the condition that the sum square of these values be between 0 and 1:

$$S_{\tilde{A}_s}(x_i) = \mu^2_{\tilde{A}_s}(x_i) + \vartheta^2_{\tilde{A}_s}(x_i) + h^2_{\tilde{A}_s}(x_i) \to 0 \le S_{\tilde{A}_s}(x_i) \le 1, \tag{2}$$

then refusal degree $R_{\tilde{A}_s}$ of u in the spherical fuzzy set $\tilde{A}_s$ will be as follows:

$$R_{\tilde{A}_s}(x_i) = \sqrt{1 - S_{\tilde{A}_s}(x_i)}, \tag{3}$$

Definition 2. [8] Score (Sc) and accuracy (Ac) functions of sorting a SFN are defined as follows, respectively:

$$Sc\left(\tilde{A}\right) = \left(\mu_{\tilde{A}} - \frac{h_{\tilde{A}}}{2}\right)^2 - \left(\vartheta_{\tilde{A}} - \frac{h_{\tilde{A}}}{2}\right)^2, \tag{4}$$

$$Ac\left(\tilde{A}\right) = \mu^2_{\tilde{A}} + \vartheta^2_{\tilde{A}} + h^2_{\tilde{A}}, \tag{5}$$

Remark 1. Note if $\tilde{A}$ and $\tilde{B}$ are SFNs then $\tilde{A} < \tilde{B}$ if and only if:

i. $Sc\left(\tilde{A}\right) < Sc\left(\tilde{B}\right)$,

or

ii. $Sc\left(\tilde{A}\right) = Sc\left(\tilde{B}\right)$ and $Ac\left(\tilde{A}\right) < Ac\left(\tilde{B}\right)$.

Definition 3. [24] Let X be the universal set and $x_i \in X; \forall i = 1, 2, \ldots n$ then $\tilde{A} = \{\langle x_i; \mu_{\tilde{A}}(x_i), \vartheta_{\tilde{A}}(x_i), h_{\tilde{A}}(x_i)\rangle | x_i \in X\}$ be a spherical fuzzy set (SFS) with corresponding weight vector $w = \{w_1, w_2, \ldots, w_n\}$; $w_i \in [0, 1]$; $\sum_{i=1}^{n} w_i = 1$. Spherical Fuzzy Weighted Arithmetic Mean ($SFWAM$) and Spherical Fuzzy Weighted Geometric Mean ($SFWGM$) are defined as follows respectively:

$$SFWAM_w\left(\tilde{A}\right) = \underset{i=1}{\overset{n}{\oplus}} w_i\tilde{A}(x_i) = w_1\tilde{A}(x_1) + w_2\tilde{A}(x_2) + \ldots + w_n\tilde{A}(x_n) = \left\{\left(1 - \prod_{i=1}^{n}\left(1 - \mu_{\tilde{A}}^2(x_i)\right)^{w_i}\right)^{0.5}, \prod_{i=1}^{n}\left(\vartheta_{\tilde{A}}(x_i)\right)^{w_i}, \left(\prod_{i=1}^{n}\left(1 - \mu_{\tilde{A}}^2(x_i)\right)^{w_i} - \prod_{i=1}^{n}\left(1 - \mu_{\tilde{A}}^2(x_i) - h_{\tilde{A}}^2(x_i)\right)^{w_i}\right)^{0.5}\right\}, \tag{6}$$

$$SFWGM_w\left(\tilde{A}\right) = \underset{i=1}{\overset{n}{\otimes}} \left(\tilde{A}(x_i)\right)^{w_i} = \left(\tilde{A}(x_1)\right)^{w_1} \times \left(\tilde{A}(x_2)\right)^{w_2} \times \ldots \times \left(\tilde{A}(x_n)\right)^{w_n} = \left\{\prod_{i=1}^{n}\left(\mu_{\tilde{A}}(x_i)\right)^{w_i}, \left(1 - \prod_{i=1}^{n}\left(1 - \vartheta_{\tilde{A}}^2(x_i)\right)^{w_i}\right)^{0.5}, \left(\prod_{i=1}^{n}\left(1 - \vartheta_{\tilde{A}}^2(x_i)\right)^{w_i} - \prod_{i=1}^{n}\left(1 - \vartheta_{\tilde{A}}^2(x_i) - h_{\tilde{A}}^2(x_i)\right)^{w_i}\right)^{0.5}\right\}, \tag{7}$$

In the following, we will introduce an entropy measure for spherical fuzzy sets (SFSs), which are proposed by Aydogdu and Gul in [25].

Definition 4. [25] Let X be the universal set and $x_i \in X; \forall i = 1, 2, \ldots n$ then $\tilde{A} = \{\langle x_i; \mu_{\tilde{A}}(x_i), \vartheta_{\tilde{A}}(x_i), h_{\tilde{A}}(x_i)\rangle | x_i \in X\rangle\}$ be a spherical fuzzy set (SFS). Spherical Fuzzy entropy measure $E_{SFS}\left(\tilde{A}\right)$ is defined as:

$$E_{SFS}\left(\tilde{A}\right) = \frac{1}{n}\sum_{i=1}^{n}\left[1 - \frac{4}{5}\left(\left|\mu_{\tilde{A}}^2(x_i) - \vartheta_{\tilde{A}}^2(x_i)\right| + \left|h_{\tilde{A}}^2(x_i) - 0.25\right|\right)\right], \tag{8}$$

3 Methodology: Spherical Fuzzy Linear Assignment (SF-LAM) in MAGDM Algorithm

In this section, we first will introduce the novel spherical fuzzy linear assignment (SF-LAM) and proposed a multi-attribute group decision making (MAGDM) model. In this way, we will be able to show the applicability of the proposed SF-LAM concept. The proposed SF-LAM is composed of several steps as given in follows. Table 1 presents the linguistic terms and their corresponding spherical fuzzy numbers.

Step 1. Collect the decision-makers' judgments by using Table 1. Consider a group of d decision-makers, $D = \{D_1, D_2, \ldots, D_d\}$ with corresponding weight vector $\tau_j = \{\tau_1, \tau_2, \ldots, \tau_d\}$ where $\sum_{j=1}^{d} \tau_j = 1$, $\tau_j \geq 0$, which participated in a group decision-making problem, where a finite set of alternatives, $A = \{A_1, A_2, \ldots, A_M\}$ are evaluated based on a finite set of criteria, $C = \{C_1, C_2, \ldots, C_n\}$, with corresponding weight vector $w_i = \{w_1, w_2, \ldots, w_n\}$ where $\sum_{i=1}^{n} w_i = 1$, $w_i \geq 0$. Each class of criteria is constructed of some sub-layer criteria. So, we will have the set of criteria $C = \{C_i | i = 1, 2, \ldots, n\}$, set of sub-criteria $C_i = \{C_{ip} | p : 1, 2, \ldots, P\}$ and set of

Table 1. Spherical fuzzy linguistic scales

Spherical Fuzzy Linguistic Scales $(\widetilde{LS}_s)$	SFNs (μ, ϑ, h)
Absolutely High Possible (AHP)	(0.9, 0.1, 0.1)
Very high Possible (VHP)	(0.8, 0.2, 0.2)
High Possible (HP)	(0.7, 0.3, 0.3)
Slightly High Possible (SHP)	(0.6, 0.4, 0.4)
Equally Possible (EP)	(0.5, 0.5, 0.5)
Slightly Low Possible (SLP)	(0.4, 0.6, 0.4)
Low Possible (LP)	(0.3, 0.7, 0.3)
Very Low Possible (VLP)	(0.2, 0.8, 0.2)
Absolutely Low Possible (ALP)	(0.1, 0.9, 0.1)

Table 2. Different suppliers of different aviation fuels

		Fuels supplied by each supplier			
		F_1	F_2	F_3	F_4
Suppliers	S_1	F_1	–	F_3	–
	S_2	–	–	F_3	F_4
	S_3	F_1	F_2	–	–
	S_4	–	–	F_3	F_4
	S_5	–	F_2	–	–

sub sub-criteria $C_{ip} = \{C_{ipq} | q : 1, 2, \ldots, Q\}$. Judgments of decision-makers are stated in a linguistic term based on Table 1. Each decision-maker d expresses his opinion about the performance of alternative A_m regard to criterion C_n using $SFS^d_{m(ipq)}$, so that $SFS^d_{m(ipq)} = \left(\mu^d_{m(ipq)}, \vartheta^d_{m(ipq)}, I^d_{m(ipq)}\right)$, $\forall m = 1, 2, \ldots, M$; therefore, the individual decision matrices are obtained as in Table 3.

Step 2. Determining the weights of DMs in spherical fuzzy decision matrices and evaluate entropy, divergence, and weights of DMs:

We used the spherical fuzzy entropy measure to determine the weights for each expert by using Eq. (8–10).

$$D_{SFS}(D_d) = 1 - E_{SFS}(D_d), \tag{9}$$

Where, $E_{SFS}(D_d)$ is showing the spherical fuzzy entropy of D_d (d[th] decision maker) and $D_{SFS}(D_d)$ is divergency of D_d.

$$W_{SFS}(D_d) = \frac{D_{SFS}(D_d)}{\sum_{j=1}^{d} D_{SFS}(D_d)}, \tag{10}$$

Step 3. Aggregate the individual decision matrices based on aggregation operators to get the GDM matrix. Naturally, decision-makers have different judgments about elements of the decision matrix. Therefore, the aggregation operators must be used in order to get the unified matrix.

Step 4. Determining the weights of criteria: Evaluate entropy, divergence, and weights of criteria in SF structure.

As mentioned in the literature and the same in step (2.1), weights of criteria, sub-criteria, and sub-sub criteria based on the practical information transmitted by each class

Table 3. Spherical fuzzy decision matrix for decision-maker d

Alternatives	Criteria	C_1					$\cdots$	C_n				
	Sub-Criteria	C_{11}		C_{12}		$\cdots$	$\cdots$	C_{n1}		C_{n2}		$\cdots$
	Sub Sub-Criteria	C_{111}	$\cdots$	C_{121}	$\cdots$	$\cdots$	$\cdots$	C_{n11}	$\cdots$	C_{n21}	$\cdots$	$\cdots$
A_1		$SFS^d_{1(111)}$	$\cdots$	$SFS^d_{1(121)}$	$\cdots$	$\cdots$	$\ddots$	$SFS^d_{1(n11)}$	$\cdots$	$SFS^d_{1(n21)}$	$\cdots$	$\cdots$
A_2		$SFS^d_{2(111)}$	$\cdots$	$SFS^d_{2(121)}$	$\cdots$	$\cdots$		$SFS^d_{2(n11)}$	$\cdots$	$SFS^d_{2(n21)}$	$\cdots$	$\cdots$
$\vdots$		$\vdots$	$\cdots$	$\vdots$	$\cdots$	$\cdots$		$\vdots$	$\cdots$	$\vdots$	$\cdots$	$\cdots$
A_M		$SFS^d_{M(111)}$	$\cdots$	$SFS^d_{M(121)}$	$\cdots$	$\cdots$		$SFS^d_{M(n11)}$	$\cdots$	$SFS^d_{M(n21)}$	$\cdots$	$\cdots$

of criterion in the process of decision making. So, it is applied spherical fuzzy entropy method to determine the weights of those criteria in SF structure same as step (2.1) by using Eq. (8–10).

Then weights of criteria are determined in step 4 means a specific criterion could carry a different amount of significance by considering its role in spherical fuzzy aggregated decision making.

Step 5. Aggregate the GDM matrix for sub-sub-criteria and then sub-criteria using determined weight vectors in step (4.1). Hence, in this step, an aggregated weighted GDM matrix is composed.
Step 6. Compute the elements of the scored decision matrix by utilizing the spherical fuzzy score function to obtained defuzzified (scored) decision matrix.
Step 7. Establish the rank frequency non-negative matrix λ_{mk} with elements that represent the frequency that A_m is ranked as the Mth criterion-wise ranking.
Step 8. Calculate and establish the weighted rank frequency matrix Π, where the Π_{mk}measures the contribution of A_m to the overall ranking. Note that each entry $\Pi_{mk}$of the weighted rank frequency matrix, $\Pi$is a measure of the concordance among all criteria in the Mth alternative and k^{th} ranking.

$$\Pi_{mk} = w_{i1} + w_{i2} + \cdots + w_{i\lambda_{MM}} \tag{11}$$

Step 9. Define the permutation matrix P as a square $(m \times m)$ matrix and set up the following linear assignment model according to the Π_{mk}value. The linear assignment model can be written in the following linear programming format:

$$Max \quad \sum_{m=1}^{M} \sum_{k=1}^{M} \Pi_{mk} \cdot P_{mk} \tag{12}$$

$$\text{Subject to.} \sum_{k=1}^{M} P_{mk} = 1, \quad \forall m = 1, 2, \ldots, M; \tag{13}$$

$$\sum_{m=1}^{M} P_{mk} = 1, \quad \forall k = 1, 2, \ldots, M; \tag{14}$$

$$P_{mk} = 0 \text{ or } 1 \text{ for all } m \text{ and } k. \tag{15}$$

Step 10. Solve the linear assignment model, and obtain the optimal permutation matrix P^* for all m and k.
Step 11. Calculate the internal multiplication of matrix $(P^*.A)$ and obtain the optimal order of alternatives.

4 Application

In this application an aviation fuel sustainable supply chain problem assumed in an uncertain environment considering all aspects and facets that could have for the criteria of the problem. The aviation fuels: Algal (F_1), Soybean based aviation fuels are type of biofuels (F_2), Aviation Gasoline (AVGAS)) (F_3) and Natural gas-based aviation fuels

(F_4) are selected from different researches in literature part [26–28]. Alternatives are aviation fuels together their suppliers. The are many suppliers all over the world, which provide the above fuels for aviation industries [29], among those companies could refer to top suppliers like British Petroleum (BP), Shell, Exxon Mobil, Chevron, Total, and Gazprom [30]. In this study, it is considered top supplier for each fuel, so according to Table 2 it is determined that which company produces and supplies which type of aviation fuels. As an example, the provider S_1 supplies algal based fuels (F_1) and AVAGAS (F_3) aviation fuel types, which could show as: $S_1 = \{F_1, F_3\}$.

It is considerable to mention that such a detailed problem has much vagueness around it and all criteria of problem contain uncertainty. This fact inspired us to consider our model in spherical fuzzy environment. As mentioned before, many essential factors needed to indicate the appropriate fuel and then relevant supplier related to that fuel, in this study these factors are called indicators. Another point of view to make our model more realistic is that indicators (criteria) are designed in a multi-level indicator model, which means that each indicator contains some lower-level indicators, are called sub-indicators of those upper indicators. Each level of indicators is self-reliant at the same time and dependent. Sub-indicators determine the indicators comprehensively, because they interact vertically and constructed indicators. Also, whole levels of criteria, including main criteria, sub-criteria, and sub-sub criteria and are selected from the literature part from different articles [31–33] and [34]:

1. Economic (C_1):
   a. Cost (C_{11}):
      (1) Capital Cost (C_{111})
      (2) Production Cost (C_{112})
   b. Lifespan (C_{12}):
      (1) Lifespan (C_{121})
2. Environmental (C_2):
   a. Ecological (C_{21}):
      (1) Green House Gas (GHG) Emission (C_{211})
      (2) PM_{10} Emission (C_{212})
      (3) $PM_{2.5}$ Emission (C_{213})
   b. Supplementary resources (C_{22}):
      (1) Water resource needed (C_{221})
      (2) Electricity resource needed (C_{222})
   c. Energy efficiency (C_{23}):
      (1) Energy Consumption to produce (C_{231})
3. Social (C_3)
   a. Acceptability rate (C_{31}):
      (1) Social acceptability rate (C_{311})
      (2) Aviation Industries acceptability rate (C_{312})
   b. Technologic maturity related to fuel (C_{32}):
      (1) Technologic maturity related to fuel (C_{321})
   c. Innovation on technology (C_{33}):
      (1) Innovation on technology (C_{331})
4. Market Reliability (C_4)
   a. Production (C_{41}):
      (1) Production Capacity (C_{411})
      (2) Availability of resources (C_{412})
   b. Supplementary resources (C_{42}):
      (1) Water resource needed (C_{421})
      (2) Electricity resource needed (C_{422})
   c. Energy efficiency (C_{43}):
      (1) Energy Consumption to produce (C_{431})

In Step 9, the linear assignment model is constructed as follows:

$$Max\ Z = \sum_{m=1}^{9} \sum_{k=1}^{9} \Pi_{mk} . P_{mk}$$

$$\text{S.t.} \begin{cases} \sum_{m=1}^{9} P_{mk} = 1;\ \forall k \\ \sum_{k=1}^{9} P_{mk} = 1;\ \forall m \\ P_{mk} = \{0,\ 1\};\ \forall m, k \end{cases}$$

For results part by solving the linear assignment model, and obtain the optimal permutation matrix P^* for all m and k. It is used an Excel solver, Microsoft office 2019, to solve the above model and found the solution presented optimal permutation matrix P^* as follows. Based on optimal permutation matrix above, $P_{31} = 1, P_{62} = 1, P_{73} = 1$, $P_{14} = 1$, $P_{55} = 1$, $P_{96} = 1$, $P_{27} = 1$, $P_{88} = 1$ and $P_{49} = 1$ together with optimal objective function value $Z = 3.6387$. Also, optimal order is interpretable based on the result of the optimal solution of P^*, means that whenever an alternative gets 1 value in only and only one rank. So, based on $P_{31} = 1$ it is clear that A_3 stands in $1st$ rank, A_6 stands in $2nd$ rank, A_7 stands in $3rd$ rank. A_3 which stands in $1st$ rank is AVAGAS fuel type ($F3$) which is supplying by a second supplier ($S2$), briefly could say $A3(S2, F3)$ is the first optimal choice then $A6(S3, F2)$ which is a Soybean based fuel type ($F2$) supplying by ($S3$) is the second choice.

$$P^* = \begin{bmatrix} 0\,0\,0\,1\,0\,0\,0\,0\,0 \\ 0\,0\,0\,0\,0\,0\,1\,0\,0 \\ 1\,0\,0\,0\,0\,0\,0\,0\,0 \\ 0\,0\,0\,0\,0\,0\,0\,0\,1 \\ 0\,0\,0\,0\,1\,0\,0\,0\,0 \\ 0\,1\,0\,0\,0\,0\,0\,0\,0 \\ 0\,0\,1\,0\,0\,0\,0\,0\,0 \\ 0\,0\,0\,0\,0\,0\,0\,1\,0 \\ 0\,0\,0\,0\,0\,1\,0\,0\,0 \end{bmatrix}$$

But based on step 11, it is more investigable the optimal order of the alternatives. The resulted vector of alternatives by step 11 shows the rank of alternatives, respectively. Thereby, it is evaluated that the optimal order is:

$$A_3 > A_6 > A_7 > A_1 > A_5 > A_9 > A_2 > A_8 > A_4$$

5 Conclusion

In this chapter, the novel concept of the spherical fuzzy linear assignment method (SF-LAM) in the spherical fuzzy environment based on the objective weighting method is defined and illustrated. A MAGDM method is proposed to show the advantages of the SF-LAM with objective weighting in providing resilience and a robust solution. It is also demonstrated the applicability of the proposed SF-LAM and weighting method based on entropy measure in the proposed MAGDM method. This study used the capabilities of SF-LAM and entropy weighting method in determining the optimal alternative in a multi-layer criteria structure. The reliability of the proposed MAGDM algorithm is based on SF-LAM and entropy weighting method in the sustainable supply chain of the aviation fuel. SSC in the aviation fuel problem is one of the hardest and complicated issues, which deliberately shows the proposed method's advantages. Based on that, four different fuel alternatives supplied by five suppliers are considered which make nine different alternatives for our proposed SSC in the aviation fuel problem. Many criteria through different levels of criteria are considered to catch a sustainable aviation fuel and sustainable supplier of that fuel. So, it could refer that many criteria are considered in different three layers for four classes of criteria, which are interacting with each other. In this method, not only could determine the sustainable fuel but also at the same time could evaluate the supplier. Alternative is AVAGAS fuel type ($F3$) which is supplying by a second supplier ($S2$), briefly say $A3(S2, F3)$, stands in the first place as sustainable fuel selection. So, A3 as an AVAGAS will give more sustainability to SSC of aviation fuel problems. For future studies, we propose various applications in different fields of study like financial, banking, social, network, health care, manufacturing, and transportation systems.

References

1. Rodger, J.A., George, J.A.: Triple bottom line accounting for optimizing natural gas sustainability: a statistical linear programming fuzzy ILOWA optimized sustainment model approach to reducing supply chain global cybersecurity vulnerability through information and communications technology. J. Clean. Prod. **142**, 1931–1949 (2017)
2. Cannibals with forks: the triple bottom line of 21st century business. Choice Rev. Online. (1999)
3. Govindan, K., Agarwal, V., Darbari, J.D., Jha, P.C.: An integrated decision making model for the selection of sustainable forward and reverse logistic providers. Ann. Oper. Res. **273**(1–2), 607–650 (2017). https://doi.org/10.1007/s10479-017-2654-5
4. Slaper, T., Hall, T.: The Triple Bottom Line : What Is It and How Does It Work? Indiana University Kelley School of Business (2011)
5. Govindan, K., Khodaverdi, R., Jafarian, A.: A fuzzy multi criteria approach for measuring sustainability performance of a supplier based on triple bottom line approach. J. Clean. Prod. **47**, 345–354 (2013)
6. Zadeh, L.A.: Fuzzy sets. Inf. Control **8**, 338–353 (1965)
7. Kutlu Gündoğdu, F., Kahraman, C.: From 1D to 3D Membership:Sphericalfuzzy Sets
8. Kutlu Gündoğdu, F., Kahraman, C.: A novel fuzzy TOPSIS method using emerging interval-valued spherical fuzzy sets. Eng. Appl. Artif. Intell. **85**, 307–323 (2019)
9. Ashraf, S., Abdullah, S., Aslam, M., Qiyas, M., Kutbi, M.A.: Spherical fuzzy sets and its representation of spherical fuzzy t-norms and t-conorms. J. Intell. Fuzzy Syst. **36**, 6089–6102 (2019)
10. Mahmood, T., Ullah, K., Khan, Q., Jan, N.: An approach toward decision-making and medical diagnosis problems using the concept of spherical fuzzy sets. Neural Comput. Appl. **31**(11), 7041–7053 (2018). https://doi.org/10.1007/s00521-018-3521-2
11. Kutlu Gündoğdu, F., Kahraman, C.: A novel VIKOR method using spherical fuzzy sets and its application to warehouse site selection. J. Intell. Fuzzy Syst. **37**, 1197–1211 (2019)
12. Kahraman, C., Kutlu Gundogdu, F., Cevik Onar, S., Oztaysi, B.: Hospital location selection using spherical fuzzy TOPSIS. In: Proceedings of the 2019 Conference of the International Fuzzy Systems Association and the European Society for Fuzzy Logic and Technology (EUSFLAT 2019). Atlantis Press, Paris (2019)
13. Donyatalab, Y., Farrokhizadeh, E., Garmroodi, S.D.S., Shishavan, S.A.S.: Harmonic mean aggregation operators in spherical fuzzy environment and their group decision making applications. J. Multi.-valued. Log. Soft Comput. (2019)
14. Kahraman, C., Onar, S.C., Oztaysi, B.: Performance measurement of debt collection firms using spherical fuzzy aggregation operators. In: Kahraman, C., Cebi, S., Cevik Onar, S., Oztaysi, B., Tolga, A.C., Sari, I.U. (eds.) INFUS 2019. AISC, vol. 1029, pp. 506–514. Springer, Cham (2020). https://doi.org/10.1007/978-3-030-23756-1_63

15. Gul, M., Yucesan, M.: Hospital preparedness assessment against COVID-19 pandemic: a case study in Turkish tertiary healthcare services. Math. Probl. Eng. **2021** (2021)
16. Bernardo, J.J., Blin, J.M.: A programming model of consumer choice among multi-attributed brands. J. Consum. Res. **4**, 111 (1977)
17. Razavi Hajiagha, S.H., Shahbazi, M., Amoozad Mahdiraji, H., Panahian, H.: A bi-objective score-variance based linear assignment method for group decision making with hesitant fuzzy linguistic term sets. Technol. Econ. Dev. Econ. **24**, 1125–1148 (2018)
18. Chen, T.Y.: A linear assignment method for multiple-criteria decision analysis with interval type-2 fuzzy sets. Appl. Soft Comput. J. **13**, 2735–2748 (2013)
19. Liang, D., Darko, A.P., Xu, Z., Zhang, Y.: Partitioned fuzzy measure-based linear assignment method for Pythagorean fuzzy multi-criteria decision-making with a new likelihood. J. Oper. Res. Soc. **71**, 1–15 (2019)
20. Bashiri, M., Badri, H., Hejazi, T.H.: Selecting optimum maintenance strategy by fuzzy interactive linear assignment method. Appl. Math. Model. **35**, 152–164 (2011)
21. Liang, D., Darko, A.P., Xu, Z., Quan, W.: The linear assignment method for multicriteria group decision making based on interval-valued Pythagorean fuzzy Bonferroni mean. Int. J. Intell. Syst. **33**, 2101–2138 (2018)
22. Donyatalab, Y., Seyfi-Shishavan, S.A., Farrokhizadeh, E., Kutlu Gündoğdu, F., Kahraman, C.: Spherical fuzzy linear assignment method for multiple criteria group decision-making problems. Informatica **31**, 707–722 (2020)
23. Seyfi-Shishavan, S.A., Kutlu Gündoğdu, F., Donyatalab, Y., Farrokhizadeh, E., Kahraman, C.: A novel spherical fuzzy bi-objective linear assignment method and its application to insurance options selection. Int. J. Inf. Technol. Decis. Mak. **20**, 1–31 (2021)
24. Gündoğdu, F.K., Kahraman, C.: Spherical fuzzy sets and spherical fuzzy TOPSIS method. J. Intell. Fuzzy Syst. **36**, 337–352 (2019)
25. Aydoğdu, A., Gül, S.: A novel entropy proposition for spherical fuzzy sets and its application in multiple attribute decision-making. Int. J. Intell. Syst. **35**, 1354–1374 (2020)
26. Zhao, S.Y., Li, W.J.: Fast asynchronous parallel stochastic gradient descent: a lock-free approach with convergence guarantee. In: 30th AAAI Conference on Artificial Intelligence. AAAI 2016 (2016)
27. Kandaramath Hari, T., Yaakob, Z., Binitha, N.N.: Aviation biofuel from renewable resources: routes, opportunities and challenges (2015)
28. Zahran, S., Iverson, T., McElmurry, S.P., Weiler, S.: The effect of leaded aviation gasoline on blood lead in children. J. Assoc. Environ. Resour. Econ. **4**, 575–610 (2017)
29. Airport Suppliers (Aviation Fuel Suppliers). https://www.airport-suppliers.com/suppliers/fuel-handling/
30. Aviation Fuel Market by Product and Geography - Forecast and Analysis 2020–2024 (Technavio). https://www.technavio.com/report/aviation-fuel-market-industry-analysis
31. Ren, J., Fedele, A., Mason, M., Manzardo, A., Scipioni, A.: Fuzzy Multi-actor Multi-criteria Decision Making for sustainability assessment of biomass-based technologies for hydrogen production. Int. J. Hydrogen Energy. **38**, 9111–9120 (2013)
32. Ren, J., Manzardo, A., Mazzi, A., Zuliani, F., Scipioni, A.: Prioritization of bioethanol production pathways in China based on life cycle sustainability assessment and multicriteria decision-making. Int. J. Life Cycle Assess. **20**(6), 842–853 (2015). https://doi.org/10.1007/s11367-015-0877-8
33. Afgan, N.H., Carvalho, M.G.: Sustainability assessment of hydrogen energy systems. Int. J. Hydrogen Energy **29**, 1327–1342 (2004)
34. Ren, J., Xu, D., Cao, H., Wei, S., Dong, L., Goodsite, M.E.: Sustainability decision support framework for industrial system prioritization. AIChE J. **62**, 108–130 (2016)

Present Worth Analysis Using Spherical Fuzzy Sets

Eda Boltürk[1(✉)] and Sukran Seker[2]

[1] Department of Industrial Engineering, Istanbul Technical University, 34367 Macka, Istanbul, Turkey
bolturk@itu.edu.tr

[2] Department of Industrial Engineering, Yildiz Technical University, 34349 Besiktaş, Istanbul, Turkey
sseker@yildiz.edu.tr

Abstract. One of the important investment techniques is present worth analysis (PWA) in engineering economics. Investment values are specified by humans and these values are based on human thoughts. It is quite hard to specify them in crisp values. In that manner, fuzzy logic could be the better alternative to calculate vagueness. Vague cash flows, uncertain life, and uncertain time value of money give rise to usage of fuzzy PWA. In this study, spherical fuzzy PWA method is developed to handle the fuzzy parameters of investments. Spherical fuzzy sets (SFSs) have better powerful tool in order to model the indefiniteness in investment analysis problems. To the best of our knowledge, there is no study about PWA based on SFSs method in literature. We give a numerical application of our proposed spherical fuzzy PWA method to show applicability of the proposed approach.

Keywords: Present worth analysis · Spherical fuzzy sets · Engineering economics

1 Introduction

Investment analysis is choosing the most appropriate investment alternative to get more profit. In order to make a profit, investors make investments taking into account many potential risks [1]. In engineering economics, different techniques or methods are presented to help investors in the evaluation different investment alternatives and determination the most suitable alternative among them. Some of these methods are Present worth analysis (PWA), future worth analysis (FWA), annual worth analysis (AWA), internal rate of return (IRR), benefit/cost (B/C) ratio analysis, and rate of return (ROR) analysis. The initial costs (IC), annual cash flows (ACF) and estimated salvage values (SV) in the useful life are used as input in investment analysis [2]. As the most frequent used engineering economics technique, PWA aims to reveal the equivalent worth of all future costs and revenues at the present time using a discount rate. Accordingly, all cash flows are transformed into the present time [3]. It is preferred because of its ease of calculation and obtaining effective results.

C. Kahraman et al. (Eds.): INFUS 2021, LNNS 308, pp. 777–788, 2022.
https://doi.org/10.1007/978-3-030-85577-2_91

Today, one of the factors of the global economy is that businesses have to make investment decisions under uncertain conditions [4]. If there is not enough information for a particular parameter, fuzzy set theory (FST) introduced by Zadeh [5] is an excellent tool to deal with this uncertainty.

In FST, only the membership degree of an item is considered. Then, Atanassov [6] generalized the notion of FST to intuitionistic fuzzy sets (IFSs) by presenting a membership degree (μ) and a non-membership degree (ϑ) with the state of $\mu + \vartheta \leq 1$. In order to strength the notion of IFSs, Yager [7] presented the notion of Pythagorean fuzzy sets (PFSs) by providing a new condition $\mu^2 + \vartheta^2 \leq 1$. Sometimes in real life, we face with many problems which cannot be handled by using IFSs or PFSs. Considering the hesitancy of decision makers (DMs) about one parameter, Spherical Fuzzy Sets (SFSs) is proposed by Gundogdu and Kahraman [8] to handle uncertainty more extensively as a generalization of PFSs. However, SFSs satisfies the condition that squared sum of membership, non-membership, and hesitancy degrees should be between 0 and 1 while each degree must be expressed in [0, 1].

The methods of engineering economics with fuzzy extension are produced extensively in the literature. Buckley [9] introduced investment techniques with FST in the mathematics of finance. Fuzzy PWA introduced by Chiu and Park [10] and they employed triangular fuzzy numbers to identify fuzzy cash flows and fuzzy interest rates. In order to evaluate investment alternatives Kuchta [11] proposed fuzzy PWA using interval-valued fuzzy numbers. Kahraman et al. [12] developed intuitionistic fuzzy AWA and hesitant fuzzy AWA using triangular hesitant fuzzy data, triangular intuitionistic fuzzy data, interval-valued hesitant data, and interval-valued intuitionistic fuzzy (IVIF) data. In addition, Kahraman et al. [13] proposed IVIF-PWA for CNC option evaluation. Kahraman et al. [14] applied PWA using PFS to deal with the fuzzy parameters of investments. Sergi and Sari [4] developed fermatean fuzzy net present worth technique using fermatean fuzzy interest rates and cash flows in evaluation investment project. Sarı and Kuchta [15] presented fuzzy global sensitivity analysis method for fuzzy net present value to show the effects of the elements on the worth of an investment project. Sari and Kahraman [16] introduced interval type-2 fuzzy engineering economics techniques named PWA, FWA and AWA by performing both triangular and trapezoidal interval type-2 fuzzy sets.

As a real life application, investment problems consist of uncertain and imprecise information. In this study, PWA using SFSs is presented for the first time in the literature for investment analysis problems. SFSs is selected since SFSs suggest a larger preference space for DMs and also, each DM can express his or her own hesitations independently.

The organization of paper is given as follows: In Sect. 2, the preliminaries of SFSs are given. In Sect. 3, the spherical PWA is introduced. In Sect. 4, an application of proposed extension is given. The conclusion is given in Sect. 5 with suggestions for future studies.

2 Spherical Fuzzy Sets

In this sub-section, we will give the preliminaries on SFSs.

Definition 1: SFSs, $\tilde{A}_S$ in the universe of discourse U is given as in Eq. (1).

$$\tilde{A}_s = \left\{ \left\langle u, (\mu_{\tilde{A}_s}(u), v_{\tilde{A}_s}(u), \pi_{\tilde{A}_s}(u)) \middle| u \in U \right\} \right. \tag{1}$$

where

$$\mu_{\tilde{A}_S} : U \rightarrow [0, 1], \;\; v_{\tilde{A}_S}(u) : U \rightarrow [0, 1], \;\; \pi_{\tilde{A}_S} : U \rightarrow [0, 1]$$

and

$$0 \leq \mu^2_{\tilde{A}_S}(u) + v^2_{\tilde{A}_S}(u) + \pi^2_{\tilde{A}_S}(u) \leq 1 \qquad \forall u \in U \tag{2}$$

For each u, the numbers $\mu_{\tilde{A}_S}(u)$, $v_{\tilde{A}_S}(u)$ and $\pi_{\tilde{A}_S}(u)$ are the degree of membership, non-membership and hesitancy of u to $\tilde{A}_S$, respectively.

Definition 2: Addition and multiplication of SFSs are given as in Eqs. (3–6):

Addition:

$$\tilde{A}_S \oplus \tilde{B}_S = \left\{ \left(\mu^2_{\tilde{A}_S} + \mu^2_{\tilde{B}_S} - \mu^2_{\tilde{A}_S}\mu^2_{\tilde{B}_S} \right)^{1/2}, v_{\tilde{A}_S} v_{\tilde{B}_S}, \left(\left(1 - \mu^2_{\tilde{B}_S}\right)\pi^2_{\tilde{A}_S} + \left(1 - \mu^2_{\tilde{A}_S}\right)\pi^2_{\tilde{B}_S} - \pi^2_{\tilde{A}_S}\pi^2_{\tilde{B}_S} \right)^{1/2} \right\} \tag{3}$$

Multiplication:

$$\tilde{A}_S \otimes \tilde{B}_S = \left\{ \mu_{\tilde{A}_S}\mu_{\tilde{B}_S}, \left(v^2_{\tilde{A}_S} + v^2_{\tilde{B}_S} - v^2_{\tilde{A}_S}v^2_{\tilde{B}_S} \right)^{1/2}, \left(\left(1 - v^2_{\tilde{B}_S}\right)\pi^2_{\tilde{A}_S} + \left(1 - v^2_{\tilde{A}_S}\right)\pi^2_{\tilde{B}_S} - \pi^2_{\tilde{A}_S}\pi^2_{\tilde{B}_S} \right)^{1/2} \right\} \tag{4}$$

Multiplication by a scalar, $\lambda > 0$:

$$\lambda \cdot \tilde{A}_S = \left\{ \left(1 - \left(1 - \mu^2_{\tilde{A}_S}\right)^{\lambda} \right)^{1/2}, v^{\lambda}_{\tilde{A}_S}, \left(\left(1 - \mu^2_{\tilde{A}_S}\right)^{\lambda} - \left(1 - \mu^2_{\tilde{A}_S} - \pi^2_{\tilde{A}_S}\right)^{\lambda} \right)^{1/2} \right\} \tag{5}$$

λ. power of $\tilde{A}_S$;$\lambda > 0$

$$\tilde{A}^{\lambda}_S = \left\{ \mu^{\lambda}_{\tilde{A}_S}, \left(1 - \left(1 - v^2_{\tilde{A}_S}\right)^{\lambda} \right)^{1/2}, \left(\left(1 - v^2_{\tilde{A}_S}\right)^{\lambda} - \left(1 - v^2_{\tilde{A}_S} - \pi^2_{\tilde{A}_S}\right)^{\lambda} \right)^{1/2} \right\} \tag{6}$$

Definition 3: Spherical Weighted Arithmetic Mean (SWAM) with respect to $w = (w_1, w_2 \ldots\ldots, w_n)$; $w_i \in [0, 1]$; $\sum_{i=1}^{n} w_i = 1$, is given as in Eq. (7):

$$SWAM_w(A_{S1}, \ldots\ldots, A_{Sn}) = w_1 A_{S1} + w_2 A_{S2} + \ldots\ldots + w_n A_{Sn}$$

$$= \left\{ \left[1 - \prod_{i=1}^{n} (1 - \mu^2_{A_{Si}})^{w_i} \right]^{1/2}, \prod_{i=1}^{n} v^{w_i}_{A_{Si}}, \left[\prod_{i=1}^{n} (1 - \mu^2_{A_{Si}})^{w_i} - \prod_{i=1}^{n} (1 - \mu^2_{A_{Si}} - \pi^2_{A_{Si}})^{w_i} \right]^{1/2} \right\} \tag{7}$$

Definition 4: Score function and accuracy function of sorting SFSs are given as in Eqs. (8–9), respectively;

$$Score(A_S) = \left(\mu_{A_S} - \pi_{A_S}\right)^2 - \left(v_{A_S} - \pi_{A_S}\right)^2 \tag{8}$$

$$Accuracy(A_S) = \mu_{A_S}^2 + v_{A_S}^2 + \pi_{A_S}^2 \tag{9}$$

Note that: $A_S < B_S$ if and only if

i. $Score(A_S) < Score(B_S)$ or
ii. $Score(A_S) = Score(B_S)$ and $Accuracy(A_S) < Accuracy(B_S)$

Definition 5: Defuzzification formula for criteria weights is given in Eq. (10) and then, this value is normalized as in Eq. (11) [17].

$$Defuzzification\left(\tilde{w}_j^s\right) = \sqrt{\left|100 * \left[\left(3\mu_{\tilde{A}_s} - \frac{\pi_{\tilde{A}_s}}{2}\right)^2\right] - \left(\frac{v_{\tilde{A}_s}}{2} - \pi_{\tilde{A}_s}\right)^2\right|} \tag{10}$$

$$\overline{w}_j^s = \frac{Defuzzification\left(\tilde{w}_j^s\right)}{\sum_{J=1}^{n} Defuzzification\left(\tilde{w}_j^s\right)} \tag{11}$$

3 Spherical Present Worth Analysis

The parameters of first cost (FC), salvage value (SV) annual benefit (AB), annual cost (AC), interest rate (i), and life (n) are given with spherical fuzzy membership numbers (SFN) in Eqs. (12–17).

$$\widetilde{FC} = \left\{ \begin{array}{c} \langle fc_1, SFN_1, \ldots, SFN_m\rangle, \langle fc_2, SFN_1, \ldots, SFN_m\rangle, \\ \ldots, \langle fc_k, SFN_1, \ldots, SFN_m\rangle \end{array} \right\} \tag{12}$$

$$\widetilde{AC} = \left\{ \begin{array}{c} \langle ac_1, SFN_1, \ldots, SFN_m\rangle, \langle ac_2, SFN_1, \ldots, SFN_m\rangle, \ldots, \\ \langle ac_k, SFN_1, \ldots, SFN_m\rangle \end{array} \right\} \tag{13}$$

$$\widetilde{AB} = \left\{ \begin{array}{c} \langle ab_1, SFN_1, \ldots, SFN_m\rangle, \langle ab_2, SFN_1, \ldots, SFN_m\rangle, \ldots, \\ \langle ab_k, SFN_1, \ldots, SFN_m\rangle \end{array} \right\} \tag{14}$$

$$\widetilde{SV} = \left\{ \begin{array}{c} \langle sv_1, SFN_1, \ldots, SFN_m\rangle, \langle sv_2, SFN_1, \ldots, SFN_m\rangle, \ldots, \\ \langle sv_k, SFN_1, \ldots, SFN_m\rangle \end{array} \right\} \tag{15}$$

$$\tilde{\imath} = \left\{ \begin{array}{c} \langle i_1, SFN_1, \ldots, SFN_m\rangle, \langle i_2, SFN_1, \ldots, SFN_m\rangle, \ldots, \\ \langle i_k, SFN_1, \ldots, SFN_m\rangle \end{array} \right\} \tag{16}$$

$$\tilde{n} = \left\{ \begin{array}{c} \langle n_1, SFN_1, \ldots, SFN_m\rangle, \langle n_2, SFN_1, \ldots, SFN_m\rangle, \ldots, \\ \langle n_k, SFN_1, \ldots, SFN_m\rangle \end{array} \right\} \tag{17}$$

The equation of PWA is given in Eq. (18):

$$\widetilde{PW}_s = -\widetilde{FC}_s - \widetilde{AC}_s\left[\frac{(1+\tilde{\imath}_s)^{\tilde{n}_s}-1}{\tilde{\imath}_s(1+\tilde{\imath}_s)^{\tilde{n}_s}}\right] + \widetilde{AB}_I\left[\frac{(1+\tilde{\imath}_s)^{\tilde{n}_s}-1}{\tilde{\imath}_s(1+\tilde{\imath}_s)^{\tilde{n}_s}}\right] + \widetilde{SV}_s(1+\tilde{\imath}_s)^{-1} \tag{18}$$

Aggregation of SFSs is performed by Eq. (7). After, parameter values can be computed by multiplying the defuzzified values for membership functions with parameter values. The defuzzification and normalization formulas are used in order to get crisp values for each parameter.

4 Application

E1, E2 and E3 are three experts who try to calculate the present worth of a summer house in Istanbul. E1, E2 and E3 have their own weight are 0.4, 0.3 and 0.3, respectively. They give three ideas for each parameter with specific weights as in Table 1. Firstly, each expert's thoughts are aggregated using Eq. (7) for each parameter (see Table 2). E1's aggregated value for first cost is calculated as follows:

$$\mu fc_{E1} = \sqrt{\left(1-\left(\left(1-0.2^2\right)^{0.5}\right)\times\left(\left(1-0.3^2\right)^{0.2}\right)\times\left(\left(1-0.2^2\right)^{0.3}\right)\right)} = 0.296$$

$$\vartheta fc_{E1} = \left(0.4^{0.5}\right) * \left(0.5^{0.2}\right) * \left(0.6^{0.3}\right) = 0.472$$

$$\pi fc_{E1} = \sqrt{\begin{array}{c}\left(\left(\left(1-0.2^2\right)^{0.5}\right)\times\left(\left(1-0.3^2\right)^{0.2}\right)\times\left(\left(1-0.4^2\right)^{0.3}\right)\right)- \\ \left(\left(\left(1-0.2^2-0.2^2\right)^{0.5}\right)\times\left(\left(1-0.3^2-0.7^2\right)^{0.2}\right)\times\left(\left(1-0.4^2-0.2^2\right)^{0.3}\right)\right)\end{array}} = 0.267$$

Then, each membership value is multiplied with respect to expert's weights (see Table 3). The result for (0.362, 0.741, 0.176) is calculated as follows:

$$\mu fc_{E1} = \sqrt{\left(1-(1-0.296)^{0.4}\right)} = 0.362$$

$$\vartheta fc_{E1} = 0.472 \times 0.4 = 0.741$$

$$\pi fc_{E1} = \sqrt{\left(\left(\left(1-0.267^2\right)^{0.4}\right)-\left(\left(1-0.267^2-0.396^2\right)^{0.4}\right)\right)} = 0.176$$

Then, each membership functions are defuzzified and normalized using Eqs. (10–11). Normalized value of $(\mu fc_{E1}, \vartheta fc_{E1}, \pi fc_{E1})$ is calculated as follows:

$$Defuzzification(fc_{E1}) = \sqrt{\left|\left(\left(3\times 0.362-\frac{0.176}{2}\right)^2-\left(\frac{0.741}{2}-0.176\right)\right)\right|} = 0.895$$

Sum of defuzzified parameters of fc_{E1}, fc_{E2} and fc_{E3} is 3.36.

$$\text{Normalization }(fc_{E1}) = \frac{0.895}{3.36} = 0.267$$

Table 1. The parameter values and spherical fuzzy membership functions.

Parameters	Experts	Experts Weights	Weights	Value	(μ, ϑ, π)	Parameters	Experts	Experts Weights	Weights	Value	(μ, ϑ, π)
$\widetilde{FC}$	E1	0.4	0.5	20000	(0.2,0.4,0.2)	$\widetilde{SV}$	E1	0.4	0.5	500	(0.2,0.4,0.2)
			0.2	30000	(0.3,0.5,0.7)				0.2	550	(0.3,0.5,0.2)
			0.3	40000	(0.4,0.6,0.5)				0.3	600	(0.4,0.6,0.2)
	E2	0.3	0.5	35000	(0.5,0.7,0.2)		E2	0.3	0.5	520	(0.5,0.7,0.2)
			0.2	45000	(0.6,0.2,0.3)				0.2	540	(0.6,0.2,0.2)
			0.3	55000	(0.7,0.4,0.2)				0.3	560	(0.75,0.4,0.2)
	E3	0.3	0.5	30000	(0.5,0.5,0.2)		E3	0.3	0.5	350	(0.5,0.5,0.2)
			0.2	45000	(0.6,0.4,0.1)				0.2	550	(0.6,0.4,0.2)
			0.3	50000	(0.5,0.5,0.2)				0.3	750	(0.5,0.5,0.2)
$\widetilde{AB}$	E1	0.4	0.5	1000	(0.2,0.4,0.2)	$\tilde{\iota}$	E1	0.4	0.5	0,8	(0.2,0.4,0.2)
			0.2	2000	(0.3,0.5,0.6)				0.2	0,9	(0.3,0.5,0.1)
			0.3	3000	(0.4,0.6,0.55)				0.3	0,8	(0.4,0.6,0.2)
	E2	0.3	0.5	1500	(0.6,0.2,0.2)		E2	0.3	0.5	0,85	(0.1,0.9,0.1)
			0.2	1700	(0.5,0.7,0.4)				0.2	0,85	(0.6,0.6,0.2)
			0.3	1900	(0.4,0.2,0.2)				0.3	0,8	(0.7,0.4,0.5)

(continued)

Table 1. *(continued)*

Parameters	Experts	Experts Weights	Weights	Value	(μ, ϑ, π)	Parameters	Experts	Experts Weights	Weights	Value	(μ, ϑ, π)
	E3	0.3	0.5	1350	(0.5,0.5,0.2)		E3	0.3	0.5	0,88	(0.5,0.3,0.8)
			0.2	1450	(0.6,0.4,0.1)				0.2	0,89	(0.6,0.1,0.7)
			0.3	1550	(0.5,0.5,0.2)				0.3	0,8	(0.5,0.5,0.2)
$\widetilde{AC}$	E1	0.4	0.5	250	(0.2,0.4,0.2)	$\tilde{n}$	E1	0.4	0.5	10	(0.5,0.4,0.2)
			0.2	300	(0.3,0.5,0.2)				0.2	12	(0.4,0.5,0.2)
			0.3	350	(0.4,0.6,0.5)				0.3	14	(0.6,0.6,0.2)
	E2	0.3	0.5	270	(0.5,0.75,0.05)		E2	0.3	0.5	10	(0.5,0.7,0.2)
			0.2	280	(0.6,0.25,0.2)				0.2	13	(0.9,0.2,0.2)
			0.3	290	(0.7,0.4,0.2)				0.3	15	(0.5,0.4,0.2)
	E3	0.3	0.5	300	(0.4,0.5,0.2)		E3	0.3	0.5	11	(0.7,0.5,0.2)
			0.2	310	(0.6,0.4,0.2)				0.2	13	(0.8,0.4,0.2)
			0.3	320	(0.5,0.5,0.2)				0.3	14	(0.5,0.5,0.2)

Table 2. Spherical aggregation membership values.

Parameters	μ	ϑ	π	Parameters	μ	ϑ	π
$\widetilde{FC}$	0.296	0.472	0.267	$\widetilde{SV}$	0.296	0.472	0.083
	0.594	0.461	0.208		0.618	0.461	0.209
	0.523	0.478	0.159		0.523	0.478	0.166
$\widetilde{AB}$	0.296	0.472	0.249	$\tilde{\imath}$	0.296	0.472	0.077
	0.532	0.257	0.195		0.506	0.651	0.266
	0.523	0.478	0.159		0.523	0.281	0.648
$\widetilde{AC}$	0.296	0.472	0.083	$\tilde{n}$	0.518	0.472	0.164
	0.594	0.499	0.182		0.656	0.461	0.235
	0.481	0.478	0.148		0.683	0.478	0.238

Table 3. Spherical weighted aggregation membership values in expert values and normalized valued of them.

Parameters	μ	ϑ	π	Normalized Value	Parameters	μ	ϑ	π	Normalized Value
$\widetilde{FC}$	0.362	0.741	0.176	0.267	SV	0.362	0.741	0.054	0.263
	0.487	0.793	0.135	0.432		0.501	0.793	0.137	0.447
	0,446	0.801	0.098	0.389		0.446	0.801	0.102	0.389
$\widetilde{AB}$	0.362	0.741	0.163	0.274	$\tilde{\imath}$	0.362	0.741	0.050	0.283
	0.451	0.665	0.121	0.403		0.437	0.879	0.164	0.370
	0.446	0.801	0.098	0.389		0.446	0.683	0.455	0.387
$\widetilde{AC}$	0.362	0.741	0.054	0.273	$\tilde{n}$	0.503	0.741	0.115	0.322
	0.487	0.812	0.117	0.432		0.523	0.793	0.159	0.469
	0.422	0.801	0.089	0.363		0.540	0.801	0.166	0.486

Final FC is calculated as follows:

$$\begin{aligned} FC = & (20000 \times 0.5 + 6000 \times 0.2 + 12000 \times 0.3) \times 0.267 \\ & + (17500 \times 0.5 + 9000 \times 0.2 + 16500 \times 0.3) \times 0.432 \\ & + (15000 \times 0.5 + 9000 \times 0.2 + 15000 \times 0.3) \times 0.389 \\ & = 41,195.6 \end{aligned}$$

Other parameters are calculated as above (see Table 4). Finally, PWA calculated by Eq. (18) and the value is \$–41,251.6.

Table 4. The final values.

Parameters	Weighted values	Sum	Normalized Weights	Final Values	Parameters	Weighted values	Sum	Normalized Weights	Final Values
FC	10000				SV	250			
	6000					110			
	12000	28000	0.267			180	540	0.263	
	17500					260			
	9000					108			
	16500	43000	0.432			168	536	0.447	
	15000					175			
	9000					110			
	15000	39000	0.389			225	510	0.389	
		41,195.6				579.9			
AB	500				i	0,4			
	400					0,18			
	900	1800	0.274			0,24	0.82	0.283	
	750					0,425			
	340					0,17			
	570	1660	0.403			0,24	0.835	0.370	
	675					0,44			
	290					0,178			

(continued)

Table 4. (*continued*)

Parameters	Weighted values	Sum	Normalized Weights	Final Values	Parameters	Weighted values	Sum	Normalized Weights	Final Values
	465	1430	0.389			0,24	0.858	0.387	
		1,717.5				0.87			
AC	125				n	5			
	60					2,4			
	105	290	0.273			4,2	11.6	0.322	
	135					5			
	56					2,6			
	87	278	0.432			4,5	12.1	0.469	
	150					5,5			
	62					2,6			
	96	308	0.363			4,2	12.3	0.486	
		310.9				15.4			

5 Conclusion

PWA is one of the most popular techniques used in the field of engineering economics when evaluating investment projects. However, investment analysis occurs under conditions of uncertainty or risk due to lack of necessary prior information. Different methods and techniques are used in the literature to evaluate investment projects under uncertainty and risk conditions. To overcome uncertainty, FST has been applied extensively in the literature. In this study, as a new extension of FST, SFSs is used for handling investment analysis problems. SFSs is selected since it provides a wider range of preferences for DMs and also each DM can express its hesitations independently. The applicability of the spherical fuzzy PWA is shown with illustrative investment analysis problem. The proposed extension generates effective results due to ease of application procedure of spherical fuzzy PWA in investment analysis problems. Further, the proposed approach can be applied for different investment analysis problems such as energy investment problems, transportation investment analysis or warehouse investment analysis problems.

References

1. Aydin, S., Kahraman, C., Kabak, M.: Evaluation of investment alternatives using present worth analysis with simplified neutrosophic sets. Eng. Econ. **29**(3), 254–263 (2018)
2. Aydın, S., Kahraman, C., Kabak, M.: Decision making for energy investments by using neutrosophic present worth analysis with interval-valued parameters. Eng. Appl. Artif. Intell. **92**, 103639 (2020)
3. Kahraman, C., Çevik, S., Öztayşi, B.: Interval-valued intuitionistic fuzzy investment analysis: application to CNC lathe selection. Int. Feder. Autom. Control **49**, 1323–1328 (2016)
4. Sergi, D., Sari, I.U.: Fuzzy capital budgeting using fermatean fuzzy sets. In: Kahraman, C., Cevik Onar, S., Oztaysi, B., Sari, I.U., Cebi, S., Tolga, A.C. (eds.) INFUS 2020. AISC, vol. 1197, pp. 448–456. Springer, Cham (2021). https://doi.org/10.1007/978-3-030-51156-2_52
5. Zadeh, L.A.: Fuzzy sets. Inf. Control **8**, 338–353 (1965)
6. Atanassov, K.T.: Intuitionistic fuzzy sets. Fuzzy Sets Syst. **20**(1), 87–96 (1986)
7. Yager, R.: Pythagorean fuzzy subsets. In: Joint IFSA World Congress and NAFIPS Annual Meeting, Edmonton, Canada, pp. 57–61 (2013)
8. Kutlu Gündoğdu, F., Kahraman, C.: Spherical fuzzy sets and spherical fuzzy TOPSIS method. J. Intell. Fuzzy Syst. **36**(1), 337–352 (2019)
9. Buckley, J.J.: The fuzzy mathematics of finance. Fuzzy Sets Syst. **21**, 257–273 (1987)
10. Chiu, C.Y., Park, C.S.: Fuzzy cash flow analysis using present worth criterion. Eng. Econ. **39**(2), 113–138 (1994)
11. Kuchta, D.: Fuzzy capital budgeting. Fuzzy Sets Syst. **111**, 367–385 (2000)
12. Kahraman, C., Onar, C.S., Oztaysi, B.: Engineering economic analyses using intuitionistic and hesitant fuzzy sets. J. Intell. Fuzzy Syst. **29**, 1151–1168 (2015)
13. Kahraman, C., Onar, S.Ç., Öztayşi, B.: Interval valued intuitionistic fuzzy investment analysis: application to CNC lathe selection. IFAC-PapersOnLine **49**(12), 1323–1328 (2016)
14. Kahraman, C., Onar, S.C., Oztaysi, B.: Present worth analysis using Pythagorean fuzzy sets. In: Kacprzyk, J., Szmidt, E., Zadrożny, S., Atanassov, K.T., Krawczak, M. (eds.) IWIFSGN/EUSFLAT -2017. AISC, vol. 642, pp. 336–342. Springer, Cham (2018). https://doi.org/10.1007/978-3-319-66824-6_30
15. Sari, I.D., Kuchta, D.: Fuzzy global sensitivity analysis of fuzzy net present value. Control Cybern. **41**(2), 481–496 (2012)

16. Sari, I.U., Kahraman, C.: Interval type-2 fuzzy capital budgeting. Int. J. Fuzzy Syst. **17**(4), 635–646 (2015)
17. Kutlu Gündoğdu, F., Kahraman, C.: A novel spherical fuzzy analytic hierarchy process and its renewable energy application. Soft. Comput. **24**(6), 4607–4621 (2019). https://doi.org/10.1007/s00500-019-04222-w

Spherical Fuzzy EXPROM Method: Wastewater Treatment Technology Selection Application

Cengiz Kahraman, Basar Oztaysi, and Sezi Cevik Onar(✉)

Department of Industrial Engineering, Istanbul Technical University, 34367 Macka, Besiktas, Istanbul, Turkey
cevikse@itu.edu.tr

Abstract. EXtension of the PROMethee (EXPROM) methods were first introduced in 1991. These methods try to find a solution for ranking the alternatives more accurately using the available information. EXPROM-I method performs a partial ranking of alternatives whereas EXPROM-II does a full ranking of the alternatives. Vague and imprecise data of multi-criteria decision making problems can be better captured by spherical fuzzy sets (SFS) than ordinary fuzzy sets. SFS are an extension of picture fuzzy sets, presenting a larger definition volume for the parameters of membership function. In this paper, spherical fuzzy EXPROM method is developed and applied to the solution of a wastewater treatment technology selection problem.

Keywords: Spherical fuzzy sets · EXPROM method · Water treatment technology · Multi-criteria decision making

1 Introduction

EXtension of PROMethee (EXPROM) methods were first introduced by Diakoulaki and Koumoutsos (1991). It tries to rank the alternatives by using widely available information. In the EXPROM I method, only the entering and leaving flows are examined and a partial ranking is done whereas in the EXPROM II method, the net flow is determined as the final value, and the full ranking of the alternatives is performed. EXPROM methods belongs to the compensatory methods. The qualitative attributes are transformed into the quantitative attributes. It does not require the independency of attributes. EXPROM methods were used in some areas such as the country market selection (Gorecka and Szalucka 2013), sustainable water resources planning (Raju et al. 2000), and material selection (Caliskan et al. 2013; Kumar and Ray 2015), host city selection decision (Gorecka 2020), material selection (Chingo et al. 2020), membrane ranking (Cardena et al. 2021),

EXPROM method has been rarely used since it was introduced in 1991. Figure 1 illustrates the usage frequencies of EXPROM method by years. As it is seen from Fig. 1, the maximum freqency that it was used within a year is only 3.

C. Kahraman et al. (Eds.): INFUS 2021, LNNS 308, pp. 789–801, 2022.
https://doi.org/10.1007/978-3-030-85577-2_92

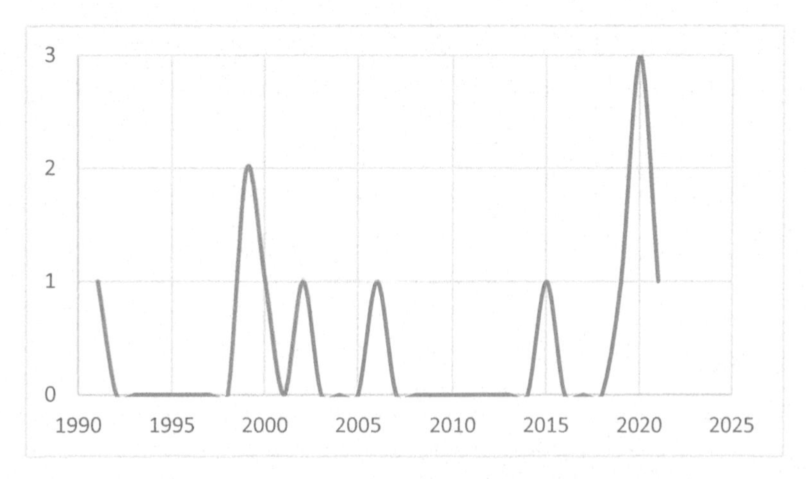

Fig. 1. Usage frequencies of EXPROM method by years

Figure 2 shows the source countries that the EXPROM method has been used in their publications. The leading country that most uses it is India. Then, France and Poland follow India.

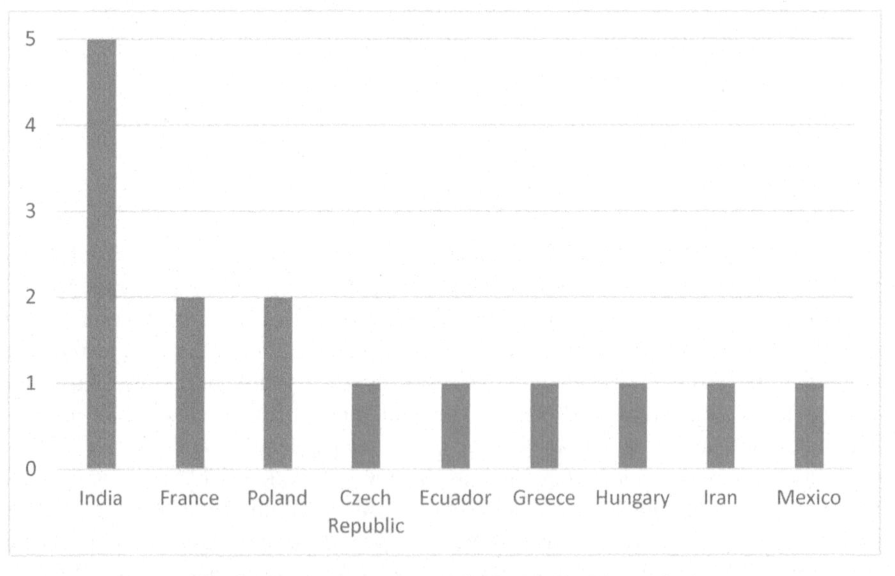

Fig. 2. Source countries publishing EXPROM method

Figure 3 gives the percentages of the subject areas of the EXPROM publications. The largest three percentages belong to computer science, decision science, and mathematics, respectively.

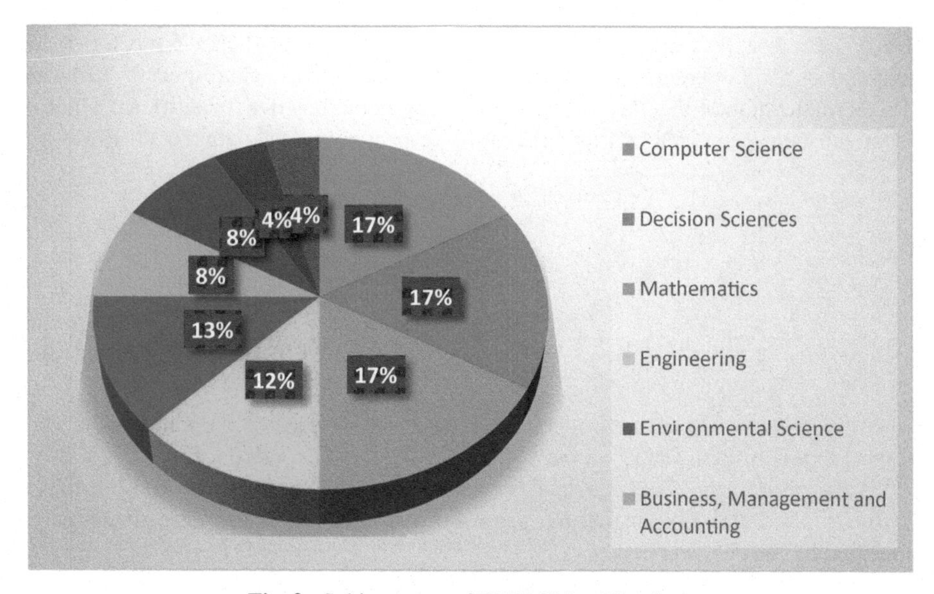

Fig. 3. Subject areas of EXPROM publications

Linguistic evaluations generally involve much more uncertainty than crisp evaluations which are composed of exact numerical values. Fuzzy set theory has been the leading theory to capture the uncertainy in decision making problems (Bolturk and Kahraman 2018; Kahraman and Tolga 1998; Kaya and Kahraman 2010;. To handle the vagueness in these linguistic evaluations, spherical fuzzy sets (SFS) are preferred in this paper. Spherical fuzzy EXPROM method first time is developed in this paper. Spherical fuzzy sets present a larger domain to assign membership, non-membership and hesitancy degrees with a constraint which their squared sum is at most one.

The remaining of this paper is organized as follows. Section 2 includes the step of the classical EXPROM method. Section 3 presents the preliminaries of single valued spherical fuzzy sets. Section 4 gives the steps of the SF EXPROM method. Section 5 presents the application of the proposed fuzzy multi-criteria decision making method. Section 6 concludes the paper.

2 Classical EXPROM Method

The decision matrix is given by Eq. (1) based on the information received from the decision.

$$\begin{vmatrix} f_1(A_1) & \cdots & f_j(A_1) & f_{j+1}(A_1) & \cdots & f_n(A_1) \\ \vdots & \ddots & \vdots & \vdots & \ddots & \vdots \\ f_1(A_i) & \cdots & f_j(A_i) & f_{j+1}(A_i) & \cdots & f_n(A_i) \end{vmatrix}$$

$$\begin{vmatrix} \vdots & \ddots & \vdots & \vdots & \ddots & \vdots \\ f_1(A_m) & \cdots & f_j(A_m) & f_{j+1}(A_m) & \cdots & f_n(A_m) \end{vmatrix} \tag{1}$$

where $i = 1, 2, \ldots, m; j = 1, 2, \ldots, n$ and $\{A_1, A_2, \cdots, A_m\}$ is a finite set of the alternatives and $f_j(*)$ represents the evaluations of alternatives with respect to attributes. The normalization of the decision matrix with respect to positive (benefit) and negative (cost) attributes is relized by Eqs. (2) and (3), respectively (Diakoulaki et al. 1991).

$$\tilde{x}_{ij} = \frac{\tilde{r}_{ij} - \tilde{r}_i^-}{\tilde{r}_i^+ - \tilde{r}_i^-} i = 1, 2, \ldots, m; j = 1, 2, \ldots, n; \text{ for positive attributes} \quad (2)$$

$$\tilde{x}_{ij} = \frac{\tilde{r}_{ij} - \tilde{r}_i^+}{\tilde{r}_i^- - \tilde{r}_i^+} i = 1, 2, \ldots, m; j = 1, 2, \ldots, n; \text{ for negative attributes} \quad (3)$$

where $\tilde{x}_{ij}$ represents a normalized value of the decision matrix for ith alternative with respect to jth attribute and $\tilde{r}_i^+ = max(\tilde{r}_1, \tilde{r}_2, \ldots, \tilde{r}_m), r_i^{-+} = min(\tilde{r}_1, \tilde{r}_2, \ldots, \tilde{r}_m)$.

Furthermore, the decision maker determines the weights of the attributes $w_1, w_2, \ldots, w_n$. In addition, all the parameters of the weak preference function are specified by the decision maker.

The Weak Preference Function

In order to determine the amount of weak preference function, the difference between the pair of alternatives is first obtained from Eq. (4) (Alinezhad and Khalili 2019a, 2019b).

$$d_j(A_i, A_{i'}) = f_j(A_i) - f_j(A_{i'}), i, i' \epsilon \{1, 2, \ldots, m\}, j = 1, 2, \ldots, n \quad (4)$$

Therefore, the value of the weak preference function is calculated based on Eq. (5).

$$P_j(A_i, A_{i'}) = f_j(d_j(A_i, A_{i'})), 0 \leq P_j(A_i, A_{i'}) \leq 1 \quad (5)$$

The type of the preference function should first be specified to determine the values of the weak preference function according to the type of attributes. There are six different types of preference functions that can be easily found in the literature (Alinezhad and Khalili 2019a).

The Weak Preference Index

With respect to the weight of attributes, the weak preference index is given by Eq. (6) (Diakoulaki and Koumoutsos 1991)

$$WP(A_i, A_{i'}) = \sum\nolimits_{j=1}^{n} P_j(A_i, A_{i'}).w_j / \sum\nolimits_{j=1}^{n} w_j, i, i' \varepsilon \{1, 2, \ldots, m\} \quad (6)$$

The Strict Preference Function

The ideal and anti-ideal solutions of each attribute are obtained by Eqs. (7) and (8), respectively (Diakoulaki and Koumoutsos 1991).

$$f_j(\dot{x}) = max\{f_j(x_1), f_j(x_1), \ldots, f_j(x_1)\}, j = 1, 2, \ldots, n \quad (7)$$

$$f_j(\ddot{x}) = min\{f_j(x_1), f_j(x_1), \ldots, f_j(x_1)\}, j = 1, 2, \ldots, n \tag{8}$$

Then, the maximum spreading of the strict preference function is calculated by using Eq. (9) (Diakoulaki and Koumoutsos 1991).

$$dm_j = f_j(\dot{x}) - f_j(\ddot{x}), j = 1, 2, \ldots, n \tag{9}$$

Equation (10) shows the strict preference function (Diakoulaki and Koumoutsos 1991).

$$P_j^{'}(A_i, A_{i'}) = max(0, (d_j(A_i, A_{i'}) - L_j))/(dm_j - L_j), \tag{10}$$

In Eq. (10), the range of the strict preference function (L) for ordinary functions is equal to zero and equal to an infinite number for the Gaussian criterion.

The Strict Preference Index

The strict preference index is calculated by Eq. (11) (Diakoulaki and Koumoutsos 1991).

$$SP(A_i, A_{i'}) = \sum_{j=1}^{n} P_j^{'}(A_i, A_{i'}).w_j / \sum_{j=1}^{n} w_j, i, i' \varepsilon \{1, 2, \ldots, m\} \tag{11}$$

The Entering and Leaving Flows

First, the values of the total preferences index are determined by Eq. (12) (Diakoulaki and Koumoutsos 1991).

$$TP(A_i, A_{i'}) = min\{1, WP(A_i, A_{i'}) + SP(A_i, A_{i'})\}, i, i' \varepsilon \{1, 2, \ldots, m\} \tag{12}$$

Then, the entering and leaving flows are obtained as shown in Eqs. (13) and (14) (Diakoulaki and Koumoutsos 1991).

$$\varphi^{+}(A_i) = \frac{1}{m-1} \sum_{A_i \varepsilon A} TP(A_i, A_{i'}), i, i' \varepsilon \{1, 2, \ldots, m\} \tag{13}$$

$$\varphi^{-}(A_i) = \frac{1}{m-1} \sum_{A_i \varepsilon A} TP(A_{i'}, A_i), i, i' \varepsilon \{1, 2, \ldots, m\} \tag{14}$$

The Net Flow

In this method, the full ranking, including (P^I, I^{II}) is obtained. The net flow values are calculated by Eq. (15) and then, the alternatives are ranked (Diakoulaki and Koumoutsos 1991).

$$\varphi(A_i) = \varphi^{+}(A_i) - \varphi^{-}(A_i), i = 1, 2, \ldots, m \tag{15}$$

The Final Ranking of Alternatives (EXPROM I Method)
Initially, Eqs. (16) to (19) are considered (Diakoulaki and Koumoutsos 1991).

$$A_i P^+ A_{i'} \; if \; \varphi^+(A_i) > \varphi^+\left(A_{i'}\right), i, i' \epsilon\{1, 2, \ldots, m\} \tag{16}$$

$$A_i I^+ A_{i'} \; if \; \varphi^+(A_i) = \varphi^+\left(A_{i'}\right), i, i' \epsilon\{1, 2, \ldots, m\} \tag{17}$$

$$A_i P^- A_{i'} \; if \; \varphi^-(A_i) < \varphi^-\left(A_{i'}\right), i, i' \epsilon\{1, 2, \ldots, m\} \tag{18}$$

$$A_i I^- A_{i'} \; if \; \varphi^-(A_i) = \varphi^-\left(A_{i'}\right), i, i' \epsilon\{1, 2, \ldots, m\} \tag{19}$$

On the other hand, the alternative A_i is better than the alternative $A_{i'}$, if:

$$A_i P A_{i'} \text{if} \begin{cases} A_i P^+ A_{i'} \; and \; A_i P^- A_{i'} \\ A_i P^+ A_{i'} \; and \; A_i I^- A_{i'} \\ A_i I^+ A_{i'} \; and \; A_i P^- A_{i'} \end{cases}, i, i' \varepsilon\{1, 2, \ldots, m\} \tag{20}$$

And, the alternatives A_i and $A_{i'}$ are indifferent to each other, if:

$$A_i I A_{i'} \text{ if } A_i I^- A_{i'} \text{ and } A_{i'} I^- A_i, \; i, \; i' \varepsilon\{1, 2, \ldots, m\} \tag{21}$$

Accordingly, all alternatives are ranked.

The Final Ranking of Alternatives (EXPROM II Method)
In this method, the alternative A_i is better than the alternative $A_{i'}$, if:

$$A_i P^{II} A_{i'} \; if \; \varphi(A_i) > \varphi\left(A_{i'}\right), i, i' \epsilon\{1, 2, \ldots, m\} \tag{22}$$

And, the alternatives A_i and $A_{i'}$ are indifferent to each other, if:

$$A_i I^{II} A_{i'} \; if \; \varphi(A_i) = \varphi\left(A_{i'}\right), i, i' \epsilon\{1, 2, \ldots, m\} \tag{23}$$

Consequently, all alternatives are ranked.

3 Preliminaries of Spherical Fuzzy Sets

Spherical fuzzy sets (SFS) were introduced by Kutlu Gundogdu and Kahraman (2019a). These sets are based on the fact that the hesitancy of a decision maker can be assigned separately satisfying the condition that the squared sum of membership, non-membership and hesitancy degrees is at most equal to 1. In the following, preliminaries of SFS are given:

Definition 1. Single valued Spherical Fuzzy Sets (SFS) $\tilde{A}_S$ of the universe of discourse U is given by

$$\tilde{A}_S = \left\{\left\langle u, (\mu_{\tilde{A}_S}(u), \nu_{\tilde{A}_S}(u), \pi_{\tilde{A}_S}(u)) \right| u \in U\right\} \tag{24}$$

where $\mu_{\tilde{A}_S}(u) : U \to [0, 1], \ \ \nu_{\tilde{A}_S}(u) : U \to [0, 1], \ \ \pi_{\tilde{A}_S}(u) : U \to [0, 1]$.
and

$$0 \leq \mu^2_{\tilde{A}_S}(u) + \nu^2_{\tilde{A}_S}(u) + \pi^2_{\tilde{A}_S}(u) \leq 1 \qquad \forall u \in U \tag{25}$$

For each u, the numbers $\mu_{\tilde{A}_S}(u)$, $\nu_{\tilde{A}_S}(u)$ and $\pi_{\tilde{A}_S}(u)$ are the degree of membership, non-membership and hesitancy of u to$\tilde{A}_S$, respectively. Spherical fuzzy sets have become very popular in a short time period (Kutlu Gundogdu and Kahraman 2019a,2019b, 2019c, 2019d, 2019e, 2019f; Oztaysi et al. 2020a; Oztaysi et al. 2020b; Cevik Onar et al. 2020;). Figure 4 illustrates the differences among intuitionistic fuzzy sets (IFS), Pythagorean fuzzy sets (PFS), neutrosophic sets (NS), and spherical fuzzy sets (SFS).

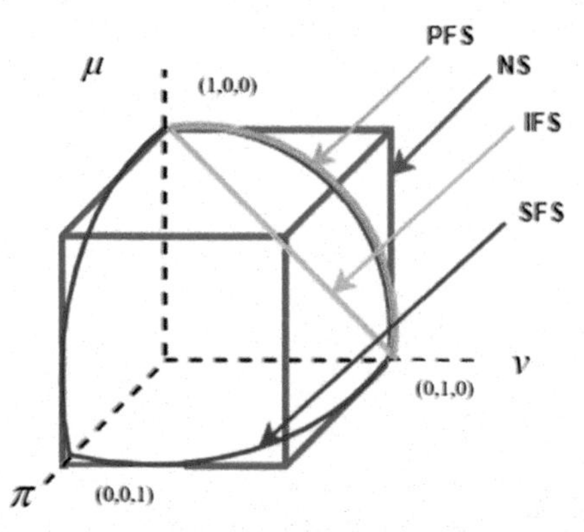

Fig. 4. Geometric representations of IFS, PFS, NS, and SFS (Kutlu Gündogdu and Kahraman 2019)

Definition 2. Let $\tilde{A} = (\mu_A, \nu_A, \pi_A)$ and $\tilde{B} = (\mu_B, \nu_B, \pi_B)$. The distance between $\tilde{A}$ and $\tilde{B}$ is calculated by Eq. (26) (Khan et al. 2020).

$$D^s\left(\tilde{A}, \tilde{B}\right) = 1 - \frac{\mu_A^2 \times \mu_B^2 + \nu_A^2 \times \nu_B^2 + \pi_A^2 \times \pi_B^2}{\mu_A^4 \bigvee \mu_B^4 + \nu_A^4 \bigvee \nu_B^4 + \pi_A^4 \bigvee \pi_B^4} \tag{26}$$

4 Spherical Fuzzy EXPROM Method

In the fuzzy decision matrix linguistic terms given in Table 1 can be used. Then, this matrix is normalized to have a crisp numerical decision matrix. In the spherical fuzzy EXPROM method, in the normalization step, by considering the positive and negative values, the best and worst values for each criterion are used in Eq. (27) for benefit attributes, and Eq. (28) for cost attributes.

For benefit attributes,

$$\tilde{x}_{ij} = \frac{1 - \dfrac{\mu_{ij}^2 \times \mu_-^2 + v_{ij}^2 \times v_-^2 + \pi_{ij}^2 \times \pi_-^2}{\mu_{ij}^4 \vee \mu_-^4 + v_{ij}^4 \vee v_-^4 + \pi_{ij}^4 \vee \pi_-^4}}{1 - \dfrac{\mu_-^2 \times \mu_+^2 + v_-^2 \times v_+^2 + \pi_-^2 \times \pi_+^2}{\mu_-^4 \vee \mu_+^4 + v_-^4 \vee v_+^4 + \pi_-^4 \vee \pi_+^4}}, \tag{27}$$

For cost attributes,

$$\tilde{x}_{ij} = \frac{1 - \dfrac{\mu_{ij}^2 \times \mu_+^2 + v_{ij}^2 \times v_+^2 + \pi_{ij}^2 \times \pi_+^2}{\mu_{ij}^4 \vee \mu_+^4 + v_{ij}^4 \vee v_+^4 + \pi_{ij}^4 \vee \pi_+^4}}{1 - \dfrac{\mu_-^2 \times \mu_+^2 + v_-^2 \times v_+^2 + \pi_-^2 \times \pi_+^2}{\mu_-^4 \vee \mu_+^4 + v_-^4 \vee v_+^4 + \pi_-^4 \vee \pi_+^4}}, \tag{28}$$

After the normalization step, the matrix becomes a crisp normalized matrix and classical EXPROM method can be applied.

Table 1. Spherical fuzzy scale

Linguistic Terms	(μ, v, π)
Absolutely High (AH)	(0.9, 0.1, 0.1)
Very High (VH)	(0.8, 0.2, 0.2)
High (H)	(0.7, 0.3, 0.3)
Slightly High (SH)	(0.6, 0.4, 0.4)
Equal (E)	(0.5, 0.5, 0.5)
Slightly Low (SL)	(0.4, 0.6, 0.4)
Low (L)	(0.3, 0.7, 0.3)
Very Low (VL)	(0.2, 0.8, 0.2)
Absolutely Low (AL)	(0.1, 0.9, 0.1)

5 Application to Waste Water Treatment Technology Selection

A municipality tries to select the most appropriate wastewater treatment technology. The available treatment technologies are filtration (A1), vertical biological reactors (A2), septic tanks (A3), and activated sludge (A4). The considered criteria are land requirement (C1), initial cost (C2), size of community (C3), and easy to operate (C4). C1 and C2 are cost criteria whereas C3 and C4 are benefit criteria. The crisp weights of the criteria are 0.3, 0.2, 0.2, and 0.3, respectively. A group of experts constructed the linguistic decision matrix given in Table 2. Using the scale in Sect. 4, the numerical values have been determined as given in Table 3.

Table 2. Decision matrix

	C1	C2	C3	C4
A1	AH	H	L	SL
A2	VL	VH	H	H
A3	H	L	VH	VH
A4	AH	AL	VH	H

Table 3. Decision matrix with fuzzy numbers

	C1	C2	C3	C4
A1	(0.9, 0.1, 0.1)	(0.7, 0.3, 0.3)	(0.3, 0.7, 0.3)	(0.4, 0.6, 0.4)
A2	(0.2, 0.8, 0.2)	(0.8, 0.2, 0.2)	(0.7, 0.3, 0.3)	(0.7, 0.3, 0.3)
A3	(0.7, 0.3, 0.3)	(0.3, 0.7, 0.3)	(0.8, 0.2, 0.2)	(0.8, 0.2, 0.2)
A4	(0.9, 0.1, 0.1)	(0.1, 0.9, 0.1)	(0.8, 0.2, 0.2)	(0.7, 0.3, 0.3)

Table 4. Normalized decision matrix

	C1	C2	C3	C4
A1	0	0,25599695	0	0
A2	1	0	0,91520292	0,87390333
A3	0,42247803	0,91061142	1	1
A4	0	1	1	0,87390333

Using, Eqs. (27) and (28), the normalized decision matrix is obtained as given in Table 4. The weak preference matrix is given in Table 5. The strict preference matrix is given in Table 6.

Table 5. Weak preference matrix

	C1	C2	C3	C4	TOTAL
WP(1,2)	−0,3	0,05119939	−0,18304058	−0,262171	−0,69401
WP(1,3)	−0,12674341	−0,13092289	−0,2	−0,3	−0,75767
WP(1,4)	0	0,2	0,4	0,9	1,5
WP(2,1)	0,3	−0,05119939	0,18304058	0,262171	0,694012
WP(2,3)	0,17325659	−0,18212228	−0,01695942	−0,037829	−0,06365
WP(2,4)	0,3	−0,2	−0,01695942	0	0,083041
WP(3,1)	0,12674341	−0,01787772	0	0,037829	0,146695
WP(3,2)	−0,17325659	0,18212228	0,01695942	0,037829	0,063654
WP(3,4)	0,12674341	−0,01787772	0	0,037829	0,146695
WP(4,1)	0	0,14880061	0,2	0,262171	0,610972
WP(4,2)	−0,3	0,2	0,01695942	0	−0,08304
WP(4,3)	−0,12674341	0,01787772	0	−0,037829	−0,14669

Table 6. Strict preference matrix

SP(1,2)	C1	C2	C3	C4	TOTAL
SP(1,3)	0	0,255997	0	0	0,051199
SP(1,4)	0	0	0	0	0
SP(2,1)	0	1	2	3	1,5
SP(2,3)	1	0	0,915203	0,873903	0,745212
SP(2,4)	0,577522	0	0	0	0,173257
SP(3,1)	1	0	0	0	0,3
SP(3,2)	0,422478	0	0	0,126097	0,164572
SP(3,4)	0	0,910611	0,084797	0,126097	0,236911
SP(4,1)	0,422478	0	0	0,126097	0,164572
SP(4,2)	0	0,744003	1	0,873903	0,610972
SP(4,3)	0	1	0,084797	0	
SP(1,2)	0	0,089389	0	0	

The total preference matrix is given in Table 7.

Table 7. Total preference matrix

TP(1,2)	−0,64281
TP(1,3)	−0,75767
TP(1,4)	1
TP(2,1)	1
TP(2,3)	0,109602
TP(2,4)	0,383041
TP(3,1)	0,311267
TP(3,2)	0,300565
TP(3,4)	0,311267
TP(4,1)	1
TP(4,2)	0,133919
TP(4,3)	−0,12882

The dominant aggregate matrix is given in Table 8. The leaving, entering and net flows are given in Table 9. The ranking of the water treatment alternatives are obtained as A2 > A3 > A4 > A1.

Table 8. Dominnt aggregate matrix

	C1	C2	C3	C4
A1	–	−0,64281	−0,75767	1
A2	1	–	0,109602	0,383041
A3	0,311267108	0,300565	–	0,311267
A4	1	0,133919	−0,12882	–

Table 9. Leaving, entering, and net flows

	Leaving flow	Entering flow	Net outranking	Ranking
A1	−0,133493036	0,770422369	−0,903915405	4
A2	0,497547688	−0,069443053	0,566990741	1
A3	0,307699676	−0,258960269	0,566659945	2
A4	0,33503395	0,564769231	−0,229735281	3

6 Conclusion

EXtension of the PROMethee methods try to find a solution for ranking the alternatives based on outranking relations more accurately using the available information. EXPROM has not been often used in the literature but it has been relatively used more frequently in the recent years. The developed spherical fuzzy EXPROM method has been first time developed and used in a multi-criteria wastewater treatment technology selection problem. Defuzzification operation has been employed in an early step since sbtruction and division operations are not well defined in fuzzy sets extensions. This can be a disadvantage of the proposed method. For further research, we suggest EXPROM method to be developed by type-2 fuzzy sets or hesitant fuzzy sets. Their results can be compared with this study.

References

Alinezhad, A., Javad, K.: New methods and applications in multiple attribute decision making (MADM). In: International Series in Operations Research & Management Science, vol. 277 (2019a). https://doi.org/10.1007/978-3-030-15009-9

Alinezhad, A., Javad, K.: EXPROM I & II Method. In: International Series in Operations Research & Management Science, Springer, vol. 277, pp. 181–191 (2019b). https://doi.org/10.1007/978-3-030-15009-9_24

Bolturk, E., Kahraman, C.: A novel interval-valued neutrosophic AHP with cosine similarity measure. Soft Comput. **22**(15), 4941–4958 (2018)

Caliskan, H., Kursuncu, B., Kurbanoglu, C., Guven, S.Y.: Material selection for the tool holder working under hard milling conditions using different multi criteria decision making methods. Mater. Des. **45**, 473–479 (2013)

Cardeña, R., et al.: Evaluation and ranking of polymeric ion exchange membranes used in microbial electrolysis cells for biohydrogen production. Bioresource Technol. **319**, 124182 (2021)

Chingo, C., Martínez-Gómez, J., Narváez, C.R.A.: Material selection using multi-criteria decision making methods for geomembranes. Int. J. Math. Oper. Res. **16**(1), 24–52 (2020)

Diakoulaki, D., Koumoutsos, N.: Cardinal ranking of alternative actions: Extension of the PROMETHEE method. Eur. J. Oper. Res. **53**(3), 337–347 (1991)

Górecka, D.: Applying multi-criteria decision aiding methods to the process of selecting a host city for sporting event. J. Phys. Educ. Sport **20**(149), 1069–1076 (2020)

Górecka, D., Szalucka, M.: Country market selection in international expansion using multicriteria decision aiding methods. Multiple Criteria Dec. Making **1**(8), 32–35 (2013)

Kahraman, C., Tolga, E.: Data envelopment analysis using fuzzy concept. In: ISMVL'98 28th IEEE International Symposium on Multiple-Valued Logic, Fukuoka/Japan, Proceedings, pp. 338–343 (1998)

Kaya, I., Kahraman, C.: Development of fuzzy process accuracy index for decision making problems. Inf. Sci. **180**(6), 861–872 (2010)

Khan, M.J., Kumam, P., Deebani, W., Kumam, W., Shah, Z.: Distance and similarity measures for spherical fuzzy sets and their applications in selecting mega projects. Mathematics **8**, 519 (2020)

Kumar, R., Ray, A.: Optimal selection of material: An eclectic decision. J. Inst. Eng. (India): Series C **96**(1), 29–33 (2015)

Kutlu Gundogdu, F., Kahraman, C.: Extension of CODAS with spherical fuzzy sets. Multiple Valued Logic Soft Comput. **33**(4–5), 481–505 (2019)

Kutlu Gündoğdu, F., Kahraman, C.: A novel VIKOR method using spherical fuzzy sets and its application to warehouse site selection. J. Intell. Fuzzy Syst. **37**, 1197–1211 (2019)

Kutlu Gündoğdu, F., Kahraman, C.: Spherical fuzzy sets and spherical fuzzy TOPSIS method. J. Intell. Fuzzy Syst. **36**(1), 337–352 (2019)

Kutlu Gündoğdu, F., Kahraman, C.: Extension of WASPAS with spherical fuzzy sets. Informatica **30**(2), 269–292 (2019)

Kutlu Gündoğdu, F., Kahraman, C.: A novel spherical fuzzy analytic hierarchy process and its renewable energy application. Soft. Comput. **24**(6), 4607–5462 (2019)

Kutlu Gündoğdu, F., Kahraman, C.: A novel fuzzy TOPSIS method using emerging interval-valued spherical fuzzy sets. Eng. Appl. Artif. Intell. **85**, 307–323 (2019)

Onar, S.C., Kahraman, C., Oztaysi, B.: Multi-criteria spherical fuzzy regret based evaluation of healthcare equipment stocks. J. Intell. Fuzzy Syst. **39**(5), 5987–6599 (2020)

Oztaysi, B., Cevik Onar, S., Kutlu Gundogdu, F., Kahraman, C.: Location based advertisement selection using spherical fuzzy AHP-VIKOR. J. Multi-Valued Logic Soft Comput. **35**, 5–23 (2020)

Oztaysi, B., Onar, S.C., Kahraman, C.: A dynamic pricing model for location based systems by using spherical fuzzy AHP scoring. J. Intell. Fuzzy Syst. **39**(5), 6293–6302 (2020)

Raju, K.S., Duckstein, L., Arondel, C.: Multicriterion analysis for sustainable water resources planning: a case study in Spain. Water Resour. Manage **14**(6), 435–456 (2000)

Tangent Similarity Measure of Cubic Spherical Fuzzy Sets and Its Application to MCDM

Ajay Devaraj(✉) and J. Aldring

Sacred Heart College (Autonomous), Tirupattur District 635601, Tamilnadu, India

Abstract. The objective of this paper is to present a new perspective of tangent similarity measure of cubic spherical fuzzy sets. First, the concept of spherical fuzzy sets (SFSs) is extended to cubic spherical fuzzy notion. Then the fundamental concepts of Cubic-Spherical fuzzy sets (Cubic-SFSs) and their operations are studied. The idea of tangent similarity measure (TAN-SM) for Cubic-SFSs is introduced. Also some of the properties of TAN-SM are investigated numerically, and these concepts have been used to develop a new decision-making technique which will have influence in the field of decision science. Further, tangent fuzzy aggregation operator (TFAO) is introduced and a medical diagnosis intelligent system has been demonstrated based on the proposed MCDM approach. Finally, a sensitive analysis is conducted to show the efficiency of TAN-SM of Cubic-SFSs.

Keywords: Cubic spherical fuzzy sets (Cubic-SFSs) · Tangent similarity measures (TAN-SM) · Aggregation operators · MCDM

1 Introduction

The concept of fuzzy set theory [1] was developed by Zadeh in 1965 by assigning membership grade values to each element of the set in the interval $[0,1]$ and it is utilized for describing situations where the results are imprecise. This traditional fuzzy set has been applied in various fields. As this traditional fuzzy set theory deals only with positive membership degree of elements, Atanassov introduced negative membership or non-membership function to cover the gaps in the fuzzy set theory and the resulting set is called an intuitionistic fuzzy set [2]. Therefore, the notion of IFS theory is an extension of FS theory. Atanassov handled both the degree of membership $\left(\xi_{\tilde{\mathbb{I}_{\mathbb{F}}}}(\dot{v})\right)$ and non-membership $\left(\psi_{\tilde{\mathbb{I}_{\mathbb{F}}}}(\dot{v})\right)$ where $\xi_{\tilde{\mathbb{I}_{\mathbb{F}}}}(\dot{v}) + \psi_{\tilde{\mathbb{I}_{\mathbb{F}}}}(\dot{v}) \leq 1$. In some cases, the decision makers may provide their preference values like $\xi_{\tilde{\mathbb{I}_{\mathbb{F}}}}(\dot{v}) = 0.7$ and $\psi_{\tilde{\mathbb{I}_{\mathbb{F}}}}(\dot{v}) = 0.6$, which violates the condition of intuitionistic fuzzy set as the sum of these values is more than 1. Therefore to deal with such issues, Yager developed the concept of Pythagorean fuzzy set [3] with the condition that $\xi^2_{\tilde{\mathbb{I}_{\mathbb{F}}}}(\dot{v}) + \xi^2_{\tilde{\mathbb{I}_{\mathbb{F}}}}(\dot{v}) \leq 1$, clearly increasing the range of membership and non-membership and hence Pythagorean fuzzy set theory

C. Kahraman et al. (Eds.): INFUS 2021, LNNS 308, pp. 802–810, 2022.
https://doi.org/10.1007/978-3-030-85577-2_93

became more important and interesting research area. Many aggregation operators have been introduced by Yager and Abbasov [4] to handle MCDM problems in pythagorean fuzzy environment. Fermatean fuzzy sets is an extension of Pythagorean fuzzy sets which was introduced by Senapati and Yager [5] and it has been implemented in MCDM [6].

Neutrosophic set [7] and neutrosophic cubic sets [8] are yet another important three dimension generalization of classical fuzzy set. Similarly, in a different perspective, Spherical fuzzy set theory was introduced by Kutlu Gundogdu et al. [9] and it is one of the newly extended notions of fuzzy set theory. SFSs handle uncertainty and vagueness more efficiently than PFSs. From the review of literature, we can observe that more studies on spherical fuzzy sets have been carried out [10–12]. However, SFSs have so far been used only with single valued information. So in order to increase the applicability of SFSs, Cubic-SFS is needed. Hence the main objective of this research is to develop cubic-SFSs and to introduce Cu-SF tangent similarity measure for MCDM model.

The remaining part of the paper is organized as follows. Section 2 describes basic definitions in spherical fuzzy sets and their operations. Cubic-SFSs are introduced in Sect. 3 and also their operations are discussed in detail. Cu-SF tangent similarity measure is also introduced. In Sect. 4, an MCDM approach is illustrated based on Cu-SF tangent similarity measure. Then, in Sect. 5, the MCDM approach is applied to medical diagnosis problem and the results are analyzed. Finally, the contribution of the paper and future work have been reviewed in conclusion section.

2 Preliminaries

Definition 1 [1]. *A fuzzy set* $\tilde{\mathbb{F}}$ *is defined on a universe of discourse* $\overset{*}{V}$ *as the form:* $\tilde{\mathbb{F}} = \left\{ \langle \dot{v}, \xi_{\tilde{\mathbb{F}}}(\dot{v}) \rangle \,|\dot{v} \in \overset{*}{V} \right\}$*, where* $\xi_{\tilde{\mathbb{F}}}(\dot{v}) :\to [0,1]$*. Here* $\xi_{\tilde{\mathbb{F}}}(\dot{v})$ *denotes membership function to each* $\dot{v}$*.*

Definition 2 [2]. *An intuitionistic fuzzy sets* $\tilde{\mathbb{I}_{\mathbb{F}}}$ *is defined as a set of ordered pairs over a universal set* $\overset{*}{V}$ *given by* $\tilde{\mathbb{I}_{\mathbb{F}}} = \left\{ \langle \dot{v}, (\xi_{\tilde{\mathbb{I}_{\mathbb{F}}}}(\dot{v}), \psi_{\tilde{\mathbb{I}_{\mathbb{F}}}}(\dot{v})) \rangle \,|\dot{v} \in \overset{*}{V} \right\}$*, where* $\xi_{\tilde{\mathbb{I}_{\mathbb{F}}}}(\dot{v}) : \overset{*}{V} \to [0,1], \psi_{\tilde{\mathbb{I}_{\mathbb{F}}}}(\dot{v}) : \overset{*}{V} \to [0,1]$ *and with the condition* $\xi_{\tilde{\mathbb{I}_{\mathbb{F}}}}(\dot{v}) + \psi_{\tilde{\mathbb{I}_{\mathbb{F}}}}(\dot{v}) \leq 1$ *for each element* $x \in \overset{*}{V}$*. Here the membership and non-membership functions are denoted as* $\xi_{\tilde{\mathbb{I}_{\mathbb{F}}}}(\dot{v})$ *and* $\psi_{\tilde{\mathbb{I}_{\mathbb{F}}}}(\dot{v})$ *respectively.*

Definition 3 [9]. *Let* $\tilde{\mathbb{S}_{\mathbb{F}}}$ *be a spherical fuzzy sets in the universal discourse* $\overset{*}{V}$ *defined as:* $\tilde{\mathbb{S}_{\mathbb{F}}} = \left\{ \langle \dot{v}, (\xi_{\tilde{\mathbb{S}_{\mathbb{F}}}}(\dot{v}), \psi_{\tilde{\mathbb{S}_{\mathbb{F}}}}(\dot{v}), \pi_{\tilde{\mathbb{S}_{\mathbb{F}}}}(\dot{v})) \rangle \,|\dot{v} \in \overset{*}{V} \right\}$*, where* $\xi_{\tilde{\mathbb{S}_{\mathbb{F}}}}(\dot{v}) : \overset{*}{V} \to [0,1], \psi_{\tilde{\mathbb{S}_{\mathbb{F}}}}(\dot{v}) : \overset{*}{V} \to [0,1], \pi_{\tilde{\mathbb{S}_{\mathbb{F}}}}(\dot{v}) : \overset{*}{V} \to [0,1]$ *and* $0 \leq \xi^2_{\tilde{\mathbb{S}_{\mathbb{F}}}}(\dot{v}) + \pi^2_{\tilde{\mathbb{S}_{\mathbb{F}}}}(\dot{v}) + \pi^2_{\tilde{\mathbb{S}_{\mathbb{F}}}}(\dot{v}) \leq 1$ *for each* $\dot{v}$*. The values* $\xi_{\tilde{\mathbb{S}_{\mathbb{F}}}}(\dot{v}), \psi_{\tilde{\mathbb{S}_{\mathbb{F}}}}(\dot{v}), \pi_{\tilde{\mathbb{S}_{\mathbb{F}}}}(\dot{v})$ *are membership, non-membership and hesitancy functions of* $\dot{v}$ *in* $\tilde{\mathbb{S}_{\mathbb{F}}}$ *respectively.*

3 Cubic-SFSs

Definition 4. *Let $\overset{*}{V}$ be a non empty universal set or universe of discourse. A cubic-SFSs $\left(Cu - \tilde{\mathbb{S}}_{\mathbb{IF}}\right)$ is constructed in the following form:*

$$Cu - \tilde{\mathbb{S}}_{\mathbb{IF}}(\dot{v}) = \left\{\dot{v}, \left[\xi_{\tilde{\mathbb{S}}_{\mathbb{F}}^{I}}(\dot{v}), \psi_{\tilde{\mathbb{S}}_{\mathbb{F}}^{I}}(\dot{v}), \pi_{\tilde{\mathbb{S}}_{\mathbb{F}}^{I}}(\dot{v})\right]; \langle \xi_{\tilde{\mathbb{S}}_{\mathbb{F}}}(\dot{v}), \psi_{\tilde{\mathbb{S}}_{\mathbb{F}}}(\dot{v}), \pi_{\tilde{\mathbb{S}}_{\mathbb{F}}}(\dot{v})\rangle \,|\dot{v} \in \overset{*}{V}\right\}$$

where $\xi_{\tilde{\mathbb{S}}_{\mathbb{F}}^{I}}(\dot{v}), \psi_{\tilde{\mathbb{S}}_{\mathbb{F}}^{I}}(\dot{v}), \pi_{\tilde{\mathbb{S}}_{\mathbb{F}}^{I}}(\dot{v})$ are interval valued spherical fuzzy sets; $\xi_{\tilde{\mathbb{S}}_{\mathbb{F}}^{I}}(\dot{v}) = \left[\xi_{\tilde{\mathbb{S}}_{\mathbb{F}}^{-}}(\dot{v}), \xi_{\tilde{\mathbb{S}}_{\mathbb{F}}^{+}}(\dot{v})\right] \subseteq [0,1]$ is the degree of membership interval values; $\psi_{\tilde{\mathbb{S}}_{\mathbb{F}}^{I}}(\dot{v}) = \left[\psi_{\tilde{\mathbb{S}}_{\mathbb{F}}^{-}}(\dot{v}), \psi_{\tilde{\mathbb{S}}_{\mathbb{F}}^{+}}(\dot{v})\right] \subseteq [0,1]$ is the degree of non-membership interval values; $\pi_{\tilde{\mathbb{S}}_{\mathbb{F}}^{I}}(\dot{v}) = \left[\pi_{\tilde{\mathbb{S}}_{\mathbb{F}}^{-}}(\dot{v}), \pi_{\tilde{\mathbb{S}}_{\mathbb{F}}^{+}}(\dot{v})\right] \subseteq [0,1]$ is the degree of hesitancy interval values; and $\langle \xi_{\tilde{\mathbb{S}}_{\mathbb{F}}}(\dot{v}), \psi_{\tilde{\mathbb{S}}_{\mathbb{F}}}(\dot{v}), \pi_{\tilde{\mathbb{S}}_{\mathbb{F}}}(\dot{v})\rangle \in [0,1]$ are membership, non-membership and hesitancy degrees respectively. For convenience, a cubic spherical fuzzy element in a $Cu - SFSs$ is simply denoted by $\left[\xi_{\tilde{\mathbb{S}}_{\mathbb{F}}^{I}}, \psi_{\tilde{\mathbb{S}}_{\mathbb{F}}^{I}}, \pi_{\tilde{\mathbb{S}}_{\mathbb{F}}^{I}}\right]; \langle \xi_{\tilde{\mathbb{S}}_{\mathbb{F}}}, \psi_{\tilde{\mathbb{S}}_{\mathbb{F}}}, \pi_{\tilde{\mathbb{S}}_{\mathbb{F}}}\rangle$ where, $\left\langle \xi_{\tilde{\mathbb{S}}_{\mathbb{F}}^{I}}, \psi_{\tilde{\mathbb{S}}_{\mathbb{F}}^{I}}, \pi_{\tilde{\mathbb{S}}_{\mathbb{F}}^{I}}\right\rangle \subseteq [0,1]$ and $\langle \xi_{\tilde{\mathbb{S}}_{\mathbb{F}}}, \psi_{\tilde{\mathbb{S}}_{\mathbb{F}}}, \pi_{\tilde{\mathbb{S}}_{\mathbb{F}}}\rangle \in [0,1]$, satisfying the condition that $0 \leq \left\langle \xi_{\tilde{\mathbb{S}}_{\mathbb{F}}^{+}}, \psi_{\tilde{\mathbb{S}}_{\mathbb{F}}^{+}}, \pi_{\tilde{\mathbb{S}}_{\mathbb{F}}^{+}}\right\rangle \leq 3, \quad 0 \leq \langle \xi_{\tilde{\mathbb{S}}_{\mathbb{F}}}, \psi_{\tilde{\mathbb{S}}_{\mathbb{F}}}, \pi_{\tilde{\mathbb{S}}_{\mathbb{F}}}\rangle \leq 3.$

Definition 5. *Let $Cu - \tilde{\mathbb{S}}_{\mathbb{IF}}(\dot{v})$ be a cubic-SFSs in $\overset{*}{V}$. Then, $Cu - \tilde{\mathbb{S}}_{\mathbb{IF}}(\dot{v})$ is said to be internal if $\xi_{\tilde{\mathbb{S}}_{\mathbb{F}}^{-}}(\dot{v}) \leq \xi_{\tilde{\mathbb{S}}_{\mathbb{F}}}(\dot{v}) \leq \xi_{\tilde{\mathbb{S}}_{\mathbb{F}}^{+}}(\dot{v})$, $\psi_{\tilde{\mathbb{S}}_{\mathbb{F}}^{-}}(\dot{v}) \leq \psi_{\tilde{\mathbb{S}}_{\mathbb{F}}}(\dot{v}) \leq \psi_{\tilde{\mathbb{S}}_{\mathbb{F}}^{+}}(\dot{v})$ and $\pi_{\tilde{\mathbb{S}}_{\mathbb{F}}^{-}}(\dot{v}) \leq \pi_{\tilde{\mathbb{S}}_{\mathbb{F}}}(\dot{v}) \leq \pi_{\tilde{\mathbb{S}}_{\mathbb{F}}^{+}}(\dot{v})$. Similarly, $Cu - \tilde{\mathbb{S}}_{\mathbb{IF}}(\dot{v})$ is said to be external if $\xi_{\tilde{\mathbb{S}}_{\mathbb{F}}}(\dot{v}) \notin \left[\xi_{\tilde{\mathbb{S}}_{\mathbb{F}}^{-}}(\dot{v}), \xi_{\tilde{\mathbb{S}}_{\mathbb{F}}^{+}}(\dot{v})\right]$, $\psi_{\tilde{\mathbb{S}}_{\mathbb{F}}}(\dot{v}) \notin \left[\psi_{\tilde{\mathbb{S}}_{\mathbb{F}}^{-}}(\dot{v}), \psi_{\tilde{\mathbb{S}}_{\mathbb{F}}^{+}}(\dot{v})\right]$ and $\pi_{\tilde{\mathbb{S}}_{\mathbb{F}}}(\dot{v}) \notin \left[\pi_{\tilde{\mathbb{S}}_{\mathbb{F}}^{-}}(\dot{v}), \pi_{\tilde{\mathbb{S}}_{\mathbb{F}}^{+}}(\dot{v})\right]$ for all $\dot{v} \in \overset{*}{V}$.*

3.1 Operations on Cubic-SFSs

Let us take two collections of cubic spherical fuzzy sets such as $Cu - SFSs(\dot{v})$ and $Cu - SFSs(\dot{w})$. Then the following operations are defined;

1. Union of two Cubic-SFSs is denoted as $\left[Cu - \tilde{\mathbb{S}}_{\mathbb{IF}}(\dot{v})\right] \cup \left[Cu - \tilde{\mathbb{S}}_{\mathbb{IF}}(\dot{w})\right]$

$$\begin{aligned}
= \Bigg\{ & \left[\min\left(\xi_{\tilde{\mathbb{S}}_{\mathbb{F}}^{-}}(\dot{v}), \xi_{\tilde{\mathbb{S}}_{\mathbb{F}}^{-}}(\dot{w})\right), \max\left(\xi_{\tilde{\mathbb{S}}_{\mathbb{F}}^{+}}(\dot{v}), \xi_{\tilde{\mathbb{S}}_{\mathbb{F}}^{+}}(\dot{w})\right)\right], \left[\max\left(\psi_{\tilde{\mathbb{S}}_{\mathbb{F}}^{-}}(\dot{v}), \psi_{\tilde{\mathbb{S}}_{\mathbb{F}}^{-}}(\dot{w})\right), \min\left(\psi_{\tilde{\mathbb{S}}_{\mathbb{F}}^{+}}(\dot{v}), \psi_{\tilde{\mathbb{S}}_{\mathbb{F}}^{+}}(\dot{w})\right)\right], \\
& \Bigg[\max\left(\left(1 - \left(\left[\min\left(\xi_{\tilde{\mathbb{S}}_{\mathbb{F}}^{-}}(\dot{v}), \xi_{\tilde{\mathbb{S}}_{\mathbb{F}}^{-}}(\dot{w})\right)\right]^2 + \left[\max\left(\psi_{\tilde{\mathbb{S}}_{\mathbb{F}}^{-}}(\dot{v}), \psi_{\tilde{\mathbb{S}}_{\mathbb{F}}^{-}}(\dot{w})\right)\right]^2\right)\right)^{1/2}, \left[\min\left(\pi_{\tilde{\mathbb{S}}_{\mathbb{F}}^{-}}(\dot{v}), \pi_{\tilde{\mathbb{S}}_{\mathbb{F}}^{-}}(\dot{w})\right)\right]\right), \\
& \min\left(\left(1 - \left(\left[\max\left(\xi_{\tilde{\mathbb{S}}_{\mathbb{F}}^{+}}(\dot{v}), \xi_{\tilde{\mathbb{S}}_{\mathbb{F}}^{+}}(\dot{w})\right)\right]^2 + \left[\min\left(\psi_{\tilde{\mathbb{S}}_{\mathbb{F}}^{+}}(\dot{v}), \psi_{\tilde{\mathbb{S}}_{\mathbb{F}}^{+}}(\dot{w})\right)\right]^2\right)\right)^{1/2}, \left[\max\left(\pi_{\tilde{\mathbb{S}}_{\mathbb{F}}^{+}}(\dot{v}), \pi_{\tilde{\mathbb{S}}_{\mathbb{F}}^{+}}(\dot{w})\right)\right]\right)\Bigg]; \\
& \langle \max\left(\xi_{\tilde{\mathbb{S}}_{\mathbb{F}}}(\dot{v}), \xi_{\tilde{\mathbb{S}}_{\mathbb{F}}}(\dot{w})\right), \min\left(\psi_{\tilde{\mathbb{S}}_{\mathbb{F}}}(\dot{v}), \psi_{\tilde{\mathbb{S}}_{\mathbb{F}}}(\dot{w})\right), \\
& \quad \min\left(\left(1 - \left(\left[\max\left(\xi_{\tilde{\mathbb{S}}_{\mathbb{F}}}(\dot{v}), \xi_{\tilde{\mathbb{S}}_{\mathbb{F}}}(\dot{w})\right)\right]^2 + \left[\min\left(\psi_{\tilde{\mathbb{S}}_{\mathbb{F}}}(\dot{v}), \psi_{\tilde{\mathbb{S}}_{\mathbb{F}}}(\dot{w})\right)\right]^2\right)\right)^{1/2}, \left[\max\left(\pi_{\tilde{\mathbb{S}}_{\mathbb{F}}}(\dot{v}), \pi_{\tilde{\mathbb{S}}_{\mathbb{F}}}(\dot{w})\right)\right]\right)\rangle . \Bigg\}
\end{aligned}$$

2. Intersection of Cubic-SFSs is denoted as $\left[Cu - \tilde{\mathbb{S}}_{\mathbb{IF}}(\dot{v})\right] \cap \left[Cu - \tilde{\mathbb{S}}_{\mathbb{IF}}(\dot{w})\right]$

$$
\begin{aligned}
&= \Big\{\Big[\max\left(\xi_{\tilde{\mathbb{S}}_{\mathbb{F}}^{-}}(\dot{v}), \xi_{\tilde{\mathbb{S}}_{\mathbb{F}}^{-}}(\dot{w})\right), \min\left(\xi_{\tilde{\mathbb{S}}_{\mathbb{F}}^{+}}(\dot{v}), \xi_{\tilde{\mathbb{S}}_{\mathbb{F}}^{+}}(\dot{w})\right)\Big], \Big[\min\left(\psi_{\tilde{\mathbb{S}}_{\mathbb{F}}^{-}}(\dot{v}), \psi_{\tilde{\mathbb{S}}_{\mathbb{F}}^{-}}(\dot{w})\right), \max\left(\psi_{\tilde{\mathbb{S}}_{\mathbb{F}}^{+}}(\dot{v}), \psi_{\tilde{\mathbb{S}}_{\mathbb{F}}^{+}}(\dot{w})\right)\Big], \\
&\Bigg[\min\left(\left(1 - \left(\left[\max\left(\xi_{\tilde{\mathbb{S}}_{\mathbb{F}}^{-}}(\dot{v}), \xi_{\tilde{\mathbb{S}}_{\mathbb{F}}^{-}}(\dot{w})\right)\right]^2 + \left[\min\left(\psi_{\tilde{\mathbb{S}}_{\mathbb{F}}^{-}}(\dot{v}), \psi_{\tilde{\mathbb{S}}_{\mathbb{F}}^{-}}(\dot{w})\right)\right]^2\right)\right)^{1/2}, \left[\max\left(\pi_{\tilde{\mathbb{S}}_{\mathbb{F}}^{-}}(\dot{v}), \pi_{\tilde{\mathbb{S}}_{\mathbb{F}}^{-}}(\dot{w})\right)\right]\right), \\
&\quad \max\left(\left(1 - \left(\left[\min\left(\xi_{\tilde{\mathbb{S}}_{\mathbb{F}}^{+}}(\dot{v}), \xi_{\tilde{\mathbb{S}}_{\mathbb{F}}^{+}}(\dot{w})\right)\right]^2 + \left[\max\left(\psi_{\tilde{\mathbb{S}}_{\mathbb{F}}^{+}}(\dot{v}), \psi_{\tilde{\mathbb{S}}_{\mathbb{F}}^{+}}(\dot{w})\right)\right]^2\right)\right)^{1/2}, \left[\min\left(\pi_{\tilde{\mathbb{S}}_{\mathbb{F}}^{+}}(\dot{v}), \pi_{\tilde{\mathbb{S}}_{\mathbb{F}}^{+}}(\dot{w})\right)\right]\right)\Bigg]; \\
&\langle \min\left(\xi_{\tilde{\mathbb{S}}_{\mathbb{F}}}(\dot{v}), \xi_{\tilde{\mathbb{S}}_{\mathbb{F}}}(\dot{w})\right), \max\left(\psi_{\tilde{\mathbb{S}}_{\mathbb{F}}}(\dot{v}), \psi_{\tilde{\mathbb{S}}_{\mathbb{F}}}(\dot{w})\right), \\
&\qquad \max\left(\left(1 - \left(\left[\min\left(\xi_{\tilde{\mathbb{S}}_{\mathbb{F}}}(\dot{v}), \xi_{\tilde{\mathbb{S}}_{\mathbb{F}}}(\dot{w})\right)\right]^2 + \left[\max\left(\psi_{\tilde{\mathbb{S}}_{\mathbb{F}}}(\dot{v}), \psi_{\tilde{\mathbb{S}}_{\mathbb{F}}}(\dot{w})\right)\right]^2\right)\right)^{1/2}, \left[\min\left(\pi_{\tilde{\mathbb{S}}_{\mathbb{F}}}(\dot{v}), \pi_{\tilde{\mathbb{S}}_{\mathbb{F}}}(\dot{w})\right)\right]\right)\Big\rangle . \Big\}
\end{aligned}
$$

3. Algebraic sum of Cubic-SFSs is denoted as $\left[Cu - \tilde{\mathbb{S}}_{\mathbb{IF}}(\dot{v})\right] \oplus \left[Cu - \tilde{\mathbb{S}}_{\mathbb{IF}}(\dot{w})\right]$

$$
\begin{aligned}
&= \Big\{\Big[\sqrt{\xi^2_{\tilde{\mathbb{S}}_{\mathbb{F}}^{-}}(\dot{v}) + \xi^2_{\tilde{\mathbb{S}}_{\mathbb{F}}^{-}}(\dot{w}) - \xi^2_{\tilde{\mathbb{S}}_{\mathbb{F}}^{-}}(\dot{v}).\xi^2_{\tilde{\mathbb{S}}_{\mathbb{F}}^{-}}(\dot{w})}, \quad \sqrt{\xi^2_{\tilde{\mathbb{S}}_{\mathbb{F}}^{+}}(\dot{v}) + \xi^2_{\tilde{\mathbb{S}}_{\mathbb{F}}^{+}}(\dot{w}) - \xi^2_{\tilde{\mathbb{S}}_{\mathbb{F}}^{+}}(\dot{v}).\xi^2_{\tilde{\mathbb{S}}_{\mathbb{F}}^{+}}(\dot{w})}\Big], \Big[\psi_{\tilde{\mathbb{S}}_{\mathbb{F}}^{-}}(\dot{v}).\psi_{\tilde{\mathbb{S}}_{\mathbb{F}}^{-}}(\dot{w}), \quad \psi_{\tilde{\mathbb{S}}_{\mathbb{F}}^{+}}(\dot{v}).\psi_{\tilde{\mathbb{S}}_{\mathbb{F}}^{+}}(\dot{w})\Big], \\
&\Big[\sqrt{\left(1 - \xi^2_{\tilde{\mathbb{S}}_{\mathbb{F}}^{-}}(\dot{w})\right)\pi^2_{\tilde{\mathbb{S}}_{\mathbb{F}}^{-}}(\dot{v}) + \left(1 - \xi^2_{\tilde{\mathbb{S}}_{\mathbb{F}}^{-}}(\dot{v})\right)\pi^2_{\tilde{\mathbb{S}}_{\mathbb{F}}^{-}}(\dot{w}) - \pi^2_{\tilde{\mathbb{S}}_{\mathbb{F}}^{-}}(\dot{v}).\pi^2_{\tilde{\mathbb{S}}_{\mathbb{F}}^{-}}(\dot{w})}, \\
&\qquad \sqrt{\left(1 - \xi^2_{\tilde{\mathbb{S}}_{\mathbb{F}}^{+}}(\dot{w})\right)\pi^2_{\tilde{\mathbb{S}}_{\mathbb{F}}^{+}}(\dot{v}) + \left(1 - \xi^2_{\tilde{\mathbb{S}}_{\mathbb{F}}^{+}}(\dot{v})\right)\pi^2_{\tilde{\mathbb{S}}_{\mathbb{F}}^{+}}(\dot{w}) - \pi^2_{\tilde{\mathbb{S}}_{\mathbb{F}}^{+}}(\dot{v}).\pi^2_{\tilde{\mathbb{S}}_{\mathbb{F}}^{+}}(\dot{w})}\Big]; \\
&\quad \Big\langle \sqrt{\xi^2_{\tilde{\mathbb{S}}_{\mathbb{F}}}(\dot{v}) + \xi^2_{\tilde{\mathbb{S}}_{\mathbb{F}}}(\dot{w}) - \xi^2_{\tilde{\mathbb{S}}_{\mathbb{F}}}(\dot{v}).\xi^2_{\tilde{\mathbb{S}}_{\mathbb{F}}}(\dot{w})}, \quad \psi_{\tilde{\mathbb{S}}_{\mathbb{F}}}(\dot{v}).\psi_{\tilde{\mathbb{S}}_{\mathbb{F}}}(\dot{w}), \quad \sqrt{\left(1 - \xi^2_{\tilde{\mathbb{S}}_{\mathbb{F}}}(\dot{w})\right)\pi^2_{\tilde{\mathbb{S}}_{\mathbb{F}}}(\dot{v}) + \left(1 - \xi^2_{\tilde{\mathbb{S}}_{\mathbb{F}}}(\dot{v})\right)\pi^2_{\tilde{\mathbb{S}}_{\mathbb{F}}}(\dot{w}) - \pi^2_{\tilde{\mathbb{S}}_{\mathbb{F}}}(\dot{v}).\pi^2_{\tilde{\mathbb{S}}_{\mathbb{F}}}(\dot{w})}\Big\rangle\Big\}
\end{aligned}
$$

4. Algebraic product of Cubic-SFSs is denoted as $\left[Cu - \tilde{\mathbb{S}}_{\mathbb{IF}}(\dot{v})\right] \otimes \left[Cu - \tilde{\mathbb{S}}_{\mathbb{IF}}(\dot{w})\right]$

$$
\begin{aligned}
&= \Big\{\Big[\xi_{\tilde{\mathbb{S}}_{\mathbb{F}}^{-}}(\dot{v}).\xi_{\tilde{\mathbb{S}}_{\mathbb{F}}^{-}}(\dot{w}), \quad \xi_{\tilde{\mathbb{S}}_{\mathbb{F}}^{+}}(\dot{v}).\xi_{\tilde{\mathbb{S}}_{\mathbb{F}}^{+}}(\dot{w})\Big], \quad \Big[\sqrt{\psi^2_{\tilde{\mathbb{S}}_{\mathbb{F}}^{-}}(\dot{v}) + \psi^2_{\tilde{\mathbb{S}}_{\mathbb{F}}^{-}}(\dot{w}) - \psi^2_{\tilde{\mathbb{S}}_{\mathbb{F}}^{-}}(\dot{v}).\psi^2_{\tilde{\mathbb{S}}_{\mathbb{F}}^{-}}(\dot{w})}, \quad \sqrt{\psi^2_{\tilde{\mathbb{S}}_{\mathbb{F}}^{+}}(\dot{v}) + \psi^2_{\tilde{\mathbb{S}}_{\mathbb{F}}^{+}}(\dot{w}) - \psi^2_{\tilde{\mathbb{S}}_{\mathbb{F}}^{+}}(\dot{v}).\psi^2_{\tilde{\mathbb{S}}_{\mathbb{F}}^{+}}(\dot{w})}\Big], \\
&\Big[\sqrt{\left(1 - \psi^2_{\tilde{\mathbb{S}}_{\mathbb{F}}^{-}}(\dot{w})\right)\pi^2_{\tilde{\mathbb{S}}_{\mathbb{F}}^{-}}(\dot{v}) + \left(1 - \psi^2_{\tilde{\mathbb{S}}_{\mathbb{F}}^{-}}(\dot{v})\right)\pi^2_{\tilde{\mathbb{S}}_{\mathbb{F}}^{-}}(\dot{w}) - \pi^2_{\tilde{\mathbb{S}}_{\mathbb{F}}^{-}}(\dot{v}).\pi^2_{\tilde{\mathbb{S}}_{\mathbb{F}}^{-}}(\dot{w})}, \\
&\qquad \sqrt{\left(1 - \psi^2_{\tilde{\mathbb{S}}_{\mathbb{F}}^{+}}(\dot{w})\right)\pi^2_{\tilde{\mathbb{S}}_{\mathbb{F}}^{+}}(\dot{v}) + \left(1 - \psi^2_{\tilde{\mathbb{S}}_{\mathbb{F}}^{+}}(\dot{v})\right)\pi^2_{\tilde{\mathbb{S}}_{\mathbb{F}}^{+}}(\dot{w}) - \pi^2_{\tilde{\mathbb{S}}_{\mathbb{F}}^{+}}(\dot{v}).\pi^2_{\tilde{\mathbb{S}}_{\mathbb{F}}^{+}}(\dot{w})}\Big]; \\
&\Big\langle \psi_{\tilde{\mathbb{S}}_{\mathbb{F}}}(\dot{v}).\psi_{\tilde{\mathbb{S}}_{\mathbb{F}}}(\dot{w}), \quad \sqrt{\xi^2_{\tilde{\mathbb{S}}_{\mathbb{F}}}(\dot{v}) + \xi^2_{\tilde{\mathbb{S}}_{\mathbb{F}}}(\dot{w}) - \xi^2_{\tilde{\mathbb{S}}_{\mathbb{F}}}(\dot{v}).\xi^2_{\tilde{\mathbb{S}}_{\mathbb{F}}}(\dot{w})}, \quad \sqrt{\left(1 - \psi^2_{\tilde{\mathbb{S}}_{\mathbb{F}}}(\dot{w})\right)\pi^2_{\tilde{\mathbb{S}}_{\mathbb{F}}}(\dot{v}) + \left(1 - \psi^2_{\tilde{\mathbb{S}}_{\mathbb{F}}}(\dot{v})\right)\pi^2_{\tilde{\mathbb{S}}_{\mathbb{F}}}(\dot{w}) - \pi^2_{\tilde{\mathbb{S}}_{\mathbb{F}}}(\dot{v}).\pi^2_{\tilde{\mathbb{S}}_{\mathbb{F}}}(\dot{w})}\Big\rangle\Big\}
\end{aligned}
$$

5. Scalar multiplication of Cubic-SFSs; $\lambda > 0$

$$
\begin{aligned}
\lambda.\left(Cu - \tilde{\mathbb{S}}_{\mathbb{IF}}(\dot{v})\right) = &\Bigg\{\Bigg[\sqrt{1 - \left(1 - \xi^2_{\tilde{\mathbb{S}}_{\mathbb{F}}^{-}}(\dot{v})\right)^{\lambda}}, \sqrt{1 - \left(1 - \xi^2_{\tilde{\mathbb{S}}_{\mathbb{F}}^{+}}(\dot{v})\right)^{\lambda}}\Bigg], \quad \Big[\psi_{\tilde{\mathbb{S}}_{\mathbb{F}}^{-}}(\dot{v}), \quad \psi_{\tilde{\mathbb{S}}_{\mathbb{F}}^{+}}(\dot{v})\Big] \\
&\Bigg[\sqrt{\left(1 - \xi^2_{\tilde{\mathbb{S}}_{\mathbb{F}}^{-}}(\dot{v})\right)^{\lambda} - \left(1 - \xi^2_{\tilde{\mathbb{S}}_{\mathbb{F}}^{-}}(\dot{v}) - \pi^2_{\tilde{\mathbb{S}}_{\mathbb{F}}^{-}}(\dot{v})\right)^{\lambda}}, \quad \sqrt{\left(1 - \xi^2_{\tilde{\mathbb{S}}_{\mathbb{F}}^{-}}(\dot{v})\right)^{\lambda} - \left(1 - \xi^2_{\tilde{\mathbb{S}}_{\mathbb{F}}^{-}}(\dot{v}) - \pi^2_{\tilde{\mathbb{S}}_{\mathbb{F}}^{-}}(\dot{v})\right)^{\lambda}}\Bigg]; \\
&\Bigg\langle \sqrt{1 - \left(1 - \xi^2_{\tilde{\mathbb{S}}_{\mathbb{F}}}(\dot{v})\right)^{\lambda}}, \quad \psi_{\tilde{\mathbb{S}}_{\mathbb{F}}}(\dot{v}), \quad \sqrt{\left(1 - \xi^2_{\tilde{\mathbb{S}}_{\mathbb{F}}^{-}}(\dot{v})\right)^{\lambda} - \left(1 - \xi^2_{\tilde{\mathbb{S}}_{\mathbb{F}}^{-}}(\dot{v}) - \pi^2_{\tilde{\mathbb{S}}_{\mathbb{F}}^{-}}(\dot{v})\right)^{\lambda}}\Bigg\rangle\Bigg\}
\end{aligned}
$$

6. Power of Cubic-SFSs; $\lambda > 0$

$$
\begin{aligned}
\left(Cu - \tilde{\mathbb{S}}_{\mathbb{IF}}(\dot{v})\right)^{\lambda} = &\Bigg\{\Big[\xi_{\tilde{\mathbb{S}}_{\mathbb{F}}^{-}}(\dot{v}), \quad \xi_{\tilde{\mathbb{S}}_{\mathbb{F}}^{+}}(\dot{v})\Big], \quad \Bigg[\sqrt{1 - \left(1 - \psi^2_{\tilde{\mathbb{S}}_{\mathbb{F}}^{-}}(\dot{v})\right)^{\lambda}}, \sqrt{1 - \left(1 - \psi^2_{\tilde{\mathbb{S}}_{\mathbb{F}}^{+}}(\dot{v})\right)^{\lambda}}\Bigg] \\
&\Bigg[\sqrt{\left(1 - \psi^2_{\tilde{\mathbb{S}}_{\mathbb{F}}^{-}}(\dot{v})\right)^{\lambda} - \left(1 - \psi^2_{\tilde{\mathbb{S}}_{\mathbb{F}}^{-}}(\dot{v}) - \pi^2_{\tilde{\mathbb{S}}_{\mathbb{F}}^{-}}(\dot{v})\right)^{\lambda}}, \quad \sqrt{\left(1 - \psi^2_{\tilde{\mathbb{S}}_{\mathbb{F}}^{-}}(\dot{v})\right)^{\lambda} - \left(1 - \psi^2_{\tilde{\mathbb{S}}_{\mathbb{F}}^{-}}(\dot{v}) - \pi^2_{\tilde{\mathbb{S}}_{\mathbb{F}}^{-}}(\dot{v})\right)^{\lambda}}\Bigg]; \\
&\Bigg\langle \xi_{\tilde{\mathbb{S}}_{\mathbb{F}}}(\dot{v}), \quad \sqrt{1 - \left(1 - \psi^2_{\tilde{\mathbb{S}}_{\mathbb{F}}}(\dot{v})\right)^{\lambda}}, \quad \sqrt{\left(1 - \xi^2_{\tilde{\mathbb{S}}_{\mathbb{F}}^{-}}(\dot{v})\right)^{\lambda} - \left(1 - \xi^2_{\tilde{\mathbb{S}}_{\mathbb{F}}^{-}}(\dot{v}) - \pi^2_{\tilde{\mathbb{S}}_{\mathbb{F}}^{-}}(\dot{v})\right)^{\lambda}}\Bigg\rangle\Bigg\}
\end{aligned}
$$

3.2 Tangent Similarity Measure of Cubic-SFSs

We discuss the tangent similarity measure of Cubic-SFSs and their properties

Definition 6. *Let $Cu - \tilde{\mathbb{S}_{\mathbb{F}}}(\dot{v}_i) \forall i = 1, 2$ be two Cubic-SFSs in the universe of discourse $\overset{*}{V}$. Then, the tangent similarity measure of the Cubic-SFSs is defined as follows:*

$$T_{Cu}\left(\tilde{\mathbb{S}_{\mathbb{F}}}(\dot{v}_1), \tilde{\mathbb{S}_{\mathbb{F}}}(\dot{v}_2)\right) = 1 - \left\{\tan\left[\frac{\pi}{16}(\wp_I) + \frac{\pi}{8}(\wp_S)\right]\right\} \tag{1}$$

where, the values of $\wp_I$ and $\wp_S$ are respectively,

$$\wp_I = \frac{1}{6}\left(\left(\xi_{\tilde{\mathbb{S}_{\mathbb{F}}}^-}(\dot{v}_1) - \xi_{\tilde{\mathbb{S}_{\mathbb{F}}}^-}(\dot{v}_2)\right)^2 + \left(\xi_{\tilde{\mathbb{S}_{\mathbb{F}}}^+}(\dot{v}_1) - \xi_{\tilde{\mathbb{S}_{\mathbb{F}}}^+}(\dot{v}_2)\right)^2 + \left(\psi_{\tilde{\mathbb{S}_{\mathbb{F}}}^-}(\dot{v}_1) - \psi_{\tilde{\mathbb{S}_{\mathbb{F}}}^-}(\dot{v}_2)\right)^2 + \left(\psi_{\tilde{\mathbb{S}_{\mathbb{F}}}^+}(\dot{v}_1) - \psi_{\tilde{\mathbb{S}_{\mathbb{F}}}^+}(\dot{v}_2)\right)^2 + \left(\pi_{\tilde{\mathbb{S}_{\mathbb{F}}}^-}(\dot{v}_1) - \pi_{\tilde{\mathbb{S}_{\mathbb{F}}}^-}(\dot{v}_2)\right)^2 + \left(\pi_{\tilde{\mathbb{S}_{\mathbb{F}}}^+}(\dot{v}_1) - \pi_{\tilde{\mathbb{S}_{\mathbb{F}}}^+}(\dot{v}_2)\right)^2\right)$$

$$\wp_S = \frac{1}{2}\left(\left(\xi_{\tilde{\mathbb{S}_{\mathbb{F}}}}(\dot{v}_1) - \xi_{\tilde{\mathbb{S}_{\mathbb{F}}}}(\dot{v}_2)\right)^2 + \left(\psi_{\tilde{\mathbb{S}_{\mathbb{F}}}}(\dot{v}_1) - \psi_{\tilde{\mathbb{S}_{\mathbb{F}}}}(\dot{v}_2)\right)^2 + \left(\pi_{\tilde{\mathbb{S}_{\mathbb{F}}}}(\dot{v}_1) - \pi_{\tilde{\mathbb{S}_{\mathbb{F}}}}(\dot{v}_2)\right)^2\right)$$

Here, $\wp_I$ and $\wp_S$ denotes the interval and individual values of Cubic-SFSs.

Example 1. Let $Cu - \tilde{\mathbb{S}_{\mathbb{F}}}(\dot{v}_1) = \{[0,0], [1,1], [1,1]; \langle 0,1,1\rangle\}$ and $Cu - \tilde{\mathbb{S}_{\mathbb{F}}}(\dot{v}_2) = \{[1,1], [0,0], [0,0]; \langle 1,0,0\rangle\}$ be two Cubic-SFSs, then the tangent similarity measure calculate as follows: The values of $\wp_I = \frac{1}{6}\left((0-1)^2 + (0-1)^2 + (1-0)^2 + (1-0)^2 + (1-0)^2 + (1-0)^2\right) = \frac{1}{6}(6)$. Similarly, the values of $\wp_S = \frac{1}{2}\left((0-1)^2 + (1-0)^2 + (1-0)^2\right) = \frac{1}{2}(3)$, Then $T_{Cu}\left(\tilde{\mathbb{S}_{\mathbb{F}}}(\dot{v}_1), \tilde{\mathbb{S}_{\mathbb{F}}}(\dot{v}_2)\right) = 1 - \left\{\tan\left[\frac{\pi}{16}\left(\frac{1}{6}(6)\right) + \frac{\pi}{8}\left(\frac{1}{2}(3)\right)\right]\right\} = 1 - 1 = 0$. Clearly we can see that the tangent similarity measure is zero. Since each Cubic-SFSs is opponent to each others.

In general, the tangent similarity measure of Cubic-SFSs of $Cu - \tilde{\mathbb{S}_{\mathbb{F}}}(\dot{v}_i)$ and $Cu - \tilde{\mathbb{S}_{\mathbb{F}}}(\dot{w}_i)$ is defined as follows:

$$T_{Cu}\left(\tilde{\mathbb{S}_{\mathbb{F}}}(\dot{v}_i), \tilde{\mathbb{S}_{\mathbb{F}}}(\dot{w}_i)\right) = 1 - \frac{1}{t}\sum_{i=1}^{t}\left\{\tan\left[\frac{\pi}{16}(\wp_I) + \frac{\pi}{8}(\wp_S)\right]\right\} \tag{2}$$

The tangent similarity measures of Cubic-SFSs satisfies the following properties:

1. $0 \leq T_{Cu}\left(\tilde{\mathbb{S}_{\mathbb{F}}}(\dot{v}_i), \tilde{\mathbb{S}_{\mathbb{F}}}(\dot{w}_i)\right) \leq 1$
2. $T_{Cu}\left(\tilde{\mathbb{S}_{\mathbb{F}}}(\dot{v}_i), \tilde{\mathbb{S}_{\mathbb{F}}}(\dot{w}_i)\right) = 1 \Leftrightarrow \tilde{\mathbb{S}_{\mathbb{F}}}(\dot{v}_i) = \tilde{\mathbb{S}_{\mathbb{F}}}(\dot{w}_i)$
3. $T_{Cu}\left(\tilde{\mathbb{S}_{\mathbb{F}}}(\dot{v}_i), \tilde{\mathbb{S}_{\mathbb{F}}}(\dot{w}_i)\right) = T_{Cu}\left(\tilde{\mathbb{S}_{\mathbb{F}}}(\dot{w}_i), \tilde{\mathbb{S}_{\mathbb{F}}}(\dot{v}_i)\right)$
4. If $Cu - \tilde{\mathbb{S}_{\mathbb{F}}}(\dot{r}_i)$ is a Cu-SFSs in $\overset{*}{V}$ and $Cu - \tilde{\mathbb{S}_{\mathbb{F}}}(\dot{v}_i) \subseteq Cu - \tilde{\mathbb{S}_{\mathbb{F}}}(\dot{w}_i) \subseteq Cu - \tilde{\mathbb{S}_{\mathbb{F}}}(\dot{r}_i)$, then $T_{Cu}\left(\tilde{\mathbb{S}_{\mathbb{F}}}(\dot{v}_i), \tilde{\mathbb{S}_{\mathbb{F}}}(\dot{r}_i)\right) \leq T_{Cu}\left(\tilde{\mathbb{S}_{\mathbb{F}}}(\dot{v}_i), \tilde{\mathbb{S}_{\mathbb{F}}}(\dot{w}_i)\right)$ and $T_{Cu}\left(\tilde{\mathbb{S}_{\mathbb{F}}}(\dot{v}_i), \tilde{\mathbb{S}_{\mathbb{F}}}(\dot{r}_i)\right) \leq T_{Cu}\left(\tilde{\mathbb{S}_{\mathbb{F}}}(\dot{w}_i), \tilde{\mathbb{S}_{\mathbb{F}}}(\dot{r}_i)\right)$

If the weight of different elements $\dot{v}_i$ is $\alpha_i = (\alpha_1, \alpha_2, \alpha_3, \ldots, \alpha_t)^T$, which satisfies the condition that $0 \leq \alpha_i \leq 1$ and $\sum_{i=1}^{t} \alpha_i = 1$, then the corresponding weighted tangent similarity measure of Cubic-SFSs is defined as follows:

$$T_{Cu}^{\alpha}\left(\tilde{\mathbb{S}_{\mathbb{F}}}(\dot{v}_i), \tilde{\mathbb{S}_{\mathbb{F}}}(\dot{w}_i)\right) = 1 - \frac{1}{t}\sum_{i=1}^{t} \alpha_i \left\{\tan\left[\frac{\pi}{16}(\wp_I) + \frac{\pi}{8}(\wp_S)\right]\right\} \tag{3}$$

Also the weighted tangent similarity measure of Cubic-SFSs satisfies all the above mentioned properties.

4 The Decision Making Model of Cubic-SFSs

Here the developed Cubic-SFSs are used to frame a new decision making model based on tangent similarity measure. In the process of decision making, consider m number of alternatives $(R_1, R_2, R_3, \ldots, R_m)$ and n number of attributes $(C_1, C_2, C_3, \ldots, C_n)$ with the weight vectors of criteria $\alpha_i (i = 1, 2, 3, \ldots, n)$. Then the decision making approach can described as follows.

Step 1. Frame the Cubic spherical fuzzy (Cubic-SF) decision matrix $DM = \left[e_{ij}^{\kappa}\right]_{m\times n}$ $\forall i = 1, 2, 3, \ldots, m; j = 1, 2, 3, \ldots, n$. where e_{ij}^{k} denotes the Cubic-SF values of alternative R_i on C_j given by experts E_κ

Step 2. Fix the ideal Cubic-SF numbers $\nabla^j\left(Cu - \tilde{\mathbb{S}_{\mathbb{F}}}(\dot{C}_j)\right)$ to each criteria

Step 3. Find the tangent similarity measure of e_{ij}^{k} and ideal Cubic-SF numbers i.e. $T_{Cu}\left(e_{ij}^{k}, \nabla^j\left(\tilde{\mathbb{S}_{\mathbb{F}}}(\dot{C}_j)\right)\right)$ and form fuzzy decision matrix (FDM)

Step 4. Calculate the aggregate values for each alternatives on criteria using tangent fuzzy aggregation operator given by

$$TFAO(x_1, x_2, x_3, \ldots, x_n) = 1 - \prod_{j=1}^{n}\left(1 - \tan\left(\frac{\pi}{4}\left(x_j^2\right)\right)\right) \tag{4}$$

Step 5. Grade the alternatives according to their aggregate values

5 Medical Diagnosis Systems Using TAN-SM of Cubic-SF

It is more uncertain to find the disease of patients based on the symptoms they have. So in order to address the need, a medical diagnosis intelligent system has been developed based on the concept of cubic-SF information. A group of $(R_i; i = 1, 2, 3, 4)$ patients have been taken and their symptoms are collected as C_1-Temperature , C_2-Cough, C_3-Throat pain, C_4-Body pain. Also, the chance of risk of disease are mentioned as follows: ∇^1-Viral fever, ∇^2-Tuberculosis, ∇^3-Typhoid, ∇^4-Throat disease. In this medical diagnosis problem, when we collect samples from the patients in certain intervals of time, the characteristic values between patients and the indicated symptoms are represented by the Cubic-SF decision matrix. The reference range of disease are converted into Cubic-SF ideal numbers. Then, we proceed to decision making process to find the risk of disease of patients as follows;

Step 1. Here, the relation between patients and their symptoms are mentioned in cubic spherical fuzzy decision matrix $DM = \left[e_{ij}^{\kappa}\right]_{m\times n}$ as shown below

$$\left\{\begin{array}{c|c|c|c|c} & C_1 & C_2 & C_3 & C_4 \\ R_1 & \begin{bmatrix}[0.5,0.6], & [0.1,0.3], \\ [0.2,0.4]; & \langle 0.6,0.2,0.3\rangle\end{bmatrix} & \begin{bmatrix}[0.5,0.6], & [0.1,0.3], \\ [0.2,0.4]; & \langle 0.6,0.2,0.3\rangle\end{bmatrix} & \begin{bmatrix}[0.4,0.6], & [0.3,0.5], \\ [0.4,0.7]; & \langle 0.5,0.6,0.4\rangle\end{bmatrix} & \begin{bmatrix}[0.2,0.4], & [0.7,0.8], \\ [0.8,0.9]; & \langle 0.3,0.8,0.9\rangle\end{bmatrix} \\ R_2 & \begin{bmatrix}[0.6,0.8], & [0.1,0.2], \\ [0.2,0.3]; & \langle 0.7,0.1,0.2\rangle\end{bmatrix} & \begin{bmatrix}[0.6,0.7], & [0.1,0.2], \\ [0.2,0.3]; & \langle 0.6,0.1,0.2\rangle\end{bmatrix} & \begin{bmatrix}[0.3,0.4], & [0.6,0.7], \\ [0.8,0.9]; & \langle 0.3,0.6,0.9\rangle\end{bmatrix} & \begin{bmatrix}[0.5,0.7], & [0.3,0.4], \\ [0.4,0.5]; & \langle 0.5,0.3,0.5\rangle\end{bmatrix} \\ R_3 & \begin{bmatrix}[0.4,0.6], & [0.2,0.3], \\ [0.1,0.3]; & \langle 0.6,0.2,0.2\rangle\end{bmatrix} & \begin{bmatrix}[0.5,0.6], & [0.2,0.3], \\ [0.3,0.4]; & \langle 0.6,0.3,0.4\rangle\end{bmatrix} & \begin{bmatrix}[0.3,0.5], & [0.7,0.8], \\ [0.6,0.7]; & \langle 0.4,0.8,0.7\rangle\end{bmatrix} & \begin{bmatrix}[0.6,0.7], & [0.2,0.4], \\ [0.4,0.5]; & \langle 0.5,0.4,0.3\rangle\end{bmatrix} \\ R_4 & \begin{bmatrix}[0.7,0.8], & [0.1,0.2], \\ [0.1,0.2]; & \langle 0.8,0.1,0.2\rangle\end{bmatrix} & \begin{bmatrix}[0.6,0.7], & [0.1,0.2], \\ [0.1,0.3]; & \langle 0.7,0.1,0.2\rangle\end{bmatrix} & \begin{bmatrix}[0.3,0.4], & [0.6,0.7], \\ [0.7,0.8]; & \langle 0.3,0.7,0.8\rangle\end{bmatrix} & \begin{bmatrix}[0.5,0.8], & [0.4,0.5], \\ [0.4,0.6]; & \langle 0.7,0.4,0.4\rangle\end{bmatrix}\end{array}\right\}$$

Step 2. The reference range of disease with regard to symptoms by the form of ideal Cubic-SF numbers to each criteria are given as follows:

$$\begin{array}{lcl} \nabla^1\left(Cu-\tilde{\mathbb{S}_{\mathrm{IF}}}(\dot{C}_1)\right) & : & [0.52,0.65],[0.19,0.48],[0.20,0.40];\langle 0.41,0.20,0.48\rangle \\ \nabla^2\left(Cu-\tilde{\mathbb{S}_{\mathrm{IF}}}(\dot{C}_2)\right) & : & [0.62,0.75],[0.19,0.35],[0.20,0.30];\langle 0.46,0.10,0.35\rangle \\ \nabla^3\left(Cu-\tilde{\mathbb{S}_{\mathrm{IF}}}(\dot{C}_3)\right) & : & [0.52,0.69],[0.35,0.48],[0.23,0.34];\langle 0.42,0.26,0.49\rangle \\ \nabla^4\left(Cu-\tilde{\mathbb{S}_{\mathrm{IF}}}(\dot{C}_4)\right) & : & [0.65,0.75],[0.19,0.35],[0.10,0.26];\langle 0.57,0.10,0.35\rangle \end{array}$$

Step 3. The tangent similarity measure of e_{ij}^{k} and the ideal Cubic-SF numbers are used to form fuzzy decision matrix ($FDM = [x_{ij}]_{m\times n}$)

$$FDM = [x_{ij}]_{m\times n} = \begin{array}{c} \\ R_1 \\ R_2 \\ R_3 \\ R_4 \end{array}\begin{array}{c} \begin{array}{cccc}\nabla^1/C_1 & \nabla^2/C_2 & \nabla^3/C_3 & \nabla^4/C_4\end{array} \\ \begin{pmatrix} 0.9851 & \mathbf{0.9918} & 0.9684 & 0.7710 \\ 0.9620 & \mathbf{0.9906} & 0.9124 & 0.9806 \\ 0.9752 & \mathbf{0.9859} & 0.9150 & 0.9758 \\ 0.9465 & \mathbf{0.9828} & 0.9180 & 0.9688 \end{pmatrix}\end{array}$$

Step 4. The aggregate values for each alternatives on criteria using Eq. 4 are calculated and are shown as; $TFAO(R_1) = 0.99995, TFAO(R_1) = 0.99996$, $TFAO(R_1) = 0.99995, TFAO(R_1) = 0.99984$
Step 5. Then, the grade of alternatives is $R_2 > R_3 > R_1 > R_4$

5.1 Sensitive Analysis

In this problem, each element of cubic spherical fuzzy decision matrix (e_{ij}^{k}) and the corresponding ideal Cubic-SFNs are more similar. Hence we can find from TAN-SM that almost every alternative is of equal importance. Also we highlight that all the patients are mostly affected by cough and Tuberculosis. For better understanding the results have been demonstrated in Figs. 1 and 2.

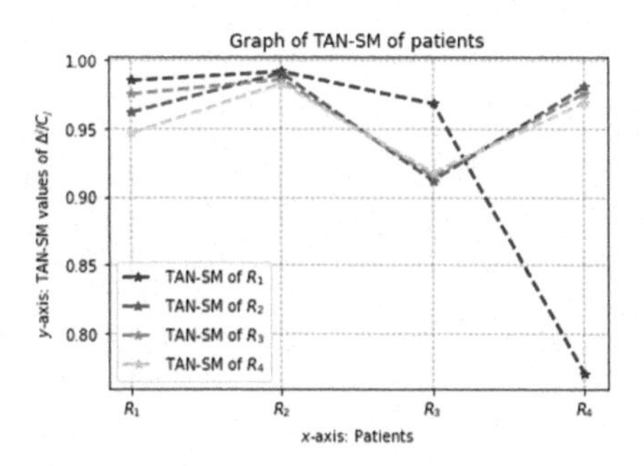

Fig. 1. TAN-SM of patients

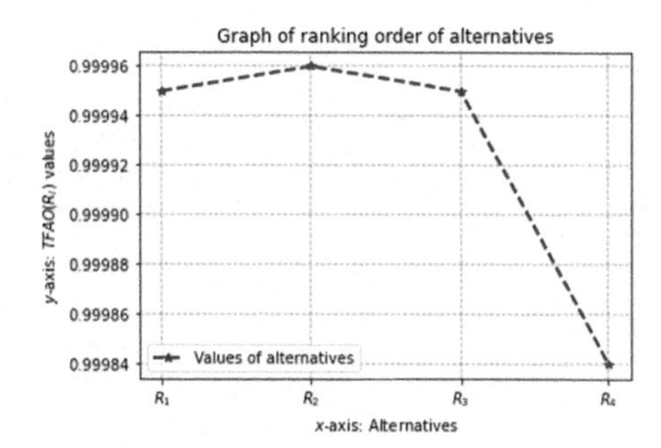

Fig. 2. The ranking order of alternatives

6 Conclusion

In this paper, the notion of Cubic-SFSs have been developed with their fundamental operations. Then, the proposed tangent similarity measure for Cubic-SFSs is investigated with a numerical example. Further, a decision making system has been introduced based on Cubic-SF tangent similarity measure and it is more effective in handling medical diagnosis problem with Cubic-SF information. So, the proposed method has been executed in order to find the disease over the symptoms in patients. And the sensitive analysis describes more about the findings under Cubic-SF medical diagnosis system. This research could be extended further in many decision models and similarity measure algorithms and could be applied to pattern recognition and fuzzy clustering analysis.

References

1. Zadeh, L.A.: Fuzzy sets. Inf. Control **8**, 338–353 (1965)
2. Atanassov, K.T.: Intuitionistic fuzzy sets. Fuzzy Sets Syst. **20**(1), 87–96 (1986)
3. Yager, R.R.: Pythagorean fuzzy subsets. In: 2013 Joint IFSA World Congress and NAFIPS Annual Meeting (IFSA/NAFIPS), Edmonton, Canada, pp. 57–61 (2013)
4. Yager, R.R., Abbasov, A.M.: Pythagorean membership grades, complex numbers, and decision making. Int. J. Intell. Syst. **28**(5), 436–452 (2013)
5. Senapati, T., Yager, R.R.: Fermatean fuzzy sets. J. Am. Intell. Hum. Comput. **11**(2), 663–674 (2020)
6. Garg, H., Shahzadi, G., Akram, M.: Decision-making analysis based on fermatean fuzzy yager aggregation operators with application in COVID-19 testing facility. Mathematical Problems in Engineering, pp. 1–16 (2020)
7. Smarandache, F.: A Unifying Field in Logics: Neutrosophic Logic. Philosophy, American Research Press, pp. 1–141 (1999)
8. Jun, Y., Smarandache, F., Kim, C.: Neutrosophic cubic sets. New Math. Nat. Comput. **13**, 41–54 (2017)
9. Kutlu Gündoßdu, F., Kahraman, C.: Spherical fuzzy sets and spherical fuzzy TOPSIS method. J. Intell. Fuzzy Syst. **36**(1), 337–352 (2019)
10. Ashraf, S., Abdullah, S.: Spherical aggregation operators and their application in multiattribute group decision-making. Int. J. Intell. Syst. **34**(3), 493–523 (2019)

11. Kutlu Gundogdu, F., Kahraman, C., Karasan, A.: Spherical fuzzy VIKOR method and its application to waste management, vol. 1029 (2019). https://doi.org/10.1007/978-3-030-23756-1118
12. Aydin, S., Kahraman, C.: A spherical fuzzy multi expert MCDM method based on the entropy and cosine similarity. In: Developments of Artificial Intelligence Technologies in Computation And Robotics - (Flins) World Scientific, vol. 12, p. 157 (2020)

Big Data-Driven in COVID-19 Pandemic Management System: Evaluation of Barriers with Spherical Fuzzy AHP Approach

Yağmur Arıöz(✉), Ibrahim Yılmaz, Abdullah Yıldızbaşı, and Cihat Öztürk

Department of Industrial Engineering, Ankara Yıldırım Beyazıt University, Ankara, Turkey
yarioz@ybu.edu.tr

Abstract. Big Data-driven management system has attracted significant attention worldwide as it provides several capabilities to improve strategic, tactical, and operational decisions to eventually create a notable impact on the COVID-19 pandemic. Monitoring, surveillance, detection, and prevention of the global pandemic cases are provided with simultaneous access and management with big data tools. Bringing the pandemic to a normalization level and taking it under control in the health supply chain could be achieved through big data-driven management which is lifesaving applications. However, the adoption of the management system in the fight against the COVID-19 pandemic includes many obstacles all over the world. Mainly, these are data-related characteristics, technological incompetence, socio-economic structure, and governmental policies. In this scope, the main motivation of this study is determining and evaluating critical criteria by using a Multi-Criteria Decision-Making (MCDM) approach. The Spherical Fuzzy AHP methodology proposed as one of the novel MCDM methods enables to obtain managerial implications by comparing to the significance of criteria. The findings can assist to understand the actual nature of the barriers and potential benefits of big data-driven in COVID-19 pandemic management system and make policy regarding curb the pandemic. Therefore, this study is a contribution to academicians, researchers, and practitioners with a different perspective on COVID-19 pandemic management.

Keywords: Big data · COVID-19 pandemic · Spherical fuzzy AHP · Multi-criteria decision making

1 Introduction

During the unprecedented coronavirus disease 2019 (COVID-19) pandemic has become a huge threat to global public health [1]. The COVID-19 pandemic requires us to precisely foresee the spread of disease and decide how to embrace managerial strategies to provide sustainable health conditions [2]. In this scope, governmental health policies based on big-data are implemented to curb the pandemic which leads to unbalanced living conditions. Monitoring, surveillance, detection, and prevention of the global pandemic cases are provided with simultaneous access and management with big data tools [3]. For

C. Kahraman et al. (Eds.): INFUS 2021, LNNS 308, pp. 811–818, 2022.
https://doi.org/10.1007/978-3-030-85577-2_94

this purpose, developing strategic solutions by capturing and analyzing data in the health supply chain can enhance the better preventative performance of the pandemic. From this aspect, big data applications are lifesaving. However, the adoption of the big-data-driven management system is encountered many barriers in the global network. These barriers consist of four main categories are the socio-economic structure, technological incompetence, data-related characteristics and governmental policies. This paper presents the determining and evaluating barrier criteria by using a Multi-Criteria Decision-Making (MCDM) approach. For this purpose, the developed a three-dimensional spherical fuzzy set enables effective handling of high uncertainty and identifying decision-makers judgments [4]. As a novelty of this paper, the Spherical Fuzzy AHP methodology proposed as one of the MCDM methods are enables to obtain managerial implications for this special case by comparing to the significance of criteria and are used to find the final rank of these. In this context, this article aims to the following Research Questions (RQ):

RQ1: What are the barriers big-data driven management in the fight of the COVID-19 pandemic?

RQ2: What is the relationship between importance weight of the barrier criteria?

RQ3: How can be improving big-data-driven management system in the struggling with the COVID-19 pandemic?

Considering all of these research questions the main purpose of this article is to determine and evaluate barriers in big-data-driven management struggle with the pandemic. The remainder of this paper is organized as follows: Sect. 2 gives a literature review. Section 3 includes methodology. Section 4 consists of a numerical case study. Section 5 presents results and discussion according to the case study. Finally, the conclusions and possible future work are included in Section 6.

2 Research Background

Combating the COVID-19, the benefits of big data-driven applications have a significant role. For this purpose, there are many significant articles in the literature. In one of these articles, Bragazzi et al. (2020) emphasized the study of the role of AI and big data applications to overcome the COVID-19 pandemic [5]. Qiu et al. (2020) mentioned that statistical analysis of pandemic process using internet search data [6]. Ienca et al. (2020) addressed the benefits of data-collection and data-processing in managing the COVID-19 pandemic [7]. Javaid et al. (2020) also stated examining the benefits of using big data from Industry 4.0 technologies to defeat the COVID-19 pandemic. Guraya (2020) highlighted that big data analytics can provide understanding and combating the COVID-19. Also, the author emphasized the procedures executed as laparoscopic endoscopic under consensus guidelines [8].

Although the literature is encompassed many articles related to the role of big data in managing the COVID-19 pandemic, there are rare articles related to barriers of big-data-driven management. One of these rare articles, Zhou et al. (2020) tackled the challenges for Geographic Information Systems (GIS) with big data in the COVID-19 pandemic [9]. In this context, they emphasized that the use of GIS and big data technologies is an important method of managing the outbreak in the fight against COVID-19. The process of regional detection and monitoring of cases provides facilities such as model instruction and mapping for data with GIS. The authors emphasized the need to develop a data-driven system in the future due to the difficulties faced by GIS usage in integrating their data. In the literature review, we determined our main research motivation starting from point of the research gap that evaluates the barriers of big data-driven in COVID-19 pandemic management system.

In this scope, we suggested Spherical Fuzzy AHP, one of the novel developed MCDM methods that can obtain more precise decisions under high uncertainty and prejudices by comparing these barriers. Kutlu Gündoğdu and Kahraman (2019) who contribute a three-dimensional spherical fuzzy set to the literature used spherical fuzzy AHP methodology for renewable energy location selection [10]. Mathew et al. (2020) suggested spherical fuzzy AHP-TOPSIS for advanced manufacturing system selection [4]. This proposed methodology is provided robust and competitive results under high uncertainty.

3 Methodology

The spherical fuzzy AHP method is consists of five steps as presented in this section. Figure 1 illustrates a flowchart related to methodology steps.

Step 1. Building the hierarchical structure Level 1 demonstrates an objective based on the score index. Level 2 substitutes that the score index is determined depend on a finite set of criteria $C = \{C_1, C_2, \ldots C_n\}$. The hierarchical structure has many sub-criteria which specified for each C.

Step 2. Building pairwise comparisons by using linguistic terms which are given in Table 1. Equations (1) and (2) obtained the score indices (SI) by using Table 2.

Table 1. Linguistic measures of importance used for pairwise comparisons

	(μ, ϑ, π)	Score Index (SI)
Absolutely more importance (AMI)	(0.9, 0.1, 0.0)	9
Very high importance (VHI)	(0.8, 0.2, 0.1)	7
High importance (HI)	(0.7, 0.3, 0.2)	5
Slightly more importance (SMI)	(0.6, 0.4, 0.3)	3
Equally importance (EI)	(0.5, 0.4, 0.4)	1
Slightly low importance (SLI)	(0.4, 0.6, 0.3)	1/3
Low importance (LI)	(0.3, 0.7, 0.2)	1/5
Very low importance (VLI)	(0.2, 0.8, 0.1)	1/7
Absolutely low importance (ALI)	(0.1, 0.9, 0.0)	1/9

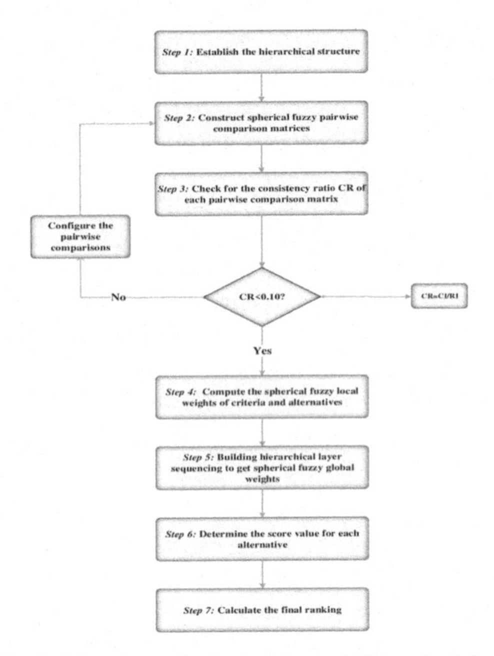

Fig. 1. Flowchart of spherical fuzzy AHP methodology

Step 3. Determine score indices by using the transformation of linguistic terms. After, check the consistency according to the threshold which is identified %1.

$$\mathrm{SI} = \sqrt{\left|100 * \left[\left(\mu_{\tilde{A}s} - \pi_{\tilde{A}s}\right)^2 - \left(v_{\tilde{A}s} - \pi_{\tilde{A}s}\right)^2\right]\right|} \tag{1}$$

for AMI, VHI, HI, SMI and EI,

$$\frac{1}{\mathrm{SI}} = \frac{1}{\sqrt{\left|100 * \left[\left(\mu_{\tilde{A}s} - \pi_{\tilde{A}s}\right)^2 - \left(v_{\tilde{A}s} - \pi_{\tilde{A}s}\right)^2\right]\right|}} \tag{2}$$

for EI, SLI, LI, VLI and ALI

Table 2. Pairwise comparison of main criteria

Criteria	C1	C2	C3	C4	$(\tilde{w}^S)$	$(\overline{w}^S)$
C1	EI	HI	AMI	SMI	(0.73, 0.26, 0.23)	0.360
C2	LI	EI	SMI	LI	(0.35, 0.63, 0.27)	0.158
C3	ALI	SLI	EI	VLI	(0.37, 0.61, 0.29)	0.169
C4	SLI	HI	VHI	EI	(0.65, 0.35, 0.25)	0.313

CR = 0.066

Step 4. Computing the spherical fuzzy local weights of the criteria and alternatives. SWAM which is given in Eq. (3) for each criterion, detects the weight of each alternative. To compute the weight of spherical fuzzy, the weighted arithmetic mean is used.

$$\mathrm{SWAM_w}(A_{s1}, \ldots, A_{sn}) = w_1 A_{s1} + w_2 A_{s2} + \ldots + w_n A_{sn}$$
$$= \left\langle \left[1 - \prod_{i=1}^{n}(1 - \mu_{\tilde{A}si}^2)^{w_i}\right]^{1/2}, \prod_{i=1}^{n} v_{\tilde{A}si}^{w_i}, \left[\prod_{i=1}^{n}(1 - \mu_{\tilde{A}si}^2)^{w_i} - \prod_{i=1}^{n}(1 - \mu_{\tilde{A}si}^2 - \pi_{\tilde{A}si}^2)^{w_i}\right]^{1/2} \right\rangle \tag{3}$$

where $w = 1/n$

Step 5. Construct the hierarchical layer sequencing to get global weights. To predict the final ranking is cumulated the spherical fuzzy weights at each level. For this purpose, there are two potential methods. Firstly, by applying score function (S) in Eq. (4) the criterion weights are defuzzified. Then, is applied the normalization in Eq. (5) and the spherical fuzzy multiplication in Eq. (6).

$$\mathrm{S}\left(\tilde{w}_j^s\right) = \sqrt{\left|100 * \left[\left(3\mu_{\tilde{A}s} - \frac{\pi_{\tilde{A}s}}{2}\right)^2 - \left(\frac{v_{\tilde{A}s}}{2} - \pi_{\tilde{A}s}\right)^2\right]\right|} \tag{4}$$

$$\overline{w}_j^s = \frac{S\left(\tilde{w}_j^s\right)}{\sum_{J=1}^{n} S\left(\tilde{w}_j^s\right)} \tag{5}$$

$$\tilde{A}_{S_{ij}} = \overline{w}_j^s.\tilde{A}_{S_i} = \left\langle \left(1 - \left(1 - \mu_{\tilde{A}s}^2\right)^{\overline{w}_j^s}\right)^{1/2}, v_{\tilde{A}s}^{\overline{w}_j^s}, \left(\left(1 - \mu_{\tilde{A}s}^2\right)^{\overline{w}_j^s} - \left(1 - \mu_{\tilde{A}s}^2 - \pi_{\tilde{A}s}^2\right)^{\overline{w}_j^s}\right)^{1/2} \right\rangle \forall i \tag{6}$$

Equation (7) illustrates the final score $(\widetilde{F})$. This score is get by applying spherical fuzzy arithmetic summation for Ai.

$$\tilde{F} = \sum\nolimits_{j=11}^{n} \tilde{A}_{S_{ij}} = \tilde{A}_{S_{i1}} \oplus \tilde{A}_{S_{i2}} \ldots \oplus \tilde{A}_{S_{in}} \forall i$$

$$\text{i.e.}\tilde{A}_{S_{11}} \oplus \tilde{A}_{S_{12}}$$
$$= \left\langle \left(\left(\mu_{\tilde{A}s_{11}}^2 + \mu_{\tilde{A}s_{12}}^2 - \mu_{\tilde{A}s_{11}}^2 \mu_{\tilde{A}s_{12}}^2\right)^{1/2}, v_{\tilde{A}s_{11}} v_{\tilde{A}s_{12}}, \left(1 - \mu_{\tilde{A}s_{12}}^2\right)\pi_{\tilde{A}s_{11}}^2 + \left(1 - \mu_{\tilde{A}s_{11}}^2\right)\pi_{\tilde{A}s_{12}}^2 - \pi_{\tilde{A}s_{11}}^2\right)^{1/2} \right\rangle \tag{7}$$

The step can follow without defuzzification. According to Eq. (8) states the calculating spherical fuzzy global preference weights.

$$\prod_{j=1}^{n} \tilde{A}_{S_{ij}} = \tilde{A}_{S_{i1}} \otimes \tilde{A}_{S_{i2}} \ldots \otimes \tilde{A}_{S_{in}} \forall i$$

$$\text{i.e.}\tilde{A}_{S_{11}} \otimes \tilde{A}_{S_{12}}$$
$$= \left\langle \mu_{\tilde{A}s_{11}} \mu_{\tilde{A}s_{12}}, \left(v_{\tilde{A}s_{11}}^2 + v_{\tilde{A}s_{12}}^2 - v_{\tilde{A}s_{11}}^2 v_{\tilde{A}s_{12}}^2\right)^{1/2}, \left(\left(1 - v_{\tilde{A}s_{12}}^2\right)\pi_{\tilde{A}s_{11}}^2 + \left(1 - v_{\tilde{A}s_{11}}^2\right)\pi_{\tilde{A}s_{12}}^2 - \pi_{\tilde{A}s_{11}}^2 \pi_{\tilde{A}s_{12}}^2\right)^{1/2} \right\rangle \tag{8}$$

Step 6. Equation (8) represents the fuzzy final score and *Step 7* is calculating the final ranking.

4 Case Study

The case study evaluates the barriers that are developed basis on spherical fuzzy sets as the expert assessment involving uncertainty and vagueness. The barrier criteria of big-data-driven management in the fight against the COVID-19 pandemic were determined via literature. The hierarchical structure of criteria is shown in Fig. 2.

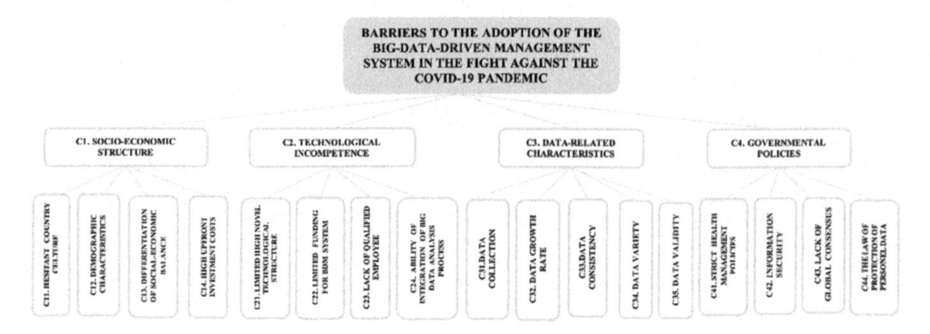

Fig. 2. The hierarchical structure of criteria

Spherical fuzzy pairwise comparison matrice of the main criteria and the CR value presented as follows (Table 3):

Table 3. Pairwise comparison of main criteria

	C1	C2	C3	C4
C1	**EI**	LI	VLI	LI
C2	VHI	**EI**	SMI	SMI
C3	AMI	SMI	**EI**	SMII
C4	SLI	LI	ALI	**EI**

CR = 0.01

The global weights of the criteria are calculated as follows and shown in Fig. 3.

The score values of the main criteria are presented in Fig. 4. According to results, score value of criteria were calculated as Data-Related Characteristics (0.3466), Governmental, Policies (0.3015), Technological Incompetence (0.2018), Socio-Economic Structure (0.1502), respectively.

5 Result and Discussion

This study has addressed and evaluated the barriers in the adoption of big-data driven approaches to combating the pandemic. Seventeen barriers were identified through a methodology using an exhaustive literature search. Barriers to the adoption of a big

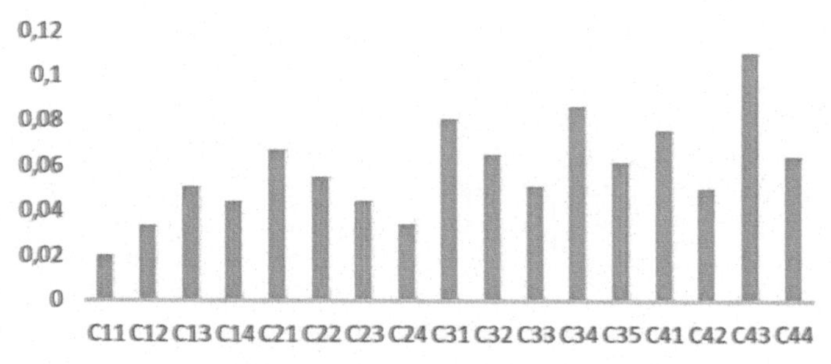

Fig. 3. Global weights of sub-criteria

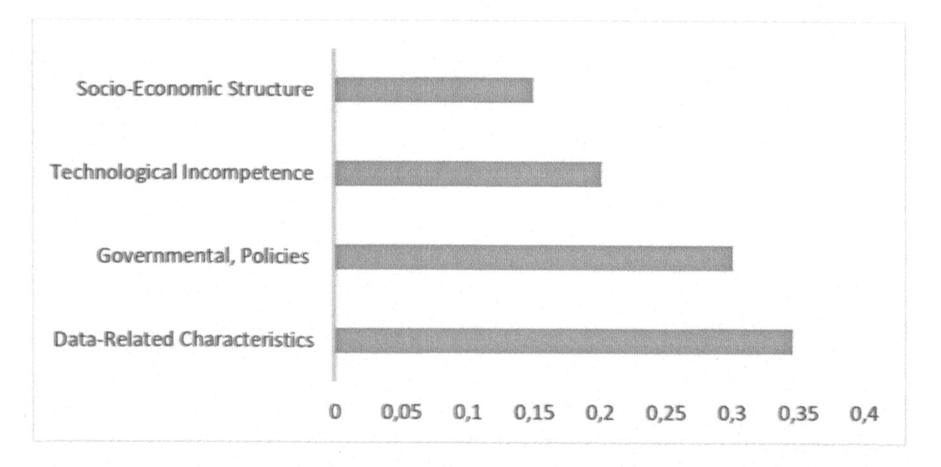

Fig. 4. Score values of the main criteria

data-based management system in the fight against the COVID-19 pandemic have been developed in spherical fuzzy sets based on uncertain expert assessment. This study contributes to the literature in terms of the actual nature of the barriers and potential benefits of big data-driven in COVID-19 pandemic management system and make policy regarding curb the pandemic. According to the score values, the criteria are ranked as C3 > C4 > C2 > C1. While data-related barriers have the highest importance, socio-economic structure barriers have the lowest importance in this rank. Developing strategic solutions by capturing and analyzing data in the health supply chain can enhance the better preventative performance of the pandemic. From this context, characteristics such as the growth rate, validity and collection of data play a noteworthy role in the management of the pandemic. Obtained results give that the importance of the validity and sustainability of government policies depends on the data. Developing reliable infrastructure and information technologies regarding technological deficiencies provides convenience in combating the pandemic. From these perspectives, big-data-driven approaches should be considered in terms of developing managerial strategies to fight against the pandemic.

References

1. Hu, S., Xiong, C., Yang, M., Younes, H., Luo, W., Zhang, L.: A big-data driven approach to analyzing and modeling human mobility trend under non-pharmaceutical interventions during COVID-19 pandemic. Transp. Res. Part C Emerg. Technol. **124**, 102955 (2021)
2. Javaid, M., Haleem, A., Vaishya, R., Bahl, S., Suman, R.: Diabetes and metabolic syndrome : clinical research & reviews industry 4.0 technologies and their applications in fighting COVID-19 pandemic. Diab. Metab. Syndr. Clin. Res. Rev. **14**(4), 419–422 (2020)
3. Shu, D., Ting, W., Carin, L., Dzau, V., Wong, T.Y.: Digital technology and COVID-19. Nat. Med. **26**(4), 459–461 (2020)
4. Mathew, M., Chakrabortty, R.K., Ryan, M.J.: A novel approach integrating AHP and TOPSIS under spherical fuzzy sets for advanced manufacturing system selection. Eng. Appl. Artif. Intell. **96**, 103988 (2020)
5. Bragazzi, N.L., Dai, H., Damiani, G., Behzadifar, M., Wu, J.: How big data and artificial intelligence can help better manage the COVID-pandemic. Int. J. Environ. Res. Publ. Health **17**(9), 4–11 (2020)
6. Qiu, H.J., et al.: Using the internet search data to investigate symptom characteristics of COVID-19: a big data study. World J. Otorhinolaryngol.-Head Neck Surg. **6**(1), 40–48 (2020)
7. Ienca, M., Vayena, E.: On the responsible use of digital data to tackle the COVID-19 pandemic. Nat. Med. **26**(4), 463–464 (2020)
8. Guraya, S.Y.: Transforming laparoendoscopic surgical protocols during the COVID-19 pandemic; big data analytics, resource allocation and operational considerations. Int. J. Surg. **80**(August), 21–25 (2020)
9. Zhou, C., et al.: COVID-19: challenges to GIS with big data. Geograph. Sustain. **1**(1), 77–87 (2020)
10. Kutlu Gündoğdu, F., Kahraman, C.: A novel spherical fuzzy analytic hierarchy process and its renewable energy application. Soft. Comput. **24**(6), 4607–4621 (2019). https://doi.org/10.1007/s00500-019-04222-w

Complex T-Spherical Fuzzy N-Soft Sets

Muhammad Akram(✉) and Maria Shabir

Department of Mathematics, University of the Punjab,
New Campus, Lahore, Pakistan
m.akram@pucit.edu.pk

Abstract. This research article broadens the literature by presenting a new model, namely, complex T-spherical fuzzy N-soft sets, which is capable to handle both aspects of two-dimensional information involved in the satisfaction, abstinence and dissatisfaction nature of human decisions. We define complex T-spherical fuzzy N-soft sets and present some of their fundamental operations. We illustrate these operations with examples. Further, we discuss some of their properties.

Keywords: N-soft set · Complex T-spherical fuzzy N-soft sets

1 Introduction

The membership function in fuzzy sets [14], restricted to $[0, 1]$, corresponds to level of satisfaction and only describes the opinion of the nature yes (positive response). Atanassov [6] extended the fuzzy sets by adding the non-membership function, restricted to $[0, 1]$, to express the degree of dissatisfaction, provided that the sum of both membership and non-membership degrees should not exceed 1 and established a new model, namely, intuitionistic fuzzy sets. The idea of picture fuzzy sets [7], a generalization of intuitionistic fuzzy set, represent the neutral part in the decisions of human beings along with the positive membership and the negative membership, provided that the sum of positive, neutral and negative memberships should be less or equal to 1. In 2019, Gündogdu and Kahraman [9] presented the idea of spherical fuzzy sets along with the spherical fuzzy TOPSIS method to tackle multi-criteria decision making problems. A spherical fuzzy set is an extension of picture fuzzy sets. Later, Kahraman et al. [10] applied the spherical fuzzy TOPSIS method to choose the most appropriate site for a hospital. Mahmood et al. [12] presented T-spherical fuzzy sets with the basic operations and developed certain aggregation operators based on T-spherical fuzzy sets. Ashraf et al. [5] provided the decision making strategies on the basis of spherical fuzzy aggregation operators. Ali et al. [4] considered complex T-spherical fuzzy aggregation operators and Guleria [11] put forward the notion of T-spherical fuzzy soft sets with aggregation operators. On the other hand, Fatimah et al. [8] introduced the concept of N-soft sets as an extension of soft set [13]. Akram et al. [1,2] studied group decision-making methods under

C. Kahraman et al. (Eds.): INFUS 2021, LNNS 308, pp. 819–834, 2022.
https://doi.org/10.1007/978-3-030-85577-2_95

complex spherical fuzzy information. Moreover, Akram et al. [3] discussed complex spherical fuzzy N-soft sets. As a continuation of this study, we introduce the concept of complex T-spherical fuzzy N-soft sets and discuss some of their fundamental operations. We illustrate the operations with examples. We also investigate some of their properties.

2 Complex T-Spherical Fuzzy N-Soft Sets

Definition 2.1. [4] A subset $\bar{g}$ of universal set U with $m_{\bar{g}}$, $\alpha_{\bar{g}} : U \rightarrow [0, 1]$; $q_{\bar{g}}, \nu_{\bar{g}} : U \rightarrow [0, 1]$ and $p_{\bar{g}}, \delta_{\bar{g}} : U \rightarrow [0, 1]$ representing the degrees of truthness, neutrality and falsity, respectively, is said to be *complex T-spherical fuzzy set* ($CTSFS$) if the following conditions hold: $m_{\bar{g}}(u)^n + p_{\bar{g}}(u)^n + q_{\bar{g}}(u)^n \leq 1$ and $\alpha_{\bar{g}}(u)^n + \nu_{\bar{g}}(u)^n + \delta_{\bar{g}}(u)^n \leq 1$, for all n (natural number) and $u \in U$. The CTSFS $\bar{g}$ can be represented as:

$$\bar{g} = \langle (u, m_{\bar{g}}(u)e^{i2\pi\alpha_{\bar{g}}(u)}, q_{\bar{g}}(u)e^{i2\pi\nu_{\bar{g}}(u)}, p_{\bar{g}}(u)e^{i2\pi\delta_{\bar{g}}(u)}) | u \in U \rangle,$$

where, $i = \sqrt{-1}$, $m_{\bar{g}}$, $q_{\bar{g}}$, $p_{\bar{g}}$, are the amplitude terms and $\alpha_{\bar{g}}$, $\nu_{\bar{g}}$, $\delta_{\bar{g}}$ are the periodic terms, which belong to unit closed interval. The degree of refusal is as follows:

$$\theta_{\bar{g}}(u) = \sqrt[n]{1 - (m_{\bar{g}}(u)^n + p_{\bar{g}}(u)^n + q_{\bar{g}}(u)^n)} e^{i2\pi \sqrt[n]{1-(\alpha_{\bar{g}}(u)^n + \nu_{\bar{g}}(u)^n + \delta_{\bar{g}}(u)^n)}}.$$

The triplet $\left(m_{\bar{g}}(u)e^{i2\pi\alpha_{\bar{g}}(u)}, q_{\bar{g}}(u)e^{i2\pi\nu_{\bar{g}}(u)}, p_{\bar{g}}(u)e^{i2\pi\delta_{\bar{g}}(u)}\right)$ is known as complex T-spherical fuzzy number ($CTSFN$).

Definition 2.2. [8] Let E be subset of parametric set Y over the universal set U with set of ordered grades $X = \{0, 1, 2, \ldots, N-1\}$ and $N \in \{2, 3, \ldots\}$. A triple $(\bar{f}, \mathrm{E}, N)$ along with mapping $\bar{f} : \mathrm{E} \rightarrow 2^{U \times X}$ is called *N-soft set* (NS_fS) over U if there exist a unique pair $(u_k, x_j^k) \in U \times \mathrm{E}$, $\forall\ u_k \in U$ and $e_j \in \mathrm{E}$.

Definition 2.3. Let E be subset of parametric set Y over the universal set U and $\mathbb{P}(CTSFNs)$ denotes the collection of CTSFNs over U. A pair $(\bar{f}, \mathrm{E})$ with set-valued function $\bar{f} : \mathrm{E} \rightarrow \mathcal{P}(CTSFNs)$ is called *complex T-spherical fuzzy soft set* ($CTSFS_fS$) over U. The $CTSFS_fS$ over universal set U can be denoted as:

$$\begin{aligned} \bar{f}(v_j) &= \{\langle u_k, m_j(u_k)e^{i2\pi\alpha_j(u_k)}, q_j(u_k)e^{i2\pi\nu_j(u_k)}, p_j(u_k)e^{i2\pi\delta_j(u_k)} \rangle | u_k \in U, e_j \in \mathrm{E}\} \\ &= \{\langle u_k, m_{kj}e^{i2\pi\alpha_{kj}}, q_{kj}e^{i2\pi\nu_{kj}}, p_{kj}e^{i2\pi\delta_{kj}} \rangle\}, \end{aligned}$$

where $m_{kj}e^{i2\pi\alpha_{kj}}$, $q_{kj}e^{i2\pi\nu_{kj}}$ and $p_{kj}e^{i2\pi\delta_{kj}}$ are complex-valued functions of truthness, neutrality and falsity, respectively, with the amplitude terms $m_{\bar{g}}$, $q_{\bar{g}}$, $p_{\bar{g}}$ and the periodic terms $\alpha_{\bar{g}}$, $\nu_{\bar{g}}$, $\delta_{\bar{g}}$ taken from unit closed interval that comply with pursuance conditions:

$$0 \leq m_{kj}^n + q_{kj}^n + p_{kj}^n \leq 1,$$

$$0 \leq \alpha_{kj}^n + \nu_{kj}^n + \delta_{kj}^n \leq 1.$$

Definition 2.4. Let E be subset of parametric set Y over the universal set U with set of grades $X = \{0, 1, 2, \ldots, N-1\}$ and $N \in \{2, 3, \ldots\}$. A triplet $(\bar{f}_{\bar{g}}, \mathrm{E}, N)$ along with mapping $\bar{f}_{\bar{g}} : \mathrm{E} \to 2^{U \times X} \times \mathcal{P}(CTSFNs)$ is called *complex T-spherical fuzzy N-soft set* ($CTSFNS_fS$) on universe U iff $\bar{f} : \mathrm{E} \to 2^{U \times X}$ be an N-soft set and $\bar{g} : \mathrm{E} \to \mathcal{P}(CTSFNs)$ be a mapping assigning CTSFNs according to the rank x_j^k of each element u_k. The $CTSFNS_fS$ can be denoted as:

$$\begin{aligned}\bar{f}_{\bar{g}}(v_j) &= \{\langle(\bar{f}(v_j), \bar{g}(v_j))\rangle : e_j \in \mathrm{E}\},\\ &= \{\langle((u_k, x_j^k), (m_j(u_k)e^{i2\pi\alpha_j(u_k)}, q_j(u_k)e^{i2\pi\nu_j(u_k)}, p_j(u_k)e^{i2\pi\delta_j(u_k)}))\rangle\},\\ &= \{\langle((u_k, x_j^k), (m_{kj}e^{i2\pi\alpha_{kj}}, q_{kj}e^{i2\pi\nu_{kj}}, p_{kj}e^{i2\pi\delta_{kj}}))\rangle\},\end{aligned}$$

where, $m_{kj}e^{i2\pi\alpha_{kj}}$, $q_{kj}e^{i2\pi\nu_{kj}}$ and $p_{kj}e^{i2\pi\delta_{kj}}$ are complex-valued functions of truthness, neutrality and falsity, respectively, with the amplitude terms $m_{\bar{g}}$, $q_{\bar{g}}$, $p_{\bar{g}}$ and the periodic terms $\alpha_{\bar{g}}$, $\nu_{\bar{g}}$, $\delta_{\bar{g}}$ taken from unit closed interval that comply with pursuance conditions:

$$0 \leq m_{kj}^n + q_{kj}^n + p_{kj}^n \leq 1,$$

$$0 \leq \alpha_{kj}^n + \nu_{kj}^n + \delta_{kj}^n \leq 1.$$

Remark 2.1. *We see that:*

1. *When $n = 2$, $CTSFNS_fS$ becomes complex spherical fuzzy N-soft set.*
2. *When $n = 1$, $CTSFNS_fS$ becomes complex picture fuzzy N-set.*
3. *When $n = 2$ and $q_{kj} = \nu_{kj} = 0$, $CTSFNS_fS$ becomes complex Pythagorean fuzzy N-soft set.*
4. *When $n = 1$ and $q_{kj} = \nu_{kj} = 0$, $CTSFNS_fS$ becomes complex intuitionistic fuzzy N-soft set.*
5. *When $N = 2$ and $\alpha_{kj} = \nu_{kj} = \delta_{kj} = 0$, $CSFNS_fS$ becomes T-spherical fuzzy soft set.*

Example 2.5. In reference with the Definition 2.4, when there is huge vagueness and uncertainty in the data and existing environments are ineffectual to confront the events then the decision-makers depict information in two dimensional $CTSFNS_fS$. For an instance, let $U = \{U_1 =$ B.S.C (Bachelor of Science), $U_2 =$ M.B.B.S (Bachelor of Medicine and Bachelor of Surgery), $U_3 =$ Doctor of Pharmacy, $U_4 =$ Biotechnology$\}$ be the set of courses which have to be chosen by the students after Fsc (Pre-medical) in Pakistan and $\mathrm{E} = \{\mathrm{E}_1 =$ Future Jobs, $\mathrm{E}_2 =$ Academic Merit, $\mathrm{E}_3 =$ Fee and Expenditures, $\mathrm{E}_4 =$ Satisfaction stability$\}$ be the most prominent subset of feature's set Y from the last 2 years. The consultants rank the choices in compliance with these conflicting features and the initial survey, summarized in the form of a 6-soft set given is in Table 1, where

0 means 'very Bad',
1 means 'Bad',

2 means 'Average',
3 means 'Good',
4 means 'Great',
5 means 'Excellent'.

Table 1. 6-soft set evaluated by experts

E/U	U_1	U_2	U_3	U_4
E$_1$	1	5	3	2
E$_2$	1	5	3	3
E$_3$	0	5	4	2
E$_4$	1	5	3	2

Table 1 can be taken as natural convention of 6- soft set model. The ratings of courses follow the criteria given in Table 2.

Table 2. Grading criteria

$x_j^k/\bar{g}$	True membership		Neutral membership		False membership	
Grades	m_{jk}	$2\pi\alpha_{jk}$	q_{jk}	$2\pi\nu_{jk}$	p_{jk}	$2\pi\delta_{jk}$
$x_j^k=0$	$[0, 0.15)$	$[0, 0.30\pi)$	$[0, 0.0170)$	$[0, 0.0340\pi)$	$(0.90, 1.00]$	$[1.80\pi, 2.00\pi]$
$x_j^k=1$	$[0.15, 0.30)$	$[0.30\pi, 0.60\pi)$	$[0.0170, 0.0819)$	$[0.0340\pi, 0.1638\pi)$	$(0.75, 0.90]$	$[1.50\pi, 1.80\pi)$
$x_j^k=2$	$[0.30, 0.50)$	$[0.60\pi, 1.00\pi)$	$[0, 0.0170)$	$[0, 0.0340\pi)$	$(0.50, 0.75]$	$[1.00\pi, 1.50\pi)$
$x_j^k=3$	$[0.50, 0.75)$	$[1.00\pi, 1.50\pi)$	$[0.0170, 0.0819)$	$[0.0340\pi, 0.1638\pi)$	$(0.30, 0.50]$	$[0.60\pi, 1.00\pi)$
$x_j^k=4$	$[0.75, 0.90)$	$[1.50\pi, 1.80\pi)$	$[0, 0.0170)$	$[0, 0.0340\pi)$	$(0.15, 0.30]$	$[0.30\pi, 0.60\pi)$
$x_j^k=5$	$[0.90, 1.00]$	$[1.80\pi, 2.00\pi]$	$[0.0170, 0.0819)$	$[0.0340\pi, 0.1638\pi)$	$[0, 0.15]$	$[0, 0.30\pi)$

Using the specified information in Table 2, the $CTSF6S_fS$, shown in Table 3, is defined as:

$$\begin{aligned}\bar{f}_{\bar{g}}(\mathrm{E}_1) = \{&((U_1, 1), (0.160e^{i0.340\pi}, 0.021e^{i0.044\pi}, 0.920e^{i1.860\pi})),\\ &((U_2, 5), (0.900e^{i1.820\pi}, 0.020e^{i0.060\pi}, 0.010e^{i0.040\pi})),\\ &((U_3, 3), (0.510e^{i1.040\pi}, 0.028e^{i0.058\pi}, 0.450e^{i0.920\pi})),\\ &((U_4, 2), (0.350e^{i0.680\pi}, 0.014e^{i0.030\pi}, 0.780e^{i1.580\pi}))\},\end{aligned}$$

$$\begin{aligned}\bar{f}_{\bar{g}}(\mathrm{E}_2) = \{&((U_1, 1), (0.180e^{i0.380\pi}, 0.023e^{i0.048\pi}, 0.940e^{i1.90\pi})),\\ &((U_2, 5), (0.910e^{i1.840\pi}, 0.018e^{i0.038\pi}, 0.010e^{i0.040\pi})),\\ &((U_3, 3), (0.520e^{i1.060\pi}, 0.029e^{i0.060\pi}, 0.460e^{i0.940\pi})),\\ &((U_4, 3), (0.530e^{i1.080\pi}, 0.031e^{i0.064\pi}, 0.470e^{i0.960\pi}))\},\end{aligned}$$

$$\begin{aligned}\bar{f}_{\bar{g}}(\mathrm{E}_3) = \{&((U_1, 0), (0.070e^{i0.120\pi}, 0.015e^{i0.028\pi}, 0.985e^{i1.972\pi})),\\ &((U_2, 5), (0.920e^{i1.860\pi}, 0.019e^{i0.040\pi}, 0.040e^{i0.060\pi})),\\ &((U_3, 4), (0.760e^{i1.540\pi}, 0.010e^{i0.022\pi}, 0.200e^{i0.420\pi})),\\ &((U_4, 2), (0.320e^{i0.660\pi}, 0.013e^{i0.028\pi}, 0.770e^{i1.560\pi}))\},\end{aligned}$$

$$\begin{aligned}\bar{f}_{\bar{g}}(\mathrm{E}_4) = \{&((U_1, 1), (0.190e^{i0.400\pi}, 0.025e^{i0.052\pi}, 0.950e^{1.92i\pi})),\\ &((U_2, 5), (0.930e^{i1.880\pi}, 0.020e^{i0.042\pi}, 0.030e^{0.080i\pi})),\\ &((U_3, 3), (0.540e^{i1.100\pi}, 0.032e^{i0.066\pi}, 0.48e^{i0.980\pi})),\\ &((U_4, 2), (0.340e^{i0.700\pi}, 0.015e^{i0.032\pi}, 0.790e^{i1.600\pi}))\}.\end{aligned}$$

This Example proves the importance and efficiencies of the proposed $CTSFNS_fS$ that has ability to handle such type of modern surveys.

Table 3. The $CTSF6S_fS$ $(\bar{f}_{\bar{g}}, \mathrm{E}, 6)$

$(\bar{f}_{\bar{g}}, \mathrm{E}, 6)$	E_1	E_2
U_1	$(1, (0.160e^{i0.340\pi}, 0.021e^{i0.044\pi}, 0.920e^{i1.860\pi}))$	$(1, (0.180e^{i0.380\pi}, 0.023e^{i0.048\pi}, 0.940e^{i1.90\pi}))$
U_2	$(5, (0.900e^{i1.820\pi}, 0.020e^{i0.060\pi}, 0.010e^{i0.040\pi}))$	$(5, (0.910e^{i1.840\pi}, 0.018e^{i0.038\pi}, 0.010e^{i0.040\pi}))$
U_3	$(3, (0.510e^{i1.040\pi}, 0.028e^{i0.058\pi}, 0.450e^{i0.920\pi}))$	$(3, (0.520e^{i1.060\pi}, 0.029e^{i0.060\pi}, 0.460e^{i0.940\pi}))$
U_4	$(2, (0.350e^{i0.680\pi}, 0.014e^{i0.030\pi}, 0.780e^{i1.580\pi}))$	$(3, (0.530e^{i1.080\pi}, 0.031e^{i0.064\pi}, 0.470e^{i0.960\pi}))$
	E_3	E_4
U_1	$(0, (0.070e^{i0.120\pi}, 0.015e^{i0.028\pi}, 0.985e^{i1.972\pi}))$	$(1, (0.190e^{i0.400\pi}, 0.025e^{i0.052\pi}, 0.950e^{1.92i\pi}))$
U_2	$(5, (0.920e^{i1.860\pi}, 0.019e^{i0.040\pi}, 0.040e^{i0.060\pi}))$	$(5, (0.930e^{i1.880\pi}, 0.020e^{i0.042\pi}, 0.030e^{0.080i\pi}))$
U_3	$(4, (0.760e^{i1.540\pi}, 0.010e^{i0.022\pi}, 0.200e^{i0.420\pi}))$	$(3, (0.540e^{i1.100\pi}, 0.032e^{i0.066\pi}, 0.48e^{i0.980\pi}))$
U_4	$(2, (0.320e^{i0.660\pi}, 0.013e^{i0.028\pi}, 0.770e^{i1.560\pi}))$	$(2, (0.340e^{i0.700\pi}, 0.015e^{i0.032\pi}, 0.790e^{i1.600\pi}))$

Definition 2.6. A $CTSFNS_fS$ $(\bar{f}_{\bar{g}}, \mathrm{E}, N)$ on universe U is known as *efficient* if $\bar{f}_{\bar{g}}(e_j) = \langle (u_k, N-1), 1, 0, 0 \rangle$ for some $e_j \in \mathrm{E}, u_k \in U$ where $(\bar{f}, \mathrm{E}, N)$ is an NS_fS.

Example 2.7. Consider Example 2.5 and let $(\bar{f}_{\bar{g}}, \mathrm{E}, 6)$ be $CTSF6S_fS$. The set shown in Table 3 is not efficient. However, the $CTSF5S_fS$ $(\tilde{f}_{\bar{g}}, \tilde{\mathcal{V}}, 5)$ in Table 6 is efficient.

Definition 2.8. Let U be a universal set and $(\bar{f}_{\bar{g}}, \mathrm{E}, N_1)$ and $(\dot{f}_{\dot{g}}, \dot{\mathrm{E}}, N_2)$ be two $CTSFN_1S_fS$ and $CTSFN_2S_fS$, respectively, on U. Both sets are known to be *equal* if and only if $\bar{f} = \dot{f}$, $\bar{g} = \dot{g}$, $\mathrm{E} = \dot{\mathrm{E}}$ and $N_1 = N_2$.

Definition 2.9. Let U be a universal set and $(\bar{f}_{\bar{g}}, \mathrm{E}, N)$ be a $CTSFNS_fS$ on U. The *weak complement* of $CTSFNS_fS$ is defined as the weak complement of the N-soft set $(\bar{f}, \mathrm{E}, N)$, that is, any N-soft set such that $\bar{f}^c(e_j) \cap \bar{f}(e_j) = \emptyset$ for all $e_j \in \mathrm{E}$. The weak complement of $CTSFNS_fS$ of $(\bar{f}_{\bar{g}}, \mathrm{E}, N)$ is represented as $(\bar{f}^c_{\bar{g}}, \mathrm{E}, N)$.

Example 2.10. Consider Example 2.5 and let $(\bar{f}_{\bar{g}}, \mathrm{E}, 6)$ be $CTSF6S_fS$. The weak complement $(\bar{f}^c_{\bar{g}}, \mathrm{E}, N)$ of $(\bar{f}_{\bar{g}}, \mathrm{E}, 6)$ is calculated in Table 4.

Table 4. Weak complement of $(\bar{f}_{\bar{g}}, \mathrm{E}, 6)$

$(\bar{f}^c_{\bar{g}}, \mathrm{E}, 6)$	E_1	E_2
U_1	$(3, (0.160e^{i0.340\pi}, 0.021e^{i0.044\pi}, 0.920e^{i1.860\pi}))$	$(0, (0.180e^{i0.380\pi}, 0.023e^{i0.048\pi}, 0.940e^{i1.90\pi}))$
U_2	$(2, (0.900e^{i1.820\pi}, 0.020e^{i0.060\pi}, 0.010e^{i0.040\pi}))$	$(3, (0.910e^{i1.840\pi}, 0.018e^{i0.038\pi}, 0.010e^{i0.040\pi}))$
U_3	$(1, (0.510e^{i1.040\pi}, 0.028e^{i0.058\pi}, 0.450e^{i0.920\pi}))$	$(4, (0.520e^{i1.060\pi}, 0.029e^{i0.060\pi}, 0.460e^{i0.940\pi}))$
U_4	$(5, (0.350e^{i0.680\pi}, 0.014e^{i0.030\pi}, 0.780e^{i1.580\pi}))$	$(5, (0.530e^{i1.080\pi}, 0.031e^{i0.064\pi}, 0.470e^{i0.960\pi}))$
	E_3	E_4
U_1	$(4, (0.070e^{i0.120\pi}, 0.015e^{i0.028\pi}, 0.985e^{i1.972\pi}))$	$(2, (0.190e^{i0.400\pi}, 0.025e^{i0.052\pi}, 0.950e^{1.92i\pi}))$
U_2	$(2, (0.920e^{i1.860\pi}, 0.019e^{i0.040\pi}, 0.040e^{i0.060\pi}))$	$(3, (0.930e^{i1.880\pi}, 0.020e^{i0.042\pi}, 0.030e^{0.080i\pi}))$
U_3	$(1, (0.760e^{i1.540\pi}, 0.010e^{i0.022\pi}, 0.200e^{i0.420\pi}))$	$(0, (0.540e^{i1.100\pi}, 0.032e^{i0.066\pi}, 0.48e^{i0.980\pi}))$
U_4	$(4, (0.320e^{i0.660\pi}, 0.013e^{i0.028\pi}, 0.770e^{i1.560\pi}))$	$(1, (0.340e^{i0.700\pi}, 0.015e^{i0.032\pi}, 0.790e^{i1.600\pi}))$

Definition 2.11. Let U be universal set and $(\bar{f}_{\bar{g}}, \mathrm{E}, N)$ be a $CTSFNS_fS$ on U. The *Strong complement* of $CTSFNS_fS$, denoted by $(\bar{f}'_{\bar{g}}, \mathrm{E}, N)$, is defined as:

$$\bar{f}'(e_j) = \begin{cases} x^k_j - 1, & \text{if } x^k_j = (N-1) - x^k_j, \\ (N-1) - x^k_j, & \text{otherwise}, \end{cases}$$

for all $e_j \in \mathrm{E}$ and $u_k \in U$, satisfying the condition $(\bar{f}_{\bar{g}}, \mathrm{E}, N) \cap (\bar{f}'_{\bar{g}}, \mathrm{E}, N) = \emptyset$.

Example 2.12. Consider Example 2.5 with $(\bar{f}_{\bar{g}}, \mathrm{E}, 6)$ be $CTSF6S_fS$ and Table 6 containing $CTSF5S_fS$ $(\tilde{f}_{\tilde{g}}, \tilde{\mathrm{V}}, 5)$, the strong complement of $(\bar{f}_{\bar{g}}, \mathrm{E}, 6)$ and $(\tilde{f}_{\tilde{g}}, \tilde{\mathrm{V}}, 5)$ are calculated in Table 5 and 7, respectively.

Table 5. Strong complement of $(\bar{f}_{\bar{g}}, \mathrm{E}, 6)$

$(\bar{f}'_{\bar{g}}, \mathrm{E}, 6)$	E_1	E_2
U_1	$(4, (0.160e^{i0.340\pi}, 0.021e^{i0.044\pi}, 0.920e^{i1.860\pi}))$	$(4, (0.180e^{i0.380\pi}, 0.023e^{i0.048\pi}, 0.940e^{i1.90\pi}))$
U_2	$(0, (0.900e^{i1.820\pi}, 0.020e^{i0.060\pi}, 0.010e^{i0.040\pi}))$	$(0, (0.910e^{i1.840\pi}, 0.018e^{i0.038\pi}, 0.010e^{i0.040\pi}))$
U_3	$(2, (0.510e^{i1.040\pi}, 0.028e^{i0.058\pi}, 0.450e^{i0.920\pi}))$	$(2, (0.520e^{i1.060\pi}, 0.029e^{i0.060\pi}, 0.460e^{i0.940\pi}))$
U_4	$(3, (0.350e^{i0.680\pi}, 0.014e^{i0.030\pi}, 0.780e^{i1.580\pi}))$	$(2, (0.530e^{i1.080\pi}, 0.031e^{i0.064\pi}, 0.470e^{i0.960\pi}))$
	E_3	E_4
U_1	$(5, (0.070e^{i0.120\pi}, 0.015e^{i0.028\pi}, 0.985e^{i1.972\pi}))$	$(1, (0.190e^{i0.400\pi}, 0.025e^{i0.052\pi}, 0.950e^{1.92i\pi}))$
U_2	$(0, (0.920e^{i1.860\pi}, 0.019e^{i0.040\pi}, 0.040e^{i0.060\pi}))$	$(0, (0.930e^{i1.880\pi}, 0.020e^{i0.042\pi}, 0.030e^{0.080i\pi}))$
U_3	$(1, (0.760e^{i1.540\pi}, 0.010e^{i0.022\pi}, 0.200e^{i0.420\pi}))$	$(2, (0.540e^{i1.100\pi}, 0.032e^{i0.066\pi}, 0.48e^{i0.980\pi}))$
U_4	$(3, (0.320e^{i0.660\pi}, 0.013e^{i0.028\pi}, 0.770e^{i1.560\pi}))$	$(3, (0.340e^{i0.700\pi}, 0.015e^{i0.032\pi}, 0.790e^{i1.600\pi}))$

Table 6. The $CSF5S_fS(\tilde{f_{\bar{g}}}, \tilde{\mathbb{V}}, 5)$

	E3	E4	E5
U_1	$(0, (0.100e^{i0.240\pi}, 0.012e^{i0.022\pi}, 0.985e^{i1.964\pi}))$	$(2, (0.400e^{i0.820\pi}, 0.017e^{i0.0356\pi}, 0.600e^{i1.220\pi}))$	$(0, (0.050e^{i0.140\pi}, 0.870e^{i1.720\pi}, 0.880e^{i1.800\pi}))$
U_2	$(0, (0.020e^{i0.060\pi}, 0.012e^{i0.026\pi}, 0.980e^{i1.962\pi}))$	$(1, (0.200e^{i0.360\pi}, 0.027e^{i0.050\pi}, 0.910e^{i1.824\pi}))$	$(1, (0.160e^{i0.340\pi}, 0.100e^{i0.204\pi}, 0.890e^{i1.784\pi}))$
U_3	$(1, (0.300e^{i0.560\pi}, 0.019e^{i0.042\pi}, 0.885e^{1.720i\pi}))$	$(3, (0.700e^{i1.420\pi}, 0.019e^{i0.040\pi}, 0.300e^{i0.560\pi}))$	$(4, (1, 0, 0))$
U_4	$(3, (0.650e^{i1.320\pi}, 0.018e^{i0.038\pi}, 0.280e^{i0.580\pi}))$	$(1, (0.200e^{i0.360\pi}, 0.027e^{i0.050\pi}, 0.910e^{i1.824\pi}))$	$(2, (0.500e^{i1.100\pi}, 0.100e^{i0.180\pi}, 0.590e^{1.280i\pi}))$

Table 7. Strong complement of $(\tilde{f_{\bar{g}}}, \tilde{\mathbb{V}}, 5)$

$(\bar{f}'_{\bar{g}}, \tilde{\mathbb{V}}, 5)$	E3	E4	E5
U_1	$(4, (0.100e^{i0.240\pi}, 0.012e^{i0.022\pi}, 0.985e^{i1.964\pi}))$	$(1, (0.400e^{i0.820\pi}, 0.017e^{i0.0356\pi}, 0.600e^{i1.220\pi}))$	$(4, (0.050e^{i0.140\pi}, 0.870e^{i1.720\pi}, 0.880e^{i1.800\pi}))$
U_2	$(4, (0.020e^{i0.060\pi}, 0.012e^{i0.026\pi}, 0.980e^{i1.962\pi}))$	$(3, (0.200e^{i0.360\pi}, 0.027e^{i0.050\pi}, 0.910e^{i1.824\pi}))$	$(3, (0.160e^{i0.340\pi}, 0.100e^{i0.204\pi}, 0.890e^{i1.784\pi}))$
U_3	$(4, (0.300e^{i0.560\pi}, 0.019e^{i0.042\pi}, 0.885e^{1.720i\pi}))$	$(1, (0.700e^{i1.420\pi}, 0.019e^{i0.040\pi}, 0.300e^{i0.560\pi}))$	$(0, (1, 0, 0))$
U_4	$(1, (0.650e^{i1.320\pi}, 0.018e^{i0.038\pi}, 0.280e^{i0.580\pi}))$	$(3, (0.200e^{i0.360\pi}, 0.027e^{i0.050\pi}, 0.910e^{i1.824\pi}))$	$(1, (0.500e^{i1.100\pi}, 0.100e^{i0.180\pi}, 0.590e^{1.280i\pi}))$

Theorem 2.13. *A strong complement of $CTSFNS_fS$ is also known as weak complement but weak complement may not be strong complement.*

Proof. The prove is straight forward from the definitions of strong complement and weak complement. Moreover, Table 4 representing the weak complement of $CTSF6S_fS$, does not satisfies the definition of strong complement therefore, weak complement may not be strong complement. □

Theorem 2.14. *Let $((\bar{f}^c_{\bar{g}}), \mathrm{E}, N)$ and $((\bar{f}'_{\bar{g}}), \mathrm{E}, N)$ be weak and strong complement of $CTSFNS_fS$ $(\bar{f}_{\bar{g}}, \mathrm{E}, N)$, respectively, then;*

1 $((\bar{f}^c)^c_{\bar{g}}, \mathrm{E}, N) \neq (\bar{f}_{\bar{g}}, \mathrm{E}, N)$,

2 $((\bar{f}')'_{\bar{g}}, \mathrm{E}, N) \begin{Bmatrix} = (\bar{f}_{\bar{g}}, \mathrm{E}, N) \text{ if } N \text{ is even} \\ \neq (\bar{f}_{\bar{g}}, \mathrm{E}, N) \text{ if } N \text{ is odd.} \end{Bmatrix}.$

Proof. The prove follows from the Definitions 2.9 and 2.11. □

Definition 2.15. Let U be universal set and let $(\bar{f}_{\bar{g}}, \mathrm{E}, N)$ be a $CTSFNS_fS$ on U. The *complex T-spherical fuzzy complement* $(\bar{f}_{\bar{g}^c}, \mathrm{E}, N)$ of $CTSFNS_fS$ is defined as

$$\bar{f}_{\bar{g}^c}(e_j) = \langle (u_k, x^k_j, (p_{kj}e^{i2\pi\delta_{kj}}, (q_{kj})e^{i2\pi(\nu_{kj})}, m_{kj}e^{i2\pi\alpha_{kj}}))\rangle.$$

Example 2.16. Consider Example 2.5 and let $(\bar{f}_{\bar{g}}, \mathrm{E}, 6)$ be $CTSF6S_fS$, the complex T-spherical fuzzy complement $(\bar{f}_{\bar{g}^c}, \mathrm{E}, N)$ is calculated in Table 8.

Table 8. The complex T-spherical fuzzy complement $(\bar{f}_{\bar{f}^c}, \mathrm{E}, N)$ of the $CTSF6S_fS$

$(\bar{f}_{\bar{g}^c}$, E, 6)	E$_1$	E$_2$
U_1	$(1, (0.920e^{i1.860\pi}, 0.021e^{i0.044\pi}, 0.160e^{i0.340\pi}))$	$(1, (0.940e^{i1.90\pi}, 0.023e^{i0.048\pi}, 0.180e^{i0.380\pi}))$
U_2	$(5, (0.010e^{i0.040\pi}, 0.020e^{i0.060\pi}, 0.900e^{i1.820\pi}))$	$(5, (0.010e^{i0.040\pi}, 0.018e^{i0.038\pi}, 0.910e^{i1.840\pi}))$
U_3	$(3, (0.450e^{i0.920\pi}, 0.028e^{i0.058\pi}, 0.510e^{i1.040\pi}))$	$(3, (0.460e^{i0.940\pi}, 0.029e^{i0.060\pi}, 0.520e^{i1.060\pi}))$
U_4	$(2, (0.780e^{i1.580\pi}, 0.014e^{i0.030\pi}, 0.350e^{i0.680\pi}))$	$(3, (0.470e^{i0.960\pi}, 0.031e^{i0.064\pi}, 0.530e^{i1.080\pi}))$
	E$_3$	E$_4$
U_1	$(0, (0.985e^{i1.972\pi}, 0.015e^{i0.028\pi}, 0.070e^{i0.120\pi}))$	$(1, (0.950e^{1.92i\pi}, 0.025e^{i0.052\pi}, 0.190e^{i0.400\pi}))$
U_2	$(5, (0.040e^{i0.060\pi}, 0.019e^{i0.040\pi}, 0.920e^{i1.860\pi}))$	$(5, (0.030e^{0.080i\pi}, 0.020e^{i0.042\pi}, 0.930e^{i1.880\pi}))$
U_3	$(4, (0.200e^{i0.420\pi}, 0.010e^{i0.022\pi}, 0.760e^{i1.540\pi}))$	$(3, (0.48e^{i0.980\pi}, 0.032e^{i0.066\pi}, 0.540e^{i1.100\pi}))$
U_4	$(2, (0.770e^{i1.560\pi}, 0.013e^{i0.028\pi}, 0.320e^{i0.660\pi}))$	$(2, (0.790e^{i1.600\pi}, 0.015e^{i0.032\pi}, 0.340e^{i0.700\pi}))$

Definition 2.17. Let U be universal set and $(\bar{f}_{\bar{g}}, \mathrm{E}, N)$ be a $CTSFNS_fS$ on U. The *weak complex T-spherical fuzzy complement* $((\bar{f}_{\bar{g}})^c, \mathrm{E}, N)$ is defined as the weak complement $(\bar{f}^c_{\bar{g}}, \mathrm{E}, N)$ and complex T-spherical fuzzy complement $(\bar{f}_{\bar{g}^c}, \mathrm{E}, N)$ of $(\bar{f}_{\bar{g}}, \mathrm{E}, N)$, respectively.

Example 2.18. Consider Example 2.5 and let $(\bar{f}_{\bar{g}}, \mathrm{E}, 6)$ be $CTSF6S_fS$, the weak complex T-spherical fuzzy complement $((\bar{f}_{\bar{g}})^c, \mathrm{E}, N)$ is calculated in Table 9.

Table 9. Weak complex T-spherical fuzzy complement $((\bar{f}_{\bar{g}})^c, \mathrm{E}, 6)$ of the $CTSF6S_fS$

$((\bar{f}_{\bar{g}})^c$, E, 6)	E$_1$	E$_2$
U_1	$(3, (0.920e^{i1.860\pi}, 0.021e^{i0.044\pi}, 0.160e^{i0.340\pi}))$	$(0, (0.940e^{i1.90\pi}, 0.023e^{i0.048\pi}, 0.180e^{i0.380\pi}))$
U_2	$(2, (0.010e^{i0.040\pi}, 0.020e^{i0.060\pi}, 0.900e^{i1.820\pi}))$	$(3, (0.010e^{i0.040\pi}, 0.018e^{i0.038\pi}, 0.910e^{i1.840\pi}))$
U_3	$(1, (0.450e^{i0.920\pi}, 0.028e^{i0.058\pi}, 0.510e^{i1.040\pi}))$	$(4, (0.460e^{i0.940\pi}, 0.029e^{i0.060\pi}, 0.520e^{i1.060\pi}))$
U_4	$(5, (0.780e^{i1.580\pi}, 0.014e^{i0.030\pi}, 0.350e^{i0.680\pi}))$	$(5, (0.470e^{i0.960\pi}, 0.031e^{i0.064\pi}, 0.530e^{i1.080\pi}))$
	E$_3$	E$_4$
U_1	$(4, (0.985e^{i1.972\pi}, 0.015e^{i0.028\pi}, 0.070e^{i0.120\pi}))$	$(2, (0.950e^{1.92i\pi}, 0.025e^{i0.052\pi}, 0.190e^{i0.400\pi}))$
U_2	$(2, (0.040e^{i0.060\pi}, 0.019e^{i0.040\pi}, 0.920e^{i1.860\pi}))$	$(3, (0.030e^{0.080i\pi}, 0.020e^{i0.042\pi}, 0.930e^{i1.880\pi}))$
U_3	$(1, (0.200e^{i0.420\pi}, 0.010e^{i0.022\pi}, 0.760e^{i1.540\pi}))$	$(0, (0.48e^{i0.980\pi}, 0.032e^{i0.066\pi}, 0.540e^{i1.100\pi}))$
U_4	$(4, (0.770e^{i1.560\pi}, 0.013e^{i0.028\pi}, 0.320e^{i0.660\pi}))$	$(1, (0.790e^{i1.600\pi}, 0.015e^{i0.032\pi}, 0.340e^{i0.700\pi}))$

Definition 2.19. Let $(\bar{f}_{\bar{g}}, \mathrm{E}, N)$ be a $CTSFNS_fS$ on U, then the *strong complex T-spherical fuzzy complement* $((\bar{f}_{\bar{g}})', \mathrm{E}, N)$ is defined as a strong complement $(\bar{f}'_{\bar{g}}, \mathrm{E}, N)$ and a complex T-spherical fuzzy complement $(\bar{f}_{\bar{g}^c}, \mathrm{E}, N)$ of $(\bar{f}_{\bar{g}}, \mathrm{E}, N)$, given by:

$$\bar{f}'_{\bar{g}^c}(u_k) = \begin{cases} (x_j^k - 1, (p_{kj}e^{i2\pi\delta_{kj}}, (q_{kj})e^{i2\pi(\nu_{kj})}, m_{kj}e^{i2\pi\alpha_{kj}})) & \text{if } x_j^k = (N-1) - x_j^k, \\ ((N-1) - x_j^k, (p_{kj}e^{i2\pi\delta_{kj}}, (q_{kj})e^{i2\pi(\nu_{kj})}, m_{kj}e^{i2\pi\alpha_{kj}})) & \text{otherwise,} \end{cases}$$

for all $e_j \in \mathrm{E}$ and $u_k \in U$.

Example 2.20. The strong T-spherical fuzzy complement $(\bar{f}'_{\bar{g}}, \mathrm{E}, N)$, of $(\bar{f}_{\bar{g}}, \mathrm{E}, 6)$, given in Table 3, is computed in Table 10.

Table 10. Strong T-spherical fuzzy complement of $(\bar{f}_{\bar{g}}, \mathrm{E}, 6)$

$(\bar{f}'_{\bar{g}}, \mathrm{E}, 6)$	E_1	E_2
U_1	$(4, (0.920e^{i1.860\pi}, 0.021e^{i0.044\pi}, 0.160e^{i0.340\pi}))$	$(4, (0.940e^{i1.90\pi}, 0.023e^{i0.048\pi}, 0.180e^{i0.380\pi}))$
U_2	$(0, (0.010e^{i0.040\pi}, 0.020e^{i0.060\pi}, 0.900e^{i1.820\pi}))$	$(0, (0.010e^{i0.040\pi}, 0.018e^{i0.038\pi}, 0.910e^{i1.840\pi}))$
U_3	$(2, (0.450e^{i0.920\pi}, 0.028e^{i0.058\pi}, 0.510e^{i1.040\pi}))$	$(2, (0.460e^{i0.940\pi}, 0.029e^{i0.060\pi}, 0.520e^{i1.060\pi}))$
U_4	$(3, (0.780e^{i1.580\pi}, 0.014e^{i0.030\pi}, 0.350e^{i0.680\pi}))$	$(2, (0.470e^{i0.960\pi}, 0.031e^{i0.064\pi}, 0.530e^{i1.080\pi}))$
	E_3	E_4
U_1	$(5, (0.985e^{i1.972\pi}, 0.015e^{i0.028\pi}, 0.070e^{i0.120\pi}))$	$(4, (0.950e^{1.92i\pi}, 0.025e^{i0.052\pi}, 0.190e^{i0.400\pi}))$
U_2	$(0, (0.040e^{i0.060\pi}, 0.019e^{i0.040\pi}, 0.920e^{i1.860\pi}))$	$(0, (0.030e^{0.080i\pi}, 0.020e^{i0.042\pi}, 0.930e^{i1.880\pi}))$
U_3	$(1, (0.200e^{i0.420\pi}, 0.010e^{i0.022\pi}, 0.760e^{i1.540\pi}))$	$(2, (0.48e^{i0.980\pi}, 0.032e^{i0.066\pi}, 0.540e^{i1.100\pi}))$
U_4	$(3, (0.770e^{i1.560\pi}, 0.013e^{i0.028\pi}, 0.320e^{i0.660\pi}))$	$(3, (0.790e^{i1.600\pi}, 0.015e^{i0.032\pi}, 0.340e^{i0.700\pi}))$

Theorem 2.21. *Let $(\bar{f}^c_{\bar{g}}, \mathrm{E}, N)$ and $(\bar{f}'_{\bar{g}}, \mathrm{E}, N)$ be weak and strong complex T-spherical fuzzy complement of $CTSFNS_fS$ $(\bar{f}_{\bar{g}}, \mathrm{E}, N)$, respectively, then*

1 $(((\bar{f}_{\bar{g}})^c)^c, \mathrm{E}, N) \neq (\bar{f}_{\bar{g}}, \mathrm{E}, N)$,

2 $([(\bar{f}_{\bar{g}})']', \mathrm{E}, N) \begin{Bmatrix} = (\bar{f}_{\bar{g}}, \mathrm{E}, N) \text{ if } N \text{ is even} \\ \neq (\bar{f}_{\bar{g}}, \mathrm{E}, N) \text{ if } N \text{ is odd.} \end{Bmatrix}$.

Proof. The prove follows from the Definitions 2.17 and 2.19. □

Definition 2.22. Let U be a non-empty set and $(\bar{f}_{\bar{g}}, \mathrm{E}, N_1)$ and let $(\tilde{f}_{\bar{g}}, \tilde{\mathrm{V}}, N_2)$ be $CTSFN_1S_fS$ and $CTSFN_2S_fS$ on U, respectively, their *restricted intersection* is defined as $(\tilde{L}_{\tilde{M}}, \tilde{G}, \tilde{O}) = (\bar{f}_{\bar{g}}, \mathrm{E}, N_1) \hat{\cap} (\tilde{f}_{\bar{g}}, \tilde{\mathrm{V}}, N_2)$, with $\tilde{L}_{\tilde{M}} = \bar{f}_{\bar{g}} \hat{\cap} \tilde{f}_{\bar{g}}$, $\tilde{G} = \mathrm{E} \cap \tilde{\mathrm{V}}$, $\tilde{O} = \min(N_1, N_2)$, i.e., $\forall e_j \in \tilde{G}$, $u_k \in U$ we have

$$\begin{aligned}\tilde{L}_{\tilde{M}}(e_j) &= \Big\langle (x^k_j, (m_{jk}e^{i2\pi\alpha_{jk}}, q_{jk}e^{i2\pi\nu_{jk}}, p_{jk}e^{i2\pi\delta_{jk}})) \Big\rangle, \\ &= \Big\langle (\min(x^{1k}_j, x^{2k}_j), \min(m^1_{kj}, m^2_{kj})e^{i2\pi\min(m^1_{kj}, m^2_{kj})}, \max(q^1_{kj}, q^2_{kj})e^{i2\pi\max(\nu^1_{kj}, \nu^2_{kj})}, \\ &\quad \max(p^1_{kj}, p^2_{kj})e^{i2\pi\max(\delta^1_{kj}, \delta^2_{kj})}) \Big\rangle,\end{aligned}$$

where, $(x^{1k}_j, (m^1_{kj}e^{i2\pi\alpha^1_{kj}}, q^1_{kj}e^{i2\pi\nu^1_{kj}}, p^1_{kj}e^{i2\pi\delta^1_{kj}})) \in \bar{f}_{\bar{g}}$ and $(x^{2k}_j, (m^2_{kj}e^{i2\pi\alpha^2_{kj}}, q^2_{kj}e^{i2\pi\nu^2_{kj}}, p^2_{kj}e^{i2\pi\delta^2_{ba}})) \in \tilde{f}_{\bar{g}}$.

Example 2.23. The restricted intersection $(\tilde{L}_{\tilde{M}}, \tilde{G}, \tilde{O})$ of $(\bar{f}_{\bar{g}}, \mathrm{E}, 6)$ and $(\tilde{f}_{\tilde{g}}, \tilde{\mathrm{V}}, 5)$, given by Table 3 and Table 6, respectively, is calculated in Table 11.

Table 11. The restricted intersection $(\tilde{L}_{\tilde{M}}, \tilde{G}, 5)$

$(\tilde{L}_{\tilde{M}}, \tilde{G}, 5)$	E_3	E_4
U_1	$(0, (0.070e^{i0.120\pi}, 0.015e^{i0.028\pi}, 0.985e^{i1.972\pi}))$	$(1, (0.190e^{i0.400\pi}, 0.025e^{i0.052\pi}, 0.950e^{1.92i\pi}))$
U_2	$(0, (0.020e^{i0.060\pi}, 0.012e^{i0.026\pi}, 0.980e^{i1.962\pi}))$	$(1, (0.200e^{i0.360\pi}, 0.027e^{i0.050\pi}, 0.910e^{i1.824\pi}))$
U_3	$(1, (0.300e^{i0.560\pi}, 0.019e^{i0.042\pi}, 0.885e^{1.720i\pi}))$	$(3, (0.540e^{i1.100\pi}, 0.032e^{i0.066\pi}, 0.48e^{i0.980\pi}))$
U_4	$(2, (0.320e^{i0.660\pi}, 0.018e^{i0.038\pi}, 0.770e^{i1.560\pi}))$	$(1, (0.200e^{i0.360\pi}, 0.027e^{i0.050\pi}, 0.910e^{i1.824\pi}))$

Theorem 2.24. *Let $(\bar{f}_{\bar{g}}, \mathrm{E}, N_1)$ and $(\tilde{f}_{\tilde{g}}, \tilde{\mathrm{V}}, N_2)$ be $CTSFN_1S_fS$ and $CTSFN_2S_fS$ on U, respectively, and $(\tilde{L}_{\tilde{M}}, \tilde{G}, \tilde{O}) = (\bar{f}_{\bar{g}}, \mathrm{E}, N_1) \hat{\cap} (\tilde{f}_{\tilde{g}}, \tilde{\mathrm{V}}, N_2)$ be their restricted intersection. Then $(\tilde{L}_{\tilde{M}}, \tilde{G}, \tilde{O})$ is the largest $CTSFNS_fS$ containing both $(\bar{f}_{\bar{g}}, \mathrm{E}, N_1)$ and $(\tilde{f}_{\tilde{g}}, \tilde{\mathrm{V}}, N_2)$.*

Proof. The prove follows from the definition of restricted intersection. □

Definition 2.25. Let $(\bar{f}_{\bar{g}}, \mathrm{E}, N_1)$ and $(\tilde{f}_{\tilde{g}}, \tilde{\mathrm{V}}, N_2)$ be $CTSFN_1S_fS$ and $CTSFN_2S_fS$ on U, respectively, their *extended intersection* is defined as $(\tilde{D}_{\tilde{Q}}, \tilde{T}, \tilde{S}) = (\bar{f}_{\bar{g}}, \mathrm{E}, N_1) \check{\cap} (\tilde{f}_{\tilde{g}}, \tilde{\mathrm{V}}, N_2)$, with $\tilde{D}_{\tilde{Q}} = \bar{f}_{\bar{g}} \check{\cap} \tilde{f}_{\tilde{g}}$, $\tilde{T} = \mathrm{E} \cup \tilde{\mathrm{V}}$, $\tilde{S} = \max(N_1, N_2)$, that is, $\forall e_j \in \tilde{T}$ and $u_k \in U$, we have

$$\tilde{D}_{\tilde{Q}}(e_j) = \begin{cases} (x_j^{1k}, m_{kj}^1 e^{i2\pi\alpha_{kj}^1}, q_{kj}^1 e^{i2\pi\nu_{kj}^1}, p_{kj}^1 e^{i2\pi\delta_{kj}^1}), & \text{if } e_j \in \mathrm{E} - \tilde{\mathrm{V}}, \\ (x_j^{2k}, m_{kj}^2 e^{i2\pi\alpha_{kj}^2}, q_{kj}^2 e^{i2\pi\nu_{kj}^2}, p_{kj}^2 e^{i2\pi\delta_{kj}^2}), & \text{if } e_j \in \tilde{\mathrm{V}} - \mathrm{E}, \\ \Big(\min(x_j^{1k}, x_j^{2k}), \min(m_{kj}^1, m_{kj}^2) e^{i2\pi\min(\alpha_{kj}^1, \alpha_{kj}^2)}, \max & \text{if } e_j \in \tilde{\mathrm{V}} \cap \mathrm{E}, \\ (q_{jk}^1, q_{jk}^2) e^{i2\pi\max(\nu_{kj}^1, \nu_{kj}^2)}, \max(p_{kj}^1, p_{kj}^2) e^{i2\pi\max(\delta_{kj}^1, \delta_{kj}^2)}\Big), & \end{cases}$$

where $(x_j^{1k}, m_{kj}^1 e^{i2\pi\alpha_{kj}^1}, q_{kj}^1 e^{i2\pi\nu_{kj}^1}, p_{kj}^1 e^{i2\pi\delta_{kj}^1}) \in \bar{f}_{\bar{g}}$ and $(x_j^{2k}, m_{kj}^2 e^{i2\pi\alpha_{kj}^2}, q_{kj}^2 e^{i2\pi\nu_{kj}^2}, p_{kj}^2 e^{i2\pi\delta_{kj}^2}) \in \tilde{f}_{\tilde{g}}$.

Example 2.26. The extended intersection $(\tilde{D}_{\tilde{Q}}, \tilde{T}, 6)$ of $(\bar{f}_{\bar{g}}, \mathrm{E}, 6)$ and $(\tilde{f}_{\tilde{g}}, \tilde{\mathrm{V}}, 5)$, given by Table 3 and Table 6, respectively, is calculated in Table 12.

Table 12. The extended intersection $(\tilde{D}_{\tilde{Q}}, \tilde{T}, \tilde{S})$

$(\tilde{D}_{\tilde{Q}}, \tilde{T}, \tilde{S})$	E1	E2	E3	E4
U_1	$(1,(0.160e^{i0.340\pi}, 0.021e^{i0.044\pi}, 0.920e^{i1.860\pi}))$	$(1,(0.180e^{i0.380\pi}, 0.023e^{i0.048\pi}, 0.940e^{i1.90\pi}))$	$(0,(0.070e^{i0.120\pi}, 0.015e^{i0.028\pi}, 0.985e^{i1.972\pi}))$	$(1,(0.190e^{i0.400\pi}, 0.025e^{i0.052\pi}, 0.950e^{1.92i\pi}))$
U_2	$(5,(0.900e^{i1.820\pi}, 0.020e^{i0.060\pi}, 0.010e^{i0.040\pi}))$	$(5,(0.910e^{i1.840\pi}, 0.018e^{i0.038\pi}, 0.010e^{i0.040\pi}))$	$(0,(0.020e^{i0.060\pi}, 0.012e^{i0.026\pi}, 0.980e^{i1.962\pi}))$	$(1,(0.200e^{i0.360\pi}, 0.027e^{i0.050\pi}, 0.910e^{i1.824\pi}))$
U_3	$(3,(0.510e^{i1.040\pi}, 0.028e^{i0.058\pi}, 0.450e^{i0.920\pi}))$	$(3,(0.520e^{i1.060\pi}, 0.029e^{i0.060\pi}, 0.460e^{i0.940\pi}))$	$(1,(0.300e^{i0.560\pi}, 0.019e^{i0.042\pi}, 0.885e^{1.720i\pi}))$	$(3,(0.540e^{i1.100\pi}, 0.032e^{i0.066\pi}, 0.48e^{i0.980\pi}))$
U_4	$(2,(0.350e^{i0.680\pi}, 0.014e^{i0.030\pi}, 0.780e^{i1.580\pi}))$	$(3,(0.530e^{i1.080\pi}, 0.031e^{i0.064\pi}, 0.470e^{i0.960\pi}))$	$(2,(0.320e^{i0.660\pi}, 0.018e^{i0.038\pi}, 0.770e^{i1.560\pi}))$	$(1,(0.200e^{i0.360\pi}, 0.027e^{i0.050\pi}, 0.910e^{i1.824\pi}))$
	E5			
U_1	$(0,(0.050e^{i0.140\pi}, 0.870e^{i1.720\pi}, 0.880e^{i1.800\pi}))$			
U_2	$(1,(0.160e^{i0.340\pi}, 0.100e^{i0.204\pi}, 0.890e^{i1.784\pi}))$			
U_3	$(4,(1,0,0))$			
U_4	$(2,(0.500e^{i1.100\pi}, 0.100e^{i0.180\pi}, 0.590e^{1.280i\pi}))$			

Definition 2.27. Let U be a non-empty set and $(\bar{f}_{\bar{g}}, \mathrm{E}, N_1)$ and $(\tilde{f}_{\tilde{g}}, \tilde{\mathbb{V}}, N_2)$ be $CTSFN_1S_fS$ and $CTSFN_2S_fS$ on U, respectively, their *restricted union* is defined as $(\tilde{\mathbb{L}}_{\tilde{\mathbb{M}}}, \tilde{\mathfrak{G}}, \tilde{\mathfrak{D}})$
$= (\bar{f}_{\bar{g}}, \mathrm{E}, N_1) \hat{\cup} (\tilde{f}_{\tilde{g}}, \tilde{\mathbb{V}}, N_2)$, with $\tilde{\mathbb{L}}_{\tilde{\mathbb{M}}} = \bar{f}_{\bar{g}} \hat{\cup} \tilde{f}_{\tilde{g}}$, $\tilde{\mathfrak{G}} = \mathrm{E} \cap \tilde{\mathbb{V}}$, $\tilde{\mathfrak{D}} = \max(N_1, N_2)$, i.e., $\forall e_j \in \tilde{\mathfrak{G}}$, $u_k \in U$ we have

$$\tilde{\mathbb{L}}_{\tilde{\mathbb{M}}}(e_j) = \left\langle (x_j^k, (m_{jk}e^{i2\pi\alpha_{jk}}, q_{jk}e^{i2\pi\nu_{jk}}, p_{jk}e^{i2\pi\delta_{jk}})) \right\rangle,$$
$$= \left\langle (\max(x_j^{1k}, x_j^{2k}), \max(m_{kj}^1, m_{kj}^2)e^{i2\pi\max(\alpha_{kj}^1, \alpha_{kj}^2)}, \min(q_{kj}^1, q_{kj}^2)e^{i2\pi\min(\nu_{kj}^1, \nu_{kj}^2)}, \min(p_{kj}^1, p_{kj}^2)e^{i2\pi\min(\delta_{kj}^1, \delta_{kj}^2)}) \right\rangle,$$

where $(x_j^{1k}, (m_{kj}^1 e^{i2\pi\alpha_{kj}^1}, q_{kj}^1 e^{i2\pi\nu_{ba}^1}, p_{kj}^1 e^{i2\pi\delta_{kj}^1})) \in \bar{f}_{\bar{g}}$ and $(x_j^{2k}, (m_{kj}^2 e^{i2\pi\alpha_{kj}^2}, q_{kj}^2 e^{i2\pi\nu_{ba}^2}, p_{kj}^2 e^{i2\pi\delta_{kj}^2})) \in \tilde{f}_{\tilde{g}}$.

Example 2.28. The restricted union of $(\bar{f}_{\bar{g}}, \mathrm{E}, 6)$ and $(\tilde{f}_{\tilde{g}}, \tilde{\mathbb{V}}, 5)$, given by Table 3 and Table 6, respectively, is calculated in Table 13.

Table 13. Restricted union $(\tilde{\mathbb{L}}_{\tilde{\mathbb{M}}}, \tilde{\mathfrak{G}}, 6)$

$(\tilde{\mathbb{L}}_{\tilde{\mathbb{M}}}, \tilde{\mathfrak{G}}, 6)$	E_3	E_4
U_1	$(0, (0.100e^{i0.240\pi}, 0.012e^{i0.022\pi}, 0.985e^{i1.964\pi}))$	$(2, (0.400e^{i0.820\pi}, 0.017e^{i0.0356\pi}, 0.600e^{i1.220\pi}))$
U_2	$(5, (0.920e^{i1.860\pi}, 0.012e^{i0.026\pi}, 0.040e^{i0.060\pi}))$	$(5, (0.930e^{i1.880\pi}, 0.020e^{i0.042\pi}, 0.030e^{0.080i\pi}))$
U_3	$(4, (0.760e^{i1.540\pi}, 0.010e^{i0.022\pi}, 0.200e^{i0.420\pi}))$	$(3, (0.700e^{i1.420\pi}, 0.019e^{i0.040\pi}, 0.300e^{i0.560\pi}))$
U_4	$(3, (0.650e^{i1.320\pi}, 0.013e^{i0.028\pi}, 0.280e^{i0.580\pi}))$	$(2, (0.340e^{i0.700\pi}, 0.015e^{i0.032\pi}, 0.790e^{i1.600\pi}))$

Definition 2.29. Let $(\bar{f}_{\bar{g}}, \mathrm{E}, N_1)$ and $(\tilde{f}_{\tilde{g}}, \tilde{\mathbb{V}}, N_2)$ be $CTSFN_1S_fS$ and $CTSFN_2S_fS$ on U, respectively, their *extended union* is defined as $(\tilde{\mathfrak{P}}_{\tilde{\mathfrak{Q}}}, \tilde{\mathfrak{T}}, \tilde{\mathfrak{B}}) = (\bar{f}_{\bar{g}}, \mathrm{E}, N_1) \breve{\cup} (\tilde{f}_{\tilde{g}}, \tilde{\mathbb{V}}, N_2)$, with $\tilde{\mathfrak{P}}_{\tilde{\mathfrak{Q}}} = \bar{f}_{\bar{g}} \breve{\cup} \tilde{f}_{\tilde{g}}$, $\tilde{\mathfrak{T}} = \mathrm{E} \cup \tilde{\mathbb{V}}$, $\tilde{\mathfrak{B}} = \max(N_1, N_2)$, that is, $\forall e_j \in \tilde{T}$ and $u_k \in U$, we have

$$\tilde{\mathfrak{P}}_{\tilde{\mathfrak{Q}}}(e_j) = \begin{cases} (x_j^{1k}, m_{kj}^1 e^{i2\pi\alpha_{kj}^1}, q_{kj}^1 e^{i2\pi\nu_{kj}^1}, p_{kj}^1 e^{i2\pi\delta_{kj}^1}), & \text{if } v_j \in \mathrm{E} - \tilde{\mathbb{V}}, \\ (x_j^{2k}, m_{kj}^2 e^{i2\pi\alpha_{kj}^2}, q_{kj}^2 e^{i2\pi\nu_{kj}^2}, p_{kj}^2 e^{i2\pi\delta_{kj}^2}), & \text{if } e_j \in \tilde{\mathbb{V}} - \mathrm{E}, \\ \Big(\max(x_j^{1k}, x_j^{2k}), \max(m_{kj}^1, m_{kj}^2) e^{i2\pi\max(\alpha_{kj}^1, \alpha_{kj}^2)}, \min & \text{if } e_j \in \tilde{\mathbb{V}} \cap \mathrm{E}, \\ (q_{kj}^1, q_{kj}^2) e^{i2\pi\min(\nu_{kj}^1, \nu_{kj}^2)}, \min(p_{kj}^1, p_{kj}^2) e^{i2\pi\min(\delta_{kj}^1, \delta_{kj}^2)} \Big), & \end{cases}$$

where $(x_j^{1k}, (m_{kj}^1 e^{i2\pi\alpha_{kj}^1}, q_{kj}^1 e^{i2\pi\nu_{kj}^1}, p_{kj}^1 e^{i2\pi\delta_{kj}^1})) \in \bar{f}_{\bar{g}}$ and $(x_j^{2k}, (m_{kj}^2 e^{i2\pi\alpha_{kj}^2}, q_{kj}^2 e^{i2\pi\nu_{kj}^2}, p_{kj}^2 e^{i2\pi\delta_{kj}^2})) \in \tilde{f}_{\tilde{g}}$.

Example 2.30. The extended union $(\tilde{\mathfrak{P}}_{\tilde{\mathfrak{Q}}}, \tilde{\mathfrak{T}}, \tilde{\mathfrak{B}})$ of $(\bar{f}_{\bar{g}}, \mathrm{E}, 6)$ and $(\tilde{f}_{\tilde{g}}, \tilde{\mathbb{V}}, 5)$, given by Table 3 and Table 6, respectively, is calculated in Table 14.

Table 14. Extended union of $(\tilde{\mathfrak{P}}_{\tilde{\mathfrak{Q}}}, \tilde{\mathfrak{T}}, \tilde{\mathfrak{B}})$

$(\tilde{\mathfrak{P}}_{\tilde{\mathfrak{Q}}}, \tilde{\mathfrak{T}}, \tilde{\mathfrak{B}})$	E_1	E_2	E_3	E_4
U_1	$(1, (0.160e^{i0.340\pi}, 0.021e^{i0.044\pi}, 0.920e^{i1.860\pi}))$	$(1, (0.180e^{i0.380\pi}, 0.023e^{i0.048\pi}, 0.940e^{i1.90\pi}))$	$(0, (0.100e^{i0.240\pi}, 0.012e^{i0.022\pi}, 0.985e^{i1.964\pi}))$	$(2, (0.400e^{i0.820\pi}, 0.017e^{i0.0356\pi}, 0.600e^{i1.220\pi}))$
U_2	$(5, (0.900e^{i1.820\pi}, 0.020e^{i0.060\pi}, 0.010e^{i0.040\pi}))$	$(5, (0.910e^{i1.840\pi}, 0.018e^{i0.038\pi}, 0.010e^{i0.040\pi}))$	$(5, (0.920e^{i1.860\pi}, 0.012e^{i0.026\pi}, 0.040e^{i0.060\pi}))$	$(5, (0.930e^{i1.880\pi}, 0.020e^{i0.042\pi}, 0.030e^{0.080i\pi}))$
U_3	$(3, (0.510e^{i1.040\pi}, 0.028e^{i0.058\pi}, 0.450e^{i0.920\pi}))$	$(3, (0.520e^{i1.060\pi}, 0.029e^{i0.060\pi}, 0.460e^{i0.940\pi}))$	$(4, (0.760e^{i1.540\pi}, 0.010e^{i0.022\pi}, 0.200e^{i0.420\pi}))$	$(3, (0.700e^{i1.420\pi}, 0.019e^{i0.040\pi}, 0.300e^{i0.560\pi}))$
U_4	$(2, (0.350e^{i0.680\pi}, 0.014e^{i0.030\pi}, 0.780e^{i1.580\pi}))$	$(3, (0.530e^{i1.080\pi}, 0.031e^{i0.064\pi}, 0.470e^{i0.960\pi}))$	$(3, (0.650e^{i1.320\pi}, 0.013e^{i0.028\pi}, 0.280e^{i0.580\pi}))$	$(2, (0.340e^{i0.700\pi}, 0.015e^{i0.032\pi}, 0.790e^{i1.600\pi}))$
	E_5			
U_1	$(0, (0.050e^{i0.140\pi}, 0.870e^{i1.720\pi}, 0.880e^{i1.800\pi}))$			
U_2	$(1, (0.160e^{i0.340\pi}, 0.100e^{i0.204\pi}, 0.890e^{i1.784\pi}))$			
U_3	$(4, (1, 0, 0))$			
U_4	$(2, (0.500e^{i1.100\pi}, 0.100e^{i0.180\pi}, 0.590e^{1.280i\pi}))$			

Theorem 2.31. *Let $(\bar{f}_{\bar{g}}, \mathrm{E}, N_1)$ and $(\tilde{f}_{\tilde{g}}, \tilde{\mathfrak{V}}, N_2)$ be CTSFN_1S_fS and CTSF N_2S_fS on U, respectively, and $(\tilde{\mathfrak{P}}_{\tilde{\mathfrak{Q}}}, \tilde{\mathfrak{T}}, \tilde{\mathfrak{B}}) = (\bar{f}_{\bar{g}}, \mathrm{E}, N_1) \check{\cup} (\tilde{f}_{\tilde{g}}, \tilde{\mathfrak{V}}, N_2)$ be their extended union. Then $(\tilde{\mathfrak{P}}_{\tilde{\mathfrak{Q}}}, \tilde{\mathfrak{T}}, \tilde{\mathfrak{B}})$ is the smallest CTSFNS_fS containing both $(\bar{f}_{\bar{g}}, \mathrm{E}, N_1)$ and $(\tilde{f}, \tilde{\mathfrak{V}}, N_2)$.*

Proof. The prove follows from the definition of extended union. □

Now we discuss some properties and their proves.

Theorem 2.32. *Let $(\bar{f}_{\bar{g}}, \mathrm{E}, N_1)$ be a CTSFNS_fS over a non-empty set U. Then,*

1 $(\bar{f}_{\bar{g}}, \mathrm{E}, N_1) \check{\cap} (\bar{f}_{\bar{g}}, \mathrm{E}, N_1) = (\bar{f}_{\bar{g}}, \mathrm{E}, N_1)$
2 $(\bar{f}_{\bar{g}}, \mathrm{E}, N_1) \hat{\cap} (\bar{f}_{\bar{g}}, \mathrm{E}, N_1) = (\bar{f}_{\bar{g}}, \mathrm{E}, N_1)$
3 $(\bar{f}_{\bar{g}}, \mathrm{E}, N_1) \check{\cup} (\bar{f}_{\bar{g}}, \mathrm{E}, N_1) = (\bar{f}_{\bar{g}}, \mathrm{E}, N_1)$
4 $(\bar{f}_{\bar{g}}, \mathrm{E}, N_1) \hat{\cup} (\bar{f}_{\bar{g}}, \mathrm{E}, N_1) = (\bar{f}_{\bar{g}}, \mathrm{E}, N_1)$

Proof. 1.

$$R.H.S = (\bar{f}_{\bar{g}}, \mathrm{E}, N_1) \check{\cap} (\bar{f}_{\bar{g}}, \mathrm{E}, N_1), \tag{1}$$

where the extended intersection of two $CTSFNS_fSs$ is calculated as:

$$(\tilde{D}_{\tilde{Q}}, \tilde{T}, \tilde{S}) = (\bar{f}_{\bar{g}}, \mathrm{E}, N_1) \check{\cap} (\bar{f}_{\bar{g}}, \mathrm{E}, N_1), \tag{2}$$

with $\tilde{T} = \mathrm{E} \cup \mathrm{E}$, $\tilde{S} = \max(N_1, N_1)$ and

$$\tilde{D}_{\tilde{Q}}(v_j) = \begin{cases} (x_j^{1k}, (m^1_{kj} e^{i2\pi\alpha^1_{kj}}, q^1_{kj} e^{i2\pi\nu^1_{kj}}, p^1_{kj} e^{i2\pi\delta^1_{kj}})), & \text{if } e_j \in \mathrm{E} - \mathrm{E}, \\ (x_j^{1k}, (m^1_{kj} e^{i2\pi\alpha^1_{kj}}, q^1_{kj} e^{i2\pi\nu^1_{kj}}, p^1_{kj} e^{i2\pi\delta^1_{kj}})), & \text{if } e_j \in \mathrm{E} - \mathrm{E}, \\ (\min(x_j^{1k}, x_j^{1k}), (\min(m^1_{kj}, m^1_{kj}) e^{i2\pi \min(\alpha^1_{kj}, \alpha^1_{kj})}, \max(q^1_{kj} & \text{if } e_j \in \mathrm{E} \cap \mathrm{E}. \\ , q^1_{kj}) e^{i2\pi \max(\nu^1_{kj}, \nu^1_{kj}}, \max(p^1_{kj}, p^1_{kj}) e^{i2\pi \max(\delta^1_{kj}, \delta^1_{kj})})), & \end{cases}$$

Case 1 : If $e_j \in \mathrm{E} - \mathrm{E} = \emptyset$,

$$\tilde{D}_{\tilde{Q}}(e_j) = \bar{f}_{\bar{g}}(e_j). \tag{3}$$

Case 2 : If $e_j \in \mathrm{E} - \mathrm{E} = \emptyset$,

$$\tilde{D}_{\tilde{Q}}(e_j) = \bar{f}_{\bar{g}}(e_j). \tag{4}$$

Case 3 : If $e_j \in \mathrm{E} \cap \mathrm{E} = \mathrm{E}$,

$$\begin{aligned} \tilde{D}_{\tilde{Q}}(e_j) &= (\min(x_j^{1k}, x_j^{1k}), (\min(m^1_{kj}, m^1_{kj}) e^{i2\pi \min(\alpha^1_{kj}, \alpha^1_{kj})}, \max(q^1_{kj}, q^1_{kj}) \\ &\quad e^{i2\pi \max(\nu^1_{kj}, \nu^1_{kj}}, \max(p^1_{kj}, p^1_{kj}) e^{i2\pi \max(\delta^1_{kj}, \delta^1_{kj})})), \\ &= (x_j^{1k}, (t^1_{kj} e^{i2\pi\alpha^1_{kj}}, q^1_{kj} e^{i2\pi\nu^1_{kj}}, p^1_{kj} e^{i2\pi\delta^1_{kj}})), \\ &= \bar{f}_{\bar{g}}(e_j). \end{aligned} \tag{5}$$

From Equations (2), (3), (4) and (5), $(\tilde{D}_{\tilde{Q}}, \tilde{T}, \tilde{S}) = (\bar{f}_{\bar{g}}, \mathrm{E}, N_1)$ and further Eq. (1) implies $(\bar{f}_{\bar{g}}, \mathrm{E}, N_1) \check{\cap} (\bar{f}_{\bar{g}}, \mathrm{E}, N_1) = (\bar{f}_{\bar{g}}, \mathrm{E}, N_1)$.
Similarly parts (2),(3) and (4) can be certify. □

Theorem 2.33. *Let* $(\bar{f}_{\bar{g}}, \mathrm{E}, N_1)$ *and* $(\tilde{f}_{\bar{g}}, \mathcal{V}, N_2)$ *be* $CTSFN_1S_fS$ *and* $CTSFN_2$ S_fS*, respectively, over the same universe* U*, then the absorption properties hold:*

1 $((\bar{f}_{\bar{g}}, \mathrm{E}, N_1) \check{\cup} (\tilde{f}_{\bar{g}}, \tilde{\mathcal{V}}, N_2)) \check{\cap} (\bar{f}_{\bar{g}}, \mathrm{E}, N_1) = (\bar{f}_{\bar{g}}, \mathrm{E}, N_1)$
2 $(\bar{f}_{\bar{g}}, \mathrm{E}, N_1) \check{\cup} ((\tilde{f}_{\bar{g}}, \tilde{\mathcal{V}}, N_2) \check{\cap} (\bar{f}_{\bar{g}}, \mathrm{E}, N_1)) = (\bar{f}_{\bar{g}}, \mathrm{E}, N_1)$
3 $((\bar{f}_{\bar{g}}, \mathrm{E}, N_1) \check{\cap} (\tilde{f}_{\bar{g}}, \tilde{\mathcal{V}}, N_2)) \check{\cup} (\bar{f}_{\bar{g}}, \mathrm{E}, N_1) = (\bar{f}_{\bar{g}}, \mathrm{E}, N_1)$
4 $(\bar{f}_{\bar{g}}, \mathrm{E}, N_1) \check{\cap} ((\tilde{f}_{\bar{g}}, \tilde{\mathcal{V}}, N_2) \check{\cup} (\bar{f}_{\bar{g}}, \mathrm{E}, N_1)) = (\bar{f}_{\bar{g}}, \mathrm{E}, N_1)$

Proof. 1. Let the extended union of $CTSFN_1S_fS$ $(\bar{f}_{\bar{g}}, \mathrm{E}, N_1)$ and $CTSFN_2S_fS$ $(\tilde{f}_{\bar{g}}, \tilde{\mathcal{V}}, N_2)$, be

$$(\tilde{\mathcal{P}}_{\tilde{\mathcal{Q}}}, \tilde{\mathfrak{T}}, \tilde{\mathfrak{B}}) = (\bar{f}_{\bar{g}}, \mathrm{E}, N_1) \check{\cup} (\tilde{f}_{\bar{g}}, \tilde{\mathcal{V}}, N_2), \tag{6}$$

with $\tilde{\mathfrak{T}} = \mathrm{E} \cup \tilde{\mathcal{V}}$, $\tilde{\mathfrak{B}} = \max(N_1, N_2)$ and
$\tilde{\mathcal{P}}_{\tilde{\mathcal{Q}}}(e_j) = (x_j^k, (t_{kj} e^{i2\pi\alpha_{kj}}, q_{kj} e^{i2\pi\nu_{kj}}, p_{kj} e^{i2\pi\delta_{kj}})),$

$$= \begin{cases} (x_j^{1k}, (t_{kj}^1 e^{i2\pi\alpha_{kj}^1}, q_{kj}^1 e^{i2\pi\nu_{kj}^1}, p_{kj}^1 e^{i2\pi\delta_{kj}^1})), & \text{if } e_j \in \mathrm{E} - \tilde{\mathcal{V}}, \\ (x_j^{2k}, (t_{kj}^2 e^{i2\pi\alpha_{kj}^2}, q_{kj}^2 e^{i2\pi\nu_{kj}^2}, p_{kj}^2 e^{i2\pi\delta_{kj}^2})), & \text{if } e_j \in \tilde{\mathcal{V}} - \mathrm{E}, \\ (\max(x_j^{1k}, x_j^{2k}), (\max(t_{kj}^1, t_{kj}^1) e^{i2\pi \max(\alpha_{kj}^1, \alpha_{kj}^1)}, & \text{if } e_j \in \mathrm{E} \cap \tilde{\mathcal{V}}. \\ \min(q_{kj}^1, q_{kj}^1) e^{i2\pi \min(\nu_{kj}^1, \nu_{kj}^1}, \min(p_{kj}^1, p_{kj}^1) e^{i2\pi \min(\delta_{kj}^1, \delta_{kj}^1)})), & \end{cases}$$

Now, consider the restricted intersection of $(\tilde{\mathfrak{P}}_{\tilde{\mathfrak{Q}}}, \tilde{\mathfrak{T}}, \tilde{\mathfrak{B}})$ and $(\bar{f}_{\bar{g}}, \mathrm{E}, N_1)$, that is defined as

$$(\tilde{L}_{\tilde{M}}, \tilde{G}, \tilde{O}) = (\tilde{\mathfrak{P}}_{\tilde{\mathfrak{Q}}}, \tilde{\mathfrak{T}}, \tilde{\mathfrak{B}}) \hat{\cap} (\bar{f}_{\bar{g}}, \mathrm{E}, N_1),$$

with $\tilde{G} = \tilde{\mathfrak{T}} \cap \mathrm{E}$, $\tilde{O} = \min(\tilde{\mathfrak{B}}, N_1) = N_1$ and

$$\begin{aligned} \tilde{L}_{\tilde{M}}(e_j) = (&\min(x_j^{1k}, x_j^{1k}), (\min(t_{kj}^1, t_{kj}^1) e^{i2\pi \min(\alpha_{kj}^1, \alpha_{kj}^1)}), \max(q_{kj}^1, q_{kj}^1) \\ &e^{i2\pi \max(\nu_{kj}^1, \nu_{kj}^1)}, \max(p_{kj}^1, p_{kj}^1) e^{i2\pi \max(\delta_{kj}^1, \delta_{kj}^1)}))), \end{aligned} \tag{7}$$

for all $e_j \in \tilde{G} = \mathrm{E} \cap \tilde{\mathcal{V}}$, so that $e_j \in \mathrm{E}$, $e_j \in \tilde{\mathcal{V}}$. If $e_j \in \mathrm{E}$, then there are three cases.

Case 1: if $e_j \in \mathrm{E} - \tilde{\mathcal{V}}$, using Eqs. (6) and (7), we get,

$$\begin{aligned} \tilde{L}_{\tilde{M}}(e_j) &= (\min(x_j^{1k}, x_j^{1k}), (\min(t_{kj}^1, t_{kj}^1) e^{i2\pi \min(\alpha_{kj}^1, \alpha_{kj}^1)}), \max(q_{kj}^1, q_{kj}^1) \\ &\quad e^{i2\pi \max(\nu_{kj}^1, \nu_{kj}^1)}, \max(p_{kj}^1, p_{kj}^1) e^{i2\pi \max(\delta_{kj}^1, \delta_{kj}^1)}))) \\ &= (x_j^{1k}, t_{kj}^1 e^{i2\pi\alpha_{kj}^1}, q_{kj}^1 e^{i2\pi\nu_{kj}^1}, p_{kj}^1 e^{i2\pi\delta_{kj}^1}) \\ &= \bar{f}_{\bar{g}}(e_j). \end{aligned} \tag{8}$$

Case 2: if $e_j \in \tilde{\mathcal{V}} - \mathrm{E}$, since $e_j \in \tilde{G} = \mathrm{E} \cap \tilde{\mathcal{V}}$ implies $e_j \in \mathrm{E}$, therefore, this case is omitted.

Case 3: if $e_j \in \tilde{\mathcal{V}} \cap \mathrm{E}$, using Eqs. (6) and (7), we get,

$$\begin{aligned} \tilde{L}_{\tilde{M}}(e_j) &= (x_j^{1k}, t_{kj}^1 e^{i2\pi\alpha_{kj}^1} q_{kj}^1 e^{i2\pi\nu_{kj}^1}, p_{kj}^1 e^{i2\pi\delta_{kj}^1}) \\ &= \bar{f}_{\bar{g}}(e_j). \end{aligned} \tag{9}$$

Thus from Eqs. (8) and (9), we get $((\bar{f}_{\bar{g}}, E, N_1) \breve{\cup} (\tilde{f}_{\tilde{g}}, \tilde{\mathcal{V}}, N_2)) \hat{\cap} (\bar{f}_{\bar{g}}, \mathrm{E}, N_1) = (\bar{f}_{\bar{g}}, \mathrm{E}, N_1)$.

2. The proves of parts (2), (3) and (4) are same as (1). □

3 Conclusion

We have defined a new type of fuzzy sets called complex T-spherical fuzzy N-soft sets. We have described some properties and operations. In future, we aim to extend our study to multi-attribute group decision making methods under the framework of complex T-spherical fuzzy N-soft sets.

References

1. Akram, M., Kahraman, C., Zahid, K.: Extension of TOPSIS model to the decision-making under complex spherical fuzzy information. Soft Comput. (2021)
2. Akram, M., Kahraman, C., Zahid, K.: Group decision-making based on complex spherical fuzzy VIKOR approach. Knowl. Based Syst. **216**, 106793 (2021)
3. Akram, M., Shabir, M., Al Kenani, A.N., Alcantud, J.C.R.: Hybrid decision making frameworks under complex spherical fuzzy N-soft sets. J. Math., 1–46 (2021)
4. Ali, Z., Mahmood, T., Yang, M.-S.: Complex T-spherical fuzzy aggregation operators with application to multi-attribute decision making. Symmetry **12**, 1311 (2020)
5. Ashraf, S., Abdullah, S., Aslam, M., Qiyas, M., Kutbi, M.A.: Spherical fuzzy sets and its representation of spherical fuzzy t-norms and t-conorms. J. Intell. Fuzzy Syst. **36**, 6089–6102 (2019)
6. Atanassov, K.: Intuitionistic fuzzy sets. Fuzzy Sets Syst. **20**(1), 87–96 (1986)
7. Cuong, B.C.: Picture fuzzy sets-first results. Neuro-Fuzzy Systems with Applications, Institute of Mathematics, Hanoi (2013)
8. Fatimah, F., Rosadi, D., Hakim, R.B.F., Alcantud, J.C.R.: N-soft sets and their decision-making algorithms. Soft Comput. **22**(12), 3829–3842 (2018)
9. Gündogdu, F.K., Kahraman, C.: Spherical fuzzy sets and spherical fuzzy TOPSIS method. J. Intell. Fuzzy Syst. **36**(1), 337–352 (2019)
10. Kahraman, C., Gündogdu, F.K., Onar, S.C., Oztaysi, B.: Hospital Location Selection Using Spherical Fuzzy TOPSIS Method. Atlantis Press (2019)
11. Guleria, A., Bajaj, R.K.: T-spherical fuzzy soft sets and its aggregation operators with application in decision making. Scientia Iranica **28**(2), 1014–1029 (2021)
12. Mahmood, T., Ullah, K., Khan, Q., Jan, N.: An approach toward decision-making and medical diagnosis problems using the concept of spherical fuzzy sets. Neural Comput. Appl. **31**(11), 7041–7053 (2019)
13. Molodtsov, D.A.: Soft set theory-first results. Comput. Mathe. Appl. **37**(4–5), 19–31 (1999)
14. Zadeh, L.A.: Fuzzy sets. Inf. Control **8**(3), 338–356 (1965)

Type-2 Fuzzy Sets

An Integration of Interval Type-2 Fuzzy Set with Equitable Linguistic Approach Based on Multi-criteria Decision Making: Flood Control Project Selection Problems

Nurnadiah Zamri[1(✉)], Syibrah Naim[2], and Zamali Tarmudi[3]

[1] Faculty of Informatics and Computing, Universiti Sultan Zainal Abidin, Besut Campus, 22200 Besut, Terengganu, Malaysia
nadiahzamri@unisza.edu.my

[2] Technology Department, Endicott College of International Studies (ECIS), Woosong University, Daejeon, South Korea

[3] Universiti Teknologi MARA, Cawangan Johor (Kampus Segamat), KM 12, Jalan Muar, 85009 Segamat, Johor, Malaysia

Abstract. Flooding is a big issue of widespread concern. Flood control project selection is needed to overcome the flooding phenomenon while achieving other objectives such as decreases the global burden of morbidity, mortality, social and economic disruptions, and stress on health services. Hence, the preferences of Decision Makers (DMs) from diverse backgrounds are needed to obtain the best project. However, flood control project selection faced high levels of conflict. Therefore, we propose a new equitable linguistic scale that provides an inclusive evaluation from integrating DMs' preferences and opinions. Besides, we also propose a hybrid averaging approach of linear orders; consists of type-reduction method, ambiguity method and (Elimination and Choice Expressing Reality) ELECTRE-based non-outranked method by the new linguistic scales for Interval Type-2 Fuzzy Technique for Order Preference by Similarity to Ideal Solution (IT2FTOPSIS) method. An actual case experiment to evaluate seven different flood control projects is carried out in Malaysia based on the evaluation from the Department of Drainage of Irrigation agencies. The result shows that Dikes (levees/embankment)/Channel improvement/Diversion schemes are the best flood control project. Besides, this proposed IT2FTOPSIS can offer a measure of the DMs' opinions and preferences. Correlation values have also proved that this proposed IT2FTOPSIS in line with the DMs' decision compared to existing IT2FTOPSIS.

Keywords: Multiple criteria decision-making · Interval Type-2 fuzzy set · Ambiguity · Type-reduction method · ELECTRE I · IT2FTOPSIS

1 Introduction

One of the most significant natural disasters in Malaysia is a flood. Floods can cause damage to nearly almost 4.9 million Malaysian people [1]. According to the Decision

C. Kahraman et al. (Eds.): INFUS 2021, LNNS 308, pp. 837–844, 2022.
https://doi.org/10.1007/978-3-030-85577-2_96

Makers' (DMs) preferences, this paper focuses on flood control project selection to overcome the flooding phenomenon. However, the DMs may have numerous preferences according to their experts and experiences. Therefore, it was recommended to discover Multiple Criteria Decision Making (MCDM) methods for solving flood control project selection problems as MCDM is known as one of the efficient methods to handle the conflicts and uncertainties among DMs' judgment [2]. Despite many MCDM methods, the Technique for Order Preference by Similarity to Ideal Solution (TOPSIS) [3] is preferred as the aim for the analysis because of its constancy and ease of use in terms of cardinal information. Fuzzy TOPSIS (FTOPSIS) is presented to handle uncertainty in linguistic judgment [4]. However, the existed T1 FTOPSIS still lacks in defining the uncertainties [5–8].

Interval Type-2 Fuzzy TOPSIS (IT2FTOPSIS) was established to overcome the uncertainties problems [5]. IT2FTOPSIS is believed to give room for more flexibility in IT2FTOPSIS because of the nature of Interval Type-2 Fuzzy Sets (IT2FSs), where it can present more uncertainties compared to Type-1 Fuzzy Sets (T1FSs). Therefore, this paper aims to build a new equitable linguistic scale based IT2FTOPSIS model. This equitable linguistic scale responds to the subjective decisions from the experts, where it offers an equitable in positive and negative scales. Besides, the scale is equally strong between the lowest of the scale and the highest of the scale [9]. Instead of proposing a new equitable linguistic scale, a hybrid averaging approach of linear order method is also developed to be included in this new equitable linguistic scale. Next, it can formulate a collective decision environment for the ELECTRE-based non-outranked method. For this hybrid averaging approach, we used the concept of ambiguity by Ban et al. [10] and type-reduction method by Wu and Mendel [11]. There are no further explorations on type-reduction in MADM yet [12, 13]. Later, a real case study on flood control project selection is applied to check the efficiency of the proposed method. Lastly, correlation coefficients are presented to show the robustness of the results. Overall, this paper is structured as follow; Sect. 2 discusses on the proposed of the IT2FTOPSIS method, focusses on experiments and results, Sect. 4 presents the analysis and results and lastly, Sect. 5 concludes.

2 The Proposed of Method

A new brand method to solve the IT2FTOPSIS method is developed in this section. The proposed IT2FTOPSIS consists of; equitable linguistic scales, hybrid averaging approach of linear order method and an ELECTRE-based non-outranked method. Table 1 are the overviews steps of the proposed IT2FTOPSIS procedures:

X is a set of alternatives, where $X = \{x_1, x_2, \ldots, x_n\}$,
F is a set of criteria, where $F = \{f_1, f_2, \ldots, f_m\}$.
k is a set DMs, where $D_1, D_2, \ldots,$ and D_k

Table 1. Steps of the proposed method

Step	Description
Step 1:	**Create a hierarchical diagram of MCDM case** Step 1 focuses on constructing the design matrix Y_p of the pth DM. Then, construct the average decision matrix, respectively $$Y_p = \left(\bar{f}^p_{ij}\right)_{m\times n} = \begin{matrix} f_1 \\ f_2 \\ \vdots \\ f_m \end{matrix} \overset{\begin{matrix} x_1 & x_2 & \cdots & x_n \end{matrix}}{\begin{bmatrix} \tilde{\tilde{f}}^p_{11} & \tilde{\tilde{f}}^p_{12} & \cdots & \tilde{\tilde{f}}^p_{1n} \\ \tilde{\tilde{f}}^p_{21} & \tilde{\tilde{f}}^p_{22} & \cdots & \tilde{\tilde{f}}^p_{2n} \\ \vdots & \vdots & \vdots & \vdots \\ \tilde{\tilde{f}}^p_{m1} & \tilde{\tilde{f}}^p_{m2} & \cdots & \tilde{\tilde{f}}^p_{mn} \end{bmatrix}}$$ $Y = \left(\tilde{\tilde{f}}_{ij}\right)_{m\times n}$, where $\tilde{\tilde{f}}_{ij} = \left(\frac{\tilde{\tilde{f}}^1_{ij} \oplus \tilde{\tilde{f}}^2_{ij} \oplus \ldots \oplus \tilde{\tilde{f}}^k_{ij}}{k}\right)$, $\tilde{\tilde{f}}_{ij}$ is a positive and negative IT2TrFN, $1 \le i \le m,\ 1 \le j \le n,\ 1 \le p \le k$, and k denotes the number of DMs
Step 2:	**Define the values of weight** The weighting matrix W_p of the criteria of the DM is constructed using Fig. 1
Step 3:	**Construct the weighted of decision matrices** The weighted decision matrix $\overline{Y}_w$, are constructed in this step $$\overline{Y}_w = \left(\tilde{\tilde{v}}_{ij}\right)_{m\times n} = \begin{matrix} f_1 \\ f_2 \\ \vdots \\ f_m \end{matrix} \overset{\begin{matrix} x_1 & x_2 & \cdots & x_n \end{matrix}}{\begin{bmatrix} \tilde{\tilde{v}}_{11} & \tilde{\tilde{v}}_{12} & \cdots & \tilde{\tilde{v}}_{1n} \\ \tilde{\tilde{v}}_{21} & \tilde{\tilde{v}}_{22} & \cdots & \tilde{\tilde{v}}_{2n} \\ \vdots & \vdots & \vdots & \vdots \\ \tilde{\tilde{v}}_{m1} & \tilde{\tilde{v}}_{m2} & \cdots & \tilde{\tilde{v}}_{mn} \end{bmatrix}},$$ where $\tilde{\tilde{v}}_{ij} = \tilde{\tilde{w}}_i \otimes \tilde{\tilde{f}}_{ij}$, $1 \le i \le m$, and $1 \le j \le n$
Step 4:	**Construct the hybrid averaging approach using the linear order method** The linear order hybrid averaging operation is employed to determine the collective evaluation value of weighted decision matrix $\tilde{\tilde{v}}^{hybrid}_{ij}$ and the collective importance weight $\tilde{\tilde{w}}^{hybrid}_{j}$ $y(x) =$ $$\frac{1}{2}\left[\left(\left(\frac{-6l+6u-x-y}{12}\right)^L \Big/ 2\right)\left(\frac{H_1\left(\tilde{A}^L_1\right)}{H_2\left(\tilde{A}^L_1\right)}\right) + \left(\left(\frac{-6l+6u-x-y}{12}\right)^U \Big/ 2\right)\left(\frac{H_1\left(\tilde{A}^U_1\right)}{H_2\left(\tilde{A}^U_1\right)}\right)\right]$$ where $\left(\frac{-6l+6u-x-y}{12}\right)^L$ represented as $\left(\underline{y_l}(x) + \overline{y_l}(x)\right)$ and $\left(\frac{-6l+6u-x-y}{12}\right)^U$ represented as $\left(\underline{y_r}(x) + \overline{y_r}(x)\right)$

(*continued*)

Table 1. (*continued*)

Step	Description
Step 5:	**Calculate the concordance matrix** Thus, the concordance matrix is defined as follows, $\tilde{\tilde{C}} = \begin{bmatrix} - & \cdots & \tilde{\tilde{c}}_{1f} & \cdots & \tilde{\tilde{c}}_{1(m-1)} & \tilde{\tilde{c}}_{1m} \\ \vdots & \ddots & \vdots & \ddots & \vdots & \vdots \\ \tilde{\tilde{c}}_{g1} & \cdots & \tilde{\tilde{c}}_{gf} & \cdots & \tilde{\tilde{c}}_{g(m-1)} & \tilde{\tilde{c}}_{gm} \\ \vdots & \ddots & \vdots & \ddots & \vdots & \vdots \\ \tilde{\tilde{c}}_{m1} & \cdots & \tilde{\tilde{c}}_{mf} & \cdots & \tilde{\tilde{c}}_{m(m-1)} & - \end{bmatrix}$ Or we can say as, the concordance level, $\tilde{\tilde{C}}$, can be defined as the average of the elements in the concordance matrix, represented by $\tilde{\tilde{c}}_{gf} = \sum_{g=1}^{m} \sum_{f=1}^{m} c_{gf} \Big/ m(m-1)$
Step 6:	**Calculate the discordance matrix** Thus, the discordance matrix is defined as $\tilde{\tilde{D}} = \begin{bmatrix} - & \cdots & \tilde{\tilde{d}}_{1f} & \cdots & \tilde{\tilde{d}}_{1(m-1)} & \tilde{\tilde{d}}_{1m} \\ \vdots & \ddots & \vdots & \ddots & \vdots & \vdots \\ \tilde{\tilde{d}}_{g1} & \cdots & \tilde{\tilde{d}}_{gf} & \cdots & \tilde{\tilde{d}}_{g(m-1)} & \tilde{\tilde{d}}_{gm} \\ \vdots & \ddots & \vdots & \ddots & \vdots & \vdots \\ \tilde{\tilde{d}}_{m1} & \cdots & \tilde{\tilde{d}}_{mf} & \cdots & \tilde{\tilde{d}}_{m(m-1)} & - \end{bmatrix}$ where $\tilde{\tilde{d}}_{gf} = \frac{\max_{j \in J_D} \left\| \tilde{\tilde{v}}_{ij}^{hybrid}(A_g) - \tilde{\tilde{v}}_{ij}^{hybrid}(A_j) \right\|}{\max_{j} \left\| \tilde{\tilde{v}}_{ij}^{hybrid}(A_g) - \tilde{\tilde{v}}_{ij}^{hybrid}(A_j) \right\|}$ $= \frac{\max_{j \in J_D} \left\| d\left(\max\left(\tilde{\tilde{v}}_{ij}^{hybrid}(A_g), \tilde{\tilde{v}}_{ij}^{hybrid}(A_j) \right), \tilde{\tilde{v}}_{ij}^{hybrid}(A_j) \right) \right\|}{\max_{j} \left\| d\left(\max\left(\tilde{\tilde{v}}_{ij}^{hybrid}(A_g), \tilde{\tilde{v}}_{ij}^{hybrid}(A_j) \right), \tilde{\tilde{v}}_{ij}^{hybrid}(A_j) \right) \right\|}$
Step 7:	**Construct the global matrix** The global matrix $\tilde{\tilde{Z}}$ is calculated by peer to peer subtraction of the elements of the matrices $\tilde{\tilde{C}}$ and $\tilde{\tilde{D}}$
Step 8:	**Selection stage** PIS and NIS are determined in the last stage, calculated the separation measures, relative closeness coefficients, and lastly rank the preference order

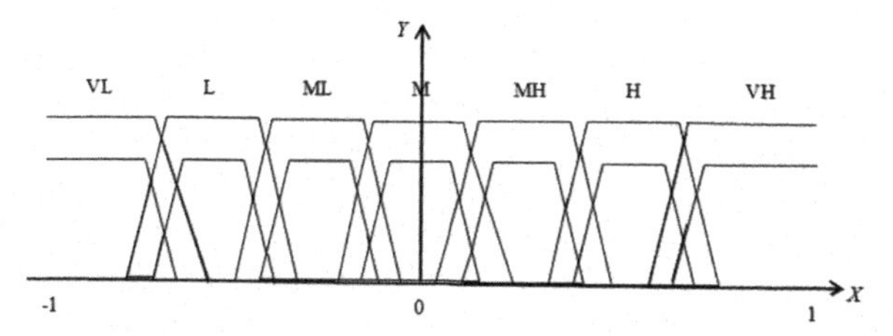

Fig. 1. The proposed equitable linguistic scale

With an attempt to consider both sides; negative and positive sides, the linear order approach, and the ELECTRE-based non-outranking method based IT2FTOPSIS, it would anticipate that the proposed IT2FTOPSIS method makes a more comprehensive look.

3 Experiment on Flood Control Project Selection

Flood management, in general plays an important role in protecting people and their property from flooding. In the face of this mounting threat, it may be necessary to approach the issue of flood management from a novel angle [14]. There are seven alternatives for flood controls are used to decrease flood damages which are the construction of:

$(A_1) =$ dam/reservoir
$(A_2) =$ dikes (levees/embankment)/channel improvement/diversion scheme
$(A_3) =$ pumping station
$(A_4) =$ flood barrier/barrage/flood gate/flood wall
$(A_5) =$ retention basins/watershed
$(A_6) =$ retention pond
$(A_7) =$ catchment areas

Then, five main criteria are selected based on seven alternatives. This study considered five main criteria with sub-criteria which were;

Economic factors (C_1) : Project costs (C_{11}), Operation and maintenance (C_{12}), Project benefit (C_{13}), Reliability economic parameter (C_{14})
Social factors (C_2): Social acceptability (C_{21}), Demographic changes (C_{22}), Effects on infrastructure (C_{23}), Recreation activity (C_{24})
Environment factors (C_3): Climate impact (C_{31}), Water quality impact (C_{34}), Nature conservation (C_{32}), Sanitary condition (C_{34})
Technical factors (C_4): Lifetime (C_{41}), Adaptability (C_{42}), Level of protection (C_{43}), Technical complexity (C_{44}), Flexibility of the project (C_{45})
Management factors (C_5): Land use (C_{51}), Area (C_{52}).

Furthermore, the face-to-face interviews were conducted based on the questionnaires designed according to the new IT2FTOPSIS requirement from the four different DMs. The linguistic data were collected based on interviewing of four authorised personnel (stated as D_1, D_2, D_3, and D_4) from Department of Drainage of Irrigation agencies using the equitable linguistic scales. Then, to evaluate the rating of alternatives with respect to each criterion in the form of a decision matrix, the equitable linguistic rating scales for rating. The result shows that, the ranking order is given by $A_2 \succ A_4 \succ A_5 \succ A_7 \succ A_6 \succ A_3 \succ A_1$ and the best alternative selection is A_2 which is Dikes (levees/embankment)/Channel Improvement/Diversion schemes. Dikes (levees/embankment)/Channel Improvement/Diversion schemes is ranked first, followed by Flood barrier/Barrage/Flood gate/Flood wall, River basins/Watershed, Catchment areas, Retention pond, and Pumping station. Dam/Reservoir is ranked last. Thus, after considered all the seven criteria, twenty sub-criteria and the opinion from the four DMs, Dikes (levees/embankment)/Channel Improvement/Diversion schemes recorded the highest closeness coefficient at 0.2067. It is important to compare the proposed method with the existing approach. Next section describes the analysis and results of comparing the output of flood control project selection and other existing MCDM methods.

4 Analysis and Results

Questionnaires were distributed randomly to ask all four DMs from the Department of Drainage of Irrigation about their opinions and judgments on flood control project selection needed to overcome the flooding problem before the experiments were conducted. Spearman Correlation is used to visualize the agreement and diagnose agreement between real output data of flood control project selections with the 1) IT2FTOPSIS [5], and 2) our proposed IT2FTOPSIS. Preferably, this Spearman Correlation was used to find the relationship correlation between the user's decisions and the output decisions. From the calculations, results are summarized as Table 2, respectively, as follows:

Table 2. Correlation values between the proposed IT2FTOPSIS and Chen and Lee's IT2FTOPSIS [8]

	Spearman Correlation for proposed IT2FTOPSIS	p-value	Spearman Correlation for IT2FTOPSIS [5]	p-value
D_1	**0.929**	0.003	0	1
D_2	**0.786**	0.036	0.036	0.939
D_3	**0.893**	0.007	0.214	0.645
D_4	**0.929**	0.003	0.071	0.879

*Bold value indicates the highest correlation value

Based on the results from Table 2, our proposed method scores the highest Spearman Correlation value compared with IT2FTOPSIS [5]. This is proved that, the hybrid IT2 fuzzy theories with the proposed method offer more nearer group of DMs' decisions than to the other fuzzy theory methods when the output reached the user's decision, meaning that it has obtained a higher correlation value.

5 Conclusion

This paper developed an IT2FTOPSIS method that considered the DMs' valuation based on the DMs' opinions and preferences. The theory of the equitable linguistic scale, the hybrid averaging approach of linear order method; consists of ambiguity method and type-reduction method, and ELECTRE-based non-outranked method are well matched when dealing with uncertainties and ambiguities. An actual case experiment to evaluate seven different flood control projects is carried out in Malaysia based on the evaluation from the Department of Drainage of Irrigation agencies. Results show that Dikes (levees/embankment)/Channel Improvement/Diversion schemes is the most suitable flood control project to solve the flood problems. Besides, it was discovered that our proposed method; the equitable linguistic scale, hybrid averaging approach and ELECTRE-based non-outranked method are capable of offering more space for the various DMs' conflicts. The proposed IT2FTOPSIS is more suitable to handle the uncertainties among DMs' decision. The experiments found that IT2FTOPSIS [5] provides lower correlation values than our proposed IT2FTOPSIS with [5]'s linguistic scales and our proposed method. Thus, the proposed IT2FTOPSIS offered a better ranking compared to other decision methods. For the upcoming work, to retrieve higher levels of uncertainties in decision making problems, we aim to integrate this proposed IT2FTOPSIS method with general type-2 membership functions. General type-2 is believed can more increase uncertainties in the DMs' agreement. The higher the uncertainties in the DM's agreement, the closer and mimic the group of human decisions.

Acknowledgment. Thanks goes to FRGS-RACER: RACER/1/2019/STG06/UNISZA//1 that fully supported this research.

References

1. Abdullah, L., Goh, C., Zamri, N., Othman, M.: Application of interval valued intuitionistic fuzzy TOPSIS for flood management. J. Intell. Fuzzy Syst. **38**(1), 873–881 (2020)
2. Triantaphyllou, E., Shu, B., Nieto Sanchez, S., Ray, T.: Multi-criteria decision making: an operations research approach. Encyclopedia of Electrical and Electronics Engineering, (J.G. Webster, Ed.), vol. 15, pp. 175–186. Wiley, New York (1998)
3. Hwang, C.L., Yoon, K.S.: Multiple Attribute Decision Making: Methods and Applications. Springer, Heidelberg (1981)
4. Chen, C.: Extensions of the TOPSIS for group decision-making under fuzzy environment. Fuzzy Sets Syst. **114**(1), 1–9 (2000)
5. Chen, S.-M., Lee, L.-W.: Fuzzy multiple attributes group decision-making based on the interval type-2 TOPSIS method. J. Expert Syst. Appl. **37**, 2790–2798 (2010)
6. Wang, W., Liu, X., Qin, Y.: Multi-attribute group decision making models under interval type-2 fuzzy environment. Knowl.-Based Syst. **30**, 121–128 (2012)
7. Hu, J., Zhang, Y., Chen, X., Liu, Y.: Multi-criteria decision-making method based on possibility degree of interval type-2 fuzzy number. Knowl.-Based Syst. **43**, 21–29 (2013)
8. Chen, S.-M., Wang, C.-Y.: Fuzzy decision-making systems based on interval type-2 fuzzy sets. Inf. Sci. **242**, 1–21 (2013)
9. Jiang, F., Preethy, A.P., Zhang, Y.-Q.: Compensating hypothesis by negative data. In: IEEE International Conference on Neural Networks and Brain, pp. 1986–1990 (2005)

10. Ban, A., Brândaş, A., Coroianu, L., Negruţiu, C., Nica, O.: Approximations of fuzzy numbers by trapezoidal fuzzy numbers preserving the ambiguity and value. Comput. Math. Appl. **61**(5), 1379–1401 (2011)
11. Wu, H., Mendel, J.M.: Uncertainty bounds and their use in the design of interval Type-2 Fuzzy logic systems. IEEE Trans. Fuzzy Syst. **10**(5), 622–639 (2002)
12. Khosravi, A., Nahavandi, S., Khosravi, R.: A new neural network-based type reduction algorithm for interval type-2 fuzzy logic systems. In: IEEE International Conference on Fuzzy Systems, pp. 1–6 (2013)
13. Chen, T.-Y.: A signed-distance-based approach to importance assessment and multi-criteria group decision analysis based on interval type-2 fuzzy set. Knowl. Inform. Syst. **35**(1), 193–231 (2013)
14. Murdoch, W.O.: The economics of flood management. A case study in the river Lossie Catchment Scotland. Master thesis, University of Edinburgh (2005)

A Comparative Study of FAHP with Type-1 and Interval Type-2 Fuzzy Sets for ICT Implementation in Smart Cities

Dušan Milošević[1](✉), Mimica Milošević[2], and Dušan Simjanović[3]

[1] Faculty of Electronic Engineering, University of Nis, Aleksandra Medvedeva 14, 18106 Nis, Serbia
dusan.milosevic@elfak.ni.ac.rs
[2] Faculty of Business Economics and Entrepreneurship, Mitropolita Petra 8, 11000 Belgrade, Serbia
[3] Faculty of Information Technology, Metropolitan University, Tadeuša Košćuška 63, 11158 Belgrade, Serbia
dusan.simjanovic@metropolitan.ac.rs

Abstract. In Type-1 Fuzzy Sets (T1FS), membership functions have not associated with uncertainty. Type-2 Fuzzy Sets (T2FS) provides an opportunity to incorporate the uncertainty of membership functions into the theory of fuzzy sets at the expense of the number of additional computational operations. Therefore, Interval Type-2 Fuzzy Sets (IT2FS) is introduced where the number of computational operations is reduced to T2FS while preserving the good properties that T2FS possesses. This paper presents a comparative study of the Fuzzy Analytic Hierarchy Process (FAHP) with T1FS and IT2FS. The proposed methods rank the criteria to find the dominant indicators for creating optimal access in the implementation of Information and Communication Technology (ICT) in the concept development of the smart city. In this paper, we measure the similarity of the ranks in decision problems, applying Spearman's rank correlation coefficient and the new Salabun's coefficient suitable for comparing rankings in the field of decision making.

Keywords: Multi-criteria decision-making · Fuzzy Analytic Hierarchy Process · Type-1 Fuzzy Sets · Interval Type-2 Fuzzy Sets · Smart cities · Information and communication technology

1 Introduction

The aspiration to achieve sustainable urban development has taken root in recent years within the concept of a smart city [1]. The smart city is resulting as a solution to the needs posed by the urbanization process but above all as an idea for using advanced technology to improve the people's lives in the city. The rapid development of ICT has significantly raised the quality of people's lives and changed the way they work, providing significant advantages in business, education, shopping, and leisure. To improve quality, react faster to changes, and overcome problems, the strives to implement ICT in all elements of the

C. Kahraman et al. (Eds.): INFUS 2021, LNNS 308, pp. 845–852, 2022.
https://doi.org/10.1007/978-3-030-85577-2_97

city such as energy, transport, infrastructure, and facilities, culture, entertainment [2–4]. ICT improves urban infrastructure by involving data collection sensors to ensure the real-time information necessary for resource management and public safety control. The trends that are estimated to be most important are the IoT, 5G, big data, and cloud services. The significant role of ICT is to drive the activities of sustainable and efficient urban services. Table 1 presents indicators of smart city development from the aspect of ICT application.

Table 1. ICT in the smart city concept.

Governance - X_1	Environment – X_2	Buildings – X_3
Digital public administration, transparency and open-data - X_{11} Participation in management, e-management, e-services – X_{12} Smart public safety through ICT solutions – X_{13}	Smart energy – X_{21} Smart water-management systems to enable sustainable water management –X_{22} Smart waste management systems strengthening the application of surveillance and collection and disposal of waste – X_{23}	Connected facility management - X_{31} Smart house - smart building management solutions, enabling users to better monitor – X_{32} Smart construction – X_{33}
Mobility – X_4	**Education – X_5**	**Healthcare – X_6**
Intelligent transport systems management of traffic flows - X_{41} Smart services for public transport - ICT solutions via built-in sensors in vehicles to detect and inform when there is a problem with the vehicle - X_{42} Smart urban logistics - GIS -based maps and visualization - X_{43}	Urban education platforms, digital learning formats – X_{51} Improving e-business and ecommerce skills – X_{52} Connectivity, content management and integration of ICT – X_{53}	Telemedicine applications, electronic medical records – X_{61} Remote diagnosis to the patient at home, integrated health information systems – X_{62} Ambient assisted living – X_{63}

Our research develops a conceptual framework of a smart city from the aspect of ICT integration, with application FAHP type -1 and interval type-2 fuzzy sets.

The paper structure has given in the sequel. In the introduction, the indicators for ICT in the smart city concept are given. The second section gives the methodology and analysis of type-1 and type-2 fuzzy numbers with algorithm. The third section presents results and discussion. The fourth section deals with the conclusion.

2 Methodology

One of the suitable tools for dealing with incomplete or partially known information is the T1FS. In this type of fuzzy set, the membership function takes values from the

[0, 1] and map each element of the universal set into this interval. It turns out that this fact can also be a limitation in many cases. Therefore, Zadeh introduced the fuzzy type-1 set extension to the type-2 fuzzy sets [5]. The uncertainty of the crisp membership degree of T1FS occurring due to human opinion causes a not easy task to operate and quantify uncertainty and determine the perfect weight of membership function, which makes T1FS not always desirable. The solution to this uneasiness in decision-making lies in type-2 fuzziness, where fuzzy sets membership function's grade is a fuzzy-1 set. The fault of the use of T2FS, as an extension of T1FS, is in operating the secondary membership function and highly challenging computation [6]. Due to the lower number of computational operations, researchers usually choose IT2FS, a particular form of T2FS, since the secondary membership function, in general, is equal to 1. Kahraman et al. [7] has introduced an Interval Type-2 FAHP method with a new ranking method for T2FS and Oztaysi in [8] using triangular and trapezoidal IT2FS preferred the uncertainty in investment projects. In this paper, group decision-making is applied on type-1 and interval type-2 fuzzy sets, with FAHP, proposed by Sari et al. [9].

2.1 Type-1 and Type-2 Fuzzy Numbers

Type-1 fuzzy number (T1FN) is a special fuzzy set $\mathcal{F} = \{(x, \mu_{\mathcal{F}}(x)), x \in \mathbb{R}\}$, and $\mu_{\mathcal{F}}(x)$: $\mathbb{R} \to [0, 1]$ is a continuous function. In this paper, triangular and trapezoidal fuzzy numbers have used. A trapezoidal and triangular fuzzy number can be respectively denoted as $\tilde{a} = \left(l, m^l, m^r, u\right)$ and $\tilde{b} = (l, m, u)$ with the membership functions [10, 11]

$$\mu_{\mathcal{F}}(x) = \begin{cases} \frac{x-l}{m^l-l}, & x \in [l, m^l] \\ 1, & x \in [m^l, m^r] \\ \frac{u-x}{u-m^r}, & x \in [m^r, u] \\ 0, & otherwise \end{cases} \quad \mu_{\mathcal{F}}(x) = \begin{cases} \frac{x-l}{m-l}, & x \in [l, m] \\ \frac{u-x}{u-m}, & x \in [m, u] \\ 0, & otherwise \end{cases} \tag{1}$$

Type-2 fuzzy number (T2FN) is defined as a fuzzy set

$$\Phi = \left\{\left((x, y), \mu_{\phi}(x, y)\right) \forall x \in X, \forall y \in J_x \subseteq [0, 1], 0 \leq \mu_{\phi}(x, y) \leq 1\right\} \tag{2}$$

where J_x denotes an interval in [0, 1]. Interval type-2 fuzzy number (IT2FN) is special case of T2FN when membership function $\mu_{\Phi}(x, y) = 1$. The trapezoidal IT2FN can be represent as follows $\tilde{\tilde{a}} = \left(\left(\tilde{a}^U; h_1(\tilde{a}^U), h_2(\tilde{a}^U)\right), \left(\tilde{a}^L; h_1(\tilde{a}^L), h_2(\tilde{a}^L)\right)\right)$, where $\tilde{a}^U = \left(l^U, \left(m^l\right)^U, (m^r)^U, u^U\right)$ and $\tilde{a}^L = \left(l^L, \left(m^l\right)^L, (m^r)^L, u^L\right)$ are trapezoidal T1FN, while $h_1\left(\tilde{a}^U\right), h_2\left(\tilde{a}^U\right), h_1\left(\tilde{a}^L\right)$ and $h_2\left(\tilde{a}^L\right)$ represent middle left and right vertex heights of the upper and the lower trapeze, respectively (see Fig. 1). Heights $h_1\left(\tilde{a}^U\right)$, $h_2\left(\tilde{a}^U\right)$, $h_1\left(\tilde{a}^L\right)$ and $h_2\left(\tilde{a}^L\right)$ belong to the interval [0, 1]. The triangular IT2FN can be represent as follows: $\tilde{\tilde{b}} = \left(\left(\tilde{b}^U; h\left(\tilde{b}^U\right)\right), \left(\tilde{b}^L; h\left(\tilde{b}^L\right)\right)\right)$, where $\tilde{b}^U = \left(l^U, m^U, u^U\right)$ and $\tilde{b}^L = \left(l^L, m^L, u^L\right)$ are triangular T1FN, while $h\left(\tilde{b}^U\right)$ and $h\left(\tilde{b}^L\right)$ represent triangle heights of the upper and the lower triangle, respectively (see Fig. 1). The heights $h\left(\tilde{b}^U\right)$ and $h\left(\tilde{b}^L\right)$ belong to the interval [0, 1].

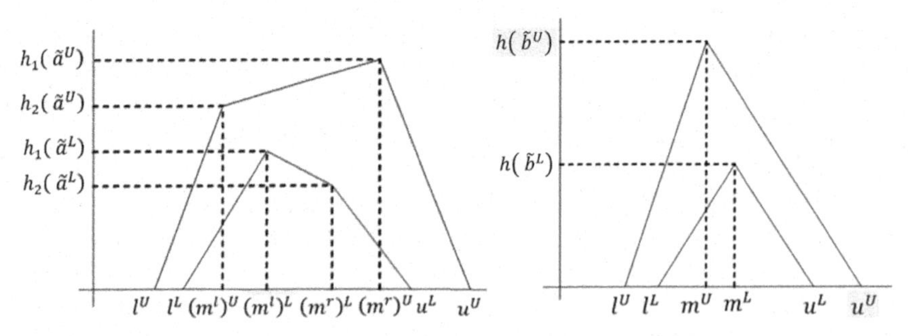

Fig. 1. Graphical representation of trapezoidal (left) and triangular (right) IT2FS

For two trapezoidal IT2FNS $\tilde{\tilde{a}}_1 = \left(\left(\tilde{a}_1^U; h_1(\tilde{a}_1^U), h_2(\tilde{a}_1^U)\right), \left(\tilde{a}_1^L; h_1(\tilde{a}_1^L), h_2(\tilde{a}_1^L)\right)\right)$ and $\tilde{\tilde{a}}_2 = \left(\left(\tilde{a}_2^U; h_1(\tilde{a}_2^U), h_2(\tilde{a}_2^U)\right), \left(\tilde{a}_2^L; h_1(\tilde{a}_2^L), h_2(\tilde{a}_2^L)\right)\right)$ arithmetic operation are defined in Table 2.

Table 2. The arithmetic operation trapezoidal IT2FNs

$$\tilde{\tilde{a}}_1 \oplus \tilde{\tilde{a}}_2 = \begin{pmatrix} \left(\tilde{a}_1^U \oplus \tilde{a}_2^U; \min\left(h_1(\tilde{a}_1^U), h_1(\tilde{a}_2^U)\right), \min\left(h_2(\tilde{a}_1^U), h_2(\tilde{a}_2^U)\right)\right), \\ \left(\tilde{a}_1^L \oplus \tilde{a}_2^L; \min\left(h(\tilde{a}_1^L), h_1(\tilde{a}_2^L)\right), \min\left(h_2(\tilde{a}_1^L), h_2(\tilde{a}_2^L)\right)\right) \end{pmatrix}$$

$$\tilde{\tilde{a}}_1 \ominus \tilde{\tilde{a}}_2 = \begin{pmatrix} \left(\tilde{a}_1^U \ominus \tilde{a}_2^U; \min\left(h_1(\tilde{a}_1^U), h_1(\tilde{a}_2^U)\right), \min\left(h_2(\tilde{a}_1^U), h_2(\tilde{a}_2^U)\right)\right), \\ \left(\tilde{a}_1^L \ominus \tilde{a}_2^L; \min\left(h_1(\tilde{a}_1^L), h_1(\tilde{a}_2^L)\right), \min\left(h_2(\tilde{a}_1^L), h_2(\tilde{a}_2^L)\right)\right) \end{pmatrix}$$

$$\tilde{\tilde{a}}_1 \odot \tilde{\tilde{a}}_2 = \begin{pmatrix} \left(\tilde{a}_1^U \odot \tilde{a}_2^U; \min\left(h_1(\tilde{a}_1^U), h_1(\tilde{a}_2^U)\right), \min\left(h_2(\tilde{a}_1^U), h_2(\tilde{a}_2^U)\right)\right), \\ \left(\tilde{a}_1^L \odot \tilde{a}_2^L; \min\left(h_1(\tilde{a}_1^L), h_1(\tilde{a}_2^L)\right), \min\left(h_2(\tilde{a}_1^L), h_2(\tilde{a}_2^L)\right)\right) \end{pmatrix}$$

$$\tilde{\tilde{a}}_1 \oslash \tilde{\tilde{a}}_2 = \begin{pmatrix} \left(\tilde{a}_1^U \oslash \tilde{a}_2^U; \min\left(h_1(\tilde{a}_1^U), h_1(\tilde{a}_2^U)\right), \min\left(h_2(\tilde{a}_1^U), h_2(\tilde{a}_2^U)\right)\right), \\ \left(\tilde{a}_1^L \oslash \tilde{a}_2^L; \min\left(h_1(\tilde{a}_1^L), h_1(\tilde{a}_2^L)\right), \min\left(h_2(\tilde{a}_1^L), h_2(\tilde{a}_2^L)\right)\right) \end{pmatrix}$$

$$k\tilde{\tilde{a}}_1 = \left(\left(k\tilde{a}_1^U; h_1(\tilde{a}_1^U), h_2(\tilde{a}_1^U)\right), \left(k\tilde{a}_1^L; h_1(\tilde{a}_1^L), h_2(\tilde{a}_1^L)\right)\right)$$

$$\sqrt[n]{\tilde{\tilde{a}}_1} = \left(\left(\sqrt[n]{\tilde{a}_1^U}; h_1(\tilde{a}_1^U), h_2(\tilde{a}_1^U)\right), \left(\sqrt[n]{\tilde{a}_1^L}; h_1(\tilde{a}_1^L), h_2(\tilde{a}_1^L)\right)\right)$$

Algorithm Step 1: Construct the evaluation matrices $A = \left(a_{ij}\right)_{n\times n}$ for all considered preference criteria where $a_{ij}, i, j = 1, 2, \ldots, n$ is equal: a_{ij} in the AHP algorithm; $\tilde{b}_{ij} = \left(l_{ij}, m_{ij}, u_{ij}\right)$ in the triangular type -1 FAHP; $\tilde{\tilde{b}}_{ij} = \left(\left(\tilde{b}_{ij}^U; h\left(\tilde{b}_{ij}^U\right)\right), \left(\tilde{b}_{ij}^L; h\left(\tilde{b}_{ij}^L\right)\right)\right)$ in the triangular interval type-2 FAHP; $\tilde{a}_{ij} = \left(l_{ij}, m_{ij}^l, m_{ij}^r, u_{ij}\right)$ in the trapezoidal type

-1 FAHP; $\tilde{\tilde{a}}_{ij} = \left(\left(\tilde{a}_{ij}^{U}; h_1\left(\tilde{a}_{ij}^{U}\right), h_2\left(\tilde{a}_{ij}^{U}\right)\right), \left(\tilde{a}_{ij}^{L}; h_1\left(\tilde{a}_{ij}^{L}\right), h_2\left(\tilde{a}_{ij}^{L}\right)\right)\right)$ in the trapezoidal interval type-2 FAHP. Corresponding crisp values (CV), linguistic variables, triangular and trapezoidal IT2FNs are in Table 3. The aggregation of different experts' opinions is calculated by the averaging method. Based on the corresponding linguistic assessments of k experts (l_i, m_i, u_i) or $\left(l_i, m_i^l, m_i^r, u_i\right)$, aggregated crisp value has obtained by $\frac{1}{k}\sum_{i=1}^{k} m_i$ or $\frac{1}{2k}\sum_{i=1}^{k}\left(m_i^l + m_i^r\right)$ rounded to the nearest integer. The corresponding linguistic value of the aggregate opinion, in both cases, is then obtained from Table 3. The meanings of linguistic variables are: equal importance (E), very weak dominance (VW), fair weak dominance (FW), slowly weak dominance (SW), slightly strong dominance (SS), fairly strong dominance (FS), very strong dominance (VS), extremely dominance (ED) and absolute dominance (A).

Table 3. Crisp values (CV), linguistic variables (LV), triangular and trapezoidal IT2FNs

CV	LV	Triangular interval type-2 fuzzy scale		Trapezoidal interval type-2 fuzzy scale	
1	E	(1, 1, 3; 1)	(1, 1, 2; 0.9)	(1, 1, 1, 3; 1, 1)	(1, 1, 1, 2; 0.9, 0.9)
2	VW	(1, 2, 3; 1)	(1.5, 2, 2.5; 0.9)	(1,1.5,2.5,3;1,1)	(1.5, 1.75, 2.25, 2.5; 0.9, 0.9)
3	FW	(1, 3, 5; 1)	(2, 3, 4; 0.9)	(1, 2, 4, 5; 1, 1)	(2, 2.5, 3.5, 4; 0.9, 0.9)
4	SW	(3, 4, 5; 1)	(3.5, 4.4, 5; 0.9)	(3, 3.5, 4.5, 5; 1, 1)	(3.5, 3.75, 4.25, 4.5; 0.9, 0.9)
5	SS	(3, 5, 7; 1)	(4, 5, 6; 0.9)	(3, 4, 6, 7; 1, 1)	(4, 4.5, 5.5, 6; 0.9, 0.9)
6	FS	(5, 6, 7; 1)	(5.5, 6, 6.5; 0.9)	(5, 5.5, 6.5, 7; 1, 1)	(5.5, 5.75, 6.25, 6.5; 0.9, 0.9)
7	VS	(5, 7, 9; 1)	(6, 7, 8; 0.9)	(5, 6, 8, 9; 1, 1)	(6, 6.5, 7.5, 8; 0.9, 0.9)
8	ED	(7, 8, 9; 1)	(7.5, 8, 8.5; 0.9)	(7, 7.5, 8.5, 9; 1, 1)	(7.5, 7.75, 8.25, 8.5; 0.9, 0.9)
9	A	(7, 9, 9; 1)	(8, 9, 9; 0.9)	(7, 9, 9, 9; 1, 1)	(8, 9, 9, 9; 0.9, 0.9)

Step 2: According to the approach like in [12], the consistency of the evaluation matrices $A = \left[a_{ij}\right]_{n\times n}$ has been examined. The matrix consistency index CI and consistency index CR are: $CI = \frac{\lambda_{max}-n}{n-1}$ and $CR = \frac{CI}{RI}$. If the consistency index CR is less than 0.1, it is considered acceptable. Then the consistency of the corresponding matrices in the case of type-1 and interval type-2 are also acceptable.

Step 3: Calculate, for each row, the geometric mean:

$$g_i = [a_{i1} \odot a_{i2} \odot \cdots \odot a_{in}]^{\frac{1}{n}}, i = 1, 2, \ldots, n, \tag{3}$$

where a_{ij} is the same as in Step 1.

Step 4: Calculate the fuzzy weights of each criterion:

$$w_i = g_i \odot \left[g_1 \oplus g_2 \oplus \cdots \oplus g_n\right]^{-1}, i = 1, 2, \ldots, n. \tag{4}$$

Step 5: The defuzzified value is obtained using the center of area method [14]:

- $\frac{1}{4}(l + 2m + u)$ in the case of triangular type-1 FAHP;
- $\frac{1}{8}\left(l^U + u^U + l^L + u^L + 2h\left(\tilde{b}_{ij}^U\right)m^U + 2h\left(\tilde{b}_{ij}^L\right)m^L\right)$ in the triangular interval type-2 FAHP;
- $\frac{1}{4}\left(l + m^l + m^r + u\right)$ in the trapezoidal type-1 FAHP
- $\frac{1}{8}\left(l^U + u^U + l^L + u^L + h\left(\tilde{a}_{ij}^U\right)\left(\left(m^l\right)^U + (m^r)^U\right) + h\left(\tilde{a}_{ij}^L\right)\left(\left(m^l\right)^L + (m^r)^L\right)\right)$ in the trapezoidal interval type-2 FAHP

3 Results

In this section, an algorithm from Sect. 2.2 has been applied. The significance of each criterion and sub-criterion have linguistically expressed in Table 3. In Table 4, a pairwise comparison of criteria and sub-criteria obtained from experts is given. Since all comparison matrices hold that CR < 0.1, they are consistent.

Table 4. Pairwise evaluation matrix of criteria and sub-criteria

	X_1	X_2	X_3	X_4	X_5	X_6
X_1	1	E	VW	FW	FW	FW
X_2	1/ E	1	VW	FW	FW	FW
X_3	1/ VW	1/ VW	1	VW	VW	VW
X_4	1/ FW	1/ FW	1/ VW	1	E	E
X_5	1/ FW	1/ FW	1/ VW	1/E	1	E
X_6	1/ FW	1/ FW	1/ VW	1/E	1/E	1

	X_{12}	X_{11}	X_{13}		X_{21}	X_{22}	X_{23}		X_{31}	X_{33}	X_{32}
X_{12}	1	VW	FW	X_{21}	1	VW	VW	X_{31}	1	FW	SS
X_{11}	1/VW	1	VW	X_{22}	1/VW	1	E	X_{33}	1/FW	1	FW
X_{13}	1/FW	1/VW	1	X_{23}	1/VW	1/ E	1	X_{32}	1/SS	1/FW	1
	X_{41}	X_{43}	X_{42}		X_{52}	X_{53}	X_{51}		X_{61}	X_{62}	X_{63}
X_{41}	1	VW	FW	X_{52}	1	FW	SS	X_{61}	1	FW	SS
X_{43}	VW	1	VW	X_{53}	1/FW	1	FW	X_{62}	1/FW	1	FW
X_{42}	1/FW	1/VW	1	X_{51}	1/SS	1/FW	1	X_{63}	1/SS	1/FW	1

Aggregation of four expert opinions was performed as specified in the Step 1 of the algorithm. The final ranking results are given in Fig. 2

Using the proposed algorithms, different rankings were obtained, one using the FAHP method and using the type-1 and interval type-2 FAHP methods. In ranking similarity analysis, the Spearman rank correlation coefficient is used the most [15]

$$r_s = 1 - \frac{6\sum_{i=1}^{n} d_i^2}{n\left(n^2 - 1\right)}, d_i = R_{x_i} - R_{y_i}. \tag{5}$$

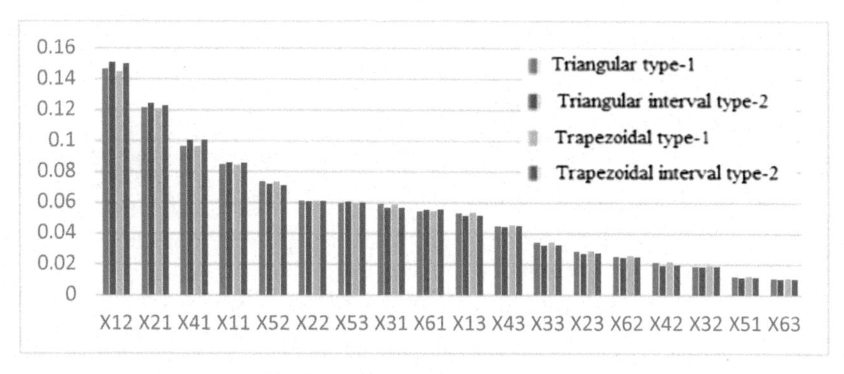

Fig. 2. Final ranking of sub-criteria

Salabun and Urbaniuk have recently introduced a WS coefficient, where changes of the ranks on the top of ranking have more influence on coefficient's value [16]

$$WS = 1 - \sum_{i=1}^{n}\left(2^{-R_{x_i}} \frac{\left|R_{x_i} - R_{y_i}\right|}{\max\left\{\left|1 - R_{x_i}\right|, \left|n - R_{x_i}\right|\right\}}\right). \tag{6}$$

In formulas (5) and (6), n is number of elements in the ranking, while R_{x_i} and R_{y_i} are ranks of the element i in the compared rankings. Using formulas (5) and (6), the ranking obtained by AHP and four FAHP algorithms have compared and the coefficients $r_s = 0.994$ and $WS = 0.997$ were obtained. Since the obtained value for $WS = 0.997 > 0.880$, one can see, according to Salabun, that the rankings have high similarity.

4 Conclusion

In this paper, the opportunities for smart city development, from the aspect of ICT application, have been explored. Given the role of ICT in smart cities, this research develops a model for priority activities in city transformation into sustainable smart cities. Methodologically, the paper contribution is the parallel application FAHP with type -1 fuzzy sets and interval type-2 fuzzy sets for ICT implementation in smart cities. The dominant measures are participation in management, e-management, e-services. The results show significant similarity in the application of these methods. Future research will relate to the application of newly developed methods spherical fuzzy analytical hierarchy process in the field of smart city development.

References

1. Angelidou, M., Psaltoglou, A., Komninos, N., Kakderi, C., Tsarchopoulos, P., Panori, A.: Enhancing sustainable urban development through smart city applications. J. Sci. Technol. Policy Manag. **9**(2), 146–169 (2018)
2. Milošević, D., Stanojević, A., Milošević, M: AHP method in the function of logistic in development of smart cities model. In: Proceeding of the 6th International Conference: Transport and logistic Til, pp. Niš (2017)

3. Milošević, M.R., Milošević, D.M., Stanojević, A.D., Stević, D.M., Simjanović, D.J.: Fuzzy and interval AHP approaches in sustainable management for the architectural heritage in smart cities. Mathematics **9**, 304 (2021)
4. Milošević, M., Milošević, D., Stanojević, A.: Managing Cultural Built Heritage in Smart Cities Using Fuzzy and Interval Multi-criteria Decision Making. In: Kahraman, C., Cevik Onar, S., Oztaysi, B., Sari, I.U., Cebi, S., Tolga, A.C. (eds.) INFUS 2020. AISC, vol. 1197, pp. 599–607. Springer, Cham (2021). https://doi.org/10.1007/978-3-030-51156-2_69
5. Zadeh, L.A.: The concept of a linguistic variable and its application to approximate reasoning – I. Inf. Sci. **8**(3), 199–249 (1975)
6. Mendel, J.M., John, R.I., Liu, F.L.: Interval type-2 fuzzy logical systems made simple. IEEE Trans. Fuzzy Syst. **14**, 808–821 (2006)
7. Kahraman, C., Oztaysi, B., Sari, I.U., Turanoglu, E.: Fuzzy analytic hierarchy process with interval type-2 fuzzy sets. Knowl. Based Syst. **59**, 48–57 (2014)
8. Oztaysi, B.: A group decision making approach using interval type-2 fuzzy AHP for enterprise information systems project selection. J. Mul. Valued Log. Soft Comput. **24**, 475–500 (2015)
9. Sari, I.U., Oztaysi, B., Kahraman, C.: Fuzzy AHP using type II fuzzy sets: an application to warehouse location selection. In Doumpos, M., Grigoroudis, E. (eds.) Multicriteria Decision Aid and Artificial Intelligence: Links, Theory and Applications, pp. 258–308. Wiley (2013)
10. Milošević, D.M., Milošević, M.R., Simjanović, D.J.: Implementation of adjusted Fuzzy AHP method in the assessment for reuse of industrial buildings. Mathematics **8**, 1697 (2020)
11. Ozkok, B.A.: Finding fuzzy optimal and approximate fuzzy optimal solution of fully fuzzy linear programming problems with trapezoidal fuzzy numbers. J. Intell. Fuzzy Syst. **36**, 1389–1400 (2019)
12. Milošević, M.R., Milošević, D.M., Stević, D.M., Stanojević, A.D.: Smart city: modeling key indicators in serbia using IT2FS. Sustainability **11**(13), 3536 (2019)
13. Milošević, A., Milošević, M., Milošević, D.,Selimi, A.: AHP multi-criteria method for sustainable development in construction. In: Proceedings of the 4th International Conference, Contemporary Achievements in Civil Engineering, pp. 929–938. Faculty of Civil Engineering, Subotica, Serbia (2016)
14. Do, Q.H., Chen, J.F., Hsieh, H.N.: Trapezoidal fuzzy AHP and fuzzy comprehensive evaluation approaches for evaluating academic library service. WSEAS Trans. Comput. **14**, 2224–2872 (2015)
15. Ceballos, B., Lamata, M.T., Pelta, D.A.: A comparative analysis of multi-criteria decision-making methods. Progress Artif. Intell. **5**(4), 315–322 (2016). https://doi.org/10.1007/s13748-016-0093-1
16. Sałabun, W., Urbaniak, K.: A new coefficient of rankings similarity in decision-making problems. In: Krzhizhanovskaya, V.V., Závodszky, G., Lees, M.H., Dongarra, J.J., Sloot, P.M.A., Brissos, S., Teixeira, J. (eds.) ICCS 2020. LNCS, vol. 12138, pp. 632–645. Springer, Cham (2020). https://doi.org/10.1007/978-3-030-50417-5_47

A New Cuckoo Search Algorithm Using Interval Type-2 Fuzzy Logic for Dynamic Parameter Adaptation

Maribel Guerrero, Fevrier Valdez(✉), and Oscar Castillo

Tijuana Institute of Technology, Tijuana, México
maribel.guerrero@tectijuana.edu.mx, {fevrier, ocastillo}@tectijuana.mx

Abstract. The main objective of this work is to select the best parameters to be dynamically adjusted in the Cuckoo Search via Lévy Flight (CS) Algorithm using interval type 2 fuzzy logic. The main objective is to dynamically change the parameters, our fuzzy will include a Mamdani type system that will include an input and output to dynamically adjust the algorithm and improve the performance of the optimized CS algorithm. In order to demonstrate the performance and results of the algorithm, the comparison is made between the original algorithm, type 1 fuzzy logic and type, tests were performed with benchmark mathematical functions. The results of the simulation shows that the proposed algorithm has good results with respect to using type-1 fuzzy logic in CS or using the original CS algorithm without parameter adaptation. Lévy flights are used in the algorithm with the intention of exploring solutions and randomness solutions in space.

Keywords: Cuckoo search · Levy flights · Interval type-2 · Fuzzy logic · Parameters

1 Introduction

Nature has its own intelligence, which is why it is a source of inspiration for various bio-inspired algorithms [1], based on evolution, intelligence of swarms of species of bees, fish, birds or based on ecology, herbage, physics or astrology, researchers analyze and study the behavior of nature to be able to carry out mathematical models in the form of optimization problems [2–4].

In the real world, there are many complex engineering problems, which is why computer science allows giving rise to the bio-inspired algorithms that we observe in nature and which can be programmed to solve various science and engineering problems [5].

There are various meta-heuristics that have the purpose of improving the results of complex problems, the main objective is to optimize costs, resources, time and contributing to the environment [6–10].

Cuckoo search with Type–1 fuzzy logic (FCS1), in this work we have implemented the algorithm with dynamic parameter adjustment using intervals type - 2 fuzzy logic

C. Kahraman et al. (Eds.): INFUS 2021, LNNS 308, pp. 853–860, 2022.
https://doi.org/10.1007/978-3-030-85577-2_98

(FCS2). In the literature, we find that several researchers have implemented type 2 to their goal heuristics to improve results [7].

With the support of type 1 fuzzy sets, it has been shown that there are various contributions to optimizing the results [9], fuzzy rules, adjusting the parameters and membership functions, perform statistical tests that show that promising results with type 2 fuzzy set [10].

Below it is indicated how this work was organized: the Sect. 2 shows the background and basic concepts, bioinspired algorithms, type 1 and type 2 fuzzy sets by intervals and on the CS algorithm (Cuckoo Search).

In Sect. 3 you can find the purpose of our method, Sect. 4 the comparisons and results obtained from the simulations and Sect. 5 the conclusions of this paper.

2 Basic Concepts

The basic concepts presented in this section help to understand the present work.

2.1 Bio-inspired Algorithms

The Bio-inspired algorithms search for optimal solutions to complex problems, these algorithms go in search of better solutions, for discover the best fitness, the fitness is compared in each iterative process, each of these optimization solutions is known as a local optimum (in this paper we are looking for a minimum fitness, however, it depends on the orientation of the problem you can go in search of maximums value). When working with stochastic algorithms we can lose promising solutions, it can also be the case of being trapped in a local optimum and not finding good results, therefore, it is important to know the algorithm and that the parameters keep their balance of exploration and exploitation.

2.2 The Type – 1 Fuzzy Set and Proposed

In classical logic there are only two values, on the contrary in fuzzy set it allows to have a set of values between [0, 1], this helps us in process optimization, implement it in fuzzy controllers, transferring the way of thinking of a human being that they are not numbers, if not on the contrary, they are linguistic variables, in order to program rules are required, which must come from an expert, we can program them of the If-Then type and be able to process information with a level of uncertainty, and transfer this knowledge into a controller or any algorithm that needs to be optimized. The first controller to use type 1 fuzzy set was in the United Kingdom for a steam engine [11].

In recent years we can observe that in everyday life we already have control systems that implement fuzzy logic, in our homes we have washing machines with the fuzzy system label, the control systems of current washing machines give the option of selecting water levels, temperature, force in the wash depends on the amount of clothes and type of fabric to be able to make better decisions about the care of the clothes, energy and water care.

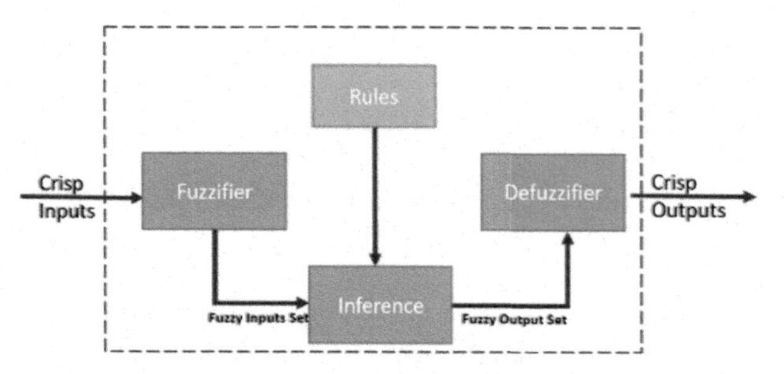

Fig. 1. Diagram description of Mamdani FCS1 FIS

In Fig. 1 describing an Mamdani system FCS1, a one crisp input and output, the fuzzy input set passes to an inference system that contains if-then rules, then, the fuzzy output set enters a defuzzifier, which has as crisp outputs (which results in a single real number).

2.3 Description the Interval Type 2 Fuzzy Sets

In this section description of the interval of Type-2 Fuzzy Sets, has vagueness and also represents a degree of uncertainty [12]. In the Fig. 2 shown the Propose of the Type-2 Fuzzy Sets.

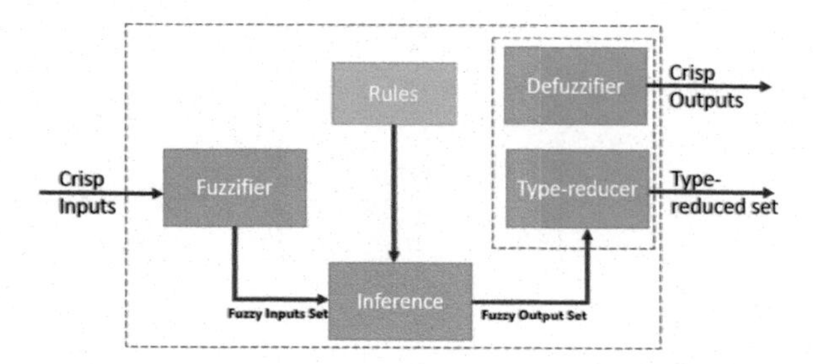

Fig. 2. Diagram description of Mamdani FCS2 and FCS3 FIS.

In the Fig. 2 presents the interval of the FCS2, which a input and two outputs are possible, in the output is a single real number [13].

FCS1 Fis includes a fuzzfier, inference system, fuzzy rules, defuzzifier and FCS2 Fis in its output includes a type reducer.

2.4 Cuckoo Search Algorithm

In 2009, the creators of the bio-inspired algorithm of Cuckoo Search via levy flight was Xin-she Yang and Suash Deb [14, 15]. It mimics the behavior of birds called Cuco, which have a very peculiar maternity approach, because they lay their eggs in nests host nests including other types of bird species, they have the ability to camouflage their eggs so that they are not discovered and have a greater probability of survival [16].

The CS algorithm uses Lévy flights to add randomness to the selection of new solutions and also to explore solution space.

The CS algorithm has the following rules [14, 16]:

1. The first rule indicates that each cuckoo bird will have a single egg, the flights of levy help to select nests to place a new egg.

$$x_i^{(t+1)} = x_i^t + \alpha \cdot \text{randn(size(D))} \otimes \boldsymbol{L\acute{e}vy}(\beta) \otimes s_i \tag{1}$$

$x_i^{(t+1)}$ The new nest positions are stored in a vector, **t**: It is the number of the iteration, x_i^t It is a vector that contains the current position of the nest, **α**: It is a constant it is recommended that it has a value greater than zero, ***randn*** is random scalar value, ***Lévy*** **(*β*)** : It is a Lévy flight and D: Dimensions.

$$\boldsymbol{Si} = (x_i^t - \boldsymbol{xbest}) \tag{2}$$

x_i^t: It contains the position of the nest that we are evaluating and is stored in this vector, ***xbest***: The vector saves the most promising solution, if $x_i^t = \boldsymbol{Xbest}$, the solution is maintained.

$$\boldsymbol{L\acute{e}vy}(\boldsymbol{\beta}) = \frac{\sigma(\beta) \cdot randn}{|randn|^{\frac{1}{\beta}}} \tag{3}$$

β: It is a constant (1 ≤ β ≤ 2). The following equation shows how to find the sigma standard deviation:

$$\boldsymbol{\sigma}(\boldsymbol{\beta}) = \left\{ \frac{\Gamma(1+\beta)\sin(\frac{\pi\beta}{2})}{\Gamma\left[\frac{1+\beta}{2}\right]\beta 2^{(\frac{\beta-1}{2})}} \right\}^{1/\beta}, \boldsymbol{\sigma}(\beta) = 1, \tag{4}$$

The gamma function is described below:

$$\boldsymbol{\Gamma}(z) = \int_0^\infty t^{z-1} e^{-t} dt \tag{5}$$

2. In the next iteration, the eggs with the best quality.
3. The number of nests is maintained, the $pa \in [0, 1]$ is the probability of discovering the intrusive egg in a communal nest. In case of discovery a new nest is generated
4. In the second rule, the quality or fitness of a solution is proportional to the objective function.

3 Proposal Method

In the Fig. 3 we see the proposal of the method in which we can see that the cuckoo search method will be analyzed with type 1 fuzzy logic (FCS1) and interval type 2 fuzzy logic (FCS2).

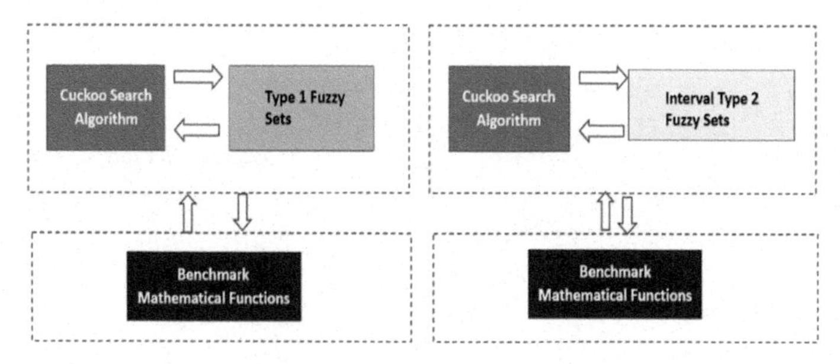

Fig. 3. . Proposal method

Table 1. Rules of the FSC1 and FCS2 method

Type system: Mamdani
3 Rules

If (Iteration is Low) then (Alpha is High)
If (Iteration is Medium) then (Alpha is Medium)
If (Iteration is High) then (Alpha is Low)

Table 2. The parameters used for simulations

Parameters	CS	FCS1	FCS2
Pa		75%	
Beta		1.5	
Alpha	5%	Dynamic	Dynamic

In the Table 1, we show at the three rules that our type 1 and 2 fuzzy inference system will have, with an input (iteration) and an output (alpha), the inference system is of type Mamdani. The parameter that we will use with dynamic adjustment is Alpha, it is the most indicated to improve the results [17].

The Table 2 shows the parameters used for simulations in CS, FCS1 and FCS2, with a population of 100 nests.

4 Simulation Results

Below are the results obtained in comparison with CS, FCS1 and FCS2.

In the Tables 3, 4, 5, 6 and 7 show the simulations with five benchmark functions, the original CS algorithms, FCS1, FCS2 are compared of average and standard deviation are displayed for each function, with 8, 16, 32, 64 and 128 dimensions.

Table 3. Results for the Rosenbrock function with the CS, FCS1 and FCS2.

Rosenbrock function							
Average				Standard deviation			
Dim.	CS	FCS1	FCS2	Dim.	CS	FCS1	FCS2
8	6.62E-08	9.67E-12	2.05E-11	8	1.17E-07	2.40E-11	4.94E-11
16	5.75E-01	5.68E-02	7.79E-02	16	6.05E-01	6.65E-02	7.55E-02
32	1.91E+01	1.75E+01	1.89E+01	32	4.81E+00	6.68E-01	3.45E-01
64	6.10E+01	5.60E+01	5.78E+01	64	3.87E+01	3.00E-01	2.21E-01
128	3.66E+02	1.25E+02	1.29E+02	128	4.23E+01	3.47E-01	9.34E-01

Table 4. Results for the Sphere function with the CS, FCS1 and FCS2

Sphere function							
Average				Standard deviation			
Dim.	CS	FCS1	FCS2	Dim.	CS	FCS1	FCS2
8	7.91E-46	3.07E-85	5.18E-92	8	1.61E-45	8.86E-85	8.04E-92
16	6.78E-27	2.65E-48	4.39E-48	16	3.13E-27	1.39E-48	2.76E-48
32	7.01E-15	2.36E-23	1.31E-20	32	2.88E-15	1.18E-23	6.57E-21
64	6.43E-08	5.22E-10	1.67E-07	64	2.01E-08	1.94E-10	4.38E-08
128	8.33E-04	1.54E-03	3.33E-02	128	1.54E-04	3.19E-04	5.79E-03

Table 5. Results for the Ackley function with the CS, FCS1 and FCS2.

Ackley function							
Average				Standard deviation			
Dim.	CS	FCS1	FCS2	Dim.	CS	FCS1	FCS2
8	7.92E-15	1.95E-15	2.31E-15	8	3.25E-15	1.64E-15	1.76E-15
16	2.63E-07	4.44E-15	4.44E-15	16	2.99E-07	0.00E+00	0.00E+00
32	1.29E-03	3.27E-11	7.74E-10	32	1.12E-03	1.09E-11	1.81E-10
64	2.11E-01	1.06E-04	1.73E-03	64	2.71E-01	1.95E-05	2.44E-04
128	2.79E+00	2.00E-01	1.55E+00	128	2.52E-01	3.20E-02	1.90E-01

Table 6. Results for the Rastrigin function with the CS, FCS1 and FCS2.

Rastrigin function							
Average				Standard deviation			
Dim.	CS	FCS1	FCS2	Dim.	CS	FCS1	FCS2
8	1.08E-02	0.00E+00	0.00E+00	8	4.06E-02	0.00E+00	0.00E+00
16	1.63E+01	9.14E+00	1.00E+01	16	2.35E+00	3.39E+00	3.17E+00
32	7.83E+01	7.66E+01	7.99E+01	32	1.02E+01	1.70E+01	1.88E+01
64	2.67E+02	3.05E+02	3.38E+02	64	2.33E+01	4.80E+01	4.58E+01
128	7.72E+02	8.91E+02	9.49E+02	128	4.16E+01	6.91E+01	6.58E+01

Table 7. Results for the Griewank function with the CS, FCS1 and FCS2

Griewank function							
Average				Standard deviation			
Dim.	CS	FCS1	FCS2	Dim.	CS	FCS1	FCS2
8	2.84E-02	1.76E-02	1.35E-02	8	9.53E-03	8.12E-03	7.88E-03
16	5.01E-07	8.66E-07	5.29E-07	16	3.39E-06	5.21E-06	8.41E-13
32	8.76E-08	3.24E-15	3.98E-09	32	1.11E-07	6.85E-15	1.80E-08
64	3.57E-04	1.98E-06	2.74E-04	64	9.93E-04	1.71E-06	1.41E-04
128	1.72E-01	2.90E-01	1.12E+00	128	3.56E-02	5.09E-02	2.51E-02

5 Conclusions

In this work we show five sets of simulations, benchmark functions are compared, the alpha variable has a value of 5% in the original CS, while the algorithm with FCS1 and interval FCS2 are dynamically adjusted.

We conclude that the proposed algorithm has good results by dynamic parameter adjustment compared to the original. The sphere function shows good results in 8 dimensions for FCS2, while with FCS1 promising results are found in 16 to 64 dimensions, the original algorithm only showed a better solution in 128 dimensions. In the Ackley function in all cases the FCS1 algorithm shows good results compared to FCS2 and the original algorithm. The Rastrigin function showed good results in 8 dimensions with FCS1 and FCS2, while CS shows good the best standard deviation from 16 to 128 dimensions. The Griewank function for 8, 16 and 128 dimensions finds the best deviation with FCS2 and for 32 and 64 dimensions it was for FCS1.

In the future work analyze the behavior of the FOU (uncertainty footprint) to see evaluate its behavior when increasing or decreasing the size.

References

1. Yang, X.-S.: Nature-Inspired Metaheuristic Algorithms. Luniver Press, Bristol (2010)

2. Caraveo, C., Valdez, F., Castillo, O.: A new optimization meta-heuristic algorithm based on self-defense mechanism of the plants with three reproduction operators. Soft. Comput. **22**(15), 4907–4920 (2018). https://doi.org/10.1007/s00500-018-3188-8
3. Perez, J., et al.: Bat algorithm with parameter adaptation using interval type-2 fuzzy logic for benchmark mathematical functions. In: 2016 IEEE 8th International Conference on Intelligent Systems (IS). IEEE (2006)
4. Yang, X.-S.: Bat algorithm for multi-objective optimisation. Int. J. Bio-Inspir. Comput. **3**(5), 267–274 (2011)
5. Olivas, F., et al.: Ant colony optimization with dynamic parameter adaptation based on interval type-2 fuzzy logic systems. Appl. Soft Comput. **53**, 74–87 (2017)
6. Amador-Angulo, L., Castillo, O.J.S.C.: A new fuzzy bee colony optimization with dynamic adaptation of parameters using interval type-2 fuzzy logic for tuning fuzzy controllers. Soft Comput. **22**(2), 571–594 (2018)
7. Olivas, F., et al.: Dynamic parameter adaptation in particle swarm optimization using interval type-2 fuzzy logic. Soft Comput. **20**(3), 1057–1070 (2016)
8. Olivas, F., et al.: Interval type-2 fuzzy logic for dynamic parameter adaptation in a modified gravitational search algorithm. Inf. Sci. **476**, 159–175 (2019)
9. Amador-Angulo, L., Castillo, O.: Optimal design of fuzzy logic systems through a chicken search optimization algorithm applied to a benchmark problem. In: Melin, P., Castillo, O., Kacprzyk, J. (eds.) Recent Advances of Hybrid Intelligent Systems Based on Soft Computing. SCI, vol. 915, pp. 229–247. Springer, Cham (2021). https://doi.org/10.1007/978-3-030-58728-4_14
10. Castillo, O., et al.: A high-speed interval type 2 fuzzy system approach for dynamic parameter adaptation in metaheuristics. Eng. Appl. Artif. Intell. **85**, 666–680 (2019)
11. Zadeh, L.A.: Fuzzy logic. Computer **21**(4), 83–93 (1988)
12. Wu, D., Mendel, J.M.J.E.A.o.A.I.: Recommendations on designing practical interval type-2 fuzzy systems. Eng. Appl. Artif. Intell. **85**, 182–193 (2019)
13. Sanchez, M.A., Castillo, O., Castro, J.R.: Method for measurement of uncertainty applied to the formation of interval type-2 fuzzy sets. In: Design of Intelligent Systems Based on Fuzzy Logic, Neural Networks and Nature-Inspired Optimization, pp. 13–25. Springer (2015). https://doi.org/10.1007/978-3-319-17747-2_2
14. Yang, X.-S., Deb, S.: Cuckoo search via Lévy flights. In: 2009 World Congress on Nature & Biologically Inspired Computing (NaBIC). IEEE (2009)
15. Shehab, M., et al.: Hybridizing cuckoo search algorithm with bat algorithm for global numerical optimization. J. Supercomput. **75**(5), 2395–2422 (2019)
16. Guerrero, M., Castillo, O., García, M.: Cuckoo search algorithm via Lévy flight with dynamic adaptation of parameter using fuzzy logic for benchmark mathematical functions. In: Design of Intelligent Systems Based on Fuzzy Logic, Neural Networks and Nature-Inspired Optimization, pp. 555–571. Springer, Cham (2015)
17. Guerrero, M., et al.: A new algorithm based on the cuckoo search with dynamic adaptation of parameters using fuzzy systems. J. Univ. Math. **1**(1), 32–61 (2018)

Z Numbers

Monitoring Stability of Plant Species to Harmful Urban Environment Under Z-information

Olga M. Poleshchuk(✉)

Moscow Bauman State Technical University, Moscow, Russia

Abstract. The paper developed a model for monitoring of plant species in urban environment under Z-information. The main task of green species monitoring in the city is to determine the types of plants that are the most adapted to the ecological conditions of the urban environment in order to plan various activities carried out at landscaping objects. When assessing the state of plantings, even a highly qualified expert cannot always be absolutely sure of assigning plants to one of the status categories. The information coming from experts is given with a certain degree of reliability, which must be taken into account. Previously, to determine the state of plant species, the reliability of the incoming information was never taken into account due to the lack of a mathematical apparatus. To formalize expert assessments of plants, the author uses Z-numbers, the components of which are the values of semantic spaces. To determine these values, a method developed by the author earlier is used. Assessments of the state of plant species are found by comparative analysis with an ideal reference state. For this, a distance between the state of the plant species, represented as a set of Z-numbers, and the ideal state is determined. A numerical example of monitoring the state of plant species on the Boulevard Ring avenue of Moscow is given.

Keywords: Z-number · Plant species · Monitoring · Semantic space

1 Introduction

The harmful effects of the cities undoubtedly have a detrimental effect on the plants growing within them. The resistance of different plant species to this effect significantly affects the quality of their functions: air purification, regulation of humidity, both air and soil, creation of recreation areas and protection from heat, etc. In order to identify the resistance of plants to the climatic conditions of cities and identify the most resistant ones for inclusion in the landscaping plan, the state of various plant species is regularly monitored. The assessment of the state of plants is carried out both by physical devices and visually by experts who cannot be absolutely sure of the complete reliability of the information they provide. However, this fact was not taken into account for many years due to the lack of an adequate mathematical apparatus. The possibility of eliminating this deficiency appeared after determining in 2011 the concept of Z-number and Z-information by Professor Lotfi Zadeh [1].

C. Kahraman et al. (Eds.): INFUS 2021, LNNS 308, pp. 863–870, 2022.
https://doi.org/10.1007/978-3-030-85577-2_99

To process Z-information (information with Z-numbers), it was necessary to develop a mathematical apparatus that includes operations on Z-numbers, ranking Z-numbers, the distance between them, etc.

In the papers [2–6], operations on Z-numbers are developed. In these papers, different approaches to operating are proposed, which are based on the probabilistic representation of the second component of Z-numbers [2, 3] and on the developed aggregating indicators of Z-numbers [4–6].

There are several different approaches to ranking Z-numbers [7]. In [8], the authors propose to rank Z-numbers based on a comparative analysis of their components. In [9], the authors propose to rank Z-numbers using an extension of the expected utility function.

To determine the distance between Z-numbers, the authors [6] use the weighted first components with the help of defuzzification the second components [10]. The authors [8] use the weighted first components of Z-numbers together with the both components of Z-numbers. In [11], to determine the distance the authors use the Jaccard similarity measure. In [5], for determination the distance between Z -numbers the author uses the aggregating segments of Z-numbers.

The developed mathematical apparatus for Z-information processing is used in [12–15] to support decision making.

For the processing of Z-information, methods have been developed that allow formalizing both components of Z-numbers and presenting them as values of linguistic variables with the properties of completeness and orthogonality [16–20].

These methods have been applied in [21] to determine the ratings of plant species based on linguistic variables. In fact, this is the only paper that proposes an approach to assessing plant species based on fuzzy set theory. The example given in it showed the effectiveness and advantages of this approach compared to the traditional scoring approach. However, the absence of a developed apparatus for taking into account the reliability of expert information did not allow to formalize the assessment of reliability when processing that information.

In this paper, this drawback is eliminated using a model that allows us to monitor the stability of plant species taking into account the reliability of expert assessments.

The numerical example given in the paper demonstrates the effectiveness of the developed model in monitoring plant species on the Boulevard Ring avenue of Moscow.

The paper contains five sections. Section 2 includes the necessary basic concepts and definitions. Section 3 presents the model for monitoring of plant species in urban environment under Z-information developed by the author. Section 4 provides a numerical example. Section 5 concludes this paper.

2 Basic Concepts and Definitions

A linguistic variable with the name X and the names of its values $T(X) = \{X_l,\ l = \overline{1, m}\}$ is called a collection $\{X, T(X), U, V, S\}$. The values of the linguistic variable are fuzzy variables. The names of the values are determined by the rule V, and the rule S defines for each fuzzy variable the corresponding fuzzy set of U[16].

In [22], the definition of weighted point for fuzzy number is given. Based on [22], in [23] the definition of weighted segment $[c_1, c_2]$ for fuzzy number $\tilde{A} = (a_1, a_2, a_L, a_R)$ is given: $c_1 = \int_0^1 \frac{2a_1-(1-\alpha)a_L}{2} 2\alpha d\alpha = a_1 - \frac{1}{6}a_L$, $c_2 = \int_0^1 \frac{2a_2+(1-\alpha)a_R}{2} 2\alpha d\alpha = a_2 + \frac{1}{6}a_R$.

In [21], a method was developed to formalize the state of plant species based on statistical information. This method is based on the frequency approach and the concept of geometric probability. The initial information is the result of an expert assessment of plants within a scale with seven linguistic levels $A_l,\ l = \overline{1,7}$: A_1 - «old dead», A_2 - «recently died», A_3 - «drying up», A_4 - «very weak», A_5 - «mean weak», A_6 - «moderately weak», A_7 - «no signs of weakness». To construct the membership function for each level, the relative number of plants that were assigned to it during the assessment is calculated.

For example, if K plants were assessed, of which $k_l,\ l = \overline{1,7}$ was assigned respectively to the l-th level of the scale, then the membership functions of levels $A_l,\ l = \overline{1,7}$ will be constructed in such a way that the area of the figure (triangle or trapezoid) bounded by the graph of the corresponding function is equal to $\frac{k_l}{K},\ l = \overline{1,7}$. Only neighboring functions have intersection points, the ordinate of the intersection point is 0.5, and the abscissas of the intersection points are $\frac{k_1}{K}, \frac{k_1+k_2}{K}, \frac{\sum_{l=1}^{3} k_l}{K}, \frac{\sum_{l=1}^{4} k_l}{K}, \frac{\sum_{l=1}^{5} k_l}{K}, \frac{1-k_7}{K}$.

A Z - number is a pair of two fuzzy numbers $Z = \left(\tilde{A}, \tilde{R}\right)$, in which the first number $\tilde{A}$ is the value of a fuzzy variable, and the second number $\tilde{R}$ is a fuzzy value of the reliability of the first number [1].

3 Monitoring Stability of Plant Species Under Z-information

To assess the state of plant species, experts use a scale with seven linguistic values A_l, $l = \overline{1,7}$: A_1 - «old dead», A_2 - «recently died», A_3 - «drying up», A_4 - «very weak», A_5 - «mean weak», A_6 - «moderately weak», A_7 - «no signs of weakness». To assess the reliability of plant health assessments, experts use a scale with five linguistic values: U - «Unlikely», NVL - «Not very likely», L - «Likely», VL - «Very likely», EL - «Extremely likely».

The scales used are formalized using linguistic variables according to methods [19, 20]. The state of the i-th plant species is represented as a linguistic variable with values $A_l,\ l = \overline{1,7}$ and membership functions $\mu_{il},\ l = \overline{1,7},\ i = \overline{1,n}$. The fuzzy numbers corresponding to the $\mu_{il},\ l = \overline{1,7},\ i = \overline{1,n}$ are denoted by $\tilde{A}_{il},\ l = \overline{1,7},\ i = \overline{1,n}$.

A linguistic variable R with name «Reliability of information» has five linguistic terms: U, NVL, L, VL, EL that correspond to the fuzzy numbers $\tilde{R}_1 = (0, 0, 0.25)$, $\tilde{R}_2 = (0.25, 0.25, 0.25)$, $\tilde{R}_3 = (0.5, 0.25, 0.25)$, $\tilde{R}_4 = (0.75, 0.25, 0.25)$, $\tilde{R}_5 = (1, 0.25, 0)$.

The state of the i-th plant species is represented as a set of seven Z-numbers $Z_{il} = \left(\tilde{A}_{il}, \tilde{R}_{il}\right),\ l = \overline{1,7},\ i = \overline{1,n}$, fuzzy number $\tilde{R}_{il} = (r_{il1}, r_{il2}, r_{il3}),\ l = \overline{1,7},\ i = \overline{1,n}$ equals to one of fuzzy number $\tilde{R}_k,\ k = \overline{1,5}$. An expert provides an assessment of each planting with one of the reliability levels U, NVL, L, VL, EL, which correspond to fuzzy numbers $\tilde{R}_k,\ k = \overline{1,5}$. If the expert estimated M plants of the i-th species by the level

A_l, $l = \overline{1,7}$ with reliability $\tilde{R}_{iml} = (r_{iml1}, r_{iml2}, r_{iml3})$, $i = \overline{1,n}$, $m = \overline{1,M}$, $l = \overline{1,7}$, where fuzzy number $\tilde{R}_{iml}$, $i = \overline{1,n}$, $m = \overline{1,M}$, $l = \overline{1,7}$ equals to one of the fuzzy number $\tilde{R}_k$, $k = \overline{1,5}$, then $r_{il1} = \max_m(r_{mil1})$, $r_{il2} = \max_m(r_{mil2})$, $r_{il3} = \max_m(r_{mil3})$ [15].

Let us define the ideal state of a plant species as a set of seven Z-numbers $\left(\tilde{A}_l^{id}, \tilde{R}_l^{id}\right)$, $l = \overline{1,7}$, where $\tilde{A}_1^{id} = \tilde{A}_2^{id} = \tilde{A}_3^{id} = \tilde{A}_4^{id} = \tilde{A}_5^{id} = \tilde{A}_6^{id} = (0, 0, 0)$, $\tilde{A}_7^{id} = (0, 1, 0, 0)$, $\tilde{R}_l^{id} = \tilde{R}_5$, $l = \overline{1,7}$. Determine the worst state of a plant species as a set of seven Z-numbers $\left(\tilde{A}_l^{w}, \tilde{R}_l^{w}\right)$, $l = \overline{1,7}$, where $\tilde{A}_1^{w} = (0, 1, 0, 0)$, $\tilde{A}_2^{w} = \tilde{A}_3^{w} = \tilde{A}_4^{w} = \tilde{A}_5^{w} = \tilde{A}_6^{w} = \tilde{A}_7^{w} = (0, 1, 0)$, $\tilde{R}_l^{w} = \tilde{R}_1$, $l = \overline{1,7}$.

In [15], the definition of weighted point c(based on the definition of weighted segment [23]) of fuzzy number $\tilde{A} = (a_1, a_2, a_L, a_R)$ is given:

$$c = \frac{a_1 + a_2}{2} + \frac{(a_R - a_L)}{12} \tag{1}$$

Weighted points of $\tilde{A}_{il}$, $\tilde{R}_{il}$, $l = \overline{1,7}$, $i = \overline{1,n}$ we denote as a_{il}, r_{il}, $l = \overline{1,7}$, $i = \overline{1,n}$ and weighted points of $\tilde{A}_l^{id}$, $\tilde{R}_l^{id}$, $l = \overline{1,7}$ as a_l^{id}, r_l^{id}, $l = \overline{1,7}$. Then the distance of the i-th plant species to the ideal state we define in the following form:

$$\rho_i = \sqrt{\sum_{l=1}^{7}\left[\left(a_{il} - a_l^{id}\right)^2 + \left(r_{il} - r_l^{id}\right)^2\right]} \tag{2}$$

The minimum distance ρ_i, $i = \overline{1,n}$ is equal to zero, in which case the state of the plant species coincides with the ideal state $\left(\tilde{A}_l^{id}, \tilde{R}_l^{id}\right)$, $l = \overline{1,7}$. The maximum distance when the state of the plant species coincides with worst state $\left(\tilde{A}_l^{w}, \tilde{R}_l^{w}\right)$, $l = \overline{1,7}$ (in this case all the plants are old dead, and the degree of reliability is «Unlikely»). Let

$$d_{\max} = \sqrt{\sum_{l=1}^{7}\left[\left(a_l^{w} - a_l^{id}\right)^2 + \left(r_l^{w} - r_l^{id}\right)^2\right]} \tag{3}$$

For the range of values of the distance ρ_i, $i = \overline{1,n}$ to be the segment [0, 1], we normalize ρ_i, $i = \overline{1,n}$ to the maximum value $d_{\max}$.

$$d_i = \frac{\rho_i}{d_{\max}}, \ i = \overline{1,n} \tag{4}$$

The smaller the distance d_i, the higher the rating of the i-th plant species.

4 Numerical Example

The state of eight plant species (European white elm, Rough-bark poplar, Witch elm, Norway maple, European ash, Large-leaved linden, Cottonwood, Common lilac) growing on the Moscow Boulevard Ring was evaluated by experts. For this purpose, the

Table 1. The relative numbers of planting and reliability.

№	Plantings	$\frac{k_1}{K}, R$	$\frac{k_2}{K}, R$	$\frac{k_3}{K}, R$	$\frac{k_4}{K}, R$	$\frac{k_5}{K}, R$	$\frac{k_6}{K}, R$	$\frac{k_7}{K}, R$
1	European white elm	0, EL	0.08, L	0.05, VL	0.19, VL	0.36, EL	0.27, VL	0.05, EL
2	Rough-bark poplar	0, EL	0.16, L	0.04, EL	0.3, L	0.4, VL	0.07, VL	0.03, VL
3	Witch elm	0, EL	0.04, NVL	0.03, VL	0.5, EL	0.3, EL	0.05, EL	0.08, EL
4	Norway maple	0.06, EL	0.05, L	0.03, VL	0.09, VL	0.46, EL	0.24, L	0.07, EL
5	European ash	0, EL	0.02, L	0.05, VL	0.5, EL	0.3, VL	0.06, EL	0.07, EL
6	Large-leaved linden	0, EL	0.08, NVL	0.06, VL	0.22, EL	0.36, VL	0.26, VL	0.02, EL
7	Cottonwood	0.02, EL	0, L	0.04, VL	0.36, VL	0.36, VL	0.18, L	0.04, EL
8	Common lilac	0.01, EL	0.02, NVL	0.06, EL	0.23, L	0.41, VL	0.26, L	0.01, VL

experts used the scale: A_1 - «old dead», A_2 - «recently died», A_3 - «drying up», A_4- «very weak», A_5 - «mean weak», A_6 - «moderately weak», A_7 - «no signs of weakness». The results are presented in Table 1.

According to Table 1, we construct linguistic variables with values $\tilde{A}_{il}$, $l = \overline{1, 7}$, $i = \overline{1, 8}$. For example, we have for the «European white elm»: $\tilde{A}_{11} = (0, 0, 0)$, $\tilde{A}_{12} = (0, 0.055, 0, 0.05)$, $\tilde{A}_{13} = (0.105, 0.05, 0.05)$, $\tilde{A}_{14} = (0.155, 0.225, 0.05, 0.19)$, $\tilde{A}_{15} = (0.415, 0.545, 0.19, 0.27)$, $\tilde{A}_{16} = (0.815, 0.925, 0.27, 0.05)$, $\tilde{A}_{17} = (0.975, 1, 0.05, 0)$.

Linguistic terms NVL, L, VL, EL of the linguistic variable R with name «Reliability of information» correspond to the fuzzy numbers $\tilde{R}_1 = (0, 0, 0.25)$, $\tilde{R}_2 = (0.25, 0.25, 0.25)$, $\tilde{R}_3 = (0.5, 0.25, 0.25)$, $\tilde{R}_4 = (0.75, 0.25, 0.25)$,

$\tilde{R}_5 = (1, 0.25, 0)$. Thus, the state of the «European white elm» is described by the following set of Z-numbers $Z_{11} = \left(\tilde{A}_{11}, \tilde{R}_5\right)$, $Z_{12} = \left(\tilde{A}_{11}, \tilde{R}_3\right)$, $Z_{13} = \left(\tilde{A}_{11}, \tilde{R}_4\right)$, $Z_{14} = \left(\tilde{A}_{14}, \tilde{R}_4\right)$, $Z_{15} = \left(\tilde{A}_{15}, \tilde{R}_5\right)$, $Z_{16} = \left(\tilde{A}_{16}, \tilde{R}_4\right)$, $Z_{17} = \left(\tilde{A}_{17}, \tilde{R}_5\right)$. The ideal state of plant species is represented as a set of Z-numbers $\left(\tilde{A}_l^{id}, \tilde{R}_5\right)$, $l = \overline{1, 7}$, where $\tilde{A}_1^{id} = \tilde{A}_2^{id} = \tilde{A}_3^{id} = \tilde{A}_4^{id} = \tilde{A}_5^{id} = \tilde{A}_6^{id}$, $(0, 0, 0)$, $\tilde{A}_7^{id} = (0, 1, 0, 0)$. For all the fuzzy numbers, weighted points are found according to the formula (1), which are used to calculate the distance d_i of the states of plant species to the ideal state according to the formulas (2)–(4), $d_{\max} = 3.0721$. The results obtained are presented in Table 2.

Table 2. Distances of states of plant species to the ideal state and rating of plant species.

№	Plantings	d_i	Rating
1	European white elm	0.4307	2
2	Rough-bark poplar	0.5946	8
3	Witch elm	0.4988	5
4	Norway maple	0.5272	7
5	European ash	0.3936	1
6	Large-leaved linden	0.4842	3
7	Cottonwood	0.4881	4
8	Common lilac	0.4894	6

According to the results obtained, the most resistant species is Norway maple, the least resistant is Rough-bark poplar. All plant species with a distance to the ideal state less than 0.5 are recommended for inclusion in the landscaping plan.

5 Conclusion

The contribution of the paper is that the model developed in it opens up new possibilities for solving environmental problems and, in particular, the problem of monitoring green spaces in large cities in terms of taking into account the reliability of the information received during the assessment.

Plant species have different resistance to the harmful effects of the urban environment. To identify the most resistant plant species in order to include them in the city greening plan, their state is monitored. Experts who assess the state of plants are not always sure of the complete reliability of these assessments. Therefore, to formalize the expert information arriving with a certain level of reliability, Z-numbers are used in the paper. Both components of Z-numbers are values of semantic spaces. A method developed by the author is used to construct these spaces.

In this paper, for the first time, a model for assessing plant species has been developed, which takes into account the reliability of expert information. The assessment of the state of plant species is carried out by comparative analysis with an ideal reference state. For this, the distance between the state of the plant species, represented as a set of Z-numbers, and the ideal state, also formalized as a set of Z-numbers, is determined. The smaller the distance between the state of the plant species and the ideal state, the better the state of the plant species and the higher its rating.

A numerical example of monitoring the state of plant species on the Boulevard Ring avenue of Moscow is given. Eight plant species were evaluated by experts with certain levels of reliability. As a result, the most resistant plant species were identified, which are recommended for inclusion in the landscaping plan.

Future research is expected to focus on solving the problems of assessing ecosystem services provided by green spaces in large cities under Z-information (mitigating the impact of heat, reducing the risk of floods in cities, the absorbency of woody plants, the concentration of phytoncides in the air, and so on).

References

1. Zadeh, L.A.: A note on Z-numbers. Inf. Sci. **14**(181), 2923–2932 (2011)
2. Aliev, R.A., Alizadeh, A.V., Huseynov, O.H.: The arithmetic of discrete Z-numbers. Inf. Sci. **1**(290), 134–155 (2015)
3. Aliev, R.A., Huseynov, O.H., Zeinalova, L.M.: The arithmetic of continuous Z-numbers. Inf. Sci. **373**, 441–460 (2016)
4. Dutta, P., Boruah, H., Ali, T.: Fuzzy arithmetic with and without α - cut method: a comparative study. Int. J. Latest Trends Comput. **1**(2), 99–107 (2011)
5. Poleshchuk, O.M.: Novel approach to multicriteria decision making under Z-information. In: Proceedings of the International Russian Automation Conference (RusAutoCon), p. 8867607 (2019). https://doi.org/10.1109/RUSAUTOCON.2019.8867607
6. Kang, B., Wei, D., Li, Y., Deng, Y.: A method of converting Z-number to classical fuzzy number. J. Inf. Comput. Sci. **9**(3), 703–709 (2012)
7. Sharghi, P., Jabbarova, K.: Hierarchical decision making on port selection in Z-environment. In: Proceedings of the Eighth International Conference on Soft Computing, Computing with Words and Perceptions in System Analysis, Decision and Control, pp. 93–104 (2015)
8. Wang, F., Mao, J.: Approach to multicriteria group decision making with Z-numbers based on Topsis and Power Aggregation Operators. Math. Probl. Eng. 1–18 (2019)
9. Aliyev, R.R., Talal Mraizid, D.A., Huseynov, O.H.: Expected utility based decision making under Z-information and its application. Comput. Intell. Neurosci. **3**, 364512 (2015)
10. Yager, R.R., Filev, D.P.: On the issue of defuzzification and selection based on a fuzzy set. Fuzzy Sets Syst. **5**(3), 255–272 (1993)
11. Aliyev, R.R.: Similarity based multi-attribute decision making under Z-information. In: Proceedings of the Eighth International Conference on Soft Computing, Computing with Words and Perceptions in System Analysis, Decision and Control, pp. 33–39 (2015)
12. Aliev, R.A., Zeinalova, L.M.: Decision-making under Z-information. In: Human-Centric Decision-Making Models for Social Sciences, pp. 233–252 (2013)
13. Kang, B., Wei, D., Li, Y., Deng, Y.: Decision making using Z-numbers under uncertain environment. J. Inf. Comput. Sci. **7**(8), 2807–2814 (2012)
14. Aliev, R.K., Huseynov, O.H., Aliyeva, K.R.: Aggregation of an expert group opinion under Z-information. In: Proceedings of the Eighth International Conference on Soft Computing, Computing with Words and Perceptions in System Analysis, Decision and Control, pp. 115–124 (2015)
15. Poleshchuk, O.M.: Object monitoring under Z-information based on rating points. In: Kahraman, C., Cevik Onar, S., Oztaysi, B., Sari, I.U., Cebi, S., Tolga, A.C. (eds.) INFUS 2020. AISC, vol. 1197, pp. 1191–1198. Springer, Cham (2021). https://doi.org/10.1007/978-3-030-51156-2_139
16. Zadeh, L.A.: The Concept of a linguistic variable and its application to approximate reasoning. Inf. Sci. **8**, 199–249 (1975)
17. Poleshchuk, O.M.: Creation of linguistic scales for expert evaluation of parameters of complex objects based on semantic scopes. In: Proceedings of the International Russian Automation Conference, (RusAutoCon – 2018), p. 8501686 (2018)
18. Poleshchuk, O., Komarov, E.: The determination of rating points of objects and groups of objects with qualitative characteristics. In: Proceedings of the Annual Conference of the North American Fuzzy Information Processing Society – NAFIPS'2009, p. 5156416 (2009)

19. Ryjov, A.P.: The concept of a full orthogonal semantic scope and the measuring of semantic uncertainty. In: Proceedings of the Fifth International Conference Information Processing and Management of Uncertainty in Knowledge-Based Systems, pp. 33–34 (1994)
20. Poleshchuk, O.M.: Formalization, prediction and recognition of expert evaluations of telemetric data of artificial satellites based on Type-II fuzzy sets. In: Hassanien, A.E., Darwish, A., El-Askary, H. (eds.) Machine Learning and Data Mining in Aerospace Technology. SCI, vol. 836, pp. 39–64. Springer, Cham (2020). https://doi.org/10.1007/978-3-030-20212-5_3
21. Darwish, A., Poleshchuk, O.: New models for monitoring and clustering of the state of plant species based on sematic spaces. J. Intell. Fuzzy Syst. **3**(26), 1089–1094 (2014)
22. Chang, Y.-H.: Hybrid fuzzy least- squares regression analysis and its reliability measures. Fuzzy Sets Syst. **119**, 225–246 (2001)
23. Domrachev, V.G., Poleshchuk, O.M.: On the construction of a regression model under fuzzy source data. Avtomatika I Telemekhanika **11**, 74–81 (2003)

Critical Path Method for Z-fuzzy Numbers

Ewa Marchwicka and Dorota Kuchta(✉)

Wroclaw University of Science and Technology, Wyb. Wyspianskiego 27, 50-370 Wroclaw, Poland
dorota.kuchta@pwr.edu.pl

Abstract. Project critical path is one of basic notions in project management, known for many years already. Its determination is straightforward and unambiguous in case of crisp activity durations, but the problem becomes more complex when the activity durations are not known precisely and given in the form of fuzzy numbers. Numerous approaches have been proposed for the determination of the critical path in case of fuzzy activity durations. In this paper the case of activity durations given in the form of Z-fuzzy numbers will be considered for the first time. The application of Z-fuzzy numbers allows to model the credibility degree of estimations. This problem (taking into account estimators who are optimistic, pessimistic, volatile or accurate) will be analyzed and a concept and method of determining the critical path based on Z-fuzzy durations proposed. The critical path determined thanks to our approach will take into account not only the lack of "hard" information, but also human features of people involved in the estimation. The result will be thus more realistic and more useful in practical applications.

Keywords: Critical Path Method (CPM) · Fuzzy numbers · Z-fuzzy numbers

1 Introduction

The critical path is a very important notion in project management, essential for project planning and control. Its definition is clear and unequivocal in the case of crisp activity durations, but the comfort of having reliable crisp estimations of activities duration is rarely given to project managers in today's uncertain world [1]. To have an accurate project network model, we need to resort to non-crisp numbers for duration estimation. Apart from probabilistic models, like the well-known PERT [2], fuzzy modelling is widely used in this context. A literature search in Mendeley, with the string to be included in the article title being *"fuzzy" and "project" and "activity" and "duration"*, rendered almost 6000 results. That is why there is a need for methods to determine critical paths or the degree of criticality of individual paths for project networks with fuzzy activity durations. Although there has been lots of research in this area, which is shown in the next section, and although this research covers numerous different types of fuzzy numbers, there is an important gap: the problem of determining the critical path in the case when project activity durations are modelled through Z-fuzzy numbers [3] has not been considered in the literature so far. Z-fuzzy numbers are important for modelling estimations of uncertain values performed by humans, as they allow us to take the features

C. Kahraman et al. (Eds.): INFUS 2021, LNNS 308, pp. 871–878, 2022.
https://doi.org/10.1007/978-3-030-85577-2_100

of the human estimator into account. What is more, Z-fuzzy numbers have already been used in project management (e.g. to express the uncertain knowledge about project activities progress [4], but not in the context of project estimations. The objective of this paper is thus to propose a critical path method for this case.

The organization of the paper is as follows: in Sect. 2 we present the relevant literature review, in Sect. 3 we describe our proposed approach, in Sect. 4 we illustrate it through a computational example. Conclusions sketch further research paths.

2 Related Literature

We conducted a literature search in Mendeley, with search string (applied to the title): "fuzzy" AND "critical" AND "path". We obtained 165 results. Then we repeated the same search in Scopus and ScienceDirect. No single work referring to critical path in the context of Z-fuzzy numbers has been identified. On the other hand, papers referring to critical path methods for fuzzy durations modeled not only by well-known types of fuzzy numbers, but also by less common types, like hexagonal, octagonal, type-2, hesitant, intuitionistic fuzzy numbers [5–15] have been identified, which means that our research fits well into the current trends of research in fuzzy project modelling.

3 Z-fuzzy Critical Path Method

3.1 Method Overview

Below a method for determining project's critical path, when activity durations are represented in the form of Z-fuzzy numbers [3] (Zadeh, 2011), is presented. To the best knowledge of the authors, based on the literature review presented above in Sect. 2, this approach has not previously been considered in project management. In case of applying classical fuzzy numbers to project's critical path, each activity duration is represented as one fuzzy number $\tilde{A}$, which is obtained from experts' estimations and, as any fuzzy number, is defined by its universe of discourse U and a membership function μ_A. The method presented in this article extends the classical fuzzy critical path approach by considering the credibility of experts' estimations, represented by another fuzzy number $\tilde{Z}$. This means that, in the case being described, activity durations are defined by a Z-fuzzy number, which is an ordered pair $\left(\tilde{A}, \tilde{Z}\right)$, where $\tilde{A}$ and $\tilde{Z}$ are fuzzy numbers and represent experts' estimations and experts' credibility, accordingly. For simplicity, it is assumed that both $\tilde{A}$ and $\tilde{Z}$ are triangular fuzzy numbers.

The method combines duration estimation, which is provided by experts as in case of standard fuzzy critical path approach, with credibility assessment, which is based on the history of experts' estimations (Fig. 1). Credibility assessment of experts' decision is based on 3-class classification. This classification determines whether experts' decisions are optimistic, pessimistic or volatile (i.e. sometimes optimistic and sometimes pessimistic). Credibility assessment enriches standard experts' estimations and changes representation from standard fuzzy number representation to the Z-fuzzy number representation. Finally, this representation is converted back to standard fuzzy number and an of the existing algorithms (e.g. [16]) for fuzzy Critical Path Method with triangular fuzzy numbers is applied.

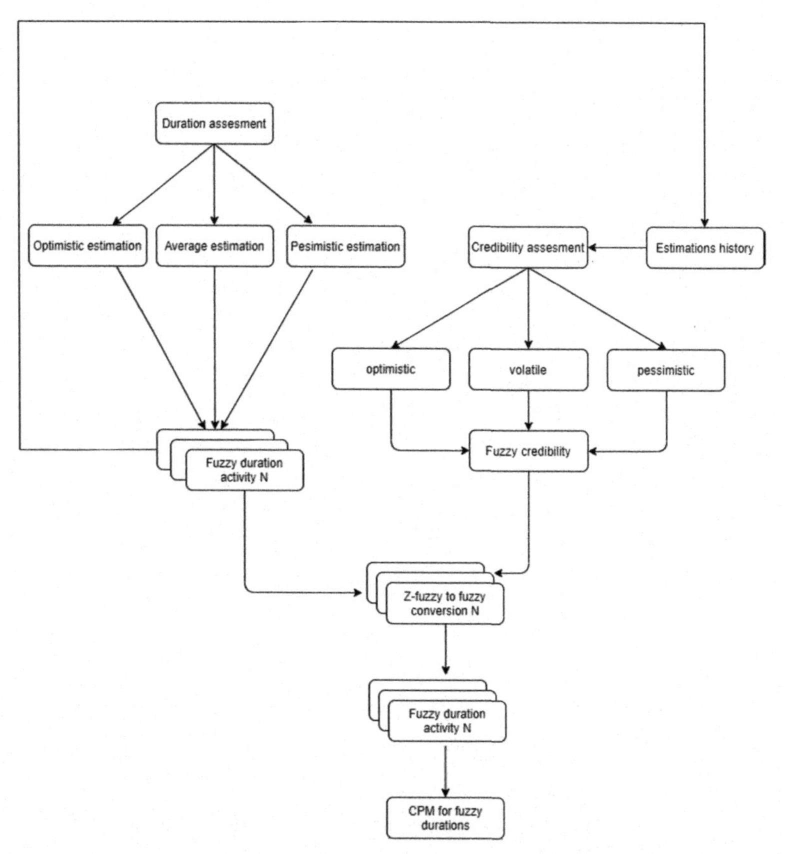

Fig. 1. Overview of CPM method described in this paper (activity durations are represented as *Z*-fuzzy numbers).

3.2 Credibility Assessment

To obtain credibility assessment, a history of previous estimations is needed. For the purpose of determining credibility, a calculation of differences between real values and previous estimations is performed. These values are then classified according to the following rules: (1) if all three estimation differences are greater than zero, it means the estimation is optimistic and label "optimistic" is assigned to it; (2) if all three estimation differences are less than zero, it means the estimation is pessimistic and label "pessimistic" is assigned to it; (3) if some of the three estimation differences are greater than zero and others are less than zero, it means that it cannot be determined whether the estimation is optimistic or pessimistic and a label "volatile" is assigned to it. These rules are listed in Table 1, where $\tilde{A} = \left(\tilde{A}_a, \tilde{A}_b, \tilde{A}_c\right)$ stands for estimation and $R = (R_a, R_b, R_c)$ stands for the real value that was observed post factum.

The labels are then counted. Final label for the whole dataset is determined according to the label that is most common. For example, having 100 previous estimations available (denoted here as $N = 100$), when 40 of them were labeled as "optimistic", 30 were

Table 1. Rules used for experts' credibility assessments (performed on each historical estimation).

Rule no.	Condition	Label applied
(1)	$R_a - \tilde{A}_a > 0$ AND $R_b - \tilde{A}_b > 0$ AND $R_c - \tilde{A}_c > 0$	"optimistic" label
(2)	$R_a - \tilde{A}_a < 0$ AND $R_b - \tilde{A}_b < 0$ AND $R_c - \tilde{A}_c < 0$	"pessimistic" label
(3)	otherwise	"volatile" label

labeled as "pessimistic" and 30 were labeled as "volatile", finally the estimations get the label "optimistic" as this was the most common label. This is a simplification used for the method, but the labeling approaches can be further analyzed as part of future research. After obtaining labels, the credibility of estimations is determined. For this purpose, another fuzzy number $\tilde{Z}$ is constructed, based on the same historical data that was already used for obtaining labels. For simplicity, $\tilde{Z}$ is approximated as a mean value from the relative deviations between real values and estimations calculated on all available data collected. The equation that is used is (1):

$$\tilde{Z} = \left(\tilde{Z}_a, \tilde{Z}_b, \tilde{Z}_c\right) = \left(1 + (\sum R_a - \tilde{A}_a)/\tilde{A}_a N, 1 + (\sum R_b - \tilde{A}_b)/\tilde{A}_b N, 1 + (\sum R_c - \tilde{A}_c)/\tilde{A}_c N\right) \quad (1)$$

Number $\tilde{Z}$ is depicted in Fig. 2. Each expert estimation is then an ordered pair of fuzzy numbers $\left(\tilde{A}, \tilde{Z}\right)$, so-called Z-fuzzy number, where the first number represents the estimation and the second one represents the credibility of this estimation.

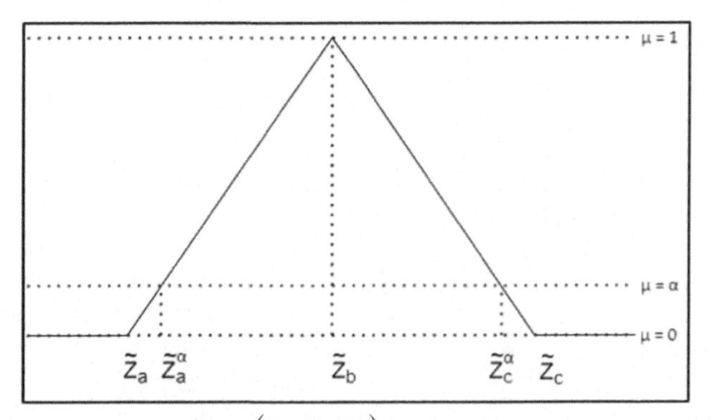

Fig. 2. Approximated number $\tilde{Z} = \left(\tilde{Z}_a, \tilde{Z}_b, \tilde{Z}_c\right)$ representing experts' estimations credibility $\left(\tilde{Z}_a^\alpha, \tilde{Z}_b, \tilde{Z}_c^\alpha\right)$ represents number $\tilde{Z}$ at α level.

3.3 Z-fuzzy to Fuzzy Conversion

As already introduced, Z-fuzzy number representation needs to be converted back to a standard fuzzy number. Both credibility of the estimations (numbers $\tilde{Z}$) and estimation labels ("optimistic", "pessimistic" or "volatile") are used for converting Z-fuzzy number ordered pair $\left(\tilde{A}, \tilde{Z}\right)$ to a standard fuzzy number. Let us denote this number as $\tilde{B}$.

This number is obtained depending on the label previously assigned, using the following rules: (3) if the label is "optimistic", we should correct the values using pessimistic estimation of $\tilde{Z}$, (4) if the label is "pessimistic", we should correct the values using optimistic estimation of $\tilde{Z}$, (5) if the label is "volatile", we should correct optimistic values using optimistic estimation of $\tilde{Z}$ and pessimistic values using pessimistic estimation of $\tilde{Z}$. This way the set of possible values will be very broad, because smallest components of number $\tilde{A}$ will be multiplied by smallest components of $\tilde{Z}$ and largest components of $\tilde{A}$ will be multiplied by largest components. These three sample rules for converting Z-fuzzy numbers are presented in a table below (Table 2).

Table 2. Rules used for Z-fuzzy number $\left(\tilde{A}, \tilde{Z}\right)$ to standard fuzzy number $\tilde{B}$ conversion.

Rule no.	Condition	Conversion applied
(3)	"optimistic" label	$\tilde{B} = \left(\tilde{B}_a, \tilde{B}_b, \tilde{B}_c\right) = \left(\tilde{A}_a\tilde{Z}_c, \tilde{A}_b\tilde{Z}_c, \tilde{A}_c\tilde{Z}_c\right)$
(4)	"pessimistic" label	$\tilde{B} = \left(\tilde{B}_a, \tilde{B}_b, \tilde{B}_c\right) = \left(\tilde{A}_a\tilde{Z}_a, \tilde{A}_b\tilde{Z}_a, \tilde{A}_c\tilde{Z}_a\right)$
(5)	"volatile" label	$\tilde{B} = \left(\tilde{B}_a, \tilde{B}_b, \tilde{B}_c\right) = \left(\tilde{A}_a\tilde{Z}_a, \tilde{A}_b\tilde{Z}_b, \tilde{A}_c\tilde{Z}_c\right)$

It is also possible that the rules from the table above will be applied using some α level. Then the conversions will have a corresponding form depicted in Table 3.

Table 3. Rules used for Z-fuzzy number $\left(\tilde{A}, \tilde{Z}\right)$ to standard fuzzy number $\tilde{B}$ conversion, at a given α level.

Rule no.	Condition	Conversion applied
(3b)	"optimistic" label	$\tilde{B} = \left(\tilde{B}_a, \tilde{B}_b, \tilde{B}_c\right) = \left(\tilde{A}_a\tilde{Z}_c^\alpha, \tilde{A}_b\tilde{Z}_c^\alpha, \tilde{A}_c\tilde{Z}_c^\alpha\right)$
(4b)	"pessimistic" label	$\tilde{B} = \left(\tilde{B}_a, \tilde{B}_b, \tilde{B}_c\right) = \left(\tilde{A}_a\tilde{Z}_a^\alpha, \tilde{A}_b\tilde{Z}_a^\alpha, \tilde{A}_c\tilde{Z}_a^\alpha\right)$
(5b)	"volatile" label	$\tilde{B} = \left(\tilde{B}_a, \tilde{B}_b, \tilde{B}_c\right) = \left(\tilde{A}_a\tilde{Z}_a^\alpha, \tilde{A}_b\tilde{Z}_b^\alpha, \tilde{A}_c\tilde{Z}_c^\alpha\right)$

A sample α level of fuzzy number $\tilde{Z}$ is illustrated in the above section in Fig. 3.

3.4 Remarks

There are a few important properties of the method that are worth mentioning. First, it should be noted that expert estimations that are described in the method are expressed in arbitrary units, which are dependent on the application. In many applications the estimations are given in days. This allows easy comparison of the values that were estimated to the values that were observed. Second, it should be noted that expert estimations can be both single-person estimations or group estimations. The history of estimations that is used for credibility assessment should consider the same group of experts, but it will not always be possible, as experts may change. Because the history of estimations changes over time, only the latest history of estimations should be used for calculation, not the whole available history.

It is worth mentioning that the method does not introduce any complexity to experts' estimations, because experts still need to provide three estimations of each activity duration (optimistic, expected and pessimistic). The complexity increases in relation to data collection. It is required that the history of both estimations and real values are being tracked for the purpose of credibility assessment.

4 Computational Example

We consider here the following example, in which, for simplification, we use the formula $(a + 2b + c/4)$ [17] for the defuzzification of the triangular fuzzy number (a, b, c), and the critical path in a network with fuzzy triangular durations is determined on the basis of these defuzzifications.

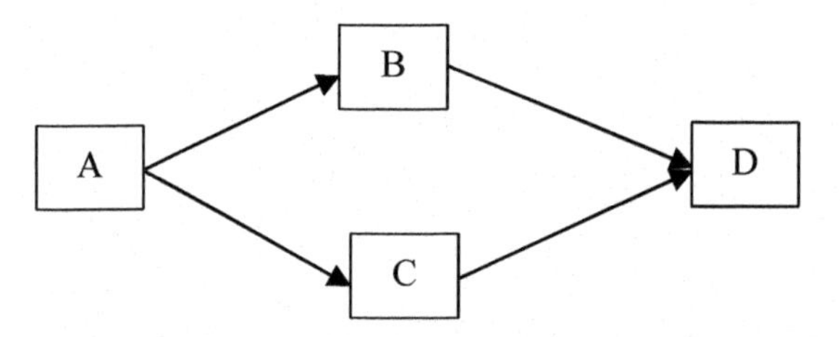

Fig. 3. Example project network.

Duration of the project activities is determined (Table 4) by means of triangular fuzzy numbers (Scenario 1) and by means of Z-fuzzy numbers (Scenario 2).

Table 4. Project activities duration.

	Scenario 1	Scenario 2
Activity A	(1,2,3)	((1,2,3),(1,1,1))
Activity B	(5,6,7)	((5,6,7),(0,7, 0,8, 1))
Activity C	(3,5,8)	((3,5,8),(1, 1,2, 1,3))
Activity D	(1,2,3)	((1,2,3),(1,1,1))
Path ABD (deffuzzified)	10	8,2 (pessimistic label)
Path ACD (deffuzzified)	9,25	10,825 (optimistic label)

In Scenario 1 we do not take into account the estimator's personal features (his or her credibility). In this case path ABD would be the critical one. In Scenario 2 we consider the estimators' features: we assume that the estimators of A and D are unbiased (accurate), but the estimator of B has been a pessimist so far and that of C an optimist. This has changed the criticality of the paths: path ACD turns out to be critical, and the slack of the non-critical activity is now higher than in case of Scenario 1. Given that Scenario 1 is probably unrealistic and the network with Scenario 1 data does not reflect well the reality, it is visible that our approach can be of utmost importance for a reliable project planning.

5 Conclusions

In this paper we propose to use Z-fuzzy numbers in modelling project activities durations and to base the critical path method on such project networks. Such an approach allows taking into account individual features of project estimators and to adjust distortions in estimations due to their biases. It is then possible to reduce too pessimistic and increase too optimistic duration estimations accordingly, so that project scheduling is based on realistic data.

Further research is needed to extend the known fuzzy critical path methods to the Z-fuzzy numbers case and to verify the approach in praxis. Cooperation with psychologists would be helpful in the identification of estimators types.

Acknowledgements. This research was supported by the National Science Centre (Poland), under Grant 394311, 2017/27/B/HS4/01881: "Selected methods supporting project management, taking into consideration various stakeholder groups and using type-2 fuzzy numbers".

References

1. Cleden, D.: Managing Project Uncertainty. Gower Publishing Ltd, UK (2017)
2. Chinneck, J.W.: Practical optimization: a gentle introduction. Carleton University. Internet: https://www.optimization101.org Accessed 06 Mar 2021
3. Zadeh, L.A.: A Note on Z-numbers. Inf. Sci. **181**(14), 2923–2932 (2011). https://doi.org/10.1016/j.ins.2011.02.022
4. Hendiani, S., Bagherpour, M., Mahmoudi, A., Liao, H.: Z-number based earned value management (ZEVM): a novel pragmatic contribution towards a possibilistic cost-duration assessment. Comput. Ind. Eng. **143** (2020). https://doi.org/10.1016/j.cie.2020.106430
5. Cheng, F., Lin, M., Yuksel, S., Dincer, H.: A hybrid hesitant 2-tuple IVSF decision making approach to analyze PERT-based critical paths of new service development process for renewable energy investment projects. IEEE Access (2020). https://doi.org/10.1109/ACCESS.2020.3048016
6. Dorfeshan, Y., Mousavi, S.M.: Soft computing based on an interval type-2 fuzzy decision model for project-critical path selection problem. Int. J. Appl. Ind. Eng. **5**(1), 1–24 (2018). https://doi.org/10.4018/ijaie.2018010101
7. Dorfeshan, Y., Mousavi, S., Mohagheghi, V., Vahdani B.: Selecting project-critical path by a new interval type-2 fuzzy decision methodology based on MULTIMOORA, MOOSRA and TPOP methods. Comput. Ind. Eng. **124** (2018). https://doi.org/10.1016/j.cie.2018.04.015

8. Begum, S.G., Praveena, J.P.N., Rajkumar, A.: Critical path through interval valued hexagonal fuzzy number. Int. J. Innovative Technol. Expl. Eng. **8**(11), 1190–1193 (2019). https://doi.org/10.35940/ijitee.J9290.0981119
9. Kaur, P., Kumar, A.: A modified ranking approach for solving fuzzy critical path problems with LR flat fuzzy numbers. Control Cybern. **41**(1), 171–190 (2012)
10. Narayanamoorthy, S., Maheswari, S.: The intelligence of octagonal fuzzy number to determine the fuzzy critical path: a new ranking method. Sci. Program. 1–8 (2016). https://doi.org/10.1155/2016/6158208
11. Pardha, B., Shankar, N.R.: Critical path in a project network using TOPSIS method and linguistic trapezoidal fuzzy numbers. Int. J. Sci. Eng. Res. **6**(11), 24–32 (2015)
12. Katenkuram, R.R., Dhodiya, J.M.: Possibilistic distribution for selection of critical path in multi objective multi-mode problem with trapezoidal fuzzy number. Int. J. Recent Technol. Eng. **8**(4), 10833–10842 (2019). https://doi.org/10.35940/ijrte.d4372.118419
13. Rameshan, N., Dinagar, D.S.: Solving fuzzy critical path with octagonal intuitionistic fuzzy number. AIP Conf. Proc. **2277**(090022), 1–8 (2020). https://doi.org/10.1063/5.0025360
14. Samayan, N., Sengottaiyan, M.: Fuzzy critical path method based on ranking methods using hexagonal fuzzy numbers for decision making. J. Intell. Fuzzy Syst. **32**(1), 157–164 (2017). https://doi.org/10.3233/JIFS-151327
15. Yogashanthi, T., Ganesan, K.: Modified critical path method to solve networking problems under an intuitionistic fuzzy environment. ARPN J. Eng. Appl. Sci. **12**(2), 398–403 (2017)
16. Chanas, S., Zieliński, P.: Critical path analysis in the network with fuzzy activity times. Fuzzy Sets Syst. **122**(2), 195–204 (2001). https://doi.org/10.1016/S0165-0114(00)00076-2
17. Bohlender, G., Kaufmann, A., Gupta, M.M.: Introduction to Fuzzy Arithmetic, Theory and Applications. Mathematics of Computation. Van Nostrand Reinhold, USA (1991). https://doi.org/10.2307/2008199

Defuzzification of Intuitionistic Z-Numbers for Fuzzy Multi Criteria Decision Making

Nik Muhammad Farhan Hakim Nik Badrul Alam[1,2], Ku Muhammad Naim Ku Khalif[1(✉)], Nor Izzati Jaini[1], Ahmad Syafadhli Abu Bakar[3], and Lazim Abdullah[4]

[1] Centre for Mathematical Sciences, Universiti Malaysia Pahang, Gambang, Malaysia
kunaim@ump.edu.my

[2] Faculty of Computer and Mathematical Sciences, Universiti Teknologi MARA Pahang, Bandar Tun Abdul Razak Jengka, Pahang, Malaysia

[3] Mathematics Division, Centre for Foundation Studies in Science, University of Malaya, Kuala Lumpur, Malaysia

[4] Management Science Research Group, Faculty of Ocean Engineering Technology and Informatics, Universiti Malaysia Terengganu, Kuala Nerus, 21030 Kuala Terengganu, Terengganu, Malaysia

Abstract. Z-numbers and intuitionistic fuzzy numbers are both important as they consider the reliability of the judgement, membership and non-membership functions of the numbers. The combination of these two numbers produce intuitionistic Z-numbers which need to be defuzzified before aggregation of multiple experts' opinions could be done in the decision making problems. This paper presents the generalised intuitionistic Z-numbers and proposes a centroid-based defuzzification of such numbers, namely intuitive multiple centroid. The proposed defuzzification is used in the decision making model and applied to the supplier selection problem. The ranking of supplier alternatives is evaluated using the ranking function based on centroid. In the present paper, the ranking is improved since the intuitionistic fuzzy numbers (IFN) are integrated within the evaluations which were initially in form of Z-numbers, considering their membership and non-membership grades. The ranking of the proposed model gives almost similar ranking to the existing model, with simplified but detailed defuzzification method.

Keywords: Intuitionistic Z-numbers · Defuzzification · Intuitive multiple centroid · Ranking function · Aggregation

1 Introduction

With the basis of uncertainty, Zadeh [1] introduced the knowledge of fuzzy set by considering the value in the interval [0, 1] for its membership value instead of crisp numbers. This concept was further extended by Dubois and Prade [2] by defining a fuzzy number (FN), which is a fuzzy subset of real number whereby its maximum membership values are surrounding the average value [2]. Some commonly used shapes of FNs are trapezoidal and triangular FNs.

C. Kahraman et al. (Eds.): INFUS 2021, LNNS 308, pp. 879–887, 2022.
https://doi.org/10.1007/978-3-030-85577-2_101

In 1986, the theory of intuitionistic fuzzy sets (IFS) was introduced by Atanassov [3] which consist of the degrees of membership and non-membership to represent the degree of confidence and non-confidence of the fuzzy sets, respectively. The intuitionistic fuzzy set basically generalizes the classical fuzzy set. Grzegrorzewski [4] then defined the intuitionistic fuzzy numbers (IFN). The definition given was general and then refined into a trapezoidal intuitionistic fuzzy numbers (TrIFNs) by [5]. Zadeh [6] in 2011 proposed the definition of Z-numbers which consists of two components; constraint and reliability. Both components are fuzzy numbers which can be considered as trapezoidal fuzzy numbers. The defuzzification of Z-numbers via intuitive multiple centroid was developed by Ku Khalif *et al.* [7]. In this method, the trapezoidal fuzzy numbers are partitioned into three areas and the centroid of each area is calculated. Connecting the sub-centroids produces a new triangle and the centroid of this triangle could further be calculated.

Recently, Sari and Kahraman [8] defined the intuitionistic fuzzy Z-numbers by refining the membership and non-membership grades of the constraint and reliability components of Z-numbers. Both of the grades are able to indicate the hesitancy of the experts for both components of Z-numbers in the decision making. However, the defuzzification methods proposed by them use a parameter τ, which is set to be a very large number, such as 100. The setting of this parameter is too subjective, thus the defuzzification output may vary depending on the τ value. Hence, a new defuzzication approach of intuitionistic Z-numbers, namely intuitive multiple centroid is proposed in this paper.

This paper is organized as follows: the introduction of the paper is explained in Sect. 1 and some preliminaries on intuitionistic fuzzy sets, Z-numbers and intuitionistic fuzzy Z-numbers are given in Sect. 2; Sect. 3 presents the defuzzification of intuitionistic Z-numbers (IZN) using intuitive multiple centroid; Sect. 4 presents the multi criteria decision making (MCDM) model based on IZNs and Sect. 5 illustrates the proposed model using supplier selection problem. Conclusion is presented in Sect. 6.

2 Preliminaries

Some preliminaries on the intuitionistic fuzzy sets, Z-numbers and the recently proposed intuitionistic fuzzy Z-numbers are presented in this section.

2.1 Intuitionistic Fuzzy Sets (IFS)

An IFS in the universe of discourse, U is of the form [3]:

$$I = \langle x, \mu_I(x), \nu_I(x) | x \in U \rangle \tag{1}$$

where $\mu_I(x)$ and $\nu_I(x)$ represent the membership and non-membership functions of the element x, respectively, satisfying $0 \leq \mu_I(x) + \nu_I(x) \leq 1$. The function

$$\pi_I(x) = 1 - [\mu_I(x) + \nu_I(x)] \tag{2}$$

denotes the degree of indeterminacy of x. A trapezoidal intuitionistic fuzzy number (TrIFN), denoted by $(a_2, a_3, a_4, a_5; a_1, a_3, a_4, a_6)$ can be illustrated in Fig. 1.

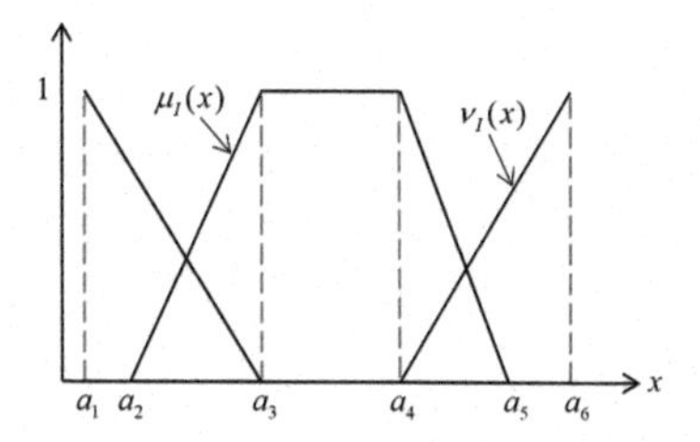

Fig. 1. A TrIFN.

2.2 Z-Numbers

A Z-number is of the form $Z = (A, R)$, where A is the fuzzy constraint and R measures the reliability of A [6]. For simplicity, both components of Z-numbers are considered as trapezoidal FNs.

The following are steps proposed by Ku Khalif *et al.* [7] for converting the Z-number into a regular FN via intuitive multiple centroid approach:

Step 1: The reliability component is converted into κ.

$$\kappa = \frac{2(r_1 + r_4) + 7(r_2 + r_3)}{18} \tag{3}$$

Step 2: The weight κ is added to the first component of Z-number.

$$Z^{\kappa} = \{\langle x, \mu_{R^{\kappa}}(x) | \mu_{R^{\kappa}}(x) = \kappa\mu_R(x), x \in [0, 1]\rangle\} \tag{4}$$

Step 3: Z^{κ} is converted into a regular FN.

$$Z' = \left\{\left\langle x, \mu_{Z'}(x) | \mu_{Z'}(x) = \mu_A\left(\sqrt{\kappa}x\right), x \in [0, 1]\right\rangle\right\} \tag{5}$$

2.3 Intuitionistic Fuzzy Z-Numbers

Sari and Kahraman [8] proposed the intuitionistic fuzzy Z-number of the form $Z = (A_I, R_I)$ where A_I and R_I are denoted by IFNs. In their literature, $A_I = (a_2, a_4, a_5, a_7, \delta_1; a_1, a_3, a_6, a_8, \eta_1)$ and $R_I = (r_2, r_4 = r_5, r_7, \delta_2; r_1, r_3 = r_6, r_8, \eta_2)$ are trapezoidal FNs for the fuzzy restriction component and and triangular fuzzy number for the reliability component, respectively.

3 Intuitionistic Z-Numbers (IZNs)

In the present paper, both components of the intuitionistic fuzzy Z-numbers are considered as trapezoidal FNs. Thus, the trapezoidal FNs for the fuzzy restriction and reliability components are defined in this section. Next, defuzzification of the proposed IZNs using intuitive multiple centroid is proposed.

3.1 Proposed Generalised IZNs

An IZN is of the form $Z = (A_I, R_I)$ where A_I and R_I are denoted by the TrIFNs as shown in Fig. 2.

Note that $A_I = (a_2, a_3, a_4, a_5, \delta_1; a_1, a_3, a_4, a_6, \eta_1)$ and $R_I = (r_2, r_3, r_4, r_5, \delta_2; r_1, r_3, r_4, r_6, \eta_2)$ are trapezoidal FNs for the fuzzy restriction and reliability components, respectively. The membership and non-membership functions of restriction component are defined in Eq. (6).

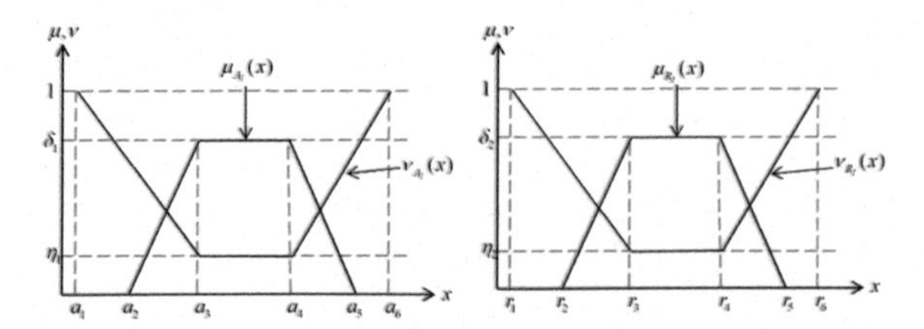

Fig. 2. A trapezoidal intuitionistic Z-number, $Z = (A_I, R_I)$

$$\mu_{A_I}(x) = \begin{cases} \frac{x-a_2}{a_3-a_2}\delta_1 & , \ a_2 \le x \le a_3 \\ \delta_1 & , \ a_3 \le x \le a_4 \\ \frac{a_5-x}{a_5-a_4}\delta_1 & , \ a_4 \le x \le a_5 \\ 0 & , \ \text{otherwise} \end{cases} \qquad \nu_{A_I}(x) = \begin{cases} \frac{\eta_1-1}{a_3-a_1} + \frac{a_3-\eta_1 a_1}{a_3-a_1} & , \ a_1 \le x \le a_3 \\ \eta_1 & , \ a_3 \le x \le a_4 \\ \frac{1-\eta_1}{a_6-a_4} + \frac{\eta_1 a_6-a_4}{a_6-a_4} & , \ a_4 \le x \le a_6 \\ 1 & , \ \text{otherwise} \end{cases} \tag{6}$$

In the following, the membership and non-membership functions of reliability component are defined in Eq. (7).

$$\mu_{R_I}(x) = \begin{cases} \frac{x-r_2}{r_3-r_2}\delta_2 & , \ r_2 \le x \le r_3 \\ \delta_2 & , \ r_3 \le x \le r_4 \\ \frac{r_5-x}{r_5-r_4}\delta_2 & , \ r_4 \le x \le r_5 \\ 0 & , \ \text{otherwise} \end{cases} \qquad \nu_{A_I}(x) = \begin{cases} \frac{\eta_2-1}{r_3-r_1}x + \frac{r_3-\eta_2 r_1}{r_3-r_1} & , \ r_1 \le x \le r_3 \\ \eta_2 & , \ r_3 \le x \le r_4 \\ \frac{1-\eta_2}{r_6-r_4}x + \frac{\eta_2 r_6-r_4}{r_6-r_4} & , \ r_4 \le x \le r_6 \\ 1 & , \ \text{otherwise} \end{cases} \tag{7}$$

Note that the main difference between the proposed intuitionistic Z-numbers as compared to the one defined in [8] is that the values a_3 and a_4, locating the "shoulders" of the trapezoid are the same for both the membership and non-membership functions, as used by many literatures involving the TrIFNs [9–12].

3.2 Defuzzification of Intuitionistic Fuzzy Z-Numbers

The methodology for the defuzzification of IZNs using intuitive multiple centroid (IMC) considers the centroid of membership and non-membership functions. The defuzzification of the membership function is adopted from [7] as follows:

$$IMC(x, y) = \left(\frac{2(r_2 + r_5) + 7(r_3 + r_4)}{18}, \frac{7\delta_2}{18} \right) \tag{8}$$

The defuzzification of the non-membership function is developed as in the following steps:

Step 1: The trapezoidal plane is divided into three areas and the sub-centroid of each area, α, β and γ is calculated. The sub-centroids obtained are given in Eqs. (9)–(11).

$$\alpha(x, y) = \left(r_1 + \frac{2}{3}(r_3 - r_1), \eta_2 + \frac{2}{3}(1 - \eta_2)\right) \tag{9}$$

$$\beta(x, y) = \left(\frac{r_3 + r_4}{2}, \eta_2 + \frac{1}{2}(1 - \eta_2)\right) \tag{10}$$

$$\gamma(x, y) = \left(r_4 + \frac{1}{3}(r_6 - r_4), \eta_2 + \frac{2}{3}(1 - \eta_2)\right) \tag{11}$$

Step 2: The sub-centroids α, β and γ are connected, creating a triangular plane as in Fig. 3.

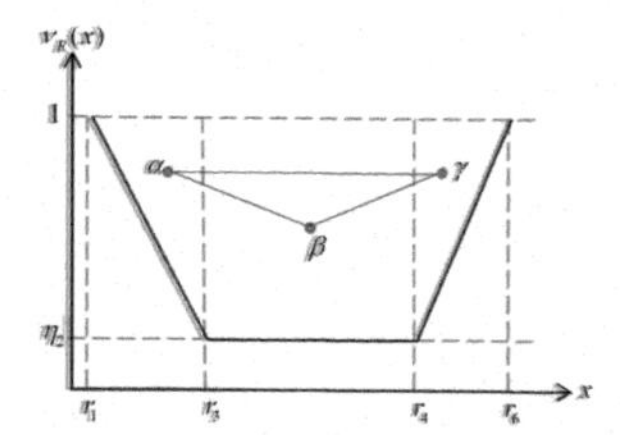

Fig. 3. Sub-centroids of non-membership functions of the reliability component

Step 3: The centroid index of multiple intuitive centroid of (x, y) are calculated as in Eq. (12).

$$\begin{aligned} IMC(x, y) &= \left(\frac{\alpha(x) + \beta(x) + \gamma(x)}{3}, \frac{\alpha(y) + \beta(y) + \gamma(y)}{3}\right) \\ &= \left(\frac{2(r_1 + r_6) + 7(r_3 + r_4)}{18}, \frac{\eta_2 + 11}{18}\right) \end{aligned} \tag{12}$$

After obtaining the centroid index, the authors are interested in developing the steps of converting the intuitionistic Z-numbers into IFNs. The proposed steps are given below:

Step 1: Convert the reliability component into a weight, κ.

$$\kappa = \frac{2(r_2 + r_5) + 7(r_3 + r_4)}{18} \times \frac{7\delta_2}{18} + \frac{2(r_1 + r_6) + 7(r_3 + r_4)}{18} \times \frac{\eta_2 + 11}{18} \tag{13}$$

Step 2: Add κ to the first component of the IZNs (Fig. 4).

$$Z^{\kappa} = \{\langle x, \mu_{R^{\kappa}}(x), \nu_{R^{\kappa}}(x)\rangle | \mu_{R^{\kappa}}(x) = \kappa\mu_R(x), \nu_{R^{\kappa}}(x) = \kappa\nu_R(x), x \in [0, 1]\} \tag{14}$$

Theorem 1: $E_{A^{\kappa}}(x) = \kappa E_A(x), x \in X$ subject to $\mu_{A^{\kappa}}(x) = \kappa \mu_A(x)$.

Proof:

$$\begin{aligned} E_{A^{\kappa}}(x) = \Big[& a_2, a_3, a_4, a_5; \tfrac{2(r_2+r_5)+7(r_3+r_4)}{18} \times \tfrac{7\delta_2}{18} + \tfrac{2(r_1+r_6)+7(r_3+r_4)}{18} \times \tfrac{11+\eta_2}{18}; \\ & a_1, a_3, a_4, a_6; \tfrac{2(r_2+r_5)+7(r_3+r_4)}{18} \times \tfrac{7\delta_2}{18} + \tfrac{2(r_1+r_6)+7(r_3+r_4)}{18} \times \tfrac{11+\eta_2}{18} \Big] \\ = & (a_2, a_3, a_4, a_5, \kappa; a_1, a_3, a_4, a_6, \kappa) = \kappa E_A(x). \end{aligned} \tag{15}$$

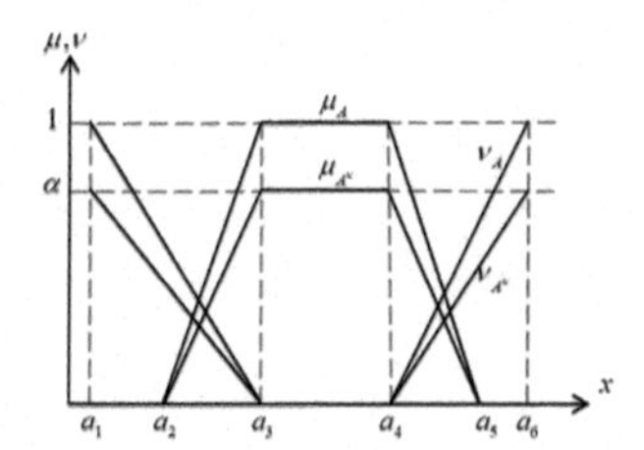

Fig. 4. Membership and non-membership functions of weighted IZN

Step 3: Convert the weighted IZN into an IFN (Fig. 5).

$$Z' = \left\{ \langle x, \mu_{Z'}(x), \nu_{Z'}(x) \rangle | \mu_{Z'}(x) = \mu_A\left(\sqrt{\kappa} x\right), \nu_{Z'}(x) = \nu_A\left(\sqrt{\kappa} x\right), x \in [0, 1] \right\} \tag{16}$$

Theorem 2: $E_{Z'}(x) = \sqrt{\kappa} E_A(x), x \in \sqrt{\kappa} X$ subject to $\mu_{Z'}(x) = \mu_A\left(\sqrt{\kappa} x\right)$.

Proof:

$$\begin{aligned} E_{A^{\kappa}}(x) = \Bigg[& a_2, a_3, a_4, a_5; \sqrt{\tfrac{2(r_2+r_5)+7(r_3+r_4)}{18} \times \tfrac{7\delta_2}{18} + \tfrac{2(r_1+r_6)+7(r_3+r_4)}{18} \times \tfrac{11+\eta_2}{18}}; \\ & a_1, a_3, a_4, a_6; \sqrt{\tfrac{2(r_2+r_5)+7(r_3+r_4)}{18} \times \tfrac{7\delta_2}{18} + \tfrac{2(r_1+r_6)+7(r_3+r_4)}{18} \times \tfrac{11+\eta_2}{18}} \Bigg] \\ = & \left(a_2, a_3, a_4, a_5, \sqrt{\kappa}; a_1, a_3, a_4, a_6, \sqrt{\kappa}\right) = \sqrt{\kappa} E_A(x). \end{aligned} \tag{17}$$

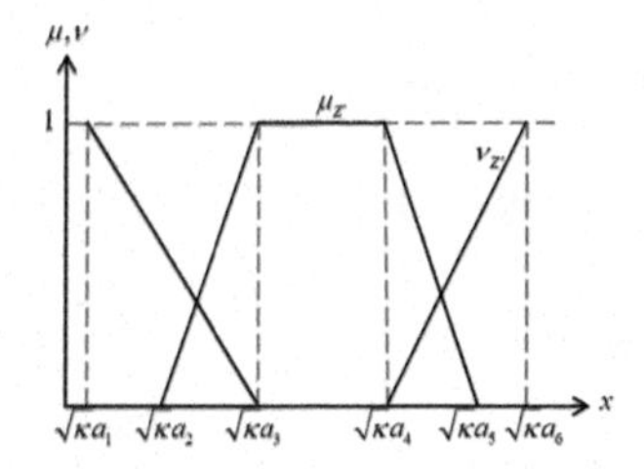

Fig. 5. Membership and non-membership functions of IFN obtained from IZN

4 MCDM Based on Intuitionistic Z-Number

In this section, the authors proposed a multi criteria decision making model in which the experts' opinions are in form of intuitionistic Z-number which will be defuzzified using intuitive multiple centroid as proposed in the previous section. The proposed model is as follows:

Step 1: The experts' opinion is obtained and translated into trapezoidal intuitionistic Z-numbers (TrIZNs). The linguistic terms for the restriction component are adopted from [9] while for the reliability component are proposed in Table 1.

Table 1. Linguistic terms and corresponding TrIZNs for reliability

Linguistic terms	TrIZNs
Not Sure (NS)	$\langle(0.0, 0.0, 0.0, 0.1; 1), (0.0, 0.0, 0.0, 0.1; 0)\rangle$
Not Very Sure (NVS)	$\langle(0.1, 0.2, 0.4, 0.5; 1), (0.1, 0.2, 0.4, 0.5; 0)\rangle$
Sure (S)	$\langle(0.5, 0.6, 0.8, 0.9; 1), (0.5, 0.6, 0.8, 0.9; 0)\rangle$
Very Sure (VS)	$\langle(0.9, 1.0, 1.0, 1.0; 1), (0.9, 1.0, 1.0, 1.0; 0)\rangle$

Step 2: The intuitionistic Z-numbers are then transformed into classical trapezoidal intuitionistic fuzzy numbers (TrIFNs) using the proposed method in Sect. 3.
Step 3: Aggregate the evaluations from all the decision makers using arithmetic averaging operator.

$$\frac{\sum_{k=1}^{n}(a_{k2},a_{k3},a_{k4},a_{k5};a_{k1},a_{k3},a_{k4},a_{k6})}{n} = \left(\frac{\sum_{k=1}^{n}a_{k2}}{n},\frac{\sum_{k=1}^{n}a_{k3}}{n},\frac{\sum_{k=1}^{n}a_{k4}}{n},\frac{\sum_{k=1}^{n}a_{k5}}{n};\frac{\sum_{k=1}^{n}a_{k1}}{n},\frac{\sum_{k=1}^{n}a_{k3}}{n},\frac{\sum_{k=1}^{n}a_{k4}}{n},\frac{\sum_{k=1}^{n}a_{k6}}{n}\right) \quad (18)$$

Step 4: Aggregate all the criteria of the ith row using the arithmetic averaging operator.
Step 5: Calculate the ranking function, $R(A)$ based on centroid [13].

$$R(I_A) = \sqrt{\frac{1}{2}\left(\left[\tilde{x}_\mu(I_A) - \tilde{y}_\mu(I_A)\right]^2 + \left[\tilde{x}_\nu(I_A) - \tilde{y}_\nu(I_A)\right]^2\right)} \quad (19)$$

where

$$\tilde{x}_{\mu}(I_A) = \frac{1}{3}\left(\frac{a_4^2 + a_5^2 - a_2^2 - a_3^2 - a_2a_3 + a_4a_5}{a_4 + a_5 - a_2 - a_3}\right)$$

$$\tilde{x}_{\nu}(I_A) = \frac{1}{3}\left(\frac{2a_6^2 - 2a_1^2 + 2a_3^2 + 2a_4^2 + a_1a_3 - a_4a_6}{a_4 + a_6 - a_1 - a_3}\right)$$

$$\tilde{y}_{\mu}(I_A) = \frac{1}{3}\left(\frac{a_2 + 2a_3 - 2a_4 - a_5}{a_2 + a_3 - a_4 - a_5}\right)$$

$$\tilde{y}_{\nu}(I_A) = \frac{1}{3}\left(\frac{2a_1 + a_3 - a_4 - 2a_6}{a_1 + a_3 - a_4 - a_6}\right).$$

5 Application in Supplier Selection Problem

Adopting the evaluations made by three decision makers from [14], the authors have the set of evaluations in form of TrIZNs. Next, the TrIZNs are transformed into TrIFNs. Using Eq. (18), the TrIFNs are aggregated for the three decision makers ($k = 3$). Next, all the criteria are integrated using Eq. (18) and the following values are obtained:

$$\begin{aligned}
I_{A_1} &= \langle (0.261, 0.321, 0.380, 0.440; 0.216, 0.321, 0.380, 0.485) \rangle \\
I_{A_2} &= \langle (0.213, 0.265, 0.317, 0.369; 0.177, 0.265, 0.317, 0.405) \rangle \\
I_{A_3} &= \langle (0.342, 0.298, 0.352, 0.407; 0.204, 0.298, 0.352, 0.446) \rangle \\
I_{A_4} &= \langle (0.263, 0.295, 0.326, 0.358; 0.241, 0.295, 0.326, 0.381) \rangle \\
I_{A_5} &= \langle (0.285, 0.318, 0.352, 0.386; 0.261, 0.318, 0.352, 0.410) \rangle \\
I_{A_6} &= \langle (0.230, 0.270, 0.310, 0.350; 0.215, 0.270, 0.310, 0.366) \rangle
\end{aligned}.$$

Finally, the ranking functions for I_{A_i}, where $i = 1, 2, \ldots, 6$ are calculated using (19). The following ranking functions are obtained: $R(I_{A_1}) = 0.123685$, $R(I_{A_2}) = 0.090099$, $R(I_{A_3}) = 0.104851$, $R(I_{A_4}) = 0.278681$, $R(I_{A_5}) = 0.338737$ and $R(I_{A_6}) = 0.168720$. From the ranking functions obtained, we have the following ranking of alternatives: $A_5 \succ A_4 \succ A_6 \succ A_1 \succ A_3 \succ A_2$. The ranking of the proposed MCDM model gives almost similar ranking when compared to the ranking obtained by [14], $A_4 \succ A_5 \succ A_3 \succ A_6 \succ A_1 \succ A_2$ and $A_4 \succ A_5 \succ A_6 \succ A_3 \succ A_1 \succ A_2$.

6 Conclusion

The combination of Z-numbers and intuitionistic fuzzy numbers produce intuitionistic Z-numbers, which are composed of membership and non-membership grades of the restriction and reliability. This present paper proposed an intuitive multiple centroid method for converting the trapezoidal intuitionistic Z-numbers (TrIZNs) to trapezoidal intuitionistic fuzzy numbers before the aggregation could be done. Hence, a new defuzzification method of TrIZNs is given in this paper. The MCDM model based on intuitionistic Z-numbers was proposed and implemented in supplier selection problem. The ranking of alternatives using the proposed model is improved since the intuitionistic fuzzy numbers are integrated within the evaluations which were initially in form of Z-numbers, considering their membership and non-membership grades. In the future, the MCDM model using TrIZNs can be improved by using different aggregation operators and ranking methods. The developed model can be applied in many other MCDM problems such as university ranking, consumer purchasing selection and portfolio selection.

Acknowledgement. This research paper is supported financially by Fundamental Research Grant Scheme under Ministry of Higher Education Malaysia FRGS/1/2019/STG06/UMP/02/9.

References

1. Zadeh, L.A.: Fuzzy sets. Inf. Control **8**(3), 338–353 (1965)
2. Dubois, D., Prade, H.: Operations on fuzzy numbers. Int. J. Syst. Sci. **9**(6), 613–626 (1978)
3. Atanassov, K.T.: Intuitionistic fuzzy set. Fuzzy Sets Syst. **20**, 87–97 (1986)
4. Grzegrorzewski, P.: The hamming distance between intuitionistic fuzzy sets. In: Proceedings of the 10th IFSA World Congress, Istanbul, Turkey, pp. 35–38 (2003)
5. Nehi, H.M., Maleki, H.R.: Intuitionistic fuzzy numbers and it's applications in fuzzy optimization problem. In: Proceedings of the 9th WSEAS International Conference on Systems, pp. 1–5. World Scientific and Engineering Academy and Society (WSEAS), Athens (2005)
6. Zadeh, L.A.: A note on Z-numbers. Inf. Sci. **181**(14), 2923–2932 (2011)
7. Ku Khalif, K.M.N., Gegov, A., Abu Bakar, A.S.: Hybrid fuzzy MCDM model for Z-numbers using intuitive vectorial centroid. J. Intell. Fuzzy Syst. **33**(2), 791–805 (2017)
8. Sari, I.U., Kahraman, C.: Intuitionistic fuzzy Z-numbers. In: International Conference on Intelligent and Fuzzy Systems 2020, pp. 1316–1324. Springer, Cham (2020)
9. Ye, J.: Expected value method for intuitionistic trapezoidal fuzzy multicriteria decision-making problems. Expert Syst. Appl. **38**(9), 11730–11734 (2011)
10. De, P.K., Das, D.: Ranking of trapezoidal intuitionistic fuzzy numbers. In: 2012 12th International Conference on Intelligent Systems Design and Applications (ISDA), pp. 184–188). IEEE, Kochi (2012)
11. Shaw, A.K., Roy, T.K.: Trapezoidal intuitionistic fuzzy number with some arithmetic operations and its application on reliability evaluation. Int. J. Math. Oper. Res. **5**(1), 55–73 (2013)
12. Li, X., Chen, X.: Multi-criteria group decision making based on trapezoidal intuitionistic fuzzy information. Appl. Soft Comput. **30**, 454–461 (2015)
13. Prakash, A.A., Suresh, M., Vengataasalam, S.: A new ranking of intuitionistic fuzzy numbers using a centroid concept. Math. Sci. **10**, 177–184 (2016)
14. Wang, F., Mao, J.: Approach to multicriteria group decision making with Z-numbers based on TOPSIS and power aggregation operators. Math. Probl. Eng. **2019**, 1–18 (2019)

Clustering Z-Information Based on Semantic Spaces

Olga M. Poleshchuk(✉)

Moscow Bauman State Technical University, Moscow, Russia

Abstract. In the paper a model for clustering Z-information is developed. Z-information is represented by the sets of Z-numbers. Each of the sets is an expert criterion model for assessing a certain characteristic. The first and the second components of Z-numbers are the values of full orthogonal semantic spaces. The first components of Z-numbers are the formalizations of the levels of the linguistic scale used by experts to assess the characteristic. The second components of Z-numbers are the formalizations of the levels of the linguistic scale used by experts to assess the reliability of their results. The distances between expert criteria (sets of Z-numbers) are found using aggregating indicators for Z-numbers. Based on pairwise distances between sets of Z-numbers (expert criteria), fuzzy binary relation of similarity is determined on a set of criteria. A fuzzy conformity relation is constructed using the transitive closure of the determined fuzzy similarity relation. The constructed fuzzy conformity relation is decomposed into equivalence relations. Depending on α-levels of fuzzy conformity relations, the set of expert criteria, represented by sets of Z-numbers, can be divided into clusters of criteria similar among themselves with α-levels of confidence.

Keywords: Z-number · Z-information · Z-clustering · Semantic space

1 Introduction

The technique used to group similar data into clusters is called clustering. For clustering fuzzy data, the well-known c-means fuzzy clustering algorithm is often used [1]. Researchers combine this algorithm with other algorithms and techniques. So, in [2] the c-means fuzzy clustering algorithm is combined with the genetic algorithm. In [3, 4], the c-means fuzzy clustering algorithm is improved by minimizing the objective function using the particle swarm optimization algorithm [5]. The use of particle swarm optimization algorithm for fuzzy information is discussed in [6–9].

Cluster analysis of expert information based on fuzzy relations, where transitive relations play a significant role, is of great interest. The use of fuzzy relations in the clustering of expert information was discussed in [10, 11]. Transitive relations have many convenient properties and provide the ability to partition a set into disjoint classes (clusters) of similarity. However, the real results of expert polls are often not transitive. Therefore, for solving problems of fuzzy cluster analysis of expert information, it is of great interest to transform the initial non-transitive relation into a transitive one. Such a

C. Kahraman et al. (Eds.): INFUS 2021, LNNS 308, pp. 888–894, 2022.
https://doi.org/10.1007/978-3-030-85577-2_102

transformation is provided by the transitive closure operation, which was first considered in [12–14]. In [12], a clustering procedure was proposed based on the transitive closure of the similarity relation obtained as a result of a survey of experts. Cluster analysis under fuzzy information and traditional cluster analysis are discussed in [15–18].

After extending in 2011 by prof. Lotfi Zadeh the concept of a fuzzy number and definition of a Z-number [19], it became necessary to develop methods for clustering Z-information (information with Z-numbers).

Since 2011, methods and models have been developed for operating on Z-numbers, determining the distance between them, ranking them and so on. [20–23]. The developed methods and models have been successfully applied to solve decision-making problems [24–28].

In [29], a fuzzy clustering algorithm is described that combines the conversion of Z-numbers into fuzzy numbers and then the application of c-means fuzzy clustering algorithm. In [30], the problem of bimodal clustering is formulated in terms of constructing Z-valued clusters. This paper explores the fundamentals of the bimodal clustering problem and proposes a comprehensive solution method. In addition, an approach is proposed based on the relation proved in [31] between type-II fuzzy sets and Z-numbers.

The objective of the paper is to develop a model for clustering Z-information represented by full orthogonal semantic spaces [32, 33], the values of which have a certain level of reliability. The model developed in the paper is new and necessary for expert information processing with a certain level of reliability in order to avoid subjectivity and the risk of errors in decision making.

The paper contains four sections. Section 2 includes the necessary basic concepts and definitions. Section 3 presents the model for clustering Z-information developed by the author. Section 4 concludes this paper.

2 Basic Concepts and Definitions

A full orthogonal semantic space X is a linguistic variable $\{X, T(X), U, V, S\}$ with a fixed term-set $T(X) = \{X_l, l = \overline{1, m}\}$ and the following properties of continuous membership functions $\mu_l(x), l = \overline{1, m}$: $U_l = \{x \in U : \mu_l(x) = 1\} \forall l = \overline{1, m}$ is an interval; $\mu_l(x), l = \overline{1, m}$ does not increase and does not decrease correspondingly to the right and to the left of U_l; $\sum_{l=1}^{m} \mu_l(x) = 1, \forall x \in U$. The values $X_l, l = \overline{1, m}$ and fuzzy values $\tilde{X}_l, l = \overline{1, m}(\mu_l(x), l = \overline{1, m})$ are determined by the rules V and S [33].

To formalize the terms of semantic spaces, the methods developed in [16, 34] will be applied. According to these methods, the formalizations of terms are trapezoidal or triangular fuzzy numbers. The construction of these numbers can be described as follows: 0.5-cut of unit square is divided into non-overlapping intervals, which correspond to the relative frequencies of the estimates. These intervals are the midlines of trapezoids or triangles, which are the graphs of the membership functions of fuzzy numbers. Membership functions of two adjacent fuzzy numbers have one intersection point when the values of the functions are equal to 0.5. To find the abscissas of the intersection points, you need to add the relative frequencies sequentially. To find the values of the arguments in which the membership functions take on unit values, you need to deviate from the intersection points by half the minimum interval of two adjacent ones. Non-adjacent membership functions have no intersection points.

In [28], for a fuzzy number $\tilde{B} = (b_1, b_2, b_L, b_R)$ an aggregating indicator ϑ is determined:

$$\vartheta = \frac{b_1 + b_2}{2} + \frac{(b_R - b_L)}{12}. \tag{1}$$

A Z - number is an ordered pair of two fuzzy numbers $Z = \left(\tilde{A}, \tilde{R}\right)$ that restricts the assessment and reliability of a person's (expert's) judgment [19].

In [28], the distance between two Z-numbers $Z_1 = \left(\tilde{B}_1, \tilde{R}_1\right), Z_2 = \left(\tilde{B}_2, \tilde{R}_2\right)$ with aggregating indicators $(b_1, r_1), (b_2, r_2)$ of fuzzy numbers $\tilde{B}_1, \tilde{R}$ and $\tilde{B}_2, \tilde{R}_2$ is determined:

$$d(Z_1, Z_2) = \sqrt{\frac{1}{3}(b_1 - b_2)^2 + \frac{1}{3}(b_1 r_1 - b_2 r_2)^2 + \frac{1}{3}(r_1 - r_2)^2}. \tag{2}$$

3 Problem Formulation and Solution

Suppose that n experts evaluate the characteristic B of a set of objects and use the scale with values $B_l, l = \overline{1, m}$. Experts give their evaluations with certain levels of reliability and to evaluate these levels they use the scale with values $R_j, j = \overline{1, k}$. To formalize the scales used in [16, 34], methods have been developed using full orthogonal semantic spaces that are used in this paper. The scale for the i-th expert is represented as a full orthogonal semantic space $B_i, i = \overline{1, n}$ with fuzzy values $\tilde{B}_{il} = \left(b_{il}^1, b_{il}^2, b_{il}^L b_{il}^R\right), l = \overline{1, m}, i = \overline{1, n}$. The scale for evaluating the reliability of expert information is represented as a full orthogonal semantic space R with fuzzy values $\tilde{R}_j = \left(r_j^1, r_j^2, r_j^L, r_j^R\right), j = \overline{1, k}$.

Criterion $Z_i, i = \overline{1, n}$ of the i-th expert for evaluating characteristic B is represented as the set of m Z-numbers $Z_i = \left(\left(\tilde{B}_{il}, \tilde{R}_{il}\right), l = \overline{1, m}\right), i = \overline{1, n}$, fuzzy number $\tilde{R}_{il} = \left(r_{il}^1, r_{il}^2, r_{il}^L, r_{il}^R\right), l = \overline{1, m}, i = \overline{1, n}$ equals to one of fuzzy number $\tilde{R}_j = \left(r_j^1, r_j^2, r_j^L, r_j^R\right), j = \overline{1, k}$.

We denote the aggregating indicators of $\tilde{B}_{il}, \tilde{R}_{il}, l = \overline{1, m}, i = \overline{1, n}$ as $b_{il}, r_{il}, l = \overline{1, m}, i = \overline{1, n}$.

Using the aggregating indicators, we determine the distance $d\left(Z_i, Z_j\right)$ between the i-th and j-th expert criteria:

$$d\left(Z_i, Z_j\right) = \frac{1}{m}\sum_{l=1}^{m}\sqrt{\frac{1}{3}\left(b_{il} - b_{jl}\right)^2 + \frac{1}{3}\left(b_{il}r_{il} - b_{jl}r_{jl}\right)^2 + \frac{1}{3}\left(r_{il} - r_{jl}\right)^2}, i = \overline{1, n}, j = \overline{1, n}. \tag{3}$$

Let

$$d_{\max} = \max_{i,j} d\left(Z_i, Z_j\right), i = \overline{1, n}, j = \overline{1, n}. \tag{4}$$

Then a difference index of two expert criteria we define as:

$$\delta_{ij} = \frac{d(Z_i, Z_j)}{d_{\max}}, i = \overline{1,n}, j = \overline{1,n}. \quad (5)$$

Based on (5), we define a similarity index of two expert criteria:

$$\rho_{ij} = 1 - \delta_{ij}, i = \overline{1,n}, j = \overline{1,n}. \quad (6)$$

Proposition. $\tilde{\Psi}$ with membership function $\mu_{\Psi}(Z_i, Z_j) = \rho_{ij}$ defines a fuzzy binary relation of similarity on the set $Z_i = \left(\left(\tilde{B}_{il}, \tilde{R}_{il}\right), l = \overline{1,m}\right), i = \overline{1,n}$.

Proof. For $\tilde{\Psi}$ to be a fuzzy binary relation of similarity, it is necessary to prove that $\tilde{\Psi}$ possesses the properties of reflexivity and symmetry. Let us prove that $\mu_{\Psi}(Z_i, Z_i) = 1, i = \overline{1,n}$ (property of reflexivity):

$$\mu_{\Psi}(Z_i, Z_i) = \rho_{ii} = 1 - \delta_{ii} = 1 - \frac{\frac{1}{m}\sum_{l=1}^{m}\sqrt{\frac{1}{3}(b_{il}-b_{il})^2+\frac{1}{3}(b_{il}r_{il}-b_{il}r_{il})^2+\frac{1}{3}(r_{il}-r_{il})^2}}{d_{\max}} = 1.$$

Let us prove that $\mu_{\Psi}(Z_i, Z_j) = \mu_{\Psi}(Z_j, Z_i), i = \overline{1,n}, j = \overline{1,n}$ (symmetry property).

$$\begin{aligned}\mu_{\Psi}(Z_i, Z_j) &= \rho_{ij} = 1 - \delta_{ij} \\ &= 1 - \frac{\frac{1}{m}\sum_{l=1}^{m}\sqrt{\frac{1}{3}(b_{il}-b_{jl})^2+\frac{1}{3}(b_{il}r_{il}-b_{jl}r_{jl})^2+\frac{1}{3}(r_{il}-r_{jl})^2}}{d_{\max}} \\ &= 1 - \frac{\frac{1}{m}\sum_{l=1}^{m}\sqrt{\frac{1}{3}(b_{jl}-b_{il})^2+\frac{1}{3}(b_{jl}r_{jl}-b_{il}r_{il})^2+\frac{1}{3}(r_{jl}-r_{il})^2}}{d_{\max}} \\ &= 1 - \delta_{ji} = \rho_{ji} = \mu_{\Psi}(Z_j, Z_i).\end{aligned}$$

Proposition is proved.

The main problem is that $\tilde{\Psi}$ is generally not transitive, that is $\mu_{\Psi}(Z_i, Z_j) \geq \max_{Z_l} \min\{\mu_{\Psi}(Z_i, Z_l), \mu_{\Psi}(Z_l, Z_j)\}, i = \overline{1,n}, j = \overline{1,n}, l = \overline{1,n}$. However, it is possible to make it transitive based on the union of compositions of $\tilde{\Psi}$ with itself $\widehat{\Psi} = \tilde{\Psi}^{n-1}$ [35].

Experts' criteria $Z_1, Z_2, ..., Z_p, p \leq n$ o will be considered similar with a confidence level $\alpha \in (0, 1)$, if $\mu_{\widehat{\Psi}}(Z_i, Z_j) \geq \alpha$ for all $Z_i, Z_j, i = \overline{1,p}, j = \overline{1,p}$.

According to the [35], $\widehat{\Psi}$ can be decomposed as follows: $\widehat{\Psi} = \max_{\alpha}\{\alpha\Psi_{\alpha}\}, 0 < \alpha \leq 1$, where Ψ_{α} is an equivalence relation in the sense of ordinary set theory.

The developed fuzzy clustering model of Z-information makes it possible to analyze expert criteria (with a certain level of reliability), formalized using Z-numbers $Z_i = \left(\left(\tilde{B}_{il}, \tilde{R}_{il}\right), l = \overline{1,m}\right), i = \overline{1,n}$ and determine clusters of similar criteria.

The possibility of such an analysis of Z-information reduces the subjectivity of evaluation procedures and the risks of errors in decision-making tasks based on expert information with a certain level of reliability.

4 Conclusion

The problem of cluster analysis of fuzzy information, given with a certain level of reliability, is relevant and requires a solution. The paper deals with expert criteria for the evaluation of some characteristic formalized with Z-numbers. The number of Z-numbers when formalizing the criteria coincides with the number of levels of the linguistic scale used to evaluate the characteristic. The formalizations of the levels of this linguistic scale, as well as the levels of the scale for evaluating the reliability of expert information, are the values of full orthogonal semantic spaces.

In the paper a difference index and a similarity index of two expert criteria based on the distance between them is defined. Based on similarity index, a fuzzy binary similarity relation on the set of the expert criteria is constructed and it is proved. Using this relation, its transitive closure is constructed, which can be decomposed into equivalence relations in the sense of ordinary set theory. This allows to analyze expert criteria with a certain level of reliability and determine clusters of similar criteria.

The possibility of such an analysis is especially relevant for problems with the active participation of experts, when only expert information is available for decision-making, and the final results significantly depend on its reliability.

Further research will focus on solving the problem of finding a group expert criterion under Z-information based on cluster analysis of individual expert criteria using the model developed in this paper.

References

1. Bezdek, J.C.: Selected applications in classifier design. Pattern Recognit. Fuzzy Object. Funct. Algorithms **2**, 203–239 (1981)
2. Bezdek, J.C., Hathaway, R.J.: Optimization of fuzzy clustering criteria using genetic algorithms. In: Proceedings of the IEEE Conference on Evolutionary Computation, vol. 2, pp. 589–594 (1994)
3. Runkler, T.A., Katz, C.: Fuzzy clustering by particle swarm optimization. In: Proceedings of the IEEE International Conference on Fuzzy Systems, pp. 34–41 (2006)
4. Liu, H.C., Yih, J.M., Wu, D.B., Liu, S.W.: Fuzzy C-mean clustering algorithms based on Picard iteration and particle swarm optimization. In: Proceedings of the International Workshop on Geoscience and Remote Sensing (ETT and GRS-2008), vol. 2, pp. 75–84 (2008)
5. Kennedy, J., Eberhart, R.C.: Particle swarm optimization. In: Proceedings of the IEEE International Joint Conference on Neural Networks, vol. 4, pp. 1942–1948 (1995)
6. Mehdizadeh, E., Tavakkoli Moghaddam, R.: Fuzzy particle swarm optimization algorithm for a supplier clustering problem. J. Ind. Eng. **1**, 17–24 (2008)
7. Melin, P., Olivas, F., Castillo, O., Valdez, F., Soria, J., Valdez, M.: Optimal design of fuzzy classification systems using PSO with dynamic parameter adaptation through fuzzy logic. Expert Syst. Appl. **40**(8), 3196–3206 (2013)
8. Chen, M., Ludwig, A.: Particle swarm optimization based fuzzy clustering approach to identify optimal number of clusters. J. Artif. Intell. Soft Comput. Res. **4**(1), 43–56 (2014)
9. Phyo, O., Chaw, E.: Comparative study of fuzzy PSO (FPSO) clustering algorithm and fuzzy c-means (FCM) clustering algorithm. Natl. J. Parallel Soft Comput. **1**(1), 62–67 (2019)
10. Ruspini, E.H.: A new approach to clustering. Inf. Control **15**, 22–32 (1969)
11. Ruspini, E.H.: Numerical methods for fuzzy clustering. Inf. Sci. **2**, 319–350 (1970)

12. Tamura, S., Higuchi, S., Tanaka, K.: Pattern classification based on fuzzy relations. IEEE Trans. Syst. Man Cybern. **1**, 61–66 (1971)
13. Zadeh, L.A.: Similarity relations and fuzzy orderings. Inf. Sci. **3**, 177–200 (1971)
14. Dunn, J.C.: A fuzzy relative of the ISODATA process and its use in detecting compact well-separated clusters. J. Cybern. **3**, 32–57 (1973)
15. Ruspini, E.H.: Recent developments in fuzzy clustering. In: Fuzzy Set and Possibility Theory. Pergamon Press, N.Y., pp. 133–146 (1982)
16. Poleshchuk, O., Komarov, E.: The determination of rating points of objects with qualitative characteristics and their usage in decision making problems. Int. J. Comput. Math. Sci. **3**(7), 360–364 (2009)
17. Darwish, A., Poleshchuk, O.: New models for monitoring and clustering of the state of plant species based on sematic spaces. J. Intell. Fuzzy Syst. **3**(26), 1089–1094 (2014)
18. Poleshchuk, O.M., Komarov, E.G., Darwish, A.: Assessment of the state of plant species in urban environment based on fuzzy information of the expert group. In: Proceedings of the XX IEEE International Conference on Soft Computing and Measurements, (SCM-2017), pp. 651–654 (2017)
19. Zadeh, L.A.: A note on Z-numbers. Inf. Sci. **14**(181), 2923–2932 (2011)
20. Kang, B., Wei, D., Li, Y., Deng, Y.: A method of converting Z-number to classical fuzzy number. J. Inf. Comput. Sci. **9**(3), 703–709 (2012)
21. Aliev, R.A., Alizadeh, A.V., Huseynov, O.H.: The arithmetic of discrete Z-numbers. Inform. Sci. **1**(290), 134–155 (2015)
22. Aliev, R.A., Huseynov, O.H., Zeinalova, L.M.: The arithmetic of continuous Z-numbers. Inform. Sci. **373**, 441–460 (2016)
23. Poleshchuk, O.M.: Novel approach to multicriteria decision making under Z-information. In: Proceedings of the International Russian Automation Conference (RusAutoCon), p. 8867607 (2019). https://doi.org/10.1109/RUSAUTOCON.2019.8867607
24. Aliyev, R.R., Talal Mraizid, D.A., Huseynov, O.H.: Expected utility based decision making under Z-information and its application. Comput. Intell. Neurosci. **3**, 364512 (2015)
25. Sharghi, P., Jabbarova, K.: Hierarchical decision making on port selection in Z-environment. In: Proceedings of the Eighth International Conference on Soft Computing, Computing with Words and Perceptions in System Analysis, Decision and Control, pp. 93–104 (2015)
26. Wang, F., Mao, J.: Approach to multicriteria group decision making with Z-numbers based on TOPSIS and power aggregation operators. Math. Probl. Eng. **2019**, 1–18 (2019)
27. Poleshchuk, O.M.: Expert group information formalization based on Z-numbers. J. Phys.: Conf. Ser. **1703**, 012010 (2020). https://doi.org/10.1088/1742-6596/1703/1/012010
28. Poleshchuk, O.M.: Object monitoring under Z-information based on rating points. Adv. Intell. Syst. Comput. **1197**, 1191–1198 (2021). https://doi.org/10.1007/978-3-030-51156-2_139
29. Jamal, M., Khalif, K., Mohamad, S.: The implementation of Z-numbers in fuzzy clustering algorithm for wellness of chronic kidney disease patients. J. Phys.: Conf. Ser. **1366**, 012058 (2018)
30. Aliev, R.A., Pedrycz, W., Guirimov, B.G., Huseynov, O.H.: Clustering method for production of Z-numbers based if-then rules. Inf. Sci. **520**, 155–176 (2020)
31. Aliev, R., Guirimov, B.: Z-number clustering based on general Type-II fuzzy sets. Adv. Intell. Syst. Comput. **896**, 270–278 (2018)
32. Zadeh, L.A.: The Concept of a linguistic variable and its application to approximate reasoning. Inf. Sci. **8**, 199–249 (1975)
33. Ryjov, A.P.: The concept of a full orthogonal semantic scope and the measuring of semantic uncertainty. In: Proceedings of the Fifth International Conference Information Processing and Management of Uncertainty in Knowledge-Based Systems, pp. 33–34 (1994)

34. Poleshchuk, O.M.: Creation of linguistic scales for expert evaluation of parameters of complex objects based on semantic scopes. In: Proceedings of the International Russian Automation Conference, (RusAutoCon – 2018), p. 8501686 (2018)
35. Averkin, A.N., Batyrshin, I.Z., Blishun, A.F., Silov, V.B., Tarasov, V.B.: Fuzzy Sets in Models of Control and Artificial Intelligence. Nauka, Moscow (1986)

Author Index

C. Kahraman et al. (Eds.): INFUS 2021, LNNS 308, pp. 895–897, 2022.
https://doi.org/10.1007/978-3-030-85577-2

Printed by Printforce, the Netherlands